Advances in Cardiovascular Technology

New Devices and Concepts

Advances in Cardiovascular Technology

New Devices and Concepts

Edited by

Jamshid H. Karimov, MD, PhD
Lerner Research Institute, Cleveland Clinic, Cleveland, OH, United States; Cleveland Clinic Lerner College of Medicine of Case Western Reserve University, Cleveland, OH, United States

Kiyotaka Fukamachi, MD, PhD
Lerner Research Institute, Cleveland Clinic, Cleveland, OH, United States; Cleveland Clinic Lerner College of Medicine of Case Western Reserve University, Cleveland, OH, United States

Marc Gillinov, MD
Department Chairman, Department of Thoracic and Cardiovascular Surgery, Cleveland Clinic, Cleveland, OH, United States; Heart, Thoracic, and Vascular Institute, Cleveland Clinic, Cleveland, OH, United States

ELSEVIER

ACADEMIC PRESS
An imprint of Elsevier

Academic Press is an imprint of Elsevier
125 London Wall, London EC2Y 5AS, United Kingdom
525 B Street, Suite 1650, San Diego, CA 92101, United States
50 Hampshire Street, 5th Floor, Cambridge, MA 02139, United States
The Boulevard, Langford Lane, Kidlington, Oxford OX5 1GB, United Kingdom

ISBN: 978-0-323-95878-3

For Information on all Academic Press publications
visit our website at https://www.elsevier.com/books-and-journals

Publisher: Mara Conner
Acquisitions Editor: Carrie Bolger
Editorial Project Manager: Rafael Guilherme Trombaco
Production Project Manager: Poulouse Joseph
Cover Designer: Matthew Limbert

Typeset by MPS Limited, Chennai, India

There's a way to do it better. Find it.

Thomas Alva Edison (1847–1931), inventor and research laboratory pioneer

Contents

Part VI
Hemodynamic patient monitoring 461

List of contributors

Jonathan Afoke, MD, FRCS
Department of Cardiothoracic Surgery,
Hammersmith Hospital,
Imperial College Healthcare NHS Trust, London, United Kingdom

Mahboob Ali, MD, MRCP (IRE), MRCP (UK), FACC
Sutter Gould Medical Foundation,
Sutter Health, Stockton, CA, United States

Cristiano Amarelli, MD
Department of Cardiac Surgery and Transplants,
Monaldi, A.O. dei Colli,
Naples, Italy

Yoshimori An, MD
Department of Cardiovascular Medicine,
Cleveland Clinic, Cleveland, OH,
United States

Balaram Anandamurthy, MD
Department of Intensive Care and Resuscitation,
Anesthesiology Institute,
Cleveland Clinic,
Cleveland, OH, United States

Amedeo Anselmi, MD, PhD
Division of Thoracic and Cardiovascular Surgery,
Pontchaillou University Hospital,
Rennes, France; Univ Rennes, CHU Rennes, Inserm,
LTSI - UMR 1099,
Rennes, France

Kedar K. Aras, MD
Department of Biomedical Engineering,
The George Washington University,
Washington, DC, United States

Imran Arif, MD
University of Cincinnati,
Cincinnati, OH, United States

Vincent Auffret, MD, PhD
Univ Rennes, CHU Rennes, Inserm, LTSI - UMR 1099,
Rennes, France; Division of Cardiology,
Pontchaillou University Hospital, Rennes, France

Michael Austriaco, BS
Cleveland Clinic, Cleveland, OH,
United States

Robert H. Bartlett, MD
Department of General Surgery,
Division of Acute Care Surgery,
University of Michigan, Ann Arbor, MI,
United States

Michael I. Brener, MD
Division of Cardiology,
Columbia University Medical Center,
New York, NY, United States

Daniel J.P. Burns, MD
Department of Thoracic & Cardiovascular Surgery,
Cleveland Clinic,
Cleveland, OH, United States

Thomas D. Callahan, MD
Extraction Center, Cardiac Electrophysiology,
Cleveland Clinic,
Cleveland, OH, United States

Adam W. Cates, PhD
AtriCure, Inc.,
Mason, OH, United States

Laura Cercenelli, PhD
eDIMES Lab, Laboratory of Bioengineering,
Department of Experimental Diagnostic and Specialty Medicine (DIMES), Alma Mater Studiorum—University of Bologna,
Bologna, Italy

Jacob Chacko, MBBS, BSc Eng, FRCS-CTh
Department of Cardiothoracic Surgery,
Hammersmith Hospital,
Imperial College Healthcare NHS Trust, London,
United Kingdom

Raphaelle A. Chemtob, MD
Department of Thoracic & Cardiovascular Surgery,
Cleveland Clinic, Cleveland, OH,
United States

Yeon Sik Choi, PhD
Center for Bio-Integrated Electronics,
Northwestern University, Evanston, IL,
United States; Querrey Simpson Institute for
Bioelectronics, Northwestern University,
Evanston, IL, United States; Department of Materials
Science and Engineering,
Northwestern University, Evanston, IL,
United States

Ivan Corazza, MSc
Medical Physics Coordination Center,
Department of Experimental Diagnostic and Specialty
Medicine (DIMES),
Alma Mater Studiorum—University of Bologna,
Bologna, Italy

Anna Corsini, MD
Cardiology Unit, Department of Experimental Diagnostic
and Specialty Medicine (DIMES),
Alma Mater Studiorum—University of Bologna,
Bologna, Italy

James L. Cox, MD
Division of Cardiac Surgery,
Department of Surgery, Bluhm Cardiovascular Institute,
Northwestern University Feinberg School of Medicine
and Northwestern Medicine,
Chicago, IL, United States

Ruggero De Paulis, MD
Cardiac Surgery Department,
European Hospital, Rome, Italy;
Unicamillus University
Rome, Italy

Erwan Donal, MD, PhD
Univ Rennes,
CHU Rennes, Inserm, LTSI - UMR 1099,
Rennes, France; Division of Cardiology,
Pontchaillou University Hospital, Rennes, France

Colin K. Drummond, PhD, MBA
Department of Biomedical Engineering,
Case Western Reserve University, Cleveland, OH,
United States

Seraina Anne Dual, PhD
Department of Radiology,
Stanford University,
Palo Alto, CA, United States

David D'Alessandro, MD
Division of Cardiac Surgery,
Transplantation and Mechanical Circulatory Support,
Massachusetts General Hospital, Boston, MA,
United States

Igor R. Efimov, PhD
Department of Biomedical Engineering,
The George Washington University,
Washington, DC, United States;
Department of Biomedical Engineering,
Northwestern University,
Evanston, IL, United States

Michael Emery S., MD, MS, FACC
Sports Cardiology Center,
Department of Cardiovascular Medicine,
Section of Clinical Cardiology, Miller Family Heart,
Vascular and Thoracic Institute,
Cleveland Clinic,
Cleveland, OH, United States

Jaime-Jürgen Eulert-Grehn, MD
Department of Cardiothoracic and Vascular Surgery,
German Heart Center Berlin, Berlin, Germany;
DZHK (German Center for Cardiovascular Research),
Partner Site Berlin, Germany

Frank M. Fago, BS
AtriCure, Inc., Mason, OH,
United States

Giulio Folino, MD
Cardiac Surgery Department,
European Hospital,
Rome, Italy

Kiyotaka Fukamachi, MD, PhD
Lerner Research Institute,
Cleveland Clinic, Cleveland, OH, United States;
Cleveland Clinic Lerner College of Medicine of
Case Western Reserve University,
Cleveland, OH, United States

Isaac George, MD
Division of Cardiac, Thoracic and Vascular Surgery,
Columbia University Medical Center and the Columbia
College of Physicians and Surgeons,
New York, NY, United States

Marc Gillinov, MD
Heart, Thoracic, and Vascular Institute,
Cleveland Clinic, Cleveland, OH, United States;
Department Chairman,
Department of Thoracic and Cardiovascular Surgery,
Cleveland Clinic,
Cleveland, OH, United States; Department of Thoracic &
Cardiovascular Surgery, Cleveland Clinic, Cleveland,
OH, United States

Patrick M. Grady, CCP
Department of Perfusion Services,
Cleveland Clinic,
Cleveland, OH, United States

Jordan Groskurth, MD
Cardiothoracic Surgery Fellow,
Spectrum Health,
Grand Rapids, MI, United States

Philip Hartley, BmBCh, MChem (Oxon), PGCE, MRCS
Department of Cardiothoracic Surgery,
Hammersmith Hospital, Imperial College Healthcare NHS Trust,
London, United Kingdom

Kevin E. Hodges, MD
Department of Thoracic & Cardiovascular Surgery,
Cleveland Clinic,
Cleveland, OH, United States

David J. Horvath, MSME
R1 Engineering,
Euclid, OH, United States

Jeffrey J. Hsu, MD, PhD, FACC
Ahmanson-UCLA Cardiomyopathy Center,
Division of Cardiology, UCLA Health,
Los Angeles, CA, United States

Ayman A. Hussein, MD
Department of Cardiovascular Medicine,
Cleveland Clinic,
Cleveland, OH, United States

Atsushi Ikeda, MD, PhD
Department of Medicine,
Nihon University School of Medicine,
Tokyo, Japan

Jamshid H. Karimov, MD, PhD, FACC, FAHA
Lerner Research Institute,
Cleveland Clinic,
Cleveland, OH, United States; Cleveland Clinic Lerner College of Medicine of Case Western Reserve University,
Cleveland, OH, United States

Helen S. Knight
Department of Biomedical Engineering,
The George Washington University,
Washington, DC, United States

Christopher K. Koehler, CCP
Department of Perfusion Services,
Cleveland Clinic,
Cleveland, OH, United States

Nakul Kumar, MD
Department of Intensive Care and Resuscitation,
Anesthesiology Institute,
Cleveland Clinic,
Cleveland, OH, United States;
Department of Cardiothoracic Anesthesiology,
Anesthesiology Institute,
Cleveland Clinic,
Cleveland, OH, United States

Shunsuke Kuroda, MD
Department of Cardiovascular Medicine,
Cleveland Clinic,
Cleveland, OH, United States

Knut Kvernebo, MD, PhD
Institute of Clinical Medicine,
University of Oslo,
Department of Thoracic Surgery,
Oslo University Hospital, Oslo, Norway

Panagiotis G. Kyriazis, BSc (Hons), MRes
Department of Cardiothoracic Surgery,
Hammersmith Hospital, Imperial College Healthcare NHS Trust,
London, United Kingdom;
National Heart and Lung Institute,
Faculty of Medicine, Imperial College London,
London, United Kingdom

Guillaume Leurent, MD
Univ Rennes,
CHU Rennes, Inserm, LTSI - UMR 1099,
Rennes, France;
Division of Cardiology,
Pontchaillou University Hospital,
Rennes, France

Guiqing Liu, BSc, MSc, MRCS
Department of Cardiothoracic Surgery,
Hammersmith Hospital, Imperial College Healthcare NHS Trust,
London, United Kingdom

Mark S. Lobosky, BSME
Lerner Research Institute,
Cleveland Clinic,
Cleveland, OH, United States

Amanda L. Mahoney, PhD, RN
Dittrick Medical History Center,
Case Western Reserve University,
Cleveland, OH, United States

Emanuela Marcelli, MSc
eDIMES Lab, Laboratory of Bioengineering,
Department of Experimental Diagnostic and Specialty Medicine (DIMES),
Alma Mater Studiorum—University of Bologna,
Bologna, Italy

Joseph Mauch, BS
Cleveland Clinic Lerner College of Medicine of
Case Western Reserve University,
Cleveland, OH, United States

Patrick M. McCarthy, MD
Division of Cardiac Surgery, Department of Surgery,
Bluhm Cardiovascular Institute, Northwestern University
Feinberg School of Medicine and Northwestern Medicine,
Chicago, IL, United States

Alana N. Miniovich
Department of Biomedical Engineering,
The George Washington University,
Washington, DC, United States

Chihiro Miyagi, MD
Lerner Research Institute,
Cleveland Clinic,
Cleveland, OH, United States

Takuma Miyamoto, MD, PhD
Lerner Research Institute,
Cleveland Clinic,
Cleveland, OH, United States

Matthew V. Monti, BS
AtriCure, Inc.,
Mason, OH, United States

Ryan A. Moore, MD, MSc
Heart Institute,
Cincinnati Children's Hospital Medical Center,
Cincinnati, OH, United States;
Department of Pediatrics, University of Cincinnati
College of Medicine, Cincinnati, OH, United States

David L.S. Morales, MD
Heart Institute,
Cincinnati Children's Hospital Medical Center,
Cincinnati, OH, United States;
Department of Surgery,
University of Cincinnati College of Medicine,
Cincinnati, OH, United States

Hani Najm, MD
Pediatric and Congenital Heart Surgery,
Cleveland Clinic, Cleveland, OH, United States;
Cleveland Clinic Lerner College of Medicine of Case
Western Reserve University,
Cleveland, OH, United States

Hiroshi Nakagawa, MD, PhD
Department of Cardiovascular Medicine,
Cleveland Clinic,
Cleveland, OH, United States

Vinci Naruka, MA Hons (Cantab), MB BChir, MRCS
Department of Cardiothoracic Surgery,
Hammersmith Hospital, Imperial College Healthcare
NHS Trust,
London, United Kingdom

Johan Nilsson, MD, PhD
Department of Translational Medicine,
Thoracic Surgery and Bioinformatics,
Lund University and Skane University Hospital,
Lund, Sweden

Carrie L. Ochocki, CCP
Department of Perfusion Services,
Cleveland Clinic,
Cleveland, OH, United States

Lucrezia Palmieri, MD
Department of Cardiac Surgery and Transplants,
Monaldi, A.O. dei Colli,
Naples, Italy

Dermot M.J. Phelan, MD, PhD, FASE, FACC
Sanger Heart & Vascular Institute—Charlotte,
Charlotte, NC, United States

Alistair Phillips, MD, FACC, FACS
Pediatric and Congenital Heart Surgery,
Cleveland Clinic, Cleveland, OH, United States;
Cleveland Clinic Lerner College of Medicine of
Case Western Reserve University,
Cleveland, OH, United States

Anthony R. Polakowski, MEng
Lerner Research Institute,
Cleveland Clinic,
Cleveland, OH, United States

Evgenij Potapov, MD
Department of Cardiothoracic and Vascular Surgery,
German Heart Center Berlin, Berlin,
Germany;
DZHK (German Center for Cardiovascular Research),
Partner Site Berlin, Germany

Shane S. Pullins, CCP
Department of Perfusion Services,
Cleveland Clinic,
Cleveland, OH, United States

Prakash P. Punjabi, FRCS, FESC, MS, MCh, FCCP, FFSTEd, Diplomate NBE
Department of Cardiothoracic Surgery,
Hammersmith Hospital, Imperial College Healthcare
NHS Trust, London, United Kingdom; National Heart and

Lung Institute,
Faculty of Medicine, Imperial College London,
London, United Kingdom

Suzanne Ratzloff, MA
Health Research International,
Lakewood, OH, United States

Rahul D. Renapurkar, MD
Thoracic and Cardiovascular Imaging,
Imaging Institute,
Cleveland Clinic,
Cleveland, OH, United States

Ellen T. Roche, PhD
Institute for Medical Engineering and Science,
Massachusetts Institute of Technology,
Cambridge, MA, United States;
Department of Mechanical Engineering,
Massachusetts Institute of Technology,
Cambridge, MA, United States

Luca Rosalia
Institute for Medical Engineering and Science,
Massachusetts Institute of Technology,
Cambridge, MA, United States;
Harvard-MIT Health Sciences and Technology Program,
Massachusetts Institute of Technology,
Cambridge, MA, United States

Mossab Saeed Y., MD
Harvard Medical School,
Boston Children's Hospital,
Boston, MA, United States

Shiva Sale, MD
Department of Cardiothoracic Anesthesiology,
Anesthesiology Institute,
Cleveland Clinic,
Cleveland, OH, United States

Andrea Salica, MD
Cardiac Surgery Department,
European Hospital,
Rome, Italy

Raffaele Scaffa, MD
Cardiac Surgery Department,
European Hospital,
Rome, Italy

Marianne Schmid Daners, PhD
Product Development Group Zurich,
Mechanical and Process Engineering,
ETH Zurich, Zurich, Switzerland

Dhruv Seshadri R., MS
Department of Biomedical Engineering,
Case Western Reserve University,
Cleveland, OH, United States

Christoph Starck, MD
Department of Cardiothoracic and Vascular Surgery,
German Heart Center Berlin, Berlin, Germany;
DZHK (German Center for Cardiovascular Research),
Partner Site Berlin, Germany

Randall C. Starling, MD, MPH, FACC, FAHA, FESC, FHFA, FHFSA
Cleveland Clinic Lerner College of Medicine of
Case Western Reserve University,
Cleveland, OH, United States;
Kaufman Center for Heart Failure, Heart, Thoracic, and
Vascular Institute, Cleveland Clinic,
Cleveland, OH, United States

Nicholas A. Szugye, MD
Heart Institute,
Cincinnati Children's Hospital Medical Center,
Cincinnati, OH, United States

W. H. Wilson Tang, MD
Heart, Vascular and Thoracic Institute,
Cleveland Clinic,
Cleveland, OH, United States

Varun Thachil
Case Western Reserve University,
Cleveland, OH, United States

Tomasz Timek, MD, PhD
Division of Cardiothoracic Surgery,
Spectrum Health, Grand Rapids,
MI, United States;
Michigan State University College of Human Medicine,
Grand Rapids, MI, United States

Niraj Varma, MD, PhD
Heart and Vascular Institute,
Cleveland Clinic,
Cleveland, OH, United States

Jean-Philippe Verhoye, MD, PhD
Division of Thoracic and Cardiovascular Surgery,
Pontchaillou University Hospital, Rennes, France;
Univ Rennes, CHU Rennes, Inserm, LTSI - UMR 1099,
Rennes, France

Brett J. Wakefield, MD
Department of Intensive Care and Resuscitation,
Anesthesiology Institute, Cleveland Clinic,

Cleveland, OH, United States;
Department of Cardiothoracic Anesthesiology,
Anesthesiology Institute, Cleveland Clinic,
Cleveland, OH, United States

Oussama M. Wazni, MD
Department of Cardiovascular Medicine,
Cleveland Clinic,
Cleveland, OH, United States

Luca Weltert, MD
Cardiac Surgery Department,
European Hospital, Rome, Italy;
Unicamillus University Rome, Italy

Bruce L. Wilkoff, MD
Extraction Center, Cardiac Electrophysiology,
Cleveland Clinic,
Cleveland, OH, United States

Rose T. Yin, PhD
Department of Biomedical Engineering,
The George Washington University,
Washington, DC, United States

Katsuaki Yokoyama
Department of Medicine,
Nihon University School of Medicine,
Tokyo, Japan

James B. Young, MD
Kaufman Center for Heart Failure and Recovery,
Heart, Vascular and Thoracic Institute,
Cleveland Clinic,
Cleveland, OH, United States;
Cleveland Clinic Lerner College of Medicine of
Case Western Reserve University,
Cleveland, OH, United States

Margherita Zecchi
Medical Physics Coordination Center,
Department of Experimental Diagnostic and Specialty
Medicine (DIMES), Alma Mater Studiorum—University
of Bologna, Bologna, Italy

Andreas Zuckermann, MD
Department of Cardiovascular Surgery, University of
Vienna, Vienna, Austria

Foreword

Innovation is the result of creative individuals who strive to improve and will never accept the *status quo*. Today, innovation is becoming a part of a culture across the world's leading centers of cardiovascular care, and technological advances are the celebration of innovation.

We are at an inflection point in the treatment of cardiac and vascular pathology. Advanced, smart, highly efficient technology platforms and opportunities for sophisticated artificial data-adopting therapeutics are being introduced and implemented at a never-before-seen rate. With a growing passion to identify, address and pursue opportunities in the area of cardiovascular care, the entire field is evolving at an unprecedentedly rapid pace.

The editors of this exciting new book are all innovators who have each independently changed the practice of cardiovascular care. We can thank these editors for all the contributions they have made in the areas of valve disease and advanced heart failure. This team has accepted the challenge to create this compendium, which relates where we have been and where we are going. This ambitious project has 39 compelling, state-of-the-art chapters that span all aspects of cardiovascular disease, including cardiac surgery, cardiology, interventional cardiology, critical care, top-notch engineering innovation efforts, and the difficult but exciting journey from bench to commercialization. Important chapters from numerous key opinion leaders, clinical and engineering champions and world experts are must reads to learn the "pearls" and "pitfalls," and to understand the technology that is in current practice and in the advanced development pipeline.

The advances in cardiovascular care in the past two decades are truly remarkable; there is truly a surge of new information emerging every year. In the United States, we are fortunate to have regulatory guidance from the Food and Drug Administration, whose responsibility it is to ensure safety and efficacy through careful evaluation of new therapies. Organizations with similar roles and authorities across Europe and the rest of the world are guardians and committed associates of the rigorous device approval process. Thus, it is important to acknowledge that innovators maintain a close and positive relationship with regulatory agencies to ensure the efficient and safe approval of innovative products.

For investigators who are early in their professional careers, this book will enlighten you and provide guidance as you embark on your path to alleviate disease and improve health through innovation. The inspiration and extensive resources this book offers will significantly aid you in understanding the theoretical, practical, and applied elements of cardiovascular technology and innovation. Learning from all of the outstanding authors who have contributed to this book will inspire you, align your mission, and guide you to your success. I remain privileged to have a professional association with many of the authors, and acknowledge each of the contributors to this edition who have helped me on my own professional journey, as I too seek to innovate and improve cardiovascular care.

Randall C. Starling, MD, MPH, FACC, FAHA, FESC, FHFA, FHFSA
Kaufman Center for Heart Failure, Heart, Thoracic, and Vascular Institute, Cleveland Clinic, Cleveland, OH, United States
Cleveland Clinic Lerner College of Medicine of Case Western Reserve University, Cleveland, OH, United States

A Tribute to Innovations in Cardiovascular Care Pioneered at Cleveland Clinic

Lars G. Svensson, MD, PhD
Chairman, Sydell and Arnold Miller Family Heart, Vascular & Thoracic Institute, Cleveland Clinic, Cleveland, OH, United States,
Delos M. Cosgrove, MD, Chair for Heart Disease Research,
Department of Thoracic and Cardiovascular Surgery and the Aorta Center, Sydell and Arnold Miller Family Heart, Vascular & Thoracic Institute, Cleveland Clinic, Cleveland, OH, United States

The Cleveland Clinic celebrated its 100th anniversary in 2021, and its past, present, and future successes are due to continuous innovation and process improvement. Cardiovascular innovations developed at Cleveland Clinic have highlighted the history of advances in cardiovascular care.

Don B. Effler and Lawrence Groves launched the pioneering era of cardiac surgery at Cleveland Clinic in 1956 as one of the first teams to perform open-heart surgery.[1] From there, the program has grown to more than 7000 heart operations at the Cleveland Clinic Healthcare System, and 5321 heart operations in northeast Ohio, in 2019. Early innovations included the "stopped heart," congenital heart and valve operations, contributions to heart-lung machines, and the total artificial heart by Willem Kolff and Tetsuzo Akutsu.

The big leap forward in managing cardiovascular disease was the accidental flipping of a catheter tip into the right coronary artery ostium by F. Mason Sones,[2,3] which caused the patient to go into cardiac arrest. Dr. Sones thumped the patient's chest and shouted at him to breathe. The catheter may have interfered with blood flow in the conal arteries of this patient and induced a ventricular arrhythmia. Fortunately, the patient recovered, and this set off the era of coronary and cardiac catheterization that continues to be the mainstay hoisting the sail of cardiac care.

Subsequently, cardiovascular pathophysiology, diagnosis, care, and follow-up were advanced by a flood of newer imaging techniques such as computed tomography, magnetic resonance imaging, echocardiography, ultrasound, and nuclear tracer studies. Many of these areas are now enhanced by supercomputing and a spectrum of processing aids from machine learning to artificial intelligence.

In 1967, René Favaloro, realizing the implications of coronary disease and blockages, expanded the nascent cardiac surgery field into the science of coronary artery bypass surgery.[4] At the same time, with the introduction of Medicare coverage for patients 65 years and older, the practice of coronary artery bypass surgery grew rapidly across the United States, as funding became available to build hospitals to perform cardiac surgery, and a bevy of cardiac surgery trainees spread over the country. Subsequently, careful studies by Floyd "Fred" Loop and Bruce Lytle[5] promulgated the benefits of one and then two internal mammary arteries on long-term survival.

Pari passu, valve and aortic surgery procedures grew exponentially. Multiple valve types and replacement grafts for the aorta were developed. As surgeons were availed of more support systems for doing more complex cardiac and aortic surgery, like better heart-lung machines and myocardial protection, surgeons, including those from Cleveland Clinic, advanced into the realm of doing both more complex and also less invasive procedures through smaller incisions (right thoracotomy,[6] J incisions,[7] robotic-assisted mitral valves), first pioneered by Delos "Toby" Cosgrove.

The innovation of cardiac catheterization by Sones and colleagues[8] also opened an era of increasingly complex procedures performed with catheters, evolving from cutdowns to radial artery access and procedures such as percutaneous transluminal coronary angioplasty (ballooning), percutaneous coronary intervention (stents), valve ballooning, percutaneous inserted valves and procedures (transcatheter aortic valve replacement, pulmonary valves, mitral and tricuspid valve clips, perivalvular leak closures). Other transcatheter procedures were also rapidly tested and utilized ventricular support devices, etc., including for electrophysiology advances such as left atrial occlusion devices, ablation techniques, and wireless pacemakers.

Heart failure treatments, such as transplants, which were based on better immunosuppression, left ventricular assist devices, and better heart failure medications, resulted in much better outcomes, both early and late—so much so that the official name for our section is the Section for Heart Failure Treatment and Recovery. A fully reimplantable total artificial heart that is a safe replacement is still a sought-after goal toward which our team, overseen for many years by Leonard Golding and Kiyotaka Fukamachi and colleagues, has made big strides.[9] A safe and effective heart xenograft, despite being one of the first heart replacement procedures, is still an unrealized dream more than half a century later.

The endovascular treatments for aortic aneurysms developed 30 years ago by Cleveland Clinic vascular surgeon Juan Parodi and colleagues[10] are now the mainstay of treatment of aortic disease. The complexity of procedures now achievable with endovascular grafts, particularly CHIMPS and branched grafts, is astounding, although safety and long-term results need to be improved.

The new, rapidly expanding area is wearable digital monitoring devices. These will grow in importance for maintaining the health of the general population and patients, albeit limited by privacy concerns, physician's time to interpret, and hacking.

In the following chapters, you will be able to explore historical, current, and future devices built on the shoulders of preceding giants of cardiovascular innovation. We trust you will find the innovations enlightening and exciting, allowing for the stimulation of further research and innovation to improve the care of all people.

References

1. Effler DB, Groves LK, Sones FM, Kolff WJ. Effective cardiac arrest in open-heart surgery. Report of three cases. *Cleve Clin. J. Med.* 1956;23:105–114.
2. Cleveland Clinic Website. 1950s: The golden age of medical innovation. Available at https://my.clevelandclinic.org/about/history/1950s.
3. Sones Jr FM. Cine-coronary arteriography. *Ohio State Med. J.* 1962;58:1018–1019.
4. Favaloro RG. Saphenous vein autograft replacement of severe segmental coronary artery occlusion: operative technique. *Ann. Thorac. Surg.* 1968;5:334.
5. Loop FD, Lytle BW, Cosgrove DM, et al. Influence of the internal-mammary-artery graft on 10-year survival and other cardiac events. *N. Engl. J. Med.* 1986;314:1–6.
6. Cosgrove 3rd DM, Sabik JF, Navia JL. Minimally invasive valve operations. *Ann. Thorac. Surg.* 1998;65:1535–1538. discussion 1538–9.
7. Svensson LG. Minimal-access "J" or "j" sternotomy for valvular, aortic, and coronary operations or reoperations. *Ann. Thorac. Surg.* 1997;64:1501–1503.
8. Sones Jr FM, Shirey EK, Proudfit WL, Westcott RN. Cine-coronary arteriography (abstract). *Circulation.* 1959;20:773.
9. Fukamachi K, Karimov J, Sunagawa G, et al. Generating pulsatility by pump speed modulation with continuous-flow total artificial heart in awake calves. *J. Artif. Organs.* 2017;20:381–385.
10. Veith FJ, Marin MM, Cynamon J, Schonholz C, Parodi J. 1992: Parodi, Montefiore, and the first abdominal aneurysm stent graft in the United States. *Ann. Vasc. Surg.* 2005;19:749–751.

Preface

The quintessential elements of biomedical science necessary for the successful replenishment, reinstitution, compensation, and adaptation of lost, weakened, or altered cardiac or vascular function lie in exceptional demand for compelling expertise, the extraordinary ability of people in art, and unique technology solutions. All these must accompany each modern-era patient, from early diagnosis to functional or structural recovery, from recovery to a long-term rehabilitation and thriving quality of life.

With improvements in health care have come an increase in longevity among the global population. Among the known pathologies, the cardiovascular diseases have maintained the largest, seemingly overwhelming burden on morbidity and mortality due to their high incidence and considerable detrimental effects. Over the past few decades, that burden has triggered a substantial spike in the exploration of the most appropriate treatment options and device solutions, thus enabling the portfolio of medical art techniques and manipulations, along with ultraspecialized training programs and sophisticated engineering tools, to assist that process as clinical representations of cardiac pathology become even more complex.

The purpose of this book was to bring together a large number of world-renowned scholars both within and outside of cardiovascular care to provide an authoritative and comprehensive definition and review of modern technology and instrumentation. This volume is an attempt to bring together the viewpoints and interests of all the clinical, engineering, and basic sciences. Among the contributors are many whose work has had a major impact in this field in recent years. The idea was that through the fusion of these diverse approaches, a better comprehension of the field will arise.

Specifically, we hope that the insights provided by basic investigations will stimulate those with clinical interests to seek more advanced, effective, and focused diagnostic and therapeutic modalities based on firm fundamental principles. By the same token, we wanted the very challenging clinical applications and unique aspects of cardiac and vascular interventions in humans to stimulate and challenge basic scientists and engineers to further refine their approaches in the applied research and device development process.

The impetus for this book arose from our aspiration to unite the perspectives of engineers, researchers, and clinicians into a single volume. Every so often, engineers lack the comprehensive understanding of a specific clinical problem and the issues related to it. Similarly, clinicians lack some fundamental knowledge related to cardiovascular technologies in development, including the possibilities and limitations of new devices. We set a goal of providing an overview that is equally valuable to experts from different areas, highlighting key cardiac disease representations and current solutions to address them, and offering insights on the future of technology and devices in treating cardiovascular diseases.

Our enthusiasm extends to the hope that others with an interest in cardiovascular medicine and new technology development and device applications in heart and vascular medicine—including graduate medical and engineering students, medical residents and clinical fellows, early career medical track, senior engineers, and clinicians—would benefit from this collaborative, multidisciplinary effort.

The book includes various aspects of evolving cardiovascular medicine where new device solutions are currently used or might offer benefits in the future. The need for improvement, potential solutions, and clinicians' and engineers' perspectives on cardiovascular diseases are also provided. Up-to-date information on cardiovascular technology, as well as limitations of the current technology and space for improvement, is extensively covered. Future insights on the role of cardiovascular devices and technology in improving outcomes in heart and vascular disease are addressed by top experts and pioneers in the field.

With the publication of this book, we would like to express our greatest appreciation to its distinguished contributors—the surgeons, cardiologists, engineers, scientists, researchers, and all the clinicians who generously accepted our invitation and extensively committed their time and talent. It is important to emphasize that the technology and clinical research data presented in this volume should be taken as a constellation of the contemporary views of these specialists.

We hope that the readers will benefit from the teaching points and profound expertise offered by the expert chapter authors from the leading centers of cardiovascular care, and will find this book a comprehensive, thoughtful, and clinically relevant collection, as well as an interesting resource.

May the outcome of this collaborative effort be widely circulated and may further conversations over the exciting cardiovascular field be stimulated. And may patients around the globe benefit from the knowledge gained and expanded by the readers.

Jamshid H. Karimov, MD, PhD
Kiyotaka Fukamachi, MD, PhD
Marc Gillinov, MD
Cleveland, OH, United States

PART I

Cardiovascular Practice

Chapter 1

The History of Cardiac Device Innovation: the Collaborative Triumph of the Kay-Cross Oxygenator

Amanda L. Mahoney
Dittrick Medical History Center, Case Western Reserve University, Cleveland, OH, United States

Chapter Outline

Introduction

The development of clinically viable heart-lung machines during the 1950s made possible groundbreaking advancements in cardiac surgery and contributed to the development of other key cardiac innovations. These included the creation of cardiac intensive care units, new clinical roles such as that of the perfusionist, as well as the development of artificial heart valves and other seminal medical devices. Now seen as vestiges of a primitive era in cardiac surgery, heart-lung machines and their components were absolutely critical to the many cardiac advancements of the 20th century that we now take for granted. Although other 20th century cardiac innovations such as the artificial heart tend to overshadow the significance of heart-lung machines, their history highlights the importance of practical, reliable solutions to clinical problems in advancing medicine. This chapter explores the role of one such workhorse device, the Kay-Cross Oxygenator (Fig. 1.1). The success of this frequently overlooked heart-lung machine component can be directly tied to the interdisciplinary team that developed it, who actively sought out feedback, supported learning and knowledge sharing, and encouraged modification or "tinkering" by other users.

Heart-lung machines and the rise of open heart surgery

Heart-lung machines allowed for the circulation and oxygenation of blood outside of the human body, a process known as extracorporeal circulation (EC). EC bypasses blood away from the heart, enabling surgeons to see cardiac structures and perform complex repairs, in other words, operate under direct vision in a bloodless field. As EC technologies and practices became increasingly safer and more reliable during the late 1950s, physicians had more time to precisely diagnose and repair the heart and surrounding vasculature in the operating room (OR). The result was a paradigm shift that not only transformed surgical techniques and cardiac care, but also created hope for thousands of heart patients and their families.

The heart-lung machines of the early 1950s were made up of multiple components, only some of which were commercially available, such as Sigmamotor and DeBakey pumps used to circulate blood within the EC circuit. Other parts of the apparatus were at first constructed in-house, including the crucial element responsible for gas exchange, the "artificial lung" or oxygenator. A clinically viable oxygenator was a tall order for medical researchers and engineers during the mid-20th century.

Advances in Cardiovascular Technology. DOI: https://doi.org/10.1016/B978-0-12-816861-5.00009-5

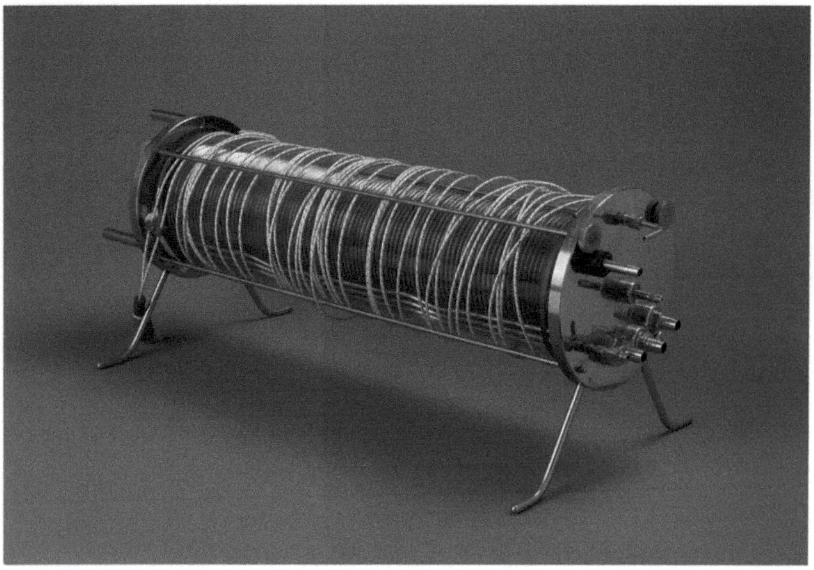

FIGURE 1.1 Kay-Cross Oxygenator, c.1955. Courtesy Dittrick Medical History Center, Case Western Reserve University.

The device needed to provide adequate gas exchange to allow the patient to receive sufficient oxygen during EC without removing too much blood volume from the body. Physiological conditions including body temperature and blood pressure needed to be maintained consistently throughout the surgery, requiring a precise rate of flow through the oxygenator to balance both pressure and the rate of oxygenation. The introduction of bacteria, large air bubbles, or particulates into the EC system could prove fatal to the patient, with the oxygenator the most vulnerable point in the circuit. Blood proteins from previous patients could cause deadly pyrogenic reactions should they remain as deposits on the internal surfaces of the oxygenator. Minimizing these risks while still meeting the physiological requirements of effective EC was no small feat.

There were three main categories of oxygenators during the mid-1950s based on the method of gas exchange used by the device: bubble, rotating screen or filming oxygenators, and membrane.[1] Each approach presented researchers with advantages and disadvantages as they sought to develop experimentally successful oxygenators and adapt them for clinical use.

Bubble oxygenators, which added bubbles of gas directly into blood within the EC system were popular due to their simplicity and affordability. The well-known example developed by Richard DeWall at the University of Minnesota was first used clinically in 1955.[2] Devices of this type provided excellent oxygenation due to the ample surface area for gas exchange created by bubbling air through the blood. However, the same bubbles of gas that made bubble oxygenators so efficient also made them dangerous and difficult to use. The bubble method was mechanically traumatic to the blood, potentially launching the clotting cascade and other undesirable physiological responses during surgery. Even with the addition of arterial reservoirs and other components that removed gas bubbles from the blood before it reentered the patient's body, the direct contact between air and blood increased the risk of air embolism. Risk for these complications increased the longer a patient was perfused using a bubble oxygenator, leading many surgical teams to limit their use to short surgical procedures, or choose a different option.

Filming oxygenators create a large surface area for gas exchange by manipulating blood into a thin film as it passes through the device. John Gibbon Jr. performed the first successful open heart surgery under EC in 1953 using a rotating screen oxygenator of his own design.[3] This category of oxygenators typically employed a sequence of disks or screens inside a reservoir and required direct contact between circulating blood and the internal surfaces of the device. As a result of this direct contact, filming oxygenators shared the same risk of air embolism and contamination with bacteria or debris. Early heart-lung machine users reported decreased fibrin formation and other signs of mechanical trauma to the blood by filming oxygenators compared to the bubble variety.[4] However, much of this evidence was anecdotal. In the early years of EC, researchers lacked a shared terminology and had yet to agree on what variables to measure when assessing oxygenator performance and comparing different methodologies.[5]

Even in the early years of EC research, clinicians and physiologists agreed that membrane oxygenators, which employed the oxygen transfer properties of thin plastic sheets for contactless gas exchange, would prove the most safe and effective approach.[6] This prediction turned out to be true as the capillary membranes currently used in extracorporeal life support are direct descendants of the experimental membrane oxygenators of the 1950s–60s.[7] Based on Willem Kolff's 1944 observation of gas exchange through passive diffusion across a cellophane surface, the advantages of membrane oxygenators were apparent and promising by the early 1950s.[8] However, researchers and manufacturers had yet to produce a plastic membrane that was resilient, thin, and flexible enough to provide sufficient oxygenation for anything beyond experimental use and very short surgeries.

Twentieth century EC researchers also correctly predicted that single-use disposable oxygenators would be the component of choice in the future.[9] It would be decades before materials science, manufacturing, and the medical supply infrastructure would make this option possible (Figs. 1.2 and 1.3). The first clinically useful disposable membrane oxygenator, known colloquially as the Bramson Lung, did not become available until shortly after 1965, over a decade after Gibbon's successful use of a heart-lung machine for open heart surgery in 1953.[10] Although the Bramson Lung and other membrane oxygenators were on the market and relatively affordable at the time, surgical teams experienced difficulty getting the devices to consistently perform even during the 1970s, relying on older heart-lung machines for longer or more complex perfusions.[11] A few different models of bubble and filming oxygenators were used across the US during this transitional period as hospitals established open heart surgery programs and manufacturers developed more sophisticated single-use disposable options. Although the DeWall-Lillehei bubble oxygenator, the Mayo-Gibbon heart-lung machine, and other early EC technologies are now museum artifacts, these technologies enabled the development of the surgical techniques and technologies that transformed cardiac care in the 20th century, not to mention saved the lives of thousands of patients.

One model of filming oxygenator, the Kay-Cross rotating disk oxygenator (KCO) was used extensively both clinically and experimentally from 1956 into the 1970s.[12] While other options were available, the KCO was a popular choice for hundreds of clinical and laboratory research groups building programs for open heart surgery and studying

FIGURE 1.2 Total Bypass Membrane Oxygenator (TMO), c.1975. Courtesy Dittrick Medical History Center, Case Western Reserve University.

FIGURE 1.3 Detail showing tightly folded membrane. Courtesy Dittrick Medical History Center, Case Western Reserve University.

FIGURE 1.4 Kay-Cross Oxygenator, detail, c.1955. Courtesy Dittrick Medical History Center, Case Western Reserve University.

the physiology of the circulatory system. Available for purchase in 1957, Pemco Inc, the device manufacturer sold over 1000 units while the KCO was in production (Fig. 1.4).[13]

Developing the Kay-Cross Oxygenator

The KCO was developed by an interdisciplinary team of surgeons, physiologists, engineers, fabricators, and technicians coordinated by cardiothoracic surgeons Earle B. Kay and Frederick S. Cross at St. Luke's Hospital in Cleveland, Ohio. Physiologists Robert M. Berne and Gerhard Brecher from the nearby Western Reserve School of Medicine, now Case Western Reserve University, were involved in the early phases of the team's EC research. Richard D. Jones, a young research assistant who joined the group, had a knack for engineering, tinkering, and problem-solving that would prove pivotal in the success of the KCO. Jones formed a decades-long research partnership with Cross and would later earn a PhD in physiology.

Kay and Cross, who first met in 1946 and discovered their shared interest in thoracic surgery, agreed to set up a clinical and surgical research partnership in Cleveland once Cross had completed a surgical fellowship and PhD at the University of Minnesota (Fig. 1.5).[14] During his time in the Minnesota program, Cross gained experience with the vanguard open heart surgical techniques developed by Richard L. Varco, John Lewis, C. Walton Lillehei, and others working under Owen Wangensteen in the early 1950s.[14] Surgical trainees in the program were required to conduct research projects alongside their surgical practice, a novel (and demanding) approach to surgical training at the time. Graduates of Wangensteen's surgical fellowships would go on to establish their own research programs, creating an influential network of elite surgeon-researchers across the US. Kay established a thoracic surgery practice and small research laboratory at St. Luke's Hospital, where Cross joined him in 1953. Using Lillehei's cross-circulation technique, Cross performed Cleveland's first open heart surgery on October 7, 1954.[14] This

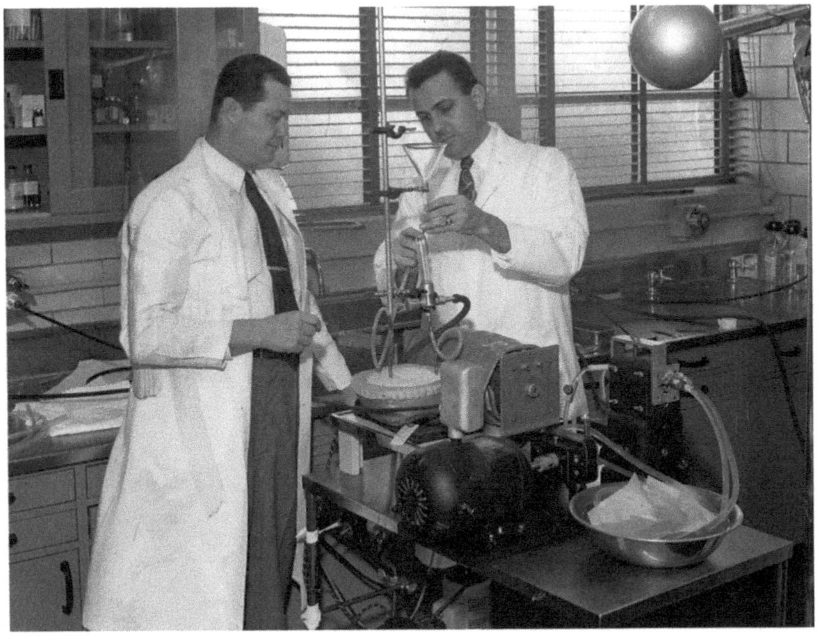

FIGURE 1.5 Earl B. Kay (left) and Frederick C. Cross (right) at St. Luke's Hospital, Cleveland, OH, 1956. Courtesy Dittrick Medical History Center, Case Western Reserve University.

approach, which used the heart and lungs of an adult to oxygenate the blood of a child, achieved good results when employed by an experienced surgical team. However cross-circulation was a viable option for a very small set of patients and presented a number of serious risks and logistical challenges. In 1955, Kay and Cross decided to work on a heart-lung machine, focusing their efforts on creating a clinically useful, relatively inexpensive oxygenator to support open heart surgery at St. Luke's Hospital. Richard D. Jones joined the team as a laboratory assistant in 1954 with additional laboratory and research staff added to the project after the group was awarded funding from the Cleveland Area Heart Association (Fig. 1.6).[15] The group partnered with Cleveland manufacturer Pemco, Inc. to create the prototype.

Kay, assisted by Cross, performed the first open heart surgery with a KCO at Cleveland's St. Vincent's Charity Hospital in January 1956, one of the first few procedures using a heart-lung machine for EC.[13] Richard Jones managed the heart-lung machine alongside other technicians, physicians, and OR nurses such as Clarabelle C. "Clara" Sittinger.[13] The group shared their results at the American Society for Artificial Internal Organs (ASAIO) meeting in April 1956, presenting on their experiences using the KCO for 12 procedures at St. Vincent's Charity Hospital and St. Luke's Hospital.[12] The KCO continued to perform well for open heart procedures with 17 surgeries completed by the time the group submitted their early clinical findings to the Journal of the American Medical Association (JAMA).[16] A total of 30 patients had undergone procedures using the KCO by July 1956, demonstrating the rapid growth of the St. Luke's Hospital open heart surgery program.[17] Cross and Jones continued to work closely with Pemco during this time to refine the KCO and experiment with design, materials, and construction.

Device design, manufacturing, and sales

Based on an experimental disk oxygenator built by Viking Björk, the KCO consisted of a series of disks mounted at precise intervals along a central shaft within a borosilicate glass cylinder.[12] The first model, constructed in 1955, used stainless steel for the disks, shaft, disk spacers, and gasketed end plates. The disks, originally Teflon-coated stainless steel, were fabricated entirely out of Teflon by 1956. Silicone coatings to prevent foaming and fibrin formation were also added to the Pyrex cylinder and endplates as these materials became more readily available. The KCO could be considered an iteration of Viking Björk's experimental disk oxygenator with a similar rotating disk design and a few key differences in materials, size, and configuration.[18] For example, Cross et al. found that the Teflon disk surfaces decreased hemolysis and eliminated fibrin formation during EC, complications that limited the clinical application of Björk's design. Shortly after its successful performance in the OR, the KCO team developed a range of Pyrex cylinder sizes and disk configurations, which made the device adaptable for experimental animals, infants, and adults.

The stainless steel construction of the KCO required precision fabrication, which could not be accomplished in-house at St. Luke's Hospital. Engineers, machinists, and sales representatives at Pemco worked closely with Kay,

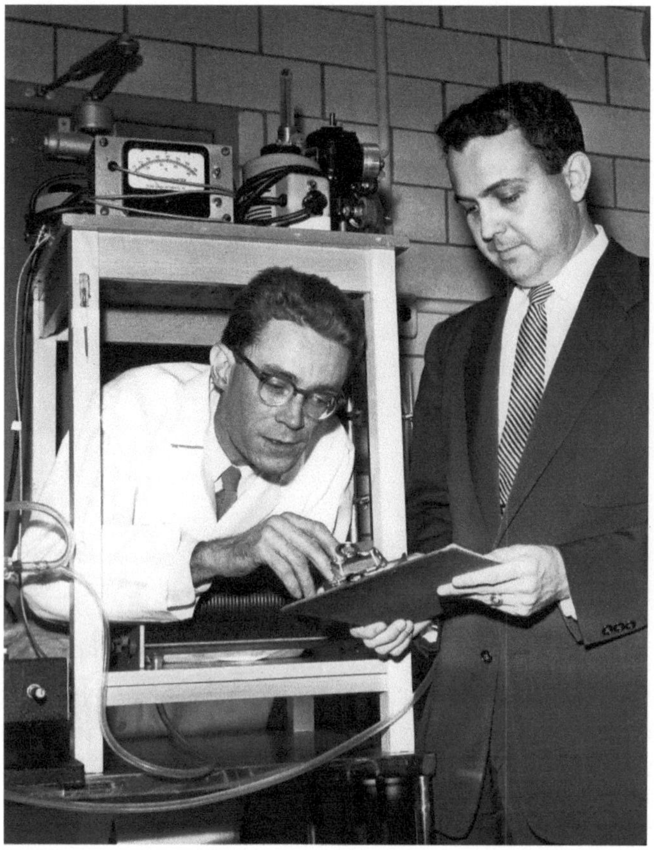

FIGURE 1.6 Richard D. Jones (left) and Frederick C. Cross (right), St. Luke's Hospital, 1956. Jones is leaning over a Kay-Cross Oxygenator prototype. Courtesy Dittrick Medical History Center, Case Western Reserve University.

Cross, and Jones to develop and perfect the oxygenator. The original design sketch used to order the prototype in June 1955 includes notations from multiple hands, and was stamped and initialed by Pemco salesmen and technicians (Fig. 1.7).[19] Working with a local manufacturer enabled Cross and Jones to experiment with elements such as silicone coatings and tweak the design and construction for experimentation with a short turnaround time. Partnering with Pemco had a number of advantages including precision-crafted stainless steel fabrication and streamlined access to new materials, such as heat-conducting coated glass, through the company's association with Corning and other manufacturers. Based on surviving correspondence and product catalogs, it is clear that the KCO team planned for Pemco to manufacture and sell the device from the beginning of their collaboration, but on a small scale. Like other groups working on oxygenators and heart-lung machines in the 1950s, Kay and Cross sought to build an oxygenator in-house for their own use in the laboratory and eventually the OR, not to create a marketable clinical technology.

Two Ohio research groups were using the KCO experimentally by March 1957, that of Jay Ankeney at University Hospitals, Cleveland and a team led by Howard D. Sirak at Ohio State University in Columbus.[15] Ankeney, who launched an open heart surgery program at University Hospitals in June 1956, used the KCO for both experimental physiological research and in his surgical practice.[20] One of many pioneering cardiac surgeons working in Cleveland during the mid-20th century, Ankeney was an early proponent of "off-pump" open heart surgical techniques, operating without the support of a heart-lung machine patients beginning in 1969.[21] During the late 1950s, however, Ankeney and other surgeons relied on dependable heart-lung machines including the KCO in their clinical practice and research.

As word of Kay and Cross's success using the KCO for open heart surgery spread, an increasing number of physicians reached out to the pair in hopes of purchasing a unit for their own labs or surgical programs. Early champions of the KCO included Robert E. Gross from the Children's Medical Center in Boston.[22] By August 1957, multiple US hospitals and universities were actively using the KCO and Pemco had sold 44 units.[23] Demand for the oxygenator surprised the pair as well as Pemco, St. Luke's Hospital administrators, and the Cleveland Area Heart Association, which partially funded the KCO research program.[12] The device had sold so well that in the spring of 1957, these stakeholders were suddenly required to agree on how the KCO should be credited as well as negotiate profit-sharing agreements.[15]

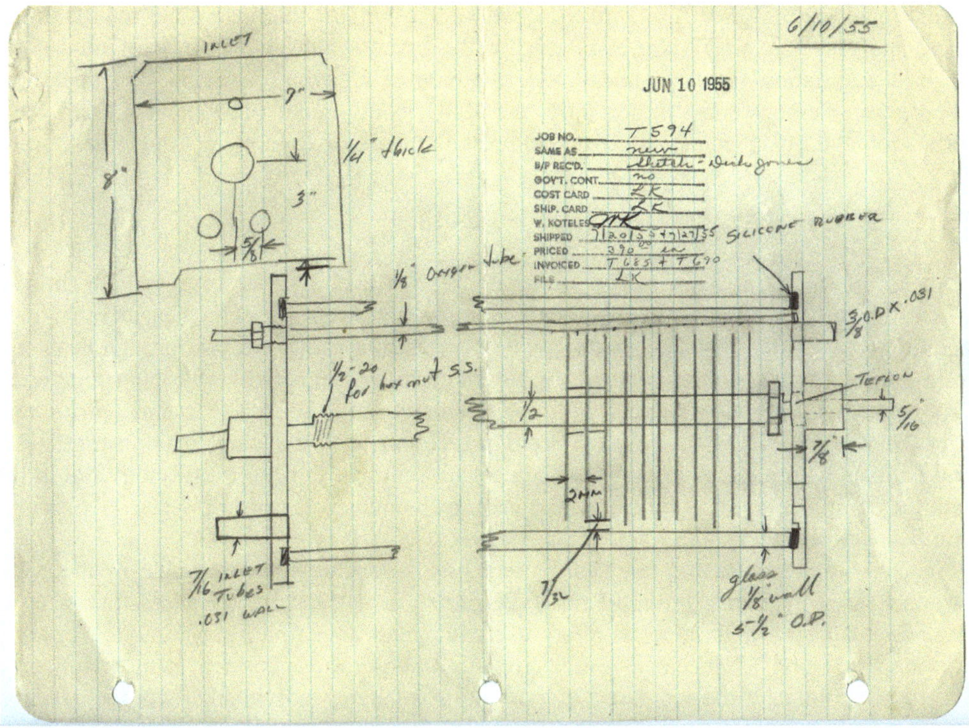

FIGURE 1.7 Kay-Cross Oxygenator design drawing, 1955. Courtesy Dittrick Medical History Center, Case Western Reserve University.

Over 100 units were sold by March 1958 as more hospitals established open heart surgery programs, a trend that continued well into the 1960s. Pemco sold over 1000 KCOs, manufacturing the devices until the 1970s and working with Kay, Cross, and other Cleveland-area cardiac surgeons on a wide range of new devices.

Other oxygenators were available by 1957 from small-scale manufacturers, including the DeWall Bubble Oxygenator, which could be purchased through Phelan Manufacturing Corporation in Minneapolis, Minnesota.[24] Research groups continued to design and manufacture oxygenators of their own design in-house, or build devices based on published designs for experimental and sometimes clinical use. This "do it yourself" approach was only practical for surgical teams housed in large, research-oriented hospitals, which had the staff, funding, and technical infrastructure to build a device. For many small, independent, or underresourced EC researchers, building their own oxygenator was not an option. A significant number of physicians and researchers interested in EC and open heart surgery couldn't or weren't interested in developing their own oxygenator or perhaps wanted to launch a cardiac surgery program at their hospitals as quickly as possible. Some groups experimenting with surgical techniques, or membrane and bubble oxygenator design were in need of a reliable "baseline" oxygenator when trialing new technologies and techniques. This created a market for a trusted, affordable oxygenator in the late 1950s. The KCO was a good option for many in the marketplace due to its reputation for reliability, ease of operation and maintenance, in addition to its commercial availability.

A practical solution

As many internationally prominent researchers studied the physiological effects of EC and developed safer, more efficient oxygenators using membrane technology, countless hospitals relied on bubble and filming oxygenators to build open heart surgery programs in the late 1950s. Three surgical groups in Ohio, those led by Joseph Sirak at Ohio State University in Columbus, Don Effler at the Cleveland Clinic, and Jay Ankeney at University Hospital in Cleveland were using the Pemco manufactured KCO in surgery as early as June of 1956, reporting their experiences back to Cross and Jones.[25] Both groups also welcomed visitors to the St. Luke's laboratory into their ORs to observe the KCO in action when Cross was unavailable to do so. The KCO was used for EC by surgeons at the Cleveland Clinic well into the 1960s, including Donald Effler and Laurence Groves, supporting the development of cardiac bypass surgery.[26]

Based on the number of experimental surgical groups that reported using the KCO in the clinical literature, the KCO was much relied-upon by surgeons and other cardiac innovators during the late 1950s through the early 1970s. The commercial availability of the device and its well-established reputation for reliability and simple operation only partially explain its popularity. Additional factors directly tied to the collaborative, interdisciplinary, and modification-friendly approach of Cross, Jones, and Kay played a significant role in the success and significance of this often-forgotten device. These included easy translation from lab to OR, relative ease of cleaning, assembly and operation, and support for new users of the device. Perhaps most importantly, the KCO team sought out feedback from other researchers and encouraged modification and tinkering as a means of improving the device and patient outcomes.

Ease of translation from lab to operating room

Kay, Cross, Jones, and other members of the KCO team held a strong conviction that in order for any new surgical technique or device to be successful in the OR, the entire operating team would need to spend considerable time together in the lab, practicing technique, becoming proficient with new devices, and learning to work through problems as a unit. Cross emphasized the importance of hands-on training as a group in correspondence with physicians interested in purchasing the KCO or seeking his advice on getting started with open heart surgery. The KCO manual, written by Cross, Jones, and technicians from Pemco and St. Luke's Hospital strongly recommended that clinicians practice using the device for EC on experimental animals noting, "Use of the pump oxygenator clinically without sufficient trial in the laboratory to become familiar with the equipment and proficient in perfusion techniques will end in disaster. Six months of intensive laboratory preparation is not excessive."[27]

Simplifying the transition from laboratory to OR and between patient populations was a core value of the KCO design team as reflected in the product range available through Pemco. Beginning in 1957, Pemco offered four sizes of the oxygenator ranging from a nine-inch canister model for laboratory use with small dogs to a 21-inch model for adult surgical patients.[19] As open-heart surgery using EC became routine for adult surgeries, a 26-inch canister for large adults was later added to the line to accommodate the increased blood volume of large individuals. Kits for adapting the 9- and 21-inch oxygenators, referred to as "conversion assemblies," provided flexible, cost-effective options for hospitals in the early stages of EC-supported surgeries (Fig. 1.8). The adaptability of the KCO also made it ideal for research groups studying the physiological effects of EC as the configuration of the device could be tailored to more

FIGURE 1.8 Pemco Heart Lung Machine c.1960. Note the Kay-Cross Oxygenator on top of the cabinet. Two DeBakey-style pumps are visible at either end of the control panel. Courtesy Dittrick Medical History Center, Case Western Reserve University.

precisely control the oxygenation, flow rate, and reservoir volume. Thus, the KCO was used extensively for physiological and surgical research well into the 1970s.[11]

(Relative) Ease of cleaning, assembly, and operation

The cleaning and sterilization of oxygenators was a challenging problem for surgical groups as the demand for open heart surgery rose, decreasing the downtime between patients and increasing the overall cost of relying on single-use disposable bubble oxygenators. As researchers experimented with using new materials in oxygenators and other heart-lung machine components, OR technicians and surgical supply managers experimented in tandem to identify safe and effective sterilization techniques for these new technologies. Many new plastics available by the late 1950s were expensive and available only in limited quantities, forcing many hospitals to reuse DeWall type oxygenators and other plastic components. Also, these materials could not withstand the extreme heat of autoclaving, which restricted technicians to using chemical solutions to eliminate bacteria, processes referred to as "cold sterilization."[28] Existing liquid sterilization products such as benzalkonium chloride (Zephiran) were insufficient for repeated use, sometimes resulting in deadly bloodstream infections caused by *Pseudomonas* and other resilient, biofilm-forming bacteria lodged inside plastic bubble oxygenators.[29]

The materials used in the KCO, stainless steel, borosilicate glass, Teflon disks, nylon bearings, and silicone-based antifoaming coatings were fully autoclavable.[27] The instruction manual, authored and regularly updated by the interdisciplinary KCO team, contained detailed instructions on cleaning, assembly, and sterilization of the device. It also included helpful tips on common related problems, such as warping of the nylon bearings due to autoclaving at temperatures above 275°F. These procedures fit easily into the existing infrastructure of US hospitals, which had well-established systems for washing surgical supplies with strong detergents to remove pyrogenic proteins, autoclaving and drying clinical equipment, and ensuring that sterile tools were on-hand near the OR. The KCO team, with their extensive experience using the device within the St. Luke's lab and surgical suite, used their practical knowledge to design an oxygenator that would work for most open heart surgery groups without the need to further invest in specialized sterilization equipment, staff, or spaces. Ease of sterilization, assembly, and storage were another major selling point for the KCO. Unlike other nondisposable oxygenators, it could be autoclaved while fully assembled, enabling hospitals to consistently have the device on-hand for emergency surgeries. This was not the case for other oxygenator options including prototype devices built in-house by other research groups and other types available for purchase, such as the Mayo-Gibbon rotating screen oxygenator.

George H. A. Clowes Jr. and William Neville at Western Reserve University School of Medicine (now Case Western Reserve University) and City Hospital in Cleveland focused their research on the membrane diffusion approach. Clowes and Neville's experimental studies of the oxygen diffusion capabilities of new plastic materials were critical to the later development of practical membrane oxygenators and other groundbreaking cardiac innovations.[30] Though a critical step toward today's gas exchange technologies, Clowes and Neville's prototype oxygenator had little success in the OR during the late 1950s through the 1960s.[31] This was in part due to the limited availability of high-quality plastic membranes at the time and the need to maintain a high blood volume within the oxygenator to ensure adequate oxygenation during EC. The elaborate, time- and resource-intensive assembly, disassembly, and sterilization process required to safely use Clowes and Neville's membrane oxygenator also significantly contributed to its impracticality for routine open heart surgery (Fig. 1.9). Presenting at the September 1957 Conference on Extracorporeal Circulation organized by the National Institutes of Health (NIH), Clowes and Neville noted that "...recent improvements in design have made it possible to set up the equipment for perfusion of an adult patient in a period of about two hours."[32] Note that this time frame did not include sterilization of the apparatus and flushing of the membrane layers to remove cleaning solution and debris. Cleaning and sterilization of the KCO was also time-consuming compared to modern standards, however the assembly and disassembly of the device was considered quick and straightforward by users, as recorded in the clinical literature of the 1950s and 1960s.[33]

Operation of the KCO was also understood as relatively simple by its developers and external users alike. In August 1957, Cross remarked to a colleague, "It is so simple to operate that even I can run it, but Dick Jones doesn't allow me to very often."[15] Surgical and research colleagues frequently commented on the ease of operation in letters to Cross and Jones, and the KCO's user-friendly design was applauded by Gross and other attendees at the Conference on Extracorporeal Circulation in 1957.[5] Kay and Cross were strong proponents of opening the St. Luke's research lab and their ORs to surgeons and technicians alike so they could observe the operation of the KCO by an experienced team firsthand. Both surgeons also encouraged feedback and new ideas on operating procedures in professional correspondence, publications, and conference proceedings.

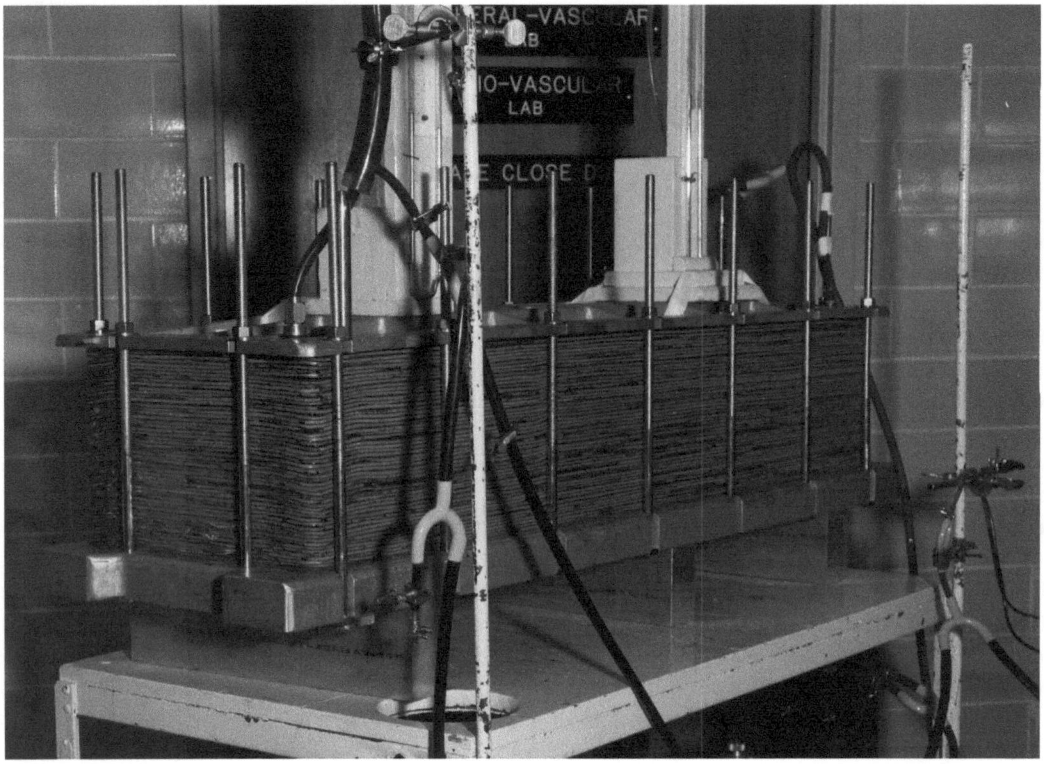

FIGURE 1.9 Clowes-Neville membrane oxygenator, c.1957. Courtesy Dittrick Medical History Center, Case Western Reserve University.

Tech support and tinkering

Kay departed St. Luke's Hospital to build a cardiac research department at St. Vincent's Charity Hospital in the late 1950s, where he also established Cleveland's first cardiac intensive care unit and developed innovative new surgical technologies. Cross and Jones continued work on the KCO at St. Luke's, and along with Koteles at Pemco, frequently answered questions and addressed concerns regarding the device from new users. Visitors to the lab and OR were frequent, with physicians and technicians sometimes remaining in Cleveland for short stints of hands-on practice in the St. Luke's lab. Most of the problem-solving took place in person in Cleveland or via correspondence, however Jones did travel to assist other users with particularly complex operational problems on a few occasions.[15] Jones, who contributed significantly to the design of the KCO as well as served as its main operator for the first few years of clinical use, was particularly well-positioned to troubleshoot operational problems. Based on the internal communications preserved in archives, Cross, Jones, and Koteles were clearly invested both in perfecting the KCO and supporting surgical teams and researchers working to get the device up and running smoothly. This "tech support" was crucial to the success of the KCO and the growth of open heart surgery programs across the US

The group's willingness to accept feedback and encouragement of tinkering and adaptation by other researchers is well-documented in the archive and clinical literature alike. Cross openly discussed and shared modifications by other users. For example, he described Sirak's use of a larger-diameter base disk to reduce blood turbulence at the NIH Extracorporeal Circulation conference, noting that this innovation was currently being investigated by the KCO group as a potential update to the original design.[5]

Kay and Cross seem to have provided nearby EC research groups and cardiac surgeons, including Sirak with preproduction KCOs in 1956 in order to further test the device and generate new ideas and feedback. As additional groups began to use the KCO in 1957, Cross routinely requested updates regarding their experiences, sharing his own findings and asking clarifying questions regarding modifications. This network of users expanded rapidly during the late 1950s, reflecting Cross's ties to other graduates of the University of Minnesota surgical program, the popularity of the KCO, and the burgeoning growth of cardiac research. Cleveland was evolving into a hub of cardiac innovation and clinical research during this era, in part due to the multiple, highly innovative EC and open heart surgery research programs active in the 1950s.

Willem Kolff, while conducting groundbreaking research on artificial organs and membrane oxygenation, also experimented with the KCO in use at the Cleveland Clinic.[9] Noting the presence of "Teflon shavings" from the oxygenator's disks and silk fragments from surgical sutures in circulating blood during open heart surgery, Kolff built a stainless steel mesh filter to use with the KCO, presenting detailed photographs and his findings, later published, at the NIH Extracorporeal Circulation conference.[5] The KCO group, who by the September 1957 conference had used the device with 104 clinical patients, had experienced no issues with debris. Shortly after returning to Cleveland after the conference, Cross reached out via telephone to Kolff to hear more about his findings as well as the filter, carefully documenting Kolff's feedback in a memo and further discussing the possible merits of related alterations with his collaborators.[15]

Jones, Cross, and in some instances Koteles continued to foster this interdisciplinary network, relying on findings, feedback, and tinkering from other researchers as they continued to perfect the KCO into the early 1960s. Cross and Jones would rely on this professional network of other cardiac innovators as they expanded their area of research to include the use of hypothermia and cardiac arrest in open heart surgery and developed new surgical equipment and devices, including heart valves.[34]

Conclusion

The professional experience of Kay and other surgeons involved in developing the KCO was certainly a factor in the group's emphasis on laboratory work and team collaboration. Cross in particular witnessed the importance of perfecting surgical technique in the laboratory, encouraging teamwork in the OR, and sharing new ideas for feedback and tinkering during his time in Wangensteen's surgical research program at the University of Minnesota. Perhaps an even stronger influence, the interdisciplinary team assembled by Kay and Cross was itself highly successful using this approach. The KCO was transformed from an experimental to a clinically viable prototype in less than a year and was commercially available with a range of useful accessories less than two years after the research program was launched at St. Luke's Hospital. Both Kay and Cross described the development of the KCO and the success of open heart surgery in Cleveland to the manufacturers, engineers, nurses, physiologists, perfusionists, anesthesiologists, and surgical technicians who directly contributed to perfecting the device. Kay and Cross frequently discussed the importance of such individuals to the success of the KCO, often crediting them by name in publications and public speeches. Many members of the research group, including anesthesiologist Ursula Kringle and laboratory technician William Goggans, shown in a 1959 photograph in Fig. 1.10, worked closely with Cross at Jones at St. Luke's Hospital for years.[12] The important role of William Koteles, the owner of Pemco, and his employees was also frequently noted by Kay and Cross, who

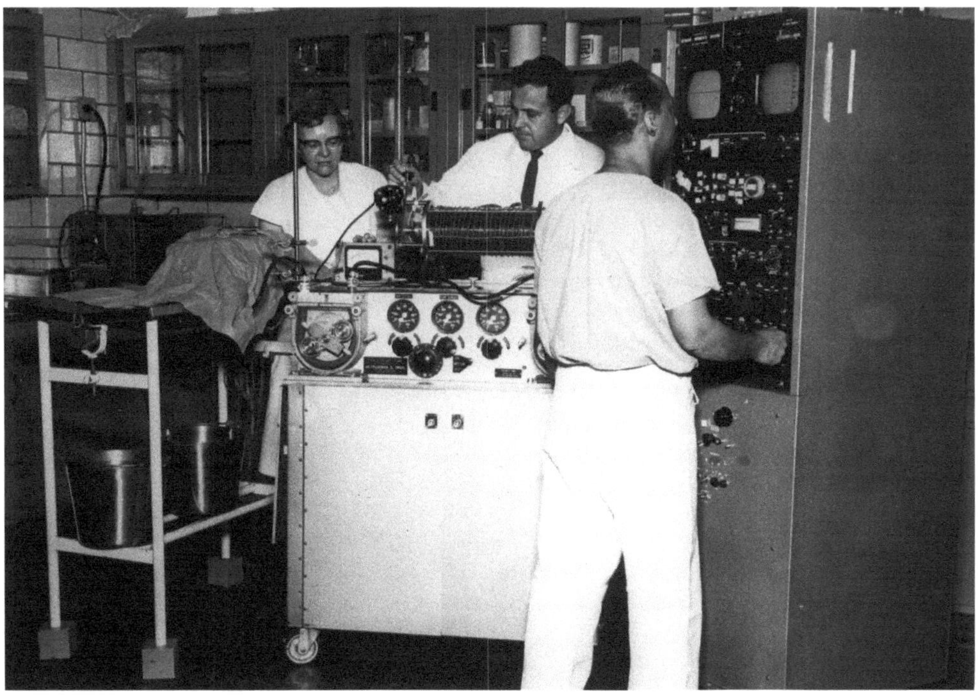

FIGURE 1.10 Ursula Kringle (left), Frederick C. Cross (center) and William Goggans (right), St. Luke's Hospital, 1959. Courtesy Dittrick Medical History Center, Case Western Reserve University.

continued to work with the company on other medical devices as late as 1970. The combination of clinical, scientific, mechanical, technical, and general problem-solving skills available in the KCO group made possible the device's success in the OR, research laboratory, and medical marketplace. These significant collaborative efforts enabled further expansion of invasive cardiovascular techniques, the development of new technologies, and the implementation of advanced cardiovascular procedures, shaping the future of cardiovascular care.

References

1. Iwahashi H, Yuri K, Nosé Y. Development of the oxygenator: past, present, and future. *J Artif Organs*. 2004;7(3):111–120.
2. Lillehei CW, DeWall RA. Design and clinical application of the helix reservoir pump-oxygenator system for extracorporeal circulation. *Postgrad Med*. 1958;23(5):561–573.
3. Gibbon JA. Application of a mechanical heart and lung apparatus to cardiac surgery. *Minn Med*. 1954;37:171.
4. Hewitt RL, Creech O. History of the pump oxygenator. *Arch Surg*. 1966;93(4):680–696.
5. Allen JG. *Extracorporeal Circulation*. Springfield, IL: Charles C Thomas, Publisher; 1958.
6. Kolff WJ, Effler DB. Disposable membrane oxygenator (heart-lung machine) and its use in experimental and clinical surgery while the heart is arrested with potassium citrate according to the Melrose technic. *ASAIO J*. 1956;2(1):13–17.
7. Kolobow T, Bowman RL. Construction and evaluation of an alveolar membrane artificial heart-lung. *ASAIO J*. 1963;9(1):238–243.
8. Kolff WJ, Berk HTJ, Welle NM, et al. The artificial kidney: a dialyser with a great area. *Acta Med Scand*. 1944;117(2):121–134.
9. Kolff WJ, Effler DB, Groves LK, Peereboom G, Moraca PP. Disposable membrane oxygenator (heart-lung machine) and its use in experimental surgery. *Cleve Clin Q*. 1956;23(2):69–97.
10. Bramson ML, Osborn JJ, Main FB, O'Brien MF, Wright JS, Gerbode F. A new disposable membrane oxygenator with integral heat exchange. *J Thorac Cardiovasc Surg*. 1965;50(3):391–400.
11. Ekert H, Montgomery D, Aberdeen E. Fibrinolysis during extracorporeal circulation: comparison of the effects of disc and membrane oxygenators. *Circ Res*. 1971;28(5):512–517.
12. Kay EB, Berne RM, Zimmerman HA, Hirose Y, Jones RD, Cross FS. Experimental and clinical experience with a rotating disk oxygenator. *ASAIO J*. 1956;2(1):94–96.
13. Kay EB. Untitled pamphlet, Earle B. Kay Papers, Dittrick Medical History Center, Case Western Reserve University.
14. Kay EB. *Thoracic and heart surgery. Medicine in Cleveland and Cuyahoga County, 1810–1976*. The Academy of Medicine of Cleveland; 1977:503–512.
15. *Research Application, Cleveland Area Heart Association*. Frederick S. Cross Papers, Dittrick Medical History Center, Case Western Reserve University; 1954.
16. Kay EB, Zimmerman HA, Berne RM, Hirose Y, Jones RD, Cross FS. Certain clinical aspects of the use of a pump oxygenator. *J Am Med Assoc*. 1956;162(7):639–641.
17. Cross FS, Berne RM, Hirose Y, Jones RD, Kay EB. Evaluation of a rotating disc type reservoir-oxygenator. *Proc Soc Exp Biol Med*. 1956;93 (2):210–214.
18. Bjork V. Brain Perfusion in dogs with artificially oxygenated blood. *Acta Chir Scand*. 1948;137(96):1–5.
19. *Pemco, Inc. Price Sheet*. Richard D. Jones Papers, Dittrick Medical History Center, Case Western Reserve University; December 10, 1957.
20. Coffin LH, Ankeney JL. The effect of extracorporeal circulation upon acid-base equilibrium in dogs. *AMA Arch Surg*. 1960;80(3):447–451.
21. Ankeney JL. Off-pump bypass surgery: The early experience, 1969–1985. *Tex Heart Inst J*. 2004;31(3):210.
22. Gross RE, Sauvage LR, Pontius RG. Cardiopulmonary bypass. *Ann Surg*. 1960;151(3).
23. Wadovick JA. Heart-lung machine pumps up confidence in cardiac surgeons. *Clevel Plain Dealer*. 1957;.
24. DeWall RA, Warden HE, Lillehei CW. *The helix reservoir bubble oxygenator and its clinical application. Extracorporeal Circulation*. Springfield, IL: Charles C Thomas, Publisher; 1958:41–56.
25. Effler DB, Groves LK. Open-heart surgery for acquired valvular heart disease: present techniques and trends. *Arch Surg*. 1962;84(2):155–169.
26. Effler DB, Favaloro RG, Groves LK, Loop FD. The simple approach to direct coronary artery surgery: Cleveland Clinic experience. *J Thorac Cardiovasc Surg*. 1971;62(4):503–510.
27. *The Kay-Cross Oxygenator Operation Manual*, Revised Edition. Richard D. Jones Papers, Dittrick Medical History Center, Case Western Reserve University; August, 1958.
28. Gibbon JH, Templeton JY. Current status of pump oxygenators in cardiac surgery and persistent problems in their use. *Prog Cardiovasc Dis*. 1958;1(1):56–65. Available from: https://doi.org/10.1016/S0033-0620(59)80036-0.
29. Keown KK, Gilman RA, Bailey CP. Open heart surgery: anesthesia and surgical experiences. *J Am Med Assoc*. 1957;165(7):781–787.
30. Clowes Jr GHA, Hopkins AL, Kolobow T. Oxygen diffusion through plastic films. *ASAIO J*. 1955;1:23–24.
31. Clowes JR, George HA. Extracorporeal maintenance of circulation and respiration. *Physiol Rev*. 1960;40(4):826–919.
32. Clowes Jr GHA, Neville WE. *Membrane oxygenator. Extracorporeal Circulation*. Springfield, IL: Charles C Thomas, Publisher; 1958:41–56.
33. Stofer RC, Lower RR, Shumway NE. The preparation, assembly, and operation of a heart-lung machine. *J Thorac Cardiovasc Surg*. 1962;43 (5):690–700.
34. Berne RM, Jones RD, Cross FS. Myocardial hypothermia in elective cardiac arrest. *J Appl Physiol*. 1958;12(3):431–436.

Chapter 2

The Critical Need for Ongoing Innovation in Cardiovascular Medicine

Patrick M. McCarthy and James L. Cox

Division of Cardiac Surgery, Department of Surgery, Bluhm Cardiovascular Institute, Northwestern University Feinberg School of Medicine and Northwestern Medicine, Chicago, IL, United States

Chapter Outline

Cardiovascular disease poses a formidable challenge. It is the leading cause of mortality in the United States with one death occurring every 36 s on average.[1] The cost of cardiovascular care is enormous and estimated to be $363.4 billion dollars per year in the United States for direct and indirect costs.[2] The psychological toll of the disease on the patient, family, and friends cannot be measured. There has been tremendous progress in the prevention and treatment of cardiovascular disease with a remarkable 56% drop in age-adjusted mortality during 1950–96, and by 22% during 1999–2013.[3] Part of these improvements are due to innovation in the care of cardiovascular patients, including prevention measures, new procedures, and new devices. But, as we witnessed through the use of extracorporeal membrane oxygenation for the sickest COVID-19 patients, the need for innovation for desperately ill patients is ever present. As Shakespeare wrote in Hamlet, "Diseases desperate grown, by desperate appliance are relieved, or not at all."

The plague of heart disease provides myriad manifestations of the disease and the challenges controlling this burden. Recent years have demonstrated the impact of innovation to treat many forms of cardiac disease, such as atherosclerosis (drug-coated stents), acquired valve disease (transcatheter aortic valve replacement for aortic stenosis, Fig. 2.1), the reduction of strokes in patients with atrial fibrillation (left atrial appendage occlusion devices), and the power of artificial intelligence (AI) to aid in the interpretation of cardiovascular imaging is just beginning. Many of these topics are covered in this outstanding book. But innovation is such a wide field in cardiovascular disease that it is not possible to encompass all of the advances in the field.

Innovation is most often equated with a new device or a new technology. Many of those are described in this textbook. But equally important are the innovative procedures which have revolutionized cardiac care, many of which were pioneered at the Cleveland Clinic.[4] Over the past several decades these include cardiac catheterization, coronary artery bypass grafting, surgical and catheter ablation of atrial fibrillation, valve repair, and minimally invasive cardiac surgery. There is an important relationship between the devices and the procedures. Medical devices may be developed to improve the procedure (e.g., AtriClip for left atrial appendage closure, valve repair rings; Fig. 2.2) or new devices may require that a procedure needs to be developed (e.g., transcatheter aortic valve replacement; ablation devices used in the Convergent Procedure). A critical aspect of many of these procedures and devices is that they were either conceived by the cardiovascular clinicians, or developed in partnership with engineers and industry. Again, Cleveland Clinic has led the way with a pioneering group of inventive surgeons and cardiologists.

Examples of innovation in cardiovascular disease prevention and treatment are all around us in our everyday practice. The Cleveland Clinic has a decades-long history of pioneering work in the laboratory to develop mechanical circulatory support, and early application of these devices in the clinical setting.[4] The sections on mechanical circulatory support and cardiovascular perfusion are particular strengths of the book. This is the most challenging target in cardiovascular medicine. The engineering demands to provide a reliable biocompatible, implantable, mechanical circulatory

Advances in Cardiovascular Technology. DOI: https://doi.org/10.1016/B978-0-12-816861-5.00025-3

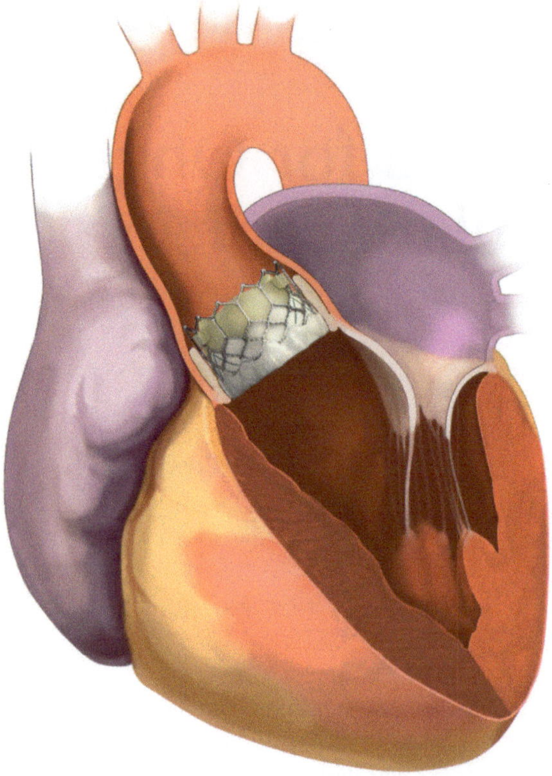

FIGURE 2.1 Transcatheter aortic valve replacement (TAVR) revolutionized the treatment of many patients with severe aortic stenosis. Many surgeons and clinicians were skeptical that this could succeed, but well performed clinical trials established the safety and effectiveness of the technology and procedure for elderly patients.

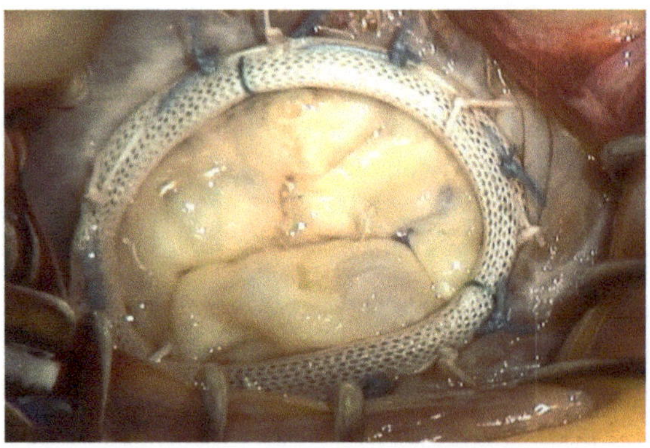

FIGURE 2.2 Innovations have been prominent in heart valve surgery. The use of mitral valve repair annuloplasty rings has led to excellent early and late durability, and for many patients has restored their life expectancy to normal.

support device are formidable. To achieve that goal seems to be a never-ending quest. Much progress has occurred since the earliest attempts in the 1960s, but much still remains to be done.

While there are many innovative surgical and interventional devices and techniques we should not forget the power of a well-coordinated, multidisciplinary team. The Cleveland Clinic motto from the days of the founders 100 years ago, "To act as a unit," has resonated over the years.[5] In cardiovascular disease this team approach became prominent as Dr. Norman Shumway and his Stanford team pioneered heart transplantation and mechanical support.[6] A more recent innovation is the development of the "heart team" to best treat complex patients with structural heart disease or even coronary artery disease.[7] The multidisciplinary team may oftentimes make the best decision about a difficult condition requiring a transcatheter or surgical approach.

The past 10 years have witnessed a disruption in the treatment of patients with severe aortic stenosis as many of those patients are now candidates for treatment with a transcatheter aortic valve replacement. Also, a subpopulation of patients with mitral regurgitation may be approached with transcatheter edge-to-edge repair, especially those with heart failure and secondary mitral regurgitation and those with degenerative mitral regurgitation who are at prohibitive or very high risk for surgery. The complexity of the mitral valve, the various pathophysiology conditions, the challenge with the annulus "landing zone," and sometimes the left ventricular cavity and outflow size have slowed the progress of transcatheter mitral valve replacement, however (Fig. 2.3). This procedure is much more challenging than transcatheter aortic valve replacement (TAVR). Look for more innovations in transcatheter and surgical devices to treat mitral disease in the next several years.

Treatment of cardiac dysrhythmias beyond medications began with creative surgical procedures using cryoablation, evolved into catheter-based radiofrequency ablation, and currently new technologies such as ultralow-temperature

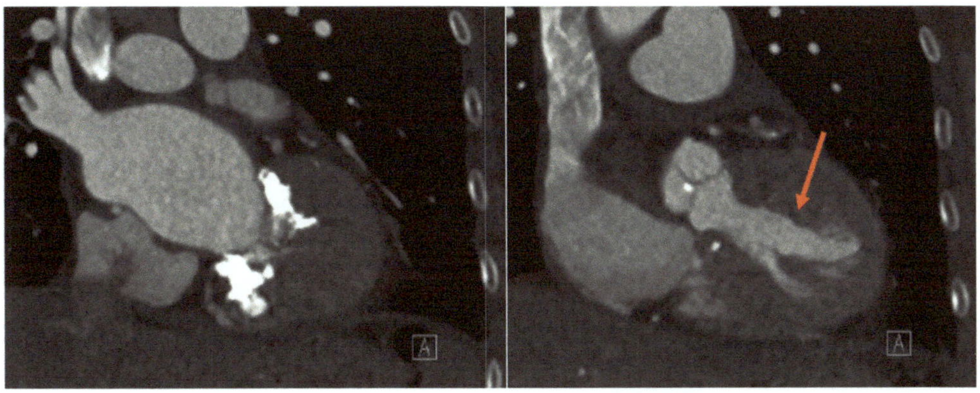

FIGURE 2.3 Transcatheter mitral valve replacement (TMVR) is conceptually a promising technology, but practical pitfalls are much more common than with TAVR. For instance, this 84-year-old frail patient with extensive mitral annular calcification and a small left ventricular cavity would ideally be treated with a less invasive approach, but TMVR technology today doesn't provide a solution. *TAVR*, Transcatheter aortic valve replacement.

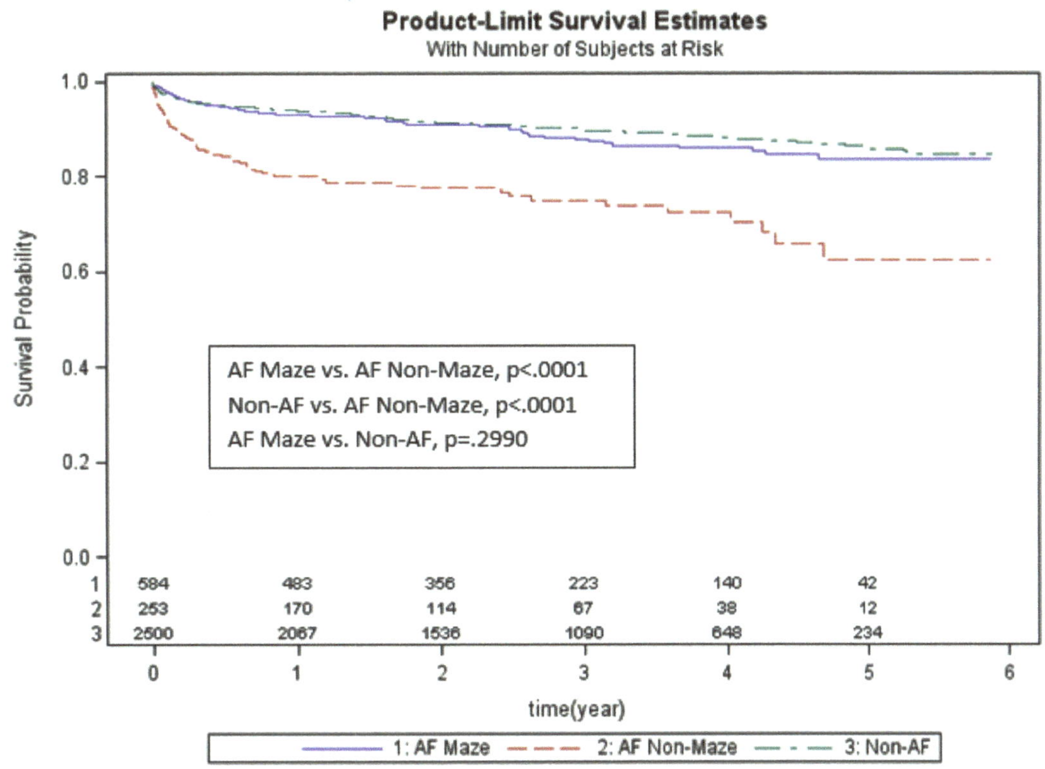

FIGURE 2.4 Surgical ablation of atrial fibrillation (AF) with concomitant cardiac surgery has been shown to be safe, reduces late stroke, and improves late survival. In this Northwestern study, after matching, survival for patients with untreated AF concomitant to other cardiac surgery was markedly worse than for those with AF ablation ($P < .0001$). AF ablation patients had similar survival to patients who had never had AF before surgery implying that survival was returned to expected in AF treated patients.[9]

cryoablation and pulsed field ablation are positioned to change our treatment paradigm yet again.[8] With the evidence of the safety and effectiveness of surgical treatment of atrial fibrillation now increasing (Fig. 2.4),[9,10] it is more commonly adopted across the world, especially with left atrial appendage closure.[11]

The future of cardiovascular innovation is bright and there are still many targets that need better therapy. At Northwestern Medicine, we have been strongly influenced by the potential of machine learning and AI using deep learning to change how we diagnose, manage, and treat cardiovascular disease. We formed the Center for Artificial Intelligence in Cardiovascular Disease within the Bluhm Cardiovascular Institute of Northwestern. Clinical projects have included the use of AI enabled "smart stethoscopes" to aid in the diagnosis of heart murmurs.[12] New AI-enabled imaging equipment has allowed not just the interpretation of echocardiographic images, but importantly also AI-directed image acquisition. Since the AI echo system directs the user to the right view, and automatically captures the best view, examinations can be performed by sparsely trained personnel with accuracy equal to that of expert sonographers.[13] This innovation will allow echocardiograms to be acquired much more widely. AI-enabled wireless biosensors are being deployed to monitor cardiovascular patients in the Northwestern cardiac intensive care units, and will be extended to the floor and home, to help reduce hospital readmissions.

Finally, there is a spirit of innovation in cardiovascular disease and it succeeds best when there is a multidisciplinary environment of learning and support. We have begun a fellowship with the Computer Science Department of Northwestern University to train three cardiovascular residents or fellows per year on machine learning and AI (Fig. 2.5). They obtain a Master of Science in Artificial Intelligence. Well-trained clinical practitioners like this will allow us to make the best use of innovation, just as practitioners have worked alongside engineers to create some of the most innovative devices that have disrupted cardiovascular care. Where this leads is not always clear as we adapt to

FIGURE 2.5 Innovation and product development in cardiovascular disease requires expertise in medicine, engineering, business and law. To prepare our trainees for the new innovations that artificial intelligence (AI) will present we began a fellowship in AI in cardiovascular medicine leading to a Master of Science in AI (MS-AI) degree. Here, the team of three MS-AI trainees are gathered together during the pandemic to develop new algorithms.

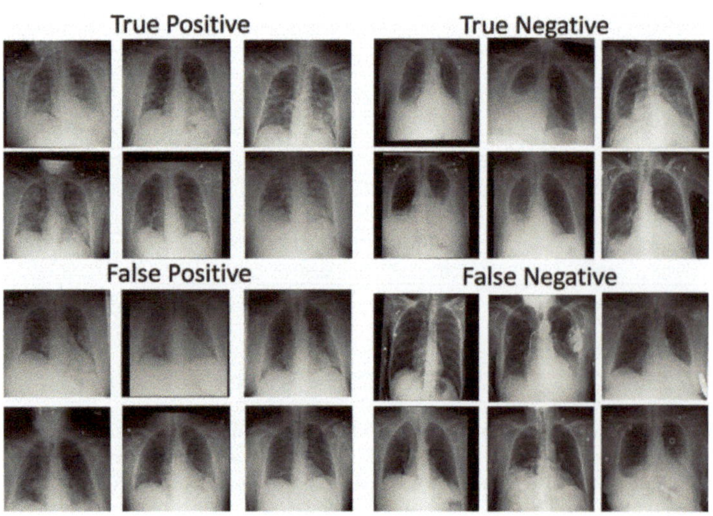

FIGURE 2.6 Using Deep Learning AI, one of the Northwestern cardiology fellows was able to develop an algorithm for the early detection of Covid-19 positive chest films. The algorithm accuracy was the same as radiologists specialized in chest CT interpretation. Early during the Covid-19 pandemic the algorithm was published and made public to foster widespread adoption.[14]

new problems. For instance, early during the Covid-19 pandemic, one of the AI cardiology fellows used deep learning to interpret chest X-ray findings to determine whether COVID-19 was present with an accuracy level that was better than expert computerized tomography (CT) radiologists (Fig. 2.6).[14]

We will continue to innovate as problems and opportunities arise. We will continue to find new ways to innovate that have not yet been conceived. The burden of cardiovascular disease is overwhelming and for those who are interested in innovative solutions to the myriad problems, this book is a good starting point.

Conflict of interest

Dr. McCarthy: Edwards Lifesciences: royalties and speaking fees; Atricure and Medtronic: speaking fees; Abbott REPAIR-Mr Trial PI (unpaid).

Dr. Cox: Consultant and Stockholder: Atricure, Inc. and SentreHEART, Inc.; Consultant, Stockholder and Board of Directors: Adagio Medical, Inc., PAVmed, Inc., Lucid Diagnostics, Inc.

References

1. *Heart Disease Facts*. https://www.cdc.gov/heartdisease/facts.htm#: ~ :text = One%20person%20dies%20every%2036,United%20States%20from%20cardiovascular%20disease.&text = About%20655%2C000%20Americans%20die%20from,1%20in%20every%204%20deaths. Accessed September 21, 2021.
2. Virani SS, Alonso A, Aparicio HJ, et al. Heart disease and stroke statistics-2021 update: a report from the American Heart Association. *Circulation*. 2021;143:e254–e743.
3. Mensah GA, Wei GS, Sorlie PD, et al. Decline in cardiovascular mortality: possible causes and implications. *Circ Res*. 2017;120:366–380.
4. *Cleveland Clinic 100 years*. https://my.clevelandclinic.org/about/history Accessed September 21, 2021.
5. Hartwell SW. *To Act as a Unit: The Story of the Cleveland Clinic*. 4th ed. Philadelphia, PA: W.B. Saunders Company; 1985.
6 Shumway N. Foreward. In: Smith JA, McCarthy PM, Sarris GE, Stinson EB, Reitz BA, eds. *The Stanford Manual of Cardiopulmonary Transplantation*. Armonk, NY: Futura Publishing Company; 1996.
7. Holmes Jr. DR, Mack M. The truly functional heart team: the devil is in the details. *J Am Heart Assoc*. 2020;9:e05035.
8. Ramirez FD, Reddy VY, Viswanathan R, Hocini M, Jais P. Emerging technologies for pulmonary vein isolation. *Circ Res*. 2020;127:170–183.
9. Lee R, McCarthy PM, Wang EC, et al. Midterm survival in patients treated for atrial fibrillation: a propensity-matched comparison to patients without a history of atrial fibrillation. *J Thorac Cardiovasc Surg*. 2012;143:1341–1351. Discussion 1350-1341.
10. Badhwar V, Rankin JS, Damiano Jr. RJ, et al. The society of thoracic surgeons 2017 clinical practice guidelines for the surgical treatment of atrial fibrillation. *Ann Thorac Surg*. 2017;103:329–341.
11. Whitlock RP, Belley-Cote EP, Paparella D, et al. Left atrial appendage occlusion during cardiac surgery to prevent stroke. *N Engl J Med*. 2021.
12. Chorba JS, Shapiro AM, Le L, et al. Deep learning algorithm for automated cardiac murmur detection via a digital stethoscope platform. *J Am Heart Assoc*. 2021;10:e019905.
13. Narang A, Bae R, Hong H, et al. Utility of a deep-learning algorithm to guide novices to acquire echocardiograms for limited diagnostic use. *JAMA Cardiol*. 2021;6:624–632.
14. Wehbe RM, Sheng J, Dutta S, et al. DeepCOVID-XR: an artificial intelligence algorithm to detect COVID-19 on chest radiographs trained and tested on a large U.S. clinical data set. *Radiology*. 2021;299:E167–E176.

PART II

Cardiovascular Perfusion

Chapter 3

Progress in Cardiovascular Perfusion and Technology

Vinci Naruka[1], Philip Hartley[1], Panagiotis G. Kyriazis[1,2], Guiqing Liu[1], Jacob Chacko[1], Jonathan Afoke[1] and Prakash P. Punjabi[1,2]

[1]*Department of Cardiothoracic Surgery, Hammersmith Hospital, Imperial College Healthcare NHS Trust, London, United Kingdom,*
[2]*National Heart and Lung Institute, Faculty of Medicine, Imperial College London, London, United Kingdom*

Chapter Outline

Introduction

Historically, several attempts were made to perfuse and aerate various organs since the late 18th and early 19th century. In 1895 Jacobi reported his experiment of blood flow through an artificial ventilated lung; in 1926 a preservation machine for both organs and complete animals was reported. In addition, the anticoagulant properties of heparin (discovered around the same period) offered the necessary stimulus for perseverance over the next 30 years. Lillehei and colleagues reported their creative controlled cross-circulation technique and reported a series of 32 patients successfully operated upon by controlled cross-circulation (Fig. 3.1). Development of the bubble oxygenator led to the early demise of this important milestone in cardiac surgery. The real breakthrough for cardiac surgery though was the development of cardiopulmonary bypass (CPB). John Gibbon heralded the start of modern-day open heart surgery by pioneering the use of the heart-lung machine. Although unsuccessful in his first patient, his second patient had his atrial septa defect (ASD) closed successfully in 1953. It was this successful operation that encouraged a number of researchers, including Kirklin and colleagues, to launch their open-heart program in 1955 using a modified Gibbon-IBM heart-lung machine.[1]

In the UK approximately 14,000/year coronary artery bypass graft (CABG) operations and just over 10,600/year valve procedures are performed. CABG surgery still remains the most common option; however, there has been observed a gradual decrease in the number of cases performed over time and this is entirely due to the drop of elective cases as the number of emergency cases is unchanged. Since 2016, the numbers of isolated aortic valve

Advances in Cardiovascular Technology. DOI: https://doi.org/10.1016/B978-0-12-816861-5.00024-1

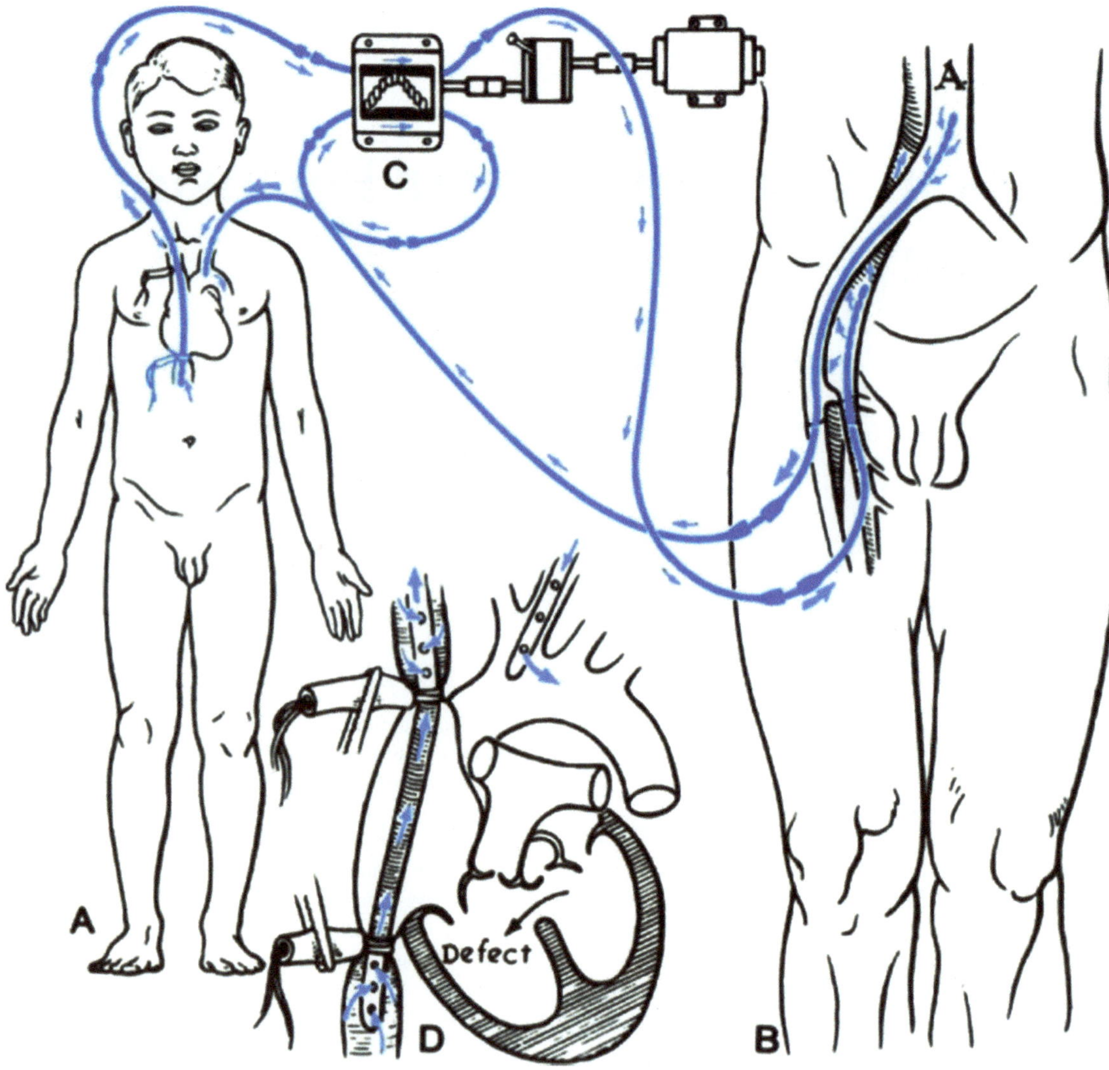

FIGURE 3.1 Classification of MiECC circuits.[2] Lillehei controlled cross circulation technique. (arrows depict direction of blood flow) (A) Patient; (B) Donor; (C) Circuit; (D) Cardiac defect. (Note that the modular type IV circuit is literally type III with a standing-by component, used only when necessary.) *X*, pump; *O*, oxygenator; *C*, cardioplegia; *T*, bubble trap/air-removing device; *V*, vent (aortic/pulmonary); *S*, soft-bag/reservoir; *H*, hard-shell/reservoir; *MiECC*, minimal invasive extracorporeal circulation.[1]

replacement (AVR) and combined AVR plus CABG operations have fallen, with the reason behind this change being the increase of transcatheter aortic valve implantation (TAVI) procedures. 2018/19 was the first time TAVI overtook surgical aortic valve replacement for the UK as whole.[3]

There is a trend toward minimally invasive procedures and mini circuits with some surgeons performing off-pump surgery. It is estimated that more than one million cardiac operations are performed on an annual basis globally with standard CPB with low operative mortality rates (approximately 1%).[1]

The CPB circuit as a whole has seen unprecedented improvements by the contribution of engineering and clinician. Expansion of the CPB to extracorporeal life support devices is a frontier that is being achieved. It is important that we continue to specifically optimize, individualize, and personalize the CPB circuit to patients, patients of all ages, patients of all weights, and patients of all body mass index (BMIs). One size does not fit all and we have to expand continuously, expand our efforts in the fields of technological improvements cannulation, and harmonize the interaction between the human and the machine in terms of blood and artificial surface (Fig. 3.2).

An ageing patient population with co-morbidities often requiring complex procedures benefits from reduction of stress by implementation of various manoeuvres and technologies. The challenge remains to minimize the secondary effects in the intensive care unit, the ward, and the long-term successful rehabilitation with minimal physiological,

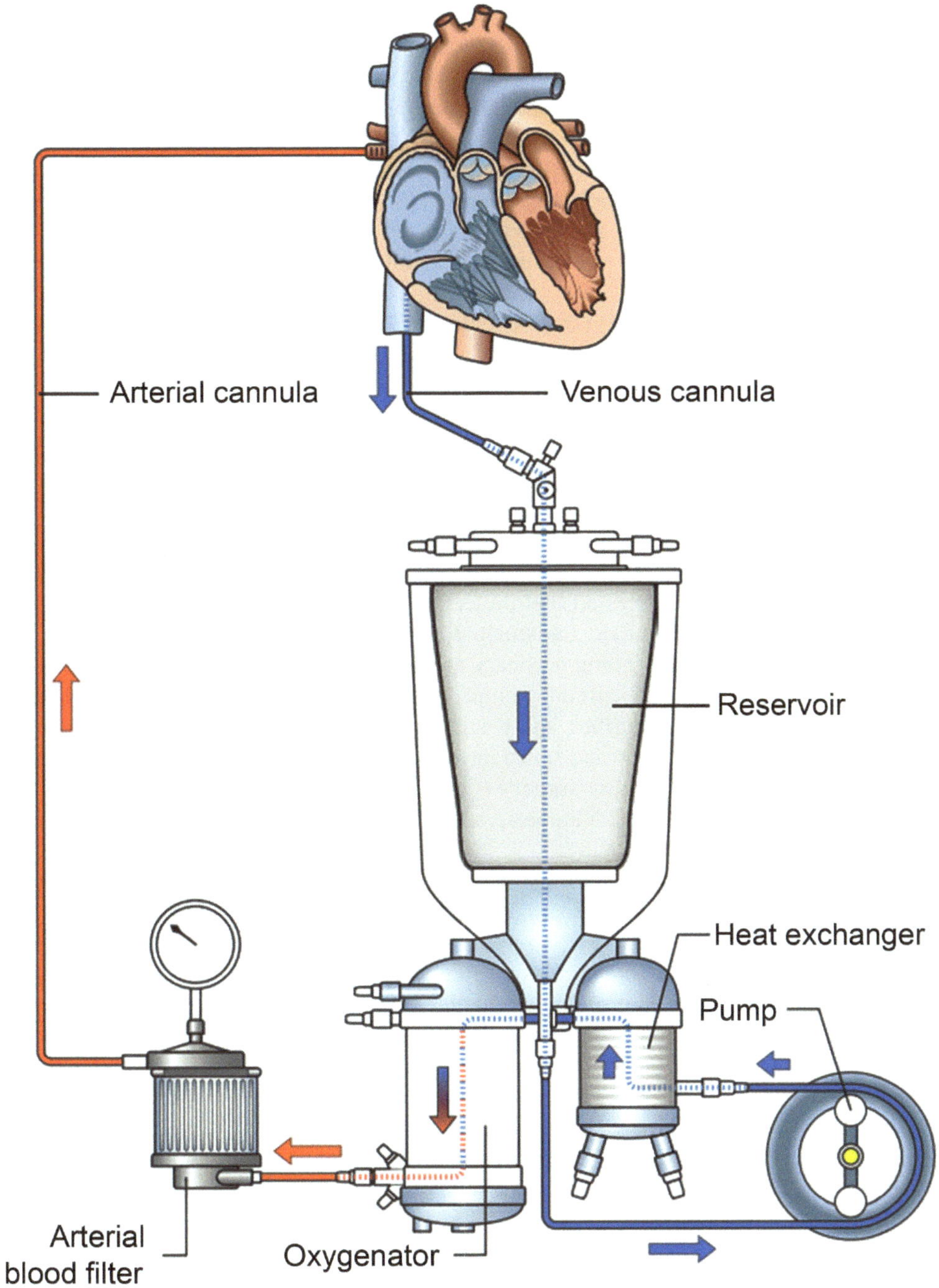

FIGURE 3.2 Overview of cardiopulmonary bypass circuit and its various components.[1]

neurological, and biochemical impairment to the patient, so that we could reach a standardization. The final goal is to truly achieve a therapeutic tool that benefits and where the risks are minimized.

Pump

Reducing hemolysis

One of the complications associated with the use of the CPB machine is hemolysis of blood components due to blood being propelled by the pump. Advances in the pump design have occurred in the last few decades to answer for this. The CPB unit is composed of a pump that is used to propel blood in the extracorporeal circuit; the two most common types are the roller pump and centrifugal pump.

The roller pump propels blood by the action of two rollers sequentially compressing a segment of tubing which causes the forward movement of blood. Its main advantages include the low cost, low prime volume and the ease of priming, it does not allow retrograde flow, and it provides a reliable constant flow rate.[4] In contrast the centrifugal pump functions by producing a constrained vortex within a polycarbonate structure, with impellers or a nest of plastic cones that are magnetically coupled with an electric motor.[5] These impellers or cones rotate rapidly generating a vortex that exerts an outward centrifugal force on the blood which exits through an outlet, resulting in a forward movement. Its main advantages include being nonocclusive and it can pump against any resistance, it is pre- and postload dependent, so any changes in resistance/pressure in the pump will alter blood flow rate so it is less likely to result in vessel dissections or pump tubing rupture. However it is more expensive, more difficult to prime, and requires higher prime volume and risk of retrograde flow.[4]

Most cardiac surgeries are carried out using a roller pump in the CPB circuit. However it is known to generate shear forces due to occlusion of the tubing between the rollers, and causes hemolysis, lipid membrane ghosts, and spallation from the tubing with increased risk of particulate embolisms, which can contribute to impaired microcirculation.[4,5] Hemolysis was examined under different flow and pressure conditions in vitro and it was noted that hemolysis was related to the duration of blood exposure to shear stress, pump pressure of the outflow, and flow rate of the pump. The centrifugal pump is associated with reduced inflammation and less trauma on the red blood cells when compared to roller pumps.[6] Further trials have demonstrated centrifugal pumps to have reduced microbubble generation, reduced complement activation, and better preservation of platelet numbers.

However, there is a multitude of studies that have attempted to demonstrate the benefits of one pump over the other. A prospective randomized trial of 1000 patients suggested that the centrifugal pump is superior for routine adult cardiac surgery with improved blood loss with reduced blood transfusions, renal function, and reduced neurological complications. A single center study of 4000 patients undergoing coronary and/or valve surgery demonstrated no difference in hospital mortality between the two pumps but centrifugal pumps are associated with reduced rate of major neurological complications in adult cardiac surgery. A further study suggested that centrifugal pumps reduce the occurrence of neurological complications such as permanent and transient neurological deficit, coma, and delirium. In contrast, a randomized trial for elective coronary artery bypass surgery (CABG) demonstrated no effect of pump type on blood loss, need for transfusion, hemoglobin or platelet numbers, and also no difference in neuropsychometric testings.[7]

A best-evidence topic analyzed 15 papers and concluded that there is no evidence for the benefit in using a centrifugal pump over a roller pump in elective CABG in terms of blood loss, clinical outcomes, or neurological problems.[8]

A recent meta-analysis of randomized controlled trials for the two pump types that included 1868 patients suggested no significant difference for hematological data (platelet counts, plasma free hemoglobin, white blood cell count, and hematocrit), postoperative blood loss, blood transfusions, neurological outcomes, or mortality.[9]

The theoretical benefit of reduced inflammatory process and hemolysis by using centrifugal pumps in prolonged CPB surgeries is yet to be proven. A randomized trial comparing the two types of pumps in patients undergoing pulmonary endoarterectomy with prolonged CPB and deep hypothermic circulatory arrest, demonstrated a reduced inflammatory response to the standard roller pumps but no significant differences in clinical outcomes.[10] Further multicenter trials are required to prove such benefits in the case of prolonged CPB surgeries.

Reducing microcirculation injury

The perfusion mode of choice for patients undergoing cardiac surgery with a CPB machine is a nonpulsatile flow. However, several studies have demonstrated nonpulsatile flow causing reduced microcirculation, decreased hemodynamics resulting in capillary collapse, microvascular shunting, and activations of inflammatory mediators.[11] The reduced microcirculatory perfusion during CPB is associated with anaerobic metabolisms, increased lactate, and multiorgan failure.[12]

Using a pulsatile flow during CPB that mimics the human blood flow may have a beneficial effect on microcirculation as the pulsatility promotes arterioles patency, which are collapsed in a nonpulsatile flow, and thus reduce the stress from the decreased release of endothelial vasodilators.[2] Pulsatile flow has been reported to improve blood flow to vital organs, reduce systemic inflammatory responses and cytokines activation, reduce endothelial damage, and generate mode hemodynamic energy, which maintains improved microcirculation of end-organs.[13]

The superiority of pulsatile flow was also demonstrated in large clinical trials with decreased incidence of myocardial infarction, reduction in the use of intraaortic balloon pumps or the need for inotropic support, and lower in-hospital mortality compared to nonpulsatile flows.

Although perfusion during CPB with a pulsatile flow could be beneficial, there are ongoing controversies about the mechanisms that favor its effects and demonstrating its benefits on the microcirculation. Nonpulsatile flow results in reduced endothelial shear stress, leading to endothelial dysfunction and increased microvascular resistance. Further experimental studies showed pulsatile shear stress induces endothelial nitric oxide synthase expression, and vasoactive substances.[14] This was also demonstrated in a clinical comparative study for patients on CPB with pulsatile flow showing increased endothelial nitric oxide release compared to nonpulsatile flow.[15]

Many studies have attempted to analyse microcirculatory perfusion in pulsatile and non-pulsatile flow. A randomised crossover study using recorded sublingual human microcirculation (recorded by sidestream dark field imaging) demonsrated no difference in microcirculation perfusion after 10 min of either pulsatile or non-pulsatile CPB flow. Another study evaluating the gastric mucosa using gastric tonomentry and gastric mucosal blood flow (using laser Doppler velocimetry) showed prevention of microvascular perfusion with pulsatile flow.[16]

More recent studies on sublingual mucosal microvascular perfusion in patients on CPB showed pulsatile flow preserving the microcirculatory perfusion throughout the early postoperative period irrespective of systemic hemodynamics. It also demonstrated an increased oxygen consumption during pulsatile flow, which favors the hypothesis that pulsatile flow CPB is beneficial for recovering from microcirculatory heterogeneity and thus reduces the risk of a reduced oxygen extraction compared to nonpulsatile flow.[17] Pulsatile perfusion has been shown to be superior by attenuating the systemic inflammatory response during CPB and reducing peak lactate levels, thus better optimizing microvascular perfusion in high-risk cardiac surgical patients[18]; however, more recently a prospective study for patients undergoing elective CABG surgery showed pulsatile flow with roller pumps having greater hemolysis with increased heme and globin levels.[19]

A pulsatile flow that maintains the human body's physiological state during CPB is theoretically logical and has shown to be beneficial in reducing microcirculatoy injury clinically; however, it is still a controversial topic and large randomized clinical trials have to address if such early postoperative benefits translate into improved outcomes in patients undergoing more complicated and high-risk cardiac surgery.

Circuit

Reducing systemic inflammatory response syndrome

Mini cardiopulmonary bypass circuit

Although the mortality rate of patients undergoing cardiac surgery has been decreasing over the past few decades, cardiac surgery patients still do experience postoperative morbidity, which occurs due to the risk of iatrogenic harm, primarily as a result of ischemia reperfusion injury (IRI) and systemic inflammatory response (SIRS). Cardiac surgery with conventional extracorporeal circulation (CECC) provokes systemic inflammatory response syndrome due to the activation of stress pathways associated with postoperative end-organ complications, and the latter are triggered by operative surgical trauma and IRI, but are further exacerbated by the interaction of air, blood, and synthetic components in the CECC apparatus.

Minimally invasive extracorporeal circulation (MiECC) systems integrate contemporary innovations in CPB technology into a strategy that aims to reduce SIRS and improve end-organ protection during cardiac surgery. They comprise a closed circuit with biologically inert blood contact surfaces and reduced priming volume; a centrifugal pump; membrane oxygenator; heat exchanger; venous bubble trap or venous air removing device; cardioplegia system and shed-blood management device. Type IV hybrid modular MiECC systems (integrating a hard-shell venous reservoir as a stand-by component for immediate conversion to an open system) overcome safety concerns and unexpected intraoperative perfusion scenarios; thus they are compatible with every cardiac surgical procedure (Fig. 3.3).

Four meta-analyses have reported substantial benefits of MIECC for 30-day mortality, stroke, myocardial infarction (MI), postoperative atrial fibrillation, and renal dysfunction. The diversity of MiECC systems, patients, and outcomes, and the poor methodological quality of many of the randomized clinical trials (RCTs) included in these reviews undermine the strength of this evidence and most hospitals continue to use CECC. Nevertheless, the potential benefit of MiECC highlights the urgent need to evaluate this technology in a large, high-quality RCT. The proposed trial should evaluate MiECC systems that conform to an established typology, recruit patients having the most common cardiac surgery operations, report important clinical outcomes, and include features to minimize bias.

Conventional vs. minimal invasive extracorporeal circulation in patients undergoing cardiac surgery (COMICS) trial is currently in progress, aiming to test the hypothesis that MiECC is effective and cost-effective compared to CECC for most cardiac surgery operations requiring ECC without circulatory arrest. In this trial patients undergoing any of the following surgical procedures (i.e., CABG, AVR, CABG + AVR) are allocated in a 1:1 ratio to either MiECC or CECC and they are followed up twice, at 30 days and 90 days postoperatively. Recruitment has commenced and a total

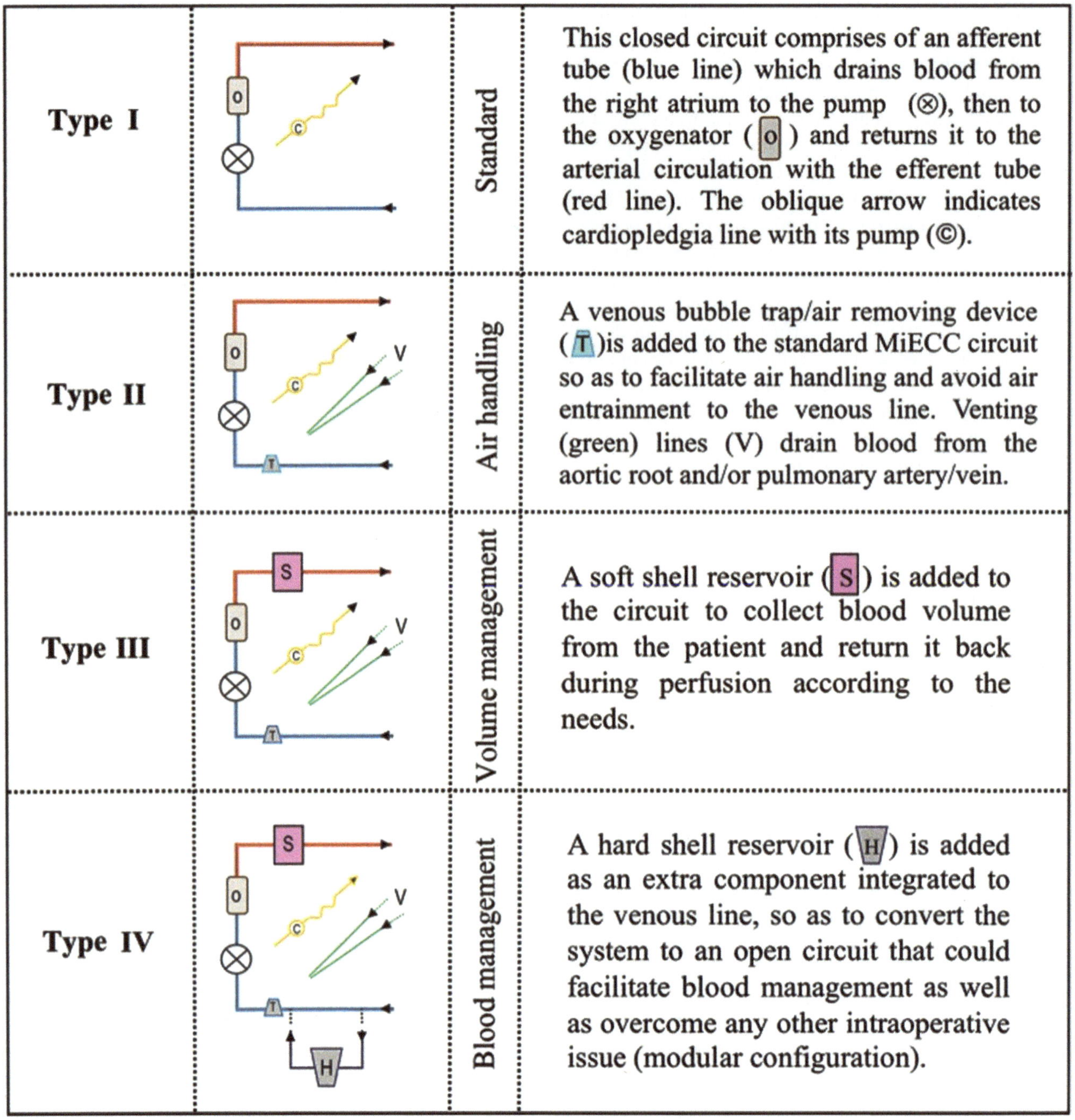

FIGURE 3.3 Classification of MiECC circuits.[2] (Note that the modular type IV circuit is literally type III with a standing-by component, used only when necessary.) *X*, Pump; *O*, oxygenator; *C*, cardioplegia; *T*, bubble trap/air-removing device; *V*, vent (aortic/pulmonary); *S*, soft-bag/reservoir; *H*, hard-shell/reservoir; *MiECC*, minimal invasive extracorporeal circulation.[20]

number between 2504 and 3500 participants will be recruited to cover any uncertainty in the power calculation. We look forward to the outcome.[21]

Reducing inflammatory response

One of the major obstacles to the development of an effective and reliable CPB circuit is the deleterious consequences associated with the interface between patient blood and the synthetic materials of the bypass circuit. This interface

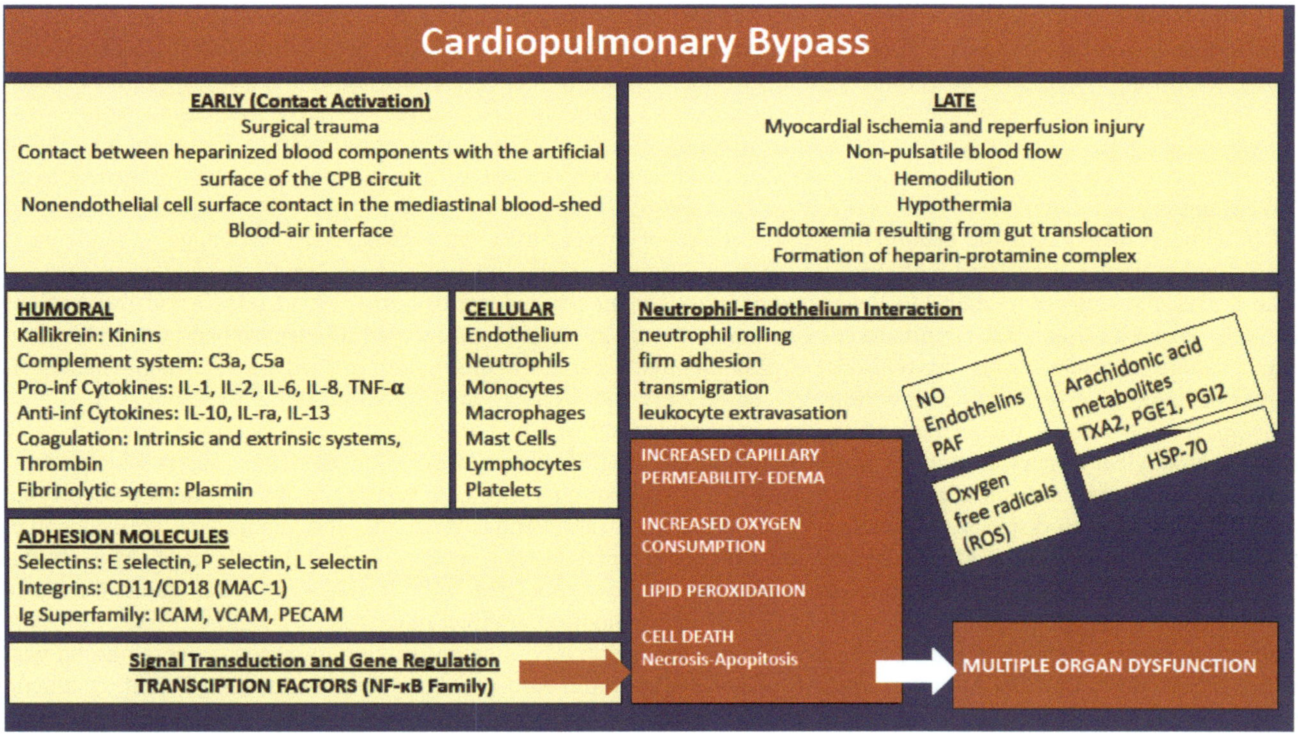

FIGURE 3.4 Cardiopulmonary bypass surgery comprehensive summary.

induces a complex inflammatory response through activation of the complement system and proinflammatory cytokines. The interaction of humoral and cellular inflammatory mediators can manifest as multiorgan dysfunction causing a coagulopathy, respiratory failure, renal insufficiency, myocardial dysfunction, or neurological deficits following cardiac surgery. The use of heparin bonded poly vinyl chloride has allowed the development of viable CPB circuits but the immunological and coagulopathic response to CPB remains a source of significant morbidity for patients undergoing cardiac surgery[22] (Fig. 3.4).

In a minimal extracorporeal circulation (MEC) circuit the venous reservoir is removed, leading to a bypass circuit of fixed volume with the patient themself acting as a reservoir of blood. The absence of a venous reservoir mandates the exclusion of air from the venous cannula as the venous return directly determines the output from the CPB circuit. Additionally, this precludes the use of conventional cell-saver devices and therefore the intraoperative transfusion of autologous blood. To some extent these disadvantages of MEC have been ameliorated by the use of suction devices activated only by direct contact with liquids or the postoperative transfusion of blood stored by a separate suction and cell-saver device. MEC circuits are currently used predominantly for isolated CABG, although they have also been used successfully for aortic valve replacement, atrial septal defect repair, and in combined valve replacement and CABG cases.[23]

The removal of the venous reservoir allows a MEC circuit to use approximately half the total length of tubing relative to a conventional bypass circuit, significantly reducing the total interface area between patient blood and synthetic material.[24] This reduction in circuit length and removal of the blood–air interface in the venous reservoir significantly reduces the coagulopathic and immunologic insults associated with CPB. The absence of cell-saver devices to transfusion pericardial bleeding may also contribute to a reduced proinflammatory stimulus, as this autologous blood is rich in proinflammatory mediators due to cellular damage and its exposure to nonendothelial surfaces. A reduced inflammatory response in MEC circuits has been supported in several studies that have shown lower concentrations of important inflammatory makers, including c-reactive protein, monocytes, and Il-6.[25]

Reducing transfusion requirements

The incidence of perioperative blood transfusion and protocol-driven triggers for transfusion remain common in cardiac surgery practice.[26] However, several large retrospective studies have demonstrated an incremental association between allogenic blood transfusion and adverse outcomes including serious infection, renal failure, postoperative cardiac dysfunction, and all-cause mortality.[27] These findings have driven a growing interest in perioperative strategies to decrease blood transfusion requirements. Optimization of the CPB circuit, to reduce the hemodilution effects of CPB and maintain the patients' hemoglobin concentration, is an important aspect in reducing perioperative transfusion requirements. Hemodilution, due to large priming volumes, additionally dilutes clotting factors and fibrinolytic proteins, thus further exacerbating the coagulopathy associated with CPB, and increases the risk of patients requiring blood transfusions following surgery.

In addition to reducing the proinflammatory stimulus of bypass, MEC circuits have the advantage of reducing the hemodilution burden associated with priming the circuit by minimizing the total circuit length. A smaller priming volume is therefore required leading to lower preoperative hematocrit and perioperative blood transfusion requirements. Patients operated using MEC circuits have been shown to require significantly fewer allogenic blood transfusions compared to those operated using conventional bypass circuits.[28] MEC circuits therefore facilitate the maintenance of a higher hematocrit during bypass, supporting oxygen delivery and reducing the risk of end-organ ischemia.

MEC is not suitable for all patients undergoing cardiac surgery and is currently used in only a relatively small number of surgical centers. Retrograde autologous priming of CPB lines, however, is a technique that is widely used to significantly reduce the hemodilution effect of CPB. Prior to commencing CPB autologous blood drains passively into the CPB pass lines to displace the priming fluid. The arterial line, venous reservoir, and venous line are sequentially drained with approximately 300 mL of priming fluid exchanged for the patients' blood at each stage. Most studies, though not all, have shown retrograde autologous priming is associated with a reduction in allogenic transfusion requirements in the perioperative period due to the smaller hemodilution effect of bypass.[29]

An alternative approach to maintaining hematocrit intraoperatively is hemofiltration to remove excess plasma water whilst the patient is on CPB. However, the role of hemofiltration in cardiac surgery, in the absence of renal dysfunction or excessive fluid overload, remains controversial. Whilst this approach may reduce transfusion requirements for the removal of fluid through hemofiltration, it has also been shown to promote hemodynamic instability, increase vasopressor requirements, and impair tissue perfusion.[30] Minimizing the use of priming fluids with retrograde autologous priming or MEC circuits is likely to be a superior option for reducing the effects of hemodilution than bypass hemofiltration.

Reducing consumption of coagulation factors

The surface coating used in CPB circuits is crucial in minimizing the activation of inflammatory and procoagulant pathways, reducing platelet activation and consumption of clotting factors. Despite extensive research, a perfectly biologically inert surface without proinflammatory or procoagulant effects has not been achieved. The most commonly used surface coating in extracorporeal circuits is unfractionated heparin. Coating bypass tubing with bonded heparin has been shown to provide an effective local anticoagulant effect with reduced d-dimer levels, platelet binding, and thrombin production. This interface also results in a strong antiinflammatory response, relative to controls, predominantly though a restriction in activation in the complement pathway.[31]

A number of alternative biocompatible coatings have been trialed in the search for a bypass circuit with an idealized biologically inert interface. Phosphorylcholine is a zwitterionic found on the surface of red cell membranes. It has been successfully used in the coating of endovascular stents to prevent stent thrombosis following percutaneous coronary intervention. When used in the coating of components of the CPB circuit, including tubing and oxygenator, it has similar anticoagulant properties to heparin-coated circuits with a possible reduction in postoperative transfusion requirements. Similarly, there is no overall difference in outcomes for patients undergoing surgery using heparin and phosphorylcholine-coated circuits.[32]

Oxygenator

Modern sheet or hollow-fiber oxygenators allow gas exchange to blood through a membrane. The sheet membrane oxygenators can be either a flat plate microporous polypropylene membrane or a spiral silicone membrane. The microporous membranes have 0.05–0.3 μm pores that first create a direct blood–gas interface and then a thin protein film forms leading to a molecular membrane. These membranes can lead to leakage of plasma after prolonged CPB use due to micropores losing their hydrophobicity from protein adsorption that leaks through the micropores. The silicone membranes are known as "true membranes" as they prevent such plasma leakage into the gas phase.[33] Hollow-fiber oxygenators have a higher surface area-to-blood volume ratio and are composed of microporous polypropylene or poly-methylpentene material. Blood flows over these fibers, while gas passes through them. Partial pressure of oxygen and of carbon dioxide is determined by adjusting the gas blender and the fresh gas flow.

Reducing hemolysis

Although most adult oxygenators are comparable, there are differences in the foreign surface area of the gas exchange membrane and on pressure drops over the oxygenators, which is considered a possible cause for hemolysis. Blood that comes in contact with a membrane oxygenator will activate and lyse cells, the oxygenator material absorbs the plasma proteins that interact with circulating platelets and cause activation of coagulation, fibrinolysis, and an inflammatory cascade. The material used on the oxygenator can cause blood trauma and to diminish this, all synthetic material surfaces in the oxygenator are coated with bioactive or biopassive coatings that reduce protein adsorption and contact activation.[34]

To reduce oxygenator material and blood trauma, modern hollow-fiber oxygenators have extraluminal flow (ELF) fiber bundles so blood does not flow inside the hollow fibers but around them, thus improving mass transfer and reducing the volume and membrane surface area.[35] More recent studies in modern oxygenators on the relationship between pressure drops and hemolysis demonstrates that this is not linear and there was no significant association between the two. The red blood cells hemolysis may not be an issue with modern oxygenators but trauma to platelets and thus activation of coagulation and inflammatory cascades persists.[36] Further research is required to understand the impact of the modern oxygenator designs on hemolysis and cell activations.

Biocompatible coating

A further issue aforementioned of membrane oxygenators is the accumulation of proteinaceous coating on the microporous membrane, reducing the area of blood–gas exchange. Biocompatible-coated oxygenators have been developed to mimic the natural endothelial lining, thus reducing plasma protein adhesion and diminishing cellular activation. They have been shown to have higher postoperative platelet count and reduced lung atelectasis but no difference in intensive care stay.

Heparin was one of the first biocompatible coatings utilized and now there are various options available, all working to mimic the natural endothelial lining. Many studies have demonstrated the advantages of using biocompatible coatings on decreased systematic inflammatory reactions, reducing activation of hemostasis, preventing platelet adhesion, and activation and preserving platelet count.[37] However, results on patient outcomes are still conflicting with either no beneficial effect in low-risk patients undergoing coronary revascularization or only minor morbidity benefits in high-risk patients.[38]

A systematic review and meta-analysis comparing biocompatible surfaces on circuits and oxygenators with conventional systems for 4360 patients demonstrated that biocompatible treatment had limited effect with reduced packed red cells transfusion and atrial fibrillation with shorter intensive care stay.[37] Researching the outcomes of biocompatible coatings would only likely be realized when other insults to blood to exposure of foreign material are minimized. Designing the optimal oxygenator should consider the balance between fiber diameters, packing fraction, membrane area, frontal area, and shunting, which can affect gas transfer, prime volume, pressure drop, and platelet activations. The careful balance for optimal performance constitutes most modern oxygenators, which are designed so they are not challenged under normal conditions of perfusion.[39]

Cardioplegia

Adequate myocardial protection plays a key role for successful outcomes in cardiac surgery. Cardioplegia is a potassium-rich solution that causes the heart to arrest in a depolarized state. It remains the gold-standard for cardioprotection.[40] There are many strategies and cardioplegia characteristics that have been shown to achieve and maintain the electromechanical quiescence required for open heart surgery. Studying and comparing these in large trials has been difficult as only a few centers share identical protocols.[41] Cardioplegia solutions can be blood-based or crystalloid-based, can be given anterograde or retrograde or both, can be warm or cold, or may involve a "hot shot." This section will explore each of these, including promising therapeutic targets for IRI.

Blood versus crystalloids

Blood cardioplegia is mixed in a ratio of 1:4 (crystalloid:blood parts) and crystalloid solutions can be intracellular (Custodiol, HTK, Bretschneider's) or extracellular (Plegisol, St. Thomas solution).[41] Blood cardioplegia has been thought to offer the advantage of delivering oxygen at cellular level via hemoglobin, providing hydrogen ion buffering, free-radical scavenging, reducing myocardial edema, improving microvascular flow, and is less associated with hemodilution.

There are several studies that have compared the metabolic response and release of cardiac enzymes, favoring blood over crystalloid cardioplegia. The clinical outcomes of the two solutions were studied in one of the largest prospective randomized clinical studies of 1440 patients undergoing on-pump CABG and on 345 aortic valve patients.[42] In both trials, all patients were gender, age, and perioperative risk matched, and there was no statistical significant difference for spontaneous sinus rhythm after aortic declamping, use of inotropic drugs or intraaortic balloon pumping, postoperative ventilatory support, bleeding and rate of allogeneic blood transfusions, perioperative myocardial infarction, episodes of atrial fibrillation, stroke or minor neurologic dysfunction, renal function, infections, physical rehabilitation, or mortality.[42] A subgroup analysis of patients with higher operative risk (age >70 years, unstable angina, diabetes, emergency operations, ejection fraction <50%, cross-clamp time >50 min, EuroSCORE II >5) in both trials also confirmed no significant difference in use of blood or crystalloid cardioplegia.[42]

A meta-analysis of 5044 patients from 34 trials also showed no difference between the two solutions for perioperative and postoperative myocardial infarction and mortality. However, they noticed a significantly lower CKMB release at 24-h postsurgery with blood cardioplegia and significantly lower incidence of low output syndrome (LOS) after reperfusing with blood cardioplegia.[43] LOS definition can be varied but it can include the requirement for inotropic or IABP support for hypotension or for >30 min to maintain blood pressure >80 mmHg; cardiac index (CI) less than 2.1 L/min/m^2 or CI < 2.01 L/min/m^2 with need for dopamine or dobutamine or adrenaline for >30 min; unusual need for inotropic or mechanical support to maintain cardiac output. A reduced LOS may lead to shorter postoperative hospital stay and it has been shown that perioperative myocardial infarction or LOS can reduce late survival.[44] The authors concluded by confirming the benefits of blood cardioplegia over crystalloid solutions.[41]

The debate about the effectiveness of crystalloid or blood cardioplegia in myocardial protection on mid- and long-term clinical outcome in aortic valve surgery is still open. Crystalloid solutions like Custodiol are attractive as they provide long myocardial protection with bloodless operative field after a single dose, and are important for valve surgery and minimally invasive procedure, compared to the repetitive application of blood cardioplegia. The use of cold blood cardioplegia seems to be preferred if there is impaired cardiac function (such as low ejection fraction) where blood provides better myocardial protection when used for aortic valve replacements.[45] The best cardioplegic solution, especially for valve surgery and in the context of impaired cardiac functions, remains still unknown and further studies are required to elucidate this.[46]

Antegrade versus retrograde versus both

Cardioplegia can be administered in two major routes to achieve cardioprotection. Anterograde follows the anatomical route of the coronary circulation by inserting the cardioplegia line into the aortic root below the aortic cross-clamp. Retrograde application is obtained by direct cannulation of the coronary sinus, which is anatomically more challenging, and it is a recognized method for myocardial protection including hypertrophied ventricles. Anterograde cardioplegia may be impaired in ventricular hypertrophy or in significant coronary artery stenosis, thus the retrograde route might be beneficial.[47]

In fact, retrograde perfusion of cardioplegia is homogenous in all the heart regions when studying human freshly explanted hearts, and the majority of this flow is through the Thebesian veins which bypass the microvasculature and are nonnutritive.

When comparing the two routes with crystalloid cardioplegia and measuring cardiomyocyte apoptosis in pigs, this was significantly elevated after ischemia-reperfusion in both groups, which proposes a suboptimal cardioprotection with either route. In particular, retrograde cardioplegia in the right ventricle was associated with a 3.4-fold higher cardiomyocyte apoptosis when compared to the anterograde one.[47] The left ventricle showed a similar difference but at a lower level. In a subsequent clinical study in 20 aortic valve surgical patients, the cardiomyocyte apoptosis was significantly increased in the left ventricle if received retrograde compared to anterograde cardioplegia.[48]

When analyzing the systolic contractile function, by analyzing the amplitude of the longitudinal systolic motion of the lateral mitral annulus, it was noted to be reduced postoperatively than before surgery in the retrograde but not in the antegrade group, suggesting less favorable left ventricular remodeling after surgery for retrograde cardioplegia. In addition, retrograde cardioplegia was associated with better protection for the left ventricle compared to the right, but it did not affect patient outcomes.[48]

A prospective randomized study of 39 patients undergoing aortic valve surgery demonstrated increased right ventricular metabolic stress with retrograde or anterograde cardioplegia compared to the left ventricle. The significantly elevated troponin I levels with both routes suggests a suboptimal myocardial protection regardless of route.[49]

A possible explanation is the variation in the venous anatomy of the heart which could affect perfusion of the right ventricular septum and free wall with retrograde cardioplegia. The differences noted for anterograde perfusion with the right ventricle compared to the left may be associated with smaller volumes of cardioplegia given into the right coronary ostia at induction. A further explanation for the reduced protection of the right ventricle with retrograde perfusion could be related to its baseline metabolic status. The left ventricle when becoming hypertrophied from overload, may undergo metabolic changes including mitochondrial function.[50] However, the exact mechanisms involved in these observations remain controversial.

When comparing combined anterograde and retrograde cardioplegia with anterograde only, there is evidence that the combined routes are favorable. In recent a prospective randomized control trial of 87 patients undergoing CABG showed that the combined cardioplegia options required significantly less inotropic support while weaning from CPB compared to anterograde only.[51]

Cold versus warm

Another important aspect for myocardial protection is the debate over the optimal temperature of cardioplegia during cardiac operations. Early techniques used cold solutions which lowered myocardial oxygen demands and the risk of ischemia damage but at the expense of inhibiting myocardial enzymes activity, thus leading to a delay in metabolic and functional recovery postcardiac surgery. Warm cardioplegia was introduced in the 1970s in the hope of maximizing the intraoperative myocardial protection and improving metabolic and functional recovery[52]; however, there is still much controversy regarding whether warm cardioplegia is superior to cold.

A prospective single-center study comparing warm/tepid and cold cardioplegia during CABG demonstrated superior outcomes with warm perfusion for perioperative mortality and myocardial infarction.[53] Another prospective randomized trial of 35 patients undergoing aortic valve surgery with hypertrophic hearts demonstrated cold blood cardioplegia is associated with less ischemic stress and reduced myocardial injury compared to warm blood.[54] The authors also suggested that both techniques still conferred suboptimal myocardial protection with higher alanine/glutamate ratio during reperfusion compared to preischemic levels in both groups.

A meta-analysis of 42 randomized trials with 5879 patients comparing warm vs. cold cardioplegia demonstrated no statistical significant difference for clinical outcomes including in-hospital mortality, length of stay, stroke incidence, atrial fibrillation, and use of balloon pumps. However, warm cardioplegia was associated with significantly better postoperative cardiac index, lower troponin concentrations on day 0, and significantly lower peak of CKMB concentrations. Overall, both warm and cold cardioplegia have been shown to be safe and effective for myocardial protection but warm is associated with better postoperative hemodynamic performance and reduced cardiocyte injury.[52]

Hot shot

Intermittent anterograde cold-blood cardioplegia followed by terminal warm blood reperfusion is known as "hot shot" and has been reported to lower ischemic/reperfusion injury.[55] However the mechanisms involved remain unclear.[56]

An early prospective clinical trial of 20 patients undergoing CABG received either cold blood cardioplegia alone or with a hot shot. This demonstrated that a hot shot accelerated myocardial metabolic recovery, preserved high-energy phosphates, improved metabolic responses to postoperative hemodynamic stresses, and reduced left atrial pressures.

A case-controlled study of 2488 patients with CABG received either cold cardioplegia only or a hot shot following cold cardioplegia. This demonstrated that patients who received a hot shot had a reduced need for direct current shocks.[57]

A prospective randomized study of 36 patients with left ventricular hypertrophy undergoing aortic valve replacement received either anterograde cold blood cardioplegia alone or with a retrograde hot shot. This demonstrated that although there were no clinical outcome differences, the retrograde hot shot did not add any extra benefit in preventing myocardial injury in this group of patients with raised troponin I levels at 4 and 12 h postsurgery in both groups.[55] This is in contrast to a randomized trial comparing 35 patients who underwent elective coronary revascularization where patients with cold blood cardioplegia alone showed significant metabolic derangements in the ischemic-reperfused hearts with higher ADP/ATP ratios and increased anaerobic metabolic activity compared to those who received a hot shot.[58]

The varied reported outcomes regarding the effectiveness of the hot shot require further biochemical and clinical studies and further trials are required to find the best contents for a hot shot and its applicability in cardiac surgery.

del Nido cardioplegia

del Nido cardioplegia solution has been used in congenital cardiac surgery for more than 20 years. It is composed of 1 part blood and 4 parts crystalloid (Plasmalyte A) solution containing potassium, lidocaine, magnesium, sodium bicarbonate, and mannitol. It is given as a single dose (20ml/kg) and repeated every 60 min if the estimated cross clamp time was over 90 min.[59]

A recent randomized controlled trial analyzed adult patients undergoing CABG or valve surgery and they received either del Nido cardioplegia or a whole blood one. They reported no significant difference in cross-clamp or CPB time in either group, but del Nido showed a higher return to spontaneous rhythm, fewer patients required inotropic support, and there was a lower increase in troponin I levels.[60]

A recent systematic review and meta-analysis of del Nido cardioplegia in adult cardiac surgery included nine studies that showed del Nido cardioplegia has significantly shorter CPB and cross-clamp time, smaller cardioplegia volumes and blood glucose levels, decreased ventilation time, and intensive care unit stay compared to conventional cardioplegia. However, there was no difference in myocardial enzymes, postoperative inotropic support, atrial fibrillation, and mortality between the two groups.[61]

There are still limited studies, which often involve healthy patients undergoing less complex procedures. A recent editorial suggested that del Nido cardioplegia provides myocardial protection that is equivalent but not superior to the current multidose blood cardioplegia.[62] Its benefits might be found in patients with multivessel coronary artery disease with LV hypertrophy, patients with reduced left ventricular ejection fraction or reduced right ventricular function and procedures that are complex requiring longer cross-clamp time.[62] Del Nido cardioplegia has been shown to be safe in routine adult cardiac surgery cases with comparable outcomes but clearly further studies and trials are required to address the other higher risk patients or procedures.

Targets for ischemia reperfusion injury

Myocardial reperfusion damage postcardioplegic ischemic arrest determines postoperative organ functional recovery, morbidity, and mortality in adult cardiac surgical patients.[41] In the last few decades, research has focused on perfecting an ideal cardioprotective solution, its delivery method, and its temperature. To provide further myocardial protection one should look at treating ischemic-reperfusion (IR) injury.[41] The following are some key developments in this area.

VE Cadherin is a key molecule for endothelial cell integrity and fluid extravasation, it interacts with fibrin β-chain and can induce capillary tube formation, cell—cell contraction, and leukocyte migration. Blocking the vascular endothelial (VE) cadherin binding sites for the fibrin β-chain by a peptide has been shown to be promising for reducing IR injury and leukocyte transmigration[63] in preclinical studies that have been designed with prolonged ischemia times, such as heart transplantation models in rodents.[64,65]

In addition to the receptors expressed by the endothelial cells barriers, there is also an upregulation of hormone secretions in IR injury, mainly endothelin (ET-1) and its receptors, which are associated with ischemic myocardial contracture. On rat hearts, ET_A selective antagonists PD155080 has been shown to decrease peak ischemic contracture, improve recovery of reperfusion left ventricular developed pressure, coronary flow, and diastolic relaxation.[66]

Mitochondrial permeability transition pores (mPTP) have been linked to ischemic injury and are sensitive to cyclosporine A which led to promising results on cyclosporine trials in reducing perfusion injury by having smaller infarcts than the placebo in patients presenting with acute ST-elevation myocardial infarction.[67,68]

A survey of practice in the UK among cardiac surgery consultants found that for on-pump CABG, 56% use cold blood cardioplegia, 14% use warm blood cardioplegia, 14% use crystalloid cardioplegia, 21% use retrograde infusion, and 16% do not use any cardioplegia (cross-clamp fibrillation).[69]

This highlights the lack of consensus in our community regarding the preparation, mode, and even the route of cardioplegia application. It is also true that most of the current cardioplegia protocols were developed for CABG patients with relatively normal morphological hearts. However, more complex and high-risk cardiac procedures are becoming more routine, for which the current cardioprotective regimens might be limited.[41] Future studies and clinical trials are required to address this issue and improve our understanding on managing IRI.

Neurological protection

There have been several developments that aim at optimizing CPB strategy to reduce central nervous system injury and therefore to confer neurological protection.

pH management

Acid–base status including pH management and CO_2 regulation are of fundamental importance. Arterial blood gas parameters are either corrected for temperature so the pH is kept constant at 7.40 (pH-stat) or the blood gas levels are uncorrected and measured at 37°C and alkalosis is permitted during cooling (alpha-stat).

While in pH-stat, blood gas increases CO_2 during cooling, which would lead cerebral blood vessels to dilate as they are dependent on PCO_2. Therefore, cerebral autoregulation would be lost and the cerebral blood flow would become pressure-dependent which may improve cooling and oxygen delivery to the brain.[70]

Alpha-stat blood gas is more appropriate in cardiac surgery and is currently recommended as it maintains autoregulation with improved control on cerebral blood flow and may be associated with improved outcomes. Further randomized trials showed these improved neurological and neuropsychological outcomes with alpha-stat pH management in patients undergoing moderate hypothermic CPB. There is evidence that suggests that in patients undergoing hypothermic circulatory arrest in cardiac surgery, the management of acid–base depends on age, with pediatric patients having better results with pH-stat and adult patients with alpha-stat.[71]

Temperature management

Cardiac surgery is performed under hypothermic (25°C–32°C) or normothermic conditions when on CPB. Hypothermia has been shown to reduce myocardial oxygen consumption, ensuring protection and increased tolerance to vital organ ischemia during CPB. However, hypothermia has also been associated with impaired myocardial cell membrane function and possible excessive postoperative bleeding. Hypothermic CPB requires a period of rewarming as patients with postoperative hypothermia (<36°C) have significant associations with adverse outcomes, such as greater mortality, prolonged ventilation, higher red blood cell transfusion, and longer intensive care unit (ICU) and hospital stay.[72]

In addition, the lack of aggressive monitoring of temperature while on CPB potentially exposes the brain to hyperthermia, which has been associated with increased morbidity and mortality in stroke patients in a meta-analysis. In CABG patients, postoperative hyperthermia is also associated with higher neuropsychological dysfunction.[73]

Similarly, there was greater neuropsychological dysfunction in patients that were randomized to a faster rate of rewarming for CABG patients. Patients that were rewarmed faster were also exposed to higher peak and mean temperatures of greater than 37°C than those rewarmed slower. The effects from cerebral hyperthermia include increased release of neurotransmitters, increased free radicals production, breakdown of the blood–brain barrier, and impaired recovery of energy metabolism.[74]

More aggressively monitoring temperature during CPB may offer the opportunity to minimize neurological injury in CABG patients. The location of where to measure temperature during CPB is also controversial. Nasopharyngeal monitoring has been reported to match well to cerebral ones. However other studies have noted that the nasopharyngeal temperatures might underestimate the jugular venous temperatures and thus expose the brain to unwanted hyperthermia.[75]

One would imagine that it would be more accurate to measure temperatures from arterial inflow in the CPB circuit. However, it has been reported that this may also be an underestimate while using conventional coupled temperature-monitoring devices, the actual temperature being significantly higher than the arterial temperature sensors in the CPB. This could be related to inadequate flow through the sensor, in fact the accuracy of the reading improved when the flow through the sensor was increased.[76]

Multiple trials have been conducted to analyze neurological outcomes with temperature management during CPB. Few randomized controlled trials concluded that there was no difference between normothermic and hypothermic CPB in terms of rates of strokes and temperature does not influence postoperative cognitive function.[77]

However, other trials have reported poorer outcomes with normothermic CPB or with hypothermic CPB. Another randomized trial noted reduced neuropsychological deficits in patients that were rewarmed to 34°C compared to those rewarmed to 37°C, but this was noted more early after surgery with no significant difference at 5-year follow-up. However, the effect of surgery on cognitive function seen early after surgery is an important predictor of cognitive performance at 5 years. Such differences can be attributed to the heterogeneity of the studies, the lack of standardized temperature monitoring locations, and in some not recognizing cerebral hyperthermia.[78]

Measuring perioperative hyperthermia is essential for neurological protection, with above 37°C associated with unnecessary neurological risk. Thus reliable measurements and avoiding perfusion temperatures exceeding 37°C might potentially assist in reducing neurological deficits. These issues have been addressed by the recent Society of Thoracic Surgeons (STS)/American Society of Extracorporeal Technology (AmSECT) guidelines on temperature management during CPB.[79] Some key recommendations include oxygenator arterial outlet temperature being the recommended surrogate for cerebral temperature but it is also assumed to be an underestimate; nasopharyngeal or pulmonary arterial temperatures are reasonable estimates of core temperature after weaning from CPB; and arterial outlet temperatures should not be higher than 37°C during CPB to prevent cerebral hyperthermia.

Minimizing return of pericardial suction blood

Reinfusing blood that has been exposed to pericardial and mediastinal surfaces from cardiotomy suctions during CPB has been associated with postoperative neurological injury. An explanation could be the increased levels of hemolysis and fat in this blood. Several studies have identified fat emboli being sourced from pericardial suctioned blood during CPB. In addition, this pericardial suctioned blood, when reinfused, exacerbated coagulation activation and fibrinolysis leading to increased blood loss.[80]

To minimize systemic inflammatory response when exposed to fat emboli, reinfusion of pericardial suction blood should be avoided, but might not be feasible in operations with large blood losses. Pericardial suctioned blood can be filtered or go through blood cell processors, such as the cell-saver, before being returned to the circuit, which has been shown to reduce reinfusion of fat emboli. Cell-savers have been shown to result in less fat being recycled during CPB compared to cardiotomy suction. Processed blood from the cardiotomy reservoir or cell-saving devices have high levels of fat emboli that are eliminated by using arterial filters which could potentially reduce neurocognitive dysfunction.[81]

Aortic assessment

The majority of type I (strokes) outcomes after CABG result from cerebral atheroembolic events, with noncalcific plaques being the greatest culprit and the least likely to be identified intraoperatively.[82]

It is well-recognized that advanced age, vascular disease, and diabetes mellitus are risk factors for perioperative strokes and predictors for aortic atherosclerosis. Up to 20% of patients undergoing CABG have been identified as having significant atherosclerosis in the aorta with >5 mm intimal thickening, and this is associated with increased risk of strokes.

In a comparative study, epiaortic ultrasonography has been shown to be more sensitive than transthoracic echocardiogram for identifying atherosclerosis of the ascending aorta and both have been shown to be superior to palpation.[82]

Both epiaortic ultrasonography and transthoracic echocardiogram have shown reduced neurological injury by identifying areas on the aorta that are free of atherosclerosis for aortic cannulation and opting for peripheral cannulation or an off-pump procedure.

Arterial line filtrations

Arterial filters have been introduced to the CPB circuits in view of microemboli and foreign material noted in the CPB. These have been shown to be beneficial. Studies have demonstrated a reduced embolic load to the brain when including arterial filters. The association between embolic load and neuropsychological outcomes have been demonstrated with arterial lines demonstrating lower neuropsychological deficits.

One randomized trial showed a reduced neuropsychological injury with a 40-μm arterial line filter but this was not demonstrated in other studies. More recently a randomized trial compared 36 patients who underwent CABG with either an integrated or nonintegrated arterial lines filter. This showed intraoperative hematocrit levels were higher, lower prime volumes were required, and the need for postoperative red blood transfusion was lower in patients who had an integrated arterial line filter.[83]

However, most studies reporting its benefits were performed with bubble oxygenators and we can question whether the existing arterial line filter design and positioning are still optimal. In fact, larger bubbles will be first fragmented by the pump, heat exchanger, and fiber bundles, so the smaller bubbles left might be smaller than the pore size of the arterial line filter.[84]

Maintaining euglycemia

There is evidence to show that hyperglycemia is associated with exacerbated neurological injury, wound infections, and mortality. Average blood glucose levels on the first day postsurgery after CABG have been correlated to adverse outcomes including stroke, myocardial infarction, septic complications, and death. An observational study of 200 diabetic patients who had undergone cardiac surgery demonstrated that poor intraoperative glycemic control was associated with increased incidence of cardiovascular, neurological, respiratory, renal, and infectious complications.[85] A further randomized trial compared nondiabetic patients undergoing CABG and, after attempting to control hyperglycemia during CPB, there was no significant effect on the combined incidence of neurologic, neuro-ophthalmologic, or neurobehavioral deficits or neurologic death. A recent study evaluated the relationship between cerebral arterial oxygen saturation with changes in blood glucose levels. It reported that hyper- or hypoglycemia during surgery on CPB did not reflect changes in cerebral oxygenation and there was no significant difference in the minimental state examination between the patients.[86]

It has been suggested that intraoperative blood glucose levels might be managed in both diabetic and nondiabetic patients by the use of an aggressive insulin dosing strategy during cardiac surgery. Hyperglycemia has been proposed to be reduced, thus leading to enhanced neurological outcomes, by avoiding glucose-containing intravenous, cardioplegic, or pump-priming solutions; increased awareness and treatment of catecholamine-induced hyperglycemia; and aggressive insulin dosing strategies.[87]

Further multicenter studies might be required to confirm the neuroprotective effect of glucose control intraoperatively but it is known that it reduces sternal wound infections and decreases morbidity and mortality in surgical ICU patients.[88]

Conclusion

Steady progress has been made in CPB techniques since modern extracorporeal circulation was first conceived by John Gibbon. Over the last several decades technological advancements and technological improvements of the devices have all ameliorated the effects of CPB. Every component has been individualized and there have been specific improvements. Patients undergoing CABG and cardiac surgery have also changed. The challenge is now in patients being elderly with co-morbidities. Cardiac surgery patients can be operated with or without CPB. On pump cardiac surgery can be conventional as any form of minimally invasive circuit. It is the constant evolution of CPB which is minimising complications.

Notwithstanding, it is essential that we as clinicians respect the physiological peculiarities of each patient and manage any untoward or unexpected complications that still continue to happen. Various components have been studied, including bubble and membrane oxygenators, improvement in oxygenators, improvement in pumps, roller pumps and centrifugal pumps, the complete circuit, and compartmentalization exclusion of venous reservoir, and these have all led to improvements and reductions in the postoperative repercussions to patients on CPB. These have a secondary influence on the duration of mechanical ventilation, as well as the length of intensive care and length in hospitals. It is important that we do not give up what John Gibbon envisaged and we continue to expand on the knowledge gathered through the years and herald our improvements by sharing knowledge and expertise to ultimately benefit our patients.

Conflicts of interest

The authors have no conflicts of interest to declare.

References

1. Punjabi PP, Taylor KM. The science and practice of cardiopulmonary bypass: from cross circulation to ECMO and SIRS. *Glob Cardiol Sci Pract.* 2013.
2. Evora PRB, Rodrigues AJ. Adult cardiopulmonary bypass in the twenty-first century—science, art or empiricism? *Card Surg—Commit Sci Technol Creat.* 2014;. Available from: https://doi.org/10.5772/57108. Published Online First.
3. National Institute for Cardiovascular Outcomes Research. National Adult Cardiac Surgery Audit (NACSA) 2020 Summary Report. Society of Cardiothoracic Surgeons in Great Britain & Ireland. Report number: 2016/17−2018/19 Data; 2020.
4. Maciej S. Roller vs Centrifugal Pumps - Perfusion I Perfusion Education; 2020. https://perfusioneducation.com/videos/roller-vs-centrifugal-pumps-perfusion/. Accessed 6 April 2021.
5. Jameel S, Colah S, Klein AA. Recent advances in cardiopulmonary bypass techniques. *Contin Educ Anaesth Crit Care Pain.* 2010;10:20−23. Available from: https://doi.org/10.1093/bjaceaccp/mkp042.
6. Just SS, Müller T, Hartrumpf M, et al. First experience with closed circuit/centrifugal pump extracorporeal circulation: cellular trauma, coagulatory, and inflammatory response. *Interact Cardiovasc Thorac Surg.* 2006;5:646−648. Available from: https://doi.org/10.1510/icvts.2005.125880.
7. Scott DA, Silbert BS, Doyle TJ, et al. Centrifugal vs roller head pumps for cardiopulmonary bypass: effect on early neuropsychologic outcomes after coronary artery surgery. *J Cardiothorac Vasc Anesth.* 2002;16:715−722. Available from: https://doi.org/10.1053/jcan.2002.128413.

8. Asante-Siaw J, Tyrrell J, Hoschtitzky A, et al. Does the use of a centrifugal pump offer any additional benefit for patients having open heart surgery? *Interact Cardiovasc Thorac Surg*. 2006;5:128–134. Available from: https://doi.org/10.1510/icvts.2005.126961.
9. Saczkowski R, Maklin M, Mesana T, et al. Centrifugal pump and roller pump in adult cardiac surgery: a meta-analysis of randomized controlled trials. *Artif Organs*. 2012;36:668–676. Available from: https://doi.org/10.1111/j.1525-1594.2012.01497.
10. Mlejnsky F, Klein AA, Lindner J, et al. A randomised controlled trial of roller vs centrifugal cardiopulmonary bypass pumps in patients undergoing pulmonary endarterectomy. *Perfusion*. 2015;30:520–528. Available from: https://doi.org/10.1177/0267659114553283.
11. Ji B, Undar A. Comparison of perfusion modes on microcirculation during acute and chronic cardiac support: is there a difference? *Perfusion*. 2007;22:115–119. Available from: https://doi.org/10.1177/0267659107080115.
12. Murphy GJ, Angelini GD. Side effects of cardiopulmonary bypass: what is the reality? *J Card Surg*. 2004;19:481–488. Available from: https://doi.org/10.1111/j.0886-0440.2004.04101.
13. Poswal P, Mehta Y, Juneja R, et al. Comparative study of pulsatile and nonpulsatile flow during cardio-pulmonary bypass. *Ann Card Anaesth*. 2004;7:44–50.
14. Li Y, Zheng J, Bird IM, et al. Effects of pulsatile shear stress on signaling mechanisms controlling nitric oxide production, endothelial nitric oxide synthase phosphorylation, and expression in ovine fetoplacental artery endothelial cells. *Endothel J Endothel Cell Res*. 2005;12:21–39. Available from: https://doi.org/10.1080/10623320590933743.
15. Lanzarone E, Gelmini F, Tessari M, et al. Preservation of endothelium nitric oxide release by pulsatile flow cardiopulmonary bypass when compared with continuous flow. *Artif Organs*. 2009;33:926–934. Available from: https://doi.org/10.1111/j.1525-1594.2009.00888.x.
16. Elbers PWG, Wijbenga J, Solinger F, et al. Direct observation of the human microcirculation during cardiopulmonary bypass: effects of pulsatile perfusion. *J Cardiothorac Vasc Anesth*. 2011;25:250–255. Available from: https://doi.org/10.1053/j.jvca.2010.06.014.
17. Koning NJ, Vonk ABA, van Barneveld LJ, et al. Pulsatile flow during cardiopulmonary bypass preserves postoperative microcirculatory perfusion irrespective of systemic hemodynamics. *J Appl Physiol*. 2012;112:1727–1734. Available from: https://doi.org/10.1152/japplphysiol.01191.2011.
18. O'Neil MP, Fleming JC, Badhwar A, et al. Pulsatile vs nonpulsatile flow during cardiopulmonary bypass: microcirculatory and systemic effects. *Ann Thorac Surg*. 2012;94:2046–2053. Available from: https://doi.org/10.1016/j.athoracsur.2012.05.065.
19. Tan Z, Besser M, Anderson S, et al. Pulsatile vs nonpulsatile flow during cardiopulmonary bypass: extent of hemolysis and clinical significance. *ASAIO J*. 2020;66:1025–1030. Available from: https://doi.org/10.1097/MAT.0000000000001154.
20. Anastasiadis K, Murkin J, Antonitsis P, et al. Use of minimal invasive extracorporeal circulation in cardiac surgery: principles, definitions and potential benefits. A position paper from the Minimal invasive Extra-Corporeal Technologies international Society (MiECTiS). *Interact Cardiovasc Thorac Surg*. 2016;22(5):647–662. Available from: https://doi.org/10.1093/icvts/ivv380. Epub January 26, 2016. PMID: 26819269; PMCID: PMC4892134.
21. The COMICS investigators. Conventional vs minimally invasive extracorporeal circulation in patients undergoing cardiac surgery: protocol for a randomised controlled trial (COMICS). *Perfusion*. 2021;36(4):388–394.
22. Remadi JP, Rakotoarivello Z, Marticho P, et al. Aortic valve replacement with the minimal extracorporeal circulation (Jostra MECC System) vs standard cardiopulmonary bypass: a randomized prospective trial. *J Thorac Cardiovasc Surg*. 2004;128:436–441. Available from: https://doi.org/10.1016/j.jtcvs.2004.01.041.
23. Yilmaz A, Sjatskig J, van Boven WJ, et al. Combined coronary artery bypass grafting and aortic valve replacement with minimal extracorporeal closed circuit circulation vs standard cardiopulmonary bypass. *Interact Cardiovasc Thorac Surg*. 2010;11:754–757. Available from: https://doi.org/10.1510/icvts.2010.241943.
24. Jarral OA, Baig K, Pettengell C, et al. National survey of UK consultant surgeons' opinions on surgeon-specific mortality data in cardiothoracic surgery. *Circ Cardiovasc Qual Outcomes*. 2016;9:414–423. Available from: https://doi.org/10.1161/CIRCOUTCOMES.116.002749.
25. Remadi JP, Rakotoarivelo Z, Marticho P, et al. Prospective randomized study comparing coronary artery bypass grafting with the new mini-extracorporeal circulation Jostra System or with a standard cardiopulmonary bypass. *Am Heart J*. 2006;151:198. Available from: https://doi.org/10.1016/j.ahj.2005.03.067.
26. Brouwers C, Hooftman B, Vonk S, et al. Benchmarking the use of blood products in cardiac surgery to stimulate awareness of transfusion behaviour: results from a four-year longitudinal study. *Neth Heart J Mon J Neth Soc Cardiol Neth Heart Found*. 2017;25:207–214. Available from: https://doi.org/10.1007/s12471-016-0936-1.
27. Santos AA, dos, Sousa AG, Piotto RF, et al. Risco de mortalidade é dose-dependente do número de unidades de concentrado de hemácias transfundidas após cirurgia de revascularização miocárdica. *Braz J Cardiovasc Surg*. 2013;28:509–517. Available from: https://doi.org/10.5935/1678-9741.20130083.
28. Abdel Aal M, ElNahal N, Bakir BM, et al. Mini-cardiopulmonary bypass impact on blood conservation strategy in coronary artery bypass grafting. *Interact Cardiovasc Thorac Surg*. 2011;12:600–604. Available from: https://doi.org/10.1510/icvts.2010.243055.
29. Murphy GS, Szokol JW, Nitsun M, et al. The failure of retrograde autologous priming of the cardiopulmonary bypass circuit to reduce blood use after cardiac surgical procedures. *Anesth Analg*. 2004;98:1201–1207. Available from: https://doi.org/10.1213/01.ane.0000112306.71113.5e. Table of contents.
30. Soliman R, Fouad E, Belghith M, et al. Conventional hemofiltration during cardiopulmonary bypass increases the serum lactate level in adult cardiac surgery. *Ann Card Anaesth*. 2016;19:45–51. Available from: https://doi.org/10.4103/0971-9784.173019.
31. Lorusso R, De Cicco G, Totaro P, et al. Effects of phosphorylcholine coating on extracorporeal circulation management and postoperative outcome: a double-blind randomized study. *Interact Cardiovasc Thorac Surg*. 2009;8:7–11. Available from: https://doi.org/10.1510/icvts.2008.176479.

32. Reser D, Seifert B, Klein M, et al. Retrospective analysis of outcome data with regards to the use of Phisio®-, Bioline®- or Softline®-coated cardiopulmonary bypass circuits in cardiac surgery. *Perfusion*. 2012;27:530–534. Available from: https://doi.org/10.1177/0267659112454558.
33. Machin D, Allsager C. Principles of cardiopulmonary bypass. *Contin Educ Anaesth Crit Care Pain*. 2006;6:176–181. Available from: https://doi.org/10.1093/bjaceaccp/mkl043.
34. De Somer F, Van Landschoot A, Van Nooten G, et al. Interaction of plasma proteins with commercial protein repellent polyvinyl chloride (PVC): a word of caution. *Perfusion*. 2008;23:215–221. Available from: https://doi.org/10.1177/0267659108099814.
35. De Somer F. Does contemporary oxygenator design influence haemolysis? *Perfusion*. 2013;28:280–285. Available from: https://doi.org/10.1177/0267659113483803.
36. Venema LH, Sharma AS, Simons AP, et al. Contemporary oxygenator design relative to hemolysis. *J Extra Corpor Technol*. 2014;46:212–216.
37. Ranucci M, Balduini A, Ditta A, et al. A systematic review of biocompatible cardiopulmonary bypass circuits and clinical outcome. *Ann Thorac Surg*. 2009;87:1311–1319. Available from: https://doi.org/10.1016/j.athoracsur.2008.09.076.
38. Karakisi SO, Kunt AG, Bozok Ş, et al. Humoral immune response and coated or uncoated oxygenators during cardiopulmonary bypass surgery. *Cardiovasc J Afr*. 2016;27:242–245. Available from: https://doi.org/10.5830/CVJA-2016-003.
39. Haworth WS. The development of the modern oxygenator. *Ann Thorac Surg*. 2003;76:S2216–S2219. Available from: https://doi.org/10.1016/j.athoracsur.2003.09.012.
40. Chambers DJ, Fallouh HB. Cardioplegia and cardiac surgery: pharmacological arrest and cardioprotection during global ischemia and reperfusion. *Pharmacol Ther*. 2010;127:41–52. Available from: https://doi.org/10.1016/j.pharmthera.2010.04.001.
41. Habertheuer A, Kocher A, Laufer G, et al. Cardioprotection: a review of current practice in global ischemia and future translational perspective. *BioMed Res Int*. 2014;2014:325725. Available from: https://doi.org/10.1155/2014/325725.
42. Øvrum E, Tangen G, Tølløfsrud S, et al. Cold blood vs cold crystalloid cardioplegia: a prospective randomised study of 345 aortic valve patients. *Eur J Cardio-Thorac Surg J Eur Assoc Cardio-Thorac Surg*. 2010;38:745–749. Available from: https://doi.org/10.1016/j.ejcts.2010.03.052.
43. Guru V, Omura J, Alghamdi AA, et al. Is blood superior to crystalloid cardioplegia? A meta-analysis of randomized clinical trials. *Circulation*. 2006;114:I331–I338. Available from: https://doi.org/10.1161/CIRCULATIONAHA.105.001644.
44. Fremes SE, Tamariz MG, Abramov D, et al. Late results of the Warm Heart Trial: the influence of nonfatal cardiac events on late survival. *Circulation*. 2000;102:III339–III345. Available from: https://doi.org/10.1161/01.cir.102.suppl_3.iii-339.
45. Hoyer A, Lehmann S, Mende M, et al. Custodiol vs cold Calafiore for elective cardiac arrest in isolated aortic valve replacement: a propensity-matched analysis of 7263 patients. *Eur J Cardio-Thorac Surg J Eur Assoc Cardio-Thorac Surg*. 2017;52:303–309. Available from: https://doi.org/10.1093/ejcts/ezx052.
46. Comentale G, Giordano R, Palma G. Comparison of the different cardioplegic strategies in cardiac valves surgery: who wins the "arm-wrestling"? *J Thorac Dis*. 2018;10:714. Available from: https://doi.org/10.21037/jtd.2018.01.133.
47. Vähäsilta T, Saraste A, Kytö V, et al. Cardiomyocyte apoptosis after antegrade and retrograde cardioplegia. *Ann Thorac Surg*. 2005;80:2229–2234. Available from: https://doi.org/10.1016/j.athoracsur.2005.05.057.
48. Vähäsilta T, Malmberg M, Saraste A, et al. Cardiomyocyte apoptosis after antegrade and retrograde cardioplegia during aortic valve surgery. *Ann Thorac Surg*. 2011;92:1351–1357. Available from: https://doi.org/10.1016/j.athoracsur.2011.05.065.
49. Lotto AA, Ascione R, Caputo M, et al. Myocardial protection with intermittent cold blood during aortic valve operation: antegrade vs retrograde delivery. *Ann Thorac Surg*. 2003;76:1227–1233. Available from: https://doi.org/10.1016/s0003-4975(03)00840-3. Discussion 1233.
50. Ruengsakulrach P, Buxton BF. Anatomic and hemodynamic considerations influencing the efficiency of retrograde cardioplegia. *Ann Thorac Surg*. 2001;71:1389–1395. Available from: https://doi.org/10.1016/s0003-4975(00)01991-3.
51. Radmehr H, Soleimani A, Tatari H, et al. Does combined antegrade-retrograde cardioplegia have any superiority over antegrade cardioplegia? *Heart Lung Circ*. 2008;17:475–477. Available from: https://doi.org/10.1016/j.hlc.2008.04.009.
52. Fan Y, Zhang A-M, Xiao Y-B, et al. Warm vs cold cardioplegia for heart surgery: a meta-analysis. *Eur J Cardio-Thorac Surg J Eur Assoc Cardio-Thorac Surg*. 2010;37:912–919. Available from: https://doi.org/10.1016/j.ejcts.2009.09.030.
53. Mallidi HR, Sever J, Tamariz M, et al. The short-term and long-term effects of warm or tepid cardioplegia. *J Thorac Cardiovasc Surg*. 2003;125:711–720. Available from: https://doi.org/10.1067/mtc.2003.105.
54. Ascione R, Caputo M, Gomes WJ, et al. Myocardial injury in hypertrophic hearts of patients undergoing aortic valve surgery using cold or warm blood cardioplegia. *Eur J Cardio-Thorac Surg J Eur Assoc Cardio-Thorac Surg*. 2002;21:440–446. Available from: https://doi.org/10.1016/s1010-7940(01)01168-x.
55. Ascione R, Suleiman SM, Angelini GD. Retrograde hot-shot cardioplegia in patients with left ventricular hypertrophy undergoing aortic valve replacement. *Ann Thorac Surg*. 2008;85:454. Available from: https://doi.org/10.1016/j.athoracsur.2007.08.039.
56. Akowuah EF, Riaz I, Shrivastava V, et al. A comparison of 250 and 500 mL of terminal warm blood cardioplegia after global myocardial ischemia: a prospective randomized study. *J Card Surg*. 2005;20:107–111. Available from: https://doi.org/10.1111/j.0886-0440.2005.200337.x.
57. Mirmohammadsadeghi P, Mirmohammadsadeghi M. Effects of single antegrade hot shot in comparison with no hot shot administration during coronary artery bypass grafting. *ARYA Atheroscler*. 2015;11:186.
58. Caputo M, Dihmis WC, Bryan AJ, et al. Warm blood hyperkalaemic reperfusion ('hot shot') prevents myocardial substrate derangement in patients undergoing coronary artery bypass surgery. *Eur J Cardiothorac Surg*. 1998;13:559–564. Available from: https://doi.org/10.1016/S1010-7940(98)00056-6.
59. Hessel EA. What's new in cardiopulmonary bypass. *J Cardiothorac Vasc Anesth*. 2019;33:2296–2326. Available from: https://doi.org/10.1053/j.jvca.2019.01.039.

60. Ad N, Holmes SD, Massimiano PS, et al. The use of del Nido cardioplegia in adult cardiac surgery: a prospective randomized trial. *J Thorac Cardiovasc Surg*. 2018;155:1011–1018. Available from: https://doi.org/10.1016/j.jtcvs.2017.09.146.
61. Li Y, Lin H, Zhao Y, et al. Del Nido cardioplegia for myocardial protection in adult cardiac surgery: a systematic review and meta-analysis. *ASAIO J Am Soc Artif Intern Organs 1992*. 2018;64:360–367. Available from: https://doi.org/10.1097/MAT.0000000000000652.
62. Lazar HL. Del Nido cardioplegia: passing fad or here to stay? *J Thorac Cardiovasc Surg*. 2018;155:1009–1010. Available from: https://doi.org/10.1016/j.jtcvs.2017.09.126.
63. Gold JP, Torres KE, Maldarelli W, et al. Improving outcomes in coronary surgery: the impact of echo-directed aortic cannulation and perioperative hemodynamic management in 500 patients. *Ann Thorac Surg*. 2004;78:1579–1585. Available from: https://doi.org/10.1016/j.athoracsur.2004.05.051.
64. Wiedemann D, Schneeberger S, Friedl P, et al. The fibrin-derived peptide Bbeta(15-42) significantly attenuates ischemia-reperfusion injury in a cardiac transplant model. *Transplantation*. 2010;89:824–829. Available from: https://doi.org/10.1097/tp.0b013e3181ccd822.
65. Zacharowski K, Zacharowski PA, Friedl P, et al. The effects of the fibrin-derived peptide Bbeta(15–42) in acute and chronic rodent models of myocardial ischemia-reperfusion. *Shock Augusta Ga*. 2007;27:631–637. Available from: https://doi.org/10.1097/SHK.0b013e31802fa038.
66. Brunner F, Opie LH. Role of endothelin-A receptors in ischemic contracture and reperfusion injury. *Circulation*. 1998;97:391–398. Available from: https://doi.org/10.1161/01.cir.97.4.391.
67. Turer AT, Hill JA. Pathogenesis of myocardial ischemia-reperfusion injury and rationale for therapy. *Am J Cardiol*. 2010;106:360–368. Available from: https://doi.org/10.1016/j.amjcard.2010.03.032.
68. Piot C, Croisille P, Staat P, et al. Effect of cyclosporine on reperfusion injury in acute myocardial infarction. *N Engl J Med*. 2008;359:473–481. Available from: https://doi.org/10.1056/NEJMoa071142.
69. Karthik S, Grayson AD, Oo AY, et al. A survey of current myocardial protection practices during coronary artery bypass grafting. *Ann R Coll Surg Engl*. 2004;86:413–415. Available from: https://doi.org/10.1308/147870804669.
70. Scallan MJH. Cerebral injury during paediatric heart surgery: perfusion issues. *Perfusion*. 2004;19:221–228. Available from: https://doi.org/10.1191/0267659104pf742oa.
71. Abdul Aziz KA, Meduoye A. Is pH-stat or alpha-stat the best technique to follow in patients undergoing deep hypothermic circulatory arrest? *Interact Cardiovasc Thorac Surg*. 2010;10:271–282. Available from: https://doi.org/10.1510/icvts.2009.214130.
72. Insler SR, O'Connor MS, Leventhal MJ, et al. Association between postoperative hypothermia and adverse outcome after coronary artery bypass surgery. *Ann Thorac Surg*. 2000;70:175–181. Available from: https://doi.org/10.1016/s0003-4975(00)01415-6.
73. Grocott HP, Mackensen GB, Grigore AM, et al. Postoperative hyperthermia is associated with cognitive dysfunction after coronary artery bypass graft surgery. *Stroke*. 2002;33:537–541. Available from: https://doi.org/10.1161/hs0202.102600.
74. Grigore AM, Grocott HP, Mathew JP, et al. The rewarming rate and increased peak temperature alter neurocognitive outcome after cardiac surgery. *Anesth Analg*. 2002;94:4–10. Available from: https://doi.org/10.1097/00000539-200201000-00002. Table of contents.
75. Kaukuntla H, Harrington D, Bilkoo I, et al. Temperature monitoring during cardiopulmonary bypass–do we undercool or overheat the brain? *Eur J Cardio-Thorac Surg J Eur Assoc Cardio-Thorac Surg*. 2004;26:580–585. Available from: https://doi.org/10.1016/j.ejcts.2004.05.004.
76. Salah M, Sutton R, Tsarovsky G, et al. Temperature inaccuracies during cardiopulmonary bypass. *J Extra Corpor Technol*. 2005;37:38–42.
77. Grigore AM, Mathew J, Grocott HP, et al. Prospective randomized trial of normothermic vs hypothermic cardiopulmonary bypass on cognitive function after coronary artery bypass graft surgery. *Anesthesiology*. 2001;95:1110–1119. Available from: https://doi.org/10.1097/00000542-200111000-00014.
78. Shann KG, Likosky DS, Murkin JM, et al. An evidence-based review of the practice of cardiopulmonary bypass in adults: a focus on neurologic injury, glycemic control, hemodilution, and the inflammatory response. *J Thorac Cardiovasc Surg*. 2006;132:283–290. Available from: https://doi.org/10.1016/j.jtcvs.2006.03.027.
79. Engelman R, Baker RA, Likosky DS, et al. The Society of Thoracic Surgeons, The Society of Cardiovascular Anesthesiologists, and The American Society of ExtraCorporeal Technology: clinical practice guidelines for cardiopulmonary bypass—temperature management during cardiopulmonary bypass. *Ann Thorac Surg*. 2015;100:748–757. Available from: https://doi.org/10.1016/j.athoracsur.2015.03.126.
80. de Vries AJ, Gu YJ, Douglas YL, et al. Clinical evaluation of a new fat removal filter during cardiac surgery. *Eur J Cardio-Thorac Surg J Eur Assoc Cardio-Thorac Surg*. 2004;25:261–266. Available from: https://doi.org/10.1016/j.ejcts.2003.11.011.
81. Kaza AK, Cope JT, Fiser SM, et al. Elimination of fat microemboli during cardiopulmonary bypass. *Ann Thorac Surg*. 2003;75:555–559. Available from: https://doi.org/10.1016/s0003-4975(02)04540-x. Discussion 559.
82. Likosky DS, Marrin CAS, Caplan LR, et al. Determination of etiologic mechanisms of strokes secondary to coronary artery bypass graft surgery. *Stroke*. 2003;34:2830–2834. Available from: https://doi.org/10.1161/01.STR.0000098650.12386.B3.
83. Gürsu Ö, Isbir S, Ak K, et al. Comparison of new technology integrated and nonintegrated arterial filters used in cardiopulmonary bypass surgery: a randomized, prospective, and single blind study. *BioMed Res Int*. 2013;2013:529087. Available from: https://doi.org/10.1155/2013/529087.
84. De Somer F. Evidence-based used, yet still controversial: the arterial filter. *J Extra Corpor Technol*. 2012;44:P27–P30.
85. Ouattara A, Lecomte P, Le Manach Y, et al. Poor intraoperative blood glucose control is associated with a worsened hospital outcome after cardiac surgery in diabetic patients. *Anesthesiology*. 2005;103:687–694. Available from: https://doi.org/10.1097/00000542-200510000-00006.
86. Kesimci E, Erkýlýc E, Alaybeyoglu F, et al. Does glycemic level affect neurologic outcomes during cardiopulmonary bypass? *J Cardiothorac Vasc Anesth*. 2016;30:S18–S19. Available from: https://doi.org/10.1053/j.jvca.2016.03.059.
87. Carvalho G, Moore A, Qizilbash B, et al. Maintenance of normoglycemia during cardiac surgery. *Anesth Analg*. 2004;99:319–324. Available from: https://doi.org/10.1213/01.ANE.0000121769.62638.EB. Table of contents.
88. van den Berghe G, Wouters P, Weekers F, et al. Intensive insulin therapy in critically ill patients. *N Engl J Med*. 2001;345:1359–1367. Available from: https://doi.org/10.1056/NEJMoa011300.

Chapter 4

Organ Perfusion and Cardiopulmonary Bypass Techniques

Christopher K. Koehler[1], Patrick M. Grady[2], Carrie L. Ochocki[1], Shane S. Pullins[1], Jamshid H. Karimov[3] and Marc Gillinov[4]

[1]*Department of Perfusion Services, Cleveland Clinic, Cleveland, OH, United States,* [2]*Department of Perfusion Services, Cleveland Clinic, Cleveland, OH, United States,* [3]*Lerner Research Institute, Cleveland Clinic, Cleveland, OH, United States,* [4]*Department Chairman, Department of Thoracic and Cardiovascular Surgery, Cleveland Clinic, Cleveland, OH, United States*

Chapter Outline

Adult perfusion circuit

The standard adult perfusion circuit consists of a hardshell venous reservoir, centrifugal pump, oxygenator with integrated filter, cardioplegia, vent suction, pump suction, and basket suction (Fig. 4.1). The custom tubing pack has many essential features that allow for standardization and adaptability. The first of which is a sealed reservoir with vacuum-assisted venous drainage (VAVD). This technique was developed at Cleveland Clinic in the mid-1990s,[1] pioneered by our colleagues Cliff Ball and Bob Foster (see Boehringer regulator below), and has been used on every operation since. This affords us the ability to decrease the priming volume by raising the reservoir height, decreasing tubing length and emptying the volume from venous lines prior to cardiopulmonary bypass (CPB). The circuit consists of two 3/8″ venous lines which allows the perfusionist to troubleshoot drainage issues, and when only using a single venous cannula, to have more operating volume in the venous reservoir. There is a recirculation line post-oxygenator that is attached to the venous reservoir and also connected to a rapid prime line which allows us to further reduce prime volume, control volume during the case, and also as a safety feature to remove air. Our custom tubing pack has a dual cardioplegia circuit which has a series of Roberts clamps on the blood and crystalloid inlets. Depending on the desired cardioplegia to be used, one set of Roberts clamps is opened and the other set closed to deliver 4:1 or 1:4 blood:crystalloid ratio.[2] For safety, all connections are color coded and have connectors in place so that connections cannot be made backwards. The pump suction, vent, and basket all have 3/8″ pump boots, which allows for lower revolutions per minute (RPMs) to decrease hemolysis and provide adequate flow for sucker bypass when needed. The custom tubing pack also has a "jump tube" from postoxygenator to the venous line which allows the perfusionist to run retrograde cerebral perfusion during circulatory arrest and to transfuse up the venous line at the end of the case so that the arterial cannula can be safely removed.

Safety has also been aided by standardizing with the LivaNova S5 heart-lung machine (HLM) with its many safety features,[3] which include a centrifugal pump head with a one-way duckbill valve, auto clamp, level, bubble, pressure

Advances in Cardiovascular Technology. DOI: https://doi.org/10.1016/B978-0-12-816861-5.00011-3

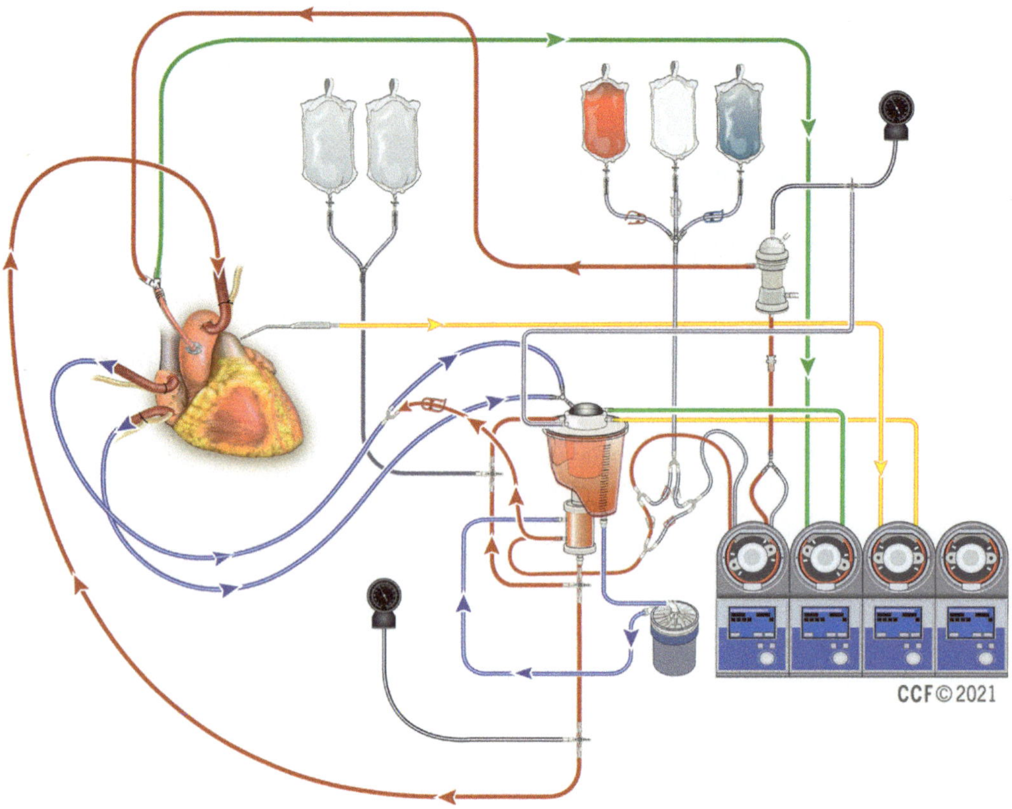

FIGURE 4.1 Adult cardiopulmonary bypass circuit.

and temperature sensors. For additional safety, we employ a Terumo Cardiovascular Devices, Inc. (CDI)[4] on every case which is an in-line trending device that actively monitors several blood gas parameters including arterial pH, O_2, CO_2, saturation, potassium, venous saturation, and hemoglobin.

At a high volume center, having equipment and supplies on hand is of paramount importance. To mitigate supply chain interruptions and ensure we have the supplies needed to care for patients, we have multiple vendors for our most important supplies. These vendors have several locations that house our custom packs and they are required to have a 60–90-day supply on hand throughout these locations. To further mitigate potential interruptions, we keep a 30-day supply at a local warehouse. We also have clinical engineers that are trained to perform preventative maintenance and repair our equipment in a timely manner.

Having a standard HLM setup that can be employed on almost every case has allowed us to keep a constant in the changing field of perfusion. This consistency has kept simplicity and safety paramount and has allowed us to reproduce consistent results in order to provide the best possible care of our patients.

Hemodilution reduction

Historically, hemodilution has been a necessary evil in CPB techniques. Priming the circuit prebypass must be achieved with either sanguineous or asanguineous (i.e., crystalloid) fluid to prevent air embolism. The priming volume of the circuit may be modified through various techniques. Numerous efforts have reduced the magnitude of hemodilution incurred when initiating CPB, including the implementation of VAVD, retrograde autologous priming (RAP), and the use of hemoconcentrators (HC). These efforts help to ameliorate anemia typically associated with cardiac surgery.[5]

Draining venous blood from the right side of the heart can be accomplished via gravity siphoning, but better drainage can be accomplished via VAVD. As described by Ogella, this requires sealing the hardshell venous reservoir and applying negative pressure to it.[1] By using negative pressure within the venous lines, more venous drainage can be achieved through smaller inner diameter venous lines. Our circuits employ two 3/8″ lines that Y together into a ½″ line prior to entering the venous reservoir. Bicaval cannulation uses both of these lines, and in the case of two-stage venous cannulation only one of the venous lines is used. This strategy reduces the priming volume and surface area of our circuit regardless of the cannulation strategy. Furthermore, by using VAVD, we are able to initiate bypass with empty (dry) venous lines by applying suction prior to unclamping them. In the absence of patent foramen ovale, all cases at Cleveland Clinic are initiated with dry venous lines. This reduces the priming volume and subsequent crystalloid solution exposure by ~140–280 mL per case depending on the cannulation strategy.

Another method by which priming volume is reduced is RAP. This technique involves exchanging the priming volume of the circuit with the patient's own blood. It requires the patient to have extra circulating blood volume and sufficient hemodynamic stability to withstand volume loss. In the case of insufficient volume or hemodynamic stability, a RAP technique deemed "RAP 2.0" is employed. This involves opening the venous lines, draining the patient, and shuttling the priming volume up the recirculation line to a plasmalyte bag. Once the fluid shuttling up to the bag begins to show blood, the recirculation line is clamped, the arterial line is opened, and bypass is evaluated and run normally.

As the name suggests, HC can also mitigate hemodilution. Cleveland Clinic puts HC into our bypass circuits as needed. On high-risk cases they may be put into the circuit prior to priming. In this case, a sanguineous priming solution may be used and hemoconcentrated prebypass. The hemoconcentrator can also be used after bypass when the perfusionist is returning aortic root vent volume. This modified ultrafiltration technique can raise the hematocrit substantially while also improving hemostasis.

Finally, hemodilution can also be reduced via another Cleveland Clinic technique: microplegia.[6] Briefly, this technique involves using a syringe pump to inject high potassium solution into whole blood in order to reduce exposure to crystalloid that typically occurs with Buckberg or del Nido cardioplegia protocols.

Goal directed perfusion

Goal directed perfusion (GDP) is a concept that has roots dating back to 1962.[7] The concept of GDP is to provide satisfactory tissue gas exchange and to allow each patient to serve as their own control. Typical measures of perfusion adequacy are keeping the hemoglobin >8 g/dL, the SvO_2 >60%–70%, PvO_2 >40 mmHg, pH = 7.35–7.45, pCO_2 = 35–45 mmHg, CI = 2.2–2.6 L/min/m^2 @ 37°C, and lactate <2 mmol/L. While these measures are useful and usually indicate adequate perfusion, basing perfusion adequacy on mixed venous sampling can overlook regional tissue malperfusion.[8]

GDP parameters include measuring consumption of oxygen (VO_2), production of CO_2 (VCO_2), and the delivery of oxygen (DO_2). Measuring and subsequently calculating these values can offer insight into the aerobic status of the patient at any given time. These GDP values are usually indexed to the patient's body surface area (BSA). While under anesthesia and at normothermia indexed VO_2 (VO_{2i}) is typically 40–50 mL O_2/min/m^2 and indexed CO_2 (VCO_{2i}) should be less than 60 mL CO_2/min/m^2.[9] Further information can be gleaned by creating ratios with these values. By dividing DO_{2i} by VCO_{2i}, the clinician can determine if the metabolic needs of the patient are currently being met, this value should be less than 5 in order to avoid a corresponding increase in lactate production.[10,11] The oxygen extraction ratio, VO_{2i}/DO_{2i}, is normally ~25% and can be used to determine the need for transfusion (~40% is a good trigger). Finally, the respiratory quotient, VCO_{2i}/VO_{2i}, is usually 0.8–1 and at values above 1 there is a corresponding increase in lactate production.

There is a large body of work supporting the need for minimum DO_{2i} during CPB in order to prevent incidence of postoperative acute kidney injury (AKI). In 2005, Ranucci and colleagues showed that increased DO_{2i} (>272 mL O_2/min/m^2) greatly reduced acute renal failure rates. These data were corroborated numerous times in the following years, and critical DO_{2i}'s range from 260[12] to >280 mL O_2/min/m^2 during the recent GIFT trial.[13] Regardless, keeping DO_{2i} in the upper 200s lowers risk of kidney injury.

The national incidence rate of all level AKI after cardiac surgery is reported to be 22%.[14] The length of stay for patients with AKI more than doubles relative to those without any AKI when matched on other variables, the postoperative costs increase from ~$18,000 to ~$37,000, and the mortality increases from 2.3% to 11.2%.[15] This gives impetus to reduce AKI rates, and one of the easiest ways to do this is to maintain adequate DO_{2i} during CPB.

There are some caveats and nuances with implementing GDP in any setting. All of the parameters and goals for GDP ratios assume normothermia. These values, especially DO_{2i}, can be significantly lowered in the setting of hypothermia. Calculation of these ratios does not include shunt flow, and if possible, a flow probe distal to all shunts should be used to measure the effective flow to the patient. Additionally, the root vent flow should be subtracted if the root vent tubing is full of blood and the clamp is off. When calculating the VCO_2, oxygenator exhaust CO_2 measurements can be obscured by sampling gas rates being lower than the waste air gas rates. Field flooding with CO_2 should be conservative (i.e., ~1 LPM of CO_2 flow). Renal injury should be assessed within 48 h after bypass and exposure to contrast dye should be noted. Finally, GDP practice is only able to be implemented with support from other disciplines (i.e., surgeons and anesthesiologists).

In our institution we have managed to increase DO_{2i} while on bypass via aggressive hemodilution reduction efforts. By using an optimized circuit to reduce prime volume, reducing further with dry venous lines and RAP methods, our patients typically have high hemoglobin levels upon initiation of CPB. While on CPB, the use of microplegia eliminates hemodilution further, and the use of 3/8″ pump boots helps to reduce the potential for mechanical destruction of cells. All of the above combine to reduce transfusion rates and ameliorate the need for excessively high flows. Typically while flowing at a 2.2–2.4 index our patient DO_{2i}'s are >300 mL O_2/min/m^2. Post CPB, the use of the Medtronic Autolog IQ autotransfusion device facilitates aggressive salvage of pump blood.

Pediatric perfusion circuit and methods

The pediatric perfusion circuit employs a combination of elements configured to suit the specific flow requirements of a patient population that ranges from 2.0 to 50 kg (Fig. 4.2; Table 4.1). Pump selection is based upon the patient BSA and flow requirements, which dictate both the size of circuit and type of prime used. Target flow (cardiac output) on bypass is calculated as 2.5 L/min × BSA. Lower flow rates are tolerated under special circumstances when cooling is employed, where the recommended minimum remains 2.0 L/min × BSA. Higher or lower CPB flows are permitted based on certain patient pathologies and at the discretion of the surgeon. Congenital anomalies often present extra flow considerations, such as the presence of shunts, collateral arteries, and MAPCAs (major aortopulmonary collateral arteries), which may necessitate much higher flows that reach one and one half to two times calculated value. Therefore, one must be careful not to underestimate the output requirements of borderline sized patients. The use of cerebral oximetry (*Masimo* SET), venous saturation (*Terumo CDI 500*), blood pressure, hematocrit and lactate levels, and oxygen delivery calculation is necessary to get a complete picture of required flow. Arterial line pressure during the conduct of CPB is not permitted to exceed 300 mmHg as measured from a port on the arterial line located distal to the oxygenator outlet. VAVD is employed only when necessary to augment inadequate venous return. A vacuum regulator (Boehringer) is used (Fig. 4.3), and reservoir pressure is limited to −40 mmHg. Neonate and infant patients rarely require negative pressure greater than −20 mmHg to achieve adequate patient drainage.

The Terumo Capiox FX-05 (Baby-FX) and Capiox FX15 oxygenators are valued for their low-prime high flow capabilities, integrated arterial filter (32-micron arterial filter screen), and ease of use. The oxygenators' self-venting feature allows for quick deairing. The FX-05 has a priming volume of 43 mL, minimum operating level of 15 mL, and can flow at rates from 0.1 to 1.5 L/min. The FX-15 has a priming volume of 144 mL, minimum operating level of 70 mL, and can flow at rates from 0.5 to 5.0 L/min. A metal bracket compatible with both oxygenator/reservoir units facilitates the quick interchanging of sizes. The pediatric tubing circuit has three ¼″ suction lines that are propelled by roller pump and color coded for easy communication with the surgical field. These lines may be used alternatively and interchangeably as cardiotomy suction, aortic root vent, LV vent, basket suction, or for any other ancillary volume retrieval as dictated by the surgical procedure. The entire perfusion circuit is coated with X-coating, a proprietary

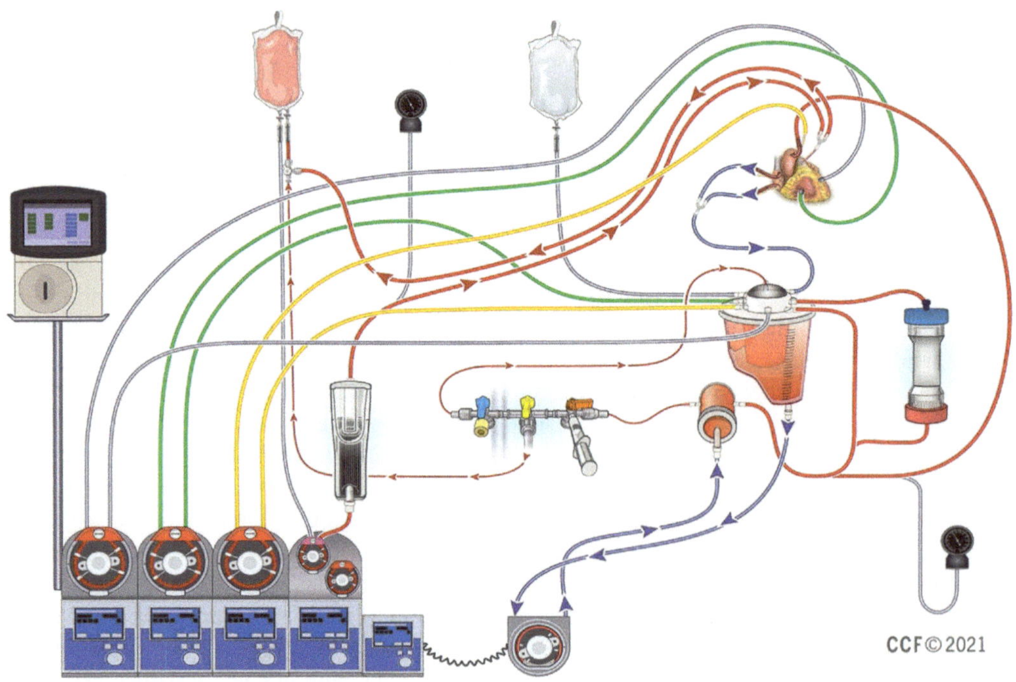

FIGURE 4.2 Pediatric cardiopulmonary bypass circuit.

TABLE 4.1 Pediatric cardiopulmonary bypass circuit sizing.

Patient	Weight (kg)	A/V loop	PUMPBOOT	Flow (mL/min)	Oxy/reservoir	Prime volume (mL)
Neonate	0–5	3/16″ × 1/4″	1/4″	<800	Terumo FX-05	173
Infant	5–12	1/4″ × 1/4″	1/4″	800–1800	Terumo FX-05	190
Child	12–25	1/4″ × 3/8″	3/8″	1800–2500	Terumo FX15–30	456
Pediatric	25–50	3/8″ × 3/8″	3/8″ or 1/2″	2500–4000	Terumo FX15–30	500

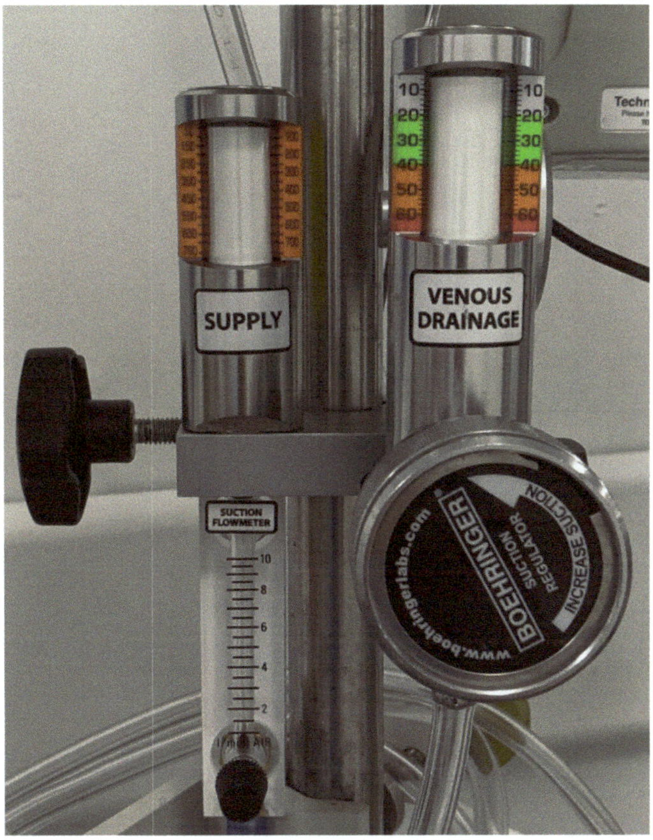

FIGURE 4.3 Boehringer vacuum-assisted venous drainage (VAVD) regulator. VAVD pioneered by Ball and Foster at Cleveland Clinic.

biopassive surface coating (Terumo) intended to reduce protein and platelet adhesion. A HEMOCHOR HPH 400 hemoconcentrator (Minntech) is placed in line for every case and used to independently control hemodilution levels of both the CPB prime and patient circulation. This helps maintain high hematocrit levels, reduce K^+ levels, maintain oncotic pressure, minimize diuretic use, and reduce the need for additional blood products on bypass. The hemoconcentrator contains biocompatible "no-rinse" fibers that do not require prepriming. The device's small size (membrane surface = 0.3 M^2, 34 mL prime volume) and high flow capability (400 mL/min) allow the perfusionist to selectively remove excess fluid introduced by priming volume and cardioplegia during CPB.

The driving force for all circuit components is the LivaNova S5 (LivaNova) roller pump. It is a positive displacement pump where tubing in a "raceway" is squeezed like a rolling pin, thereby propelling blood forward. Pump output depends upon pump speed (RPM), size of the raceway, tubing size, and occlusion of the rollers. The pump is not preload or afterload dependent which allows for precision at low flow rates. This fine control is important when performing selective regional perfusion in neonates and infants where flow rates as low as 150 mL/min may be required. Safety features of the LivaNova S5 include pressure and temperature control modules and low level and bubble detection.

Aortic and venous cannula selection is based on anatomical placement and flow requirements as determined by the operative procedure. Because the pediatric population ranges from 2.0 to 50 kg (neonate–teenager/young adult) there is a large range of sizes and styles (Figs. 4.4–4.6) Straight tip arterial cannulas are used for direct cannulation of the aorta, carotid artery, femoral artery, and innominate artery (in the case of antegrade cerebral perfusion) (Table 4.2). These cannulas are also often placed into a 3.5–6.0 mm Gore-Tex graft sewn onto a weaker vessel. All aortic cannula come with a "ring" or "bumper" (adjustable in some cases) for precise positioning and stability upon insertion. This is particularly important in pediatric patients where positional shifts of even a few millimeters can cause damage to tiny vessels. Venous cannulation is achieved by placing a single straight cannula directly into the right atrium ("single venous") or two individual cannula into the superior and inferior vena cava with snares applied ("bicaval") (Table 4.3). A right-angle bend in the venous cannula regularly used for bicaval cannulation permits them to be manipulated and turned into the vessel for optimal drainage. A metal tipped version of this is available for pediatric patients in sizes as low as 12 Fr. Its thin-walled design at small French sizes provides maximum drainage at minimal outer diameter. It should be noted

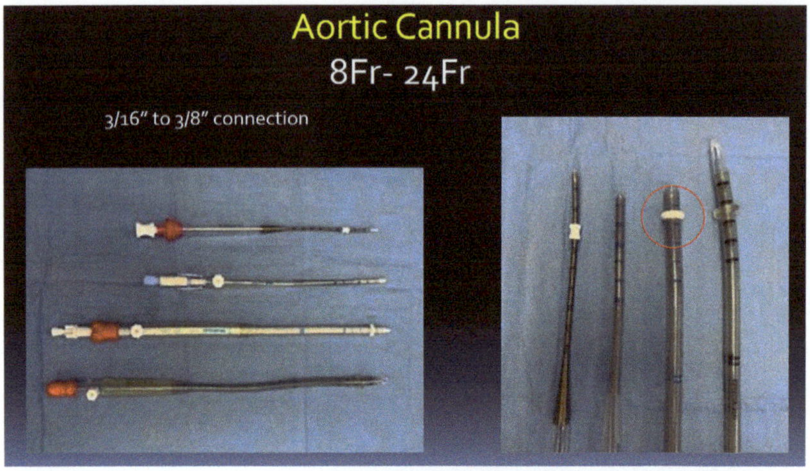

FIGURE 4.4 Aortic cannula 8 Fr–24 Fr.

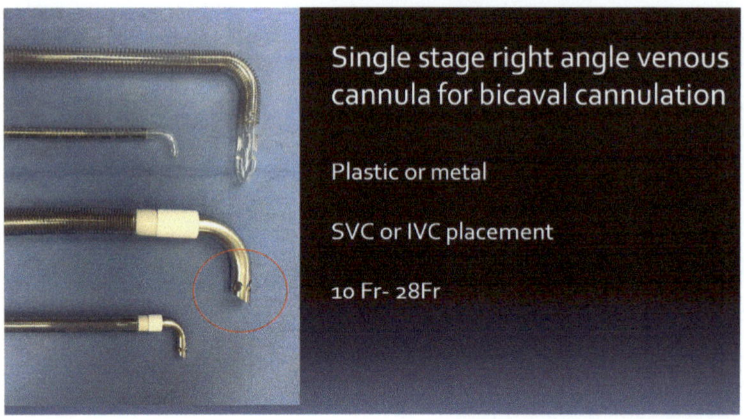

FIGURE 4.5 Right angled venous cannula.

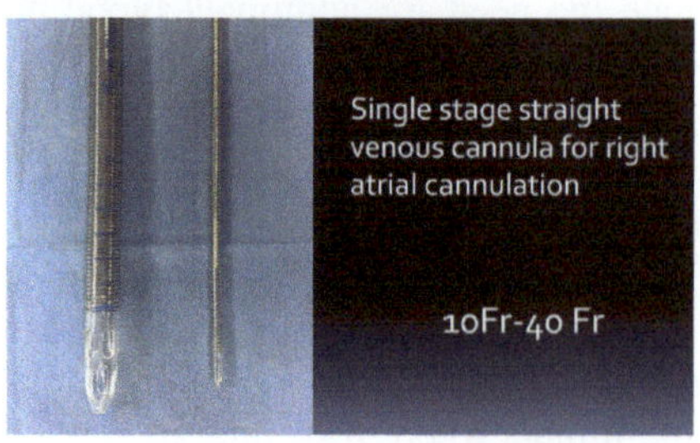

FIGURE 4.6 Single straight venous cannula.

that the metal-edged material makes this particular cannula sharper than its plastic counterparts. Great care must be used during insertion and removal to ensure that the vessels are not inadvertently punctured or torn. Improvements in cannula design and materials are constantly being developed, particularly in the pediatric arena, in an effort to solve the problem of "big hands operating in a small space." The end goal is always to provide a bloodless, immobile surgical field. Efficient and adequate cannulation is imperative to facilitate shorter time on CPB, which is critical for successful pediatric outcomes. The elements of the CPB prime are based on patient and circuit size, preoperative hematocrit of the patient, and pathophysiology. The priming volumes listed in Table 4.1 represent the minimum operating volume of the HLM (Fig. 4.7). This includes the oxygenator, all filled tubing lines, and minimum reservoir volume.

TABLE 4.2 Arterial cannula sizes.

Medtronic DLP one-piece arterial cannula	*Medtronic* Bio-Medicus arterial cannula (heparin coated)		
6 Fr	Up to 2.5 kg	8 Fr	Up to 4 kg
8 Fr	Up 5.5 kg	10 Fr	Up to 8 kg
10 Fr	Up to 11 kg	12 Fr	Up to 12 kg
12 Fr	Up to 15 kg	14 Fr	Up to 17 kg
14 Fr	Up to 25	15 Fr	Up to 40 kg
16 Fr	Up to 35 kg	17 Fr	Up to 60 kg
Medtronic EOPA			
18 Fr	Up to 50 kg		

TABLE 4.3 Venous cannula sizes.

Superior vena cava		Inferior vena cava		Single atrial cannula	
10 Fr	Up to 3 kg	10 Fr	Up to 2.5 kg	16 Fr	Up to 2.5 kg
12 Fr	Up to 7.5 kg	12 Fr	Up to 3.5 kg	18 Fr	2.5–4 kg
14 Fr	Up to 15 kg	14 Fr	Up to 5 kg	20 Fr	4–6 kg
16 Fr	Up to 25 kg	16 Fr	Up to 7.5 kg	22 Fr	6–10 kg
18 Fr	Up to 35 kg	18 Fr	Up to 12 kg	24 Fr	10–20 kg
20 Fr	Up to 45 kg	20 Fr	Up to 20 kg	28 Fr	20–30 kg
22 Fr	Up to 55 kg	22 Fr	Up to 40 kg	32 Fr	30–40 kg
		24 Fr	Up to 65 kg	36 Fr	40–75 kg

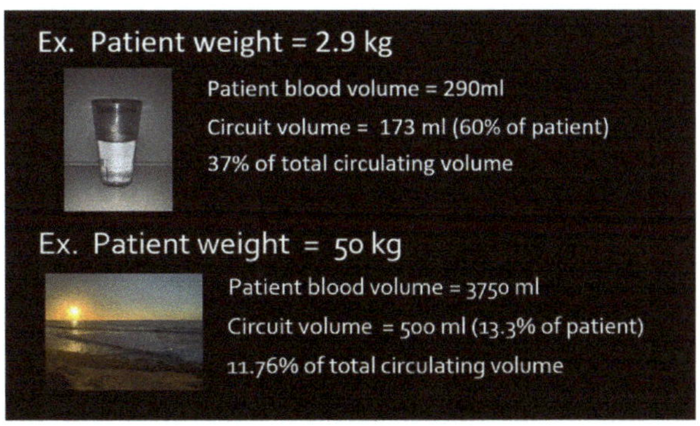

FIGURE 4.7 Cardiopulmonary bypass prime volume comparisons for pediatric population.

Efforts are made to reduce these numbers by positioning the CPB machine close to the operating table prior to initiating bypass so that the arterial–venous tubing loop (Fig. 4.2) can be cut, and the excess volume drained into a basin. Patients weighing greater than 12 kg ("Child" circuit or larger) receive a clear prime which consists of Plasmalyte 148, heparin (3.0–5.0 cc dependent on weight), and albumin (50 mL, Albumin (Human), USP, 25% solution FLEXBUMIN) to maintain oncotic pressure. An arterial blood gas sample (GEM Premier 3500 analyzer) is taken of the prime and is used to measure electrolytes (Na^+, K^+, and Ca^{2+}) and acid base balance (pH). Sodium bicarbonate (1 mEq/mL) and calcium chloride (100 mg/mL) are then incrementally added to buffer the prime and raise electrolytes to normal physiological levels. CPB circuits for patients weighing less than 12 kg ("Neonate" and "Infant" circuit) are primed with blood. A blood prime consists of one unit of washed red blood cells and either one half unit of prethawed fresh frozen plasma or 50 mL 25% albumin. The inclusion of fresh frozen plasma is at the discretion of the perfusionist and anesthetist who will be managing the patient postoperatively. When utilized, the remainder of the plasma unit is refrigerated (4-h shelf life) and administered pre- or postbypass by the anesthesiologist to replace clotting factors while minimizing donor blood exposure. Patients ≤ 5 kg always have plasma included in their prime, as well as those undergoing reoperation or where unusual bleeding may be expected postoperatively due to extreme hemodilution or pathological lack of clotting factors. When using a blood prime, the circuit is first wet and deaired with Plasmalyte 148. Once the prime has been recirculated and air removed, a portion of the Plasmalyte volume is displaced by blood at a level which still maintains the airless integrity of the circuit. Heparin is added upon addition of blood to the tubing (3 cc $\approx$ 3000 units). Following these constituents, sodium bicarbonate and calcium chloride are added until the acid–base balance (pH) is normalized and normal values of 135–145 mmol/L Na^+ and 1.2 mmol/L Ca^{2+} are achieved (as close as possible) as measured by an arterial blood gas sample. The pO_2, pCO_2, K^+, lactate, and glucose levels are also corrected. Once any blood products have been introduced to the circuit, constant low flow recirculation of the prime is maintained at 37°C. Oxygenator FiO_2 is set at 0.21 and sweep gas at 400 cc min to maintain physiological $PO_2 \approx 120–150$ mmHg and pCO_2 35–45 mmHg. The recirculating CPB prime may be hemoconcentrated using a Minntech HEMOCOR HPH 400 hemoconcentrator to reach normal K^+ (3.5–5.0 mmol/L) and a hematocrit $\geq 20\%$. Blood prime considerations are always made based upon the patient's preoperative arterial blood gas values, the goal hematocrit on bypass, and the postoperative hematocrit goal which will provide a stable separation from the HLM. In the case where an unusually high preoperative hematocrit is present due to cyanosis, whole blood may be withdrawn from the patient by anesthesia IV and reinfused later. This facilitates the conduction of bypass at a safe hematocrit while providing an autologous source of volume replacement for later use. Upon initiation of bypass, mannitol and Amicar are commonly added to the circulation (except where contraindicated). Mannitol (0.75 mg/kg) is used to stimulate diuresis and for its protective effect on renal function. Amicar (25–75 mg/kg dependent on weight) is used for reoperations and complex congenital repairs to control bleeding.

Pediatric autotransfusion

Pediatric perfusionists manage the Fresenius CATSmart cell saver to retrieve and process shed blood from the surgical field as well as wash all banked blood prior to transfusion. Shed blood is anticoagulated, collected in a sterile reservoir, and processed in a continuous flow centrifuge to obtain washed packed red cells that are reinfused to the patient or added to the CPB machine. This process removes nonerythrocytic cellular components, activated clotting factors, products of fibrinolysis and cell trauma, as well as anticoagulants.[16] All red cell units obtained from the blood bank are also washed prior to use. The washing of banked blood units removes preservative solutions which may be acidic in nature and reduces volume to simply red blood cells resuspended in 0.9% NaCl. The CATSmart system is ideally suited for the pediatric population since it operates as a continuous flow centrifuge which is preprimed and therefore demands a very minimal volume to function. Blood is withdrawn continuously from the reservoir to the washing chamber where it is washed and pumped to an attached reinfusion bag. (Figs. 4.8–4.10), The CATSmart uses ultrasonic sensors to monitor both the hematocrit (%) of the blood to be washed as well as the hematocrit (%) of the final product. The hematocrit values of washed cells are routinely $\geq 50\%$.[17] There are three wash programs which allow the operator to customize the PRC flow and wash factor according to specific scenarios. The "smart wash" and "low volume wash" programs have a lower processing speed which allows for the best possible treatment of the blood cells that can be initiated at volumes as low as 35–40 mL. If an increased processing speed is required, there is an "emergency wash" program that invokes a shorter processing time with faster blood flows. Prior to reinfusion, blood is filtered using a Haemonetics RC1VE high efficiency leukocyte removal filter. In the case of larger blood loss where volume needs to be transfused rapidly, an SQ40 Microaggregate Blood Transfusion filter may be substituted. Leukocyte filtration is preferred as it protects compromised immune systems from leukocyte associated transfusion complications such as

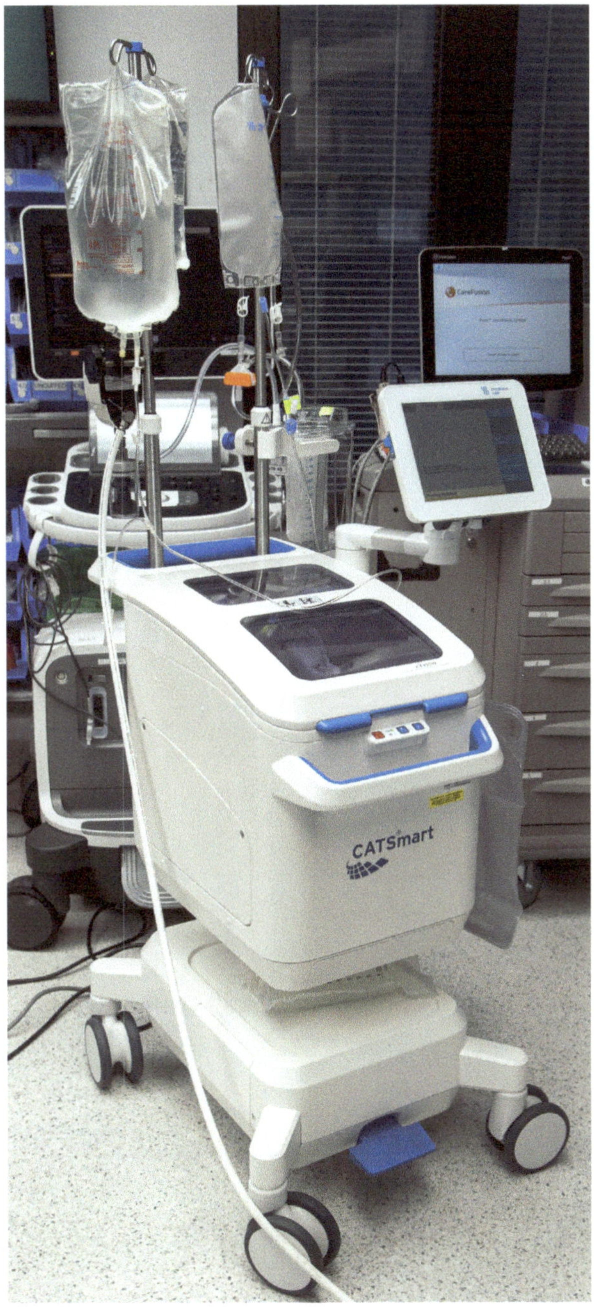

FIGURE 4.8 Fresenius Kabi CATSmart cell-saver.

Cytomegalovirus and alloimmunization. Washed blood is returned to the patient via the reinfusion bag. The continuous nature of the CATSmart system allows blood to be drawn from the reinfusion bag, in increments of 30–50 cc via syringe, and returned to the smallest patients immediately upon availability. The wash process is interrupted automatically and the operator alerted by a tone when the saline bag or reservoir is empty. At the conclusion of a case, any remaining volume in the CPB circuit after separation from the patient is emptied into the cell saver and processed in the same manner as above. The perfusionist must be careful of certain relative contraindications to the reinfusion of cell-saved blood which include infection, malignancy, sickle cell disease, thalassemia, or the aspiration of incompatible topical agents.

Installation of the CATSmart disposable system is made simple by the sterile preconnection of most of its components. Quality control of the CATSmart includes a machine self-test on initiation and routine independent monitoring of

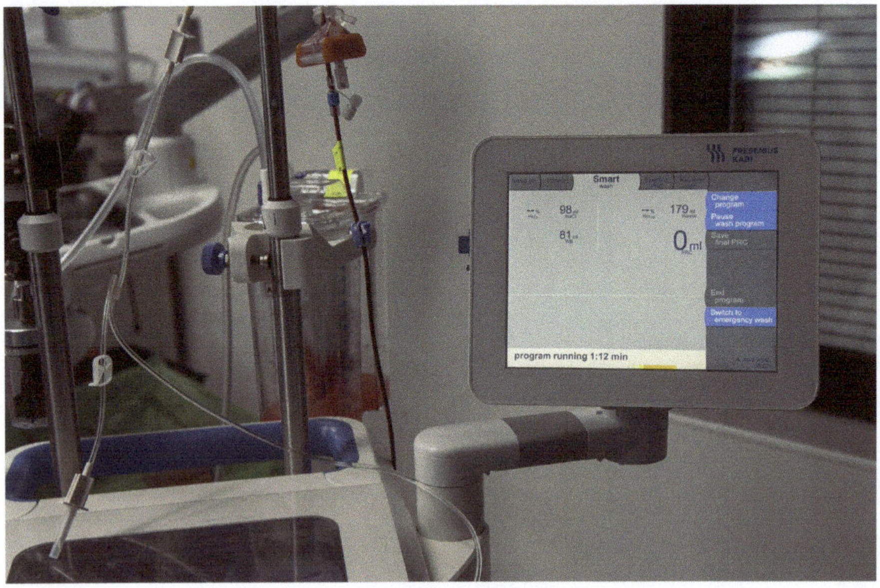

FIGURE 4.9 Smart wash program.

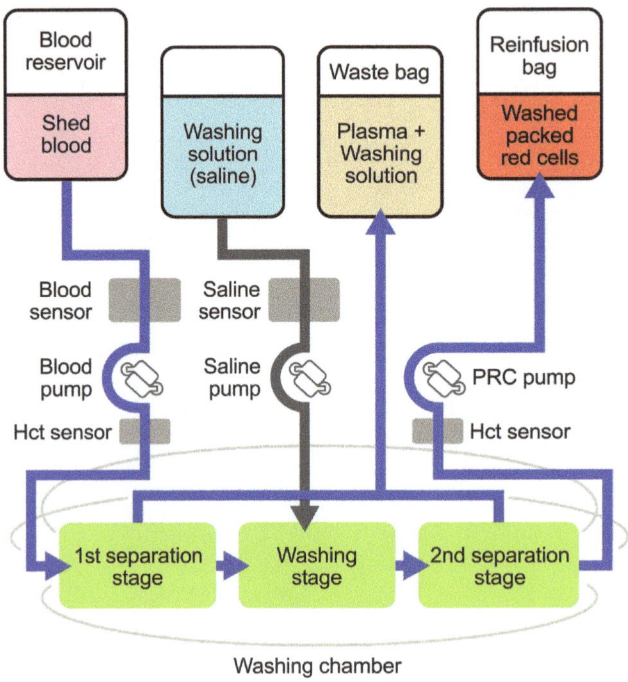

FIGURE 4.10 Fresenius washing process.[4]

possible residual heparin, as well as hematocrit and K^+ levels. This is based upon individual institution and Association for the Advancement of Blood and Biotherapies (AABB) recommendations.

GEM premier 3500

The GEM Premier 3500 blood gas analyzer (Figs. 4.11 and 4.12) can be directly mounted to the mast of the CPB pump and is therefore situated to provide arterial and venous blood gas, electrolyte, glucose, lactate, and hematocrit values to the perfusionist within minutes. The low sample volume (150 μL for BG/Hct/Lytes/Glu/Lac) quick result time (120 s), and customizable features (automatic temperature correction, critical value ranges, etc.) make it an ideal choice for the

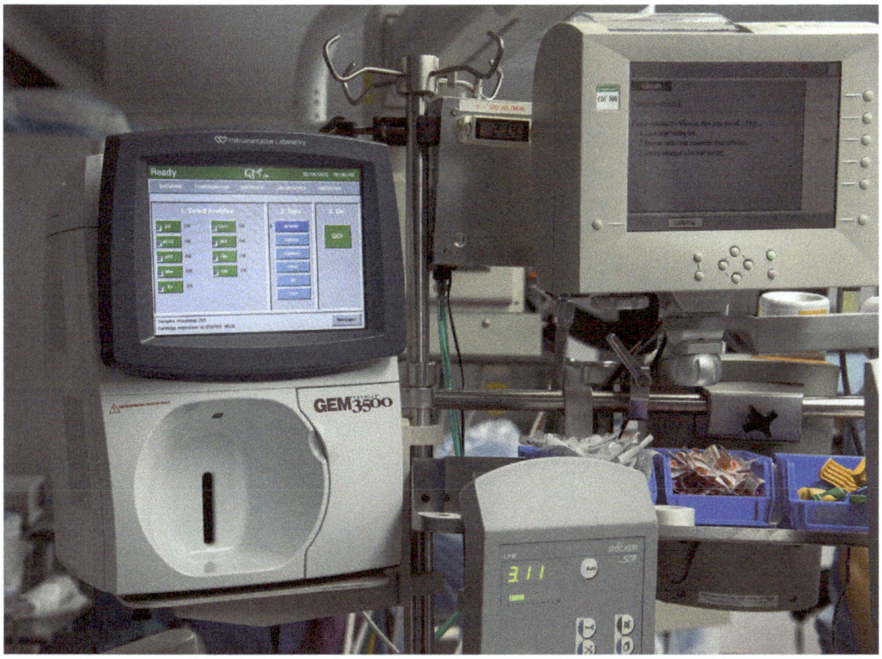

FIGURE 4.11 GEM Premier 3500 mounted on cardiopulmonary bypass pump.

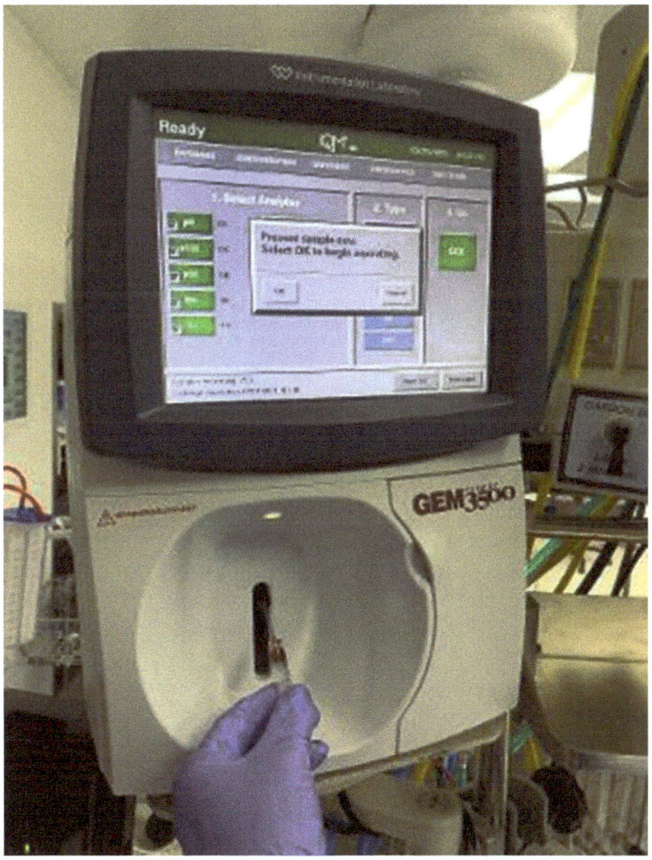

FIGURE 4.12 GEM Premier 3500 sampling.

conduct of neonatal and infant perfusion which requires careful attention to the exaggerated effects on the circulation of both temperature and dilution. Real time arterial and venous samples can be obtained directly from the manifold of the extracorporeal circuit and immediately run without leaving the hand of the perfusionist. The HLM operator can then make rapid adjustments according to current physiological conditions. This is extremely useful during operative procedures that employ hypothermic circulatory arrest where blood gases are run almost continuously while the patient is being cooled and rewarmed. Certain complex repairs require that the circulating blood temperature be lowered below 28°C so that all or parts of the circulation may be safely temporarily stopped for short periods of time. At nasal temperatures below 28°C, pH stat blood gas management is used to increase cerebral blood flow. During pH stat acid–base management, the patient's pH is maintained at a constant level by correcting pH to the patient's temperature. Arterial blood gases are corrected to the patient's actual temperature, not the standard 37°C. Compared to alpha-stat, pH stat (which aims for a pCO_2 of 40 and pH of 7.40 at the patient's actual temperature) leads to higher pCO_2 (respiratory acidosis), and increased cerebral blood flow. Often CO_2 is deliberately added to maintain a pCO_2 of 40 mmHg during hypothermia.[18] Facilitating this technique, the GEM Premier 3500 allows the operator to manually enter a patient temperature with each sample. The analyzer will then automatically calculate values at both measured (37°C) and temperature corrected values for direct comparison (Fig. 4.13). This provides the perfusionist with an efficient tool to conduct pH stat blood gas strategy at a rapid pace.

When separated from bypass, patient samples are continued to be run without leaving the operating room to monitor postoperative progress such as lung function, acidosis, saturations, and electrolyte deficiencies. Dry heparinized syringes are preferred because they ensure sufficient heparin to prevent clotting yet avoid the dilutional effect of liquid heparin.[19]

The GEM Premier 3500 allows the operator to customize analyte panels to control what parameters/analytes are measured and decide what measurements are deemed "critical values." Any results that are outside of normal range are flagged to attract the operator's attention. GEM Premier 3500 outputs are compatible with multiple electronic medical

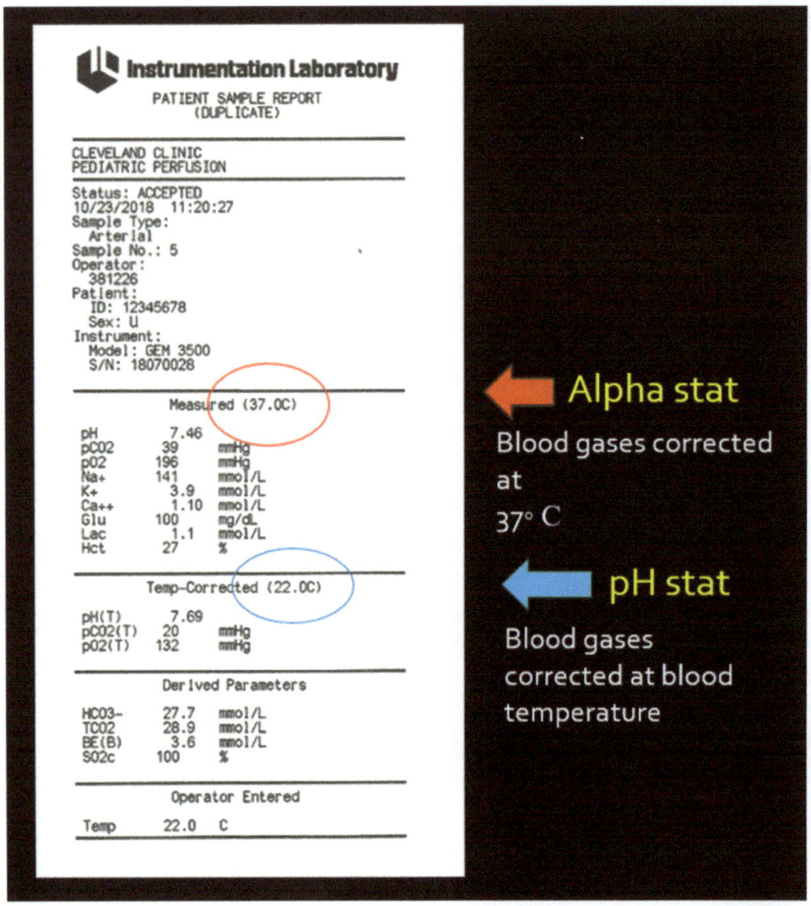

FIGURE 4.13 Autotemperature correction.

record programs for ease of record keeping. Patient and quality control data can also be easily managed with the GEMWeb custom program.

Point of Care Testing inspection is a vital safety component of any program that has personnel performing laboratory testing outside of a centralized laboratory. The GEM 3500 is designed in a way that allows it to routinely meet the accreditation standards of both the College of American Pathology (CAP) and Clinical Laboratory Improvement Amendments (CLIA) with its unique "IQM technology." The *GEM 3500* performs self-initiated quality control and assessment, automated trouble-shooting, and clot and interference detection by virtue of internal software. This automation makes the instrument safely available to the perfusionist 24 h/day with little intervention required while relieving concerns regarding the validity of its results. The instrument is relatively compact (17.5 × 11.8 × 13 in, 31.2 lbs,) portable, has a wide operating temperature (15°C–35°C), and uses self-contained, nonrefrigerated cartridges that may be ordered in increments according to amount of use. All these considerations make this analyzer a cost-effective and valuable tool for the pediatric perfusionist.

Adult autotransfusion

Autotransfusion helps to reduce the need for blood transfusion. In 2019 the Medtronic Autolog IQ was chosen as the standard autotransfusion system for the adult cardiac surgery environment, replacing the previous Medtronic Autolog which had been in service since 2007. In 2020, the Autolog IQ delivered an average of 580 mL per case, for almost 4000 cases, which equates to nearly 600 gallons of autologous blood per year.

The Medtronic Autolog IQ device is easy to set up, easy to troubleshoot, and it is easy for new staff or students to understand quickly. We have modified the device to have a longer IV pole, and we use the hidden storage compartment to store cleaning wipes for rapid room turnover. We have also added small orange zip ties to the power cord and vacuum tubing plugs to simplify the process of moving or swapping out cell savers for quality control checks. The standard gray power cord color is otherwise found on many devices in the operating room, so this helps to avoid confusion.

Arterial line filter technology

Arterial filters on the extracorporeal circuit serve two purposes: (1) microaggregate filter; and (2) bubble trap for gaseous microemboli (GME). In 1992, Pall debuted the LG6 (LeukoGuard-6) arterial leukocyte filter for extracorporeal service, thus adding leukofiltration as a third function (Fig. 4.14). The fundamental premise of the Pall LG6 is that a special coating on the filter can be used to attenuate systemic inflammatory response while on bypass despite the additional surface area required.

The Pall LG6 was evaluated in a laboratory setting in 1992.[20] At a flow rate of 4.5 liters per minute (LPM) the filter was observed to deplete polymorphonuclear leukocytes (PMNs) during the first 30–40 min of perfusion.[1] Over the next 16 years, researchers across the globe failed to identify a significant impact on patient outcomes. In 2005, a group in

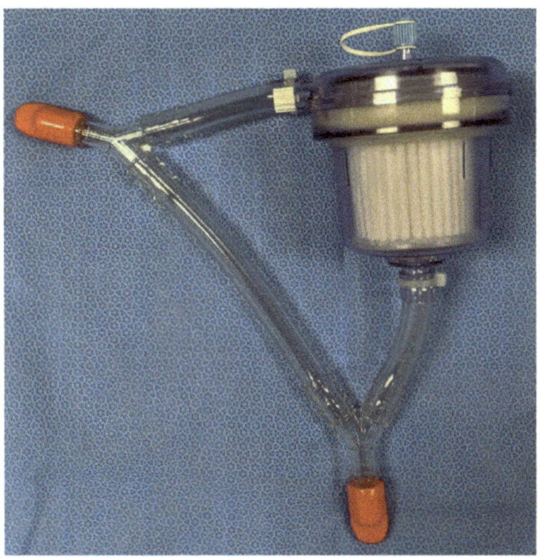

FIGURE 4.14 Pall LG6 arterial leukocyte filter with bypass line.

Finland established that Pall LG6 was ineffective and possibly harmful if deployed continuously.[21] This sparked more research on optimal deployment strategies for shorter durations (such as prior to crossclamp off, or prior to weaning of bypass), however, the effect of these strategies on patient outcomes could not be confirmed.

In 2008 the STS Blood Conservation Guidelines recommended against the use of leukocyte depletion filters for extracorporeal circulation because of the potential for harm.[22] Also in 2008, Terumo debuted the integrated arterial filter on the Capiox hollow fiber membrane oxygenator, which promised an overall reduction in surface area and prime volume while still retaining the ability to manage GME.[23] The advantages granted by integrating the arterial filter into the oxygenator increased the relative costs associated with use of the Pall LG6 (see chart below with all costs identified), including redundancy for two of its functions (microaggregate filter and bubble trap).

Though study of the Pall LG6 arterial filter continued beyond 2008, it was at a reduced pace, and still failed to identify a significant difference in patient outcomes. In 2013 a group in China established 60 min as an upper limit for continuous deployment using a similar device.[24] In 2014 a group in Brazil established that Pall LG6 was effective for neutrophil sequestration for up to 50 min at flows of 5 LPM.[25]

The Pall LG6 may be useful as one avenue to attenuate systemic inflammatory response so long as its upper limit for use is not exceeded. Based on the findings above we utilize the Pall LG6 in a limited capacity for ex vivo lung perfusion circuit (which typically operates at lower flow rates with STEEN solution, thus being safer for longer duration), and lung-ECMO (ExtraCorporeal membrane oxygenation) bypass for lung transplants (utilized if conversion is necessary to full bypass), and pulmonary embolectomies. The costs associated with its use are considered briefly in the Table 4.4.

Oxygenators—current state-of-the-art

At the Cleveland Clinic we currently use oxygenators by Terumo (adult and pediatric), and Medtronic (adult). The Terumo Capiox FX Advance (Fig. 4.15) is the standard adult oxygenator used at Cleveland Clinic. It has established a track record for reliability, and it is easy to prime quickly. The Advance was introduced in 2016 and featured several minor improvements over the regular FX, such as vortex reduction at low operating volumes. The FX Advance also lowered the minimum recommended operating volume in the reservoir from 200 to 150 mL, which improves the margin of safety and helps to reduce hemodilution.

The Medtronic Affinity Fusion (Fig. 4.16) is a backup oxygenator in case there are interruptions to the Terumo supply line. The Medtronic Affinity Fusion oxygenator is based on the same hollow fiber membrane technology, but the oxygenator is rotated 90 degrees (relative to Terumo) to appear like a barrel that stands on end.

TABLE 4.4 Pall LG6—cost versus benefit analysis.

Costs	Absolute	Relative	Reference
Prime volume increase	+250 mL	+31% approx.	Cleveland Clinic Foundation (CCF) Terumo circuit prime volume 800 mL
Surface area increase	+360 cm^2	+14% approx.	Terumo oxygenator surface area ~2500 cm^2
Setup time increase	+5–10 min	+25% approx.	Terumo circuit 15–20 min to build and prime
Material cost increase	+$130–150	Unknown	CCF Terumo and Medtronic circuit costs
Blood conservation decrease	−20–250 mL	Variable	Inhibits postbypass blood conservation
Limits arterial flow to 6 LPM	−1.0 LPM	−15% approx.	Terumo Capiox FX Advance otherwise capable of 7 LPM

Additional costs

- redundant function as a microaggregate filter if used with an oxygenator that has an integrated arterial filter
- redundant function as a bubble trap if used with an oxygenator that has an integrated arterial filter
- potential for contamination if cutting into a prebuilt circuit
- potential for harm if deployed greater than 50–60 min at 4–5 LPM (ref)

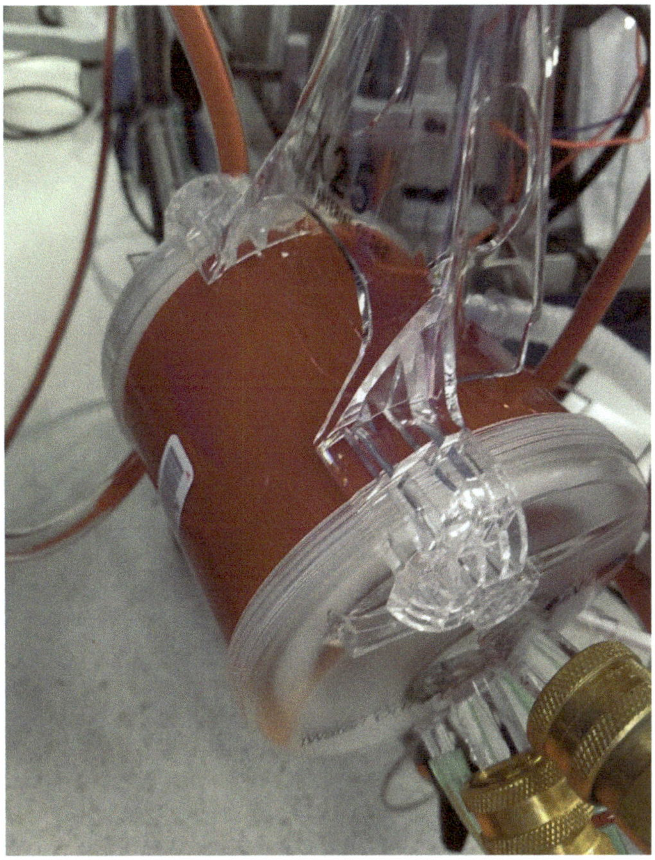

FIGURE 4.15 Terumo CAPIOX FX Advance oxygenator.

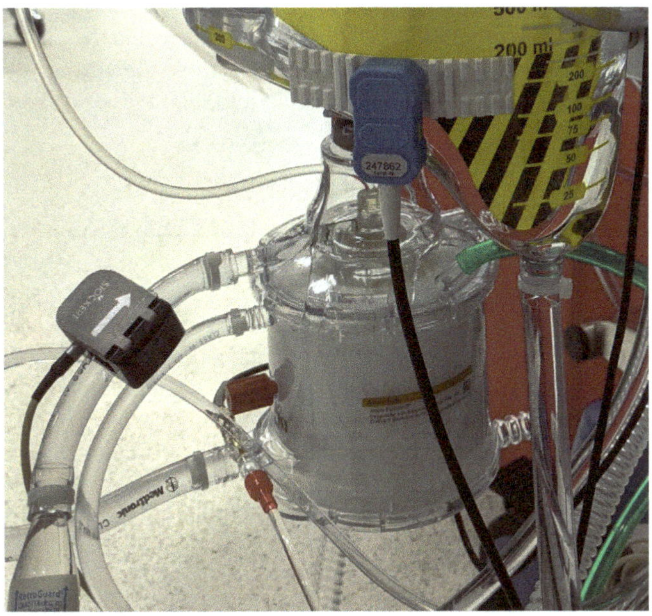

FIGURE 4.16 The Affinity Fusion with Cortiva surface coating.

TABLE 4.5 Oxygenators—comparison chart.

Oxygenator	Terumo FX25, FX25R	Medtronic Affinity Fusion	Terumo NX19
Fiber material	Microporous polypropylene	Microporous polypropylene	Microporous polypropylene
Fiber bundle SA	~2.5 m^2	2.5 m^2	~1.9 m^2
HE material	Stainless steel	Polyethylene terephthalate	Polypropylene terephthalate
HE SA	~0.2 m^2	0.4 m^2	~0.43 m^2
Arterial filter SA	600 cm^2	Not specified	360 cm^2
Priming volume	260 mL	260 mL	~185 mL
Blood Flow range	0.5–7.0 LPM	1.0–7.0 LPM	2.0–8.0 LPM
Min operating level	150 mL	200 mL	150 mL
Coating	X-Coating	Balance or Cortiva	X-Coating

HE, Heat exchanger; *SA*, surface area.

The Medtronic oxygenator has five characteristics that may affect its selection in practice. First, the Affinity Fusion often requires an extra step to prime because it can accumulate micro air near the top edges of the barrel-shaped oxygenator. To clear the micro air it is necessary to clamp out the A–V loop and direct all flow out through the cardioplegia line at high pressure. Second, the Affinity Fusion reservoir has a funnel shape (rather than Terumo's tapered rectangular shape) and a vortex is often visible with 500 mL in the reservoir. To operate safely above the level sensor it is often necessary to have more volume in the reservoir, which limits the ability to aggressively reduce hemodilution. Third, the Medtronic has a gooseneck-shaped venous port to the reservoir that can make it more difficult to prevent kinking of venous lines. Fourth, the Medtronic oxygenator also requires active purging to manage air (small shunt from top of oxygenator to venous line), which adds an extra step when going on and off bypass. Finally, the Medtronic oxygenator has Cortiva surface coating, which is heparin-based (whereas Terumo X-coating is not heparin-based), so it should be avoided in heparin-induced thrombocytopenia (HITT)-positive cases (Table 4.5).

Oxygenators—room for improvement

The Terumo Capiox hollow fiber membrane oxygenator with integrated heat exchanger made its debut in 1983.[26] Hollow fiber membrane oxygenator technology has gradually evolved and continues to be used today. In January 2019 Terumo revealed their latest hollow fiber membrane oxygenator, the Terumo NX 19 (Table 4.5), featuring a further reduction in prime volume and surface area, as well as increased flow rates, and additional heat exchange surface area.[27] Hypobaric oxygenators are under development that enable a radical reduction in GME.[28–30]

Heat exchangers

The primary heat exchanger for systemic perfusion used at Cleveland Clinic is composed of stainless steel and is integrated into the Terumo Capiox FX Advance oxygenator. The backup Medtronic Affinity Fusion oxygenator integrated heat exchanger is composed of polyethylene terephthalate (PET), the same material found in blood collection tubes.

For cardioplegia delivery, the heat exchangers of both Terumo (Fig. 4.17) and Medtronic devices (Fig. 4.18) are composed of stainless steel in a polycarbonate housing. The Terumo heat exchanger is shaped like a fire hydrant and has a priming purge located at the top center of the device, with cardioplegia delivery line extending from the top edge of the dome. The Medtronic MYOtherm XP heat exchanger is shaped like a hockey puck and has a priming purge off-center, with cardioplegia delivery line extending downwards through a 150-micron microaggregate filter.

The Medtronic Biotherm heat exchanger is larger than the MYOtherm XP and has been used in a 3/8″ line when an oxygenator is not necessary, such as for liver bypass or left heart bypass. However, production of the Biotherm has been discontinued, so the MYOtherm XP is now being used for this purpose, despite having to reduce from 3/8″ tubing to 1/4″ inflow and outflow.

The standard heater-cooler used at Cleveland Clinic is the LivaNova (formerly Sorin, originally Stockert) 3T. The device is favored because it supports the speed of the surgeons and permits reduced times on CPB. At 120 volts (down

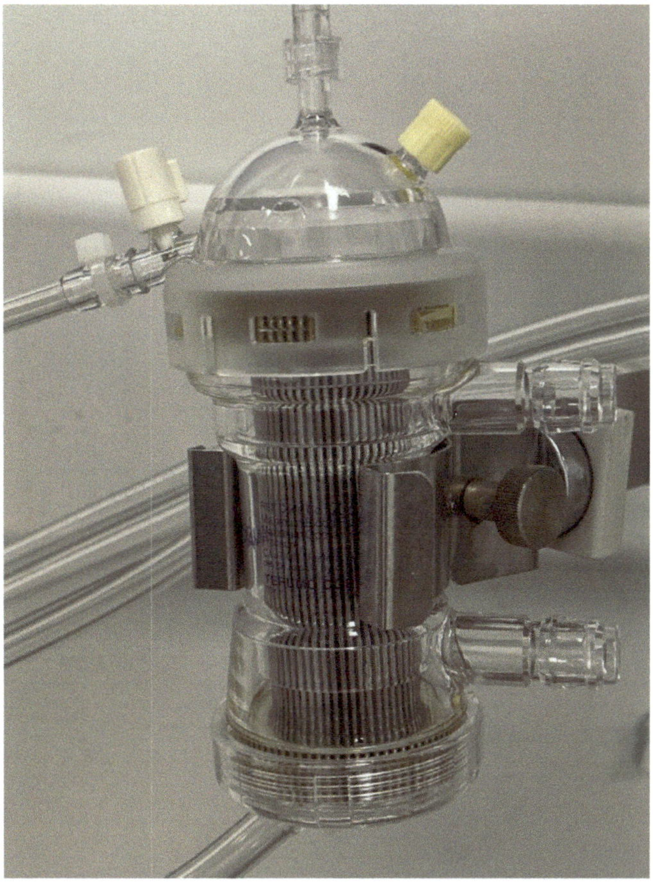

FIGURE 4.17 Terumo Capiox cardioplegia heat exchanger.

from European 240-volt model), the device gives approximately 11 LPM through the systemic circuit and approximately 8 LPM through the cardioplegia circuit.

There has been concern in the past with these devices about the potential for production and aerosolization of bacteria.[31] We have developed a rigorous cleaning and testing program in accordance with LivaNova's revised instructions for use.[32,33] The water is tested daily for concentration of hydrogen peroxide for a target of 100–500 ppm. The water is changed weekly, and the devices are disinfected with bleach every other week. Water from the devices is tested monthly (pre- and postdisinfection) to confirm that overall process is working. We have also modified the devices to have an internal vacuum system to prevent aerosolization, and modified our practice by leaving water lines connected to the bypass circuit, and not drained until the patient leaves the room. After the lines are drained and disconnected from the oxygenator, the connections are wiped with CaviWipes (CAVI) antibacterial wipes. By attacking every stage of the problem we have eliminated the risk associated with these devices.

Heater-coolers—field modifications

In order to satisfy the requirements of the 3T device, several field modifications have become necessary. To mitigate the logistical challenge involved in testing, rotation, and cleaning, we have built H_2O_2 test strip container holders out of 1-L normal saline bottles and zip-tied them to the side of the device, reinforced with heavy plastic used from Medtronic pump packs to make them durable. From the same heavy plastic we have also made H_2O_2 test sheet holders for daily testing, secured to the side panel of the device with a screw. The plastic is very durable, and in this configuration does not require much intervention or repair.

We have also found the default worm drive hose clamps to be problematic because they leave a sharp metal tail that can puncture gloves and skin. In order to eliminate the sharp edges, we use zip ties to secure the hose to the tail, and 1.5″ polyvinyl chloride (PVC) sleeves that can be fit over the worm drive clamps and zip-tied in place. The PVC

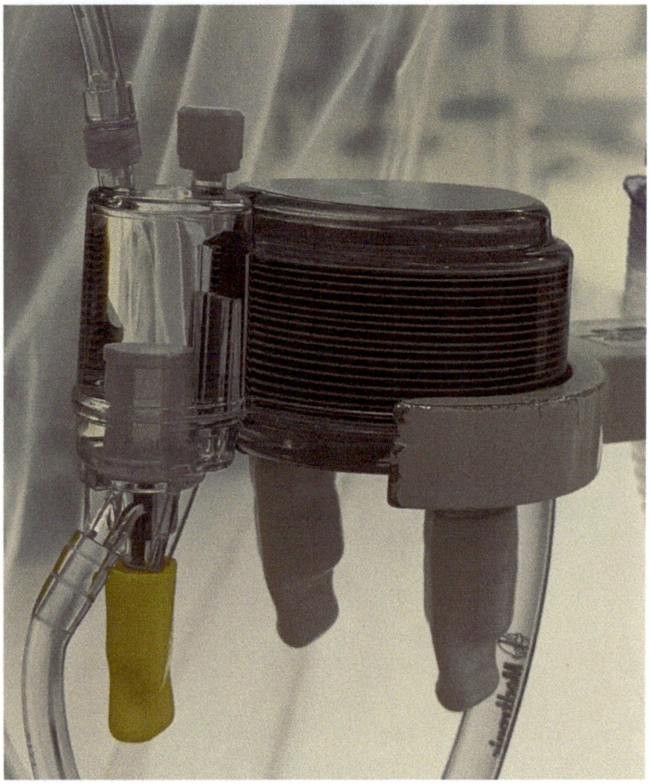

FIGURE 4.18 The Medtronic MYOtherm XP cardioplegia heat exchanger.

sleeves are very durable and typically last for one year of daily use before they are punctured and rendered useless. We have also considered heat-shrinking plastic for ease of installation and removal, but it is more expensive and thought to be less durable than PVC. A cheaper alternative is radial-seal stainless steel zip ties (to replace the worm drive clamps completely).

Heater-coolers—room for improvement

In practice we think that there is absolutely room for improvement with the 3T device or its successor. Foremost, the device is labor-intensive. The rigorous cleaning process and all related testing, supplies, and replacement of vacuum canisters and tubing, hydrogen peroxide, test strips, CaviWipes, etc. for multiple devices creates a disproportionate burden that reduces the availability of perfusionists and supply chain managers for other tasks. Second, in order to make the devices available for cleaning, it is necessary to rotate them out of service. This means coordinating their rotation (with timing constraints around existing cases) and having more devices than cases so that you are not limited by availability.

A good device would be one that simplifies the maintenance process and related costs. In the future we may see a closed-loop disposable, or a method of integration with the heat exchanger, for example, such that the burden of water draining, cleaning, and testing can be eliminated. The development of hypobaric oxygenators and their ability to radically reduce GME could offer a completely new approach.

Conclusions

At the Cleveland Clinic we have had the opportunity to refine our processes to meet the demands associated with a moral imperative to provide care for as many patients as possible. The lessons learned by our founders in time of war are still relevant today as we celebrate our 100th year (1921–2021). We expect that we will continue to grow and refine to meet the demands of tomorrow, and we look forward to the introduction of novel devices and innovative solutions that will help us to do our job better.

References

1. Ogella DA. Advances in perfusion technology—an overview. *J Indian Med Assoc*. 1999;97(10):436–437. 441. PMID: 10638107.
2. Kim K, Ball C, Grady P, Mick S. Use of del Nido Cardioplegia for Adult Cardiac Surgery at the Cleveland Clinic: Perfusion Implications. *J Extra Corpor Technol*. 2014;46(4):317–323.
3. Assets.livanova.com; 2021. [online] Available at: https://assets.livanova.com/media/com/documents/therapeutic%20areas/cardiopulmonary/brochures/im-7300030-cp_b-s5_product_catalogue_2020_rev5.pdf?ext = .pdf. Accessed 19 April 2021.
4. 550, C. *CDI® Blood Parameter Monitoring System 550: Perfusion Products: Products: Terumo Cardiovascular Systems*. [online] Terumo-cvs.com; 2021. Available at: https://www.terumo-cvs.com/products/ProductDetail.aspx?groupId = 1&familyID = 906&country = 1. Accessed 19 April 2021.
5. Loor G, Koch CG, Sabik 3rd JF, Li L, Blackstone EH. Implications and management of anemia in cardiac surgery: current state of knowledge. *J Thorac Cardiovasc Surg*. 2012;144(3):538–546.
6. Borden 2nd RA, Ball C, Grady PM, et al. Microplegia vs 4:1 blood cardioplegia: effectiveness and cost savings in complex cardiac operations. *Ann Thorac Surg*. 2020;110(4):1216–1224.
7. Galletti P, Brecher G. *Heart-Lung Bypass, Principles and Techniques of Extracorporeal Circulation*. New York: *Grune & Stratton*; 1962:258.
8. McDaniel L, Zwischenberger J, Vertrees R, et al. Mixed venous oxygen saturation during cardiopulmonary bypass poorly predicts regional venous saturation. *Anesth Analg*. 2010;80:466–472.
9. Justison G. Is timing everything? *J Extra Corpor Technol*. 2017;49(2):P13–P18.
10. Ranucci M, Carboni G, Cotza M, et al. Carbon dioxide production during cardiopulmonary bypass: pathophysiology, measure and clinical relevance. *Perfusion*. 2017;32:4–12.
11. Ranucci M, Isgro G, Romitti F, et al. Anaerobic metabolism during cardiopulmonary bypass: predictive value of carbon dioxide derived parameters. *Ann Thorac Surg*. 2006;81:2189–2195.
12. de Somer F, Mullholland J, Bryan M, Alosio T, Van Nooten G, Ranucci M. O_2 delivery and CO_2 production during cardiopulmonary bypass as determinants of acute kidney injury: time for a goal-directed perfusion management? *Crit Care*. 2011;15:R192.
13. Ranucci M, Johnson I, Willcox T, Baker R, et al. Goal-directed perfusion to reduce acute kidney injury: a randomized trial. *J Thorac Cardio Surg*. 2018;156(5):1918–1927.
14. Elmistekaway E, McDonald B, Hudson C, et al. Clinical impact of mild acute kidney injury after cardiac surgery. *Ann Thorac Surg*. 2014;98:815–822.
15. Dasta J, Kane-Gill S, Durtschi A, Pathak D, et al. Costs and outcomes of acute kidney injury (AKI) following cardiac surgery. *Nephrol Dialysis Transplant*. 2008;23(6):1970–1974.
16. Fresenius Kabi, Setup, Operation protocol, Baird Kevin. CCP, Ali Cornell, CCP. 2020.
17. CATSmart Operating Instructions, Fresenius Kabi, edition 06/09.16. Part no. M70 682, p. 11.
18. Hypothermia: pH stat management. Openanesthesia.org. Accessed February 5, 2022. Available at: https://www.openanesthesia.org/hypothermia_ph_stat_management/.
19. GEM Premier 3500 Operator's Manual, Rev 00, January 2009, Instrumentation Laboratory Company, Lexington, MA.
20. Gourlay T, Fleming J, Taylor KM. Laboratory evaluation of the Pall LG6 leukocyte depleting arterial line filter. *Perfusion*. 1992;7(2):131–140. Available from: https://doi.org/10.1177/026765919200700209.
21. Ilmakunnas M, Pesonen EJ, Ahonen J, Rämö J, Siitonen S, Repo H. Activation of neutrophils and monocytes by a leukocyte-depleting filter used throughout cardiopulmonary bypass. *J Thorac Cardiovasc Surg*. 2005;129(4):851–859. Available from: https://doi.org/10.1016/j.jtcvs.2004.07.061.
22. Gourlay T, Olivencia-Yurvati AH, Gunaydin S. STS blood conservation guidelines: the role of leukocyte filtration. *Ann Thorac Surg*. 2008;85(3):1138–1139. Available from: https://doi.org/10.1016/j.athoracsur.2007.06.017.
23. Deptula J, Valleley M, Glogowski K, Detwiler J, Hammel J, Duncan K. Clinical evaluation of the Terumo Capiox FX05 hollow fiber oxygenator with integrated arterial line filter. *J Extra Corpor Technol*. 2009;41(4):220–225.
24. Tang J, Tao K, Zhou J, et al. Long-term leukocyte filtration should be avoided during extracorporeal circulation. *Mediators Inflamm*. 2013;2013:612848. Available from: https://doi.org/10.1155/2013/612848.
25. de Amorim CG, Malbouisson LM, da Silva Jr. FC, Fiorelli AI, Murakami CK, Carmona MJ. Leukocyte depletion during CPB: effects on inflammation and lung function. *Inflammation*. 2014;37(1):196–204. Available from: https://doi.org/10.1007/s10753-013-9730-z.
26. Ennema JJ, Mook PH, Elstrodt JM, Wildevuur CR. A new hollow fiber membrane oxygenator with an integral heat exchanger: a hematological evaluation in dogs. *Thorac Cardiovasc Surg*. 1983;31(6):359–364. Available from: https://doi.org/10.1055/s-2007-1022019.
27. Terumo introduces its smallest, lowest prime volume, full-size adult* capiox® oxygenator. Terumo-europe.com. Accessed February 5, 2022. Available at: https://www.terumo-europe.com/en-emea/news/terumo-introduces-its-smallest-lowest-prime-volume-full-size-adult-capiox%C2%AE-oxygenator.
28. Gipson KE, Rosinski DJ, Schonberger RB, et al. Elimination of gaseous microemboli from cardiopulmonary bypass using hypobaric oxygenation. *Ann Thorac Surg*. 2014;97(3):879–886. Available from: https://doi.org/10.1016/j.athoracsur.2013.08.074.
29. Schuldes M, Riley JB, Francis SG, Clingan S. Effect of Normobaric versus Hypobaric Oxygenation on Gaseous Microemboli Removal in a Diffusion Membrane Oxygenator: An In Vitro Comparison. *J Extra Corpor Technol*. 2016;48(3):129–136.
30. Clingan S, Schuldes M, Francis S, Hoerr Jr. H, Riley J. In vitro elimination of gaseous microemboli utilizing hypobaric oxygenation in the Terumo® FX15 oxygenator. *Perfusion*. 2016;31(7):552–559. Available from: https://doi.org/10.1177/0267659116638148.

31. Kuehl R, Banderet F, Egli A, et al. Different Types of Heater-Cooler Units and Their Risk of Transmission of Mycobacterium chimaera During Open-Heart Surgery: Clues From Device Design. *Infect Control Hosp Epidemiol.* 2018;39(7):834–840. Available from: https://doi.org/10.1017/ice.2018.102.
32. Blakey AK, Holt DW. Improving Decreased Heater-Cooler Efficiency as a Result of Heater-Cooler Infection Control Strategy. *J Extra Corpor Technol.* 2019;51(2):73–77.
33. LivaNova Heater-Cooler System 3T Operating Instructions. Version 02/2020 - CP_IFU_16-XX-XX_USA_021.

PART III

Mechanical Circulatory Support and Organ Replacement

Chapter 5

Demand for Mechanical Circulatory Support

Kiyotaka Fukamachi[1,2], Jamshid H. Karimov[1,2] and Randall C. Starling[2,3]

[1]*Lerner Research Institute, Cleveland Clinic, Cleveland, OH, United States,* [2]*Cleveland Clinic Lerner College of Medicine of Case Western Reserve University, Cleveland, OH, United States,* [3]*Kaufman Center for Heart Failure, Heart, Thoracic, and Vascular Institute, Cleveland Clinic, Cleveland, OH, United States*

Chapter Outline

Introduction

Heart failure (HF) is a major public health concern and one of the most frequent causes of hospitalization. An estimated 6.2 million Americans ≥20 years of age have HF; projections show that the prevalence will increase 46% from 2012 to 2030, resulting in >8 million people ≥18 years of age with HF.[1] Additionally, the total percentage of the population with HF is predicted to increase from 2.42% in 2012 to 2.97% in 2030. In 2012, the total cost for HF was estimated to be $30.7 billion, of which more than two-thirds was attributable to direct medical costs. Projections suggest that by 2030, the total cost of HF will increase by 127%, to $69.8 billion.[2]

Pharmacological therapy and treatment with a resynchronization device can provide symptomatic and survival benefits; however, many patients develop progressive symptoms refractory to further medical therapy even with intravenous inotropic therapy. These end-stage HF patients have a poor quality of life (QOL) with recurrent hospitalizations and a high mortality rate. The gold standard for the treatment of end-stage HF remains heart transplantation, which is performed in only ~3600 cases per year in the United States because of the chronic shortage of organ donors. With the ever-growing population of patients with end-stage HF, however, an estimated 50,000–100,000 patients are in need of heart transplantation or mechanical circulatory support (MCS) in the United States. Therefore, a high demand for MCS is obvious. In addition, cardiogenic shock due to acute HF and decompensated chronic HF is not an indication for heart transplant, and short-term temporary MCS is also necessary.

Currently available MCS devices can support the left, right, or both ventricles, offering varying levels of circulatory support. One of the most difficult challenges in MCS is to select a proper candidate with a proper device and ideal timing. In this chapter, we discuss MCS devices for acute HF and chronic HF, along with their indications and complications. Short-term temporary MCS devices are primarily used for acute HF patients in cardiogenic shock, postcardiac arrest, and high-risk percutaneous coronary intervention (PCI), while durable MCS devices are increasingly utilized for chronic HF patients as a bridge-to-transplant (BTT) or as a destination therapy (DT).

Advances in Cardiovascular Technology. DOI: https://doi.org/10.1016/B978-0-12-816861-5.00030-7

Temporary MCS

In many patients with cardiogenic shock, medical therapy alone (inotropic therapy and intravenous vasodilators) is insufficient to optimize hemodynamics and improve end-organ function. In these cases, temporary MCS is increasingly being utilized in the management of cardiogenic shock as a bridge to recovery (BTR), bridge to decision (BTD), or BTT.[3] Especially, a BTT indication has markedly increased since the change in the United Network for Organ Sharing (UNOS) allocation policy for donor hearts in October 2018. The new system assigns high priority to patients supported with temporary MCS, while patients supported with inotropes alone are assigned a lower status. The new heart allocation system also highlighted the inadequacy of the preimplantation intent strategies, making it much more difficult to transplant a stable patient with a durable left ventricular assist device (LVAD).

There are multiple MCS devices to support left ventricle (LV) and/or right ventricle (RV) that can offer strong short-term hemodynamic support. The specific device chosen largely depends upon the etiology of cardiogenic shock, the patient's unique physiology, and the cardiac output augmentation required.

The technology has advanced so that it is miniaturized, more rapidly deployable, and can be placed percutaneously, which is preferable in unstable patients. Temporary MCS is also used in patients undergoing high-risk PCI. Currently available temporary MCS devices are:

1. Intraaortic balloon pump (IABP)
2. Percutaneous ventricular assist device (VAD)
   a. Impella (Abiomed, Danvers, Massachusetts)
   b. HeartMate PHP (percutaneous heart pump) (Abbott, St. Paul, MN)
   c. TandemHeart (LivaNova, London, United Kingdom)
3. Extracorporeal membrane oxygenation (ECMO)
4. Extracorporeal pumps (CentriMag, Abbott)

All the devices can be deployed percutaneously (except for CentriMag, which is usually used in the open chest situation). These devices can be used in various configurations, in isolation or in conjunction with other devices, providing RV and/or LV support. An important challenge is how best to pair the right patient with the right device at the right time. The relationship between the degree of invasiveness and the degree of circulatory support for each device is shown in Fig. 5.1. A detailed discussion of each device is provided in the following sections.

Intraaortic balloon pump

The IABP is a percutaneous device that was introduced more than 50 years ago,[4] but it is still the most commonly used percutaneous device in patients undergoing high-risk PCI, in cases of cardiogenic shock (cardiac index <2.2 L/min/m^2 and systolic blood pressure <90 mm Hg[5]), and as bridge therapy to other forms of more durable LVADs and ultimately heart transplant. This common usage is due to its low cost, wide availability, near universal expertise, and relatively low risk of insertion. From pre- to post-UNOS allocation policy implementation, there was a statistically significant increase in the use of IABP (7.48% vs 27.0%, $P < .001$).[6]

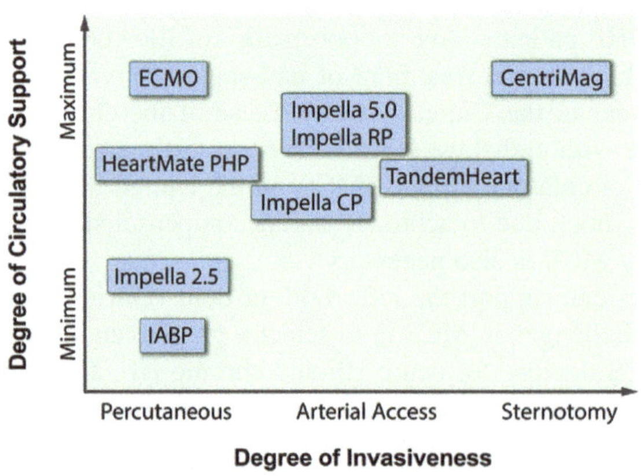

FIGURE 5.1 The relationship between the degree of invasiveness and the degree of circulatory support for each device. *ECMO*, Extracorporeal membrane oxygenation; *IABP*, intraaortic balloon pump.

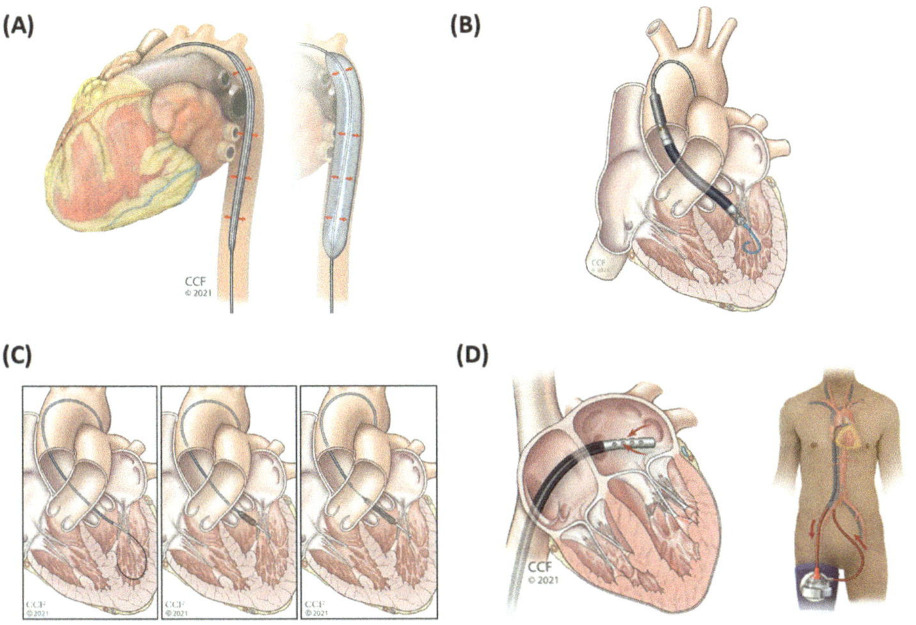

FIGURE 5.2 Short-term mechanical circulatory support devices. (A) Intraaortic balloon pump (IABP). (B) Impella. (C) HeartMate PHP. (D) TandemHeart.

The IABP can be implanted via the femoral artery or by means of percutaneous or axillary artery cut-down in a retrograde configuration (Fig. 5.2A). The catheter-based balloon (1) inflates during diastole, augmenting coronary perfusion (myocardial oxygen supply) by increasing the aortic root pressure, and (2) deflates during systole to reduce LV afterload, and thus myocardial work, thereby improving the myocardial oxygen demand/supply ratio. However, the expected increase in the cardiac output is only about 1.0 L/min,[7] and it is suggested that the more severe the LV dysfunction, the less benefit from IABP support.[8]

Complications include vascular access complications such as limb ischemia and major bleeding, platelet consumption, and balloon rupture. However, the incidence of these complications is generally low.[9]

Percutaneous VAD (Impella, Heart Mate (HM) PHP, TandemHeart)

The use of catheter-based, percutaneous VADs as a therapeutic modality in refractory cardiogenic shock has rapidly expanded. The Impella and HeartMate PHP can be placed percutaneously via arterial vascular access across the aortic valve, and draw blood from the LV and eject it into the ascending aorta. The TandemHeart is a bypass system with an extracorporeal centrifugal blood pump. It draws blood from the left atrium (LA) through a trans-septal cannula inserted from the femoral vein and ejects blood to the femoral artery. Therefore, similar to regular durable VADs, these devices can decrease LV volume and myocardial wall stress (thus myocardial oxygen demand) and increase coronary perfusion and cardiac output.

Impella

The Impella is a continuous-flow (CF) axial pump that is placed fluoroscopically into the LV in a retrograde manner across the aortic valve. It provides active support by expelling blood from the LV into the ascending aorta (Fig. 5.2B), so it can unload the LV, increase cardiac output, and improve coronary perfusion. The pump can be implanted percutaneously, via a cut-down or in the operating room. The distal portion of the Impella is a pigtail configuration to allow the device to reside in the mid-LV, approximately 3.5 cm below the aortic annulus.

The Impella family of devices includes (1) three transfemoral artery devices (2.5, CP, 5.0/5.5) and one transaortic device (LD) for support of the failing LV, and (2) one transfemoral vein device (Impella RP) for support of the failing RV. The Impella devices have been used in diverse clinical scenarios, depending on the conditions of the patients. A couple of small, randomized control trials comparing the Impella 2.5 with IABP support have failed to demonstrate a survival benefit, despite a more favorable hemodynamic profile.[10,11] Therefore, clinicians tend to use the Impella CP or Impella 5.0 for patients requiring greater hemodynamic support than the IABP or Impella 2.5.[3]

Complications include hemolysis, vascular access complications such as limb ischemia and major bleeding, pericardial tamponade due to LV perforation, mitral valve impingement with mitral regurgitation, and arrhythmia due to the presence of the device with in the LV.[12]

HeartMate PHP

The HeartMate PHP is a second-generation transcatheter axial flow circulatory support system.[13] Similar to the Impella, the HeartMate PHP is placed fluoroscopically into the LV in a retrograde manner across the aortic valve and provides active support by expelling blood from the LV into the ascending aorta (Fig. 5.2C). Unlike the Impella, the HeartMate PHP is a collapsible catheter pump that is inserted through a small 14 Fr sheath, deployed across the aortic valve, expanding to 24 Fr and able to deliver up to 5 L/min blood flow at reduced risk of vascular access complications due to its smaller catheter size.

The HeartMate PHP was approved for commercial sale in Europe and is in the advanced clinical testing stage in the United States (Supporting Patients Undergoing HIgh-Risk PCI Using a High-Flow PErcutaneous Left Ventricular Support Device (SHIELD II)). However, SHIELD II was suspended in January 2017 after receiving eight reports (1.9%) of pumps stopping during high-risk procedures, with one patient's death associated with sepsis several days after the intervention. SHIELD II resumed in January 2019 with a new composite primary endpoint and a two-part design.

TandemHeart

The TandemHeart is a bypass system with an extracorporeal, nondurable centrifugal blood pump (Fig. 5.2D). It draws blood from the LA through a trans-septal cannula inserted from the femoral vein and ejects blood to the femoral artery. The pump can deliver flow rates up to 5.0 L/min at a maximum speed of 7500 rpm.

Placement of the percutaneous LA cannula requires fluoroscopic and/or echocardiographic guidance, femoral venous advancement, and trans-septal puncture, which is technically more demanding than other modalities of percutaneous mechanical support. It provides short-term cardiac support ranging from a few hours up to 14 days. The system is highly versatile and may be used for either right, left, or biventricular support with different cannulation methods, depending upon the clinical need. An oxygenator can also be added to the system to function as an ECMO circuit,[14] albeit at lower maximum flows.

The most common complications include bleeding, dislodging of the TandemHeart cannula (due to the technical expertise required to position the device), a femoral arteriovenous fistula, thromboembolism, atrial-septal defect, limb ischemia, wound infection, and lymphocele.[15]

Extracorporeal membrane oxygenation

The general concept underlying venoarterial ECMO (VA-ECMO) is that venous blood is drained from the right heart, passed through an oxygenator, and returned to the arterial circulation to support both the circulatory and respiratory systems (Fig. 5.3A). Detailed descriptions of ECMO are provided in Chapter 11, but basically, a VA-ECMO circuit is equivalent to the cardiopulmonary bypass circuit routinely used for cardiac operations, except that it lacks a reservoir and cardiotomy suckers. It has been increasingly used in refractory cardiogenic shock as bridge to heart replacement therapy, BTD, or BTR.[16] According to the Extracorporeal Life Support Organization registry, >15,000 adult patients have been supported with VA-ECMO.

Treatment options for COVID-19 are still evolving, but ECMO may be beneficial in select cases. The mortality rate in COVID-19 patients who require mechanical ventilation is high. ECMO can be lifesaving for patients with severe forms of acute respiratory distress syndrome.

In many centers, VA-ECMO is now frequently performed by interventional cardiologists, critical care physicians, and even emergency room personnel. The two most commonly used cannulation strategies are (1) peripheral cannulation with a femoral venous cannula and femoral artery return cannula; and (2) central cannulation with a venous cannula directly in the right atrium and a return cannula in the ascending aorta.[17]

The hemodynamic improvements associated with VA-ECMO commonly restore end-organ perfusion. However, this configuration leads to an increased afterload to the LV (resulting in increased LV distention that manifests as pulmonary edema with worsening oxygenation), increases the risk of LV thrombus formation as a result of stasis in the nonejecting LV, and is associated with lower rates of myocardial recovery.[18] Therefore, if LV distention develops, the LV can be unloaded percutaneously with the placement of an Impella, or surgically with placement of a vent. Other common complications of VA-ECMO include acute limb ischemia (in the case of peripheral cannulation), stroke, bleeding, and infection.

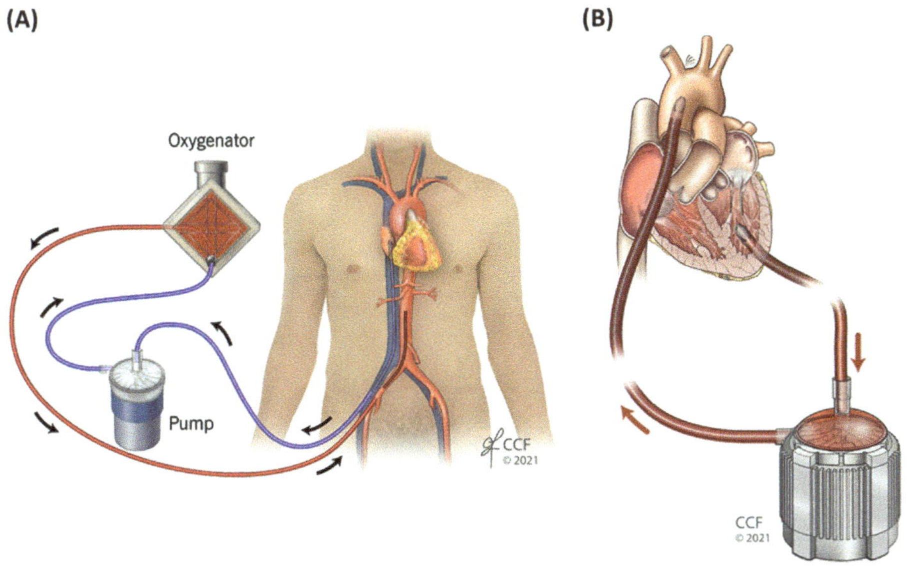

FIGURE 5.3 Short-term mechanical circulatory support devices. (A) Venoarterial extracorporeal membrane oxygenation (VA-ECMO). (B) CentriMag.

Extracorporeal pumps (CentriMag)

The CentriMag (Fig. 5.3B) is an extracorporeal, CF centrifugal type rotary blood pump that utilizes magnetically levitated, bearingless motor technology to actuate and support a centrifugal pump rotor to pump blood up to 9.9 L/min.[19] The CentriMag generates as much as 10 L/min flow and is capable of supporting one or both sides of the heart for as long as 30 days, or as part of an ECMO circuit with low risk of thromboembolic complications. To date, the device has been used successfully in more 3700 patients worldwide for such varied indications as primary graft failure after heart transplantation, neonatal life support, and postcardiotomy cardiogenic shock.[20]

By eliminating bearings, shafts, and seals typically associated with a conventional centrifugal pump, the level of hemolysis and risk of thrombus accumulation within this pump are greatly reduced, along with the potential for adverse clinical sequelae. This pump is different from other temporary MCS devices in that it must be inserted centrally when used as LVAD or right ventricular assist device (RVAD), and is intended for longer use than other short-term MCS devices.

Complications included bleeding (21%), infection (5%), respiratory failure (3%), hemolysis (5%), and neurologic dysfunction (11%).[20]

Durable MCS

Durable MCS gained public awareness when the Jarvik-7 total artificial heart (TAH) was successfully implanted in a patient in 1982.[21] LVADs have emerged as the standard of care for patients with advanced HF refractory to maximal medical therapy awaiting heart transplantation (BTT) or as a DT. LVADs were approved by the US Food and Drug Administration (FDA) as a BTT in 1994.[22] The landmark Randomized Evaluation of Mechanical Assistance for the Treatment of Congestive Heart Failure (REMATCH) study was published in 2001 and opened the era of DT in patients with end-stage HF who are not eligible for a heart transplant.[23]

In addition to LVADs, RVADs, and TAHs, MCS can also be used for patients with HF with preserved ejection fraction (HFpEF). Clinical demand for HFpEF treatment options are now urgent because there are no effective treatment options for HFpEF.[24]

In this section, indications and complications of durable MCS devices, such as LVAD, RVAD, TAH, and devices for patients with HFpEF are discussed.

Left ventricular assist device

Initial LVAD technology utilized large pulsatile pumps to provide adequate circulatory support. Pulsatile LVAD support was shown to improve survival and QOL as DT in HF patients[23]; however, the large size, limited durability, and complications hindered long-term success. As a result, CF LVAD technology was developed as an alternative to the

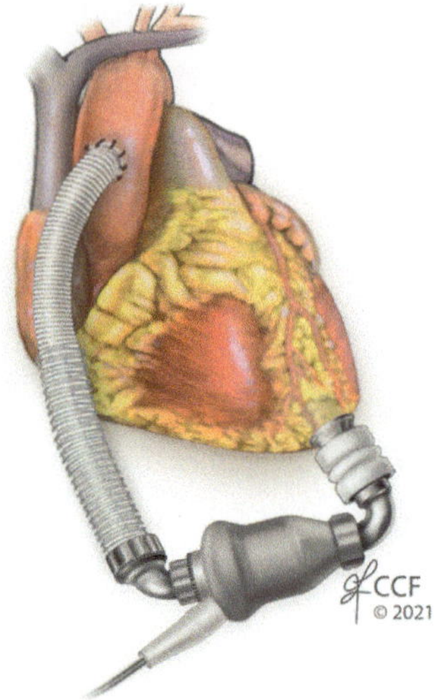

FIGURE 5.4 The HeartMate II.

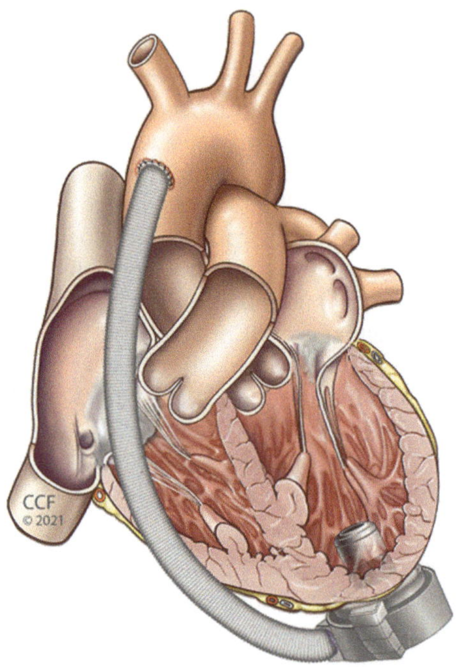

FIGURE 5.5 The HeartMate 3.

pulsatile-flow devices.[25] The clinically available CF LVADs include HeartMate II (Abbott) (Fig. 5.4), HeartMate 3 (Abbott) (Fig. 5.5), and HVAD (HeartWare Inc., Framingham, MA) (Fig. 5.6). The detailed descriptions of these devices are provided in Chapter 6.

Indications

In this section, indications are described for each of the BTT, DT, BTC (or BTD), and BTR applications; however, designations of BTT and DT are no longer clinically distinctive in the current medical era and they are dated terms, because they change over time according to the patient's clinical condition. The new coverage decision by the Centers for Medicare & Medicaid Services (CMS) did away with BTT and DT designations.

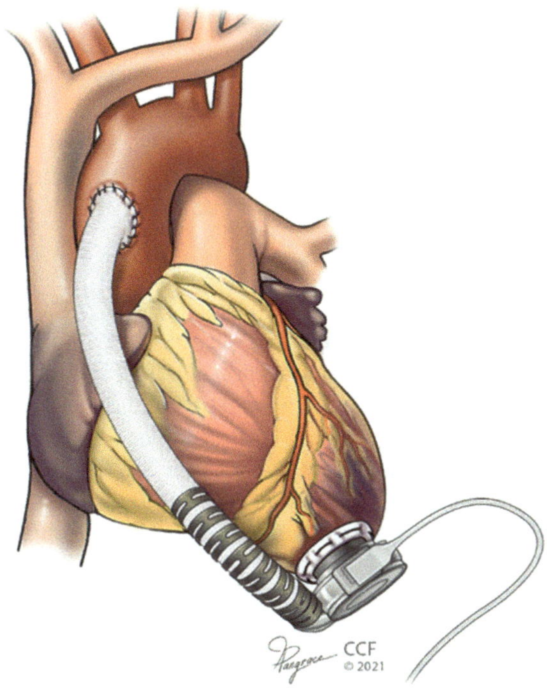

FIGURE 5.6 The HVAD.

Bridge to heart transplant

For HF patients who are awaiting a heart transplant but have developed decompensated end-stage HF refractory to medical therapy, LVADs are used as a BTT to maintain hemodynamics and improve end-organ functions. Increased waiting time for a transplant has led to the use of LVADs as a standard therapy for decompensated end-stage HF to bridge the patient to transplant.[26] The most appropriate timing for the LVAD is still not completely clear and is based on an individual balance between risk of death and the development of multiorgan system failure while on the waiting list vs. the risk of surgery and subsequent adverse effects from the LVAD.

Destination therapy

For HF patients who are not eligible to receive a heart transplant due to nonreversible comorbidities or age, LVADs are used as DT. As heart transplantation rates are inadequate to meet demand, the percentage of LVADs implanted as DT increased from 2% in 2008 to 49% in 2017 for CF LVADs. There was also an overall decline in those in whom the LVAD was implanted as a bridge to candidacy (BTC) or transplantation over this time period.[27] In addition, HF patients who are transplant-eligible can also receive LVADs as DT. The improved durability of second-generation CF LVADs led to an overall greater use of LVADs for DT; currently all implanted devices are CF LVADs.[28]

Bridge to candidacy

Patients with potentially modifiable medical or psychosocial risk factors that preclude immediate transplant listing are assigned the classification BTC or BTD. Significant heterogeneity and uncertainty associated with the end-stage HF patient population makes deciding on intention to treat between DT and BTT difficult in situations when modifiable risk factors (contraindications for heart transplant) are present but may be reversible, thus making them potential future transplant candidates. Some studies have shown that the 1-year survival of BTC patients is somewhere between that of BTT and DT patient populations, and between 30% and 44% of BTC patients are eventually listed for a donor heart.[29]

Bridge to recovery

Although it is not very common, an LVAD is also used as a BTR for patients who are sick enough to require LVAD implantation but are expected to recover their cardiac function to a point at which their LVAD can be explanted. Although the beneficial physiologic effects of LVAD support, such as unloading of LV and reduction of pulmonary congestion, are recognized, the greatest challenge with BTR is to properly identify which patients meet the BTR indication. Another challenge is how to evaluate myocardial recovery to determine if the LVAD can be explanted. Explant of devices for myocardial recovery continues to be rare in the Interagency Registry for Mechanically Assisted Circulatory Support (INTERMACS) population, occurring in less than 5% of the implants by 5 years.[30] Methods to evaluate myocardial recovery, along with aggressive medical regimens after LVAD explanation, are still in development. However, Birks et al.[31] reported that their strategy of LVAD support combined with a standardized pharmacological and cardiac function monitoring protocol resulted in an exceptionally high rate (52.3%) of LVAD explantation.

INTERMACS profiles

Timely referral and accurate patient selection are critical to a successful MCS application. The patient selection criteria for MCS are complex, as it requires a comprehensive evaluation of HF severity, preoperative risk, psychosocial factors, and compliance with the self-care regimen. Selecting appropriate candidates and appropriate timing are important to prevent long-term injury to other organs. Renal and hepatic dysfunctions are the two organ systems that are especially critical to assess when determining if a patient is an acceptable candidate for MCS.

In general, patients with severe HF with poor LV systolic function (ejection fraction <25%) and markers of adverse prognosis (persistently elevated brain natriuretic peptide, exercise limitation with peak VO_2 <10–12 mL/kg/min, and end organ dysfunction) should be considered for LVAD therapy. In 2018 alone, LVADs were implanted in more than 3000 patients. More than 3000 patients received heart transplants in the United States, with an additional 3500 patients awaiting transplantation.[32,33]

The New York Heart Association (NYHA) classification does not provide an adequate prognostic assessment to allow optimal patient selection for MCS devices. To more accurately define this situation, the INTERMACS Profiles were developed in 2009 with the first INTERMACS report. In the INTERMACS profile, patients are categorized according to clinical condition into seven progressive clinical profiles that are largely sub-classifications of NYHA class III–IV functional status (Table 5.1). Profile 1 is defined as "crash and burn" and the patient is on maximal inotropic support, with survival defined in hours. Profile 3 is defined as "stable on inotropes," either in or out of the hospital. "Resting symptoms" daily despite high doses of diuretic therapy without any intravenous medications defined Profile 4, which are essentially chronic NYHA Class IV symptoms. The designations of "exertion limited" but comfortable at rest defined Profile 5, and "Walking wounded" with limited activity outside the home defined Profiles 6 and 7. The least compromised class was Profile 7, which was added as a placeholder for patients comfortable with some limitation, without frequent exacerbations.

Inotrope dependence has traditionally been considered the threshold that prompts consideration of MCS candidacy. When these profiles were introduced, 80% of durable MCS device implantations were performed in patients in Profiles 1 and 2.[34] However, the percentage of implants in Profile 1 decreased from 41% in 2006 to 18% between 2015 and 2019,[35] because outcomes were significantly worse, with >50% mortality at 1 year (compared with 35% for patients in Profile 2).[36]

TABLE 5.1 INTERMACS profile.

Profile	Profile description	NYHA class	Time frame for intervention	LVADs implanted (2015–19)[35]	Survival at 1 year after LVAD implant[35]
1.	Critical cardiogenic shock ("crash and burn")	IV	Hours	18.0%	74.2%
2.	Progressive decline during inotropic support ("crash and burn")	IV	Hours to days	34.1%	82.2%
3.	Stable but inotrope dependent ("dependent stability")	IV	Weeks	35.5%	85.4%
4.	Resting symptoms	IV	Weeks to months	12.4%	85.0%
5.	Exertion intolerant ("exertion limited")	IIIb/IV	Variable		
6.	Profile 6: Exertion is limited ("walking wounded")	IIIb	Variable		
7.	Advanced NYHA III	III/IIIb	Not indicated		

NYHA, New York Heart Association.

Between 2015 and 2019, ambulatory patients in Profiles 4–7 accounted for 12.4% of all MCS implants included in the Society of Thoracic Surgeons (STS) INTERMACS database.[35] For those patients, we should consider the balance between the benefits by LVAD support and the incidence and severity of the complications of LVAD. The Risk Assessment and Comparative Effectiveness of Left Ventricular Assist Device and Medical Management in Ambulatory Heart Failure Patients (ROADMAP) study evaluated ambulatory, noninotrope-dependent HF patients (Profiles 4–7) who were not candidates for heart transplantation.[37] Among 195 patients with advanced ambulatory noninotrope-dependent HF, only those with higher severity of HF (Profile 4) benefitted from LVAD implantation compared with optimal medical management, despite increased complications. In patients with Profiles 5–7, no benefit of LVADs was noted.

Complications

The two-year LVAD survival rates of the most recent devices (HeartMate 3, 83%[38] and HVAD, 87%[39]) are now comparable to transplant survival (82% at 2 years).[40] However, there are still many complications, including bleeding, infection, stroke, pump thrombosis and hemolysis, RV failure, aortic insufficiency, and gastrointestinal (GI) bleeding.

The most common complication is bleeding. Up to 70% of LVAD patients have a bleeding complication over the lifetime of the LVAD, although bleeding that requires surgical intervention is not common.[41] The second common complication is infection (22%–50% of patients); driveline infection was the most prevalent type of infection (85%).[42] Stroke (ischemic > hemorrhagic) is also common, as 10% of LVAD patients are affected by stroke in the first year of support alone.[43]

Pump thrombosis and hemolysis are also common complications, and hemolysis with an elevated lactate dehydrogenase (LDH) level is usually a first sign of pump thrombosis. Initial trials of the HeartMate II reported low rates of pump thrombosis, ranging from 1.4% at 18 months[44] to 6% at 2 years[45]; however, Starling et al.[46] reported a 12.3% incidence at 2 years of confirmed or suspected pump thrombosis in a multicenter cohort of 837 patients.

The incidence of RV failure requiring an RVAD after LVAD implantation has been reported between 5% and 20%,[32] and RV failure remains a serious complication.[47] For patients who require an additional RVAD, the 1-year survival rate is only 58% and the median survival is 31%.[27]

Aortic insufficiency and GI bleeding have not been seen frequently with pulsatile LVADs, so these complications may be related to reduced pulsatility and/or continuous pumping with CF LVADs. Up to 30% of patients will develop at least moderate aortic insufficiency within 2 years after CF LVAD implantation.[48,49] The incidence of GI bleeding was found to be 6.8 events/100 patient years, or 10% in those with pulsatile LVADs[50]; however, patients with CF LVADs have an increased incidence of GI bleeding, estimated at 63 events/100 patient years or 20%–30%,[50,51] with recurrent bleeding at 30%–40%.[52] One explanation for the high GI bleeding rate is an abnormal von Willebrand factor degradation with supraphysiologic shear stress from CF LVADs that depletes functional von Willebrand factor multimers.[53]

Hospital readmissions to address these complications remain a significant limitation of device therapy, which explains why LVAD therapy has not become even more popular. Readmission after CF LVAD implant occurs in 38.6% (70% CI: 38.1%, 39.0%) of patients at 90 days and in 72.2% (70% CI: 71.8%, 72.6%) at 12 months.[35] The leading causes for unplanned readmission were GI bleeding, HF, stroke, and infection.[54,55]

Right ventricular assist device

As described in the "Left Ventricular Assist Device" section, RV failure is a serious complication in patients with LVAD support. RV failure frequently occurs because most patients with advanced LV dysfunction who are assessed for an LVAD also have some degree of RV dysfunction. Although an LVAD may improve RV function due to an increase in RV compliance and a decrease in RV afterload,[56] it may exacerbate RV dysfunction due to excessive LV unloading, resulting in a leftward shift of the interventricular septum and reducing its contribution to RV contraction and output.[57] In addition, augmentation of systemic blood flow by the LVAD increases the amount of venous return to the RV that cannot be handled well by a dysfunctional RV.

Current treatment options for RV support are limited to short-term devices such as CentriMag, Impella RP, and TandemHeart, and there is no dedicated, durable RVAD, other than using an LVAD as an RVAD. Therefore, there is a need for a dedicated, durable RVAD, and so we worked on developing the DexAide RVAD.[58] We are also working on a VAD that can be used universally as an LVAD and RVAD.[59–61] The detailed description is presented in Chapter 7.

Patients with persistent RV failure of decreased cardiac index less than 2.0 L/min/m^2 and increased central venous pressure greater than 18 mm Hg, even with maximal inotropic support (Dobutamine >7.5–15 μg/kg/min, Milrinone >0.75 μg/kg/min), are candidates for RVAD support.

The main complications of RVAD implantation are the higher incidence of bleeding and infection when the chest remains unclosed, and the necessity for resternotomy if the sternum is closed at the time of the primary operation. Several new approaches, designed to avoid these complications, have been described for temporary RVAD implantation.[62]

Total artificial heart

For patients who suffer from severe biventricular failure, a biventricular assist device using an LVAD and RVAD or a TAH is required. The only clinically available TAH in the United States is the SynCardia TAH (Fig. 5.7), which is approved as a BTT only. From 2006 to April 2017, 450 individuals in INTERMACS underwent a TAH implantation. Among those, 266 underwent transplantation and 162 died on support. At 12 months, 52% of the patients had undergone transplantation, 34% had died, and 13% were still alive and with the device.[63] These results are worse than the results with LVADs; however, it should be noted that most TAH recipients are INTERMACS Profile 1 (43%) or 2 (37%),[63] who also have poor outcomes with LVADs.

Potential candidates for TAH implantation must be transplant-eligible with end-stage restrictive cardiomyopathy, irreversible biventricular failure, allograft failure, rejection or heart transplant vasculopathy, massive myocardial infarction, refractory ventricular tachycardia, postinfarction ventricular septal defect, or end-stage congenital heart disease (CHD).

Similar to LVADs, complications of the SynCardia TAH include bleeding, infection, stroke, and acute kidney injury in this critically ill population.[63] The likelihood of major infection approached 70% within 6 months. The incidence of stroke was 23.7% within the first 6 months (13.5% ischemic and 10.2% hemorrhagic); 12% of these events resulted in moderate-severe disability and 39% in severe disability. The incidences of major and minor device malfunction were 7.1% and 28.2%, respectively, at 6 months, and the incidence of thrombosis was 1.6% during this same interval.

FIGURE 5.7 The SynCardia total artificial heart.

For the 399 patients who did not require dialysis before implantation, 114 (29%) began dialysis after SynCardia TAH implantation.

Several new TAHs are under development in the US to address these complication risks. Similar to LVADs that moved from pulsatile devices to CF devices, TAHs will also move from pulsatile devices to CF devices for the same reasons. These new devices include the Cleveland Clinic TAH (CFTAH) and the BiVACOR TAH. Both involve a single moving part, a rotor with impeller blades on each side, with each side driving the flow in the left and right centrifugal CF pumps that make up the device. These are considerably smaller than positive displacement, pulsatile TAHs. Detailed descriptions of the Cleveland Clinic CFTAH are found in Chapter 9.

New devices for HFpEF

Among HF patients, more than half appear to have preserved systolic function,[64,65] which is categorized as HFpEF. This has become an increasingly predominant form of HF, and will soon reach epidemic proportions.[66] Unlike its counterpart HFrEF, there are no effective treatment guidelines for HFpEF.[24] Traditionally available therapies for HF have entirely failed to improve the mortality or morbidity of patients with HFpEF, even though they had been convincingly effective for patients with HFrEF. For example, LVADs that are established, standard therapies for patients with HFrEF do not work well for patients with HFpEF, as their LV cavity is too small and the LV volume is insufficient for an LVAD to work effectively.[67] Therefore, device-based therapy for HFpEF is now urgently needed.

Several devices for HFpEF patients are currently under development. The detailed descriptions of these devices are provided in Chapter 12.

MCS for a pediatric population

CHD, a major cause of infant death during the first year, accounts for the largest percentage of deaths (30%–50%) resulting from birth defects.[68] Cardiac surgery for neonates and infants with CHD usually addresses complex cardiac anomalies such as hypoplastic left heart syndrome and transposition of the great artery. The surgical procedures (e.g., Norwood procedure or Jatene procedure) can be complicated by severe HF postoperatively, with failure to thrive. Persistent residual cyanosis and incomplete correction by these palliative procedures often result in unsatisfactory hemodynamic status after surgery.

Options for MCS modalities are limited to external devices: temporary ECMO, a temporary extracorporeal centrifugal VAD, and the EXCOR VAD (Berlin Heart GmbH, Berlin, Germany) (Fig. 5.8). The EXCOR VAD, a paracorporeal pneumatic pump, is currently the only VAD available to provide long-term support to small children awaiting heart transplantation. The Infant Jarvik 2000 (Jarvik Heart, Inc., New York, NY),[69] a downsized version of the adult Jarvik 2000, was used in a 5.7 kg baby in Italy. In October 2016, the US FDA granted conditional approval to conduct an Investigational Device Exemption (IDE) study of the Jarvik 2015 (Fig. 5.9), a 40% larger version of the Infant

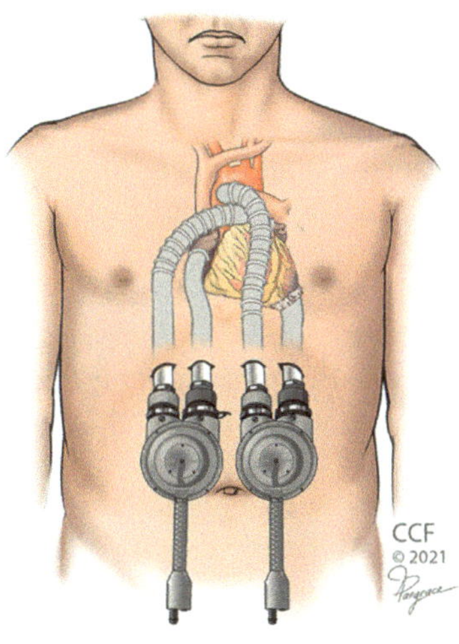

FIGURE 5.8 The EXCOR.

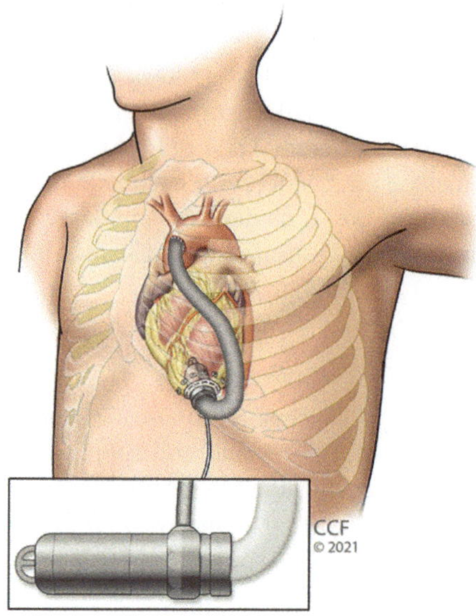

FIGURE 5.9 The Jarvik 2015.

Jarvik 2000. This study (Clinical Trials number NCT02954497) will compare the implanted, portable electrically powered Jarvik 2015 LVAD System with the Berlin Heart EXCOR Pediatric System.

CHD often involves both ventricles; in such cases, heart transplantation, a biventricular assist device, or a TAH is needed. TAHs are more appropriate than biventricular assist devices for patients with a single ventricle, thrombosed ventricles, restrictive cardiomyopathy, cardiac tumors, severe ventricular septal defects, frail ventricles for difficult cannulation, and intractable arrhythmias.

The only pediatric TAH under IDE clinical study is SynCardia's 50cc TAH-t. This 50cc model is smaller than its 70cc model but still too large for newborns or infants. Penn State is developing a TAH for infants and small children, using two pneumatically driven infant VADs.[70] We are developing a pediatric version of the Cleveland Clinic CFTAH by downsizing the adult CFTAH.[71,72]

Contraindications

There are many contraindications to MCS that are related to cardiac conditions, noncardiac but medical conditions, and nonmedical conditions. Absolute contraindications to MCS devices in general include irreversible neurologic injury, moribund state or systemic illness limiting survival, disseminated malignancy, and severe compliance barriers, such as financial, cognitive, and/or psychosocial factors.[73] Patients with coagulopathy disorders should also be considered higher-risk for postoperative complications, as hemorrhagic and thrombotic complications are common following MCS implantation.

There are many other contraindications, depending on the types of MCS device. For example, RV failure is a contraindication for LVADs, as RV failure that required an additional RVAD implantation showed higher mortality than the LVAD alone.[27] Ventricular arrhythmia that affects RV function, some cardiac anatomy (complex CHD or hypertrophic, infiltrative, or restrictive cardiomyopathy), prior surgery (Coronary artery bypass grafting (CABG) or valve replacement), and valvular diseases can also be contraindications for LVADs.

Renal dysfunction is one of the most common comorbidities in patients with HF, and has consistently been one of the strongest predictors of morbidity and mortality with ongoing medical management, but also for poor outcome after LVAD. Therefore, pre-LVAD renal dysfunction of almost any type should be factored into decision-making and expected outcomes.

Severe peripheral vascular disease and the presence of aortic stents are common contraindications for percutaneous delivery MCS devices, such as IABP, Impella, and TandemHeart. A mechanical aortic prosthesis is a contraindication for the Impella.

A small thoracic cavity (and the inability to accommodate the rigid device and its various components) is another contraindication; for example, contraindications for the SynCardia TAH include a body surface area $<1.7\ m^2$ or

<10 cm between the posterior cortex of the sternum and the anterior cortex of the T10 vertebra on computed tomography (CT) images.[74]

Conclusions

There is a high demand for MCS for patients with both acute and chronic HF as described above. Although the 2-year LVAD survival rates of the most recent devices are now comparable to transplant survival, MCS devices still suffer from frequent complications, such as bleeding, infection, stroke, pump thrombosis and hemolysis, etc.

Further technological developments and clinical research to improve MCS devices are warranted. Miniaturization of the durable device has a number of advantages, such as fewer size limitations, potential for biventricular lifelong support, and less invasive techniques for implantation. In addition, a truly internalized powered system, without the need for a driveline, is believed to reduce the incidence of device infection and improve patients' QOL postdevice implantation.

While options for biventricular support are currently limited to two implantable LVADs or temporary device support, dedicated, universal VADs suitable for right ventricular support will be developed to address end-organ failure and patient outcomes in biventricular HF.

References

1. Virani SS, et al. Heart disease and stroke statistics-2020 update: a report from the American Heart Association. *Circulation*. 2020;141(9): e139–e596.
2. Heidenreich PA, et al. Forecasting the impact of heart failure in the United States: a policy statement from the American Heart Association. *Circ Heart Fail*. 2013;6(3):606–619.
3. Nagpal AD, et al. Temporary mechanical circulatory support in cardiac critical care: a state of the art review and algorithm for device selection. *Can J Cardiol*. 2017;33(1):110–118.
4. Moulopoulos SD, Topaz S, Kolff WJ. Diastolic balloon pumping (with carbon dioxide) in the aorta—a mechanical assistance to the failing circulation. *Am Heart J*. 1962;63:669–675.
5. van Diepen S, et al. Contemporary management of cardiogenic shock: a scientific statement from the American Heart Association. *Circulation*. 2017;136(16):e232–e268.
6. Goff RR, et al. A change of heart: preliminary results of the US 2018 adult heart allocation revision. *Am J Transplant*. 2020;20(10):2781–2790.
7. Werdan K, et al. Mechanical circulatory support in cardiogenic shock. *Eur Heart J*. 2014;35(3):156–167.
8. Sintek MA, et al. Intra-aortic balloon counterpulsation in patients with chronic heart failure and cardiogenic shock: clinical response and predictors of stabilization. *J Card Fail*. 2015;21(11):868–876.
9. de Jong MM, et al. Vascular complications following intra-aortic balloon pump implantation: an updated review. *Perfusion*. 2018;33(2):96–104.
10. Seyfarth M, et al. A randomized clinical trial to evaluate the safety and efficacy of a percutaneous left ventricular assist device vs intra-aortic balloon pumping for treatment of cardiogenic shock caused by myocardial infarction. *J Am Coll Cardiol*. 2008;52(19):1584–1588.
11. O'Neill WW, et al. A prospective, randomized clinical trial of hemodynamic support with Impella 2.5 vs intra-aortic balloon pump in patients undergoing high-risk percutaneous coronary intervention: the PROTECT II study. *Circulation*. 2012;126(14):1717–1727.
12. Ali JM, Abu-Omar Y. Complications associated with mechanical circulatory support. *Ann Transl Med*. 2020;8(13):835.
13. Van Mieghem NM, et al. Design and principle of operation of the HeartMate PHP (percutaneous heart pump). *EuroIntervention*. 2018;13 (14):1662–1666.
14. DiVita M, et al. Emergent TandemHeart-ECMO for acute severe mitral regurgitation with cardiogenic shock and hypoxaemia: a case series. *Eur Heart J Case Rep*. 2020;4(1):1–6.
15. Gregoric ID, et al. Techniques and complications of TandemHeart ventricular assist device insertion during cardiac procedures. *ASAIO J*. 2009;55(3):251–254.
16. Abrams D, Combes A, Brodie D. Extracorporeal membrane oxygenation in cardiopulmonary disease in adults. *J Am Coll Cardiol*. 2014;63(25 Pt A): 2769–2778.
17. Guglin M, et al. Venoarterial ECMO for adults: JACC scientific expert panel. *J Am Coll Cardiol*. 2019;73(6):698–716.
18. Truby LK, et al. Incidence and implications of left ventricular distention during venoarterial extracorporeal membrane oxygenation support. *ASAIO J*. 2017;63(3):257–265.
19. Zhang J, et al. Computational and experimental evaluation of the fluid dynamics and hemocompatibility of the CentriMag blood pump. *Artif Organs*. 2006;30(3):168–177.
20. John R, et al. Outcomes of a multicenter trial of the Levitronix CentriMag ventricular assist system for short-term circulatory support. *J Thorac Cardiovasc Surg*. 2011;141(4):932–939.
21. Cai AW, et al. Mechanical circulatory support in the treatment of advanced heart failure. *Am J Transplant*. 2017;17(12):3020–3032.
22. Goldstein DJ, Oz MC, Rose EA. Implantable left ventricular assist devices. *N Engl J Med*. 1998;339(21):1522–1533.
23. Rose EA, et al. Long-term use of a left ventricular assist device for end-stage heart failure. *N Engl J Med*. 2001;345(20):1435–1443.

24. Ponikowski P, et al. 2016 ESC Guidelines for the diagnosis and treatment of acute and chronic heart failure: the task force for the diagnosis and treatment of acute and chronic heart failure of the European Society of Cardiology (ESC) developed with the special contribution of the Heart Failure Association (HFA) of the ESC. *Eur Heart J.* 2016;37(27):2129–2200.
25. Slaughter MS, et al. Clinical management of continuous-flow left ventricular assist devices in advanced heart failure. *J Heart Lung Transplant.* 2010;29(4 Suppl.):S1–S39.
26. Cohn L. Cardiac Surgery in the Adult. 4th ed. McGraw-Hill Education; 2012.
27. Kormos RL, et al. The Society of Thoracic Surgeons Intermacs database annual report: evolving indications, outcomes, and scientific partnerships. *J Heart Lung Transplant.* 2019;38(2):114–126.
28. Strueber M, et al. Multicenter evaluation of an intrapericardial left ventricular assist system. *J Am Coll Cardiol.* 2011;57(12):1375–1382.
29. Teuteberg JJ, et al. Implant strategies change over time and impact outcomes: insights from the INTERMACS (Interagency Registry for Mechanically Assisted Circulatory Support). *JACC Heart Fail.* 2013;1(5):369–378.
30. Teuteberg JJ, et al. The society of thoracic surgeons intermacs 2019 annual report: the changing landscape of devices and indications. *Ann Thorac Surg.* 2020;109(3):649–660.
31. Birks EJ, et al. Prospective multicenter study of myocardial recovery using left ventricular assist devices (RESTAGE-HF [Remission from Stage D Heart Failure]): medium-term and primary end point results. *Circulation.* 2020;142(21):2016–2028.
32. Kirklin JK, et al. Eighth annual INTERMACS report: special focus on framing the impact of adverse events. *J Heart Lung Transplant.* 2017;36 (10):1080–1086.
33. Colvin M, et al. OPTN/SRTR 2016 annual data report: heart. *Am J Transplant.* 2018;18(Suppl 1):291–362.
34. Stevenson LW, et al. INTERMACS profiles of advanced heart failure: the current picture. *J Heart Lung Transplant.* 2009;28(6):535–541.
35. Molina EJ, et al. The society of thoracic surgeons Intermacs 2020 annual report. *Ann Thorac Surg.* 2021;111(3):778–792.
36. Kirklin JK, et al. INTERMACS database for durable devices for circulatory support: first annual report. *J Heart Lung Transplant.* 2008;27 (10):1065–1072.
37. Shah KB, et al. Left ventricular assist devices vs medical management in ambulatory heart failure patients: an analysis of INTERMACS Profiles 4 and 5 to 7 from the ROADMAP study. *J Heart Lung Transplant.* 2018;37(6):706–714.
38. Mehra MR, et al. Two-year outcomes with a magnetically levitated cardiac pump in heart failure. *N Engl J Med.* 2018;378(15):1386–1395.
39. McGee Jr. E, et al. Evaluation of a lateral thoracotomy implant approach for a centrifugal-flow left ventricular assist device: the LATERAL clinical trial. *J Heart Lung Transplant.* 2019;38(4):344–351.
40. Lund LH, et al. The registry of the international society for heart and lung transplantation: thirty-fourth adult heart transplantation report-2017; Focus theme: allograft ischemic time. *J Heart Lung Transplant.* 2017;36(10):1037–1046.
41. Sen A, et al. Mechanical circulatory assist devices: a primer for critical care and emergency physicians. *Crit Care.* 2016;20(1):153.
42. Gordon RJ, et al. Prospective, multicenter study of ventricular assist device infections. *Circulation.* 2013;127(6):691–702.
43. Acharya D, et al. INTERMACS analysis of stroke during support with continuous-flow left ventricular assist devices: risk factors and outcomes. *JACC Heart Fail.* 2017;5(10):703–711.
44. Miller LW, et al. Use of a continuous-flow device in patients awaiting heart transplantation. *N Engl J Med.* 2007;357(9):885–896.
45. Pagani FD, et al. Extended mechanical circulatory support with a continuous-flow rotary left ventricular assist device. *J Am Coll Cardiol.* 2009;54(4):312–321.
46. Starling RC, et al. Unexpected abrupt increase in left ventricular assist device thrombosis. *N Engl J Med.* 2014;370(1):33–40.
47. Kilic A. The future of left ventricular assist devices. *J Thorac Dis.* 2015;7(12):2188–2193.
48. Jorde UP, et al. Prevalence, significance, and management of aortic insufficiency in continuous flow left ventricular assist device recipients. *Circ Heart Fail.* 2014;7(2):310–319.
49. Cowger JA, et al. Consequences of aortic insufficiency during long-term axial continuous-flow left ventricular assist device support. *J Heart Lung Transplant.* 2014;33(12):1233–1240.
50. Crow S, et al. Gastrointestinal bleeding rates in recipients of nonpulsatile and pulsatile left ventricular assist devices. *J Thorac Cardiovasc Surg.* 2009;137(1):2008–2015.
51. Draper KV, Huang RJ, Gerson LB. GI bleeding in patients with continuous-flow left ventricular assist devices: a systematic review and meta-analysis. *Gastrointest Endosc.* 2014;80(3):435–446. e1.
52. Aggarwal A, et al. Incidence and management of gastrointestinal bleeding with continuous flow assist devices. *Ann Thorac Surg.* 2012;93 (5):1534–1540.
53. Bartoli CR, et al. Pathologic von Willebrand factor degradation with a left ventricular assist device occurs via two distinct mechanisms: mechanical demolition and enzymatic cleavage. *J Thorac Cardiovasc Surg.* 2015;149(1):281–289.
54. Hasin T, et al. Readmissions after implantation of axial flow left ventricular assist device. *J Am Coll Cardiol.* 2013;61(2):153–163.
55. Akhter SA, et al. Hospital readmissions after continuous-flow left ventricular assist device implantation: incidence, causes, and cost analysis. *Ann Thorac Surg.* 2015;100(3):884–889.
56. Fukamachi K, et al. Effects of left heart bypass on right ventricular performance. Evaluation of the right ventricular end-systolic and end-diastolic pressure-volume relation in the in situ normal canine heart. *J Thorac Cardiovasc Surg.* 1990;99(4):725–734.
57. Takeda K, et al. Timing of temporary right ventricular assist device insertion for severe right heart failure after left ventricular assist device implantation. *ASAIO J.* 2013;59(6):564–569.

58. Fukamachi K, et al. Implantable continuous-flow right ventricular assist device: lessons learned in the development of a cleveland clinic device. *Ann Thorac Surg*. 2012;93(5):1746–1752. PMCID: PMC3447008.
59. Fukamachi K, et al. Advanced ventricular assist device with pulse augmentation and automatic regurgitant-flow shut-off. *J Heart Lung Transplant*. 2016;35(12):1519–1521.
60. Miyamoto T, et al. Effects of blood pump orientation on performance: In vitro assessment of universal advanced ventricular assist device. *Artif Organs*. 2020;44(10):1055–1060.
61. Miyamoto T, et al. An advanced universal circulatory assist device for left and right ventricular support: first report of an acute in vivo implant. *J Thorac Cardiovasc Surg Open*. 2020; (in press).
62. Saeed D, et al. Alternative right ventricular assist device implantation technique for patients with perioperative right ventricular failure. *J Thorac Cardiovasc Surg*. 2015;149(3):927–932.
63. Arabia FA, et al. Interagency registry for mechanically assisted circulatory support report on the total artificial heart. *J Heart Lung Transplant*. 2018;37(11):1304–1312.
64. Pfeffer MA, Shah AM, Borlaug BA. Heart failure with preserved ejection fraction in perspective. *Circ Res*. 2019;124(11):1598–1617.
65. Borlaug BA, Paulus WJ. Heart failure with preserved ejection fraction: pathophysiology, diagnosis, and treatment. *Eur Heart J*. 2011;32 (6):670–679.
66. Sharma K, Kass DA. Heart failure with preserved ejection fraction: mechanisms, clinical features, and therapies. *Circ Res*. 2014;115(1):79–96.
67. Moscato F, et al. Use of continuous flow ventricular assist devices in patients with heart failure and a normal ejection fraction: a computer-simulation study. *J Thorac Cardiovasc Surg*. 2013;145(5):1352–1358.
68. Gilboa SM, et al. Mortality resulting from congenital heart disease among children and adults in the United States, 1999 to 2006. *Circulation*. 2010;122(22):2254–2263.
69. Kirklin JK. Advances in mechanical assist devices and artificial hearts for children. *Curr Opin Pediatr*. 2015;27(5):597–603.
70. Weiss WJ, et al. The Penn State pediatric total artificial heart. In: Karimov JH, Fukamachi K, Starling RC, eds. *Mechanical Support for Heart Failure: Current Solutions and New Technologies*. Springer; 2020:577–585.
71. Fukamachi K, et al. Initial in vitro testing of a paediatric continuous-flow total artificial heart. *Interact Cardiovasc Thorac Surg*. 2018;26 (6):897–901.
72. Karimov JH, et al. Early in vivo experience with the pediatric continuous-flow total artificial heart. *J Heart Lung Transplant*. 2018;37 (8):1029–1034.
73. Combes A, et al. Temporary circulatory support for cardiogenic shock. *Lancet*. 2020;396(10245):199–212.
74. Friedline K, Hassinger P. Total artificial heart freedom driver in a patient with end-stage biventricular heart failure. *AANA J*. 2012;80(2):105–112.

Chapter 6

Continuous-Flow Ventricular Assist Devices

Takuma Miyamoto[1], Kiyotaka Fukamachi[1,2] and Jamshid H. Karimov[1,2]

[1]*Lerner Research Institute, Cleveland Clinic, Cleveland, OH, United States,* [2]*Cleveland Clinic Lerner College of Medicine of Case Western Reserve University, Cleveland, OH, United States*

Chapter Outline

Abbreviations

AEs	adverse events
AI	aortic insufficiency
BTT	bridge-to-transplant
CFD	computational fluid dynamics
CAP	continued access protocols
CF	continuous-flow
CF-VADs	continuous-flow ventricular assist devices
DT	destination therapy
DTE	device thrombogenicity emulation
DCT	double-cuff tipless
DLI	driveline infections
ELEVATE	Evaluating the HeartMate 3 with Full MagLev Technology in a Post-Market Approval Setting
EPPY	events per patient-year
ECMO	extracorporeal membrane oxygenation
FDA	Food and Drug Administration
GI	gastrointestinal
HMW	high molecular weight
INTERMACS	Interagency Registry for Mechanically Assisted Circulatory Support
ILS	intermittent low speed
INR	international normalized ratio
LATERAL	Left Anterior Thoracotomy with Upper Hemisternotomy or Right Anterior Thoracotomy
LVAD	left ventricular assist device
MCS	mechanical circulatory support
MOMENTUM	Multicenter Study of MagLev Technology in Patients Undergoing Mechanical Circulatory Support Therapy with HeartMate 3

Advances in Cardiovascular Technology. DOI: https://doi.org/10.1016/B978-0-12-816861-5.00005-8

PREVENT	PREVENtion of HeartMate II Pump Thrombosis through Clinical Management
REVIVE-IT	Randomized Evaluation of VAD InterVEntion before Inotropic Therapy
RA	right atrium
RHF	right heart failure
RVAD	right ventricular assist device
ROADMAP	Risk Assessment and Comparative Effectiveness of Left Ventricular Assist Device and Medical Management in Ambulatory Heart Failure Patients
SSS	scalar shear stress
SS	shear stress
SSI	silicone-skin interface
TAH	total artificial heart
US	United States
VAD	ventricular assist device
vWF	von Willebrand factor

Introduction

Current trend of continuous-flow ventricular assist device therapy

The continuous-flow (CF) ventricular assist device (CF-VAD) is an effective therapy for patients with advanced (stage D) heart failure. Advanced heart failure is further classified into Interagency Registry for Mechanically Assisted Circulatory Support (INTERMACS) 1–7 profiles. According to ACC/AHA guidelines in 2013, durable mechanical circulatory support (MCS) has Class IIa indication for carefully selected patients with stage D heart failure. Currently, the majority of patients' implanted CF-VAD support is under inotrope support.[1]

In 2021, The Society of Thoracic Surgeons INTERMACS 2020 Annual Report showed the current status of VAD therapy.[2] The report included all adult (aged >19 years) patients who underwent implantation with a Food and Drug Administration (FDA)-approved durable MCS device from January 1, 2010, to December 31, 2019, with follow-up through June 30, 2020. A total of 27,298 adult patients received an FDA-approved durable MCS device with STS-INTERMACS registration. For the reported 10-year period, 26,604 (97.5%) patients underwent implantation with a durable CF- left ventricular assist device (LVAD) (96%) or CF biventricular VAD (4%) support during the index LVAD operation. In 2019, 3198 primary LVADs were implanted in the United States (US), which is the largest annual volume in INTERMACS history. They evaluated the changing trend of patient characteristics, indication, device selection, outcome, and adverse events (AEs).

Patient characteristics changed slightly. Compared with patients from 2010 to 2014, patients from 2015 to 2019 were more unstable, and more patients were treated with temporary MCS before durable LVAD implant in the recent era versus the earlier era (36.8% vs 26.0%; $P<.001$), including extracorporeal membrane oxygenation support and an intraaortic balloon pump. According to the INTERMACS Profile at the time of implantation, in the recent era, 52.1% of the patients were Profile 1 or 2, and 12.4% were Profiles 4–7, a decrease from 18.6% in the earlier era.

Device strategy changed significantly in 2019, after implementation of a new heart allocation system. From January 2010 to December 2019, 21.9% of devices were implanted as a bridge-to-transplant (BTT), 26.9% were for bridge-to-candidate, and 50.4% were for destination therapy (DT). In 2019, 73.1% of patients were implanted CF-LVADs as DT.

There are three CF-LVADs approved by FDA as BTT and DT:

- the HeartMate II, an axial-flow pump, approved in 2008 as BT, 2010 as DT;
- the HVAD, a centrifugal-flow pump, approved 2012 as BTT, 2017 as DT, and
- the HeartMate 3, a centrifugal-flow pump, approved 2017 as BTT, 2018 as DT.

A dramatic shift also occurred from 2010 to 2019 in device selection as new devices showed improved clinical results and were approved for use. Axial-flow devices were the predominant device type implanted during 2010–14. However, the use of HeartMate II declined after 2015. In 2019, the HeartMate 3, HVAD, HeartMate II were implanted to 77.7%, 20.5%, and 1.8% of CF-LVAD patients, respectively.

Among patients implanted as DT:

- The 1- and 2-year survivals in the recent era are higher (82.3% and 73.1%, respectively) than those (80.5% and 69.1%) in the earlier era.
- The median patient survival has increased from 46.5 months in the earlier era to 54.6 months in the recent era.

- The hazard for AEs was high in the early postoperative period (90 days), except for device specific infection, which consists predominantly of driveline infections (DLI).

The most common AEs in the early and late periods after CF-LVAD implant are major infection [early, 1.349; late, 0.440 events per patient-year (EPPY)] and major bleeding (early, 1.433; late, 0.347 EPPY). Only 59% and 67% of patients are free from infection and major bleeding, respectively, at 1 year, and approximately half of bleeding episodes are gastrointestinal (GI) bleeding. Freedom from stroke was 87% at 1 year. Hospital readmission rate is still high and is a significant limitation of this therapy; readmission occurs in 38.6% of patients at 90 days and in 72.2% at 12 months.

Teuteberg et al.[3] compared clinical results of these three devices using the same cohort in 2019. Among 13,016 adult patients who were implanted CF-LVADs from January 2014 to December 2018, there was no significant difference in survival between the HeartMate II and HVAD devices, but the HeartMate 3 device had a 1-year survival of 87%, which was significantly higher in an unadjusted comparison with other devices. For AEs, the unadjusted freedom from GI bleeding, stroke, and infection was superior in the HeartMate 3 device. GI bleeding remains common in the first year after implant with 25%, 20%, and 12% of patients having a GI bleed for HeartMate II, HVAD, and HeartMate 3 devices, respectively. The freedom from any first stroke at 1 year was 88% for HeartMate II, 84% for HVAD, and 93% for HeartMate 3 devices. Major infection was the most common AE, and freedom from major infection at 1 year was 60% for HeartMate II, 57% for HVAD and 67% for the HeartMate 3 devices.

Axial- and centrifugal-flow pumps

The two major types of CF-VAD pump are axial- and centrifugal-flow pumps. A CF-VAD pumps blood from the ventricle or atrium to the ascending or descending artery to support the failed ventricle. A CF-VAD consists of inlet and outlet ports and a single moving part, which is a rotor. The rotor is supported or suspended within a pump housing containing a motor to spin the rotor at the desired speed.

The primary difference between axial- and centrifugal-flow pumps lies in the design of their rotating elements. In an axial-flow pump, the inlet and outlet blood paths are positioned parallel to the axis of rotation, and the rotor acts like a propeller in a pipe that pushes fluid forward. In a centrifugal-flow pump, the outlet path is positioned 90 degrees relative to the axis of rotation, and the rotor acts as a spinning disk with blades, and throws the blood from the impeller blades to the outflow cannula along a tangential course.[4]

The flow of CF-VADs is determined by pressure differences between the inlet and outlet (head pressure), rotating speed, and blood viscosity. A pump head curve (pressure—flow curve) shows the relationship between pump flow and pressure difference (head pressure) across the pump at different operating pump speed and the pressure—flow curve is characteristic to each pump.

In general, centrifugal-flow pumps work with lower operational speeds and have a flatter pressure—flow curve than axial-flow pumps. Thus centrifugal-flow pumps are typically more efficient, resulting in lower power consumption compared with axial-flow pumps. Higher pressure head sensitivity in centrifugal-flow pumps results in a lower likelihood of ventricular suction events, and greater flow pulsatility than axial-flow pumps. Stanfield et al.[5] compared the ability of both axial- and centrifugal-flow pumps regarding their capability for pulsatility in vitro and showed that centrifugal-flow pumps created more pulsatile flow than axial-flow pumps.

Pump bearing system

Another distinguishing characteristic of CF pumps is the bearing system, including contact bearings (mechanical bearings), and noncontact bearings (hydrodynamic bearings, composite hydrodynamic and magnetic bearings, and active and passive magnetic suspension).[6]

There are many shapes of mechanical bearings (ball-cup bearing, cone-bearing etc.). The bearings can stabilize axial and radial movement of a rotor. Most mechanical bearings, such as the HeartMate II, have proven to be robust and reliable. However, a major limitation of this mechanism is heat generation around bearings that represents potential sites for fibrin deposition with thrombus formation.

Noncontact-bearing designs incorporate either hydrodynamic or magnetic forces or a combination of both. Hydrodynamic bearings, such as the HVAD, use pressure generated from a fluid film that acts to separate the rotating surfaces of the impeller from the stationary base of the pump. Hydrodynamic bearings don't need a position sensor, so the pump can be smaller than a magnetic bearing pump. However, in order to generate a hydrodynamic force, the surface of the impeller must be moving within a certain speed range. If the impeller is stopped or operated in slow rotational speed, the impeller may contact the housing or become unstable.

Another limitation is that hydrodynamic bearings have higher heat stress at the fluid film. In a magnet bearing, such as the HeartMate 3, magnetic forces may be a passive permanent magnet or active electrical magnet in design. With a magnetic bearing, the pump can have a larger flow gap inside the pump, and lower shear stress (SS). However, for stable suspension of magnetic bearing, the pump needs complex electromagnetic systems, requiring position sensors, electromagnets and extra conductors, connector pins, and electronics to execute the dedicated position control algorithm, and these complex systems are susceptible to failure.

Axial-flow pump

HeartMate II

Device description

The HeartMate II is an axial-flow rotary VAD composed of a blood pump, percutaneous lead, external power source, and system driver (Fig. 6.1).[8] The dimensions are: weight, 350 g; size, 7.0 cm long, and 4.0 cm in diameter at its largest diameter.[9] The operating range is 6000–15,000 rpm, and the flow is up to 10 L/min (the pressure–flow curves are shown in Fig. 6.2). The pump rotor spins on two bearings located at either end; both bearings react against radial and axial thrust loads. Each bearing's stationary element is located in the hub of the respective inlet and outlet stators. The bearings are the ball-and-socket hydrodynamic composite bearing, which is similar to standard ball-cup jewel bearings. These are mechanical bearings, so they take high load during operation, and are susceptible to wear. However, according to Sundareswaran et al.,[7] the HeartMate II bearing is reported to be robust, with the highest bearing wear rate was 1.46 μm/year, and the bearing had an estimated life of 17 years or longer, so it is claimed not to be a limiting factor for long-term support.

The motor is contained in the pump housing and creates a spinning magnetic field that spins the rotor to pump blood. The rotor contains a permanent magnet located inside a 12 mm diameter titanium duct. Blood entering the HeartMate II flows across the inlet stator, which has three blades that structurally support the inlet stator and straighten the flow field before entering the rotor (straightener). The rotor also has three blades and creates radial velocity.

At the exit, the flow field encounters the outlet stator. Outlet stator vanes have a twisted shape, designed to convert the radial velocity to the axial direction. As the flow is being turned, the exit hub narrows, diffusing the flow and converting kinetic energy to static pressure (diffuser). Inflow conduits and outflow grafts are sealed graft. Textured surfaces

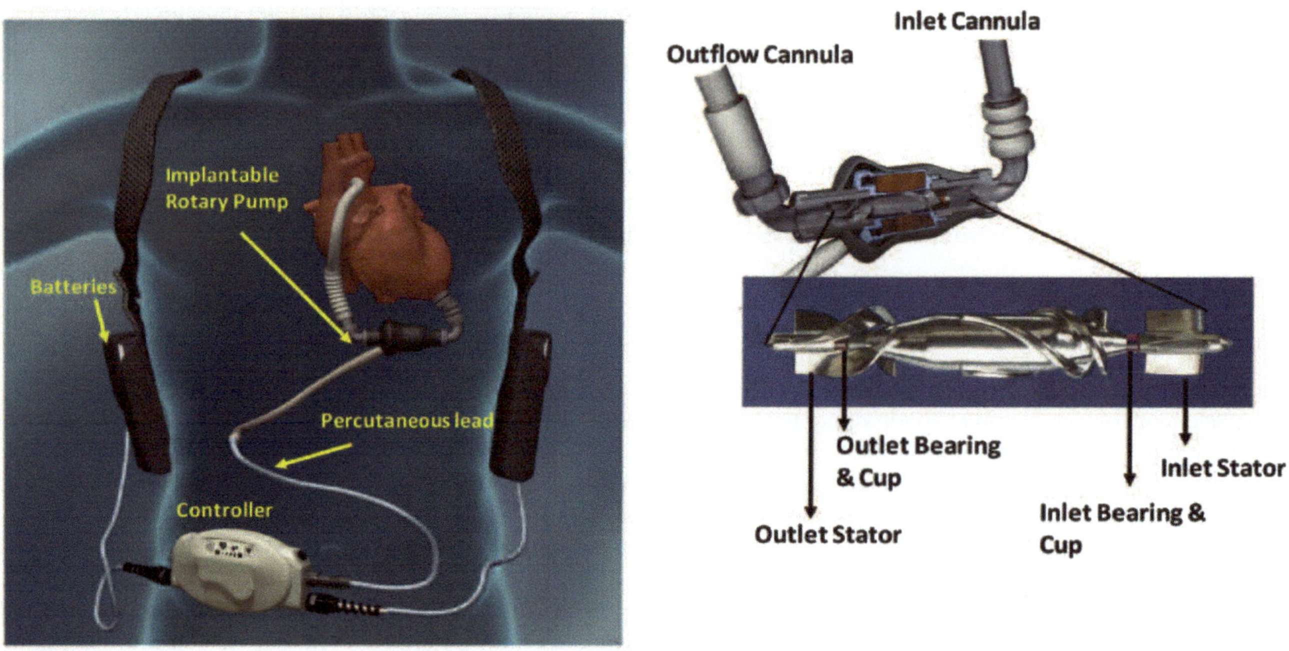

FIGURE 6.1 HeartMate II left ventricular assist system. *Used with permission from Sundareswaran KS, Reichenbach SH, Masterson KB, Butler KC, Farrar DJ. Low bearing wear in explanted HeartMate II left ventricular assist devices after chronic clinical support. ASAIO J. 2013;59(1):41–45.*[7]

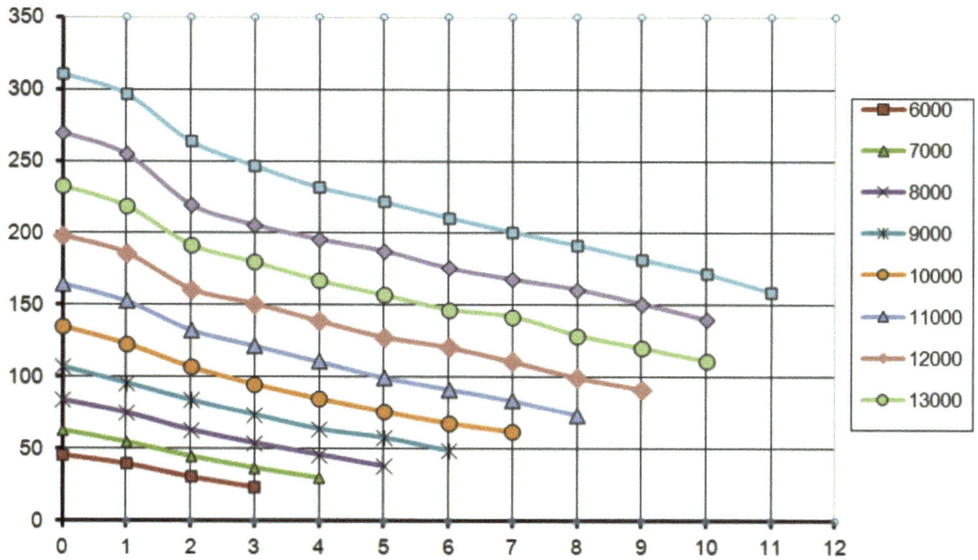

FIGURE 6.2 Pressure–flow curves of the HeartMate II. *Used with permission from Abbott (Abbott Park, IL).*

coat the outside and inner lumen of the inflow conduit and the outflow elbow to stimulate tissue growth and create a natural lining. The HeartMate II is cradled between the inlet and outlet elbow connectors. The implantation requires pump pocket creation at the preperitoneal space; that is, below the left rectus muscle, above the posterior rectus sheath, to position the pump parallel to the diaphragm.[9]

The HeartMate II motor is commutated using electromagnetic field sensing that maintains pump speed independent of rotor torque. This results in a linear current-pump flow relationship. Using this in combination with the pressure–-flow relationship, the pump flow can be estimated. The estimated flow values are also used to define relative levels of "flow pulse" delivered during the systolic–diastolic phases of the cardiac cycle.

In general, flow estimation in the axial-flow pump is less reliable than that of the centrifugal-flow pump. Slaughter et al.[10] compared intraoperative estimated flow values with an ultrasonic flow probe placed around the outflow graft. Flow estimation demonstrated a moderate linear correlation with ultrasonic flow probe measurements [$y = 0.74x + 0.99$ ($R^2 = 0.56$)]. A root mean square error of 0.8 L/min and a 15%–20% difference at flows of 4 of 6 L/min was observed between flow estimation and direct flow measurement.

Clinical results

The development of the HeartMate II began in the early 1990s as a collaborative project between Nimbus, Inc. (Rancho Cordova, California, USA) and the University of Pittsburgh.[9] On July 27, 2000, the first human implantation was conducted at the Sheba Medical Center in Israel. A European study was terminated early due to thrombus formation at the inlet and outlet stators. Engineers reexamined the textured surface and flow paths, and shifted the start of the smooth surface slightly upstream so that it ended slightly downstream of its original location.

In November 2003, the newly designed HeartMate II was implanted into an 18-year-old man at the Texas Heart Institute.[11] The first clinical trial (43 cases) in the United States was reported by Frazier et al.[12] in 2007. In this report, the average duration of support was 258 days and the total duration of support was more than 31 patient-years. Of the 43 implanted patients, 35 (81%) survived the operation without significant complications and were discharged from the hospital in NYHA functional class I.

The same year, a prospective, multicenter study was reported by Miller et al.[13] In this study, 133 BTT patients underwent implantation of the HeartMate II. The survival rate during support was 75% at 6 months and 68% at 12 months with significant improvement in functional status and in quality of life. As a result, in 2008, the HeartMate II received FDA approval for BTT.

In 2009, Pagani et al.[14] reported the 18 month-results of a prospective, multicenter study, in which 281 patients underwent implantation of the HeartMate II for BTT and had the same good results. In the same year, Slaughter et al.[15] reported a multicenter randomized trial, comparing outcomes in DT patients who received either a pulsatile-flow LVAD (HeartMate XVE, $n = 66$), or the Heart Mate II ($n = 134$). Results showed superior results with the HeartMate II compared with the HeartMate XVE; the primary composite end point (survival free from disabling stroke and reoperation to repair or replace the device) was achieved in more patients with the HeartMate II than with the HeartMate XVE (46% vs 11%), and patients with HeartMate II had better actuarial survival rates at 2 years (58% vs 24%). In 2010, the HeartMate II was approved by the FDA for DT.

The HeartMate II is approved by the FDA for NYHA class IIIb (recent dyspnea at rest) and IV (symptoms of heart failure in rest or small secrecies) patients, which is equivalent to INTERMACS profiles 1–5. However, the majority of the LVAD implantations were performed in patients who were hospitalized and dependent on intravenous inotropic support.

In 2015 and 2017, Estep and Starling et al.[16,17] reported an observational study of 1- and 2-year results of data for the HeartMate II ($n = 97$) compared with optimal medical management (OMM, $n = 103$) in patients with noninotrope-dependent heart failure [Risk Assessment and Comparative Effectiveness of Left Ventricular Assist Device and Medical Management in Ambulatory Heart Failure Patients (ROADMAP) trial]. At 2 years, the authors reported: (1) higher survival with improved functional status, improved quality of life, and reduced depression in the HeartMate II group; (2) no increased mortality with delaying LVAD implant with close monitoring; (3) more hospitalizations in the LVAD than in the OMM group; and (4) a greater rate of major AEs in LVAD than OMM subjects in year 1, but a reduction in LVAD AEs in year 2.

This study showed a potential benefit of the HeartMate II implant for patients with less severe (noninotropic support) heart failure, but this benefit is limited by the high rate of major AEs, especially in the first year of implant. The National Institutes of Health's National Heart, Lung, and Blood Institute convened a working group in March 2008 to discuss a trial to assess the use of long-term VAD therapy in patients who are less ill than patients currently eligible for DT. This resulted in the Randomized Evaluation of VAD InterVEntion before Inotropic Therapy (REVIVE-IT) pilot study.[18] The specific objective of REVIVE-IT was to compare LVAD therapy with OMM in patients with less advanced heart failure than current LVAD indications to determine if wider permanent LVAD use in less ill patients would be associated with improved survival, quality of life, or functional capacity. However, REVIVE-IT failed to accomplish its goal with the HeartMate II due to concerns about pump thrombosis.

Adverse events

AEs are limiting factor of LVAD therapy, especially complications related to hemocompatibility, such as bleeding, hemolysis, and thrombus formation. In the ROADMAP study, the composite event rate was significantly higher for LVAD versus medical management (1.74 vs 0.98 EPPY; $P < .001$) and GI bleeding was the most common AE for LVAD patients.[16] Freedom from rehospitalization at 2 years was very low (16% for medical management patients and 8% for LVAD patients, $P = .005$). The main reason for readmission of OMM patients was worsening heart failure. Readmissions for LVAD patients were largely due to AEs.

Michelis et al.[19] described dynamic estimates of event-free survival with conditional survival probabilities in 8245 adult patients implanted with a CF-LVAD as DT. For patients who experienced a nonfatal AE within the first year, subsequent 1-year conditional survival was lower than in the absence of that event for stroke (65% vs 75%; $P < .001$), device-related infection (64% vs 76%; $P < .001$), and pump thrombosis or malfunction (64% vs 76%; $P < .001$).

Seese et al.[20] evaluated the sequences of AEs preceding mortality attributed to multisystem organ failure in patients with a LVAD by analyzing 3765 AEs after 536 patients with LVAD experienced multisystem organ failure death. They identified two distinct multisystem organ failure clusters based on their proportion of AE types and survival time. The early-death cluster (renal failure to respiratory failure to death) occurred in 418 patients, who had 2304 AEs and a median survival of 1 month. The late-death cluster (bleeding to infection, to respiratory failure to death) occurred in 118 patients, with 1461 AEs and a median survival of 11 months.

von Willebrand factor deficiency and gastrointestinal bleeding

Multifactorial bleeding is a major AE in patients supported with a CF-LVAD, brought on by the requirement of anticoagulation, acquired von Willebrand disease and impaired platelet aggregation. Data from INTERMACS suggest that the most frequent locations of the first bleeding episode after implantation are mediastinal (45%), the thoracic pleural space (12%), lower GI tract (10%), chest wall (8%), and upper GI tract (8%).[21]

Acquired von Willebrand disease is unique complication among patients exposed to high SS, such as aortic stenosis and CF-LVAD support. The von Willebrand factor (vWF) is a large multimeric glycoprotein involved in hemostasis. The protein is present in plasma and stored as multimers in the Weibel-Palade bodies of endothelial cells, megakaryocytes and subendothelial connective tissue. The vWF plays a critical role in primary hemostasis. At sites of vascular injury, turbulent blood flow and altered SS trigger conformational activation of vWF multimers. Subsequently, vWF binds to subendothelial collagen and platelets to produce the primary hemostatic plug. VADs cause turbulent flow and alter SS profiles. With these supraphysiological SS, more than 9 Pa activate vWF, resulting in downsizing of the high molecular weight (HMW) multimers by vWF-cleaving metalloprotease, ADAMTS13. The resultant loss of vWF HMW multimers causes impaired vWF function and platelet aggregation.

Most patients supported with CF-LVAD have some degree of vWF degradation. Geisen et al.[22] reported missing biologically active large multimers in all 10 tested VAD patients. A pathologic collagen binding ratio was found in all 10 tested VAD patients, and a reduced ristocetin cofactor activity ratio was found in 10 of 12 VAD patients. Both are important indicators of vWF activity.

Klovaite et al.[23] reported that 11 of 16 patients treated with the HeartMate II had decreased ristocetin cofactor activity ratio, and fewer large vWF multimers than healthy control patients. Uriel et al.[24] reported the prevalence of coagulation abnormalities, including laboratory assessment for von Willebrand syndrome, and related bleeding events during device support for 79 patients supported by HeartMate II. HMW vWF multimers were measured in 31 patients and were reduced in all patients, and 58% of patients had bleeding.

Kang et al.[25] evaluated the effect of the HeartMate II on vWF in vitro, focusing on pump speed and considering SS created by pump speed. They observed decreased vWF multimers and significantly increased vWF degradation fragments at 11,400, 10,000, and 8600 rpm at the same degree, and concluded that the HeartMate II caused vWF degradation regardless of its speed (within the clinical operational range). Acquired von Willebrand disease is a characteristic complication of CF-VAD support and is related to bleeding AEs.

The GI tract is common nonsurgical bleeding spot in patients with CF-LVAD. Demirozu et al.[26] investigated GI bleeding and its relationship to arteriovenous malformations (AVMs) in 172 patients with the HeartMate II. Thirty-two patients (19%) had GI bleeding after 63 ± 62 days of HeartMate II support. Ten patients had GI bleeding from an AVM. Sixteen patients had upper GI bleeding and 15 had lower GI bleeding and 1 had upper and lower GI bleeding.

Aggarwal et al.[27] conducted a retrospective analysis of 101 patients implanted with the HeartMate II. GI bleeding occurred 22.8% of patients; 57% of the patients with bleeding episodes bled from the upper GI tract, and 35% of patients bled from the lower GI tract. Risk factors of GI bleeding were previous history of GI bleeding, elevated international normalized ratio (INR), and low platelet count.

Draper et al.[28] conducted a meta-analysis to determine prevalence and risk factors for GI bleeding with 17 case control and cohort studies [1839 LVAD patients, of whom 1697 (92%) had CF-LVADs]. The prevalence of GI bleeding in LVAD patients was 23%. Subgroup analysis demonstrated that older age and elevated creatinine were associated with GI bleeding.

Treatment of GI bleeding typically requires reduction or elimination of anticoagulation or antiplatelet therapy. However, Stulak et al.[29] reported the association of GI bleeding and subsequent thromboembolic events defined as stroke, transient ischemic attack, hemolysis or suspected or confirmed pump thrombosis in 389 patients supported by CF-LVAD. They observed GI bleeding in 116 patients (30%), and patients with GI bleed were at sevenfold increased risk for having a subsequent thromboembolic event. GI bleeding is also associated with lower rate of cardiac transplantation.[30]

Some studies have shown a relationship between pulsatility and GI bleeding or vWF deficiency. Wever-Pinzon et al.[31] assessed the effects of different degrees of pulsatility on the incidence of nonsurgical bleeding with 134 patients supported by the HeartMate II. Major bleeding occurred in 33 (25%) patients (70% GI, 21% epistaxis, 3% genitourinary, and 6% intracranial). In multivariable analysis, reduced pulsatility was associated with an increased risk of bleeding.

Muthiah et al.[32] prospectively studied 28 patients implanted with the HVAD for up to 360 days. A higher vWF collagen binding activity/Antigen ratio (CBAS/Ag) was associated with aortic valve opening ($P = .02$). In patients with a closed aortic valve at day 30, the vWF CBA/Ag fell significantly by 0.49 ± 0.1 below baseline ($P < .0001$), whereas in those who had an open or intermittently open aortic valve, the fall from baseline was nonsignificant (0.16 ± 0.2, $P = .47$). This pattern persisted to day 180.

Vincent et al.[33] investigated the relationship between pulsatility and vWF under CF support in vitro and in vivo. They demonstrated the proteolytic degradation of vWF multimers by high shear CF pump in a circulatory loop without pulsatility. In the in vivo study in swine models, they observed that the magnitude of the VWF degradation is modulated by the pulsatility level in the high SS level condition, and that the restoration of pulsatility is a trigger for the endothelial release of vWF.

Shear stress and blood damage

Extra physiological SS inside CF-LVADs causes blood damage, such as platelet activation, von Willebrand multimer degradation, or hemolysis and some studies were investigated intra-pump environment, such as SS distribution or flow streamlines. Significant amount of AEs may relate to blood damage and the nonphysiological flow conditions inside CF-VADs. Wall SS in the normal circulation is mostly under 2 Pa, ranging up to 15 Pa in the smallest vessels.

Fraser et al.[34] demonstrated computational fluid dynamics (CFD) analysis of five different CF pumps (three axial-flow and two centrifugal-flow) to flow fields within VADs and to determine the range of SS and residence times experienced by the blood inside them and to evaluate their contribution to blood damage. They defined SS that cause blood damage as follows: (1) less than approximately 1 Pa have been linked with thrombosis [low scalar shear stress (SSS)]; (2) higher than 9 Pa is a potential indicator of vWF degradation (high SSS), higher than 50 Pa is an indicator of potential platelet activation (higher SSS), and higher than 150 Pa is associated with hemolysis (very high SSS).

Results showed that the low SSS percentage (low SSS volume as a percentage of the total VAD volume) was larger for the axial-flow VADs (25%–30%) than for the centrifugal-flow VADs (5%–15%). The high SSS volume generally increases with the pressure head in the HeartMate-II like axial-flow pump, in which high SSS ranged from around 17% at 50 mm Hg to around 25% at 150 mm Hg. Higher SSS volume increases with the pressure head and was approximately 3%–5% of the pump. The very high SSS regions in the HeartMate II-like pump was approximately 0.2%–1.3%.

Kang et al.[25] demonstrated that the reduced HeartMate II speed (within the clinical operational range) did not significantly decrease vWF degradation in a mock circulatory loop with human blood. They also showed the relationship between HeartMate II speed and wall sheer stress SS at different rpms (Fig. 6.3). This shows that the instantaneous SS at a single location between the impeller and pump housing was modeled by computational fluid modeling. Reduced LVAD speed produced only a modest decrease in SS, so with clinical operational range, HeartMate II did not approach physiologic levels of SS.

Khoo et al.[35] performed CFD analyses on the two VADs, the HVAD and the HeartMate II, and then used the velocity gradients to calculate the magnitudes of normal and SS components; the normal stresses act perpendicular to the faces, whereas the shear forces act parallel to the faces. The results reveal that while the magnitude of the normal stress is lower than the SS, there are substantial regions of normal stress in both VADs. In the HeartMate II, the highest SS were found in the blade tip gaps (up to 400 Pa), whereas the negative, low-pressure side of the blade experienced the highest normal stress (up to 100 Pa) along the curvature of the blade. High-velocity regions (between 5 and 8 m/s) at the mid-section of the impeller and housing also experienced SS. In the diffuser section, the highest normal stresses were between the hub and the blades (up to 50 Pa). The results showed that there are large regions experiencing high SS magnitudes and high normal stress sufficient to cause RBC and vWF deformation in both the HeartMate II and the HVAD.

Flow pattern and thrombus formation

Irregular flow pattern within the blood pump, such as flow stagnation and recirculation is considered to be associated with thrombus formation. In 1999, Wu et al.[36] reported flow characteristics in the HeartMate II under development using fluorescent particle imaging. The flow within a sector of the inlet stator and outlet stator were studied by observing planes from the blade root to the tip. At the inlet section of the pump, the particle path lines were straight and parallel, except within a small annular region of backflow observed at the rotor blade tip and near the inlet housing wall, and

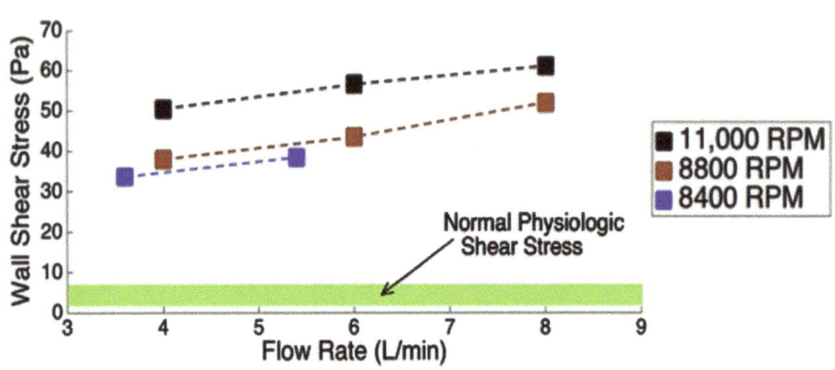

FIGURE 6.3 The relationship between HeartMate II speed and wall shear stress. *Used with permission from Kang J, Zhang DM, Restle DJ, et al. Reduced continuous-flow left ventricular assist device speed does not decrease von Willebrand factor degradation. J Thorac Cardiovasc Surg. 2016;151(6):1747–1754. e1741.*[25]

a small recirculation was observed behind the inlet stator blade under nominal conditions. At the outlet section of the pump, the flow features in the outlet duct demonstrated complex, strong three-dimensional characteristics.

Fraser et al.[34] observed the same kind of flow field with CFD analysis of the HeartMate II-like axial-flow pump. At the higher speed and lower flow rates, there was some preswirl before the impeller. The flow through the impeller itself was smooth for all operating conditions. Disturbances developed in the diffuser blade passages and the helical flow were enhanced by the bend in the outlet.

Chiu et al.[37] simulated the HeartMate II by CFD to predict the stress exposure and flow trajectories of flowing platelets. They found that the trajectories depicted the formation of well-defined recirculation zones that spanned approximately 1–2 mm across at the downstream of the inlet stator blades, while other platelet trajectories formed stagnant flow patterns, attaching to the rear hub (Fig. 6.4). Some platelet trajectories were observed forming entrapped circular patterns at the entry of the impeller blades, closely following the rotational motion of the impeller. These flow patterns correspond geographically to the sites of the thrombus formation in analysis of malfunctioning explanted HeartMate II devices.

Yang et al.[38] extended investigations to include pulsatile conditions. As the flow rate was reduced to below 3.8 L/min, a rapid transition to a chaotic velocity field occurred, exhibiting a large toroidal vortex adjacent to the inlet bearing. The pathlines in the downstream stator section were consistently chaotic for all hemodynamic conditions

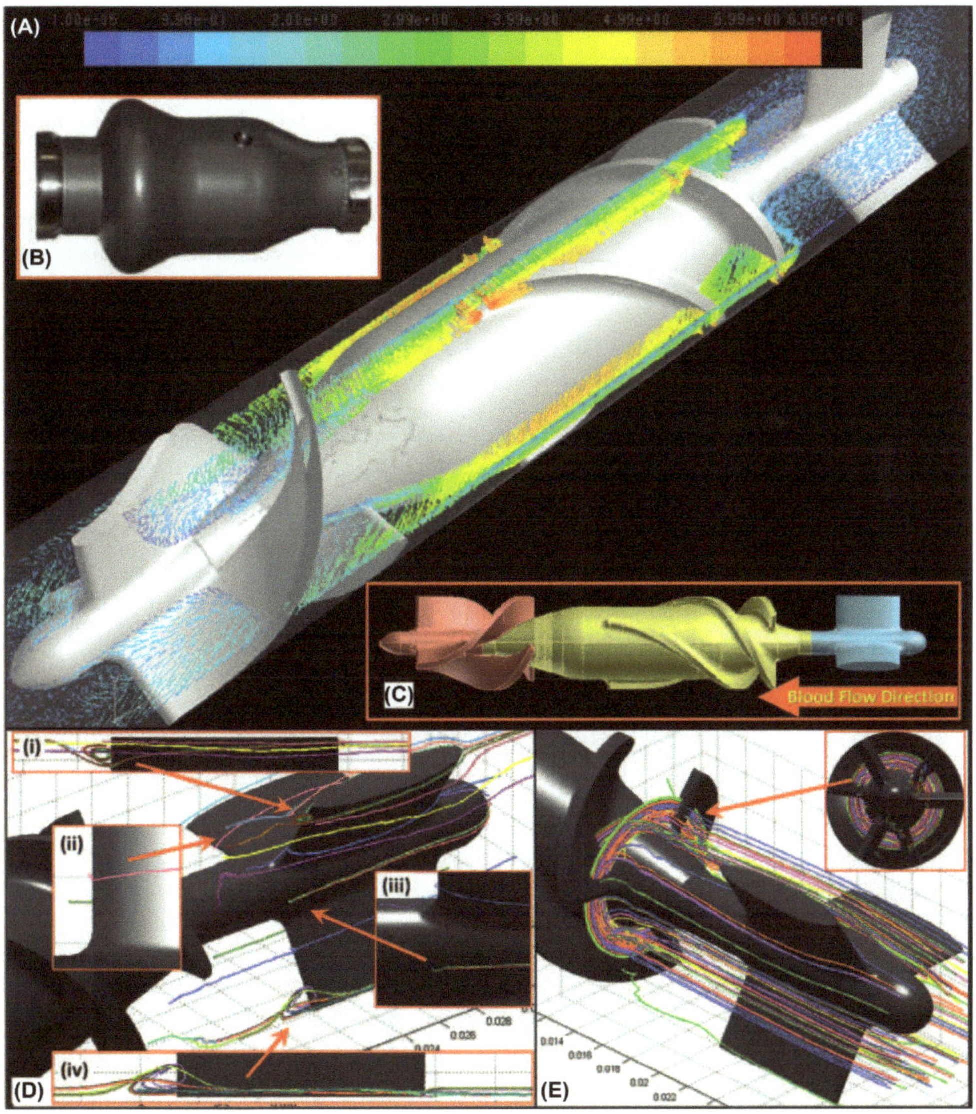

FIGURE 6.4 Numerical prediction of thrombus formation patterns in the HeartMate II. (A) Freeze frame depiction of the velocity vector field of the HeartMate II simulation (4 L/min of cardiac output with the impeller speed of 10,000 rpm). (B and C) Exterior and interior of the HeartMate II: flow straightener, impeller, and diffuser (*blue*, *yellow*, and *pink*, respectively). (D and E) Stagnant platelet trajectories are observed along the posterior side of the flow straightener, and the entry of the impeller blades is characterized by stagnant flow patterns. Recirculation zones were observed at the downstream of the flow-straightener blades spanning approximately 1–2 mm (D: inset i, ii, and iii). Platelet trajectories indicate a stagnation zone at the rear hub (D: inset iv) and entrapped circular flow patterns at the entry toward the impeller blades (E). *Used with permission from Chiu WC, Slepian MJ, Bluestein D. Thrombus formation patterns in the HeartMate II ventricular assist device: clinical observations can be predicted by numerical simulations. ASAIO J. 2014;60(2):237–240.*[37]

investigated. The addition of pulsatility caused the flow field to become periodically chaotic during the diastolic portion of the cardiac cycle, depending on the instantaneous flow rate and acceleration.

Wu et al.[39] reported the mathematical model to predict patterns of thrombus deposition that have been observed clinically in the HeartMate II. Their model showed the regions where the thrombus preferentially accumulates, including the majority of the upstream flow straightener hub, the rear of the straightener vanes and the upstream bearing region. This finding is in good agreement with the regions in which the thrombus has been reported clinically.

Pump thrombosis

The biggest limitation of the HeartMate II would be pump thrombosis. In early 2011, clinicians began to notice a spike in the incidence of pump thrombosis with the HeartMate II. Investigators at Cleveland Clinic initially sounded the alert, which led to a focused analysis nested within the INTERMACS database.[40]

In the analysis, 6910 adult patients from 132 institutions who received a HeartMate II LVAD were analyzed. Freedom from device exchange or death due to thrombosis decreased from 99% at 6 months in 2009 to 94% in 2012 ($P = .0001$). Multivariable hazard function analysis showed risk factors for pump thrombosis included later implant year, younger age, higher creatinine, larger body mass index, white race, left ventricular ejection fraction above 20%, and higher lactate dehydrogenase level at 1 month. The investigators concluded that pump exchange or death due to pump thrombosis increased during 2011 and 2012, but the magnitude of the increase remained relatively small.

At the same time, Starling et al.[41] also reported an apparent increase in the rate of device thrombosis among 837 patients at three major institutions who received the HeartMate II, where 895 devices were implanted from 2004 through mid-2013; the mean age of the patients was 55 ± 14 years. A total of 72 pump thromboses were confirmed in 66 patients. Starting in March 2011, the occurrence of confirmed pump thrombosis at three months after implantation increased from 2.2% to 8.4% by January 1, 2013. Before March 1, 2011, the median time from implantation to thrombosis was 18.6 months; from March 2011 onward, it was 2.7 months.

Saeed et al.[42] conducted a retrospective review of 80 patients undergoing HeartMate II implantation to evaluate the relationship between aortic valve opening and thrombotic events. Patients with a thrombotic event had significantly higher aortic valve opening scores compared to those without. Also, there was a trend for subjects with thromboembolic events to be maintained on lower LVAD speeds. This suggested that lower LVAD speed and more frequent aortic valve opening are associated with thromboembolism.

Uriel et al.[43] performed a root-cause analysis in all device thrombosis cases confirmed upon explant among 177 patients implanted with a HeartMate II. Among them, 19 (11%) were found to have device thrombosis of various etiologies after a mean of 351 ± 311 days, representing 0.12 EPPY. Of the five mechanically induced thromboses, three patients had abnormal inflow cannula position, two had bend relief disconnect with deformed outflow graft, and one had a hypercoagulable disorder. In the remaining 13 patients, device thrombosis occurred after 357 ± 383 days; INR at the time of diagnosis was 1.81, and mean device speed was 8855 ± 359 rpm. Five of 13 patients (38%) had an infection during the month leading up to device thrombosis. Of note, lactate dehydrogenase was already elevated at the time of discharge in patients who would later develop nonmechanical device thrombosis and so it can be an early risk marker of device thrombosis.

In 2015, Smedira et al.[44] followed up on an abrupt increase in pump thromboses. A total of 995 pumps thrombosed, with risk peaking within weeks of implant. The risk-adjusted increase in pump thrombosis began in 2010, reached a maximum in 2012, and then plateaued at a level that was 3.3-times higher than pre-2010. Pump exchange, younger age, and larger body mass index were important predictors. The probability of death within 3 months after pump thrombosis was 24%.

Soltani et al.[45] reported a possible association between design modification and pump thrombosis in the HeartMate II. In 2010, the cable strain relief of the HeartMate II device was improved and sealed grafts were introduced. After introduction of the new cable design, incidence of cable damage in HeartMate II patients dropped from 0.06 to 0 EPPY ($P = .03$), while pump thrombosis increased from 0.02 to 0.14 EPPY ($P < .001$) after the sealed graft was introduced.

Maltais et al.[46] conducted PREVENT (PREVENtion of HeartMate II Pump Thrombosis Through Clinical Management) recommendations study. They prospectively assessed the HeartMate II pump thrombosis rate among 300 patients implanted with the HeartMate II at 24 participating sites in the current era when participating centers adhered to the PREVENT recommendations. The primary end-point (confirmed pump thrombosis at 3 months) analysis showed a confirmed pump thrombosis of 2.9% at 3 months and 4.8% at 6 months. Full adherence to implant techniques, heparin bridging, and pump speeds >9000 rpms resulted in a significantly lower risk of pump thrombosis (1.9% vs 8.9%;

$P = .01$) and lower composite risk of suspected thrombosis, hemolysis, and ischemic stroke (5.7% vs 17.7%; $P = .01$) at 6 months.

Baumann et al.[47] pathologically and immunohistochemically examined 28 thrombi removed from 17 HeartMate II devices. Two groups of thrombi were found: those formed in the inlet/outlet and those on the rotor. The majority of the thrombi (20/28) were located on the rotor: nine at the inlet bearing, five on the rotor vanes, and six at the outlet bearing. Laminated thrombi formed around the inlet bearing in rings, an area of blood recirculation. The presence of distinct rings suggests development of the clot over time.

Rowlands et al.[48] conducted a retrospective analysis of 29 explanted HeartMate II devices due to suspected pump thrombosis. The outlet bearing/stator region had the highest frequency of deposition ($\approx$72%), and the preponderance of thrombi appeared white and unlaminated in their microstructure. Conversely, 41.3% of VADs had thrombus on the fore bearing of the pump, and these formations tended to be red and laminated, indicating they formed in layers over time. Fourteen VADs presented pump thrombosis in multiple regions; analysis of the data showed a statistically significant correlation between deposition in the fore bearing and subsequent thrombosis in the outlet bearing/stator.

Although there is no clear evidence, many factors were considered to be associated with the increase in pump thrombosis, such as a trend to reduce the anticoagulation regimen in response to the high bleeding complication rate, and a decreased pump speed to keep the aortic valve open during pump support and prevent aortic insufficiency (AI). The HeartMate II pump has high SS inside the pump, and also has irregular flow, as shown in many studies, so the pump could have narrow safety margin to prevent AEs related to hemocompatibility.

Infection

Infection is an inevitable AE when using artificial material, especially with VADs, since the pump is connected to the outside of the body by a driveline. In the original INTERMACS definition, four broad categories were included: (1) localized nondevice infection; (2) percutaneous site and/or pocket infection; (3) an infection involving any internal pump component and inflow or outflow tract, including the infection of blood-contacting surfaces of the LVAD; and (4) sepsis.[21]

Topkara et al.[49] reported incidence of infection, types of infection, microbiologic profile, and association of infections with clinical endpoints in 81 patients supported by CF-LVAD. Forty-two patients (51.9%) had at least one type of infection with a mean follow-up period of 9.2 ± 9.2 months. Subgroup analysis revealed that sepsis was significantly associated with increased mortality in the CF LVAD cohort. Resistant *Staphylococcus* and *Pseudomonas* species were the most common pathogens leading to device- and nondevice-related local infections. Development of driveline or pocket infection had no effect on survival in patients with CF-LVAD support.

Sharma et al.[50] examined the incidence and predictors of DLI in 143 patients who underwent HeartMate II implantation. DLI developed in 18 patients (12%) at a median of 182 days after implantation:

- Infections were superficial in 15 (83%) and deep in three (17%).
- Trauma was documented in six patients (33%).
- Seven patients (39%) needed readmission for DLI.
- Three patients (17%) needed surgical debridement.

All patients were managed successfully, without the need for device explantation or urgent cardiac transplantation. No patient required continuous antibiotic prophylaxis after the infection subsided. Risk factor analysis identified duration of support as the only independent predictor of infection. The odds of having a DLI rose by 4% for every month of support.

Gordon et al.[51] reported the microbiology and epidemiology of infections in 150 patients at 11 US cardiac centers who were followed prospectively until transplantation, explantation for recovery, death, or for 1 year. Eighty-six patients (57%) received HeartMate II devices. Thirty-three patients (22%) developed 34 VAD-related infections with an incidence rate of 0.10 per 100 person-days. The median time to infection was 68 days. The DLI was the most common ($n = 28$); 18 (64%) were associated with invasive disease. *Staphylococcus* was the most common pathogen (47%), but *Pseudomonas* or other Gram-negative bacteria caused 32% of infections. A history of depression and elevated baseline serum creatinine were independent predictors of VAD infection.

Investigating a DLI prevention strategy, Dean et al.[52] evaluated a surgical tunneling technique to keep the entire driveline velour portion in the subcutaneous tunnel, resulting in a silicone-skin interface (SSI) at the exit site. To assess long-term freedom from DLI with this technique, a multicenter SSI registry was initiated. A total of 200 patients who were implanted with the SSI tunneling method and on HeartMate II support for at least 10 months were compared with a control group of 201 patients also on HeartMate II support for at least 10 months with the traditional method (in which a small section of the velour portion of the driveline was externalized). The 1-year and 2-year DLI prevalence rates of were significantly lower in the SSI group [9% and 19%, compared with 23% and 35% in the control group (hazard ratio 0.49, 95% confidence interval 0.33–0.73, $P < .001$)]. The EPPY was 0.11 and 0.22 for the SSI and control groups, respectively ($P < .001$).

Based on a multivariate analysis, age and DL exit side were the only independent variables associated with DLI. The reason for the improved DLI rate with SSI group is not clear, but Singh et al.[53] suggested that with the silicone externalized technique, a sinus is formed around the exit site and has little to no skin adhesion. Minor trauma to the driveline did not disrupt the skin–drive line interface. Furthermore, the internalized velour portion can act as an additional anchor within the preperitoneal pocket and the subcutaneous plane to bolster driveline stability.

Cagliostro et al.[54] reported the effects of care of the driveline to DLI. DLI data were collected prospectively on all LVAD patients implanted between 2009 and 2013 at Columbia University Medical Center. Driveline care was changed from a dry sterile dressing without a standard anchoring device to a standardized kit, which included silver gauze dressing and a standard anchoring device. During the study period, 107 patients were implanted with LVADs before implementation of a standardized kit and 159 thereafter. DLI event rate improved from 0.18 to 0.07 event per patient-year, corresponding to a relative risk reduction of 62.5%. In addition, the 1-year freedom from infection was significantly higher in patients with the standardized kit.

Right heart failure

Right heart failure (RHF) has been a major limitation of LVAD treatment. According to INTERMACS, the definition of RHF is as follows[21]:

1. Early acute RHF: Need for implantation of a temporary or durable right ventricular assist device (RVAD), including extracorporeal membrane oxygenation (ECMO) concomitant with LVAD implantation (RVAD implanted before the patient leaves the operating room).
2. Early postimplant RHF: Need for implantation of a temporary or durable right VAD (including ECMO) within 30 days following LVAD implantation for any duration of time; or, failure to wean from inotropic or vasopressor support or inhaled nitric oxide within 14 days following LVAD implantation, or having to initiate this support within 30 days of implant for at least 14 days.
3. Death occurring in patients within 14 days of LVAD implant who have not received an RVAD but who remain on inotropes or vasopressors at the time of death and meet criteria for the diagnosis of RHF
4. Late RHF: Need for implantation of an RVAD (including ECMO) more than 30 days after an LVAD implantation. This may occur within the index hospitalization for LVAD implant or during subsequent rehospitalization for any diagnosis which results in a need for temporary or permanent right-sided mechanical assist devices. Hospitalization that occurs more than 30 days postimplant and which requires intravenous diuretics or inotropic support for at least 72 h.

Kormos et al.[55] evaluated the incidence, risk factors, and effect on outcomes of early RHF in a large population of patients implanted with the HeartMate II in patients ($n = 484$) enrolled in the bridge-to-transplantation clinical trial. Overall, 30 (6%) patients required an RVAD, 35 (7%) required extended inotropes, and 33 (7%) required late inotropes. A significantly greater percentage of patients without RHF survived to transplantation, recovery, or ongoing device support at 180 days compared with patients with RHF (89% vs 71%, $P < .001$). Multivariate analysis revealed that a central venous pressure/pulmonary capillary wedge pressure ratio of greater than 0.63, need for preoperative ventilator support, and blood urea nitrogen level of greater than 39 mg/dL were independent predictors of RHF after LVAD implantation. There are many prediction models or risk factor of RHF, but their performance is still limited.[56–60]

Takeda et al.[61] reported the incidence and clinical significance of late RHF in 336 patients who underwent CF-LVAD implantation. Of these, 293 patients (87%) discharged with isolated LVAD support were included in the study. Late RHF occurred in 33 patients (11%) at a median of 99 days after discharge (range 19–1357 days). Freedom from late RHF rates were 87%, 84%, and 79% at 1, 2, and 3 years, respectively. RHF recurred in 15 patients. Three patients required RVAD insertion. A univariable Cox proportional hazards regression model revealed that diabetes mellitus, body mass index >29 and blood urea nitrogen level >41 mg/dL were significant predictors of late RHF. Overall survival at 2 years was significantly worse in patients who developed late RHF (60% vs 85%, $P = .016$).

Aortic insufficiency

AI was an unexpected, VAD-specific AE. Cowger et al.[62] examined AI development following LVAD implantation by reviewing echocardiograms from 78 subjects undergoing HeartMate-XVE or HeartMate-II. Overall, at 6 months, 89% ± 4% of subjects were free from moderate or worse AI, but this dropped to 74% ± 7% and 49% ± 13% by 12 and 18 months, respectively. Correlates of AI progression included female sex, smaller body surface area, and the HeartMate-II device. Correlates of progressive AI on postoperative echocardiograms included increasing aortic sinus diameter, an aortic valve that remained closed or only intermittently opened, and lower left ventricular diastolic and systolic volumes.

Pak et al.[63] compared the prevalence of AI after HeartMate II versus HeartMate XVE support and assessed the role of aortic root diameter and aortic valve opening in the development of AI in 67 HeartMate XVE and 63 HeartMate II patients. Freedom from AI at 6 and 12 months was 94.5% and 88.9% in HeartMate XVE patients and 83.6% and 75.2% in HeartMate II patients, respectively. Aortic root diameters tended to be larger at baseline and follow-up compared with those who did not have AI. Lastly, AI was more common in patients whose aortic valve did not open. Aortic diameter, aortic valve opening, and HeartMate II support were associated with AI development.

Cowger et al.[64] evaluated the effect of AI on patient morbidity and mortality in 930 echocardiograms completed in 166 patients supported by the HeartMate II. During 291 person-years of follow-up, mild-moderate or worse AI developed in 70 (0.38 EPPY), moderate or worse AI in 36 (0.17 EPPY), moderate-severe AI in 11 (0.039 EPPY), and severe AI in 2 (0.0069 EPPY). Overall 2-year survival and 2-year survival after onset of moderate or worse AI was 87% ± 6.2% and 65% ± 11%, respectively, compared with 76% ± 5.1% and 76% ± 5.1%, respectively, in those with less AI ($P = .57$).

Jorde et al.[65] studied the prevalence and management of AI in 232 patients with CF-LVAD. Kaplan–Meier analysis revealed that freedom from greater-than-mild de novo AI at 1 year was 77.6% ± 4.2%, and that at least moderate AI is expected to develop in 37.6% ± 13.3% after 3 years. Nonopening of the AV was strongly associated with de novo AI development in patients without prospective discharge speed optimization. Seven of 21 patients with at least moderate AI developed symptomatic heart failure requiring surgical intervention.

Holley et al.[66] evaluated the impact on clinical outcomes and mortality in CF-LVAD recipients in 237 patients implanted with the HeartMate II. Moderate or severe AI occurred in 32 (13.5%) patients. Risk factors that significantly affected the development of AI included older age at the time of implantation, female gender, longer duration of LVAD support, and DT strategy. Freedom from moderate or severe AI was 94%, 76%, and 65% of patients at 1, 3, and 5 years, respectively. There was no difference in survival between recipients who developed significant AI and those who did not.

Sayer et al.[67] assessed the hemodynamics of LVAD patients with at least mild AI, at baseline and in response to device speed changes. They found that LVAD patients with at least mild AI have increased filling pressures and a lower pulmonary artery pulsatile index. Normalization of filling pressures can be achieved by increasing LVAD speed; however, this concomitantly worsens AI severity.

Truby et al.[68] conducted a retrospective analysis of patients with CF-LVAD in the INTERMACS study. Among 10,603 eligible patients, 1399 patients on CF-LVAD support developed moderate-to-severe AI. Prevalence of significant AI progressively increased over time. Predictors of worsening AI included older age, female sex, smaller body mass index, mild preimplantation AI, and DT strategy. Moderate-to-severe AI was associated with significantly higher left ventricular end-diastolic diameter, reduced cardiac output, and higher levels of brain natriuretic peptide. Significant AI was associated with higher rates of rehospitalization (32.1% vs 26.6%, respectively, at 2 years; $P = .015$) and mortality (77.2% vs 71.4%, respectively, at 2 years; $P = .005$), conditional upon survival to 1 year.

Auvil et al.[69] assessed outcomes in 221 patients who received a HeartMate II or HeartWare. Patients were stratified by AI severity at 6 months. Moderate AI was a significant predictor of 2-year mortality in a multivariable model ($P = .024$). However, functional exercise capacity (measured by a 6-minute walk test) and incidence of RHF at 1 year were not significantly different between groups ($P = .1421$; $P = .2189$).

Yehya et al.[70] reported nine cases of transcatheter aortic valve replacement in patients implanted with a CF-LVAD for treatment of symptomatic AI. All nine patients had one transcatheter aortic valve replacement procedure with resolution of AI and were discharged. Procedural complications include valve migration warranting a second valve for stabilization, retroperitoneal and groin hematoma, and pseudoaneurysm requiring thrombin injection. A significant improvement in the New York Heart Association classification was noted from the time of implant to 6 months.

Summary

The HeartMate II is a mechanical-bearing axial-flow CF pump. According to the INTERMACS report, the market share of the HeartMate II is approximately 2%. However, the HeartMate II is the first FDA-approved CF-LVAD, implanted in more than 27,000 patients, and the most studied device in CF-VAD history. After clinical application of the HeartMate II or transitioning from pulsatile flow pump, we experienced unanticipated complications which (AI and GI bleeding).[71] Those complications are considered to be related to the unphysiological nature of CF-VAD support, such as supraphysiological SS, nonpulsatile or reduced pulsatile circulation, or alternative flow path from the left ventricular apex to the ascending aorta.

Risk factors for AI are closed aortic valve, and nonpulsatile flow circulation, so there was a trend to reduce pump speed to keep the aortic valve open. For GI bleeding, physicians tried to reduce SS by reducing the pump speed and lowering the anticoagulation regimen in response to occurrence of GI bleeding or to prevent it. There is no clear evidence that these trends caused it, but then we observed an increased incidence of pump thrombosis with HeartMate II support.

As discussed later in this chapter, pump thrombosis is the most concerning complication of the HeartMate II. For most axial-flow devices, we observed a high rate of pump thrombosis, hemolysis, and GI bleeding. As CFD analysis shows, the HeartMate II has flow recirculation around the inlet bearing and a chaotic flow pattern around the outlet bearing, which are points of pump thrombosis, and creates high SS inside the pump, which causes vWF degradation. The HeartMate II has a relatively narrow safety operational margin for hemocompatibility-related AEs.

ReliantHeart aVAD

The MicroMed DeBakey VAD (ReliantHeart Inc., Houston, TX) is an earlier version of the aVAD that was the first implanted CF-VAD in a human. Beginning in 1988, Michael E. DeBakey, George P. Noon, and engineers at the National Aeronautics and Space Administration/Johnson Space Center started the project to develop a CF-VAD.[72] The pump is a titanium, electromagnetically actuated axial-flow pump. The dimensions are: 30.5 mm (diameter), 76.2 mm (length), and 93 g (weight). The priming volume of the pump, including the inflow cannula, is 25 cm^3, and it pumps in excess of 10 L/min. The internal pump system consists of a titanium inflow cannula and apical ring, MicroMed DeBakey VAD pump, flow probe, Dacron outflow conduit graft, and percutaneous cable assembly with controller connector (Fig. 6.5).

Clinical trials of the MicroMed DeBakey VAD began in Europe in November 1998 and in the US in June 2000. In 2001, Noon et al.[73] reported 51 patients implanted with the MicroMed DeBakey VAD; the probability of survival at 30 days after VAD implant was 81%. In 2003, Goldstein et al.[74] reported worldwide results of 150 patients who underwent placement of the Micromed DeBakey VAD as a BTT. Follow-up was 100%, with 30.4 patient-years of cumulative support time. Mean age was 48 ± 4 years, and 18% were female. Mean support time was 75 ± 81 days. AE rates (EPPY) were: reoperation for bleeding, 2.03; hemolysis, 0.61; device infection, 0.16; thromboembolic events, 0.61; pump thrombus, 0.61; and pump failure, 0.13. Eight-two patients (55%) were either bridged to transplantation, recovery or are ongoing and 68 (45%) died.

In 2010, Noon et al.[75] reported on a revised version of the MicroMed DeBakey VAD, the HeartAssist 5 (ReliantHeart Inc.), chiefly in regard to the prevention of pump thrombus and power fluctuation. These revisions have

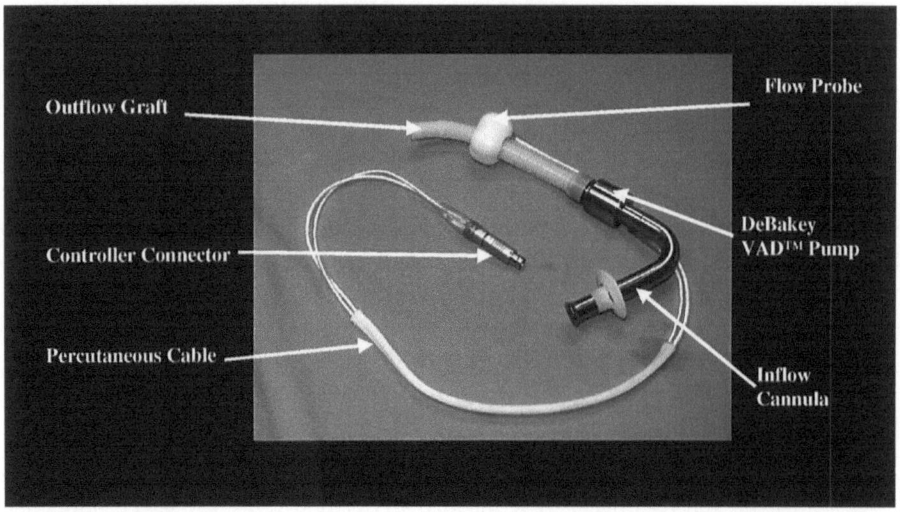

FIGURE 6.5 MicroMed DeBakey VAD axial-flow pump with titanium inflow cannula and apical ring, Dacron outflow conduit and pump flow probe, and cable assembly with controller connector. *Used with permission from Noon GP, Morley D, Irwin S, Benkowski R. Development and clinical application of the MicroMed DeBakey VAD. Curr Opin Cardiol. 2000;15(3):166–171.*[72]

included, for example, a change from the original shaft-and-ring bearing to a silicon carbide cup-and-ball bearing, and changes to the inflow cannula, to the front and rear gaps, and to the inducer and flow-straightener configurations. Stronger magnets were implanted in the impeller.

Girdhar et al.[76] conducted biocompatibility analysis of the HeartAssist 5 by device thrombogenicity emulation (DTE). DTE combines in silico numerical simulations with in vitro measurements by correlating device hemodynamics with platelet activity coagulation markers. They compared MicroMed DeBakey VAD with the HeartAssist 5. The results showed an improved streamline and platelet dispersion pattern, and low stress accumulation to platelets inside the pump with the HeartAssist 5.

Chiu et al.[77] compared the HeartMate II VAD with the HeartAssist 5 using the same method, with the following results:

1. Platelets flowing through the HeartAssist 5 are exposed to significantly lower stress accumulation that leads to platelet activation than the HeartMate II, especially at the impeller-shroud gap regions.
2. The recirculation zones and stagnant platelet trajectories that were observed at the downstream of the flow straightener of HeartMate II were reduced with the HeartAssist 5.
3. Platelet activation rates measured in vitro with the VADs mounted in recirculation flow-loops had a 2.5-fold significantly higher platelet activation rate value with the HeartMate II.

There have been no clinical trials with the HeartAssist 5, so these results have not yet been confirmed with patients.

The HeartAssist 5 has been modified to a new aVAD (ReliantHeart Inc.), axial-flow intrapericardial device (Fig. 6.6).[78] The design is based on that of the predicate axial-flow HeartAssist 5 device. Because the rotors and bearings are similar, the HeartAssist 5 and the aVAD have very similar pressure–flow characteristics. With speeds ranging between 6000 and 12,000 rpm and mean inflow-to-outflow pressure gradients between 40 and 90 mm Hg, the mean device flow is between approximately 2 and 7 L/min.

Novel features of the aVAD include the FastConnect system (ReliantHeart Inc.), which enables the adjustment of the pump depth within the ventricle and rapid securement of the pump into the sewing ring via a single screw to prevent a suction event, and the TrueFlow (ReliantHeart Inc.) ultrasonic flow probe, which is placed around the outflow graft and provides direct, real-time measurement of the LVAD flow that can be used to track device performance. The system controller also incorporates VADLink (ReliantHeart Inc.), which is a web-based remote monitoring portal. A subscriber identity module card is inserted into the system controller, which enables real-time monitoring of device flow, rpm, and power consumption via mobile device or any computer.

In 2017, Schmitto et al.[78] reported the first human study in a 61-year-old woman. Her postoperative course was uneventful, and 6 months after implantation, pump parameters were stable, but lactate dehydrogenase was 735 units per liter.

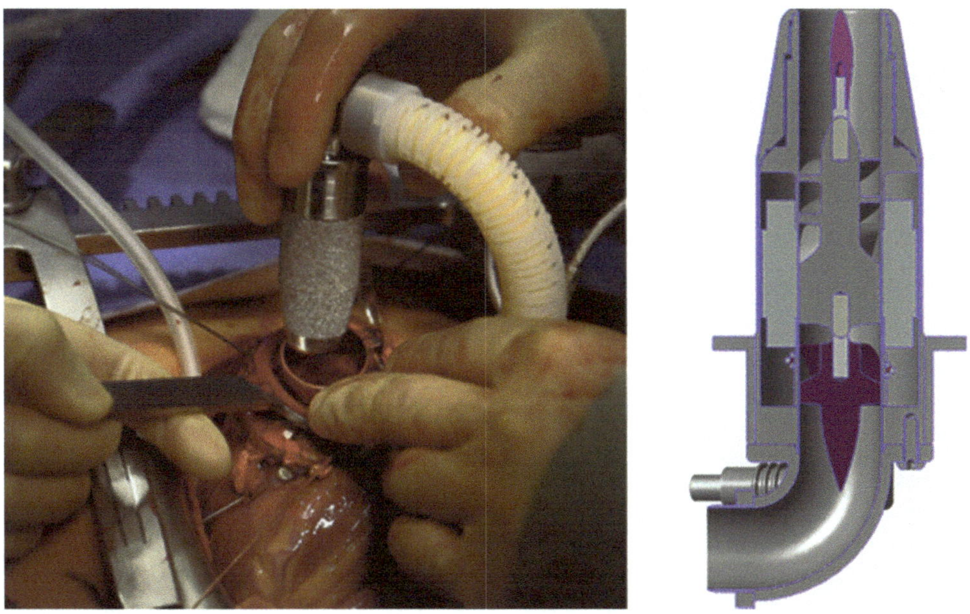

FIGURE 6.6 The aVAD blood pump at the time of ventricular insertion (left). Schematic cutaway showing the aVAD blood pump (right). *Used with permission from Schmitto JD, Hanke JS, Dogan G, et al. First implantation of a novel left ventricular assist device: the ReliantHeart aVAD. Ann Thorac Surg. 2017;104(4):e311–e313.*[78]

Hohmann et al.[79] reported their experience with 11 patients (9 males) who received an aVAD for a telemonitoring cohort. Mean age at LVAD implantation was 59 ± 5.1 years and the total follow-up time was 2438 patient-days. A total of 6216 alarm messages were generated in 11 patients, mostly low-flow alarms due to hypovolemia or suspected pump thrombosis. During follow-up, one patient died, one received a cardiac transplant and one had the VAD explanted because of pump thrombosis. Pump thrombosis was suspected in five patients, with eight episodes of sudden flow decreases and laboratory signs of hemolysis.

INCOR

Incor (Berlin Heart GmbH, Berlin, Germany) is a magnetically levitated axial-flow pump for support. The speed is from 5000 to 10,000 rpm and can provide a flow of up to 6 L/min against a pressure of 100 mm Hg (Fig. 6.7).[80] The pump is 123 mm in length, 30 mm in diameter, and weighs 200 g. It was designed with the aims of minimum wear, low energy consumption, and reduced damage to the blood components.

Like the HeartMate II, the inflow cannula is angled. At the pump inlet, the flow straightener straightens the blood stream; at the outlet, the flow passes the stationary diffuser. Magnetic suspension bearings stabilize the impeller actively in the axial direction and passively in the radial direction. Thus the impeller has no mechanical contact with the pump housing.

The impeller is driven by an "air gap" motor with an efficiency of more than 90%, which ensures that the pump does not generate any significant amount of heat. The blood is then transported to the aorta through the outflow cannula. Both cannula are made of silicon; the pump is of titanium alloy. All blood-contacting titanium surfaces are heparin-coated by the Carmedaw process.

In 2004, Hetzer et al.[80] reported the first clinical results in 24 patients. The 30-day mortality rate was 8%, and 79% of patients were able to be discharged home. The cumulative time on the device is 6.9 years; the longest individual time was 12.6 months. There were no structural defects or failures in the system. Cardiac output ranged between 4 and 6 L/min in all instances. There were no cases of infection and no hemolysis. There were three instances of transient ischemic attack, two cases developed late cardiac tamponade, and one case of GI bleeding.

Schmid et al.[81] also reported clinical experience with 15 consecutive patients (10 men, 5 women), 24–59 years of age. The patients had support duration ranging from 9 to 436 days and cumulative mechanical support was 6.2 years. Nine patients remain alive, but six patients died during assist. No early bleeding complications were seen, but late bleeding occurred in four patients. Minor cerebral thromboembolic events with transient neurologic symptoms occurred in three patients; severe stroke had to be treated in one patient. Systemic emboli were seen in two patients. Thrombus-related pump dysfunction was suspected in three patients, and managed by intensifying anticoagulation.

As shown in Fig. 6.8, Incor has a unique inflow cannula shape. Schmid et al.[82] investigated the relationship between neurologic AE rate and length of the inflow cannula in 216 consecutive patients [138 short (24 mm) inflow cannula and 79 long (34 mm) inflow cannula]. At the end of the observation period, patients with a long inflow cannula demonstrated a better survival rate than those with a short inflow cannula. The thromboembolic AE rate was also significantly lower. Only three patients (3.8%) with a long cannula had a thromboembolic AE (0.11 EPPY), compared with 32 (23.2%) of short cannula patients (0.50 EPPY, $P < .001$). This is an interesting study which showed the relationship between inflow cannula length and thromboembolic event.

Jarvik 2000

The Jarvik 2000 Heart (Jarvik Heart, Inc., New York, NY) is a small axial-flow impeller pump weighing 90 g, with a displacement volume of 25 cm^3.[83] The pump is implanted into the apex of the ventricle, and a 16 mm vascular graft conveys blood to the descending thoracic aorta or ascending aorta. The impeller spins within an electromagnetic field at between 8000 and 12,000 rpm, delivering a systemic blood flow of 3–8 L/min.

Westaby et al.[84] reported the first human implant in 2000. Frazier et al.[85] reported initial clinical results with 10 patients and showed good hemodynamic improvement. Then in 2003, Frazier et al.[86] reported 22 cases of Jarvik 2000 support as LVAD as a BTT (in the United States) and in 4 patients as DT (in the United Kingdom). The DT patients were supported for an average of 502 days, while support in the BTT group averaged 67 days (range 13–214). In the DT group, two patients died (at 382 and 95 days). Two patients have lived for extended periods; both were alive and well at 642 and 889 days after implant. Thirteen patients (59%) in the BTT group underwent heart transplants, and 12 remained alive with an average posttransplant follow-up of 15 months (range 0.8–29 months). Seven patients (32%) died waiting for heart transplant.

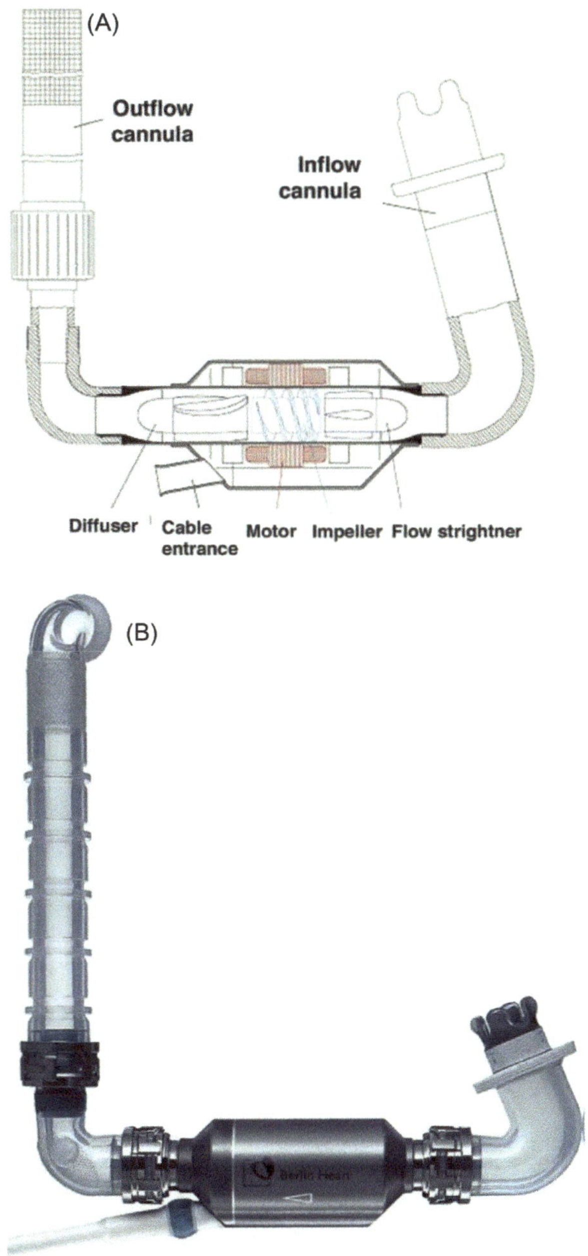

FIGURE 6.7 (A) Schematic cross-section of the INCOR pump. (B) The INCOR axial-flow pump with the inflow and outflow cannulae and the power cable. *Used with permission from Hetzer R, Weng Y, Potapov EV, et al. First experiences with a novel magnetically suspended axial flow left ventricular assist device. Eur J Cardiothorac Surg. 2004;25(6):964–970.*[80]

In 2010, the Jarvik 2000 bearing design was changed from a pin and sleeve to a unique cone, because ring thrombus was observed in the crevice between the rotating and stationary surfaces with the pin and sleeve bearing (Fig. 6.9).[83] Jarvik et al.[83] reported successful animal implant with miniaturized Jarvik pumps with cone bearing. Stanfield et al.[88] conducted an in vitro study to compare hydrodynamic performance of those two bearings and found that the hydrodynamic performance of the cone bearing pump was superior to that of the pin bearing pump.

Tarzia et al.[89] investigated the long-term outcomes of patients who received the Jarvik 2000 LVAD, depending on bearing design, in 99 patients. There was a significant reduction in both hemorrhagic and ischemic stroke risk with the cone bearing. Moreover, the cone bearing provides a hemodynamic improvement with consequent reduction in the rate of RHF.

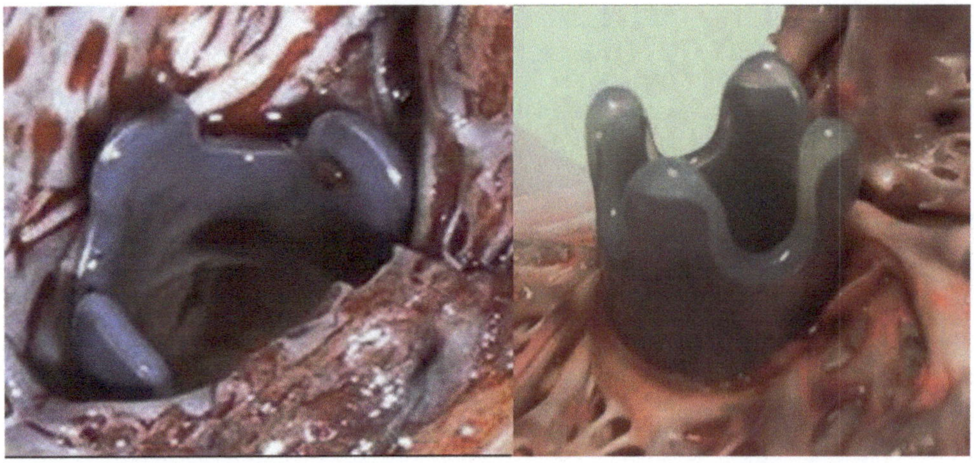

FIGURE 6.8 Posttransplant presentation of the short cannula on the left and the long cannula on the right. *Used with permission from Schmid C, Jurmann M, Birnbaum D, et al. Influence of inflow cannula length in axial-flow pumps on neurologic adverse event rate: results from a multi-center analysis. J Heart Lung Transpl. 2008;27(3):253–260.*[82]

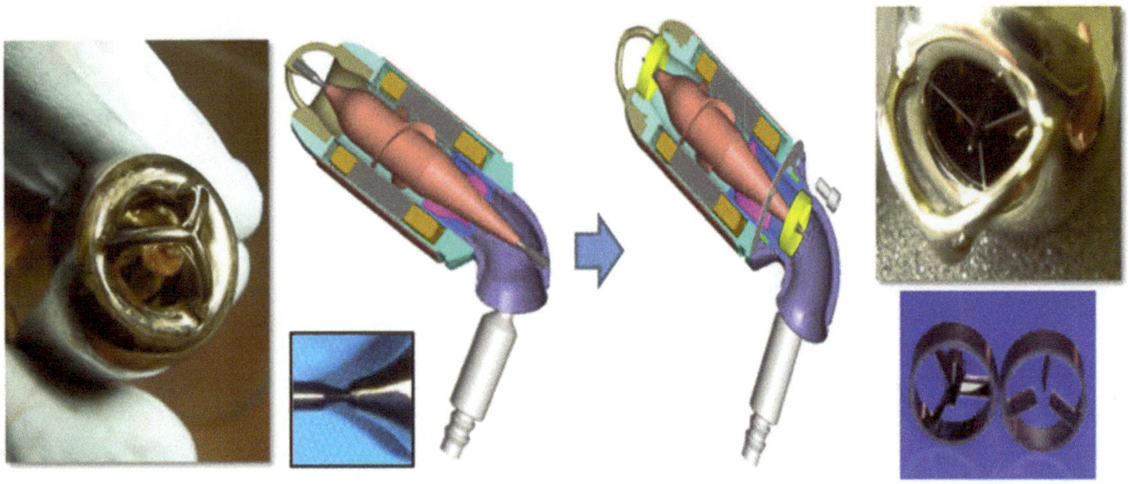

FIGURE 6.9 Conversion from pin- to cone-bearing pumps. The pin-bearing pump demonstrates the typical "O ring" of thrombus, which is less apparent in the cone-bearing pumps. *Used with permission from Selzman CH, Koliopoulou A, Glotzbach JP, McKellar SH. Evolutionary improvements in the Jarvik 2000 left ventricular assist device. ASAIO J. 2018;64(6):827–830.*[87]

Kohno et al.[90] reported a multicenter prospective study of the Jarvik 2000 using a Japanese registry with 83 patients. Median support duration was 191 days (maximum 758). Overall survival proportions at 1 and 2 years were 85.0% and 79.3%, respectively. Common AEs included major bleeding (27.7%), new infection (31.3%), stroke (20.5%) and device malfunction (20.5%). De novo aortic regurgitation was observed in 17 patients, six of whom developed at least moderate regurgitation.

Selzman et al.[87] summarized device improvement history. To improve thromboresistance, three main design changes were implemented. First was the intermittent low speed (ILS) controller. To washout the aortic root and prevent progress of AI, the controller was redesigned to automatically reduce pump speed to 7000 rpm for 8 s each minute, regardless of the speed setting of the device. Kohno et al.[91] collected data on 85 heart transplant-eligible Jarvik 2000 devices with ILS controllers. The clinical benefit of ILS controller was not evident that freedom from moderate or greater AI was 84.4%, 66.1%, and 60.2% at 1, 2, and 3 years, respectively.

The second improvement was the microsphere coating. The angle between the intraventricular pump and the surrounding left ventricular endocardium is an area of relative stasis that is prone to develop thrombus (wedge thrombus). To prevent thrombus here, the outer pump surface was modified by adding a titanium microsphere coating (Fig. 6.10). The third improvement was bearing (described before).

High SS is still a problem with this pump. Akiyama et al.[92] investigated the incidence, etiology, and outcome of GI bleeding in patients with a CF-LVAD. Fifty-four patients received one of the following: a CF-LVAD [HeartMate II ($n = 29$), Jarvik 2000 ($n = 9$), DuraHeart (Terumo Heart Inc., Ann Arbor, MI) ($n = 9$), EVAHEART (Sun Medical

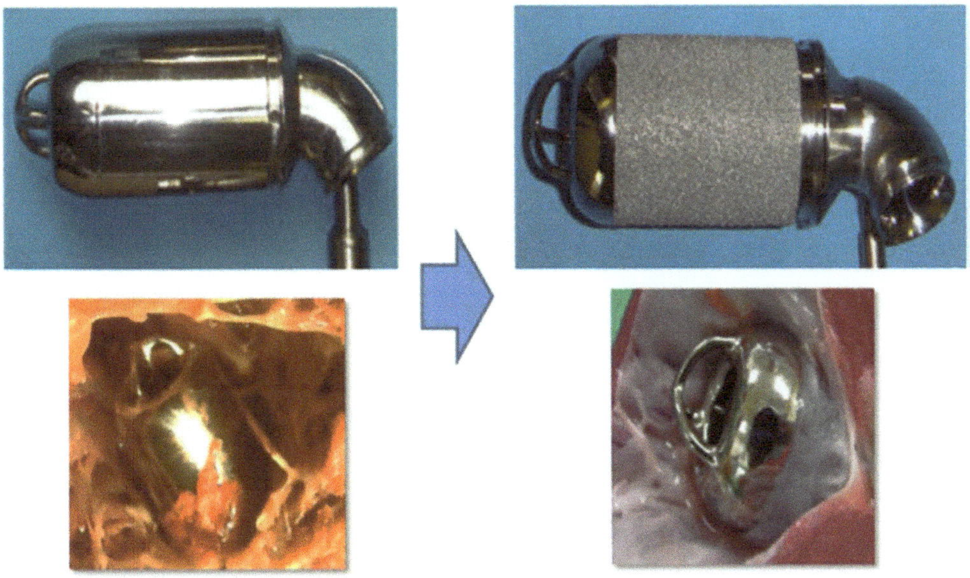

FIGURE 6.10 Titanium microsphere coating of the intraventicular pump. Representative photos of pre- (left) and post- (right) coating, with thrombus at the left ventricular pump junction in the noncoated pump. *Used with permission from Selzman CH, Koliopoulou A, Glotzbach JP, McKellar SH. Evolutionary improvements in the Jarvik 2000 left ventricular assist device.* ASAIO J. *2018;64 (6):827–830.*[87]

Technology Research Corp., Nagano, Japan) ($n = 6$), and HeartWare HVAD ($n = 1$)]. Eight (15%) presented with overt GI bleeding (12 events). GIB patients were significantly older ($P = .04$) and their preoperative inferior vena cava diameter was larger ($P = .02$). Multivariate analysis revealed that the use of the Jarvik 2000 ($P = .003$) was a risk factor for GI bleeding.

Another characteristic feature of the Jarvik 2000 pump is that it is the only device that was designed to anastomose the outflow graft to the descending aorta. However, in a multicenter study, Tarzia et al.[93] investigated short- and long-term outcomes of patients implanted with the Jarvik 2000 LVAD, according to the ascending versus descending outflow graft connection to the aorta, and demonstrated that outflow graft anastomosis to the ascending aorta was associated with better long-term survival, independent of age and perfusion techniques.

Centrifugal-flow pump

HVAD

Device description

The HeartWare System consists of an implantable centrifugal-flow pump (HVAD), an external controller, and external power sources.[94] A wide-blade impeller with a hybrid suspension system uses passive magnetic and hydrodynamic thrust bearings to create a contact-free rotation of the impeller. The HVAD system consists of a blood pump with an integrated, partially sintered inflow cannula; a 10 mm diameter gel-impregnated polyester outflow graft, and a percutaneous driveline (Fig. 6.11).

The pump can be implanted in the pericardial space. A strain relief is used on the outflow graft to prevent kinking and secures the outflow graft to the pump. The pump has a displaced volume of 50 mL, weighs 160 g, and has one moving part—an impeller, which spins blood to generate up to 10.0 L/min of flow (Fig. 6.12). There are two motors in the pump housing, with one motor providing redundancy.[95]

The HeartWare System pump has three main parts: (1) a front housing with an integrated inflow cannula; (2) a rear housing with a magnetic center-post; and (3) the rotating impeller (Fig. 6.13). Front and rear housings are hybrid titanium—ceramic assemblies; each contains a hermetically sealed motor stator. Dual motor stators enhance system efficiency and provide power redundancy to rotate the impeller.

The wide-bladed impeller is designed to accommodate four large rare earth motor magnets, which allow for a short axial height and high motor efficiency. The impeller also contains a separate stack of three rare earth magnets with like poles together as part of the impeller suspension system.

A corresponding magnetic stack contained within the center post develops strong repulsive magnetic forces to maintain impeller radial support. Vertical alignment of the magnetic stack within the center post is shifted downward relative to the impeller magnetic stack to develop an axial magnetic force to keep the impeller toward the front housing. When

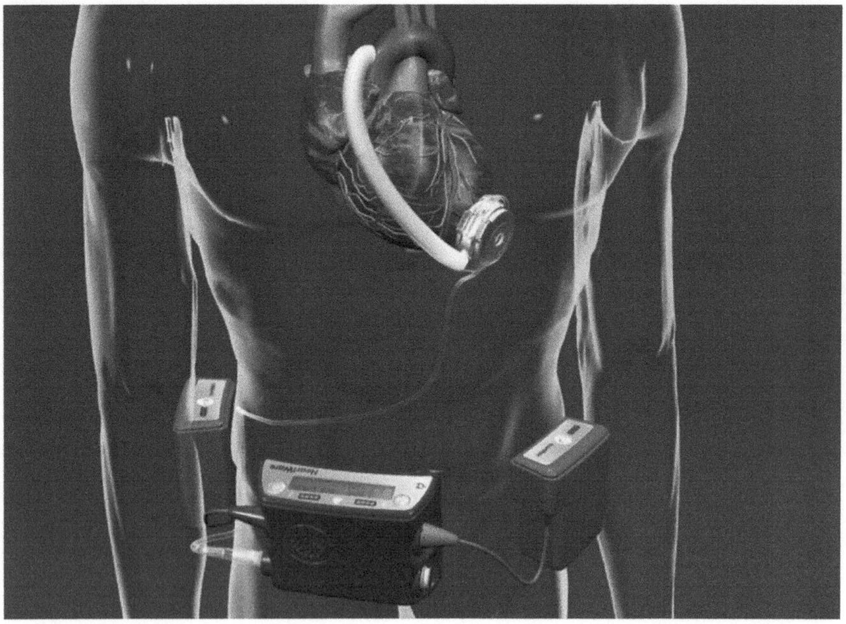

FIGURE 6.11 Internal components of the HeartWare system. *Used with permission from Larose JA, Tamez D, Ashenuga M, Reyes C. Design concepts and principle of operation of the HeartWare ventricular assist system. ASAIO J. 2010;56(4):285–289.*[95]

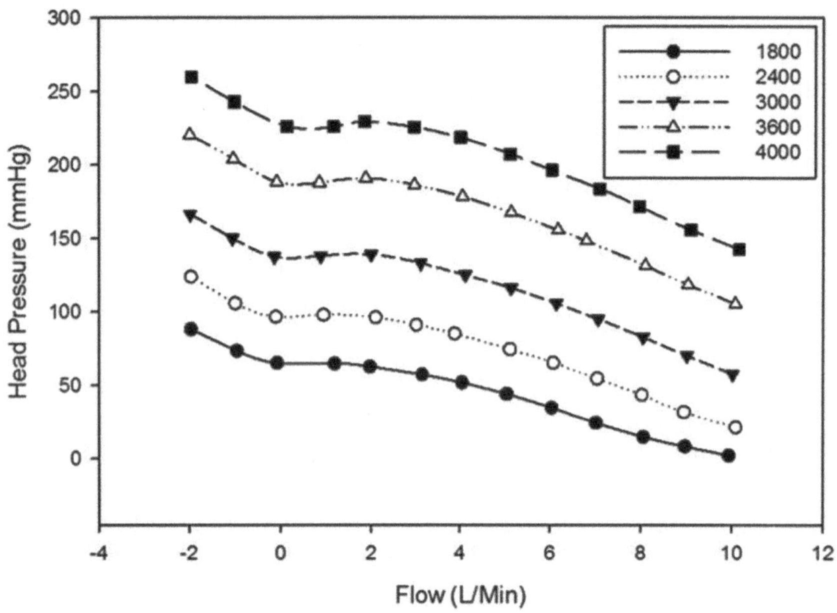

FIGURE 6.12 Pressure flow curve for the HVAD. *Used with permission from Larose JA, Tamez D, Ashenuga M, Reyes C. Design concepts and principle of operation of the HeartWare ventricular assist system.* ASAIO J. *2010;56(4):285–289.*[95]

the pump is turned on, the impeller rotates, and lift is created by the hydrodynamic thrust bearings pushing the impeller away from the front housing. A fluid (blood) barrier maintains a gap between the impeller and the front housing. The pump is very reliable; no device failure was observed during a 6-year reliability test using eight HVAD pumps.[96]

Energy to rotate the impeller is provided through an electromagnetic coupling between the impeller and the motor stators located in the front and rear pump housings. The pump can adjust the pump speed, ranging from 1800 to 4000 rpm (the recommended clinical operating speed range is 2400–3200 rpm). Speeds below 2400 rpm should only be used during the implant procedure when weaning from cardiopulmonary bypass. Speeds above 3200 rpm are seldom needed and may increase the risk of suction events.[94]

CFD analysis demonstrated three flow paths: (1) through the inflow cannula and into four impeller flow channels, after which blood is collected in the housing and exits the pump through the outflow graft; (2) starting under the impeller and going upward through the annular gap between the impeller and the center post, and then reentering the primary

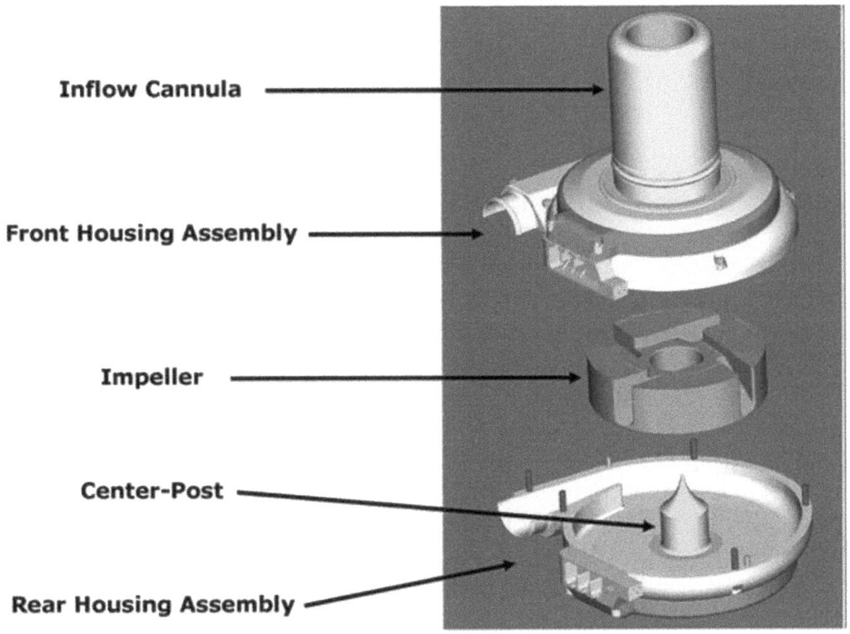

FIGURE 6.13 Three main assemblies of the HVAD pump: (1) front housing with integrated inflow cannula; (2) rear housing and centerpost; and (3) wide-blade suspended impeller. HVAD. *Used with permission from Larose JA, Tamez D, Ashenuga M, Reyes C. Design concepts and principle of operation of the HeartWare ventricular assist system.* ASAIO J. *2010;56(4):285–289.*[95]

flow path; (3) starting at the impeller slots and leading to the hydrodynamic thrust bearings at the top of impeller, where blood reenters the primary flow path and exits the pump outlet.[95]

The monitor is a touch screen tablet that uses proprietary software to display system performance and to permit adjustment of selected controller parameters. When connected to a controller, the monitor receives continuous data from the controller and displays real-time and historical pump information. The monitor also displays alarm conditions and can provide notification of available controller software updates. The monitor waveforms display important information about pump status to clinicians.

Rich et al.[97] provided the theoretical foundation for understanding the determinants of HVAD waveform characteristics and provided practical examples of how to interpret and integrate changes in HVAD waveforms into clinical practice. Gross et al.[98] investigated occurrences and durations of suction events and their associations with tachycardia in stable outpatients and showed further improvement of suction detection with high-resolution LVAD monitoring.[94,95]

A current-flow (IQ) curve is used to estimate instantaneous and average flow through the pump. Estimated flow is computed with values for pump speed and motor current. The default blood viscosity setting is 2.7 cP, and the adjustable range is 2.2–4.3 cP, which is adjusted by measured hematocrit. Viscosity and motor stator configuration determine the active IQ curve used to calculate estimated flow.

Speed and current are exponentially averaged. Exponentially averaged speed is used to determine the closest reference speeds. Reference speeds and the exponentially averaged current are used to determine the current and flow values from the active IQ curve. Reyes et al.[99] assessed the algorithm accuracy in in vitro and in vivo studies and showed strong agreement with measured flow in the in vivo study ($y = 0.96x + 0.54$, $R^2 = 0.92$).

A cyclic-controlled speed change function (Lavare Cycle) allows for changes in left ventricular filling and flow rate through the LVAD once per minute during a 3-s cycle. The impeller rotational speed change is fixed at 200 rpm, which is adequate to allow some increase of ventricular volume but not low enough to cause left ventricular distension and low cardiac output. The impeller rotational speed decreases below the set speed by 200 rpm for 2 s, then increases 400 rpm (200 rpm above the set speed) for 1 s, and then returns to the set point. The cyclic-controlled speed change will only operate between 2000 and 3800 rpm, with low and high limits of speed change at 1800 and 4000 rpm, respectively.

Zimpfer et al.[100] reported in vitro and clinical findings on the effects of the Lavare cycle. The effect of pump speed changes on the intraventricular flow field was examined with an in vitro particle image velocimetry model. The clinical impact of the Lavare cycle was also evaluated through a retrospective review of the ReVOLVE study. Particle image velocimetry showed increased ventricular washout with an active Lavare cycle. With the Lavare cycle on, there was also a 22% decrease in the stagnation index compared with when the Lavare cycle was off.

In the ReVOLVE registry, patients with the Lavare cycle turned on ($n = 215$) were supported for 497 patient-years, whereas patients who did not use the speed modulation ($n = 33$) were supported for 39.3 patient-years. The Lavare cycle did not significantly affect patient survival, as both groups had approximately an 80% survival after 1 year. Patients using the Lavare cycle had significantly fewer rates of stroke (0.06 vs 0.20 EPPY, $P = .0008$), sepsis (0.03 vs 0.15 EPPY, $P = .0003$) and RHF (0.03 vs 0.18 EPPY, $P < .0001$).

Clinical results

In 2010, Wieselthaler et al.[101] reported initial clinical results of the HVAD in Europe and Australia in 23 patients (mean age, 47.9 years) in end-stage heart failure. The primary end point was survival to heart transplant or survival to 180 days on the device, whichever occurred first. Mean duration of support was 167 ± 143 days (range, 13–425 days), and mean pump flow was 6.1 ± 1.1 L/min. At the 180-day end point, two patients had undergone successful transplant at 157 and 175 days, two patients died while on the device, and 19 patients continued pump support for more than 180 days. Actuarial survival rate was 91% and 86% at 6 months and 1 year respectively.

Strueber et al.[102] expanded these first clinical results to 50 patients for a minimum of 2 years. Of the 50 patients, 20 (40%) received transplants, 4 (8%) had the pump explanted after myocardial recovery, and 17 (34%) continued support at 2 years. Nine (18%) patients died during support from sepsis ($n = 3$), multiple organ failure ($n = 3$), or hemorrhagic stroke ($n = 3$). The most frequent AEs were infection and bleeding. In 2012, Aaronson et al.[103] reported multicenter BTT trial in 140 patients who received the HVAD, comparing with 499 patients who received the HeartMate II in US. Success occurred in 90.7% of HVAD patients and 90.1% of HeartMate II, establishing the noninferiority of the investigational pump ($P < .001$; 15% noninferiority margin). The HVAD was approved as BTT indication by FDA in 2012.

Slaughter et al.[104] extended this pivotal BTT trial, adding a continued access protocols (CAP) trial. The primary outcome, success, was defined as survival on the originally implanted device, transplant, or explant for ventricular recovery at 180 days. Secondary outcomes included an evaluation of survival, functional and quality of life outcomes, and AEs. A total of 332 patients in the BTT and CAP trial have completed their assessment. Survival in patients receiving the HVAD was 91% at 180 days and 84% at 360 days. Quality of life scores improved significantly, and AE rates remain low.

In 2014, Strueber et al.[105] reported the postmarket Registry to Evaluate the HeartWare Left Ventricular Assist System (ReVOLVE) to collect post–CE Mark Trial clinical data with a total of 254 commercial HVAD implants. Duration of support ranged from 1 to 1057 days, with a mean of 363 ± 280 days (median 299.5 days). Success in patients with the HeartWare system was 87% at 6 months, 85% at 1 year, 79% at 2 years, and 73% at 3 years. The most common AE was bleeding. There were 101 bleeding events (0.40 EPPY) occurring in 28% of patients, and an additional 5% of patients had GI bleeding (0.06 EPPY). DLI occurred in only 6% of patients, at a rate of 0.07 EPPY. Twelve patients (5%, 0.05 EPPY) developed sepsis. Stroke occurred in 8% of patients (0.08 EPPY). RHF occurred in 9% of patients (0.10 EPPY), whereas renal failure was reported in 4% (0.04 EPPY).

Rogers et al.[106] compared the HVAD and HeartMate II in patients who were DT candidates in a multicenter randomized trial involving 446 patients who were assigned, in a 2:1 ratio, to the study device or the control device in 48 US sites (ENDURANCE trial). The primary endpoint was survival at 2 years free from disabling stroke, or device removal for malfunction or failure. The intention-to-treat-population included 297 participants assigned to the HVAD and 148 participants assigned to the HeartMate II.

The primary end point was achieved in 164 patients in the study group and 85 patients in the control group, and showed noninferiority of the study device relative to the control device. Regarding AEs, more patients in the HeartMate II group than in the HVAD group had device malfunction or device failure requiring replacement (16.2% vs 8.8%), and more patients in the HVAD group had strokes (29.7% vs 12.1%). In a post hoc analysis, the authors found that a recorded mean arterial blood pressure of higher than 90 mm Hg was associated with a risk of stroke—predominantly hemorrhagic strokes—that was 34% higher than that among patients whose blood pressure was 90 mm Hg or lower.

In response to the result, Milano et al.[107] reported on the ENDURANCE Supplemental Trial, a prospective, multicenter evaluation of 465 patients with DT indication, randomized 2:1 to HVAD ($n = 308$) with blood pressure management protocol, or the HeartMate II ($n = 157$). The primary endpoint was the 12-month incidence of transient ischemic attack or stroke with residual deficit 24 weeks postevent. Secondary endpoints included the composite of freedom from death, disabling stroke, and need for device replacement or urgent transplantation, as well as comparisons of stroke or transient ischemic attack rates in HVAD cohorts in ENDURANCE Supplemental and ENDURANCE.

Following discharge from the index hospitalization, all HVAD subjects were required to adhere to a blood pressure management protocol in addition to receiving the standard of care. The recommended blood pressure range was defined as mean arterial pressure <85 mm Hg for the automated cuff method and an opening pressure <90 mm Hg for the Doppler cuff method. Results showed that the enhanced blood pressure protocol significantly reduced mean arterial blood pressure. The primary endpoint was not achieved (14.7% with HVAD vs 12.1% with HeartMate II). However, the secondary composite endpoint demonstrated superiority of the HVAD (76.1%) versus control (66.9%) ($P = .04$).

Supplemental Trial compared with the HVAD cohort of the ENDURANCE trial, although superiority was not shown (chi-square superiority $P = .10$). However, an analysis of the overall strokes in HVAD patients in ENDURANCE compared with ENDURANCE Supplemental revealed an improvement of the neurological event profile in general, and hemorrhagic strokes in particular were reduced by 50.5% ($P = .02$). The overall stroke rates for the control device were not significantly different between the control arms in the two trials, including the rate of disabling strokes, which occurred in 4.7% of patients (7 of 149) in the ENDURANCE trial and 6.4% of those (10 of 157) enrolled in the ENDURANCE Supplemental Trial at 12 months.

Adverse events

Blood damage

Thamsen et al.[108] investigated and compared the flow features of the HeartMate II and the HVAD associated with blood damage. The results were compared regarding different SS levels. Also, residence times were computed as a rough measure for exposure times to these SS. There was a difference for SS beyond the 9 Pa threshold, so that the HVAD seemed to have larger areas with vWF damage (Fig. 6.14). For the higher SS, which corresponded to potential hemolysis and platelet activation, both pumps had a very similar volume fraction. The sum of the cell residence times at these SS was higher for the HVAD for all thresholds. In the HVAD pump, SS was high at the hydrodynamic bearing.

Thamsen et al.[109] assessed the axial gap size between impeller and upper housing and its sensitivity to speed, flow, and pressure difference of HVAD using X-ray radiography. The gap size increased with rotational speeds, dependent on the viscosity (2.8, 6.9, and 9.4 mm/1000 rpm for 1, 4, and 8 mPas, respectively), but was independent from the volume flow and the pressure head at constant speeds.

Chen et al.[110] evaluated the flow dynamics of the HVAD and the effect of pump speed modulation on blood trauma. The highest WSS occurred in the narrow hydrodynamic bearing gap between the top wedged surfaces of the impeller and the upper housing. Under the constant speed operation, about 75%–77% of the volume inside the pump was exposed to SSS below 10 Pa, while almost 22%–24% of volume was exposed to SSS from 10 to 100 Pa. Only a small fraction of the volume (around 0.6%–0.7%) was exposed to SSS above 100 Pa. Under the pulse-modulated speed operation, the percentage of the volumes exposed to the three levels of SSS varied at different instants due to variation in pump speed. At the middle of acceleration and deceleration, about 75% of the volume was exposed to SSS below 10 Pa, 23% of the volume to SSS from 10 to 100 Pa, and about 0.75% of the volume to SSS above 100 Pa.

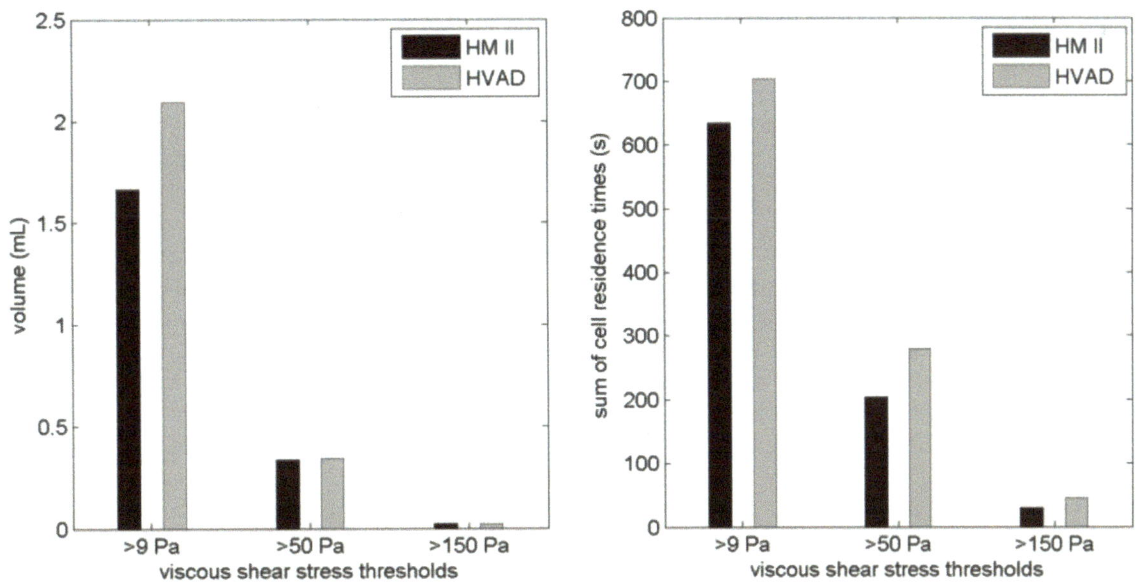

FIGURE 6.14 Volume fractions (left) subjected to specified viscous SS thresholds for vWF damage (9 Pa), platelet activation (50 Pa), and hemolysis (150 Pa), and corresponding sums of cell residence times (right). *Used with permission from Thamsen B, Blümel B, Schaller J, et al. Numerical analysis of blood damage potential of the HeartMate II and HeartWare HVAD rotary blood pumps. Artif Organs. 2015;39(8):651–659.*[108]

While the pump speed was at 4000 rpm, the volume exposed to higher SSS levels increased. About 68% of the volume was exposed to SSS below 10 Pa and 30% of the volume to SSS from 10 to 100 Pa, while about 1% of the volume was exposed to SSS above 100 Pa. During the low-speed rotation at 2400 rpm, around 80% of pump volume was exposed to SSS below 10 Pa, 17%–22% of the volume to SSS from 10 to 100 Pa, and a smaller percentage (<0.5%) of the volume to SSS above 100 Pa. The computational model did not demonstrate any differences in the time average hemolysis index between constant and modulated pump speed operations, thereby predicting that pulse-modulated speed operation may help to restore vascular pulsatility without any further increased risk of blood trauma.

Birschmann et al.[111] investigated 20 patients supported with the HeartMate II or HVAD and measured plasma coagulation, full blood count, and clinical chemistry parameters. Acquired von Willebrand syndrome could be detected in all patients. Signs of hemolysis, as measured by lactate dehydrogenase levels (mean, 470 U/L HeartMate II, 250 U/L HVAD; $P < .001$), were more pronounced in the HeartMate II patients. In contrast, D-dimer analysis indicated a significantly higher activation of the coagulation system in HVAD patients (mean, 0.94 mg/L HeartMate II, 2.01 mg/L HVAD; $P < .01$).

Meyer et al.[112] also showed the presence of acquired vWF disease with HeartMate II and HVAD patients. vWF testing was performed in 102 of 175 consecutive patients after LVAD implantation. Acquired vWS was found in all patients, demonstrated by a decrease in the HMW multimers of vWF to 30% ± 14% in HeartMate II patients and 34% ± 13% in patients with an HVAD.

Mondal et al.[113] compared device-specific platelet functionality in 26 heart failure patients supported with three CF LVADs: the HeartMate II ($n = 8$), the Jarvik 2000 ($n = 9$), and the HVAD ($n = 9$). Intraplatelet reactive oxygen species (ROS) generation, mitochondrial damage, and platelet apoptosis were compared among the device types before and after implantation at every week up to 1 month. Intraplatelet ROS, mitochondrial damage, and platelet apoptosis were significantly elevated in the HVAD group compared with the other two device groups after implantation. Major bleeding, infections, systemic inflammatory response syndrome, and RHF were found to be more common among the HVAD group than others. Intraplatelet ROS and platelet damage levels returned to baseline in both the HeartMate II and the Jarvik groups, while in the HeartWare group they remained elevated.

Pump thrombosis and stroke

As shown in the ENDURANCE trial, the rate of pump thrombosis is lower with the HVAD than with the HeartMate II. Potapov et al.[114] reported results of 225 patients (median age 55.4 years, range 7–82 years, 40 women) supported with the HVAD. Cumulative follow-up in all 225 patients was 151.9 patient/years. There were 12 incidences of pump thrombosis. In six patients, early pump thrombosis (after <30 days) requiring pump exchange occurred after a median time of seven days, with an incidence rate of 2.7%. In six patients, late pump thrombosis requiring pump exchange occurred after a median of 380 days, with an incidence rate of 2.7%. The overall incidence was 5.33%, with 0.079 thromboses per year. All but one thrombotic event occurred in pumps without sintering of the inflow cannula. In 69 patients with sintered inflow cannulae, there was only one incident of pump thrombosis (incidence rate: 1.5%). In the original design of the HVAD, the external and internal surfaces of the inflow cannula consisted of a polished titanium alloy. However, as tissue ingrowth encircling the external surface of the polished inflow cannula was observed, the external surface of the inflow cannula was modified to include sintered titanium microspheres that covered approximately half the length of the inflow cannula, extending from the base to the midsection circumferentially.

Najjar et al.[115] reported the analysis of a study that included 382 patients who underwent implantation of the HVAD as part of the BTT and subsequent CAP trial. There were 34 pump thrombus events in 31 patients (8.1%, 0.08 EPPY rate). Medical management of pump thrombus was attempted in 30 cases, and was successful in 15 (50%). Sixteen patients underwent pump exchange, and two underwent urgent transplantation. Five patients with a pump thrombus died after medical therapy failed, four of whom also underwent a pump exchange. According to a multivariable analysis, significant risk factors for pump thrombus included a mean arterial pressure >90 mm Hg, aspirin dose ≤81 mg, INR ≤2, and an INTERMACS profile level of ≥3 at implant. The authors also compared sintered versus nonsintered inflow cannula. Of these, 272 (71.2%) received a nonsintered HVAD, and 110 (28.8%) received an HVAD with a sintered inflow cannula. The pump thrombus event rate was similar (0.08 EPPY) in these two groups.

Glass et al.[116] observed tissue growth around the sintered inflow cannula of the HVAD and evaluated thrombus complications. Cases of HVAD devices removed at the time of heart transplant were identified, then the gross and microscopic findings were reviewed along with 22 patients. Gross pathologic examination revealed thrombi on the outer aspect of the HVAD inflow cannula in 23 of 24 devices (96%). Complications during the period of device support included nine thromboembolic events (41%), including six ischemic strokes (27%), two intracoronary thromboembolic events, and one splenic infarct. The nidus for thrombus formation appears to be the smooth-sintered interface of the HVAD inflow cannula.

Stulak et al.[117] evaluated treatment strategies for HVAD pump thrombosis. Among 175 patients, there were 36 pump thromboses in 21 patients, for a total event rate of 0.22 events/patient-year of support. Initial medical treatment was used in 29 episodes (tissue plasminogen activator in 24, heparin alone in four, and heparin plus glycoprotein IIb/IIIa in one) and surgical (device exchange) in seven. Medical treatment was successful in 14 of 29 episodes (48%). Complications of medical treatment included hemorrhagic stroke in six patients (21%), the need for urgent device exchange/transplant in six (21%), and death in three (10%). Surgical treatment was successful in all seven patients (100%). No significant early complications or early deaths occurred after device exchange.

Aortic insufficiency

There are only a few articles investigating AI of the HVAD. Bhagra et al.[118] investigated the development of AI after HVAD implantation and found that 24.5% of patients had AI more than mild, and only one patient had AI at 1 year (1.9%). da Rocha et al.[119] examined the prevalence of significant AI in 102 patients who received the HVAD ($n = 77$) or the HeartMate II ($n = 25$). Significant AI (more than mild) was found in 32 patients (31.4%). De novo AI occurred after 183 ± 246 days of support. A permanently closed aortic valve corresponds with a greater prevalence of AI when compared with intermittent or complete opening of the aortic valve ($P = .004$). Patients with the HeartMate II or the HVAD did not show any differences in terms of AI prevalence.

Gastrointestinal bleeding

Some studies showed equivalent or a greater prevalence of GI bleeding events in HVAD patients compared with the HeartMate II. Lalonde et al.[120] compared the devices in 46 consecutive patients who underwent implantation of either an HVAD ($n = 13$) or a HeartMate II ($n = 33$). Survival and incidence of perioperative bleeding, renal dysfunction, liver dysfunction, and infection were similar between the groups. However, HVAD devices had a significantly higher incidence of GI bleeding (31% vs 0% in HeartMate II patients, $P < .01$) and stroke (44% vs 10% in HeartMate II patients, at 1 year $P = .04$).

Gaffey et al.[121] also compared bleeding complications among patients supported with the HeartMate II (HeartMate II, $n = 105$) and the HeartWare (HVAD, $n = 34$). There were no significant differences in demographics or comorbidities. The incidence of GI bleeding was as low as 3.9% in the HeartMate II and 2.9% in the HVAD ($P = .78$). Within the postoperative period, there was no difference in product transfusion: red blood cells ($P = .10$), fresh frozen plasma ($P = .19$), and platelets ($P = .89$). Postoperatively, a higher but not significantly different number of HeartMate II patients returned to the operating room for bleeding ($n = 27$) compared to HVAD ($n = 6$, $P = .35$).

Saeed et al.[122] investigated the prevalence of GI bleeding events in patients supported with the HVAD and activated Lavare cycle. Among 105 HVAD patients, 32 GI bleeding events were documented in 22 patients (21%). with a mean age of 58 ± 11 years after median duration of 22 days (0.34 EPPY).

Kawabori et el. retrospectively analyzed the location, incidence, and survival effect of GI bleeding events in recipients in 526 patients who underwent primary implantation of an HeartMate II ($n = 403$) or HVAD ($n = 123$). Of the 526 patients, 140 (26.6%) had a GI bleeding event (HeartMate II: $n = 100$; HVAD: $n = 40$), 92 (17.5%) had a single GI bleeding event, and 48 (9.1%) had multiple GI bleeding events (range: 2–9 events). HVAD recipients had a higher incidence of both upper and lower GI bleeding events ($P < .001$ and $P = .002$, respectively) than HeartMate II recipients. AVM was the most common etiology for GI bleeding (50 patients/72 events); for this group, the average time-to-event was 300.4 days, the recurrence rate was 34%, and the 90-day and 1-year survival rates were 88.3% and 66.7%, respectively.

Goldstein et al.[123] evaluated GI bleeding in patients receiving the HVAD in the pivotal BTT trial and under the CAP. Overall, 59 of 382 (15.4%) patients experienced 108 GI bleeding events (0.27 EPPY). Mean time to first bleed was 273.1 days and 86.1% of events occurred beyond 30 days. Freedom from GI bleeding was 84.1% at 1 year. The most common etiology of bleeding was AVM, and the most common site was the small intestine. Repeat bleeding was infrequent, though patients with GI bleeding required more readmissions and developed nondevice infections more frequently.

Biventricular support

The HVAD is small intrapericardial device, so it is more suitable as an RVAD than the HeartMate II. Krabatsch et al.[124] discussed the problems of using the HVAD as an RVAD, and developed a method that allows the use of HVAD for right ventricular assistance.

There are three major problems in using the HVAD as an RVAD:

1. In the pulmonary circulation (low afterload), the flow would be too high even at the lowest adjustable speed, theoretically resulting in severe pulmonary edema.
2. The inflow cannula of the HVAD is 35 mm long—too long to be inserted into right ventricle or atrium.
3. The optimal anatomical site for this connection had to be determined.

They then did an in vitro study to examine the flow characteristics of the system with different outflow diameters (10–4 mm). They also looked for a material for safe and satisfactory reduction of the effective length of the inflow cannula for better fitting to the RV dimensions. Reduction of the outflow graft to an inner diameter of ~5 mm adds enough resistance to the system that in patients with a normal pulmonary resistance, the pump would deliver between 3.5 and 7 L/min. They added two 5 mm silicon suture rings to the system's "apical" sewing ring to reduce the effective length of the inflow cannula. Connection of the pump to the anterior free wall of the right ventricle ensures good orientation of the inflow cannula within the RV cavity, with sufficient space to prevent the inflow cannula from suction to the opposite interventricular septum.

Tran et al.[125] described a successful series of biventricular support procedures in patients utilizing an HVAD in a left ventricular and right atrial configuration. For surgical procedures, the inflow cannula was inserted to the right atrium (RA) with a 7–8 mm high polytetrafluoroethylene felt ring stack for inflow cannula to penetrate approximately 1 cm into the right atrial cavity. Outflow graft length was approximately 15 cm for increased afterload. They reported on 11 patients with evidence of biventricular failure who underwent implantation with CF-LVAD and RA-HVAD pumps. A total of 4314 biventricular VAD support days were accumulated. Seven patients have undergone orthotopic heart transplant, while three are ambulatory and are either waiting transplant or reconsideration for transplantation. There was one mortality in this case series, which was due to an intracranial bleed from supratherapeutic anticoagulation. Two other patients experienced hemorrhagic strokes, but without neurologic sequelae, whereas no patients have experienced ischemic strokes. There were two episodes of GI bleeding.

Marasco et al.[126] also reported the outcomes of using the HVAD in a biventricular configuration in 93 patients. The average age was 47.4 ± 12.93 years, with 75.3% being male, and a total follow-up of 104.9 patient-years. Seventeen of the 93 patients (18.3%) underwent orthotopic heart transplantation during the 2-year follow-up period. The most common AEs seen in the study cohort were bleeding ($n = 33$, 35.5%), infection ($n = 24$, 25.8%), and respiratory failure ($n = 19$, 20.4%). Seventeen patients (18.3%) experienced device malfunction events: nine thrombus events in eight patients (all RVADs), seven patients who underwent device exchanges (six RVADs, one LVAD), one patient with a controller fault, and one patient with an RVAD controller exchange owing to controller temperature.

The AE profile compared patients who had their HVAD RVAD implanted in the RA ($n = 32$) with patients who had their HVAD RVAD implanted in the RV ($n = 56$). This study showed no statistically significant difference in the incidence of AEs between the groups, although there were statistically higher event rates in the RA-implanted RVADs than in the RV in terms of bleeding (0.82 vs 0.30, $P = .05$), cardiac arrhythmia (0.19 vs 0.01, $P = .004$), device malfunction/failure (0.39 vs 0.13, $P = .04$), pericardial fluid collection (0.35 vs 0.01, $P < .0001$), and respiratory failure (0.58 vs 0.14, $P = .003$). There was no significant difference in survival at 1 year. Overall Kaplan – Meier survival for all HVAD biventricular VADs was estimated to be 56% at 1 year and 47% at 2 years.

Maynes et al.[127] reported on a meta-analysis to compare the outcomes of patients undergoing SynCardia total artificial heart (TAH) and biventricular HVAD support for biventricular HF. Twelve studies, including 512 patients in the TAH group versus 38 patients in the Bi-HVAD group, were pooled. There was a comparable incidence of stroke, acute kidney injury, overall infection, and DLI. Postoperative bleeding was present in 42% of TAH versus 23% of Bi-HVAD ($P = .22$). Patients in the TAH group had shorter duration of support. At the mean follow-up time of 120 days, patients in both groups had similar overall mortality (TAH 36% vs Bi-HVAD 26%, $P = .44$). Discharge home on support was achieved in 6% of TAH patients versus 73% of Bi-HVAD ($P < .01$).

They also conducted a meta-analysis to compare the outcomes of right-sided HVAD using the RA versus RV approach in 13 articles with 56 cases.[128] Overall, 21/56 patients (37.5%) had RA HVAD, while 35/56 (62.5%) had RV HVAD. The median time of support was 351 days (IQR 136–626) for RA HVAD and 135 days (IQR 61–244) for RV HVAD ($P = .02$). Pump thrombosis occurred at a similar rate [RA: 3/10 (30.0%) vs RV: 6/20 (30.0%), $P = 1$], as did GI bleeding [RA: 10/35 (28.6%) vs RV: 5/21 (23.8%), $P = .94$] during the follow-up time period. Kaplan–Meier analysis when censored for transplant showed higher survival with the RA HVAD compared to RV HVAD ($P = .036$), with an estimated survival at 1 year of 91.7% (95% CI 77.3–100.0) in RA HVAD versus 66.2% (95% CI 48.9–89.6) for RV HVAD.

Minimally invasive implantation

HVAD is the only CF-VAD that is approved by the FDA for minimally invasive implantation. There are some variations in technique. Cheung et al.[129] reported a case of sternum-sparing implantation of the HVAD through small incisions, avoiding cardiopulmonary bypass, through an upper ministernotomy and left minithoracotomy. The apical puncture was performed under rapid ventricular pacing, followed by insertion of the inflow of the pump. Popov et al.[130] reported their technique with right and left thoracotomy with cardiopulmonary bypass. Schmitto et al.[131] also described a case report with a detailed surgical procedure. Cardiopulmonary bypass was established by venous cannulation into the right femoral vein and arterial cannulation into the ascending aorta via an upper hemisternotomy.

Maltais et al.[132] reported their experience with 283 patients implanted with the HVAD. A minimally invasive left lateral thoracotomy implant strategy was used in 53 patients (19%), of which 22 (42%) were implanted off-pump. Median age at implant was 59 years, and 183 patients (65%) were implanted as BTT. Follow-up was 100% complete for a total of 302 patient years of support.

In-hospital mortality was 4.9% (14 patients), and was comparable between the conventional sternotomy and minimally invasive cohorts ($P = .15$). Patients in the minimally invasive cohort had a shorter median length of stay ($P < .01$) and had a lower incidence of postimplant pump thrombus ($P = .02$).

The Left Anterior Thoracotomy with Upper Hemisternotomy or Right Anterior Thoracotomy (LATERAL) trial is a multicenter, prospective, nonrandomized, single-arm trial that utilized data from 144 patients enrolled in the INTERMACS database at 26 centers in the US and Canada.[133] They compared midterm results and complications using the LATERAL approach versus BTT and CAP trial patients. The primary composite end point was success at 180 days, defined as alive on the originally implanted device and free from disabling stroke, transplanted or explanted for recovery. The key secondary end point was mean length of initial hospital stay.

The primary end point was successfully achieved in 88.1% of patients and was significantly greater than the predefined performance goal of 77.5% set from historical sternotomy data ($P = .0012$). The key secondary end point—mean length of initial hospital stay—was 18 days and was significantly shorter than the predefined performance goal of 26.1 days obtained from historical sternotomy data ($P < .0001$). The incidence of bleeding requiring reoperation was 3.5% at 30 days, 5.6% at 6 months, and 5.6% at 1 year.

Summary

The HVAD is a hybrid of magnetic and hydrodynamic bearing intrapericardial centrifugal-flow pumps. It is the first FDA-approved centrifugal-flow pump. Clinical trials showed lower rates of pump thrombosis compared to the HeartMate II, but failed to demonstrate survival benefits, mainly due to the high stroke rate. The reason for this high stroke rate is not clear, but there may be two possible sites of thrombus formation– around the inflow cannula (wedge thrombus) or at the hydrodynamic bearing.

The HVAD had a smooth titanium inflow cannula at first, which was then modified to a half-textured inflow cannula. The half-textured inflow cannula showed some improvement of clinical results, but clinicians still observe thrombus formation around it, especially at the transition site. It is not clear, and not well studied, how tissue grows around the textured surface, and what shape of inflow cannula is the most effective to prevent wedge thrombus formation.

The HVAD had the same occurrence rate of shear-mediated complications, such as vWF disease and GI bleeding. According to CFD results, this may be due to the high shear rate of the hydrodynamic bearing. Another novel feature of the HVAD is Lavarge cycle to improve pump washout and create some pulsatility and ensure an aortic valve opening, but there is no clear evidence that this feature reduced GI bleeding or AI.

The HVAD is small intrapericardial pump so it does not need pump pocket creation, but the infection rate is almost the same or slightly improved compared to the HeartMate II.[134] The great advantage of the pump is that it can be used as biventricular support and implanted with a minimally invasive approach. We still need more evidence to show clinical benefits of these applications.

HeartMate 3

Device description

The HeartMate 3 system is an intrapericardial centrifugal-flow device, with an inflow cannula inserted into the left ventricular apex and an outflow graft anastomosed to the ascending aorta (Fig. 6.15).[135] The implanted components include the inflow cannula, pump housing, motor, control electronics, outflow graft with a bend relief, and percutaneous driveline.

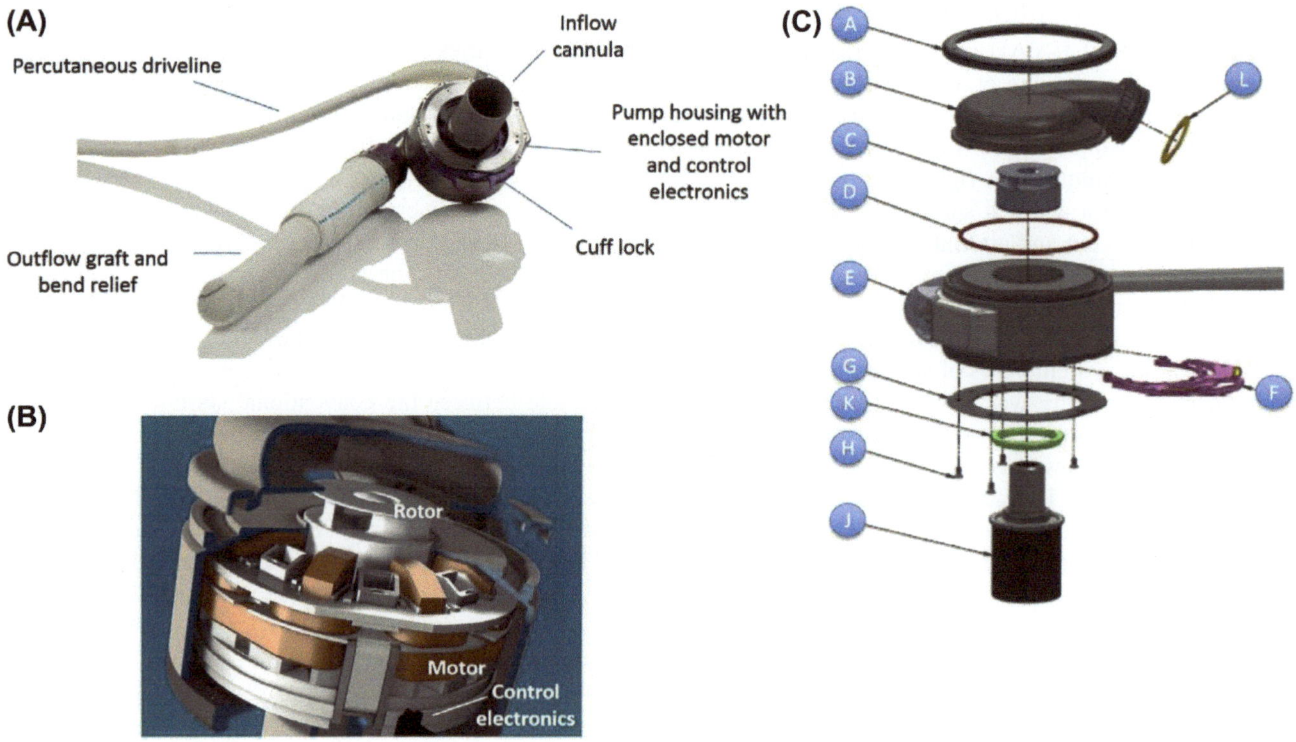

FIGURE 6.15 (A) Implanted components of the HeartMate 3 include the pump with an integral inflow cannula, cuff lock for quickly securing the pump, outflow graft and bend relief, and percutaneous driveline. (B) Enclosed in the pump housing is a rotor and, hermetically sealed from the blood flow, a motor and control electronics. (C) An exploded view of the HeartMate 3 expose: the pump cover screw ring (a), pump cover (b), rotor (c), o-ring (d), motor assembly (e), cuff lock (f), cuff lock cover (g), cuff lock cover screws (h), inflow cannula (i), and sealing washers (j and k). *Used with permission from Bourque K, Cotter C, Dague C, et al. Design rationale and preclinical evaluation of the HeartMate 3 left ventricular assist system for hemocompatibility.* ASAIO J. *2016;62(4):375–383.*[135]

The HeartMate 3 pumps blood up to 10 L/min, and the operating speed ranges from 3000 to 9000 rpm (Fig. 6.16). The characteristic features of the HeartMate 3 are (1) a magnetically levitated rotor (Full MagLev, St. Jude Medical, Inc., Pleasanton, CA), (2) wide blood flow gaps, and (3) artificial pulse capability. The pump rotor is fully supported by contactless magnetic levitation. Both the drive and levitation of the rotor are accomplished using a single stator comprising iron pole pieces, a back iron, copper coils, and position sensors. Measuring the position of a permanent magnet in the rotor and controlling the current in the drive and levitation coils enables active control of the radial position and rotational speed of the rotor. Because the permanent magnet is attracted to the iron pole pieces, the rotor passively resists excursion in the axial direction, whether translating or tilting.

The electronics and software necessary to control the motor drive and levitation are integrated into the lower housing with the stator; these components plus the rotor comprise the motor. The 20 mm diameter inflow cannula is rigidly affixed to the lower housing with the stator; these components plus the rotor comprise the motor.

The inflow cannula is a sintered inflow cannula, the same as the HeartMate II. The outflow cannula assembly includes a 14 mm sealed woven polyester graft. A reinforced tube around the graft serves as a bend relief to prevent kinking and abrasion. The artificial pulse mode is always used, which varies the rotor speed from the user-set speed every 2 s to produce changes in blood flow and arterial blood pressure (Fig. 6.17).

Clinical results

The HeartMate 3 received its CE mark in October 2015. In 2015, Schmitto et al.[136] reported the first clinical case in a 55-year-old man. Hanke et al.[137] reported outcomes and AEs of a single-center cohort 6 months after HeartMate 3 implantation with 27 patients. Within 1 year after implantation, one patient received a heart transplant and three patients died. Thirty-day survival was 88.9% and 6 months, 85.2%. No pump thrombosis and no strokes were observed within 6 months. RHF was diagnosed in one patient. No technical complications and no pump exchanges were documented.

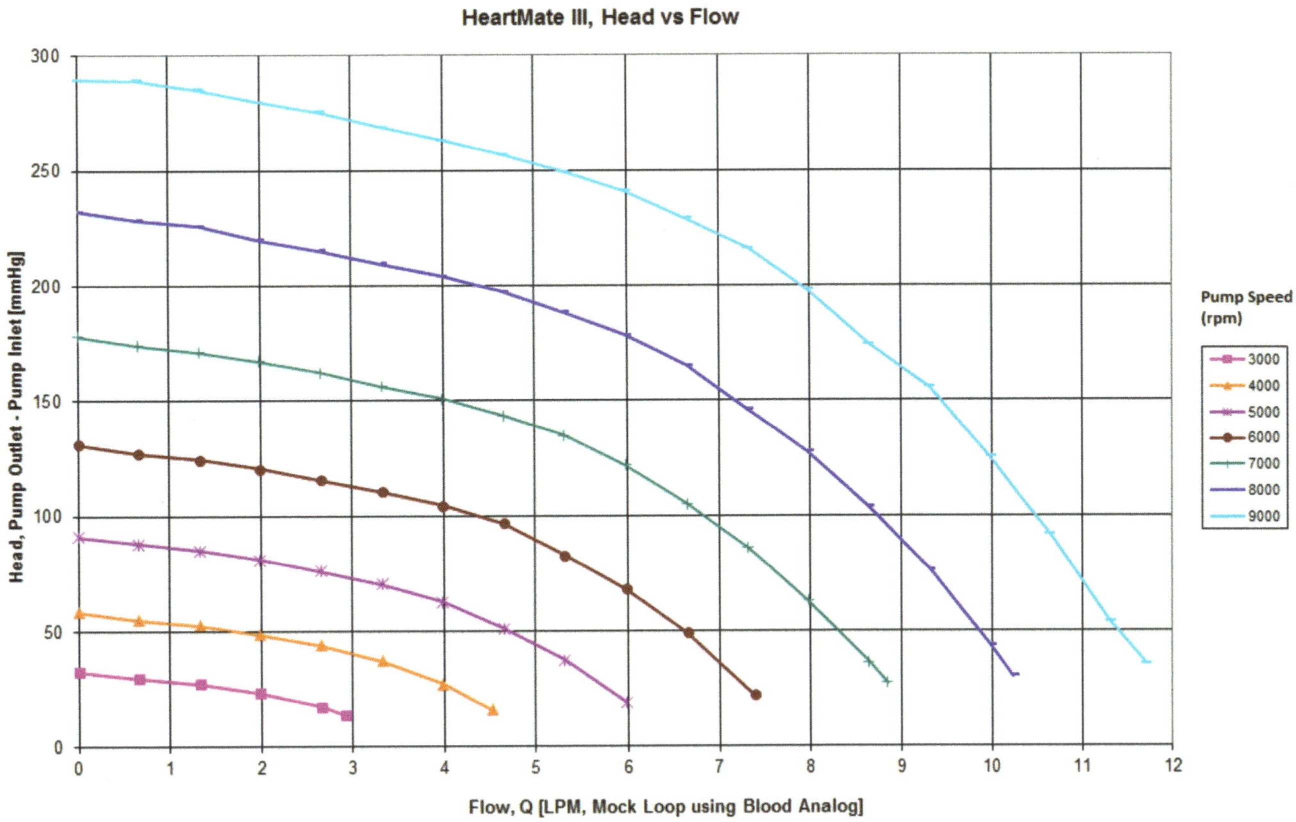

FIGURE 6.16 Pressure flow relationship of the HeartMate 3. *Used with permission from Abbott (Abbott Park, IL).*

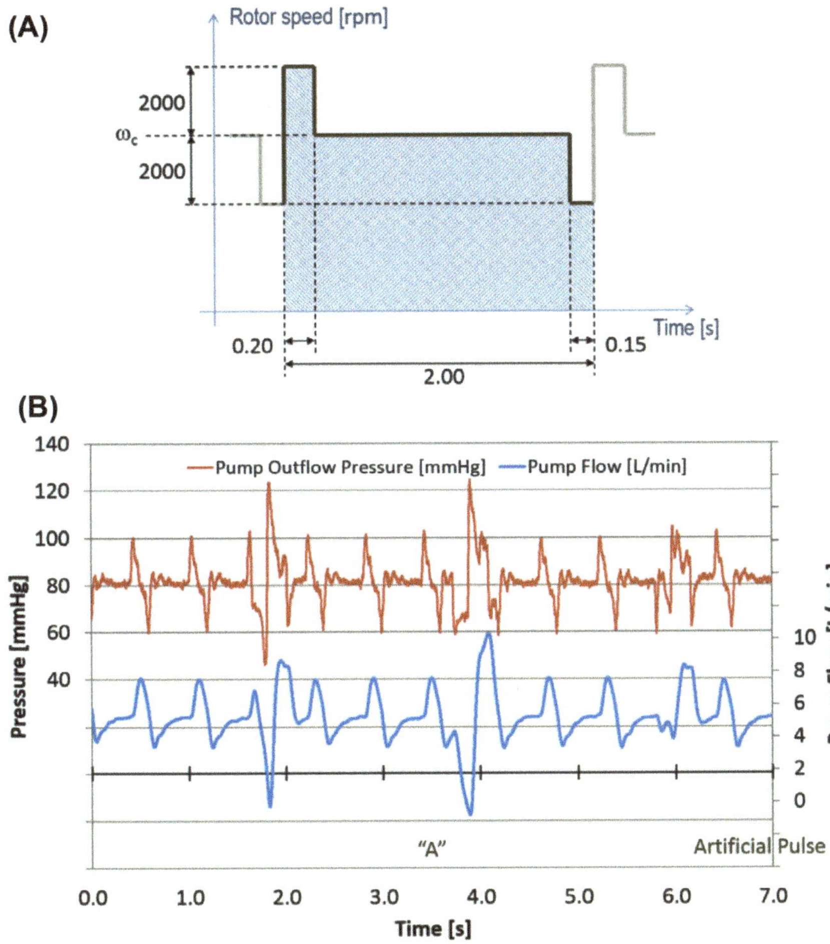

FIGURE 6.17 (A) Schematic diagram of the artificial pulse. (B) The effect of the artificial pulse can be seen superimposed upon the simulated native pulse in an in vitro mock circulatory loop. *Used with permission from Bourque K, Cotter C, Dague C, et al. Design rationale and preclinical evaluation of the HeartMate 3 left ventricular assist system for hemocompatibility. ASAIO J. 2016;62(4):375–383.*[135]

Krabatsch et al.[138] followed this cohort for 2 years and reported the 1-year results in 50 patients (54% BTT and 46% DT). At 1 year, 74% of the patients remained on support, 18% expired, 6% were transplanted, and 2% were explanted. The AEs include GI bleeding (12%), DLI (16%), strokes (18%), and outflow graft thrombosis (2%). There was no hemolysis, pump thrombosis or pump malfunction through 1 year.

Gustafsson et al.[139] reported the results of the Evaluating the HeartMate 3 with Full MagLev Technology in a Post-Market Approval Setting (ELEVATE) registry, which was planned to study postapproval outcomes with the HeartMate 3. They reported the 6-month outcomes of the patients who received HeartMate 3 as their primary implant ($n = 463$) and the survival of the 540 patients in the full cohort. Baseline characteristics included a mean age of 55.6 ± 11.7 years, 89% male, 48% ischemic etiology, and 70% on inotropes. The majority of patients (66%) were BTT and 32% were INTERMACS profile 1–2. Full cohort survival at 6 months was 82% ± 2%. In the enrolled primary implant patients, there was no incidence of pump thrombosis, the major bleeding rate was 25%, major infection 35%, and any stroke type 5%. Functional capacity improved significantly. Freedom from unplanned rehospitalizations at 6 months was 68% ± 2%.

In 2018, Mehra et al.[140] reported 2-year results of the Multicenter Study of MagLev Technology in Patients Undergoing Mechanical Circulatory Support Therapy with HeartMate 3 (MOMENTUM 3) trial, which was a randomized noninferiority and superiority trial, comparing the HeartMate III with the HeartMate II pump in patients with advanced heart failure, irrespective of the intended goal of support (BTT or DT). The composite primary end point was survival at 2 years free of disabling stroke (with disabling stroke indicated by a modified Rankin score of >3; scores range from 0 to 6, with higher scores indicating more severe disability) or survival free of reoperation to replace or remove a malfunctioning device. Of 366 patients, 190 were assigned to the HeartMate 3 group and 176 to the HeartMate II group.

In the intention-to-treat population, the primary end point occurred in 151 patients (79.5%) in the HeartMate 3 pump group, as compared with 106 (60.2%) in the HeartMate II group [absolute difference, 19.2 percentage points; 95% lower confidence boundary, 9.8 percentage points ($P < .001$ for noninferiority); hazard ratio, 0.46; 95% confidence interval (CI), 0.31–0.69 ($P < .001$ for superiority)]. Reoperation for pump malfunction was less frequent in the HeartMate 3 group than in the HeartMate II group [3 patients (1.6%) vs 30 patients (17.0%); hazard ratio, 0.08; 95% CI, 0.03–0.27; $P < .001$]. The rates of death and disabling stroke were similar in the two groups, but the overall stroke rate was lower in the HeartMate 3 group (10.1% vs 19.2%; hazard ratio, 0.47; 95% CI, 0.27–0.84, $P = .02$).

Schmitto et al.[141] reported on a European multicenter, prospective, nonrandomized, single-arm study to assess safety and outcomes of patients 2 years after implantation with the HeartMate 3 in 50 adults. The median support duration was 694 days (range: 19–833 days). At 2 years, Kaplan–Meier survival was 74% ± 6%, 5 (10%) patients were transplanted, and 32 patients (64%) remained with support. AE rates include bleeding requiring surgery (16%), GI bleeding (20%), DLI (24%), ischemic stroke (16%), hemorrhagic stroke (8%), RHF (14%), and outflow graft thrombosis (2%). Notably, no hemolysis, pump thrombosis, or pump malfunction events occurred.

Adverse events

Hemocompatibility

Bourque et al.[135] reported CFD analysis of the HeartMate 3. CFD-based hemocompatibility of the HeartMate 3 was analyzed by reporting volumetric SS distribution, shear-exposed areas, and the Hemolysis Index, and compared with the corresponding values for the HeartMate II. Overall, SS distribution was relatively homogenous and smooth across the flow passages. Typical high-shear areas for the axial impeller were confined at the blade tip, whereas the HeartMate 3 impeller blades exhibited SS at the concave side of the blade leading edge and the convex side of the blade trailing edge.

The SS histogram for the nominal operating condition indicated that more than 99% of the blood volume inside the HeartMate 3 was exposed to low SS (<50 Pa), compared with 97% of pump volume in the HeartMate II. Less than 0.03% of the HeartMate 3 was exposed to high shear values >150 Pa, compared with 0.09% of the HeartMate II. Similarly, the surface area with wall SS exceeding 150 Pa was 7.7% of the impeller in the HeartMate 3, compared with 26% in the HeartMate II.

In agreement with SS loading comparisons, CFD estimation of the hemolysis index for the HeartMate 3 was 14% that of HeartMate II. Surface washing through the secondary flow paths of the HeartMate 3 was evaluated by estimating the local residence time inside gaps and comparing to the mean residence time of blood traveling through the pump primary flow path. Simulations showed that secondary flow residence time was comparable with primary flow residence time.

Wiegmann et al.[142] investigated the effect of the artificial pulse mode on the flow fields and stresses using CFD. The speed decrease of the artificial pulse severely destabilized the flow, leading to multiple vortices. The subsequent speed increase led to an overall stabilization of the flow. The final speed change from top speed back to nominal led to the anticipated overall reduction of velocities, but not to the formation of major vortices.

For SS, in the "artificial pulse" scenario, viscous stress levels in the rotor decreased during the period of low flow and increased during high rotational speed. Total stresses in the rotor and volute rose considerably during the period of low flow, and zones of elevated stresses were also located towards the inlet. At the time of the increase in rotor speed, large parts of the rotor experienced total stresses >100 Pa. Overall levels of total stress decreased only as the system returned to nominal conditions. Total stresses were considerably higher than viscous ones, resulting in a shift to higher total stress values and longer exposure times in the artificial pulse case.

Boraschi et al.[143] also evaluated the thrombogenic profile of the HVAD Lavare Cycle and HeartMate 3 artificial pulse via comprehensive numerical evaluation of (1) pump washout, (2) stagnation zones, (3) SS regimens, and (4) modeling of platelet activation status via the platelet activity state model. They did not observe a significant effect of the Lavare Cycle or Artificial pulse on washout time, and improved flow stagnation. However, in the Artificial pulse mode of the HeartMate 3, the increased SS during the acceleration and high-speed phase, and increased platelet activity, state value compared with static environment.

Stroke

The HeartMate 3 had significantly fewer pump thrombosis events compared with the HeartMate II, but the question of whether the HeartMate 3 has a lower stroke rate compared with the HeartMate II or HVAD is still unanswered. Colombo et al.[144] performed a comprehensive analysis of stroke events to evaluate their longitudinal occurrence, clinical correlates, patterns, and impact on outcome across the 2-year duration of support by using the cohort of the MOMENTUM 3 trial. In 361 patients with the intended implant (189 HeartMate 3 and 172 HeartMate II), 65 strokes (40 ischemic strokes and 25 hemorrhagic strokes) occurred in 52 patients at a median of 131 (range, 1−733) days. No difference in stroke rate was noted between 0 and 180 days of follow-up between devices. However, long-term (181−730 days after LVAD) stroke incidence was 3.3 times lower for the HeartMate 3 group (HeartMate 3: 0.04 vs HeartMate II: 0.13 EPPY; odds ratio, 0.23; 95% CI, 0.08−0.63; $P = .01$). Treatment with the HeartMate 3 pump was the only independent predictor of lower stroke events.

Li et al.[145] systematically identified and analyzed key clinical trials and registries evaluating the HeartMate II, HeartMate 3, and HVAD devices. Reported neurologic events were nonexclusively categorized into ischemic stroke, hemorrhagic stroke, disabling stroke, fatal stroke, and other neurologic events per the studies' definitions. Event rates were standardized to EPPY and freedom-from-event formats, and seven key clinical trials and registries were included in the analysis. There was significant variation and overlap in neurologic event rates for the three VAD platforms across clinical trials [all neurologic events (EPPY): HeartMate 3 0.17−0.21; HeartMate II 0.19−0.26; HVAD 0.16−0.28]. The authors found that no pump performed consistently better for all types of neurologic events.

Bleeding complications postimplant

Bansal et al.[146] evaluated the structure and functional characteristics of vWF HMWMs in patients with the HeartMate 3 ($n = 60$) and the HeartMate II ($n = 30$). Of 51 HeartMate 3 and 29 HeartMate II evaluable patients, those implanted with the HeartMate 3 pump exhibited greater preservation of the vWF HMWM ratio than those with the HeartMate II pump at 90 days after implantation (54.1% vs 42.4%, $P < .0001$). Laboratory values for all vWF assays (antigen, activity, and coagulation factor VIII activity) remained within the normal functional range, with no significant differences observed between the pumps at 90 days after implantation. Multivariable modeling identified the HeartMate 3 pump as the only independent variable that determined postimplantation preservation of the structural integrity of vWF HMWMs.

Klaeske et al.[147] analyzed whether the HeartMate 3 is superior to the HVAD in preserving the multimeric structure of vWF in 70 patients with an implanted HeartMate 3 ($n = 35$) or HVAD ($n = 35$). Acquired vWF syndrome, which is characterized by a decrease in vWF HMWMs, was found in 97.1% of patients in the HeartMate 3 group and 100% of patients in the HVAD group. vWF activity was elevated in patients with an implanted HeartMate 3 device (153.7% ± 54.4%) compared with those with an HVAD device (126.3% ± 39.7%, $P = .02$). However, no differences in incidence of surgical and nonsurgical bleeding were observed in MOMENTUM trial comparing the HeartMate 3 and HeartMate II. In the CE Mark study, bleeding requiring surgery and GI bleeding occurred 16% and 20% of patients, respectively.

In response to low pump thrombosis events, compatible or low stroke events and compatible bleeding events in HeartMate 3 support compared to the other two devices, there are some studies to reduce anticoagulation regimen in patients with HeartMate 3. Saeed et al.[148] evaluated the effect of aspirin dose on hemocompatibility outcomes with the HeartMate 3. Overall, 321 HeartMate 3 patients [usual-dose (325 mg): n = 141, low-dose (81 mg): $n = 180$] were included in this analysis. Usual-dose group patients were younger (57 ± 13 vs 60 ± 12 years, $P = .035$) and were assigned DT less often (55% vs 67%, $P = .029$) than the low-dose group. At 2 years, a similar proportion of patients in the usual- and low-dose groups (43.4% vs 45.3%, $P = .94$) met the primary end point. There were no differences in survival free from hemorrhagic (usual-dose: 54.4% vs low-dose: 51.7%, $P = .42$) or thrombotic (usual-dose: 76.8% vs low-dose: 75.7%, $P = .92$) events.

Netuka et al.[149] investigated the safety of low-intensity anticoagulation in patients with the HeartMate 3. The Minimal AnticoaGulation EvaluatioN To aUgment heMocompatibility (MAGENTUM 1) pilot study was a prospective, single-arm study of low-intensity warfarin anticoagulation in patients implanted with the HeartMate 3. After standard warfarin anticoagulation (INR 2.0–3.0) and aspirin for 6 weeks postimplant, patients were transitioned to a lower INR target range of 1.5–1.9. The primary end point was a composite of survival free of pump thrombosis, disabling stroke, or major bleeding (excluding perioperative bleeding) with at least a 6-month postimplant follow-up. They enrolled 15 patients (mean age 57.3 ± 13.3 years), including 13 men with advanced heart failure. The primary end point was met in 14 of 15 (93 ± 6%) patients; one patient developed recurrent GI bleeding. No thrombotic events occurred.

Biventricular support

There are a few case reports or case studies of the HeartMate 3 as an RVAD.[150,151] Theoretically, the HeartMate 3 is more suitable as an RVAD than as an HVAD, since the rotor can be stable in wider range of rotational speeds with MagLev technology. Lavee et al.[152] reported on 14 cases of biventricular VAD support with a dual HeartMate 3 worldwide. Nine of these patients (64%) were alive as of January 1, 2018. Eight of the 9 have continued on BiVAD support for 95 to 636 (mean 266) days: seven at home, and one successfully transplanted after 98 days of support.

Minimally invasive implantation

The HeartMate 3 is also suitable for less invasive implant.[141,153] Wood et al.[154] reported the results of 10 patients who underwent complete sternal-sparing HeartMate 3 implantation using bilateral thoracotomies. The median length of stay in the intensive care unit was 3.5 days. No patients required reexploration for postoperative bleeding. There was no incidence of right ventricle failure, stroke, renal failure, hepatic failure, or myocardial infarction. The median length of hospitalization after LVAD implantation was 14.5 days (interquartile range, 12–17 days).

Gosev et al.[155] reported a retrospective, single-center study that included consecutive patients implanted with the HeartMate 3, in a comparison of surgical approaches: sternal sparing or traditional sternotomy. The primary outcome was overall survival at 6 months. Secondary outcomes included perioperative complications, blood product utilization, and hospital readmissions. Of 105 patients, 41 (39%) were implanted via the sternal-sparing approach. There were no intraoperative conversions. The sternal-sparing patients were younger; otherwise, all other characteristics were similar between cohorts. The sternal-sparing cohort demonstrated a significantly lower incidence of severe RHF (7% vs 28%, $P = .012$), fewer blood-product transfusions (41% vs 86%, $P < .001$), and a shorter index hospital length of stay (15.5 vs 21 days, $P = .018$). Six-month survival was 93% for the SS cohort.

Riebandt et al.[156] reported the outcomes of a propensity score analysis designed to compare less invasive and full sternotomy LVAD implantation in 159 consecutive patients who received the HVAD or the HeartMate 3. Seventy-five patients were matched and compared (HVAD 83%; less invasive approach 43%; mean age 60 ± 12 years; 89%, male; 48%, ischemic cardiomyopathy; 37%). Patient groups had comparable preoperative characteristics. Less invasive LVAD implantation was successful in all patients, with no intraoperative conversions. In-hospital mortality was 16% in both groups. Patients in the less invasive group had lower rates of blood product usage, but otherwise, outcomes were comparable.

Summary

The HeartMate 3 is a full magnetic-bearing centrifugal-flow pump with a wide flow gap and low SS. Magnetic bearing is considered to be weak because it needs a complex control system of rotor stability, but the HeartMate 3 seems robust and no major pump dysfunction has occurred.

The HeartMate 3 is the newest FDA-approved pump, and the market share is approximately 80%. In clinical trials, the HeartMate 3 exhibited better survival rates, lower rates of pump thrombosis, and the same or lower stroke rate compared with the HeartMate II. According CFD analysis, HeartMate 3 had lower SS and better flow patterns compared with the HeartMate II or HVAD; however, clinically, patients with HeartMate 3 still presented with vWF disease, and there is no clear evidence that HeartMate 3 support contributes to a lower rate of GI bleeding.

The HeartMate 3 has the artificial pulse mode, but some reports emphasized the increased blood damage of this mode. The device is a small intrapericardial pump, but the infection rate is about the same as the HeartMate II. Overall, the HeartMate 3 has shown favorable clinical results, but we still need more clinical experience with this device.

EVAHEART 2

EVAHEART (Evaheart, Inc., Houston, TX) is a centrifugal-flow CF-LVAD with a hydraulically levitated impeller that has been used in Japan since 2005 as a BTT (Fig. 6.18).[158] This pump has a unique bearing system in which sterile water for injection is continuously circulated through the driveline and around the rotating impeller shaft and bearing interface to apply a hydraulic levitation force (cool seal unit). The cool seal unit has demonstrated durability of up to 10 years in clinical use, without any adverse infectious events. The impeller features a unique open-vane design with large gaps in the pump housing wall (700–1000 μm), minimizing wall SS. The pump's normal operating speed is 1800–2400 rpm, providing a maximum flow rate of 10–15 L/min.

The EVAHEART 1 was miniaturized and developed as the EVAHEART 2 (Fig. 6.19).[157] The EVAHEART 1 required thick pump housing walls for a screw bolt assembly. To miniaturize the pump size without compromising the pump flow characteristics, the pump's external walls were made thinner and were assembled through a press-fit. All of the surfaces that contact blood (including the impeller) and the blood pathway dimensions remained unchanged. Compared with the EVAHEART 1, the EVAHEART 2 was miniaturized by 30% in weight and 26% in displacement volume. The driveline diameter decreased from 9.8 to 7.8 mm. Because the blood pathway dimensions are identical, the pump performance curves reported previously remain the same.

The pressure flow curves of the EVAHEART 2 are relatively "flat," meaning the flow is sensitive to head pressure.[157] Additionally, the EVAHEART 2 is capable of high end-systolic peak flow, which contributes to the preservation of arterial pulsatility.

Another unique feature of the EVAHEART 2 is the "double-cuff tipless" (DCT) inflow cannula, which does not protrude into the LV chamber. Instead, the DCT cannula is secured to the endocardial surface through a mesh suture cuff. The EVAHEART 2 adopted the DCT inflow cannula specifically to mitigate the risk of stroke and inflow suction events.

Fig. 6.20 shows the following components of the DCT inflow design: (1) a polished, methacryloyloxyethyl phosphorylcholine–coated titanium conduit for blood drainage; (2) a polyester surgical mesh for approximating cored apical myocardium; (3) a polytetrafluorethylene suture cuff; and (4) an O-ring sealed connector.

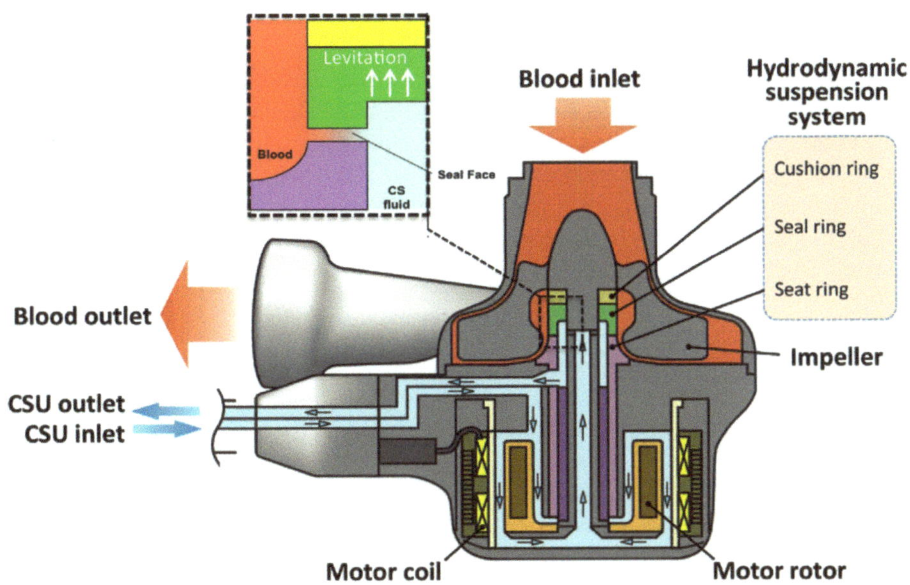

FIGURE 6.18 Mechanical seal cross-sectional schematic view of the EVAHEART 2. *Used with permission from Motomura T, Tuzun E, Yamazaki K, Tatsumi E, Benkowski R, Yamazaki S. Preclinical evaluation of the EVAHEART 2 centrifugal left ventricular assist device in bovines. ASAIO J. 2019;65 (8):845–854.*[157]

FIGURE 6.19 Miniaturization of the pump and driveline from the EVAHEART 1 (left) to the EVAHEART 2 (right). *Used with permission from Motomura T, Tuzun E, Yamazaki K, Tatsumi E, Benkowski R, Yamazaki S. Preclinical evaluation of the EVAHEART 2 centrifugal left ventricular assist device in bovines.* ASAIO J. *2019;65(8):845–854.*[157]

Saito et al.[158] reported clinical results of the EVAHEART 1 in 96 consecutive patients enrolled in the Japanese Registry for Mechanically Assisted Circulatory Support trial from 2011 to 2013 at 14 Japanese centers. Patients' median age was 43 years (85% male). The mean support duration was 384.7 days, with a cumulative duration of 101.2 years. The Kaplan–Meier survival rate during support was 93.4% at 6 months, 87.4% at 1 year and 87.4% at 2 years. Seventy-seven patients (80.2%) remained on support, seven received a transplant, and 10 died during support. Major AEs included driveline infection (14.6%) and neurologic events such as ischemic stroke (17.7%), hemorrhage (13.5%), transient ischemic attack (3.1%), pump thrombosis (1%) and hemolysis (1%). There was no GI bleeding or RHF requiring RVAD. There was no pump exchange due to mechanical failure.

Imamura et al.[160] evaluated the characteristic performance of the EVAHEART 1 and occurrence of AI related to it in 45 patients implanted with a CF-LVAD (EVAHEART, 14; DuraHeart, 9; HeartMate II, 14; Jarvik 2000, 7; HeartWare, 1) between 2006 and 2015. There were no significant differences in preoperative patient characteristics between any of the groups. Among preoperative data, EVAHEART usage was the only significant predictor of being AI-free ($P = .012$, odds ratio 15.79). After the surgery, most of the patients (93%) avoided AI during EVAHEART treatment. Pulse pressure was higher in the EVAHEART group, and the aortic root was less dilated in the EVAHEART group compared with the other groups.

Bartoli et al.[161] tested the hypothesis that the EVAHEART 1 (Evaheart Inc.), which was designed to minimize SS, reduced vWF degradation versus the HeartMate II in vitro. Human blood was circulated for 12 h in mock circulatory loops through a HeartMate II or an EVAHEART. The HeartMate II eliminated large vWF multimers and significantly ($P < .05$) increased 10 of 11 vWF degradation fragments at 6 and 12 h. The increase was approximately twofold at 6 h and 2.2-fold at 12 h.

In contrast, the EVAHEART LVAS modestly reduced large vWF multimers and significantly increased 5 of 11 and 8 of 11 vWF degradation fragments at 6 and 12 h, respectively. The increase was approximately 1.5-fold at 6 h and 1.7-fold at 12 h. The EVAHEART caused significantly less degradation ($P < .01$) than the HeartMate II of the 140 kDa vWF fragment (cleavage product of ADAMTS-13, the vWF protease).

They also investigated the effect of reduced pump speed in the EVAHEART to vWF with human blood in vitro. Blood was circulated in ex vivo mock circulatory loops for 6 h with an EVAHEART LVAS at 2300 ($n = 12$), 2100 ($n = 8$), or 1800 rpm ($n = 8$). Pump speed reduction protected vWF from pathologic degradation. At lower rpms, significantly greater levels of vWF multimers were observed. Similarly, at lower rpms, significantly fewer vWF fragments, a product of vWF degradation, were observed.

Zayat et al.[162] evaluated blood damage induced by the HeartMate 3 and the EVAHEART. After 6 h of mock-loop circulation under both settings, the free plasma hemoglobin concentration was significantly lower in the EVAHEART compared with the HeartMate 3. Loss of vWF HMWM occurred in both LVADs and settings, but the

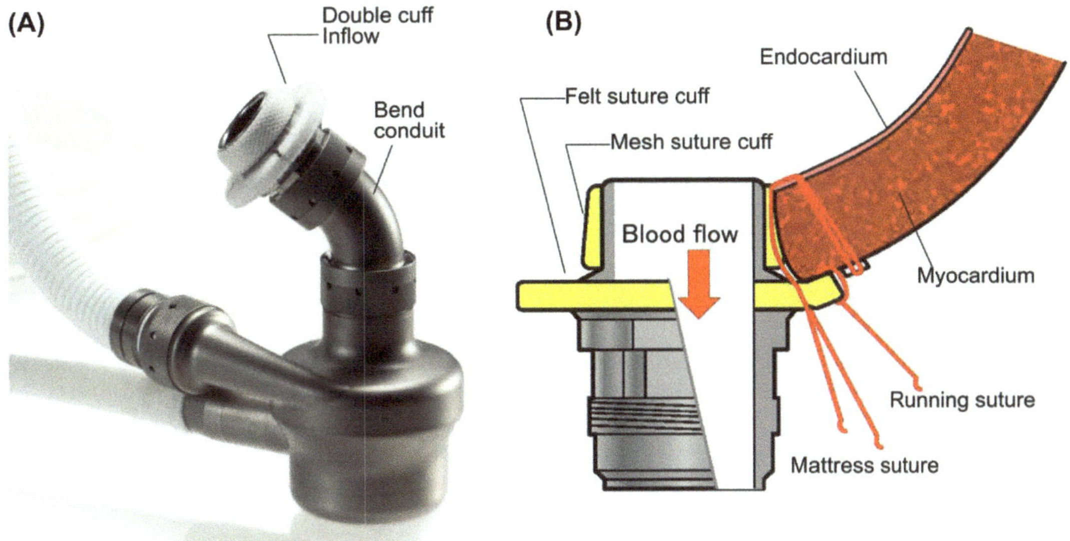

FIGURE 6.20 EVAHEART 2 with tipless inflow cannula (A) and cross-sectional fitting schematic of the DCT inflow cannula (B) sutured to the apical myocardium with appropriate suture technique. *DCT*, Double-cuff tipless. *Used with permission from Motomura T, Kelly B. Surgical techniques for implanting the EVAHEART 2 double cuff tipless inflow cannula. ASAIO J. 2019;65(8):e86–e89.*[159]

vWF:activity/vWF:antigen ratio after 6 h was significantly lower in the HeartMate 3 ($P = .009$). The thrombin–antithrombin complex level was significantly lower with the EVAHEART for both settings ($P < .0001$).

Ichihara et al.[163] reported on a clinical study showing that incidence of GI bleeding was relatively low in patients with an EVAHEART, and there was a modest degradation of HMW vWF multimers. Akiyama et al.[92] investigated the incidence, etiology, and outcome of GI bleeding in 54 patients with a CF-LVAD [HeartMate II ($n = 29$), Jarvik 2000 ($n = 9$), DuraHeart (Terumo Heart Inc., Ann Arbor, MI, USA, $n = 9$), EVAHEART ($n = 6$), and HVAD ($n = 1$)]. Among them, eight (15%) presented with overt GI bleeding (12 events). Patients with GI bleeding were significantly older ($P = .04$), and their preoperative inferior vena cava diameter was larger ($P = .02$). Multivariate analysis revealed that the use of Jarvik 2000 ($P = .003$) was a risk factor for GI bleeding. In total, 85.8% of patients were free from GI bleeding at 1 year. No patient with EVAHEART had GI bleeding.

Stroke events were a major limitation of the EVAHEART 1. Matsumoto et al.[164] compared the hemodynamic performance and clinical results of the EVAHEART and HeartMate II in 14 patients who received the EVAHEART and 28 who received the HeartMate II. Mean follow-up was 35.5 ± 14.8 months for the EVAHEART and 29.8 ± 6.5 months for the HeartMate II. Survival rates were comparable between the two groups. After 24 months, freedom from DLI was 28% with the EVAHEART, and 85% with HeartMate II; freedom from neurologic events was 21% with the EVAHEART, and 89% with the HeartMate II. In response to clinical results in Japan, many efforts were made to reduce wedge thrombus formation with the EVAHEART.

May-Newman et al.[165] analyzed velocity fields inside the left ventricle with varying length of inflow cannula, and found that the shortest (1 cm) inflow cannula insertion depth produced the flow pattern that exhibited the least apical flow stasis and greatest pulsatility, and should correlate to a lower risk of thrombus formation. They then assessed the "tipless" inflow cannula in a mock circulatory loop compared with the usual protruding inflow cannula.[166] The protruding inflow cannula interrupted the growth and restricted the movement of the vortex, and produced areas of low velocity and pulsatility adjacent to the cannula. The tipless cannula exhibited an uninterrupted pattern of the mitral jet toward the LV apex, which allowed the diastolic vortex to grow and aid in the washout of this region. In addition, the tipless cannula increased aortic valve flow, which reduced stasis in the left ventricular outflow tract.

Motomura et al.[157] reported on an in vivo study of the EVAHEART 2 with the new cannula, which demonstrated good biocompatibility of the system. In 2020, Kawamura et al.[167] reported the first clinical experience with the EVAHEART 2.

Conclusions

Since the first human implant of a CF-VAD in 1998, the technology has greatly expanded. Today, CF-VAD support is the most effective therapy for patients with advanced heart failure, especially those who are not candidates for heart transplantation. As the technology has developed, indications for CF-VAD therapy are changing and expanding, and clinical results are improved.

However, the major limitation of this therapy is the high rate of AEs, especially those related to hemocompatibility. As seen in this chapter, many engineering efforts, preclinical and clinical studies have been performed to evaluate the causes, risk factors, and pump characteristics of CF-LVADs. Much progress has been made, but we have not yet overcome the problems. Therefore, more technological innovation and in-depth evaluation of biological device interaction will be needed to render the circulatory support solutions even more effective and safe.

References

1. Yancy CW, Jessup M, Bozkurt B, et al. ACCF/AHA guideline for the management of heart failure: a report of the American College of Cardiology Foundation/American Heart Association Task Force on Practice Guidelines. *J Am Coll Cardiol.* 2013;62(16):e147–e239.
2. Molina EJ, Shah P, Kiernan MS, et al. The Society of Thoracic Surgeons Intermacs 2020 Annual Report. *Ann Thorac Surg.* 2021;.
3. Teuteberg JJ, Cleveland Jr. JC, Cowger J, et al. The Society of Thoracic Surgeons Intermacs 2019 Annual Report: the changing landscape of devices and indications. *Ann Thorac Surg.* 2020;109(3):649–660.
4. Moazami N, Fukamachi K, Kobayashi M, et al. Axial and centrifugal continuous-flow rotary pumps: a translation from pump mechanics to clinical practice. *J Heart Lung Transpl.* 2013;32(1):1–11.
5. Stanfield JR, Selzman CH. In vitro pulsatility analysis of axial-flow and centrifugal-flow left ventricular assist devices. *J Biomech Eng.* 2013;135(3):34505.
6. Tang PC, Pagani FD. Continuous-flow device engineering and pump technology. *Cardiol Clin.* 2018;36(4):451–463.
7. Sundareswaran KS, Reichenbach SH, Masterson KB, Butler KC, Farrar DJ. Low bearing wear in explanted HeartMate II left ventricular assist devices after chronic clinical support. *ASAIO J.* 2013;59(1):41–45.
8 MANUALS AND. RESOURCES for HeartMate II® Left Ventricular Assist System. Available from: https://manuals.sjm.com.
9. Griffith BP, Kormos RL, Borovetz HS, et al. HeartMate II left ventricular assist system: from concept to first clinical use. *Ann Thorac Surg.* 2001;71(3 Suppl.):S116–S120. Discussion S114–S116.
10. Slaughter MS, Bartoli CR, Sobieski MA, et al. Intraoperative evaluation of the HeartMate II flow estimator. *J Heart Lung Transpl.* 2009;28 (1):39–43.
11. Frazier OH, Delgado 3rd RM, Kar B, Patel V, Gregoric ID, Myers TJ. First clinical use of the redesigned HeartMate II left ventricular assist system in the United States: a case report. *Tex Heart Inst J.* 2004;31(2):157–159.
12. Frazier OH, Gemmato C, Myers TJ, et al. Initial clinical experience with the HeartMate II axial-flow left ventricular assist device. *Tex Heart Inst J.* 2007;34(3):275–281.
13. Miller LW, Pagani FD, Russell SD, et al. Use of a continuous-flow device in patients awaiting heart transplantation. *N Engl J Med.* 2007;357 (9):885–896.
14. Pagani FD, Miller LW, Russell SD, et al. Extended mechanical circulatory support with a continuous-flow rotary left ventricular assist device. *J Am Coll Cardiol.* 2009;54(4):312–321.
15. Slaughter MS, Rogers JG, Milano CA, et al. Advanced heart failure treated with continuous-flow left ventricular assist device. *N Engl J Med.* 2009;361(23):2241–2251.
16. Starling RC, Estep JD, Horstmanshof DA, et al. Risk assessment and comparative effectiveness of left ventricular assist device and medical management in ambulatory heart failure patients: the ROADMAP study 2-year results. *JACC Heart Fail.* 2017;5(7):518–527.
17. Estep JD, Starling RC, Horstmanshof DA, et al. Risk assessment and comparative effectiveness of left ventricular assist device and medical management in ambulatory heart failure patients results from the ROADMAP study. *J Am Coll Cardiol.* 2015;66(16):1747–1761.
18. Pagani FD, Aaronson KD, Kormos R, et al. The NHLBI REVIVE-IT study: understanding its discontinuation in the context of current left ventricular assist device therapy. *J Heart Lung Transpl.* 2016;35(11):1277–1283.
19. Michelis KC, Zhong L, Peltz M, et al. Dynamic forecasts of survival for patients living with destination left ventricular assist devices: insights from INTERMACS. *J Am Heart Assoc.* 2020;9(14):e016203.
20. Seese L, Movahedi F, Antaki J, et al. Delineating pathways to death by multisystem organ failure in patients with a left ventricular assist device. *Ann Thorac Surg.* 2020;.
21. Kormos RL, Antonides CFJ, Goldstein DJ, et al. Updated definitions of adverse events for trials and registries of mechanical circulatory support: a consensus statement of the mechanical circulatory support academic research consortium. *J Heart Lung Transpl.* 2020;39(8):735–750.
22. Geisen U, Heilmann C, Beyersdorf F, et al. Non-surgical bleeding in patients with ventricular assist devices could be explained by acquired von Willebrand disease. *Eur J Cardiothorac Surg.* 2008;33(4):679–684.
23. Klovaite J, Gustafsson F, Mortensen SA, Sander K, Nielsen LB. Severely impaired von Willebrand factor-dependent platelet aggregation in patients with a continuous-flow left ventricular assist device (HeartMate II). *J Am Coll Cardiol.* 2009;53(23):2162–2167.
24. Uriel N, Pak SW, Jorde UP, et al. Acquired von Willebrand syndrome after continuous-flow mechanical device support contributes to a high prevalence of bleeding during long-term support and at the time of transplantation. *J Am Coll Cardiol.* 2010;56(15):1207–1213.
25. Kang J, Zhang DM, Restle DJ, et al. Reduced continuous-flow left ventricular assist device speed does not decrease von Willebrand factor degradation. *J Thorac Cardiovasc Surg.* 2016;151(6):1747–1754. e1741.

26. Demirozu ZT, Radovancevic R, Hochman LF, et al. Arteriovenous malformation and gastrointestinal bleeding in patients with the HeartMate II left ventricular assist device. *J Heart Lung Transpl*. 2011;30(8):849–853.
27. Aggarwal A, Pant R, Kumar S, et al. Incidence and management of gastrointestinal bleeding with continuous flow assist devices. *Ann Thorac Surg*. 2012;93(5):1534–1540.
28. Draper KV, Huang RJ, Gerson LB. GI bleeding in patients with continuous-flow left ventricular assist devices: a systematic review and meta-analysis. *Gastrointest Endosc*. 2014;80(3):435–446. e431.
29. Stulak JM, Lee D, Haft JW, et al. Gastrointestinal bleeding and subsequent risk of thromboembolic events during support with a left ventricular assist device. *J Heart Lung Transpl*. 2014;33(1):60–64.
30. Holley CT, Harvey L, Roy SS, et al. Gastrointestinal bleeding during continuous-flow left ventricular assist device support is associated with lower rates of cardiac transplantation. *ASAIO J*. 2015;61(6):635–639.
31. Wever-Pinzon O, Selzman CH, Drakos SG, et al. Pulsatility and the risk of nonsurgical bleeding in patients supported with the continuous-flow left ventricular assist device HeartMate II. *Circ Heart Fail*. 2013;6(3):517–526.
32. Muthiah K, Connor D, Ly K, et al. Longitudinal changes in hemostatic parameters and reduced pulsatility contribute to non-surgical bleeding in patients with centrifugal continuous-flow left ventricular assist devices. *J Heart Lung Transpl*. 2016;35(6):743–751.
33. Vincent F, Rauch A, Loobuyck V, et al. Arterial pulsatility and circulating von willebrand factor in patients on mechanical circulatory support. *J Am Coll Cardiol*. 2018;71(19):2106–2118.
34. Fraser KH, Zhang T, Taskin ME, Griffith BP, Wu ZJ. A quantitative comparison of mechanical blood damage parameters in rotary ventricular assist devices: shear stress, exposure time and hemolysis index. *J Biomech Eng*. 2012;134(8):081002.
35. Khoo DP, Cookson AN, Gill HS, Fraser KH. Normal fluid stresses are prevalent in rotary ventricular assist devices: a computational fluid dynamics analysis. *Int J Artif Organs*. 2018;41(11):738–751.
36. Wu ZJ, Antaki JF, Burgreen GW, Butler KC, Thomas DC, Griffith BP. Fluid dynamic characterization of operating conditions for continuous flow blood pumps. *ASAIO J*. 1999;45(5):442–449.
37. Chiu WC, Slepian MJ, Bluestein D. Thrombus formation patterns in the HeartMate II ventricular assist device: clinical observations can be predicted by numerical simulations. *ASAIO J*. 2014;60(2):237–240.
38. Yang F, Kormos RL, Antaki JF. High-speed visualization of disturbed pathlines in axial flow ventricular assist device under pulsatile conditions. *J Thorac Cardiovasc Surg*. 2015;150(4):938–944.
39. Wu WT, Yang F, Wu J, Aubry N, Massoudi M, Antaki JF. High fidelity computational simulation of thrombus formation in Thoratec HeartMate II continuous flow ventricular assist device. *Sci Rep*. 2016;6:38025.
40. Kirklin JK, Naftel DC, Kormos RL, et al. Interagency Registry for Mechanically Assisted Circulatory Support (INTERMACS) analysis of pump thrombosis in the HeartMate II left ventricular assist device. *J Heart Lung Transpl*. 2014;33(1):12–22.
41. Starling RC, Moazami N, Silvestry SC, et al. Unexpected abrupt increase in left ventricular assist device thrombosis. *N Engl J Med*. 2014;370 (1):33–40.
42. Saeed O, Maybaum S, Alessandro DD, Goldstein DJ, Patel SR. Aortic valve opening and thrombotic events with continuous-flow left ventricular assist devices. *J Heart Lung Transpl*. 2014;33(1):109–112.
43. Uriel N, Han J, Morrison KA, et al. Device thrombosis in HeartMate II continuous-flow left ventricular assist devices: a multifactorial phenomenon. *J Heart Lung Transpl*. 2014;33(1):51–59.
44. Smedira NG, Blackstone EH, Ehrlinger J, et al. Current risks of HeartMate II pump thrombosis: non-parametric analysis of Interagency Registry for Mechanically Assisted Circulatory Support data. *J Heart Lung Transpl*. 2015;34(12):1527–1534.
45. Soltani S, Kaufmann F, Vierecke J, et al. Design changes in continuous-flow left ventricular assist devices and life-threatening pump malfunctions. *Eur J Cardiothorac Surg*. 2015;47(6):984–989.
46. Maltais S, Kilic A, Nathan S, et al. PREVENtion of HeartMate II pump thrombosis through clinical management: the PREVENT multi-center study. *J Heart Lung Transpl*. 2017;36(1):1–12.
47. Baumann Kreuziger L, Slaughter MS, Sundareswaran K, Mast AE. Clinical relevance of histopathologic analysis of HeartMate II thrombi. *ASAIO J*. 2018;64(6):754–759.
48. Rowlands GW, Pagani FD, Antaki JF. Classification of the frequency, severity, and propagation of thrombi in the HeartMate II left ventricular assist device. *ASAIO J*. 2020;66(9):992–999.
49. Topkara VK, Kondareddy S, Malik F, et al. Infectious complications in patients with left ventricular assist device: etiology and outcomes in the continuous-flow era. *Ann Thorac Surg*. 2010;90(4):1270–1277.
50. Sharma V, Deo SV, Stulak JM, et al. Driveline infections in left ventricular assist devices: implications for destination therapy. *Ann Thorac Surg*. 2012;94(5):1381–1386.
51. Gordon RJ, Weinberg AD, Pagani FD, et al. Prospective, multicenter study of ventricular assist device infections. *Circulation*. 2013;127 (6):691–702.
52. Dean D, Kallel F, Ewald GA, et al. Reduction in driveline infection rates: results from the HeartMate II Multicenter Driveline Silicone Skin Interface (SSI) Registry. *J Heart Lung Transpl*. 2015;34(6):781–789.
53. Singh A, Russo MJ, Valeroso TB, et al. Modified heartmate II driveline externalization technique significantly decreases incidence of infection and improves long-term survival. *ASAIO J*. 2014;60(6):613–616.
54. Cagliostro B, Levin AP, Fried J, et al. Continuous-flow left ventricular assist devices and usefulness of a standardized strategy to reduce driveline infections. *J Heart Lung Transpl*. 2016;35(1):108–114.

55. Kormos RL, Teuteberg JJ, Pagani FD, et al. Right ventricular failure in patients with the HeartMate II continuous-flow left ventricular assist device: incidence, risk factors, and effect on outcomes. *J Thorac Cardiovasc Surg*. 2010;139(5):1316–1324.
56. Aissaoui N, Salem JE, Paluszkiewicz L, et al. Assessment of right ventricular dysfunction predictors before the implantation of a left ventricular assist device in end-stage heart failure patients using echocardiographic measures (ARVADE): combination of left and right ventricular echocardiographic variables. *Arch Cardiovasc Dis*. 2015;108(5):300–309.
57. Atluri P, Goldstone AB, Fairman AS, et al. Predicting right ventricular failure in the modern, continuous flow left ventricular assist device era. *Ann Thorac Surg*. 2013;96(3):857–864.
58. Kato TS, Farr M, Schulze PC, et al. Usefulness of two-dimensional echocardiographic parameters of the left side of the heart to predict right ventricular failure after left ventricular assist device implantation. *Am J Cardiol*. 2012;109(2):246–251.
59. Drakos SG, Janicki L, Horne BD, et al. Risk factors predictive of right ventricular failure after left ventricular assist device implantation. *Am J Cardiol*. 2010;105(7):1030–1035.
60. Matthews JC, Koelling TM, Pagani FD, Aaronson KD. The right ventricular failure risk score. A pre-operative tool for assessing the risk of right ventricular failure in left ventricular assist device candidates. *J Am Coll Cardiol*. 2008;51(22):2163–2172.
61. Takeda K, Takayama H, Colombo PC, et al. Incidence and clinical significance of late right heart failure during continuous-flow left ventricular assist device support. *J Heart Lung Transpl*. 2015;34(8):1024–1032.
62. Cowger J, Pagani FD, Haft JW, Romano MA, Aaronson KD, Kolias TJ. The development of aortic insufficiency in left ventricular assist device-supported patients. *Circ Heart Fail*. 2010;3(6):668–674.
63. Pak SW, Uriel N, Takayama H, et al. Prevalence of de novo aortic insufficiency during long-term support with left ventricular assist devices. *J Heart Lung Transpl*. 2010;29(10):1172–1176.
64. Cowger JA, Aaronson KD, Romano MA, Haft J, Pagani FD. Consequences of aortic insufficiency during long-term axial continuous-flow left ventricular assist device support. *J Heart Lung Transpl*. 2014;33(12):1233–1240.
65. Jorde UP, Uriel N, Nahumi N, et al. Prevalence, significance, and management of aortic insufficiency in continuous flow left ventricular assist device recipients. *Circ Heart Fail*. 2014;7(2):310–319.
66. Holley CT, Fitzpatrick M, Roy SS, et al. Aortic insufficiency in continuous-flow left ventricular assist device support patients is common but does not impact long-term mortality. *J Heart Lung Transpl*. 2017;36(1):91–96.
67. Sayer G, Sarswat N, Kim GH, et al. The hemodynamic effects of aortic insufficiency in patients supported with continuous-flow left ventricular assist devices. *J Card Fail*. 2017;23(7):545–551.
68. Truby LK, Garan AR, Givens RC, et al. Aortic insufficiency during contemporary left ventricular assist device support: analysis of the INTERMACS registry. *JACC Heart Fail*. 2018;6(11):951–960.
69. Auvil B, Chung J, Ameer A, et al. Moderate aortic insufficiency with a left ventricular assist device portends a worse long-term survival. *ASAIO J*. 2020;66(7):780–785.
70. Yehya A, Rajagopal V, Meduri C, et al. Short-term results with transcatheter aortic valve replacement for treatment of left ventricular assist device patients with symptomatic aortic insufficiency. *J Heart Lung Transpl*. 2019;38(9):920–926.
71. Frazier OH. Unforeseen consequences of therapy with continuous-flow pumps. *Circ Heart Fail*. 2010;3(6):647–649.
72. Noon GP, Morley D, Irwin S, Benkowski R. Development and clinical application of the MicroMed DeBakey VAD. *Curr Opin Cardiol*. 2000;15(3):166–171.
73. Noon GP, Morley DL, Irwin S, Abdelsayed SV, Benkowski RJ, Lynch BE. Clinical experience with the MicroMed DeBakey ventricular assist device. *Ann Thorac Surg*. 2001;71(3 Suppl.):S133–S138. Discussion S144-136.
74. Goldstein DJ. Worldwide experience with the MicroMed DeBakey ventricular assist device as a bridge to transplantation. *Circulation*. 2003;108 (Suppl. 1):Ii272–Ii277.
75. Noon GP, Loebe M. Current status of the MicroMed DeBakey noon ventricular assist device. *Tex Heart Inst J*. 2010;37(6):652–653.
76. Girdhar G, Xenos M, Alemu Y, et al. Device thrombogenicity emulation: a novel method for optimizing mechanical circulatory support device thromboresistance. *PLoS One*. 2012;7(3):e32463.
77. Chiu WC, Girdhar G, Xenos M, et al. Thromboresistance comparison of the HeartMate II ventricular assist device with the device thrombogenicity emulation- optimized HeartAssist 5 VAD. *J Biomech Eng*. 2014;136(2):021014.
78. Schmitto JD, Hanke JS, Dogan G, et al. First implantation of a novel left ventricular assist device: the ReliantHeart aVAD. *Ann Thorac Surg*. 2017;104(4):e311–e313.
79. Hohmann S, Veltmann C, Duncker D, et al. Initial experience with telemonitoring in left ventricular assist device patients. *J Thorac Dis*. 2019;11(Suppl. 6):S853–S863.
80. Hetzer R, Weng Y, Potapov EV, et al. First experiences with a novel magnetically suspended axial flow left ventricular assist device. *Eur J Cardiothorac Surg*. 2004;25(6):964–970.
81. Schmid C, Tjan TD, Etz C, et al. First clinical experience with the Incor left ventricular assist device. *J Heart Lung Transpl*. 2005;24 (9):1188–1194.
82. Schmid C, Jurmann M, Birnbaum D, et al. Influence of inflow cannula length in axial-flow pumps on neurologic adverse event rate: results from a multi-center analysis. *J Heart Lung Transpl*. 2008;27(3):253–260.
83. Jarvik R. Jarvik 2000 pump technology and miniaturization. *Heart Fail Clin*. 2014;10(1 Suppl.):S27–S38.
84. Westaby S, Banning AP, Jarvik R, et al. First permanent implant of the Jarvik 2000 Heart. *Lancet*. 2000;356(9233):900–903.

85. Frazier OH, Myers TJ, Gregoric ID, et al. Initial clinical experience with the Jarvik 2000 implantable axial-flow left ventricular assist system. *Circulation*. 2002;105(24):2855–2860.
86. Frazier OH, Myers TJ, Westaby S, Gregoric ID. Use of the Jarvik 2000 left ventricular assist system as a bridge to heart transplantation or as destination therapy for patients with chronic heart failure. *Ann Surg*. 2003;237(5):631–636. Discussion 636–637.
87. Selzman CH, Koliopoulou A, Glotzbach JP, McKellar SH. Evolutionary improvements in the Jarvik 2000 left ventricular assist device. *ASAIO J*. 2018;64(6):827–830.
88. Stanfield JR, Selzman CH. In vitro hydrodynamic analysis of pin and cone bearing designs of the Jarvik 2000 adult ventricular assist device. *Artif Organs*. 2013;37(9):825–833.
89. Tarzia V, Di Giammarco G, Di Mauro M, et al. From bench to bedside: can the improvements in left ventricular assist device design mitigate adverse events and increase survival? *J Thorac Cardiovasc Surg*. 2016;151(1):213–217.
90. Kohno H, Matsumiya G, Sawa Y, et al. The Jarvik 2000 left ventricular assist device as a bridge to transplantation: Japanese registry for mechanically assisted circulatory support. *J Heart Lung Transpl*. 2018;37(1):71–78.
91. Kohno H, Matsumiya G, Sawa Y, et al. Can the intermittent low-speed function of left ventricular assist device prevent aortic insufficiency? *J Artif Organs*. 2021;.
92. Akiyama M, Sakatsume K, Sasaki K, et al. The incidence, risk factors, and outcomes of gastrointestinal bleeding in patients with a left ventricular assist device: a Japanese single-center cohort study. *J Artif Organs*. 2020;23(1):27–35.
93. Tarzia V, Di Giammarco G, Bagozzi L, et al. From bench to bedside: impact of left ventricular assist device outflow conduit anastomosis position on outcome. *Artif Organs*. 2020;.
94 MANUALS AND. RESOURCES for Medtronic HVADTM ventricular assist device. Available from: https://global.medtronic.com/xg-en/e/response/heartware-hvad-system.html.
95. Larose JA, Tamez D, Ashenuga M, Reyes C. Design concepts and principle of operation of the HeartWare ventricular assist system. *ASAIO J*. 2010;56(4):285–289.
96. Reyes C, Chorpenning K, LaRose JA, Gomez R, Tamez D. Six-year in-vitro reliability results of the HeartWare HVAD pump. *ASAIO J*. 2014;60(5):541–544.
97. Rich JD, Burkhoff D. HVAD flow waveform morphologies: theoretical foundation and implications for clinical practice. *ASAIO J*. 2017;63(5):526–535.
98. Gross C, Schima H, Schlöglhofer T, et al. Continuous LVAD monitoring reveals high suction rates in clinically stable outpatients. *Artif Organs*. 2020;44(7):E251–E262.
99. Reyes C, Voskoboynikov N, Chorpenning K, et al. Accuracy of the HVAD pump flow estimation algorithm. *ASAIO J*. 2016;62(1):15–19.
100. Zimpfer D, Strueber M, Aigner P, et al. Evaluation of the HeartWare ventricular assist device Lavare cycle in a particle image velocimetry model and in clinical practice. *Eur J Cardiothorac Surg*. 2016;50(5):839–848.
101. Wieselthaler GM, Driscoll GO, Jansz P, Khaghani A, Strueber M. Initial clinical experience with a novel left ventricular assist device with a magnetically levitated rotor in a multi-institutional trial. *J Heart Lung Transpl*. 2010;29(11):1218–1225.
102. Strueber M, O'Driscoll G, Jansz P, Khaghani A, Levy WC, Wieselthaler GM. Multicenter evaluation of an intrapericardial left ventricular assist system. *J Am Coll Cardiol*. 2011;57(12):1375–1382.
103. Aaronson KD, Slaughter MS, Miller LW, et al. Use of an intrapericardial, continuous-flow, centrifugal pump in patients awaiting heart transplantation. *Circulation*. 2012;125(25):3191–3200.
104. Slaughter MS, Pagani FD, McGee EC, et al. HeartWare ventricular assist system for bridge to transplant: combined results of the bridge to transplant and continued access protocol trial. *J Heart Lung Transpl*. 2013;32(7):675–683.
105. Strueber M, Larbalestier R, Jansz P, et al. Results of the post-market Registry to Evaluate the HeartWare Left Ventricular Assist System (ReVOLVE). *J Heart Lung Transpl*. 2014;33(5):486–491.
106. Rogers JG, Pagani FD, Tatooles AJ, et al. Intrapericardial left ventricular assist device for advanced heart failure. *N Engl J Med*. 2017;376(5):451–460.
107. Milano CA, Rogers JG, Tatooles AJ, et al. HVAD: the ENDURANCE supplemental trial. *JACC Heart Fail*. 2018;6(9):792–802.
108. Thamsen B, Blümel B, Schaller J, et al. Numerical analysis of blood damage potential of the HeartMate II and HeartWare HVAD rotary blood pumps. *Artif Organs*. 2015;39(8):651–659.
109. Thamsen B, Plamondon M, Granegger M, et al. Investigation of the axial gap clearance in a hydrodynamic-passive magnetically levitated rotary blood pump using X-Ray radiography. *Artif Organs*. 2018;42(5):510–515.
110. Chen Z, Jena SK, Giridharan GA, et al. Shear stress and blood trauma under constant and pulse-modulated speed CF-VAD operations: CFD analysis of the HVAD. *Med Biol Eng Comput*. 2019;57(4):807–818.
111. Birschmann I, Dittrich M, Eller T, et al. Ambient hemolysis and activation of coagulation is different between HeartMate II and HeartWare left ventricular assist devices. *J Heart Lung Transpl*. 2014;33(1):80–87.
112. Meyer AL, Malehsa D, Budde U, Bara C, Haverich A, Strueber M. Acquired von Willebrand syndrome in patients with a centrifugal or axial continuous flow left ventricular assist device. *JACC Heart Fail*. 2014;2(2):141–145.
113. Mondal NK, Sorensen EN, Feller ED, Pham SM, Griffith BP, Wu ZJ. Comparison of intraplatelet reactive oxygen species, mitochondrial damage, and platelet apoptosis after implantation of three continuous flow left ventricular assist devices: HeartMate II, Jarvik 2000, and HeartWare. *ASAIO J*. 2015;61(3):244–252.

114. Potapov EV, Stepanenko A, Kaufmann F, et al. Thrombosis and cable damage in the HeartWare pump: clinical decisions and surgical technique. *ASAIO J*. 2013;59(1):37–40.
115. Najjar SS, Slaughter MS, Pagani FD, et al. An analysis of pump thrombus events in patients in the HeartWare ADVANCE bridge to transplant and continued access protocol trial. *J Heart Lung Transpl*. 2014;33(1):23–34.
116. Glass CH, Christakis A, Fishbein GA, et al. Thrombus on the inflow cannula of the HeartWare HVAD: an update. *Cardiovasc Pathol*. 2019;38:14–20.
117. Stulak JM, Dunlay SM, Sharma S, et al. Treatment of device thrombus in the HeartWare HVAD: success and outcomes depend significantly on the initial treatment strategy. *J Heart Lung Transpl*. 2015;34(12):1535–1541.
118. Bhagra S, Bhagra C, Özalp F, et al. Development of de novo aortic valve incompetence in patients with the continuous-flow HeartWare ventricular assist device. *J Heart Lung Transpl*. 2016;35(3):312–319.
119. da Rocha ESJG, Meyer AL, Eifert S, Garbade J, Mohr FW, Strueber M. Influence of aortic valve opening in patients with aortic insufficiency after left ventricular assist device implantation. *Eur J Cardiothorac Surg*. 2016;49(3):784–787.
120. Lalonde SD, Alba AC, Rigobon A, et al. Clinical differences between continuous flow ventricular assist devices: a comparison between HeartMate II and HeartWare HVAD. *J Card Surg*. 2013;28(5):604–610.
121. Gaffey AC, Chen CW, Chung JJ, et al. Is there a difference in bleeding after left ventricular assist device implant: centrifugal versus axial? *J Cardiothorac Surg*. 2018;13(1):22.
122. Saeed D, Maxhera B, Sadat N, Petrov G, Albert A, Lichtenberg A. Gastrointestinal bleeding in patients with heartware ventricular assist device: does the activation of the lavare cycle make a difference? *ASAIO J*. 2018;64(1):126–128.
123. Goldstein DJ, Aaronson KD, Tatooles AJ, et al. Gastrointestinal bleeding in recipients of the HeartWare ventricular assist system. *JACC: Heart Fail*. 2015;3(4):303–313.
124. Krabatsch T, Hennig E, Stepanenko A, et al. Evaluation of the HeartWare HVAD centrifugal pump for right ventricular assistance in an in vitro model. *ASAIO J*. 2011;57(3):183–187.
125. Tran HA, Pollema TL, Silva Enciso J, et al. Durable biventricular support using right atrial placement of the HeartWare HVAD. *ASAIO J*. 2018;64(3):323–327.
126. Marasco S, Simon AR, Tsui S, et al. International experience using a durable, centrifugal-flow ventricular assist device for biventricular support. *J Heart Lung Transpl*. 2020;39(12):1372–1379.
127. Maynes EJ, O'Malley TJ, Luc JGY, et al. Comparison of SynCardia total artificial heart and HeartWare HVAD biventricular support for management of biventricular heart failure: a systematic review and meta-analysis. *Ann Cardiothorac Surg*. 2020;9(2):69–80.
128. Maynes EJ, O'Malley TJ, Patel P, et al. Right atrial versus right ventricular HeartWare HVAD position in patients on biventricular HeartWare HVAD support: a systematic review. *Artif Organs*. 2020;44(9):926–934.
129. Cheung A, Lamarche Y, Kaan A, et al. Off-pump implantation of the heartware HVAD left ventricular assist device through minimally invasive incisions. *Ann Thorac Surg*. 2011;91(4):1294–1296.
130. Popov AF, Hosseini MT, Zych B, Simon AR, Bahrami T. HeartWare left ventricular assist device implantation through bilateral anterior thoracotomy. *Ann Thorac Surg*. 2012;93(2):674–676.
131. Schmitto JD, Molitoris U, Haverich A, Strueber M. Implantation of a centrifugal pump as a left ventricular assist device through a novel, minimized approach: upper hemisternotomy combined with anterolateral thoracotomy. *J Thorac Cardiovasc Surg*. 2012;143(2):511–513.
132. Maltais S, Anwer LA, Tchantchaleishvili V, et al. Left lateral thoracotomy for centrifugal continuous-flow left ventricular assist device placement: an analysis from the mechanical circulatory support research network. *ASAIO J*. 2018;64(6):715–720.
133. McGee Jr. E, Danter M, Strueber M, et al. Evaluation of a lateral thoracotomy implant approach for a centrifugal-flow left ventricular assist device: the LATERAL clinical trial. *J Heart Lung Transpl*. 2019;38(4):344–351.
134. Olmsted RZ, Critsinelis A, Kurihara C, et al. Severe LVAD-related infections requiring surgical treatment: incidence, predictors, effect on survival, and impact of device selection. *J Card Surg*. 2019;34(2):82–91.
135. Bourque K, Cotter C, Dague C, et al. Design rationale and preclinical evaluation of the HeartMate 3 left ventricular assist system for hemocompatibility. *ASAIO J*. 2016;62(4):375–383.
136. Schmitto JD, Hanke JS, Rojas SV, Avsar M, Haverich A. First implantation in man of a new magnetically levitated left ventricular assist device (HeartMate III). *J Heart Lung Transpl*. 2015;34(6):858–860.
137. Hanke JS, Dogan G, Rojas SV, et al. First experiences with HeartMate 3 follow-up and adverse events. *J Thorac Cardiovasc Surg*. 2017;154(1):173–178.
138. Krabatsch T, Netuka I, Schmitto JD, et al. Heartmate 3 fully magnetically levitated left ventricular assist device for the treatment of advanced heart failure - 1 year results from the CE mark trial. *J Cardiothorac Surg*. 2017;12(1):23.
139. Gustafsson F, Shaw S, Lavee J, et al. Six-month outcomes after treatment of advanced heart failure with a full magnetically levitated continuous flow left ventricular assist device: report from the ELEVATE registry. *Eur Heart J*. 2018;39(37):3454–3460.
140. Mehra MR, Goldstein DJ, Uriel N, et al. Two-year outcomes with a magnetically levitated cardiac pump in heart failure. *N Engl J Med*. 2018;378(15):1386–1395.
141. Schmitto JD, Krabatsch T, Damme L, Netuka I. Less invasive HeartMate 3 left ventricular assist device implantation. *J Thorac Dis*. 2018;10(Suppl. 15):S1692–S1695.
142. Wiegmann L, Thamsen B, de Zélicourt D, et al. Fluid dynamics in the HeartMate 3: influence of the artificial pulse feature and residual cardiac pulsation. *Artif Organs*. 2019;43(4):363–376.

143. Boraschi A, Bozzi S, Thamsen B, et al. Thrombotic risk of rotor speed modulation regimes of contemporary centrifugal continuous-flow left ventricular assist devices. *ASAIO J*. 2020;.
144. Colombo PC, Mehra MR, Goldstein DJ, et al. Comprehensive analysis of stroke in the long-term cohort of the MOMENTUM 3 study. *Circulation*. 2019;139(2):155–168.
145. Li S, Beckman JA, Cheng R, et al. Comparison of neurologic event rates among HeartMate II, HeartMate 3, and HVAD. *ASAIO J*. 2020;66 (6):620–624.
146. Bansal A, Uriel N, Colombo PC, et al. Effects of a fully magnetically levitated centrifugal-flow or axial-flow left ventricular assist device on von Willebrand factor: a prospective multicenter clinical trial. *J Heart Lung Transpl*. 2019;38(8):806–816.
147. Klaeske K, Dieterlen MT, Scholz U, et al. Acquired von Willebrand factor deficiency is reduced in HeartMate 3 patients†. *Eur J Cardiothorac Surg*. 2019;56(3):444–450.
148. Saeed O, Colombo PC, Mehra MR, et al. Effect of aspirin dose on hemocompatibility-related outcomes with a magnetically levitated left ventricular assist device: an analysis from the MOMENTUM 3 study. *J Heart Lung Transpl*. 2020;39(6):518–525.
149. Netuka I, Iválk P, Tučanová Z, et al. Evaluation of low-intensity anti-coagulation with a fully magnetically levitated centrifugal-flow circulatory pump-the MAGENTUM 1 study. *J Heart Lung Transpl*. 2018;37(5):579–586.
150. Bansal A, Akhtar F, Zwintscher NP, Ostrow S, Desai S, Fary D. Use of the Heartmate 3 for biventricular support as a bridge to heart transplant-first US implant. *J Card Surg*. 2019;34(12):1629–1631.
151. Potapov EV, Kukucka M, Falk V, Krabatsch T. Biventricular support using 2 HeartMate 3 pumps. *J Heart Lung Transpl*. 2016;35 (10):1268–1270.
152. Lavee J, Mulzer J, Krabatsch T, et al. An international multicenter experience of biventricular support with HeartMate 3 ventricular assist systems. *J Heart Lung Transpl*. 2018;37(12):1399–1402.
153. Carrozzini M, Bejko J, Gerosa G, Bottio T. Bilateral mini-thoracotomy approach for minimally invasive implantation of HeartMate 3. *Artif Organs*. 2019;43(6):593–595.
154. Wood KL, Ayers BC, Sagebin F, et al. Complete sternal-sparing HeartMate 3 implantation: a case series of 10 consecutive patients. *Ann Thorac Surg*. 2019;107(4):1160–1165.
155. Gosev I, Wood K, Ayers B, et al. Implantation of a fully magnetically levitated left ventricular assist device using a sternal-sparing surgical technique. *J Heart Lung Transpl*. 2020;39(1):37–44.
156. Riebandt J, Schlöglhofer T, Moayedifar R, et al. Less invasive left ventricular assist device implantation is safe and reduces intraoperative blood product use: a propensity score analysis VAD implantation techniques and blood product use. *ASAIO J*. 2020;.
157. Motomura T, Tuzun E, Yamazaki K, Tatsumi E, Benkowski R, Yamazaki S. Preclinical evaluation of the EVAHEART 2 centrifugal left ventricular assist device in bovines. *ASAIO J*. 2019;65(8):845–854.
158. Saito S, Yamazaki K, Nishinaka T, et al. Post-approval study of a highly pulsed, low-shear-rate, continuous-flow, left ventricular assist device, EVAHEART: a Japanese multicenter study using J-MACS. *J Heart Lung Transpl*. 2014;33(6):599–608.
159. Motomura T, Kelly B. Surgical techniques for implanting the EVAHEART 2 double cuff tipless inflow cannula. *ASAIO J*. 2019;65(8): e86–e89.
160. Imamura T, Kinugawa K. Centrifugal pump EVAHEART prevents development of aortic insufficiency preserving pulse pressure. *Int Heart J*. 2016;57(1):127–128.
161. Bartoli CR, Kang J, Zhang D, et al. Left ventricular assist device design reduces von willebrand factor degradation: a comparative study between the HeartMate II and the EVAHEART left ventricular assist system. *Ann Thorac Surg*. 2017;103(4):1239–1244.
162. Zayat R, Moza A, Grottke O, et al. In vitro comparison of the hemocompatibility of two centrifugal left ventricular assist devices. *J Thorac Cardiovasc Surg*. 2019;157(2):591–599. e594.
163. Ichihara Y, Nishinaka T, Komagamine M, Yamada Y, Yamazaki K. Preservation of von Willebrand factor multimers and function in patients with an EVAHEART centrifugal-type, continuous-flow left ventricular assist device. *J Heart Lung Transpl*. 2017;36(7):814–817.
164. Matsumoto Y, Fujita T, Fukushima S, et al. Comparison of hemodynamic performance and clinical results with EVAHEART versus HeartMate II. *ASAIO J*. 2017;63(5):562–567.
165. May-Newman K, Moon J, Ramesh V, et al. The effect of inflow cannula length on the intraventricular flow field: an in vitro flow visualization study using the evaheart left ventricular assist device. *ASAIO J*. 2017;63(5):592–603.
166. May-Newman K, Montes R, Campos J, et al. Reducing regional flow stasis and improving intraventricular hemodynamics with a tipless inflow cannula design: an in vitro flow visualization study using the EVAHEART LVAD. *Artif Organs*. 2019;43(9):834–848.
167. Kawamura A, Toda K, Sawa Y. First clinical experience with the double cuff tipless inflow cannula in the EVAHEART left ventricular assist system: case report. *Artif Organs*. 2020;44(4):436–437.

Chapter 7

Research and Development for Creating a Universal Ventricular Assist Device

Anthony R. Polakowski[1], Takuma Miyamoto[1], Kiyotaka Fukamachi[1,2] and Jamshid H. Karimov[1,2]

[1]*Lerner Research Institute, Cleveland Clinic, Cleveland, OH, United States,* [2]*Cleveland Clinic Lerner College of Medicine of Case Western Reserve University, Cleveland, OH, United States*

Chapter Outline

Background

Heart failure is a serious health issue and a primary contributor to cardiovascular mortality that affects approximately 26 million people worldwide.[1] The incidence of heart failure is also rising rapidly in developing nations. Heart transplantation can provide a remarkable improvement in quality of life and survival in selected patients with end-stage heart disease,[2,3] but a shortage of donor hearts always limits this option.[4] In the United States, an estimated 50,000–100,000 patients are in need of heart transplantation or mechanical circulatory support (MCS), such as ventricular assist devices (VADs), but transplantation is performed in approximately 3600 cases per year. Instead, such support is usually provided by left ventricular assist devices (LVADs), the most widely used means for improving the hemodynamic status of both transplant-eligible and noneligible patients.[3] The rate of patient enrollment has grown to exceed 2500 implants per year for durable MCS devices such as LVADs and total artificial hearts. Device strategy is changing such that the number of LVADs implanted as destination therapy has continued to grow. However, as support duration increases, clinical limitations and specific complications of continuous-flow LVADs have become apparent. These include a loss of arterial pulsatility,[4,5] development of aortic regurgitation,[6] pump thrombosis or hemolysis,[7,8] severe backflow during pump stoppage, difficulty evaluating native heart function and weanability,[9,10] and acute or late right heart failure.[11,12]

Biventricular failure represents an extremely ill population of patients who have end-stage heart disease that involves both ventricles (even if its initial cause was limited to left-sided heart disease).[13] There is a high incidence of adverse events and poor 1-year survival in the patients with biventricular assist devices.[14,15] With no linkage to any specific pump type, this condition continues to limit the long-term success of durable MCS therapies, and so demands timely intervention to sustain deteriorating hemodynamics and to facilitate symptomatic improvement.[15]

Despite major breakthrough with continuous-flow MCS technology in recent years, major efforts were made to deliver LVADs to clinical practice. However, there are still no durable right ventricular assist devices (RVADs) available to address right heart failure. Existing VADs are not specifically designed for right ventricular support, which is limited by LVAD operational range.[16] We are developing an innovative, automatic, universal ventricular assist device (UVAD), intended for biventricular support, or for individual ventricular use as an RVAD or LVAD (suitable for either right or left heart assist), which is designed to address performance limitations by introducing an automatic and

Advances in Cardiovascular Technology. DOI: https://doi.org/10.1016/B978-0-12-816861-5.00039-3

dynamic pump regulation property also found in our continuous-flow total artificial heart.[17] This pump's novel feature is that the rotor can move axially to regulate the impeller discharge aperture size, depending on the pressure difference between the inflow and outflow and pump speed. This feature makes the pump more sensitive to pressure differences in the pump, allowing the performance to dynamically change during the cardiac cycle.

In this chapter, we describe UVAD development and report the in vitro and in vivo device performance evaluations, the results of computational fluid dynamics (CFD). and provide an overview of the research and engineering processes that have been instrumental in the fabrication of advanced device prototype and iterations, testing the biocompatibility of unique UVAD architecture in a range of simulated hemodynamic conditions.

Technology overview

UVAD is a continuous-flow centrifugal pump (Fig. 7.1). The rotor is free to axially move in and out of the housing cavity in response to system pressures, thereby opening and closing an aperture at the impeller discharge.[1] The strong magnetic force provides torque and spins the rotor, and a weaker magnetic force acts on the rotating assembly in the axial direction to close the aperture at the impeller outside diameter when power is interrupted. The hydraulic forces created by the pump rotation act in the opposite direction to open it. Thus, if the pump rotation stops, the hydraulic forces disappear, and the axial magnetic force causes the impeller discharge aperture to automatically close, inhibiting backflow of blood through the pump. Through passive modulation of the pump aperture opening, the dynamic coupling of pump capacity is automatically and precisely timed to the action of the ventricle.[2]

A cross-section of the pump is shown in Fig. 7.2. The pump is 44 mm in diameter and 57 mm long, with a total pump displacement volume of 70 mL and a priming volume of 15 mL. This magnetically and hydrodynamically levitated centrifugal pump has a performance range of 2000–3500 rpm and produces flows up to 10 L/min. As shown in Fig. 7.2, the rotor can move axially, depending on the pressure difference in the pump (pump differential pressure, dP) and rotational speed. A weak magnetic force acts to close the aperture, and the hydraulic forces during rotor rotation act to open it. This small but dynamic axial movement of the rotor changes the performance in response to the cardiac cycle. During diastole (pump dP is high), the rotor moves toward the aperture and decreases the flow. During systole (pump dP is low), the rotor moves against the aperture and increases the flow. In this way, the UVAD is expected to provide increased pulsatility of the pump flow and pressure (pulse augmentation). When the rotor stops, the hydraulic forces diminish, and the magnetic force immediately moves the rotor toward the aperture (regurgitant flow shut-off). If the pump runs in a lower-pressure environment such as the pulmonary circulation, the aperture opening is automatically reduced, and the entire pump performance is decreased to provide adequate flow to the pulmonary circulation (RVAD use).

Although the overall design of the UVAD is uncomplicated, refinements do exist and will be incorporated in the next phases of device development. This chapter will describe future design improvements for the UVAD, which will be discussed in detail and grouped as they relate to the three subassemblies: Volute Housing, Rotor, and Stator. Preliminary results from CFD analyses and in vivo studies will also be covered.

Volute housing enhancements

The Volute Housing envelops the other two subassemblies, provides the basic shape of the UVAD, and provides inflow and outflow connecting ports.

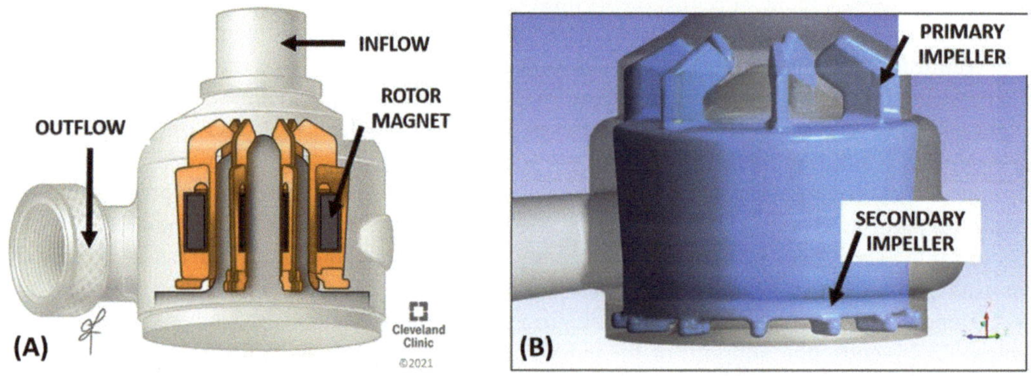

FIGURE 7.1 Universal ventricular assist device basics. Universal ventricular assist device basics. (A) Cross section. (B) Impeller regions.

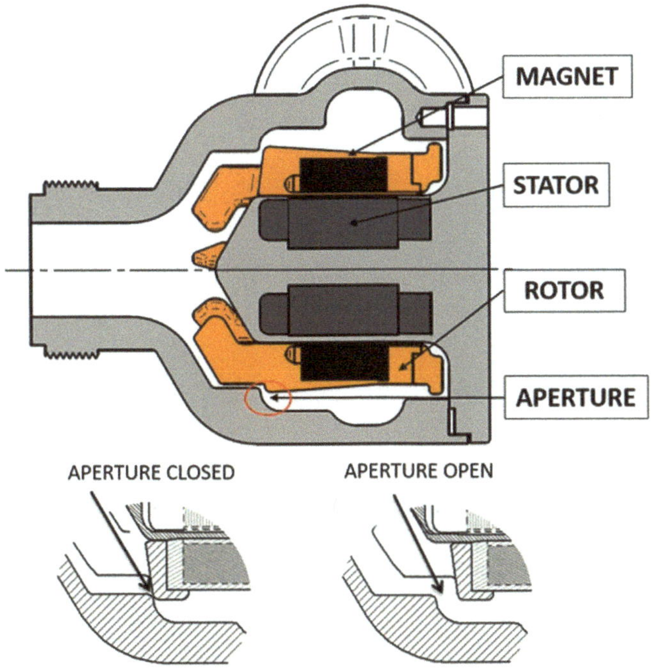

FIGURE 7.2 Cross section and aperture details.

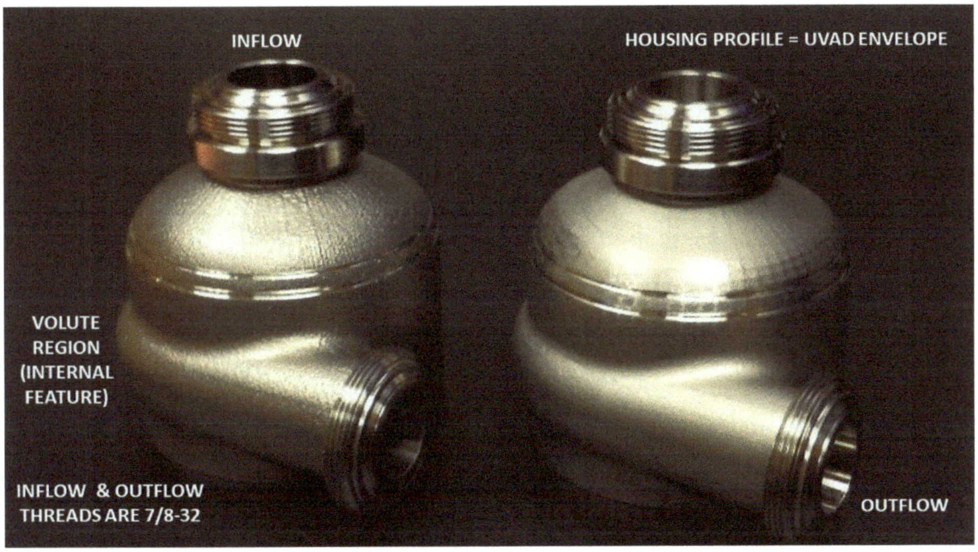

FIGURE 7.3 Latest volute housing.

To prevent and minimize tissue damage, overall device weight should be reduced when possible. Minimizing wall thicknesses and shortening the protrusions for the inflow and the outflow ports will be prioritized. Some areas of this housing currently have wall thicknesses exceeding 0.165 in (4.2 mm), which can be reduced to a safe, nominal thickness of 0.050 in (1.3 mm). This design improvement will result in a significant weight reduction.

Because of its complex exterior and internal volute surfaces, traditional machining methods were not used to produce the overall geometric shape of the Volute Housing. Instead, 3D-printing of this component in titanium was performed, which resulted in acceptable parts. In Fig. 7.3, the left-hand housing was printed during an earlier build, while the right-hand housing was printed more recently. Both were printed by Protolabs (Maple Plain, MN) using the same high-resolution, 20 μm layer option. With improvements in 3D-printing technology over the years, utilizing the normal-resolution, 30-μm layer option will be explored to reduce cost while producing new housings, which should prove satisfactory for development work. Fig. 7.3 shows the improved smoother surface finishes of the newer (right-hand) housing's outer diameter and the outflow protrusion feature. These two features are shown as 3D-printed and not machined.

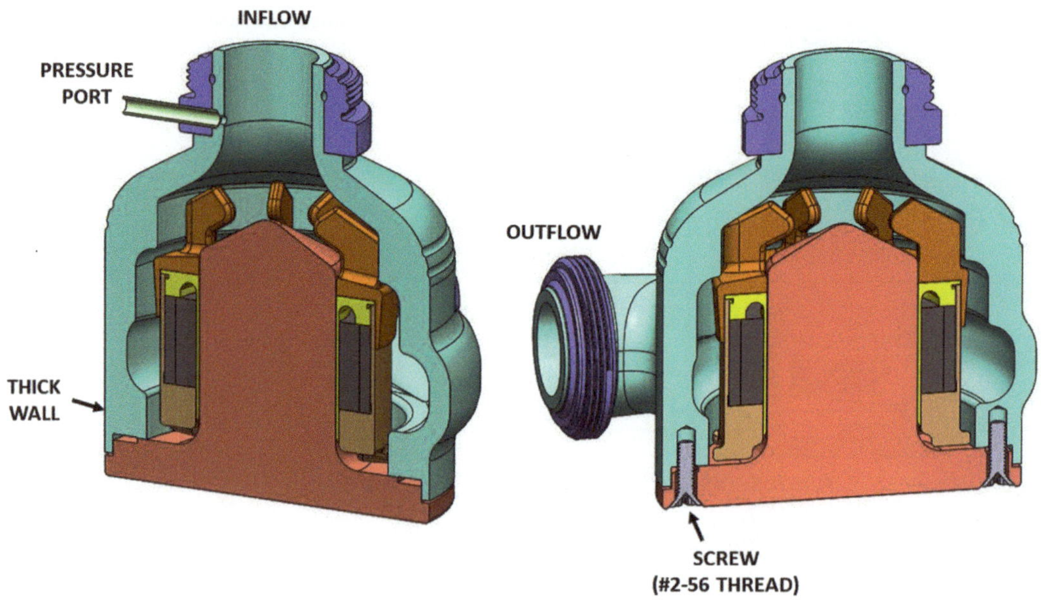

FIGURE 7.4 Assembly cross sections.

To reduce the possibility of blood clots, special attention is required to create a smooth surface finish for the internal volute feature. Being a volute, this surface is a spiral in nature and doesn't have a constant radius. In the past, polishing the volute was found to be time-consuming and expensive, and was accomplished using smaller, powered hand tools. For the next batch of Volute Housings, a proprietary micromachining process will be employed by MicroTek Finishing (Cincinnati, OH) to produce a mirror-like finish on this difficult and challenging internal feature.

During pump development, pressure readings at points along the flow path can be taken and monitored to determine pump performance. In Fig. 7.4, a pressure port is shown in the inflow area upstream of the rotor/stator cavity. Because of the simplicity of the UVAD and a project goal to minimize the overall size, numerous pressure ports are not feasible, but pressure ports at the inflow and outflow will be considered for inclusion. After rudimentary CFD analysis of this device, another point or two in the vicinity of the rotor cavity may be identified as points of interest, in which to add ports for pressure monitoring.

Another area for improvement involves the attachment of the Stator to the Volute Housing. In the current design iteration, six screws are employed to join these two subassemblies together, along with a dowel pin for alignment. Though the screws are small (# 2–56 threads), it may be observed that the radial wall thickness becomes fairly thick (0.225″, 5.7 mm) in this region of the Volute Housing (Fig. 7.4). With the current fastener orientation (screws are parallel to the axis of the UVAD), this thicker wall is translated over an axial length in excess of 0.205″ (5.2 mm). This juncture will be reviewed to see if an off-the-shelf retaining ring can be incorporated. Or, another approach is to include mechanical "catch" features to align and join the Volute Housing and Stator. With rapid prototyping (3D-printing), multiple technical solutions may be trialed, which will result in substantial device optimization in a short time.

Improvements to rotating assembly

The latest rotors produced consisted of a magnet ring encased by three plastic 3D-printed components with an option for shim washers for performance adjustment (Fig. 7.5).

In the next phase of this device's development, titanium components will replace the printed plastic material in order to improve product durability. Fig. 7.6 shows that combining components numbered 3 and 4 may be possible, or some other combination of design features that result in the magnet ring being "sandwiched" between a top and bottom rotor piece. It may be possible to employ traditional machining methods for one component, and produce the other component by 3D-printing with titanium depending on the complexity of each part.

As mentioned previously, shim washers (placed on the top face of component # 3 and below component # 4; Fig. 7.6) may be used to adjust pump performance. By adjusting the length of the rotor, the aperture size (Fig. 7.2) is influenced and controlled as the rotor operates axially. This aperture feature acts as a metering mechanism, increasing

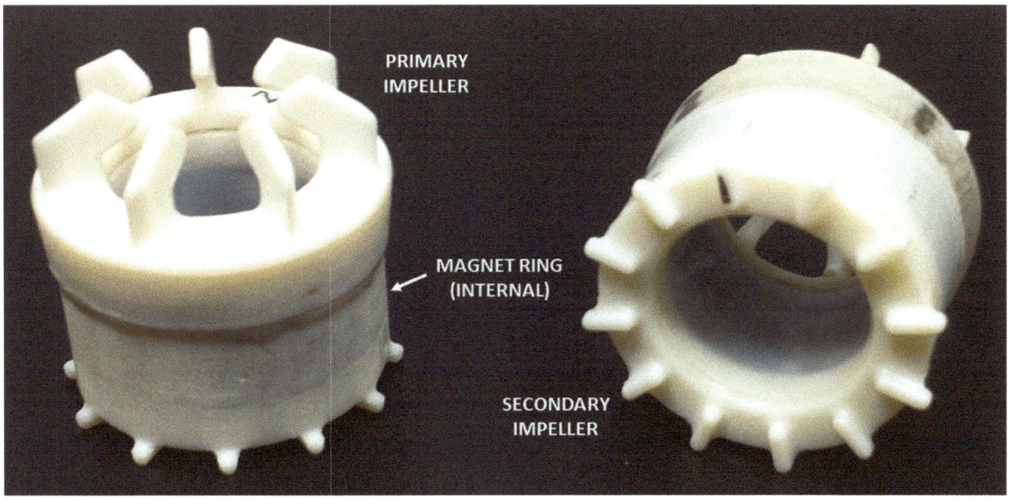

FIGURE 7.5 Latest rotor.

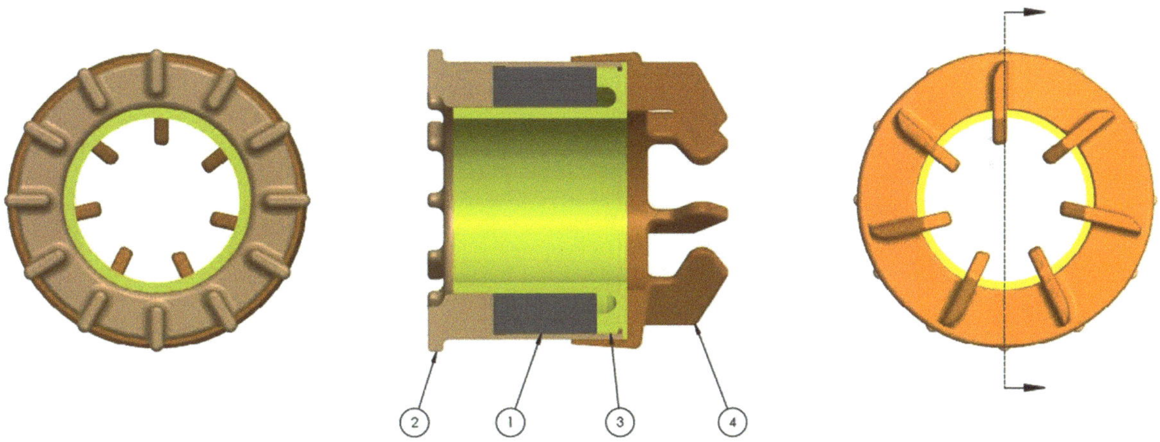

FIGURE 7.6 Rotor cross section.

or decreasing the flow through the UVAD as needed by the patient. With a comprehensive tolerance stackup analysis, a dimensional scheme may be determined to eliminate the need for this shimming process.

One more topic for future rotor fabrication is the technique when joining the top and bottom rotor pieces. In the current rendition the various components are fastened together using epoxy, which has proven reliable during UVAD operation. However, if a different impeller design (at either end) were to be tested, or if some feature needed to be reworked, the epoxy does not allow for easy disassembly or rework of the subassembly. A possible solution is having threads incorporated into the top and bottom shells, and other simple fastening methods can be considered.

Engineering and optimization of motor stator

The Stator is a fairly simple and straightforward structure (Fig. 7.7) consisting of motor coils occupying the core of a central protrusion feature, a structural flange for attachment to the Volute Housing, and a driveline that provides device power and control. Due to this simplicity, the probable mechanical changes are limited to a couple of areas.

First, as discussed in the Volute Housing section, a different attachment strategy would alter the general appearance of this subassembly. If mechanical "catch" features, or a retaining ring, were used, the fastener holes would be designed out, and the flange could be thinner (to eliminate extra material and reduce weight). As stated previously, many design solutions can be explored.

Second, the driveline's size and location will be reviewed. Reducing the diameter of the driveline (and number of electrical signals) to its bare minimum is a final optimizing step. As for driveline location, the axial orientation (being

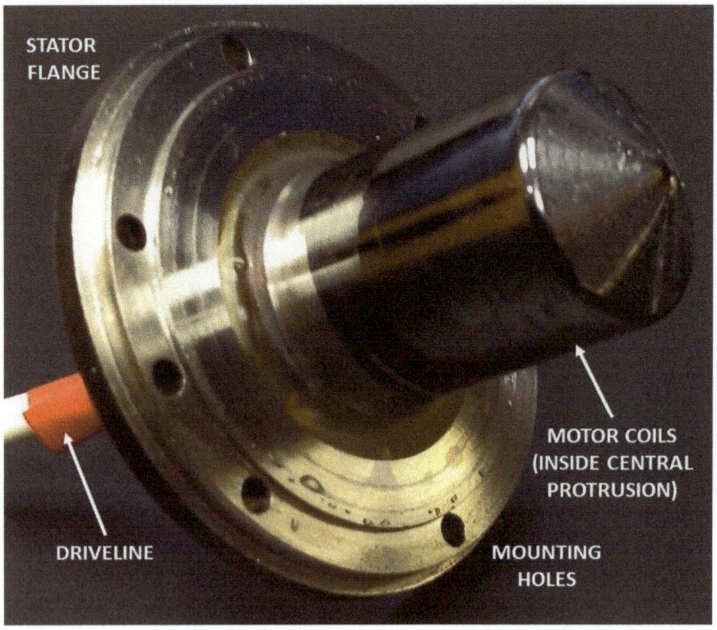

FIGURE 7.7 Latest stator.

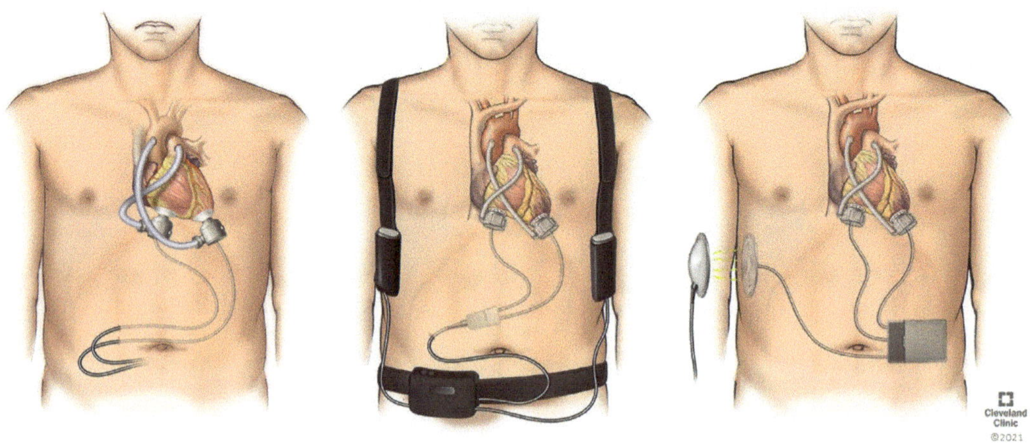

FIGURE 7.8 Drive line exteriorization paths.

positioned away from the inflow and outflow conduits) is arguably an ideal location for testing this device, but is it the best configuration when implanted in a patient's chest? As a continuous improvement effort, our surgical and engineering teams will review the positions and orientations of the inflow, outflow, and driveline to ensure a proper, and hopefully the least obtrusive, fit inside the chest cavity of the patient. In Fig. 7.8, three system configurations are illustrated. Depending on the hardware involved, there could be one or two exteriorization paths leading to the power source and user interface. Ultimately, if a wireless charging option is implanted, then no exteriorization would be needed.

Computational fluid dynamics analysis

The initial focus of the CFD study was to create and validate a CFD model of the UVAD design by comparing predicted pump performance with in vitro measurements under steady state flow conditions. ANSYS R18.2 (Canonsburg, PA) software was used for these simulations; specific ANSYS software tools included DesignModeler, Meshing, CFX, and CFD-Post. To reduce the influence of the applied boundary conditions, 6-inch long inflow and outflow extensions were added to the pump CFD model (Fig. 7.9). Leveraging the cylindrical symmetry of the pump housing surrounding both impellers, a multiframes of reference "frozen rotor" model was applied for these steady state simulations.

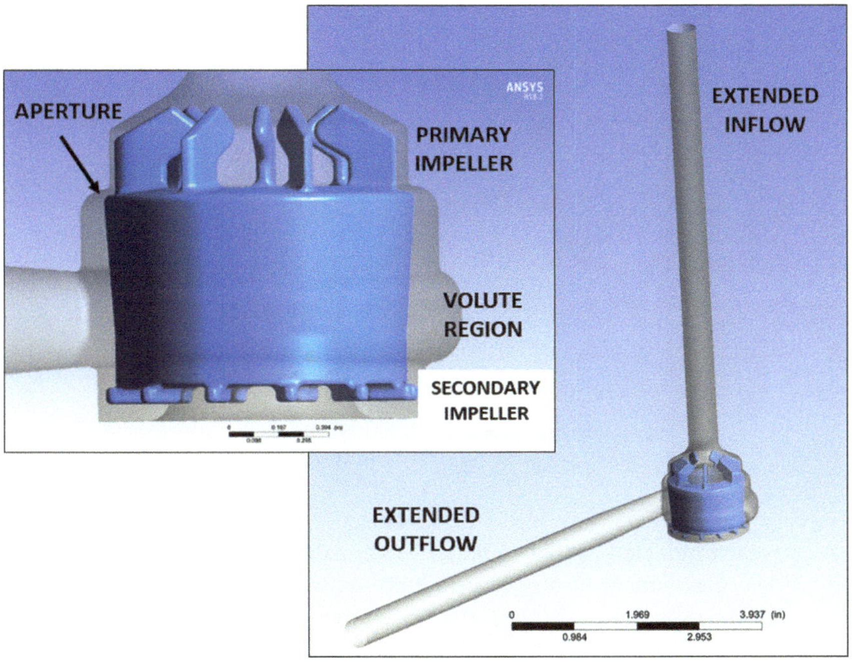

FIGURE 7.9 Computational fluid dynamics model with extended inflow and outflow.

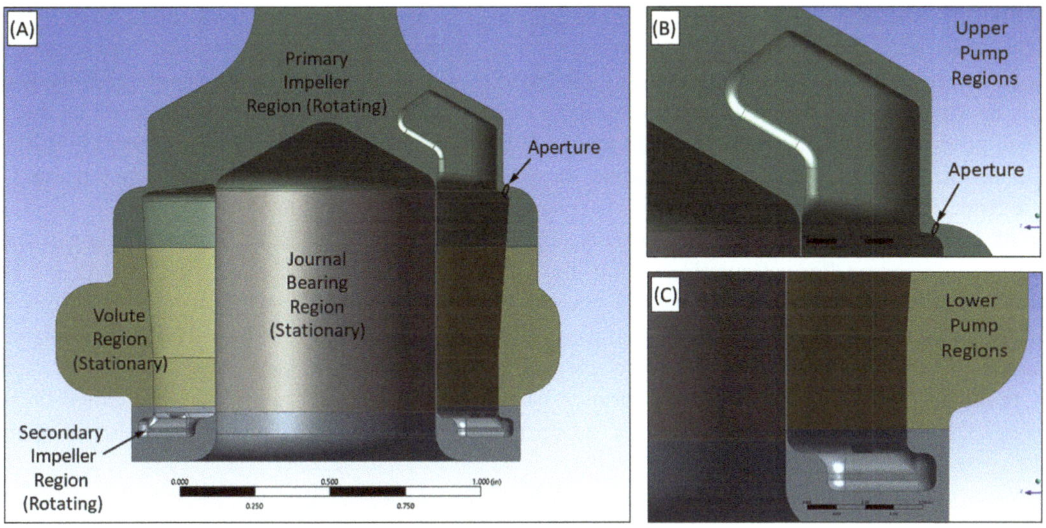

FIGURE 7.10 Cross section of different flow regions with close-up views of primary and secondary impeller regions. Cross section of different flow regions (A) with close-up views of primary (B) and secondary (C) impeller regions.

The pump fluid volume was extracted from the three-dimensional UVAD model and separated into rotating and nonrotating flow domains (Fig. 7.10A–C). Key to this effort was developing a CFD model where the different flow domains would automatically adjust and shift in response to changes in rotor axial position and aperture size. Tetrahedral mesh elements with eight prism inflation layers along all the walls were used in the volute and both impeller regions. Hexahedral mesh elements were used for the cylindrical-shaped journal bearing flow path and for the inflow and outflow extensions. Images of this refined mesh model near the primary and secondary impellers are shown in Fig. 7.11A and B. The area averaged mesh y + value, which is a dimensionless distance from the wall based upon the local wall shear stress, on the impeller surfaces was 1.2, with a maximum value of 2.8. These y + values were for highest shear condition, with a flow rate of 8 L/min and impeller rotational speed of 3600 rpm (Fig. 7.11C). Such near-wall mesh resolution is intended to accurately capture the flow field, velocity gradients, and hence shear stress, along the walls.[18]

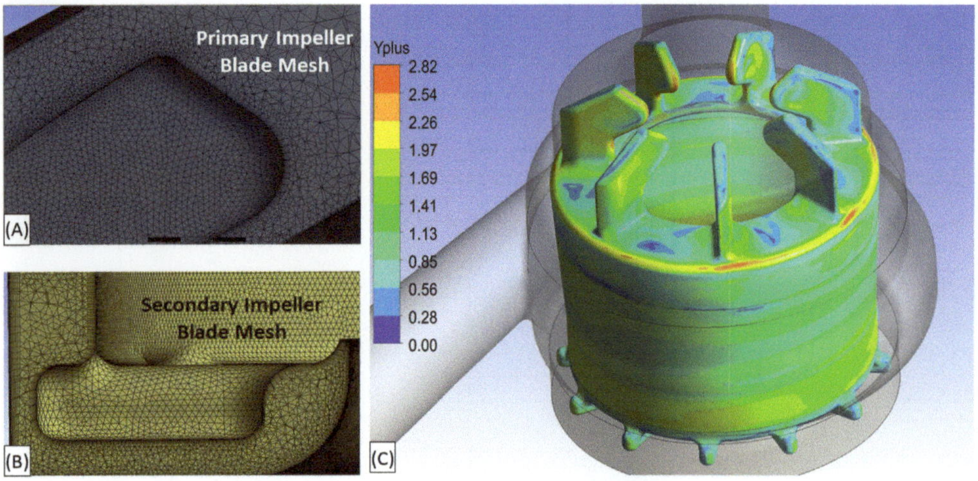

FIGURE 7.11 Mesh in primary and secondary impeller regions and contour plot. Mesh in primary (A) and secondary (B) impeller regions and contour plot (C).

For the validation study, three flow rates were each modeled at four impeller rotational speeds spanning the pump's intended range of use under both RVAD and LVAD operation. A water/glycerin mixture, with fluid properties matching the in vitro test conditions ($\rho = 1060\ \text{kg/m}^3$; $\mu = 2.03$ centipoise), was used for this study. The effects of turbulent flow within the pump were captured using the Reynolds Averaged Navier–Stokes shear stress transport $k-\omega$ turbulence model.[19] An average inflow static pressure and an outflow mass flow rate, matching the in vitro test conditions, were used as boundary conditions. The axial location of the rotating assembly, which was estimated photographically through a clear pump housing, varied by only ±0.003″ across the range of flow conditions measured. As this degree of variation was comparable with the axial measurement resolution, a position was chosen with the rotating assembly at the average axial location for the data collected. A concentric radial location for the rotating assembly was used due to the very small bearing clearance with limited potential radial displacement of the rotor relative to the overall pump dimensions. The predicted hydraulic pressure head and wall static pressures, at six positions throughout the pump, were compared with in vitro experimental data to validate the CFD model. The CFD solutions were run on 12- or 16-core workstations at SimuTech Group (Hudson, OH).

A density of $1060\ \text{kg/m}^3$ and a Cross non-Newtonian viscosity model[20,21] with an infinite shear viscosity of 3.5 centipoise, was used for blood flow simulations. First, the pump's flow fields under nominal RVAD and LVAD operating conditions were compared. Then, the change in pump performance with varying rotor axial positions was explored. The intent was to evaluate the sensitivity of the pump's hydraulic performance to the aperture size. Lastly, as the pump housing was intentionally designed with a large clearance to ensure sufficient room for rotor axial movement, the impact of reducing the secondary impeller clearance was evaluated. An additional variable called "residence time," calculated using a volumetric user scalar with a source term, was included in the secondary impeller clearance simulations.

CFD model validation

The CFD-predicted hydraulic pressure rise agreed well with the in vitro measured data, within 6.5% at 2300 rpm and within 3.5% for the higher rotor speeds. A plot comparing the hydraulic pressure rise test data with the CFD results is shown in Fig. 7.12. Typical ranges for the flow rates and pressure increases under both RVAD and LVAD operation are also shown in this Fig. 7.12. The wall static pressures, taken at six positions throughout the pump (Fig. 7.13A), also agreed well with experimental data. The largest differences occurred in the primary impeller region with the CFD results under-predicting the experimental values by less than 7% for all flow/speed conditions other than at 7 L/min and 3100 rpm. At that condition, the CFD-predicted primary impeller region pressures were 10% lower than the measured values. Fig. 7.13B–D show the good agreement between the CFD results and test data at the pressure tap locations over the wide range of flow/speed conditions studied. For the pressure taps in the primary and secondary impeller regions, the static pressure values from the CFD model were average values around the perimeter of the pump at the axial location of the pressure tap. The intent was to "average out" the impact of the impeller blade position relative to the pressure tap location and, in doing so, better mimic the measured "time-averaged" pressure values at those locations.

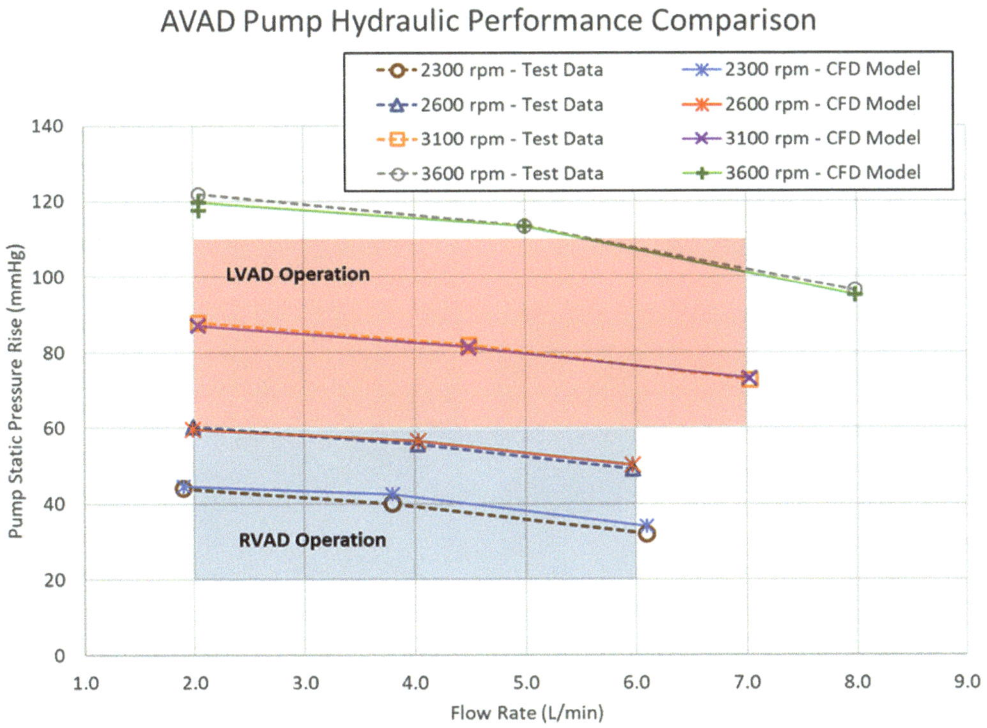

FIGURE 7.12 Comparison of hydraulic performance curves.

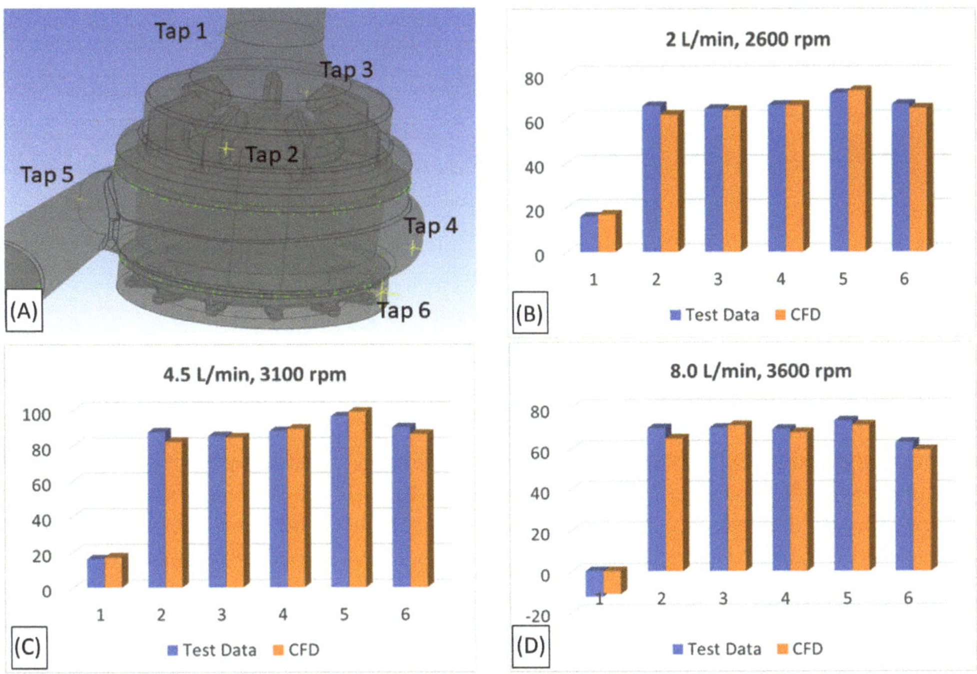

FIGURE 7.13 Pressure tap locations and static pressures at six locations with different flow / speed conditions. Pressure tap locations (A) and static pressures (B-D) at six locations with different flow / speed conditions.

RVAD and LVAD operation—similarity of flow patterns

The intent of the aperture and the offset volute are to enable the operation of the UVAD under both RVAD and LVAD conditions. This is accomplished by providing similar flow patterns through each of the primary impeller blade passages, achieving uniformly distributed static pressures circumferentially on the impeller surfaces, and reducing flow recirculation within the volute over the wide range of RVAD/LVAD operating conditions. The images in Fig. 7.14A–C show the variation in velocity in the stationary frame on a plane cutting through the primary impeller blade for the

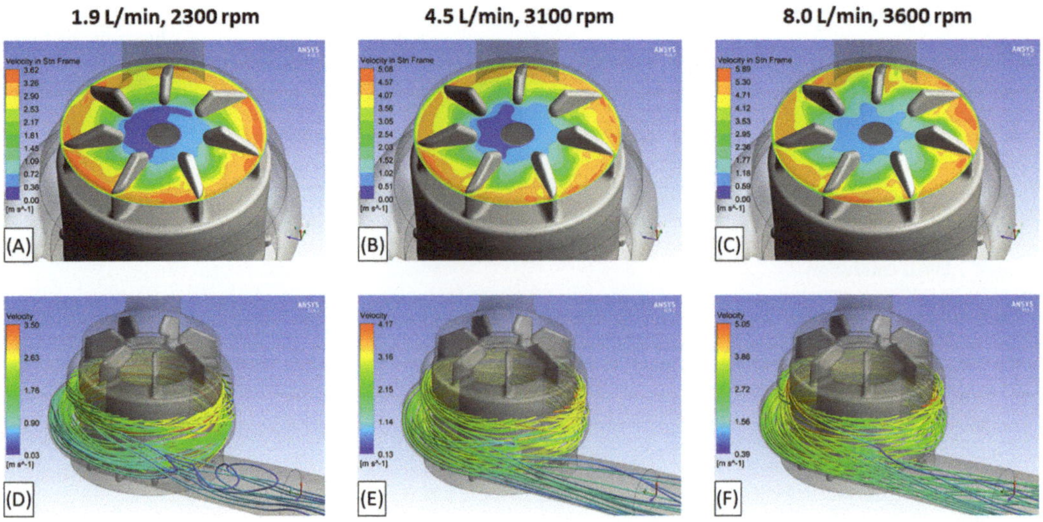

FIGURE 7.14 Contour plots of velocity (A–C) and flow streamlines (D–F).

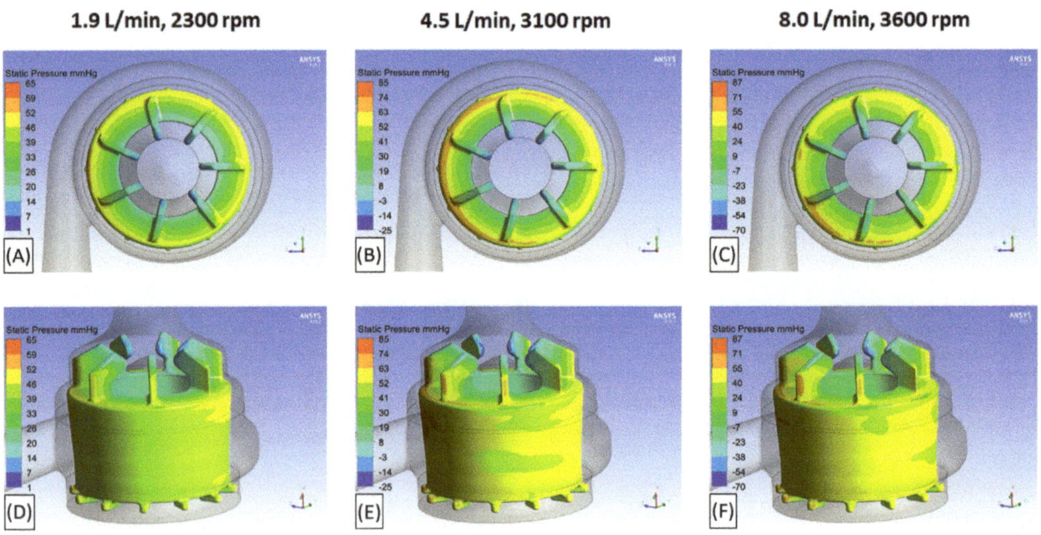

FIGURE 7.15 Contour plots of static pressure at different flow/speed conditions. Contour plots of static pressure at different flow / speed conditions. (A–C) For primary impeller. (D–F) For rotating assembly.

lowest, middle, and maximum operating conditions. The velocity contours show very good circumferential similarity for the blood flowing through the impeller blade passages. The images in Fig. 7.14D–F show the downstream path for streamlines released at the interface between the primary impeller and volute regions. The flow patterns within the volute are again quite similar over this wide range of flow conditions. In addition, the absence of separated, recirculating flow regions within the volute regions is quite promising.

Static pressure contour plots on the primary impeller surfaces (Fig. 7.15A–C) and the rotating assembly surfaces (Fig. 7.15D–F) show that some variation in static pressure occurs on the surfaces nearest to the outflow diffuser. However, these pressure variations are small and localized, helping to create a more uniform flow circumferentially around the pump. The more circumferentially uniform flow field should reduce the extent of isolated, recirculating flow regions within the offset volute region that could lead to thrombus formation.[22]

RVAD versus LVAD comparison—nominal flow conditions

In addition to comparing the flow fields within the UVAD at a nominal LVAD and RVAD flow rate of 4.5 L/min, the effect of using blood vs. the water/glycerin mixture was also explored. The CFD results presented in Fig. 7.16 summarize the findings from this study.

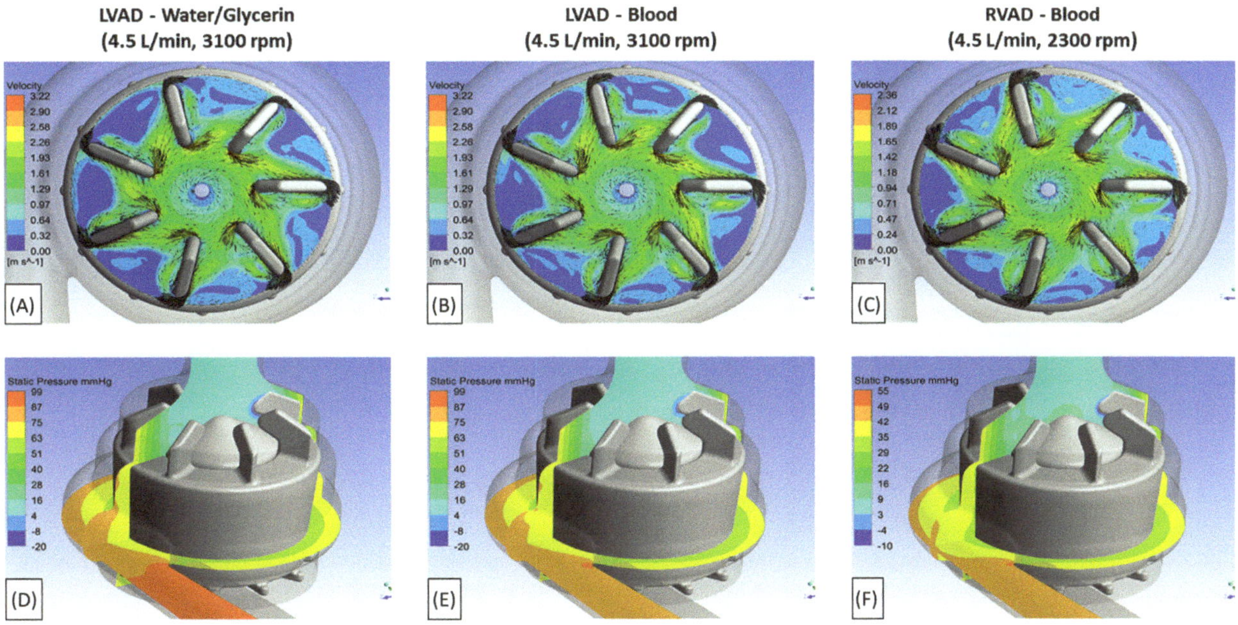

FIGURE 7.16 Contour plots of relative velocity (A–C) and static pressure (D–F) with different fluid and ventricular assist device conditions.

The upper set of images in Fig. 7.16A–C present velocity contours and tangential velocity vectors on a plane cutting through the primary impeller region. The velocity contours and vectors are relative to the rotating primary impeller. The velocity contours for the LVAD operation, at 3100 rpm, are extremely similar between the water/glycerin and blood solutions. Both solutions show higher relative speeds near the center cone, where the flow enters, and decreasing relative flow velocity within the blade passages, where the fluid is rotating with the blades. The flow patterns at the lower RVAD rotational speed of 2300 rpm are very similar to the LVAD results. For all these simulations, significant blade leading edge-induced flow separation occurs and so reducing the blade leading edge flow incidence angle was identified as a design improvement for the next-generation UVAD design.

The lower set of images in Fig. 7.16D–F show the variation in static pressure throughout the UVAD and exiting through the diffuser. The water/glycerin solution provided a slightly higher outflow static pressure than that for the blood simulation, 89.2 mm Hg versus 87.5 mm Hg. The increased blood viscosity only caused a small reduction in outflow pressure because the flow within the pump is largely turbulent, and the inertial forces (velocity times density) are characteristically dominant over the viscous forces. Here again, a strong circumferential similarity and uniformity of the pressure fields exists for the UVAD operating as both an LVAD and an RVAD.

Impact of rotor axial position

The axial position of the rotating assembly was moved from 0.0″ to 0.089″, opening the aperture as shown in Fig. 7.17A and B. The goal for this work was to better understand the sensitivity of the pump's performance to varying the rotor axial location and aperture size. A blood flow rate of 5 L/min and rotor speed of 3400 rpm was used for this study. Fig. 7.17C shows the variation in pump total pressure rise and net rotor axial force with rotor axial position. The pump total pressure rise was relatively insensitive (92–98 mm Hg) to rotor axial position over a wide range of rotor axial positions (0.015″–0.089″). As the aperture closes, the total pressure losses through the aperture increase causing the outflow pressure to decrease as intended. The net axial force acting on the rotor is also relatively stable, with values varying between −0.15 and −0.49 N, across a wide range of rotor axial positions. These results suggest that only when the aperture begins to close, in response to system pressures, will the UVAD's hydraulic performance and axial position change significantly. Such consistency in performance and rotor axial stability, with the aperture open, was one of the initial UVAD design goals.

Reduced secondary impeller clearance

In reviewing the CFD results from the nominal flow conditions study, a recirculating flow region was observed near the secondary impeller. A design modification was therefore proposed with the axial clearance in the secondary impeller

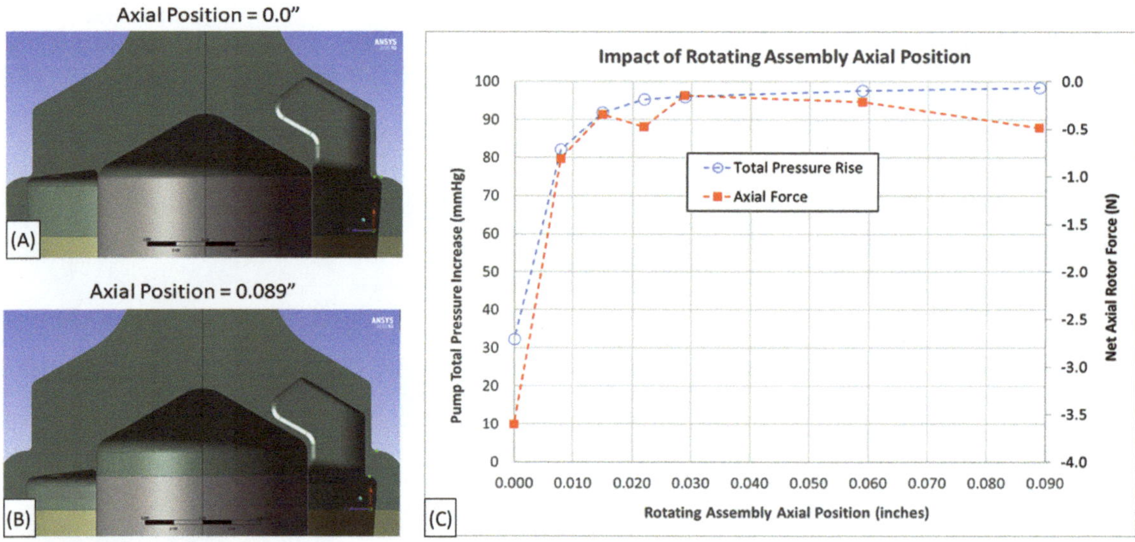

FIGURE 7.17 Cross sections with rotating assembly at two axial positions and plot of pressure rise and axial rotor force versus axial position. Cross sections (A-B) with rotating assembly at two axial positions and plot (C) of pressure rise and axial rotor force versus axial position.

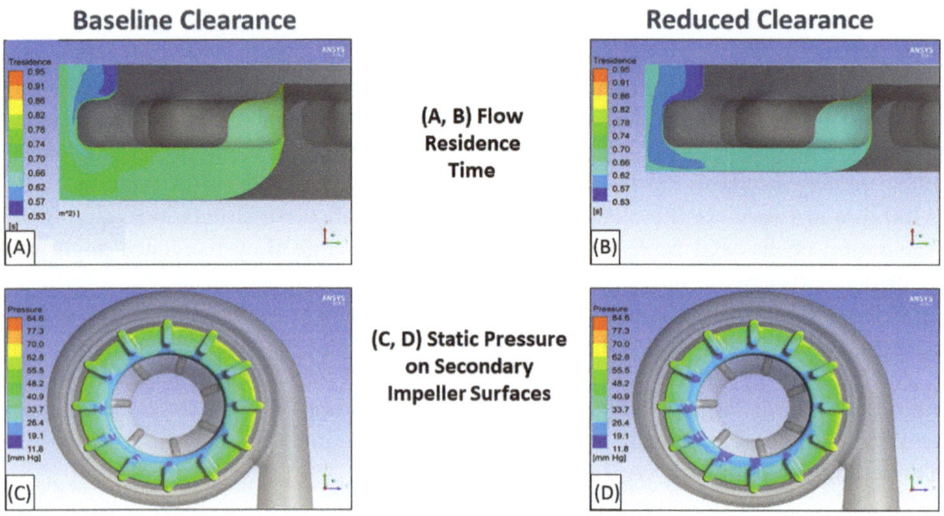

FIGURE 7.18 Contour plots of residence time (A, B) and static pressure (C, D).

region reduced by 0.394″ (1 mm). The intent was twofold: (1) to remove excess flow volume in that region, to reduce the extent of the observed flow recirculation; and (2) to reduce the secondary impeller pressure, encouraging increased rotor axial movement. Such increased sensitivity in axial movement with rotor speed would help differentiate LVAD from RVAD operation. A blood flow rate of 4.5 L/min, rotor speed of 3100 rpm, and a rotor axial location of 0.022″ was used for this study. The smaller secondary clearance reduced the volume averaged residence time in the secondary impeller region by 10%. A lower average flow residence time indicates improved wash-out with less flow recirculation near the secondary impeller. Such improved flow patterns should reduce the likelihood of thrombus forming near the secondary impeller.[22] Contour plots comparing flow residence times within the secondary impeller regions are shown in Fig. 7.18A and B. With the aperture isolating the two impeller regions, the reduced secondary clearance had negligible impact on the pump's hydraulic pressure rise (−0.2 mm Hg). Reducing the clearance increases the local speed along the inner surfaces of the secondary impeller, specifically near the journal bearing connection. This localized speed increase is accompanied by a reduction in the static pressure near the inside of the secondary impeller as shown in Fig. 7.18C and D. Lower secondary impeller pressures reduce the rotor net axial force from −0.16 to −0.30 N, providing increased axial movement of the rotor in response to varying pressure differences. This increased responsiveness is desired and could improve the responsiveness and pulsatility in the pump's operation.

In vivo testing

The in vivo study was approved by the Cleveland Clinic's Institutional Animal Care and Use Committee, and all animals received humane care in compliance with the "Principles of Laboratory Animal Care" formulated by the National Society for Medical Research and the Guide for the Care and Use of Laboratory Animals. The pump components, including inflow cannulae (left cannula: straight, length = 69 mm, diameter = 14 mm; right cannula: angled, length = 125 mm, diameter = 14 mm) are shown in Fig. 7.19. A total of five acute in vivo studies were performed using male Jersey calves (mean body weight = 83.2 kg, range = 72.4–91.2 kg). With the calves under general anesthesia, full median sternotomy were performed. UVAD implantation was conducted without cardiopulmonary bypass (CPB) support. The outflow graft was anastomosed to the ascending aorta (for RVAD, to the pulmonary trunk) and the inflow cannula was inserted to the ventricular apex (for RVAD, through the right ventricular diaphragmatic surface). The device was activated at low speed and performance data was collected in various hemodynamic conditions.

The measured hemodynamic and pump performance data included the following: (1) heart rate; (2) central venous pressure (fluid-filled catheter at the jugular vein); (3) aortic pressure (AoP; fluid-filled catheter in the peripheral artery and on the outflow graft); (4) pulmonary artery pressure (PAP; fluid-filled catheter to the outflow graft, only for the RVAD study); (5) left ventricular pressure or pump inflow pressure (fluid-filled catheter); (6) pump flow (inline flow probe in the outflow conduit); (7) motor current (analog output from the control module); and (8) motor speed (analog output from the control module).

Hemodynamic and pump performance data were collected in the following manner:

1. *Speed range test.* Data were obtained at a baseline (without pump support) and at 2000, 2500, 3000, and 3500 rpm.
2. *Pump stoppage test.* The pump was stopped to evaluate the shut-off feature. The pump was run at 3000 rpm for the LVAD configuration and at 2000 or 2500 rpm for the RVAD configuration. The pump was stopped for ≥5 s using the controller, and then the outflow conduit was clamped. The hemodynamic parameters and regurgitation flow were recorded a total of 10 times.

Acute hemodynamic and pump performance parameters were digitized in real time at a sampling rate of 200 Hz with a PowerLab (ADInstruments Inc., Colorado Springs, CO) data acquisition system. The data were analyzed using a custom-made visual basic program in Excel software (Microsoft Corporation, Redmond, WA). The pulse pressure was calculated as (maximum AoP—minimum AoP).

The UVAD was successfully implanted as an LVAD ($n = 3$, A01, A02, and A05) and as an RVAD ($n = 2$, A03 and A04) without CPB support (Fig. 7.20A and B). There were no complications during the procedure, and anatomical fitting was acceptable. The speed range of the UVAD was 2000–3500 rpm as an LVAD and 2000–3000 rpm as an RVAD, and up to 4.3 L/min (1.1–4.3 L/min) was achieved for both configurations. The pump performance was compared with a bench test (Fig. 7.21). The pump performance was observed to be lower in vivo at high speed (3000 or 3500 rpm) due to some degree of inflow suction; the overall pump performance was shown to be consistent throughout the experiments with the values observed during preoperative in vitro device testing. Fig. 7.22 shows a representative pressure flow loop of the UVAD implanted as an LVAD. The widest pressure flow loop was observed at 2500 rpm, indicating the greatest pulsatility. The smallest loops were observed at 3000 and 3500 rpm.[23]

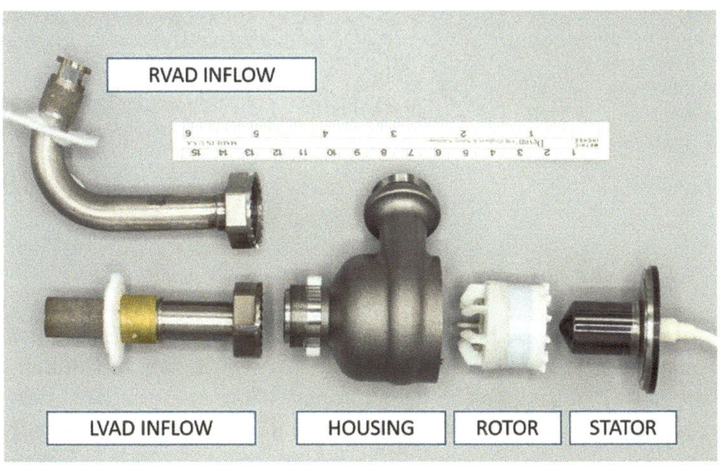

FIGURE 7.19 In vivo pump components.

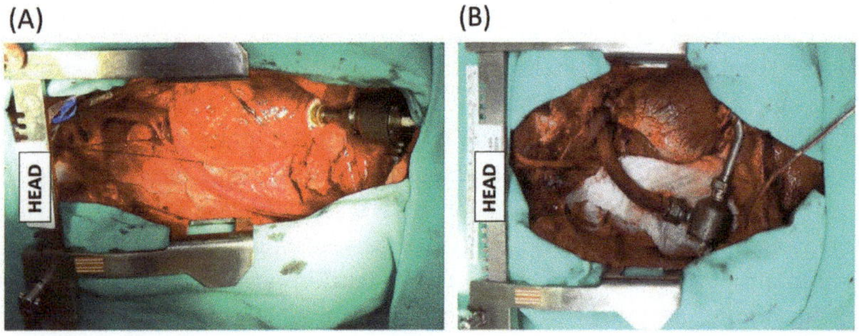

FIGURE 7.20 Implanted devices. (A) Left ventricular assist device. (B) Right ventricular assist device.

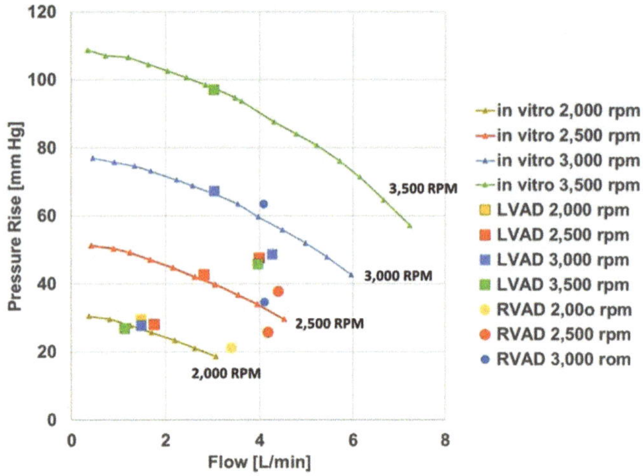

FIGURE 7.21 Pump performance test results.

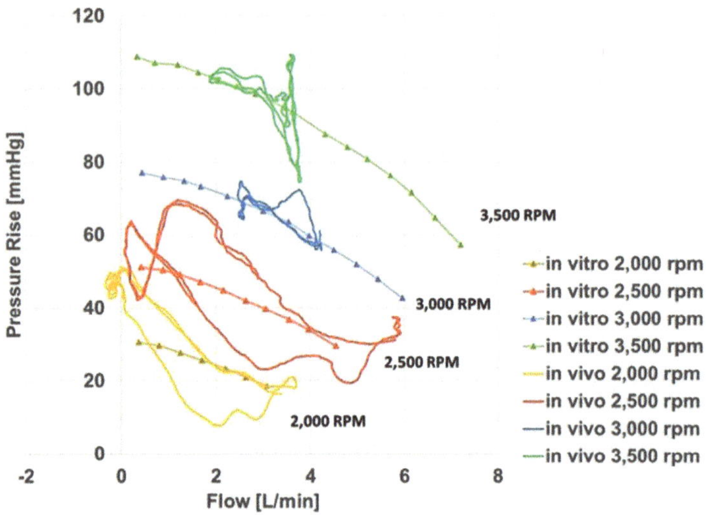

FIGURE 7.22 Pressure flow loops.

Summary

Because it is the largest component, significant weight reduction will be realized after optimizing the design and wall thicknesses of the Volute Housing. Polishing techniques and installing pressure ports as needed will be key to achieving exceptional pump performance values and monitoring our results during future development. Since it is the only moving part, developing a titanium Rotor will result in a more durable, longer-lasting device. Determining a better attachment strategy between the Volute Housing and Stator will require some thought and creativity, but will result in a more compact device envelope. In addition, a critical review of the UVAD's inflow, outflow, and driveline positions will be ongoing. With several exteriorization options available, how the device is implanted and handled during clinical use will be scrutinized. With a nice amount of CFD study completed, more in vivo testing can be planned with the intention that theoretical and clinical results can be compared and contrasted.

References

1. Ambrosy AP, Fonarow GC, Butler J, et al. The global health and economic burden of hospitalizations for heart failure: lessons learned from hospitalized heart failure registries. *J Am Coll Cardiol*. 2014;63:1123–1133.
2. Garbade J, Bittner HB, Barten MJ, Mohr FW. Current trends in implantable left ventricular assist devices. *Cardiol Res Pract*. 2011;2011:290561.
3. Wever-Pinzon O, Drakos SG, Kfoury AG, et al. Morbidity and mortality in heart transplant candidates supported with mechanical circulatory support: is reappraisal of the current United network for organ sharing thoracic organ allocation policy justified? *Circulation*. 2013;127: 452–462.
4. Moazami N, Hoercher KJ, Fukamachi K, et al. Mechanical circulatory support for heart failure: past, present and a look at the future. *Expert Rev Med Dev*. 2013;10:55–71.
5. Travis AR, Giridharan GA, Pantalos GM, et al. Vascular pulsatility in patients with a pulsatile- or continuous-flow ventricular assist device. *J Thorac Cardiovasc Surg*. 2007;133:517–524.
6. Pak SW, Uriel N, Takayama H, et al. Prevalence of de novo aortic insufficiency during long-term support with left ventricular assist devices. *J Heart Lung Transplant*. 2010;29:1172–1176.
7. Bartoli CR, Zhang D, Kang J, et al. Clinical and In vitro evidence that subclinical hemolysis contributes to LVAD thrombosis. *Ann Thorac Surg*. 2018;105:807–814.
8. Starling RC, Moazami N, Silvestry SC, et al. Unexpected abrupt increase in left ventricular assist device thrombosis. *N Engl J Med*. 2014;370:33–40.
9. Sunagawa G, Byram N, Karimov JH, et al. The contribution to hemodynamics even at very low pump speeds in the HVAD. *Ann Thorac Surg*. 2016;101:2260–2264.
10. Sunagawa G, Byram N, Karimov JH, et al. In vitro hemodynamic characterization of HeartMate II at 6000 rpm: implications for weaning and recovery. *J Thorac Cardiovasc Surg*. 2015;150:343–348.
11. Anderson M, Morris DL, Tang D, et al. Outcomes of patients with right ventricular failure requiring short-term hemodynamic support with the Impella RP device. *J Heart Lung Transplant*. 2018;37:1448–1458.
12. Takeda K, Takayama H, Colombo PC, et al. Late right heart failure during support with continuous-flow left ventricular assist devices adversely affects post-transplant outcome. *J Heart Lung Transplant*. 2015;34:667–674.
13. Dandel M, Krabatsch T, Falk V. Left ventricular vs. biventricular mechanical support: decision making and strategies for avoidance of right heart failure after left ventricular assist device implantation. *Int J Cardiol*. 2015;198:241–250.
14. Shah P, Ha R, Singh R, et al. Multicenter experience with durable biventricular assist devices. *J Heart Lung Transplant*. 2018;37:1093–1101.
15. Vierecke J, Gahl B, de By T, et al. Results of primary biventricular support: an analysis of data from the EUROMACS registry. *Eur J Cardiothorac Surg*. 2019;.
16. Karimov JH, Sunagawa G, Horvath D, Fukamachi K, Starling RC, Moazami N. Limitations to chronic right ventricular assist device support. *Ann Thorac Surg*. 2016;102:651–658.
17. Horvath DJ, Karimov JH, Byram NA, et al. Advantages of integrating pressure-regulating devices into mechanical circulatory support pumps. *ASAIO J*. 2019;65:e1–e3.
18. ANSYS-CFX Modeling Guide. Chapter 4, *Turbulence and Near-Wall Modeling, Modeling Flow Near the Wall*, Version R18.2.
19. Menter FR. Two-equation eddy-viscosity turbulence models for engineering applications. *AIAA J*. 1994;32(8).
20. Cross MM. Rheology on non-Newtonian fluids: a new flow equation for pseudoplastic systems. *J Colloid Sci*. 1965;20:417–437.
21. Banerjee RK, Cho YI, Kensey K. Effect of non-Newtonian viscosity of blood on steady and pulsatile flow in stenosed arteries. *Adv Bioeng*. 1991;20:103–106.
22. Chiu WC, Slepian M, Bluestein D. Thrombus formation patterns in the HeartMate II ventricular assist device: clinical observations can be predicted by numerical simulation. *ASAIO J*. 2014;60:237–240.
23 Miyamoto T, Kado Y, Horvath DJ, et al. An advanced universal circulatory assist device for left and right ventricular support: first report of an acute in vivo implant. *J Thorac Cardiovasc Surg Open*. 2020;3:140–148.

Chapter 8

Right Ventricular Assist Device

Jaime-Jürgen Eulert-Grehn[1,2], Christoph Starck[1,2] and Evgenij Potapov[1,2]

[1]Department of Cardiothoracic and Vascular Surgery, German Heart Center Berlin, Berlin, Germany, [2]DZHK (German Center for Cardiovascular Research), Partner Site Berlin, Germany

Chapter Outline

Background

Heart failure is a complex entity. The focus is mostly on the left ventricle.[1] This is reflected in the development of long-term ventricular assist devices (VADs), which are developed for high performance at the systemic level. Well-accepted definitions of right heart failure (RHF) are scarce, and so quantifying right ventricular ejection fractions is difficult.[2] Isolated RHF can have many causes, such as right ventricular involvement in myocardial infarction or RHF in acute pulmonary embolism. In addition, chronic pulmonary hypertension leads to remodeling of the right ventricle and ultimately can lead to RHF.

On the other hand, RHF can be a part of left ventricular failure-dominated heart failure. One treatment is left ventricular mechanical support. Although RHF post-continuous flow left ventricular assist device (cfLVAD) implantation can occur in approximately 14%–34% of patients, two randomized control studies (ENDURANCE and MOMENTUM) reported right ventricular assist device (RVAD) implantation rates of under <5%.[3,4] Kormos et al. reported a rate of 3.6% for simultaneous RVAD during isolated cfLVAD implantation. In this cohort of cfLVAD patients, 2% required a RVAD within 14 days and less than 1% required an RVAD after 14 days.[5]

The immediate postoperative period is the most crucial with regard to RHF post-cfLVAD implantation. The 2019 Society of Thoracic Surgeons (STS) INTERMACS Annual Report showed that RHF occurred predominantly in the first month, with a freedom from RHF of 62%–71% at 1 year. RHF represents not only a challenge regarding diagnosis and treatment, but is also the cause of death in 11% of patients with cfLVAD support.[6,7] Patients at INTERMACS 1 and 2 have two to three times the risk of RHF, with approximately 10% of INTERMACS 1 patients requiring an RVAD. Other risk factors are prior coronary artery bypass or heart valve surgery, INTERMACS 1 and 2, and extracorporeal membrane oxygenation (ECMO) within 48 h prior to cfLVAD implantation.[8]

Currently no device is specially tailored for support of the right ventricle and perfusion of the low-resistance pulmonary bed. Therefore, in the last 10 years, many groups around the world have tried to tailor cfLVAD devices to support the right ventricle. This combination is referred to as a continuous flow biventricular assist device (cfBVAD or cfBiVAD). With current pump designs, four main factors have to be kept in mind while implanting cfLVAD for right heart support:

1. the highly trabeculated right ventricle;
2. the smaller cavity and thinner muscle wall of the right ventricle;
3. the difference in vascular resistance; and

Advances in Cardiovascular Technology. DOI: https://doi.org/10.1016/B978-0-12-816861-5.00033-2

4. the pulmonary vascular bed as scavenger for thrombi.

In sum, we as clinicians face two challenges regarding patients with RHF: definition and treatment. For patients with RHF post-cfLVAD implantation, most registries and publications refer to the revised INTERMACS definition of RHF[9]:

Symptoms or findings of persistent right ventricular failure are characterized by *both* of the following:

1. Documentation of elevated central venous pressure (CVP) by:
    - Direct measurement (e.g., right heart catheterization) with evidence of a CVP or right atrial pressure (RAP) >16 mmHg;
      OR
    - Findings of a significantly dilated inferior vena cava with absence of inspiratory variation by echocardiography;
      OR
    - Clinical findings of elevated jugular venous distension at least halfway up the neck in an upright patient.
2. Manifestations of elevated CVP characterized by:
    - Clinical findings of peripheral edema (grade >2 + either new or unresolved);
      OR
    - Presence of ascites or palpable hepatomegaly on physical examination (unmistakable abdominal contour) or by diagnostic imaging;
      OR
    - Laboratory evidence of worsening hepatic (total bilirubin > 2.0 mg/dL) or renal dysfunction (creatinine > 2.0 mg/dL).

Right heart failure severity grades

Mild right heart failure

1. VAD Implant Admission
   Patient meets both criteria for RHF plus:
    - Postimplant inotropes, inhaled nitric oxide or intravenous vasodilators not continued beyond post-op day 7 following VAD implant;
      AND
    - No inotropes continued beyond post-op day 7 following VAD implant.
2. Surveillance periods (3 months, 6 months, 12 months, and every 6 months thereafter) following VAD implant
   Patient meets both criteria for RHF plus:
    - No readmissions for RHF since last surveillance period;
      AND
    - No inotropes since last surveillance period.

Moderate right heart failure

1. VAD Implant Admission
   Patient meets both criteria for RHF plus:
    - Postimplant inotropes, inhaled nitric oxide or intravenous vasodilators continued beyond post-op day 7 and up to post-op day 14 following VAD implant;
2. Surveillance periods (3 months, 6 months, 12 months, and every 6 months thereafter) following VAD implant.
   Patient meets both criteria for RHF plus:
    - Limited to one (1) readmission for intravenous diuretics/vasodilators to treat RHF since last surveillance period;
      AND
    - No inotropes since last surveillance period.

Severe right heart failure

1. VAD Implant Admission
   Patient meets both criteria for RHF plus:

- CVP or RAP greater than 16 mm Hg;
  AND
- Prolonged postimplant inotropes, inhaled nitric oxide or intravenous vasodilators continued beyond post-op day 14 following VAD implant.

2. Surveillance periods (3 months, 6 months, 12 months and every 6 months thereafter) following VAD implant
   Patient meets both criteria for RHF plus:
   - Need for inotropes at any time since last surveillance period;
     OR
   - Two (2) or more readmissions for intravenous diuretics/vasodilators to treat RHF since last surveillance period;
     OR
   - Requiring RVAD support at any time after hospital discharge;
     OR
   - Death at any time following discharge from the VAD implant hospitalization with RHF as the primary cause.

Severe-acute right heart failure

1. VAD Implant Admission
   Patient meets both criteria for RHF plus:
   - CVP or RAP greater than 16 mmHg;
     AND
   - Need for RVAD at any time following VAD implant;
     OR
   - Death during the VAD implants hospitalization with RHF as the primary cause.

As current risk models don't accurately predict post-cfLVAD RHF, it is important to take into account the chronicity.[10] The chronicity and reversibility of right ventricular dysfunction can preoperatively be judged by grade of chronic congestive dermatosis. Regarding RVAD implantation, the recent INTERMACS report shows that less than 0.1% of all implants are isolated RVADs.[5,6] Therefore, we will focus on implant strategies and outcomes of durable cfLVADs as RVADs for biventricular support. Of all the implanted continuous flow assist devices in the INTERMACS registry, approximately 3.5% have adjunct RVAD support.[5,6] The current continuous flow device available, the HeartMate 3, can be used for right ventricular support. Preoperative cardiac CT is required for procedural and access planning, especially in redo procedures.

Implant site and ring augmentation

The inflow cannula of continuous flow RVADs can be implanted either into the right atrium or ventricle. The right atrium has become the implantation side of choice.[11] Irrespective of whether the right atrium or ventricle is chosen, ring augmentation is most often necessary because of the thinner wall and the smaller cavities on the right side.

Furthermore, after implantation, a various amount of remodeling leads to a substantial decrease in dimensions, especially of the dilated right ventricle after the acute setting of right ventricular failure.[12] Especially because pump thrombosis is a major concern with continuous flow pumps on the right side, a careful implantation, considering implant site and intracavitary length of the inflow cannula, is of major importance. The inflow ring and cannula have to be adapted to the thin wall of the right atrium by augmentation with Teflon rings for extraluminal height reduction, with the goal of leaving around 5 mm of the inflow cannula protruding into the lumen.

The length of the inflow cannula of the HeartMate 3 is roughly 32 mm. This length is ideal for implantation into the thicker-walled and larger left ventricle, but not for the right side. With the height of the implantation ring of about 5 mm, and about 4 mm of thickness of the right ventricular free wall, the intracavitary length of the 32 mm long cannula would be about 21 mm. In this case, the cannula may touch the septum, resulting in partial or complete obstruction with consequent suction. Adding 4–6 layers of Teflon plates to the implantation ring reduces the intracavitary length of the cannula (Fig. 8.1A and B). This ring augmentation can be omitted in the case of a very big right atrium or ventricle, bearing in mind that the cavitary dimensions can decrease substantially post-RVAD implantation. This can increase the risk for suction events.

A hole has to be made in the glued Teflon layers (Fig. 8.1C). On the other hand, too little protrusion risks tissue overgrowth (Fig. 8.1D).

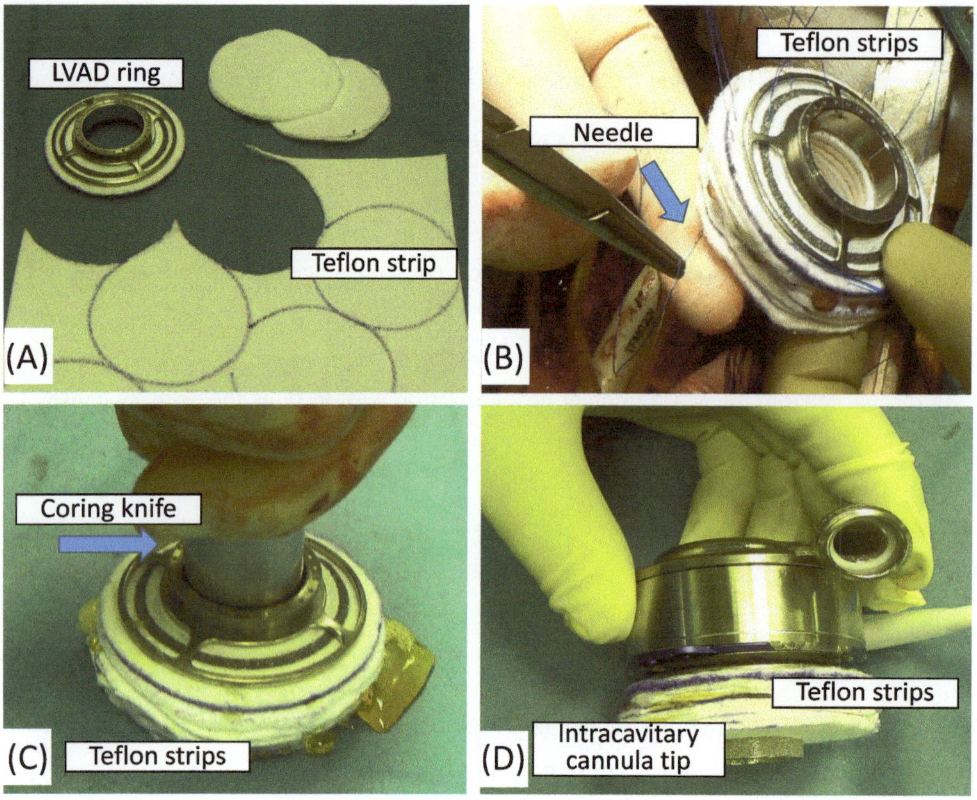

FIGURE 8.1 (A) Teflon stripes for ring augmentation. (B) Intraoperative picture of ring augmentation: the needle has to go through all 5–8 layers of Teflon stripes and the ring. (C) HeartMate 3 cutting knife (in this case used for a HeartMate 3 pump). (D) Inflow cannula.

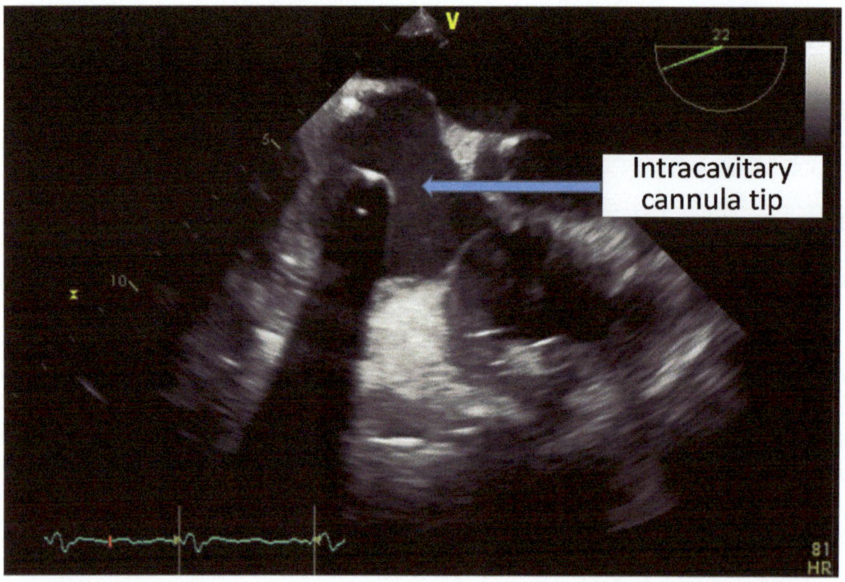

FIGURE 8.2 Transesophageal echocardiography showing a right atrial placement of the inflow cannula. Important to have sufficient distance to the neighboring structures (as seen in this figure) toward the interatrial septum and the tricuspid valve. The tip of the cannula is above the middle of the tricuspid valve.

The implant is done under transesophageal echocardiography guidance (Fig. 8.2) for two reasons:

1. sufficient distance to the inferior vena cava;
2. heading above the middle of the tricuspid valve.

The extracavitary section becomes larger, with a need of enough space between the chest wall and the right ventricle. This is one of the main reasons why the right atrial approach, as opposed to the ventricular implantation side, has gained more interest in the last years (Fig. 8.3).

The lack of space can be tackled by placing the pump extrapericardially towards the right pleural cavity. A pericardial patch is wrapped around the pump to avoid direct adhesions to the pump (Fig. 8.4). The pericardium adjacent to the right atrial wall can be left in place. The sutures have to be placed carefully to avoid damage to the phrenic nerve. This strategy is especially helpful in redo cases with intrapericardial adhesions. When appropriate, a temporary RVAD can be implanted and replaced with a continuous-flow RVAD later (Fig. 8.5).

Outflow graft narrowing

Outflow graft narrowing in length and diameter has been adopted from the pulmonary banding procedure in congenital heart surgery and was used for adopting the HeartWare device for right ventricular support. The fully magnetically levitated HeartMate 3 pump can be run safely at a low rotation speed. For this pump, no outflow graft narrowing is implemented.

Results

Goldstein et al. (ISHLT Mechanically Assisted Circulatory Support Registry) analysis reports a 12-month survival of isolated left ventricular assist devices (LVAD) of 82.9% and of simultaneous RVAD and LVAD of 58.5%. Interestingly, in the second year of support, about 10% of patients with isolated LVAD die, while about 5% of simultaneous RVAD and LVAD patients die. Looking in to various single and multicenter reports, three points are worth considering: (1) which device (HW or HeartMate 3); (2) implantation site; and (3) rate of pump thrombosis.[5]

One of the earliest larger reports of 13 patients with HeartWare Ventricula Assist Device (HVAD) as RVAD reported a rate of 50% pump thrombosis with the right ventricle as implant site. They then changed to the right atrial approach and observed one pump thrombosis in seven patients. An overall rate of 30% of pump thrombosis was observed.[11] Tran et al. observed in a right atrium-only cohort with HVAD as RVAD a pump thrombosis rate of 36%.[13] Shah et al., a recent multicenter report with a 50:50 distribution between atrial and ventricular implantation sites, reported a pump thrombosis rate of 37%.[14]

In the largest single center report to date from our center with 33 HVAD and six HeartMate 3, the pump thrombosis rate was 30%.[15] In most reports, HVAD was the predominant VAD. Lavee et al. examined in a multicenter report exclusively the results with HeartMate 3, in 12 out of 14 with a right atrial approach. One pump thrombosis was observed.[16] McGiffin et al. recently published their experience with HeartMate 3 as RVAD. The right atrium was used

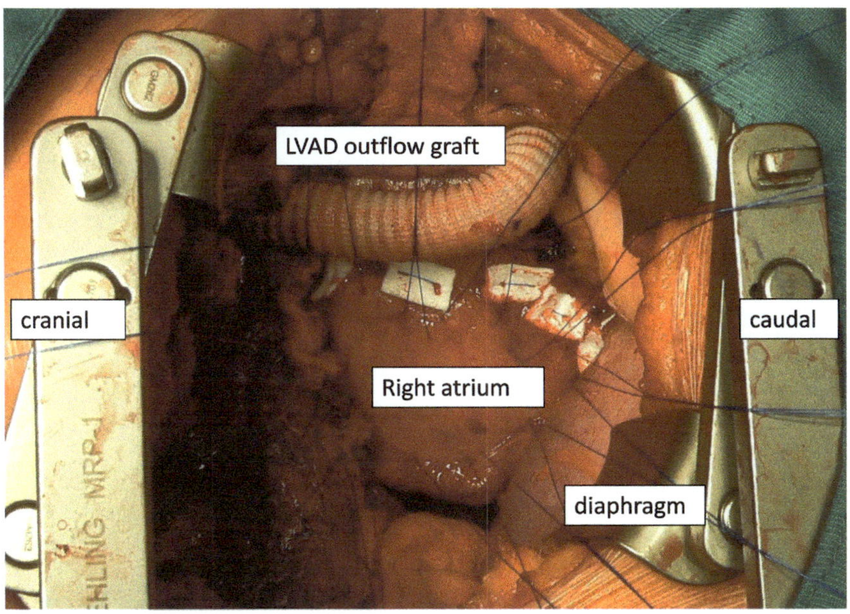

FIGURE 8.3 Placement of the right atrial sutures (access through right lateral thoracotomy).

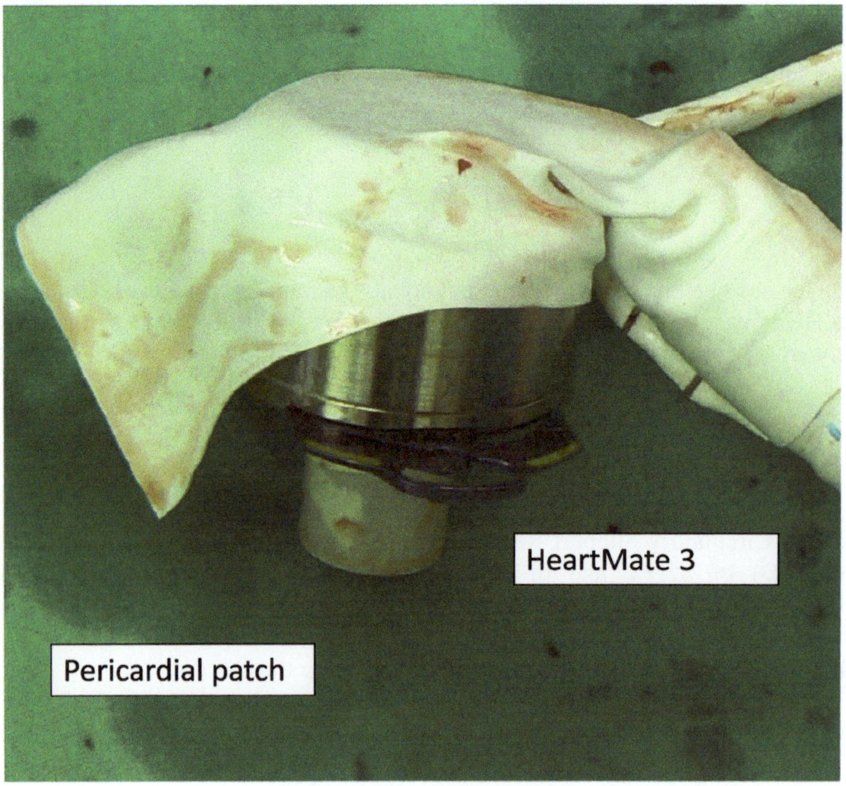

FIGURE 8.4 Pericardial covering of HM3.

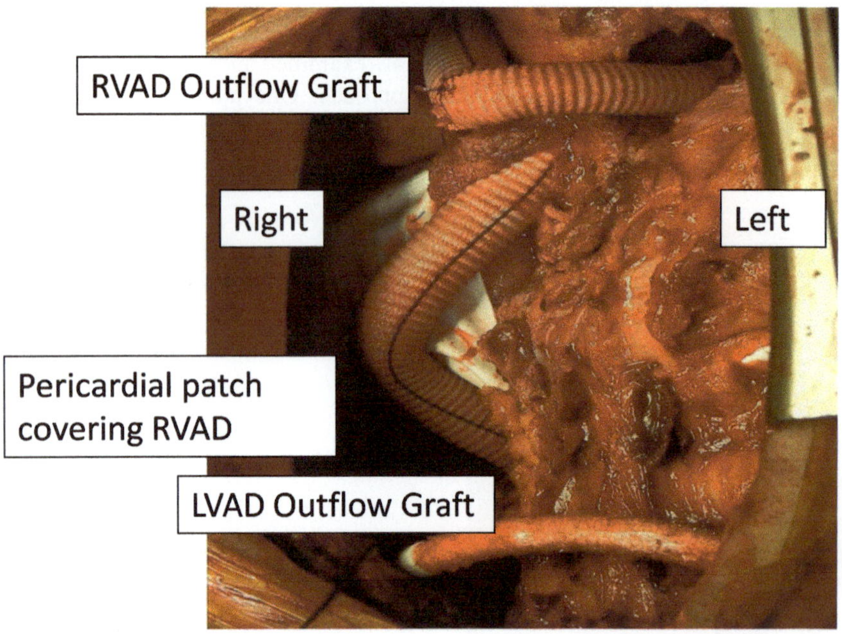

FIGURE 8.5 Intraoperative photo of the chest after explanting a temporary RVAD and implanting a cfRVAD after primary cfLVAD implantation.

in all cases, with a pump thrombosis rate of 20%.[17] Patients with biventricular failure can be supported with the contemporary continuous flow assist device HeartMate 3 as biventricular assist device (BVAD) (Fig. 8.6).

The above modifications are not used by all groups. The main decisions have to be made regarding implantation site (atrial or ventricular) and device. This is a tailored approach; especially in small chests, not all devices fit all.

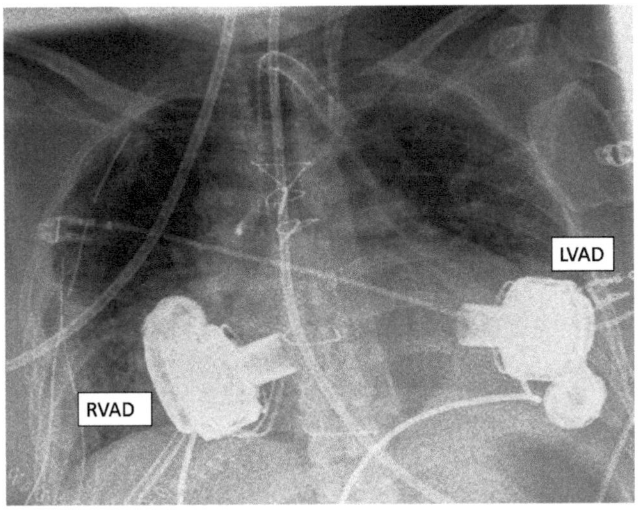

FIGURE 8.6 Postoperative X-ray after BIVAD implantation with two HeartMate 3.

Discussion

From the actual point of view, and considering the efforts of the last years, the chances of developing a VAD tailored to support the right ventricle are slim. The low number of patients worldwide with right/biventricular assist devices do not support meaningful financial investments in this kind of device.

Second, a lack of concurrence from device manufacturers has emerged during the last couple of years, with the withdrawal of the HVAD aggravating the current situation. Therefore, the next years will focus on intra-, pre-, and postprocedural improvements of the current available approach, especially in addressing RVAD pump thrombosis.

Conclusion

Although there is currently no approved device for right ventricular support, the experiences and outcomes with cfLVAD show that with some little modifications, biventricular support with two continuous flow ventricular assist devices (cfVADs) is feasible and apparently noninferior to the current available alternatives of mechanical circulatory support.

References

1. Vijiiac A, Onciul S, Guzu C, et al. Forgotten no more-the role of right ventricular dysfunction in heart failure with reduced ejection fraction: an echocardiographic perspective. *Diagnostics (Basel)*. 2021;11(3):548.
2. Sheehan F, Redington A. The right ventricle: anatomy, physiology and clinical imaging. *Heart*. 2008;94:1510–1515.
3. Rogers JG, Pagani FD, Tatooles AJ, et al. Intrapericardial left ventricular assist device for advanced heart failure. *N Engl J Med*. 2017;376 (5):451–460.
4. Mehra MR, Uriel N, Naka Y, et al. A fully magnetically levitated left ventricular assist device—final report. *N Engl J Med*. 2019;380 (17):1618–1627.
5. Kormos RL, Cowger J, Pagani FD, et al. The Society of Thoracic Surgeons Intermacs database annual report: evolving indications, outcomes, and scientific partnerships. *J Heart Lung Transplant*. 2019;38:114–126.
6. Teuteberg JJ, Cleveland Jr JC, Jennifer C, et al. The Society of Thoracic Surgeons Intermacs 2019 annual report: the changing landscape of devices and indications. *Ann Thorac Surg*. 2020;109:649–660.
7. Goldstein DJ, Meyns B, Xie R, et al. Third annual report from the ISHLT mechanically assisted circulatory support registry: a comprarison of centrifugal and axial continuous flow left ventricular assist devices. *J Heart Lung Transplant*. 2019;38:352–363.
8. Kiernan MS, Wilson Grandin E, Brinkley Jr M, et al. Early right ventricular assist device utilization in patients undergoing continuous-flow left ventricular assist device implantation: incidence and risk factors from INTERMACS. *Circ Heart Fail*. 2017;10(10):e003863.
9. https://www.uab.edu/medicine/intermacs/intermacs-documents.
10. Potapov E, Schoenrath F, Falk V. Clinical signs of right ventricular failure following implantation of a left ventricular assist device. *Eur J Heart Fail*. 2020;22(2):383–384.

11. Shehab S, Macdonald PS, Keogh AM, et al. Long-term biventricular HeartWare ventricular assist device support—case series of right atrial and right 30 ventricular implantation outcomes. *J Heart Lung Transplant.* 2016;35(4):466–473.
12. Venkatachalam S, Wu G, Ahmad M. Echocardiographic assessment of the right ventricle in the current era: application in clinical practice. *Echocardiography.* 2017;34(12):1930–1947.
13. Tran HA, Pollema TL, Enciso JS, et al. Durable biventricular support using right atrial placement of the HeartWare HVAD. *ASAIO J.* 2017;.
14. Shah P, Ha R, Ramesh S, et al. Multicenter experience with durable biventricular assist devices. *J Heart Lung Transplant.* 2018;37(9):1093–1101.
15. Eulert-Grehn J-J, Lanmüller P, Schönrath F, et al. Two implantable continuous-flow ventricular assist devices in a biventricular configuration: technique and results. *Interact Cardiovasc Thorac Surg.* 2018;27(6):938–942.
16. Lavee J, Mulzer J, Krabatsch T, et al. An international multicenter experience of biventricular support with HeartMate 3 ventricular assist systems. *J Heart Lung Transplant.* 2018;37(12):1399–1402.
17. McGiffin D, Kure C, McLean J, et al. The results of a single-center experience with HeartMate 3 in a biventricular configuration. *J Heart Lung Transplant.* 2021;40(3):193–200.

Chapter 9

Advanced Approaches for Total Artificial Heart Development

Jamshid H. Karimov[1,2], Anthony R. Polakowski[1], Kiyotaka Fukamachi[1,2] and David J. Horvath[3]

[1]Lerner Research Institute, Cleveland Clinic, Cleveland, OH, United States, [2]Cleveland Clinic Lerner College of Medicine of Case Western Reserve University, Cleveland, OH, United States, [3]R1 Engineering, Euclid, OH, United States

Chapter Outline

Introduction

Heart failure (HF) is a serious healthcare issue, affecting more than 20 million people worldwide,[1,2] and is a primary contributor to cardiovascular mortality. In the United States alone, by 2030, >8 million people (nearly 1 in every 33) are projected to have HF. The prevalence of advanced stage heart disease is rising rapidly, with more elderly Americans hospitalized for HF than for any other medical condition.[3] Heart transplantation can provide a remarkable improvement in quality of life in patients with end-stage HF,[4,5] but a shortage of donor hearts is a major limit for this option.[6]

In the United States, an estimated 50,000–100,000 patients are in need of heart transplantation or mechanical circulatory support (MCS), but transplantation is performed in only ~3600 cases per year (2020). Instead, such support is usually provided by left ventricular assist devices (LVADs), the most widely used means for improving the hemodynamic status of both transplant-eligible and noneligible patients.[4]

Initially employed as a bridge to transplantation, MCS was most frequently used to prolong survival while patients were on the waiting list until another therapy could be proposed or a donor organ became available. Today, an increasing number of patients with end-stage HF are being placed on long-term ventricular support as "destination therapy," with better survival and quality of life.

During the last decade, the LVAD landscape has moved from volume-displacement pulsatile-flow technology toward continuous-flow (CF) rotary pumps. CF pumps provide greater mechanical reliability, smaller size, improved durability, and better patient outcomes that approach those of heart transplantation. Since its approval by the US Food and Drug Administration (FDA) in April 2008, the HeartMate II (HMII) (Abbott, Lake Bluff, IL), an axial pump, has seen 25% annual growth in the number of LVAD implants throughout the world, exceeding 26,600 total implants as of 2018. The HVAD (Medtronic, Minneapolis, MN), a centrifugal pump, has been implanted in more than 17,000 patients worldwide. The newer HeartMate 3 CF LVAD (Abbott) was compared to its axial-flow predecessor (HMII) in more than 1000 patients in the MOMENTUM trial and demonstrated a lower incidence of pump thrombosis and disabling stroke.[7] According to the eighth annual report of the Interagency Registry for Mechanically Assisted Circulatory

Advances in Cardiovascular Technology. DOI: https://doi.org/10.1016/B978-0-12-816861-5.00037-X

Support, more than 95% of implants in the current era are CF devices. This is a pace of more than 2500 implants per year,[8] which clearly indicates the future direction of this device therapy.

Technologies intended for durable MCS are considered as replacement therapy for advanced-stage cardiac disease because of the lack of organ donation. The breakthrough development of CF technology in currently available ventricular assist devices (VAD) has afforded a wealth of engineering and design advantages in the development of a series of novel total artificial heart (TAH) platforms.[9] The recent success of CF MCS devices has led to the growing acceptance of these devices as a viable therapeutic option for end-stage HF patients who are not responsive to any existing drug or electrophysiologic treatment.[10]

The modern CF MCS-based therapy is realized as an implantation of a blood pump with a rotor bearing necessary to ensure its reliability during extended operation.[11] Due to their simple structure and low power consumption, blood pumps with hydrodynamic bearings are making their way into clinical applications. Such hydrodynamically suspended rotary blood pumps with noncontact bearings are particularly effective at enhancing blood compatibility.[6,11] There are multiple new generation artificial hearts in both preclinical development and in clinical trials for both bridge-to-transplant and destination therapy.[12–16] The TAH technology looks promising and has been shown to be effective for circulatory support in select patients with end-stage HF, and novel CF TAH designs under development are believed to have great potential to expand the role of TAH.[17] However, a number of considerable technological barriers still remain before a bioartificial heart can move toward advanced testing and eventually to broad clinical application.[18]

The development of CF rotary pump technology represents an important landmark centered on a novel pump design concept. These newer pumps have almost completely supplanted pulsatile, volume-displacement devices. CF rotary blood pumps are typically designed with an "axial" or "centrifugal" blood flow path, featuring an internal rotor within that path.[5] However, clinical application of CF has created a unique physiologic state, the consequences of which remain largely unknown and are being explored.[9] By eliminating a moving membrane and the multiple valves needed by pulsatile pumps, CF TAHs have the benefits of a smaller design and the potential for superior long-term mechanical dependability.

The TAHs are usually made up of an implantable rotary blood pump, along with components located outside the patient's body: a wearable controller connected to the pump via a percutaneous cable; wearable rechargeable batteries; battery charger; alternating- and direct-current (AC/DC) power supplies; and a hospital device for system control and monitoring.

The TAH controller drives the pump's electrical motor; controls the pump speed or flow based on user commands or feedback signals; collects, processes, and stores data; performs self-diagnostics; transmits information to and receives data from other system components, that is, hospital monitor and batteries; and provides various types of user interface (audible, visual, and tactile).[19] The TAH, in response to its hemodynamic environment, can act as a virtual sensor, providing feedback to indicate and track relative changes in flow and differential pressure. TAHs are connected to their controller units through an exterior driveline. All of the devices' left and right inflows originate from the native atrial chamber to provide blood flow to the pump, and outflows connected to the patient's native vessels.

Cleveland Clinic's CF TAH (CFTAH) is a double-ended centrifugal blood pump and is composed of a single rotating assembly with an embedded magnet that is shorter than the motor's steel laminations.[20] This difference in length allows a degree of free axial movement in the direction of net axial force that is primarily caused by an imbalance of pump inlet (atrial) pressures. The range of relatively free axial movement is bounded by increasing magnetic axial stiffness, which acts to restrain excess movement of the rotor. An aperture that opens and closes with axial movement of the rotor connects the right pump impeller section with the right volute. The aperture allows the rotating assembly to move axially to a magnetic/hydraulic "force-balanced" axial position.[21] This self-pressure regulation of the pump through automatic adjustment of the aperture has been reported previously.[22]

In this chapter, we focus on the research and development efforts at Cleveland Clinic to address the demands for end-stage HF treatment using total heart replacement technology; specifically, on the portfolio of CFTAH devices currently under development.

Cleveland Clinic total artificial heart design

The effort to create an advanced TAH pump in our institution originated with a creative challenge to craft an ultrasimple TAH (Fig. 9.1). Our novel CFTAH concept has undergone several iterations since the device conception as an alpha prototype development, resulting in an in vivo laboratory system capable of chronic support and automatic control. Although the system requires no flow or pressure sensors to function, a proposal was made to introduce a rotor position sensor in our research to demonstrate the optional benefits for hemodynamic monitoring.

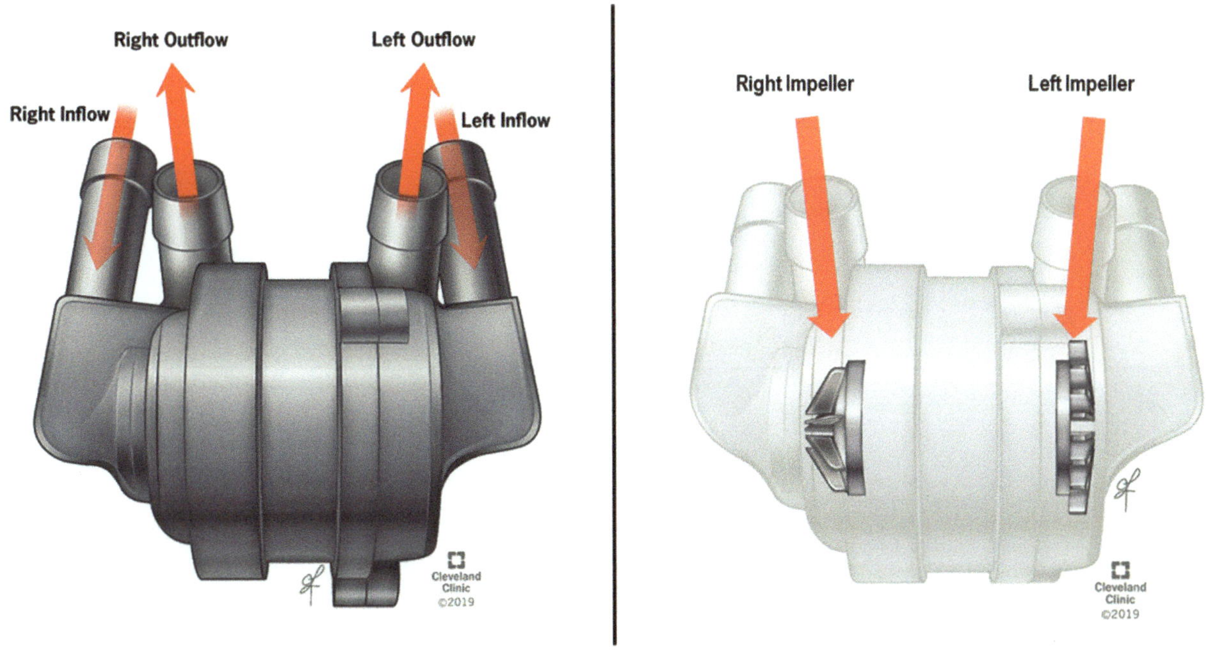

FIGURE 9.1 Cleveland Clinic continuous-flow artificial heart design.

The CFTAH design was unique, exploring radical new concepts in TAH design. The CFTAH design comprised a single, valveless CF pump with passive inherent hydraulic flow and pressure regulation to take over the function of the native heart. This "double-ended centrifugal pump" concept had a single, continuously rotating, brushless DC motor and pump assembly with a centrifugal pump on both ends (Fig. 9.2). Pump speed was modulated to create pulsatile flow and pressure. The electrical power response to speed modulation was monitored via controller and used to indicate incipient suction and appropriate speed reduction. This design differentiated itself from current clinical TAHs in that the speed-modulated, single-shaft, dual centrifugal pump allows for a significant reduction in size while eliminating the use of valves and pressure or flow sensors for control, which is believed to improve the overall system reliability.

The core breakthrough of this elegant, innovative pump concept was that the hemodynamic function of balancing systemic and pulmonary flows while minimizing atrial pressure differences is incorporated directly into the pump's hydromechanical design. This inherent function was engineered as an automatic, immediate, passively regular feature, and required no electronic intervention. The CFTAH control scheme was based on this passive self-balancing of pump flows to maintain atrial pressure balance and prevent atrial wall suction without the use of sensors for control feedback.

The relatively high number of complex components in existing pulsatile TAHs, as well as their large size, potential for failure, and limited durability create an obvious need for a small, reliable TAH. The availability of a small, simplified alternative with a single moving component will address many problems with existing technologies and provide better management for more patients. Ultimately, such a device would be less costly, as it needs only a single motor controller to balance left and right flows, while eliminating valves, compliance chambers, actuator mechanisms, and pressure or flow sensors. The Cleveland Clinic CFTAH accomplishes passive, sensorless autoregulation of pump flows and filling pressures by varying relative left and right pump efficiencies, based on the position of its single moving part, to balance the left and right pump-filling pressures and pump outputs.[20,23]

Preliminary data have shown that these design features do regulate flows and pressures over wide ranges of systemic and pulmonary vascular resistance.[20,24] Fig. 9.3 illustrates how the rotating assembly position affects pump geometry to correct any imbalance in atrial pressure. Note that the rotor magnet is shorter than the steel stator laminations, so that there is a window of axial movement that is relatively free of magnetic restoring forces, allowing rotor movement to respond primarily to pressure forces.

Inlet pressures are the most dominant terms in force balancing. If right atrial pressure (RAP) becomes higher than left atrial pressure (LAP), the rotating assembly is pushed to the left (Fig. 9.3A), opening the aperture at the outer diameter of the right pump, increasing right pump output, decreasing RAP and increasing LAP. Conversely, if LAP is higher than RAP, the rotating assembly is pushed to the right (Fig. 9.3B), closing the aperture of the right pump and reducing right pump output, again driving the system to balance. The narrow blood passage between the rotating assembly

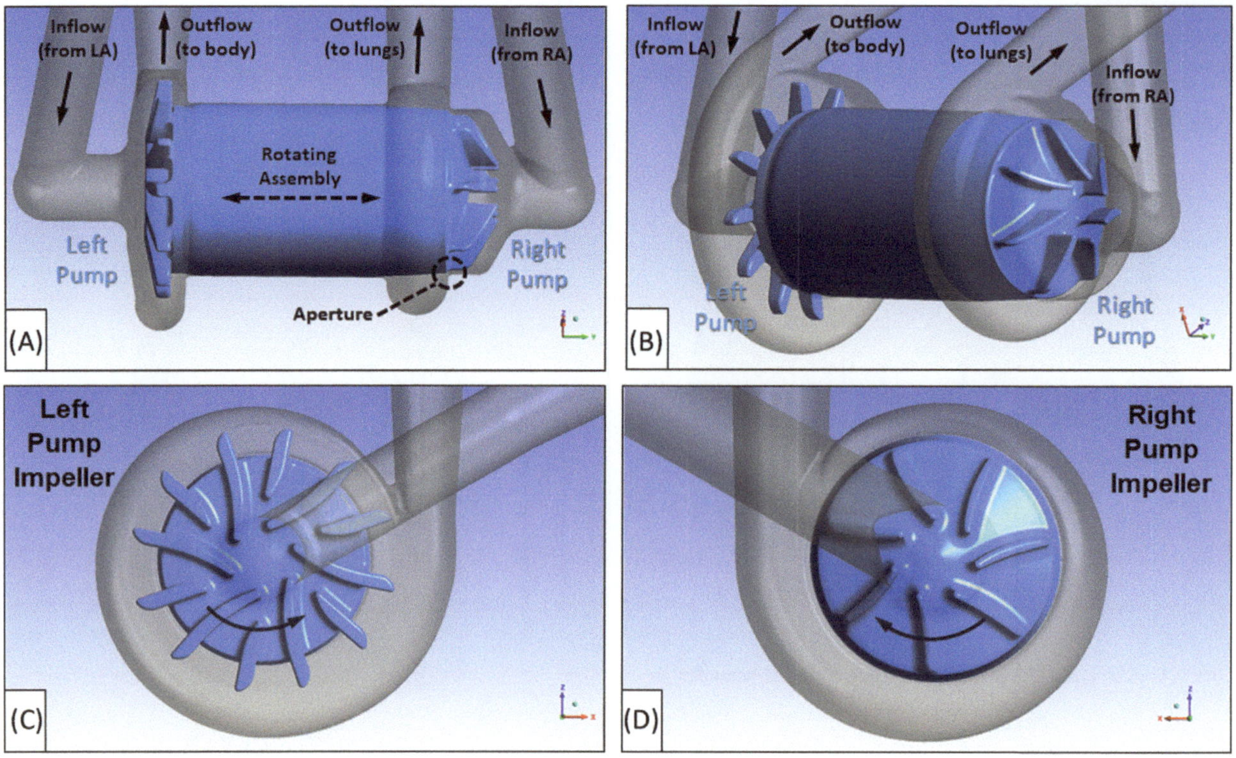

FIGURE 9.2 The CFTAH080 design showing the rotating assembly with left and right impeller regions, the thin connecting journal bearing blood flow path, and the right pump aperture. The right pump moves the blood returning from the body, via the right atrium (RA), to the lungs. The left pump receives the blood returning from the lungs, via the left atrium (LA), and pumps that blood throughout the body. (A) Side view of rotating assembly and blood flow path. (B) Isometric view of rotating assembly and blood flow path. (C) Left pump impeller with six primary blades and six splitter blades. (D) Right pump impeller with seven primary impeller blades. The direction of rotation is indicated by arrows.

functions as a connection between left and right chambers of the pump, and also serves as a fluid bearing (Fig. 9.3C) as axial movement occurs during device pressure adjustments (self-regulation) through opening and closure of the right pump aperture (Fig. 9.3D).

The CFTAH platform is currently represented through three different devices, intended for an adult population (Fig. 9.4A), pediatric patients (Fig. 9.4B) and infants (Fig. 9.4C). Each of these devices has specific design requirements related to its performance, size, weight, and displacement (Fig. 9.5); therefore, each is undergoing a different stage of development and will be described in below sections of this Chapter.

Computational fluid dynamics for biocompatibility analysis

Stator bore design and journal-bearing

The computational fluid dynamics (CFD) analysis has been extensively employed during CFTAH development with the focus on the radial suspension of the rotating assembly and blood bearing. The rotating assembly derives lift from its own motion, creating a hydrodynamic bearing force that requires stabilization. The key to the radial stabilization is a radial offset between the stator-bearing bore and the motor winding's steel laminations. This offset applies a radial magnetic force that counteracts the hydrodynamic force as the rotating assembly moves to a "force-balanced" radial position.

The journal-bearing blood passage is a thin, cylindrical region with a nominal radial clearance of 0.004 inches that connects the left and right impeller regions. Its role is to support the rotating assembly and to provide the hydraulic radial force for the hydrodynamic bearing[25] (Fig. 9.6).

The earlier design (CFTAH080) was found to be within an acceptable range in terms of hydrodynamic stability and biocompatibility.[22] However, the CFTAH080 had a lower-than-desired axial stiffness and, correspondingly, higher-than-desired axial travel. Therefore, a new design (CFTAH100) has been proposed, which incorporates a new stacked

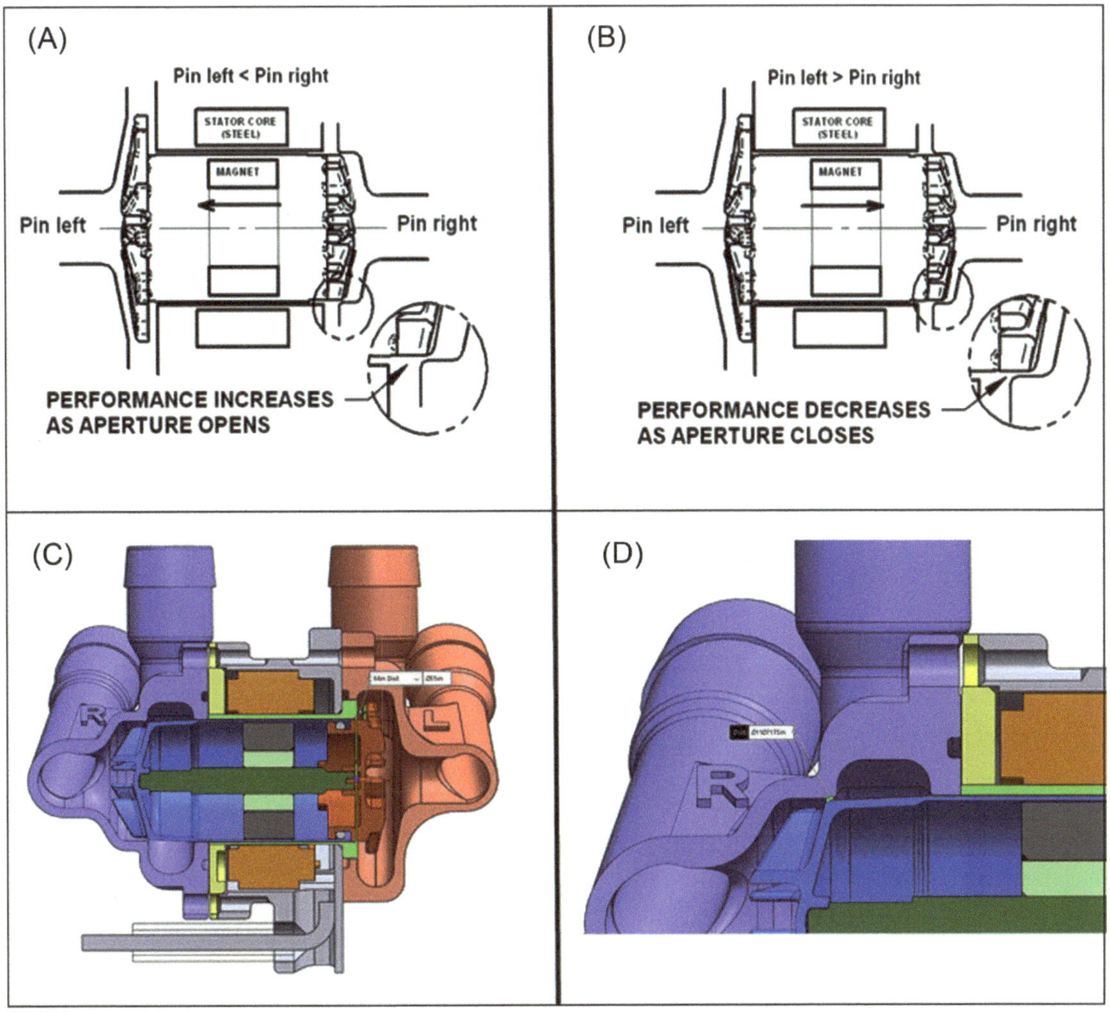

FIGURE 9.3 Atrial pressure difference effects on rotor position and right pump performance (A, B) and right aperture design (C, D).

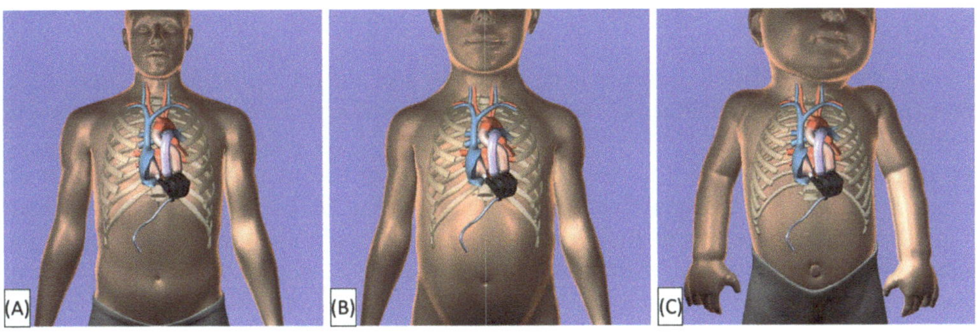

FIGURE 9.4 Schematic illustration of device portfolio of continuous-flow total artificial heart intended for adult (A), pediatric (B) and infant use (C).

magnet that offers an increased axial magnetic stiffness. However, the radial magnetic stiffness of the CFTAH100 is reduced due to the gap between the split magnets and an overall shorter total magnet length.[26] A lower radial magnetic stiffness increases the risk of the rotor instability (the rotor bouncing around within the bearing clearance), causing reliability issues and potentially reduced biocompatibility.

To improve the stability of the CFTAH100, two parameters were identified: lamination-stator offset and bearing radial clearance. A larger lamination-stator offset resulted in a higher magnetic force, and therefore an improved

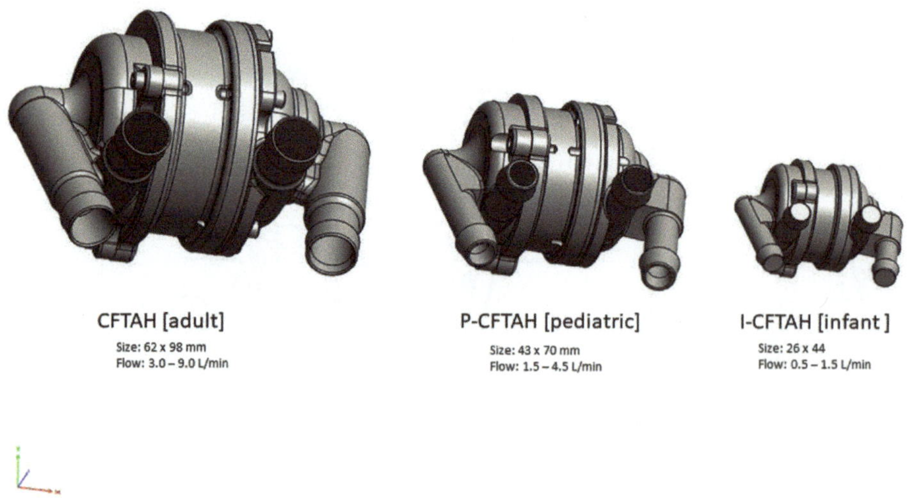

FIGURE 9.5 Cleveland Clinic CFTAH lineup and basic performance specifications.

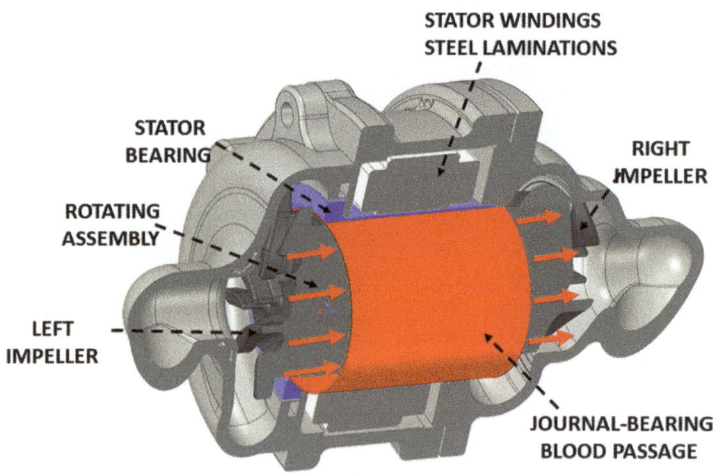

FIGURE 9.6 CFTAH journal-bearing blood passage.

stability. Moreover, higher radial clearance made the rotor assembly more stable by lowering the potential of the hydrodynamic lift surpassing the magnetic load.

During optimization, the rotor was displaced; therefore, the magnet field was also shifted. Sufficient air (free space) outside the pump was also modeled to account for the effect of the magnetic field leakage at all the faces of the pump that could most affect the radial stiffness calculation.

Fig. 9.7A shows the magnetic flux density of the motor for the centered position of the rotor. A detailed view of the field for the rotor at the maximum 0.014 inches radial displacement is also shown in Fig. 9.7B (the range of flux density was adjusted in order to clearly illustrate the differences of flux density between the right and the left side of the figure). The radial displacement allows the field on the side of the rotor with the decreased gap to be increased, with an associated increase of force, while the opposite side experiences a decreased field with a lower force. Such a differential results in a net force pulling the rotor in the direction to further decrease the gap. The radial displacement does not significantly change the direction of the net force, but does result in an increase in net force. Since the magnetic properties remain basically linear, the net force is also essentially linear. Using a radial displacement of 0.014 inches, the resulting magnetic radial stiffness is determined to be 680 N/inch (Fig. 9.8).

Hydrodynamic bearings are low-clearance, purely passive-type, and their gap is designed to be dependent on the bearing load. The bearing film requires a dynamic pressure buildup that is usually generated by means of a decreasing gap. Different types of bearings provide this pressure buildup in different ways. Disadvantages are the very small gaps, which lead to high fluid friction and high shear stresses that may initiate hemolysis. Also, due to reduced gaps, high-precision machining is required.[27]

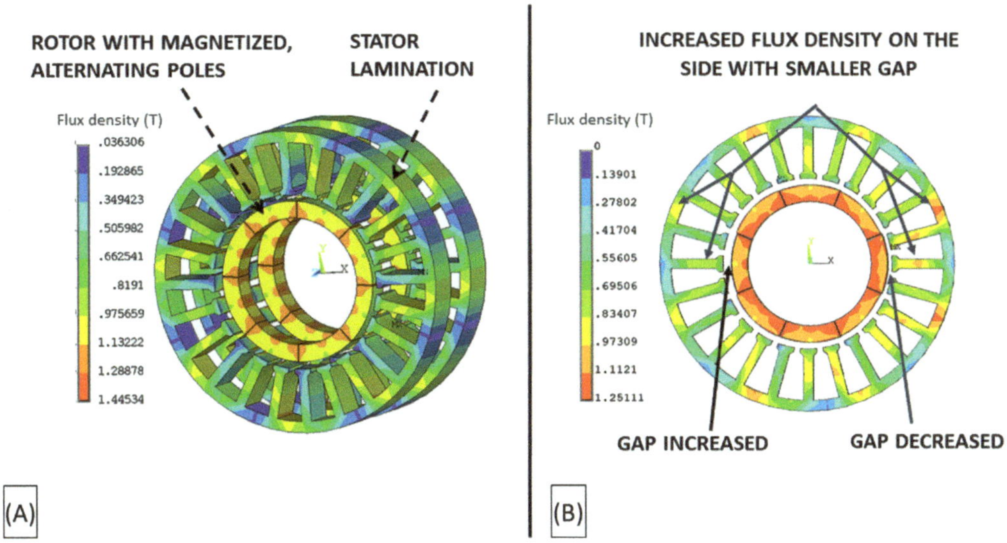

FIGURE 9.7 Magnetic field of the motor: (A) rotor at the centered position with respect to the stator; (B) rotor moved radially from the centered position by 0.014″.

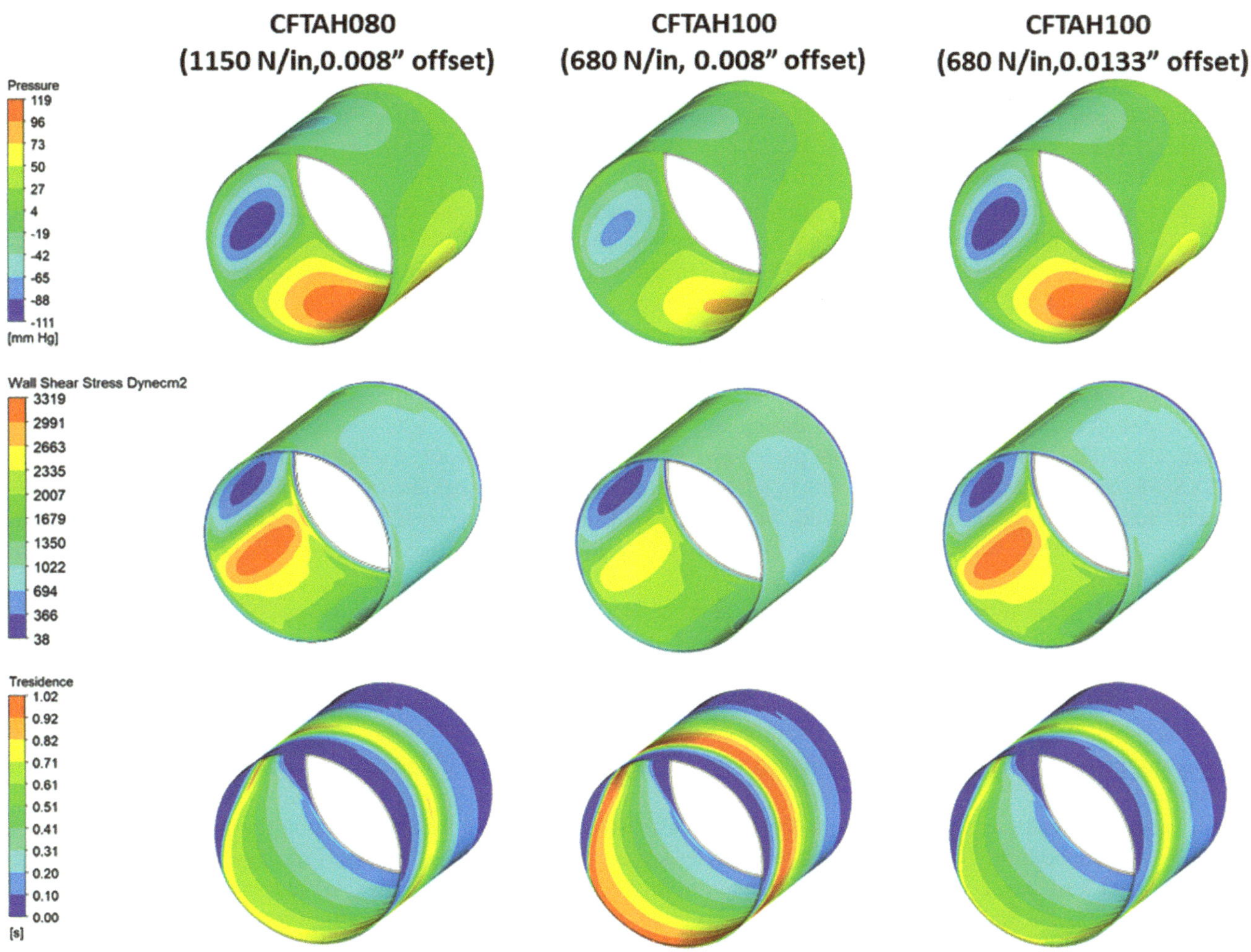

FIGURE 9.8 Lateral force (N) versus radial displacement (in).

Minor manufacturing inconsistencies can play a large role in unstable journal-bearing operations.[28] Unlike their rolling element counterparts (i.e., ball bearings), there is an optimal size for a journal-bearing under a given application. If the bearing is too small for its application, the journal may experience bearing rub caused by inadequate radial clearance between the journal and the bearing. If the bearing is too large for its application, unstable orbital motion can cause the journal to move progressively off-center until it contacts the bearing.

The CFD analysis of the journal-bearing was primarily focused on the suspension performance under CF conditions, where hydrodynamic bearing models were tested with given constant flow rate, speed, and bearing differential pressure representing the expected nominal for clinical operation. However, in the actual clinical application, the inlet and outlet pressures of the CFTAH would vary according to the periodic pressure change of the entire atrial balance and pressure distributions within the device-vasculature system, and would also be affected by compliances of the connected aorta and pulmonary tree. Thus, the characteristics of the suspension performance of the proposed journal-bearing will be dynamically adjusted. Therefore, the scales of lubrication films and groove depths and blood in the gap should be updated continuously, and the corresponding flow rate should be as large as possible in order to improve washout.[29,30]

A set of CFD simulations was carried out to compare a range of journal-bearing designs with respect to different performance parameters such as power loss, blood residence time, flow rate and shear stress. A final design that offers the most suitable compromise between different performance parameters and fabrication constraints was selected. However, the effects of hemolysis and thrombotic formations within the pump gaps and/or blood-contacting surfaces are still to be verified using the advanced CFD techniques, and validated in vivo.

Right pump impeller design and modeling

The flow path and geometry for the baseline CFTAH080 design are shown in Fig. 9.2A and B. For this design, the rotor is free to move both axially and radially until it reaches a magnetic/hydrodynamic force-balanced position. Another unique feature of this pump is that the right pump includes a narrow fluid path, an aperture that connects the impeller region with the volute. This aperture functions as a differential pressure-regulating valve to automatically balance the right and left pump inlet pressures. The differential pressure-regulating feature keeps the right and left blood inlet pressures safely balanced.

The left pump, composed of six primary impeller blades and six smaller splitter blades, provides the hydraulic pressure head (60–110 mm Hg) to supply the body with oxygenated blood over its operating range (3–8 L/min) (Fig. 9.2C). By contrast, the right pump has seven larger primary impeller blades without splitters, designed to supply the lower hydraulic pressure head required (20–40 mm Hg) to provide blood flow to and from the lungs[31] (Fig. 9.2D).

The primary goal of CFD analyses performed at typical conditions, similar to those in vivo flow conditions, was to identify differences in the flow patterns around the left and right impellers. Such differences could reveal potential reasons for the thrombus formations on the right impeller surfaces. Using the insight gained by comparing the left and right impeller flow fields, the CFD modeling would then be used to identify an improved, next-generation right impeller design aimed at reducing the likelihood of thrombus formation.

The CFTAH CFD simulations were performed using ANSYS software (ANSYS Inc., Canonsburg, PA). The ANSYS-Meshing software was used to create a hybrid (tetrahedral, prism, hexahedral) mesh of typically 18–19 million elements for the flow domains. Prism layers were used on all internal pump surfaces except for the journal-bearing and inlet/outlet extensions, where hexahedral mesh elements were used to capture the flow field and velocity gradients along the walls.

The CFTAH CFD model was previously validated by comparison with rotor torque and in vitro pressure mapping data over the intended range of operating conditions.[21] For that study, steady state, frozen rotor CFD solutions were conducted using the ANSYS-CFX software with fluid properties matching the in vitro test conditions ($\rho = 1078$ kg/m^3; $\mu = 2.63$ centipoise). The shear stress transport (SST) $k-\omega$ turbulence model was applied in all the fluid regions, except for the very thin journal-bearing region, where laminar flow conditions were used.[32] For each flow condition modeled, the rotating assembly axial position was moved to a location that provided a right pump pressure increase equal to the in vitro test data.

The area-averaged mesh y + values (a dimensionless distance from the wall based upon the local wall shear stress) were below 1 for both the left and right pump impeller surfaces. This indicates that the first layer of elements, adjacent to the impeller surfaces, resides within the viscous sublayer and an appropriate level of near-wall mesh refinement to accurately predict the wall shear stress values. The CFD simulation results agreed with experimental values to within 5%–10% for rotor torque, hydraulic pressure heads, and 10 static pressure taps placed at different locations throughout the pump.

The validated CFTAH CFD model was then used to compare and evaluate the flow fields near the right and left impeller surfaces using blood. A Cross non-Newtonian blood viscosity model[33,34] was included to help identify low-shear stress regions. Flow residence time (a volumetric passive scalar variable) was also predicted in the simulations. For the design comparison simulations, the blood flow rate was set at 8 L/min and the impeller rotational speed at 3100 rpm to mimic the in vivo test conditions. For the baseline and recommended next-generation right impeller designs, the low flow rate/low impeller rotational speed condition of 3 L/min and 2200 rpm was also modeled.

With a goal of identifying a new right impeller design with the potential to eliminate thrombus formation, numerous design variations were explored, including the impeller surface cone angle, blade width, blade shape, number of blades, inlet configuration, and center region design. These changes were intended to remove regions with sustained low wall shear stress on the impeller surfaces, where thrombus could form, and reduce the extent of inward-flowing blood returning from the higher shear aperture region, where platelet activation is more likely to occur.[35–37] In addition to improving the right pump's flow characteristics, it was important to maintain the hydraulic performance of the baseline right pump design.

The first step in this study was to identify flow-related differences between the left and right impeller regions that could lead to thrombus deposition. As a starting point, the wall shear stress acting on the impeller surfaces was compared at different angular positions to capture the transient nature of the flow fields. Fig. 9.9A presents the wall shear stress acting on the impeller surfaces after 720 degrees of rotations. The values are limited to an upper value of 500 dynes/cm^2 to better identify lower shear stress regions. The right impeller has larger regions of low shear stress along the downstream (suction) side of the blades and in the center region opposite the inlet port.

To capture the combined effect of low shear stress and increased residence time, a flow variable defined as fluid shear stress divided by residence time was calculated. Regions where this combined variable was low, corresponding to low fluid shear stress and long residence times, would be more susceptible to thrombus formation. The right impeller exhibits lower values of this variable, particularly between the impeller blades and in the center region of the impeller (Fig. 9.9B). For these images, the values for fluid shear stress/residence time were computed on surfaces offset 0.003″ from the impeller surfaces, as residence time is undefined on the actual impeller surfaces. Similar trends, concerning the locations with low values for both quantities, were found for the other impeller rotational positions. The low wall

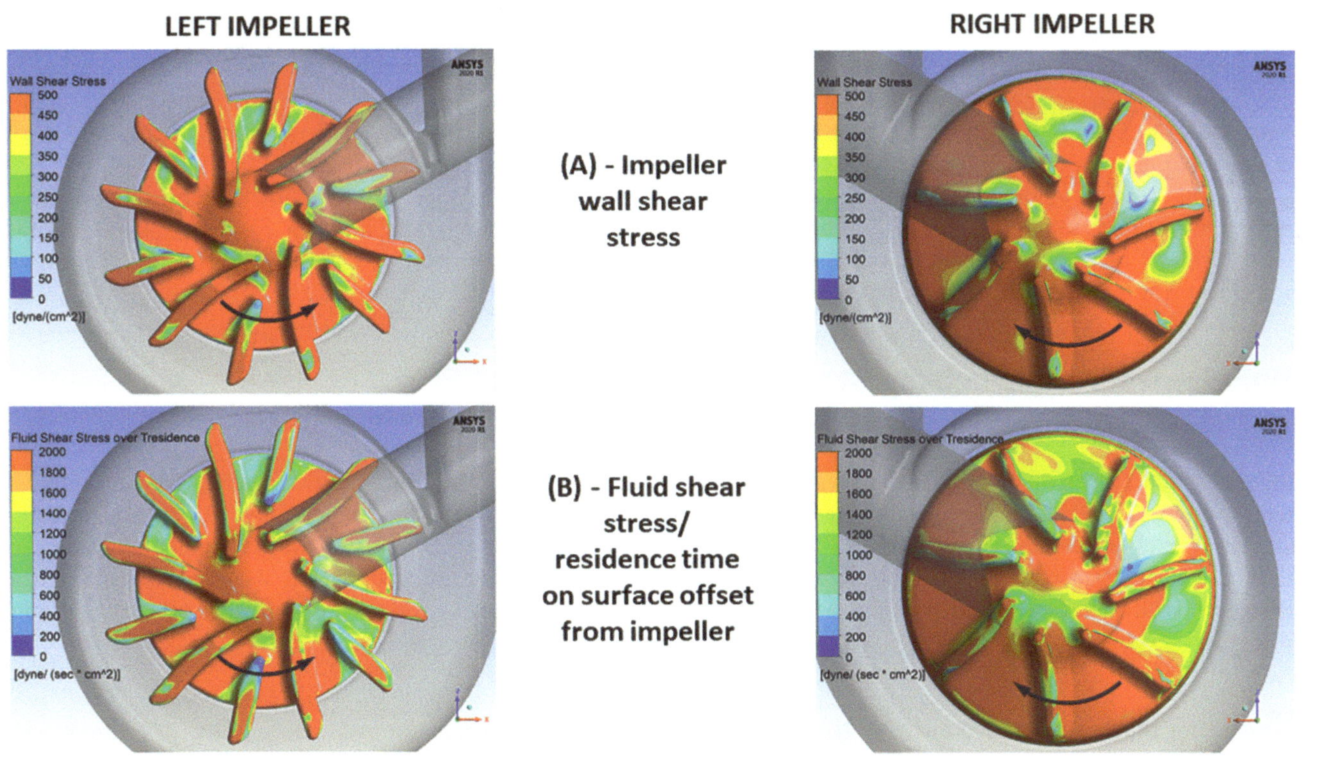

FIGURE 9.9 (A) Wall shear stress on the impeller surfaces. (B) Fluid shear stress/residence time on surface offset 0.003″ from impeller surfaces. All images at rotation of 720 degrees. Ranges limited to identify regions with lower values, which indicate increased potential for thrombus formation. The direction of rotation is indicated by arrows.

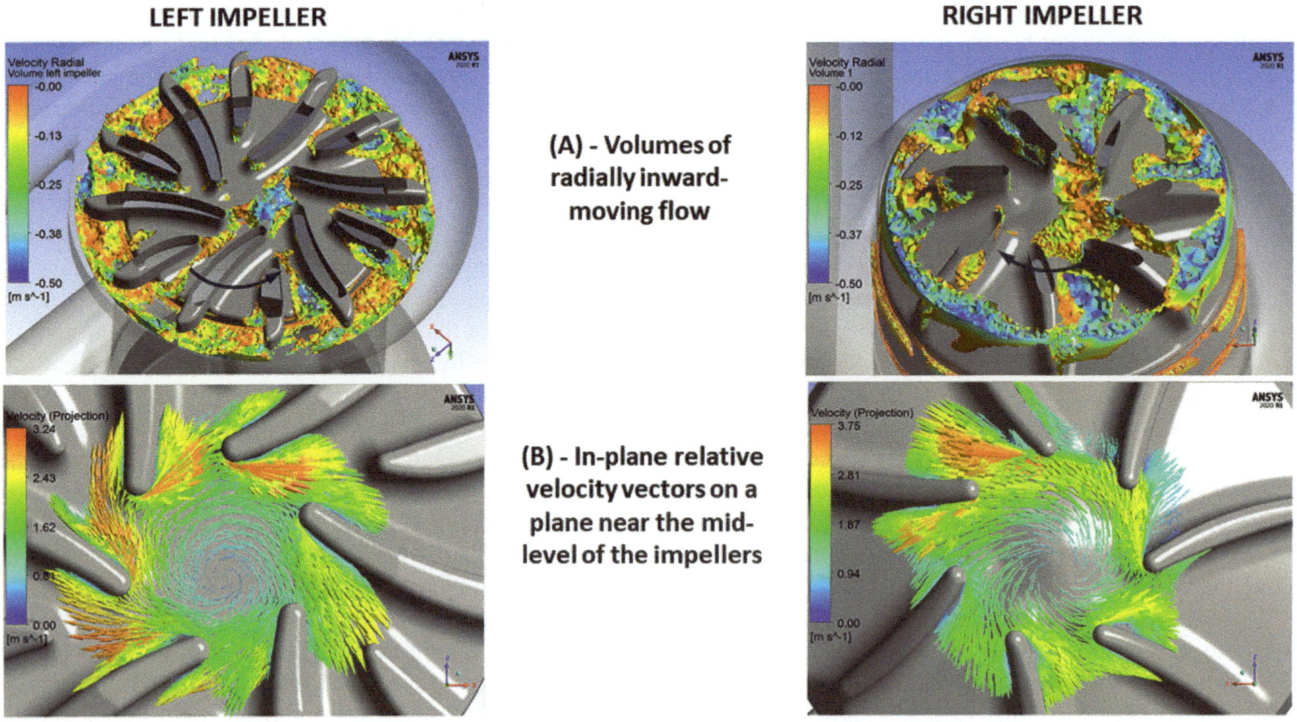

FIGURE 9.10 (A) Fluid volumes of radially inward-moving flow. (B) In-plane, relative velocity vectors on planes near the mid-level of the impellers. All images at rotations of 720 degrees. Regions with inward-moving flow between the impellers or flow separation off the blade-leading edges, as seen for the right impeller, have increased potential for thrombus formation. The direction of rotation is indicated by arrows.

shear stress and low fluid shear stress/residence time regions, near the center and between the blades of the right impeller, corresponded well with the location of the star-shaped thrombus formed during the final long-term in vivo study.

Identifying regions of radially inward-moving flow and flow separation near the impeller blades were two other key areas of focus. The images in Fig. 9.10A present fluid volumes of radially inward-moving flow. These images show regions with radially inward velocities between 0.0 and −0.50 m/s. The inward-moving flow occurs near the rotor center and along the downstream, suction side of the right impeller blades.

In-plane relative velocity vectors on planes near the mid-level of the impellers are shown in Fig. 9.10B. These images reveal flow separation originating at the blade tips for several of the right impeller blades (12:00–3:00 position). Minimal blade-leading edge-initiated flow separation was observed for the left impeller design, compared with significant flow separation in several of the right impeller blades.

Right impeller design changes

Reducing the flow area between impeller blades

To reduce the extent of radially inward-moving flow, several designs with reduced flow area between the impeller blades were modeled. The first concept changed the number of primary impeller blades from seven to six and added six smaller splitter blades in between the primary blades. This configuration, resembling the left impeller design, has increased blade surface area intended to increase the pressure head. Another approach was to use wedge-shaped impeller blades with increased width moving radially outwards. A third design increased the impeller surface cone angle from 15 to 25 degrees (Fig. 9.11).

Evaluating blade curvature and center region design

Several design concepts aimed at improving the alignment between the incoming flow and the impeller blade tips, with a goal of reducing the leading edge-induced flow separation, were also modeled. For the first concept, the blade curvature was increased by reducing the leading-edge blade angle from 50 to 38 degrees, relative to the horizontal, to better

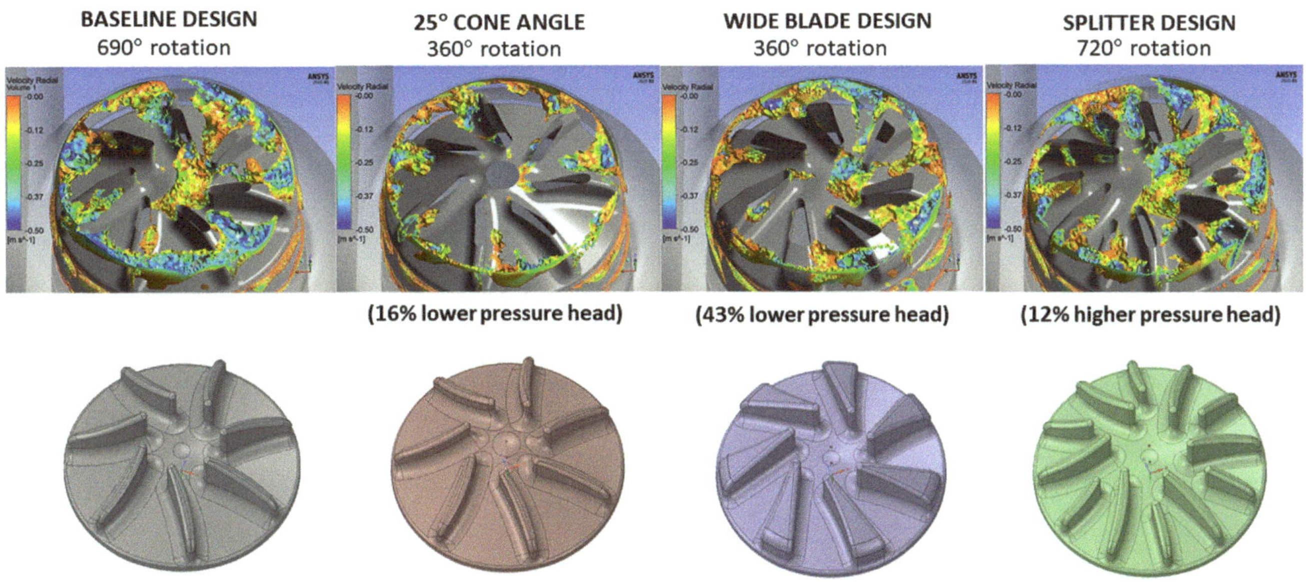

FIGURE 9.11 Fluid volumes with radially inward-moving blood flow near the right pump impeller for different designs intended to reduce the fluid volume between the impeller blades. The volume of inward-moving flow was significantly reduced by increasing the impeller base cone angle from 15 to 25 degrees. Pressure head increased with the splitter design and significantly decreased with the wide blade design. The direction of impeller rotation is clockwise for these images.

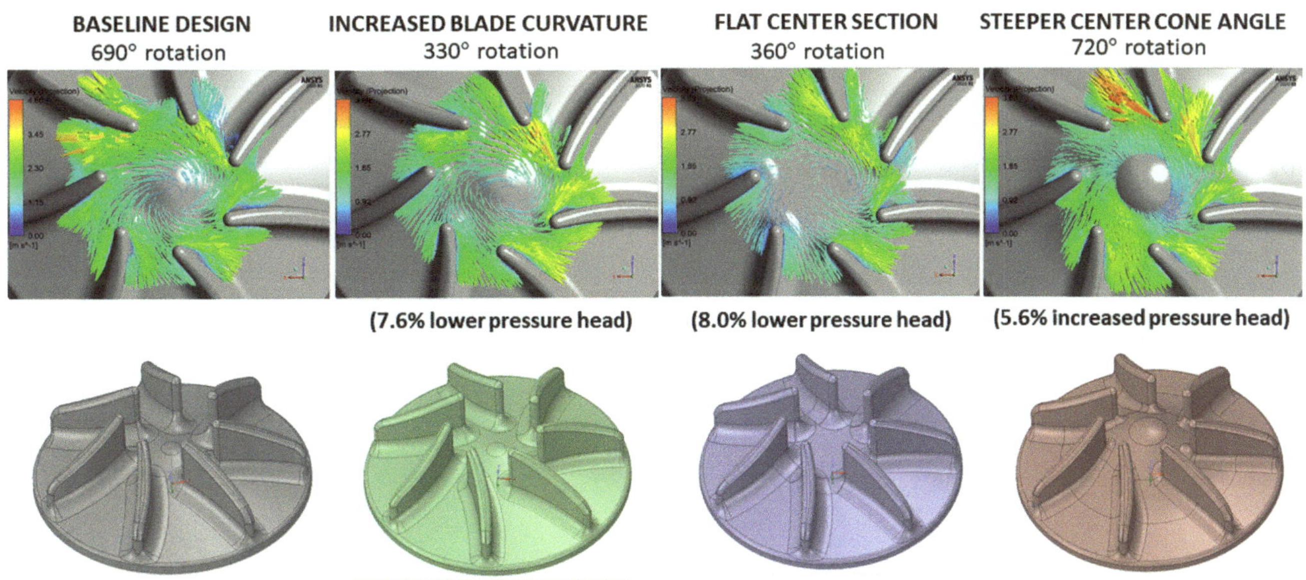

FIGURE 9.12 In-plane, relative velocity vectors on planes near the mid-level of the impellers for different impeller designs. Increasing the blade curvature and including a steeper center cone region improved the alignment of the incoming flow with the impeller blade tips. The increased blade curvature and flat center designs reduced the pressure head by 8%, whereas the steeper cone angle model increased the pressure head by 6%. The direction of impeller rotation is clockwise for these images.

align the incoming flow with the impeller blade tips. Two other concepts, a flat center and a steeper cone center, were also explored.

A comparison of in-plane relative velocity vectors on a plane near the mid-level of the impellers for the baseline and design variations is presented in Fig. 9.12. Reducing the leading-edge blade angle and increasing the center cone angle improved the alignment of the incoming flow with the impeller blade tips. This improved flow/blade tip alignment should reduce the extent of flow separation downstream of the blade-leading edges.

Next-generation right impeller design

Leveraging the knowledge gained through the CFD results for the design variations modeled, a configuration was selected for the next-generation CFTAH right impeller design. This new design combined several features from the simulated designs, including a 25 degrees impeller surface cone angle, six primary/six splitter blades, and a reduction in the leading-edge blade angle. An additional feature, 2:1 elliptical-shaped leading edges on all the impeller blades, was also included in the new design. This feature was intended to further reduce the extent of blade-leading, edge-initiated flow separation.[38] All the previously modeled designs had spherically shaped, fully rounded impeller blade tips.

Fig. 9.13 shows the changes made between the baseline and next-generation right impeller designs. The changes made in the next-generation right impeller design significantly reduced the extent of radially inward-moving flow near the primary impeller blades and in the center region of the impeller. Fig. 9.13 presents the volume of inwards moving flow, within the range of 0–0.5 m/s, for both impeller designs after 690 and 720 degrees of rotation. The 30 degrees angle variation was selected to capture the two extremes in angular position for the impeller blades. For the next-generation design, inward-moving flow was still observed around the downstream regions near the splitter blades. These regions occurring along the right impeller housing walls, away from the impeller surfaces and near the outer perimeter of the impeller housing, would be less likely to promote thrombus formation on the impeller.

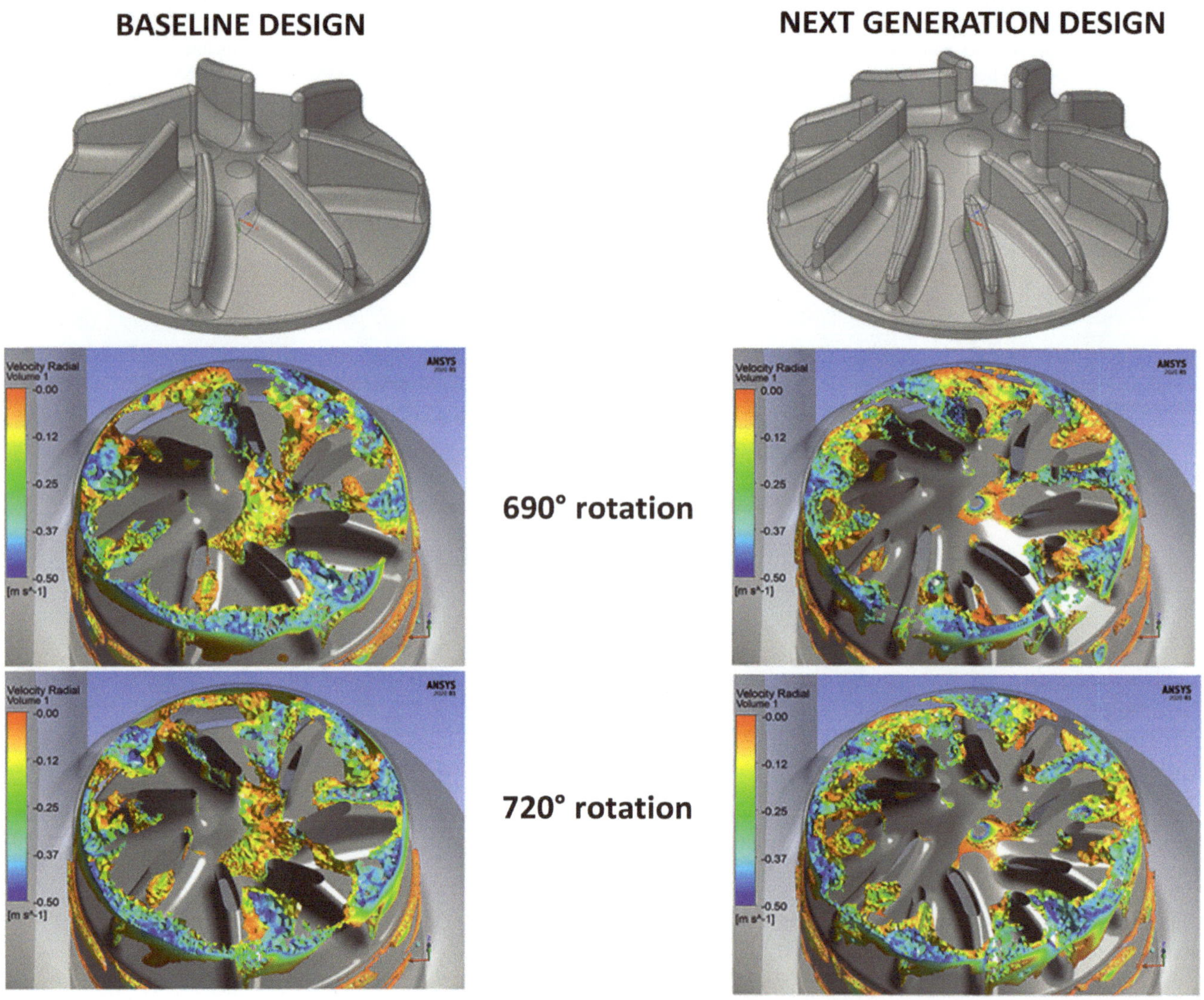

FIGURE 9.13 Fluid volumes with radially inward-moving blood flow near the right pump impeller for the baseline and next-generation design at two angles of rotation. The next-generation design significantly reduced the extent of inward-moving flow around the primary impeller blades and in the center region of the impeller. The direction of impeller rotation is clockwise for these images.

Significant improvement in the alignment of the incoming flow with the leading edges of the primary blades was also achieved with the next-generation design. In-plane, relative velocity vectors on planes near the mid-level of the impellers, after 690 and 720 degrees of impeller rotation, are shown in Fig. 9.14. The number of impeller blades with leading edge-induced flow separation is reduced from three (12:00–3:00 positions) to one (2:00 position) with the next-generation design. This reduction in the leading edge-induced flow separation should reduce the extent of flow separation along the suction side of the impeller blades, and as a result, help to reduce the likelihood of thrombus formation along the blade surfaces.

The inward-moving flow between the blades had just experienced higher shear passing through the impeller blade passages and returning from the aperture region. It was important to keep the volume of this blood, which was potentially at a higher state of platelet activation, minimized and further apart from the impeller surfaces.[39] Likewise, improving the alignment between the incoming blood flow and the impeller blades would reduce the extent of flow separation, recirculation, and potential thrombus formation on the suction side of the impeller blades.

The design of CFTAH self-regulating response also acts to mitigate atrial suction.[23,40,41] If the left inlet is occluded due to suction, the left inlet pressure drops, pulling the rotating assembly to the left, thereby opening the aperture of the right pump, increasing LAP and automatically recovering from the suction condition. Similarly, if the right inlet is

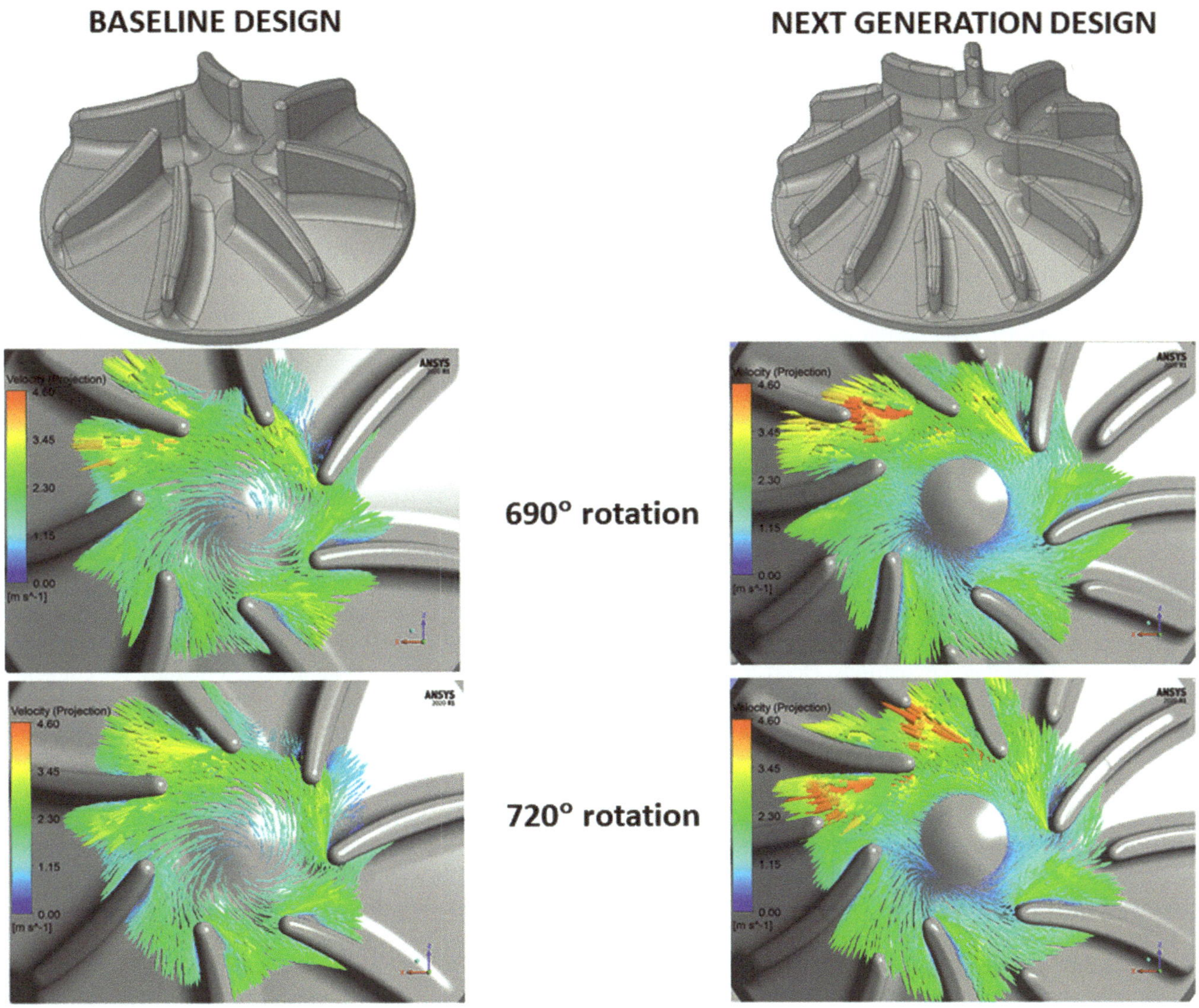

FIGURE 9.14 In-plane, relative velocity vectors on planes near the mid-level of the impellers for the baseline and next-generation right impeller designs. The next-generation design significantly reduced the degree of primary impeller leading edge blade-induced flow separation. The direction of impeller rotation is clockwise for these images.

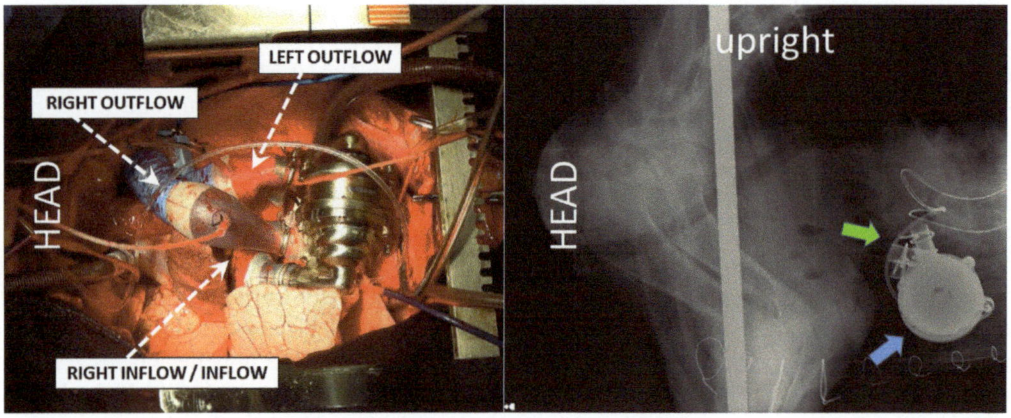

FIGURE 9.15 Adult CFTAH implantation in chronic animal model.

occluded due to suction, the right inlet pressure drops, pulling the rotating assembly to the right, thereby closing the aperture of the right pump and decreasing the suction on the right atrium.

Our earlier series of in vivo single-piece CFTAH experiments has demonstrated desired hemodynamic performance and pump operation within design specifications. The Cleveland Clinic CFTAH was implanted Jersey calves (*n* = *17*; weight range, 77.0–93.9 kg) through a median sternotomy (*n* = *9*) or a right thoracotomy (*n* = *8*)[42] for elective chronic implantation of 14, 30, or 90 days. The improved pump design in six animals brought about no chronic hemolysis. Our most recent (2020) CFTAH implant (Fig. 9.15), using a double-stacked motor design and improved right impeller design, has achieved the intended 30-day duration in the chronic animal (weight, 88.0 kg) without any perioperative and postoperative complications.

Our 90-day biocompatibility studies in calves *(n* = *2)* were successfully completed at the planned durations and showed remarkable pump performance and biocompatibility without any postoperative anticoagulation therapy.[22] The CFTAH passively self-balances left and right circulations without electronic intervention. The nominal external dimensions of the latest CFTAH design used in these studies were 98.4 mm (length), 62 mm (diameter), 160 cc (volume displacement), and 486 g (device weight with cable and connector off the scale). All chronic experiments demonstrated good hemodynamic performance of the CFTAH without organ thromboembolism. The newer CFTAH design device generation, with a revised right impeller, is currently undergoing validation in vivo.

Total artificial heart for pediatric use

MCS has become standard therapy for adult patients with end-stage HF; however, in pediatric patients with congenital heart disease, the options for chronic MCS are limited to paracorporeal devices or off-label use of devices intended for implantation in adults. Congenital heart disease and cardiomyopathy often involve both the left and right ventricles; in such cases, heart transplantation or a biventricular assist device (BVAD) or a TAH is needed to adequately sustain both pulmonary and systemic circulations. Heart transplantation in infants and children is an accepted therapy for congenital heart disease, but the availability of suitable donor hearts varies and is never enough to meet the clinical need. MCS is also defined as a standard option, but its use in infants and children is limited due to size and functional range restrictions.

At Cleveland Clinic, we are developing a miniature pediatric version of the CFTAH described above (the P-CFTAH) with a single motor and single rotating assembly[43,44] (Fig. 9.16). The Cleveland Clinic team was the first to report successful implantation of the prototype P-CFTAH in an acute animal study (Fig. 9.17).[44] The results, recently published, are from a series of implantations of our P-CFTAH in four healthy lambs.[45] The working model prototype P-CFTAH pump was derived from the adult CFTAH configuration by downsizing it by a scale factor of 0.70 (one-third of its total volume) to enable implantation in the infant chest. The pump flow range (1.5–4.5 L/min) is intended to provide support to patients weighing up to 50 kg (average weight of 14-year-olds).

The pressure-flow curves demonstrated a broad range of left pump flow. At a delta pressure of 100 mm Hg, the left pump flow range was 0.4–4.7 L/min with a pump speed of 3500 and 5000 rpm. The difference in left minus right atrial pressures (LAP-RAP) was maintained within ±5 mm Hg with a wide range of systemic-to-pulmonary vascular resistance ratios (between 1.4 and 35), with and without pump speed modulation. Our findings met the proposed design

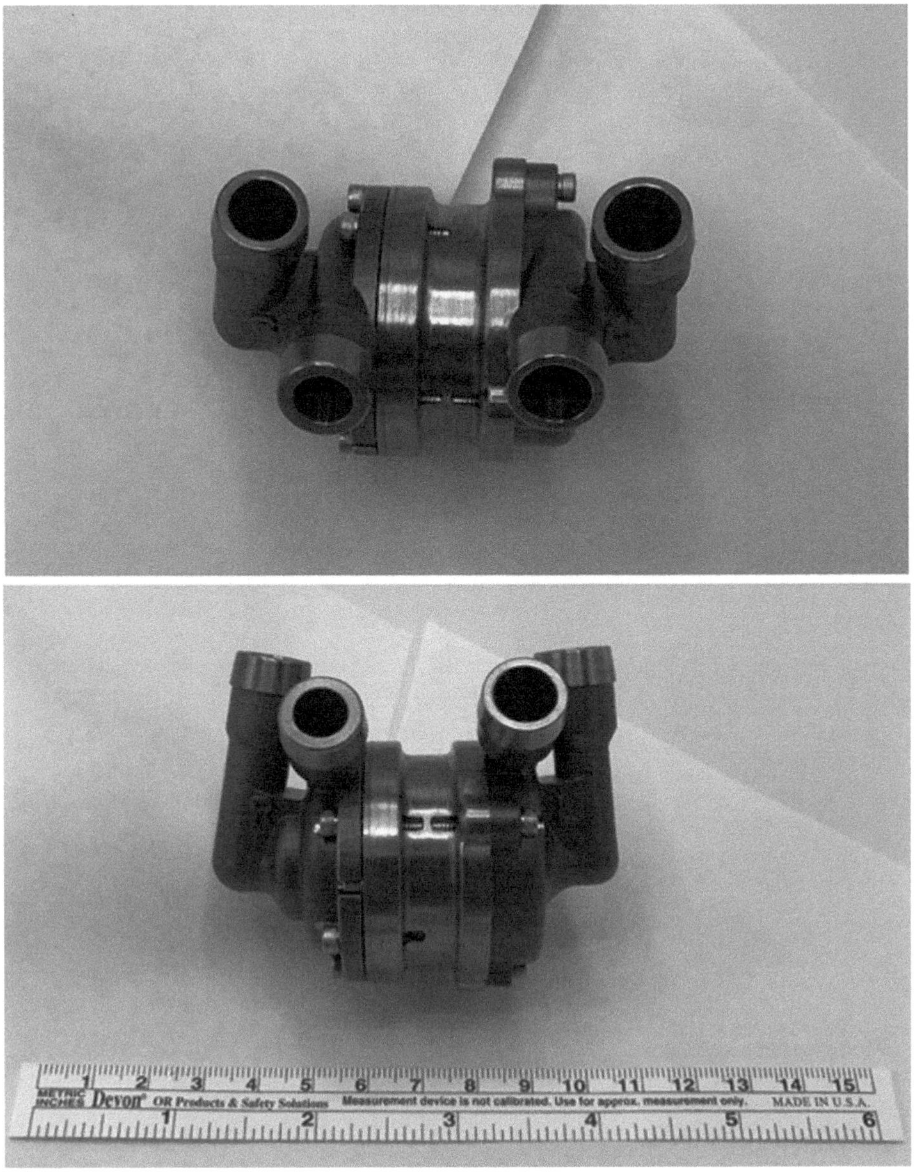

FIGURE 9.16 Pediatric continuous-flow total artificial heart (P-CFTAH) prototype.

requirements for pump flow range (1.5–4.5 L/min) and self-regulation (LAP-RAP within −5 to +10 mm Hg). The ongoing in vivo experiments will help us evaluate the hydraulic performance and self-regulating mechanical design of the P-CFTAH under selected hemodynamic conditions. Multiple challenges to pediatric MCS devices exist, and these difficulties need to be overcome to advance the field and improve quality of life, safety and survival of pediatric patients in need of MCS.[46]

Total artificial heart for very small patients

In order to explore the possibilities of using the CFTAH platform and designing further downsizing of the pump, we recently tested in silico the performance of the initial device model of an infant continuous-flow total artificial heart (I-CFTAH) with one motor and one rotating assembly supported by a hydrodynamic bearing.[47] The I-CFTAH (Fig. 9.18) (flange diameter, 26.9 mm; length, 44 mm) was envisioned to provide hemodynamic support to children ranging from 3 to 9 kg, with nominal design point matched to a 6 kg child, requiring 1.0 L/min of flow. I-CFTAH was tested using an in silico virtual simulation mock loop (VML; MATLAB®; MathWorks) and simulated the hemodynamics as they surge through the system implanted with a blood pump (run at 0.5 L/min for 3 kg infant).

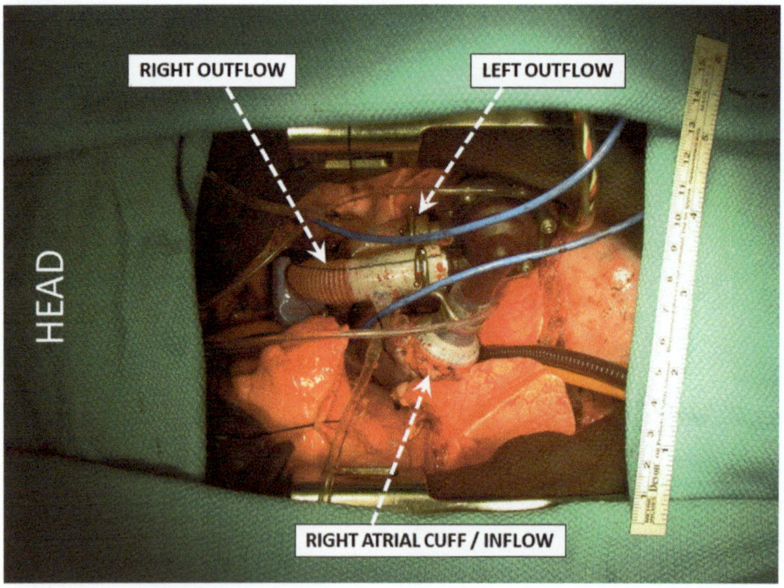

FIGURE 9.17 P-CFTAH prototype implantation in animal model.

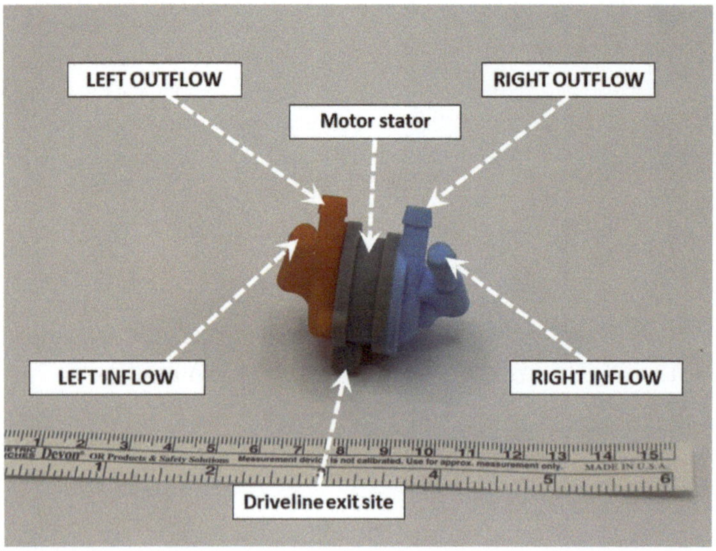

FIGURE 9.18 Infant continuous-flow total artificial heart (I-CFTAH) device model.

A lumped-parameter model was used, including systemic/pulmonary circulation, values for impedance, beat rate, and blood volume. The early assessment during in silico experiments demonstrated the intended performance of the scaled down pump (left and right pump flows, 0.5 L/min at 4380 rpm). The difference between the left and right atrial pressures was maintained within ± 5–15 mm Hg. Simulation of I-CFTAH was successfully performed and allowed the creation of arterial pulsation with this device. The development of the I-CFTAH is ongoing, and we will be looking to investigate the performance of this device in an animal model.

Biventricular heart failure support using a dual centrifugal pump

The use of CF (nonpulsatile) hemodynamic support using two CF pumps was first performed in our institution as biventricular support in a 75-kg calf supported for 34 days.[48] Today, the efforts are ongoing at Cleveland Clinic, with implantable devices to develop better options for mechanical support of a failing heart using CF devices.[49]

When left ventricular filling remains inadequate despite pharmacotherapy to enhance right ventricular (RV) function and lower pulmonary vascular resistance, a right ventricular assist device (RVAD) may be required.[50] The primary reason that VADs have not been designed specifically for RVAD applications is that there is much less clinical need (and therefore a limited market) than for LVADs.[51,52] Although clinically approved implantable LVADs are not designed for right-sided support, LVAD implantation is being considered for right heart support simply because an effective, easily implantable RVAD is not yet available. The absence of a dedicated, clinically evaluated, long-term RVAD significantly diminishes the treatment options for patients with isolated and post-LVAD RV failure, as well as for those with biventricular HF requiring two pumps (BVAD).[53,54]

The first use of an implantable CF pump for RV support was reported in 2004. A 21-year-old patient with worsened RV failure and elevated peripheral vascular resistance after implantation of a Jarvik 2000 FlowMaker was treated with implantation of a second one.[50] By directing the inlet toward the inferior vena cava–right atrial junction and keeping right atrial pressures at 5–10 cm H_2O, surgeons achieved stable RV unloading and satisfactory blood flow to the left-sided pump. Other groups consecutively reported on limited, off-label clinical use of two LVADs. Saito et al.[54] converted a 6-month survival patient from a Toyobo paracorporeal device (Nipro, Osaka, Japan) to a Jarvik 2000 as an RVAD and a DuraHeart (Terumo Heart, Inc., Ann Arbor, MI) for LVAD support.

Successful application of two HVADs (Medtronic, Minneapolis, MN) to treat severe biventricular failure was previously reported.[55–57] A group from the German Heart Institute (Deutsches Herzzentrum) in Berlin reported on 10 patients with idiopathic dilated cardiomyopathy who received biventricular HVADs.[58] In three of these patients, physicians were able to turn the RVAD off following long-term BVAD support, having confirmed that the patients had achieved stable hemodynamics and that no thromboembolic events and no regurgitation through the right pump had occurred.

Currently, temporary support via RVADs/BVADs is available by CF extracorporeal or paracorporeal circulation support systems[52,59] such as the CentriMag (Thoratec, Pleasanton, CA), the Bio-Pump (Medtronic, Minneapolis, MN), and the TandemHeart (CardiacAssist, Inc., Pittsburgh, PA). Progress in the development of clinically available rotary pumps now permits physicians to choose a hybrid approach for BVAD placement, using both an implantable LVAD and a temporary extracorporeal RVAD.[60–63] The adaptation of an LVAD to function as an RVAD remains a technical challenge because of both ventricular distinctions and pump characteristics. BVAD support can be effective, but continues to be associated with lower survival and a worse quality of life than left ventricular support alone.[64]

Despite progress in therapies for left HF, the biventricular failure represents an extremely ill population with end-stage cardiac disease that often involves both ventricles, even if its initial cause was limited to left-sided heart disease. There is a high incidence of adverse events and poor 1-year survival in the BVAD population.[65,66] With no linkage to any specific pump type, this condition demands timely intervention to sustain deteriorating hemodynamics and to facilitate symptomatic improvement, while continuing to limit the long-term success of durable MCS therapies.[66] TAHs are more appropriate than BVADs for patients with thrombosed ventricles, restrictive cardiomyopathy, cardiac tumors, and severe ventricular septal defects, frail ventricles for difficult cannulation, congenital abnormalities, and intractable arrhythmias.

We recently performed a feasibility assessment of MCS using a miniature CF BVAD based on a P-CFTAH for biventricular support (Fig. 9.19).[67] With the introduction of a unique cannulation approach, we were able to test

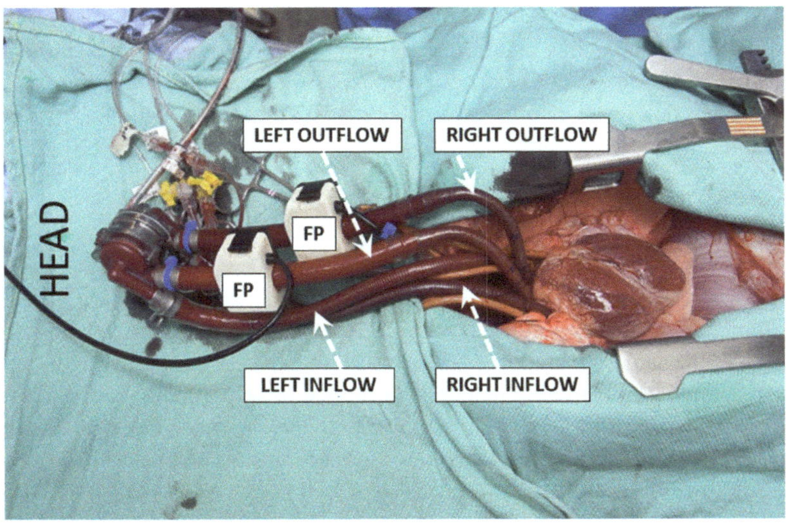

FIGURE 9.19 Intraoperative view of biventricular assist device (BVAD) in acute animal model. *FP*, Ultrasonic flow probe.

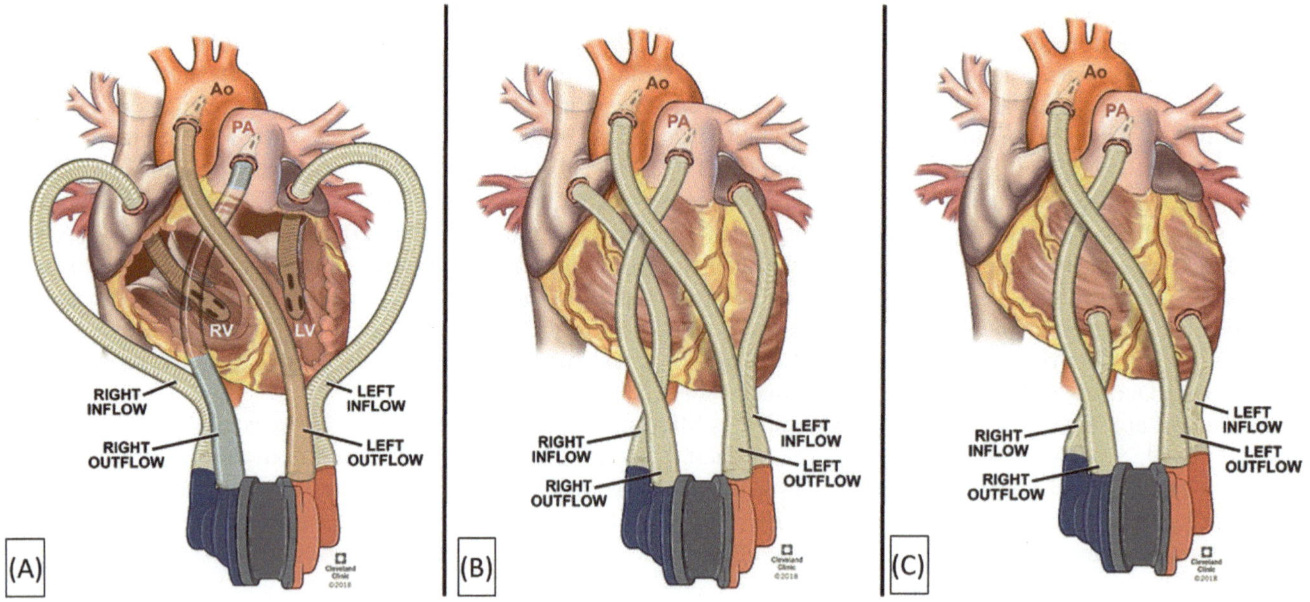

FIGURE 9.20 Schematic illustration of biventricular support using single device approach. *Ao*, Aorta; *LV*, ventricular cannulation; *PA*, pulmonary artery; *right*, atrial cannulation. Experimental setup (A), atrial (B) and ventricular cannulation (C).

different options for pump cannulations (Fig. 9.20A). In its downscaled configuration, the dimensions of the present blood pump have been shown to be implantable for TAH support in patients whose BSA is ≥ 0.3 m^2.[68] However, the device has never been tested in the presence of the native heart ventricles. The BVAD used in this study is a single-piece device (P-CFTAH) that is valveless and does not employ flow or pressure sensors for its operation or control.[20,69] All acute experiments in lambs ($n = 4$) showed adequate device performance and stability of device operation.

The in vivo data results compared favorably with previously collected in vitro pump performance. In both cannulation approaches [atrial (Fig. 9.20B) and ventricular (Fig. 9.20C)], the BVAD performance was maintained adequately within the 3000–6000 rpm range after implantation. The device self-regulating performance was demonstrated by the left and right atrial pressure difference (LAP-RAP) falling predominantly within the range of −5 to 10 mm Hg in left and right HF conditions with either atrial and ventricular cannulation types.

This early series of acute in vivo experiments with the single-piece BVAD support has demonstrated the intended hemodynamic output, device features, and feasibility of BVAD support using one device and one controller unit. More development will be needed to investigate true hemodynamic requirements for BVAD support using a single device, degree or self-regulation, and potential advantages or drawbacks in comparison with the dual-VAD approach for biventricular HF treatment. This will be the required technology mapping step to indicate the most appropriate patient population for whatever approach will become available in the future.

As of today, the adult version of the Cleveland Clinic CFTAH has undergone multiple design optimizations (including changes to right impeller and stacked motor configuration), and is currently undergoing animal testing for pump performance and physiologic and hemodynamic parameters. It will also be assessed for its use as a BVAD in select HF populations where higher hemodynamic output will be required.

Summary

There has been a continuous stream of new devices and modifications to existing technologies in recent years. These therapies have improved outcomes for patients in both morbidity and mortality. Significant technological and therapeutic advances in TAHs have been achieved in recent years, which is a major breakthrough in the field. However, issues with blood pumps continue to be the focus of increasingly sophisticated research in device development. Investigators are sharing insights into the physiological characteristics of mechanical circulation, and newer, more refined device technologies are continually emerging. Biocompatibility of the device components in direct contact with native tissues and blood is improving. However, the biocompatibility of devices, as well as current methods of device control, are still far from ideal, and merit continuing investigation.

The theoretical advantages of centrifugal-flow pumps over axial-flow pumps could move the field more toward centrifugal pump design in the future; however, with new device designs and better materials being used, there is certainly much room for improvements in pulsatile blood pump technologies, as well. Centrifugal pumps have better flow characteristics that can minimize hemolysis and perhaps also lead to less need for anticoagulation therapy.

In addition, such pumps have a greater degree of "pulsatility" inherent in their design, which may translate into fewer problems with aortic insufficiency, aortic valve fusion, and gastrointestinal bleeding. Finally, the particular pressure/flow characteristics of these pumps and their inherently more accurate flow estimation will likely lead to the development of automated speed-control algorithms. These may provide a much safer, more effective way of operating these pumps compared with the fixed-speed mode of operation that is a current design feature.

The studies performed recently at Cleveland Clinic marked a significant milestone in proving that CFTAH technologies can support life for a long duration with a single device. Today, device development and biocompatibility studies at Cleveland Clinic are ongoing under active funding from the National Institutes of Health and institutional support. The implementation of this project and achievement of this important milestone suggests that single-piece, autoregulated and sensorless devices have significant potential for further development into a valid clinical option in HF patients.

Issues surrounding how best to manage RV failure are still at the crux of every decision that is made for LVAD implantation. Better techniques need to be developed to identify those patients who will or will not benefit from LVAD therapy alone. Developments such as a dedicated, permanently implanted RVAD; safe, effective conduit and pump speed modifications to current LVAD technology; or single-device solutions suitable for BVAD support will greatly expand applications for chronic biventricular failure. Complete cardiac replacement with CFTAHs, while currently in early development and application, will play a key role in addressing a subset of patients with chronic biventricular failure. Smaller, simpler CFTAHs that are proven reliable may even move the field completely in this direction.

It is expected that advances in the field of MCS will expand the use of support devices to the point where they may offer a long-term substitute in terms of quality of life, or at least provide a feasible alternative to heart transplantation for a greater number of terminally ill HF patients.[70]

Notably, invasive options to treat end-stage HF continue to be updated and expanded as more innovative devices emerge. The most effective, durable, reliable, and fully autoregulated devices will undoubtedly save lives. However, until the ideal pump has been created and brought into accepted clinical practice, patient survival will continue to be primarily determined by an early decision to implant univentricular or biventricular devices, and not by choice of device type.

In addition to early diagnosis and treatment with MCS devices, primary efforts and attention should continue to be provided for better prevention of HF by treating risk factors at an early stage so that the device therapies can be more focused and available when HF is not further manageable by more conservative means. Further research will be needed to boost these endeavors and to benefit the demanding clinical need of the growing HF patient population.

Our early bench evaluation and feasibility studies have demonstrated promising results of the CFTAH device platform, showing that it is capable of delivering full systemic and pulmonary circulatory support over a wide range of hemodynamic conditions in adults, pediatric patients, and infants. Research and development on biventricular support using a single-device approach is ongoing. Ultimately, more insights will be provided on the best suitable TAH platform and device configurations for biventricular circulatory support at earlier stages of HF.

References

1. Benjamin EJ, Virani SS, Callaway CW, et al. American Heart Association Council on E, Prevention Statistics C, Stroke Statistics S. Heart disease and stroke statistics-2018 update: a report from the American Heart Association. *Circulation*. 2018;137:e67–e492.
2. Benjamin EJ, Muntner P, Alonso A, et al. Heart disease and stroke statistics-2019 update: a report from the American Heart Association. *Circulation*. 2019;139:e56–e528.
3. Heidenreich PA, Albert NM, Allen LA, et al. American Heart Association Advocacy Coordinating C, Council on Arteriosclerosis T, Vascular B, Council on Cardiovascular R, Intervention, Council on Clinical C, Council on E, Prevention, Stroke C. Forecasting the impact of heart failure in the United States: a policy statement from the American Heart Association. *Circ Heart Fail*. 2013;6:606–619.
4. Wever-Pinzon O, Drakos SG, Kfoury AG, et al. Morbidity and mortality in heart transplant candidates supported with mechanical circulatory support: is reappraisal of the current United network for organ sharing thoracic organ allocation policy justified? *Circulation*. 2013;127:452–462.
5. Garbade J, Bittner HB, Barten MJ, Mohr FW. Current trends in implantable left ventricular assist devices. *Cardiol Res Pract*. 2011;2011:290561.
6. Moazami N, Hoercher KJ, Fukamachi K, et al. Mechanical circulatory support for heart failure: past, present and a look at the future. *Expert Rev Med Dev*. 2013;10:55–71.
7. Mehra MR, Uriel N, Naka Y, et al. A fully magnetically levitated left ventricular assist device—final report. *N Engl J Med*. 2019.

8. Kirklin JK, Pagani FD, Kormos RL, et al. Eighth annual INTERMACS report: special focus on framing the impact of adverse events. *J Heart Lung Transplant*. 2017;36:1080–1086.
9. Baldwin AC, Gemmato CJ, Cohn WE, Frazier OH. Feasibility of long-term continuous flow total heart replacement in calves. *Int J Artif Organs*. 2021; 391398820987851.
10. Moazami N, Fukamachi K, Kobayashi M, et al. Axial and centrifugal continuous-flow rotary pumps: a translation from pump mechanics to clinical practice. *J Heart Lung Transplant*. 2013;32:1–11.
11. Fu Y, Hu Y, Huang F, Zhou M. The impact of pulsatile flow on suspension force for hydrodynamically levitated blood pump. *J Healthcare Eng*. 2019;2019:8065920.
12. Glynn J, Song H, Hull B, et al. The OregonHeart total artificial heart: design and performance on a mock circulatory loop. *Artif Organs*. 2017;41:904–910.
13. Jurney PL, Glynn JJ, Dykan IV, et al. Characterization of a pulsatile rotary total artificial heart. *Artif Organs*. 2020;.
14. Timms D, Fraser J, Hayne M, Dunning J, McNeil K, Pearcy M. The BiVACOR rotary biventricular assist device: concept and in vitro investigation. *Artif Organs*. 2008;32:816–819.
15. Carpentier A, Latrémouille C, Cholley B, et al. First clinical use of a bioprosthetic total artificial heart: report of two cases. *Lancet (London, England)*. 2015;386:1556–1563.
16. Pelletier B, Spiliopoulos S, Finocchiaro T, et al. System overview of the fully implantable destination therapy—ReinHeart-total artificial heart. *Eur J Cardiothorac Surg*. 2015;47:80–86.
17. Melton N, Soleimani B, Dowling R. Current role of the total artificial heart in the management of advanced heart failure. *Curr Cardiol Rep*. 2019;21:142.
18. Lundberg MS, Baldwin JT, Buxton DB. Building a bioartificial heart: obstacles and opportunities. *J Thorac Cardiovasc Surg*. 2017;153:748–750.
19. Medvedev AL, Karimov JH, Kuban BD, Horvath DJ, Moazami N, Fukamachi K. Unlocking the box: basic requirements for an ideal ventricular assist device controller. *Expert Rev Med Dev*. 2017;14:393–400.
20. Fukamachi K, Horvath DJ, Massiello AL, et al. An innovative, sensorless, pulsatile, continuous-flow total artificial heart: device design and initial in vitro study. *J Heart Lung Transplant*. 2010;29:13–20.
21. Kobayashi M, Horvath DJ, Mielke N, et al. Progress on the design and development of the continuous-flow total artificial heart. *Artif Organs*. 2012;36:705–713.
22. Karimov JH, Moazami N, Kobayashi M, et al. First report of 90-day support of 2 calves with a continuous-flow total artificial heart. *J Thorac Cardiovasc Surg*. 2015;150:687–693. e681.
23. Horvath D, Byram N, Karimov JH, et al. Mechanism of self-regulation and in vivo performance of the Cleveland Clinic continuous-flow total artificial heart. *Artif Organs*. 2017;41:411–417.
24. Fumoto H, Horvath DJ, Rao S, et al. In vivo acute performance of the Cleveland Clinic self-regulating, continuous-flow total artificial heart. *J Heart Lung Transplant*. 2010;29:21–26.
25. Fukamachi KKJ, Horvath D, Byram N, Goodin M. *Impeller for Artificial Heart Blood Pumps*. WIPO; 2019.
26. Kuban BD, Karimov JH, Miyamoto T, Kado Y, Horvath DJ, Fukamachi K. Total artificial heart incorporating novel stacked motor; first in vivo results. *J Heart Lung Transplant*. 2020;39:S413.
27. Kink T, Reul H. Concept for a new hydrodynamic blood bearing for miniature blood pumps. *Artif Organs*. 2004;28:916–920.
28. Ranganath NK, Rashidi M, Antaki JF, et al. Mechanical blood-immersed bearings in continuous-flow rotary blood pumps. *ASAIO J*. 2020;66:343–347.
29. Han Q, Zou J, Ruan X, Fu X, Yang H. A novel design of spiral groove bearing in a hydrodynamically levitated centrifugal rotary blood pump. *Artif Organs*. 2012;36:739–746.
30. Murashige T, Sakota D, Kosaka R, et al. Plasma skimming in a spiral groove bearing of a centrifugal blood pump. *Artif Organs*. 2016;40:856–866.
31. Sunagawa G, Horvath DJ, Karimov JH, Moazami N, Fukamachi K. Future prospects for the total artificial heart. *Expert Rev Med Dev*. 2016;13:191–201.
32. Menter FR. Two-equation eddy-viscosity turbulence models for engineering applications. *AIAA J*. 1994;32:1598–1605.
33. Cho Y, Kensey KR. Effects of the non-Newtonian viscosity of blood on flows in a diseased arterial vessel. Part 1: Steady flows. *Biorheology*. 1991;241–262.
34. Banerjee RK, Cho Y, Kensey KR. Effect of non-Newtonian viscosity of blood on steady and pulsatile flow in stenosed arteries. *Adv Bioeng*. 1991;20:103–106.
35. Leverett LB, Hellums JD, Alfrey CP, Lynch EC. Red blood cell damage by shear stress. *Biophys J*. 1972;12:257–273.
36. Bluestein D, Niu L, Schoephoerster RT, Dewanjee MK. Fluid mechanics of arterial stenosis: relationship to the development of mural thrombus. *Ann Biomed Eng*. 1997;25:344–356.
37. Einav S, Bluestein D. Dynamics of blood flow and platelet transport in pathological vessels. *Ann N Y Acad Sci*. 2004;1015:351–366.
38. Davis MJAJ. Design of flat plate leading edges to avoid flow separation. *AIAA J*. 1980;18:598–600.
39. Sheriff J, Bluestein D, Girdhar G, Jesty J. High-shear stress sensitizes platelets to subsequent low-shear conditions. *Ann Biomed Eng*. 2010;38:1442–1450.
40. Horvath D, Karimov JH, Byram N, et al. Sensorless suction recognition in the self-regulating Cleveland Clinic continuous-flow total artificial heart. *ASAIO J*. 2015;61:726–728.

41. Horvath DJ, Karimov JH, Byram NA, et al. Advantages of integrating pressure-regulating devices into mechanical circulatory support pumps. *ASAIO J*. 2019;65:e1–e3.
42. Karimov JH, Moazami N, Sunagawa G, et al. Median sternotomy or right thoracotomy techniques for total artificial heart implantation in calves. *Artif Organs*. 2016;40:1022–1027.
43. Fukamachi K, Karimov JH, Horvath DJ, et al. Initial in vitro testing of a paediatric continuous-flow total artificial heart. *Interact Cardiovasc Thorac Surg*. 2018;26:897–901.
44. Karimov Jamshid H, Horvath D, Byram N, et al. Abstract 16894: initial in vivo results of Cleveland Clinic pediatric continuous-flow total artificial heart implant. *Circulation*. 2017;136:A16894.
45. Karimov JH, Horvath DJ, Byram N, et al. Early in vivo experience with the pediatric continuous-flow total artificial heart. *J Heart Lung Transplant*. 2018;37:1029–1034.
46. Fukamachi K, Karimov JH, Miyamoto T. Challenges in pediatric mechanical circulatory support devices. *Artif Organs*. 2019;43:441–443.
47. Karimov JH, Polakowski A, Horvath D, et al. Development of continuous-flow total artificial heart for use in infants. *J Heart Lung Transplant*. 2019;38:S482.
48. Golding LR, Jacobs G, Murakami T, et al. Chronic nonpulsatile blood flow in an alive, awake animal 34-day survival. *Trans Am Soc Artif Intern Organs*. 1980;26:251–255.
49. Karimov JH, Horvath D, Byram N, et al. Mechanical circulatory support for biventricular heart failure using continuous-flow total artificial heart. *J Heart Lung Transplant*. 2019;38:S343.
50. Frazier OH, Myers TJ, Gregoric I. Biventricular assistance with the Jarvik FlowMaker: a case report. *J Thorac Cardiovasc Surg*. 2004;128:625–626.
51. Fukamachi K, Shiose A, Massiello AL, et al. Implantable continuous-flow right ventricular assist device: lessons learned in the development of a Cleveland Clinic device. *Ann Thorac Surg*. 2012;93:1746–1752.
52. Karimov JH, Sunagawa G, Horvath D, Fukamachi K, Starling RC, Moazami N. Limitations to chronic right ventricular assist device support. *Ann Thorac Surg*. 2016;102:651–658.
53. Loforte A, Stepanenko A, Potapov EV, et al. Temporary right ventricular mechanical support in high-risk left ventricular assist device recipients vs permanent biventricular or total artificial heart support. *Artif Organs*. 2013;37:523–530.
54. Saito S, Sakaguchi T, Miyagawa S, et al. Biventricular support using implantable continuous-flow ventricular assist devices. *J Heart Lung Transplant*. 2011;30:475–478.
55. Strueber M, Meyer AL, Malehsa D, Haverich A. Successful use of the HeartWare HVAD rotary blood pump for biventricular support. *J Thorac Cardiovasc Surg*. 2010;140:936–937.
56. Hetzer R, Krabatsch T, Stepanenko A, Hennig E, Potapov EV. Long-term biventricular support with the heartware implantable continuous flow pump. *J Heart Lung Transplant*. 2010;29:822–824.
57. Shehab S, Macdonald PS, Keogh AM, et al. Long-term biventricular HeartWare ventricular assist device support—case series of right atrial and right ventricular implantation outcomes. *J Heart Lung Transplant*. 2016;35:466–473.
58. Potapov E, Schweiger M, Vierecke J, et al. Discontinuation of HeartWare RVAD support without device removal in chronic BIVAD patients. *ASAIO J*. 2012;58:15–18.
59. Kaczorowski DJ, Woo YJ. Who needs an RVAD in addition to an LVAD? *Cardiol Clin*. 2011;29:599–605.
60. Chen JM, Levin HR, Rose EA, et al. Experience with right ventricular assist devices for perioperative right-sided circulatory failure. *Ann Thorac Surg*. 1996;61:305–310. Discussion 311–303.
61. Lazar JF, Swartz MF, Schiralli MP, et al. Survival after left ventricular assist device with and without temporary right ventricular support. *Ann Thorac Surg*. 2013;96:2155–2159.
62. Meineri M, Van Rensburg AE, Vegas A. Right ventricular failure after LVAD implantation: prevention and treatment. *Best Pract Res Clin Anaesthesiol*. 2012;26:217–229.
63. Raina A, Patarroyo-Aponte M. Prevention and treatment of right ventricular failure during left ventricular assist device therapy. *Crit Care Clin*. 2018;34:439–452.
64. Aissaoui N, Morshuis M, Schoenbrodt M, et al. Temporary right ventricular mechanical circulatory support for the management of right ventricular failure in critically ill patients. *J Thorac Cardiovasc Surg*. 2013;146:186–191.
65. Shah P, Ha R, Singh R, et al. Multicenter experience with durable biventricular assist devices. *J Heart Lung Transplant*. 2018;37:1093–1101.
66. Vierecke J, Gahl B, de By T, et al. Results of primary biventricular support: an analysis of data from the EUROMACS registry. *Eur J Cardiothorac Surg*. 2019;.
67. Karimov JH, Horvath DJ, Miyamoto T, et al. First in vivo experience with biventricular circulatory assistance using a single continuous flow pump. *Semin Thorac Cardiovasc Surg*. 2020;32:456–465.
68. Fukamachi K, Karimov JH, Byram NA, et al. Anatomical study of the Cleveland Clinic continuous-flow total artificial heart in adult and pediatric configurations. *J Artif Organs*. 2018;21:383–386.
69. Fukamachi K, Horvath DJ, Byram N, Sunagawa G, Karimov JH, Moazami N. Advanced ventricular assist device with pulse augmentation and automatic regurgitant-flow shut-off. *J Heart Lung Transplant*. 2016;35:1519–1521.
70. Karimov JH, Polakowski AR, Fukamachi K, Miyamoto T. Progress in mechanical circulatory support: challenges and opportunities. *Artif Organs*. 2019; (in press).

Chapter 10

Cardiac Transplantation and Organ Preservation

Cristiano Amarelli[1], Lucrezia Palmieri[1], Jamshid H. Karimov[2], Andreas Zuckermann[3], David D'Alessandro[4], Johan Nilsson[5] and Robert H. Bartlett[6]

[1]*Department of Cardiac Surgery and Transplants, Monaldi, A.O. dei Colli, Naples, Italy,* [2]*Lerner Research Institute, Cleveland Clinic, Cleveland, OH, United States,* [3]*Department of Cardiovascular Surgery, University of Vienna, Vienna, Austria,* [4]*Division of Cardiac Surgery, Transplantation and Mechanical Circulatory Support, Massachusetts General Hospital, Boston, MA, United States,* [5]*Department of Translational Medicine, Thoracic Surgery and Bioinformatics, Lund University and Skane University Hospital, Lund, Sweden,* [6]*Department of General Surgery, Division of Acute Care Surgery, University of Michigan, Ann Arbor, MI, United States*

Chapter Outline

Introduction

The transplantation landscape after 2022

During the first 50 years of its history, heart transplantation has been the foremost innovation in cardiac surgery due to its experimental nature, and the continued challenges due to organ shortage and the changes in patients' clinical profiles benefiting from heart replacement therapy.

Facing a continuous improvement of mid-term and long-term outcomes, clinicians have dealt much more with managing immunosuppressive issues and their common complications, paving the way for the relative stability of perioperative mortality from the introduction of CNI up to the present.

During the past few years, due to the changes of allocation policies and the more complicated clinical and surgical profile of the cardiac recipients and to the still-high incidence of primary graft failure (PGF), graft preservation in organ transplantation has returned as a research priority. On the one hand, the improvements in medical management of outpatients suffering from chronic heart failure and the availability of a valuable answer to patients' clinical needs represented by left ventricular assist devices (LVADs) has shifted the indication toward the allocation of organs to urgent candidates.[1] The consequence has been the increase in ischemic times to bring the organs to the point where the urgent recipient stays and an increased chance of PGF due to the rise of surgical complexity and addition of donor and recipient risk factors.[2]

The dreadful course of PGF, its effects on postoperative ICU stays, and the need for expensive treatments (ECMO, temporary circulatory support) are the main reasons to push toward the research of alternative paradigms to improve myocardial preservation. The opportunity to recondition the organs and possibly enlarge the acceptability criteria for heart donation has been a further call for new technologies (Fig. 10.1).

Advances in Cardiovascular Technology. DOI: https://doi.org/10.1016/B978-0-12-816861-5.00016-2

A large amount of research emerging from abdominal organs to reduce graft failure and delayed functioning has led to rediscussing the previously accepted and almost standardized graft preservation based on cold static storage (CSS).[3–5]

Intriguingly, until now, no precise data have existed regarding the right cardioplegic solution for CSS preservation and the right temperature, and few data are broadly accepted to reduce the burden of PGF.

Ex situ organ perfusion (ESOP; better than ex vivo since it is used predominantly after a brain or circulatory death) has been advocated as the more physiologic way to perfuse the organs during the transport time. Still, little disrupting evidence clearly supported the advantage of this approach. Recently, the data coming from a nonrandomized trial on ex situ hypothermic perfusion of the heart and from cold hypothermic blood perfusion of the liver supported the benefit of cold perfusion as a possible alternative strategy.[6–8]

This chapter will deal with the advances and the innovative paradigms leading to reengineering the preservation of the myocardium and the potential of new upcoming reconditioning strategies to reduce BDD organ shortage and move toward DCD donation.

The weight of graft preservation on primary graft failure

The first heart transplantation, performed by Christiaan Barnard in Cape Town, South Africa, was a DCD donation with a very short ischemic time since the two teams worked in parallel due to the fear of PGF.

During the first 40 years of heart transplantation, the preservation of the organ has been continuously and rarely changed without a systematic focus on it, thus leading to a lack of clear demonstrations about the superiority of one CSS solution over another. As many as 167 different types of preservation fluids are claimed to be available for cold storage[9]; Celsior is safe and effective in myocardial preservation.[10] Our retrospective single-center analysis of 200 transplant procedures demonstrated that Celsior[11] provides comparably effective myocardial preservation in both standard and suboptimal or even marginal donor grafts. However, the addition of mechanical arrest, ventricular unloading,

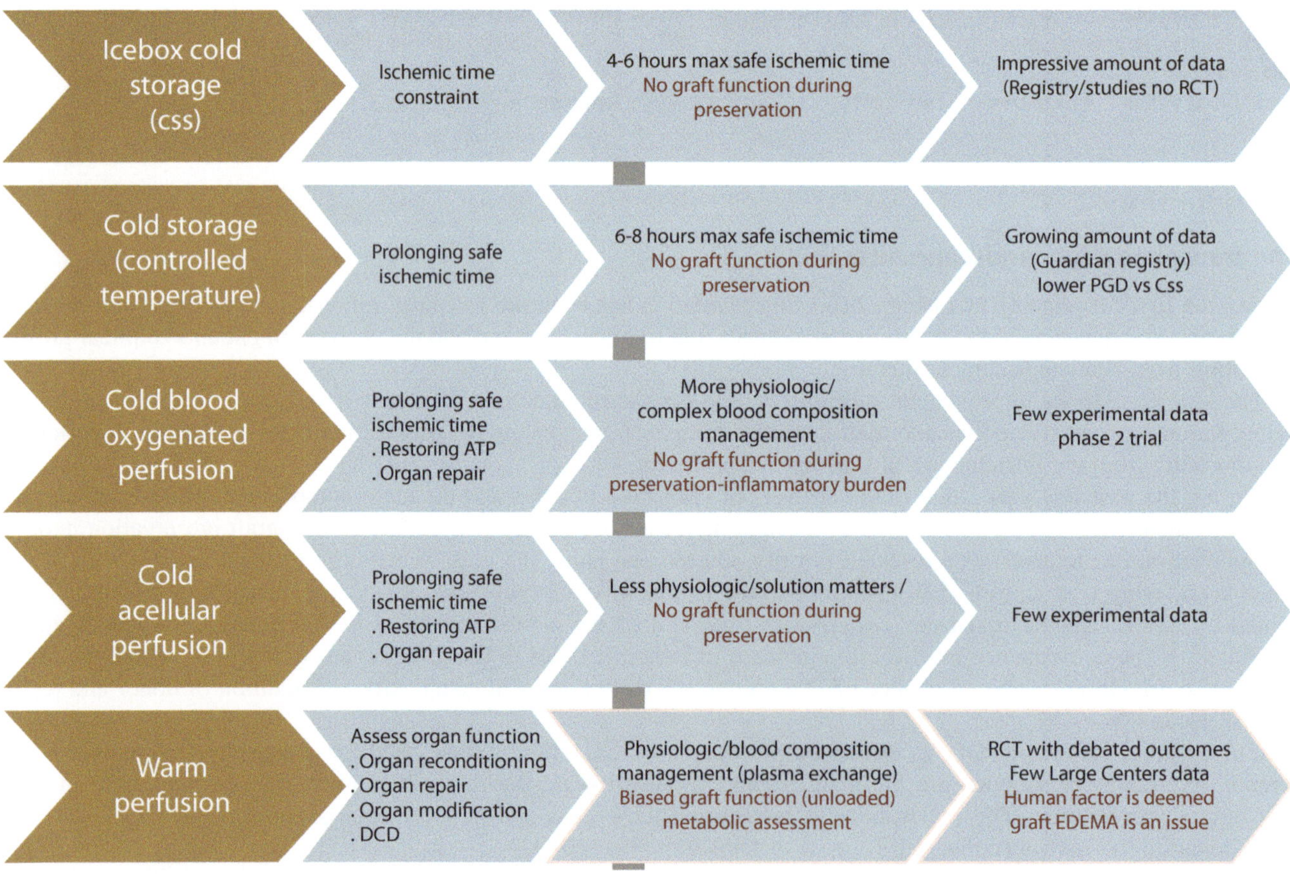

FIGURE 10.1 Strengths and drawbacks of new approaches to myocardial preservation.

and hypothermic preservation, combined with other clinical challenges, fail to protect all the patients from primary graft dysfunction (PGD), which still accounts for most of the deaths after heart transplant within 30 days, along with paramount effects in terms of costs and prolongation of ICU stay.[12] The risk of PGD is strictly related to the recipient pool with the inclusion of higher risk candidates. It is further exacerbated by the extension of donor criteria to include the use of previously unacceptable organs. The right ventricle is the Achilles heel of graft preservation, since it is more vulnerable to injury from ischemia and benefits the most from reliable cardioplegic protection.[13] Thus, benchmarking protection standards and transplantation outcomes using available cardiac preservation solutions is critical for developing future advances in myocardial preservation technology.

During these years, the advantages of one composition over another and new intracellular solutions have been periodically claimed to extend the safe ischemic time. Many centers continue to use St. Thomas cardioplegia as a standard of care for CSS in real-world practice. Some evidence was produced regarding the superiority of intracellular solutions when the ischemic time is deemed longer or the donor is deemed at higher risk.[14] Older donors carry a lower tolerance to cold ischemic preservation; in Europe, due to the donor shortage, there has been an impressive progressive increase of mean donor age. The absence of a granular definition of organ function has led to the need to design scores that could characterize the risk brought by the donor in a numerical value (ETS or adapted score from Zuckermann).[15,16] These scores have been proposed to ease the process of allocating the organs from the perspective of the OPOs, thus facilitating the process of defining the acceptability criteria and the donor pool of different centers[17]

The issue of organ preservation in heart transplantation has been flawed by assessing donor quality and the possible modifications due to brain death and its management. It appears clear that the graft function after 24–48 h from reperfusion is quite different from that seen during the evaluation of the graft during the retrieval (Fig. 10.2). Within these changes, there are:

- the intrinsic quality and function of the graft as assessed during retrieval;
- the amount of ischemic damage;
- the amount of damage due to the freezing;
- the rewarming and reoxygenating injury;

and in the end, the amount of reperfusion injury could be related to the amount of ischemia but also to immunologic reasons.

Looking at this list of factors, it appears clear why many of the studies that aimed to assess the preservation solution were not capable of achieving definitive results. Indeed, the complex puzzle of factors that led to PGF has been the

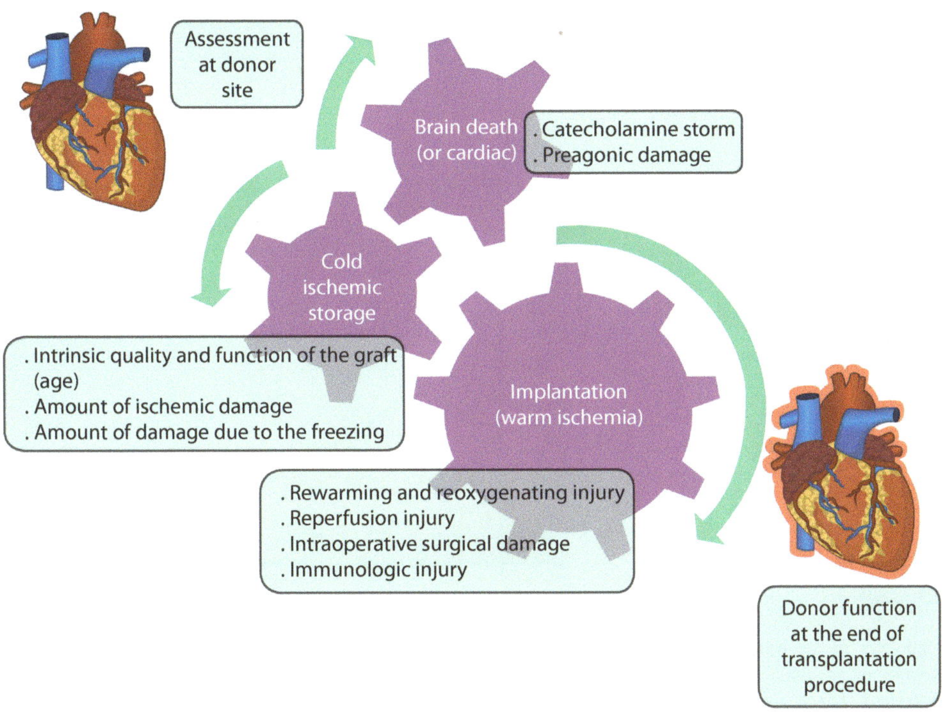

FIGURE 10.2 Determinants of graft performance at the end of a heart transplant procedure.

object of investigations that amplify the donor pool more than obtaining more physiologic graft preservation during transport.

Furthermore, an indefinite amount of evidence is emerging regarding the possible incremental effect of the poor quality of preservation on the graft immunogenicity, thus supporting a potential role of the quality of graft preservation on its long-term functioning.[18,19]

Equally, due to the organ shortage constraint, the current availability of a device achieving normothermic graft perfusion has suddenly becomes the platform to expand the donation frontiers toward circulatory death (DCD) more than just to improve and standardize the quality of organ preservation.[20]

Which solution to preserve the heart?

Adequacy of organ preservation is essential for successful heart transplantation. Suboptimal graft preservation is the basis of subsequent ischemia–reperfusion injury (IRI). It plays a critical role in the pathogenesis of PGD and acute right heart failure, significantly affecting early postoperative morbidity and mortality.[21]

With the spread of solid organ transplantation in the 1970s, plenty of preservation solutions have been developed explicitly for organ transplantation, including the Euro-Collins (EC), Celsior (CE), University of Wisconsin (UW), and Histidine-Tryptophan-Ketoglutarate (HTK) solutions.

Besides electrical silence, the other critical component of modern donor graft preservation techniques is the emptying and the use of hypothermia to inhibit metabolic activity (Fig. 10.3).

Blood-containing solutions or crystalloid solutions have populated the landscape of organ protective solutions. The latter are divided into "extracellular" and "intracellular" solutions. Their Na content solely determines this difference.

Intracellular solutions (e.g., UW solution) are known to cause chemical damage from potassium-induced injury to the coronary endothelium, leading to allograft vasculopathy, contraction band necrosis,[22] and coronary vasoconstriction from the release of endothelin.[23]

Therefore, the ideal preserving fluid has not yet been established. A variety of commercial preservation solutions or noncommercial, locally compounded solutions are used today by each heart transplant center, based on historical and regional preferences.

Although the composition differs, their preservation principles are analogous.

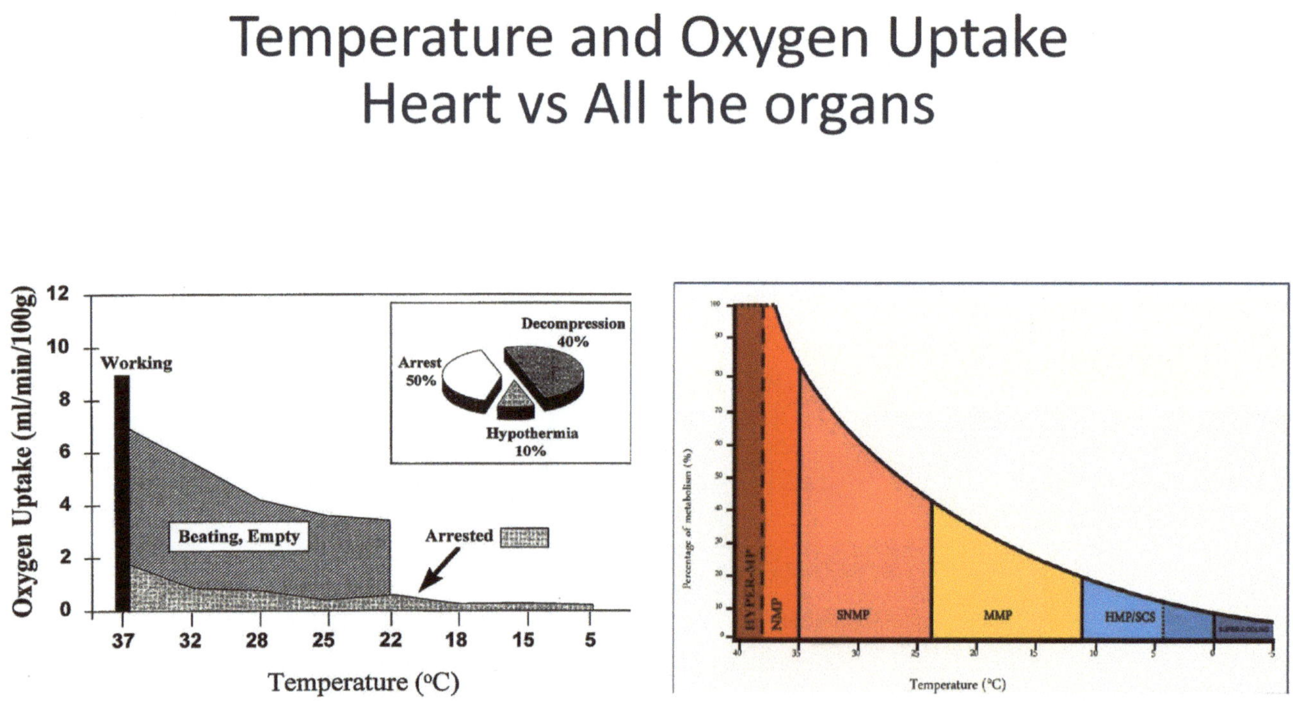

FIGURE 10.3 Determinants of graft metabolism in the heart versus other organs. *Modified from (left) Gravlee GP.* Cardiopulmonary Bypass: Principles and Practice; *and (right) Machine perfusion: cold versus warm, versus neither. Update on Clinical Trials.* Semin Liver Dis. *2020;40:264–281. https://doi.org/10.1055/s-0040-1713118.*

Insights into the physiopathology of cardiac energetics have led to three guidelines for the formulation of preservation fluids to:

1. warrant hypothermic arrest of metabolism;
2. recreate a physical and biochemical environment to maintain the viability of the ultrastructural components of the tissue during hypothermic metabolic slowing; and
3. minimize the burden of reperfusion injury.

Little comparative, real-world data on the preservation solutions exist[13]; the UNOS-STAR database was incomplete with 27.1% of patients with missing data (2.5%); "unknown" or "other" solution used (11.2%), unspecified cardioplegia solutions (12.1%), no flush used (0.3%), crystalloid flush (saline/Ringer solution, 0.2%), or other preservation solutions. UW solution (41.7%), and Celsior (22.6%; Genzyme Corp Cambridge, MA) were the most commonly used solutions. Only 8.7% reported HTK or HTK/Custodiol (Essential Pharmaceuticals, LLC, Durham, NC) use.

Newly designed solutions like the UW solution are claimed to achieve the recovering of electrical activity more rapidly with excellent myocardial protection.[24] These solutions are designed to:

1. immediately arrest the heart;
2. warrant effective membrane stabilization;
3. mitigate hypothermic cellular swelling;
4. reduce the presence of free radical scavengers; and
5. provide precursors for the repletion of high phosphate stores during reperfusion.[25]

Extracellular solutions (e.g., HTK), on the other hand, have a few key advantages:

1. reducing extracellular sodium concentration, being isotonic with cytoplasmic levels to minimize cellular swelling;
2. lowering extracellular calcium level to facilitate electromechanical silencing of cardiomyocytes;
3. excellent buffering capacity over a broad pH range from the presence of histidine; and
4. less viscous, allowing a more rapid flush with the thorough evacuation of intravascular contents and facilitating rapid cardiac cooling.[26]

Molecules with osmotic properties (e.g., glucose, mannitol, lactobionate, rabinose) aim to prevent graft edema due to sodium and water retention during the ischemic hypothermic preservation.

Antioxidants (e.g., allopurinol, glutathione, tryptophan, mannitol, histidine) may reduce the burden of oxygen free radicals.

Metabolic acidosis due to anaerobic metabolism and adenosine triphosphate (ATP) hydrolysis is counteracted by buffer molecules (e.g., phosphate, bicarbonate, histidine). Finally, adenosine, glutamic acid, or glutamate induce higher levels of ATP and preserve mitochondrial function.

In 2008, Rauen et al. showed that marked toxicity was observed during cold incubation, with histidine being the most likely responsible agent.[27] Adenosine 5′-triphosphate hydrolysis is deemed to lead to hydrogen ion production (ongoing at a low level even at temperatures close to 0°C) and histidine has been found to increase intracellular levels of redox-active chelatable iron.[28] Indeed, improvements of the composition should be based on replacing such effects.

In the attempt to achieve optimal composition in heart transplantation models, HTK-N has been designed to remove harmful components.[29] Further cold-induced, iron-independent injuries are mediated by chloride.[27] Thus, the modifications in experiments with HTK consisted of:

- a chloride-poor HTK composition in which histidine was replaced with *N*-acetyl-L-histidine and added with L-arginin.
- glycine, alanine (inhibition of sodium influx), arginine (NO synthase substrate), aspartate (additional energy substrate), and iron chelators (deferoxamine and LK 614) were added to the novel HTK-solution.

Iron chelators decreased the generation of cardiac F2-isoprostanes, a highly specific and sensitive biomarker of oxidative stress and improved parameters of cell function (e.g., enzyme release of creatine kinase [CK] and lactate dehydrogenase and decreased parameters of cell injury). Nevertheless, iron chelators in high concentrations have a negative effect on heart preservation due to toxicity.

In rats, left ventricular systolic pressure and coronary blood flow were improved, whereas the number of apoptotic cells was clearly reduced in the HTK-N group through the better-preserved endothelial function, accounting for the fast myocardial recovery.[30]

HTK-N for heart transplantation now needs to be confirmed in nonrodent animal models and in humans to demonstrate its superiority.

Until now, the "safe" ischemic time is expected to relate to the type of storage method. However, due to the prevalence of the single-center volume studies, with a low number of organs with prolonged ischemic time and the connatural clinical tendency to accept a risk factor mitigating other risk factors (i.e., donor age, D/R matching), risk factor analyses tend not to reach an adequate statistical power.

However, the ischemia reperfusion injury (IRI) remains the inevitable complex pathophysiological process during organ transplantation. Ischemia is unavoidable in solid organ transplantation with CSS, thus leading to IRI. The only real solution to IRI is to avoid ischemia through continuous cold or warm perfusion to preserve cell integrity and metabolism. The first period of ischemia-induced cellular damage based on oxygen and nutrient deprivation during the ischemic phase is followed by a second injury phase during reperfusion based on the inflammatory response, ROS production, calcium overload, correction of pH, and, in the end, mitochondrial dysfunction. When evaluating a donor during retrieval, an IRI-interdependent two-phase process must be kept in mind since it interferes both with the acceptability criteria into the donor pool as well as the endurance of transplanted organs. Marginal organs are known to be more prone to IRI, adding the additional risk of graft failure and preventing the extension of the donor pool. Therefore, IRI and organ preservation represents an ideal target to overcome the hallmark problem of transplant medicine, that is, organ shortage.

However, the new concept of dynamic organ preservation will have to deal with one of the key elements in organ preservation that remains, that is, seeking the optimal preservation solution.

Today, following the most updated evidence, HTK was claimed to be superior in myocardial preservation for heart transplantation when compared to other cardioplegic solutions. Nevertheless, the protection of donor hearts after CSS is still limited to 3–6 h in contrast to most abdominal organs.

However, Koerner and colleagues[31] clearly demonstrated that superior results were achieved for heart transplantation if the same solution is administered with a perfusion rate of approximately 3500 mL within 10 min. This perfusion protocol is different than the one everywhere used for intraoperative cardioplegia (1500–2000 mL within 6–8 min).

Warm or cold?

The impressive promise of the ESOP is to warmly perfuse the organ so as to maintain a physiologic environment and a normal composition of the blood perfusing the graft. Although this promise is attractive and fascinating, it appears evident looking at the data coming from the literature that there are many drawbacks and possible "fake friends" on the road toward prolonged perfusion of the graft. Furthermore, freezing risk has been recently highlighted, thus stimulating the adoption of controlled temperature systems to reduce the burden of freezing, allowing better organ preservation. The donor heart can also sustain an injury during the warm phase during implantation and rewarming, accounting for up to 50% of total ischemic injury.

One point that has been highlighted as a possible weakness of the organ care system (OCS) perfusion system called Transmedics is that the blood flow and the amount of oxygen needed from an empty beating heart at subnormothermic temperature are far lower than the blood flow actually delivered .

Furthermore, many articles from the group of Robert Bartlett have disclosed how there is a need for an exchange of the plasma composition during a prolonged perfusion. The challenges proceeding from the experience with graft cross-circulation up to 72 h represents a clear call to create machines or bioreactors capable of preserving the integrity of the organ during a prolonged perfusion.[32]

In fact, if the original aim of the OCS was to safely protect the graft during the transport to the OR, during the years after its market availability it has rapidly become clear that the costs and the risks of a suboptimal preservation should have been counterbalanced from the upcoming possibility to widen the indication for use toward the DCD donation and the preconditioning. Indeed, 4–6 h, the upfront of the ischemic time with the CSS paradigm, are deemed to be a minimal safe time. The extension over 24 h and possibly toward 72 h may be considered a valuable objective.[33] On the other hand, as always happens when the standard of care is losing its superiority, many new solutions have been postulated to relaunch the safety and simplicity of cold preservation.

Sherpapak represents the first temporal solution to achieve an extension of the safe ischemic time within a paradigm of cold static saline preservation. The device strengths are the simplicity and the capacity to verify the temperature, avoiding both excessive freezing and suboptimal temperature. The data from the Guardian Registry claimed a lower grade of PGF and lower costs due to long ICU stays.[34] It remains a static system and its indication remains mainly the reduction of the damage due to graft preservation in the new allocative landscape privileging urgency.

Recently, a new system of nonischemic heart preservation (NIHP) has undergone a first-in-human phase 2 trial.[35] The system is based on a low temperature, pressure, and flow perfusion with a nonbeating floating heart. The data on the first six patients showed excellent results with optimal heart preservation, a lower burden of edema, and no cases of PGD requiring postoperative ECMO.

The system was based entirely on the Steen perfusion solution. Further studies have been deemed necessary to launch this new device in the clinical field.[36]

Recently, due to the obstacles encountered with warm perfusion, many different solutions based on cold perfusion have emerged:

- (VPS) of cold perfusion with oxygenated acellular preservation solution has been proposed to FDA
  - Cold, oxygenated, acellular preservation solution
  - Small, lightweight, and portable
  - Intuitive use/handling by one person
  - Exploring possibility of analyzing perfusate other than lactate
- Bridge to Life (LifeCradle)
  - Cold, oxygenated, acellular preservation solution (Beltzer)
  - Retrograde flushing of air before starting perfusion
  - Small, lightweight, and portable
  - Continuous display/recording of preservation conditions
  - Intuitive use/handling by one person
  - Completely self-contained
  - Patented device technology uses a perfusion solution to support organs at a controlled temperature in transport, providing oxygen and nutrients.
  - Circulates an oxygenated, hypothermic, nutrient solution in a highly controlled and monitored environment.
- Paragonix Sherpa Perfusion
  - Cold, oxygenated, acellular preservation solution
- Organ Recovery System
  - Oxygenated, acellular preservation solution

Hypothermia (<10°C) during perfusion offers the same protective benefits as CS. However, lowering the temperature below 15°C–20°C does not significantly reduce oxygen uptake during continuous oxygenated ex situ heart perfusion (ESHP).[37–39] Also, deep hypothermia itself could lead to cell death, as it results in disruption of ion regulatory mechanisms and imbalances in ATP supply and demand. To date, the majority of studies describing different temperatures during oxygenated ESHP focus on profound hypothermia (<10°C) and normothermia (>35°C).

Company name	Device name	Studies	Stage of development
Paragonix Technologies (Cambridge, MA, USA)	Sherpa Perfusion Cardiac Transport System	Animal studies reported	Unknown
Bridge to Life (Northbrook, IL, USA)	LIFECRADDLE Heart Perfusion Organ Transport System	Animal studies Unused human hearts	Preclinical testing
Vascular Perfusion Solutions, Inc. (San Antonio, TX, USA)	VP.S ENCORE Cardiac preservation device	Animal studies	Preclinical testing FDA Breakthrough Device Designation
Organ Recovery Systems (Itasca, IL, USA)	HeartPort System	Animal studies	Unknown

Blood or crystalloid? Which carrier for oxygen?

Blood-warm perfusion is unanimously agreed to be the most accurately physiological way to perfuse the graft during the transport. The OCS Transmedics device utilizes this principle[40]; however, there are many concerns about the perfusion pressure, the flows, and the composition of the blood coming from a patient in brain death, so recirculating the brainstorm's catecholamines and the inotropes since the reduced oxygen consumption expected with a fully unloaded left ventricle (LV) (Fig. 10.4). Moreover, the need for trained personnel and the risk of loss of the organ due to a technical failure are the major constraints to the rise of machine perfusion in the field of heart transplantation.[41] The possibility of using blood solutions in a philosophy of cold storage has also been proposed; some concerns exist about the inflammatory burden due to the blood component activation and the need to remove cytokines to preserve the endothelium. It is quite sparsely used to repeat cardioplegia on arrival at the OR before completing the anastomosis to remove the metabolic products of the myocardial metabolism. Some centers tend to privilege blood reperfusion to provide nutrients and buffer the unavoidable acidosis due to the long ischemic time. A preliminary series of patients recently underwent heart transplantation after receiving a complimentary dose of 4:1 modified DNP (blood:CP) during donor heart implantation. Del Nido (DN) cardioplegia is an extracellular cardioplegic solution developed initially for the pediatric population and, in the last years, has increased in popularity in the adult population with a modified composition. The DN solution shows, in the experimental settings, the potential to offer superior myocardial protection for older patients undergoing cardiac surgery.[42] In these series, during donor heart harvesting, a cold crystalloid heart preservation solution (UW) was used for myocardial protection.

During donor heart implantation in the recipient, a modified cold DN dose was administered according to the different ischemic times of the donor heart to achieve additional myocardial protection. The modified DN dose in our preliminary series was given by anterograde. Myocardial protection with a complimentary dose of cardioplegia during heart transplantation is still not standardized, and several different techniques were reported. In many centers, complimentary doses of cold blood cardioplegia are used with repeated infusions every 30 min as soon as the graft is removed from the cold storage solution in the operating room. Subsequently, a warm blood cardioplegic reperfusion (hot shot) is also administered before removing the aortic clamp. Otherwise, a continuous retrograde hyperkalemic perfusion using warm blood was similarly proposed as ideal myocardial protection during donor heart implantation.[43] Tevaearai et al. recently proposed administering an additional dose of crystalloid cardioplegia containing procaine immediately before implantation.[44]

In a recent preliminary series, a modified DN (4:1) cardioplegia was utilized as a complementary dose during donor heart implantation after a previous first dose of UW solution during donor heart harvesting. DN cardioplegia shows a decreased calcium accumulation during myocardial ischemia in a setting of a depolarized arrest. The utility of modified DN (4:1) in clinical practice remains under evaluation, but preliminary studies are encouraging. The modified DN solution in this preliminary group of patients was used safely and effectively, paving the way again in the knowledge of myocardial protection during heart transplantation.[45]

Timing of infusion, method of administration (anterograde and retrograde), and amount of total complementary dose have still not been widely investigated. Furthermore, the long-term survival and benefit of using this solution might be confirmed in a randomized trial evaluating the amount of myocardial enzymes' release, the incidences of graft failure, and of comorbidities.

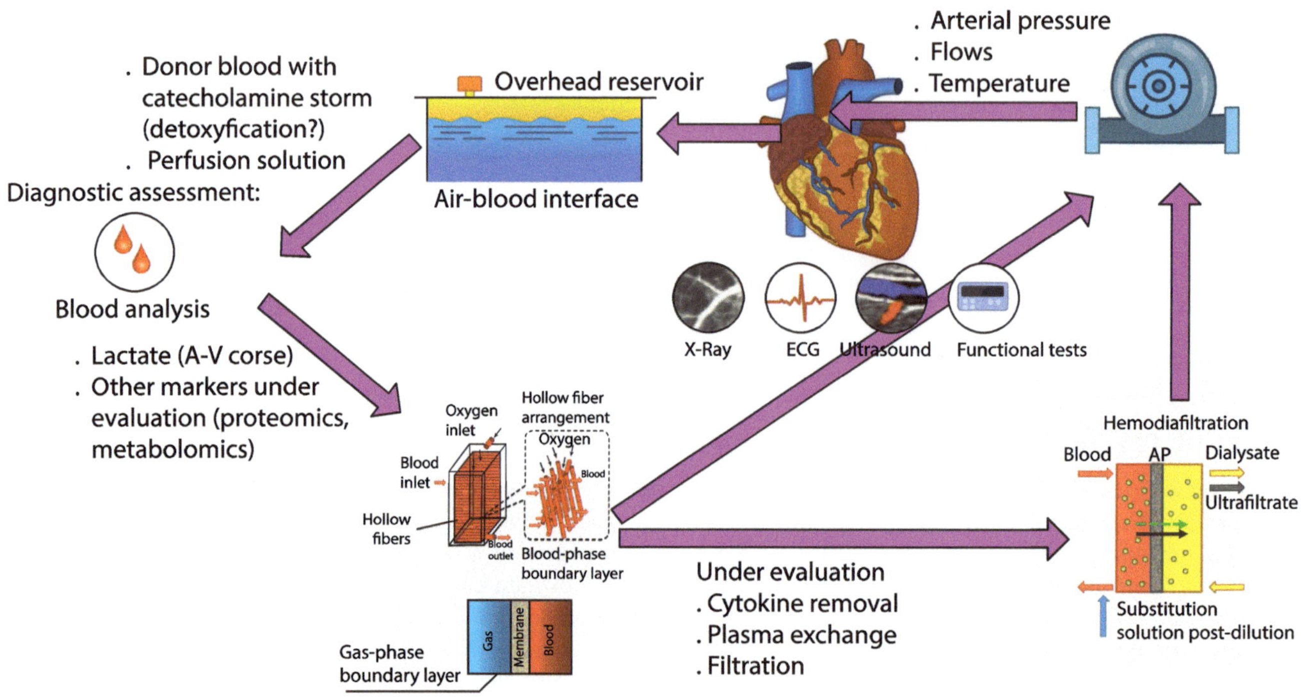

FIGURE 10.4 Insights into ex situ perfusion devices and potential improvements.

Ischemic or nonischemic?

Prolonged ischemic time, especially over 4 h, is associated with PGD and reduced recipient survival. Rather than relying on CSS, which is the current standard of care during transport, organ preservation techniques have been developed that allow for ex vivo cardiac perfusion in order to preserve donor organ function and expand the donor pool by enabling procurement from distant locations.[46] Different methods of ex vivo cardiac perfusion have been described, including subnormothermic and hypothermic ex vivo perfusion.

The OCS (Transmedics) is the only normothermic ex vivo perfusion platform available clinically.[20] It is based on a portable system warranting temperature-regulated and oxygenated blood to perfuse the donor heart, which continues beating unloaded and remains metabolically active during transport, as assessed by checking the lactate level. In 2015, the ex vivo perfusion of donor's hearts for human heart transplantation (PROCEED II) trial evaluated clinical outcomes when using the OCS compared to static cold storage for donor heart transport.[7] The PROCEED II trial was a prospective, open-label, multicenter randomized trial that enrolled 130 donors, randomly assigned either to the OCS arm (67 grafts) or (the remaining) to CSS. Thirty-day recipient and graft survival, adverse events, and severe rejection rates were found to be similar between the OCS and CSS groups. The mean cold ischemic time for the OCS group was 113 min, compared to 195 min in the control group.[20] However, a follow-up study of a subset of the PROCEED II cohort failed to demonstrate any significant difference in 2-year survival, CAV development, and incidence of acute rejection between the two groups. The OCS was also validated in the single-arm OCS Heart EXPAND Trial, focused on marginal donors with an anticipated total ischemic time of ≥4 h or ≥2 h with LVH, EF 40%–50%, downtime ≥20 min, and/or age under 55 years. This study showed that, with a mean perfusion time of 6.35 h, recipient short-term survival (95% 30-day and 88% at 6 months) and incidence of graft dysfunction at 24 h after transplant (10.7%) for the 75 donor hearts perfused with the OCS were acceptable. However, the most feared aspects of OCS are related to the risk of accidents during perfusion in a setting of warm perfusion and with the need for trained personnel and the risk of human errors during the perfusion course.

The supporters of these drawbacks faced this concern by trying to develop a system using ex situ hypothermic perfusion. This concept, despite appealing to the heart, prevents the possibility of a continuous functional assessment that has been claimed as the fundamental strength of the OCS. Hypothermic ex vivo perfusion infuses cold preservation solution into the donor coronary arteries to deliver oxygen and facilitate metabolites' washout.

Nilsson recently reported the first-in-human experience of their nonischemic heart preservation (NIHP) system in which the donor heart[34] is perfused with a cold (8°C) blood-based oxygenated cardioplegic solution. In opposite to the OCS the investigators used bank blood. The system has been named nonischemic heart preservation so as to underline the absence of ischemia during organ preservation. The group chose to study the NIHP system in a nonrandomized phase II study and applied this system to six donor hearts. The median preservation time in the NIHP group was 223 min as compared to 194 min in the static cold storage group (similar to the mean time of the control group in the PROCEED II). Recipients of the hearts transported with NIHP had 100% event-free survival. Intriguingly, there were no cases of PGD within 24 h, ECMO use within 7 days, or acute cellular rejection greater than 2R within 180 days and no deaths at 6 months, as compared to 72% of patients within the CSS group. This exciting finding was deemed to be related to lower endothelial damage and the prevention of ischemia/reperfusion injury associated with ischemic avoidance. At the European Society for Organ Transplantation (ESOT) Congress in Milan 2021, the group presented an interim analysis of their randomized trial, including 36 patient, showing a significant decrease in CK-MB leak at 12 hours, and improved cardiac index at 24 hour post x-clamp release compared to the control group. It was also proposed that, in the event of an interruption to the ex vivo perfusion system, the hypothermic NIHP system provides additional protection to the donor's heart as compared both to a normothermic system and to CSS. Nevertheless, the advantage over CSS is the potential of recharging the metabolic stores of the graft.

These technologies support the development of ex vivo cardiac perfusion systems as a viable way to expand donor heart utilization by allowing procurement from remote locations, reducing ischemia/reperfusion injury, and enabling further assessment of "marginal" donor hearts. Blood cardioplegia induction included XHS cardioplegia with fresh donor blood collected just prior to administering cardioplegia at 25% of total cardioplegia volume (similar to above)[40]. After procuring the heart, it may be connected to the XVIVO perfusion system (XVIVO Perfusion, Gothenburg, Sweden) and continuously perfused with oxygenated XHS at 8°C, with pressure maintained at 20 mm Hg at a physiological pH (7.2–7.6) until transplantation. An XVIVO dual lumen cannula (central inflow, axially oriented outflow, and quick connecting for machine circuit connection) is placed in the transected ascending aorta, with the tip of the cannula in the distal ascendent aorta. Aortic valve incompetence must be prevented during perfusion. The mitral valve is made incompetent by transvalvular placement of silicon tubing sutured to the left atrium, positioned in order to prevent LV distention, as in an OCS device. The XVIVO perfusion system is used in the randomized study performed by the LUND group, the multi-center study in Europe, and the Australian and New Zealand NIHP trial. The Australian and New Zealand NIHP trial recently presented a case using the heart box for almost 6 hours (total ischemic time 7,5 hours) with excellent results. To further support the benefits of nonischemic cardiac perfusion (NIHP), mitigating myocardial damage and inflammation has resulted in up to 6 months of survival in pig-to-baboon cardiac xenotransplants, using a nonischemic continuous perfusion preservation technique (NIHP) and antiinflammatory agents, among other adjuncts.[47] NIHP prevents I/R injury and has been shown to ease removal from cardiopulmonary bypass and reduce the need for catecholamine support. However, NIHP is expensive, and the components of the oxygenated perfusate are complex with components of unknown significance.[48]

Traditional CSS is well-tolerated in allotransplantation and has been used since the inception of heart transplantation, resulting in PGD at 8%–10%.[12,34,49] In contrast to traditional static preservation (i.e., slush), the NIHP strategy utilizes a hyperoncotic cardioplegic perfusate containing packed red blood cells, human serum albumin (HSA), dextran 40, inotropes, hormones, antibiotics, and cocaine to minimize ischemia–reperfusion phenomena and prevent myocardial edema.[48]

There is literature to support our observations. It is known that during solid organ procurement, preservation and subsequent reperfusion result in oxygen radical formation, edema, microvascular injury, and compromised microcapillary circulation. The underlying mechanism is likely twofold: a continued ischemic phenomenon, despite reperfusion, and a paradoxical damage progression due to the reperfusion of ischemic areas (i.e., IRI). There are definitive multifocal, patchy ischemic areas that result upon reperfusion of ischemic striated muscle due to constriction of afferent arterioles, obstruction of the capillaries themselves, or interstitial edema.

Secondly, reperfusion injury occurs upon restoring flow in ischemic areas due to oxygen-free radical formation from NADPH and xanthine oxidase, causing endothelial cell damage. Most relevant to xenotransplantation, oxidative stress also causes leukocyte adherence to endothelial cells, activation, and further endothelial damage.[50,51] These mechanisms of graft injury are largely augmented by recipient inflammation at the graft endothelium. Since immunological insults of much higher intensity are seen in cross-species transplantation, the differences seen in PGD in allotransplantation versus xenotransplantation can be explained by augmented IRI by inflammation.[52] Perhaps by minimizing ischemia, using either blood cardioplegia induction or NIHP, downstream effects of ischemia reperfusion are reduced.

Furthermore, serum albumin serves to increase the oncotic pressure intravascularly, abolishing some of the microcapillary dysfunction, while dextran 40 protects the endothelial lining from leukocyte interaction. To what extent each component contributes to mitigating further damage by ischemia reperfusion is mostly unknown.

Goerlich hypothesized that combining HSA and dextran 40 with blood cardioplegia induction strategies is equally effective as NIHP, but no data exist in support.[47] Moreover, it was observed that there was significantly less vasopressor requirement in recipients who received NIHP, suggesting that vasoplegia after transplantation is reduced. This observation should be interpreted with caution, as the evidence is limited, but it does provide potential evidence that NIHP mitigates inflammation-induced vasoplegia not seen by other preservation strategies.

How to assess organ preservation?

An essential component of the future systems for graft preservation (long-term ex vivo heart warm or cold perfusion or updated CSS) will be assessing cardiac function to ascertain the preservation quality until surgical implantation. Current clinical methods for ESHP lack this function. The OCS system, which perfuses a normothermic heart that beats, uses serial measurements of lactate consumption to assess the metabolic preservation. Lactate consumption provides an idea of the metabolic state which, in combination with previous cardiac function assessment, enables a safe transplant. The NIHP system, which perfuses a hypothermic cardioplegic heart, does not have the same need for this type of evaluation. However, serial measurements of CK-MB and TnI have been postulated as a means to validate the usage of the graft since they may represent a valuable marker of organ preservation. In the clinical arena, increased leak means increased risk of PGD[53,54] More sensitive heart-damaged markers are being developed and will probably be used in the future with the increased knowledge of proteomics and the upcoming techniques of basic research (Fig. 10.5). Another approach may be ex situ perfusion of the heart after recovery, similar to EVLP for lungs. The functional assessment can then take place at the transplant center with a machine intended for the purpose. Whether the whole heart or only a certain part of the heart's function should be assessed is still open for debate. An evaluation of the whole heart can be valuable if the goal is to achieve more than 6 h of preservation, but the machine then becomes more complex.[55]

During preservation, perfusate samples may be analyzed during ESHP before connecting the heart.

Furthermore, hearts may be weighed before preservation, after preservation, and at the end of normothermic ex situ evaluation. In experimental works, biopsies from standardized locations on the apex of the LV were taken and stored in formaldehyde at baseline in vivo, prior to preservation, after preservation, and after normothermic ex situ evaluation. Five-micrometer-thick

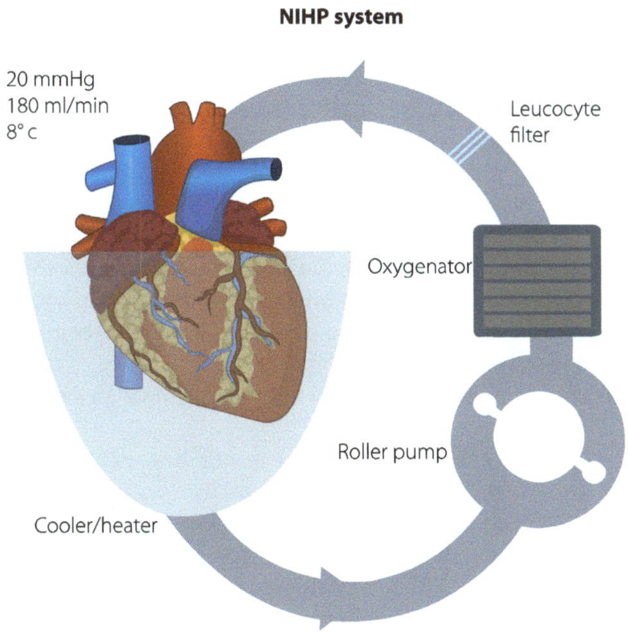

FIGURE 10.5 Nonischemic cardiac perfusion (NIHP) scheme.

cross-sections were prepared, embedded in paraffin, and stained with Sirius Red. Samples were scored 0 (none) to 3 (severe) for edema, contraction bands, cytoplasmic coagulation necrosis, congestion, and the influx of neutrophils.[56]

Alternatively, to judge functional preservation, a conically shaped balloon was inserted into the LV through the mitral valve annulus for operational evaluation. The balloon was connected to a pressure transducer for continuous pressure recording.

Intraventricular balloon diastolic pressure was kept at 0–10 mm Hg, to avoid impairments of endocardial flow due to high intraventricular pressure.

Volume can be gradually added to the balloon (maximum of 40 mL). Systolic function may be represented by contractility, defined as the top rate of intraventricular pressure rise (dP/dtmax), and elastance. Diastolic function was represented by lusitropy, defined as the maximum rate of intraventricular pressure fall (dP/dtmin), and stiffness.[56]

Hurdles and challenges of prolonged cardiac perfusion

Recently the group of Robert Bartlett published seminal work dealing with a significant lengthening of ex situ perfusion by modifying the milieu of the perfusion.[57–61] This group addressed this challenge as a moving target by attempting to reach a reproducible way to sustain the organs during the ESHP. The first work[58] explored the possibility of adding to the circuit a cross-circulation with an animal, thus exploiting the metabolic system of the animal as a host while diluting the toxic metabolites during prolonged ESHP.[59] In the following experiences, the authors verify that the benefit was mainly due to plasma exchange from cross-circulated animals more than to whole blood.[60,61] The subsequent step was to define the amount of fresh plasma needed to maintain a physiologic milieu and a working graft. This hypothesis-generating experiment showed that modification of blood composition is the metabolic and biological basis of graft failure over the prolonged time of perfusion. Air–blood interfaces, relation with epicardial fat, and biomaterials have relevant roles in this complex interplay. Interestingly, no metabolomics, nor proteomics, nor other analysis was performed to mechanistically understand the physiological basis of the graft failure. Evaluation of lactates, weight, and histologic changes of the graft have been used to evaluate the quality of preservation. Prolonged graft perfusion of more than 24 h—up to 3 days—is expected to redraw the landscape of heart transplantation, moving toward the reconditioning of DCD donors and their allocation based on systematic algorithms also encompassing immunologic compatibility, thus reducing the high burden of complications due to immunosuppressive therapies needed, until now, in the clinical field of heart transplantation.

From cardiac preservation to organ banks

Prolonged organ ex vivo perfusion will lead to organ banks when it is demonstrated to be safe. Organ banks could be centralized, a few per country, or as an alternative, organ perfusion machines will travel, bringing the organs where they are needed. An expanded brain-dead pool and cardiac-dead donors are the targets for procurement teams. Procurement surgeons from the recipient hospitals will no longer need to go to the donor if the quality of the organs could be assessed before implantation. Organs will be stored on ice and transported to the central organ bank facility. This assumes that organs stored on ice for 12–24 h can be routinely restored to normal function by normothermic ex vivo perfusion. Stored cold organs could be resuscitated using normothermia to determine functional status, treat the organ if needed, and perhaps undertake immune modulation. Tissue typing, blood typing, etc. will be initiated at the time of procurement. The organ bank could match the organs to potential recipients, and the organ could be transported to the transplant hospital for elective transplantation.

The development of full organ banks could be a commercial enterprise or an organ procurement organization system component.

With the actual policies of myocardial preservation (CSS) in our historical data, the level of HLA matching was more than 4[62]; once the ischemic time with ESHP and organ scarcity (organ reconditioning in banks) is reduced, the crucial challenge will be to put together all the factors related to donor and recipient characteristics in a platform of machine learning or even artificial intelligence to predict or even govern a fair allocation of the few organs between the large number of possible candidates aiming for life-changing organ transplantation.

Therefore, organ perfusion and preservation technologies could carry a highly disruptive impact on the traditional rapid procurement and short preservation model of delivery of transplantation. Building an assessment and repair center exponentially increases the acquisition of knowledge through a seamless integration with regenerative medicine in clinical research–industry partnerships, thus allowing a more rapid translation of therapies into clinical practice.

The development of organ banking could reshape the global delivery of transplantation (Fig. 10.6), and if undertaken under clear and transparent ethical principles, could make a significant impact in reducing the global inequities in access to transplantation and technologies.

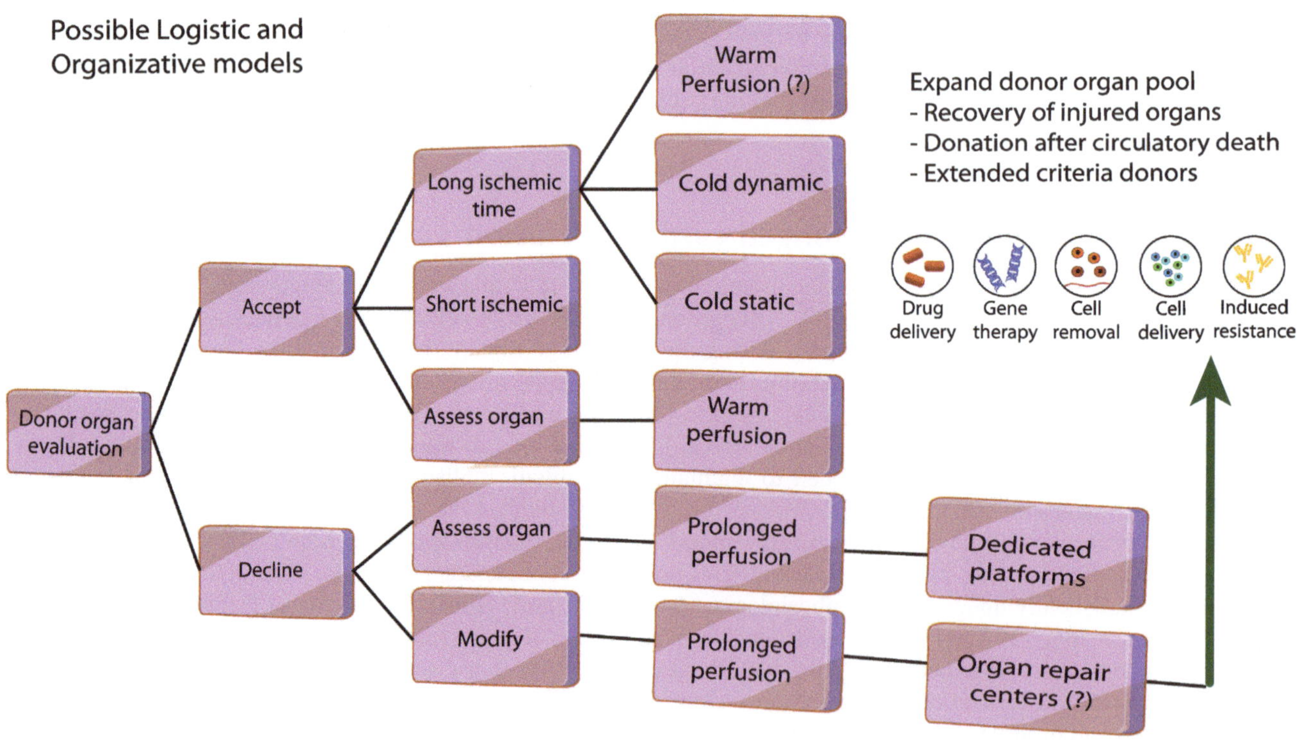

FIGURE 10.6 Future logistics of new approaches of myocardial preservation.

Conclusions

PGD is the leading cause of mortality in patients undergoing heart transplantation.[12] Despite advances in allocation methods, cardiopulmonary bypass, and postoperative care, the incidence is still around 8%–10%.[34,49]

The research is currently moving in three different directions:

- Improve the actual standard of care (CSS) through a certified, temperature-controlled device.
- Move toward prolonged warm perfusion to extend the safe travel time of the organ and change the paradigms of organ preservation. Keeping the organ beating and evaluating its functional state beyond the mere metabolic evaluation appear to be the crucial endpoints.
- Move toward prolonged cold perfusion with or without blood to extend the safety of CSS at the price of a simple portable console without too much demand for its working.

References

1. Cogswell R, John R, Estep JD, et al. An early investigation of outcomes with the new 2018 donor heart allocation system in the United States. *J Heart Lung Transpl.* 2020;39(1):1–4. Available from: https://doi.org/10.1016/j.healun.2019.11.002. Epub 2019 Nov 20. PMID: 31810767.
2. Benck L, Kransdorf EP, Emerson DA, et al. Recipient and surgical factors trigger severe primary graft dysfunction after heart transplant. *J Heart Lung Transpl.* 2021;40(9):970–980. Available from: https://doi.org/10.1016/j.healun.2021.06.002. Epub 2021 Jun 10. PMID: 34272125.
3. Heylen L, Pirenne J, Samuel U, et al. Effect of donor nephrectomy time during circulatory-dead donor kidney retrieval on transplant graft failure. *Br J Surg.* 2020;107(1):87–95. Available from: https://doi.org/10.1002/bjs.11316. Epub 2019 Oct 1. PMID: 31573084.
4. Weissenbacher A, Vrakas G, Nasralla D, Ceresa CDL. The future of organ perfusion and re-conditioning. *Transpl Int.* 2019;32(6):586–597. Available from: https://doi.org/10.1111/tri.13441. Epub 2019 May 8. PMID: 30980772; PMCID: PMC6850430.
5. Dengu F, Abbas SH, Ebeling G, Nasralla D. Normothermic machine perfusion (NMP) of the liver as a platform for therapeutic interventions during ex-vivo liver preservation: a review. *J Clin Med.* 2020;9(4):1046. Available from: https://doi.org/10.3390/jcm9041046. PMID: 32272760; PMCID: PMC7231144.
6. Ravaioli M, Maroni L, Angeletti A, et al. Hypothermic oxygenated perfusion vs static cold storage for expanded criteria donors in liver and kidney transplantation: protocol for a single-center randomized controlled trial. *JMIR Res Protoc.* 2020;9(3):e13922. Available from: https://doi.org/10.2196/13922. Published 2020 Mar 19.
7. Ardehali A, Esmailian F, Deng M, et al. PROCEED II trial investigators. Ex-vivo perfusion of donor hearts for human heart transplantation (PROCEED II): a prospective, open-label, multicentre, randomised non-inferiority trial. *Lancet.* 2015;385(9987):2577–2584. Available from: https://doi.org/10.1016/S0140-6736(15)60261-6. Epub 2015 Apr 14. PMID: 25888086.
8. Nasralla D, Coussios CC, Mergental H, et al. A randomized trial of normothermic preservation in liver transplantation. *Nature.* 2018;557(7703):50–56. Available from: https://doi.org/10.1038/s41586-018-0047-9. Epub 2018 Apr 18. PMID: 29670285.
9. Demmy TL, Biddle JS, Bennett LE, Walls JT, Schmaltz RA, Curtis JJ. Organ preservation solutions in heart transplantation—patterns of usage and related survival. *Transplantation.* 1997;63(2):262–269. Available from: https://doi.org/10.1097/00007890-199701270-00015. PMID: 9020328.
10. Hernández A, Borrego JM, Gomez S, et al. Myocardial preservation using Celsior: clinical results in high-risk cardiac transplantation. *Transpl Proc.* 2005;37(3):1543–1545. Available from: https://doi.org/10.1016/j.transproceed.2005.02.030. PMID: 15866667.
11. De Santo LS, Amarelli C, Romano G, et al. High-risk heart grafts: effective preservation with Celsior solution. *Heart Vessel.* 2006;21(2):89–94. Available from: https://doi.org/10.1007/s00380-005-0867-z. PMID: 16550309.
12. Singh SSA, Dalzell JR, Berry C, Al-Attar N. Primary graft dysfunction after heart transplantation: a thorn amongst the roses. *Heart Fail Rev.* 2019;24(5):805–820. Available from: https://doi.org/10.1007/s10741-019-09794-1.
13. Mokbel M, Zamani H, Lei I, et al. Histidine-tryptophan-ketoglutarate solution for donor heart preservation is safe for transplantation. *Ann Thorac Surg.* 2020;109(3):763–770. Available from: https://doi.org/10.1016/j.athoracsur.2019.07.012. Epub 2019 Aug 27. PMID: 31470011.
14. Hendry PJ, Labow RS, Keon WJ. A comparison of intracellular solutions for donor heart preservation. *J Thorac Cardiovasc Surg.* 1993;105(4):667–673. PMID: 8469000.
15. Smits JM, De Pauw M, de Vries E, et al. Donor scoring system for heart transplantation and the impact on patient survival. *J Heart Lung Transpl.* 2012;31(4):387–397. Available from: https://doi.org/10.1016/j.healun.2011.11.005. Epub 2011 Dec 16. PMID: 22177692.
16. Angleitner P, Kaider A, Smits JM, et al. The adapted heart donor score. *Transpl Int.* 2021;34(3):546–560. Available from: https://doi.org/10.1111/tri.13822. Epub 2021 Feb 22. PMID: 33455049; PMCID: PMC7986732.
17. Khush KK, Ball RL. Great variability in donor heart acceptance practices across the United States. *Am J Transpl.* 2020;20(6):1582–1596. Available from: https://doi.org/10.1111/ajt.15760. Epub 2020 Jan 20. PMID: 31883229; PMCID: PMC7261633.
18. Critchley WR, Stone JP, Liao Q, et al. Non-ischemic heart preservation via hypothermic cardioplegic perfusion induces immunodepletion of donor hearts resulting in diminished graft infiltration following transplantation. *Front Immunol.* 2020;11:1621. Available from: https://doi.org/10.3389/fimmu.2020.01621. PMID: 32849549; PMCID: PMC7399062.
19. Kvietkauskas M, Leber B, Strupas K, Stiegler P, Schemmer P. Machine perfusion of extended criteria donor organs: immunological aspects. *Front Immunol.* 2020;11:192. Available from: https://doi.org/10.3389/fimmu.2020.00192. PMID: 32180769; PMCID: PMC7057848.

20. Tong CKW, Khush KK. New approaches to donor selection and preparation in heart transplantation. *Curr Treat Options Cardiovasc Med.* 2021;23(5):28. Available from: https://doi.org/10.1007/s11936-021-00906-5. Epub 2021 Mar 23. PMID: 33776401; PMCID: PMC7985579.
21. Avtaar Singh SS, Banner NR, Rushton S, Simon AR, Berry C, Al-Attar N. ISHLT primary graft dysfunction incidence, risk factors, and outcome: a UK National Study. *Transplantation.* 2019;103(2):336–343. Available from: https://doi.org/10.1097/TP.0000000000002220. PMID: 29757910.
22. Drinkwater DC, Rudis E, Laks H, et al. University of Wisconsin solution vs Stanford cardioplegic solution and the development of cardiac allograft vasculopathy. *J Heart Lung Transpl.* 1995;14(5):891–896. PMID: 8800725.
23. Gharagozloo F, Bulkley BH, Hutchins GM, et al. Potassium-induced cardioplegia during normothermic cardiac arrest. Morphologic study of the effect of varying concentrations of potassium on myocardial anoxic injury. *J Thorac Cardiovasc Surg.* 1979;77(4):602–607. PMID: 423594.
24. Karck M, Vivi A, Tassini M, et al. The effectiveness of University of Wisconsin solution on prolonged myocardial protection as assessed by phosphorus 31-nuclear magnetic resonance spectroscopy and functional recovery. *J Thorac Cardiovasc Surg.* 1992;104(5):1356–1364. PMID: 1434717.
25. Salvadori M, Rosso G, Bertoni E. Update on ischemia-reperfusion injury in kidney transplantation: pathogenesis and treatment. *World J Transpl.* 2015;5:52–67.
26. Choong YS, Gavin JB. Functional recovery of hearts after cardioplegia and storage in University of Wisconsin and in St. Thomas' Hospital solutions. *J Heart Lung Transpl.* 1991;10(4):537–546. PMID: 1911796.
27. Rauen U, Klempt S, de Groot H. Histidine-induced injury to cultured liver cells, e_ects of histidine derivatives and of iron chelators. *Cell Mol Life Sci.* 2007;64:192–205.
28. Mohr A, Brockmann JG, Becker F. HTK-N: modified histidine-tryptophan-ketoglutarate solution-A promising new tool in solid organ preservation. *Int J Mol Sci.* 2020;21(18):6468. Available from: https://doi.org/10.3390/ijms21186468. PMID: 328997720; PMCID: PMC7555843.
29. Bahde R, Palmes D, Gemsa O, et al. Attenuated cold storage injury of rat livers using a modified HTK solution. *J Surg Res.* 2008;146:49–56.
30. Loganathan S, Radovits T, Hirschberg K, et al. Effects of Custodiol-N, a novel organ preservation solution, on ischemia/reperfusion injury. *J Thorac Cardiovasc Surg.* 2010;139(4):1048–1056. Available from: https://doi.org/10.1016/j.jtcvs.2009.09.034. Epub 2009 Nov 27. PMID: 19945120.
31. Koerner MM, Tenderich G, Posival H, Wlost S, Gromzik H, Koerfer R. Cardiac transplant survival in relation to preservation. In: Collins GM, Dubernard JM, Land W, Persijn GG, eds. *Procurement, Preservation and Allocation of Vascularized Organs.* Dordrecht: Springer; 1997. Available from: https://doi.org/10.1007/978-94-011-5422-2_35.
32. García Sáez D, Zych B, Sabashnikov A, et al. Evaluation of the organ care system in heart transplantation with an adverse donor/recipient profile. *Ann Thorac Surg.* 2014;98(6):2099–2105. Available from: https://doi.org/10.1016/j.athoracsur.2014.06.098. Discussion 2105-6. Epub 2014 Oct 23. PMID: 25443013.
33. D'Alessandro D, Philpott J, Boeve T, Pham S, Zuckermann A. (285). First report of the GUARDIAN Registry: an international consortium examining the effect of controlled hypothermic preservation in heart transplantation. *J Heart Lung Transplant.* 2021;40(4, Suppl.):S127. Available from: https://doi.org/10.1016/j.healun.2021.01.398.
34. Nilsson J, Jernryd V, Qin G, et al. A nonrandomized open-label phase 2 trial of nonischemic heart preservation for human heart transplantation. *Nat Commun.* 2020;11(1):2976. Available from: https://doi.org/10.1038/s41467-020-16782-9. PMID: 32532991; PMCID: PMC7293246.
35. Rojas SV, Ius F, Schibilsky D. Cardiac transplantation in higher risk patients: is ex vivo heart perfusion a safe preservation technique? *A Two Cent Experience.* 2019;38(4, Suppl.):S43. Available from: https://doi.org/10.1016/j.healun.2019.01.091.
36. Messer S, Ardehali A, Tsui S. Normothermic donor heart perfusion: current clinical experience and the future. *Transpl Int.* 2015;28(6):634–642. Available from: https://doi.org/10.1111/tri.12361. Epub 2014 Jul 7. PMID: 24853906.
37. Buckberg GD. Myocardial temperature management during aortic clamping for cardiac surgery: protection, preoccupation, and perspective. *J Thorac Cardiovasc Surg.* 1991;102(6):895–903.
38. Buckberg G. Studies of the effects of hypothermia on regional myocardial blood flow and metabolism during cardiopulmonary bypass. IV. Topical atrial hypothermia in normothermic beating hearts. *J Thorac Cardiovasc Surg.* 1977;73(2):195–200.
39. Cleveland JC, Meldrum DR, Rowland RT, Banerjee A, Harken AH. Optimal myocardial preservation: cooling, cardioplegia, and conditioning. *Ann Thorac Surg.* 1996;61(2):760–768.
40. Schroder JN, D'Alessandro D, Esmailian F. Successful utilization of extended criteria donor (ECD) hearts for transplantation - results of the OCS™ heart EXPAND trial to evaluate the effectiveness and safety of the OCS heart system to preserve and assess ECD hearts for transplantation. *J Heart Lung Transplant.* 2019;38(4):S42. Available from: https://doi.org/10.1016/j.healun.2019.01.088. Supplement, ISSN 1053-2498.
41. Valooran GJ, Nair SK, Chandrasekharan K, Simon R, Dominic C. Del Nido cardioplegia in adult cardiac surgery - scopes and concerns. *Perfusion.* 2016;31(1):6–14. Available from: https://doi.org/10.1177/0267659115608936. Epub 2015 Oct 7. PMID: 26445810.
42. Matte GS, del Nido PJ. History and use of del Nido cardioplegia solution at Boston Children's Hospital. *J Extra Corpor Technol.* 2012;44(3):98–103. Erratum in: J Extra Corpor Technol. 2013 Dec;45(4):262. PMID: 23198389; PMCID: PMC4557532.
43. Ribeiro RV, Friedrich JO, Ouzounian M, et al. Supplemental cardioplegia during donor heart implantation: a systematic review and meta-analysis. *Ann Thorac Surg.* 2020;110(2):545–552. Available from: https://doi.org/10.1016/j.athoracsur.2019.10.094. Epub 2020 Jan 21. PMID: 31972127.
44. Tevaearai Stahel HT, Unger D, Schmidli J, et al. Supplemental cardioplegia immediately before graft implantation may improve early post-transplantation outcome. *Front Surg.* 2014;1:46. Available from: https://doi.org/10.3389/fsurg.2014.00046. Nov 28 PMID: 25593970; PMCID: PMC4286961.

45. Gallo M, Trivedi JR, Slaughter MS. Myocardial protection with complementary dose of modified Del Nido cardioplegia during heart transplantation. *J Card Surg*. 2019;34(11):1387–1389. Available from: https://doi.org/10.1111/jocs.14223. Epub 2019 Aug 26. PMID: 31449689.
46. Sponga S, et al. Ex-vivo perfusion on marginal donors in heart transplantation: clinical resuts and pathological findings. *J Heart Lung Transplant*. 2019;38(4):S42–S43.
47. Goerlich CE, Griffith B, Singh AK, et al. Blood cardioplegia induction, perfusion storage and graft dysfunction in cardiac xenotransplantation. *Front Immunol*. 2021;12:667093. Available from: https://doi.org/10.3389/fimmu.2021.667093. Published 2021 Jun 9.
48. Steen S, Paskevicius A, Liao Q, Sjöberg T. Safe orthotopic transplantation of hearts harvested 24 hours after brain death and preserved for 24 hours. *Scand Cardiovasc J SCJ*. 2016;50:193–200. Available from: https://doi.org/10.3109/14017431.2016.1154598.
49. Nicoara A, Ruffin D, Cooter M, et al. Primary graft dysfunction after heart transplantation: incidence, trends, and associated risk factors. *Am J Transpl*. 2018;18(6):1461–1470. Available from: https://doi.org/10.1111/ajt.14588. Epub 2017 Dec 14. PMID: 29136325; Menger MD. Microcirculatory disturbances secondary to ischemia-reperfusion. *Transplant Proc*. 1995;27:2863–2865.
50. Menger MD. Microcirculatory disturbances secondary to ischemia-reperfusion. *Transpl Proc*. 1995;27:2863–2865.
51. Yang Q, He GW, Underwood MJ, Yu CM. Cellular and molecular mechanisms of endothelial ischemia/reperfusion injury: perspectives and implications for postischemic myocardial protection. *Am J Transl Res*. 2016;8(2):765–777. PMID: 27158368; PMCID: PMC4846925.
52. Toldo S, Mauro AG, Cutter Z, Abbate A. Inflammasome, pyroptosis, and cytokines in myocardial ischemia-reperfusion injury. *Am J Physiol Heart Circ Physiol*. 2018;315:H1553–H1568. Available from: https://doi.org/10.1152/ajpheart.00158.2018.
53. Chen RJ, Yu WH. Marker games: seeking better biomarkers? *Am J Respir Crit Care Med*. 2016;194(1):11–13. Available from: https://doi.org/10.1164/rccm.201602-0211ED. PMID: 27367885.
54. De Santo LS, Torella M, Romano G, et al. Perioperative myocardial injury after adult heart transplant: determinants and prognostic value. *PLoS One*. 2015;10(5):e0120813. Available from: https://doi.org/10.1371/journal.pone.0120813. PMID: 25942400; PMCID: PMC4420471.
55. Bona M, Wyss RK, Arnold M, et al. Cardiac graft assessment in the era of machine perfusion: current and future biomarkers. *J Am Heart Assoc*. 2021;10(4):e018966. Available from: https://doi.org/10.1161/JAHA.120.018966. Epub 2021 Jan 30. Erratum in: *J Am Heart Assoc*. 2021;10(12): e014673. PMID: 33522248; PMCID: PMC7955334.
56 van Suylen V, Vandendriessche K, Neyrinck A, et al. Oxygenated machine perfusion at room temperature as an alternative for static cold storage in porcine donor hearts. *Artif Organs*. 2022 Feb;46(2):246–258. Available from: https://doi.org/10.1111/aor.14085.
57. Tchouta L, Drake D, Hoenerhoff M, et al. Twenty-four-hour normothermic perfusion of isolated ex vivo hearts using plasma exchange. *J Thorac Cardiovasc Surg*. 2020;. Available from: https://doi.org/10.1016/j.jtcvs.2020.11.158. S0022–5223(20)33335-3. Epub ahead of print. PMID: 33485659.
58. Trahanas JM, Witer LJ, Alghanem F, et al. Achieving 12 hour normothermic ex situ heart perfusion: an experience of 40 porcine hearts. *ASAIO J*. 2016;62(4):470–476. Available from: https://doi.org/10.1097/MAT.0000000000000382. PMID: 27164040; PMCID: PMC4925299.
59. Church JT, Alghanem F, Deatrick KB, et al. Normothermic ex vivo heart perfusion: effects of live animal blood and plasma cross circulation. *ASAIO J*. 2017;63(6):766–773. Available from: https://doi.org/10.1097/MAT.0000000000000583. PMID: 28394815; PMCID: PMC5630486.
60. McLeod JS, Poling C, Church JT, et al. Ex vivo heart perfusion for 72 hours using plasma cross circulation. *ASAIO J*. 2020;66(7):753–759. Available from: https://doi.org/10.1097/MAT.0000000000001061. PMID: 31453833.
61. Tchouta LN, Alghanem F, Rojas-Pena A, Bartlett RH. Prolonged (≥24 hours) normothermic (≥32°C) ex vivo organ perfusion: lessons from the literature. *Transplantation*. 2020;. Available from: https://doi.org/10.1097/TP.0000000000003475. Epub ahead of print. PMID: 33031222.
62. Crudele V, Cacciatore F, Grimaldi V, et al. Human leukocyte antigen-DR mismatch is associated with increased in-hospital mortality after a heart transplant. *Exp Clin Transpl*. 2013;11(4):346–351. Available from: https://doi.org/10.6002/ect.2012.0276. Epub 2013 May 29. PMID: 23734843.

Chapter 11

Extracorporeal Membrane Oxygenation in Adults

Brett J. Wakefield[1,2], Nakul Kumar[1,2], Shiva Sale[2] and Balaram Anandamurthy[1]

[1]*Department of Intensive Care and Resuscitation, Anesthesiology Institute, Cleveland Clinic, Cleveland, OH, United States,*
[2]*Department of Cardiothoracic Anesthesiology, Anesthesiology Institute, Cleveland Clinic, Cleveland, OH, United States*

Chapter Outline

Introduction

Extracorporeal membrane oxygenation (ECMO) is a form of cardiopulmonary bypass used to rescue patients with severe cardiorespiratory failure refractory to conventional treatments in the critical care units. The term ECMO is used interchangeably with extracorporeal life support (ECLS), albeit ECLS is a more appropriate term. The first successful use of ECMO in an adult was published more than four decades ago.[1] However, for the first two to three decades it was mainly used in neonates and pediatrics. Only in the last 20 years has there been an exponential growth in the use of ECLS in adults. This is due to both improvements in technology (innovation in cannula, pump, oxygenator design,[2] tubing biocoating, and equipment miniaturization) and evolution of intensive care medicine. There is also an increase in demand since the 2009 H1N1 pandemic (Fig. 11.1).[2–4]

The basic design of the ECMO circuit involves a large cannula draining blood from the right atrium via inferior vena cava (IVC) and/or superior vena cava (SVC) and returning it to the patient after adding oxygen and removing carbon dioxide with the help of an oxygenator (membrane lung) and a centrifugal pump. If the blood is returned to the right atrium, the pump supports the lungs (gas exchange) and is called venovenous (VV) ECMO. If the blood is returned to the arterial side (either femoral artery, axillary artery, or ascending aorta), the pump supports gas exchange

Advances in Cardiovascular Technology. DOI: https://doi.org/10.1016/B978-0-12-816861-5.00036-8

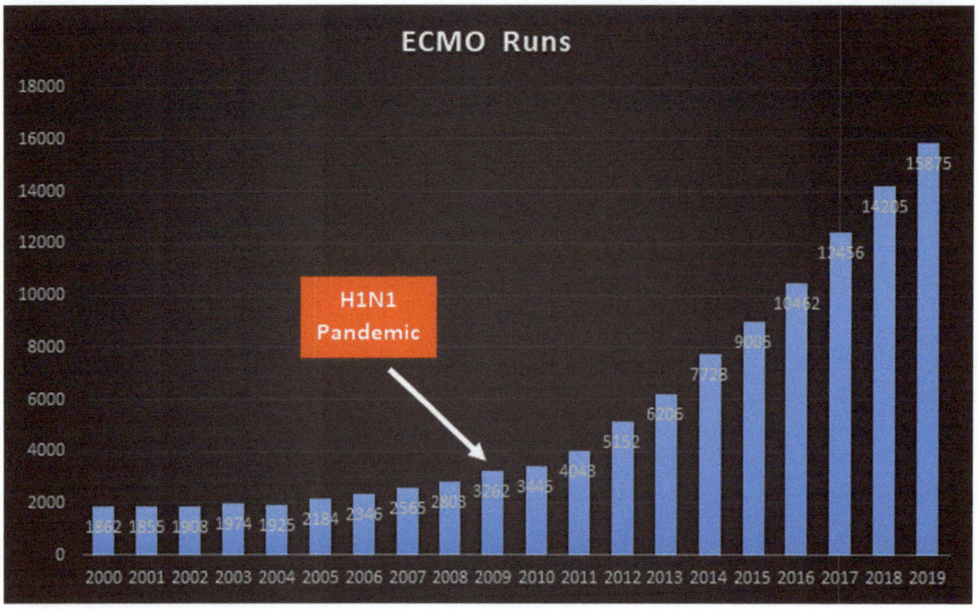

FIGURE 11.1 Chart demonstrating increase in ECMO runs after the 2009 H1N1 pandemic. *Data from ELSO 2020 International Summary.*

and circulation and is called venoarterial (VA) ECMO (Figs. 11.2 and 11.3). In certain situations, both VV and VA are combined to support an injured lung and heart, resulting in V-AV ECMO. The main goal of ECMO therapy is to support the injured organ and reverse or prevent multiorgan dysfunction so the patient can either recover or be eligible for advanced therapies. Unfortunately, about 30%–50% of patients do not recover or qualify for advanced therapies or transplantation leading to the withdrawal of support resulting in mortality.

Venovenous extracorporeal membrane oxygenation for respiratory failure

Introduction

ECMO is used to rescue patients from severe respiratory failure when the risk of death is high with conventional therapies and they are unable to provide adequate oxygenation or ventilation safely. VV ECMO is the configuration most commonly used to provide respiratory support, although VA ECMO or V-AV ECMO may be used in special circumstances.

Physiology of VV ECMO

Understanding VV ECMO physiology is essential for optimal patient management.[5] VV ECMO improves gas exchange but does not provide direct circulatory support as it is in series with the native circulation. In VV ECMO, blood is drained from the IVC, SVC, or both, and after adding oxygen and removing carbon dioxide, it is returned to the right atrium. A fraction of the blood reinfused from the ECMO circuit is drained back into the circuit before entering the right ventricle, resulting in recirculation which decreases the efficacy of ECMO support. Recirculation is detailed further in the chapter. The highly oxygenated blood returning from the ECMO circuit mixes with the poorly oxygenated blood from the native circulation resulting in oxygen enriched mixed venous saturation. The right ventricle pumps this blood into the lungs where further gas exchange can occur depending on the native lung function (Fig. 11.4). However, in the majority of patients at the initiation of VV ECMO support, the native lungs do not make any additional contribution to gas exchange and the patient is completely dependent on the ECMO support.[6]

Oxygenation

Oxygenation depends on the ratio of ECMO blood flow to the patient's intrinsic cardiac output, the recirculation fraction, and the native lung function. Higher circuit blood flows lead to better oxygenation. In order to achieve an oxygen saturation greater than 90%, circuit flows need to capture at least 60% of the patient's cardiac output.[7,8] Recirculation should be expected if augmentation in the ECMO blood flow rate does not result in the expected increase in patient's

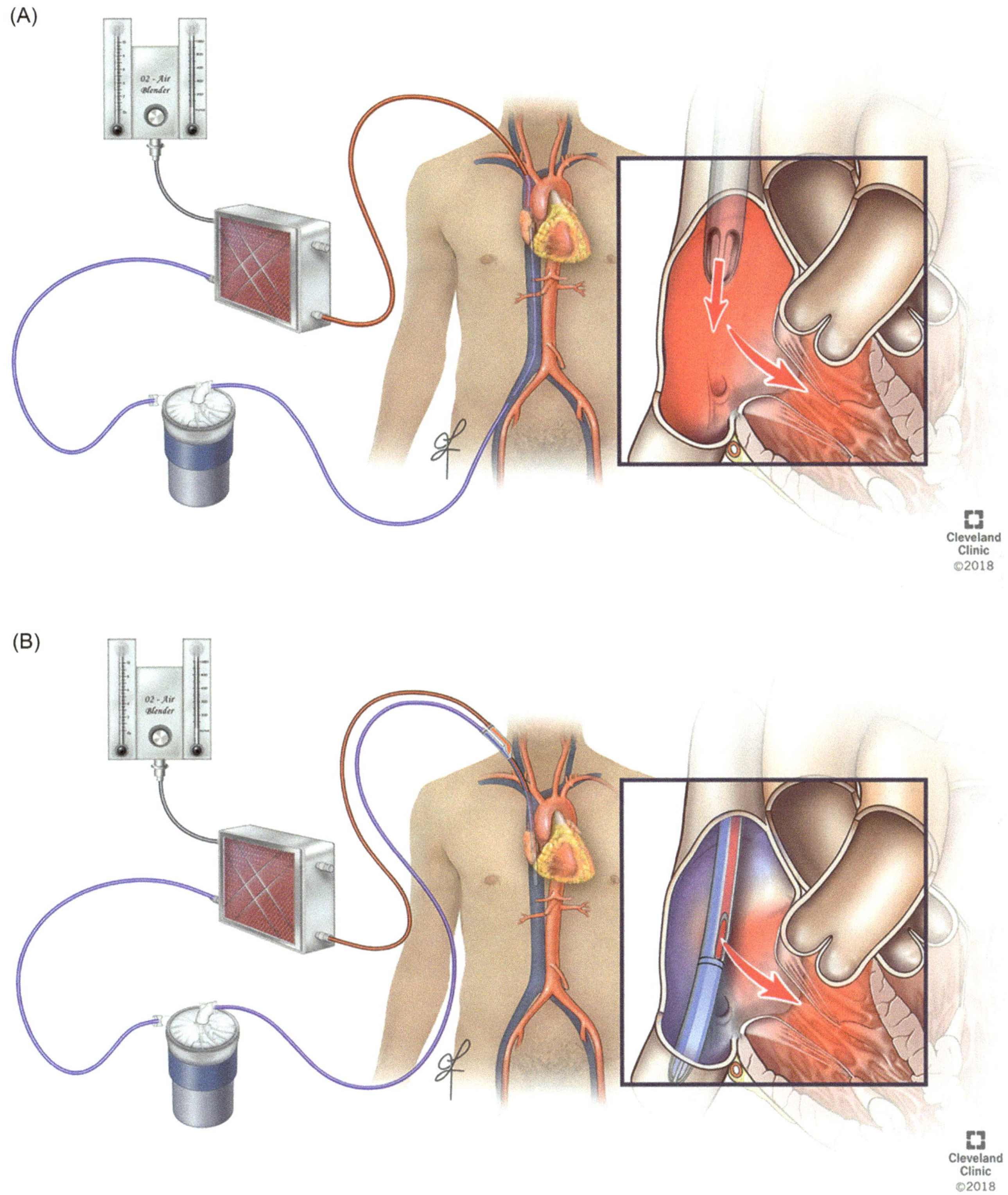

FIGURE 11.2 Venovenous ECMO with right femoral venous drainage and right internal jugular reinfusion.

oxygen saturation. Physical examination of the cannulas may reveal the presence of oxygenated blood in both cannulas. Recirculation can be minimized by optimizing cannula position or using a well-placed double lumen cannula. If both the SVC and IVC are cannulated, ensure the cannula tips are at least 10 cm apart.

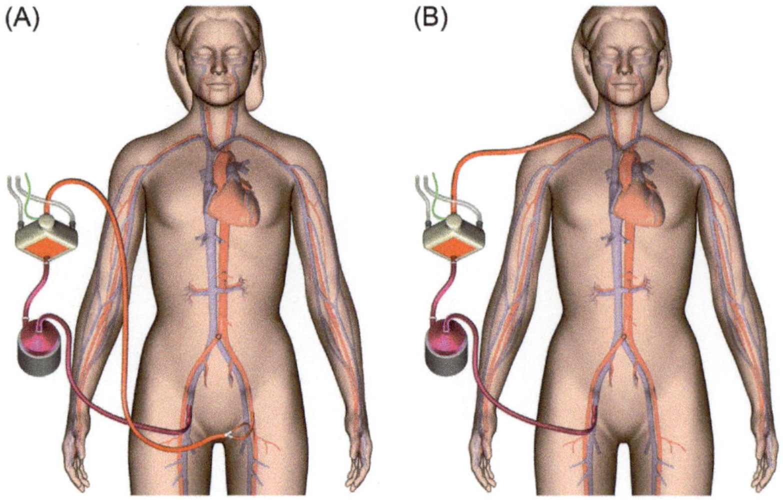

FIGURE 11.3 Peripheral VA ECMO configuration with right femoral venous inflow and right femoral arterial outflow.

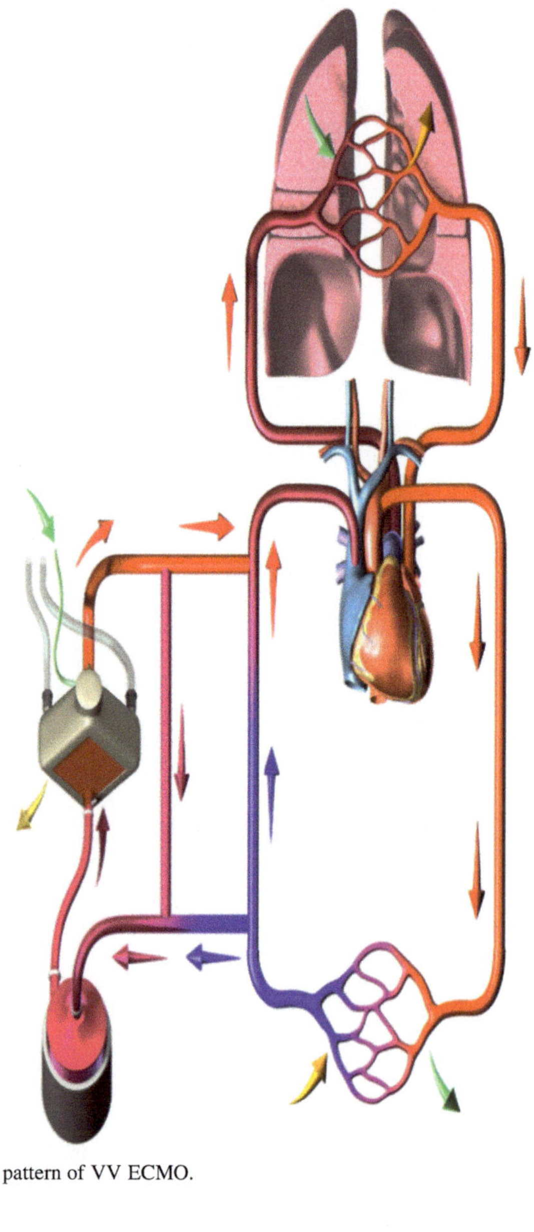

FIGURE 11.4 Physiology and blood flow pattern of VV ECMO.

Lung protective ventilation should be employed with the initiation of VV ECMO support. As the lung injury resolves and compliance improves, ventilator support can be slowly increased to aid gas exchange. Hypoxemia is common on VV ECMO and is typically well-tolerated. If oxygen delivery (DO_2) is greater than twice the oxygen consumption (VO_2), patients do not develop tissue hypoxia. However, refractory hypoxemia may occur in the setting of increasing oxygen consumption in patients with hyperdynamic cardiac function, such as uncontrolled pain, anxiety, or fever. Circuit flows may be unable to maintain the >60% fraction (ECMO flow/Total CO) required for optimal oxygenation. In these circumstances ensuring adequate sedation, analgesia, neuromuscular blockade, cooling with targeted temperature management, and blood transfusion to raise the hemoglobin above 10 g/dL may be required. Rarely beta-blockers are used to reduce native cardiac output and increase the fraction of blood captured by the ECMO circuit. Careful monitoring is required because there is a risk of decreasing DO_2 by decreasing cardiac output.[9]

Carbon dioxide removal

VV ECMO is more efficient for carbon dioxide removal than oxygen transfer. Carbon dioxide removal is a function of fresh gas flow (sweep) and requires significantly less circuit blood flow when compared to oxygen transfer. The fresh gas flow contains little or no carbon dioxide and hence allows rapid transfer and removal of carbon dioxide along the concentration gradient. Furthermore, carbon dioxide is significantly more soluble in plasma than oxygen. Circuit flow as low as 0.5–1.0 L/min can allow the removal of a patient's entire minute production of carbon dioxide (VCO_2).

Indications and contraindications

VV ECMO is only indicated in patients with severe respiratory failure and refractory hypoxemia or hypercarbia with high risk of demise despite maximal medical therapy or requiring potentially injurious mechanical ventilation settings to provide acceptable gas exchange (Table 11.1). Various protocols may have different triggers for VV ECMO support. Table 11.2 presents the criteria for VV ECMO in the EOLIA (ECMO to Rescue Lung Injury in Severe ARDS) trial.[10] Other indications include complex airway surgery, bronchopleural fistula, severe pulmonary edema, and primary graft dysfunction after lung transplantation, pulmonary embolus, or other causes of respiratory failure and hypoxemia. The EOLIA criteria is used to select patients for ECMO even in the recent COVID-19 pandemic. In crisis situations where hospital systems are overwhelmed, many hospitals resort to triaging only young patients with the highest chance of survival to ECMO. ELSO recommended against starting new ECMO programs during the pandemic.[11]

TABLE 11.1 Deleterious mechanical ventilation.

Deleterious mechanical ventilation
• Plateau pressure >30 cm H_2O
• Tidal volume >6 mL/kg
• Driving pressure[a] >14 cm H_2O

[a]Driving pressure (ΔP) = Peak inspiratory pressure (PIP) − positive end expiratory pressure (PEEP).

TABLE 11.2 Indications for VV ECMO support from the EOLIA trial.[10]

Indications for VV ECMO support
• PaO_2:FiO_2 Ratio <50 mmHg for >3 h despite adjunctive therapies[a] and optimization of MV for hypoxemia[b]
• PaO_2:FiO_2 Ratio <80 mmHg for >6 h despite adjunctive therapies and optimization of MV for hypoxemia
• Arterial blood pH <7.25 with P_aCO_2 ≥60 mmHg for >6 h despite optimization of MV for hypercarbia[c]

[a]Adjunctive therapies: inhaled nitric oxide, recruitment maneuvers, prone position, neuromuscular blockade.
[b]FiO_2 ≥ 80%, Tidal volume = 6 mL/kg, PEEP ≥ 10 cm H_2O.
[c]Respiratory rate 35/min, Plateau pressure ≤ 32 cm H_2O, PEEP ≥ 8 cm H_2O.

Any patient can be placed on VV ECMO support as long as large-bore venous cannulas can be safely placed (for adults single lumen dual site, usually 20–28 Fr for drainage, 16–20 Fr for return; and for double lumen, usually 28–32 Fr). However, VV ECMO support may be futile or not advisable in some settings. See absolute and relative contraindications in Table 11.3.

Patient selection

Patient selection is critical to improve outcomes and optimize resources. There are several validated scoring systems to predict outcomes, such as the RESP (Respiratory ECMO survival Prediction) score and PRESRVE (Predicting death for Severe ARDS on V-V ECMO) score.[12,13] However, none of these scoring systems replace clinical judgment, and utilizing an experienced multidisciplinary team to make these complex decisions is very helpful. Adequate cardiac function is a prerequisite for VV ECMO. VA ECMO should be considered in patients with decreased cardiac function. In the absence of major contraindications and markers of poor outcome, like old age and multiorgan failure, the clinical trajectory of the patient should be considered before deciding to initiate ECMO.[14]

ECMO management

Critical care management involves the management of the patient and the circuit (Fig. 11.5). Patient management can encompass all systems and involves all aspects of critical care medicine. From a neurological standpoint, when ECLS is initiated, patients may require heavy sedation and potentially neuromuscular blockade. The eventual goal, however, is to get patients awake, either tracheostomy as early as feasible or extubated and spontaneously breathing. The patients should be comfortable, well-nourished, and able to participate in their care, including respiratory and physical therapy. This goal is especially imperative for potential transplant candidates to improve their physical strength and optimize their chances for a successful lung transplant. A primary goal of VV ECMO is to prevent and limit iatrogenic injury from injurious mechanical ventilation. The patients should be on "lung rest" [tidal volume <4 mL/kg, plateau pressure

TABLE 11.3 Absolute and relative contraindications to VV ECMO support.

Absolute contraindication	Relative contraindication
• Severe or irreversible neurologic injury • Irreversible lung injury in a patient unsuitable for lung transplantation • Life expectancy <6 months (terminal illness) • Advanced multiorgan failure • Prolonged cardiopulmonary resuscitation (>60 min) • Prolonged mechanical ventilation (7–10 days) • Inability to gain adequate access	• Refusal to accept blood products (Jehovah's witness) • Coagulopathy • Advanced age • Morbid obesity • Inability to anticoagulate

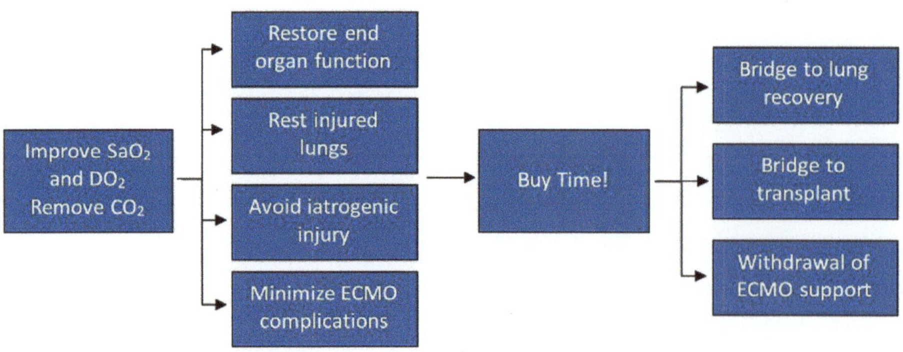

FIGURE 11.5 Goals of VV ECMO. The primary goal of VV ECMO is to protect the injured lungs.

<24 cm H_2O, driving pressure <14 cm H_2O, and optimal positive end expiratory pressure (PEEP)] as soon as feasible after initiation of ECMO.[15]

VV ECMO depends on the native circulation for the delivery of oxygen. Therefore, it is essential to maintain adequate cardiac output and blood pressure. Patients should be euvolemic, and transfusion thresholds vary depending on whether the ECMO is a bridge to transplant or a bridge to recovery. As long as oxygen delivery is adequate, a conservative red blood cell transfusion threshold of 7 g/dL is acceptable. However, if oxygen delivery is inadequate after optimizing the blood flow, increasing the hemoglobin to greater than 10 g/dL or higher is acceptable in patients on a bridge to recovery pathway. The ECMO circuit can cause consumption coagulopathy and disseminated intravascular coagulation (DIC), therefore, daily monitoring of all coagulation parameters (prothrombin time (PT), activated partial thromboplastin time (aPTT), fibrinogen, and platelets) is required and should be corrected as clinically indicated. Thromboelastography (TEG) or thromboelastometry (ROTEM) is useful when managing bleeding or thrombotic complications. ECMO complications may be either related to the circuit or the patient. Table 11.4 presents common complications associated with ECMO therapy.

With regards to the circuit, there are only four possible manipulations. In the setting of hypoxemia, circuit flow should be increased as allowed by the cannula size and rated flow (ability to raise saturation from 75% to 95% for a given EBF) of the oxygenator. Second, in patients with hypercarbia, the fresh gas flow (sweep) should be increased (max up to 15 L/min) to allow for increased carbon dioxide removal. Third, in the setting of reduced mixed venous oxygenation, oxygen consumption may need to be reduced with targeted temperature management (mild hypothermia <36°C) or neuromuscular blockade. Finally, in the setting of refractory hypoxemia despite ECMO circuit and patient optimization, a second venous drainage cannula may be required in order to allow increased flow to improve oxygenation.

Weaning

Patients can stay on ECMO for days to weeks (generally 2–4 weeks). Once the lungs start to recover, there is an increase in compliance with increased tidal volume and reduced airway pressures. The patient's chest X-ray may improve, and arterial blood gases will stabilize. As the lungs begin to take over oxygenation and carbon dioxide removal, the fresh gas flow (sweep) can be reduced. When the patient's compliance is improved significantly (tidal volume 6 mL/kg, plateau pressure <30 cm H_2O) and the sweep decreased to <2 L/min, a weaning trial should be undertaken. During weaning trial, the sweep is turned off and the ventilator is adjusted to optimal settings for gas exchange. It is vital to maintain lung-protective ventilation during the weaning process and after decannulation. If the patient can tolerate 0 L/min fresh gas flow for more than 6–24 h, the cannulas can be removed at the bedside. Usually, there is no need to wean blood flow or F_DO_2 (fraction of ECMO oxygen delivery—similar to the fraction of inspired oxygen (FiO_2) of a ventilator) of the circuit (wean sweep not flow!). However, a patient may desaturate on the odd occasion when the sweep is turned entirely off from as little as 0.5 L/min. In this case, a slower wean may be required and the

TABLE 11.4 ECMO complications.

Patient-associated complications	Circuit-associated complications

• AKI requiring RRT (28.2%)
• Bleeding
  ◦ Surgical/Cannula site (12.5%)
  ◦ Gastrointestinal (5%)
  ◦ ICH (3.5%)
• Cardiac arrhythmia (7%)
• Sepsis/Infection (6.4%)
• Pneumothorax (5.4%)
• Cardiac arrest (4.2%)
• DIC (1.2%)
• Thrombosis (DVT)
• Hemolysis
	• Circuit thrombosis (9.2%)
	• Oxygenator failure (6.3%)
	• Cannula issues (4.7%)
	  ◦ Cannula dislodgement
	  ◦ Tubing rupture
	• Air embolism (1.1%)
	• Pump failure (0.9%)

Percentages represent the incidence of the complication in patients requiring VV ECMO for respiratory failure. Data from the ELSO registry 2021.
AKI, Acute kidney injury; *RRT*, renal replacement therapy; *ICH*, intracerebral hemorrhage; *DIC*, disseminated intravascular coagulation; *DVT*, deep venous thrombosis.

F_DO_2 is gradually turned down (from 100% to 21%). Once at an F_DO_2 of 21%, the fresh gas flow can likely be discontinued. Again, if the patient can tolerate 0 L/min fresh gas flow for more than 6–24 h, the cannulas can be removed.

Evidence and outcomes

Many case series and observational studies demonstrate survival with VV ECMO between 50% and 70%.[16] However, only four randomized controlled trials have been completed, two of which were done before 2000,[17,18] reflecting old ECMO technology and intensive care unit (ICU) management. In both of these studies, mortality and complications were unacceptably high in the ECMO group, and survival was similar to the conventional arm of the trial. The CESAR (Conventional ventilatory support vs Extracorporeal membrane oxygenation for Severe Adult Respiratory failure) trial was the first trial to show a mortality benefit of ECMO treatment with a 16% higher survival in the ECMO group.[19] Nonetheless, there were several methodological issues with this trial. EOLIA[5] tried to address those issues, but unfortunately, this study was stopped by the Data Safety Monitoring Board before completion for futility.[10] Subsequent secondary analysis and Bayesian analysis of the trial did show a statistically significant survival benefit with ECMO. ELSO registry reports a survival of over 70% for VV ECMO for respiratory failure, which is probably close to real-life experience. Outcomes may be better in high-volume centers compared to low-volume centers.

In the last two decades, we have made great strides in improving circuit components and patient management. However, ECLS remains a high-risk and expensive intervention where outcomes depend on appropriate patient selection and management.

Venoarterial extracorporeal membrane oxygenation for cardiac failure

Introduction

Venoarterial extracorporeal membrane oxygenation (VA ECMO) is a form of temporary mechanical circulatory support used in cases of refractory cardiogenic shock, cardiac arrest, or postcardiotomy shock. A precise definition of cardiogenic shock does not exist, and the definition varies from trial to trial. However, in general, cardiogenic shock involves a state of hypoperfusion despite adequate fluid resuscitation and signs of end-organ dysfunction, such as oliguria, altered mental status, lactic acidosis, and mottled or cool extremities in the setting of ventricular dysfunction.[20]

Refractory cardiogenic shock requires escalation to mechanical circulatory support. The type of mechanical circulatory support depends on the patient. Post myocardial infarction ventricular septal defects and mitral valve papillary muscle ruptures may only require an intraaortic balloon pump (IABP). Isolated left ventricular failure can be managed with a microaxial left ventricular assist device (Impella, Abiomed, Danvers, MA) or a left atrial to arterial device (TandemHeart). However, in cases of severe hemodynamic instability or right ventricular failure, VA ECMO may be required. VA ECMO drains venous blood from the central veins and returns oxygenated blood to the arterial system, bypassing the left and right ventricles and the lungs. The circuit significantly reduces cardiac preload while increasing LV (Left ventricle) afterload. VA ECMO can also provide pulmonary support in addition to cardiac support.

Indications and contraindications

Any cause of refractory cardiogenic shock may be an indication for VA ECMO. The most common indications for VA ECMO are postcardiotomy cardiogenic shock (postcardiopulmonary bypass cardiogenic shock following cardiotomy during cardiac surgery) and acute myocardial infarction.[21] VA ECMO is a frequent requirement in patients undergoing heart transplantation and durable ventricular assist device implantation. Medical indications include myocarditis, peripartum cardiomyopathy, refractory arrhythmias, pulmonary embolism, pulmonary hypertension, drug intoxication, and hypothermia.

There are few contraindications to VA ECMO support. VA ECMO should only be placed in patients with potential for recovery or as a bridge to heart transplantation or ventricular assist device implantation. Relative contraindications include advanced age, obesity, inability to anticoagulate, and chronic organ dysfunction such as advanced chronic obstructive lung disease, metastatic cancer, cirrhosis, or renal failure. Aortic insufficiency can lead to left ventricular distension on VA ECMO support, and valve replacement or repair should be considered if VA ECMO is required and insufficiency is more than moderate. Mild to moderate aortic insufficiency may be tolerated with the addition of left ventricular venting (addressed further in the chapter).

Outcomes

Survival following VA ECMO is around 30%–40%, with slightly higher survival rates seen in patients with myocarditis (64%) or following heart transplantation (57%).[21,22] Two meta-analyses of VA ECMO patients with postcardiotomy cardiogenic shock have demonstrated survival to discharge rates of 30%–34%.[23,24] The duration of VA ECMO treatment varies with mortality in a nonlinear manner. A registry study evaluating VA ECMO patients found that mortality increased each day for the first 4 days and decreased each day until day 12. After the 12th day, there was no change in mortality.[21] Another study in patients with postcardiotomy shock found improved survival in patients undergoing 4–7 days of VA ECMO treatment compared to shorter or longer runs.[25] Failure to correct organ dysfunction with VA ECMO likely represents the poor survival seen in patients with short VA ECMO runs. In fact, improvement in lactate levels in the first 24–48 h has been correlated with improved survival.[22] Similar to other procedures, mortality following VA ECMO is higher at low-volume centers. A German registry study found that centers performing more than 100 procedures per year had a mortality of 53.4%, while those who performed less than 50 procedures per year had 65% mortality.[26] Risk factors for mortality following VA ECMO include age, cardiac arrest, chronic liver disease, elevated total bilirubin, and the presence of a left ventricular vent.[27]

Cannulation strategies

Cannulation for VA ECMO support can be either peripheral or central. The femoral or axillary arteries can be used for peripheral cannulation. Surgical femoral cut-down or percutaneous access offers access to the femoral vasculature for quick initiation of peripheral VA ECMO (Fig. 11.3). A 6–8 French distal reperfusion catheter should be placed into the superficial femoral artery to ensure adequate perfusion to the distal lower extremity. Primary disadvantages of peripheral femoral VA ECMO are upper body hypoxemia and lower extremity ischemia. Axillary cannulation provides easier ambulation, antegrade flow, and decreased differential hypoxemia; however, it requires an end-side surgical graft anastomosis to avoid critical arm ischemia.

Central cannulation typically occurs via sternotomy in the postcardiac surgical period when the ascending aorta can be directly cannulated. Venous drainage can be direct atrial cannulation or peripheral drainage using the right internal jugular or femoral veins. Disadvantages to central cannulation include an increased risk of bleeding and the requirement for an open chest with an increased risk of mediastinitis. Central cannulation also typically precludes extubation and patient mobilization. If prolonged support is anticipated, central cannulas can be tunneled to promote chest closure or cannulas moved to peripheral location. Cannula sizes are chosen to provide adequate circulatory support (60 mL/kg/min blood flow at our institution). The size of the venous cannula is the primary factor limiting the total blood flow provided by the circuit. In contrast, the size of the arterial cannula determines the pressure of the system. Periodically, optimal flows are unachievable with one venous drainage cannula and a second venous cannula is required.

There are no randomized controlled trials examining peripheral versus central cannulation strategies in VA ECMO, and there is no consensus regarding the superiority of one method over the other. The method of cannulation is mostly determined by clinical circumstances. Central cannulation avoids the issue of differential hypoxia (detailed further in the chapter) and offers a larger venous drainage cannula which can improve cardiac unloading. On the other hand, peripheral access may reduce infections and improve survival at the expense of increased vascular complications.[28] A large 2019 meta-analysis of almost 1700 VA ECMO patients demonstrated reduced bleeding, continuous VV hemofiltration, and blood product transfusions in peripheral VA ECMO compared to central.[29] However, there was no difference with regards to sepsis, limb complications, stroke, or mortality. Further, a retrospective evaluation of 800 patients undergoing postcardiotomy VA ECMO support found higher mortality in patients with a central cannulation strategy.[30]

Hemodynamic goals

The primary aim of VA ECMO support is to restore adequate organ perfusion. Maintaining an appropriate flow and cardiac index ($CI \geq 2.2$ L/min/m^2) with the ECMO circuit should restore blood flow to the major organs and potentially allow improvement in lactate clearance, mentation, urine output, and liver function tests. The mean arterial pressure should be at least 60 mm Hg to ensure adequate perfusion and may need to be supported with vasoactive medications. Adequate flow is also supported with a mixed venous oxygen saturation greater than 60%. Considering VA ECMO bypasses the cardiopulmonary structures, it is important to promote some degree of cardiac output and arterial line pulsatility (goal pulse pressure ≥ 10–20 mm Hg). This allows blood flow through the cardiopulmonary tree and decreases areas or regions of stasis which could contribute to intracardiac thrombus formation. Inotropes or an IABP may be required to achieve pulsatility and blood flow. If unable to generate adequate arterial pulsatility, special attention must be paid to anticoagulation to avoid disastrous intracardiac thrombosis. Intermittent states of low ECMO flow, termed chugging, can occur when venous drainage cannot meet the desired flow. Chugging is evident when quick and substantial ($>20\%$) changes in flows or pump speed (RPM). Typically, this is due to a malpositioned cannula, high pump speeds, or hypovolemia. Cannula position should be confirmed with a chest or abdominal radiograph. Mixed venous oxygenation should be measured to determine if flows are too high. If flows are appropriate and the cannulas are well-positioned, a fluid challenge is likely to improve consistent flows.

Differential hypoxia

A unique disadvantage to peripheral femoral VA ECMO occurs in patients with improving cardiac function and concomitant respiratory failure when poorly oxygenated blood enters the left heart and is ejected into the ascending aorta and upper body. This phenomenon, referred to as north–south syndrome, Harlequin syndrome, or differential hypoxia, results in differential oxygenation to the body with upper extremity, cerebral, and myocardial hypoxemia. In fact, myocardial hypoxemia may be the first problem to develop and can hinder the recovery of cardiac function.[31] Mixing of poorly oxygenated blood from the ascending aorta and well-oxygenated blood from the femoral arterial cannula causes mixing in the descending aorta (Fig. 11.6). The level of mixing depends on the patient's intrinsic cardiac function and the VA ECMO flows. This syndrome only develops with respiratory failure resulting in poorly oxygenated blood entering the left heart. Rarely this syndrome will present with obvious physical exam characteristics such as upper body cyanosis. It is very important to monitor arterial blood gases via a right upper extremity arterial line, right upper extremity pulse oximetry, and cerebral oximetry to detect this problem early.

The management of deferential hypoxemia may require changing the arterial cannula from femoral to axillary or central location. Another option is adding a second venous reinfusion cannula to the RA (Right atrium) resulting in V-VA ECMO. It is vital to note that conversion to V-AV ECMO reduces arterial flow and reduces the VA ECMO support component. Increasing the ECMO flow may work transiently but more durable solution is to either change the cannula location or configuration.

Left ventricular decompression

While VA ECMO is used to allow for cardiac recovery, arterial inflow increases left ventricular afterload. The increased afterload can decrease cardiac output allowing blood to collect in the left ventricle (via Thebesian, bronchial, and systemic venous return), leading to elevated ventricular pressures. Increased left ventricular pressures increase wall tension and increase myocardial oxygen demand resulting in myocardial ischemia and delaying left ventricular recovery.[32] Furthermore, increased left ventricular pressures increase pulmonary vascular congestion leading to pulmonary edema and reduced pulmonary function. This phenomenon does not occur in every VA ECMO patient. If the aortic valve opens with each heartbeat, a pulse pressure of at least 10 mm Hg is maintained in the arterial tree, and the

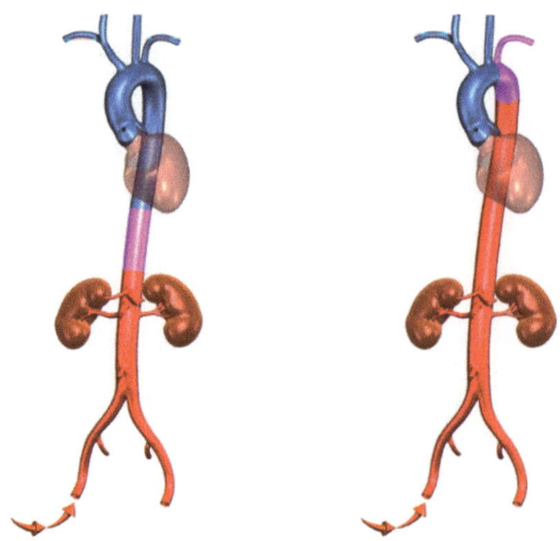

FIGURE 11.6 Harlequin syndrome (mixing of blood in Fem-Fem VA ECMO).

pulmonary capillary wedge pressure remains low, the heart is likely adequately decompressed. If not, inotropes should be initiated to optimize forward output and decompress the left ventricle. VA ECMO flows may be decreased; however, flows must still be high enough to maintain peripheral perfusion. If these maneuvers fail to decompress the left ventricle, left ventricular venting must be employed.

Retrospective studies and meta-analyses have demonstrated improved mortality in patients on VA ECMO receiving a broad range of left ventricular decompression modalities.[33,34] Multiple devices exist to help vent the left ventricle while on VA ECMO (Table 11.5). Some institutions use percutaneous IABP due to their ability to reduce left ventricular afterload while improving coronary perfusion. A 2019 meta-analysis of over 4000 patients found VA ECMO plus an IABP was associated with decreased hospital mortality in patients with cardiogenic shock compared to VA ECMO alone.[35] In addition, there were no differences in neurologic, gastrointestinal, or limb-related complications. However, an IABP does not directly vent the left ventricle or reduce ventricular preload. The TandemHeart (CardiacAssist, Inc, LivaNova, UK) uses a percutaneous transeptal catheter to drain the left atrium. Similar cannulas can be placed percutaneously and connected to the venous inflow cannula offering left ventricular preload reduction. The primary disadvantages of these cannulas include the requirement for a transeptal puncture, the potential for atrial suction events, and the potential for left ventricular thrombus formation in the setting of a competent mitral valve.[36] Specialized pulmonary vents can be placed via the upper central venous system to vent blood via the distal pulmonary arteries; however, these are not frequently used. Vents can be placed surgically via the atrium, pulmonary veins, or directly through the left ventricular apex. These allow ventricular offloading but require surgical implantation, an open chest, and may increase the risk of bleeding complications. Microaxial left ventricular assist devices, such as the Impella, offer an ideal solution to left ventricular distension (Fig. 11.7). This device is inserted either percutaneously through the femoral artery or via a surgical graft of the axillary artery. It directly unloads the left ventricle while propelling blood into the ascending aorta. The action reduces left ventricular distention and promotes blood flow through the heart thus reducing myocardial ischemia, pulmonary edema, and cardiac thrombus formation. Recent small studies have shown improved mortality in patients with VA ECMO and Impella (ECPELLA)-mediated left ventricular venting.[37,38]

Lower extremity ischemia

Acute lower extremity limb ischemia occurs in about 20% of patients with femoral VA ECMO and is associated with increased mortality.[39,40] Distal reperfusion cannulas are commonly placed to aid in distal perfusion and reduce the risk of limb ischemia.[41] It is essential to monitor distal flow with Doppler or tissue oximetry as pulses will likely not be

TABLE 11.5 Left ventricular decompression modalities.

Device	Advantages	Disadvantages
Intraaortic balloon pump	Easy placement	Not a true vent
TandemHeart	Directly drains left atrium	Requirement for septal puncture Potential for left ventricular thrombus Atrial suction
Pulmonary vents	Easy placement	Potential for left heart thrombus Not frequently used
Surgical vent	Direct left heart drainage	Surgical implantation Open chest Infection Bleeding
Impella	Direct left ventricular venting Low risk of cardiac thrombus	Potentially difficult placement Mitral valve damage Cost

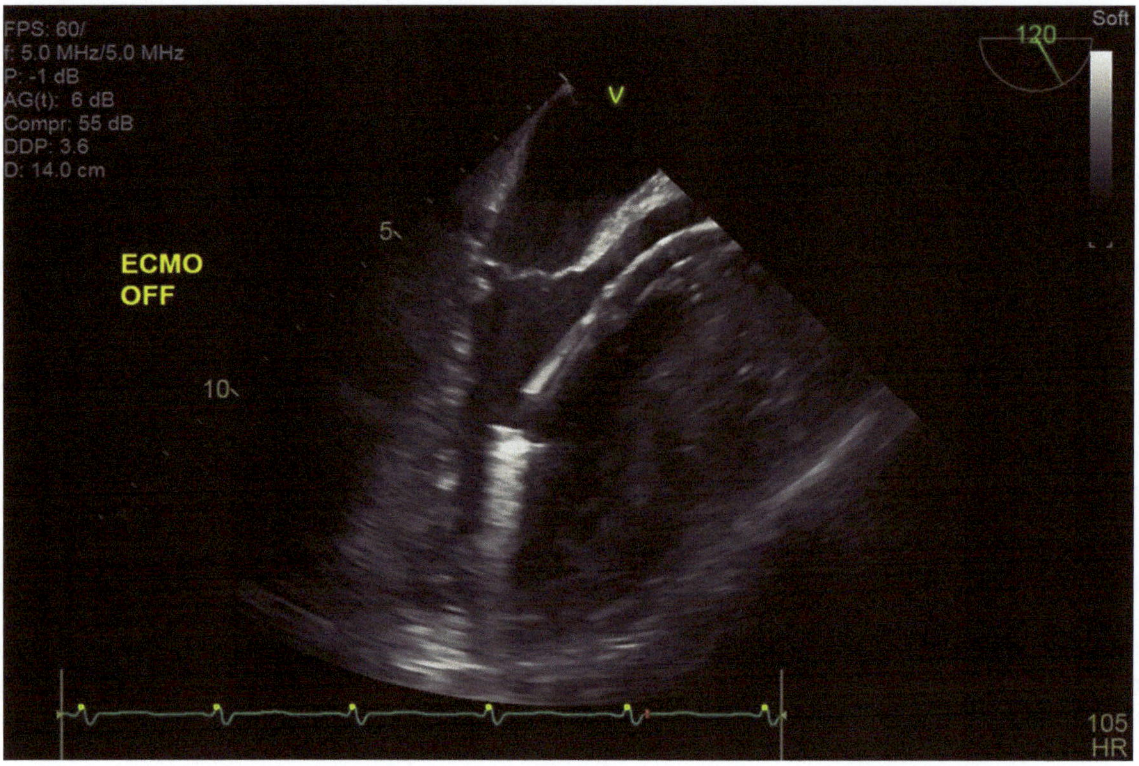

FIGURE 11.7 Midesophageal long axis view demonstrating a microaxial percutaneous left ventricular assist device (Impella).

palpable given the continuous flow of the ECMO circuit and distal reperfusion cannula.[42] Extremities should be evaluated daily for signs of compartment syndrome or ischemia. Any new onset hyperkalemia should initiate a thorough investigation for ischemia, including serial creatinine kinase measurements, Doppler blood flow evaluation, and physical exam. Involving the vascular surgical service early may be helpful in quick and efficient intervention should ischemia arise.

Weaning

The requirements for weaning are controversial, and protocols vary from institution to institution.[43] VA ECMO should not be weaned too early because the risk of weaning failure is high. Typically, 48 h of VA ECMO is required before consideration for weaning trials is entertained. After 48 h and upon return of cardiac function (Table 11.6), weaning may be possible. Lowering VA ECMO flows can increase the risk of thrombus formation and catastrophic embolization. Anticoagulation is encouraged to mitigate this problem. Heparin can be used with a goal activated PTT greater than 60 s. Pulmonary function should be evaluated. If pulmonary function is poor, conversion of VA ECMO to VV ECMO may be required. Once the return of cardiac function has been established, a weaning trial may be performed. With echocardiographic guidance, either transthoracic or transesophageal, ECMO flows can slowly be decreased in a step-wise fashion. While decreasing flows, left and right ventricular function and size and mitral and tricuspid regurgitation should be monitored to evaluate the response to the reduced ECMO support. Flows should not be decreased to less than 1 L/min due to concern for thrombosis formation. Inotrope and vasopressor doses may need to be increased slightly for successful weaning. With minimal flows, left ventricular ejection fraction should be greater than 20%–25%, aortic velocity time integral should be greater than 10–12 cm, and tissue Doppler of the lateral mitral annulus peak systolic velocity should be greater than or equal to 6 cm/s.

Extracorporeal cardiopulmonary resuscitation

Extracorporeal cardiopulmonary resuscitation (ECPR) is a newer technique for advanced cardiovascular rescue that utilizes an extracorporeal circuit to support blood flow in cardiac arrest refractory to traditional CPR (Cardio Pulmonary

TABLE 11.6 Signs of cardiac recovery.

Signs of cardiac recovery

• Hemodynamic improvement
  • Decreasing inotrope support
  • Decrease vasopressor support
• Improved pulsatility of both arterial and pulmonary arterial catheters
• Echocardiographic evidence of improved ventricular function, without significant ventricular distension of valvular disease
• Correction of metabolic derangements

Resuscitation) methods. This method allows organ perfusion to be maintained while addressing potentially reversible conditions.[44] ECPR specifically refers to rapid deployment of VA ECMO to assist with the return of spontaneous circulation (ROSC) after cardiac arrest.[45,46] This terminology is considered by some to be appropriate when VA ECMO is instituted both during manual/mechanical CPR and also within 20 min of ROSC.

As opposed to conventional CPR (CCPR), ECPR can be employed as a bridge to decision making, organ recovery, or transplantation, and has also been used as a bridge to palliative care.[47] However, significant resources need to be in place to allow for the successful application of ECPR. These resources often begin with a focus on high-quality CPR before and during the deployment of ECMO, with minimal interruptions between chest compressions, such as those available in an ICU setting. Additionally, a dedicated and well-coordinated ECMO team must be available for rapid cannulation and successful ECMO circulatory support.

Currently, there is insufficient evidence to suggest using ECPR routinely for all patients suffering from cardiac arrest.[44,45,48] However, select patients may benefit from ECPR if it is rapidly implemented in the setting of a potentially reversible etiology for cardiac arrest given a limited period of extracorporeal cardiopulmonary mechanical support.[45,49–51] A recent meta-analysis also reported a 13% absolute increase in survival and a 14% increase in favorable neurological outcome at 30-days for patients treated with ECPR compared to CCPR during refractory cardiac arrest.[52] Additionally, there is also mounting evidence for more favorable neurologic outcomes when targeted temperature management is utilized via ECPR and ECMO postresuscitation via a heat exchanger connected to the ECMO circuit.[51,53]

For in-hospital cardiac arrest, the use of ECPR has been associated with improved survival compared to manual CPR.[49–51,54] Additionally, the duration of CCPR before ECPR initiation along with earlier initiation of ECPR is a vital factor for success.[46,49,54,55] Unfortunately, the majority of studies involving ECPR are nonrandomized or retrospective in nature, making them subject to bias. More research is needed to help identify patients who would benefit most from ECPR compared to conventional techniques.

ECMO anticoagulation

The hemostatic system undergoes complex changes under ECMO support. The contact of blood with the nonendothelial surfaces of the ECMO circuit leads to activation of the coagulation cascade, coagulation factor consumption, platelet activation, and fibrinolysis.[56] Furthermore, ECMO patients may experience an acquired von Willebrand syndrome and factor XIII deficiency.[57] Bleeding and thrombotic complications represent a significant proportion of ECMO morbidity and mortality. Circuit thrombosis may occur in 10%–30% of patients, while bleeding complications may occur in more than 50% of patients.[58,59] Finding the balance of reducing thrombotic complications while simultaneously avoiding an increase in bleeding complications has proved difficult. Traditionally, systemic anticoagulation has been utilized for VV and VA ECMO to combat the increased thrombogenicity inherent in the ECMO circuit; however, the degree of anticoagulation, method of anticoagulation, and whether anticoagulation should even be utilized are controversial.

Heparin is the first line anticoagulant used for ECMO therapy. A 50–100 units/kg bolus dose is administered immediately before cannulation, and a heparin infusion is initiated in the absence of any overt clinical bleeding. Therapeutic heparin is continued until ECMO decannulation unless significant bleeding develops. Heparin administration can be monitored with activated clotting time (ACT), aPTT, anti-Xa levels, TEG, or ROTEM. Classically, the aPTT has been used due to its widespread availability, minimal cost, and quick turnaround time. However, diurnal variation, liver disease, acute phase reactants, factor deficiencies and inhibitors, lupus anticoagulant, and vitamin K abnormalities can alter the aPTT response to heparin.[60] As a result, anti-Xa monitoring has gained an appreciation for patients requiring heparin infusions. A large 2020 meta-analysis of over 2000 patients found that an anti-Xa-based anticoagulation strategy was associated with fewer bleeding events, lower mortality, and no change in thrombotic events when compared to time-based methods (ACT, aPTT, TEG, ROTEM).[61] ACT monitoring results in extremely fast results; however, it has not been shown to correlate with anti-Xa levels and is likely an inferior method of anticoagulation monitoring in ECMO patients.[62] A direct comparison between aPTT and anti-Xa demonstrated that aPTT monitoring may result in a longer time to therapeutic levels and higher heparin doses.[63] The ideal method of heparin monitoring is currently unknown; however, aPTT and anti-Xa monitoring remain the most commonly used strategies. Given the limitations of aPTT, an anti-Xa-based anticoagulation strategy may become the standard in the future.

Due to the challenges associated with heparin monitoring, heparin resistance, and the development of heparin-induced thrombocytopenia (HIT), alternative anticoagulants, such as direct thrombin inhibitors bivalirudin or argatroban, have been trialed in ECMO patients. Bivalirudin has a rapid onset, a half-life of 25 min, and requires dose adjustment in patients with renal dysfunction.[64] It is metabolized by proteolytic cleavage, which restores thrombin activity and can lead to clotting in the setting of stasis. Thus patients requiring VA ECMO support and bivalirudin anticoagulation must have pulsatility and flow through the cardiopulmonary tree to avoid catastrophic cardiac thrombus formation. This effect of bivalirudin metabolism may be particularly important during weaning trials as flows are decreased and, in fact, beneficial in the setting of soft tissue hematoma or hemothorax. On the other hand, argatroban has an onset of 30 min and a half-life of 45 min. Hepatic metabolism requires dose adjustment in liver failure, while no dose adjustment is required for renal failure. Due to the different metabolism, argatroban does not result in clot formation in the setting of stasis. Data regarding direct thrombin inhibitor use during ECMO therapy is limited, but current retrospective studies suggest these medications may be viable alternatives to heparin but not superior.[65]

Anticoagulation strategies may differ when managing VV versus VA ECMO. The immediate continuity of the oxygenator and cannula with the systemic circulation increases the morbidity and mortality associated with thrombus formation. Thrombus can embolize to the systemic circulation causing stroke, mesenteric ischemia, and extremity ischemia. aPTT goals of 40–50 s and anti-Xa levels 0.2–0.4 offer adequate anticoagulation for the VV ECMO patient, while aPTT goals of 50–65 s and anti-Xa levels of 0.2–0.7 may be required for VA ECMO. Patients requiring VA ECMO support following cardiac surgery may require 24–48 h without anticoagulation to allow for improvement in periprocedural bleeding. Due to the high incidence of bleeding complications and the associated morbidity and mortality, controversy has recently arisen over the anticoagulation requirements in ECMO patients. Two retrospective evaluations demonstrated no difference in overall morality in patients whether or not anticoagulation was used.[66,67] While there are a small number of studies suggesting the safety of anticoagulation avoidance in patients undergoing ECMO support, more research is required, and the current standard of care is to administer systemic anticoagulation in patients without evidence of significant bleeding.

Bleeding complications are common on ECMO support. Coagulation parameters should be routinely monitored, and blood products may be required. In the setting of active bleeding, platelets should be administered to keep the platelet count above 50 K/μL, and cryoprecipitate or fibrinogen concentrate should be administered to keep the fibrinogen level above 200 mg/dL. Fresh frozen plasma may be required in the setting of massive transfusion or abnormalities in the PT or TEG/ROTEM. In addition, an acquired von Willebrand syndrome may occur in up to 80%–100% of patients undergoing ECMO support.[68] ECMO-induced increased shear stress can destroy high-molecular-weight multimers leading to an overall decrease in von Willebrand factor activity. Furthermore, reduced pulsatility may contribute to decreased endothelial cell release of von Willebrand factor. This syndrome typically develops within 24 h of cannulation and is reversible upon decannulation. Theoretically, 1-deamino-8-D-arginine vasopressin (DDAVP) may help improve hemostatic parameters in patients with acquired von Willebrand disease. Furthermore, von Willebrand factor concentrate may improve bleeding complications as well, however, potentially at an increased thrombosis risk.[69]

Equipment

As previously discussed, extracorporeal mechanical equipment includes cannulas secured to a tubing circuit, a pump, and an oxygenator (Fig. 11.8). Several important considerations are necessary with regard to materials and equipment for successful ECMO function.

Cannula size

Higher blood flow rates (as described by Poiseuille's Law) require short cannulas with larger internal diameters. Generally, an ECLS arterial return cannula size of 18–20 Fr can permit flows >4 L/min. Typical adult drainage cannula

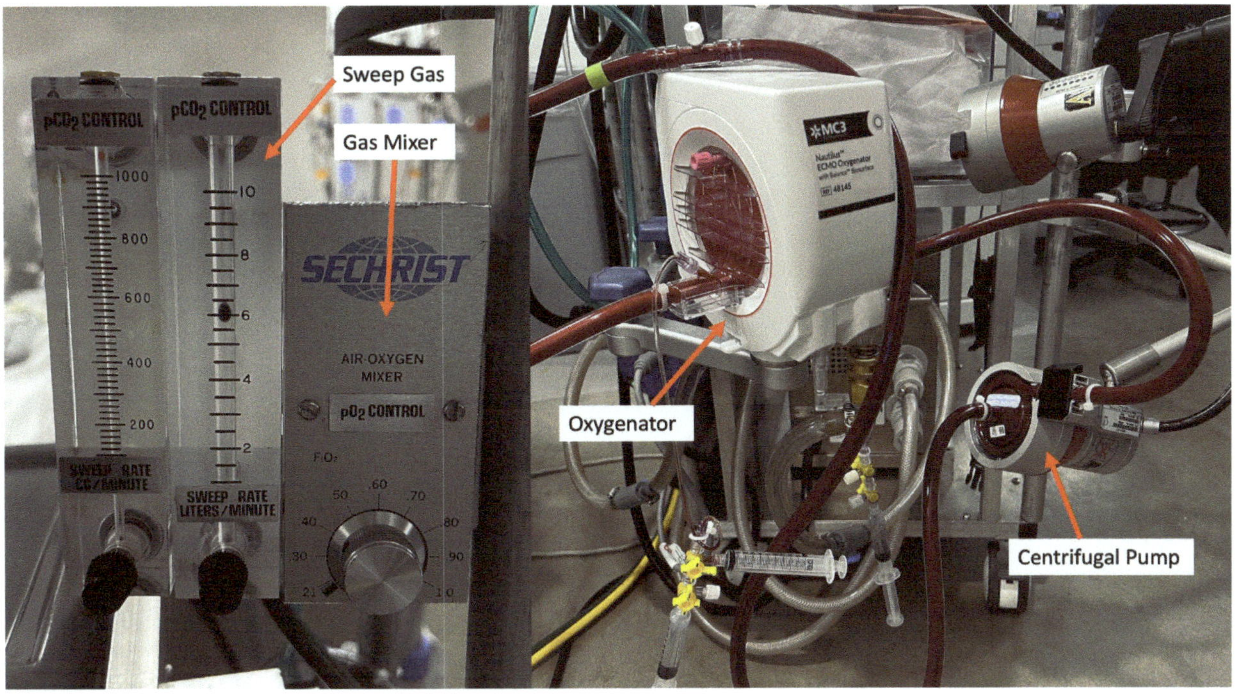

FIGURE 11.8 Rotaflow seen in its pump head housing, connected to a quadrox oxygenator. (Left) Fresh gas flow (sweep gas) and gas mixer. (Right) ECMO circuit with oxygenator and centrifugal pump.

sizes range from 20 to 28 Fr, with larger patients requiring larger cannulas (Fig. 11.9). ECMO cannulas are often wire-reinforced which allows for greater durability and prevents luminal occlusion. Some cannulas may include a malleable wire along the cannula body, allowing for customizability with their angle of insertion and cardiac position.[70] Importantly, cannula sizing based on patient size and ECMO support needs is vital to achieve optimal flow and avoid hemolysis. A pressure gradient of <100 mm Hg for arterial return and a gradient <50 mm Hg for venous drainage is ideal.[70] Dual-lumen, single-site cannulas are a recent advance in cannula technology that can be employed for VV ECMO support, most often placed via the right internal jugular vein or left subclavian vein (Figs. 11.8B and 11.9). These cannulas allow drainage and return via a single site, often improving patient comfort, facilitating extubation, and allowing greater mobility.[71] Some of these dual-lumen cannulas offer a bicaval design, with drainage lumens placed in both the SVC and IVC, with a return lumen placed in the right atrium. To reduce the risk of blood recirculation, imaging (such as transesophageal echocardiography) is often necessary to ensure the appropriate placement of these cannulas. For optimal oxygen delivery, the reinfusion port (middle port) is directed toward the tricuspid valve in the right atrium.[70,71]

Circuit and pump

The ECLS circuit is primed with a balanced crystalloid, such as plasmalyte (Baxter corporation, Deerfield, IL, USA). It is often connected to polyvinyl chloride tubing with a biocompatible surface coating (e.g., heparin) to reduce the inflammatory cascade and platelet activation that is classically associated with blood contact with foreign materials.[72] These circuits are connected to either a roller or centrifugal pump (Fig. 11.10). Roller pumps create direct suction on the venous circuit through physical pressure to generate forward flow. These pumps are now rarely used in ECMO due to drawbacks such as heavier motors limiting portability, tubing wear and tear, risk of tubing rupture, tubing fragmentation, red cell injury, and insensitivity to preload and afterload.[73]

Most ECMO circuits have now transitioned to centrifugal, magnetically driven impeller pumps with heat resistant bearings (2nd generation) or magnetically levitated bearing-less impeller pumps (3rd generation). Centrifugal pumps are dependent on preload, afterload, and rotational speed. Unlike roller pumps, centrifugal pumps are lighter, more portable, undergo more minor wear and tear of components, and rarely cause circuit rupture.

Initial centrifugal pumps suffered issues with blood stagnation, inappropriate pump heating, and thrombosis. The newer centrifugal pumps have improved upon these deficiencies and now allow greater RPM (Revolutions Per Minute) variability and reduced thrombosis. Despite the enhancements in pump technology and ECMO portability using

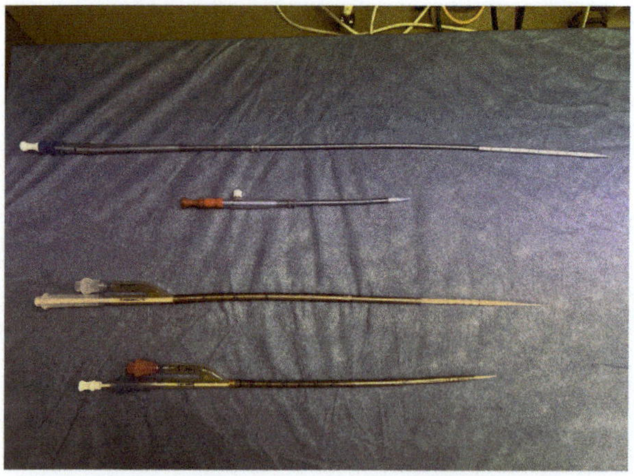

FIGURE 11.9 (A) Multistage access cannula for venous drainage. The distal end contains multiple access ports to allow efficient drainage. These cannulas are often used for internal jugular and femoral venous drainage. (B) Fem-Flex (Edwards Lifesciences, Irvine, CA, USA) arterial cannula. These cannulas are frequently used for blood return in either VA or VV ECMO. (C) Dual-Lumen Avalon (Maquet Getinge Group, Rastatt, Germany) cannula shown with inner dilator. (D) Crescent Dual Lumen (Medtronic, Dublin, Ireland) cannula shown with inner dilator.

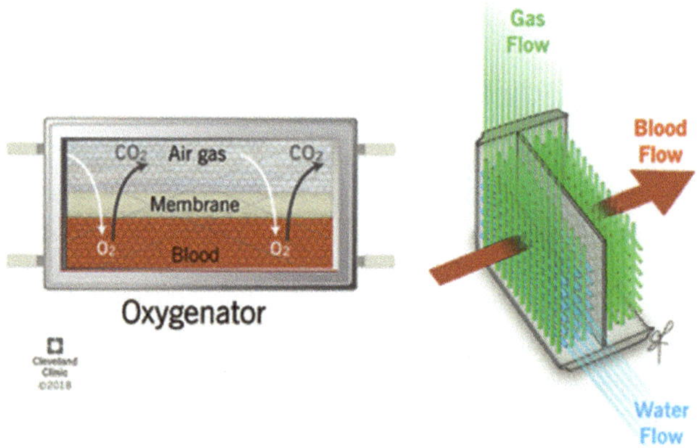

FIGURE 11.10 Illustrations representing blood, gas flow, and heated water through the oxygenator.

centrifugal pumps, significant complications are associated with usage. These include the ability for retrograde flow in the setting of pump failure and micro air bubble formation from heat generation. Additionally, if there is an occlusion on the return line, cavitation and hemolysis can ensue at the inlet. The risk of microemboli, gaseous embolism, hemolysis, and ultimately coagulopathy is a significant risk inherent with current centrifugal pumps and requires careful attention with prolonged use. The Rotaflow (Maquet, Hirrlingen, Germany), Cardiohelp (Maquet, Getinge group, Rastatt, Germany), and CentriMag (Thoratec Corporation, Pleasanton, CA) are examples of FDA-approved centrifugal pumps currently used in the United States.

Oxygenator

Modern membrane oxygenators attempt to recreate the blood:air interface by using low resistance capillary fibers made of poly-methyl-pentene (PMP) that allow gas permeability and facilitate intraluminal gas transfer across the concurrent blood flow around the fiber bundle (Fig. 11.9). Compared to older silicone-based membrane oxygenators that required much larger priming volumes and suffered from micropore leakage, the newest PMP oxygenators tend to require lower priming volume and lead to lower pressure drops across the oxygenator membrane while maintaining high gas exchange efficiency. Additionally, these newer PMP oxygenators, such as the Quadrox iD (Maquet, Hirrlingen, Germany), are more resistant to plasma leaking, allowing greater longevity and longer-term use before a significant decline in membrane permeability occurs requiring an oxygenator exchange.

Oxygen delivery is dependent on multiple variables aside from the oxygenator, including hemoglobin content, the fraction of oxygen delivery (F_DO_2), and oxygen saturation. Surface area is also limited compared to lung tissue (2 m^2 vs 70 m^2, respectively), thus flow rates can affect oxygenation as blood dwell time is affected. Internal flow resistance creating higher pressure drops inside the oxygenator will lead to greater shear stress and hemolysis, along with reducing oxygenator efficiency. Oxygenator clotting is often detected with rising D-dimer levels and trans oxygenator pressure gradients. Monitoring pre- and postoxygenator pressures in real time can allow for prompt recognition of impending oxygenator failure and allow for timely exchange. Accumulation of condensation will impair carbon dioxide exchange by decreasing carbon dioxide permeability. High gas flow for seconds to minutes can allow drying and elimination of condensate to restore carbon dioxide transfer.

Heater/cooler exchanger

The Quadrox iD also supports heat transfer via a polyurethane heat exchanger by providing a separate channel surrounding the inner blood and gas chamber for heated nonsterile water to flow through, perpendicular to blood flow (Fig. 11.5). Heat exchangers can also be installed separately into the ECMO circuit for patient heating and cooling.

Monitoring

Vigilance in monitoring patients on ECMO support is vital to their success. Numerous safety monitoring devices are present for each patient, assessing both the components of ECMO circuitry and patient stability. Circuit-related monitoring begins with full circuit checks to identify any issues with power supply, tubing, connections, oxygenator, or heat exchanger.

Blood flow and RPM

An electronic pump controller will digitally reflect the current centrifugal pump RPMs and the corresponding blood flow through the circuit at those RPMs using noninvasive flow probes attached to the pump and the circuit. A set RPM may not always deliver the same blood flow in centrifugal pumps because blood flow is dependent on preload, afterload, and cannula size. In the event of retrograde flow, circuit censors will alarm. Some pumps are designed to counteract retrograde flow by matching RPM speed to create 0 mL/min blood flow or clamp the tubing altogether. A backup battery and hand crank for manual pump operation are some additional safety features often incorporated on pumps.

Circuit pressure monitoring

Several pressure monitoring sites can be utilized to assess appropriate ECMO flow. The venous line pressure is measured prepump and monitored for excess negative pressure (should not exceed -300 mm Hg) which can create significant negative suction, cavitation, and hemolysis in the circuit. The postoxygenator pressure measurement is a combination of the pump RPMs, tubing/cannula size, and the patient's arterial pressure. Very high pressures, such as caused by occlusion of the arterial line, indicate impending circuit interruption. A rising delta pressure gradient between pre- and postoxygenator sites indicates worsening oxygenator resistance (potential thrombosis). The sweep gas pressure should not exceed blood pressure to avoid air embolism. Although the risk is low with newer PMP membranes, it is recommended to keep the oxygenator below the level of the patient.

Hemolysis monitoring

In addition to anticoagulation monitoring as discussed previously in the chapter, plasma-free hemoglobin monitoring has been associated with increased hemolysis and levels >50 mg/dL may be a potential indicator of increased mortality in ECMO patients.[74] Increased plasma-free hemoglobin has also been associated with worsening organ function, such as acute renal injury.[75]

Neuromonitoring

Brain perfusion and preservation is a core end point during ECMO support, yet assessment can be challenging depending on the level of support and the sedation that is often required to maintain patient safety (especially during VA

ECMO). When a neurological exam is difficult to perform (e.g., during heavy sedation), techniques such as intermittent or continuous electroencephalogram monitoring can assist with ruling out seizure activity, computed tomography imaging can help to rule out intracranial ischemia or hemorrhage, and cerebral oximetry/transcranial Doppler can ensure adequate blood flow during ECMO support.

Conclusion

ECMO support is a life-saving intervention which is increasing in utilization year after year. Knowledge of the physiology, management, and components of the ECMO circuit is of paramount importance in caring for these patients for best outcomes.

References

1. Hill JD, O'Brien TG, Murray JJ, et al. Prolonged extracorporeal oxygenation for acute post-traumatic respiratory failure (shock-lung syndrome). Use of the Bramson membrane lung. *N Engl J Med*. 1972;286(12):629–634.
2. Bartlett RH, Deatrick KB. Current and future status of extracorporeal life support for respiratory failure in adults. *Curr Opin Crit Care*. 2016;22(1):80–85.
3. Wang D, Zhou X, Liu X, Sidor B, Lynch J, Zwischenberger JB. Wang-Zwische double lumen cannula-toward a percutaneous and ambulatory paracorporeal artificial lung. *ASAIO J*. 2008;54(6):606–611.
4. Australia, New Zealand Extracorporeal Membrane Oxygenation Influenza I, Davies A, et al. Extracorporeal membrane oxygenation for 2009 influenza A(H1N1) acute respiratory distress syndrome. *JAMA*. 2009;302(17):1888–1895.
5. Combes A, Schmidt M, Hodgson CL, et al. Extracorporeal life support for adults with acute respiratory distress syndrome. *Intensive Care Med*. 2020;46(12):2464–2476.
6. Abrams D, Schmidt M, Pham T, et al. Mechanical ventilation for acute respiratory distress syndrome during extracorporeal life support. research and practice. *Am J Respir Crit Care Med*. 2020;201(5):514–525.
7. Bartlett RH. Physiology of gas exchange during ECMO for respiratory failure. *J Intensive Care Med*. 2017;32(4):243–248.
8. Schmidt M, Tachon G, Devilliers C, et al. Blood oxygenation and decarboxylation determinants during venovenous ECMO for respiratory failure in adults. *Intensive Care Med*. 2013;39(5):838–846.
9. Patel B, Arcaro M, Chatterjee S. Bedside troubleshooting during venovenous extracorporeal membrane oxygenation (ECMO). *J Thorac Dis*. 2019;11(Suppl. 14):S1698–S1707.
10. Combes A, Hajage D, Capellier G, et al. Extracorporeal membrane oxygenation for severe acute respiratory distress syndrome. *N Engl J Med*. 2018;378(21):1965–1975.
11. Bartlett RH, Ogino MT, Brodie D, et al. Initial ELSO guidance document: ECMO for COVID-19 patients with severe cardiopulmonary failure. *ASAIO J*. 2020;66(5):472–474.
12. Schmidt M, Bailey M, Sheldrake J, et al. Predicting survival after extracorporeal membrane oxygenation for severe acute respiratory failure. The Respiratory Extracorporeal Membrane Oxygenation Survival Prediction (RESP) score. *Am J Respir Crit Care Med*. 2014;189(11):1374–1382.
13. Schmidt M, Zogheib E, Roze H, et al. The PRESERVE mortality risk score and analysis of long-term outcomes after extracorporeal membrane oxygenation for severe acute respiratory distress syndrome. *Intensive Care Med*. 2013;39(10):1704–1713.
14. Ahuja A, Shekar K. Patient selection for VV ECMO: have we found the crystal ball? *J Thorac Dis*. 2018;10(Suppl. 17):S1979–S1981.
15. Del Sorbo L, Goffi A, Goligher E, Fan E, Slutsky AS. Setting mechanical ventilation in ARDS patients during VV-ECMO: where are we? *Minerva Anestesiol*. 2015;81(12):1369–1376.
16. Munshi L, Walkey A, Goligher E, Pham T, Uleryk EM, Fan E. Venovenous extracorporeal membrane oxygenation for acute respiratory distress syndrome: a systematic review and meta-analysis. *Lancet Respir Med*. 2019;7(2):163–172.
17. Morris AH, Wallace CJ, Menlove RL, et al. Randomized clinical trial of pressure-controlled inverse ratio ventilation and extracorporeal CO_2 removal for adult respiratory distress syndrome. *Am J Respir Crit Care Med*. 1994;149(2 Pt 1):295–305.
18. Zapol WM, Snider MT, Hill JD, et al. Extracorporeal membrane oxygenation in severe acute respiratory failure. A randomized prospective study. *JAMA*. 1979;242(20):2193–2196.
19. Peek GJ, Mugford M, Tiruvoipati R, et al. Efficacy and economic assessment of conventional ventilatory support vs extracorporeal membrane oxygenation for severe adult respiratory failure (CESAR): a multicentre randomised controlled trial. *Lancet*. 2009;374(9698):1351–1363.
20. Vahdatpour C, Collins D, Goldberg S. Cardiogenic shock. *J Am Heart Assoc*. 2019;8(8):e011991.
21. Smith M, Vukomanovic A, Brodie D, Thiagarajan R, Rycus P, Buscher H. Duration of veno-arterial extracorporeal life support (VA ECMO) and outcome: an analysis of the Extracorporeal Life Support Organization (ELSO) registry. *Crit Care*. 2017;21(1):45.
22. Hu RTC, Broad JD, Osawa EA, et al. 30-Day outcomes post veno-arterial extra corporeal membrane oxygenation (VA-ECMO) after cardiac surgery and predictors of survival. *Heart Lung Circ*. 2020;29(8):1217–1225.
23. Khorsandi M, Dougherty S, Bouamra O, et al. Extra-corporeal membrane oxygenation for refractory cardiogenic shock after adult cardiac surgery: a systematic review and meta-analysis. *J Cardiothorac Surg*. 2017;12(1):55.

24. Wang L, Wang H, Hou X. Clinical outcomes of adult patients who receive extracorporeal membrane oxygenation for postcardiotomy cardiogenic shock: a systematic review and meta-analysis. *J Cardiothorac Vasc Anesth*. 2018;32(5):2087–2093.
25. Mariscalco G, El-Dean Z, Yusuff H, et al. Duration of venoarterial extracorporeal membrane oxygenation and mortality in postcardiotomy cardiogenic shock. *J Cardiothorac Vasc Anesth*. 2020;.
26. Becher PM, Gossling A, Schrage B, et al. Procedural volume and outcomes in patients undergoing VA-ECMO support. *Crit Care*. 2020;24(1):291.
27. Vigneshwar NG, Kohtz PD, Lucas MT, et al. Clinical predictors of in-hospital mortality in venoarterial extracorporeal membrane oxygenation. *J Card Surg*. 2020;35(10):2512–2521.
28. Danial P, Hajage D, Nguyen LS, et al. Percutaneous vs surgical femoro-femoral veno-arterial ECMO: a propensity score matched study. *Intensive Care Med*. 2018;44(12):2153–2161.
29. Raffa GM, Kowalewski M, Brodie D, et al. Meta-analysis of peripheral or central extracorporeal membrane oxygenation in postcardiotomy and non-postcardiotomy shock. *Ann Thorac Surg*. 2019;107(1):311–321.
30. Mariscalco G, Salsano A, Fiore A, et al. Peripheral vs central extracorporeal membrane oxygenation for postcardiotomy shock: multicenter registry, systematic review, and meta-analysis. *J Thorac Cardiovasc Surg*. 2020;160(5):1207–1216. e1244.
31. Prisco AR, Aguado-Sierra J, Butakoff C, et al. Concomitant respiratory failure can impair myocardial oxygenation in patients with acute cardiogenic shock supported by VA-ECMO. *J Cardiovasc Transl Res*. 2021;.
32. Cevasco M, Takayama H, Ando M, Garan AR, Naka Y, Takeda K. Left ventricular distension and venting strategies for patients on venoarterial extracorporeal membrane oxygenation. *J Thorac Dis*. 2019;11(4):1676–1683.
33. Al-Fares AA, Randhawa VK, Englesakis M, et al. Optimal strategy and timing of left ventricular venting during veno-arterial extracorporeal life support for adults in cardiogenic shock: a systematic review and meta-analysis. *Circ Heart Fail*. 2019;12(11):e006486.
34. Baldetti L, Gramegna M, Beneduce A, et al. Strategies of left ventricular unloading during VA-ECMO support: a network meta-analysis. *Int J Cardiol*. 2020;312:16–21.
35. Li Y, Yan S, Gao S, et al. Effect of an intra-aortic balloon pump with venoarterial extracorporeal membrane oxygenation on mortality of patients with cardiogenic shock: a systematic review and meta-analysisdagger. *Eur J Cardiothorac Surg*. 2019;55(3):395–404.
36. Kim AR, Park H, Lee SE, et al. Outcomes of left ventricular unloading with a transseptal cannula during extracorporeal membrane oxygenation in adults. *Artif Organs*. 2021;45(4):390–398.
37. Schrage B, Becher PM, Bernhardt A, et al. Left ventricular unloading is associated with lower mortality in patients with cardiogenic shock treated with venoarterial extracorporeal membrane oxygenation: results from an international, multicenter cohort study. *Circulation*. 2020;142(22):2095–2106.
38. Patel SM, Lipinski J, Al-Kindi SG, et al. Simultaneous venoarterial extracorporeal membrane oxygenation and percutaneous left ventricular decompression therapy with impella is associated with improved outcomes in refractory cardiogenic shock. *ASAIO J*. 2019;65(1):21–28.
39. Son AY, Khanh LN, Joung HS, et al. Limb ischemia and bleeding in patients requiring venoarterial extracorporeal membrane oxygenation. *J Vasc Surg*. 2021;73(2):593–600.
40. Gulkarov I, Bobka T, Elmously A, et al. The effect of acute limb ischemia on mortality in patients undergoing femoral venoarterial extracorporeal membrane oxygenation. *Ann Vasc Surg*. 2020;62:318–325.
41. Hanley SC, Melikian R, Mackey WC, Salehi P, Iafrati MD, Suarez L. Distal perfusion cannulae reduce extracorporeal membrane oxygenation-related limb ischemia. *Int Angiol*. 2021;40(1):77–82.
42. Steffen RJ, Sale S, Anandamurthy B, et al. Using near-infrared spectroscopy to monitor lower extremities in patients on venoarterial extracorporeal membrane oxygenation. *Ann Thorac Surg*. 2014;98(5):1853–1854.
43. Lusebrink E, Stremmel C, Stark K, et al. Update on weaning from veno-arterial extracorporeal membrane oxygenation. *J Clin Med*. 2020;9(4).
44. Holmberg MJ, Geri G, Wiberg S, et al. Extracorporeal cardiopulmonary resuscitation for cardiac arrest: a systematic review. *Resuscitation*. 2018;131:91–100.
45. Link MS, Berkow LC, Kudenchuk PJ, et al. Part 7: Adult advanced cardiovascular life support: 2015 American Heart Association guidelines update for cardiopulmonary resuscitation and emergency cardiovascular care. *Circulation*. 2015;132(18 Suppl. 2):S444–S464.
46. Inoue A, Hifumi T, Sakamoto T, Kuroda Y. Extracorporeal cardiopulmonary resuscitation for out-of-hospital cardiac arrest in adult patients. *J Am Heart Assoc*. 2020;9(7):e015291.
47. Abrams DC, Prager K, Blinderman CD, Burkart KM, Brodie D. Ethical dilemmas encountered with the use of extracorporeal membrane oxygenation in adults. *Chest*. 2014;145(4):876–882.
48. Kim SJ, Kim HJ, Lee HY, Ahn HS, Lee SW. Comparing extracorporeal cardiopulmonary resuscitation with conventional cardiopulmonary resuscitation: a meta-analysis. *Resuscitation*. 2016;103:106–116.
49. Soar J, Nolan JP, Bottiger BW, et al. European resuscitation council guidelines for resuscitation 2015: Section 3. Adult advanced life support. *Resuscitation*. 2015;95:100–147.
50. Twohig CJ, Singer B, Grier G, Finney SJ. A systematic literature review and meta-analysis of the effectiveness of extracorporeal-CPR vs conventional-CPR for adult patients in cardiac arrest. *J Intensive Care Soc*. 2019;20(4):347–357.
51. Chen X, Zhen Z, Na J, Wang Q, Gao L, Yuan Y. Associations of therapeutic hypothermia with clinical outcomes in patients receiving ECPR after cardiac arrest: systematic review with meta-analysis. *Scand J Trauma Resusc Emerg Med*. 2020;28(1):3.
52. Ouweneel DM, Schotborgh JV, Limpens J, et al. Extracorporeal life support during cardiac arrest and cardiogenic shock: a systematic review and meta-analysis. *Intensive Care Med*. 2016;42(12):1922–1934.

53. Blumenstein J, Leick J, Liebetrau C, et al. Extracorporeal life support in cardiovascular patients with observed refractory in-hospital cardiac arrest is associated with favourable short and long-term outcomes: a propensity-matched analysis. *Eur Heart J Acute Cardiovasc Care*. 2016;5 (7):13–22.
54. Shin TG, Choi JH, Jo IJ, et al. Extracorporeal cardiopulmonary resuscitation in patients with inhospital cardiac arrest: a comparison with conventional cardiopulmonary resuscitation. *Crit Care Med*. 2011;39(1):1–7.
55. Bartos JA, Grunau B, Carlson C, et al. Improved survival with extracorporeal cardiopulmonary resuscitation despite progressive metabolic derangement associated with prolonged resuscitation. *Circulation*. 2020;141(11):877–886.
56. Andrews J, Winkler AM. Challenges with navigating the precarious hemostatic balance during extracorporeal life support: implications for coagulation and transfusion management. *Transfus Med Rev*. 2016;30(4):223–229.
57. Fang ZA, Navaei AH, Hensch L, Hui SR, Teruya J. Hemostatic management of extracorporeal circuits including cardiopulmonary bypass and extracorporeal membrane oxygenation. *Semin Thromb Hemost*. 2020;46(1):62–72.
58. Murphy DA, Hockings LE, Andrews RK, et al. Extracorporeal membrane oxygenation-hemostatic complications. *Transfus Med Rev*. 2015;29 (2):90–101.
59. Dalton HJ, Reeder R, Garcia-Filion P, et al. Factors associated with bleeding and thrombosis in children receiving extracorporeal membrane oxygenation. *Am J Respir Crit Care Med*. 2017;196(6):762–771.
60. Byun JH, Jang IS, Kim JW, Koh EH. Establishing the heparin therapeutic range using aPTT and anti-Xa measurements for monitoring unfractionated heparin therapy. *Blood Res*. 2016;51(3):171–174.
61. Willems A, Roeleveld PP, Labarinas S, et al. Anti-Xa vs time-guided anticoagulation strategies in extracorporeal membrane oxygenation: a systematic review and meta-analysis. *Perfusion*. 2020;36(5):501–512.
62. Delmas C, Jacquemin A, Vardon-Bounes F, et al. Anticoagulation monitoring under ECMO support: a comparative study between the activated coagulation time and the anti-Xa activity assay. *J Intensive Care Med*. 2020;35(7):679–686.
63. Kulig CE, Schomer KJ, Black HB, Dager WE. Activated partial thromboplastin time vs anti-factor Xa monitoring of heparin anticoagulation in adult venoarterial extracorporeal membrane oxygenation patients. *ASAIO J*. 2021;67(4):411–415.
64. Burstein B, Wieruszewski PM, Zhao YJ, Smischney N. Anticoagulation with direct thrombin inhibitors during extracorporeal membrane oxygenation. *World J Crit Care Med*. 2019;8(6):87–98.
65. Berei TJ, Lillyblad MP, Wilson KJ, Garberich RF, Hryniewicz KM. Evaluation of systemic heparin versus bivalirudin in adult patients supported by extracorporeal membrane oxygenation. *ASAIO J*. 2018;64(5):623–629.
66. Kurihara C, Walter JM, Karim A, et al. Feasibility of venovenous extracorporeal membrane oxygenation without systemic anticoagulation. *Ann Thorac Surg*. 2020;110(4):1209–1215.
67. Wood KL, Ayers B, Gosev I, et al. Venoarterial-extracorporeal membrane oxygenation without routine systemic anticoagulation decreases adverse events. *Ann Thorac Surg*. 2020;109(5):1458–1466.
68. Kalbhenn J, Schmidt R, Nakamura L, Schelling J, Rosenfelder S, Zieger B. Early diagnosis of acquired von Willebrand Syndrome (AVWS) is elementary for clinical practice in patients treated with ECMO therapy. *J Atheroscler Thromb*. 2015;22(3):265–271.
69. Mazzeffi M, Bathula A, Tabatabai A, et al. Von Willebrand factor concentrate administration for acquired von willebrand syndrome- related bleeding during adult extracorporeal membrane oxygenation. *J Cardiothorac Vasc Anesth*. 2021;35(3):882–887.
70. Lequier L, Horton SB, McMullan DM, Bartlett RH. Extracorporeal membrane oxygenation circuitry. *Pediatr Crit Care Med*. 2013;14(5 Suppl. 1):S7–S12.
71. Tipograf Y, Gannon WD, Foley NM, et al. A dual-lumen bicaval cannula for venovenous extracorporeal membrane oxygenation. *Ann Thorac Surg*. 2020;109(4):1047–1053.
72. Kopp R, Mottaghy K, Kirschfink M. Mechanism of complement activation during extracorporeal blood-biomaterial interaction: effects of heparin coated and uncoated surfaces. *ASAIO J*. 2002;48(6):598–605.
73. Kaluza M, May B, Doenst T. Using a roller pump for establishing extra-corporal membrane oxygenation (ECMO)—technical considerations for times of crisis. *Perfusion*. 2021;. 267659121996182.
74. Omar HR, Mirsaeidi M, Socias S, et al. Plasma free hemoglobin is an independent predictor of mortality among patients on extracorporeal membrane oxygenation support. *PLoS One*. 2015;10(4):e0124034.
75. Lyu L, Long C, Hei F, et al. Plasma free hemoglobin is a predictor of acute renal failure during adult venous-arterial extracorporeal membrane oxygenation support. *J Cardiothorac Vasc Anesth*. 2016;30(4):891–895.

Chapter 12

Device-Based Circulatory Support Therapy for Heart Failure with Preserved Ejection Fraction

Chihiro Miyagi[1], **Jamshid H. Karimov**[1,2], **Randall C. Starling**[2,3] **and Kiyotaka Fukamachi**[1,2]

[1]*Lerner Research Institute, Cleveland Clinic, Cleveland, OH, United States,* [2]*Cleveland Clinic Lerner College of Medicine of Case Western Reserve University, Cleveland, OH, United States,* [3]*Kaufman Center for Heart Failure, Heart, Thoracic, and Vascular Institute, Cleveland Clinic, Cleveland, OH, United States*

Chapter Outline

Introduction

There are many therapies for heart failure (HF) patients. However, HF includes a wide range of etiological and pathological aspects, so we should consider the application of each device and/or pharmacological therapy by the type of HF. One of the most accepted ways of categorizing HF is to divide it by left ventricular (LV) ejection fraction (EF): HF with preserved EF (HFpEF) and HF with reduced EF (HFrEF). The dividing line for these two groups differs between an EF of 40% (European guidelines) and 50% (US guidelines), but HFpEF is widely defined as an EF of more than 40%.

In this chapter, we present the device-based options for HFpEF patients. While many conservative therapies, including pharmacological treatments, have shown promising efficacy in HFrEF patients, they have failed entirely to improve outcomes for HFpEF patients. This is alarming, because the number of HFpEF patients is growing, and their prognosis is as devastating as that of HFrEF patients. Therefore, device treatments have become a much more attractive option for HFpEF patients. Many novel attempts have been made to apply a current device to HFpEF, or to develop a new device tailored for HFpEF features. We have summarized each device's unique features, development process, safety and efficacy, and the advantages and disadvantages, in detail.

Before going into detail about the devices, let us explain the clinical features of HFpEF. Two important and typical clinical characteristics of HFpEF are exercise intolerance and exertional dyspnea due to high left atrial (LA) pressure (LAP) and pulmonary capillary wedge pressure (PCWP). These are caused by a progressive decrease of LV relaxation

Advances in Cardiovascular Technology. DOI: https://doi.org/10.1016/B978-0-12-816861-5.00020-4

and LA compliance, especially during exercise. Although the etiology arises from a large variety of conditions, LV hypertrophy or concentric remodeling is known to be present in 59% of HFpEF cases, and diastolic dysfunction in 69%.[1] Also, from previous reports, we know that a larger LA correlates with a higher risk of hospitalization, and LA size is independently associated with the morbidity and mortality of HFpEF patients. In summary, LV hypertrophy or concentric remodeling, diastolic dysfunction, and LA enlargement could be present in most patients with HFpEF, and LA size and function can be potential therapeutic targets.

Interatrial shunt device

For HFpEF patients, higher pressure and larger LA size are important targets for intervention because they are strongly related to symptoms and hospitalization. Some devices have been developed with the goal of LAP reduction and they have become candidates for therapy for HFpEF patients. One example is an artificial interatrial shunt, which makes a shunt between the LA and right atrium (RA) and reduces LAP by introducing left-to-right flow. This concept originated from the Lutembacher syndrome; a patient with mitral stenosis and interatrial septal defect manifests fewer symptoms than a patient with mitral stenosis only.[2,3] Although the interatrial shunt devices (IASDs) are palliative treatment for HFpEF patients, they do reduce LAP and relieve symptoms. Since most of the devices can be inserted percutaneously, easy delivery and minimal interference with other devices[2] would make them very attractive alternatives as a treatment for HFpEF patients.

Compared to other percutaneous devices, IASDs have proven the most reliable in terms of safety, as seen in a number of studies for human use. An important point in evaluating these devices is long-term patency, as the patency of the shunt is related to long-term efficacy for HF patients, as reported by Rajeshkumar et al.[4] The demands for anticoagulation therapy and/or antiplatelet therapy also affect safety and long-term patency. Also, which unidirectional or bidirectional flow has better efficacy for HFpEF patients is unknown, as is whether patency or preventing right-to-left shunt have a greater effect on the prognosis. Since there is not enough data about the efficacy of all of the devices in HFpEF patients, further evaluation is required.

Currently, there are three IASDs and two devices with similar concepts: (1) the IASD; (2) the V-Wave device, (3) the Atrial Flow Regulator (AFR), (4) the Transcatheter Atrial Shunt System, and (5) the NoYA system. The detailed features of each device are illustrated below.

Interatrial shunt device (IASD)

The IASD (IASD II, Corvia Medical Inc., Tewksbury, MA) is a bare-metal nitinol self-expanding mesh frame with a disk diameter of 19 mm which can create an 8 mm shunt (Fig. 12.1). Patients who have the IASD implanted are treated with lifelong aspirin and short-term (usually 3–6 months) clopidogrel.[6] The progress of investigation of the clinical utility of IASDs is summarized in Table 12.1.

The main highlights of this IASD are the two human clinical trials: Reduce Elevated Left Atrial Pressure in Patients With Heart Failure (REDUCE LAP-HF) and REDUCE LAP-HF Randomized Trial I (REDUCE LAP-HF I). In the REDUCE LAP-HF trial, which included 64 patients, a significant decrease in PCWP at peak exercise from baseline to 6 months after implantation (34.7 ± 7.0 mm Hg to 31.6 ± 8.0 mm Hg, $P = .025$) and a slight decrease in PCWP at rest (17.4 ± 5.2 mm Hg to 16.5 ± 6.7 mm Hg) were reported.[12]

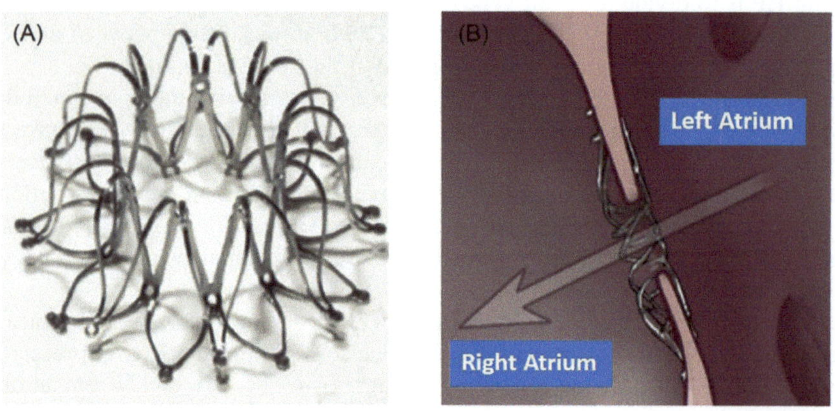

FIGURE 12.1 (A) Interatrial shunt device (IASD). (B) Illustration of an IASD attached to interatrial septum. *From Emani S, Burkhoff D, Lilly SM. Interatrial shunt devices for the treatment of heart failure.* Trends Cardiovasc Med. *2020.[5] Published permitted to reuse by ELSEVIER.*

TABLE 12.1 IASD development.

Study	Study design	All Pt.	IASD Pt.	HFpEF/HFrEF	Author	F/u time	Anticoagulant/ antiplatelet	Patency	Result
Computer simulation	Simulation study	39 (data)	—	39/0	Kaye et al.[7]	—	—	—	PCWP at rest: 10 to 7 mm Hg PCWP at exercise: 28 to 17 mm Hg LV CO at rest: 5.3 to 4.7 L/min LV CO at exercise: 9.9 to 8.8 L/min
Feasibility	Single arm	11	11/11	11/0	Søndergaard et al.[8]	1 months	Aspirin (lifelong), clopidogrel (institutional)	11/11	PCWP: 19 ± 5 to 14 ± 3 mm Hg 6-MWD: 322 ± 151 to 368 ± 123 m
					Malek et al.[9]	12 months		11/11	NYHA class III/IV: 45%/0% to 82%/18%
REDUCE LAP-HF (NCT01913613)	Multicenter, open-label, single arm, Phase I	66	64/66	66/0	Hasenfuß et al.[10]	6 months	Aspirin (75–325 mg, indefinitely), clopidogrel (75 mg, 6 months)	60/60	PCWP at exercise: 34 to 10 mm Hg Exercise duration: 7.3 to 8.2 min Qp/Qs: 1.06 to 1.27
					Kaye et al.[11]	1 year		48/48	PCWP at exercise: N.A ($n = 12$) 6-MWD: 331 ± 90 to 363 ± 93
					Wessler et al.[12]	6 months/1 year		No data	PCWP at exercise: (Base) 34.1 ± 7.6, (6 months) 31.6 ± 8.0, (1 year) 33.2 ± 10.4 mm Hg
					Kaye et al.[13]	3 years		No data	Mortality: 33% lower in IASD patients than predicted cohort
					Hanff et al.[14]	6 months		No data	Higher LA compliance and RA reservoir strain were related to greater LA volume reduction after IASD implantation

(Continued)

TABLE 12.1 (Continued)

Study	Study design	All Pt.	IASD Pt.	HFpEF/HFrEF	Author	F/u time	Anticoagulant/ antiplatelet	Patency	Result
REDUCE LAP-HF I (NCT02600234)	Randomized, sham-controlled, double-blind, Phase II	44	20/22	44/0	Feldman et al.[15]	1 month	Aspirin, clopidogrel (if the patient was on clopidogrel at baseline)	19/19	Significant difference in the decrease of PCWP at exercise
					Shah et al.[16]	6 months/1 year		20/20	Lower hospitalization rate in IASD arm, RV size in IASD arm increased at 6 months but stay the same at 1 year
	Pooled analysis	79	79	79/0	Obokata et al.[17]	1 month/ 6 months	—	78/79	60 patients from REDUCE LAP-HF at 6 months, 19 patients from REDUCE LAP-HF I at 1 months PCWP at exercise: -2 ± 7 mm Hg
REDUCE LAP-HF II (NCT03088033)	Randomized, triple-blind	608	—	608/0	June 2017-				
REDUCE LAP-HF III (NCT03191656)	Observational	100	—	Both	July 2017-				

IASD, Interatrial shunt device; *Pt*, patients; *F/u*, follow up; *PCWP*, pulmonary capillary wedge pressure; *LV*, left ventricle; *CO*, cardiac output; *6-MWD*, 6-minute walk distance; *NYHA*, New York Heart Association; *Qp/Qs*, pulmonary/systemic flow ratio; *LA*, left atrial; *RA*, right atrial; *RV*, right ventricular.

Keye et al.[13] calculated 1- and 3-year predicted cohort outcomes applying the Meta-analysis Global Group in Chronic Heart Failure (MAGGIC) prognostic model (an estimated non-IASD group) to the REDUCE LAP-HF cohort and compared its estimated Kaplan–Meier survival curves with the observed actual outcomes (IASD-treated group). The actual all-cause mortality of IASD-treated patients 3 years after the IASD implantation was 33% lower than that of predicted patients ($P = .02$). This is the longest interval result among all the published data of the IASDs.

The latest report from this cohort is by Hanff et al.,[14] who investigated predictors for the efficiency of IASD therapy. From this report, higher LA compliance appeared to be more important than LA strain in determining volume offloading of LA. Also, a decrease from baseline in the LA volume index at 6 months after IASD implantation was independently associated with a higher baseline in the LA compliance and the RA reservoir strain.[14] Therefore, higher LA compliance and RA reservoir strain at baseline would be possible predictors of greater efficacy of IASD.

From the REDUCE LAP-HF I study, long-term results at 1 year are available. All the IASD shunts were patent after 1 year, and the major adverse events did not differ significantly between the IASD group and the control group.[16] As for the right ventricular sizes, those of the IASD group were larger than the controls at 6 months, though further increase was not observed at 12 months. The quality of life and exercise tolerance were similar in both groups 1 year after the implantation.[16]

Obokata et al.[17] performed a pooled analysis from these two trials (REDUCE LAP-HF and REDUCE LAP-HF I) evaluating the effect of the shunt flow to the pulmonary artery function. Data from 79 patients (60 patients from REDUCE LAP-HF, 19 from REDUCE LAP-HF I) were obtained by right heart catheter at 1 month and/or 6 months. The pulmonary vascular function at rest and during exercise showed improvement without compromising systemic perfusion. However, in this study, patients with high pulmonary vascular resistance or right ventricular dysfunction were excluded.

Based on these results, which indicated a high level of safety, patency, and efficacy, a pivotal multicenter randomized clinical trial (REDUCE LAP-HF II, NCT03088033) was started. This is a randomized, controlled trial, which compares the device efficacy to optimal medical treatment (with a sham procedure) among HFpEF patients (HF >40%). Also, the REDUCE LAP-HF III trial (NCT03191656) is currently planning to enroll 100 patients (both HFpEF and HFrEF) to evaluate the more definite clinical utility of IASDs.

V-Wave device

The first generation of the V-Wave Interatrial Shunt (V-Wave, Caesarea, Israel) (Fig. 12.2A) is an hourglass-shaped nitinol self-expanding prosthesis covered with a polytetrafluoroethylene polymer on the left side and a tri-leaflet porcine pericardium valve on the right side to prevent right-to-left flow and paradoxical embolism and early closure.[19] After implantation, anticoagulation is administered (warfarin for 3 months, and aspirin, 81–100 mg/day, indefinitely).[20] This device underwent a prospective, multicenter, single-arm, open-label trial with both HFpEF and HFrEF patients that clarified its safety and feasible outcomes.[18]

It is worth noting that five of 36 patients (14%) had total shunt occlusions, and 13 (36%) had shunt restenosis due to pannus formation 1 year after device insertion. Although patients who maintained patent shunts ($n = 18$) exhibited a

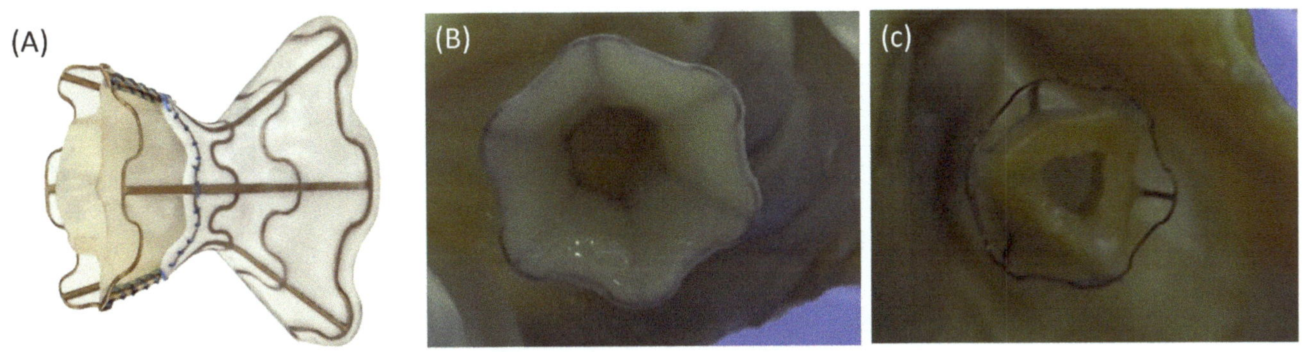

FIGURE 12.2 (A) V-Wave (first generation); (B) left side of the V-Wave (first generation) without any thickness of leaflet; (C) right side of the V-Wave (first generation) with thickened and stenotic leaflet (the part made of porcine pericardium) 1 year after the implantation. *From Rodés-Cabau J, et al. Interatrial shunting for heart failure: early and late results from the first-in-human experience with the V-Wave system.* JACC Cardiovasc Interv. *2018;11(22):2300–2310.[18] Published permitted to reuse by ELSEVIER.*

significant improvement in PCWP and a rise in LV EF at 12 months,[18] the rate of shunt narrowing/occlusion up to 50% was much higher compared to other devices.

The potential cause of stenosis or occlusion was investigated when one of the shunts was explanted during cardiac transplantation approximately 2 years after the procedure. Intrashunt valve deterioration was suggested as the main mechanism, as the bioprosthetic leaflets were thickened and stenotic with pannus (Fig. 12.2B and C).[18] From this observation, along with the lack of thrombus, modifications were implemented, including removal of the valve, and newer generations of the V-Wave shunt were produced. The second-generation (valveless) V-Wave is an hourglass-shaped device made of nitinol with expanded polytetrafluoroethylene encapsulation (Fig. 12.3). The first-in-human experience with 10 patients was recently reported by Guimaraes et al.,[21] showing better patency with only antiplatelet therapy (the same regimen as the IASD).

A new trial, Reducing Lung Congestion Symptoms in Advanced Heart Failure (RELIEVE-HR, NCT03499236), including 500 patients regardless of their EF and using the second-generation (valveless) V-Wave device, began in September 2018, and will be completed in 2021. Table 12.2 is a summary of V-Wave devices.

Atrial flow regulator

The AFR (Occlutech, Helsingborg, Sweden) is a self-expanding, double-disk wire mesh made of nitinol with a central penetration that enables interatrial bidirectional flow (Fig. 12.4). The center waist section connects the two disks and works as a shunt. Shunt sizes of 6, 8, and 10 mm fenestration diameters are available, with 18, 24, and 30 mm total device diameters, respectively.[3,23] The main procedural difference from the IASD and V-Wave devices is the necessity to perform a balloon atrial septostomy prior to device implantation.[24] Antiplatelets are usually given to patients for 6 months after implantation.[4]

In the beginning, the AFR was intended to be a therapy for pulmonary artery hypertension (PAH) to decrease the RA and right ventricular pressure by introducing right-to-left flow. Rajeshkumar et al.[4] conducted a nonrandomized pilot study of 12 patients with severe irreversible PAH. All procedures were successful without any major complications, and all patients had relief of syncope, improved 6-min walking distance, and improved cardiac index. The device was patent in all patients at a median follow-up of 189 days.[4]

After the success of IASD usage for HF patients, the AFR also started to be used for the HF treatment. An international nonrandomized clinical trial (PRELIEVE, NCT03030274) was completed in September 2020 with 36 patients with symptomatic HF including both patients with HFpEF ($n = 20$) or HFrEF ($n = 16$). The 3-month results were published by Paitazoglou et al.[24] recently, reporting that all implantation succeeded with 8 mm device ($n = 28$) or 10 mm device ($n = 8$) and the significant PCWP decrease in HFpEF group (-5.2 ± 8.8 mm Hg, $P = .03$) and New York Heart Association (NYHA) class ranking improvement in both group (HFrEF: -1.4 ± 0.2, $P = .0005$, HFpEF: -1.1 ± 0.2, $P = .0003$) at 3 months had been observed (Table 12.3).

Transcatheter atrial shunt system

The Transcatheter Atrial Shunt System (Edwards Life-sciences, Irvine, CA) has a unique approach to the left-to-right shunt by approaching LA via the coronary sinus (CS). This CS-to-LA shunt is a bare-nitinol implant flanked by four

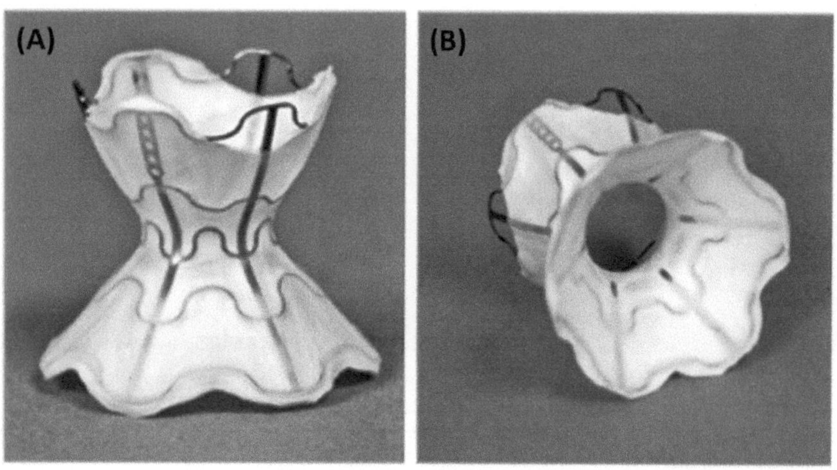

FIGURE 12.3 (A) Hourglass-shaped V-Wave (second generation); (B) valveless design of V-Wave (second generation). *From Guimarães L, et al. Interatrial shunt with the second-generation V-Wave system for patients with advanced chronic heart failure.* EuroIntervention. *2020;15 (16):1426–1428.*[21] *Published and permitted to reuse by Europa Group.*

TABLE 12.2 V-Wave device development.

Study	Study design	All Pt.	V-Wave Pt.	HFpEF/HFrEF	Author	F/u time	Anticoagulant/antiplatelet	Patency	Result
Feasibility	Single-arm	10	10/10	0/10	Del Trigo et al.[22]	3 months	Warfarin (or DOAC), aspirin (81–100 mg, 3 months)	10/10	PCWP: 23 ± 5 to 17 ± 8 mm Hg 6-MWD: 244 ± 112 to 318 ± 134 m
Feasibility	Multicenter, open-label, single-arm	38	38/38	8/30	Rodés-Cabau et al.[18]	3 months/1 year	Warfarin (3 months), low-dose aspirin	19/36	NYHA class III/IV: (Base) 100%, (3 months) 22%, (1 year) 39% 6-MDW: (Base) 290 ± 112, (3 months) 340 ± 92, (1 year) 324 ± 105 m
Second generation (valveless) feasibility	Single-arm	10	9/10	1/5	Guimaraes et al.[21]	1 year	Aspirin (indefinitely), clopidogrel (3 months)	6/6	NYHA class III/IV: 100% to 16.7%
Second generation (valveless) RELIEVE-HF (NCT03499236)	Randomized, sham-controlled	500	—	Both	October 2018-				

Pt, Patients; *F/u*, follow up; *PCWP*, pulmonary capillary wedge pressure; *6-MWD*, 6-minute walk distance; *NYHA*, New York Heart Association.

FIGURE 12.4 Atrial flow regulator. *From Emani S, Burkhoff D, Lilly SM. Interatrial shunt devices for the treatment of heart failure.* Trends Cardiovasc Med. *2020.*[5] *Published and permitted to reuse by ELSEVIER.*

arms with an internal shunting diameter of 7 mm (Fig. 12.5). Two arms are situated on the LA wall, while the other two lie within the CS, with the arms surrounding a central shunting lumen. The interatrial septum is increasingly used by septal occluding devices for closure of the patent foramen ovale or iatrogenic atrial septal defect following transseptal procedures.

Both aspirin and clopidogrel were introduced, but in those with prior anticoagulation, an oral anticoagulant plus clopidogrel monotherapy were introduced.[25] The first-in-human implantation was performed on 11 patients (7 HFpEF patients and 4 HFrEF patients), and successful percutaneous atriotomy was achieved in eight. In the other three patients, the unsuccessful percutaneous atriotomy was caused by an inability to track the guidewire delivery catheter into the CS due to the anatomic constraints. The median follow-up interval was 201 days (156–260 days), and seven patients showed improvement in NYHA class and a significant reduction in PCWP (−9 mm Hg, interquartile range: −9.5 to −8 mm Hg) at median 100 days' (81–174 days) assessment.[25] The shunts sustained patency, and no significant procedural complications were reported.

NoYA interatrial shunt system

The NoYA Interatrial Shunt System (DiNova Medtech, Shanghai, China) is a self-expanding, flowerlike nitinol stent used as a radioablation catheter to create a persistent hole in the interatrial septum (http://en.dinovamedtech.com/Products/Content.aspx?Id = 36). Since the catheter is removed after the ablation and no foreign device is left implanted, there is no requirement for anticoagulation or antiplatelet therapy after the procedure. The diameter of the hole can be adjusted from 4 to 12 mm. Although a pilot study of 10 patients has done, the results have not been published yet.

LV expander

LV expanders are devices that expand the LV from the inside or outside, as the name implies. There are currently two such devices, but neither have been fully documented so far, and their safety and efficacy, influence on cardiac function, and the need for an anticoagulant, are still unknown. The idea of expanding the LV could be a radical and reasonable cure considering the anatomical feature of HFpEF, but the effect might be temporary if further hypertrophic changes occur to the LV chamber.

CORolla assist device

The CORolla (CorAssist Inc., Haifa, Israel) is an elastic self-expanding device (Fig. 12.6) that is attached to the endocardium and enhances LV diastolic expansion from inside, improves LV filling, and lowers LV end-diastolic pressure and LAP.[2,3] At first, the CORolla was implanted by minimally invasive trans-apical approach off-pump for a patient

TABLE 12.3 AFR device development.

Study	Study design	All Pt.	AFR Pt.	HFpEF/HFrEF	Author	F/u time	Anticoagulant/antiplatelet	Patency	Result
Feasibility	Single arm	12	12/12	-(PAH)	Rajeshkumar et al.[4]	10–296 days	Aspirin and clopidogrel (both 6 months)	12/12	6-MWD: 377.7 ± 33.2 to 423 ± 31.3 m
AFR-PRELIEVE, (NCT03030274)	Multicenter, open-label, single arm	36	36/36	20/16	Paitazoglou et al.[24]	3 months	No data	36/36	PCWP of HFpEF: −5.2 ± 8.8 mm Hg

AFR, Atrial flow regulator; *Pt.*, patients; *F/u*, follow up; *PAH*, pulmonary artery hypertension; *6-MWD*, 6-minute walk distance; *PCWP*, pulmonary capillary wedge pressure.

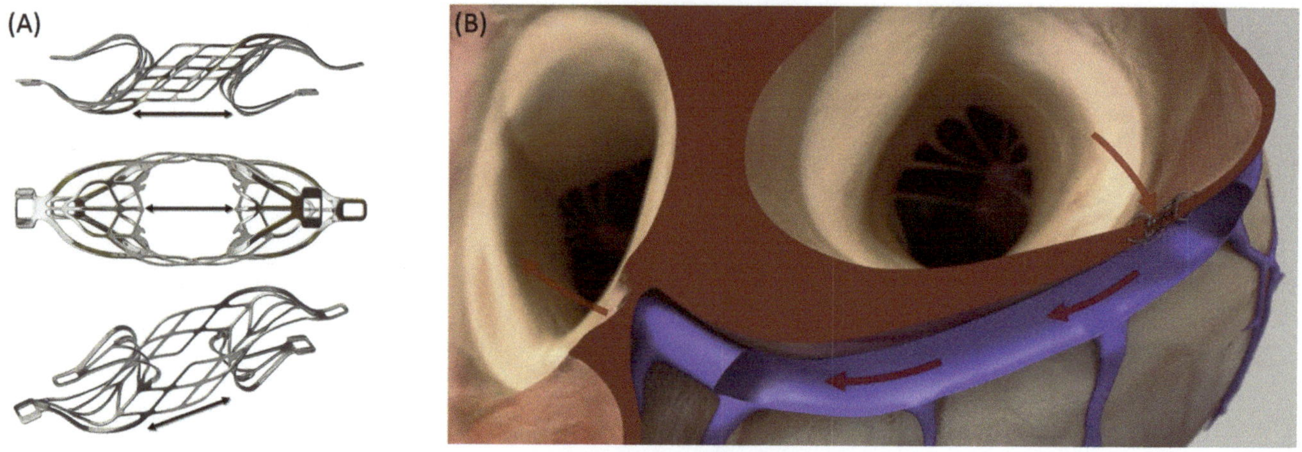

FIGURE 12.5 (A) Transcatheter atrial shunt system; (B) the view of the shunt device attached between the left atrium and coronary sinus. *From Simard T, et al. Percutaneous atriotomy for levoatrial-to-coronary sinus shunting in symptomatic heart failure: first-in-human experience.* JACC Cardiovasc Interv. *2020;13(10):1236–1247.*[25] *Published permitted to reuse by ELSEVIER.*

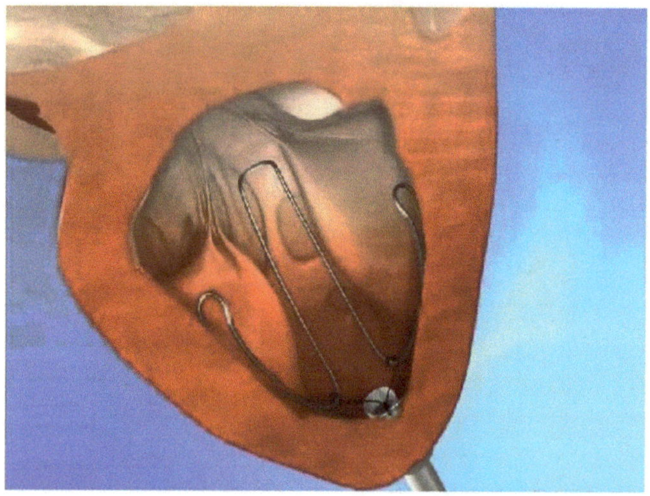

FIGURE 12.6 CORolla assist device. *From Burlacu A, et al. Novel percutaneous interventional therapies in heart failure with preserved ejection fraction: an integrative review.* Heart Fail Rev. *2019;24(5):793–803.*[2] *Published and permitted to reuse by Springer.*

with aortic stenosis and severe diastolic dysfunction.[2] The percutaneous trans-apical insertion of this device was developed in 2017, and both approaches are used today.

The company's website is the main source of information on clinical outcomes achieved with this device (https://corassist.com/news-resources/). The implanted cases exhibited an improvement in LV mass, LA indexed volume, and consistent LV remodeling at 6 months. This might suggest a certain efficacy and safety of the device, but we look forward to published results. Also, an ongoing open-label trial enrolling 10 patients aims to highlight the safety and feasibility of the device in a 12-month follow-up in HFpEF patients (NCT02499601).

ImCardia

The ImCardia (CorAssist Inc., Haifa, Israel) is an elastic self-expanding device attached to the epicardial surface of the LV. The device is inserted through an off-pump procedure with a median sternotomy during surgery for aortic valvular disease. As reported in a conference presentation, 10 patients with severe aortic stenosis underwent a surgical treatment with ImCardia and showed LV mass regression and LA area decrease.[26] However, there were no published results; moreover, the trial was terminated, as it was decided that the device needs further improvement.

Electrical therapy for HFpEF

Cardiac contraction modulation

Cardiac contractility modulation (CCM) is a pacemaker-like therapy for HF that applies relatively high-voltage (up to 7.5 V), long duration (up to 20 msec), biphasic electric signals to the right ventricular septum within the absolute myocardial refractory period.[27] CCM delivers nonexcitatory electrical signals (does not elicit a new contraction) and improves the LV contraction without increasing myocardial oxygen consumption by affecting the biology of the failing myocardium (Fig. 12.7). There are nearly 3000 patients currently implanted worldwide,[29,30] with a series of Optimizer devices (Optimizer Smart; Optimizer III; Optimizer IV, Impulse Dynamics, Marlton, NJ) that employ three leads: one to the RA and two to the right ventricular septum.

The CCM was developed for symptomatic HFrEF patients with slightly prolonged QRS interval who were not eligible for cardiac resynchronization therapy. The benefits of CCM for HFrEF patients' quality of life and exercise tolerance has been illustrated in three randomized studies: the FIX-HF-4 study in European Union,[31] the FIX-HF-5 trial in the United States (including 428 patients),[32] and the FIX-HF-5C study, conducted in sites in the United States and European Union with 160 patients.[28] Notably, the subgroup analysis of FIX-HF-5 revealed that CCM was more effective when the baseline EF is higher (between 35% and 45%, rather than <35%).[33]

Based on these results, investigations into the efficacy of CCM on HFpEF patients were begun. The first case report of two symptomatic HFpEF patients (both 59 year-old females with full standard HF therapy) by Tschöpe et al.,[29] showed early improvement in diastolic function with CCM, with an increase in EF reserve for 3 months. Frequent endomyocardial biopsies were performed, and an early and continuous increase in the expression of key cardiac contractile proteins was revealed. Improvements in collagen composition were also noted. In addition, an increase in cardiac protein expression was observed remotely, not just near the CCM probes. This may indicate that the CCM led to a global beneficial effect of cardiac reverse remodeling in HFpEF patients, equal to that observed in HFrEF patients.

An ongoing trial (CCM-HFpEF, NCT03240237) to evaluate the safety and efficacy of CCM therapy in now enrolling up to 60 HFpEF patients who are symptomatic despite appropriate medications. Also, a new CCM delivery algorithm was recently developed to eliminate the need for an atrial sensing lead—the two-lead CCM device (2-Lead Optimizer Smart System, Impulse Dynamics, Marlton, NJ). The FIX-HF-5C2 study (NCT03339310) was conducted to test the two-lead device's performance, clinical effect, and safety in 60 patients with LV EF 25%–45%. The device showed comparable efficacy and fewer device-related adverse effects at 24 weeks.[34] The clinical trial with HFpEF patients using this new two-lead device is expected to begin soon.

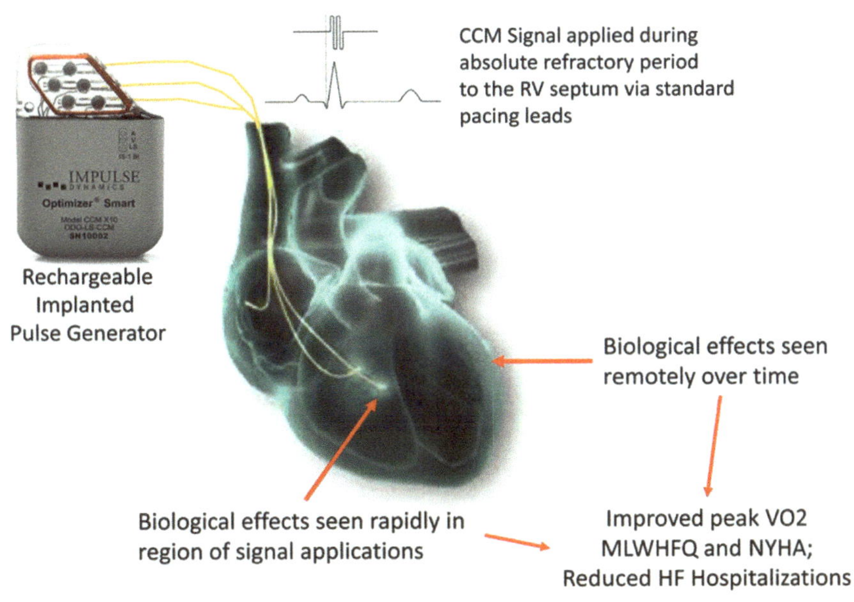

FIGURE 12.7 Cardiac contractility modulation (CCM) and its therapeutic mechanism. *MLWHFQ*, Minnesota Living With Heart Failure questionnaire; *NYHA*, New York Heart Association. *From Abraham WT, et al. A randomized controlled trial to evaluate the safety and efficacy of cardiac contractility modulation.* JACC Heart Fail. *2018;6 (10):874–883.[28] Published and permitted to reuse by ELSEVIER.*

Left atrial permanent pacing

The concept of atrial electrical dyssynchrony following reduced LA function has been an important therapeutic target for HFpEF patients,[35,36] as an independent correlation has been established between the severity of the intraatrial dyssynchrony and reduced LA function in HFpEF patients. LA pacing (CS pacing) has recently been adapted for atrial resynchronization function, although its original target was atrial tachyarrhythmias. Since the prolonged intraatrial dyssynchrony was possibly associated with the deteriorated clinical symptoms of HFpEF patients,[37] the attempt to use LA pacing for HFpEF patients began to receive priority.

During the implantation procedure, a specially designed CS lead is used to stimulate the middle-to-distal area of the CS.[38] Laurent et al.[39] conducted a pilot study with six HFpEF patients; after 3 months' activation of LA pacing, a 2-week randomized, double-blind, crossover phase started with the active CS pacing group and the inactive pacing group. The 3 months activation showed significant improvements in subjects' 6-min walk (190 ± 15 m to 240 ± 25 m, $P < .05$), mitral A wave duration (104 ± 8 to 158 ± 25 msec, $P = .002$), and E/A and E/e' ratio (3.4 ± 1.3 to 1.8 ± 0.9, $P = .009$, and 22.6 ± 4.6 to 15.3 ± 4.3, $P = .006$, respectively). Of note, after the pacing inactivation, a significant decrease of the 6-minute walk was observed, implying an on/off response.

Despite the potential of this device with a new perspective for electrical therapy in HFpEF, the long-term effects and safety are still unknown, as there have been few trials. The Left Atrial Pacing in Diastolic Heart Failure (LEAD) study (NCT01618981), a French single-blind, randomized, controlled crossover study involving 48 patients, began in 2012 and concluded in 2016, but the results have not been reported yet.

Baroreceptor activation therapy

Baroreceptor activation therapy (BAT), which was developed as a treatment for resistant hypertension,[40] is also a pacemaker-like device with an electrical technology that stimulates the carotid sinuses. Arterial baroreceptors are located at the aortic arch and carotid sinuses and induce an arterial baroreflex, which is a strong regulator of intravascular stressed blood volume and provides regulation of blood pressure and heart rate. The reduced baroreceptor-heart rate reflex sensitivity is known as an independent predictor of outcome in HF,[41] thus BAT is now considered a beneficial device for treating autonomic dysregulation in HF patients.

In rat models, the baroreflex failure had been reported as a potentially important pathogenesis in acute HF, especially in HFpEF.[42] The first generation BAT that was clinically applied to HFpEF patients was the Rheos System (CVRx Inc., Minneapolis, Minnesota), which delivers chronic therapy with a pulse generator implanted in the pectoral region and one lead in contact with the carotid sinus. The Rheos HOPE4HF (Health Outcomes Prospective Evaluation for Heart Failure with Ejection Fraction >40%) trial (NCT00957073) was a randomized, multicenter, open-label trial that began in 2009, and ended in 2018.

The second-generation BAT device, the BAROSTIM NEO (CVRx Inc., Minneapolis, Minnesota) was intended to be a treatment for HFrEF patients.[43] The largest clinical randomized study of BAROSTIM NEO, by Abraham et al.,[43] compared two groups of HFrEF patients: one with BAT ($n = 70$) and the other without BAT ($n = 76$). The major adverse event-free rate for BAT was 97.2%, and there were also significant improvements in 6-minute walk, NYHA class ranking, and quality of life score in the BAT group.

As for HFpEF patients, there is a prospective, ongoing registry (BAROSTIM THERAPY in Heart Failure With Preserved Ejection Fraction: A Post-Market Registry with the CE-Marked BAROSTIM NEO system, NCT02876042). This observational prospective study will enroll up to 70 subjects after the BAROSYSTEM NEO implantation. The feasibility and efficacy of BAT in HFpEF patients is expected to be much clearer by this trial.

Left ventricular assist devices for diastolic dysfunction

As documented in other chapters of this book, the left ventricular assist device (LVAD) is a standard and essential option for end-stage HF patients with dilated or ischemic cardiomyopathy, and has been demonstrated to improve morbidity and mortality in HFrEF patients.[44] However, the situation for HFpEF patients is completely different, since there are limited reports of an LVAD being implanted in these patients.[45–47] This is because their typical remarkably small LV tends to be incompatible for LVAD therapy.

When we investigate any reports of LVAD implantation for HFpEF patients, we find only very limited results. However, as regards LVAD patients with end-stage restrictive cardiomyopathy (RCM) or hypertrophic cardiomyopathy (HCM), some HFpEF patients seem to have been included in this phenotype. RCM and HCM patients have many common features with HFpEF patients such as a narrow LV cavity and diastolic dysfunction. Therefore, the experience of LVAD implantation for RCM/HCM patients can be useful when considering LVAD for HFpEF. It is still extremely rare for RCM/HCM patients to receive LVADs prior to heart transplantations; only 2% of LVAD patients have RCM/HCM.[48]

HeartMate II

The outcome of eight patients with RCM or HCM who were implanted with a continuous-flow axial LVAD (HeartMate II, Abbott, Abbott Park, IL) was reported by Topilski et al.,[45] who compared the results with 75 patients with dilated or ischemic cardiomyopathy. They performed an additional LV myectomy for HFpEF patients to create enough space in the LV to insert the inflow cannula. The 1-year survival rate of RCM or HCM patients was no different from that of the dilated or ischemic cardiomyopathy patients.

It is noteworthy that this report mentions the "suck-down event of the LA," which was observed in one of the patients. Usually, the HeartMate II has a built-in system to detect suck-down events of the LV, and it can automatically drop the pump speed to the low limit. However, the LV wall in RCM and HCM patients is sometimes too stiff and thick to be deflated even with the negative pressure inside the LV, and this may cause LA collapse as a result.

Even worse, the pump is unable to detect the LA collapse because it is not a direct sucking with inflow cannula and tissue. The authors explained this phenomenon using a transesophageal echocardiogram view. They explained that the LA pressure was estimated to be low, and the impaired right ventricular function of the HCM patient could be one reason for the decrease in LAP and the collapse of the LA. Judicious increments of LVAD rpm, with strict echocardiographic guidance, should be done to prevent this unique but devastating suck-down event.[45]

According to previous reports, HCM patients have often been categorized as contraindications or exclusions for an LVAD implantation because of the reduced LV end-diastolic dimensions.[49] However, a lack of experience with HCM patients who have undergone LVAD implantation is likely affecting this rationale, and some patients with end-stage HF by RCM or HCM may benefit from the HeartMate II.

HVAD

With the results of the HeartMate II studies as a basis, Muthiah et al.[47] reported outcomes of three HCM patients implanted with a centrifugal continuous-flow LVAD (HeartWare HVAD, Medtronic, Minneapolis, MN). Data from these patients were compared with 36 dilated cardiomyopathy patients who received the HVAD. In both groups, the mean pulmonary arterial and RA pressures improved, and even though myomectomy was not performed in any HCM patients, suction events were not observed. One patient died due to the inlet obstruction with an intraventricular clot, after the antiplatelet therapy had been modified because the patient had had a transient ischemic attack a week before. The authors concluded that patients with HCM may benefit from HVAD implantation, at least in the short-to-medium term, as a bridge to transplant, so these patients should not be excluded from consideration of mechanical circulatory support.

A case report of applying HVAD with an LA cannulation to a 57-year old HCM patient was recently reported.[50] This "LA-to-aorta LVAD technique" had only been used in pediatric cases before,[51] but since pediatric HF patients have the same problem (small LV cavity) as HFpEF patients, the authors tried implanting the HVAD with an LA inflow cannula via transseptal RA approach (Fig. 12.8A). This made good use of the markedly dilated atrium of the patient avoiding suction events, but the pump speed had to be adjusted as low as 2600 rpm to achieve the partial opening of the aortic valve with every cardiac cycle.

The patient's postoperative course was smooth, and there were no hemocompatibility issues over 4 months until the subsequent heart transplantation was performed (Fig. 12.8B). The potential risks of this concept are, narrow range of pump speed availability, LV thrombus formation, and difficulty with right ventricular assist device placement when needed. Therefore, the authors emphasized that the application of this configuration should be limited to those who otherwise ineligible and who have severe LA dilation to accommodate a LA cannula.[50]

LVAD for HFpEF (Computer modeling)

As a good lesson for us to predict the efficacy of an LVAD for HFpEF patients, a simulation study reported by Moscato et al.[52] used a computer model of continuous-flow LVAD applied to HFpEF patients' hemodynamics, and showed that it can offer hemodynamic benefits for HFpEF patients. Both at rest and exercise conditions were simulated

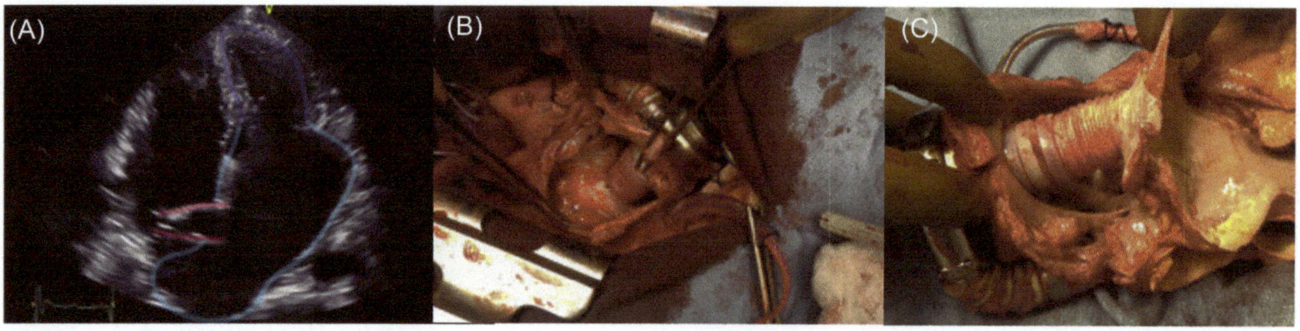

FIGURE 12.8 (A) Four-chamber echocardiography view of the left atrial inflow cannula; (B) in situ view of the Gore-Tex graft anastomosed to the interatrial septum and run through the right atrium; (C) explanted heart at heart transplantation 4 months after the HVAD implantation. *From Kiamanesh O, et al. Left ventricular assist device with a left atrial inflow cannula for hypertrophic cardiomyopathy.* JACC: Case Rep. *2020;2 (13):2090–2094.*[50] *Published and permitted to reuse by ELSEVIER.*

with different pump settings, and it became clear that the LVAD assistance could benefit the cardiac output (CO) during physical activity, unload the LV, and decrease the pulmonary circulation. This means that LVADs have the potential to relieve the symptoms of marked intolerance of HFpEF patients and improve their quality of life.

However, the simulation also revealed a need to adjust the pump setting commensurate with the level of exertion; that is, a low pump speed should be maintained at rest to reduce a risk of suction, and a high pump speed during exercise to avoid ineffective pulmonary unloading. Therefore, a constant pump speed for both rest and exercise conditions is not advisable, and strict control of pump speed and fluid balance after LVAD implantation is needed for HFpEF patients. This simulation study does not take into consideration the narrowness and thickness of the LV wall in HFpEF patients, so further research is required.

Devices in development (preclinical)

As described above, LVADs and pharmacological therapies that have yielded favorable results in HFrEF do not work as well in HFpEF patients. Therefore, new devices or techniques are strongly needed, tailored to the treatment of HFpEF patients in consideration of their features and hemodynamics. In this section, we present some new, pump-based mechanical cardiac support devices in development as future possible treatments for HFpEF patients. Each device has a novel and unique concept, and is trying to accumulate enough reliable data to be used in humans.

CoPulse: valveless pulsatile pump

The CoPulse is a valveless, pulsatile pump with a single cannula directly connected to the LV from the apex, and adds approximately 30 mL to the LV stroke volume. The pump is separated from a pneumatic chamber and blood chamber by a membrane that is acquired from a Berlin Heart Excor pump (BerlinHeart GmbH, Berlin, Germany).[53] The reciprocating membrane movement is driven by positive and vacuum pressures of the pneumatic chamber via a pneumatic line connected to the system, and the ejection and filling of the CoPulse pump are synchronized with the LV cardiac cycle by utilizing the heart's LV pressure as an input signal.

To date, in silico,[54] in vitro (hybrid mock circulatory loop), and ex vivo (porcine heart model)[53] experiments have been published. In the hemodynamic evaluation using the in vitro test bench in which HFpEF patients with significant comorbidities and hypertrophy were simulated, CoPulse decreased mean LAP (from 26 to 18 mm Hg) and the stroke volume of LV plus pump (87.0–99.4 mL) and increased mean CO (6.2–7.1 L/min).

The characteristic features of this device are its pulsatility and valveless design to reduce shear stress and blood clot generation. The design and flow distribution were analyzed by the computational fluid dynamics data and 4D-flow magnetic resonance imaging. The results demonstrated that the asymmetric CoPulse pump served as low shear stresses and comparable hemocompatibility for pump washout and blood residence time, as previously reported for the pulsatile ventricular assist device called Symphony (Abiomed Inc., MA, USA).[55] The Symphony has a similarly designed single port, but the number of valveless cannulas is different (two rather than one), and the connection site is the subclavian artery, not the LV.

As for a suction event, Escher et al.[53] estimated that the CoPulse drains blood synchronously from the LV only while the mitral valve is open, which should reduce the risk of sucking. However, because the position of the cannula is at the thick LV wall of HFpEF patients, further verification is needed, as with typical LVADs.

Synergy system

The Synergy System (Medtronic, Minneapolis, MN) is a micropump-based form of circulatory support, and its unique feature is that the pump inflow can be derived from either the LA or LV.[56] The pump size is roughly the same as an AA battery, and it can be implanted through a right mini-thoracotomy in an off-pump procedure. The outflow graft is anastomosed to the subclavian artery, and the micropump is then placed in a subcutaneous pocket like a pacemaker.

Although the pump development and the related project was completely abandoned about 5 years ago, the concept of LA drainage and its developing process is worth sharing. It was originally designed as a partial mechanical support for HFrEF patients. The hemodynamic favorable effects on HFrEF patients were simulated in 2007[57] and the pumps were successfully implanted in 17 HFrEF patients.[58] Subsequently, the analysis of the Synergy System in HFpEF patients had been done by Burkhoff et al.[56] in the same way, and it was suggested that this system may improve the hemodynamics of advanced HFpEF patients. This simulation study also highlighted the possibility that the Synergy System could overcome the limitation of an LV sucking event by means of LA drainage. However, thromboembolism in LV could be a potential issue because of blood stagnation in the LV, and LV remodeling would not be likely to occur.

Left atrial assist device

A left atrial assist device (LAAD) is a mixed-flow pump that is implanted in the mitral position and pumps blood from the LA to fill the LV (Fig. 12.9), with a flow of up to 10 L/min and up to 180 mm Hg. It is 33.5 mm in diameter and 33.5 mm high. This is a mixed-flow pump using a hydrodynamic bearing for radial support of the rotating assembly and a passive magnetic bearing to fix its axial position (Fig. 12.10). The pump is driven by a custom three-phase brushless, sensorless direct current. motor that was designed and built in-house. It is able to pump over a wide operating range (up to 10 L/min and up to 180 mm Hg) to maintain optimal flow in support of cardiac output.

Recently, an initial in vitro results of this concept were reported, stating that the LAAD increases CO and decreases LAP under diastolic HF conditions while maintaining the arterial pulsatility and full function of the aortic valve.[59] The various levels of diastolic HF conditions were simulated with three different speeds each. Interestingly, the diastolic HF conditions largely affected the CO at the baseline (without the LAAD), but almost maximal CO recovery was achieved

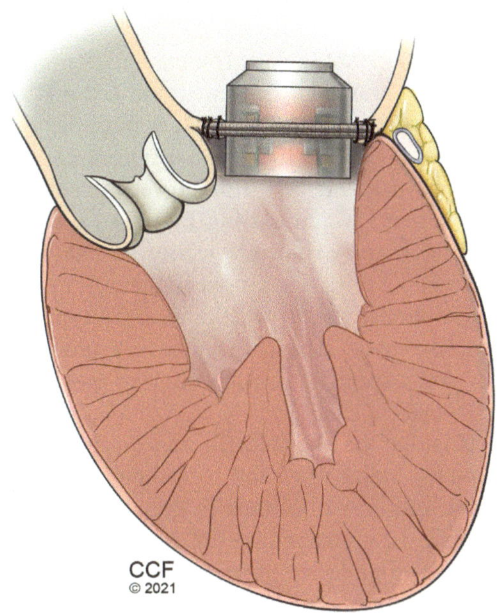

FIGURE 12.9 The left atrial assist device.

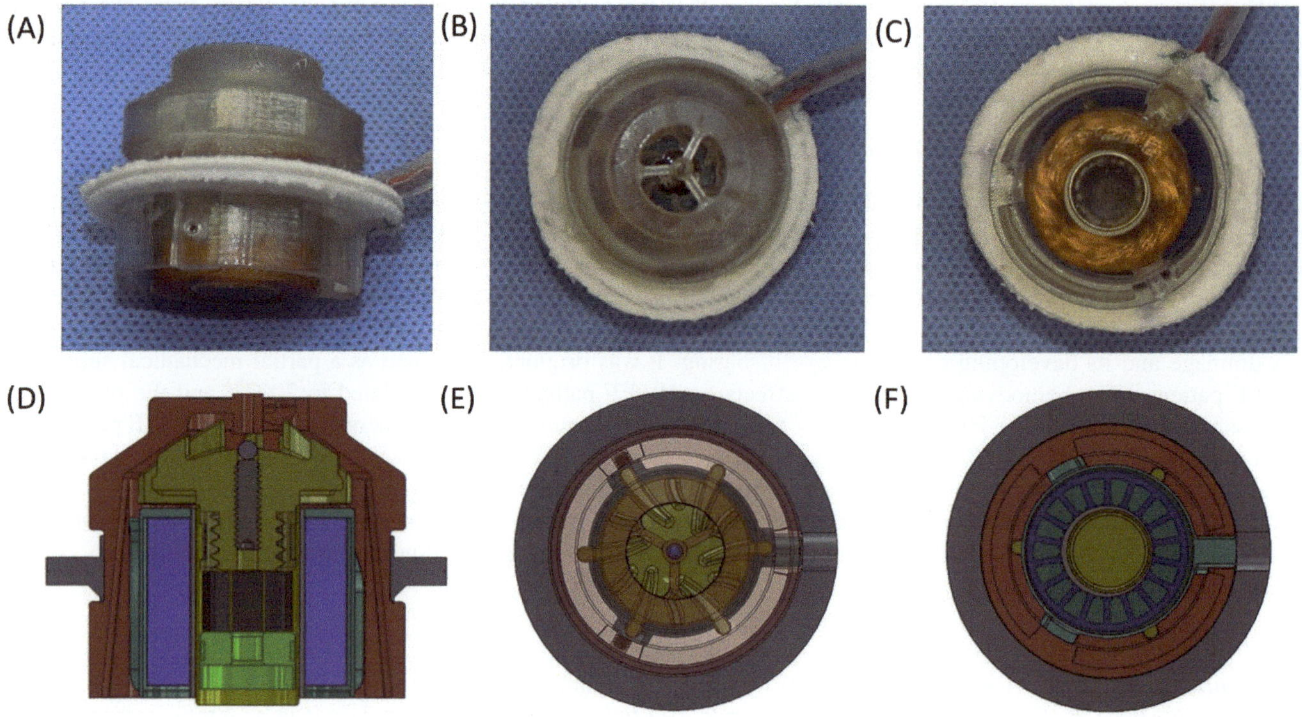

FIGURE 12.10 (A) The left atrial assist device (LAAD) seen from the front. (B) The LAAD seen from the LA side. (C) The LAAD seen from the LV side. (D) A cross-sectional view of the LAAD. (E) A schematic illustration of the LAAD inlet (LA side); (F) A schematic illustration of the LAAD outlet (LV side).

with the LAAD in all diastolic HF conditions. Similar trends were observed when increasing the mean aortic pressure and decreasing LAP.[59]

The interesting aspect of this in vitro study is that, in addition to the LAAD, two different configurations were analyzed under diastolic HF conditions: the LA-to-aorta configuration and LVAD configuration. The LA-to-aorta configuration is the similar circulation as the Synergy System or the HVAD with an LA inflow cannulation that has been described earlier. In this configuration, the arterial pulse pressure decreased as the pump speed increased, and the aortic valve was always closed.

This phenomenon was also observed in the HVAD with LA cannulation case while running at over 2800 rpm. An unopened aortic valve will raise concerns of thromboembolism in the LV and aortic root, a fusion of the aortic valve leaflets, and the development of aortic valve regurgitation, so low pulsatility and an unopened aortic valve at high pump speed became the critical limitations of the LA-to-aorta configuration. In the LVAD configuration, the limited LV volume under the diastolic HF condition made the total flow lower than that of the LAAD configuration.

A potential advantage of the LAAD is that LV remodeling may be expected because the entire LAAD flow goes through the LV to the aortic valve, and dilatation of the LV cavity and an increase in LV compliance can occur, as well. This means that the LAAD may provide a bridge-to-recovery therapy for HFpEF patients. At the beginning of the treatment, the probability of suction event seems to be low, because the typical LA of HFpEF patients is dilated,[60,61] and their LAP is significantly elevated. However, during the recovery process of HFpEF, the size of the LA could become smaller; careful maintenance of pump speed or consideration of pump removal is then needed to prevent a suction event.

Another concern with LAAD support is that the elevated LV end-diastolic pressure could be a problem, since it can have an unfavorable effect on the coronary blood flow and right ventricular function. Further in vivo studies using calves conducted so far (a typical surgical setting is illustrated in Fig. 12.11A), showed good performance and positioning of the LAAD at the mitral position (Fig. 12.11B), and no LV outflow tract obstruction (Fig. 12.11C). In an attempt to simulate a diastolic dysfunction circulation under in vivo settings, a balloon was inserted from the LV apex and inflated 50–70 mL to reduce the LV capacitance (Fig. 12.11). More information related to these experiments will be forthcoming.

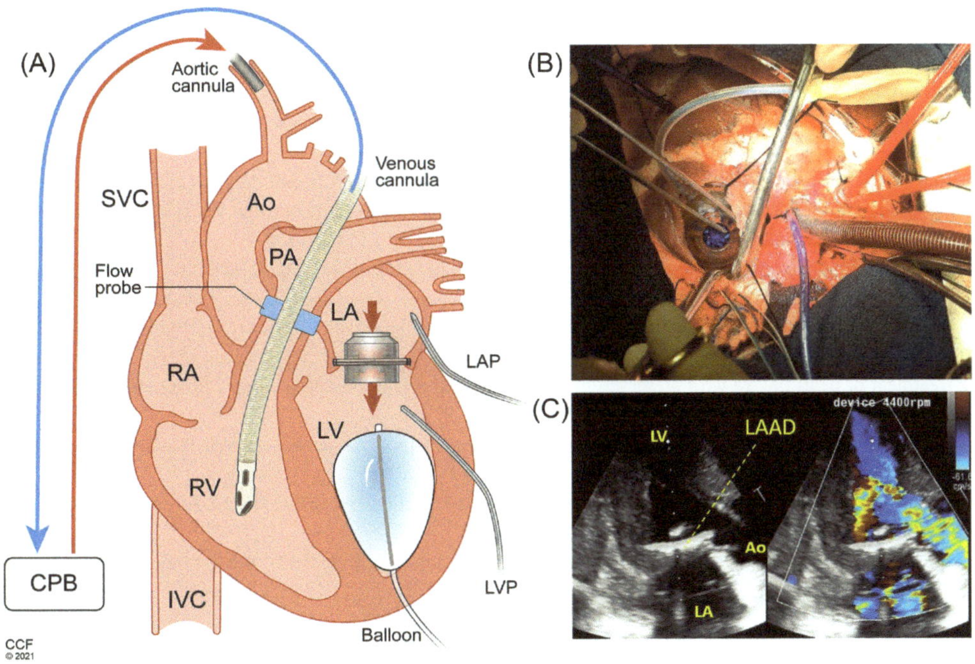

FIGURE 12.11 (A) A schematic diagram of the surgical setting for an in vivo LAAD experiment with diastolic heart failure configuration. (B) Surgical view of the LAAD implanted at the mitral position of a calf. (C) Echo view of the implanted LAAD without any obstruction of LV outflow tract.

PulseVAD

The PulseVAD (Northern Development, Strandhaugen, Oslo, Norway) is another pulsatile heart assist device, which pumps blood from the LA to the descending aorta. It can be implanted with a minimally invasive procedure without using cardiopulmonary bypass, can provide pulsatile flow, and is adaptable for both HFpEF and HFrEF patients (https://www.pulsevad.com/). Since the outflow is anastomosed to the descending aorta, and due to the pulsatile flow, the risk for cerebral emboli could be reduced, but there are no published reports on this device, and the concept has not yet been clarified.

VADovations

The VADovations cardiac assist system (VADovations, Oklahoma City, OK) is a small pump approximately the size of an AAA battery, and the procedure can be performed percutaneously. Surprisingly, after being delivered endovascularly, the pump is placed between the LA and the ascending aorta without inflow cannula or outflow conduit. The indications for this device have been expanded to systolic and diastolic dysfunction including HFpEF patients, but the safety and effectiveness have not yet been demonstrated (https://vadovations.com/technology/advantage/).

Summary

In conclusion, numerous attempts to improve the outcome of HFpEF patients have been initiated and studied with pharmacologic and now device therapies in both experimental settings and preclinical use. While each of the device solutions reviewed demonstrates a different stage of development and readiness for use, the most available choice of device-based therapies currently is the IASD. However, this is a palliative therapy and would be less effective after long-term use.

Although all the pump-based therapies are still in the preclinical phase, they are approaching first-in-human implantation as an invasive surgical therapy. Nevertheless, the variety of possible treatment methods for HFpEF patients is still limited. Further evaluation of each device and methodology, including long-term efficacy and safety, requirements for titrated anticoagulation, and aspects of interdevice interaction, will be required. In the meantime, research and development toward effective therapies will proceed until the most promising device solutions and innovative therapeutic concepts shed more light on current HFpEF understanding and treatments.

References

1. Zile MR, et al. Prevalence and significance of alterations in cardiac structure and function in patients with heart failure and a preserved ejection fraction. *Circulation.* 2011;124(23):2491–2501.
2. Burlacu A, et al. Novel percutaneous interventional therapies in heart failure with preserved ejection fraction: an integrative review. *Heart Fail Rev.* 2019;24(5):793–803.
3. Gupta A, Bailey SR. Update on devices for diastolic dysfunction: options for a no option condition? *Curr Cardiol Rep.* 2018;20(10):85.
4. Rajeshkumar R, et al. Atrial septostomy with a predefined diameter using a novel occlutech atrial flow regulator improves symptoms and cardiac index in patients with severe pulmonary arterial hypertension. *Catheter Cardiovasc Interv.* 2017;90(7):1145–1153.
5. Emani S, Burkhoff D, Lilly SM. Interatrial shunt devices for the treatment of heart failure. *Trends Cardiovasc Med.* 2020;.
6. Kaye DM, Nanayakkara S. Interatrial shunt device for heart failure with preserved ejection fraction. *Front Cardiovasc Med.* 2019;6:143.
7. Kaye D, et al. Effects of an interatrial shunt on rest and exercise hemodynamics: results of a computer simulation in heart failure. *J Card Fail.* 2014;20(3):212–221.
8. Søndergaard L, et al. Transcatheter treatment of heart failure with preserved or mildly reduced ejection fraction using a novel interatrial implant to lower left atrial pressure. *Eur J Heart Fail.* 2014;16(7):796–801.
9. Malek F, et al. Clinical outcome of transcatheter treatment of heart failure with preserved or mildly reduced ejection fraction using a novel implant. *Int J Cardiol.* 2015;187:227–228.
10. Hasenfuß G, et al. A transcatheter intracardiac shunt device for heart failure with preserved ejection fraction (reduce LAP-HF): a multicentre, open-label, single-arm, phase 1 trial. *Lancet.* 2016;387(10025):1298–1304.
11. Kaye DM, et al. One-year outcomes after transcatheter insertion of an interatrial shunt device for the management of heart failure with preserved ejection fraction. *Circ Heart Fail.* 2016;9(12):e003662.
12. Wessler J, et al. Impact of baseline hemodynamics on the effects of a transcatheter interatrial shunt device in heart failure with preserved ejection fraction. *Circ Heart Fail.* 2018;11(8):e004540.
13. Kaye DM, et al. Impact of an interatrial shunt device on survival and heart failure hospitalization in patients with preserved ejection fraction. *ESC Heart Fail.* 2019;6(1):62–69.
14. Hanff TC, et al. Assessment of predictors of left atrial volume response to a transcatheter interatrial shunt device (from the reduce LAP-HF trial). *Am J Cardiol.* 2019;124(12):1912–1917.
15. Feldman T, et al. Transcatheter interatrial shunt device for the treatment of heart failure with preserved ejection fraction (reduce LAP-HF I [reduce elevated left atrial pressure in patients with heart failure]): a phase 2, randomized, sham-controlled trial. *Circulation.* 2018;137(4):364–375.
16. Shah SJ, et al. One-year safety and clinical outcomes of a transcatheter interatrial shunt device for the treatment of heart failure with preserved ejection fraction in the reduce elevated left atrial pressure in patients with heart failure (reduce LAP-HF I) trial: a randomized clinical trial. *JAMA Cardiol.* 2018;3(10):968–977.
17. Obokata M, et al. Effects of interatrial shunt on pulmonary vascular function in heart failure with preserved ejection fraction. *J Am Coll Cardiol.* 2019;74(21):2539–2550.
18. Rodés-Cabau J, et al. Interatrial shunting for heart failure: early and late results from the first-in-human experience with the V-Wave system. *JACC Cardiovasc Interv.* 2018;11(22):2300–2310.
19. Amat-Santos IJ, et al. The V-Wave device for the treatment of heart failure. Initial experience in europe. *Rev Esp Cardiol (Engl Ed).* 2015;68 (9):808–810.
20. Amat-Santos IJ, et al. Left atrial decompression through unidirectional left-to-right interatrial shunt for the treatment of left heart failure: first-in-man experience with the V-Wave device. *EuroIntervention.* 2015;10(9):1127–1131.
21. Guimarães L, et al. Interatrial shunt with the second-generation V-Wave system for patients with advanced chronic heart failure. *EuroIntervention.* 2020;15(16):1426–1428.
22. Del Trigo M, et al. Unidirectional left-to-right interatrial shunting for treatment of patients with heart failure with reduced ejection fraction: a safety and proof-of-principle cohort study. *Lancet.* 2016;387(10025):1290–1297.
23. Patel MB, et al. Implantable atrial flow regulator for severe, irreversible pulmonary arterial hypertension. *EuroIntervention.* 2015;11 (6):706–709.
24. Paitazoglou C, et al. The AFR-prelieve trial: a prospective, non-randomised, pilot study to assess the atrial flow regulator (AFR) in heart failure patients with either preserved or reduced ejection fraction. *EuroIntervention.* 2019;15(5):403–410.
25. Simard T, et al. Percutaneous atriotomy for levoatrial-to-coronary sinus shunting in symptomatic heart failure: first-in-human experience. *JACC Cardiovasc Interv.* 2020;13(10):1236–1247.
26. Bateman RM, et al. 36th International symposium on intensive care and emergency medicine: Brussels, belgium. 15–18 march 2016. *Crit Care.* 2016;20(Suppl 2):94.
27. Borggrefe M, Mann DL. Cardiac contractility modulation in 2018. *Circulation.* 2018;138(24):2738–2740.
28. Abraham WT, et al. A randomized controlled trial to evaluate the safety and efficacy of cardiac contractility modulation. *JACC Heart Fail.* 2018;6(10):874–883.
29. Tschöpe C, et al. Cardiac contractility modulation signals improve exercise intolerance and maladaptive regulation of cardiac key proteins for systolic and diastolic function in HFpEF. *Int J Cardiol.* 2016;203:1061–1066.
30. Abi-Samra F, Gutterman D. Cardiac contractility modulation: a novel approach for the treatment of heart failure. *Heart Fail Rev.* 2016;21 (6):645–660.

31. Borggrefe MM, et al. Randomized, double blind study of non-excitatory, cardiac contractility modulation electrical impulses for symptomatic heart failure. *Eur Heart J*. 2008;29(8):1019–1028.
32. Kadish A, et al. A randomized controlled trial evaluating the safety and efficacy of cardiac contractility modulation in advanced heart failure. *Am Heart J*. 2011;161(2):329–337. e1–2.
33. Borggrefe M, Burkhoff D. Clinical effects of cardiac contractility modulation (CCM) as a treatment for chronic heart failure. *Eur J Heart Fail*. 2012;14(7):703–712.
34. Wiegn P, et al. Safety, performance, and efficacy of cardiac contractility modulation delivered by the 2-lead optimizer smart system: the FIX-HF-5C2 study. *Circ Heart Fail*. 2020;13(4):e006512.
35. Morris DA, et al. Left atrial systolic and diastolic dysfunction in heart failure with normal left ventricular ejection fraction. *J Am Soc Echocardiogr*. 2011;24(6):651–662.
36. Maass AH, Van Gelder IC. Atrial resynchronization therapy: a new concept for treatment of heart failure with preserved ejection fraction and prevention of atrial fibrillation? *Eur J Heart Fail*. 2012;14(3):227–229.
37. Liu S, et al. Impaired left atrial systolic function and inter-atrial dyssynchrony may contribute to symptoms of heart failure with preserved left ventricular ejection fraction: a comprehensive assessment by echocardiography. *Int J Cardiol*. 2018;257:177–181.
38. Daubert C, et al. Permanent left atrial pacing with a specifically designed coronary sinus lead. *Pacing Clin Electrophysiol*. 1997;20(11):2755–2764.
39. Laurent G, et al. Permanent left atrial pacing therapy may improve symptoms in heart failure patients with preserved ejection fraction and atrial dyssynchrony: a pilot study prior to a national clinical research programme. *Eur J Heart Fail*. 2013;15(1):85–93.
40. Ng MM, Sica DA, Frishman WH. Rheos: an implantable carotid sinus stimulation device for the nonpharmacologic treatment of resistant hypertension. *Cardiol Rev*. 2011;19(2):52–57.
41. La Rovere MT, et al. Prognostic implications of baroreflex sensitivity in heart failure patients in the beta-blocking era. *J Am Coll Cardiol*. 2009;53(2):193–199.
42. Funakoshi K, et al. Striking volume intolerance is induced by mimicking arterial baroreflex failure in normal left ventricular function. *J Card Fail*. 2014;20(1):53–59.
43. Abraham WT, et al. Baroreflex activation therapy for the treatment of heart failure with a reduced ejection fraction. *JACC Heart Fail*. 2015;3(6):487–496.
44. Kirklin JK, et al. Eighth annual intermacs report: special focus on framing the impact of adverse events. *J Heart Lung Transpl*. 2017;36(10):1080–1086.
45. Topilsky Y, et al. Left ventricular assist device therapy in patients with restrictive and hypertrophic cardiomyopathy. *Circ Heart Fail*. 2011;4(3):266–275.
46. Wynne E, et al. Use of a left ventricular assist device in hypertrophic cardiomyopathy. *J Card Surg*. 2011;26(6):663–665.
47. Muthiah K, et al. Centrifugal continuous-flow left ventricular assist device in patients with hypertrophic cardiomyopathy: a case series. *ASAIO J*. 2013;59(2):183–187.
48. Patel SR, et al. Outcomes of restrictive and hypertrophic cardiomyopathies after LVAD: an intermacs analysis. *J Card Fail*. 2017;23(12):859–867.
49. Lund LH, Matthews J, Aaronson K. Patient selection for left ventricular assist devices. *Eur J Heart Fail*. 2010;12(5):434–443.
50. Kiamanesh O, et al. Left ventricular assist device with a left atrial inflow cannula for hypertrophic cardiomyopathy. *JACC: Case Rep*. 2020;2(13):2090–2094.
51. Ma M, et al. A novel inflow cannulation strategy for pediatric mechanical circulatory support in small left ventricles. *J Thorac Cardiovasc Surg*. 2017;154(3):e47–e48.
52. Moscato F, et al. Use of continuous flow ventricular assist devices in patients with heart failure and a normal ejection fraction: a computer-simulation study. *J Thorac Cardiovasc Surg*. 2013;145(5):1352–1358.
53. Escher A, et al. A valveless pulsatile pump for heart failure with preserved ejection fraction: hemo- and fluid dynamic feasibility. *Ann Biomed Eng*. 2020;48(6):1821–1836.
54. Granegger M, et al. A valveless pulsatile pump for the treatment of heart failure with preserved ejection fraction: a simulation study. *Cardiovasc Eng Technol*. 2019;10(1):69–79.
55. Giridharan GA, et al. Flow dynamics of a novel counterpulsation device characterized by CFD and PIV modeling. *Med Eng Phys*. 2011;33(10):1193–1202.
56. Burkhoff D, et al. Left atrial decompression pump for severe heart failure with preserved ejection fraction: theoretical and clinical considerations. *JACC Heart Fail*. 2015;3(4):275–282.
57. Morley D, et al. Hemodynamic effects of partial ventricular support in chronic heart failure: results of simulation validated with in vivo data. *J Thorac Cardiovasc Surg*. 2007;133(1):21–28.
58. Meyns B, et al. Proof of concept: hemodynamic response to long-term partial ventricular support with the synergy pocket micro-pump. *J Am Coll Cardiol*. 2009;54(1):79–86.
59. Fukamachi K, et al. Left atrial assist device to treat patients with heart failure with preserved ejection fraction: initial in vitro study. *J Thorac Cardiovasc Surg*. 2020.
60. Persson H, et al. Diastolic dysfunction in heart failure with preserved systolic function: need for objective evidence: results from the charm echocardiographic substudy-charmes. *J Am Coll Cardiol*. 2007;49(6):687–694.
61. Santos AB, et al. Impaired left atrial function in heart failure with preserved ejection fraction. *Eur J Heart Fail*. 2014;16(10):1096–1103.

PART IV

Innovative Surgical Approaches and Interventional Techniques

Chapter 13

Cardiac Implantable Devices in Heart Failure: Role of Imaging

Rahul D. Renapurkar

Thoracic and Cardiovascular Imaging, Imaging Institute, Cleveland Clinic, Cleveland, OH, United States

Chapter Outline

Introduction

Heart failure (HF) continues to be a rising public health concern currently affecting more than 6 million people in the United States and is projected to exceed 8 million people by 2030.[1] The incidence of HF is also rising rapidly in developing nations. These trends are also seen globally and is expected to rise with success of medical therapies and increasing life expectancy.[2] Management of HF is a complex process and necessitates a multidisciplinary evaluation taking into account several factors such as acuity of HF, potential reversibility, age of the patient, and other medical comorbidities. Often, despite adequate medical therapies, HF progresses to a refractory state, and is associated with a high mortality rate.[3] Implantable cardiac devices form a group of ever-expanding devices that are used to support and treat HF. Depending on the duration and state of the HF, the intended goals of these devices vary, with the ultimate goal in acute HF being to reinstate the cardiac function towards recovery, and in chronic HF to support the ailing heart through assisted cardiac circulatory function and if applicable, act as a destination therapy (DT) tool in cases of end-stage organ failure. The success of minimally invasive devices has proved to be a game changer and has expanded the availability of these devices to a larger patient population. Imaging plays a pivotal role in the successful implantation of devices, and is used for preprocedural evaluation, selection of appropriate candidates, intraprocedural guidance, and postprocedural evaluation. Often, chest radiography (CXR) is the first used imaging tool as part of baseline evaluation. Advanced imaging options commonly used include echocardiography and computed tomography (CT). Other imaging techniques such as positron emission tomography/computed tomography (PET/CT) and magnetic resonance imaging (MRI) are used as problem solving tools; the role of MRI is limited as it is contraindicated in certain devices such as ventricular assist devices (VADs).

In this chapter, we will focus on the commonly used cardiac implantable devices and the role of imaging and imaging techniques, their normal and abnormal imaging appearances, as well as imaging evaluation of potential complications.

Advances in Cardiovascular Technology. DOI: https://doi.org/10.1016/B978-0-12-816861-5.00023-X

TABLE 13.1 Summary of commonly used cardiac implantable devices.

Device	Goal	Types
Cardiac resynchronization therapy	To improve ventricular dyssynchrony and thus LV systolic function	Can be used as a CRT-pacemaker (CRT-P) or by a combined CRT-implantable cardioverter-defibrillator (CRT-D). LV pacing done by a coronary sinus lead, or in select cases by an epicardial lead (in patients with poor access)
LV remodeling devices	To improve the LV shape and geometry, and reduce wall shear stress	*Surgical *Percutaneous
Mechanical cardiac assist devices		
IABP	To augment LV pumping and facilitate forward flow; used as a short-term therapy	
ECMO	To improve tissue oxygen deliver	Veno-venous-for respiratory failure Veno-arterial-for cardiopulmonary support
Ventricular assist devices	Used either as a bridge to transplantation or destination therapy to augment LV and/or RV pumping	First generation (pulsatile pumps)-HeartMate I, Thoratec and Novacor. Second generation (axial flow pumps)-HeartMate II, Jarvik and Berlin Heart INCOR Third generation (centrifugal pumps)-HeartMate III and HeartWare (recently withdrawn).
Total artificial heart	Used in biventricular failure or patients with failed cardiac transplantation.	SynCardia TAH the only FDA approved device.

Types of devices

A plethora of devices have been studied in the nonpharmacological management of HF; the list continues to evolve with more minimally invasive devices gaining foothold. The aims of these therapies are diverse, and seek to improve ventricular synchrony, reduce wall shear stress, and improve left ventricle (LV) contractility. The commonly used devices are summarized in Table 13.1. Surgical and nonsurgical valvular repair (such as transcatheter mitral valve repair), and investigational devices that are not yet available or approved for clinical use are also part of the therapeutic interventions, although are beyond the scope of this chapter and will not be discussed further.

LV remodeling devices

Surgical approaches to remodeling of LV following an ischemic event have been tried over the years, although the conclusive benefit of these is still uncertain. In the Surgical Treatment for Ischemic Heart Failure (STICH) trial, which compared the outcomes of coronary bypass graft (CABG) alone versus CABG with surgical anterior ventricular endocardial restoration in 1000 patients with ischemic cardiomyopathy, the investigators found no significant difference in the rate of the primary outcome of death from any cause or hospitalization for cardiac causes at a median follow-up of 48 months (58% vs 59% respectively).[4] Other approaches such as partial LV resection (Batista procedure) and cardiomyoplasty have also fallen out of favor. Surgical and nonsurgical LV restoration devices that improve the LV geometry and improve wall stress have been explored and developed over the years. Conclusive data on the benefits of these devices is lacking although are attractive alternatives in principle. Earlier surgical devices such as Myosplint (Fig. 13.1) and Acord CorCap have given way for more minimally invasive and percutaneous approaches. One such device is a parachute shaped partitioning device (Cardiokinetix, Menlo Park, CA) that is made of a fluoropolymer membrane mounted on a nitinol frame. Using a percutaneous approach, this device is placed in the infarcted LV apex with the goal of isolating dysfunctional myocardium and thereby improving LV shape and geometry. Clinical studies are ongoing, although the results from PARACHUTE III trial were encouraging.[5]

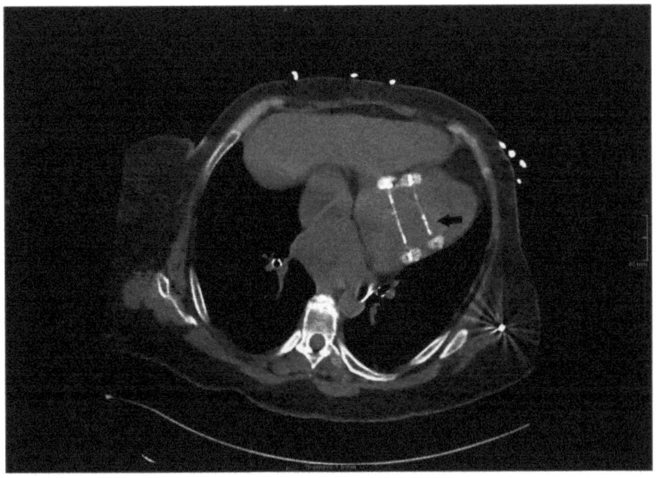

FIGURE 13.1 Normal CT imaging appearance of Myosplint partitioning device demonstrating the transventricular splints in the left ventricular cavity (arrow).

Role of imaging

Beyond intraprocedural guidance, imaging is required to ascertain satisfactory positioning of the devices, and to assess for complications. CXR is often useful as a baseline tool for confirming good positioning. These devices often have characteristic appearances and familiarity with these is essential. For example, with the newer Cardiokinetix partitioning device, the device should look like an inverted parachute or an umbrella, with the foot resting against the LV apex.[6] CT is not routinely indicated, although can be performed for evaluation of suspected complications. Possible complications include device malposition or migration and thromboembolic events.[7]

Cardiac resynchronization therapy

The cardiac resynchronization therapy (CRT) is a standard electrical treatment device for advanced systolic HF patients, which reduces mortality and reliefs symptoms in patients with HF and electrical and/or mechanical dyssynchrony. Initially, the CRT has been selectively used in patients with heart failure with reduced ejection fraction (HFrEF) with associated bundle branch block.[8] The rationale for using this therapy is that improving the electrical dyssynchrony can improve ventricular contraction and thereby mitigate adverse events such as reverse remodeling. In select patients with LVEF $\leq 35\%$ and wide QRS, CRT can improve LV contractility with immediate hemodynamic benefits.[9,10] It is also considered in select patients with left ventricular ejection fraction (LVEF) between 35% and 50% who need frequent ventricular pacing or who have a wide QRS with left bundle branch block and refractory HF.[9,10] While the CRT had been considered to be beneficial only in HFrEF patients with severe symptoms and the entry criterion for CRT were usually limited to patients with reduced LVEF there are some reports implying on potential efficacy of CRT on HFpEF patient cohort as well.[11] Typical CRT configuration includes a right atrial, right ventricular, and coronary sinus lead. In some cases, epicardial LV lead can be used. The leads are connected to the pulse generator usually placed in the subcutaneous soft tissues of the upper chest. The electrode is proximally introduced into the generator via a terminal connector pin, which should be engaged. These can be used as a CRT-pacemaker (CRT-P) or can be upgraded to CRT-defibrillator (CRT-D) in patients with suitable indication for Implantable cardioverter defibrillator (ICD).

Role of imaging

CXR is often used to evaluate optimal position of the device and check for integrity of the components. With an immediate postprocedural CXR, the position of the leads is confirmed, as well as optimal seating of the connector pin. Incomplete engagement of the lead into the connector pin can be a cause of faulty rhythm capture.[12] Often baseline CXR is done to check for the device manufacturer and MRI compatibility. Each manufacturer has its own logo that is seen on the casing of the device.[13] Newer MRI compatible devices have a radiopaque marker that is inserted at the

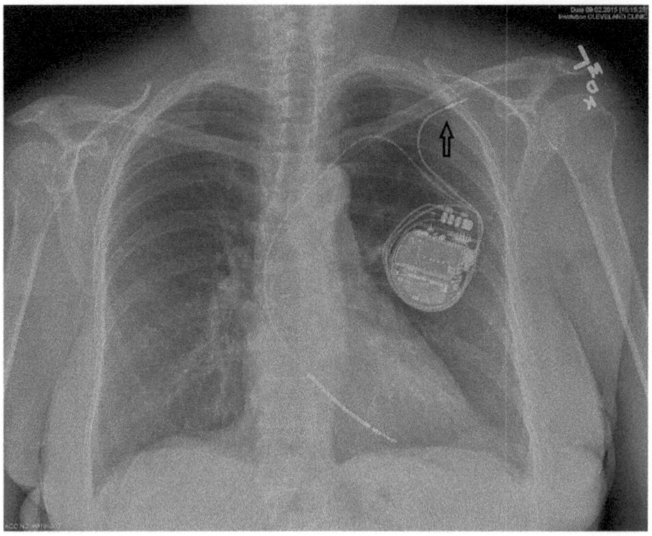

FIGURE 13.2 Fifty eight-year-old female with ICD placement and Twiddler syndrome. The right atrial lead has dislodged and is now within the expected location of left subclavian vein (arrow).

proximal end of the electrode, close to the casing. Also, in some cases, the logo itself denotes MRI compatibility. CXR is also used to evaluate for immediate and delayed complications with CT used as a problem-solving tool. Commonly seen immediate complications include pneumothorax, hemothorax, or myocardial perforation. Myocardial perforation is a rare complication seen in $<1\%$ of cases. CXR can suspect myocardial perforation based on unusual position of the lead tip, although CT can be confirmatory. On CT, perforation is suspected if the lead tip is within 3 mm of epicardial fat and is considered likely if it is conclusively seen within fat.[14] Delayed complications include lead dislodgement and fractures, and Twiddler syndrome. Lead fracture and damage to insulation are chronic complications typically seen at the fulcrum of a freely moving lead with a stationary point such as the subclavian vein–first rib junction and at the site of excessively tight fixation sutures.[12,13] Lead displacement can also be occasionally seen over time. Twiddler syndrome is a rare but unique complication resulting from unintentional or deliberate manipulation of the pulse generator within its subcutaneous pocket by the patient. This causes coiling of the leads in the long axis and can result in lead fracture or retraction[12,13] (Fig. 13.2). Reel syndrome is a variation of the same, termed when the impulse generator rotates along its transverse axis. The prevalence of this syndrome is 0.07% and is frequently seen in obese women.[13] Often, serial follow-up CXR can help in assessment of the changes in the generator position and associated coiling of the leads. Other systemic and long-term complications include vein thrombosis that may be seen in 30%–50% of patients, either acutely or over a long time period.[13] Device-related infections are on the rise and can range from pocket infections to florid lead/infective endocarditis.[15] CT can be useful in looking for fluid collections at the generator pocket site as well as evaluating lung parenchyma for cavitary nodules. In equivocal cases, Fluorodeoxyglucose (FDG)-PET/CT can be helpful in diagnosis device infections.[16]

Mechanical circulatory support devices

Mechanical circulatory support devices constitute a range of devices that augment the function of a failing ventricle to reestablish normal hemodynamics and end-organ blood flow. Since the first successful cardiopulmonary bypass in 1954, this field has come a long way and continues to evolve.[17] Currently, a vast number of devices are available that can support either the LV or right ventricle (RV) or both; these devices can further be subgrouped into short-term and long-term assist devices. Broadly, the indications for support devices can be grouped into bridge to recovery (BTR), bridge to candidacy, bridge to transplantation (BTT), and DT. Short-term devices are usually used as BTR in acute settings such as cardiogenic shock (CS) or cardiopulmonary arrest. Intermediate to long-term devices are usually used as BTT, until a donor heart becomes available. With improved survival rates and durability of long-term assist devices, use as a DT is now a reality.[17]

Short-term assist devices

Short-term Mechanical circulatory support (MCS) devices inserted percutaneously can be used as a bridge to bridge or BTR strategy. Bridge to bridge, a common strategy, especially in patients with acute on chronic HF complicated by CS, is the case when patients have a temporary device inserted with a provisional plan to transition to a more definitive intervention like a durable left ventricular assist device (LVAD) or as a BTT after clinical stabilization.

Currently available short-term MCS devices include the intraaortic balloon pump (IABP) (Getinge), the **microaxial Impella 2.5 CP** and 5.0 and 5.5 systems (Abiomed Europe, Aachen, Germany), TandemHeart (Cardiac Assist, Inc., Pittsburgh, PA), and extracorporeal membrane oxygenation (ECMO). RV support devices include the Impella Right Peripheral (RP) (Abiomed, Danvers, MA), and the TandemHeart RA-PA (LivaNova, London, UK) with blood delivery from the right atrium (RA) or inferior vena cava to the pulmonary artery.

As noted earlier, short-term devices are used to aid the recovery of a failing ventricle in an acute setting. These devices can be placed either surgically or percutaneously (Table 13.1). Percutaneously inserted devices are particularly useful in acute settings since they allow rapid attainment of circulatory support. We will review the commonly used short-term devices, namely IABP, ECMO, and Impella device.

IABP devices

CS is a deadly syndrome characterized by hypotension with evidence of tissue hypoperfusion. Historically the number one cause of CS has been acute myocardial infarction (AMI).[18]

An IABP is one of the most widely used support devices that is composed of a balloon mounted on a dual lumen flexible catheter. Helium is used to rhythmically inflate and deflate the balloon; this rhythmic pumping is controlled by a mobile console (Fig. 13.3). The desired effect of this counterpulsation mechanism is to reduced afterload and augment forward blood flow.[19] Also, during diastole, the balloon inflates, thereby increasing diastolic pressure and augmenting coronary and cerebral flow. The device is inserted percutaneously, commonly through a common femoral artery approach although alternative access sites such as subclavian and axillary arteries are also used. The catheter has a radiopaque marker at its distal end, and is advanced fluoroscopically to a position of 1 cm distal to the left subclavian artery.[20]

Imaging

CXR is routinely obtained following IABP placement to confirm satisfactory positioning and is also used for regular monitoring. On CXR, aortic knob has traditionally been used a radiographic reference landmark, although some have suggested carina to be a better reference point, with the desired radiopaque tip approximately 2 cm above it.[19] Imaging is also used to assess for complications. Vascular complications are relatively common seen in about 6%–25% of cases and include hematomas, limb and visceral ischemia, vascular perforation or dissection.[19,20] These can be well evaluated with CT. Hematomas at the cannula insertion sites are seen as high attenuation collections on noncontrast CT. Nonvascular complications are lesser in frequency and include incorrect positioning, balloon rupture, sepsis, and

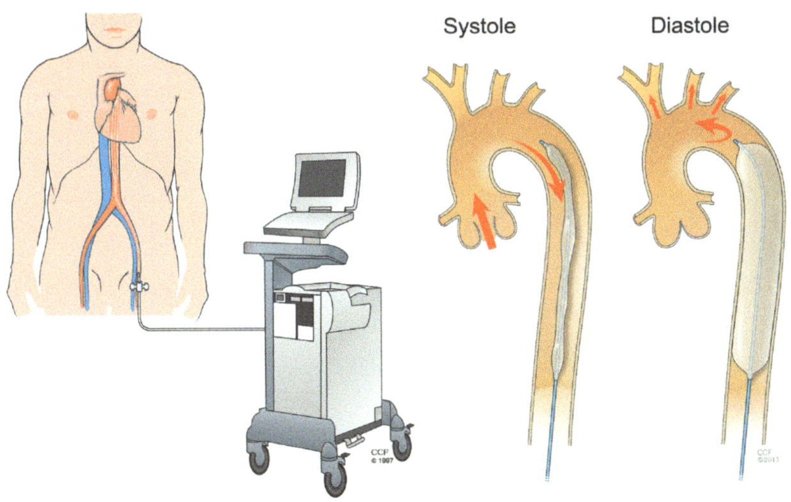

FIGURE 13.3 Schematic showing basic functioning of an intraaortic balloon pump. *Reprinted with permission, Cleveland Clinic Center for Medical Art & Photography.*

general complications related to anticoagulation. Incorrect position of IABP is however more common than expected and in one study of 63 patients, the balloon was correctly positioned in only 62% of the patients.[21] Also, this CT based study found a high incidence of abdominal visceral arterial compression; this was attributed in 68% of the cases to discrepancy of the patient size versus the length of the balloon.[22]

ECMO devices

ECMO support is a prolonged form of circulatory support that is initiated in the setting of unresponsive but potentially reversible acute respiratory and/or cardiac failure. Several clinical trials and meta-analyses have shown improvement in mortality rate in patients with severe acute respiratory failure following institution of ECMO therapy.[23–25] Data for cardiac arrest is less robust although observational studies and case series have shown improved survival rates compared to conventional cardiopulmonary resuscitation.[26,27] The basic goal of ECMO therapy is to increase systemic oxygen delivery, which is achieved by removing native venous blood, passing it through an oxygenator, and returning it to the body. Broadly, there are two types of vascular access: veno-venous (VV) and veno-arterial (VA). With a VV-ECMO, the blood is drained from the vena cava or RA and returned to the RA (Fig. 13.4A). With a VA-ECMO, the heart is completely bypassed and the blood is drained from either the vena cava or RA and returned to the arterial system (Fig. 13.4B). VV-ECMO allows respiratory support while VA-ECMO allows both cardiac and respiratory support. Also, the VA-ECMO cannulation may be central or peripheral; with a central VA-ECMO, the cannulae are placed directly in the RA and the aorta. The magnitude of hemodynamic support provided by IABP is related to the magnitude of blood volume displacement. The size of the descending aorta is related to patient weight, height, and age.

Imaging

As with IABP, CXR is often performed following ECMO placement to confirm appropriate positioning of the cannulae and for excluding immediate complications such as malpositioning and pneumothorax[28] (Fig. 13.5). On a CXR, with a VV-ECMO, the tip of the femoral cannula should be near the junction of the Inferior vena cava (IVC) and RA, and the tip of the internal jugular cannula should be near the junction of the superior vena cava (SVC) and RA. The positions of cannulae for VA-ECMO vary depending on the access sites used and if it is peripheral vs. central. Echocardiography is often used to monitor recovery of LV function following placement of ECMO. CXR and CT are often used in evaluation of complications. Bleeding and thromboembolism constitute the most common complications. Hemorrhagic complications can be seen in about 30%–50% of patients and result from platelet dysfunction and continuous anticoagulation.[29] Serious bleeding can result in body cavities such as retroperitoneal and pleural spaces; these are well evaluated with noncontrast CT. Systemic thromboembolism has been reported with varying frequency with some studies demonstrating the incidence of deep vein thrombosis as high as 70%.[30] The impact of thromboembolism is more serious with VA-ECMO as the blood is returned into the systemic circulation. Similarly, high incidence of pulmonary embolism (PE) up to 16% has been reported.[30] Mechanical complications such as arterial perforation and dissection are relatively rare (<5%).[31] These can be evaluated with ultrasound and/or CT. Local complications such as vessel obstruction and occlusion can occur at the cannulation site and are again more common with VA-ECMO.[32]

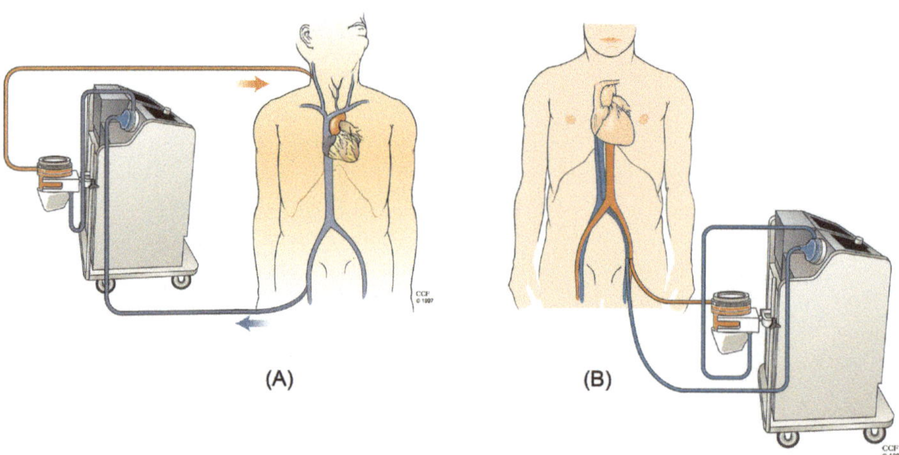

FIGURE 13.4 Schematic demonstrating basic mechanisms of veno-venous ECMO (A) and veno-arterial ECMO (B). *Reprinted with permission, Cleveland Clinic Center for Medical Art & Photography.*

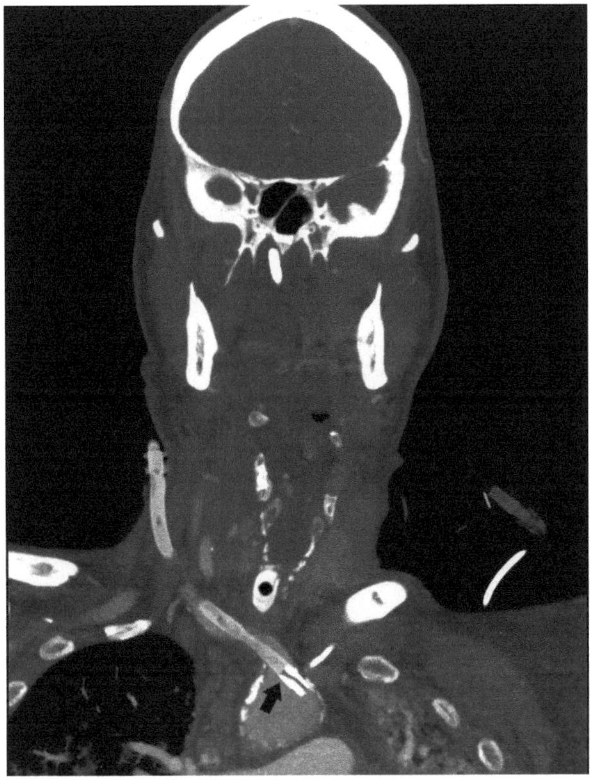

FIGURE 13.5 Maximum Intensity Projection (MIP) coronal CT image demonstrating malpositioned ECMO cannula with the venous limb terminating in the aortic arch (arrow).

Doppler ultrasound may show lack of flow as well as absent waveforms in patients with arterial occlusions.[33] In venous obstruction, color Doppler may show a lack of venous flow or absence of normal phasicity on spectral Doppler imaging.[33] Additional complications are also noted that are specific for VA-ECMO. With a VA-ECMO on full flow settings, the LV is completely bypassed. This can cause pooling of blood in the LV, aortic root, and eventually, thrombus formation. Also with VA-ECMO, when a contrast enhanced CT is performed, the injected contrast is siphoned away to the ECMO oxygenator and is returned to the arterial circulation, with the aorta getting opacified in a retrograde fashion. This creates unusual opacification patterns on Contrast enhanced computed tomography (CECT), and can mimic dissection[28,33,34] (Fig. 13.6). Imaging specialists need to be aware of this phenomenon and need to tailor the CT protocol accordingly; some of the strategies include minimizing ECMO flow rate (if clinically feasible), injecting contrast into the "arterial" ECMO tubing if possible, manually triggering the CT scanning, and if needed, performed delayed imaging.[33] Also, when the LV function starts to recover, it can lead to differential pumping of the blood. Deoxygenated blood gets pumped through the LV into the upper half of the body, while the lower body gets oxygenated blood returning from the ECMO circuit. This leads to differential cyanosis, called Harlequin or North-South syndrome of VA-ECMO.[35]

Impella devices

Impella is a continuous flow microaxial pump contained within a pigtail catheter. It consists of an inlet area that is retrogradely placed in the LV. Using a pump that rotates an Archimedes screw, the inlet portion ejects blood from the LV through the outlet into the ascending aorta[36] (Fig. 13.7). Various sizes are available with percutaneous options consisting of Impella 2.5 and Impella CP. The Impella 5.0/LD and Impella 5.5 require surgical cutdown of femoral or axillary artery but can achieve a flow rate of up to 5.5 L/min.[36,37] The Impella RP is designed for temporary RV support. While attractive in principle, so far, the existing data supporting its use are underwhelming. This is partly complicated by lack of large randomized controlled trials, existing data based on underpowered trials, and inherent difficulties in conducting trials in this patient population.[36]

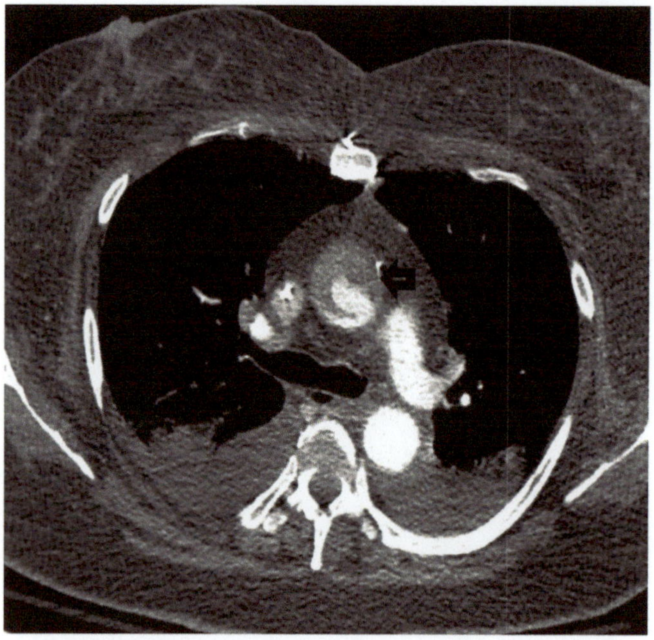

FIGURE 13.6 Axial CT image of a contrast enhanced CT in a patient with VA ECMO demonstrating mixing of opacified and nonopacified blood in the ascending thoracic aorta that can be mistaken for a dissection (arrow).

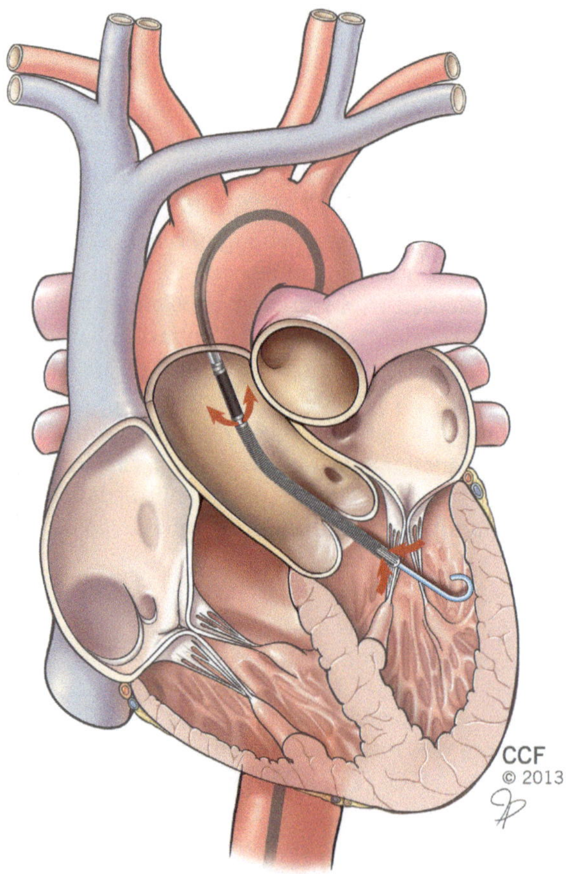

FIGURE 13.7 Schematic showing basic functioning of an Impella device. *Reprinted with permission, Cleveland Clinic Center for Medical Art & Photography.*

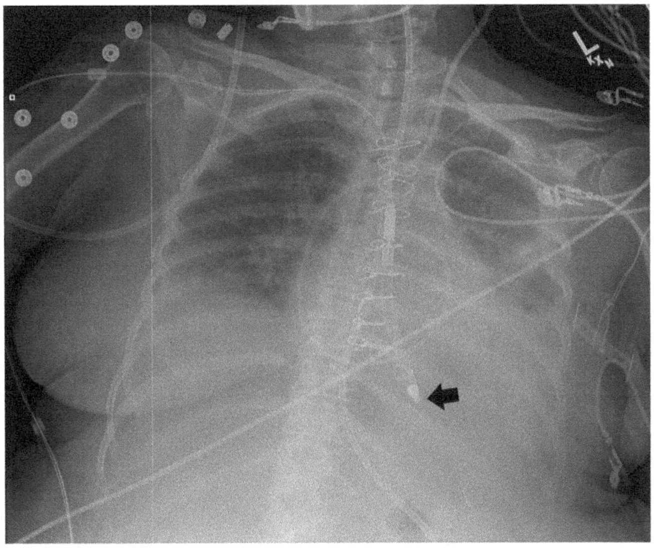

FIGURE 13.8 Portable radiograph in a patient with Impella device placement showing coiled and kinked inflow portion of the device (arrow).

Imaging

CXR is used to confirm satisfactory positioning of the device. The distal tip of the device is coiled and should lie within the left ventricular cavity.[28] The ideal location of the distal tip is 3–4 cm across the aortic valve. Inaccurate positioning deep in the LV can lead to ventricular arrhythmias. Catheter kinking may be identified on chest radiographs[38] (Fig. 13.8). Routine use of CT is not needed unless specific complications such as device migration are suspected. Other complications that are commonly seen include hemolysis, and noncontrast CT can be used to evaluate for suspected bleeding.

Long-term circulatory support devices

LVAD constitutes an important therapeutic option in patients with HFrEF who fail to respond to medical therapy. These devices have evolved considerably over the years, with successive newer generations demonstrating better outcomes and decreased rate of complications. Consequently, the indications for their use have expanded, with LVADs often used now for BTT as well as DT.[17] The first-generation devices used a pulsatile volume-displacement pump, and these were bulky and associated with limited durability and a lot of complications such as infection.[39] To overcome these limitations, second-generation devices were devised using a continuous flow rotary pump; significant advantages with these included fewer moving parts and hence smaller sizes, more durability, and decreased complications.[39] Third-generation devices were devised to eliminate all contact between impeller and drive mechanism and use centrifugal flow pumps. Key advantages include compact sizes, greater durability (5–10 years), and better blood flow to minimize hemolysis and thrombotic complications.[39] These devices are summarized in Table 13.1. Currently, most of the first-generation devices are obsolete with a trend toward increased usage of third-generation devices such as HeartMate III (HMIII). However, it should be noted that HeartWare device was recently discontinued due to the high incidence of complications such as stroke.

Imaging in patients with VAD

Imaging plays a paramount role in successful implantation of VAD. It is used in preoperative planning, intraoperative guidance, postoperative follow-up, and evaluation of complications. CXR is often used as a baseline tool for postoperative evaluation and for routine follow-up. Its role in assessment of complications is fairly limited. Echocardiography and CT play complementary roles in comprehensive evaluation of LVAD. MRI is contraindicated in patients with LVAD. FDG-PET is being increasingly used as a problem-solving tool in assessment of LVAD infections. A summary of the imaging tools and their relative strengths and limitations is highlighted in Table 13.2.

TABLE 13.2 Summary of role of different imaging modalities in assessment of complications related to commonly used implantable devices.

	Chest radiography	Echocardiography	Ultrasound and Doppler	CT	PET/CT and other modalities
Cardiac resynchronization therapy		–	–		–
Pneumothorax	+++	–	–	–	–
Twiddler syndrome	+++	–	–	–	–
Infection	+	–	–	++	In indeterminate cases
IABP				–	
Hemorrhage	++	+ (if pericardial)	+	++	–
Malposition	++	–	–	++	–
Vascular complications	–	–	++	++	–
ECMO					
Pneumothorax	+++	–	–	–	–
Hemorrhage	++	+ (if pericardial)	+	++	–
Malposition	++	–	–	++	–
Vascular complications	–	–	++	++	–
Thrombosis and stroke	–	–	+	+++	–, MRI
Ventricular assist devices					
Infection	+	–	–	++	–
LVADs					
Hemorrhage	++	+ (if pericardial)	+	++	–, 99mTc-RBC scan in GI bleed
Infection	+	++ (if bloodstream infection)	–	+++	PET/CT as a problem solving tool
Inflow cannula malposition	+	++	–	+++	
Inflow cannula thrombus	–	++	–	+++	
Outflow graft kinking	–	+ (not well seen with TTE)	–	+++	
Pump housing thrombus	–	++ (indirect predictor)	–	–	
LV thrombus	–	++	–	+++	
Bend relief disconnection	+	–	–	+++	
Driveline fracture	+++	–	–	++	
Aortic valve stenosis/regurgitation	–	+++	–	++	
RV failure	–	+++	–	+	
Stroke	–	–	–	+++	MRI of brain

Echocardiography

Detailed description of the role of echocardiography in LVAD evaluation is beyond the scope of this review. Echocardiography has multiple roles including baseline assessment and preoperative planning, intraoperative guidance (with transesophageal echocardiography), and immediate postoperative assessment of the device and valvular function. Also, echocardiography is well suited for routine postoperative evaluation and for assessment of complications.[40] However, echocardiography is not without its disadvantages, and key drawbacks include artifacts related to acoustic windowing and shadowing from the VAD and surgical drains, artifacts secondary to postoperative mediastinal and pleural air, and reduced patient mobility.[40]

Computed tomography

CT plays a pivotal role in assessment of LVAD and its associated complications. Key advantages of CT include high spatial resolution, ability for multiplanar evaluation, and ability to comprehensively evaluate the heart, lungs, mediastinum, and chest wall simultaneously. CT protocol is tailored to answer the specific clinical question. Ideally, for evaluation of LVADs, a retrospectively ECG-gated contrast enhanced CT is preferred since it allows anatomic as well as functional assessment of the LVAD and heart chambers. However, this increases the radiation dose to the patient and hence is reserved for certain situations, such as inflow cannula malposition, whereby cine imaging is critical to make a diagnosis. For infections of LVAD and its components, a regular contrast-enhanced CT can be adequate; however, care should be taken to image the abdominal wall as well, especially in cases of driveline infection. Unenhanced CT can suffice for evaluating hemorrhagic complications in the chest wall, mediastinum, pleura, or retroperitoneum.

Components and normal appearance of LVAD

LVAD has three basic components; an inflow cannula, pump, and outflow graft. In the case of second-generation devices such as HeartMate II (HMII), the pump is placed preperitoneally, between the rectus muscle and the posterior rectus sheath[28,41] (Fig. 13.9). The inflow cannula is placed in the LV cavity, while the outflow graft is anastomosed to the ascending thoracic aorta. In the case of third-generation devices such as HMIII and HeartWare, the pump is smaller and is integrated with the inflow cannula; this integrated inflow cannula is placed in the pericardial space. A driveline extends from the pump, which is tunneled through the anterior abdominal wall and connects to the electronic controller. The controller is powered by electricity or battery pack. With HeartMate II, there is an additional rigid polytetrafluoroethylene tube that is slid over the outflow graft and is secured by a snap ring. This bend relief portion prevents kinking of the outflow graft.[28]

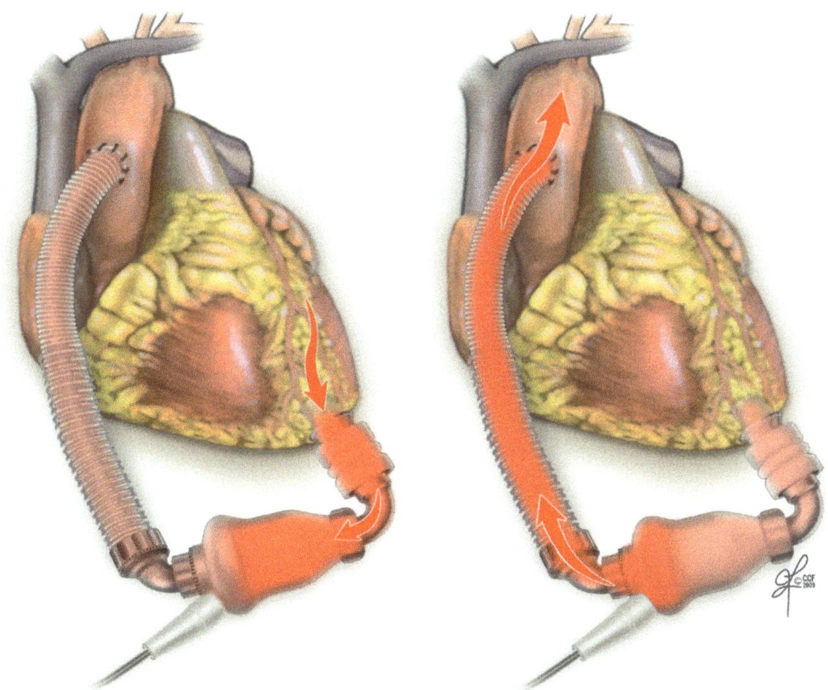

FIGURE 13.9 Schematic of a second generation left ventricular assist device (HeartMate II). *Reprinted with permission, Cleveland Clinic Center for Medical Art & Photography.*

On imaging, the inflow cannula should be directed towards the mitral valve. Angulation toward the interventricular septum or other ventricular walls can result in suction of the myocardium in the inflow cannula resulting in partial or complete obstruction. On cine imaging, the interventricular septum should be in a neutral middle position and there should be no aortic regurgitation. The outflow graft connects to the ascending thoracic aorta; there should be no kinking or obstruction. Often, a small amount of low-attenuation material between the outflow graft and overlying Gore-Tex tube graft is noted. This is actually blood that has seeped into this space and should not be mistaken for outflow graft thrombus.[41]

Imaging of complications

Bleeding

Bleeding is one of the most common complications following LVAD placement, occurring in both early postoperative and later periods. In the early postoperative period, this is attributed mostly to cardiopulmonary bypass related coagulopathic effect, hepatic congestion, and external administration of anticoagulants. Late complications are related to anticoagulation therapy as well as development of acquired von Willebrand disease.[42] Bleeding continues to be a major problem even with the newer continuous flow pumps. Common sites of bleeding include near the implantation site and close to areas of anastomoses and conduit grafts. Mucosal surfaces, including gastrointestinal, bleeding is also common, with an incidence of 14%–35% in continuous flow pumps.[43] Bleeding into the chest wall and cavities such as pleural, pericardial, or peritoneal spaces is also seen. A noncontrast CT is sufficient for diagnosis with acute hemorrhage seen as high attenuation fluid collection; its density decreases over time due to hemoglobin degradation.

Infection

Infections are one of the major causes of morbidity and mortality in patients with LVAD. A range of infectious complications have been noted that can involve the various components of the device, the body surfaces that contain them as well as systemic bloodstream. With regards to the hardware-related infections, these can occur in the pump, pump pocket, or around the driveline. Percutaneous driveline infections are the most common occurring in about 10.7%–35%,[44] with staphylococci and Gram-negative bacilli being the most commonly identified pathogens.[45] Continuous flow pumps by virtue of smaller designs and smaller drivelines have lesser infection rates, although still constitute a major long-term hazard.[46] Chest radiography has a limited role and can demonstrate parenchymal consolidation and nodules, which can be cavitary. CT can be immensely helpful and infections are seen as rim-enhancing fluid collections, soft tissue stranding adjacent to device components, or gas pockets or locules. In the assessment of a possible infection, the clinical context needs to be factored in and some of these findings can be normally seen in early postoperative state. FDG PET/CT has shown to be a good problem-solving tool in challenging case scenarios and demonstrates high sensitivity, although with variable specificity (Fig. 13.10). Some common conditions that can cause false positive FDG PET/CT include postoperative states, myocarditis, and sarcoidosis.[47]

Thrombosis and thromboembolic disease

Pump thrombosis, one of the most common and serious complications of blood pump technology (with the true contributing factors still being explored), is a multifactorial phenomenon arising from mechanical or biologic etiologies, or a

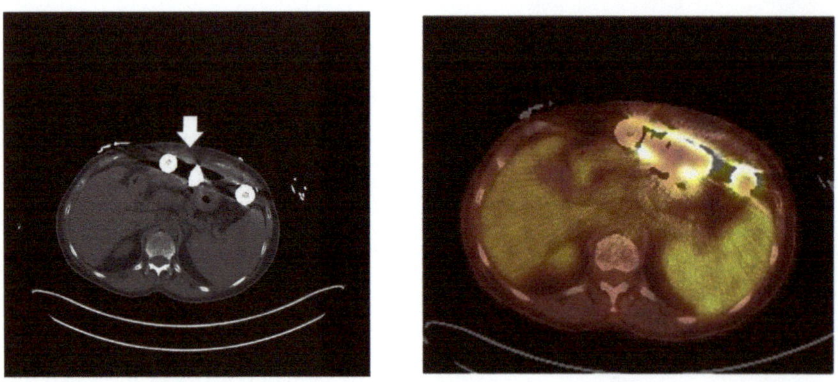

(A) (B)

FIGURE 13.10 (A) Axial CT image in a patient with HMII device demonstrating loculated fluid and gas collection close to the pump pocket (arrow). (B) Corresponding PET/CT image demonstrates FDG avidity in this region consistent with infection.

combination of these factors. Pump thrombosis refers to development of a clot within any of the components of the pump, including the inflow cannula, outflow graft, and pump housing.[28] The thrombus may originate in the pump or may be embolized from one of the heart chambers[48] (Cine Video 13.1). The causes are multifactorial and one of the postulated mechanisms includes fibrin and denatured protein deposition adjacent to the inflow bearing. Overall, the incidence has decreased with continuous flow devices with a rate of 0.01–0.11 per patient-year.[49] Thrombosis is suspected when there is development of hemolysis and/or deterioration of pump performance. CT is excellent for evaluation of thrombus in the inflow and outflow limbs and in the ventricular chambers. Delayed scan (with a delay of 70–75 s) may be used to optimally highlight the distinction between enhancing myocardium and nonenhancing thrombus.[50] However, due to streak artifacts, CT is limited in the evaluation of thrombus in the pump housing. Echocardiography with a speed-ramp test (measurement of left ventricular end-diastolic diameters at increasing VAD speeds) is a good alternative method to demonstrate device thrombosis and the two studies are often complementary in evaluation.[51] Similarly, complications such as stroke are common with reported incidence of ischemic and hemorrhagic stroke after LVAD placement of 8%–25%.[52] The causes of stroke are multifactorial but are related to increased thrombotic tendency. CT or Mr imaging of the brain can be used to assess this complication. Stroke is also common with newer continuous flow devices and is especially concerning with HeartWare device[53]; this has proven to be a major factor in discontinuation of this device.

Cine Video 13.1 3D echocardiogram loop demonstrating a mobile thrombus at the orifice of the inflow cannula. The video can be found on online at https://doi.org/10.1016/B978-0-12-816861-5.00023-X.

Mechanical complications and device failure

The geometric aspect of device–body interaction postimplant is multifactorial and may have a unique variability of configuration in each patient. Device failures are usually seen as a late complication and can involve either the external or internal components. It is critically important that the device inflow cannula be positioned with no obstruction and that graft kinking or major device dislodgement does not occur; these factors remain key requirements for correct LVAD placement. Possible LVAD technical problems can include malposition of the inflow cannula, kinking of outflow grafts, bend relief disconnection, driveline fracture, and lead fatigue.[54] Inflow cannula malfunction can happen due to thrombus formation, partial obstruction by myocardial trabeculation, or left ventricular underfilling secondary to hypovolemia or right ventricular dilatation.[28] Paradoxically, myocardial recovery leading to reverse LV remodeling can precipitate inflow cannula obstruction and malfunction due to myocardial apposition in a left ventricle that has decreased in size since the time of LVAD implantation. For evaluation of inflow cannula, echocardiography and CT are often used in conjunction. Typical findings with cannula obstruction are high-velocity aliased flow at Doppler echocardiography (peak inflow velocity >2.5 m/s with a pulsatile pump and >2 m/s with a continuous-flow device).[40] Cannula kinking is associated with loss of Doppler signal on all echocardiographic views. Cine CT is well suited for evaluation for cannula malposition and can demonstrate angulation and malposition effectively. CT is excellent for complete evaluation for outflow graft and associated narrowing or kinking, which is often not possible by echocardiography due to limited acoustic windows. Multiplanar reconstruction helps in better evaluation of outflow graft kinks. Bend relief disconnection can happen as an early complication although usually seen to develop gradually over time. Chest and abdominal radiographs especially when compared to serial prior imaging can help identify this complication. The disconnection can be partial when the snap ring appears to be disconnected (i.e., the snap ring is out of plane and/or there is lucency between the snap ring and underlying graft although the outflow graft remains covered) or there can be a complete disconnection between the bend relief conduit and the graft nut[55] (Fig. 13.11). CT is not required but is excellent at diagnosing bend relief disconnection. Driveline fractures are usually evaluated with radiography.

Other complications

Aortic regurgitation is a frequent complication seen in about 25% of patients treated with continuous flow device.[56] This is well evaluated with transthoracic echocardiography (TTE) and can be treated with increase in pump speeds. Rarely, surgery may be necessary. Right heart dysfunction is fairly common, both seen in early postoperative state and as a delayed complication. The incidence varies from 5%–50% in early perioperative period, depending on the clinical definition of RV failure (RVF) used.[57,58] Mechanical unloading of the LV by the LVAD results in bowing of the interventricular septum to the left and altered geometry of the RV, which is thought to be one of the precipitating causes of RVF. Device malfunction, tricuspid regurgitation, and progression of intrinsic RV dysfunction also contribute to late onset RVF.[59] TTE is the test of choice to assess right ventricular function and to optimize LVAD settings to minimize septal bowing. ECG-gated CT can be used to assess the systolic and diastolic flow volumes in the RV.[60] Other complications such as pericardial effusion and ventricular arrhythmias have also been noted.[61]

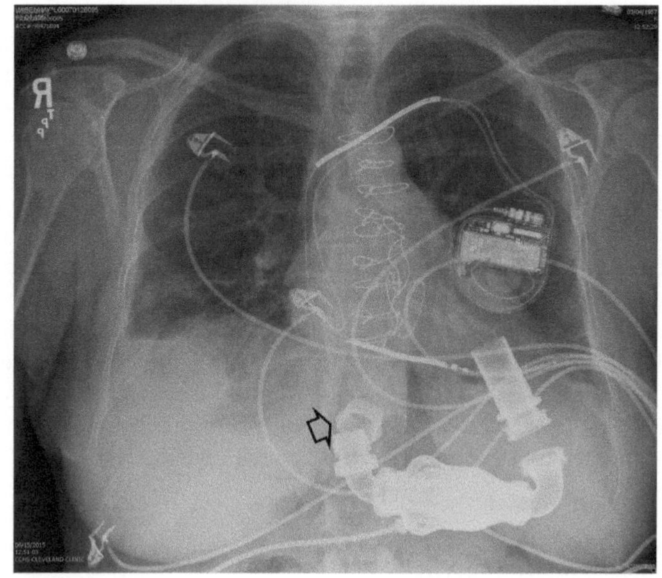

FIGURE 13.11 Frontal radiograph in a 67-year-old female with HMII device placement. There is malalignment of the snap ring of the outflow graft consistent with partial bend-relief disconnection (arrow).

Total artificial heart

Total artificial heart (TAH) constitutes an alternative long-term solution for patients with end-stage HF especially those with extensive myocardial damage and biventricular failure. While several devices have been developed over the years, SynCardia TAH (SynCardia Systems, Inc., Tucson, AZ) is currently the only commercially available TAH in the United States approved by the FDA as a bridge to heart transplantation.[62] The device system consists of an implantable TAH, which is connected by pneumatic drivelines to an external console (Fig. 13.12). The consoles control the pneumatic pumping and also monitor the device rates and functioning status. Recently, the FDA has approved a 5.5 kg portable Freedom driver that allows increased patient mobility. The implantable TAH consists of two artificial ventricles, each made of blood-containing and air-containing chambers separated by a polyurethane diaphragm. Each chamber incorporates two mechanical single leaflet tilting disk valves (SynHall (formerly Medtronic Hall)) to regulate direction of flow. The pneumatic drivers inflate the air-containing chambers, forcing blood to flow outward from the ventricles; blood passively returns into the ventricles. Since the pump occupies a volume of 400 cc in the chest cavity, an anterior–posterior chest diameter of at least 10 cm is required by CT from the anterior border of T10 vertebra to the posterior table of the sternum for successful implantation.[62,63]

Imaging

Chest radiography is often used as the initial test for immediate postoperative assessment, and for regular follow-up. The appearance of TAH is unusual with two air-filled radiolucencies overlying the cardiac silhouette corresponding to the pneumatic chambers; the chamber sizes change during different phases of cardiac cycle.[28,64] Also, the TAH mechanical valves do not lie in their expected anatomic locations. Care should be taken not to mistake the air-filled pneumatic chambers for intracavitary air, pneumomediastinum, pneumopericardium, and abscess.[64] CXR has limited role in assessment of complications; nonvisualization or visualization of only one pneumatic chamber on CXR can suggested underlying fluid or hematoma causing extrinsic compression. Spectrum of complications is similar to those with LVAD and can be evaluated with CT. A problem unique to TAH is that there is no recordable Electrocardiogram (ECG), and hence ECG-synchronized CT is not feasible. Fortunately, there is no external motion of the TAH implant, unlike that of a normal human heart (i.e., only the inside pneumatic chambers pulsate) and hence gating is not necessary.[65] MRI is contraindicated.[65]

FIGURE 13.12 Schematic of SynCardia Total Artificial Heart device. *Reprinted with permission, Cleveland Clinic Center for Medical Art & Photography.*

Conclusions

Despite advances in medical management, HF constitutes a major public health issue in the United States and throughout the world. Over the past two decades, an increasing number of surgically and nonsurgically implantable devices have emerged that can slow the downhill course of HF, and augment the failing heart; the list continues to grow. Radiologists and imaging specialists need to be aware of these devices, their functioning and complications that can develop in short- and long-term periods. Also, different imaging modalities play a multifaceted role in preoperative patient selection, intraprocedural guidance, and postoperative follow-up.

References

1. Benjamin EJ, Muntner P, Alonso A, et al. American heart association council on E, prevention statistics C and stroke statistics S. Heart disease and stroke statistics-2019 update: a report from the American heart association. *Circulation.* 2019;139:e56–e528.
2. Senni M, Tribouilloy CM, Rodeheffer RJ, et al. Congestive heart failure in the community: trends in incidence and survival in a 10-year period. *Arch Intern Med.* 1999;159(1):29–34.
3. Ho KK, Pinsky JL, Kannel WB, Levy D. The epidemiology of heart failure: the Framingham Study. *J Am Coll Cardiol.* 1993;22(4 Suppl. A):6A–13A.
4. Jones RH, Velazquez EJ, Michler RE, et al. Coronary bypass surgery with or without surgical ventricular reconstruction. *N Engl J Med.* 2009;360 (17):1705–1717.
5. Thomas M, Nienaber CA, Ince H, et al. Percutaneous ventricular restoration (PVR) therapy using the Parachute device in 100 subjects with ischaemic dilated heart failure: one-year primary endpoint results of PARACHUTE III, a European trial. *EuroIntervention.* 2015;11(6):710–717.

6. Costa MA, Pencina M, Nikolic S, et al. The PARACHUTE IV trial design and rationale: percutaneous ventricular restoration using the parachute device in patients with ischemic heart failure and dilated left ventricles. *Am Heart J.* 2013;165:531–536.
7. Costa MA, Mazzaferri Jr EL, Sievert H, et al. Percutaneous ventricular restoration using the parachute device in patients with ischemic heart failure: three-year outcomes of the PARACHUTE first-in-human study. *Circ Heart Fail.* 2014;7:752–758.
8. Jarcho JA. Resynchronizing ventricular contraction in heart failure. *N Engl J Med.* 2005;352(15):1594–1597.
9. Epstein AE, DiMarco JP, Ellenbogen KA, et al. 2012 ACCF/AHA/HRS focused update incorporated into the ACCF/AHA/HRS 2008 guidelines for device-based therapy of cardiac rhythm abnormalities: a report of the American College of Cardiology Foundation/American Heart Association Task Force on Practice Guidelines and the Heart Rhythm Society. *J Am Coll Cardiol.* 2013;61(3):e6–e75.
10. Ponikowski P, Voors AA, Anker SD, et al. 2016 ESC Guidelines for the diagnosis and treatment of acute and chronic heart failure: The Task Force for the diagnosis and treatment of acute and chronic heart failure of the European Society of Cardiology (ESC) developed with the special contribution of the Heart Failure Association (HFA) of the ESC [published correction appears in Eur Heart J. 2016]. *Eur Heart J.* 2016;37 (27):2129–2200.
11. Maass AH, van Veldhuisen DJ. Device therapy in patients with heart failure and preserved ejection fraction (HFPEF): a new frontier? *Eur J Heart Fail.* 2010;12(6):527–529.
12. Aguilera AL, Volokhina YV, Fisher KL. Radiography of cardiac conduction devices: a comprehensive review. *Radiographics.* 2011;31 (6):1669–1682.
13. Alandete Germán SP, Isarria Vidal S, Domingo Montañana ML, De la Vía Oraá E, Vilar Samper J. Pacemakers and implantable cardioverter defibrillators, unknown to chest radiography: review, complications and systematic reading. *Eur J Radiol.* 2015;84(3):499–508.
14. Steiner RM, Tegtmeyer CJ, Morse D, et al. The radiology of cardiac pacemakers. *RadioGraphics.* 1986;6(3):373–399.
15. Nery PB, Fernandes R, Nair GM, et al. Device-related infection among patients with pacemakers and implantable defibrillators: incidence, risk factors, and consequences. *J Cardiovasc Electrophysiol.* 2010;21:786–790.
16. Juneau D, Golfam M, Hazra S, et al. Positron emission tomography and single-photon emission computed tomography imaging in the diagnosis of cardiac implantable electronic device infection: a systematic review and meta-analysis. *Circ Cardiovasc Imaging.* 2017;10(4):e005772.
17. Kirklin JK, Naftel DC, Pagani FD, et al. Sixth INTERMACS annual report: a 10,000-patient database [published correction appears in J Heart Lung Transplant. 2015 Oct;34(10):1356. Timothy Baldwin, J [corrected to Baldwin, J T]]. *J Heart Lung Transpl.* 2014;33(6):555–564.
18. Trost JC, Hillis LD. Intra-aortic balloon counterpulsation. *Am J Cardiol.* 2006;97(9):1391–1398.
19. Wayangankar SA, Bangalore S, McCoy LA, et al. Temporal trends and outcomes of patients undergoing percutaneous coronary interventions for cardiogenic shock in the setting of acute myocardial infarction: a report from the CathPCI registry. *JACC Cardiovasc Interv.* 2016;9 (4):341–351.
20. Ferguson 3rd JJ, Cohen M, Freedman Jr RJ, et al. The current practice of intra-aortic balloon counterpulsation: results from the Benchmark Registry. *J Am Coll Cardiol.* 2001;38(5):1456–1462.
21. Kim JT, Lee JR, Kim JK, et al. The carina as a useful radiographic landmark for positioning the intraaortic balloon pump. *Anesth Analg.* 2007;105(3):735–738.
22. Rastan AJ, Tillmann E, Subramanian S, et al. Visceral arterial compromise during intra-aortic balloon counterpulsation therapy. *Circulation.* 2010;122(11 Suppl.):S92–S99.
23. Peek GJ, Mugford M, Tiruvoipati R, et al. Efficacy and economic assessment of conventional ventilatory support versus extracorporeal membrane oxygenation for severe adult respiratory failure (CESAR): a multicentre randomised controlled trial. *Lancet.* 2009;374:1351.
24. Combes A, Hajage D, Capellier G, et al. Extracorporeal membrane oxygenation for severe acute respiratory distress syndrome. *N Engl J Med.* 2018;378:1965.
25. Combes A, Peek GJ, Hajage D, et al. ECMO for severe ARDS: systematic review and individual patient data meta-analysis. *Intensive Care Med.* 2020;46:2048.
26. Shin TG, Choi JH, Jo IJ, et al. Extracorporeal cardiopulmonary resuscitation in patients with inhospital cardiac arrest: a comparison with conventional cardiopulmonary resuscitation. *Crit Care Med.* 2011;39:1.
27. Chen YS, Lin JW, Yu HY, et al. Cardiopulmonary resuscitation with assisted extracorporeal life-support versus conventional cardiopulmonary resuscitation in adults with in-hospital cardiac arrest: an observational study and propensity analysis. *Lancet.* 2008;372:554.
28. Mohamed I, Lau CT, Bolen MA, et al. Building a bridge to save a failing ventricle: radiologic evaluation of short- and long-term cardiac assist devices. *Radiographics.* 2015;35(2):327–356.
29. Mazzeffi M, Greenwood J, Tanaka K, et al. Bleeding, transfusion, and mortality on extracorporeal life support: ECLS working group on thrombosis and hemostasis. *Ann Thorac Surg.* 2016;101(2):682–689.
30. Parzy G, Daviet F, Persico N, et al. Prevalence and risk factors for thrombotic complications following venovenous extracorporeal membrane oxygenation: a CT scan study. *Crit Care Med.* 2020;48(2):192–199.
31. Cheng R, Hachamovitch R, Kittleson M, et al. Complications of extracorporeal membrane oxygenation for treatment of cardiogenic shock and cardiac arrest: a metaanalysis of 1,866 adult patients. *Ann Thorac Surg.* 2014;97(2):610–616.
32. Bisdas T, Beutel G, Warnecke G, et al. Vascular complications in patients undergoing femoral cannulation for extracorporeal membrane oxygenation support. *Ann Thorac Surg.* 2011;92(2):626–631.
33. Lee S, Chaturvedi A. Imaging adults on extracorporeal membrane oxygenation (ECMO). *Insights Imaging.* 2014;5(6):731–742.

34. Al-Ogaili Z, Foulner D, Passage J, et al. CT pulmonary angiography during veno-arterial extracorporeal membrane oxygenation in an adult. *J Med Imaging Radiat Oncol*. 2013;57(3):345–347.
35. Rao P, Khalpey Z, Smith R, Burkhoff D, Kociol RD. Venoarterial extracorporeal membrane oxygenation for cardiogenic shock and cardiac arrest. *Circ Heart Fail*. 2018;11(9):e004905.
36. McGovern L, Cosgrave J. Axial flow ventricular assist devices in cardiogenic shock complicating acute myocardial infarction [published online ahead of print, 2021 Apr 2]. *Heart*. 2021;. heartjnl-2020–318226.
37. Lemor A, Ya'qoub L, Basir MB. Mechanical circulatory support in acute myocardial infarction and cardiogenic shock. *Interv Cardiol Clin*. 2021;10(2):169–184.
38. Naidu SS. Novel percutaneous cardiac assist devices: the science of and indications for hemodynamic support. *Circulation*. 2011;123 (5):533–543.
39. Capdeville M, Smedira NG. Advances and future directions for mechanical circulatory support. *Anesthesiol Clin*. 2013;31(2):321–353.
40. Estep JD, Stainback RF, Little SH, Torre G, Zoghbi WA. The role of echocardiography and other imaging modalities in patients with left ventricular assist devices. *JACC Cardiovasc Imaging*. 2010;3(10):1049–1064.
41. Shroff GS, Ocazionez D, Akkanti B, et al. CT imaging of complications associated with continuous-flow left ventricular assist devices (LVADs). *Semin Ultrasound CT MR*. 2017;38(6):616–628.
42. Crow S, Chen D, Milano C, et al. Acquired von Willebrand syndrome in continuous-flow ventricular assist device recipients. *Ann Thorac Surg*. 2010;90(4):1263–1269. Discussion 1269.
43. Guha A, Eshelbrenner CL, Richards DM, Monsour Jr HP. Gastrointestinal bleeding after continuous-flow left ventricular device implantation: review of pathophysiology and management. *Methodist Debakey Cardiovasc J*. 2015;11(1):24–27.
44. Maniar S, Kondareddy S, Topkara VK. Left ventricular assist device-related infections: past, present and future. *Expert Rev Med Devices*. 2011;8(5):627–634.
45. Nienaber JJ, Kusne S, Riaz T, et al. Clinical manifestations and management of left ventricular assist device-associated infections. *Clin Infect Dis*. 2013;57(10):1438–1448.
46. Slaughter MS, Rogers JG, Milano CA, et al. Advanced heart failure treated with continuous-flow left ventricular assist device [published correction appears in N Engl J Med. 2018 Aug 16;379(7):697]. *N Engl J Med*. 2009;361(23):2241–2251.
47. Tam MC, Patel VN, Weinberg RL, et al. Diagnostic accuracy of FDG PET/CT in suspected LVAD infections: a case series, systematic review, and meta-analysis. *JACC Cardiovasc Imaging*. 2020;13(5):1191–1202.
48. Goldstein DJ, John R, Salerno C, et al. Algorithm for the diagnosis and management of suspected pump thrombus. *J Heart Lung Transpl*. 2013;32(7):667–670.
49. Eckman PM, John R. Bleeding and thrombosis in patients with continuous-flow ventricular assist devices. *Circulation*. 2012;125(24): 3038–3047.
50. Mellnick VM, Raptis DA, Raptis C, Bhalla S. Imaging of left ventricular device complications. *J Thorac Imaging*. 2013;28(2):W35–W41.
51. Uriel N, Morrison KA, Garan AR, et al. Development of a novel echocardiography ramp test for speed optimization and diagnosis of device thrombosis in continuous-flow left ventricular assist devices: the Columbia ramp study. *J Am Coll Cardiol*. 2012;60(18):1764–1775.
52. Kato TS, Schulze PC, Yang J, et al. Pre-operative and post-operative risk factors associated with neurologic complications in patients with advanced heart failure supported by a left ventricular assist device. *J Heart Lung Transpl*. 2012;31(1):1–8.
53. Rogers JG, Pagani FD, Tatooles AJ, et al. Intrapericardial left ventricular assist device for advanced heart failure. *N Engl J Med*. 2017; 376(5):451–460.
54. Potapov EV, Stepanenko A, Krabatsch T, Hetzer R. Managing long-term complications of left ventricular assist device therapy. *Curr Opin Cardiol*. 2011;26(3):237–244.
55. Yuzefpolskaya M, Uriel N, Chow DS, et al. Prevalence and timing of bend relief disconnection in patients supported by the late version HeartMate II left ventricular assist device. *J Heart Lung Transpl*. 2013;32(3):320–325.
56. Pak SW, Uriel N, Takayama H, et al. Prevalence of de novo aortic insufficiency during long-term support with left ventricular assist devices. *J Heart Lung Transpl*. 2010;29(10):1172–1176.
57. Dang NC, Topkara VK, Mercando M, et al. Right heart failure after left ventricular assist device implantation in patients with chronic congestive heart failure. *J Heart Lung Transpl*. 2006;25(1):1–6.
58. Slaughter MS, Pagani FD, Rogers JG, et al. Clinical management of continuous-flow left ventricular assist devices in advanced heart failure. *J Heart Lung Transpl*. 2010;29(4 Suppl.):S1–S39.
59. Takeda K, Takayama H, Colombo PC, et al. Late right heart failure during support with continuous-flow left ventricular assist devices adversely affects post-transplant outcome. *J Heart Lung Transpl*. 2015;34(5):667–674.
60. Remy-Jardin M, Delhaye D, Teisseire A, Hossein-Foucher C, Duhamel A, Remy J. MDCT of right ventricular function: impact of methodologic approach in estimation of right ventricular ejection fraction, part 2. *AJR Am J Roentgenol*. 2006;187(6):1605–1609.
61. Garan AR, Yuzefpolskaya M, Colombo PC, et al. Ventricular arrhythmias and implantable cardioverter-defibrillator therapy in patients with continuous-flow left ventricular assist devices: need for primary prevention? *J Am Coll Cardiol*. 2013;61(25):2542–2550.
62. Copeland JG, Smith RG, Arabia FA, et al. Cardiac replacement with a total artificial heart as a bridge to transplantation. *N Engl J Med*. 2004;351(9):859–867.

63. Roussel JC, Sénage T, Baron O, et al. CardioWest (Jarvik) total artificial heart: a single-center experience with 42 patients. *Ann Thorac Surg*. 2009;87(1):124–129.
64. Parker MS, Fahrner LJ, Deuell BP, et al. Total artificial heart implantation: clinical indications, expected postoperative imaging findings, and recognition of complications [published correction appears in AJR Am J Roentgenol. 2014 Jun;202(6):1396. de Groot, Patti Ann [corrected to de Groot, Patricia M]]. *AJR Am J Roentgenol*. 2014;202(3):W191–W201.
65. U.S. Food and Drug Administration website. SynCardia Systems Inc. CardioWest total artificial heart (TAH): directions for use. http://www.fda.gov/ohrms/dockets/ac/04/briefing/4029b1_final.pdf.

Chapter 14

Imaging Technologies and Virtual Planning for Congenital Heart Repairs

Nicholas A. Szugye[1], Ryan A. Moore[1,2] and David L.S. Morales[1,3]

[1]*Heart Institute, Cincinnati Children's Hospital Medical Center, Cincinnati, OH, United States,* [2]*Department of Pediatrics, University of Cincinnati College of Medicine, Cincinnati, OH, United States,* [3]*Department of Surgery, University of Cincinnati College of Medicine, Cincinnati, OH, United States*

Chapter Outline

Introduction

Improvements in the management of congenital heart disease (CHD) have led to improved long-term survival and an increased number of CHD patients.[1] Currently the number of adults living with CHD exceeds the number of children.[2] Furthermore, CHD survivors with increasing complexity of disease need noninvasive imaging evaluation for both lifelong surveillance and preoperative decision making. Cardiac magnetic resonance imaging (CMR) and cardiac computed tomography (CCT) are two cross-sectional imaging (CSI) modalities that have been utilized more extensively over the past decade.[3] This chapter will discuss the basics of advanced cardiac imaging, techniques for three-dimensional (3D) visualization of imaging data, and review recent innovations in surgical simulation.

Cross-sectional imaging overview

Although transthoracic echocardiography is the first-line imaging modality for CHD surveillance, CSI offers several imaging resolution and 3D reformatting benefits beyond conventional echocardiography that are advantageous for preoperative planning. CSI allows determination of relationships between cardiac and extracardiac structures. The concepts of spatial, temporal, and contrast resolutions will not be covered in extensive detail; however, will be frequently noted in the forthcoming CCT and CMR sections and should be clearly stated as the primary reason for the advancement in the quality of CSI-derived 3D imaging over the past decade.[4] Technological achievements in each type of resolution have led to improvement in submillimeter isotropic voxel acquisitions (i.e., 0.5–0.625 mm voxels that are equal on all sides; x-, y-, and z-axes), enhanced 3D multiplanar reformatting algorithms, and numerous 3D rendering and medical modeling solutions.[5,6] These advances have improved the processing of CSI data into highly accurate 3D renders that allow for visualization and manipulation of the true anatomic 3D form needed for advanced simulation.[7] CCT has been the primary imaging modality for 3D modeling and rendering due to rapid advancements in resolution techniques;

Advances in Cardiovascular Technology. DOI: https://doi.org/10.1016/B978-0-12-816861-5.00010-1

however, newer 3D-CMR sequences have started to rival CCT multiplanar reformats. Each modality will be described below related to their impact in cardiac imaging.

Role of cardiac computed tomography

CCT uses a tube construction with an X-ray emitter and detector that travel circumferentially along a gantry rotation. X-ray attenuation values recorded along the rotation are reconstructed to produce an image. Initial computed tomography (CT) scanners were unsuitable for cardiac imaging due to low temporal resolution. However, current CCT technology with multidetector scanners and dual-source CT scanners allows for rapid acquisition a full volumetric data set within a single heartbeat.[8,9] Electrocardiogram (ECG)-triggering enables data acquisition during a targeted portion of the cardiac cycle (diastole or systole), limiting artifacts from cardiac motion to produce high-quality images. For patients with CMR contraindications, a retrospectively gated CCT can produce datasets at multiple cardiac phases. This enables a functional assessment of ventricular volumes and ejection fraction, though at lower temporal resolution than CMR.[8] CCT has largely replaced cardiac catheterization for evaluation of coronary artery origins and courses.[10]

Several limitations should be noted for CCT. First, intravenous iodinated contrast is necessary to provide distinction between tissue and blood pool due to poor intrinsic contrast resolution. A well-recognized CCT health risk is the need for ionizing radiation exposure which imparts risk of malignancy in a dose-dependent manner.[11] Previous CCT technology reported radiation dosages up to 18 millisieverts (mSV) which is approximately twice the dose of a standard cardiac catheterization. However, the required radiation dose for CCT imaging has decreased over time due to multiple technological advancements. A recent multiinstitutional prospective registry assessed clinical performance of CCT in CHD patients with a median effective radiation dose of 1 mSV, demonstrating a significant decrease in radiation dosage for newer scanners.[12] It is worth mentioning that retrospectively-gated CCT (i.e., scanning over multiple cardiac cycles) still carries a higher ionizing radiation exposure than other types of CCT, typically in the range 2.5–8 mSv;[13] however, this is often not necessary for the vast majority of 3D virtual surgical planning cases where well-timed, contrast-enhanced CT angiograms are more than adequate for 3D rendering.

Role of cardiac magnetic resonance

Functional and volumetric flow CMR data is obtained from two-dimensional (2D) static and cine images in multiple cardiac imaging planes and phase contrast sequences. Cine imaging of contiguous slices through the ventricles are post-processed to quantify the ventricular end-diastolic and end-systolic volumes, stroke volumes, and ejection fractions. Phase contrast images produce a quantitative assessment of blood flow velocity, directionality, and volume. Phase contrast allows for quantification of valve regurgitation, systemic blood flow, and pulmonary blood flow, which can then be used to calculate intra- and extracardiac shunting.[14–16] Myocardial tissue characteristics (tissue edema, hyperemia, necrosis, and fibrosis) can also be assessed through various CMR sequences.[17] All of this highly valuable diagnostic data has led CMR to becoming a critical tool for the physiologic surveillance of cardiac disease and the need for surgical intervention. However, the rapid advancement in the temporal, spatial, and contrast resolution of several newer CMR 3D acquisitions have led to its potential to become the primary CSI for imaging-based surgical planning.

Contrast-enhanced 3D acquisitions during a CMR study have become part of the standard imaging protocol for cardiac patients and are nearly identical in resolution to CCT. Static 3D sequences acquire volume element (i.e., voxel) data (each voxel has equal x-, y-, and z-dimension) that can be used for multiplanar reconstruction and 3D reconstruction.[18] 3D sequences suitable for 3D reconstruction often require intravenous gadolinium contrast, although newer non-contrast enhanced acquisitions can obtain similar anatomic accuracy without the need for a contrast agent.[18]

The main limitations of CMR imaging stem from the need to acquire and compile data over multiple heartbeats, so it is necessary to account for both cardiac and respiratory motion. Respiratory motion is accounted for with breath-holds (patient holds breath for 6–12 s), respiratory triggering (acquiring data only during a particular respiratory phase),[19] or respiratory gating (acquiring data only within a narrow range of diaphragmatic excursion).[16] Cardiac motion is accounted for by ECG-gating. Multiple sequences are acquired to obtain functional and anatomic data, so studies last 30–90 min. The duration of scan time and need for breath-holds limits the use of CMR in younger patients, patients with developmental delay, claustrophobia, and patients with respiratory disease. Given the nuance of CHD anatomy, high-quality imaging study requires a cardiac imaging specialist directly involved at the time of image acquisition. Lastly, patients with metallic implants or devices (i.e., pacemakers) can cause susceptibility artifacts and may have contraindications to magnetic fields.

Imaging segmentation

Imaging segmentation is a necessary step to convert raw 3D imaging data into the 3D render used for the production of both physical 3D models and virtual models.[20] Fig. 14.1 is a schematic demonstrating a simplified workflow where segmentation serves as the fundamental image processing step to produce a 3D render.[21] The 3D models that may be produced include, but are not limited to, 3D printed, virtual reality (VR), and augmented reality (AR) models. The anatomic accuracy of 3D models is reliant on detailed and careful imaging segmentation and the tools that are available to optimize the models.[22]

Segmentation is the process of partitioning a digital image into multiple component "segments." In medical imaging, these components are typically anatomic features (i.e., organs) or pathologic features (i.e., tumor). In cardiac imaging, the complex cardiac structures are partitioned into individual segments such as systemic veins, atria, ventricles, pulmonary arteries, pulmonary veins, systemic arteries, and myocardium. Other intrathoracic structures such as bone, trachea, and lung are also segmented to visualize anatomic relationships.

The most common type of segmentation technique in cardiac imaging is threshold-based segmentation, where a distinct range of signal intensity, measured in Hounsfield units (HU) is used to separate anatomic features.[23] Structures with HU within the preselected range are first identified and those within the same region are aggregated to create a segment. These components can be further subdivided as needed and further modifications can be made manually.

In complex CHD, anatomic segmentation allows cardiologists and cardiac surgeons to share a common conception of the anatomic model to appropriate define heart defects and discuss surgical repairs. The use of 3D visualization has been shown to enhance physician communication, improve caregiver counseling, and aide in presurgical planning and decision-making.[20]

CCT versus CMR for segmentation

Image data for segmentation is most commonly taken from either CCT or CMR 3D acquisitions. In general, contrast-enhanced CCT allows for high contrast resolution between the blood pool and surrounding tissue, giving overall better

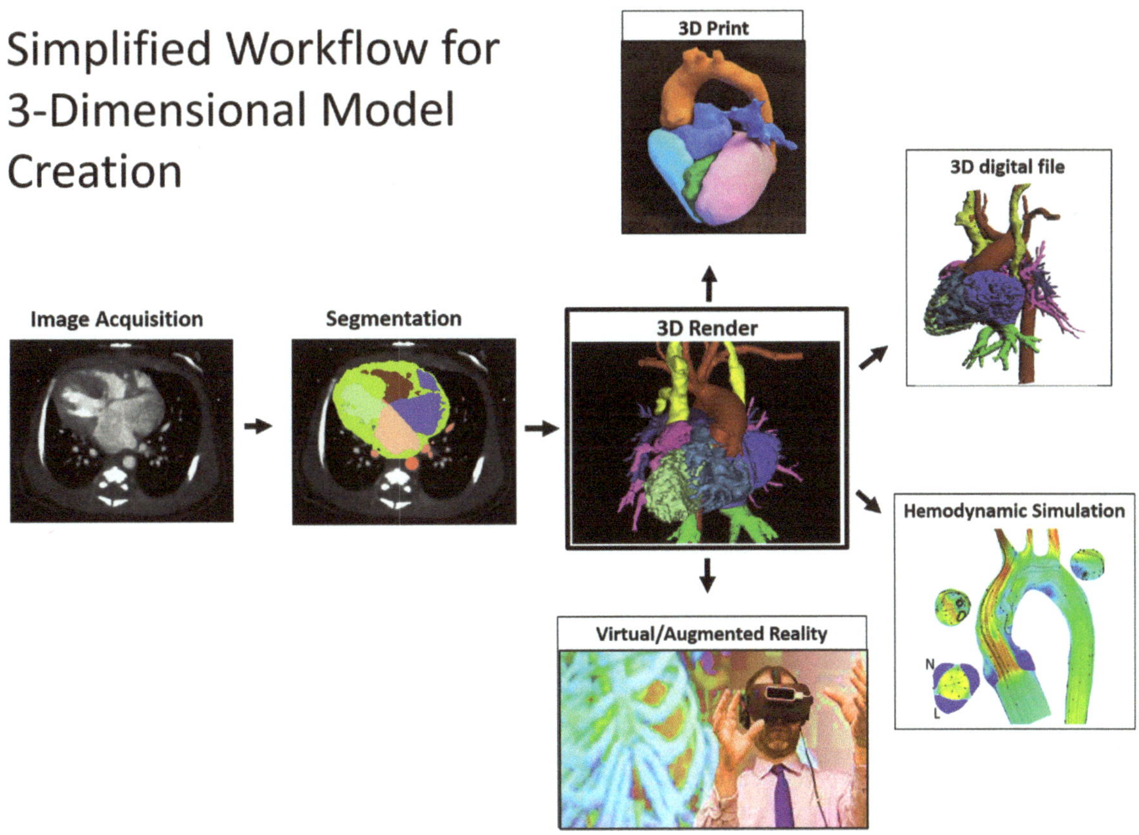

FIGURE 14.1 Simplified workflow of 3D model creation: Acquired CCT and CMR images undergo segmentation prior to the creation of a 3D render. The 3D render can be used to produce multiple 3D model types.

edge definition than CMR. The enhanced blood pool signal allows for easier anatomic segmentation which allows for high-quality volume renders, 3D surface renders, and 3D printed models.[24] A limitation of CCT for segmentation is that contrast is dependent on the timing between contrast administration and image acquisition, so either left or right heart may be more densely opacified. CMR can also be used for blood pool segmentation with good overall contrast,[25] however the signal intensity of the blood pool is typically less homogenous, making segmentation more difficult. For certain applications such as coronary imaging, specific CMR protocols can achieve high similarity to CCT in image resolution.[26]

Image rendering: surface versus volume

Image rendering refers to the transformation of source 2D imaging data (e.g., CCT or CMR) into 3D. Before discussing the various applications of 3D rendered models, it is important to highlight two distinct classes of 3D rendering techniques: surface-based and volume-based. Surface-based rendering is one of the more common techniques used in interactive digital model creation used in 3D printing, virtual surgical planning, and virtual and AR-based experiences. This type of rendering produces a dataset in the form of a surface "mesh" surrounding the outer surface of the 3D object. Surface rending is advantageous because it allows for easier correction of imaging artifacts and manipulation of the data in the segmentation process. The surface mesh can be simplified without loss of important anatomic detail. The primary limitations of surface rendered models is the substantial time and care it takes to process a 3D dataset to maintain anatomic accuracy and detail. At this time, it is critically dependent on the detailed manual segmentation processes described in the above section. With advancement in artificial intelligence-enhanced autosegmentation algorithms, there is hope that this highly effective rendering technique can rapidly produce high-performance models commonly used for 3D printing, VR, and AR.[27]

Volume rendering is a powerful image visualization technique where a 3D data set is displayed as a 2D projection. It offers the ability to visualize the multiple tissue subtypes in high definition making it a preferred method for displaying anatomic detail.[28] Volume rendering is available instantaneously in nearly all medical imaging pictures and archiving systems (PACS), with additional platforms available to allow for 3D visualization optimization through cropping, lighting (i.e., photorealism), and 3D overlay with maximum intensity projections that can mimic fluoroscopy views that are common in cardiac catheterization.[29] Volume rendering is limited by an inability to correct artifacts and requirement of higher-quality images. Volume rendering has been primarily used for 3D visualization since the datasets are too large to be 3D printed or easily manipulated for virtual surgical planning (Fig. 14.2).

3D printing

3D printing, or rapid-prototyping, is the construction of a 3D object where polymer material is deposited and cured to create a physical 3D printed object. 3D printing of heart models has long been the mainstay of 3D visualization and is

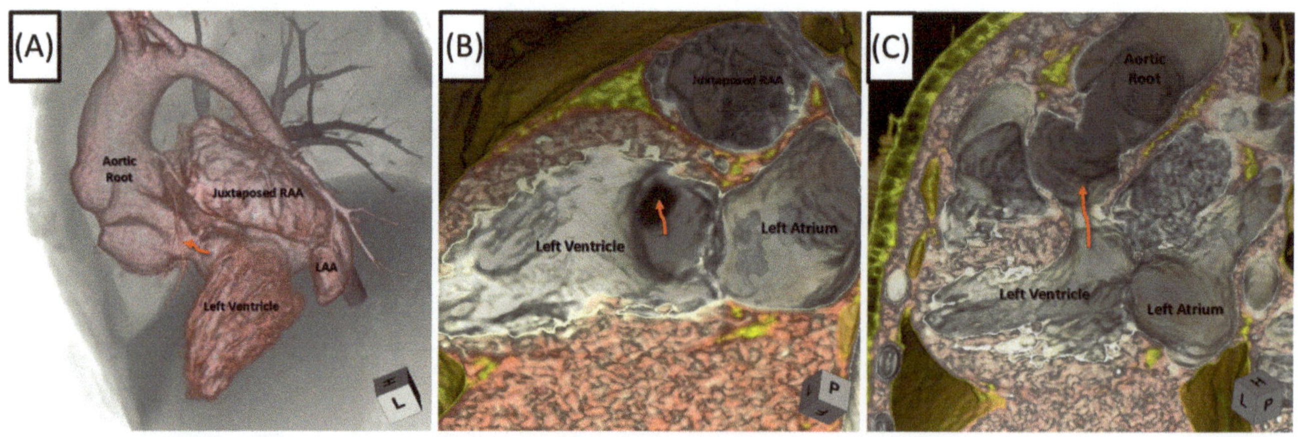

FIGURE 14.2 Volume rendering: This 3-year-old child with history of double outlet right ventricle, pulmonary atresia, and noncommitted ventricular septal defect with juxtaposed right atrial appendage (RAA) status post biventricular repair with left ventricle-to-aortic baffle and right ventricle-to-pulmonary artery conduit subsequently developed left ventricle-to-aortic baffle obstruction. (A) An endocast reconstruction with the right ventricle and pulmonary trunk removed and overlaid over a fluoroscopy-like projection demonstrates the baffle narrowing (*red arrow*). (B and C) Virtual dissection reconstructions in a two-chamber and three-chamber cut plane additionally demonstrates this focal narrowing (*red arrow*) as the baffle passes between the crest of the interventricular septum and ventriculoinfundibular fold. *LAA*, Left atrial appendage. *Courtesy Justin T. Tretter, Cincinnati Children's Hospital Medical Center.*

preferred by many surgeons because of the tactile nature of a 3D printed model.[30–32] Prior works have shown that the review of a high-quality 3D printing before an operation is useful for complex surgeries.[33,34] 3D printing has been transformative in some centers, enabling the planning of complex operations that may have not been attempted without the 3D printed model.[35] Several groups have recently shown that soft material 3D printed hearts of patient-specific data can be used for hands-on practice of operations in a 3D print surgical suite.[36–38]

3D printing has been used in preparation of a multitude of different cardiac surgeries. A common 3D printing application is planning double-outlet right ventricle repair since the complex relationship between the ventricular septal defect (VSD) and outflow tracts can be well visualized in a 3D printed model.[39,40] Other examples of 3D printing applications include major aortopulmonary collateral unifocalization, pulmonary artery pseudo-aneurysm, and cardiac tumor debulking.[41–43] 3D printing has also been used to practice catheter-based interventions on patient-specific models for pulmonary valve replacement and patent ductus arteriosus stenting.[44,45] Apart from its role in surgical planning, 3D printing has proven useful in medical education and training, as well as, assisting doctor–patient communication.[46–48]

3D printing can be accomplished via either blood pool printing or myocardial printing. Blood pool printing involves the segmentation of cardiac chambers and extracardiac vessels prior to model the intravascular space. Alternatively, the myocardium can be printed, leaving a hollow intravascular space. Each method has unique applications depending on the intended operation or procedure[49] (Fig. 14.3).

3D printing is limited by the costs of the printer (ranging from consumer grade \$500–\$10,000) to production grade (\$15,000–\$500,000), printing material (\$50–\$500 per print), maintenance costs, and time required by imaging specialists and expert technicians for model production. Depending on print volume and material, 3D printing can take 2–12 h of active time (for preprocessing and postprocessing) and up to an additional 72 h for printing, curing, and cleaning.[50] Another significant disadvantage of 3D printing for complex surgical planning is that only a single surgical strategy (incision, patch, conduit placements) can be attempted per 3D print, requiring a new 3D print to be made for each new strategy. Additionally, one does not get an understanding how the strategy functions in relationship to surrounding structures such as the chest wall, diaphragm, innominate vein, extracardiac pulmonary veins, and lungs. Future direction in 3D printing include the use of lower cost 3D print material, dynamic moving heart models, and investigation of clinical outcomes prediction.[47]

3D digital models

Given the limitations of cost and personnel, 3D printing has not been universally adopted. Predating the widespread use of 3D printing, 3D digital models viewed on a computer screen were used to examine anatomy.[51–53] 3D digital models viewed on a computer screen have been shown to be equally valuable as a 3D printed heart[30] with the additional advantages of decreased cost, decreased time, and capacity to produce multiple iterations of the same model.

There have been several unique applications of evaluating 3D digital files. Computer screen analysis of 3D digital files has been used to make the SynCardia Total Artificial Heart available to patients whose weight and height fit below the standard body surface area criteria (Fig. 14.4A).[54,55] Likewise, a "fit test" was developed to test out the location of an intracorporeal ventricular assist device (VAD) in smaller patients or those with CHD (Fig. 14.4B).[56] This type of analysis has also been used to evaluate donor-recipient heart size compatibility for heart transplantation using 3D digital models for total cardiac volume (Fig. 14.4C).[57–60]

In addition to patient-specific 3D digital models, defect-specific 3D digital heart models are used frequently for patient and trainee education. The Heartpedia mobile application, available for mobile devices, uses labeled 3D digital heart models to demonstrate anatomic abnormalities as well as surgical repairs in an interactive format (Fig. 14.4D).[61]

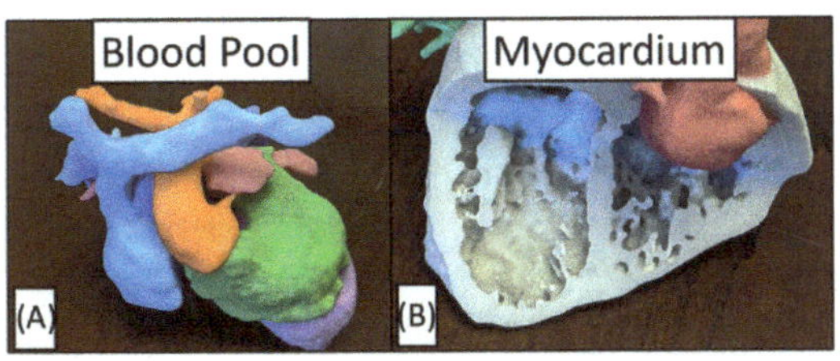

FIGURE 14.3 3D printed models: (A) A Blood Pool model with multiple colors comprising the intravascular volumes and cardiac chamber volumes. (B) A myocardial 3D render is printed in gray outlining trabeculations, papillary muscles, and intraventricular septum.

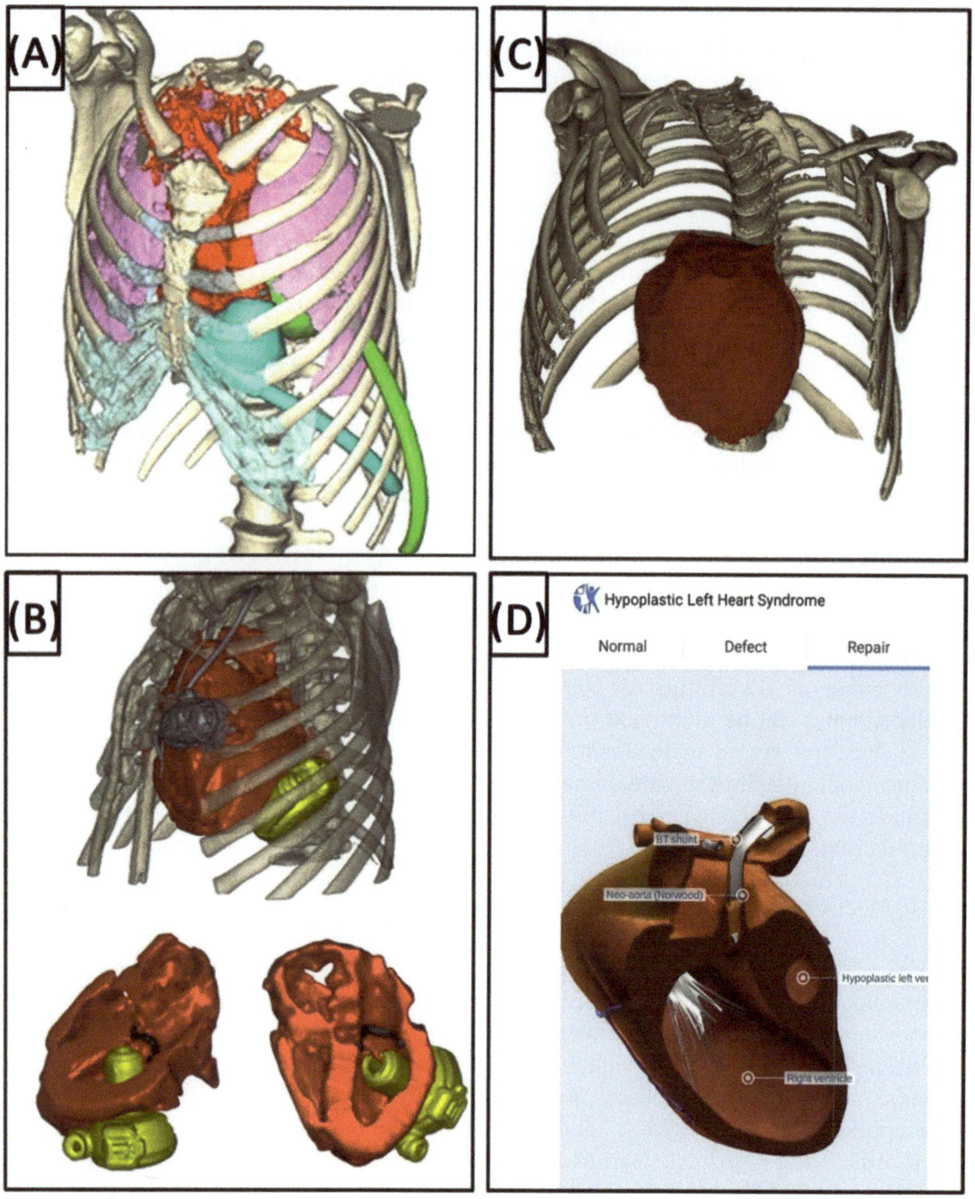

FIGURE 14.4 3D digital models: (A) Virtual fit testing of the SynCardia Total Artificial Heart enables device use in patients with smaller body size or atypical thoracic shape. (B) 3D digital models enable the evaluation of multiple ventricular assist device positions. (C) Donor-Recipient size matching in heart transplantation can be accomplished by comparison of donor and recipient 3D digital models, increasing donor availability. (D) Screenshot from the HeartPedia digital application with a labeled 3D model of the surgical repair of hypoplastic left heart syndrome.

Virtual reality

VR is an artificial digital environment through which a user is fully immersed and can manipulate objects by touch. VR offers the advantage of 3D digital models viewed on a computer screen because it gives the user depth perception and increased realism. Also, VR models enable better user-specific manipulation of objects and the user can move their own body to navigate within the model.

VR models offer an advantages of viewing from the "surgeon's perspective" as well as augmented by views not able to be seen in the operating room. VR offers the ability to view proximity of the operative field to "unseen" structures. VR model can therefore create a better shared mental model between imager and surgeon and account for the cardiac structures and major vessels in relationship to the surrounding mediastinal/chest structures. Kim et al. developed a novel VR medical image display system for group discussion where both leader and participants wore VR headsets enabling a realistic and interactive visualization of CHD anatomy via interior viewing, rotating, and zooming in and out. In this study, discussions aided by VR produced higher diagnostic accuracy compared with conventional imaging.[62]

Beyond visualization, VR possesses several important advantages over 3D printing for surgical simulation. (1) Whereas only a single surgical approach can be attempted for a 3D print, VR enables multiple surgical approaches to be attempted. VR models can be edited and revisualized as often as needed. As an early example, Sorensen et al. showed that incision simulation can be performed in preoperative planning for complex CHD, evaluating different surgical strategies.[63] (2) VR models enable an appreciation between a planned cardiac repair and the surrounding structures. For example, a right ventricular to pulmonary artery (RV-PA) conduit's relationship to the anterior chest wall can be evaluated in VR, but this is not possible with a 3D cardiac print. (3) VR can be implemented for less cost than 3D printing when accounting for initial purchase cost and maintenance costs.[64]

Our group has recently created a 3D virtual surgical suite where CCT or CMR 3D renders are downloaded into a VR system powered by video game-engine technology. Leveraging game engine technology for this purpose enables a surgeon to virtually manipulate myocardial tissue by cutting, sewing, and folding tissue. Additionally, surgical patches, baffles, and conduits can be sewed into the heart to practice a proposed surgical repair. Common surgical components that can be planned in VR include VAD and total artificial heart positioning, left ventricular outflow tract baffle, intracardiac venous baffling. This has been most useful for complex repairs such as biventricular repairs of heterotaxy syndrome, truncus arteriosus with atrioventricular septal defect, and Fontan to biventricular conversion. Fig. 14.5 shows several examples of the VR surgical suite experience. The VR surgical suite enables binocular visualization to enable the user to appreciate depth perception. An "X-ray ball" tool can be placed over anatomy to make it temporarily transparent to visualize internal structures. We also have the capability to remove different parts of the heart and enlarge it so that the surgeon can enter the heart and gain a perspective of a VSD's relationship to the aorta or of the pathway of a complex baffle that he cannot do even in the operating room. Conduits, patches, and baffles can be placed into the virtual environment to simulate patching and cutting.

In our attempt to share our technology and create a greater level of collaborations when consulting on difficult cases from other surgical programs, we have created a multiuser platform where we can virtually review patient virtual model to plan complex repairs with surgeons from other programs in the US and internationally. The platform will even have the capability to translate multiple languages so our collaborations will not be limited by language.

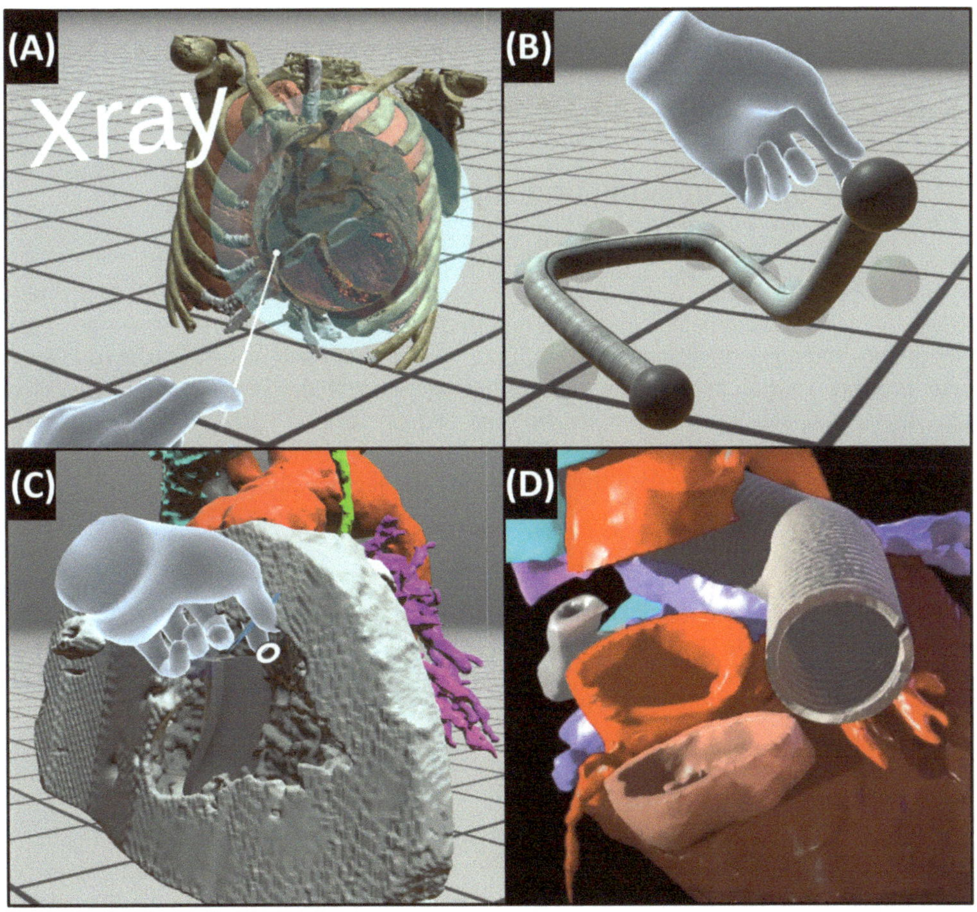

FIGURE 14.5 Virtual reality (VR) surgical suite: (A) Depth perceptive anatomic visualization is aided by an "X-Ray Ball" to swiftly cut through curvilinear anatomic planes. (B) 3D objects such as conduits can be bent in the 3D space by tactile manipulation. (C) Surgical patches are placed into the VSD planes to simulate surgical VSD closure and baffling. (D) A conduit is placed onto a transected pulmonary artery.

Lastly, VR offers several opportunities to enhance medical education. VR has been shown to be helpful in teaching and evaluating surgical planning.[65–67] VR has been used as a way to catalog and compare anatomy of cadaveric specimens.[68] Kiraly et al. have digitized a set of cadaveric specimens creating a virtual museum of digital heart defects.[69]

Augmented reality and mixed reality

In contrast to VR where the user experience is completely immersive, AR and mixed reality (MR) provide an interrelationship between both virtual and real world objects. In AR, virtual objects overlay the real-world environment with spatial registration, creating the impression that virtual and real objects coexist indistinguishably in the same space. In mixed reality (MR), virtual objects are not only overlaid, but also anchored onto real-world objects so a user can interact with both virtual and real objects (i.e., the virtual objects will "follow" the real world objects to which they are anchored). Stated differently, physical objects are integrated into the virtual world and can dynamically interact in real time. The first report of CHD surgeries using MR showed that MR help surgeons to understand complex CHD.[70] A recent advancement in MR is Echopixel's True3D (EchoPixel Inc., Santa Clara, CA, USA), an intraoperative viewing system where bookmarked 3D CT images are projected over the operating table to appear to the surgeon as a hologram over the patient.[71] In future applications, the MR environment of imaging data may be overlaid a surgical field reacting to the events of the surgery in real time.

Hemodynamic simulation

3D renders of cardiovascular anatomy can be used to simulate blood flow though the intravascular space. Patient-specific models can be used to evaluate postoperative fluid dynamics via computational fluid dynamics for hemodynamic simulation.[72] Fluid–structure interaction simulations have been used to study the influence of aortic root position on wall shear stress.[73] In postoperative Fontan patients, computation fluid dynamics can be used to simulate the efficiency of patient-specific cavopulmonary circulations in various levels of cardiac output.[74] Eventually, it may be possible to test the hemodynamic efficiency of an intended cardiac repair performed in VR, enabling the comparison of multiple surgical approaches. This would be particularly helpful to anticipate the hemodynamic efficiency for complex biventricular repairs, intracardiac baffles, or extracardiac rerouting of venous structures as well as optimizing the placement of VADs in a particular ventricle.

Addressing limitations of advanced 3D imaging and surgical simulation

Automatic segmentation

Before more widespread implementation of advanced 3D visualizations and surgical simulation is possible, the field must address the various limitations. A major impediment is the time required for preprocessing prior to the construction of a 3D digital model. The length of time needed for segmentation limits its use in all patients and takes time away from other important tasks for an imaging specialist. A recent review has shown that the segmentation process of the whole heart takes, on average, between 2 and 3 h to complete, but can take as long as 120 h for highly detailed models.[23,75] Automatic segmentation via artificial intelligence deep learning models and the production of more efficient segmentation tools (semiautomatic tools) can address the time requirement for segmentation.[27,76,77]

Affordability

Much effort has been placed on making advanced 3D visualization more cost-effective. Newer 3D printing polymers have been developed that enable 3D printing at lower cost with similar performance.[78,79] Virtual dissection can be accomplished with open source software.[29] Prior work in medical education and surgical training has shown that VR can be implemented at low cost through judicious selection of hardware and software.[80–83]

Conclusion

CSI technology has advanced to enable high-fidelity 3D renders suitable for many surgical simulation applications. Though 3D printing has long been the standard for 3D surgical planning, 3D digital models and VR methodologies are burgeoning avenues of current development. As in the authors' practice, we believe that 3D printing for complex surgical planning will be supplanted by 3D surgical planning in the virtual space. Anticipated developments include cost reduction, automation of image procession, and prospective comparison of surgical strategies with hemodynamic simulation.

References

1. Stout KK, Daniels CJ, Aboulhosn JA, et al. AHA/ACC Guideline for the management of adults with congenital heart disease: executive summary: a report of the American College of Cardiology/American Heart Association Task Force on Clinical Practice Guidelines. *J Am Coll Cardiol*. 2018;2019(73):1494–1563.
2. Stout KK, Daniels CJ, Aboulhosn JA, et al. AHA/ACC Guideline for the management of adults with congenital heart disease: a report of the American College of Cardiology/American Heart Association Task Force on Clinical Practice Guidelines. *J Am Coll Cardiol*. 2018;2019(73): e81–e192.
3. Anderson S, Figueroa J, McCracken CE, et al. Factors influencing temporal trends in pediatric inpatient imaging utilization. *J Am Soc Echocardiogr*. 2020;33:1517–1525.
4. Lin E, Alessio A. What are the basic concepts of temporal, contrast, and spatial resolution in cardiac CT? *J Cardiovasc Comput Tomogr*. 2009;3:403–408.
5. Sorensen TS, Beerbaum P, Korperich H, Pedersen EM. Three-dimensional, isotropic MRI: a unified approach to quantification and visualization in congenital heart disease. *Int J Cardiovasc Imaging*. 2005;21:283–292.
6. Sorensen TS, Korperich H, Greil GF, et al. Operator-independent isotropic three-dimensional magnetic resonance imaging for morphology in congenital heart disease: a validation study. *Circulation*. 2004;110:163–169.
7. Okuda S, Kikinis R, Geva T, Chung T, Dumanil H, Powell AJ. 3D-shaded surface rendering of gadolinium-enhanced MR angiography in congenital heart disease. *Pediatr Radiol*. 2000;30:540–545.
8. Han BK, Rigsby CK, Leipsic J, et al. Computed tomography imaging in patients with congenital heart disease, Part 2: Technical recommendations. An expert consensus document of the Society of Cardiovascular Computed Tomography (SCCT): endorsed by the Society of Pediatric Radiology (SPR) and the North American Society of Cardiac Imaging (NASCI). *J Cardiovasc Comput Tomogr*. 2015;9:493–513.
9. Han BK, Rigsby CK, Hlavacek A, et al. Computed tomography imaging in patients with congenital heart disease, Part I: Rationale and utility. An expert consensus document of the Society of Cardiovascular Computed Tomography (SCCT): endorsed by the Society of Pediatric Radiology (SPR) and the North American Society of Cardiac Imaging (NASCI). *J Cardiovasc Comput Tomogr*. 2015;9:475–492.
10. Frommelt P, Lopez L, Dimas VV, et al. Recommendations for multimodality assessment of congenital coronary anomalies: a guide from the American Society of Echocardiography: developed in collaboration with the Society for Cardiovascular Angiography and Interventions, Japanese Society of Echocardiography, and Society for Cardiovascular Magnetic Resonance. *J Am Soc Echocardiogr*. 2020;33:259–294.
11. Einstein AJ, Moser KW, Thompson RC, Cerqueira MD, Henzlova MJ. Radiation dose to patients from cardiac diagnostic imaging. *Circulation*. 2007;116:1290–1305.
12. De Oliveira Nunes M, Witt DR, Casey SA, et al. Multi-institution assessment of the use and risk of cardiovascular computed tomography in pediatric patients with congenital heart disease. *J Cardiovasc Comput Tomogr*. 2021.
13. Schmermund A, Marwan M, Hausleiter J, et al. Declining radiation dose of coronary computed tomography angiography: German cardiac CT registry experience 2009–2014. *Clin Res Cardiol*. 2017;106:905–912.
14. Ridgway JP. Cardiovascular magnetic resonance physics for clinicians: Part I. *J Cardiovasc Magn Reson*. 2010;12:71.
15. Biglands JD, Radjenovic A, Ridgway JP. Cardiovascular magnetic resonance physics for clinicians: Part II. *J Cardiovasc Magn Reson*. 2012;14:66.
16. Fratz S, Chung T, Greil GF, et al. Guidelines and protocols for cardiovascular magnetic resonance in children and adults with congenital heart disease: SCMR expert consensus group on congenital heart disease. *J Cardiovasc Magn Reson*. 2013;15:51.
17. Haaf P, Garg P, Messroghli DR, Broadbent DA, Greenwood JP, Plein S. Cardiac T1 mapping and extracellular volume (ECV) in clinical practice: a comprehensive review. *J Cardiovasc Magn Reson*. 2016;18:89.
18. Kourtidou S, Jones MR, Moore RA, et al. mDixon ECG-gated 3-dimensional cardiovascular magnetic resonance angiography in patients with congenital cardiovascular disease. *J Cardiovasc Magn Reson*. 2019;21:52.
19. Pednekar AS, Wang H, Flamm S, Cheong BY, Muthupillai R. Two-center clinical validation and quantitative assessment of respiratory triggered retrospectively cardiac gated balanced-SSFP cine cardiovascular magnetic resonance imaging in adults. *J Cardiovasc Magn Reson*. 2018;20:44.
20. Goo HW, Park SJ, Yoo SJ. Advanced medical use of three-dimensional imaging in congenital heart disease: augmented reality, mixed reality, virtual reality, and three-dimensional printing. *Korean J Radiol*. 2020;21:133–145.
21. Maier A, Syben C, Lasser T, Riess C. A gentle introduction to deep learning in medical image processing. *Z Med Phys*. 2019;29:86–101.
22. Lau IWW, Liu D, Xu L, Fan Z, Sun Z. Clinical value of patient-specific three-dimensional printing of congenital heart disease: quantitative and qualitative assessments. *PLoS One*. 2018;13:e0194333.
23. Byrne N, Velasco Forte M, Tandon A, Valverde I, Hussain T. A systematic review of image segmentation methodology, used in the additive manufacture of patient-specific 3D printed models of the cardiovascular system. *JRSM Cardiovasc Dis*. 2016;5:2048004016645467.
24. Moore RA, Riggs KW, Kourtidou S, et al. Three-dimensional printing and virtual surgery for congenital heart procedural planning. *Birth Defects Res*. 2018;110:1082–1090.

25. Farooqi KM, Lengua CG, Weinberg AD, Nielsen JC, Sanz J. Blood pool segmentation results in superior virtual cardiac models than myocardial segmentation for 3D printing. *Pediatr Cardiol.* 2016;37:1028–1036.
26. Hajhosseiny R, Rashid I, Bustin A, et al. Clinical comparison of sub-mm high-resolution non-contrast coronary CMR angiography against coronary CT angiography in patients with low-intermediate risk of coronary artery disease: a single center trial. *J Cardiovasc Magn Reson.* 2021;23:57.
27. Zhuang X. Challenges and methodologies of fully automatic whole heart segmentation: a review. *J Healthc Eng.* 2013;4:371–408.
28. Tretter JT, Gupta SK, Izawa Y, Nishii T, Mori S. Virtual dissection: emerging as the gold standard of analyzing living heart anatomy. *J Cardiovasc Dev Dis.* 2020;7.
29. Gupta SK, Spicer DE, Anderson RH. A new low-cost method of virtual cardiac dissection of computed tomographic datasets. *Ann Pediatr Cardiol.* 2019;12:110–116.
30. Batteux C, Haidar MA, Bonnet D. 3D-printed models for surgical planning in complex congenital heart diseases: a systematic review. *Front Pediatr.* 2019;7:23.
31. Cantinotti M, Valverde I, Kutty S. Three-dimensional printed models in congenital heart disease. *Int J Cardiovasc Imaging.* 2017;33:137–144.
32. Hermsen JL, Roldan-Alzate A, Anagnostopoulos PV. Three-dimensional printing in congenital heart disease. *J Thorac Dis.* 2020;12:1194–1203.
33. Vettukattil JJ. Three dimensional echocardiography in congenital heart disease. *Heart.* 2012;98:79–88.
34. Vettukattil JJ, Mohammad Nijres B, Gosnell JM, Samuel BP, Haw MP. Three-dimensional printing for surgical planning in complex congenital heart disease. *J Card Surg.* 2019;34:1363–1369.
35. Garcia A. OpHeart commentary: three-dimensional printing for surgical planning in complex congenital heart disease. *J Card Surg.* 2019;34:753.
36. Hussein N, Honjo O, Haller C, et al. Quantitative assessment of technical performance during hands-on surgical training of the arterial switch operation using 3-dimensional printed heart models. *J Thorac Cardiovasc Surg.* 2020;160:1035–1042.
37. Yoo SJ, Spray T, Austin 3rd EH, Yun TJ, van Arsdell GS. Hands-on surgical training of congenital heart surgery using 3-dimensional print models. *J Thorac Cardiovasc Surg.* 2017;153:1530–1540.
38. Yoo SJ, Hussein N, Peel B, et al. 3D modeling and printing in congenital heart surgery: entering the stage of maturation. *Front Pediatr.* 2021;9:621672.
39. Yoo SJ, van Arsdell GS. 3D printing in surgical management of double outlet right ventricle. *Front Pediatr.* 2017;5:289.
40. Bhatla P, Tretter JT, Chikkabyrappa S, Chakravarti S, Mosca RS. Surgical planning for a complex double-outlet right ventricle using 3D printing. *Echocardiography.* 2017;34:802–804.
41. Elmaghraby A, Ghareep AN, Alkuwari M, Hijazi ZM, Salustri A. Patient-specific 3-dimensional printing of tetralogy of fallot with major aortopulmonary collaterals. *JACC: Case Rep.* 2019;1:535–537.
42. Riggs KW, Dsouza G, Broderick JT, Moore RA, Morales DLS. 3D-printed models optimize preoperative planning for pediatric cardiac tumor debulking. *Transl Pediatr.* 2018;7:196–202.
43. Zhu Y, Zhang XE, Li Q, Yao H. Three-dimensional printing in a patient with pulmonary artery pseudoaneurysm and complex congenital heart disease—a case report. *Clin Case Rep.* 2020;8:2107–2110.
44. Chamberlain RC, Ezekian JE, Sturgeon GM, Barker PCA, Hill KD, Fleming GA. Preprocedural three-dimensional planning aids in transcatheter ductal stent placement: a single-center experience. *Catheter Cardiovasc Interv.* 2020;95:1141–1148.
45. Poterucha JT, Foley TA, Taggart NW. Percutaneous pulmonary valve implantation in a native outflow tract: 3-dimensional DynaCT rotational angiographic reconstruction and 3-dimensional printed model. *JACC Cardiovasc Interv.* 2014;7:e151–e152.
46. Su W, Xiao Y, He S, Huang P, Deng X. Three-dimensional printing models in congenital heart disease education for medical students: a controlled comparative study. *BMC Med Educ.* 2018;18:178.
47. Sun Z, Lau I, Wong YH, Yeong CH. Personalized three-dimensional printed models in congenital heart disease. *J Clin Med.* 2019;8.
48. Biglino G, Capelli C, Leaver LK, Schievano S, Taylor AM, Wray J. Involving patients, families and medical staff in the evaluation of 3D printing models of congenital heart disease. *Commun Med.* 2015;12:157–169.
49. Anwar S, Singh GK, Miller J, et al. 3D printing is a transformative technology in congenital heart disease. *JACC Basic Transl Sci.* 2018;3:294–312.
50. Bartel T, Müller S. 3-Dimensional printing in personalized interventional cardiology and cardiac surgery. *JACC: Case Rep.* 2019;1:538–539.
51. Lanning C, Chen SY, Hansgen A, Chang D, Chan KC, Shandas R. Dynamic three-dimensional reconstruction and modeling of cardiovascular anatomy in children with congenital heart disease using biplane angiography. *Biomed Sci Instrum.* 2004;40:200–205.
52. Sodian R, Weber S, Markert M, et al. Stereolithographic models for surgical planning in congenital heart surgery. *Ann Thorac Surg.* 2007;83:1854–1857.
53. Laschinger JC, Vannier MW, Gutierrez F, et al. Preoperative three-dimensional reconstruction of the heart and great vessels in patients with congenital heart disease. Technique and initial results. *J Thorac Cardiovasc Surg.* 1988;96:464–473.
54. Moore RA, Madueme PC, Lorts A, Morales DL, Taylor MD. Virtual implantation evaluation of the total artificial heart and compatibility: beyond standard fit criteria. *J Heart Lung Transpl.* 2014;33:1180–1183.
55. Moore RA, Lorts A, Madueme PC, Taylor MD, Morales DL. Virtual implantation of the 50cc SynCardia total artificial heart. *J Heart Lung Transpl.* 2016;35:824–827.
56. Moore RA, D'Souza GA, Villa C, Taylor MD, Morales DLS, Lorts A. Optimizing surgical placement of the HeartWare ventricular assist device in children and adolescents by virtual implantation. *Prog Pediatric Cardiol.* 2017;47:11–13.

57. Shugh SB, Szugye NA, Zafar F, et al. Expanding the donor pool for congenital heart disease transplant candidates by implementing 3D imaging-derived total cardiac volumes. *Pediatr Transpl.* 2020;24:e13639.
58. Szugye NA, Lorts A, Zafar F, Taylor M, Morales DLS, Moore RA. Can virtual heart transplantation via 3-dimensional imaging increase the maximum acceptable donor size? *J Heart Lung Transpl.* 2019;38:331–333.
59. Szugye NA, Zafar F, Ollberding NJ, et al. A novel method of donorrecipient size matching in pediatric heart transplantation: a total cardiac volumepredictive model. *J Heart Lung Transpl.* 2020.
60. Plasencia JD, Kamarianakis Y, Ryan JR, et al. Alternative methods for virtual heart transplant-size matching for pediatric heart transplantation with and without donor medical images available. *Pediatr Transpl.* 2018;22:e13290.
61. Heart Encyclopedia; 2021. Available at: https://www.cincinnatichildrens.org/patients/child/encyclopedia/heartpedia. Accessed May 28, 2021.
62. Kim B, Loke YH, Mass P, et al. A novel virtual reality medical image display system for group discussions of congenital heart disease: development and usability testing. *JMIR Cardio.* 2020;4:e20633.
63. Sorensen TS, Greil GF, Hansen OK, Mosegaard J. Surgical simulation—a new tool to evaluate surgical incisions in congenital heart disease? *Interact Cardiovasc Thorac Surg.* 2006;5:536–539.
64. Ong CS, Krishnan A, Huang CY, et al. Role of virtual reality in congenital heart disease. *Congenit Heart Dis.* 2018;13:357–361.
65. Rizzetto F, Bernareggi A, Rantas S, Vanzulli A, Vertemati M. Immersive virtual reality in surgery and medical education: diving into the future. *Am J Surg.* 2020;220:856–857.
66. Hussein N, Honjo O, Barron DJ, Haller C, Coles JG, Yoo SJ. The incorporation of hands-on surgical training in a congenital heart surgery training curriculum. *Ann Thorac Surg.* 2020.
67. Hussein N, Lim A, Honjo O, et al. Development and validation of a procedure-specific assessment tool for hands-on surgical training in congenital heart surgery. *J Thorac Cardiovasc Surg.* 2020;160:229–240. e1.
68. Seslar SP, Shepard CW, Giroud JM, et al. Lost treasures: a plea for the systematic preservation of cadaveric heart specimens through three-dimensional digital imaging. *Cardiol Young.* 2015;25:1457–1459.
69. Kiraly L, Kiraly B, Szigeti K, Tamas CZ, Daranyi S. Virtual museum of congenital heart defects: digitization and establishment of a database for cardiac specimens. *Quant Imaging Med Surg.* 2019;9:115–126.
70. Brun H, Bugge RAB, Suther LKR, et al. Mixed reality holograms for heart surgery planning: first user experience in congenital heart disease. *Eur Heart J Cardiovasc Imaging.* 2019;20:883–888.
71. EchoPixel Introduces First-Ever Intraoperative Software to Provide 3D Holographic Experience in the Operating Suite to Support Structural Heart Procedures; 2019. Available at: https://www.prnewswire.com/news-releases/echopixel-introduces-first-ever-intraoperative-software-to-provide-3d-holographic-experience-in-the-operating-suite-to-support-structural-heart-procedures-300926643.html.
72. Sundareswaran KS, de Zelicourt D, Pekkan K, et al. Anatomically realistic patient-specific surgical planning of complex congenital heart defects using MRI and CFD. *Annu Int Conf IEEE Eng Med Biol Soc.* 2007;2007:202–205.
73. Sundstrom E, Jonnagiri R, Gutmark-Little I, et al. Effects of normal variation in the rotational position of the aortic root on hemodynamics and tissue biomechanics of the thoracic aorta. *Cardiovasc Eng Technol.* 2020;11:47–58.
74. Bossers SS, Cibis M, Gijsen FJ, et al. Computational fluid dynamics in Fontan patients to evaluate power loss during simulated exercise. *Heart.* 2014;100:696–701.
75. Kim MS, Hansgen AR, Wink O, Quaife RA, Carroll JD. Rapid prototyping: a new tool in understanding and treating structural heart disease. *Circulation.* 2008;117:2388–2394.
76. Zhuang X, Rhode K, Arridge S, et al. An atlas-based segmentation propagation framework locally affine registration—application to automatic whole heart segmentation. *Med Image Comput Comput Assist Interv.* 2008;11:425–433.
77. Zhuang X, Rhode KS, Razavi RS, Hawkes DJ, Ourselin S. A registration-based propagation framework for automatic whole heart segmentation of cardiac MRI. *IEEE Trans Med Imaging.* 2010;29:1612–1625.
78. Lau I, Wong YH, Yeong CH, et al. Quantitative and qualitative comparison of low- and high-cost 3D-printed heart models. *Quant Imaging Med Surg.* 2019;9:107–114.
79. Abdelkarim A, Hageman A, Levi DS, Aboulhosn J. Operationalizing low-cost three-dimensional printing in planning for complex congenital cardiac interventions. *Mater Today Commun.* 2018;15:171–174.
80. Alvarez-Lopez F, Maina MF, Saigi-Rubio F. Use of a low-cost portable 3D virtual reality gesture-mediated simulator for training and learning basic psychomotor skills in minimally invasive surgery: development and content validity study. *J Med Internet Res.* 2020;22:e17491.
81. Parham G, Bing EG, Cuevas A, et al. Creating a low-cost virtual reality surgical simulation to increase surgical oncology capacity and capability. *Ecancermedicalscienc.* 2019;13.
82. Chang P, Chen B, Parikh M, et al. Virtual reality supplemental teaching at low-cost (VRSTL): a model and review for developing low-cost VR for medical education. *J Gen Intern Med.* 2017;32:S722.
83. Park CH, Wilson KL, Howard AM. Pilot study: supplementing surgical training for medical students using a low-cost virtual reality simulator. *Comp Med Sys.* 2014;125–127.

Chapter 15

Evolution of Device Therapies for Cardiovascular Surgical Care in Children

Alistair Phillips[1,2] and Hani Najm[1,2]

[1]Pediatric and Congenital Heart Surgery, Cleveland Clinic, Cleveland, OH, United States, [2]Cleveland Clinic Lerner College of Medicine of Case Western Reserve University, Cleveland, OH, United States

Chapter Outline

History of device therapies in children for cardiovascular surgery

Congenital heart surgery has been in the forefront of the development of cardiac surgery. Cardiac surgery was always thought to be an untannable goal because the heart was seen as the home of one's soul. There have been many attempts at open heart surgery but the initial reports were related to penetrating trauma. In 1944, the landscaped changed forever with the treatment of "blue babies" by the Hopkins Team, led by Vivian Thomas. Helen Taussig, MD followed "blue babies" and realized that babies with patent ductus arteriosus and tetralogy of Fallot had better outcomes. She approached Alfred Blalock, MD who was Chief of Surgery at The Johns Hopkins University, and asked if there was a way to make an artificial patent ductus arteriosus. Dr. Blalock tasked Vivian Thomas to investigate. Dr. Thomas developed a procedure that connected the left subclavian artery to the left pulmonary artery. This procedure was successfully completed on a dog named Anna who went on to live to the age of 15.[1] Vivian Thomas stood behind Dr. Blalock as he performed the first successful Thomas shunt on 15-month-old Eileen Saxon on November 29, 1944.[2] The Thomas shunt has successfully been performed on thousands of children worldwide and its modification is still being performed today. Most importantly, this procedure ushered in the modern era of cardiac surgery. Vivian Thomas used his ingenuity and skills as a carpenter and machinist to both conceive of the operation and develop the necessary instruments to perform the surgery. This procedure and his tutelage inspired many young surgeons, many who went on to be leaders in medicine. Most notable of these young surgeons were Denton Cooley, MD who became a pioneer in cardiac surgery, and William Longmire, MD, Dr. Blalock's first assistant during the first operation.

John Gibbons, MD performed the first successful open-heart surgery with the closure of an atrial septal defect using the heart-lung machine on May 6, 1953.[3] Dr. Gibbons worked tirelessly over 23 years to develop a heart-lung machine. After completing this first successful application Dr. Gibbons had three patients die. He never did open-heart surgery again.

C. Walton Lillehei, MD at the University of Minnesota used cross-circulation, using a donor to act as the heart-lung machine. He would connect the patient to the femoral vessels of the donor patient. His initial experience with open heart was on patients with congenital heart disease. The initial results were excellent with an overall survival of 60%, with the first procedure completed successfully on March 26, 1954. During this time Dr. Lillehei was working on improving the Gibbons' heart-lung machine[4] as were several other groups, namely Dr. Clarence Dennis of Downstate University in Brooklyn[5] and John Kirklin of the Mayo Clinic.[6] Dr. Lillehei also trained several notable surgeons that

Advances in Cardiovascular Technology. DOI: https://doi.org/10.1016/B978-0-12-816861-5.00013-7

went on to be pioneers in heart surgery, most notably Norman Shumway, MD and Christiaan Barnard, MD, who performed the first successful heart transplant.[7]

These were very important times for the blossoming of cardiac surgery and set the stage for the amazing life-changing developments. It also sets the tone for the continued device development needed for this growing medical field. However, most of this development is centered around adult acquired heart disease, and cardiac surgery for children is a secondary consideration. There are several critical areas for evolution of device therapies for children requiring congenital heart surgery that are outlined in this chapter.

Mechanical circulatory support

Extracorporeal membrane oxygenation

Extracorporeal membrane oxygenation (ECMO) was first successfully used for the treatment of adult respiratory failure in 1972.[8] There were several other initial reports using ECMO to treat adult respiratory failure. The majority of ECMO, in the early years, was used in neonates and children for respiratory failure. The sentinel work was done by Robert Bartlett, MD when he reported the use of ECMO to treat 100 neonates with respiratory failure.[9] Bartlett et al. reported three phases of development of neonatal ECMO with an initial survival of 54% and by phase 3 a 90%

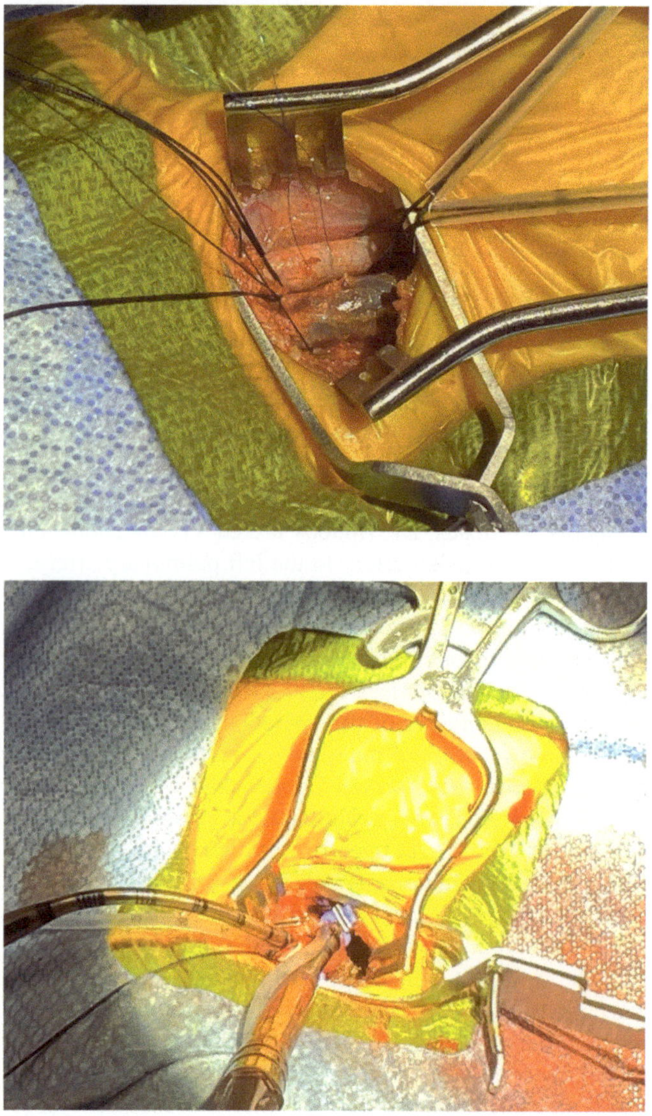

FIGURE 15.1 Cut-down for placement of veno-arterial ECMO in a 5-month-old, with placement of the arterial and venous cannulas.

survival. Over the three phases of development there was a decline in complications.[9] ECMO has been used very successful for neonatal patients with respiratory failure and in postoperative congenital heart surgery patients. With improved management of these patients in specialized neonatal and congenital heart units the requirement for ECMO in children is decreasing. ECMOs roles have increased dramatically in adult over the last 10 years, increasing by 1180%, and the numbers of ECMO centers increased by 133% from 2006 to 2016.[10] There has been evolving criteria for the use of veno-arterial ECMO over the last decade in adults from surgeon-driven for postcardiotomy shock to now multifactorial cardiogenic shock be it from sepsis or cardiac arrest or severe respiratory failure.[10] Recently with the COVID-19 pandemic, ECMO has been used successfully. There was better 90-day survival in more experienced centers, which was attributed to earlier activation and deployment of ECMO, and the earlier use of veno-veno ECMO.[11] ECMO use in adults has increased with improved survival and outcomes. The increased utilization of ECMO in adult centers has many reasons: (1) multidisciplinary teams with expertise in technologies, (2) rapid deployment of smaller more portable ECMO systems, and (3) ease of cannulation with percutaneous systems.[10,11]

Similar systems need to be developed for children. One critical area for development is improved cannulation strategies. Presently for neonates and small children we are required to use a cut-down (Fig. 15.1). As with the adults we need to have rapid deployment and improved systems to ensure ease of cannulation. For adult cannulation, ultrasonography has been exceptionally helpful; however, the probes are large and very bulky, and challenging to use in very small patients. Developing smaller more user-friendly systems that can identify and help guide placement of the cannulas would be very beneficial. Having improved delivery systems will expand the number of operators that can place a patient on ECMO and will have the added benefit of expanding the indications for the use of ECMO in children as has occurred in adults over the last 10 years.

For pediatric patients, the ECMO circuit to patient body size are exponentially larger when compared to adults (Fig. 15.2). A lot of this is related to the nominal size of the tubing to maintain adequate flow while not causing hemolysis. The consequence of having an increased circuit to patient volume ratio was increased hemolysis and resultant mortality.[12] Novel approaches are needed to reduce circuit size and develop functional oxygenators with reduced prime

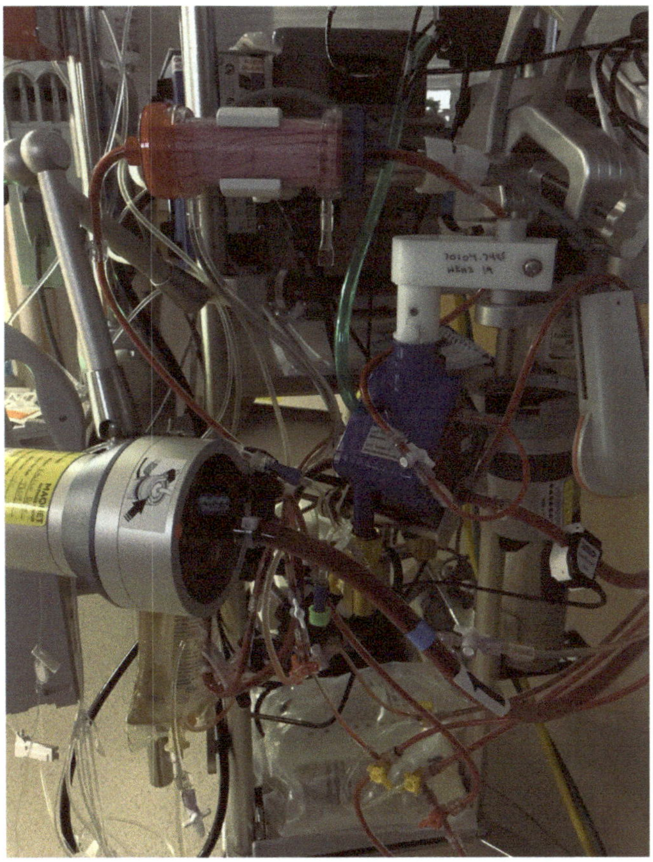

FIGURE 15.2 ECMO circuit with the centrifugal pump, oxygenator, ultrafiltration, and necessary tubing.

volumes. It will be important to develop new materials that are compatible with blood interface, and materials that can allow removal of carbon dioxide and the addition of oxygen. It is critical, as Maul et al. described, to produce a circuit for all children that has a circuit to patient volume ratio greater than 50.[12]

New devices

Presently the only FDA-approved ventricular assist device (VADs) for children less than 40 kg is the Berlin Heart EXCOR, which has been used since 1986. It is a paracorporeal pneumatically driven pulsatile pump. The outcomes are very good with 75% 1-year survival, with median duration of support of 40 days, and 29% had neurological dysfunction.[13] There clearly is a need for new devices. The NIH has spent millions of dollars on the PumpKIN trial to develop new devices but presently there are no new devices on the horizon.[14] A lot of the issues are related to high thrombosis rate and hemolysis. Developing partial flow devices for adults will have the benefit of augmenting cardiac output for marginal adult patients. These devices will operate in series with the native heart. The will be easier to implant and explant. With enhanced cardiac function, and reduced work of the heart remodeling of the diseased heart may occur. These partial flow devices for adults will provide full support for children.

3D printing

3D printing began for rapid prototyping in 1984 by Charles Hull with the development of sterolithography (SLA), which used ultraviolet laser to cure photopolymers. During the 1980s there was development of three main types of 3D printing. These printers are the mainstay of the technology used today. The three types of printing are (1) selective laser sintering (SLS), (2) SLA, and (3) fused deposition modeling (FDM). All three of these techniques are used generally for rapid prototyping. With improved imagining modalities in medicine like high-resolution CT scans and MRIs, 3D printing for humans was made possible, including high-resolution replications of the heart.

3D printing has three major areas of use in medicine:

1. Anatomic 3D prints
   a. Allows of planning for procedure with simulation of the procedure[15]
   b. Better three-dimensional understanding for education and teaching
2. Personalized medical devices being used for implantation or external use
3. Development of biological parts and systems that will be completely biocompatible

These basic printer setups are the basis for biological printing.[16] With 3D biological printing there are three general ways to build 3D printed materials: (1) biomimicry—build scaffolding and cells together; (2) autonomous self-assembly—similar to fetal development; and (3) minitissue building blocks—cell spheres.[16]

Using 3D printing will allow for better planning for procedures by greatly enhancing clinical understanding. 3D printing was instrumental in developing the surgical approach to replace a pulmonary valve via a minimally invasive method.[15] By using 3D, Phillips et al. were able to simulate the surgical replacement of the pulmonary valve using the expertise of a cardiac surgeon, interventional cardiologist, and an echocardiographer, otherwise the patient would have required open heart surgery.[15]

3D printing can also be used for the development of personalized plugs (Fig. 15.3), for occlusion of patent ductus arteriosus. There are higher fidelity printers and printers that are capable of printing with different materials. Properties of these materials could be bioabsorbed or have the feel of real tissue.

There is an expanding role of anatomical 3D models for medical simulation of complex surgeries, teaching, and education. Using 3D printing, education will be greatly expanded for the patient, family, and the whole clinical team by having a better anatomical understanding for each patient (Fig. 15.4).

There will be an expanding role of 3D printing within medicine in two critical areas: (1) developing implantable biocompatible devices for personalized medical use, like ECMO circuits, and (2) developing biological replacement organs and other human tissues.

Material science and novel devices

Mounting a bovine jugular vein into a vascular stent and then implanting into the native pulmonary valve of a lamb via a percutaneous delivery was first reported by Bonhoeffer et al. in 2000.[17] The transcatheter pulmonary valve has led the

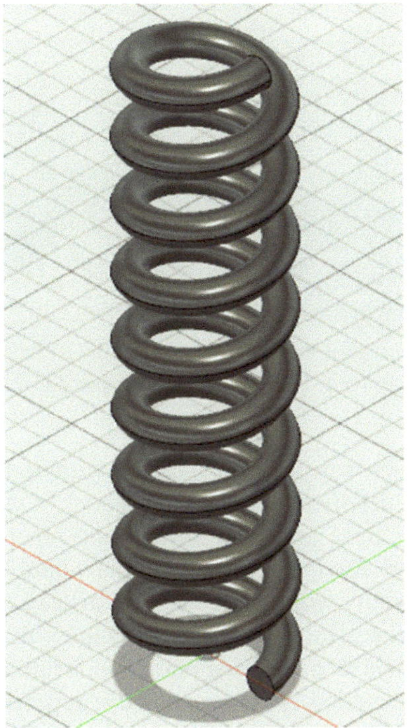

FIGURE 15.3 A personalized vascular plug developed from measurement of the pulmonary artery length and diameter.

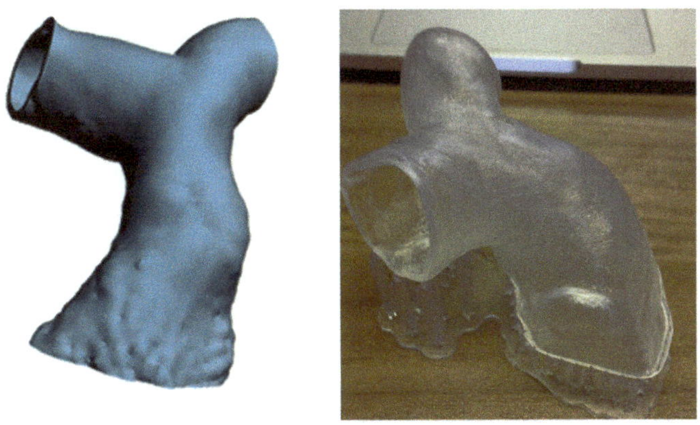

FIGURE 15.4 Image on the left shows the computer aided design (CAD) and the final 3D printing of a patient's pulmonary artery on the left used to simulate surgery.

way to transcatheter valve technology, especially for aortic valve in adults with aortic stenosis. The initial experience was the treatment of congenital heart patients.

There are several benefits to the transcatheter valves in adults. They have a larger effective orifice area (EOA), as seen in Fig. 15.5. In the Nordic Aortic Valve Intervention trial, patients with low surgical risk were randomized to surgical (SAVR) or transcatheter aortic valve replacement (TAVR), at 6 years there was significantly larger EOA and lower mean gradient for TAVR patients.[18]

Congenital heart disease was the initial focus for transcatheter valve therapy, as was cardiac surgery. These novel devices have clearly helped many thousands of people, but have limited utility for children because of lack of growth, limited sizes, and more rapid deterioration of the valve over time, mainly from calcification. There is a statistical significantly higher rate of reoperation secondary to structural valve deterioration for patients less than 60 years of age for patients having a surgical aortic valve.[19] A similar fate would be expected for TAVR valves.

Polymorphic TAVR valves are being developed that have the benefits of utilizing polymers that have improved biologic properties including hemocompatibility, reduced risk of thrombosis, don't calcify, and are very durable.

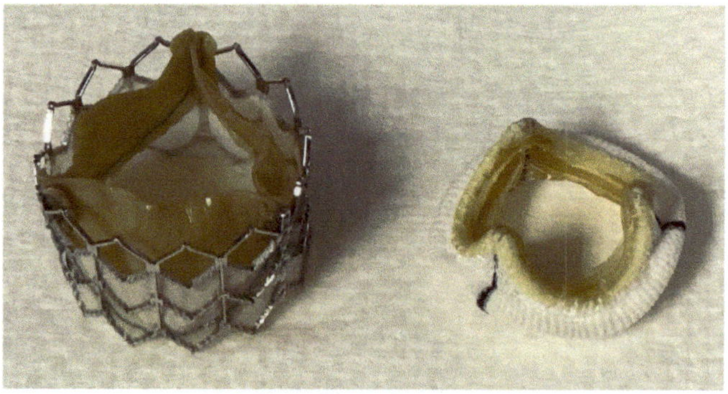

FIGURE 15.5 On the left is a 29 mm Sapien valve from Edwards Life Science compared to a 23 mm Biocor valve from St Jude Medical.

In experimental models, polymorphic valves were found to have favorable hemodynamics that are very durable, to be resistant to platelet activation, to have low calcification, and to preserve EOA.[20]

The ideal valve would grow with time, have excellent physiologic function, have excellent hemocompatibility and not be thrombogenic, and would be durable. Having a valve like this would be exceptionally beneficial for children but would have even greater application for adults with acquired heart disease. Ultimately, the new valve would integrate with the patient's own tissue. Developing a scaffolding that allows ingrowth of the patients own tissue can be produced by either using a decellularized biologic valve or developing a bioabsorbable material. Porcine aortic roots showed complete decellularization over a 16-day period, macroscopically appeared intact, and had an intact extracellular matrix.[21] The decellularized cusps had similar mechanical properties as the native cusps.[19] The decellularized aortic roots had similar hemodynamics as the native roots.[19] These decellularized porcine aortic roots can be used for aortic root replacement, maintaining excellent hemodynamic and mechanical properties with the possibility of progenitor cell ingrowth with repopulation of the patient's own tissue.

Decellularized tissues can be cultured with the patient's own stem cells to allow them to grow into the scaffolding. The new cellularized valved conduit can be then used as a surgical valve with the ability for it to grow, having excellent hemodynamics and mechanical properties. Ovine pulmonary valves were harvested as well as adipose-derived stem cells that were then separated into endothelial cell lineage, ECs, and fibroblast cell linage, FBs.[22] The ovine pulmonary valve was decellularized and the cusps and adventitia were seeded with FBs, and the cusps were seeded with ECs, the valves were reimplanted, and animals were followed for 6 months with excellent valve function.[20]

Tissue engineered heart valves were constructed with autologous cell seeding, 20×10^6 LDL−cells (myofibroblasts) onto a polymer mesh (polyglactin woven mesh sandwich). After 7 days the mesh was seeded with 2 million endothelial cells per cm^2 and cultured for another week.[23] The new leaflets were implanted into pulmonary artery of animals, a control group had implantation of only the polymer mesh, which was designed to degrade in 6−8 weeks. At follow-up, all of the control leaflets had degraded and the treatment group had functioning valve leaflets.[21]

Development of bioabsorbable materials to act as a scaffold, allows the ingrowth of the patient's own cells, while maintaining strength, flexibility, and thus to remain competent to act as a vessel. The Xeltis pulmonary valve (XPV) conduit is made from the RestoreX polymer platform and is fully synthetic.[24] The XPV was implanted in 12 children with a 12-month follow-up (group 1), and another six children had an implant of a newly designed valve XPV-2 with improved valve function (group 2). For patients in group 1, three patients had severe stenosis, gradient >40 mm Hg, and five patients developed severe pulmonary valve regurgitation due to valve prolapse. For patients in group 2, one patient developed rapidly progressive stenosis requiring explant and one patient developed mild pulmonary insufficiency at 12 months follow-up. These results are very encouraging and there are many applications for this material such as developing novel conduits, implantable valves for all patients, and developing other tissue that needs to grow, like aortic conduits.

Synthetic materials, especially polymers have several important advantages. They are cheap, can be easily molded into different forms, and can be used to produce highly reproducible products with excellent physiochemical and mechanical proprieties. A lot of these materials are not biocompatible, are thrombogenic, and are not hemocompatible. Decellularized biological tissues, like porcine aorta, as described above, have a lot of favorable physiologic properties; however, they have limited longevity and structural integrity. By combining decellularized biological tissue with polymers, hybrid materials that are biocompatible with a nonthrombogenic surface and hemocompatible with excellent physiochemical and mechanical proprieties can be produced.[25] The hybrid material was shown to have excellent tensile strength, with excellent adhesion of the two layers, as seen by two-photon microscopy.[23] These hybrid materials will be beneficial for the linings of VADs and other biologic conduits, for example, components in extracorporeal circuits, including developing hybrid materials that can "breathe," that is, gas exchange.

Using the combination of 3D printing and hybrid materials, novel scaffolding can be produced that can mimic biologic tissues. Using modern imaging, personalized computer aided designs (CADs) can be created; using a combination of 3D printers a bioabsorbable scaffolding can be created with biologic printing into the scaffolding with stem cells. Tracheal grafts were created using 3D printing with polycaprolactone and extracellular matrix sutured to the scaffolding.[26] The grafts were placed in seven animals and at the end of the 3-month study there was mucosal coverage and vascularity, with all grafts intact.[24]

Conclusion

The future for medicine is very bright. From developing novel, innovative therapies like Vivian Thomas did to developing new technologies. We do need to develop novel hybrid materials that will allow for somatic growth, and will be biocompatible. Combining these materials with 3D printing, new tissues can be produced, like tracheas, valves, and ultimately new organs. Medicine is an ever-changing exciting field. One of greatest advances has been the development of Team of Teams and development of multidisciplinary teams working to improve medical delivery on a personal level.[27]

References

1. Brogan TV, Alfieris GM. Has the time come to rename the Blalock-Taussig shunt? *Pediatric Crit Care Med.* 2003;4(4):450–453. Available from: https://doi.org/10.1097/01.PCC.0000090014.95775.3A.
2. Cooley DA. The first Blalock-Taussig shunt. *J Thorac Cardiovasc Surg.* 2010;140(4):750–751. Available from: https://doi.org/10.1016/j.jtcvs.2010.06.033. PMID: 20850653.
3. Cohn LH. Fifty years of open-heart surgery. *Circulation.* 2003;107(17):2168–2170. Available from: https://doi.org/10.1161/01.CIR.0000071746.50876.E2. PMID: 12732590.
4. Lillehei CW. Controlled cross circulation for direct-vision intracardiac surgery; correction of ventricular septal defects, atrioventricularis communis, and tetralogy of Fallot. *Postgrad Med.* 1955;17(5):388–396. Available from: https://doi.org/10.1080/00325481.1955.11708211. PMID: 14371276.
5. Dennis C, Spreng Jr DS, Nelson GE, et al. Development of a pump-oxygenator to replace the heart and lungs; an apparatus applicable to human patients, and application to one case. *Ann Surg.* 1951;134(4):709–721. Available from: https://doi.org/10.1097/00000658-195113440-00017. PMID: 14878382; PMCID: PMC1802968.
6. Kirklin JW, Dushane JW, Patrick RT, et al. Intracardiac surgery with the aid of a mechanical pump-oxygenator system (gibbon type): report of eight cases. *Proc Staff Meet Mayo Clin.* 1955;30(10):201–206. PMID: 14371757.
7. Baumgartner WA, Reitz BA, Gott VL, Shumway SJ, Norman E, Shumway MD. PhD: visionary, innovator, humorist. *J Thorac Cardiovasc Surg.* 2009;137(2):269–277. Available from: https://doi.org/10.1016/j.jtcvs.2008.11.008. PMID: 19185135.
8. Hill JD, O'Brien TG, Murray JJ, et al. Prolonged extracorporeal oxygenation for acute post-traumatic respiratory failure (shock-lung syndrome). Use of the Bramson membrane lung. *N Engl J Med.* 1972;286(12):629–634. Available from: https://doi.org/10.1056/NEJM197203232861204. PMID: 5060491.
9. Bartlett RH, Gazzaniga AB, Toomasian J, Coran AG, Roloff D, Rucker R. Extracorporeal membrane oxygenation (ECMO) in neonatal respiratory failure. 100 cases [published correction appears in Ann Surg 1987 Jan;205(1):11. Corwin AG [corrected to Coran AG]]. *Ann Surg.* 1986;204(3):236–245. Available from: https://doi.org/10.1097/00000658-198609000-00003.
10. Guglin M, Zucker MJ, Bazan VM, et al. Venoarterial ECMO for adults: JACC scientific expert panel. *J Am Coll Cardiol.* 2019;73(6):698–716. Available from: https://doi.org/10.1016/j.jacc.2018.11.038. PMID: 30765037.
11. Lebreton G, Schmidt M, Ponnaiah M, et al. Paris ECMO-COVID-19 investigators. Extracorporeal membrane oxygenation network organisation and clinical outcomes during the COVID-19 pandemic in Greater Paris, France: a multicentre cohort study. *Lancet Respir Med.* 2021;9(8):851–862. Available from: https://doi.org/10.1016/S2213-2600(21)00096-5. Epub 2021 Apr 19. Erratum in: Lancet Respir Med. 2021 Jun;9(6):e55. Erratum in: Lancet Respir Med. 2021 Jul;9(7):e62. PMID: 33887246; PMCID: PMC8055207.
12. Maul TM, Aspenleiter M, Palmer D, Sharma MS, Viegas ML, Wearden PD. Impact of circuit size on coagulation and hemolysis complications in pediatric extracorporeal membrane oxygenation. *ASAIO J.* 2020;66(9):1048–1053. Available from: https://doi.org/10.1097/MAT.0000000000001121.
13. Almond CS, Morales DL, Blackstone EH, et al. Berlin Heart EXCOR pediatric ventricular assist device for bridge to heart transplantation in US children. *Circulation.* 2013;127(16):1702–1711. Available from: https://doi.org/10.1161/CIRCULATIONAHA.112.000685. Epub 2013 Mar 28. PMID: 23538380.
14. Adachi I. Current status and future perspectives of the PumpKIN trial. *Transl Pediatr.* 2018;7(2):162–168. Available from: https://doi.org/10.21037/tp.2018.02.04.

15. Phillips AB, Nevin P, Shah A, Olshove V, Garg R, Zahn EM. Development of a novel hybrid strategy for transcatheter pulmonary valve placement in patients following transannular patch repair of tetralogy of fallot. *Catheter Cardiovasc Interv*. 2016;87(3):403–410. Available from: https://doi.org/10.1002/ccd.26315. Epub 2015 Nov 3. PMID: 26527499.
16. Murphy SV, Atala A. 3D bioprinting of tissues and organs. *Nat Biotechnol*. 2014;32(8):773–785. Available from: https://doi.org/10.1038/nbt.2958. PMID: 25093879.
17. Bonhoeffer P, Boudjemline Y, Saliba Z, et al. Transcatheter implantation of a bovine valve in pulmonary position: a lamb study. *Circulation*. 2000;102(7):813–816. Available from: https://doi.org/10.1161/01.cir.102.7.813. PMID: 10942752.
18. Søndergaard L, Ihlemann N, Capodanno D, et al. Durability of transcatheter and surgical bioprosthetic aortic valves in patients at lower surgical risk. *J Am Coll Cardiol*. 2019;73(5):546–553. Available from: https://doi.org/10.1016/j.jacc.2018.10.083. PMID: 30732707.
19. Bourguignon T, Bouquiaux-Stablo AL, Candolfi P, et al. Very long-term outcomes of the Carpentier-Edwards Perimount valve in aortic position. *Ann Thorac Surg*. 2015;99(3):831–837. Available from: https://doi.org/10.1016/j.athoracsur.2014.09.030. Epub 2015 Jan 9. PMID: 25583467.
20. Rotman OM, Kovarovic B, Chiu WC, et al. Novel polymeric valve for transcatheter aortic valve replacement applications: in vitro hemodynamic study. *Ann Biomed Eng*. 2019;47(1):113–125. Available from: https://doi.org/10.1007/s10439-018-02119-7.
21. Sierad LN, Shaw EL, Bina A, et al. Functional heart valve scaffolds obtained by complete decellularization of porcine aortic roots in a novel differential pressure gradient perfusion system. *Tissue Eng Part C Methods*. 2015;21(12):1284–1296. Available from: https://doi.org/10.1089/ten.TEC.2015.0170. PMID: 26467108; PMCID: PMC4663650.
22. Movileanu I, Harpa M, Al Hussein H, et al. Preclinical testing of living tissue-engineered heart valves for pediatric patients, challenges and opportunities. *Front Cardiovasc Med*. 2021;19(8):707892. Available from: https://doi.org/10.3389/fcvm.2021.707892. PMID: 34490371; PMCID: PMC8416773.
23. Shinoka T. Tissue engineered heart valves: autologous cell seeding on biodegradable polymer scaffold. *Artif Organs*. 2002;26(5):402–406. Available from: https://doi.org/10.1046/j.1525-1594.2002.07004.x. PMID: 12000435.
24. Morales DL, Herrington C, Bacha EA, et al. A novel restorative pulmonary valve conduit: early outcomes of two clinical trials. *Front Cardiovasc Med*. 2021;7:583360. Available from: https://doi.org/10.3389/fcvm.2020.583360. PMID: 33748192; PMCID: PMC7969645.
25. Todesco M, Zardin C, Iop L, et al. Hybrid membranes for the production of blood contacting surfaces: physicochemical, structural and biomechanical characterization. *Biomater Res*. 2021;25(1):26. Available from: https://doi.org/10.1186/s40824-021-00227-5. PMID: 34376256; PMCID: PMC8353781.
26. Rehmani SS, Al-Ayoubi AM, Ayub A, et al. Three-dimensional-printed bioengineered tracheal grafts: preclinical results and potential for human use. *Ann Thorac Surg*. 2017;104(3):998–1004. Available from: https://doi.org/10.1016/j.athoracsur.2017.03.051. Epub 2017 Jun 11. PMID: 28610885.
27. Lee TH, Duckworth AL. Organizational grit: turning passion and perseverance into performance: the view from the health care industry. *Harv Bus Rev*. 2018;.

Chapter 16

Robotic Mitral Valve Surgery: New Techniques and Solutions

Raphaelle A. Chemtob, Kevin E. Hodges, Daniel J.P. Burns and Marc Gillinov
Department of Thoracic & Cardiovascular Surgery, Cleveland Clinic, Cleveland, OH, United States

Chapter Outline

Introduction

Degenerative disease is the most common etiology causing mitral valve dysfunction, followed by ischemia, rheumatic disease, and endocarditis.[1] Mitral valve repair is the procedure of choice to correct mitral regurgitation caused by degenerative mitral valve disease, due to its well-documented superiority over valve replacement.[2,3] At reference centers, mitral valve repair is feasible in more than 95% of such patients and has been associated with improved long-term survival, greater freedom from valve-related adverse events, and better preservation of left ventricular function.[4–9]

Over the past decades, the operative techniques to repair the mitral valve have evolved with an increasing interest in the application of minimally invasive approaches.[10–15] Developments in robotic technology, instrumentation, and cardiopulmonary bypass techniques together with refinement of valve repair techniques enabled gradual introduction of robotic technology in cardiac surgery.[16–27] Furthermore, minimally invasive approaches have become popular among patients, increasing the demand for robotically assisted surgery. Today, robotic mitral valve surgery is the least invasive surgical approach to the mitral valve. It has proven to be a safe, reproducible technique with late outcomes similar to conventional sternotomy-based approaches with several advantages including reduction of surgical trauma, diminished blood loss, shorter hospital stay, and more rapid return to work.[21,22,28–33] Recent further advances in technology and surgical methods have expanded its application to the repair of complex valves. It has become a cornerstone in the management of degenerative mitral valve disease and is being performed at several dedicated surgical centers across the world with excellent outcomes.[34]

History of robotics in cardiac surgery

The era of robotics in cardiac surgery began in 1998 and represented a separate new technology within minimally invasive surgery. The Zeus model developed by Computer Motions, was the first system that allowed to perform fully

Advances in Cardiovascular Technology. DOI: https://doi.org/10.1016/B978-0-12-816861-5.00034-4

robotic cardiac surgery. The Zeus system was equipped with a 2D imaging system, a console for the operator, and three mechanical arms. The third arm was provided by a voice-controlled Automated Endoscopic System for Optimal Positioning (AESOP) robot holding the video endoscope and enabled "solo mitral surgery."[35] The initial robotic cardiac surgery cases were documented in Europe using the Zeus system. In 1998, Dr. Damiano at Pennsylvania State University performed the first robot-assisted cardiac surgery in the United States.[36] Although the Zeus system allowed for a new era in cardiac surgery, there were important limitations to be addressed, that is, the lack of 3D view, the limited degrees of freedom of the tools. In 1999, Intuitive Surgery, based in Mountain View, CA, further developed the Zeus robot into the DA VINCI Surgical System, which was FDA-approved in 2001 and is currently the most widely employed robotic system. The second generation of robots (i.e., the telemanipulation robots, Da Vinci and Zeus) represented the next step in minimally invasive surgery advancement.[36] The Da Vinci System (Intuitive Surgical, Sunnyvale, CA) overcame many disadvantages of traditional long-shafted endoscopic instruments such as limited dexterity, depth perception, and muscle strain. It enabled seven degrees of freedom, free orientation in a confined space, 3D visualization, tremor filtering, and motion scaling, as well as telemanipulation from an ergonomic surgical console. The newest Da Vinci Xi is the fourth generation Da Vinci System available, with improvements including a dual console for teaching, further enhanced high-definition (HD) vision, superior ergonomics, and increased video and audio control. The benefits derived from Zeus and the Da Vinci System contributed to the immediate application and success of robots in cardiac surgery.

In 1998, Carpentier and his colleagues in France performed the first robotically assisted mitral valve repair using an early prototype of the Da Vinci.[16] Concurrently, Mohr et al. performed five mitral repairs in Leipzig, Germany.[17] In 2000, Grossi performed partial mitral repairs using the Zeus at New York University.[18] In the same year, a first complete robotically assisted mitral valve repair in the USA was performed by Chitwood at East Carolina University using the Da Vinci System, as part of a trial on safety and efficacy of robotics in mitral surgery, leading to final FDA approval in 2003.[19,20] Meanwhile, the Zeus was discontinued after a merger of Computer Motion and Intuitive Surgery, the two main companies producing operating surgical robots, making Da Vinci the only FDA-approved robotic system in cardiac surgery. Since then, robotic mitral valve repair has been performed at several dedicated surgical centers across the world with excellent results.[21–24,28,37–44] Robotic mitral valve repair is the least invasive approach to the mitral valve and yields comparable safety and efficacy to conventional sternotomy-based approaches with several advantages.[21,25] Furthermore, robotic mitral valve surgery is associated with a high likelihood of valve repair and low operative mortality and morbidity.[23] Over the course of experience, operative techniques have evolved, facilitating robotic mitral valve repair and allowing robotic correction of more complex degenerative mitral valve pathology.[37,45] The combination of algorithm-driven patient selection, increased surgical experience, and improved technology has enhanced clinical outcomes and procedural efficiency.[23,46]

Clinical presentation

Degenerative disease is the most common etiology causing mitral valve dysfunction, followed by ischemia, rheumatic disease, and endocarditis. Degenerative (myxomatous) mitral valve disease is currently most amenable to robotic repair. Robotic repair of rheumatic valve disease or a severely calcified annulus can be very challenging and is therefore not frequently performed. Indications for mitral valve surgery depend on most recent knowledge of disease progression and survival as described in the American College of Cardiology (ACC) and the American Heart Association (AHA) guidelines.[47] Current guidelines recommend mitral valve surgery for symptomatic patients with chronic primary mitral regurgitation (Mr) irrespective of left ventricle systolic function, and asymptomatic patients with chronic severe primary Mr and with left ventricle systolic dysfunction, or progressive increase in left ventricle size or decrease in ejection fraction. Mitral valve repair is recommended in preference to mitral valve replacement when the anatomic cause of Mr is degenerative disease and if a successful and durable repair is possible. A robotically assisted approach should be considered in all patients with isolated degenerative mitral valve disease in the absence of contraindications (Table 16.1).

Preoperative testing

A screening algorithm was introduced at the Cleveland Clinic in 2011 to facilitate the selection of candidates for robotic mitral valve surgery.[23,46,48] The preoperative workup is depicted in Fig. 16.1 and includes routine transthoracic or transesophageal echocardiogram and computed tomography (CT) of the chest, abdomen and pelvis. A sternal approach is indicated when echocardiography shows more than mild aortic regurgitation, severe mitral annular calcification, left ventricular dysfunction or dilatation, or severe pulmonary hypertension. CT results are important in assessing patients for aortoiliac atherosclerosis, femoral artery diameter <7 mm or pectus excavatum, in which a sternotomy-based

TABLE 16.1 Absolute and relative contraindications for robotically assisted mitral valve repair.

Absolute contraindications:	Relative contraindications:
Severe mitral annular calcification	Mitral valve stenosis
Acute infectious endocarditis	Coronary artery disease
History of chest radiation	More than 1 + aortic regurgitation
Hemodialysis	Aortic atherosclerosis
Small femoral arteries	Previous cardiac surgery
	Peripheral arterial disease
	Chest deformities
	Severe chronic obstructive pulmonary disease
	History of chest trauma

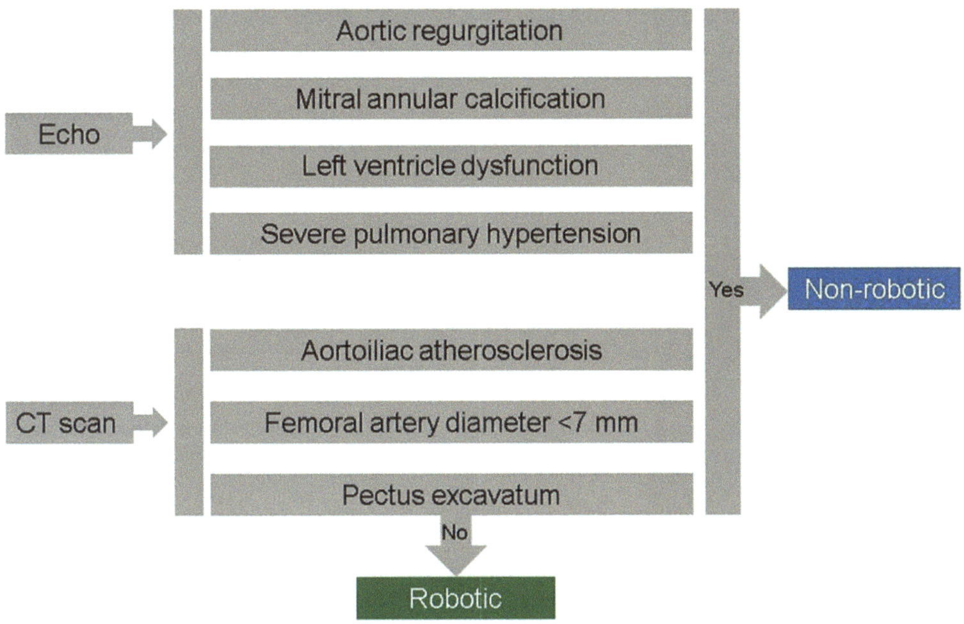

FIGURE 16.1 Screening algorithm, including routine echocardiography and computed tomography scans, to facilitate the selection of candidacy for robotic mitral valve surgery. *CT*, Computed tomography.

approach is preferred. Furthermore, our preoperative screening strategy includes routine laboratory work, chest radiography, and coronary angiography. In all female patients >45 years of age and male patients >40 years, cardiac catheterization is performed to determine the presence of coronary artery disease. If coronary artery disease is present and requires surgical treatment, a robotic approach is generally not employed in our Institution. While this screening algorithm is intentionally quite conservative, it has been associated with excellent results in our hands.[23,46] That said, it is possible to extend robotic mitral valve surgery to selected patients who do not satisfy the screening algorithm if successful surgery can be accomplished.

Surgical technique

Anesthesia and patient preparation

Anesthesia requires placement of a double lumen endotracheal tube or a single lumen tube with a bronchial blocker, to allow for single lung ventilation. A transesophageal echo probe is introduced, followed by placement of two central lines via the right internal jugular vein. The patient's right side is elevated 30 degrees with the right arm off the table and secured with a sling. The patient is prepped and draped as for sternotomy to allow conversion if necessary. External defibrillator patch electrodes are placed (away from the site of port placement) since use of internal paddles is not possible.

Cannulation

The femoral vessels are exposed prior to port placement to ensure adequate femoral vessel size for a robotic procedure (Fig. 16.2). A 3–4 cm right groin incision is made, and the arterial and venous cannulas are placed using Seldinger's technique and transesophageal echocardiogram guidance. The tip of the venous cannula (21–27 F) is positioned 2–3 cm into the superior vena cava. In patients with a body surface area greater than or equal to 2.0 cm, a second venous cannula is placed by upsizing one of the central venous lines placed by anesthesia. The arterial cannula (17–21 F) is positioned into the distal abdominal aorta or iliac artery. In cases of femoral atherosclerosis or small femoral vessels, cannulation of the axillary artery can be performed, although this is not ideal and a sternotomy should be considered.

Port placement

Proper port placement is crucial to avoid instrument conflicts and reduction in range of robotic arm motion. The right lung is deflated and a 4 cm working port is made about 3 cm lateral to the right nipple in the fourth intercostal space in the anterior axillary line. The robotic camera is introduced through the working port to assess position, good visualization, and need for a diaphragm retraction suture. The trocars for robotic arms are inserted through the sixth intercostal space in the anterior axillary line, and the second intercostal space 3–5 cm anterior to the anterior axillary line for the right and left arms, respectively (Figs. 16.3 and 16.4). The transthoracic atrial retractor port is placed in the fourth

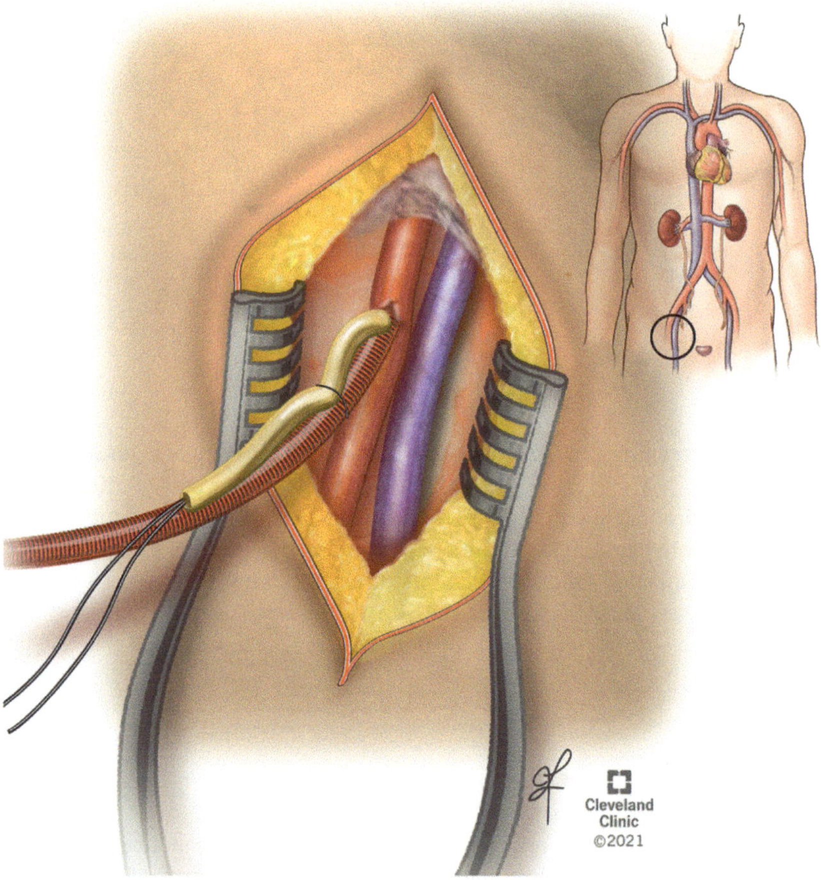

FIGURE 16.2 Femoral artery and vein cannulation for cardiopulmonary bypass. A 3–4 cm right groin incision is made for the femoral arterial and venous cannulas which are placed using Seldinger's technique and transesophageal echocardiogram guidance. The tip of the venous cannula is positioned 2–3 cm into the superior vena cava. The arterial cannula is positioned into the distal abdominal aorta or iliac artery.

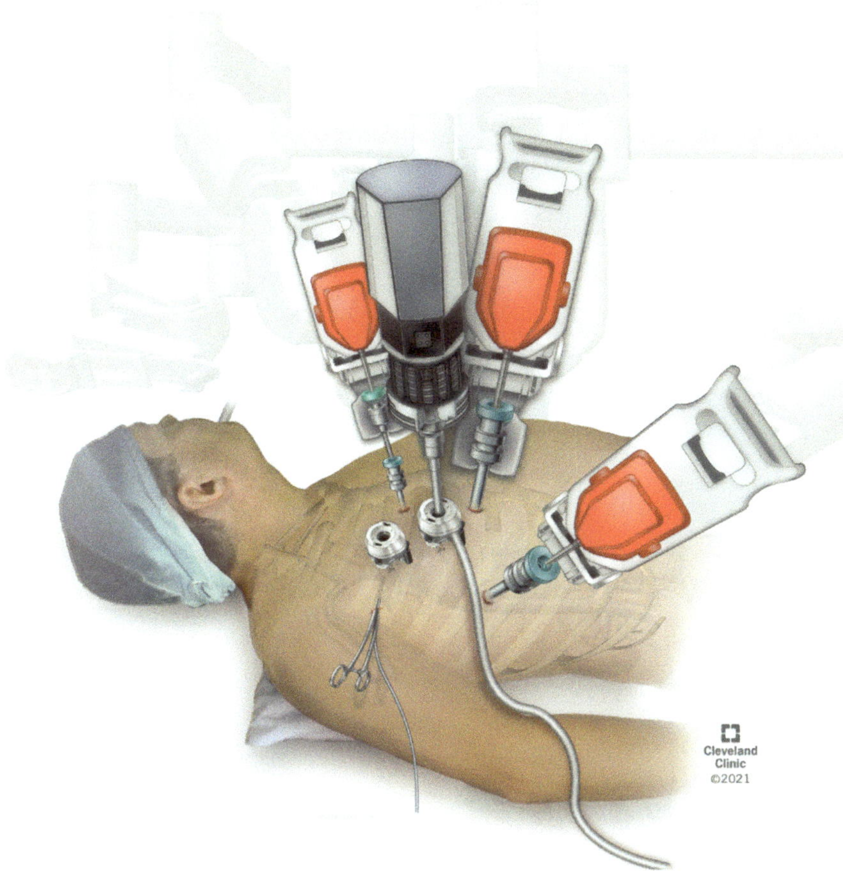

FIGURE 16.3 **Port placement and docking of the robot.** A 4 cm working port about 3 cm lateral to the right nipple in the fourth intercostal space in the anterior axillary line; three trochars for robotic arms are inserted through the sixth intercostal space in the anterior axillary line, and the second intercostal space 3–5 cm anterior to the anterior axillary line for the right and left arms, respectively; a transthoracic atrial retractor port is placed in the fourth intercostal space at the midclavicular line, medial to the working port.

intercostal space at the midclavicular line, medial to the working port. During placement of the atrial retractor port proximity to internal mammary artery should be taken into consideration.

Cardiopulmonary bypass and myocardial protection

A transthoracic aortic cross clamp (Chitwood clamp, Scanland International Inc.) is introduced through the second intercostal space in the midaxillary line. The cross clamp is positioned with the concave side cranially oriented. Cardiopulmonary bypass is initiated and the heart is decompressed. The patient is cooled to 30°C. The pericardium is opened 3–4 cm anterior to the phrenic nerve to prevent phrenic nerve traction injury. The aorta is exposed and a cardioplegia catheter is placed through the chest wall and into the aortic root and connected to an antegrade cardioplegia delivery system. The aorta is then cross clamped and a single dose of antegrade del Nido cardioplegia is delivered. We have previously demonstrated the del Nido solution to be a safe alternative to conventional cardioplegia in adult isolated valve surgery.[49]

Mitral valve repair

The mitral valve is approached through a standard left atriotomy, and the robotic atrial retractor is used for optimal mitral valve exposure. After initial inspection of the valve, standard mitral valve repair techniques are employed using robotic instruments.

For repair of the posterior leaflet, both resectional techniques and artificial chordae have excellent long-term results. We generally favor creation of artificial chordae for patients with posterior leaflet prolapse, as the procedure is simple and quick. The neochordae are created using 5–0 or 4–0 polytetrafluoroethylene (ePTFE) monofilament sutures usually made of Gore-Tex (W.L. Gore & Associates, Inc., Medical Products Division, Flagstaff, AZ).[50] An ePTFE suture is passed through the fibrous region of a papillary muscle on the same side of the valve (i.e., medial or lateral) as the

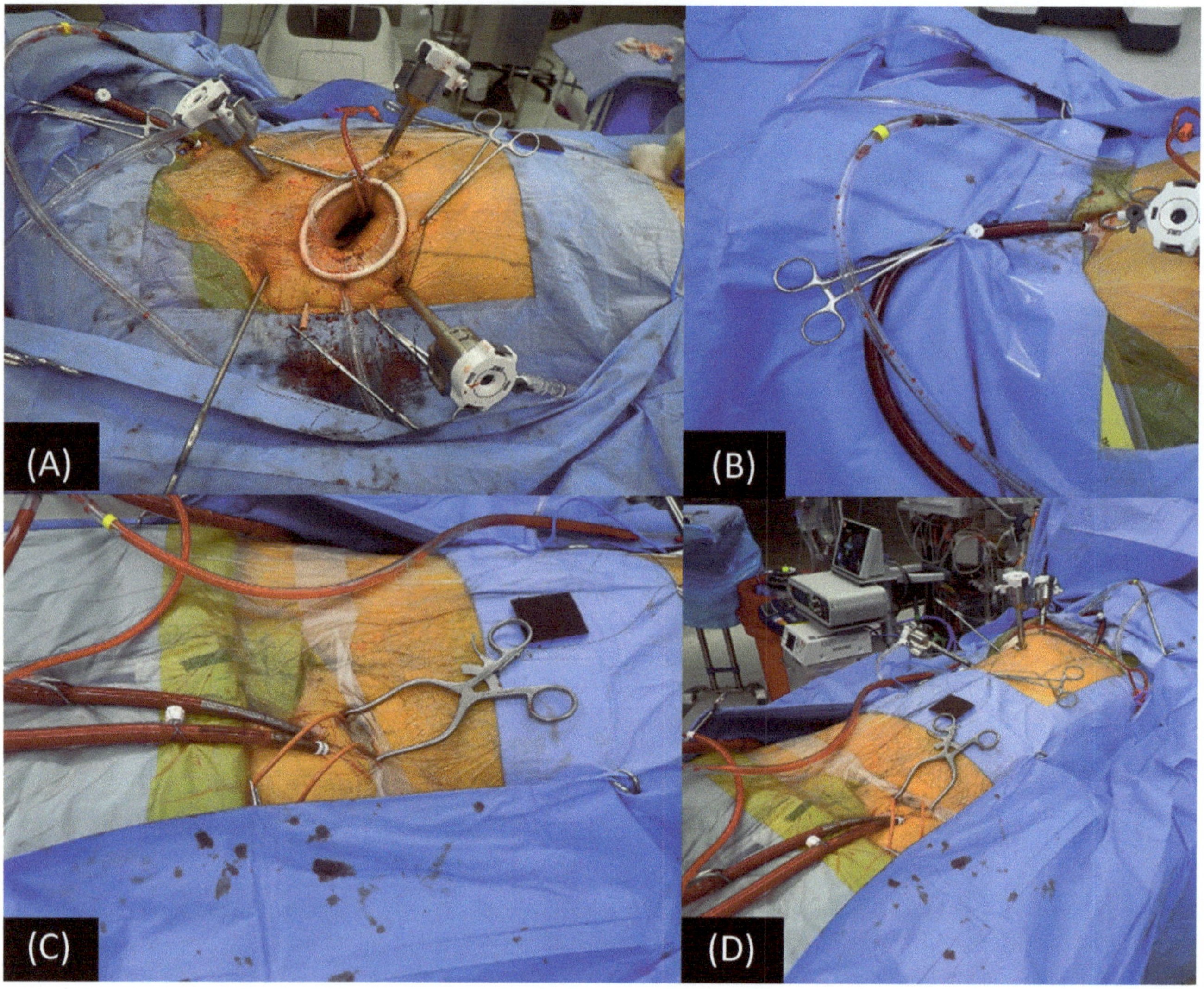

FIGURE 16.4 **Cannulation and port placement for robotic mitral valve surgery.** (A and B) A 4 cm working port about 3 cm lateral to the right nipple in the fourth intercostal space in the anterior axillary line; three trochars for robotic arms are inserted through the sixth intercostal space in the anterior axillary line, and the second intercostal space 3–5 cm anterior to the anterior axillary line for the right and left arms, respectively; a transthoracic atrial retractor port is placed in the fourth intercostal space at the midclavicular line, medial to the working port. (C) A 3–4 cm right groin incision is made for the femoral arterial and venous cannulas which are placed using Seldinger's technique and transesophageal echocardiogram guidance. The tip of the venous cannula is positioned 2–3 cm into the superior vena cava. The arterial cannula is positioned into the distal abdominal aorta or iliac artery. (D) View of the left femoral arterial and venous cannulas, and the right chest port placements.

region of prolapse. Each end of the suture is brought up to the leaflet edge and passed twice through the leaflet tissue in the region of prolapse; the needles pass from the ventricular to the atrial aspect and finish adjacent to each other on the atrial aspect (Fig. 16.5). The chordal length is adjusted to a level to prevent prolapse, using the region of P1 as a reference point. The suture is tied on the atrial side of the mitral valve leaflet. In cases when the posterior leaflet is exceedingly tall and bulky, a resectional technique (i.e., quadrangular resection and sliding repair) is an effective approach to prevent systolic anterior motion (SAM). In cases of a narrow P2 prolapse, triangular resection is used (Fig. 16.6).

Currently, creation of artificial chordae is the procedure of choice to correct isolated anterior leaflet prolapse of the mitral valve and has been associated with improved results.[51,52] A1 and A3 isolated prolapse are repaired with commisuroplasty technique.

Bileaflet prolapse, a characteristic of Barlow's disease, often presents a challenge in mitral valve repair. We favor creation of artificial chordae to the anterior and posterior leaflets, and placement of a large annuloplasty band. If there is posterior leaflet prolapse and anterior leaflet billowing with an anteriorly directed regurgitant jet, a posterior leaflet procedure will suffice (e.g., artificial chords or a resection).[53]

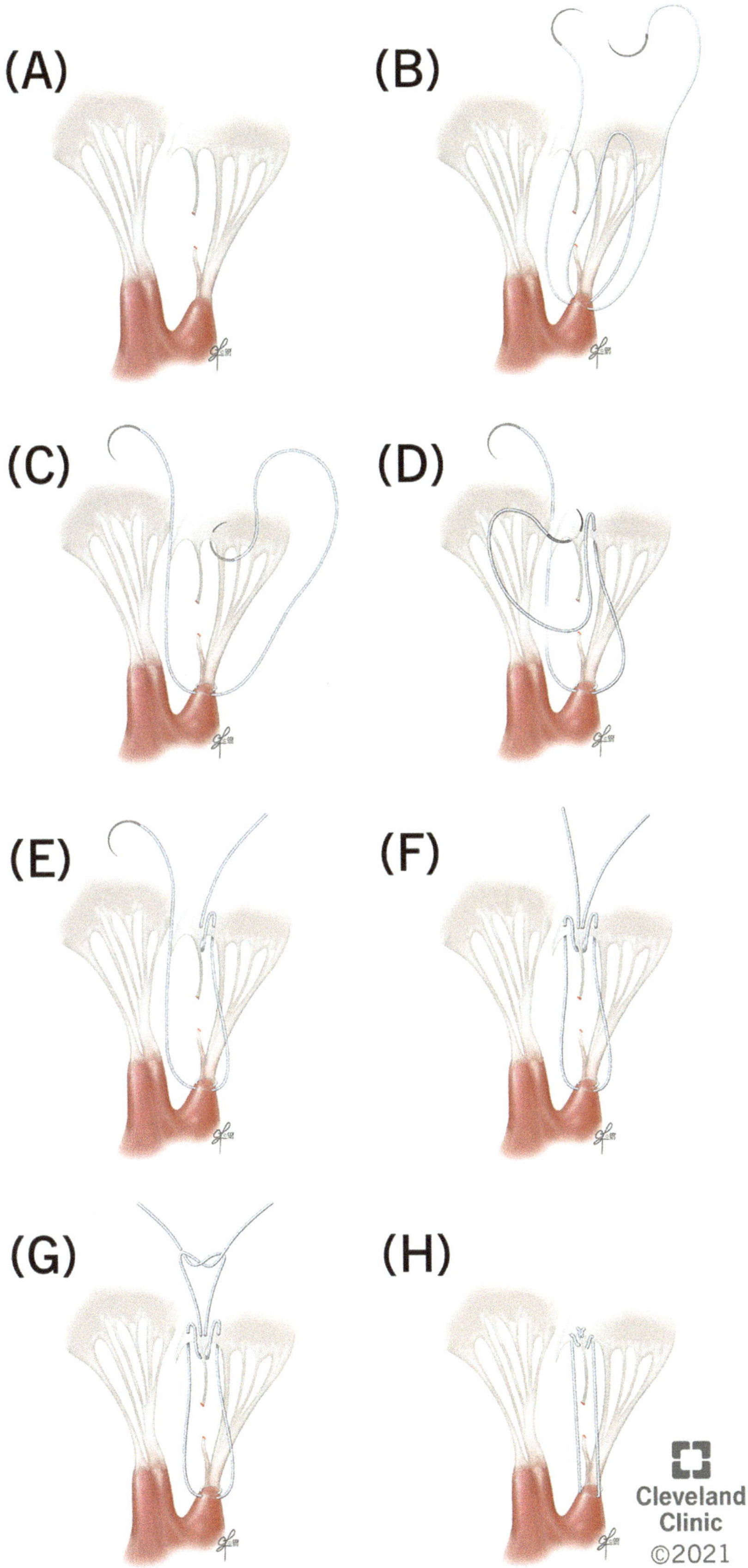

FIGURE 16.5 **Nonresectional mitral valve repair using free hand chordae.** The prolapsing segment is identified (A) and the neochordae is passed through the fibrous region of a papillary muscle on the same side on the valve (i.e., medial or lateral) as the region of prolapse (B). Each end of the neochordae is brought up to the leaflet edge (C) and passed twice through the leaflet tissue in the region of the prolapse (D–E); the needles pass from the ventricular to the atrial aspect and finish adjacent to each other (F). The chordal length is adjusted and the neochordae is tied on the atrial side of the mitral valve (G–H).

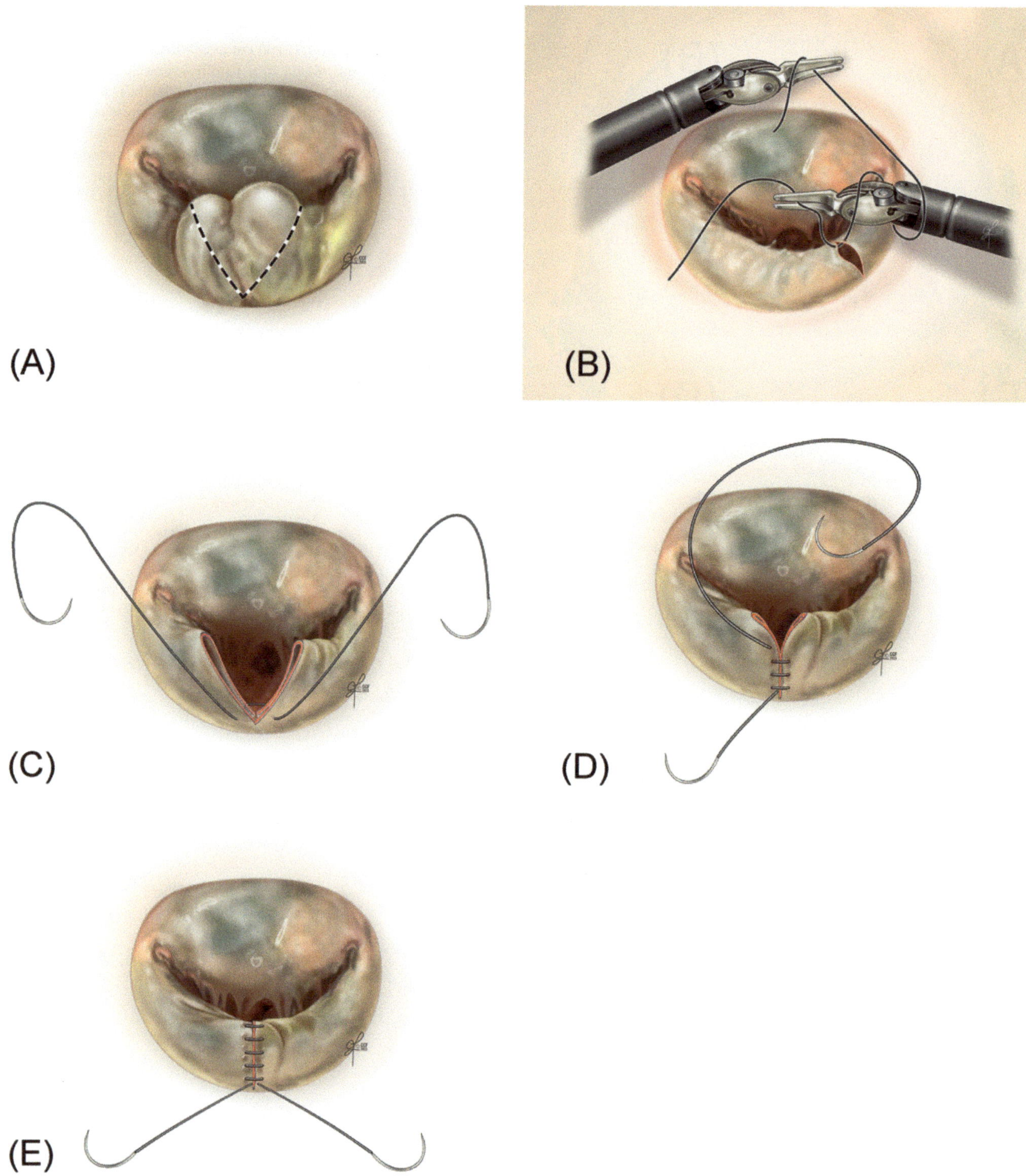

FIGURE 16.6 **Triangular resection for mitral valve repair.** A triangular shaped segment of the prolapsing leaflet tissue is excised (A). Leaflet continuity is restored (B–D).

An annuloplasty should be used in nearly all patients undergoing repair of a degenerative mitral valve to increase leaflet coaptation, restore annular size and shape, reduce tension on suture lines, and prevent future annular dilation. Mitral valvuloplasty improves repair durability regardless of the type of annuloplasty device.[54–58] We employ a running annuloplasty suture technique when applying the annuloplasty device. This method decreases operative time needed for

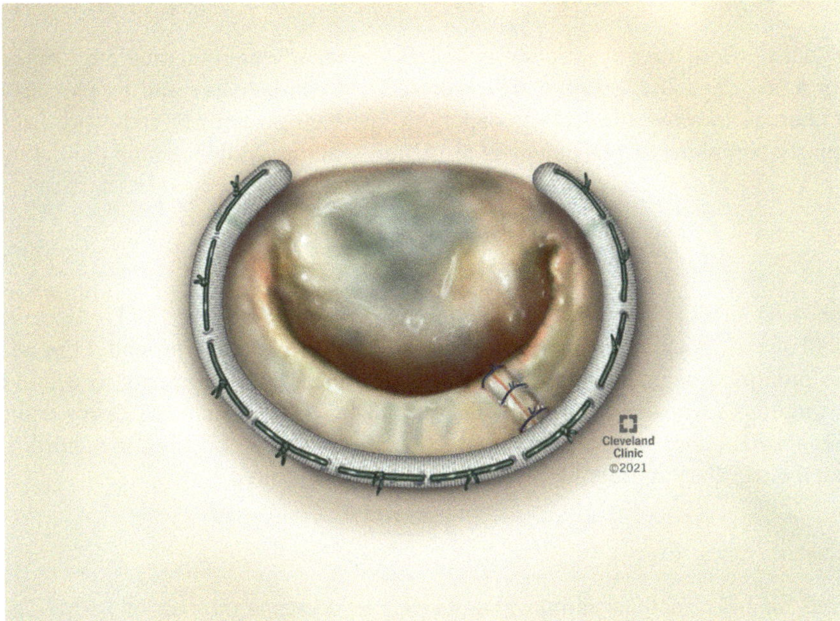

FIGURE 16.7 **Mitral annuloplasty.** An annuloplasty should be used in nearly all patients undergoing repair of a degenerative mitral valve to improve repair durability. A number of annuloplasty devices are available. The most frequently applied device at our institution is the partial Cosgrove–Edwards ring annuloplasty.

band placement. It entails placement of three 2–0 braided sutures, the first 16, the second 14, and the third 9 cm in length.[41] The first suture is tied at the right trigone and run clockwise to the midportion of the band. The second suture is tied at the midportion of the annulus, the tail is tied to the first suture and the suture is run clockwise. A third suture is tied at the left trigone and the tail is tied to the second suture (Fig. 16.7).

The atriotomy is closed with a running CV-2 ePTFE suture. The heart is filled and deaired through the atriotomy before the suture is tied. The antegrade cardioplegia catheter is then connected to suction for air removal and the cross-clamp is removed. Once transesophageal echocardiogram confirms successful repair, and normal heart rhythm is established, the cardioplegia catheter is removed and the puncture hole is closed with a pledged mattress suture. Instruments and ports are removed and cardiopulmonary bypass is discontinued.

Tips and pitfalls

Conversions

The robotic approach can be converted to a mini-anterolateral thoracotomy or a partial or complete sternotomy in cases when a robotic approach cannot provide safe and effective execution of the procedure. Although preoperative precautionary diagnostic measures are taken to avoid intraoperative conversions, the team must be prepared for the possibility of a conversion.

CO_2 insufflation

CO_2 should be continuously insufflated into the chest cavity to reduce the risk of neurologic injury due to gas embolism.[59] It is usually insufflated directly with the camera port.

Transthoracic cross clamp

Placement of the transthoracic aortic cross clamp carries risk of left atrial appendage and pulmonary artery injury.

Bleeding

Following protamine administration and decannulation, the right lung is again deflated and an endoscope introduced into the chest cavity to inspect possible bleeding sites. Hemostasis is achieved using electrocautery and additional sutures.

Systolic anterior motion

SAM of the mitral valve causing left ventricular outflow tract obstruction is a well-known adverse outcome after mitral valve repair. It may occur intraoperatively or a few days after surgery. The risk of SAM can be decreased by placing a large annuloplasty band and using artificial chordae to move the leaflet coaptation line well posteriorly and away from the left ventricular outflow tract. A sliding plasty technique or placement of an Alfieri stitch may also reduce the risk of SAM.[53,60,61]

Artificial chordae length

The neochordae are created using 5–0 or 4–0 ePTFE monofilament sutures made of Gore-Tex (W.L. Gore & Associates, Inc, Medical Products Division, Flagstaff, AZ).[50] The usual length for posterior chords is around 1 cm and is established by lowering the height of the prolapsed segment to the level of the nonprolapsed segments to achieve good leaflet coaptation. Additionally, some surgeons favor direct measurement and construction of chordal loops using a caliper.[62,63] They measure the length of the normal chord, adjacent to the area of prolapse, and construct the chord. If there is no reference chord, the distance from the papillary muscle head to the annulus is measured.[64]

Annuloplasty band size

A large annuloplasty ring or band is preferred for myxomatous mitral valve disease to decrease the risk of SAM. The ring or band size is determined by measuring the height of the anterior leaflet with appropriate sizers.[41]

Advantages and disadvantages of robotic mitral valve repair

The initial goal is to provide safe and effective repair of the mitral valve. Longer bypass and cross clamp times and high initial conversion rates have been important arguments against the robotic approach. However, increasing surgeon and team experience has led to decreases in bypass and cross clamp times which are now comparable to sternotomy-based approaches.[23,34] Similarly, experience and more careful patient selection brought down the conversion rates. In addition, a robotically assisted approach provides benefits such as better cosmesis, a lower transfusion requirement, less postoperative pain and infectious complications, shorter Intensive Care Unit and hospital stays, and faster return to work.[12,20,23,28,34,43,65,66] For the surgeon, it provides excellent visualization of structures and better spatial orientation, increased dexterity, motion scaling, and tremor reduction. Robotic mitral valve surgery can be performed with similar safety and efficacy as conventional sternotomy-based approached, and has excellent short and midterm results.[21,26–28,32,34,43,66–69] Hospital mortality is generally less than 1% and stroke rate is similarly low. On the other hand, disadvantages include lack of tactile sense and increased initial cost due to high initial investment and maintenance. It is suggested that the higher cost could be offset by shorter Intensive Care Unit and hospital stay, and societal benefit from a faster return to work.

Summary and future aspects

Over the past decades, the operative techniques to repair the mitral valve have evolved with an increasing interest in the application of minimally invasive approaches. Today, robotic mitral valve repair is the least invasive approach to the mitral valve, and is as safe and reproducible as other approaches across all types of mitral valve repair with a hospital mortality of generally less than 1% and stroke rate similarly low. Robotic mitral valve surgery has proven to be a safe, reproducible technique with late outcomes similar to conventional sternotomy-based approaches with several advantages. Furthermore, it provides the surgeon with excellent visualization of structures and better spatial orientation, increased dexterity, motion scaling and tremor reduction. As a result of improved technology and surgical techniques, feasibility of robotic mitral valve repair has been demonstrated in both simple and complex repairs. A wider application can be expected in the future with further advancement in robotic technology and procedures. Broader application of this technology is currently only feasible in high robotic volume centers due to the long learning curve and costs however, the accessibility to robotic treatment is increasing as the technology becomes more affordable while offered in a multidisciplinary setting. Furthermore, training and mentoring opportunities are becoming more available worldwide. Case volume, careful patient selection, and a team approach is vital to robotic surgery to navigate the significant learning curve.

Robotic mitral valve surgery has evolved into a safe and effective surgical option and presents a significant paradigm shift in modern cardiac surgery. Application of robotic technology has resulted in reduced recovery time while maintaining safety and outcome standards in mitral valve surgery.

References

1. Desai MY, Grigioni F, di Eusanio M, et al. Outcomes in degenerative mitral regurgitation: current state-of-the art and future directions. W.B. Saunders *Prog Cardiovasc Dis*. 2017;370–385.
2. Nishimura R, Otto CM, Bonow RO, et al. AHA/ACC focused update of the 2014 AHA/ACC guideline for the management of patients with valvular heart disease. *Circulation*. 2017;135:e1159–e1195.
3. Lazam S, Vanoverschelde JL, Tribouilloy C, et al. Twenty-year outcome after mitral repair vs replacement for severe degenerative mitral regurgitation: analysis of a large, prospective, multicenter, international registry. Lippincott Williams and Wilkins *Circulation*. 2017;135:410–422.
4. Mick SL, Keshavamurthy S, Gillinov AM. Mitral valve repair vs replacement. *Ann Cardiothorac Surg*. 2015;4:230–237.
5. David TE, Armstrong S, McCrindle BW, Manlhiot C. Late outcomes of mitral valve repair for mitral regurgitation due to degenerative disease. *Circulation*. 2013;127:1485–1492.
6. Suri RM, Schaff HV, Dearani JA, et al. Survival advantage and improved durability of mitral repair for leaflet prolapse subsets in the current era. *Ann Thorac Surg*. 2006;82:819–826.
7. Daneshmand MA, Milano CA, Rankin JS, et al. Mitral valve repair for degenerative disease: a 20-year experience. *Ann Thorac Surg*. 2009;88:1828–1837.
8. Anyanwu AC, Bridgewater B, Adams DH. The lottery of mitral valve repair surgery. *Heart*. 2010;1964–1967.
9. Castillo JG, Anyanwu AC, Fuster V, Adams DH. A near 100% repair rate for mitral valve prolapse is achievable in a reference center: implications for future guidelines. *J Thorac Cardiovasc Surg*. 2012;144:308–312. Mosby Inc.
10. Seeburger J, Borger MA, Falk V, et al. Minimal invasive mitral valve repair for mitral regurgitation: results of 1339 consecutive patients. *Eur J Cardiothorac Surg*. 2008;34:760–765.
11. Lapenna E, Torracca L, de Bonis M, la Canna G, Crescenzi G, Alfieri O. Minimally invasive mitral valve repair in the context of Barlow's disease. Elsevier USA *Ann Thorac Surg*. 2005;79:1496–1499.
12. Folliguet T, Vanhuyse F, Constantino X, Realli M, Laborde F. Mitral valve repair robotic vs sternotomy. *Eur J Cardiothorac Surg*. 2006;362–366.
13. Jones B, Krueger S, Howell D, Meinecke B, Dunn S. Robotic mitral valve repair. *Tex Heart Inst J*. 2005;32:143–146.
14. Chitwood WJ, Kypson A, Nifong L. Robotic mitral valve surgery: a technologic and economic revolution for heart centers. *Am Heart Hosp J*. 2003;1:30–39.
15. Chitwood WR, Nifong LW, Elbeery JE, et al. Robotic mitral valve repair: trapezoidal resection and prosthetic annuloplasty with the da Vinci Surgical System. *J Thorac Cardiovasc Surg*. 2000;120:1171–1172. Mosby Inc.
16. Morgan JA, Thornton BA, Peacock JC, et al. Does robotic technology make minimally invasive cardiac surgery too expensive? A hospital cost analysis of robotic and conventional techniques. *J Card Surg*. 2005;20(3):246–251.
17. Mohr FW, Falk V, Diegeler A, et al. Minimally invasive port-access mitral valve surgery. *J Thorac Cardiovasc Surg*. 1998;115(3):567–574.
18. Grossi EA, LaPietra A, Applebaum RM, et al. Case report of robotic instrument-enhanced mitral valve surgery. *J Thorac Cardiovasc Surg*. 2000;120:1169–1171. Mosby Inc.
19. Nifong LW, Chu VF, Marcus Bailey B, et al. Robotic mitral valve repair: experience with the da Vinci system. *Ann Thorac Surg*. 2003;75(2): 438–442.
20. Randolph Chitwood Jr. W, Wiley Nifong L. Minimally Invasive videoscopic mitral valve surgery: the current role of surgical robotics. *J Card Surg*. 2000;15(1):61–75.
21. Chitwood WR, Rodriguez E, Chu MWA, et al. Robotic mitral valve repairs in 300 patients: a single-center experience. *J Thorac Cardiovasc Surg*. 2008;136:436–441.
22. Nifong LW, Chitwood WR, Pappas PS, et al. Robotic mitral valve surgery: a United States multicenter trial. *J Thorac Cardiovasc Surg*. 2005;129:1395–1404.
23. Gillinov AM, Mihaljevic T, Javadikasgari H, et al. Early results of robotically assisted mitral valve surgery: analysis of the first 1000 cases. *J Thorac Cardiovasc Surg*. 2018;155:82–91. e2, Mosby Inc.
24. Suri RM, Dearani JA, Mihaljevic T, et al. Mitral valve repair using robotic technology: safe, effective, and durable. *J Thorac Cardiovasc Surg*. 2016;1450–1454. Mosby Inc.
25. Chitwood WR. Robotic mitral valve surgery: overview, methodology, results, and perspective. *Ann Cardiothorac Surg*. 2016;5:544–555. AME Publishing Company.
26. Gillinov AM, Suri R, Mick S, Mihaljevic T. Robotic mitral valve surgery: current limitations and future directions. *Ann Cardiothorac Surg*. 2016;5:573–576. AME Publishing Company.
27. Hawkins RB, Mehaffey JH, Mullen MM, et al. A propensity matched analysis of robotic, minimally invasive, and conventional mitral valve surgery. *Heart*. 2018; BMJ Publishing Group.
28. Murphy DA, Miller JS, Langford DA, Snyder AB. Endoscopic robotic mitral valve surgery. *J Thorac Cardiovasc Surg*. 2006;132:776–781.

29. Akpinar B, Guden M, Sagbas E, Sanisoglu I, Caynak B, Bayramoglu Z. Robotic-enhanced totally endoscopic mitral valve repair and ablative therapy. *Ann Thorac Surg*. 2006;81:1095–1098.
30. Smith JM, Stein H, Engel AM, McDonough S, Lonneman L. Totally endoscopic mitral valve repair using a robotic-controlled atrial retractor. *Ann Thorac Surg*. 2007;84:633–637.
31. Reade CC, Johnson JO, Bolotin G, et al. Combining robotic mitral valve repair and microwave atrial fibrillation ablation: techniques and initial results. *Ann Thorac Surg*. 2005;79:480–484.
32. Woo YJ, Nacke EA. Robotic minimally invasive mitral valve reconstruction yields less blood product transfusion and shorter length of stay. *Surgery*. 2006;140:263–267.
33. Maselli D, Pizio R, Borelli G, Musumeci F. Institutional report—valves. Endovascular balloon vs transthoracic aortic clamping for minimally invasive mitral valve surgery: impact on cerebral microemboli. *Interact Cardiovasc Thorac Surg*. 2006;5:183–186.
34. Mihaljevic T, Jarrett CM, Gillinov AM, et al. Robotic repair of posterior mitral valve prolapse vs conventional approaches: potential realized. *J Thorac Cardiovasc Surg*. 2011;141. Mosby Inc.
35. Mohr FW, Onnasch JF, Falk V, et al. The evolution of minimally invasive valve surgery - 2 year experience. *Eur J Cardiothorac Surg*. 1999; 15(3):233–238.
36. Shah J, Vyas A, Vyas D. The history of robotics in surgical specialties. *Am J Robot Surg*. 2015;1:12–20. American Scientific Publishers.
37. Loulmet DF, Ranganath NK, Neuburger PJ, Nampiaparampil RG, Galloway AC, Grossi EA. Can complex mitral valve repair be performed with robotics? An institution's experience utilizing a dedicated team approach in 500 patients. *Eur J Cardiothorac Surg*. 2019;470–478.
38. Murphy DA, Moss E, Binongo J, et al. The expanding role of endoscopic robotics in mitral valve surgery: 1,257 consecutive procedures. Elsevier USA *Ann Thorac Surg*. 2015;100:1675–1682.
39. Rodríguez E, Kypson AP, Moten SC, Nifong LW, Chitwood WR. Robotic mitral surgery at East Carolina University: a 6 year experience. *Int J Med Robot Computer Assist Surg*. 2006;211–215.
40. Cheng W, Fontana GP, de Robertis MA, et al. Is robotic mitral valve repair a reproducible approach? *J Thorac Cardiovasc Surg*. 2010;139:628–633.
41. Mihaljevic T, Jarrett CM, Gillinov AM, Blackstone EH. A novel running annuloplasty suture technique for robotically assisted mitral valve repair. *J Thorac Cardiovasc Surg*. 2010;139:1343–1344.
42. Mihaljevic T, Pattakos G, Gillinov AM, et al. Robotic posterior mitral leaflet repair: neochordal vs resectional techniques. *Ann Thorac Surg*. 2013;95:787–794.
43. Nifong LW, Rodriguez E, Chitwood WR. 540 Consecutive robotic mitral valve repairs including concomitant atrial fibrillation cryoablation. *Ann Thorac Surg*. 2012;94:38–43.
44. Suri RM, Burkhart HM, Daly RC, et al. Robotic mitral valve repair for all prolapse subsets using techniques identical to open valvuloplasty: establishing the benchmark against which percutaneous interventions should be judged. *J Thorac Cardiovasc Surg*. 2011;142:970–979.
45. Malas T, Mick S, Wierup P, Gillinov M. Five maneuvers to facilitate faster robotic mitral valve repair. *Semin Thorac Cardiovasc Surg*. 2019;31:48–50. W.B. Saunders.
46. Chemtob RA, Wierup P, Mick SL, et al. A conservative screening algorithm to determine candidacy for robotic mitral valve surgery. *J Thorac Cardiovasc Surg*. 2021; Mosby Inc.
47. Otto CM, Nishimura RA, Bonow RO, et al. ACC/AHA guideline for the management of patients with valvular heart disease: a report of the American College of Cardiology/American Heart Association Joint Committee on Clinical Practice Guidelines. *J Thorac Cardiovasc Surg*. 2020;2021. Mosby Inc.
48. Javadikasgari H, Suri RM, Gillinov AM. Robotic mitral valve repair: algorithmic approach in degenerative mitral valve disease. *Ann Cardiothorac Surg*. 2016;5:586–588. AME Publishing Company.
49. Mick SL, Robich MP, Houghtaling PL, et al. Del Nido vs Buckberg cardioplegia in adult isolated valve surgery. *J Thorac Cardiovasc Surg*. 2015;626–636. e5, Mosby Inc.
50. David TE. Artificial chordae. W.B. Saunders *Semin Thorac Cardiovasc Surg*. 2004;161–168.
51. Phillips MR, Daly RC, Schaff HV, Dearani JA, Mullany CJ, Orszulak TA. Repair of anterior leaflet mitral valve prolapse: chordal replacement vs chordal shortening. *Ann Thorac Surg*. 2000;25–29.
52. Zussa Z, Polesel E, Rocco F. Artificial chordae in the treatment of anterior mitral leaflet pathology. *Cardiovasc Surg*. 1997;5:125–128.
53. Chemtob RA, Mick S, Gillinov M, Hodges K, Wierup P. Repair of bileaflet prolapse in Barlow syndrome: the 4-chord technique. *J Card Surg*. 2019;34:605–609. Blackwell Publishing Inc.
54. Gillinov AM, Tantiwongkosri K, Blackstone EH, et al. Is prosthetic anuloplasty necessary for durable mitral valve repair? *Ann Thorac Surg*. 2009;88:76–82.
55. Pierce EL, Bloodworth CH, Imai A, et al. Mitral annuloplasty ring flexibility preferentially reduces posterior suture forces. *J Biomech*. 2018;75:58–66. Elsevier Ltd.
56. Chang BC, Youn YN, Ha JW, Lim SH, Hong YS, Chung N. Long-term clinical results of mitral valvuloplasty using flexible and rigid rings: a prospective and randomized study. *J Thorac Cardiovasc Surg*. 2007;133:995–1003.
57. Kunzelman K, Reimink M. Flexible vs regid ring annuloplasty for mitral valve annular dilatation: a finite element model. *J Hear Valve Dis*. 1998;7:108–116.
58. Chee T, Haston R, Togo A, Raja SG. Is a flexible mitral annuloplasty ring superior to a semi-rigid or rigid ring in terms of improvement in symptoms and survival? *Interact Cardiovasc Thorac Surg*. 2008;7:477–484.

59. Gallego-Poveda J, Guerra NC, Carvalheiro C, et al. Use of CO_2 in video assisted thoracic surgery and single-lumen endotracheal tube-a new less invasive approach. *J Thorac Dis*. 2017;903–906. AME Publishing Company.
60. George KM, Gillinov AM. Posterior leaflet shortening to correct systolic anterior motion after mitral valve repair. *Ann Thorac Surg*. 2008;86:1699–1700.
61. Gillinov AM, Smedira NG, Shiota T. Use of the alfieri edge-to-edge technique to eliminate left ventricular outflow tract obstruction caused by mitral systolic anterior motion. *Ann Thorac Surg*. 2004;78.
62. Gillinov M, Quinn R, Kerendi F, et al. Premeasured chordal loops for mitral valve repair. Elsevier USA *Ann Thorac Surg*. 2016;102:e269–e271.
63. Otto Von Oppell U, Mohr FW. Chordal replacement for both minimally invasive and conventional mitral valve surgery using premeasured gore-tex loops. *Ann Thorac Surg*. 2000.
64. Chemtob RA, Wierup P, Mick S, Gillinov M. Choosing the "Best" surgical techniques for mitral valve repair: lessons from the literature. *J Card Surg*. 2019;717–727. Blackwell Publishing Inc.
65. Nifong LW, Chitwood WR. Building a surgical robotics program. *Am J Surg*. 2004;188:16–18. Elsevier Inc.
66. Suri RM, Antiel RM, Burkhart HM, et al. Quality of life after early mitral valve repair using conventional and robotic approaches. *Ann Thorac Surg*. 2012;93:761–769.
67. Stevens LM, Rodriguez E, Lehr EJ, et al. Impact of timing and surgical approach on outcomes after mitral valve regurgitation operations. *Ann Thorac Surg*. 2012;93:1462–1468.
68. Ramzy D, Trento A, Cheng W, et al. Three hundred robotic-assisted mitral valve repairs: the Cedars-Sinai experience. *J Thorac Cardiovasc Surg*. 2014;147:228–235.
69. Suri RM, Taggarse A, Burkhart HM, et al. Robotic mitral valve repair for simple and complex degenerative disease. Lippincott Williams and Wilkins *Circulation*. 2015;132:1961–1968.

Chapter 17

Transcatheter Aortic Valve Procedures: Technology Update

Amedeo Anselmi[1,2], Vincent Auffret[2,3] and Jean-Philippe Verhoye[1,2]
[1]Division of Thoracic and Cardiovascular Surgery, Pontchaillou University Hospital, Rennes, France, [2]Univ Rennes, CHU Rennes, Inserm, LTSI - UMR 1099, Rennes, France, [3]Division of Cardiology, Pontchaillou University Hospital, Rennes, France

Chapter Outline

Introduction

The first-in-man with a balloon-expandable transcatheter heart valve (THV) in the aortic position was pioneered by Cribier,[1] although the first experimental works date back to 1965 and 1971.[2] Since then, Transcatheter aortic valve implantation (TAVI) has incessantly grown as a technique and a practice. It is one of the fields in the recent cardiovascular landscape where the accumulation of knowledge and synergy between clinical science and engineering have progressed at the greatest pace, and achieved a major breakthrough for the radical transformation of clinical practice in a relatively short time frame. The standardization of TAVI satisfied unanswered clinical needs in the treatment of patients deemed too severe or fragile to undergo surgery, and subsequently in the reduction of invasiveness of the curative treatment of severe aortic stenosis in patients at high and intermediate operative risk.[3] Data are currently accumulating about its use in patients deemed to be at low operative risk.[4–6] Developments in technology and design of THVs have been fundamental in such evolution, which has nonetheless entailed failures and dead ends as well.

Multiple THV concepts have been proposed over the last two decades, and numerous factors can determine their success or failure. Our purpose herein will be to provide an overview of design concepts for THVs intended for the management of aortic valve disease, with a description of their peculiar technological and design features, and we discuss their impact on the clinical results. Such information will be put in perspective with the current trends in the TAVI practice.

Advances in Cardiovascular Technology. DOI: https://doi.org/10.1016/B978-0-12-816861-5.00001-0

Anatomical and clinical requirements for TAVI

Severe aortic valve stenosis is currently the valve disturbance most frequently treated by surgical or transcatheter means. Its prevalence has been estimated to be 3.4% in the elderly (≥75 years) population.[7] Given the greater prevalence of calcified aortic stenosis than aortic regurgitation, and as the calcifications constitute a landing zone to achieve THV anchoring, the vast majority of THVs described so far (and all THVs which are beyond the investigational phase) are designed and labeled for employment in calcified aortic stenosis only. On the other hand, the success of TAVI might be at least in part ascribed to the anatomical features of the aortic valve and root in aortic stenosis, which present several favorable characteristics.

The aortic root constitutes a morphofunctional unit (aorta–valve complex) including three rings (the sinotubular junction, the ventriculo–arterial junction, and the virtual basal ring joining the hinge point of each leaflet) and one crown-like ring (corresponding to the valve leaflet insertion) (Fig. 17.1).[8,9] The illustration also depicts the anatomical continuity with the aortomitral curtain, the right and left fibrous trigones, the membranous septum continuous to the right/noncoronary interleaflet triangle intimately related to the bundle of His, and the coronary ostia in their most frequent situation. The aortic annulus is a dynamic structure, appearing more circular in systole and more elliptical in diastole[10]; nonetheless, in aortic stenosis the calcified leaflets and annulus induce modified tissue material properties and little or no changes in annular geometry between systole and diastole. This aspect, together with the bulky calcifications, favors the achievement of THV stability across the cardiac cycle. On the other hand, calcifications can be extremely variable in their volume and distribution; they show variable extension within the native annulus and toward the left ventricular outflow tract/aortomitral continuity (Fig. 17.2). These factors induce irregular landing zones explaining the challenges encountered to obtain sealing and consequently the occurrence of perivalvular leaks (PVLs) with first- and second-generation THVs. Optimal sealing is of major importance given the detrimental prognostic effect of more-than-mild aortic regurgitation due to PVL.[11] THV stability can be based on the annular/leaflet anchoring alone, or in association with supravalvular support structures. The above anatomical elements also illustrate the risk of conduction disturbance potentially leading to permanent pacemaker (PPM) implantation, which has also been recognized to have detrimental prognostic impact in the mid- to long term,[12] as well as the obvious importance of maintaining clearance of the coronary ostia both immediately and at follow-up (delayed coronary obstruction[13]). The design of any THVs, therefore, needs to comply with such anatomical and pathophysiological constraints. The depth of implantation

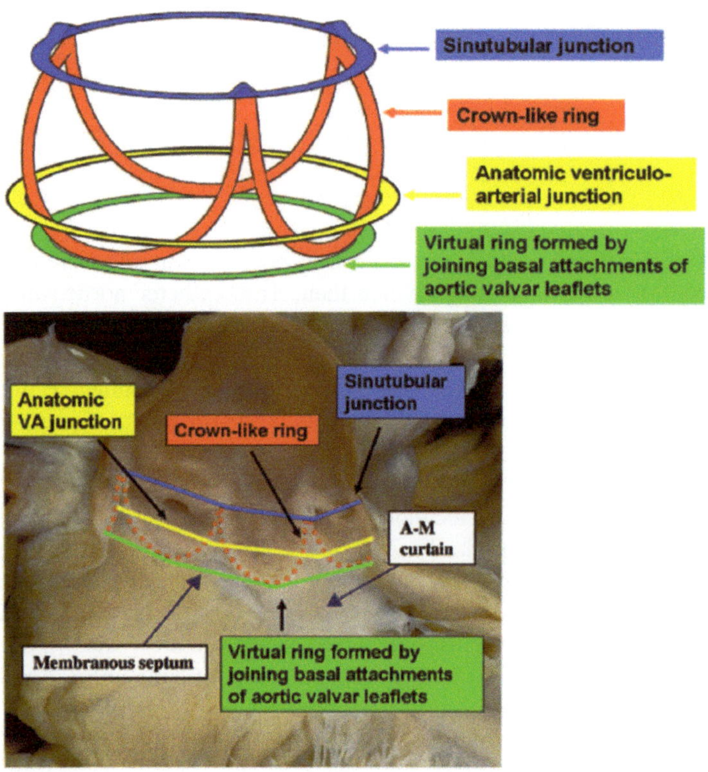

FIGURE 17.1 The three-dimensional anatomy of the aortic root, including notable elements to be accounted for in the design of THVs. *Reproduced with permission from Piazza N, de Jaegere P, Schultz C, Becker AE, Serruys PW, Anderson RH. Anatomy of the aortic valvar complex and its implications for transcatheter implantation of the aortic valve. Circ Cardiovasc Interv. 2008;1:74–81.*[8]

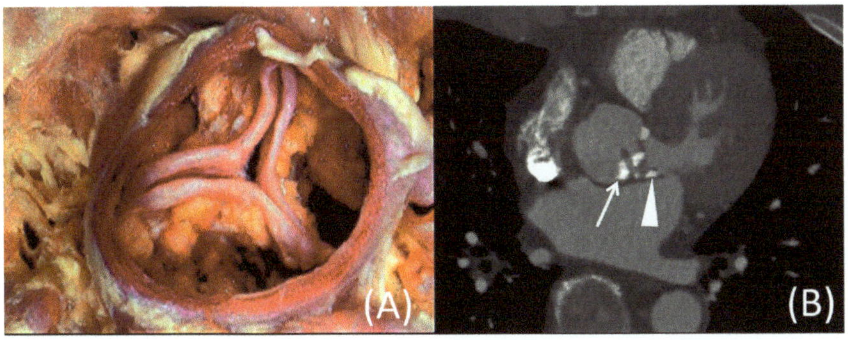

FIGURE 17.2 (A) Pathoanatomical aspect of calcified aortic stenosis in a trileaflet valve, aortic-side view. (B) CT scan imaging showing calcifications of the aortic valve in severe aortic stenosis (white arrow), associated with calcification of the aortomitral fibrous body and anterior mitral annulus (white arrowhead).

emerged as a determinant factor to be precisely controlled in order to minimize the risk of PVL and/or PPM. As well, the persistence of mitral regurgitation after TAVI is prognostically unfavorable, while the relative role of THV depth (continuity with the mitral apparatus) and the degree of intrinsic myocardial disease (secondary mitral regurgitation) in determining persistence of mitral regurgitation after TAVI is an object of debate. Concordantly, the main domains where technology/design efforts have to be provided include obtaining effective stability and sealing, limiting the occurrence of conduction disturbances, avoiding immediate or delayed obstruction of the coronary ostia, and potential to recapture and reposition the THV should the implantation depth/valve positioning be considered unsatisfactory. Overall, the landscape entails two main technological strategies to achieve stability and sealing in the context of calcified aortic stenosis. A self-expanding nitinol frame has been employed in the majority of the proposed concepts; nitinol has superelasticity and shape memory properties allowing collapsed frames into miniaturized delivery systems. Additionally, nitinol frames tend to continue radial expansion when exposed to body temperature, which is of potential interest for the reduction of PVL. One balloon-expandable concept has been employed for first-in-man implantation and is now largely diffused in the global practice. Other design pathways have been proposed, but they did not develop into clinically accepted/generalized employment, as detailed below.

The percutaneous transfemoral arterial access represents the ideal route for THV delivery: as a percutaneous access, it allows procedure performance under local anesthesia, minimal aggressiveness and faster recovery. Concordantly, miniaturization of sheaths and delivery systems has represented a main axis of development in order to allow compatibility with challenging iliofemoral anatomies. Not only absolute size of introducers and delivery systems but also gliding properties and flexibility have been addressed to various degrees and combination. The most largely diffused third-generation systems have a 14 Fr-to-16 Fr sheath internal diameter with expansion to 21 Fr during passage of the valve (balloon-expandable), and a 14 Fr-to-16 Fr profile (self-expandable). Characteristics of native iliofemoral arteries to be considered in decision-making are the inner diameter, the tortuosity, the degree and distribution of calcification, and luminal stenosis. The occurrence of access-related complications is associated with increased mortality/morbidity and prolonged hospital stay after TAVI.[14,15] In that sense, the adequacy of THV delivery systems in addressing the above requirements is one determinant of device-related outcomes and procedural success. Despite the use of current-generation devices, in major national registries the employment of a nontransfemoral access for THV delivery is variable and comprises between 10% and 15%.[16,17] THV sizing and choice of the arterial access are based on the evaluation of preoperative contrast-enhanced CT (computed tomography)-scan imaging which are currently performed through dedicated software in a semiautomatic fashion (Figs. 17.3 and 17.4). The adequacy of such evaluation, partially depending on operators' experience, remains a major determinant of final success.

Technological foundations of contemporary TAVI practice: valve devices

The SAPIEN 3 transcatheter heart valve

The Edwards SAPIEN 3 (S3) (Fig. 17.5), currently used in routine practice, represents the fifth generation of this balloon-expandable THV (Fig. 17.6). Building upon the success of previous generations, it was designed to circumvent their limitations including access-site related injury, device malposition, paravalvular leak, or periprocedural strokes. The original Cribier–Edwards prosthesis consisted of a laser-cut stainless steel tubular frame with equine pericardium leaflets. It quickly evolved, with the SAPIEN and SAPIEN XT platforms, toward a more refined cobalt–chromium stent within which are sewn bovine pericardium leaflets undergoing anticalcification treatment (Thermafix) intended to reproduce the durability of the surgically proven Perimount Magna Ease bioprosthesis (Edwards Lifesciences). The

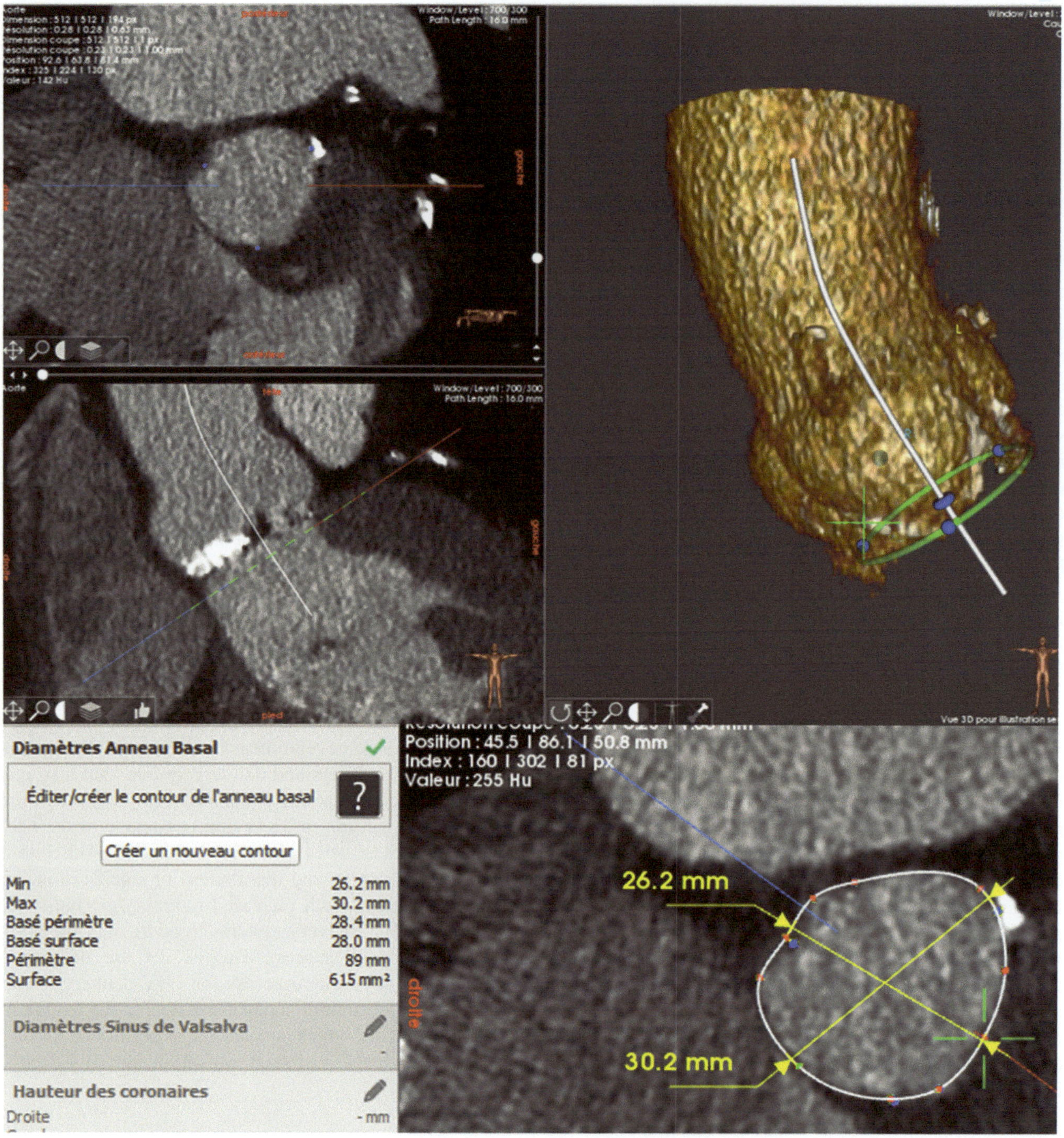

FIGURE 17.3 Example of CT-scan based semiautomatic THV sizing (generated through the EndoSize software). Identification of the virtual basal ring, of the annular perimeter and surface.

optimized cobalt–chromium frame of the S3 allows for a very low-crimped profile while maintaining a high radial strength for circularity and optimal hemodynamics.[18,19] The stent also incorporates a large open cell geometry on its outflow side and has a limited expanded height (15.5, 18, 20, and 22.5 mm for the 20, 23, 26, and 29 mm valves, respectively) to facilitate coronary access (Fig. 17.7). The pericardial leaflets are mounted inside the stent and solidarized with the inner fabric skirt. Compared with the SAPIEN XT, which only incorporated this inner skirt, the S3 adds an adaptive external polyethylene terephthalate outer skirt designed to reduce the rate of paravalvular leak (PVL) (Fig. 17.8). As its predecessors, the S3 is based on an intraannular deployment concept. Depending on the prosthesis

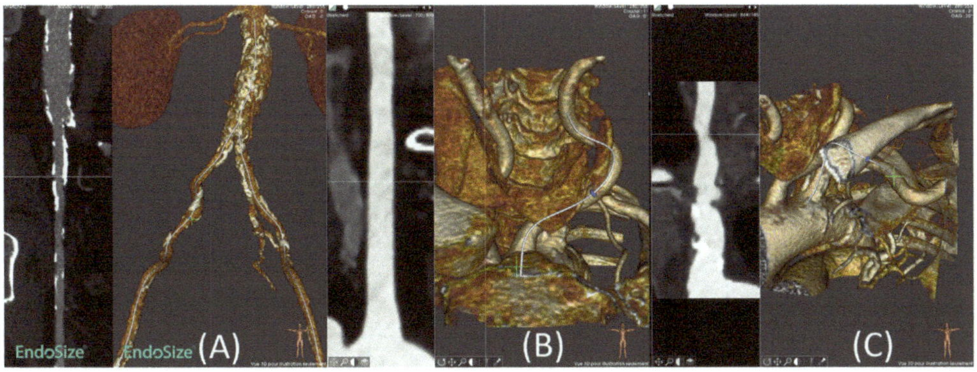

FIGURE 17.4 Example of CT-scan based semiautomatic evaluation of arterial access for TAVI (generated through the EndoSize software). (A) Evaluation of the iliofemoral axis. (B) Evaluation of the left common carotid artery. (C) Evaluation of the left subclavian artery.

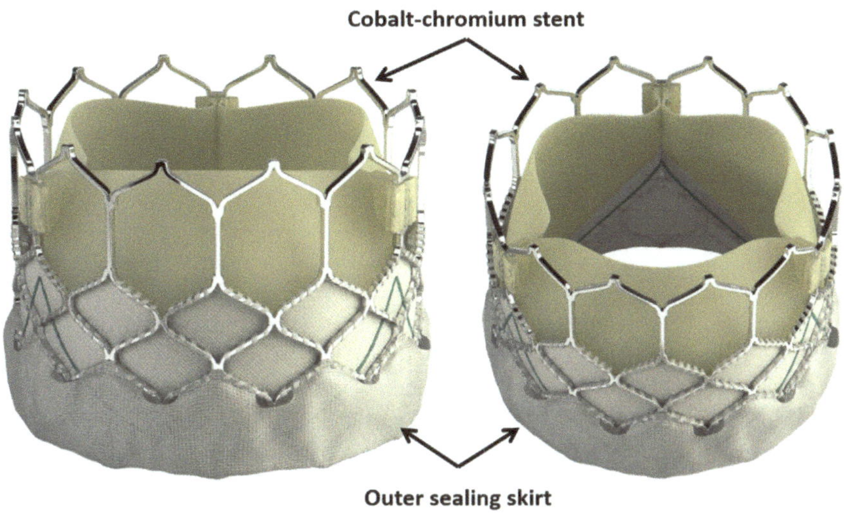

FIGURE 17.5 The Sapien 3 valve, side and side upper views. *Courtesy Edwards Lifesciences. Reproduced with permission.*

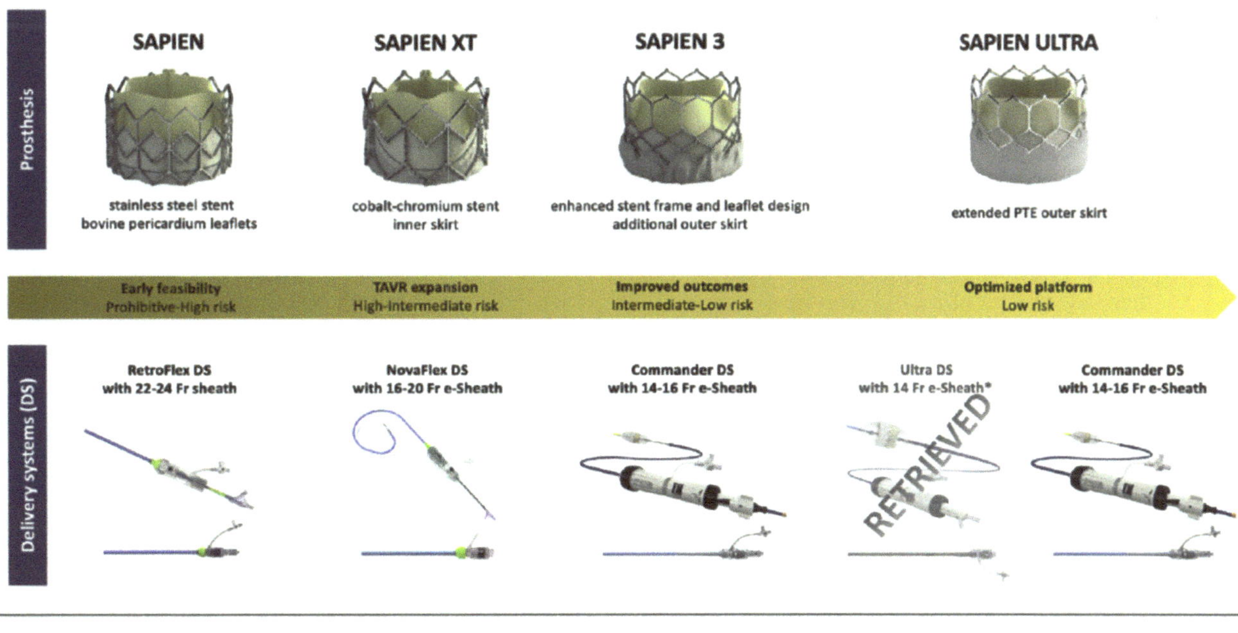

TAVR = transcatheter aortic valve replacement. *Retrieved August 2019.

FIGURE 17.6 Evolution of the Sapien THV valve models and delivery systems. *Reproduced with permission from Barbanti M, Costa G. SAPIEN 3 Ultra transcatheter aortic valve device: two sides of the moon. JACC Cardiovasc Interv. 2020;13:2639–2641.*[26]

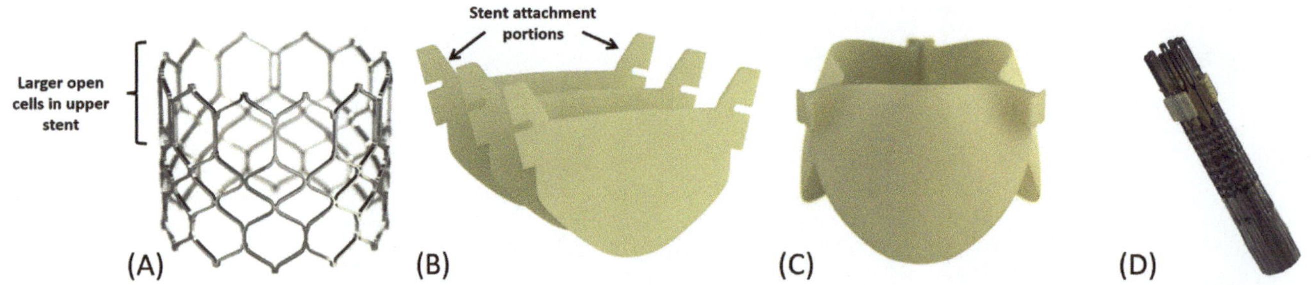

FIGURE 17.7 Components of the Sapien 3 valve. (A) Cobalt–chromium stent. (B) Pericardial valve leaflets. (C) Valve leaflets in their final position after mounting inside the stent. (D) Crimped S3 valve. *Courtesy Edwards Lifesciences. Reproduced with permission.*

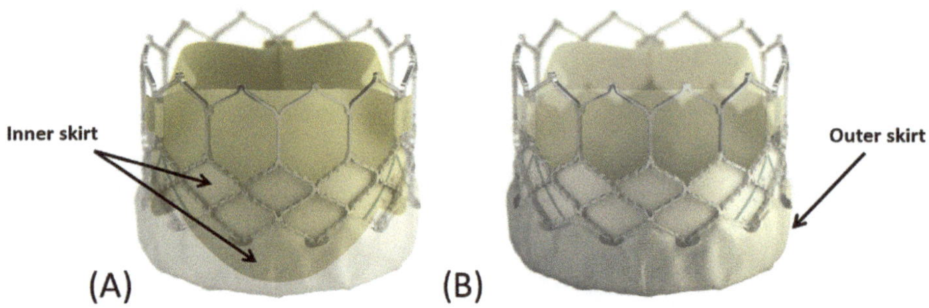

FIGURE 17.8 Components of the Sapien 3 valve. (A) Ghosted outer sealing skirt to demonstrate the inner skirt and the leaflet and skirt attachment to the stent. (B) Ghosted leaflets to demonstrate the superposition of the inner and outer sealing skirts. *Courtesy Edwards Lifesciences. Reproduced with permission.*

diameter, the height of the inner skirt ranges from 7.9 to 11.6 mm while the outer skirt measures from 5.2 to 8.1 mm.[20,21] Along with increased operators' experience and lower-risk patients, the addition of this outer skirt led to 30 days rates of moderate or severe PVL <2% with the S3 compared with pooled rates of approximately 7.0% with first-generation devices in meta-analyses.[4,22–24] The theoretical treatable range of aortic annuli with the S3 ranges from 18.6 to 29.5 mm. However, it has been demonstrated that the S3 prosthesis can be over expanded by adding 2–4 extra mL to the nominal inflation volume of the balloon depending on valve size, which is an interesting property in case of borderline annulus size or to treat very large annuli of up to 740 mm^2 corresponding to an area-derived annulus diameter of 31 mm[25] (Fig. 17.9). One limitation inherent the balloon-expandable design lies in the impossibility to recapture the valve once the balloon has been retracted in the crimped valve in the descending aorta.

The transfemoral Commander delivery system (Fig. 17.10) includes a nose cone tipped inner balloon catheter and an outer deflectable flex catheter. The attached handle has a rotating wheel to control the flex of the catheter tip, a flex indicator that indicates the degree of tip deflection, a fine adjustment wheel for precise positioning of the THV during valve alignment, a balloon lock knob to secure the balloon catheter to the flex catheter, and a flush port.[19] These elements were introduced to facilitate fine adjustment of deployment height (rotation of fine alignment wheel), recognized to impact the risk of valve malposition and the occurrence of PVL or LBBB (left bundle branch block). Two radiopaque markers at the proximal and distal ends of the balloon define the valve alignment position, while the central radiopaque marker in the balloon assists in valve positioning. Valve implantation is a two-operators procedure. Aside from the stent frame, another key feature of the S3 platform, which allows a low insertion profile, is the ingenious loading process of the THV on its catheter. Indeed, the valve is mounted on the balloon catheter distal to the balloon, which is then retracted within the crimped prosthesis in the descending aorta. Therefore, the S3 is compatible with a 36 cm working length, 14-French (Fr) (for the 20, 23, or 26 mm THVs) or 16-Fr (for the 29 mm THV) expandable sheath (eSheath) (Fig. 17.11). Nevertheless, it should be kept in mind that the outer sheath diameter is 18-Fr (for 20–26 mm S3) and 20-Fr (for 29 mm S3), transiently expanding to 24-Fr (for 20–26 mm S3) and 27-Fr (for 29 mm S3) to accommodate the THV passage. Given the compliance of femoral arteries, minimal diameters of 5.5 mm (for 14-Fr e-sheath) and 6 mm (for 16-Fr e-sheath) are recommended for a transfemoral approach. It has been demonstrated that the modifications of the delivery system resulted in a reduction in the mean sheath size used with the S3 (14.3-Fr) compared with the SAPIEN XT (18.1-Fr).[27]

The transapical/transaortic Certitude delivery system (Fig. 17.12) includes a handle with a Flex Wheel for articulation of the Balloon Catheter and a Loader. The loader allows for the delivery of the crimped THV through the hemostatic valves of the sheath. The THV is crimped between the two radiopaque shoulders on the distal and proximal ends

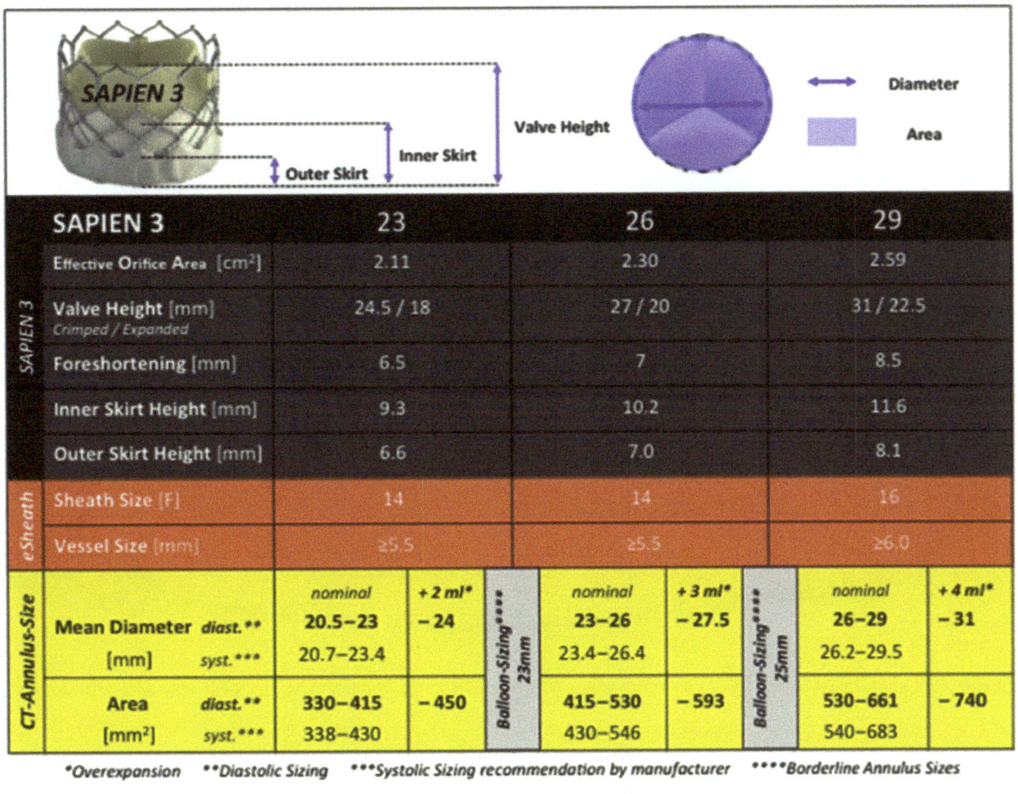

SAPIEN 3		23			26			29	
SAPIEN 3	Effective Orifice Area [cm²]	2.11			2.30			2.59	
	Valve Height [mm] Crimped / Expanded	24.5 / 18			27 / 20			31 / 22.5	
	Foreshortening [mm]	6.5			7			8.5	
	Inner Skirt Height [mm]	9.3			10.2			11.6	
	Outer Skirt Height [mm]	6.6			7.0			8.1	
eSheath	Sheath Size [F]	14			14			16	
	Vessel Size [mm]	≥5.5			≥5.5			≥6.0	
CT-Annulus-Size		nominal	+2 ml*	Balloon-Sizing**** 23mm	nominal	+3 ml*	Balloon-Sizing**** 25mm	nominal	+4 ml*
	Mean Diameter [mm] diast.**	20.5–23	–24		23–26	–27.5		26–29	–31
	syst.***	20.7–23.4			23.4–26.4			26.2–29.5	
	Area [mm²] diast.**	330–415	–450		415–530	–593		530–661	–740
	syst.***	338–430			430–546			540–683	

*Overexpansion **Diastolic Sizing ***Systolic Sizing recommendation by manufacturer ****Borderline Annulus Sizes

Sizing chart for the Edwards SAPIEN 3 transcatheter heart valve (THV) based on multidetector computed tomography (MDCT)–derived mean diameter and cross-sectional area of the aortic valve annulus obtained in the systolic and diastolic phases of the imaging. The MDCT annulus parameters in the diastolic phase of overexpanded 23-, 26-, and 29-mm S3 THVs with the addition of 2, 3, and 4 ml of volume, respectively, are also provided. A = area; D = diameter; diast. = diastole; P = perimeter; syst. = systole.

FIGURE 17.9 Sizing chart of the Sapien 3 THV. *Reproduced with permission from Shivaraju A, Kodali S, Thilo C, et al. Overexpansion of the SAPIEN 3 transcatheter heart valve: a feasibility study. JACC Cardiovasc Interv. 2015;8:2041–2043.*[25]

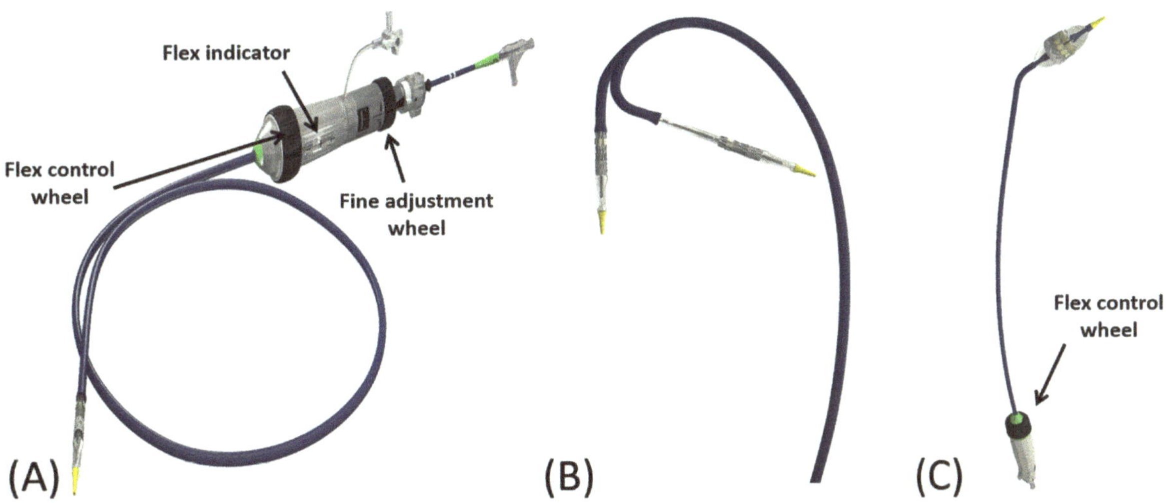

FIGURE 17.10 Delivery systems for the Sapien 3 valve. (A) Transfemoral Commander delivery system. (B) Demonstration of maximum flex with the Commander delivery system. (C) Transapical Certitude delivery system. *Courtesy Edwards Lifesciences. Reproduced with permission.*

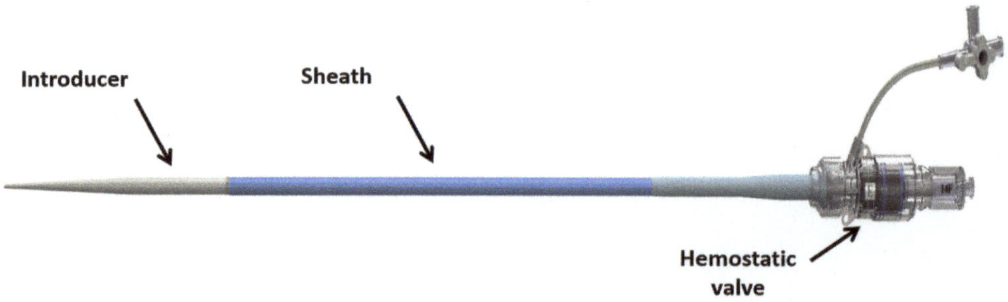

FIGURE 17.11 Dedicated transfemoral introducer (eSheath) for the Sapien 3 valve. The sheath is transiently expandable to accommodate the passage of the THV. *Courtesy Edwards Lifesciences. Reproduced with permission.*

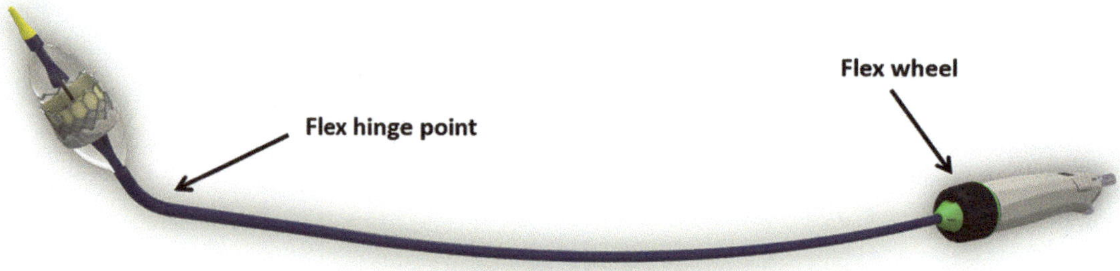

FIGURE 17.12 Dedicated transapical/transaortic delivery system (Certitude) for the Sapien 3 valve. *Courtesy Edwards Lifesciences. Reproduced with permission.*

of the balloon. A radiopaque central marker in the balloon is provided to help with valve positioning. The ergonomic design of the catheter allows for single-operator implantation of the S3 using and 18-Fr (≤26 mm THV) or 21-Fr sheath (29 mm THV). Although conceived and labeled for the transapical/transaortic access, the Certitude is currently employed in many centers worldwide for the transcarotid access, which clinical results are interesting and has been proposed by several groups as the preferred nontransfemoral access route for TAVI.[28,29] Nonetheless, and despite the diffusion of transcarotid and transsubclavian accesses for TAVI, presently no dedicated and specifically labeled delivery system has been proposed for these approaches.

The CoreValve Evolut Pro transcatheter heart valve

This device represents the third generation of the most frequently employed self-expandable THV platform. It maintains several design features of the second generation valve (the CoreValve Evolut R),[5,30] such as the cell frame guaranteeing homogeneous radial force along the axis of the device; the possibility to recapture and reposition the device until almost complete deployment; the flaring shaft tip (intended at improving the valve stability during the early deployment as well as the centering of the device inside the native valve orifice); and a 14 Fr-equivalent profile (sizes 29–31 mm) with a 18 Fr external diameter. The prevention of PVL was considered a major issue, as the lower portion of the stent was redesigned (geometry and radial force distribution) at this stage in order to improve the sealing and the rate of conduction disturbances. With the Evolut Pro model, this issue was further addressed by the addition of a porcine pericardial sheet at the lower cell frame, aimed at improving sealing (Fig. 17.13). Nonetheless, such modification obliged to the use of a 16-Fr introducing shaft profile. One expected advantage of the self-expandable valve concept lies in its capability to adapt to more irregular annular (landing zone) anatomies, or annuli in the lower/upper extremities of surface range. The 23-, 26- and 29-mm valve sizes cover (with the Evolut R 34-mm valve) a range of 18–26 mm in annular diameter and a 56.5–94.2 mm in annular perimeter. The system is based on a both intra- and supraannular support to obtain stability, with intraannular sealing. On the other hand, the height of the valve stent needs to be considered with respect to access to coronary ostia in patients with increased likelihood of percutaneous coronary disease treatment during follow-up. The self-expandable strategy is considered interesting in cases with major annular calcifications involving the left ventricular outflow tract and aortomitral continuity, purposely in order to optimize the adaptation to irregular anatomies and minimize the risk of annular rupture. The nitinol stent is expected to achieve its maximum radial force over time after deployment, potentially reducing PVL during the hours/days after the procedure.

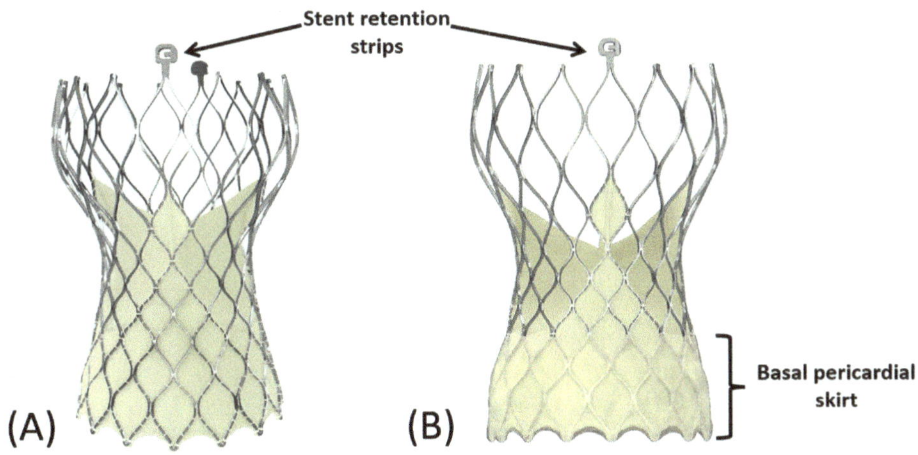

FIGURE 17.13 (A) The CoreValve Evolut R device. (B) The CoreValve Evolut Pro device. *Courtesy Medtronic. Reproduced with permission. Medtronic France, 2020.*

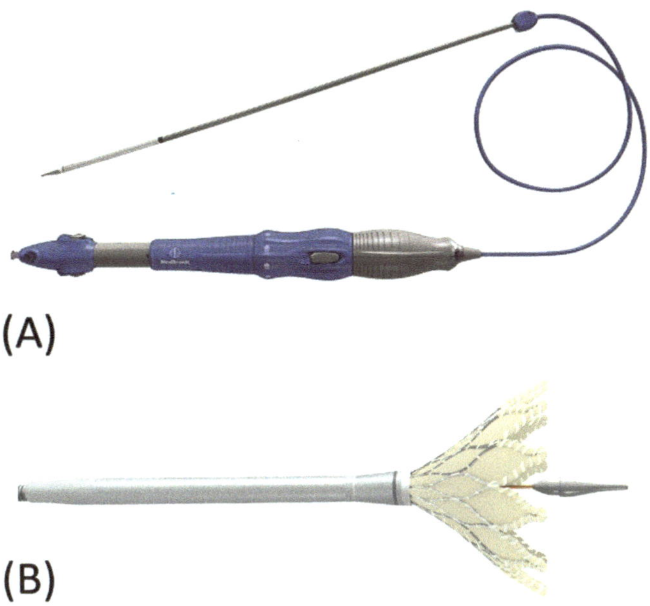

FIGURE 17.14 (A) The Enveo R delivery system. (B) Early deployment of a CoreValve Evolut R device; the flaring catheter tip is demonstrated. *Courtesy Medtronic. Reproduced with permission. Medtronic France, 2020.*

The dedicated delivery system (ENVEO Pro, 14-16 Fr equivalent for the 23−29 mm size range, Fig. 17.14) is distinguished from its predecessor by an improved nose-cone aimed at minimizing vascular trauma during its advancement throughout the arterial system, and at avoiding the uncoupling of the nosecone and the capsule while crossing the native aortic valve. The latter feature should improve alignment and device preservation until delivery. There is a 1:1 response of the capsule while actuating the deployment knob, and a tactile indicator ("rumble strip") facilitates the approach to the point of no recapture. Valve height can be adjusted during most of the deployment through management of system and guidewire tension. The atraumatic and low profile of the delivery system facilitate its employment not only for the transfemoral, but also for the transsubclavian access and possibly for the transcarotid route (although it is not labeled for these approaches). Miniaturization has been pursued also by a sheathless strategy for insertion of the delivery system.

The Portico transcatheter heart valve

This THV device based on a self-expanding concept, is composed by a nitinol stent mounting an internal porcine pericardial sheet (basal sealing skirt) and a trileaflet bovine pericardial valve (Fig. 17.15). Tissues are treated with the Linx

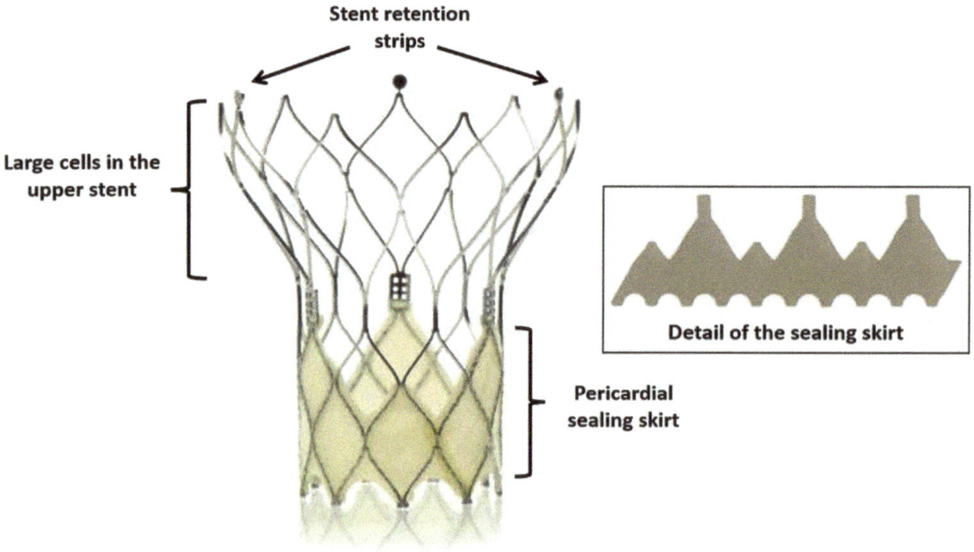

FIGURE 17.15 The Portico transcatheter heart valve. The large cell design of the upper stent is displayed. Insert: design of the basal pericardial sealing skirt. *Courtesy Abbott Cardiovascular. Reproduced with permission.*

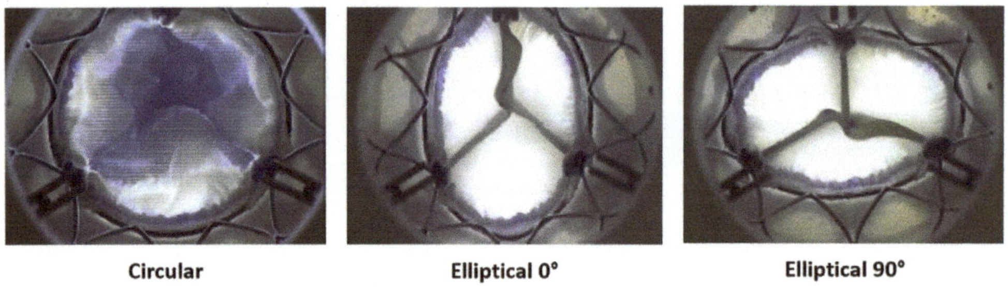

FIGURE 17.16 In vitro testing of the Portico THV deployed inside circular and elliptical mock aortic annuli. *Courtesy Abbott Cardiovascular. Reproduced with permission.*

anticalcification treatment, the same employed in an array of surgical bioprostheses.[31] The stability is given by the radial force and the sealing is supported by the basal pericardial skirt; the stent is conceived in order to provide a stable outward annular radial force within the recommended diameter range for each valve size. The nitinol frame features a large-cells structure in its upper portion (measuring between 13.5 and 20.8 Fr) intended at facilitating the coronary access during the follow-up. In in vitro experiments, the leaflet coaptation was also maintained in "extreme" elliptical configurations of the valve (Fig. 17.16). Such feature is expected to facilitate adaptation of the device into irregular and bicuspid valve anatomies. This THV is conceived for intraannular deployment, requires no rapid pacing, and is recapturable/repositionable until about 80% of stent deployment.

The transfemoral delivery system has evolved to its most recent generation (FlexNav), intended at addressing the main delivery-related issues. The FlexNav sizes 14–15 Fr equivalent. It associates an integrated sheath, a hydrophilic coating, flexibility properties and a smooth transition between the introducer and the valve capsule, as strategies to facilitate navigation of the iliofemoral anatomies. An additional layer, located at the proximal third of the catheter, is included as the strategy to optimize the stability of the system during deployment and purposedly facilitate the deployment at the intended level.

Following the initial 1-year safety/effectiveness investigations,[32] and due to a more recent appearance than competitor THVs, Portico has been the object of randomized trials vs. commercially available devices (both self- and balloon-expandable). Such evaluation was limited on one side by the inclusion of all-generations commercial valves, and on the other side by the most frequent employment of the previous-generation Portico delivery system. Conclusions suggested similar rates of the primary efficacy endpoint at 1 year and similar rates of death/stroke at 2 years, but a higher rate of the primary safety endpoint at 30 days in the Portico group.[33] This has been associated with vascular access complications, and the impact of a systematic employment of the last-generation delivery had to be addressed in a separated cohort. PPM implantation rates were similar than other self-expandable valves. Of note, this THV is also labeled for the transsubclavian/axillary access.

The Lotus Edge and Lotus Mantra transcatheter heart valve

These devices are the expression of design iterations based on the initial version of the Lotus THV. This platform consists in a self-expandable braided nitinol frame mounting a bovine pericardial trileaflet valve, conceived for intraannular position deployment. Its designed strategy to obtain stability and sealing is based on the association of the radial force of a small-cell braided stent (expected to conform to the native annular and calcified leaflet anatomy) and of an adaptive seal (expected to help in PVL prevention by filling the gaps between the stent and the landing zone) (Fig. 17.17). The adaptive seal consists of a polymer membrane covering the lower half of the device.

This THV is self-expandable, its release being mechanically-driven through the push of the delivery system applied to the upper stent buckles by rotation of the wheel at the handle. The rotation can be inversed at every moment and the valve is recapturable even after full stent deployment and before its release ("Controlled expansion"), one feature which has been reportedly associated with no instances of valve malposition in prospective investigations.[34] Consistently, the valve is pushed out of its sheath as the operators concomitantly retrieve the entire system to adjust the level of deployment. The movements are transmitted through the buckles and collars system solidarized with the valve posts. As the device is pushed outside the sheath, the nitinol stent transitions from its "relaxed" crimped form to the final shape by increasing its diameter and reducing its height (Fig. 17.18). The polymer membrane seal folds while the frame reduces its height, the resulting redundancy of the polymer membrane is expected to fill the gaps at the annulus/native leaflets until the final state of the frame is achieved. Therefore, the seal is obtained with both annular and leaflet contact. These elements yielded low rates of moderate-severe PVL in prospective studies.[35,36] While no rapid pacing is required, care must be paid under fluoroscopic surveillance until adequate locking before final release. This THV covers a 20- to 27-

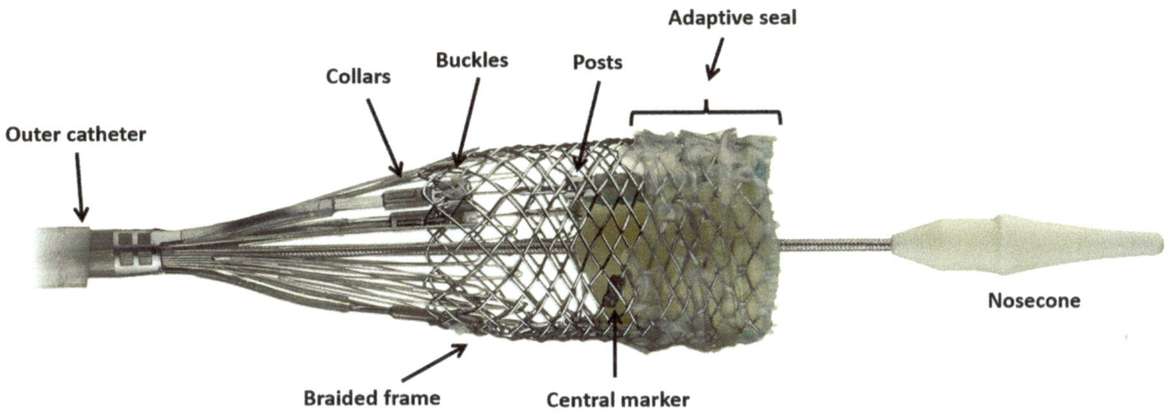

FIGURE 17.17 The Lotus Edge THV. The braided nitinol frame, the continuity between the collars/buckles system and the valve posts, and the adaptive seal are showed. *Courtesy Boston Scientific (modified). Reproduced with permission.*

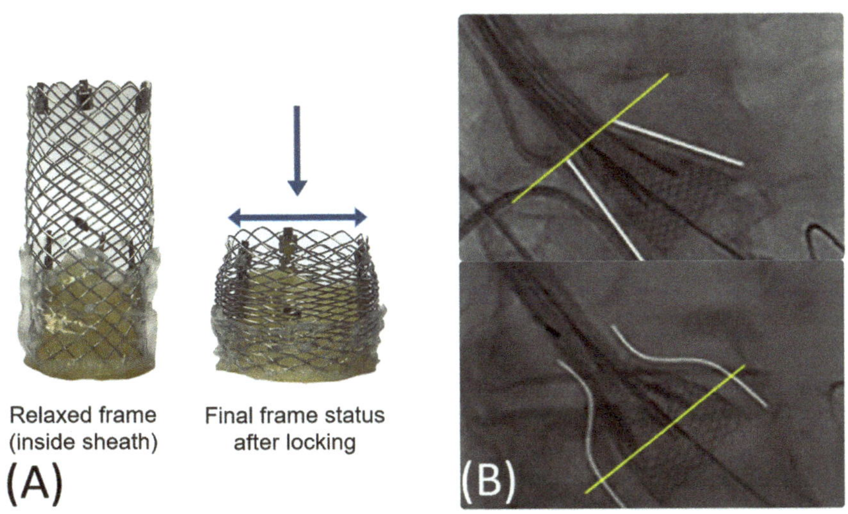

FIGURE 17.18 Release of the Lotus Edge THV. (A) transition from the relaxed crimped form to its final form after delivery. (B) The last-generation device (lower panel) favors earlier anchoring and less left ventricular outflow tract (LVOT) protrusion through a flared shape during release, not existing in the previous generation valve (upper panel). *Courtesy Boston Scientific (modified). Reproduced with permission.*

mm annular size range over three device sizes (23, 25, and 27 mm); the final device height after full deployment is 19-mm for all sizes.

One issue observed with the previous-generation Lotus valve was the considerable PPM implantation rate (up to 32%,[37,38]). Such observation was attributed in part to the radial force of the protruding lower stent portion (which is greater than other self-expanding valves), and possibly to the deployment technique itself (the valve being pushed out of the tip of the sheath may result in a deeper dive of the inflow portion of the braid in the left ventricular outflow tract (LVOT) than the recommended 4-mm below the annulus). Consequently, the Edge valve version has been equipped with a Depth Guard system, consisting in a redesigned lower nitinol stent into a flared shape, in order to favor earlier valve-annulus contact and improved stability during the last phases of deployment (Fig. 17.18). Such adaptation led to a 15.2% rate of new PPM implantation in a limited early series.[39] This THV was also labeled for the transfemoral and the direct aortic access. It is to be noted that in late 2020, this THV was recalled and the project discontinued by the manufacturer. This outcome was essentially due to the limited development of specific indications to its use and the complexity of the delivery technique and system, alongside the important investments required to develop an alternative delivery systems which were considered to be nonviable.

The Acurate Neo2 transcatheter heart valve

This THV represents the second-generation of the initial valve concept Acurate Neo, which failed to achieve noninferiority versus established competitor THVs, mainly due to PVL rates.[40,41] Acurate Neo2 consists in a self-expanding nitinol frame mounting a bovine pericardial trileaflet valve (Fig. 17.19). The upper stent portion essentially consists of three "stabilization arches" continuous with the valve posts and intended at optimizing the alignment through contact with the aortic wall. Furthermore, the wide opening of these arches might facilitate later coronary access. Such mechanism introduces some influence of the ascending aortic morphology on the final valve stability and position, as underlined by the height of the sinotubular junction as an independent predictor of more-than-mild PVL in initial experiences.[42] Such design feature needs to be considered in patients' selection. The stability is obtained through the outer radial force of the stent (which is maximal at the midportion of the lower crown,[43] where the prosthesis is expected to interact with the native annulus). Improved sealing with this second-generation device has been pursued with both an outer 60% larger pericardial skirt covering the entire extent of the lower crown, and an upper crown intended to capture the native leaflets and constrain the upper edge of the outer skirt to embrace the free leaflet edges during diastole. These are also intended at improving anchoring. Recent evidence indicates significant reduction of PVL rate with this strategy vs. the Acurate Neo valve.[44]

This THV exists in three sizes (23, 25, and 27 mm) covering a 21- to 27-mm annular diameter range. A dedicated transapical delivery system has been developed.

The Hydra THV System

This recently developed THV concept (Fig. 17.20) features a nitinol self-expanding frame, a supraannular bovine pericardial valve, and the possibility to recapture and reposition the valve until 85%–90% of full deployment. The inflow portion of the frame is nonflared, with the intended purpose to minimize the PPM implantation rate. Only the

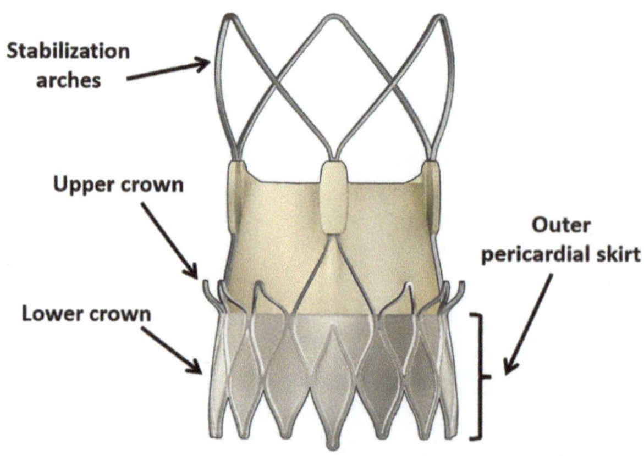

FIGURE 17.19 The Acurate Neo2 THV (scheme). *Courtesy Boston Scientific (modified). Reproduced with permission.*

transfemoral access is proposed. In clinical evaluation trials in high-risk patients with severe aortic stenosis, this device showed a 92.5% Valve Academic Research Consortium (VARC)-defined device success rate, a 7.5% PPM implantation rate, no more than mild PVL, and a 6-month 17.5% mortality.[45] These evaluations included nonetheless limited patients' series and follow-up duration, and heterogeneous implanting teams with respect to experience with THVs. The place of the Hydra in the THV landscape will be determined by future evidence over safety/effectiveness.

Technological development pathways for clinical outcome improvement

Cerebral protection

Cerebrovascular events (CVE) are one of the most dreadful complications of TAVI being associated with an increased risk of early and late mortality as well as a significant morbidity and detrimental effect upon quality of life.[46–48] The median 30 days CVE rate in a large meta-analysis of 64 studies involving >70,000 patients was 4%.[49] Embolic protection devices (EPD) may be associated with a reduced rate of early CVE potentially leading to a lower risk of in-hospital mortality.[50–53] Thus, these devices unsurprisingly generated, and still generate, a fair amount of clinical research.[54] Limitations to the diffusion of such devices include: increasing technical complexity of the TAVI procedure and the need for an additional arterial access; potential for additional embolic risk associated with aortic arch navigation; increased costs. Additionally, currently available evidence globally suggest that the addition of EPD is effective in reducing the volume and size of periprocedural silent brain lesions, but definite evidence about their potential to reduce the rate of new clinically evident neurological events is still lacking. Due to the above factors, cerebral protection devices are unfrequently employed in routine TAVI clinical practice.

Embrella Embolic Deflector

Among these devices, the first clinically evaluated was the Embrella Embolic Deflector (EED).[55–58] The system consists of two petals forming an oval-shaped nitinol frame of 59-mm length and 25.5-mm width covered with a heparin-coated porous polyurethane membrane (100-μm pore size representing the smallest among available EPD) (Fig. 17.21). It can be inserted through a 6-French right radial or brachial approach. Three markers localized at the outer point of each petal and on the distal shaft help positioning the device along the greater curvature of the aortic arch covering the brachiocephalic trunk and the left common carotid. Thus, the system is intended to deflect rather capture embolic material. Published clinical results are limited to a first-in-man study including four patients and two small nonrandomized trials.[55–57] In the PROTAVI-C Trial, 41 TAVI patients treated with the Embrella system were compared with 11 patients who underwent TAVI without cerebral protection.[56] The deployment of the device was successful in all attempted cases without procedural complications. However, the EED associated with a higher number of cerebral

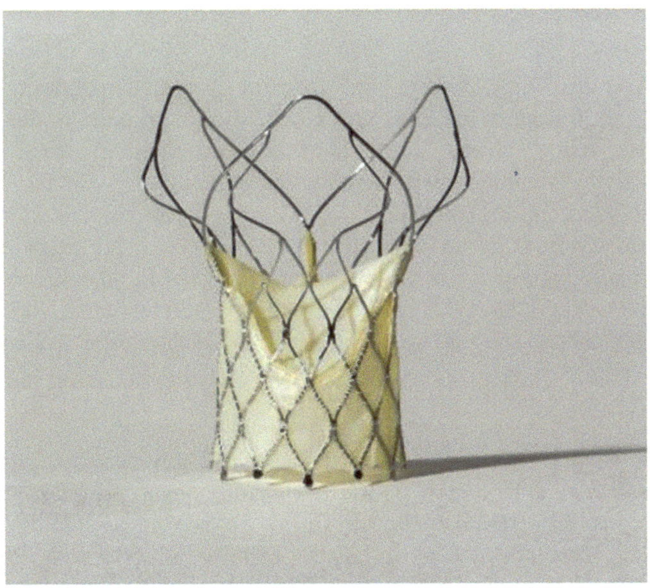

FIGURE 17.20 The Hydra THV.

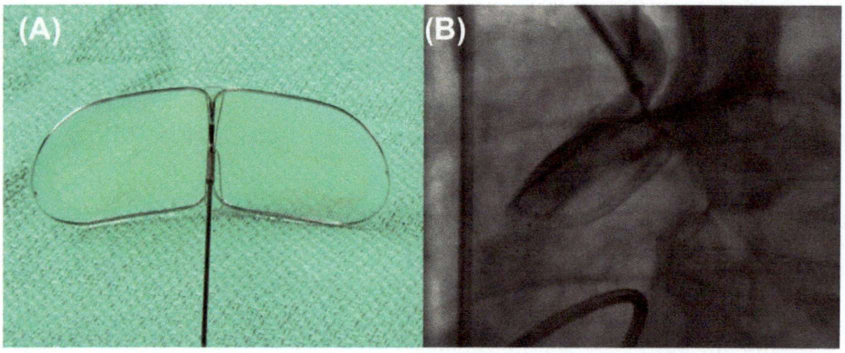

FIGURE 17.21 The Embrella Embolic Deflector System. (A) photo of the device as deployed. (B) Angiographic view of the device after deployment at the greater curvature of the aortic arch. *Reproduced with permission from Rodés-Cabau J, Kahlert P, Neumann F-J, et al. Feasibility and exploratory efficacy evaluation of the Embrella Embolic Deflector system for the prevention of cerebral emboli in patients undergoing transcatheter aortic valve replacement: the PROTAVI-C pilot study. JACC Cardiovasc Interv. 2014;7:1146–1155.*[56]

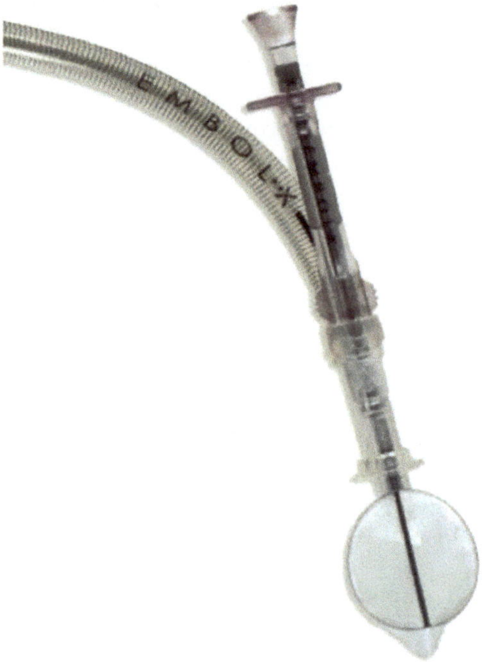

FIGURE 17.22 The Embol-X protection device.

embolic events as assessed by transcranial Doppler and diffusion-weighted magnetic resonance imaging (DW-MRI) while three patients had a CVE at 30 days in the EED group compared with none in the control group. Nonetheless, the EED was associated with a lower lesion volume. Samim et al. also compared 15 patients receiving TAVI with the EED with a historical cohort of 37 TAVI recipients without cerebral protection.[57] An increased number of new ischemic lesions was observed in the EED group on DW-MRI along with a smaller single-lesion volume. A greater number of right-sided lesions was found in the EED group suggesting a risk of healthcare-associated lesions with the use of the device. Overall, the Embrella may be protective regarding embolization of large debris to the brain, which did not translate in a reduced risk of clinically significant CVE in the limited available evidence. It can be hypothesized that deflection could be effective for larger debris, and that the navigation inside the aortic arch/supraaortic vessels is a potential trigger of nonvalvular sources of emboli.

Embol-X

The Embol-X (Fig. 17.22) was initially intended for use during open-heart surgery, yet it was used off-label during transaortic TAVI. It consists of a self-expandable filter sitting across the ascending aorta, which can be delivered through a short 17-French sheath. It includes a heparin-coated, 120-μm polyester mesh in a flexible nitinol frame, available in five sizes covering a large range of aortic diameters (22–40 mm). The Embol-X was evaluated in a randomized, single-blinded, single center trial,[58] which assigned 14 patients to transaortic TAVI with the Embol-X and 16 patients

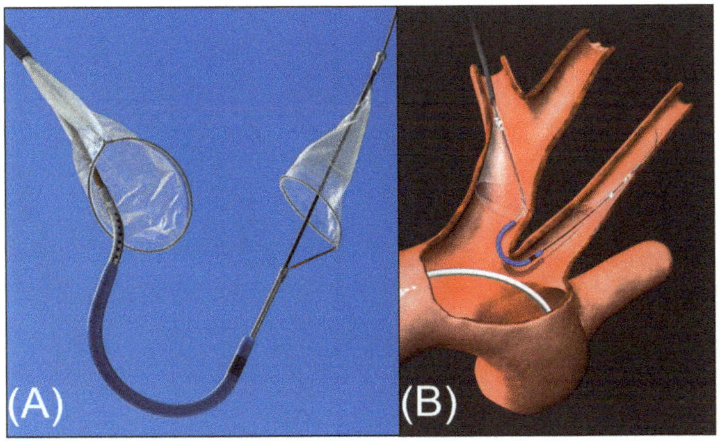

FIGURE 17.23 The Sentinel Cerebral Protection System. (A) Detail of the two independent polyurethane filters. (B) Scheme of the correctly deployed device. *Courtesy Boston Scientific. Reproduced with permission.*

to transaortic TAVI without the device. DW-MRI assessed new cerebral lesions were found in 69% of the control group and 50% of the Embol-X group. Lesion size was numerically lower, and significantly lower in the territory of the middle cerebral artery, in the Embol-X group. No clinically apparent events occurred in this study. This device is limited by its direct transaortic insertion, which condemns its employment to the very rare cases of transaortic TAVI.

Sentinel

This concept consists of two independent polyurethane filters (pore size: 140 μm) which are deployed inside the brachiocephalic trunk and the left common carotid artery (Fig. 17.23). The 6-Fr delivery catheter is steerable and has minimal protrusion inside the aortic arch, intended at minimizing interactions with both the aortic endothelium and the TAVI delivery system. It is designed not only to capture, but also to retrieve and exteriorize the embolic material. In evaluation studies, this device has been suggested effective in reducing the rate of symptomatic stroke early after TAVI in multiple investigations,[53,59] and yielded interest over its employment.

Procedural assistance through augmented imaging; machine-learning

Computer-assisted medical interventions, as a transversal discipline, integrate various technologies and sources of data, such as pre- and intraoperative imaging, virtual reality, augmented reality, interactive simulation and micro-technologies.[60] Exploratory investigations have been conducted for TAVI with respect to options for improved intraprocedural guidance. Fusion imaging applications have been proposed, overlaying CT scan-derived 3D mesh over intraprocedural fluoroscopy imaging.[61] This requires 2D/3D registration based on landmark positioning, and dynamic tracking to maintain the superimposition onto the live fluoroscopic images. Also, transesophageal echocardiography/fluoroscopy fusion imaging is possible.[62] The calculation capacity of computers equipping current hybrid ORs allows real-time tracking over live fluoroscopy images. The rationale lies in the usefulness of achieving optimal THV deployment depth, in order to minimize the risk of both coronary obstruction and PPM implantation. Particular clinical scenarios such as the THV implantation for the treatment of a degenerating surgical aortic bioprosthesis (Valve-in-Valve) facilitate the performance of intraprocedural registration and dynamic tracking due to the existence of radiopaque markers within the failing prosthesis (Fig. 17.24).[63] These approaches are potentially useful, but they are still scarcely diffused. Proof of feasibility has been provided so far, while dedicated clinical investigations are required to define their field of application and measurable benefits.

An increasingly explored and transversal approach in numerous activities, artificial intelligence applications have been proposed for risk stratification in candidates to TAVI. The limitations of established risk evaluation scores have been historically discussed; accordingly, some groups have proposed machine-learning algorithms based on an array of patient-derived variables. Such approaches have been suggested to outperform conventional risk scores for the prediction of early and 1-year mortality,[64] but their methodological reliability depends on the included variables and on the number of observations. They have nonetheless the potential to consider a far larger number of factors and patient-derived variables than conventional risk evaluation scores, including multiple anatomical factors.[65] This axis of research is still in its infancy and large multi-Institutional collaborations are required to achieve significant advancements at the current stage.

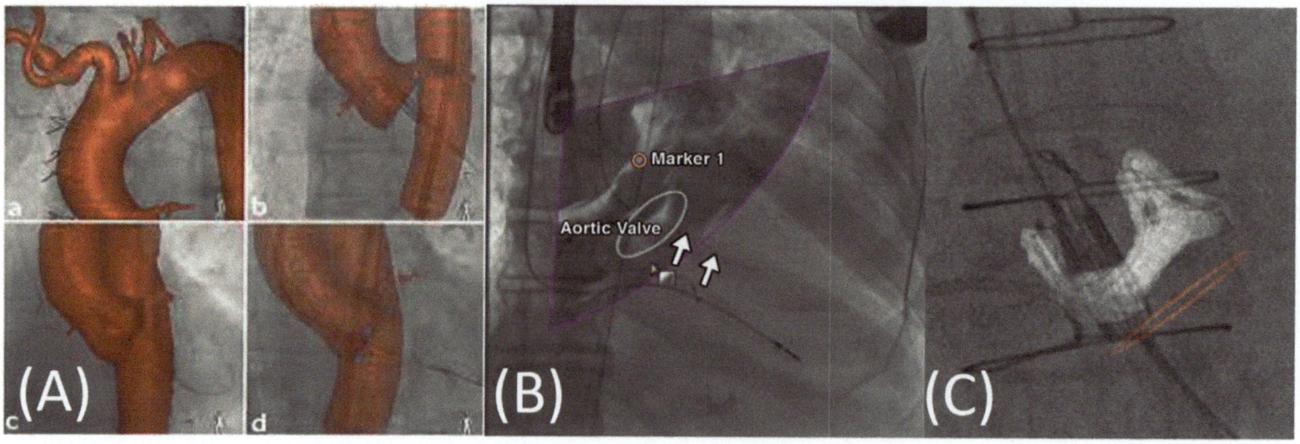

FIGURE 17.24 Augmented intraprocedural imaging in TAVI. (A) CT scan and fluoroscopy fusion imaging. A 3D mesh models of the aortic anatomy is overlayed on fluoroscopy images; introperative registration and dynamic tracking. (B) Transesophageal echocardiography and fluoroscopy fusion imaging. (C) Virtual 3D reconstruction of a failing aortic bioprosthesis and addition of a virtual plane representing the optimal deployment level for a Valve-in-Valve THV. Intraoperative registration and dynamic tracking. *Reproduced with permission from (A) Vernikouskaya I, Rottbauer W, Gonska B, et al. Image-guidance for transcatheter aortic valve implantation (TAVI) and cerebral embolic protection.* Int J Cardiol. *2017;249:90–95.*[61] *(B) Kronzon I, Jelnin V, Ruiz CE, et al. Optimal imaging for guiding TAVR: transesophageal or transthoracic echocardiography, or just fluoroscopy?* JACC Cardiovasc Imaging. *2015;8:361–370.*[62] *(C) Belhaj Soulami R, Verhoye JP, Nguyen Duc H, et al. Computer-assisted transcatheter heart valve implantation in valve-in-valve procedures.* Innovations (Phila). *2016;11:193–200.*[63]

Pathways for improved THV reproducibility: commissural alignment

With the expansion of THV use in lower-risk and younger patients' strata, predictability of THV function and durability, as well as reliability of coronary access, emerge as key requirements. Moderate-severe commissural misalignment (CMA) has been defined as a 30–45 degrees shift between the position of bioprosthetic post and the native aortic commissure. Unsurprisingly, postimplantation multidetector CT studies described no instances of CMA in a cohort of SAVR patients, and a 53% rate among randomly implanted THV. CMA had no impact on transprosthetic gradient or coronary filling, but was associated with higher rates of mild central regurgitation.[66] It can be hypothesized that misaligned THV/native aortic ring/root might induce a deformation in the THV frame, and potentially impact durability (the asymmetry of mechanical stress applied to leaflet being historically associated with increased risk of structural valve deterioration (SVD) in SAVR bioprostheses). It has been proposed that an orientation-specific deployment of commercially available THVs can be pursued through the association of (1) coregistration of the desired THV posts position with overlay of the THV frame on baseline CT scan and determination of the fluoro valve orientation on the coplanar view; and (2) intraprocedural employment of THV radiopaque markers to obtain the desired circular orientation (Fig. 17.25).[67,68] Dedicated research with adequate follow-up is required to demonstrate the effectiveness of this sophisticated approach to improve THV performance/durability.

Other technology pathways

The Centera Valve

The self-expanding CENTERA THV (Edwards Lifesciences) had several interesting features including a short height nitinol frame, a unique, low-profile (14-French compatible), motorized delivery catheter also enhancing coaxial annular alignment, a contoured frame to anchor in the annulus with a polyethylene terephthalate skirt to reduce PVL, and the possibility to recapture up to 80% of the deployment.[69,70] Despite excellent clinical outcomes, matching those of other new-generation devices,[71] the CENTERA program was stopped by Edwards Lifesciences in July 2019 because of the significant amount of resources required to optimize deliverability and expand the indications to match the S3, the latter being on the verge of US Food and Drug Administration (FDA) approval for the management of low-risk patients at the time.

The DirectFlow Valve

This THV can be comprised in the group of second-generation THVs (CE mark received in 2013), and was unique in its design concept (Fig. 17.26). It featured a bovine pericardial valve mounted inside a nonmetallic structure made of

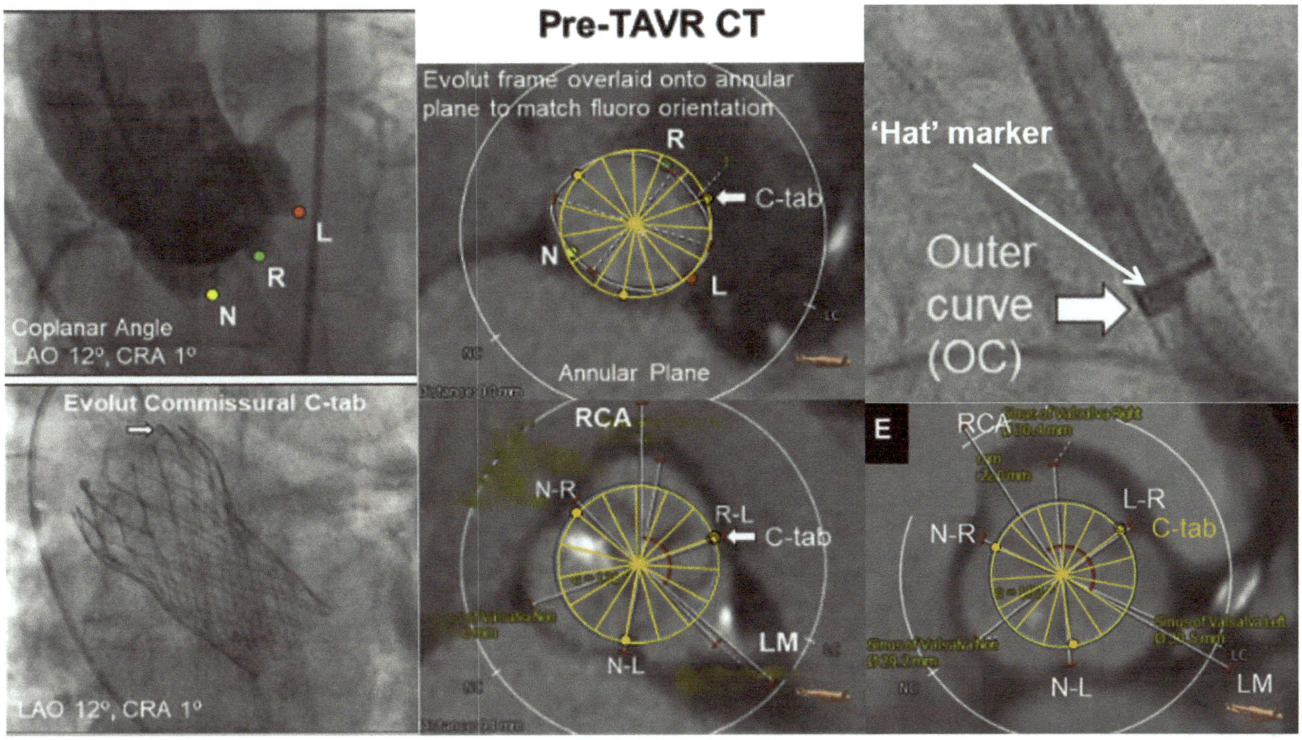

FIGURE 17.25 Proposed algorithm to improve commissural alignment in THV implantation. From left to right: coregistration of THV frame on baseline CT scan with fluoro coplanar view; tracking the "Hat" marker in the outer aortic curve as the desired orientation for improved alignment; verification of the "C-tab" commissural marker lying at the inner aortic curve corresponding to the L–R native commissure. The example of a CoreValve Evolut THV is provided. *Reproduced with permission. Modified from Tang GHL, Zaid S, Fuchs A, et al. Alignment of transcatheter aortic-valve neo-commissures (ALIGN TAVR): impact on final valve orientation and coronary artery overlap.* JACC Cardiovasc Interv. *2020;13:1030–1042.*[67]

dacron/polyester, one upper (aortic) and one lower (infraannular) ring which could be individually controlled and unfolded by saline injection at a 12 atm pressure; their position could be adjusted indefinitely until injection of a polymer hardening the structure into its final position.

Theoretically, this device comprised multiple characteristics expected to be important for effective/safe use: ability to conform to the native anatomy including asymmetric/irregular landing zones and reduce PVLs; possibility to reposition and retrieve; controlled deployment; limited rates of PPM implantation and PVL in the landscape of second-generation THVs; an 18 Fr profile; possibility to position with minimal or no contrast injection with transesophageal echocardiography (TOE) guidance.[72,73] Reasons for the discontinuation of this project may be found in the complexity of implanting procedure and of its monitoring from the point of view of the operators. Additionally, the development of a THV requires major research, development and marketing efforts with associated funding. The impossibility to secure funding in the extensive process of obtaining scientific proof of safety/efficacy and identifying the appropriate indications/positioning for the specific device, remain a major issue and a determinant factor in THV development.

The SAPIEN 3 Ultra Valve

This device was foreseen as an iteration of the S3 valve aiming at further improving device-related outcomes (Fig. 17.27). The main design objectives with this project were to simplify the procedure, decrease vascular complications, and reduce PVL. The outer skirt was replaced by a 40% higher textured sealing polyethylene terephthalate structure promoting healing and endothelization.[74–77] The delivery system was modified to reduce the number of procedural steps allowing the S3 Ultra to be crimped directly on the balloon, and eliminating the need to align it in the descending aorta or to retract the flex catheter before implantation. Finally, during transfemoral procedures, the THV is intended to be used with the Axela sheath, a 14-French expandable, and self-collapsible system with a hydrophilic coating. The outer elastomeric jacket of the sheath allows its transient expansion by forcing its active contraction after the THV passage.[74] So far, clinical data are encouraging showing devices success, major complications, and short-term survival

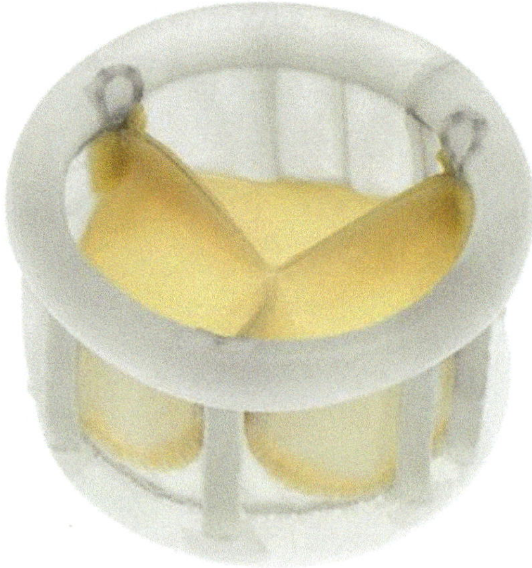

FIGURE 17.26 The DirectFlow Medical THV.

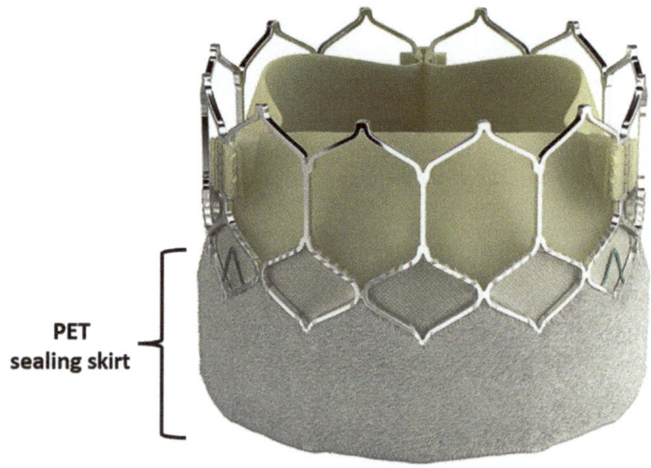

FIGURE 17.27 The Sapien 3 Ultra valve. *PET*, Polyethylene terephthalate. *Courtesy of Edwards Lifesciences. Reproduced with permission.*

rates consistent with those of the S3 THV while also demonstrating lower rates of more than mild PVL with the S3 Ultra.[21,75–77] Conversely, there were reports of burst balloons during some procedures with the new delivery system. This resulted in a few cases of vascular injury, bleeding, or need for surgical intervention, and led the FDA to issue a class I recall for the S3 Ultra delivery system in August 2019. Therefore, the previous-generation Commander delivery system and the 14-French expandable eSheath were recommended for delivery of the S3 Ultra valve. This example is a perfect illustration of the technological challenges in the evolving TAVI field, constantly balancing the drive toward potentially beneficial advancements with the need for reliability and safety in clinical practice.

The Myval THV

As the most recently developed balloon-expandable THV concept, this device consists of a Nickel-Cobalt alloy frame mounting a bovine pericardial trileaflet valve; it also features an internal Polyethylene terephthalate cuff and an external Polyethylene terephthalate sealing skirt, aimed at minimizing PVLs (Fig. 17.28). The frame height of this THV (comprised between 17.35 and 21.14 mm across the various sizes) is occupied by smaller, closed cells in its lower 47% (to provide stability through radial force and sealing) and by larger, open cells in its upper 53% (for clearance of coronary ostia). The largest circumscribable diameter in the upper cells equals 6 mm. As a specificity, 9 valve sizes can be provided for this THV: 20, 21.5, 23, 24.5, 26, 27.5, 29, 30.5, and 32 mm ("Traditional," "Intermediate" and "XL" THV

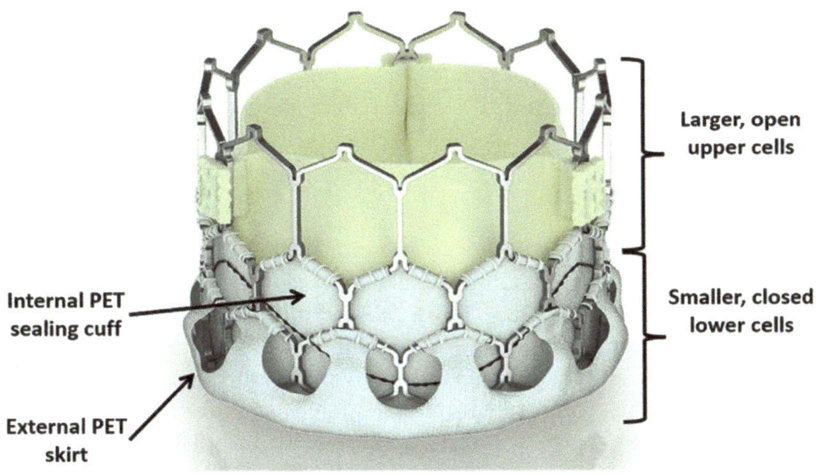

FIGURE 17.28 The Meril Myval THV. Meril Lifesciences. *Reproduced with permission. PET, Polyethylene terephthalate.*

sizes), covering a 270–804 mm^2 range in CT-derived native annular area. Such large array of sizes is intended at achieving greater granularity in valve choice, better adequacy of the valve dimensions and the annular surface in order to minimize PVLs as well as, potentially, other issues such as conduction disturbances and trauma to the native annulus in severely calcified anatomies. The effectiveness of this strategy in improving early valve function and late durability is still to be demonstrated. All sizes are compatible with a dedicated 14 Fr-equivalent steerable delivery system and expandable sheath.

In the initial prospective evaluations for early postimplantation results, Myval has been associated with a 2% mortality, a 1% stroke rate and a 2% rate of PPM.[78] Retrospective matched comparison versus established balloon-expandable valve have reported similar rates of clinical efficacy and early safety, with significantly lower rates of PPM implantation.[79] Such finding (which has been associated with the limited infraannular protrusion, theoretically comprised between 3.05 and 3.55 mm in bench tests) needs to be confirmed through the ongoing LANDMARK randomized trial designed to compare Myval with established balloon-expandable and self-expanding THVs.[80] This THV has obtained the CE (European Community mark) Mark in 2019 and has been also tested in nontransfemoral approaches (including transcarotid, transsubclavian and transapical). The above features suggest that this THV platform has significant potential for diffusion.

Tissue-engineered THVs

Tissue engineering has been long explored for bioprosthetic aortic valve replacement as a mean expected to obviate SVD, through improved biocompatibility, reduced antigenicity, potential for growth/remodeling in vivo, and resistance to infection.[81] Currently, no bioengineered bioprosthesis for surgical aortic valve replacement (SAVR) has achieved successful clinical translation or durable preclinical testing. Multiple issues persist, including: immunogenicity; reliability of in vitro expansion of cell lineages obtained from the valve recipient and of scaffolds for in vitro growth guaranteeing adequate mechanical performance; economic and regulatory challenges. In recent years, concepts based on acellular off-the-shelf constructs (cell-derived extracellular matrix) expected to recellularize in situ after implantation have been proposed for use in a self-expanding THV.[82] This concept has been tested in the ovine model. Such approach needs to address both the issues with reliability of tissue-engineered constructs, and the generic safety/reproducibility requirements of THV platforms (stability, prevention of PVL and of conduction disturbances). Despite such challenges, this research has major potential, provided that the research efforts can be consistently pursued until demonstration of improved durability in the long-term vs. nontissue-engineered THVs.

THVs for the treatment of aortic regurgitation

The vast majority of proposed THVs aim at treating calcified aortic stenosis. Native pure aortic regurgitation (PAR) constitutes a completely different anatomical scenario due to the absence of leaflet calcifications constituting the anchoring structure for THVs. Additionally, systolic/diastolic changes of the aortic annulus are possible due to absence of calcifications. Therefore, treating noncalcified aortic regurgitation requires dedicated THV design and concepts.

Concordantly, it has been reported in analytical investigation that THVs labeled for PAR yielded significantly better procedural success rates than off-label TAVI devices originally developed for the treatment of aortic stenosis.[83] The association of PAR with root and ascending aortic disease, its lower prevalence than severe aortic stenosis in elderly and high-risk patients, as well as its association with infective endocarditis, might reduce the population of potential candidates and therefore influence the accumulation of relevant evidence over THV effectiveness.

The JenaValve System

As a THV specifically designed for the management of PAR,[84] and still an investigational device, JenaValve consists of a self-expanding nitinol frame with larger cells in its upper portion, diamond-shaped smaller cells in its lower portion, a bovine pericardial valve, and a set of three radiopaque "locators." These are intended at capturing the native valve leaflets by diving at the nadir of each of them, and at constraining them against the sealing portion of the frame (Fig. 17.29). The stability and the sealing are therefore obtained at the supraannular level, with expected favorable impact of the rate of PPM implantation. More specifically, this stability and sealing strategy requires three accessible leaflets for clipping. Although CE mark studies are ongoing for label in severe aortic stenosis also, it is likely that its main point of strength lies in its ability to treat PAR.

Initially implanted through the transapical access, a transfemoral approach is currently preferred. Continued evaluation is needed for proof of reproducibility and refinement of indications, as no longitudinal cohort data have been made available so far (a CE mark study is ongoing).

The J-Valve

First implanted in 2014 and still an investigational device, this THV has been designed for use in both aortic stenosis ad aortic regurgitation. It consists of a two-stage mechanism (Fig. 17.30) including: a three-prong clasper which is deployed to capture the native aortic leaflets by seating into the aortic sinuses, and serves as the landing zone for the proper THV; a self-expanding frame mounting the valve which is subsequently deployed inside the clasper. These two elements are connected by three sutures, allowing a movable connection. Therefore, both stability and seal are obtained by capturing the native aortic leaflets against the valve frame with the combination of frame radial force and clasper rigidity. In a multicenter series of 43 implants in PAR, procedural success was 97.7%, 1-year mortality was 4.7%, new PPM implantation was 4.7% and there were no cases of more-than-mild aortic regurgitation, but the rate of valve-related reintervention was 7%.[85] The 2-years results from another series were also encouraging, with a 91.5% device success rate and no cases of more-than-mild aortic regurgitation among survivors at 2 years.[86] Such results warrant continued clinical evaluation especially regarding about durability, as the longest available follow-up (4-years) reported important rates of SVD.[87] While a transapical approach has been employed for the vast majority of clinical implantations, a transfemoral delivery system has been recently employed for the first time.

Conclusions

As one of the most significant paradigm shifts in the cardiovascular landscape, the success of TAVI is highly dependent on its technological content. Several anatomical, pathophysiological, and clinical requirements need to be satisfied, each corresponding to multiple design solutions the individual THV concepts have to propose. Therefore, the

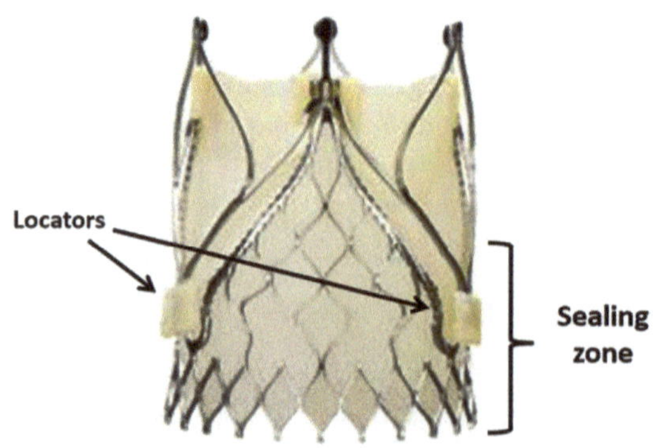

FIGURE 17.29 The JenaValve System THV.

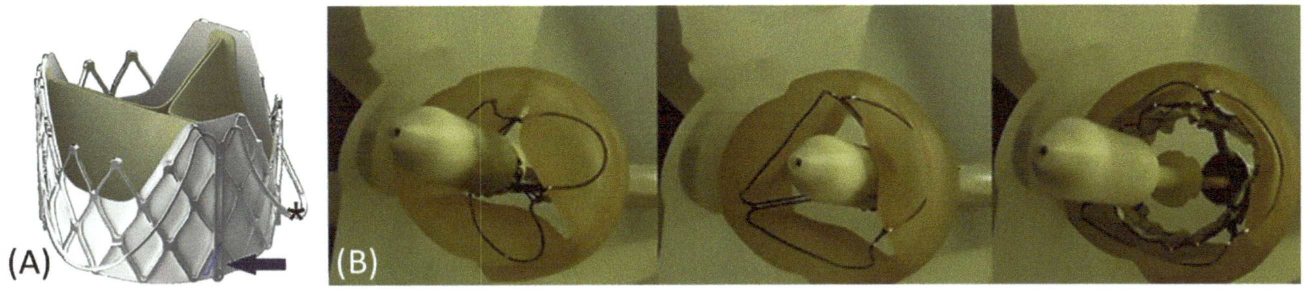

FIGURE 17.30 The J-Valve THV. (A) Scheme of the device. Asterisk indicates the three-prong clasper, black arrow indicates the connecting suture between the clasper and the self-expanding valve frame. (B) In vitro deployment sequence (clasper engagement in the aortic sinuses followed by frame deployment). *Reproduced with permission from Liu H, Yang Y, Wang W, et al. Transapical transcatheter aortic valve replacement for aortic regurgitation with a second-generation heart valve.* J Thorac Cardiovasc Surg. *2018;156:106–116.*[85]

development and establishment of a THV platform requires major efforts in terms of research, development, and investment. After the first-generation devices, the second-generation THVs have seen the appearance of numerous concepts: among these, several were discontinued or did not attain significant diffusion in the clinical practice. The continued development and numerous iterations in devices and delivery system has yielded the current clinical success, and a large body of literature has been produced with regard to clinical results and device-specific advantages/disadvantages. At the same time, there is a lack of direct comparison studies among alternative devices to support decision-making. Also, this rapidly evolving context challenges the capacities of academic medicine to provide at an adequate pace both independent clinical research and pedagogy to the use of new devices/techniques. Such a situation should elicit a reflection within academic institutions and scientific societies, as well as regulatory bodies.

In calcified aortic stenosis, radial force has emerged as the solution to obtain THV stability, with its declinations into the balloon-expandable and the self-expandable strategies. PAR represents a completely distinct pathophysiological and anatomical scenario, where radial force alone is inadequate and a leaflet capturing mechanism is required. Despite several THV concepts having been developed for PAR, their diffusion is still limited due to lower number of potential candidates, association of aortic regurgitation with thoracic aortic disease, and less demonstrated reproducibility than THV for calcified aortic stenosis. In recent years, optimized sealing through the addition of an outer skirt of different materials and extension at the THV sealing body, has determined the success of third-generation THVs and the migration of TAVI toward the lowest-risk patients' strata. Possibility to recapture and reposition (available only for self-expanding devices) represents a useful feature; although it is not essential for clinical success, it can orient decision-making in specific clinical settings (i.e., in case of increased risk of coronary obstruction). It should be also noted that the simplicity and reproducibility of the implant technique plays a role in the success and diffusion of THV concepts. Issues which are expected to be addressed in the next future through other "disruptive" technological solutions include, among others, the prevention of atrioventricular block (AVB) and PPM implantation. Miniaturization of the delivery system has also represented a major axis of development, yet its impact and other factors' on THV durability in the long term is still to be appreciated. The long-term evaluation of TAVI as a therapy should analyze separately the outcomes of the valve (durability) and the outcomes of the patients (survival, freedom from heart failure, etc.). The above elements suggest, at the current stage of knowledge, a careful extension of TAVI to patients' subgroups with greatest life expectancy before such evaluations are available. This would guarantee a viable and durable expansion of TAVI use. High-quality imaging (both for procedure planning and intraoperative guidance) is one of the key factors for the success of TAVI.

Despite miniaturization and technical experience leading to application in increasingly challenging iliofemoral anatomies, the nontransfemoral accesses continue to prove necessary in a significant proportion of cases. They allow adaptation to patients' characteristics and prevent access-related complications which carry significant morbidity burden.[88] Peripheral arterial nontransfemoral accesses (transcarotid and transsubclavian) have been suggested to be superior to the transapical/transaortic routes in terms of early morbidity related to vascular access[89,90] (patient outcome should prevail over device outcome) and their diffusion can be encouraged over previous-generation nontransfemoral routes. Nonetheless, no dedicated and specifically labeled delivery system has been proposed so far for these approaches.

The discussion of medico-economic aspects of TAVI is beyond the scopes of the present document. Nonetheless, it can be underlined that despite considerable investments for research/development and technological content, the affordability of TAVI procedures in routine use likely is an increasingly important requirement in a context of aging of the population and progressive limitations in the budgets of healthcare systems.

In conclusion, TAVI is a clear example of major healthcare progress based on adaptive technological advancements coupled with anatomical/pathophysiological knowledge. Its future is bright, yet still the boundaries of its applications are evolving, remain to be defined during the next years, and need to be evidence based.

References

1. Cribier A, Eltchaninoff H, Bash A, et al. Percutaneous transcatheter implantation of an aortic valve prosthesis for calcific aortic stenosis: first human case description. *Circulation*. 2002;106:3006–3008.
2. Moulopoulos SD, Anthopoulos L, Stamatelopoulos S, Stefadouros M. Catheter-mounted aortic valves. *Ann Thorac Surg*. 1971;5:423–430.
3. 2017 ESC/EACTS. Guidelines for the management of valvular heart disease. *Eur Heart J*. 2017;38:2739–2786.
4. Mack MJ, Leon MB, Thourani VH, et al. Transcatheter aortic-valve replacement with a balloon-expandable valve in low-risk patients. *N Engl J Med*. 2019;380:1695–1705.
5. Popma JJ, Deeb GM, Yakubov SJ, et al. Transcatheter aortic-valve replacement with a self-expanding valve in low-risk patients. *N Engl J Med*. 2019;380:1706–1715.
6. 2020 ACC/AHA guideline for the management of patients with valvular heart disease: a report of the American College of Cardiology/American Heart Association Joint Committee on Clinical Practice Guidelines. *J Thorac Cardiovasc Surg*. 2021;162:e183–e353.
7. Osnabrugge RL, Mylotte D, Head SJ, et al. Aortic stenosis in the elderly: disease prevalence and number of candidates for transcatheter aortic valve replacement: a meta-analysis and modeling study. *J Am Coll Cardiol*. 2013;62:1002–1012.
8. Piazza N, de Jaegere P, Schultz C, Becker AE, Serruys PW, Anderson RH. Anatomy of the aortic valvar complex and its implications for transcatheter implantation of the aortic valve. *Circ Cardiovasc Interv*. 2008;1:74–81.
9. Anderson RH. Clinical anatomy of the aortic root. *Heart*. 2000;84:670–673.
10. Hamdan A, Guetta V, Konen E, et al. Deformation dynamics and mechanical properties of the aortic annulus by 4-dimensional computed tomography: insights into the functional anatomy of the aortic valve complex and implications for transcatheter aortic valve therapy. *J Am Coll Cardiol*. 2012;59:119–127.
11. Takagi H, Umemoto T. ALICE (All-Literature Investigation of Cardiovascular Evidence) Group. Impact of paravalvular aortic regurgitation after transcatheter aortic valve implantation on survival. *Int J Cardiol*. 2016;221:46–51.
12. Auffret V, Puri R, Urena M, et al. Conduction disturbances after transcatheter aortic valve replacement: current status and future perspectives. *Circulation*. 2017;136:1049–1069.
13. Jabbour RJ, Tanaka A, Finkelstein A, et al. Delayed coronary obstruction after transcatheter aortic valve replacement. *J Am Coll Cardiol*. 2018;71:1513–1524.
14. Sinclair N, Mordhorst A, Yang GK, MacDonald PS, Sidhu R, Reid J. Vascular access complications and clinical outcomes of vascular surgical repairs following transcatheter aortic valve implantation (TAVI). *Ann Vasc Surg*. 2021;74:258–263. Available from: https://doi.org/10.1016/j.avsg.2020.12.032.
15. Mohammadi S, Rodés-Cabau J, Kalavrouziotis D. TAVR access: fitting the approach to the patient and not the patient to the approach. *J Am Coll Cardiol*. 2019;74:2740–2742.
16. Patel JS, Krishnaswamy A, Svensson LG, Tuzcu EM, Mick S, Kapadia SR. Access options for transcatheter aortic valve replacement in patients with unfavorable aortoiliofemoral anatomy. *Curr Cardiol Rep*. 2016;18:110.
17. Auffret V, Lefevre T, Van Belle E, et al. FRANCE TAVI Investigators. Temporal trends in transcatheter aortic valve replacement in France: FRANCE 2 to FRANCE TAVI. *J Am Coll Cardiol*. 2017;70:42–55.
18. Webb J, Gerosa G, Lefèvre T, et al. Multicenter evaluation of a next-generation balloon-expandable transcatheter aortic valve. *J Am Coll Cardiol*. 2014;64:2235–2243.
19. Binder RK, Rodes-Cabau J, Wood DA, et al. Transcatheter aortic valve replacement with the SAPIEN 3: a new balloon-expandable transcatheter heart valve. *JACC Cardiovasc Interv*. 2013;6:293–300.
20. Yudi MB, Sharma SK, Tang GHL, Kini A. Coronary angiography and percutaneous coronary intervention after transcatheter aortic valve replacement. *J Am Coll Cardiol*. 2018;71:1360–1378.
21. Moriyama N, Lehtola H, Miyashita H, Piuhola J, Niemelä M, Laine M. Hemodynamic comparison of transcatheter aortic valve replacement with the SAPIEN 3 Ultra vs SAPIEN 3: the HomoSAPIEN registry. *Catheter Cardiovasc Interv*. 2021;97:E982–E991.
22. Genereux P, Head SJ, Van Mieghem NM, et al. Clinical outcomes after transcatheter aortic valve replacement using valve academic research consortium definitions: a weighted meta-analysis of 3,519 patients from 16 studies. *J Am Coll Cardiol*. 2012;59:2317–2326.
23. Ando T, Briasoulis A, Holmes AA, Taub CC, Takagi H, Afonso L. Sapien 3 vs Sapien XT prosthetic valves in transcatheter aortic valve implantation: a meta-analysis. *Int J Cardiol*. 2016;220:472–478.
24. Rheude T, Blumenstein J, Möllmann H, Husser O. Spotlight on the SAPIEN 3 transcatheter heart valve. *Med Devices (Auckl)*. 2018;11:353–360.
25. Shivaraju A, Kodali S, Thilo C, et al. Overexpansion of the SAPIEN 3 transcatheter heart valve: a feasibility study. *JACC Cardiovasc Interv*. 2015;8:2041–2043.
26. Barbanti M, Costa G. SAPIEN 3 Ultra transcatheter aortic valve device: two sides of the moon. *JACC Cardiovasc Interv*. 2020;13:2639–2641.
27. Arai T, Lefèvre T, Hovasse T, et al. Comparison of Edwards SAPIEN 3 vs SAPIEN XT in transfemoral transcatheter aortic valve implantation: difference of valve selection in the real world. *J Cardiol*. 2017;69:565–569.

28. Folliguet TA, Teiger E, Beurtheret S, et al. Carotid vs femoral access for transcatheter aortic valve implantation: a propensity score inverse probability weighting study. *Eur J Cardiothorac Surg*. 2019;56:1140–1146.
29. Verhoye JP, Belhaj Soulami R, Tomasi J, et al. Early outcomes of transcarotid access for transcatheter aortic valve implantation. *EuroIntervention*. 2020;15:e1332–e1334.
30. Reardon MJ, Van Mieghem NM, Popma JJ, et al. SURTAVI Investigators. Surgical or transcatheter aortic-valve replacement in intermediate-risk patients. *N Engl J Med*. 2017;376:1321–1331.
31. Lehmann S, Merk DR, Etz CD, et al. Porcine xenograft for aortic, mitral and double valve replacement: long-term results of 2544 consecutive patients. *Eur J Cardiothorac Surg*. 2016;49:1150–1156.
32. Linke A, Holzhey D, Möllmann H, et al. Treatment of aortic stenosis with a self-expanding, resheathable transcatheter valve: one-year results of the international multicenter portico transcatheter aortic valve implantation system study. *Circ Cardiovasc Interv*. 2018;11:e005206.
33. Makkar RR, Cheng W, Waksman R, et al. Self-expanding intra-annular vs commercially available transcatheter heart valves in high and extreme risk patients with severe aortic stenosis (PORTICO IDE): a randomised, controlled, non-inferiority trial. *Lancet*. 2020;396:669–683.
34. Feldman TE, Reardon MJ, Rajagopal V, et al. Effect of mechanically expanded vs self-expanding transcatheter aortic valve replacement on mortality and major adverse clinical events in high-risk patients with aortic stenosis: the REPRISE III randomized clinical trial. *JAMA*. 2018;319:27–37.
35. Reardon MJ, Feldman TE, Meduri CU, et al. Reprise III Investigators. Two-year outcomes after transcatheter aortic valve replacement with mechanical vs self-expanding valves: the REPRISE III randomized clinical trial. *JAMA Cardiol*. 2019;4:223–229.
36. Meredith IT, Walters DL, Dumonteil N, et al. 1-Year outcomes with the fully repositionable and retrievable lotus transcatheter aortic replacement valve in 120 high-risk surgical patients with severe aortic stenosis: results of the REPRISE II study. *JACC Cardiovasc Interv*. 2016;9:376–384.
37. Rampat R, Khawaja MZ, Hilling-Smith R, et al. Conduction abnormalities and permanent pacemaker implantation after transcatheter aortic valve replacement using the repositionable LOTUS device: the United Kingdom experience. *JACC Cardiovasc Interv*. 2017;10:1247–1253.
38. Pilgrim T, Stortecky S, Nietlispach F, Heg D, Tueller D, Toggweiler S. Repositionable vs balloon-expandable devices for transcatheter aortic valve implantation in patients with aortic stenosis. *J Am Heart Assoc*. 2016;5:e004088.
39. Göteborg, et al. *PCR London Valves*; 2018.
40. Lanz J, Kim WK, Walther T, et al. SCOPE I investigators. Safety and efficacy of a self-expanding vs a balloon-expandable bioprosthesis for transcatheter aortic valve replacement in patients with symptomatic severe aortic stenosis: a randomised non-inferiority trial. *Lancet*. 2019;394:1619–1628.
41. Tamburino C, Bleiziffer S, Thiele H, et al. Comparison of self-expanding bioprostheses for transcatheter aortic valve replacement in patients with symptomatic severe aortic stenosis: SCOPE 2 randomized clinical trial. *Circulation*. 2020;142:2431–2442.
42. Kim WK, Möllmann H, Liebetrau C, et al. The ACURATE neo transcatheter heart valve: a comprehensive analysis of predictors of procedural outcome. *JACC Cardiovasc Interv*. 2018;11:1721–1729.
43. Kim WK, Hengstenberg C, Hilker M, et al. Transcatheter aortic valve implantation with the ACURATE neo valve: indications, procedural aspects and clinical outcomes. *EuroIntervention*. 2020;15:e1571–e1579.
44. Rück A. Results from the Early neo2 registry Acurate neo2 TAVI valve. In: *Euro PCR 2021 Conference*; 2021.
45. Chandra P, Jose J, Mattummal S, et al. Clinical evaluation of the Hydra self-expanding transcatheter aortic valve: 6 month results from the GENESIS trial. *Catheter Cardiovasc Interv*. 2021;98:371–379. Available from: https://doi.org/10.1002/ccd.29733.
46. Kapadia SR, Huded CP, Kodali SK, et al. Stroke after surgical vs transfemoral transcatheter aortic valve replacement in the PARTNER trial. *J Am Coll Cardiol*. 2018;72:2415–2426.
47. Kapadia S, Agarwal S, Miller DC, et al. Insights into timing, risk factors, and outcomes of stroke and transient ischemic attack after transcatheter aortic valve replacement in the PARTNER trial (placement of aortic transcatheter valves). *Circ Cardiovasc Interv*. 2016;9(9).
48. Nombela-Franco L, Webb JG, de Jaegere PP, et al. Timing, predictive factors, and prognostic value of cerebrovascular events in a large cohort of patients undergoing transcatheter aortic valve implantation. *Circulation*. 2012;126:3041–3053.
49. Auffret V, Regueiro A, del Trigo M, et al. Predictors of early cerebrovascular events in patients with severe aortic stenosis undergoing transcatheter aortic valve replacement: a systematic review and meta-analysis. *J Am Coll Cardiol*. 2016;68:673–684.
50. Giustino G, Sorrentino S, Mehran R, Faggioni M, Dangas G. Cerebral embolic protection during TAVR: a clinical event meta-analysis. *J Am Coll Cardiol*. 2017;69:465–466.
51. Giustino G, Mehran R, Veltkamp R, Faggioni M, Baber U, Dangas GD. Neurological outcomes with embolic protection devices in patients undergoing transcatheter aortic valve replacement: a systematic review and meta-analysis of randomized controlled trials. *JACC Cardiovasc Interv*. 2016;9:2124–2133.
52. Stachon P, Kaier K, Heidt T, et al. The use and outcomes of cerebral protection devices for patients undergoing transfemoral transcatheter aortic valve replacement in clinical practice. *JACC Cardiovasc Interv*. 2021;14:161–168.
53. Megaly M, Sorajja P, Cavalcante JL, et al. Ischemic stroke with cerebral protection system during transcatheter aortic valve replacement. *JACC Cardiovasc Interv*. 2020;13:2149–2155.
54. Gallo M, Putzu A, Conti M, Pedrazzini G, Demertzis S, Ferrari E. Embolic protection devices for transcatheter aortic valve replacement. *Eur J Cardiothorac Surg*. 2018;53:1118–1126.
55. Nietlispach F, Wijesinghe N, Gurvitch R, et al. An embolic deflection device for aortic valve interventions. *JACC: Cardiovasc Interv*. 2010;3:1133–1138.

56. Rodés-Cabau J, Kahlert P, Neumann F-J, et al. Feasibility and exploratory efficacy evaluation of the Embrella Embolic Deflector system for the prevention of cerebral emboli in patients undergoing transcatheter aortic valve replacement: the PROTAVI-C pilot study. *JACC Cardiovasc Interv*. 2014;7:1146–1155.
57. Samim M, Agostoni P, Hendrikse J, et al. Embrella embolic deflection device for cerebral protection during transcatheter aortic valve replacement. *J Thorac Cardiovasc Surg*. 2015;149:799–805. e2.
58. Wendt D, Kleinbongard P, Knipp S, et al. Intraaortic protection from embolization in patients undergoing transaortic transcatheter aortic valve implantation. *Ann Thorac Surg*. 2015;100:686–691.
59. Seeger J, Gonska B, Otto M, Rottbauer W, Wöhrle J. Cerebral embolic protection during transcatheter aortic valve replacement significantly reduces death and stroke compared with unprotected procedures. *JACC Cardiovasc Interv*. 2017;10:2297–2303.
60. Haigron P, Dillenseger JL, Luo L, Coatrieux JL. Image-guided therapy: evolution and breakthrough. *IEEE Eng Med Biol Mag*. 2010;29:100–104.
61. Vernikouskaya I, Rottbauer W, Gonska B, et al. Image-guidance for transcatheter aortic valve implantation (TAVI) and cerebral embolic protection. *Int J Cardiol*. 2017;249:90–95.
62. Kronzon I, Jelnin V, Ruiz CE, et al. Optimal imaging for guiding TAVR: transesophageal or transthoracic echocardiography, or just fluoroscopy. *JACC Cardiovasc Imaging*. 2015;8:361–370.
63. Belhaj Soulami R, Verhoye JP, Nguyen Duc H, et al. Computer-assisted transcatheter heart valve implantation in valve-in-valve procedures. *Innovations (Phila)*. 2016;11:193–200.
64. Agasthi P, Ashraf H, Pujari SH, et al. Artificial intelligence trumps TAVI2-SCORE and CoreValve score in predicting 1-year mortality post-transcatheter aortic valve replacement. *Cardiovasc Revasc Med*. 2021;24:33–41.
65. Pollari F, Hitzl W, Claes M, Grossmann I, Fischlein T. Machine learning for making aortic valve interventions complementary and not competitive. *JACC Cardiovasc Interv*. 2019;12:2112.
66. Fuchs A, Kofoed KF, Yoon SH, et al. Commissural alignment of bioprosthetic aortic valve and native aortic valve following surgical and transcatheter aortic valve replacement and its impact on valvular function and coronary filling. *JACC Cardiovasc Interv*. 2018;11:1733–1743.
67. Tang GHL, Zaid S, Fuchs A, et al. Alignment of transcatheter aortic-valve neo-commissures (ALIGN TAVR): impact on final valve orientation and coronary artery overlap. *JACC Cardiovasc Interv*. 2020;13:1030–1042.
68. Sengupta A, Alexis SL, Kovacic JC, Tang GHL. Current challenges in TAVI: neo-commissural alignment to mimic more physiologic valve implantation. *Vessel Plus*. 2020;4:40.
69. Reichenspurner H, Schaefer A, Schäfer U, et al. Self-expanding transcatheter aortic valve system for symptomatic high-risk patients with severe aortic stenosis. *J Am Coll Cardiol*. 2017;70:3127–3136.
70. Brennan PF, Spence MS. Self-expanding CENTERA valve for the treatment of severe, symptomatic aortic stenosis. *Future Cardiol*. 2019;15:79–84.
71. Tchétché D, Windecker S, Kasel AM, et al. 1-Year outcomes of the CENTERA-EU trial assessing a novel self-expanding transcatheter heart valve. *JACC Cardiovasc Interv*. 2019;12:673–680.
72. Naber CK, Pyxaras SA, Ince H, et al. A multicentre European registry to evaluate the Direct Flow Medical transcatheter aortic valve system for the treatment of patients with severe aortic stenosis. *EuroIntervention*. 2016;12:e1413–e1419.
73. Latib A, Maisano F, Colombo A, et al. Transcatheter aortic valve implantation of the direct flow medical aortic valve with minimal or no contrast. *Cardiovasc Revasc Med*. 2014;15:252–257.
74. Parma R, Hudziak D, Smolka G, Gocoł R, Ochała A, Wojakowski W. SAPIEN 3 Ultra—design and procedural features of a new balloon-expandable valve. *Cardiol J*. 2020;27:194–196.
75. Saia F, Gandolfo C, Palmerini T, et al. In-hospital and thirty-day outcomes of the SAPIEN 3 Ultra balloon-expandable transcatheter aortic valve: the S3U registry. *EuroIntervention*. 2020;15:1240–1247.
76. Rheude T, Pellegrini C, Lutz J, et al. Transcatheter aortic valve replacement with balloon-expandable valves: comparison of SAPIEN 3 Ultra vs SAPIEN 3. *JACC Cardiovasc Interv*. 2020;13:2631–2638.
77. Welle GA, El-Sabawi B, Thaden JJ, et al. Effect of a fourth-generation transcatheter valve enhanced skirt on paravalvular leak. *Catheter Cardiovasc Interv*. 2021;97:895–902.
78. Sharma SK, Rao RS, Chandra P, et al. First-in-human evaluation of a novel balloon-expandable transcatheter heart valve in patients with severe symptomatic native aortic stenosis: the MyVal-1 study. *EuroIntervention*. 2020;16:421–429.
79. Delgado-Arana JR, Gordillo-Monge MX, Halim J, et al. Early clinical and haemodynamic matched comparison of balloon-expandable valves. *Heart*. 2021;. Available from: https://doi.org/10.1136/heartjnl-2021-319349.
80. Kawashima H, Soliman O, Wang R, et al. Rationale and design of a randomized clinical trial comparing safety and efficacy of myval transcatheter heart valve vs contemporary transcatheter heart valves in patients with severe symptomatic aortic valve stenosis: The LANDMARK trial. *Am Heart J*. 2021;232:23–38.
81. Fioretta ES, Motta SE, Lintas V, et al. Next-generation tissue-engineered heart valves with repair, remodelling and regeneration capacity. *Nat Rev Cardiol*. 2021;18:92–116.
82. Lintas V, Fioretta ES, Motta SE, et al. Development of a novel human cell-derived tissue-engineered heart valve for transcatheter aortic valve replacement: an in vitro and in vivo feasibility study. *J Cardiovasc Transl Res*. 2018;11:470–482.
83. Wernly B, Eder S, Navarese EP, et al. Transcatheter aortic valve replacement for pure aortic valve regurgitation: "on-label" vs "off-label" use of TAVR devices. *Clin Res Cardiol*. 2019;108:921–930.

84. Schäfer U, Schirmer J, Niklas S, Harmel E, Deuschl F, Conradi L. First-in-human implantation of a novel transfemoral selfexpanding transcatheter heart valve to treat pure aortic regurgitation. *EuroIntervention*. 2017;13:1296–1299.
85. Liu H, Yang Y, Wang W, et al. Transapical transcatheter aortic valve replacement for aortic regurgitation with a second-generation heart valve. *J Thorac Cardiovasc Surg*. 2018;156:106–116.
86. Shi J, Wei L, Chen Y, et al. Transcatheter aortic valve implantation with J-Valve: 2-year outcomes from a multicenter study. *Ann Thorac Surg*. 2021;111:1530–1536.
87. Li F, Wang X, Wang Y, et al. Structural valve deterioration after transcatheter aortic valve implantation using J-Valve: a long-term follow-up. *Ann Thorac Cardiovasc Surg*. 2020;26:158–165.
88. Laakso T, Moriyama N, Raivio P, et al. Impact of major vascular complication access site status on mortality after transfemoral transcatheter aortic valve replacement—results from the finnvalve registry. *Circ Rep*. 2020;2:182–191.
89. Beurtheret S, Karam N, Resseguier N, et al. Femoral vs nonfemoral peripheral access for transcatheter aortic valve replacement. *J Am Coll Cardiol*. 2019;74:2728–2739.
90. De Carlo M, Giannini C, Petronio AS. *Alternative transcatheter approaches. A Practical Guide to Prevention and Treatment of Complications During Transcatheter Aortic Valve Implantation*. John Wiley & Sons Ltd; 2021.

Chapter 18

Technology Advancement for Percutaneous Interventions for Mitral Regurgitation

Mahboob Ali[1] and Imran Arif[2]

[1]Sutter Gould Medical Foundation, Sutter Health, Stockton, CA, United States, [2]University of Cincinnati, Cincinnati, OH, United States

Chapter Outline

Introduction

Mitral regurgitation (MR) is the leading cause of valvular heart disease in the United States and affects an estimated 4 million people with a drastic increase expected by 2030 due to both an aging population and population growth.[1,2] The estimated prevalence of MR in the United States is ~1.7% but this prevalence increases with age and MR affects ~9%–10% of people aged older than 75 years.[3–5] Worldwide, rheumatic heart disease is the leading cause of mitral regurgitation. In the United States, functional MR, usually associated with ischemic heart disease, is the leading cause of mitral regurgitation.[6] Untreated MR carries a poor prognosis. Reported mortality rates for secondary MR are 20% at 1 year, and about 50% at 5 years in patients with heart failure.[7,8] MR is an independent predictor of mortality in HF patients.[3–5] While the treatment of MR have been limited to optimizing medical therapy and surgical therapies, the spectrum of mitral valve percutaneous interventions is rapidly evolving, with several recent CE (Conformité Européene) and FDA (US Food and Drug Administration) approved options, and many others undergoing investigational trials. In this chapter we summarize the approved and investigational percutaneous therapeutic interventional devices available for the treatment of MR.

Anatomy and pathophysiology

The healthy mitral valve is a complex apparatus that functions as a one-way conduit from the left atrium to the left ventricle. The mitral apparatus consists of mitral annulus, leaflets, commissures, chordae tendinae, posterior left atrium, left ventricular free wall, and papillary muscles,[9] which works in a coordinated fashion to allow frictionless passage of blood through the left side of the heart.[9,10] Dysfunction of any component can lead to mitral valve pathology, including MR (Fig. 18.1).[9]

Advances in Cardiovascular Technology. DOI: https://doi.org/10.1016/B978-0-12-816861-5.00031-9

MR is the retrograde flow through the mitral valve from the left ventricle back into the left atrium during systole. Pathologic mitral regurgitation is when the amount of flow exceeds trivial mitral regurgitation on echocardiography. Most patients will tolerate mild and moderate mitral regurgitation without symptoms or need for intervention. However, severe mitral regurgitation is not well tolerated, leading to left ventricular remodeling, poor cardiac output, neurohormonal activation, irreversible myocardial damage with fibrosis, heart failure, and ultimately death.[9–11]

Patient symptoms are due to hemodynamic changes that happen due to severe MR. Secondary MR has different physiology than primary MR, as the left ventricle and left atrium are dilated in excess relative to the amount of regurgitant volume, and left atrial pressures are excessively elevated despite lower regurgitant volumes than primary MR.[12] Invasive studies reveal elevated left atrial mean pressure (LAmP), left atrial V wave pressure (LAvP), left atrial mean pressure index (LAmPI), left atrial V wave pressure index (LAvPI), increased pulmonary capillary wedge pressure (PCWP), pulmonary arterial pressures and right atrial and tricuspid annular dilatation, and increased rates for recurrent hospitalization.[12,13]

Classification of mitral regurgitation

The diagnosis, pathophysiology, natural history, and management of mitral regurgitation is intimately connected to the acuity and underlying mechanism of the valve pathology. Mitral regurgitation is classified into acute and chronic MR. The mechanics of chronic mitral regurgitation is further subclassified into primary/organic mitral regurgitation and secondary/functional mitral regurgitation. Transthoracic echocardiography remains the cornerstone for the initial assessment and severity of mitral regurgitation. Once the structure and movement of the mitral valve have been determined, mitral valve pathology can be further subclassified using the Carpentier's classification: type I, normal leaflet movement (annular dilatation or leaflet perforation); type II, excessive leaflet movement; type III, restrictive leaflet movement—IIIa, diastolic restriction (rheumatic disease), and IIIb, systolic restriction (functional disease) (Table 18.1).[6,9,14–16]

Acute mitral regurgitation

Acute mitral regurgitation can result from dysfunction of any component of the mitral valve apparatus. Acute MR has been relatively rare, but the incidence is rising secondary to the growing rates of infectious endocarditis from intravenous drug abuse. The common causes for acute MR include infectious endocarditis, chordal rupture from degenerative mitral valve disease, and papillary muscle rupture from ST-segment elevation myocardial infarction. The pathophysiology of acute mitral regurgitation involves acute volume and pressure overload of the left ventricle and left atrium, resulting in pulmonary congestion and poor cardiac output. In comparison to chronic mitral regurgitation, the left

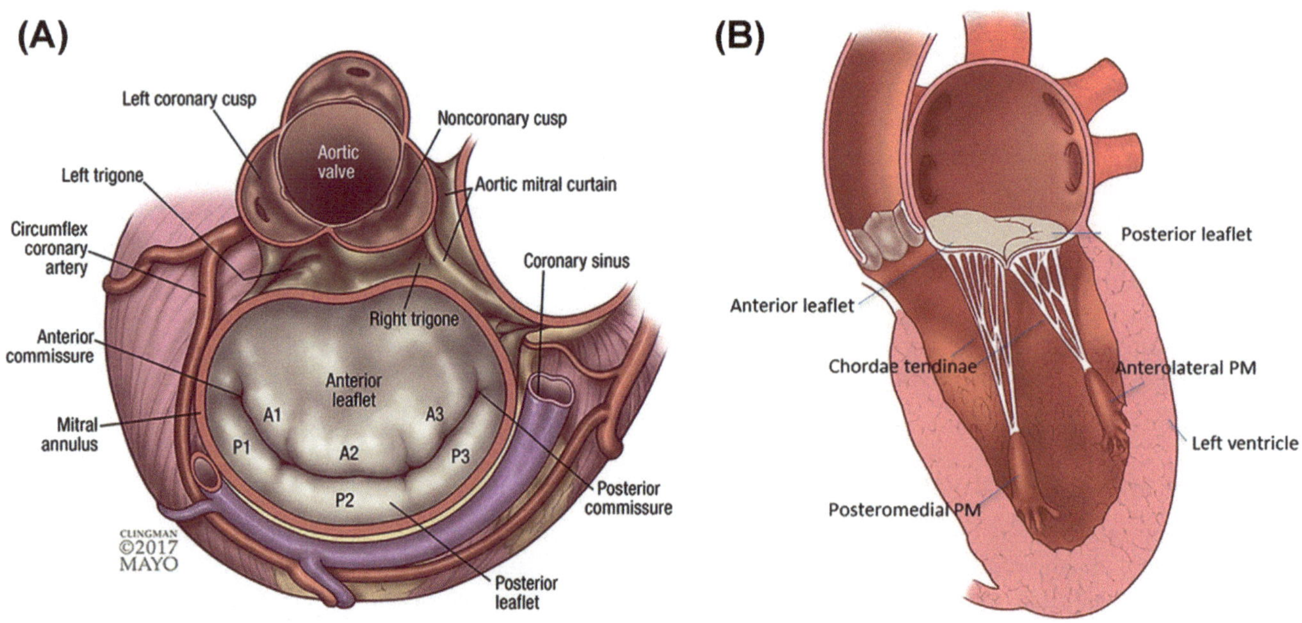

FIGURE 18.1 (A) Schematic representation of mitral valve from atrial view. (B) Sagittal view of mitral valve. *A*, Anterior; *P*, posterior; *PM*, papillary muscle. *Reproduced with permission from Marechaux et al.* Prog Cardiovasc Dis. *2017;60(3):289–304.*

TABLE 18.1 Carpentier's classification.

Carpentier's classification				
	Organic			**Functional**
	Type I (normal leaflet movement)	Type II (excessive leaflet movement)	Type IIIa (restrictive leaflet movement in diastole)	Type I/Type IIIb (restrictive leaflet movement in systole)
Nonischemic	Endocarditis (perforation); degenerative (annular calcification); congenital (cleft leaflet)	Degenerative (billowing/flail leaflets); endocarditis (ruptured chordae); traumatic (ruptured Chord/PM); rheumatic (acute RF)	Rheumatic (chronic RF); iatrogenic (radiation/drug); inflammatory (lupus/ anticardiolipin, eosinophilic endocardial disease, endomyocardial fibrosis)	Cardiomyopathy; myocarditis; left-ventricular; dysfunction (any cause)
Ischemic	**Organic**	Ruptured PM		Functional ischemia
		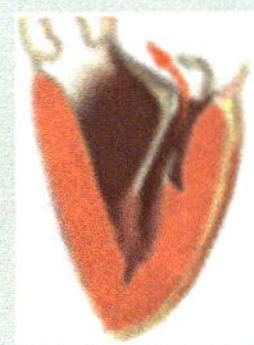		

Mr, Mitral regurgitation; *PM*, papillary muscle; *RF*, rheumatic fever.

ventricle has not had adequate time to dilate and compensate for the regurgitant volume. Aortic pressure exceeds left atrial pressure, and therefore systolic flow preferentially favors the left atrium, which results in reduced cardiac output, hypotension, and even shock. Transthoracic echocardiography can be helpful in diagnosis, by demonstrating valve morphology, severe MR jet, hyperdynamic left ventricle function, and normal left ventricular size (assuming no baseline left ventricular dilatation).[14] Acute MR carries a relatively high mortality rate, and urgent surgical intervention is often required.[14]

Prompt surgical evaluation is the mainstay of therapy for symptomatic acute, severe primary mitral regurgitation. Medical therapy can be utilized as a temporizing/supportive measure and includes diuretics to reduce pulmonary congestion and increase venous capacitance as well as vasodilator therapy to reduce afterload and aortic impedance to forward flow which can preferentially shift flow from retrograde to antegrade through the aortic valve. Patients with acute MR are often hypotensive which limits the ability to use vasodilators. Intraaortic balloon counterpulsation (IABP) is another option for nonsurgical management of acute MR. IABP reduces aortic systolic pressure (afterload), while simultaneously increasing mean and diastolic aortic pressure. In acute MR associated with hypotension, the decrease in afterload in combination with diastolic pressure augmentation, can enhance forward output without worsening hypotension and is a valuable therapy.[14,17,18]

Chronic primary/organic mitral regurgitation

In contrast to acute mitral regurgitation, chronic primary mitral regurgitation has a more indolent course and can be well compensated for long periods of time. The underlying pathology involves the mitral valve or valve apparatus, and most commonly includes degenerative (primary myxomatous disease, primary flail leaflet, annular calcification) in 60%–70% of cases, followed by ischemic MR due to obstructive coronary disease (20%), endocarditis (2%–5%), and rheumatic heart disease (2%–5%).

Chronic secondary/functional mitral regurgitation

Chronic functional mitral regurgitation differs from primary mitral regurgitation in that the mitral valve apparatus is normal, valve dysfunction arises from left ventricular remodeling. In contrast to chronic primary mitral regurgitation where valve dysfunction precedes left ventricular remodeling; the reverse order is true in functional mitral regurgitation. Left ventricular remodeling and dilatation of any etiology (ischemic or nonischemic cardiomyopathy) and left atrial dilatation due to AF (leads to annular dilatation) can result in functional mitral regurgitation and a consequent increase in left atrial pressure which may lead to pulmonary hypertension and progression of heart failure symptoms. Functional mitral regurgitation which is secondary to left ventricular remodeling is an independent predictor of adverse prognosis and mortality.[9,14–16]

Mitral regurgitation is associated with excessive mortality in patients with effective regurgitant orifice (EROA) of $>2.0\ mm^2$ irrespective of patient symptoms. It is reported to have 22% ± 3% 5-year mortality in asymptomatic patients with EROA of $40 \pm 27\ mm^2$, when treated with medical management only.[19,20] Although chronic mitral regurgitation is categorized clinically into compensated or decompensated mitral regurgitation, there are three stages of chronic MR. In stage 1 (compensated phase) long-standing primary mitral regurgitation results in left ventricular volume overload/enlargement, eccentric hypertrophy with normal systolic function. Enlargement of left atrium due to compliance keeps the left atrial pressures and thus pulmonary arterial pressures within normal limits in the initial compensatory phase.[20] That is in fact the first sign that MR is worsening. In stage 2 (transitional phase) there is mild LV dysfunction, however, it is reversible following correction of regurgitant lesion. In stage 3 (decompensated phase) LV dilates, wall stress increases and may cause irreversible myocardial damage. As this occurs, LV contractility may decline, with a reduction in LV stroke volume and ejection fraction (Fig. 18.2).[9,14,15]

Diagnosis

Diagnosis of MR is usually confirmed by imaging modalities. Transthoracic echocardiogram (TTE) and transesophageal echocardiogram (TEE) imaging tools are the commonest modalities used to make the diagnosis. Although a diagnosis of MR can be made with TTE, TEE is usually required to define etiology of the mitral regurgitation and better evaluate anatomy for the various percutaneous therapies.[3] TEE better elucidates leaflet characteristics including leaflet length, coaptation, calcification, degree of prolapse and tenting area, assessment of subvalvular apparatus, annulus diameter, and direction of regurgitant jet, which are essential for planning percutaneous intervention.[3] Severity of MR can also be assessed by additional echocardiography parameters including vena contracta (VC), proximal isovelocity surface area

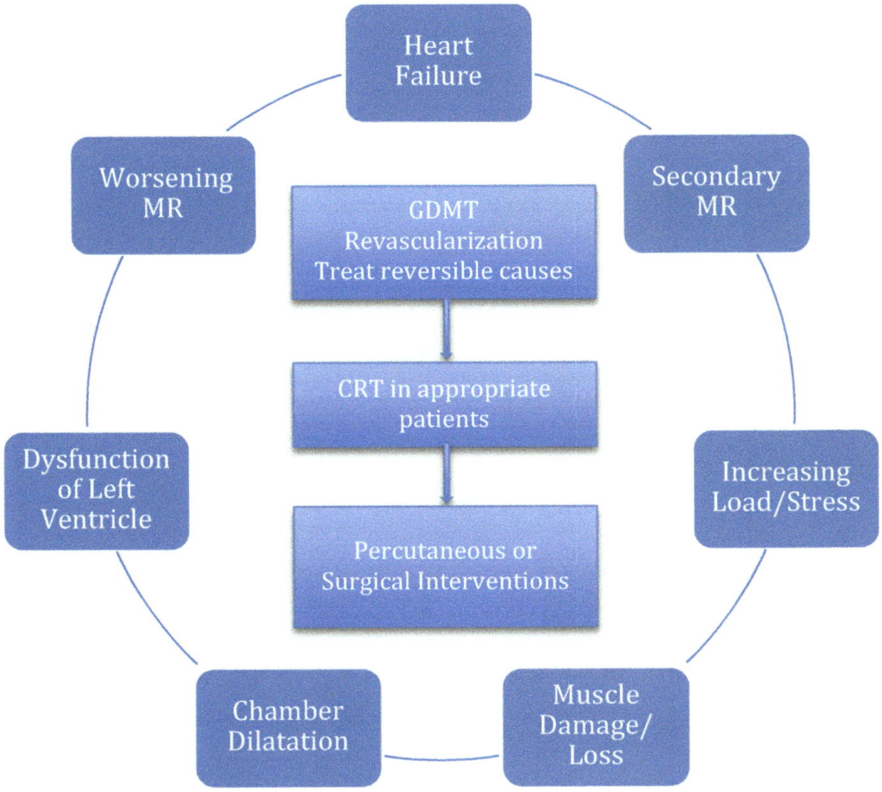

FIGURE 18.2 Vicious cycle for HF and MR wherein MR begets more HF which leads to more MR.

(PISA), effective regurgitant orifice area (EROA) and by calculating regurgitant volume (RVol) and regurgitant fraction (RF) by various methods.[3] A VC of >0.7 cm, RVol of >60 mL, RF >50%, EROA >0.4 cm^2 and dilated LV are consistent with severe MR per American guidelines (Fig. 18.3). However, per European guidelines an EROA of >0.2 cm^2 and RVol of 30 mL is consistent with severe secondary MR. Therefore, an accurate diagnosis of MR as well as a suitable therapeutic plan can be developed with the help of TTE and TEE based on the above-mentioned characteristics.[3] EROA >13 mm^2 during exercise, or an EROA >20 mm^2 at rest, and a shorter mitral deceleration time have been identified as independent predictors of cardiac mortality.[3,21]

Cardiac magnetic resonance (CMR) imaging is highly accurate in quantifying the degree of MR, LV dimensions, volumes and function, LA volumes, and can be a valuable adjunct to echocardiography.[22] There is limited data on regurgitant volume and RF thresholds for severity grading.[12] Multidetector row computed tomography (MDCT) is playing an increasing role in the management of MR. MDCT can accurately depict mitral apparatus structure and morphology, can demonstrate the relationship of the mitral apparatus to other cardiac structures, and can be valuable in planning therapeutic (particularly catheter based) interventions.[22–24]

Current treatment options for MR

Medical therapy options for chronic primary mitral regurgitation are aimed toward treating the underlying cause of valve dysfunction and toward minimizing progression of MR but the data supporting a1ny medical therapy specifically for mitral regurgitation is limited. The mainstay of treatment for chronic primary MR is surgical or percutaneous mitral valve repair or replacement. The ideal window of opportunity is while the patient is compensated, as the degree of myocardial damage is minimal. Current practice guidelines recommend surgical intervention in asymptomatic patients with primary MR and LVEF <60% or LV end-systolic dimension >40 mm.[14]

The 2014 ACC/AHA guidelines for valvular heart disease gives a Level IIa recommendation for beta-adrenergic blockade and angiotensin-aldosterone pathway inhibition. Although small studies have suggested reverse remodeling and inhibition of fibrosis with these therapies, large scale clinical trials are lacking. Vasodilator therapy in normotensive patients carries a Class III/No Benefit recommendation.[14]

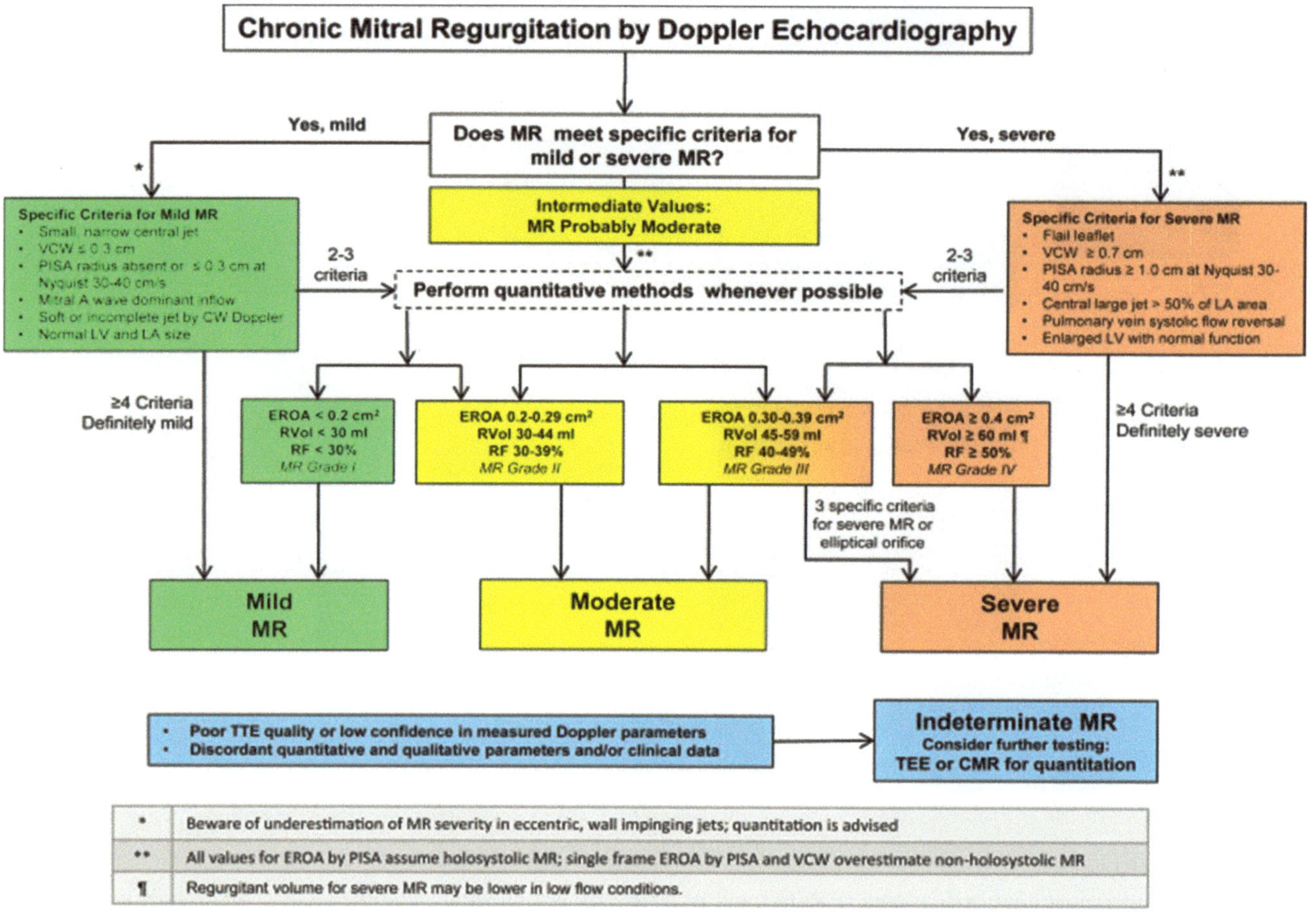

FIGURE 18.3 Algorithm for assessment MR severity by American society of Echocardiography. *Reprinted with permission from Zoghbi et al.* J AM Soc Echocardiogr. *2007;30(4):303–371.*

Medical treatment of functional mitral regurgitation includes guideline directed medical therapy (GDMT) for underlying left ventricular dysfunction. Beta-adrenergic blockade, angiotensin aldosterone pathway blockade, and cardiac resynchronization therapy (CRT) (in appropriate patients) are Class I recommendations in the 2014 ACC/AHA Guidelines for Valvular Heart Disease.[14–16] None of these therapies target mitral regurgitation, thus severity of MR may or may not improve even if the LV function improves with above therapies.

Surgery has been the definitive treatment option for low-risk patients with primary MR, and per current AHA/ACC guidelines surgery has a IIb recommendation for secondary MR.[14,15] Surgery is also class IIb recommendation per European society of cardiology guidelines in low surgical risk patients.[25] Surgical interventions did not produce encouraging results in secondary MR patients, and were associated with 60% recurrent MR and 17.6% 1-year mortality rates.[5] Thus high surgical risk patients with chronic primary or secondary MR have been treated medically until recently.

Percutaneous interventions

Percutaneous therapeutic interventions which target leaflet, annulus and valve apparatus have been developed over the last two decades for patients with MR and/or high surgical risk. Although multiple options are available, only Mitraclip is approved by FDA for percutaneous intervention in patients with high or prohibitive surgical risk primary and secondary MR. Following review of data suggesting that Mitraclip improved quality of life, New York Heart Association (NYHA) functional class and reduced hospitalizations related to heart failure due to reduction in secondary MR, it was approved for treatment of secondary MR in Japan long before it's approval in the United States based on the unmet need for secondary MR patients resistant to medical therapy, but at prohibitive risk for mitral valve surgery.[26]

Percutaneous mitral valve therapies include transcatheter mitral valve repair, direct and indirect annuloplasty, and transcatheter mitral valve replacement (TMVR) (Table 18.2).

TABLE 18.2 Transcatheter mitral devices.

Mechanism of action		Device name	Manufacturer	Characteristics	Access site	Indications	Strengths	Weakness	Development stage	Clinical trials/ studies
Leaflet repair	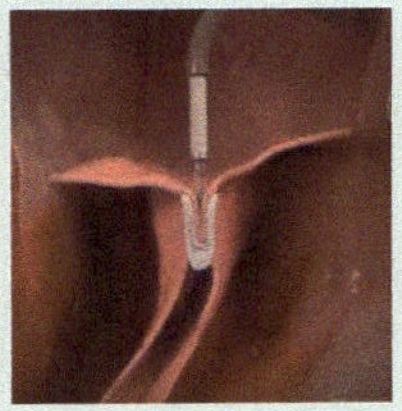	MitraClip	Abbott Vascular	Edge-to-Edge leaflet repair	24 F TF/TS	DMR FMR	Minimal invasiveness	Lack of annuloplasty	FDA approved CE mark approved	EVEREST EVEREST II REALSM ACCESS EU TRAMI GRASP European Sentinel MITRA-FR COAPT
	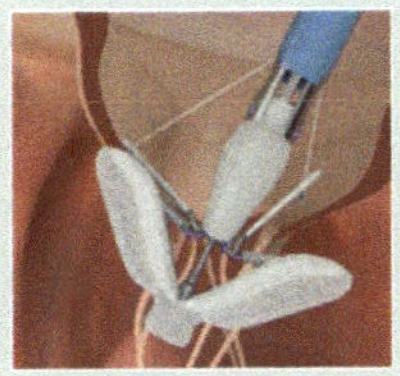	Pascal	Edwards Lifesciences	Edge-to-Edge leaflet repair	22 F TF/TS	DMR FMR	Larger paddles, independent grippers, advanced steerability	Lack of annuloplasty	CE mark approved	Compassi-onate use in 23 patients. CLASP trial ongoing
Chordal Apparatus		NeoChord	NeoChord INC	Off-pump artificial chord implantation to PML	28 F TA	PML flail or prolapse	Minimally invasive Off Pump valve repair	TA approach	CE mark approved	TOP-MINI TACT trial TACT registry ReChord AcChord
	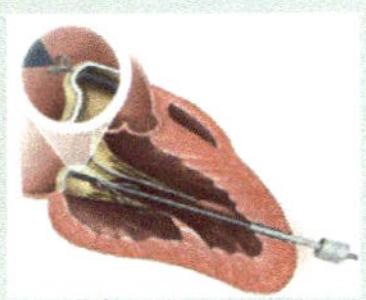	Harpoon	Harpoon Medical	Off-pump artificial chord implantation	TA	DMR PML prolapse	Minimally invasive Off Pump valve repair	TA approach	CE mark approved	Early Feasibility and CE Mark study 65 patients. TRACER (ongoing)
Direct Annuloplasty	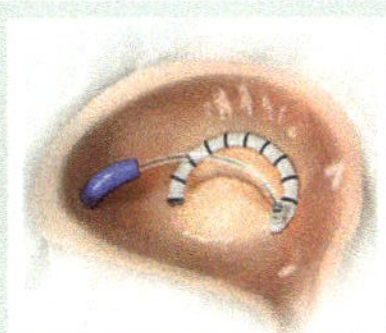	Cardioband	Edwards Lifesciences	Sutureless anchors over atrial side posterior mitral annulus	TF/TS	FMR		Complex advanced imaging Not suitable for patients with heavily calcified annulus	CE mark approved	Feasibility study- 31 patients. CE mark study- 61 patients

(*Continued*)

TABLE 18.2 (Continued)

Mechanism of action		Device name	Manufacturer	Characteristics	Access site	Indications	Strengths	Weakness	Development stage	Clinical trials/ studies
		Mitralign	Mitralign INC	Pairs of Pledgeted sutures over atrial side posterior mitral annulus	TF/TS TF/LV	FMR		Limited efficacy Not suitable for patients with EF <30% or LEDD >7 cm.	CE mark approved	Feasibility study ongoing (71 patients)
Indirect Annuloplasty	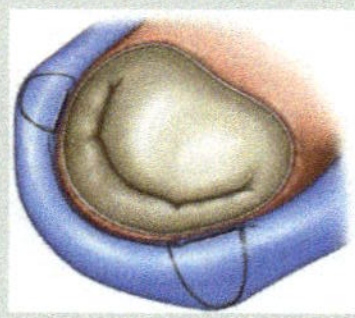	Carillon	Cardiac Dimension INC	Coronary sinus cinching with Nitinol wire and anchors	10 F TJ	FMR	Minimal invasive, simplicity	Wire fractures with older version Coronary compromise	CE mark approved	AMEDEUS TITAN TITAN II REDUCE-FMR
Mitral Valve Replacement		Tendyne	Abbott Vascular	Retrievable, nitinol frame with atrial and apical fixation	36 F TA	DMR FMR without calcification or MS	Less invasive, versatile and several sizes	TA only		Early feasibility trial- 110 patient SUMMIT trial ongoing
		Intrepid	Medtronic	Radial force and subannular cleats	35 F TA	DMR FMR	Less invasive, simple design	Sealing and durability		Early feasibility trial- 48 patient APOLLO trial ongoing
	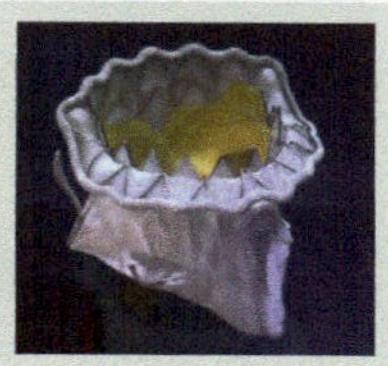	Tiara	Neovasc	Nitinol frame with ventricular anchors and sealing skirt, with native leaflet engagement	32 F TA TF/TS	DMR FMR without calcification or MS	D shaped- similar to MV Effective in reducing MR	Fractures		Early feasibility and CE mark studies- TIARA I TIARA II

	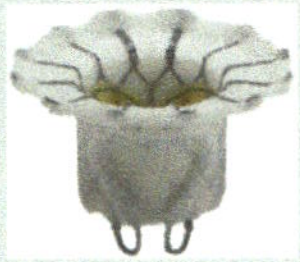	Fortis	Edwards Lifesciences	29-mm nitinol system with paddles to capture native leaflets and secure position	42 F TA TF/TS	FMR without calcification or MS	Simple design	Valve Thrombosis Sealing		US feasibility trial halted for evaluation of valve thrombosis events
	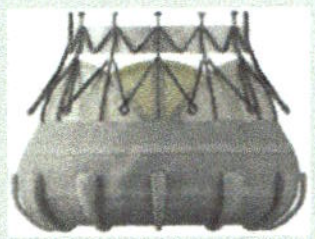	CardiAQ	Edwards Lifesciences	Self-positioning, nitinol frame with ventricular anchors with leaflet engagement	33 F TA TF/TS	DMR FMR without calcification or MS	TA and TF approach	Atrial Protrusion Thrombosis		Early feasibility trial (RELIEF trial)
	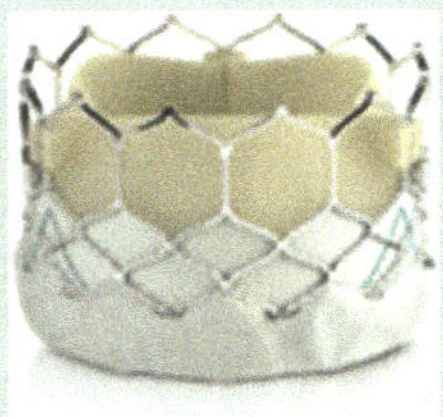	Sapien XT/ Sapien 3	Edwards Lifesciences	Balloon-expandable TAVR implanted in mitral position (severe MAC, ¢valve in valve, valve in ring)	33 F-TA 16 F TF/TS	DMR FMR Bioprosthetic Ms/MR Native calcific MS				MITRAL trials (PARTNER 3 and SITRAL) currently enrolling
	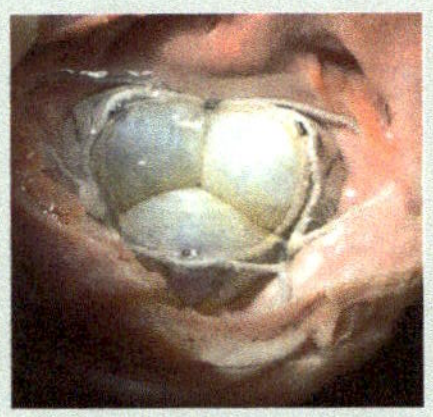	Caisson	Caisson Interventional LLC	Mitral annulus capture with engagement fibrous groove	31 F TF/TS	DMR FMR	Minimally invasive TS approach	2 Step procedure		Early feasibility study- 11 patients PRELUDE and INTERLUDE trials
	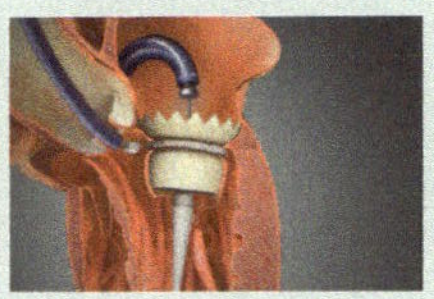	High Life	High-Life SAS	External subannular ring with valve in ring	30 F TA TF/TS		TA and TS approach	2 step procedure One size		
Ventricular Restoration	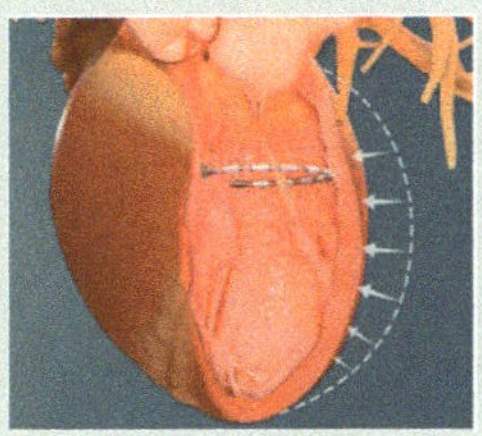	Accucinch	ANCORA Heart	Cinching left ventricles	Trans-arterial	FMR	Targets Left Ventricle. Aims to reduce LV size, prevent LV enlargement and cause reverse remodeling.			Early Feasibility study- 18 patients CINCH-2 and CorCinch studies ongoing

Transcatheter mitral valve repair

Leaflet solutions

The mechanism of **Mitraclip** edge to edge repair is based on the surgical Alfieri double orifice technique (Alfieri stitch) proposed by Dr. Alfieri in the 1990s.[27,28] Although initially approved in the United States for primary MR, following the COAPT (Cardiovascular Outcomes Assessment of the MitraClip Percutaneous Therapy for Heart) Failure Patients trial, Mitraclip has recently gained FDA approval for secondary MR.

Everest II (Endovascular Valve Edge-to-Edge REpair Study) trial randomly assigned 258 patients with grade 3 + or 4 + in a 2:1 ratio (178:80) to percutaneous mitral valve repair with Mitraclip versus conventional surgery.[28] Although there was no apparent mortality benefit for either group, Mitraclip was safer with fewer side effects, but less effective in reducing the severity of MR.[28] Mitraclip treated patients more commonly required surgery for residual MR during the first year after treatment, but between 1- and 5-year follow-up, comparably low rates of surgery for residual MV dysfunction were observed for both treatments consistent with durable MR reduction with both repair techniques.[28] Catheter-based outcomes were influenced by an initial learning curve: lack of prior experience in left atrial navigation and mitral interventions as well as the absence of 3D echocardiography. Improved operator experience and technical progress have led to improved clinical outcomes, and to application of the technology beyond classic EVEREST Trial inclusion criteria to clinically challenging cases (emergency, papillary ruptures, extreme left ventricular dysfunction, and many others).[27,28]

The multicenter German TRAMI (Transcatheter Mitral Valve Interventions) registry enrolled 828 real world patients, the majority of whom had functional MR (70%) with NYHA grade III or more symptoms (89%). This registry revealed a correlation between procedural safety and higher center volumes.[29] The major complication of bleeding requiring blood transfusion was likely related to large bore venous access.[29] The impact of learning curve on procedural outcomes was also proven in a Two-Phase observational ACCESSEU (MitraClip Therapy Economic and Clinical Outcomes Study Europe) European study, which evaluated high-risk elderly patients, mainly affected by functional MR. Mitraclip was safe (3.4% 30-day mortality) and effective in significantly improving MR grade at 12 months.[30] A significant improvement of 6-minute walk from baseline to 6 months with continued trend at 12 months was observed.[30] The HAMBURG (Transcatheter Mitral Valve Repair in Surgical High- Risk Patients: Gender-Specific Acute and Long-Term Outcomes) study analyzed gender specific differences in 592 patients, who underwent the Mitraclip procedure. Men had significantly higher prevalence of cardiovascular comorbidities, renal failure, and adverse echocardiographic parameters which resulted in longer device time and higher numbers of implanted clips but with equal procedural success. Survival was relatively higher in women, and renal dysfunction was associated with higher mortality.[31]

Two landmark trials evaluated the efficacy of Mitraclip in symptomatic heart failure patients with secondary MR. MITRA-FR Trial (Percutaneous Repair with the MitraClip Device for Severe Functional/Secondary Mitral Regurgitation) was a multicenter, open label, randomized, controlled phase 3 trial which compared Mitraclip + guideline driven medical therapy (GDMT) versus GDMT alone. A total of 307 patients were randomized, of whom 152 were assigned to Mitraclip + GDMT. Although the Mitraclip group had significant reduction of MR, there was no statistically significant difference in 1-year mortality and need for hospitalization among the two groups.[32] The COAPT Trial (Cardiovascular Outcomes Assessment of the MitraClip Percutaneous Therapy for Heart Failure Patients with Functional Mitral Regurgitation) was a multicenter, randomized, controlled, parallel-group trial that enrolled 614 patients with heart failure (majority with NYHA III-IV symptoms) and secondary MR (302 patients were randomized to Mitraclip + GDMT group vs 312 patients to GDMT alone). The Mitraclip + GDMT group had a statistically significant reduction in all-cause mortality, hospitalization, reduction in MR, improvement in quality of life and functional capacity at 24 months follow-up.[33]

There are several possible explanations for the apparent discordant clinical outcomes between these studies. The patients in COAPT trial were on intense medical therapy. The onsite heart team for COAPT trial included a heart failure specialist. Patients' heart failure medications were up-titrated to maximum tolerated doses. All those patients whose symptoms abated, or degree of MR improved were excluded. Only symptomatic patients, refractory to maximum tolerated medical therapy were included in the COAPT trial compared to the MITRA-FR trial, which allowed variable medication adjustments in each group during follow-up. The second possible explanation is related to the degree of MR in context of LV volumes. Patients in COAPT trial had larger EROA than MITRA-FR trial ($41 + 15\ mm^2$ vs $31 + 10\ mm^2$), thus had more severe MR, and relatively smaller left ventricle end diastolic volumes ($101 + 34\ mL/m^2$ vs $135 + 35\ mL/m^2$). Thus, the degree of MR was disproportionately greater than would have been expected from the

LV chamber enlargement (disproportionate MR). Reduction in MR produced significant reduction in regurgitant volumes in COAPT trial patients, with subsequent significant impact on mortality and morbidity benefit due to the complex interplay between MR and LV dysfunction. Procedural performance success rates differed and probably influenced outcomes as well. COAPT trial group had more Mitraclips per patient, less procedural complications (8.5% vs 14.6%) and more sustained reduction of MR at 1 year (5% vs 17%) compared with MITRA-FR patients respectively.[32–36] Based on the COAPT trial results, FDA has approved Mitraclip for symptomatic heart failure due to secondary MR. A meta-analysis including 2121 patients with heart failure and severe secondary MR comparing Mitraclip + GDMT ($n = 833$) versus GDMT alone ($n = 1288$) also demonstrated that Mitraclip (plus GDMT) was superior to GDMT alone, and was associated with significant survival benefit, particularly in patients with secondary MR across all subgroups.[37]

The Edwards **PASCAL** transcatheter mitral valve repair system is similar to Mitraclip and targets the mitral leaflet. In comparison to Mitraclip, it has longer paddles for larger grabbing surface, independent grippers for independent articulation of both arms of the device and has an independent steerable system. A multicenter, prospective, observational, first-in-man study involving 23 patients established feasibility of PASCAL device with a high rate of technical success and reduction of MR.[38] The CLASP trial is enrolling patients and is designed to assess safety as well as major adverse events defined as cardiovascular mortality, stroke, myocardial infarction, need for renal replacement therapy, severe bleeding and reintervention for study device-related complications at 30 days.

Chordal solutions

Neochord involves off-pump artificial chordal implantation via the transapical approach. Initial safety and efficacy experience were evaluated in the TACT (Transapical Artificial Chordae Tendinae) trial. Thirty patients with severe MR due to isolated Carpentier type II prolapse of the posterior MV leaflet and without annulus dilation were included in this study, and patients with secondary MR were excluded.[39–41] TACT concluded that off-pump transapical implantation of artificial chordae to correct MR is technically safe and feasible. The TOP-MINI (Transapical off-pump mitral valve repair with Neochord Implantation) early feasibility trial evaluated 49 patients with severe symptomatic degenerative MR (89.8% with posterior leaflet prolapse [P], 8.2% with anterior LP and 2% with combined disease) who underwent the procedure. Early results from TOP-MINI suggest that Neochord is feasible and safe with acute procedure success (defined as successful placement of at least three neochords with reduction of residual MR to less than grade 2 +) achieved in all patients and in-hospital mortality observed in 2%. At 3-month follow-up survival was 98% and MR was absent in 16 patients (33.4%), grade 1 + in 15 (31.2%), and was grade 2 + in 12 (25%) patients. Five patients (10.4%) developed recurrent severe MR due to anterior native chordae upture.[27,39,40] A prospective randomized trial of NeoChord versus open surgical repair (ReChord trial) is ongoing, plans to enroll 585 patients, and is expected to finish in 2020.

The **Harpoon** system is a polytetrafluoroethylene chordal implantation device similar to Neochord, which has completed an early feasibility study in 11 patients with PML prolapse and normal ejection fraction. Procedural success was 100% with trace MR observed immediately after the procedure and mild MR at 1 month.[42] The safety profile of the Harpoon system was confirmed in a prospective trial of 30 patients where 27 of 30 patients (90%) met the primary end point, and 10% (3/30) required conversion to open mitral surgery. There were no deaths, strokes, or permanent pacemaker implantations. MR was mild or less in 89% (24 of 27) at 1 month, and was moderate in 11% (3 of 27).[27,42,43]

Direct annuloplasty

Although conventional mitral annuloplasty can be done surgically, surgical intervention is a class IIb recommendation in patients with secondary MR and may be associated with increased mortality and morbidity.[14,15] Patients who are undergoing cardiac surgery for other reasons, for example, CABG or aortic valve disease, may have mitral annuloplasty performed as well. Percutaneous direct annuloplasty concepts are inspired of surgical annuloplasty. Several direct annuloplasty systems are currently under development and include the Cardioband, Mitralign, and Millipede systems.

Cardioband uses transeptal access to the annulus and is implanted on the posterior annulus from the anterior-lateral to the posterior-medial mitral valve commissure on the atrial side of the valve.[43–45] The early feasibility study enrolled 31 patients with secondary MR and reduced LV function (left ventricular ejection fraction 34% + 11%). At 6-month follow-up Cardioband was effective in reducing MR, improving heart failure symptoms and had a favorable safety profile.[45] The Cardioband Mitral system CE mark trial in 61 patients demonstrated a mortality rate of 21% with 96% of patients having an MR grade of 2 or less at 2 years follow-up. A 9% reduction in annular area was observed and 83% of patients were NYHA class I or II with improved 6-minute walk tests at 2-year follow-up.[46]

Mitralign is a trans-arterial system that reduces annulus size by delivering paired pledgeted sutures to the posterior mitral annulus coming from the ventricular to the atrial side of the valve. Early feasibility evaluation in 71 patients demonstrated a 70.4% device success rate with 8.9% of patients developing cardiac tamponade. At 30 days ($n = 45$) and 6-months ($n = 41$), rates for all-cause mortality, stroke, and myocardial infarction were 4.4%, 4.4%, 0.0%, and 12.2%, 4.9%, 0%, respectively. Further, a 50% reduction in MR and a significant reduction in annular size were noted at 6 months.[46–48]

Indirect annuloplasty

Indirect mitral annuloplasty may be accomplished through the coronary sinus or by altering the dimensions of the LV chamber (ventriculoplasty). At least two systems are being developed to address the coronary sinus approach: **Carillon** Mitral Contour System and Mitral Valve cerclage. The Carillon device was evaluated in the AMADEUS (CARILLON Mitral Annuloplasty Device European Union Study) trial which enrolled 48 patients with symptomatic heart failure, LVEF <40% and moderate to severe MR. Eighteen patients did not receive the device because of access issues, insufficient FMR reduction, or coronary artery compromise. The rate of major adverse events (death, MI, CS dissection/perforation, device embolization and surgery or PCI related to device) was 13% at 30 days. These included one reported death, three myocardial infarctions, and three CS dissection/perforation. At 6 months follow-up, FMR reduction ranged from 22% to 32% and 6-minute walk distance improved from 307 + 87 m (baseline) to 403 + 137 m.[49,50] Carillon was also evaluated in the TITAN (Transcatheter Implantation of Carillon Mitral Annuloplasty Device) trial. Although 36 patients initially received an implant, the device was recaptured in 17 patients (who subsequently served as control) for various reasons including coronary compromise. The successfully implanted group demonstrated significant reductions in FMR as well as LV diastolic and systolic volumes, compared with progressive LV dilation in the "control," unsuccessful implant group. Further, 6 MWD markedly improved in the implanted patients.[50,51]

However, wire fractures were observed next to the proximal anchor locking mechanism, at high strain locations on the Carillon devices, in both AMADEUS and TITAN which prompted device iteration,[51] and a modified Carillion system was evaluated in the subsequent TITAN II trial which enrolled 36 patients with symptomatic CHF, LVEF <40, and moderate or greater FMR. TITAN II demonstrated reduction in FMR, improvement in functional class and 6-minute walk, one nondevice related death 17 days post device implant. and no device wire fractures.[51,52] The REDUCEFMR Trial (Safety and Efficacy of the CARILLON Mitral Contour System in Reducing Functional Mitral Regurgitation Associated with Heart Failure trial) was a sham-controlled randomized trial that demonstrated the Carillon device to significantly reduce the degree of MR and LV volumes at 1 year follow-up.

Coronary sinus annuloplasty "**Cerclage**" is an investigational device undergoing preclinical evaluation for indirect annuloplasty and also involves the coronary sinus. Animal studies have been carried out, but human feasibility studies are yet to be done, and are planned for late 2019.[53,54]

Transcatheter mitral valve replacement

TMVR is an investigation procedure for high surgical risk in-operable patients. There are many TMVR devices available in preclinical phase, and some of these have finished feasibility studies. We will discuss some of them. These are discussed in full detail as a separate chapter in this book.

Tendyne transcatheter mitral valve has the highest number of implants to date. It is a D-shaped porcine pericardial trileaflet valve with two nitinol stents, adjustable tether and apical fixation/sealing pad, that is delivered transapically.[55] So far >100 valves have been implanted.[56] The Global feasibility trial reported 30 days outcomes on the first 30 high surgical risk patients with grade 3 or 4 MR. The device was successfully implanted in 28 patients (93.3%). One patient died on the 13th day because of pneumonia, and another patient developed leaflet thrombosis, which resolved on treatment with warfarin. At 30 days, only one patient mild MR (grade 1 +), and no MR was observed in the remaining 26 patients.

An improvement of left ventricular end-diastolic volume index (90.1 + 28.2 mL/m^2 at baseline vs 72.1 + 19.3 mL/m^2 at follow-up) and left ventricular end-systolic volume index (48.4 + 19.7 mL/m^2 vs 43.1 + 16.2 mL/m^2) was noted. Most patients (75%) reported mild or no symptoms at follow-up (NYHA class I or II).[56–58] Results of the first 100 high surgical risk patients, with STS predicted mortality of 7.9% and predominant severe MR, enrolled in the international CE mark study reveal a high procedural success rate without significant complications. There was a 97% procedural success rate, no procedural mortality, surgical conversion, or need for mechanical circulatory support and 98.7% had no or trace MR at 30 days among 76 patients with complete 30-day data.[56] Ongoing SUMMIT trial is comparing Tendyne TMVR versus conventional surgical MVR.

Intrepid TMVR system is a self-expanding bovine pericardial trileaflet valve, which is placed using a transapical delivery system. A Global multicenter, multinational study enrolled 50 high surgical risk patients with mean STS score of 6.4 + 5.5%, 84% with secondary MR, 86% with NYHA class III or IV symptoms and mean EF 43 + 12% who underwent device implant. The device was successfully implanted in 48 patients and 30 days mortality was 14%. There were no disabling strokes or repeat interventions. At median follow-up 173 days, there was no or mild MR and significant improvement in NYHA functional class in 79% of patients. No valve thrombosis events were observed, and 4 (8%) patients were admitted with HF within 30 days.[59–61] The ongoing APOLLO trial consisting of two arms, with one-arm randomizing Intrepid TMVR versus traditional surgery ($n = 650$), and a second arm that will treat patients with Intrepid TMVR who are ineligible for surgery as a single arm cohort ($n = 550$), will provide further insight into the safety and efficacy of Intrepid valve.[60]

TIARA TMVR system was implanted for compassionate use in two patients in 2014, with severe MR, depressed LV function, and NHYA class IV symptoms. The procedure was technically successful with hemodynamic improvement and improvement in NYHA class in both patients without major complication.[62] An early feasibility study enrolled eight patients, with procedural success in seven patients, and one patient was converted to open surgery. There was no mortality or major complications at 30 days in successfully implanted patients.[63] TIARA-I, an early feasibility, and TIARA-II, European CE Mark Study, are ongoing trials. So far more than 50 high-risk patients with moderate to severe left ventricular systolic dysfunction and secondary MR have received TIARA valve with 95% device implant success rate, and 8.5% mortality rate.[64]

Other TMVR systems are either in early feasibility clinical trials or yet to start feasibility trials. These include CardiAQ-Edwards, Sapien XT or Sapien 3, HighLife TMVR, CAISSON TMVR, and NCSI NaviGate Mitral Valve.

Ventricular restoration

AccuCinch is a new concept of treating MR which targets the left ventricle freewall. The technique attempts to do a "ventriculoplasty" to reduce septal free wall and LV dimensions. The exact mechanism is still unknown. The initial and updated experience were presented at TCT 2018, it has subsequently been attempted in over 30 patients. AccuCinch reduced LV diameters, increased LVEF and reduced MR besides improving functional capacity of patients. The early feasibility and CE mark studies are underway.

Conclusion

Mitral regurgitation is not a benign condition and it is associated with high mortality rates. MR is an independent predictor of mortality in HF patients. The relationship of the LV and MR is complex. Both dilated LV with reduced systolic function leads to worsening MR, whereas worsening MR leads to further volume overload and dilatation of LV in a vicious cycle. Current treatment for secondary MR is guideline directed medical treatment of heart failure, and CRT in eligible patients. There is fairly large group of patients who remain symptomatic despite maximum GDMT. Many of these patients have high surgical risk and surgery has a class IIb indication. Percutaneous catheter-based interventions in high-surgical-risk patients to improve MR, in particular those who have "disproportionate" MR to LV volumes, may improve patient outcomes by breaking this vicious cycle. Percutaneous interventions are a rapidly evolving field to improve morbidity and mortality in patients with secondary MR. With the exception of Mitraclip, percutaneous interventions for mitral regurgitation remain investigational. Mitraclip has been in use for nearly two decades, and was initially approved for primary MR in high surgical risk patient following EVERSET trials. COAPT trial was resulted in late 2018, which revealed significant mortality and morbidity benefits in severely symptomatic heart failure patients with severe secondary MR treated with Mitraclip + GDMT versus GDMT alone. This led to FDA approval of Mitraclip for treatment of secondary MR. There have been reports of use annuloplasty and Mitraclip together with significant long-term improvement in patient symptoms and outcomes. Besides, many of the above discussed percutaneous interventional devices, including annuloplasty and TMVR systems, have improved MR grade and patient's quality of life, even though they are investigational devices and in early phase. We hope that some of these devices would be standard of care for these high-risk patients in future to improve morbidity and mortality.

References

1. Chatzistergiou KT, Papanastasiou CA, Kokkinidis DG, Ziakas AG, Karvounis HI, Karamitsos TD. MitraClip device for patients with functional mitral valve regurgitation: a systematic review. *Hellenic J Cardiol*. 2019;.

2. Nkomo VT, Gardin JM, Skelton TN, Gottdiener JS, Scott CG, Enriquez-Sarano M. Burden of valvular heart diseases: a population-based study. *Lancet (London, England)*. 2006;368(9540):1005–1011.
3. Asgar AW, Mack MJ, Stone GW. Secondary mitral regurgitation in heart failure: pathophysiology, prognosis, and therapeutic considerations. *J Am Coll Cardiol*. 2015;65(12):1231–1248.
4. Nishimura RA, Vahanian A, Eleid MF, Mack MJ. Mitral valve disease—current management and future challenges. *Lancet (London, England)*. 2016;387(10025):1324–1334.
5. Eng MH, Wang DD. Transcatheter mitral valve therapy: defining the patient who will benefit. *Curr Cardiol Rep*. 2018;20(11):107.
6. Enriquez-Sarano M, Akins CW, Vahanian A. Mitral regurgitation. *Lancet (London, England)*. 2009;373(9672):1382–1394.
7. Goel SS, Bajaj N, Aggarwal B, et al. Prevalence and outcomes of unoperated patients with severe symptomatic mitral regurgitation and heart failure: comprehensive analysis to determine the potential role of MitraClip for this unmet need. *J Am Coll Cardiol*. 2014;63(2):185–186.
8. Dziadzko V, Clavel M-A, Dziadzko M, et al. Outcome and undertreatment of mitral regurgitation: a community cohort study. *Lancet (London, England)*. 2018;391(10124):960–969.
9. Fuster V, Harrington RA, Narula J, Eapen ZJ. Two Volume Set *Hurst's the Heart*. 14th ed. McGraw Hill Professional; 2017.
10. Quill JL, Hill AJ, Laske TG, Alfieri O, Iaizzo PA. Mitral leaflet anatomy revisited. *J Thorac Cardiovasc Surg*. 2009;137(5):1077–1081.
11. Carabello BA. The current therapy for mitral regurgitation. *J Am Coll Cardiol*. 2008;52(5):319–326.
12. Lancellotti P, Dulgheru R, Marchetta S, Oury C, Garbi M. Valve disease in heart failure: secondary but not irrelevant. *Heart Fail Clin*. 2019;15 (2):219–227.
13. Kuwata S, Taramasso M, Czopak A, et al. Continuous direct left atrial pressure: intraprocedural measurement predicts clinical response following MitraClip therapy. *JACC Cardiovasc Interv*. 2019;12(2):127–136.
14. Nishimura RA, Otto CM, Bonow RO, et al. ACC/AHA task force members. 2014 AHA/ACC guideline for the management of patients with valvular heart disease: executive summary: a report of the American College of Cardiology/American Heart Association Task Force on Practice Guidelines. 2014;129:2440–2492.
15. Nishimura RA, Otto CM, Bonow RO, et al. 2017 AHA/ACC focused update of the 2014 AHA/ACC guideline for the management of patients with valvular heart disease: a report of the American College of Cardiology/American Heart Association Task Force on Clinical Practice Guidelines. *Circulation*. 2017;135(25):e1159–e1195.
16. O'Gara PT, Grayburn PA, Badhwar V, et al. 2017 ACC expert consensus decision pathway on the management of mitral regurgitation: a report of the American College of Cardiology Task Force on Expert Consensus Decision Pathways. *J Am Coll Cardiol*. 2017;70(19):2421–2449.
17. Horstkotte D, Schulte HD, Niehues R, Klein RM, Piper C, Strauer BE. Diagnostic and therapeutic considerations in acute, severe mitral regurgitation: experience in 42 consecutive patients entering the intensive care unit with pulmonary edema. *J Heart Valve Dis*. 1993;2(5): 512–522.
18. Yoran C, Yellin EL, Becker RM, Gabbay S, Frater RWM, Sonnenblick EH. Mechanism of reduction of mitral regurgitation with vasodilator therapy. *Am J Cardiol*. 1979;43(4):773–777.
19. Antoine C, Benfari G, Michelena HI, et al. Clinical outcome of degenerative mitral regurgitation. *Circulation*. 2018;138(13):1317–1326.
20. Enriquez-Sarano M, Avierinos J-F, Messika-Zeitoun D, et al. Quantitative determinants of the outcome of asymptomatic mitral regurgitation. *N Engl J Med*. 2005;352(9):875–883.
21. Lancellotti P, Troisfontaines P, Toussaint A-C, Pierard LA. Prognostic importance of exercise-induced changes in mitral regurgitation in patients with chronic ischemic left ventricular dysfunction. *Circulation*. 2003;108(14):1713–1717.
22. Thavendiranathan P, Phelan D, Thomas JD, Flamm SD, Marwick TH. Quantitative assessment of mitral regurgitation: validation of new methods. *J Am Coll Cardiol*. 2012;60(16):1470–1483.
23. Delgado V, Tops LF, Schuijf JD, et al. Assessment of mitral valve anatomy and geometry with multislice computed tomography. *JACC Cardiovasc Imaging*. 2009;2(5):556–565.
24. Tops LF, Van de Veire NR, Schuijf JD, et al. Noninvasive evaluation of coronary sinus anatomy and its relation to the mitral valve annulus: implications for percutaneous mitral annuloplasty. *Circulation*. 2007;115(11):1426–1432.
25. Baumgartner H, Falk V, Bax JJ, et al. 2017 ESC/EACTS Guidelines for the management of valvular heart disease. *Eur Heart J*. 2017;38 (36):2739–2791.
26. Konishi A, Ho M, Ouchi T, Mitsutake Y, Shirato H. Regulatory approval review of transcatheter mitral valve repair—difference in the indication between the USA and Japan. *J Cardiol*. 2019;.
27. Saccocci M, Taramasso M, Maisano F. Mitral valve interventions in structural heart disease. *Curr Cardiol Rep*. 2018;20(6):49.
28. Feldman T, Kar S, Elmariah S, et al. Randomized comparison of percutaneous repair and surgery for mitral regurgitation: 5-year results of EVEREST II. *J Am Coll Cardiol*. 2015;66(25):2844–2854.
29. Eggebrecht H, Schelle S, Puls M, et al. Risk and outcomes of complications during and after MitraClip implantation: experience in 828 patients from the German TRAnscatheter mitral valve interventions (TRAMI) registry. *Catheter Cardiovasc Interv*. 2015;86(4):728–735.
30. Maisano F, Franzen O, Baldus S, et al. Percutaneous mitral valve interventions in the real world: early and 1-year results from the ACCESS-EU, a prospective, multicenter, nonrandomized post-approval study of the MitraClip therapy in Europe. *J Am Coll Cardiol*. 2013;62(12):1052–1061.
31. Tigges E, Kalbacher D, Thomas C, et al. Transcatheter mitral valve repair in surgical high-risk patients: gender-specific acute and long-term outcomes. *BioMed Res Int*. 2016;2016:3934842.
32. Obadia J-F, Messika-Zeitoun D, Leurent G, et al. Percutaneous repair or medical treatment for secondary mitral regurgitation. *N Engl J Med*. 2018;379(24):2297–2306.

33. Stone GW, Lindenfeld J, Abraham WT, et al. Transcatheter mitral-valve repair in patients with heart failure. *N Engl J Med*. 2018;379(24):2307–2318.
34. Nishimura RA, Bonow RO. Percutaneous repair of secondary mitral regurgitation—a tale of two trials. *N Engl J Med*. 2018;379(24):2374–2376.
35. Mack MJ, Abraham WT, Lindenfeld J, et al. Cardiovascular outcomes assessment of the MitraClip in patients with heart failure and secondary mitral regurgitation: design and rationale of the COAPT trial. *Am Heart J*. 2018;205:1–11.
36. Grayburn PA, Sannino A, Packer M. Proportionate and disproportionate functional mitral regurgitation: a new conceptual framework that reconciles the results of the MITRA-FR and COAPT trials. *JACC Cardiovasc Imaging*. 2019;12(2):353–362.
37. Giannini C, D'ascenzo F, Fiorelli F, et al. A meta-analysis of MitraClip combined with medical therapy vs. medical therapy alone for treatment of mitral regurgitation in heart failure patients. *ESC Heart Fail*. 2018;5(6):1150–1158.
38. Praz F, Spargias K, Chrissoheris M, et al. Compassionate use of the PASCAL transcatheter mitral valve repair system for patients with severe mitral regurgitation: a multicentre, prospective, observational, first-in-man study. *Lancet*. 2017;390(10096):773–780.
39. Colli A, Manzan E, Zucchetta F, et al. Transapical off-pump mitral valve repair with Neochord implantation: early clinical results. *Int J Cardiol*. 2016;204:23–28.
40. Colli A, Zucchetta F, Torregrossa G, et al. Transapical off-pump mitral valve repair with Neochord Implantation (TOP-MINI): step-by-step guide. *Ann Cardiothorac Surg*. 2015;4(3):295–297.
41. Seeburger J, Rinaldi M, Nielsen SL, et al. Off-pump transapical implantation of artificial neo-chordae to correct mitral regurgitation: the TACT Trial (Transapical Artificial Chordae Tendinae) proof of concept. *J Am Coll Cardiol*. 2014;63(9):914–919.
42. Gammie JS, Bartus K, Gackowski A, et al. Beating-heart mitral valve repair using a novel ePTFE cordal implantation device: a prospective trial. *J Am Coll Cardiol*. 2018;71(1):25–36.
43. Kuwata S, Taramasso M, Guidotti A, Nietlispach F, Maisano F. Ongoing and future directions in percutaneous treatment of mitral regurgitation. *Expert Rev Cardiovasc Ther*. 2017;15(6):441–446.
44. Taramasso M, Inderbitzin DT, Guidotti A, et al. Transcatheter direct mitral valve annuloplasty with the Cardioband system for the treatment of functional mitral regurgitation. *Multimed Man Cardiothorac Surg: MMCTS*. 2016;2016. mmw004.
45. Nickenig G, Hammerstingl C, Schueler R, et al. Transcatheter mitral annuloplasty in chronic functional mitral regurgitation: 6-month results with the cardioband percutaneous mitral repair system. *JACC Cardiovasc Interv*. 2016;9(19):2039–2047.
46. Guadagnoli AF, De Carlo C, Maisano F, et al. Cardioband system as a treatment for functional mitral regurgitation. *Expert Rev Med Dev*. 2018;15(6):415–421.
47. Schueler R, Nickenig G. The Mitralign: strategies for optimal patient selection and optimised results. *EuroIntervention*. 2016;12(Y):Y67–Y69.
48. Nickenig G, Schueler R, Dager A, et al. Treatment of chronic functional mitral valve regurgitation with a percutaneous annuloplasty system. *J Am Coll Cardiol*. 2016;67(25):2927–2936.
49. Schofer J, Siminiak T, Haude M, et al. Percutaneous mitral annuloplasty for functional mitral regurgitation: results of the CARILLON Mitral Annuloplasty Device European Union Study. *Circulation*. 2009;120(4):326–333.
50. Siminiak T, Wu JC, Haude M, et al. Treatment of functional mitral regurgitation by percutaneous annuloplasty: results of the TITAN Trial. *Eur J Heart Fail*. 2012;14(8):931–938.
51. Lipiecki J, Siminiak T, Sievert H, et al. Coronary sinus-based percutaneous annuloplasty as treatment for functional mitral regurgitation: the TITAN II trial. *Open Heart*. 2016;3(2):e000411.
52. Bail DHL. Treatment of functional mitral regurgitation by percutaneous annuloplasty using the Carillon Mitral Contour System-Currently available data state. *J Interv Cardiol*. 2017;30(2):156–162.
53. Kim J-H, Kocaturk O, Ozturk C, et al. Mitral cerclage annuloplasty, a novel transcatheter treatment for secondary mitral valve regurgitation: initial results in swine. *J Am Coll Cardiol*. 2009;54(7):638–651.
54. Kim JH, Sung SC, Chon MK, et al. Mitral loop cerclage as a variant form of mitral cerclage annuloplasty that adds a device (CSTV) for preventing potential complications: a preclinical proof of concept and feasibility study. *EuroIntervention*. 2016;11(14):e1669–e1679.
55. Beller JP, Rogers JH, Thourani VH, Ailawadi G. Early clinical results with the Tendyne transcatheter mitral valve replacement system. *Ann Cardiothorac Surg*. 2018;7(6):776–779.
56. Muller DWM, Farivar RS, Jansz P, et al. Transcatheter mitral valve replacement for patients with symptomatic mitral regurgitation: a global feasibility trial. *J Am Coll Cardiol*. 2017;69(4):381–391.
57. Perpetua EM, Reisman M. The Tendyne transcatheter mitral valve implantation system. *EuroIntervention*. 2015;11 Suppl(W):W78–9. Available from: https://doi.org/10.4244/EIJV11SWA23.
58. Quarto C, Davies S, Duncan A, et al. Transcatheter mitral valve implantation: 30-day outcome of first-in-man experience with an apically tethered device. *Innovations*. 2016;11(3):174–178.
59. Rodrigo Mendirichaga VS. Intrepid steps towards clinical prime time. *J Thoracic Dis*. 2018;10(Suppl. 9):S1040–S1043.
60. Paul Sorajja VB. Early experience with the Intrepid system for transcatheter mitral valve replacement. *Ann Cardiothorac Surg*. 2018;7(6):792–798.
61. Qasim Stanazai MA. The intrepid adventure of early transcatheter mitral valve replacement. *J Thoracic Dis*. 2018;10(Suppl. 9):S999–S1002.
62. Cheung A, Webb J, Verheye S, et al. Short-term results of transapical transcatheter mitral valve implantation for mitral regurgitation. *J Am Coll Cardiol*. 2014;64(17):1814–1819.
63. Verheye S, Cheung A, Leon M, Banai S. The Tiara transcatheter mitral valve implantation system. *EuroIntervention*. 2015;11(Suppl. W(W)):W71–W72.
64. Cheung A. Early experience of TIARA transcatheter mitral valve replacement system. *Ann Cardiothorac Surg*. 2018;7(6):787–791.

Chapter 19

Transcatheter Mitral Valve Replacement: Technology Update

Amedeo Anselmi[1,2], Guillaume Leurent[2,3] and Erwan Donal[2,3]

[1]*Division of Thoracic and Cardiovascular Surgery, Pontchaillou University Hospital, Rennes, France,* [2]*Univ Rennes, CHU Rennes, Inserm, LTSI - UMR 1099, Rennes, France,* [3]*Division of Cardiology, Pontchaillou University Hospital, Rennes, France*

Chapter Outline

Introduction

Transcatheter mitral valve replacement (TMVR) is among the fields of the current cardiovascular landscape yielding major evolution and eliciting the greatest enthusiasm in the medical community. Mitral regurgitation (MR) is the most frequent heart valve disorder; it can be calculated that significant (moderate or severe) MR is fourfold more prevalent than significant aortic stenosis,[1,2] but also that only a minor percentage of these patients receives interventional treatment for MR, despite its association with increased mortality. The prevalence of MR rises with age, hence a sustained increase in numbers of affected patients is expected to occur in Western-type populations. Despite correction of MR represents the greatest-growing subset of cardiac surgical procedures during the last decade in the US according to the Society of Thoraci Surgeons (STS) Database,[3] surgery is clearly insufficient to address the needs of this vast, complex and composite population. Transcatheter mitral repair has also markedly developed in the recent years and percutaneous techniques based on a "edge-to-edge" principle are now a consolidating clinical reality. Nonetheless, these techniques have limits associated with either anatomical contraindications or risk of recurrence of MR over time, that is, in secondary MR,[4,5] Recurrence of regurgitation in secondary MR is clearly more frequent after surgical mitral repair than replacement, although its prognostic impact needs to be declined in various clinical scenarios.[6] TMVR is expected to be an interesting option in all cases where the effectiveness and durability of either percutaneous or surgical mitral valve repair is considered to be suboptimal with respect to the individual patients' life expectancy.[7] In fact, despite the relatively limited number of recipients and of patient-years available for follow-up analysis, one feature associated with the clinically available devices for TMVR is the effectiveness of correction of MR, and its stability over time (1-year in some experiences[8,9]).

Numerous approaches and solutions under the engineering profile have been proposed to cope with the many anatomical requirements inherent to TMVR. In this domain, technological advancements and dedicated research are essential to meet such requirements and pursue clinically sound devices. In this chapter, we will provide an overview of the various TMVR concepts at different stages of development, with the focus on their design characteristics. These will be discussed in the perspective of their fallout on applicative results. As transcatheter mitral therapy will continue to

Advances in Cardiovascular Technology. DOI: https://doi.org/10.1016/B978-0-12-816861-5.00012-5

evolve dramatically, such considerations will hopefully be useful to clinicians for decision-making in their present and future patients, as well as to foreseeing the incoming features of what can be anticipated to be the "TMVR revolution."

The challenges inherent to TMVR

TMVR therapy is currently considered for patients with symptomatic moderate-to-severe mitral regurgitation despite optimal medical treatment, and who are noneligible or at high risk for surgical interventions. The anatomical and (patho)physiological complexity of the mitral apparatus explains the unique challenges associated with TMVR. Concordantly, a considerable number of TMVR concepts have not achieved widespread clinical application yet, and important research and development work is required to overcome multiple issues. Such specific challenges can be reduced according to generic features of transcatheter therapies. Valve stability refers to the ability of the device to consistently maintain the intended position during and after the implantation procedure. Sealing refers to the ability of the device to effectively correct MR and prevent the appearance of perivalvular leaks (PVL). Stability and sealing of TMVR devices must be achieved inside a D-shaped annulus presenting variable mechanical properties at its different segments: an anterior portion with greater rigidity (the anterior annulus and aortomitral curtain), a posterior less supported portion at the atrioventricular junction, and two fibrous trigones. The mitral annulus is a dynamic, three-dimensional structure; pathologic changes can induce a great variability in annular dimensions and in combinations of anterior–posterior and intercommissural diameters; the system is exposed to a high-pressure environment and to considerable pressure gradients. Such features complicate the achievement of uniform intraannular stability and expose the risk of device migration and PVL. Of note, mitral PVL is associated with a particularly negative prognostic impact under the profiles of both survival and valve-related morbidity.[10] The anatomical continuity with the left ventricular outflow tract (LVOT) exposes the risk of obstruction (LVOTO, LVOT obstruction) either from the device itself or from the native anterior mitral leaflet which is going to be fixed in its open position. Dynamic LVOTO is also possible. Circular, balloon-expandable transcatheter valves employed for the aortic position have been occasionally tested for the native mitral position in specific settings (severe mitral annular calcifications, MAC), with unsatisfying results[11,12] except when coupled with surgical valve fixation and removal of the native anterior leaflet.[13] This further underlines the crucial importance of properly addressing both valve stability and LVOT issues with dedicated concepts. To date, a significant proportion of candidates to TMVR is screen failed due to anatomical reasons (predicted LVOTO risk and excessive/too small annular dimensions being the most frequent instances). Hence, the challenge with imaging and procedural planning is related to the reliable prediction of the LVOTO risk in the working conditions. Baseline imaging also plays a major role in patients' selection and sizing of the device, when applicable. The capability to comply with both requirements is based on imaging quality, and spatial and temporal resolution of computed tomography (CT) scan acquisition. As a common point to all TMVR devices, the screening encompasses transthoracic and transesophageal echo and Electrocardiogram (ECG)-gated CT scan, assesses the feasibility of the procedure, the matching of the device with the patient's anatomy, and the risk of LVOTO or other complications. Additionally, valve delivery systems should allow confident endocavitary navigation and controlled valve deployment, while the possibility to reposition and/or retrieve the TMVR valve represents an undoubtedly favorable feature in this context. If each device is characterized by different strategies to obtain stability, sealing, and LVOT preservation, nearly all of them entail self-expandable nitinol frames, which shape memory, and superelasticity properties allow collapsed structures into delivery systems and adaptation to anatomical structures with controlled radial and axial forces. Nitinol is also highly biocompatible.

Devices and technologies

The Tendyne Valve

Tendyne is a pericardial trileaflet valve (Fig. 19.1) consisting of two shape-memory frames, a tether, and an apical pad. The outer frame has a D-shaped morphology, is intended to assure sealing at the native mitral annulus and at the supravalvular level, and is covered with Dacron fabric. The outer frame supports an internal, circular frame into which the proper bioprosthetic valve is mounted (Fig. 19.2A). Overall, 13 valve sizes exist, resulting from adaptation to different combination of dimensions (anterior–posterior diameter, intercommissural distance, annular perimeter), and two profiles (Standard and Low profiles for the outer frame). The available sizes can accommodate a 25–42 mm range in anteroposterior (septolateral) dimension, a 35–48 mm range in the intercommissural dimension, and a 100–145 mm range in perimeter of the native mitral orifice, the desired oversize being ~10% in the intercommissural dimension and ~5%–10% in the antero–posterior dimension. Sizing is based on CT-scan (Fig. 19.3A); a virtual fit assessment

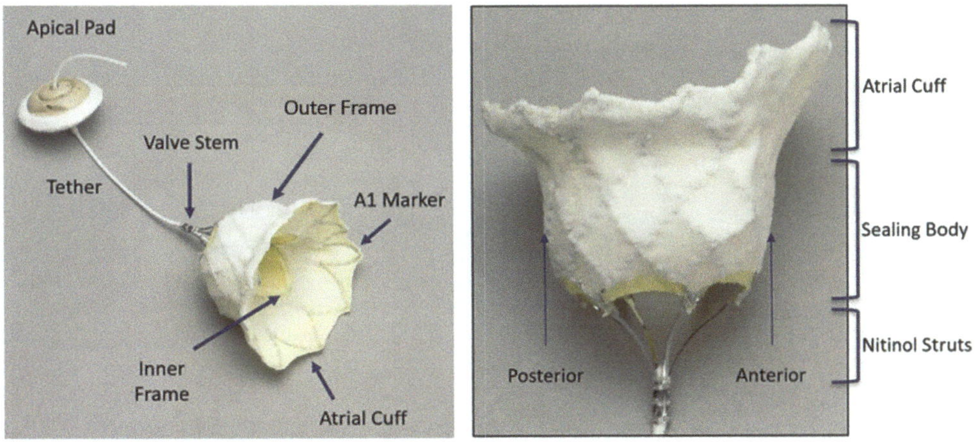

FIGURE 19.1 The Tendyne TMVR system (Standard-Profile system). *Courtesy Abbott Cardiovascular—reproduced with permission.*

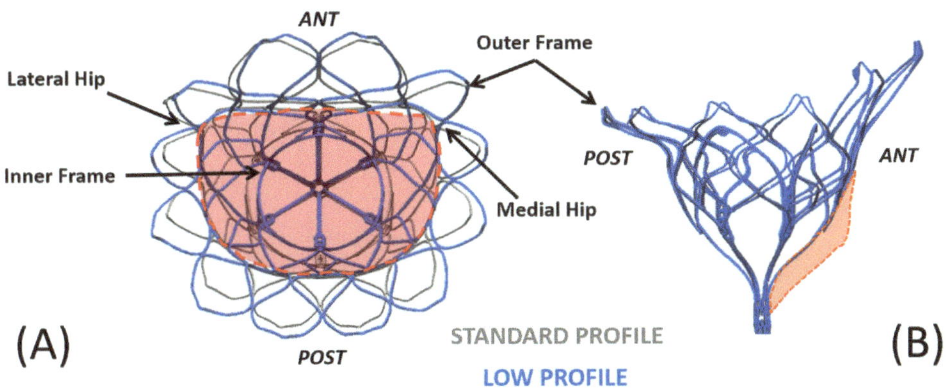

FIGURE 19.2 (A) Axial scheme of the dual-frame structure. Both Standard- and Low-Profile versions are shown. Red shaded area: D-shaped section for intraannular engagement. (B) Illustration of minimized LVOT protrusion with the Low-Profile version (*red* shaded area). *ANT*, Anterior; *POST*, posterior. *Courtesy Abbott Cardiovascular—reproduced with permission.*

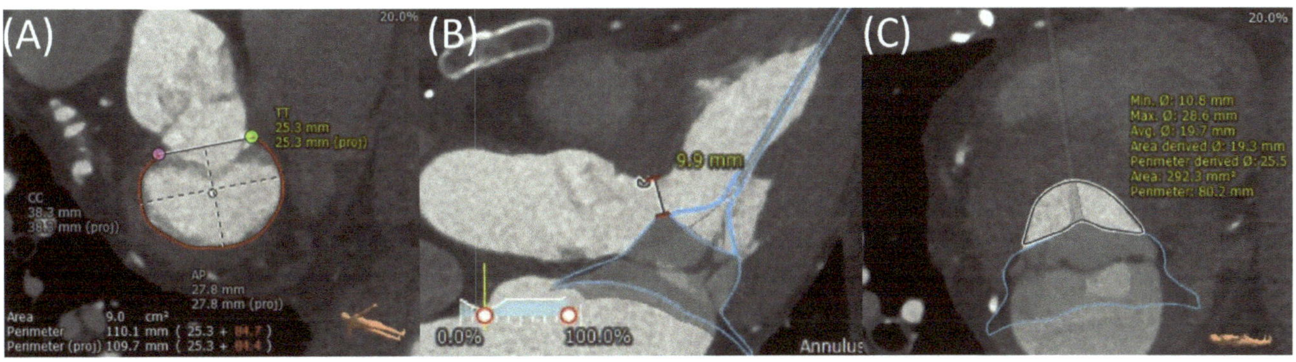

FIGURE 19.3 CT-scan based sizing, end-systole measurements. (A) The annular diameters and perimeter are measured. (B) Virtual fit of the valve in end-systole, minimum distance to interventricular septum. (C) Area of the nLVOT is measured.

(Fig. 19.3B) and a calculation of the expected nLVOT surface (Fig. 19.3C) are performed both in the systolic and diastolic phases. LVOTO prevention is based on the anterior portion of the valve structure tethering the anterior mitral leaflet, which needs nonetheless to be not excessively elongated. The Low-profile outer frame (Fig. 19.2B) retains annular dimensions and is intended to minimize the protrusion into the LVOT, for use in patients with borderline predicted neo-LVOT or increased prospected risk of LVOTO. The inner frame is circular; its effective orifice area (EOA) is 3.0 cm^2 for the Standard-Profile valves and 2.2 cm^2 for the Low-Profile valves. The tether and epicardial apical pad system ensures the anchoring (stability) of the device through counteracting axial forces; the sealing is both intraannular (radial force of the appropriately oversized outer stent), and supraannular (pull of the cuff on the atrial aspect of the native annulus and on the atrial walls). The atrial cuff is asymmetric in the anteroposterior dimension, aimed at adapting

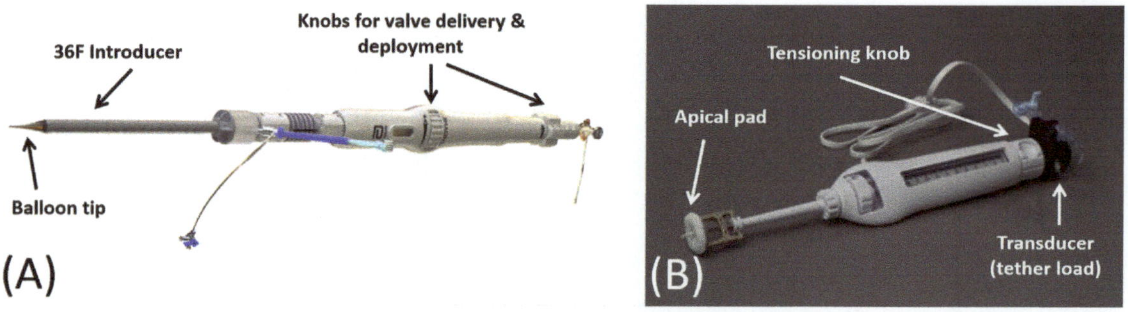

FIGURE 19.4 (A) Tendyne Generation II transapical delivery system. (B) Pad positioning system with integrated tether load transducer. *Courtesy Abbott Cardiovascular—reproduced with permission.*

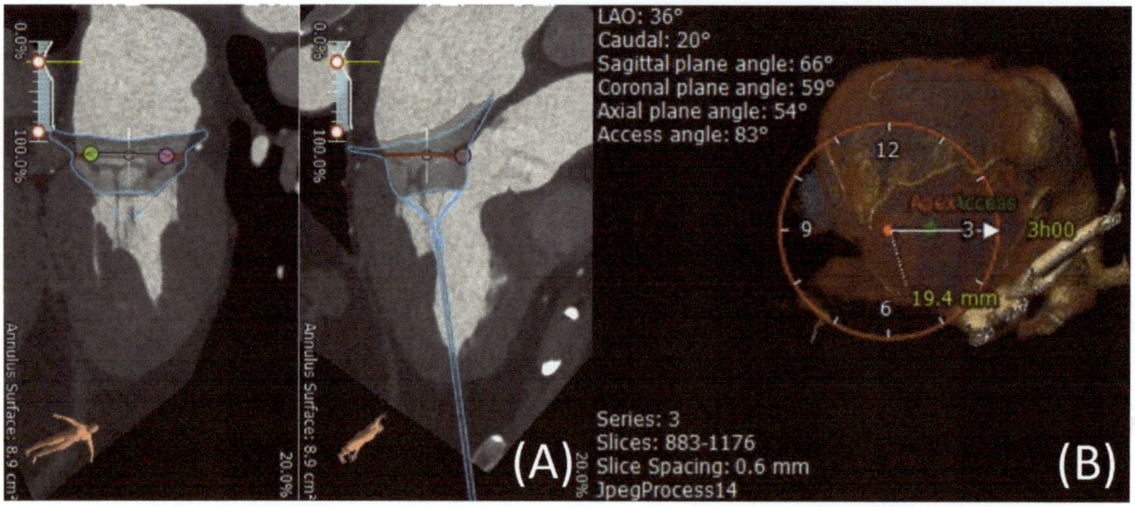

FIGURE 19.5 (A) Virtual fit of the valve in the CT-derived X-plan view (end-systole) for identification of optimal LV access at the epicardium. (B) 3D reconstruction showing the desired apical puncture site (access—*green dot*) for optimal valve seating. *Courtesy Abbott Cardiovascular—reproduced with permission.*

to the atrial anatomy. The tether and apical pad system allows full retrievability and repositionability even after full deployment and decoupling from the delivery system.

During delivery, the valve is crimped into a 36F transapical delivery system (Fig. 19.4A) which is a second generation device. The accurate choice of the site of apical puncture is of paramount importance for the success of the procedure: in fact, this will determine the perpendicularity of the tether to the mitral orifice and the effectiveness of valve seating. Such a site is planned preoperatively based on ECG-gated CT-scan with systolic and diastolic acquisition (Fig. 19.5A and B). Minor adjustments in the access site are possible case-by-case (e.g., slight posterior bias in order to minimize the LVOTO risk in predisposed patients). Such adaptive management is based on the specific design of this device.

After crossing the mitral orifice with a guidewire, the delivery system is advanced until 1–2 cm above the mitral annulus and allows controlled valve deployment and adequate rotational alignment (clocking) of the device. Deployment is performed intraannular or immediately supraannular with subsequent pull in the intraannular position, under continuous transesophageal echocardiography (TOE) and fluoroscopy control, without the need for rapid pacing. TOE provides the essential of guidance during valve deployment using both the 3D "surgical" view and the X-plane view; fluoroscopy is complementary and used to confirm the opening of the stent structure and the valve clocking through the radiopaque A1 marker (Fig. 19.6). Fine-tuning delivery and recapture is possible at this stage. In the case of LVOTO, the valve can be partially recaptured and slightly off-clocked counterclockwise (3D *en face* echo view) in order to drive away the outer frame lateral hip (Fig. 19.2A) from the LVOT and increase its surface. After retrieval of the delivery system, the tether exiting the cardiac apex is controlled and a pad positioning system is advanced over it (Fig. 19.4B) until the pad comes into contact with the myocardium. The final tensioning of the tether is another critical step, and is performed through continuous evaluation of tether load (a pressure transducer is connected to the pad

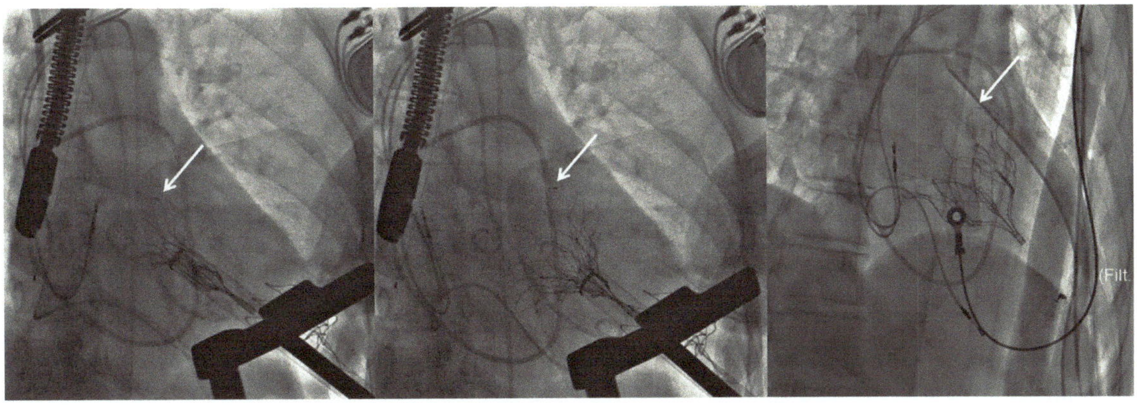

FIGURE 19.6 Intraoperative fluoroscopy views. Progressive steps of deployment of a Tendyne TMVI. White arrows: A1 marker.

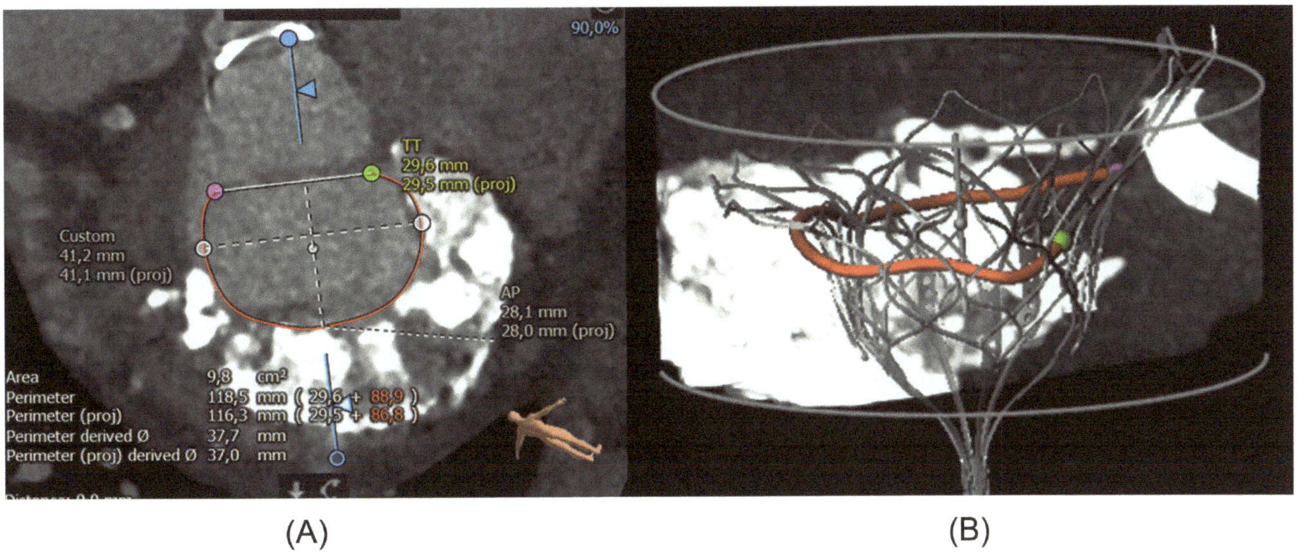

FIGURE 19.7 (A) Annular Sizing in a MAC case. (B) Virtual fit of a Tendyne valve inside the MAC. *Courtesy Abbott Cardiovascular—reproduced with permission.*

positioning system). Valve instability and PVL can be managed through increased tether tension, while excessive intraannular position can be corrected through reduced tether tension (continuous TOE surveillance); such steps are reversible until optimal tension is adjusted. The diastolic tether tension is required to exceed the diastolic left ventricular end-diastolic pressure by at least 10 mm Hg, in order to achieve valve stability across the entire cardiac cycle. Once tensioned, the anterior body of the outer frame tethers itself to the anterior mitral leaflet and subvalvular apparatus, preventing its extremity from dynamically obstructing the LVOT unless excessively long (i.e., Barlow disease). Once the pad is fixed at the cardiac apex and appropriate tensioning is finalised, the tether exiting the myocardium at the left ventricular apex is cut 5-10 cm above the epicardium. Nonetheless, this instance was rare in clinical experience. As an additional guidance to valve deployment, intraoperative fusion imaging (overlay of CT-scan 3D reconstructions over the live TOE images—semiautomatic registration and tracking based on reference points at the mitral annulus) has been proposed.[14]

Tendyne TMVR can be indicated for moderate-to-severe MR in various clinical situations: primary MR with high or prohibitive surgical risk with complex anatomy, symptomatic secondary MR despite guideline-directed medical therapy (GDMT), failed transcatheter repair, mixed-etiology mitral disease, mild-to-moderate MAC. Additionally, the valve design, the supraannular sealing, the capability for intraannular deployment, and the separation of sealing from stability likely render this device safe and effective in MR associated with severe MAC and sufficient annular diameter (Fig. 19.7), although this indication is at the present stage considered as off-label. A dedicated MAC arm exists in the

SUMMIT trial (ClinicalTrials.gov Identifier: NCT03433274) to address this issue. The outer stent can adapt to moderately irregular annular/subannular anatomies without deformation of the inner stent. As a limitation, this device requires a feasible transapical access and no contraindication to anticoagulation (the latter limitation is shared with virtually all TMVR devices). Severe mitral stenosis and extensive calcifications of the anterior leaflet are most often contraindications; nonetheless, selected cases with concomitant moderate stenosis, no leaflet calcifications, and potentially eligible for balloon predilatation can be considered.

Since the first human use in 2014,[15] about 600 implantations of Tendyne have been performed (at time of writing), and it is the only CE-marked TMVR device. Investigations show early safety and predictability of technical success (96%) since the first 100 treated patients[8]; technical success rate has increased over the more recent periods of the experience. Longitudinal follow-up shows the effectiveness and stability of MR correction at 1 year (98% none/trace MR) as well as improved patients' functional status, with no need for percutaneous PVL closure. Predictors of greater late clinical benefits are being evidenced, such as more baseline severe MR or geometry of LV/valve interaction.[16,17] Overall, the current TMVR device is unique in the number of cases treated to date and in the amount of available clinical data.

The Intrepid Valve

Intrepid consists of a self-expanding nitinol frame with a dual structure: an outer circular fixation ring and an inner stent housing a 27-mm trileaflet bovine pericardial valve.[18] The stability is obtained through the combination of the radial force of the outer frame, which needs to be oversized with respect to the native annulus (annular anchor), and the sealing is achieved through the intraannular and subannular fixation (adhesion to the native leaflets through circumferential frictional elements; Fig. 19.8). There is no need for rotational alignment (circular design) or leaflet capture, while deployment is conducted under rapid pacing. Of note, the radial stiffness of the outer frame is variable across the longitudinal axis of the valve: the atrial segment is more flexible while the intraannular and ventricular portions have greater stiffness, producing a narrow intraannular neck and a wider subvalvular body after deployment. Stability is based on annular oversizing; Intrepid is available in three sizes: 43, 46, and 50 mm, while the 54 mm valve is currently in development. Initially, only a 35F transapical delivery system was available (Fig. 19.9A), while a 35F transfemoral transseptal delivery system is currently under investigation (Fig. 19.9B) with a 29F transfemoral system in the pipeline. Nonetheless, a 42F delivery system would be necessary to deliver the 54-mm valve. The transfemoral system entails two flex zones to facilitate endocavitary navigation and alignment with the mitral orifice, and is currently the object of a nonrandomized dedicated Early Feasibility Study aimed at verifying the reproducibility of results seen with the transapical approach.

This device has the advantage of being repositionable during deployment, although full retrievability is not possible after decoupling from the delivery system. The deployment entails three steps: initial valve expansion with supraannular

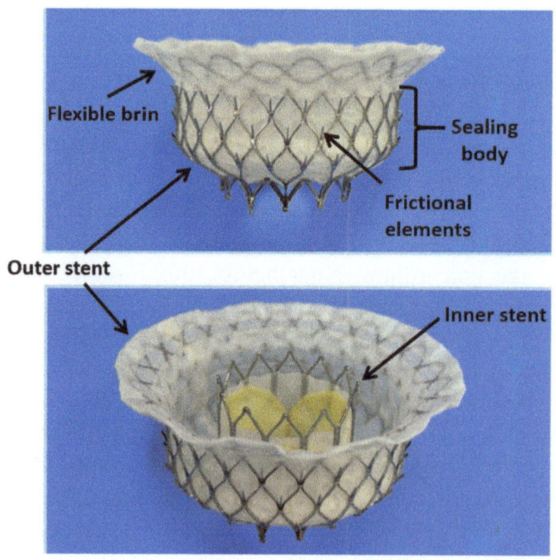

FIGURE 19.8 Design of the Intrepid valve. *Courtesy Medtronic France 2020—reproduced with permission.*

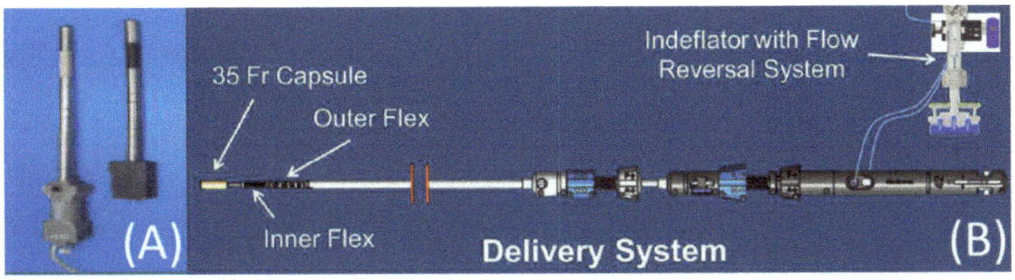

FIGURE 19.9 Transapical introducer (A) and transfemoral transseptal delivery system (B) for the Intrepid valve. *Courtesy Medtronic France 2020—reproduced with permission.*

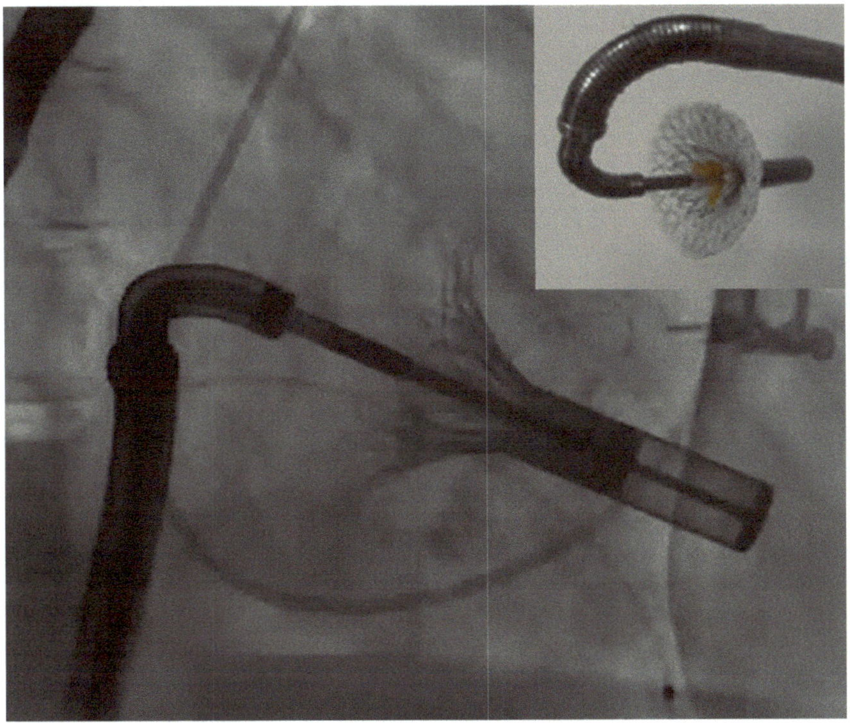

FIGURE 19.10 Transfemoral transseptal delivery of an Intrepid valve. Insert: photo of the steerable extremity of the system. *Courtesy Medtronic France 2020—reproduced with permission.*

opening of the radiopaque brin; intraannular engagement with expansion of the outer frame for fixation, and valve release. These steps are performed under TOE and fluoroscopy guidance. Irrespective to the transapical retrograde or transfemoral antegrade approach, it appears that an adequate alignment of the deploying valve with the mitral orifice is important to achieve controlled seating at the appropriate depth and, therefore, stability. Due to the principles of design and anchoring, deployment must proceed from the atrial toward the ventricular aspect even if the transseptal approach is used (Fig. 19.10). Due to the intraannular design being based on radial force and no need for leaflet capture, Intrepid can be considered for MAC cases.

The APOLLO trial is currently investigating TMVR with Intrepid in patients with severe MR and is deemed not suitable for either surgery or approved transcatheter mitral therapy.[19] APOLLO consists of two separate arms, a Primary cohort including non-MAC MR patients, and a MAC cohort. At time of writing, more than 275 Intrepid implants have been performed globally, with a reported 98% procedural success and 14% 30-day mortality in the first 50 patients in a Global Feasibility Study.[20] CE-mark approval is pending (transapical system only).

The HighLife Valve

The HighLife (HighLife SAS (Société par Actions Simplifiée), Paris, France) (Fig. 19.11A) is a self-expandable valve. It consists of a single nitinol frame covered in Dacron fabric, and a trileaflet glutaraldehyde cross-linked bovine pericardial valve. The stability is based on the interaction of the prosthesis (radial force) with a ring-shaped subannular implant (SubAnnular Implant (SAI) or dock) which is preliminary deployed outside the mitral leaflet in the subannular position. The SAI represents the landing zone of the device.[21] The valve stent is circular and roughly hourglass-shaped, the

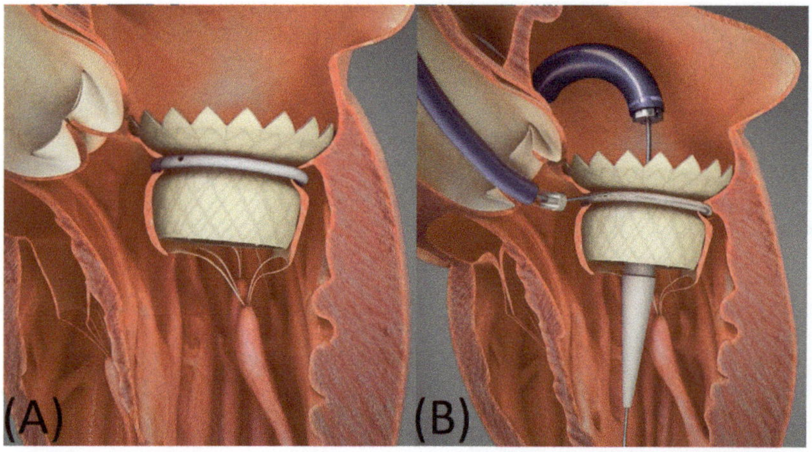

FIGURE 19.11 (A) Illustration of the HighLife valve. (B) Illustration of the transseptal implantation. *Courtesy HighLife SAS—reproduced with permission.*

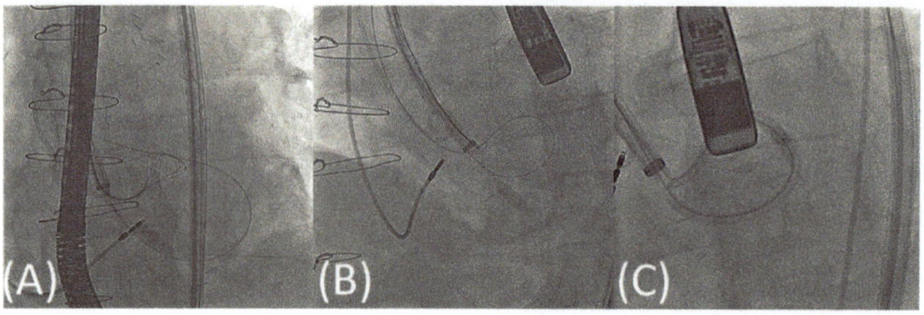

FIGURE 19.12 X-ray images of the SAI implantation. Once advanced in the left ventricle (retrograde approach through the aortic valve), a guidewire is caught with a snare (panel A) and externalized from the femoral access, thereby creating a loop around the subannular apparatus (panel B). Threaded over the two guidewires loops ends up, the SAI is then implemented and irreversibly closed (panel C).

preformed groove in the annular region aimed at interacting with the SAI to obtain stability. The sealing is obtained through an atrial cuff to be deployed above the native annulus, and through the interaction of the valve stent with both the SAI and the native leaflets. After valve deployment, these are constrained between the SAI and the valve frame, thereby preventing PVL.

This valve is currently an investigational device and available in one size only (28 mm). In the initial part of the experience, a 31-mm-sized valve was also available[22]; currently, the 28-mm valve is proposed to fit all screen-passed candidates, possibly minimizing the LVOTO risk with a less bulky device. Although initially proposed with a hybrid approach (retrograde transfemoral for SAI deployment and transapical for valve delivery), with the current new-generation valve a totally percutaneous strategy is proposed (transseptal approach for valve delivery, HighLife TSMVR (transseptal mitral valve replacement) system). The first step of the procedure consists of setting-up the SAI. Through an 18 Fr femoral artery access, a specially designed loop placement catheter (LPC) is positioned inside the left ventricle using a retrograde transaortic approach. A guidewire is advanced from this LPC in the left ventricle around the subannular apparatus, caught with a snare and externalized from the femoral access, thereby creating a loop (Fig. 19.12A). The consistency of the loop trajectory is assessed through fluoroscopy and TOE: all the chordae tendineae need to be circumscribed within the loop. The two externalized wire extremities are pulled to harmoniously tighten the loop, which allows confirming the position of the wire (Fig. 19.12B). Then, the LPC is replaced by an 18 Fr SAI delivery catheter (SDC) which permits the implementation of the SAI around the subannular apparatus. The SAI is a polymer tube with two distinct extremities: the enlarged ("female") end hosts the tapered other end ("male"), thereby closing and irreversibly latching the SAI (Fig. 19.12C). Based on such design, this device requires sufficient space between the native leaflets and the ventricular myocardium in the infraannular region. Implantation in MAC cases has not been attempted and its feasibility is potentially hindered by anatomical alterations of the leaflets and of the annulus in such circumstances. In the second step the 30 Fr (18 Fr Shaft; 30 Fr capsule) transseptal mitral valve (TSMV) delivery catheter (TSDC) is inserted through a femoral venous transseptal access. From the delivery catheter, the outflow end of the TSMV is deployed distal to the SAI in the left ventricle, before the inflow end is deployed proximally to the SAI into the atrium (Fig. 19.11B).

Self-centering and self-alignment are the theoretical key benefits of the HighLife device: the SAI prevents the prosthesis from slipping into the atrium, whereas the inflow shape of the stent prevents it from slipping into the left ventricle. The stability of the valve at the intended position results from the opposite effects of these two interactions. It also depends on maintaining the optimal deployment level of the SAI at the infraannular level and with respect to the subvalvular apparatus, as well as on the dimensions of the native annulus with respect to the diameter of the inflow prosthesis' cuff. In the screening phase, such design characteristics require careful consideration of the dimensions of the native annulus, the geometry of the native leaflets/subvalvular space, and the tethering forces applied to the native apparatus. The possibility for later valve-in-valve procedure is an additional advantage with this device. Conversely, the valve is not retrievable after full deployment, and is repositionable before full deployment.

Promising early hemodynamic performance and no paravalvular leak were reported in early experiences with the transapical approach.[22,23] An Early Feasibility Study (the HighLife (TSMVR) System trial) is currently recruiting (more than 20 human implantations to date).

The Tiara Valve

This device was implanted for the first time in a patient in 2014. Tiara is a trileaflet bovine pericardial valve mounted into a self-expanding, D-shaped nitinol frame. The stability is obtained by the intraannular radial force and sealing is achieved through capture of native mitral leaflets by dedicated anterior and posterior anchors (or tabs). These represent distinctive features of the device (Fig. 19.13). A 32F or 36F transapical delivery system was developed first. After crossing the mitral orifice with the delivery system, the atrial skirt is deployed and a rotational alignment is performed under 3D TOE monitoring (the atrial skirt being intended to seal the anterior portion of the mitral inflow); the annulus is then engaged. Resheathing, repositioning, and retrieval are possible until this stage. Subsequently, the anchors are released with progressive single-knob unsheathing. Two anterior anchors are intended to capture the anterior leaflet until the aortomitral fibrous body, and at the same time at constraining the anterior mitral leaflet adherent to the valve body, preventing dynamic LVOTO. One posterior anchor is intended at capturing the posterior leaflet over its entire height until the posterior annulus. The ventricular skirt is then completely deployed. This device requires accessible anterior/posterior leaflet tissue for effective anchoring. Cases with compromised leaflet anatomy (i.e., severely retracted or obliterated posterior leaflet such as in severe MAC) are likely unsuitable candidates. Conversely, the technique may be interesting in cases with increased risk of dynamic LVOTO (i.e., long, scarcely-tethered anterior leaflet). The valve is not retrievable after full deployment.

Tiara exists in two sizes, 35 and 40 mm, being capable of accommodating a 9.3–12.5 cm range in annular perimeter.[24] A transseptal delivery system is currently under development, with a prospected ≤30F profile. Over 83 implantations have been performed globally; the reported outcomes[25] include a 94% procedural success, an 88% none/trace MR at discharge, and 12.7% 30-day mortality.

The Edwards Evoque Valve

This device was originally conceived for delivery through a transfemoral transseptal approach. Evoque consists of circular-shaped self-expandable nitinol frame supporting a trileaflet bovine pericardial valve (Fig. 19.14A). This device currently exists in two sizes (44 and 48 mm). The stability is achieved through a system of anchors designed to capture the native mitral leaflets and subvalvular apparatus, plus the radial force of the stent. The sealing is achieved through the captured native leaflets and apparatus adhering to the valve body, as well as through a skirt with intraannular position. Evoque aims at addressing the LVOTO risk with a low intraventricular profile: the atrial protrusion is visible after full valve delivery (Fig. 19.15), while the ventricular protrusion can be minimized by the captured leaflets maintaining

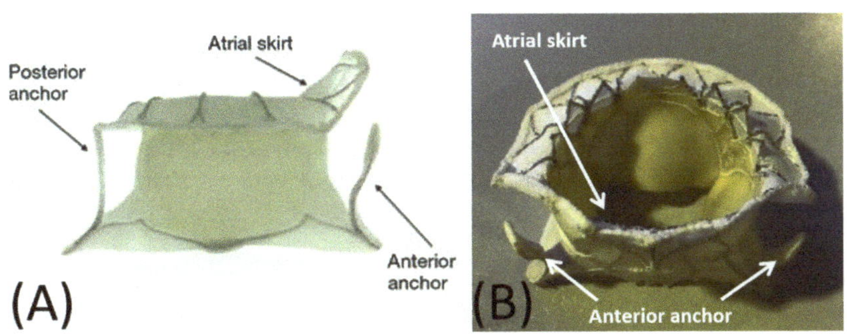

FIGURE 19.13 Illustration of the Neovasc Tiara valve. (A) Side (medial) aspect. (B) Atrial view. *Reproduced with permission from Ferrari E.* Semin Thorac Cardiovasc Surg. *2018;30:160–163; Cheung A.* Ann Cardiothorac Surg. *2018;7:787–791.*

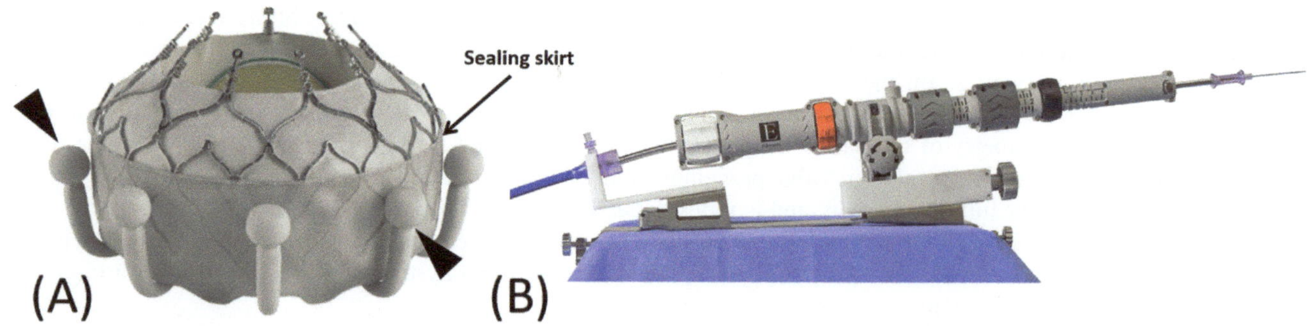

FIGURE 19.14 (A) Illustration of the Edwards Evoque valve. Arrowheads indicate the anchors intended at capturing the mitral leaflets and apparatus. (B) Delivery system for transseptal approach, allowing navigation into three planes. *Courtesy Edwards Lifesciences—reproduced with permission.*

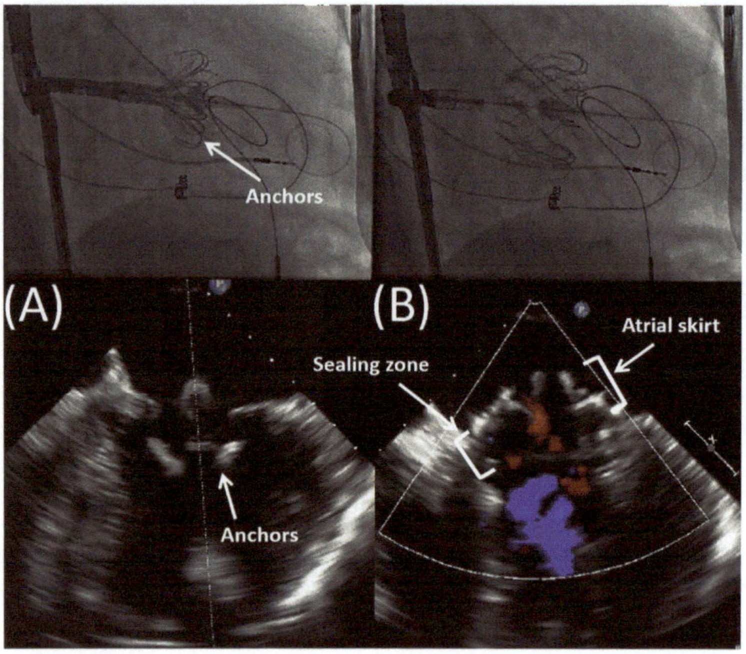

FIGURE 19.15 Implantation of the Edwards Evoque valve. (A) Fluoro and echo view of subvalvular deployment; anchors are visualized to confirm capture of the mitral leaflets and apparatus. (B) Fluoro and echo view of deployment of the inflow (atrial skirt) portion. The atrial protrusion is visualized. *Courtesy Edwards Lifesciences—reproduced with permission.*

the valve in the intraannular position. In fact, the Evoque has been developed as an evolution of its predecessor CardiAQ valve device. Lessons learned with this initial version of the device[26] prompted a design with smaller atrial and ventricular profile to minimize LVOTO risk. Additionally, the anchors-based system constraining the anterior leaflet adherent to the valve body potentially minimizes the risk of dynamic LVOTO. This might be considered in patients with excessively elongated or scarcely tethered anterior leaflet. On the other hand, and similar to other systems based on leaflet capture, effectiveness requires sufficient and mobile leaflet tissue. Cases with obliterated leaflets or bulky calcifications (i.e., severe MAC) may not be good candidates for this technique. This valve is not retrievable after full deployment.

Although no circular orientation is necessary, an effective capture of native leaflets and apparatus throughout the entire valve perimeter is essential to obtain stability and minimize PVL. TOE guidance is pivotal to such aim, as is the capacity to precisely navigate the left atrium and the inflow portion of the left ventricle (coaxiality of the system with the mitral orifice, depth control, possibility to dynamically capture the free edge of the leaflets). Therefore, the delivery system (Fig. 19.14B) allows three planes of motion for its extremity containing the crimped valve (lateral/medial, anterior/posterior and depth control motions). One further advantage of this transseptal delivery system is represented by its limited size compared to others (28F for both valve sizes). The implantation technique entails transseptal puncture, left atrial navigation, and coaxial crossing of the mitral orifice with the tip of the delivery system. The ventricular portion of the valve is deployed first, including the anchors which are visualized through fluoroscopy and echo, used to confirm the capture of the native leaflets and apparatus (Fig. 19.15A). When satisfactory, the inflow portion (atrial skirt) is released (Fig. 19.15B). In the initial feasibility experience with 14 patients, technical success was 93%, one patient required conversion to open surgery, two required percutaneous PVL closure and one secondary septal alcohol ablation due to LVOTO,[27] the remainder showing an echocardiographically good result (residual MR ≤ 1/4). The occurrence of PVL underlines the importance of effectively capturing the leaflets, and of adequate tissue amount/quality, over the entire perimeter with this system. Of note, this valve platform has been tested as well in the tricuspid position with favorable initial results.[28]

The M3 Valve

This device combines a dock-based strategy and the unique dedicated balloon-expandable prosthesis proposed for TMVR to date. The system is conceived for a transfemoral transseptal delivery with a 20F profile; the dock consists of a nitinol structure entailing an encircling turn, three functional turns, and an atrial turn (Fig. 19.16). The dedicated transseptal dock delivery system is steerable and allows navigation of the left atrium; its tip is guided down to the medial commissure where the dock is progressively released under TOE and fluoroscopy surveillance externally to the mitral leaflets and apparatus, where it progressively recovers its circular morphology. This technology, like others based on similar strategies, requires sufficient leaflet tissue and adequate space externally to the leaflets in the infraannular space. This system aims at addressing the stability of the dock in the desired position (the leaflet tethering forces potentially drawing it in the apical direction) through an atrial turn which is intended to maintain the supraannular position and counteract the pull in the apical direction. The procedure is reversible until the final release of the dock; retrievability is not possible afterwards. The valve itself is a modified Sapien 3 transcatheter valve, equipped with an outer knitted positron-emission tomography (PET) skirt (Fig. 19.16), which is deployed inside the dock using a steerable Commander delivery system. A unique size (29-mm) is provided. The stability is achieved by the infraannular dock and the radial force of the valve stent; the sealing is improved by the radial force and the knitted skirt adhering to the native leaflets against the dock. The anterior mitral leaflet is constrained by the dock in close contact with the valve body, thus the risk of dynamic LVOTO can be reduced. Conversely, with the landing zone of the valve being located at the infraannular level (the dock position), the deployed valve frame can be pushed in the apical direction and potentially toward the LVOT. Therefore, accurate preimplantation planning is important to evaluate the neo-LVOT. Similar to other dock-based strategies, this approach entails a circularization of the mitral orifice, requiring sufficiently pliable mitral leaflets.

The clinical experience with the M3 valve system consists of a first-in-man series of 10 patients and in an early feasibility study involving 35 subjects.[29] In these series, the procedural success rate was between 88% and 90% and the

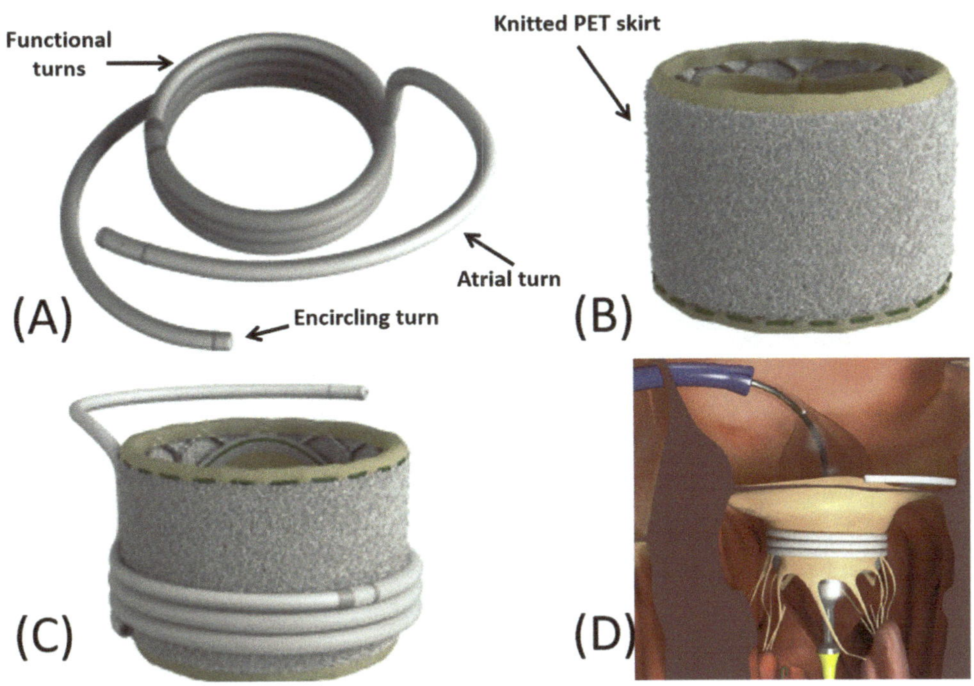

FIGURE 19.16 The Edwards M3 valve system. (A) Nitinol dock entailing three segments: encircling, functional and atrial turns. (B) The balloon-expandable valve equipped with an outer knitted skirt. (C) Valve deployed inside the dock. (D) Illustration of the deployed valve capturing the native mitral leaflets against the dock. *Modified from Webb J, et al.* J Am Coll Cardiol *2019;73:1239–1246. Reproduced with permission.*

30-day mortality was between 2.9% and 0%; one patient in the first-in-man series required percutaneous PVL closure. More information needs to be collected to further clarify the safety/effectiveness profile of this strategy.

The MitralTech CardioValve

This device consists of a circular, self-expandable frame supporting a bovine pericardial valve; it is designed for transfemoral transseptal delivery through a dedicated 28F system. Three sizes are available (40, 45, and 50 mm) for this valve concept based on annular anchoring to obtain stability and sealing. A low ventricular profile is adopted as the strategy to minimize LVOTO risk. As no specific leaflet capture system is featured, radial force and circularization of the native mitral annulus are pivotal to achieve stability. Pliability of the annulus (no severe, circular MAC) seems to be a requirement. The published experience to date is quite limited (five patients), with a reported 100% procedural success but a 60% 30-day mortality. Of note, this device is proposed for use in the tricuspid position as well: separate early feasibility studies for the mitral and tricuspid positions have been initiated; these will allow clarification of whether the design features satisfactorily comply with the anatomical/functional requirements in both positions.

The Caisson Valve

Featuring a D-shaped nitinol frame and a porcine trileaflet valve, this device was designed with a system of annular anchors intended to clamp the native annulus through counteracting forces from both its atrial and ventricular aspects under TOE control (Fig. 19.17). The deployed device was to be atrial-based as a means to manage LVOTO risk. Retrievable and repositionable until final release, a transfemoral transseptal approach with a 31F profile delivery system was employed. In 2019, and despite an Early Feasibility Study having recruited five patients, the development of this concept was discontinued, allegedly due to funding issues.

The Cephea Valve

This device, which is primarily intended for transfemoral transseptal delivery, relies on an annular and leaflet clamping strategy to ensure both stability and sealing (Fig. 19.18). It consists of a dual-frame circular self-expanding structure, entailing a ventricular disk which is deployed first in the infraannular position. Traction is applied to fully engage the ventricular aspect of the annulus. Subsequently, the atrial disk is deployed and axial forces are intended to stabilize the annular capture. The inner frame supporting the valve is conceived to maintain its circular shape despite geometrical deformation of the outer frame portions due to individual native valve anatomy. In a reported clinical case, such characteristic of the valve successfully adapted to asymmetric annular morphology associated with nonsevere MAC.[30] In experimental animal studies, it appeared that the ventricular disk successfully captured and rolled up the native anterior leaflet.[31] Together with the low ventricular protrusion, this device is expected to minimize the risk of both dynamic and nondynamic LVOTO. Concordantly, in the initial human experience in three patients, no clinically significant LVOTO was noted, with a satisfactory echocardiographic result.[32] At the current stage of development, retrievability is partial and possible until full delivery of the ventricular disk. Prospective and larger studies are expected with this device.

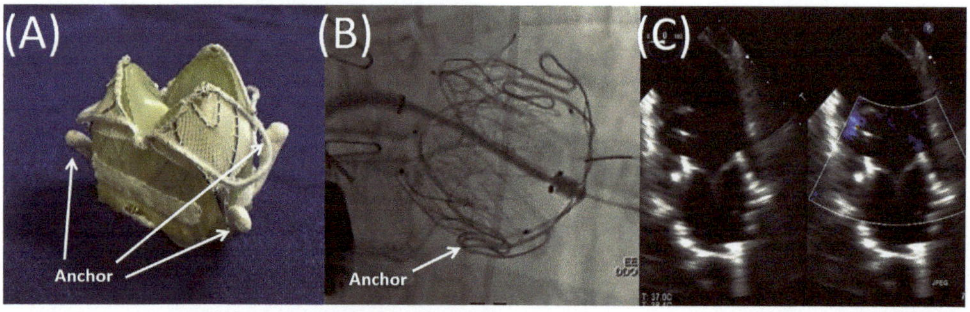

FIGURE 19.17 The Caisson valve system. (A) Photo of the anterior-lateral aspect of the device (atrial view) illustrating the anchors intended at capturing the native mitral annulus. (B) Fluoroscopy view of the deployed valve. (C) TOE view of the deployed valve, showing intraannular positioning and limited ventricular protrusion. *Modified from Regueiro A, et al.* J Am Coll Cardiol. *2019;73:1239–1246. Reproduced with permission.*

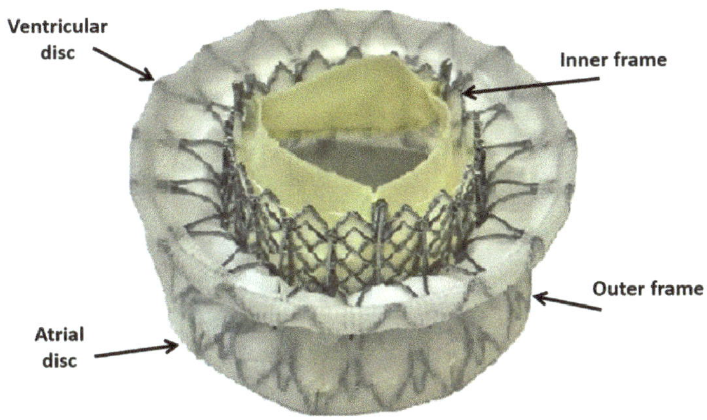

FIGURE 19.18 The Cephea valve system. The dual-frame, dual-disk system is demonstrated.

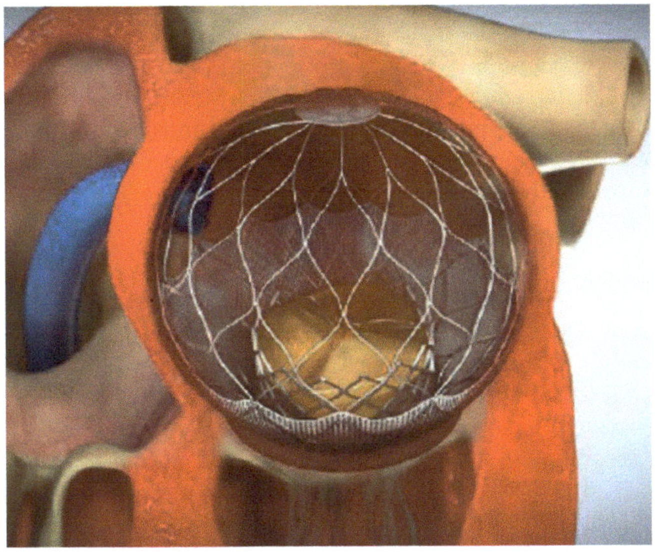

FIGURE 19.19 The AltaValve system. The atrial anchor system and supraannular valve position are illustrated.

The AltaValve

Although the initial results with this device are limited (the first-in-man case has been recently published),[33] it is remarkable due to its unique design, based on an atrial anchor concept to obtain stability (Fig. 19.19). A spherical, oversized large-cells frame mounting a bovine pericardial valve is released (transapical access) inside the left atrium, the prosthetic valve acquiring a supraannular position. This strategy is reminiscent of the atrially-implanted surgical prostheses in cases with native annulus inaccessible to stitch placement. This strategy virtually eliminates LVOTO risk, but needs to be evaluated in its capacity to obtain effective sealing between the prosthetic valve and the native annulus. Additionally, the morphology and dimensions of left atrium are greatly variable among patients in both secondary and primary MR settings, constituting a potentially limiting factor in the screening phase. It is likely that multiple sizes of the atrial frame should be available to comply with variable atrial dimensions and shapes. A transfemoral transseptal delivery system is being developed.

The Saturn Valve

This device shares with others the combination of an infraannular dock for stability and a separate trileaflet valve implant (Fig. 19.20). Nonetheless, and different to the others, this device entails locking of the valve frame with the dock elements, which are definitively solidarized. An annular structure (dock) composed of two segments and encircling the native leaflets is initially deployed from the transapical access. The self-expandable valve is subsequently

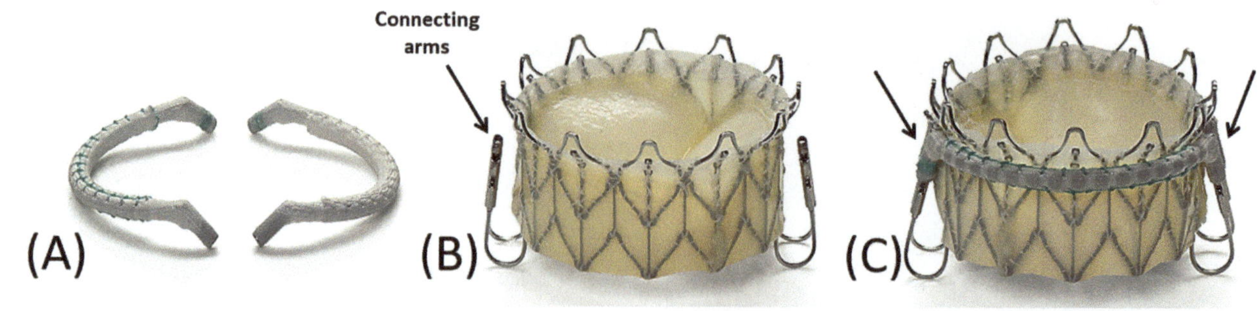

FIGURE 19.20 The Saturn valve system. (A) Two-components annular structure ("dock") for deployment externally the native mitral leaflets. (B) Self-expandable valve. Arrows indicate the locking mechanism ("connecting arms") designed to connect with the annular segments. (C) Assembled system. Arrows indicate the locked connecting arms and annular segments.

advanced inside its delivery system, connected to the annular segments (dock) through the "connecting arms," and finally deployed. Stability is obtained through interacting forces between the annular structure, the valve frame, and the native annulus. Sealing is obtained through the radial force of the valve frame capturing the native leaflets against the annular structure. Of note, the valve frame is guided and oriented in order to lock with the annular structure, which might improve valve stability by counteracting the tethering forces of the subvalvular apparatus. The ventricular protrusion measured from the annular plane is 13 mm; such a low profile is intended to minimize the LVOTO risk. A transseptal system is currently under development and initial clinical data are expected from the ongoing early feasibility study (the first-in-man case was performed in August 2020).

Conclusions

TMVR currently represents one of the most dynamic and promising domains in cardiovascular therapies; at the same time, it presents unique and multiple challenges. The success of any TMVR solution greatly depends on its design and technological content. Despite a large number of proposed devices, transapical solutions have displayed greater reproducibility to date. Nearly all strategies employ self-expandable nitinol frames. Only one transapical concept with apical tether-based fixation, allowing separation of stability from sealing, has entered practice to date after obtainment of the European Community mark (CE) Mark, and is corroborated by the largest clinical experience. Nonetheless, several other concepts display promising results and a drive toward standardization of transfemoral transseptal approaches is expected during the next years. Both safety and surgical-like results with respect to the correction of MR will need to be demonstrated nonetheless. Miniaturization of delivery systems seems to be important to such scope, and it is challenged by the size of TMVR valve structures. Systems based on annular and/or subvalvular apparatus anchoring/docking may be effective in prevention of LVOTO due to excessively elongated/nontethered anterior leaflet. Nonetheless, their application will require attentive evaluation and management strategies for both structural and functional (tethering forces) pathological changes in individual candidates. Therefore, they can be associated with specific anatomical requirements (sufficient bileaflet tissue and accessibility of the infraannular region). Systems based on annular clamping or intra/supraannular stability can obviate such limitations, but need to feature strategies for LVOTO prevention. It is likely that an array of several devices will be made available to clinicians during the next years. The decision-making between the alternative options will be based not only on experience, but also on the appreciation of anatomical and pathophysiological features in individual patients for customized valve choice.[9] These elements suggest that TMVR will probably remain for many years in the hands of multidisciplinary, highly specialized Heart Valve Teams.

Additionally, with increasing application and diffusion, the need for continued assessment of valve function and stability over time during late follow-up, as well as of valve/ventricle interaction[17] compared to state-of-the-art results of surgical bioprostheses[34] will acquire increasing scientific value to allow ongoing diffusion of TMVR.

Device development in the TMVR field requires major development and clinical research efforts. Only the cooccurrence of optimal conceptual design (addressing the multiple challenged of mitral anatomy and pathophysiology) and adequate funding/project development capabilities can lead to success. In a highly competitive context where some devices have already proven to be clinically effective and reproducible, several factors might impact to an unknown extent the fundraising for systems/devices at earlier stages of development. Such factors include the increasing amount of investments required for development and clinical investigations in this domain, the evolution of the TMVR market with respect to healthcare reimbursement policies, and the impact of the Covid-19 crisis on global investment strategies.

References

1. Nkomo V, Gardin J, Skelton T, Gottdiener J, Scott C, Enriquez-Sarano M. Burden of valvular heart diseases: a population-based study. *Lancet.* 2006;368:1005–1011.
2. United States Census Bureau. Age and sex composition. *Issued May.* 2010;2011:1–16.
3. Gammie JS, Chikwe J, Badhwar V, et al. Isolated mitral valve surgery: the society of thoracic surgeons adult cardiac surgery database analysis. Isolated mitral valve surgery: the society of thoracic surgeons adult cardiac surgery database analysis. *Ann Thorac Surg.* 2018;106:716–727.
4. Obadia JF, Messika-Zeitoun D, Leurent G, et al. Percutaneous repair or medical treatment for secondary mitral regurgitation. *N Engl J Med.* 2018;379:2297–2306.
5. Stone GW, Lindenfeld J, Abraham WT, et al. Transcatheter mitral-valve repair in patients with heart failure. *N Engl J Med.* 2018;379:2307–2318.
6. Acker MA, Parides MK, Perrault LP, et al. Mitral-valve repair vs replacement for severe ischemic mitral regurgitation. *N Engl J Med.* 2014;370:23–32.
7. Coutinho GF, Correia PM, Branco C, Antunes MJ. Long-term results of mitral valve surgery for degenerative anterior leaflet or bileaflet prolapse: analysis of negative factors for repair, early and late failures, and survival. *Eur J Cardiothorac Surg.* 2016;50:66–74.
8. Sorajja P, Moat N, Badhwar V, et al. Initial feasibility study of a new transcatheter mitral prosthesis: the first 100 patients. *J Am Coll Cardiol.* 2019;73:1250–1260.
9. Hensey M, Brown RA, Lal S, et al. Transcatheter mitral valve replacement: an update on current techniques, technologies, and future directions. *JACC Cardiovasc Interv.* 2021;14:489–500.
10. Cho IJ, Moon J, Shim CY, et al. Different clinical outcome of paravalvular leakage after aortic or mitral valve replacement. *Am J Cardiol.* 2011;107:280–284.
11. Yoon SH, Whisenant BK, Bleiziffer S, et al. Outcomes of transcatheter mitral valve replacement for degenerated bioprostheses, failed annuloplasty rings, and mitral annular calcification. *Eur Heart J.* 2019;40:441–451.
12. Alexis SL, Malik AH, El-Eshmawi A, George I, Sengupta A, et al. Surgical and transcatheter mitral valve replacement in mitral annular calcification: a systematic review. *J Am Heart Assoc.* 2021;10:e018514.
13. Russell H, Guerrero M, Salinger M, et al. Open atrial transcatheter mitral valve replacement in patients with mitral annular calcification. *J Am Coll Cardiol.* 2018;72:1437–1448.
14. Anselmi A, Corbineau H, Verhoye JP, Donal E. Transapical transcatheter mitral valve implantation in mitral annular calcification: role of fusion imaging. *JTCVS Tech.* 2020;3:146–149.
15. Quarto C, Davies S, Duncan A, et al. Transcatheter mitral valve implantation: 30-day outcome of first-in-man experience with an apically tethered device. *Innovations (Phila).* 2016;11:174–178.
16. Badhwar V, Sorajja P, Duncan A, et al. Mitral regurgitation severity predicts one-year therapeutic benefit of Tendyne transcatheter mitral valve implantation. *EuroIntervention.* 2019;15:e1065–e1071.
17. Fukui M, Sorajja P, Gössl M, et al. Left ventricular remodeling after transcatheter mitral valve replacement with tendyne: new insights from computed tomography. *JACC Cardiovasc Interv.* 2020;13:2038–2048.
18. Meredith I, Bapat V, Morriss J, McLean M, Prendergast B. Intrepid transcatheter mitral valve replacement system: technical and product description. *EuroIntervention.* 2016;12(Y):Y78–Y80.
19. McCarthy P, Kislitsina O, Malaisrie S, Davidson C. Transcatheter mitral valve replacement with intrepid. *Interv Cardiol Clin.* 2019;8:287–294.
20. TCT. Transcatheter cardiovascular therapeutics 2019 congress; 2019. https://www.tctmd.com/TCT2019.
21. Lange R, Piazza N. HighLife transcatheter mitral valve replacement. *EuroIntervention.* 2016;12(Y):Y81–Y83.
22. Regueiro A, Granada JF, Dagenais F, Rodés-Cabau J. Transcatheter mitral valve replacement: insights from early clinical experience and future challenges. *J Am Coll Cardiol.* 2017;69:2175–2192.
23. Barbanti M, Piazza N, Mangiafico S, et al. Transcatheter mitral valve implantation using the highlife system. *JACC Cardiovasc Interv.* 2017;10:1662–1670.
24. Cheung A. Early experience of TIARA transcatheter mitral valve replacement system. *Ann Cardiothorac Surg.* 2018;7:787–791.
25. PCR eCourse. PCR e-Course (pcronline.com); 2020.
26. Sondergaard L, Ussia GP, Dumonteil N, Quadri A. The CardiAQ transcatheter mitral valve implantation system. *EuroIntervention.* 2015;11 (Suppl. W):W76–W77.
27. Webb J, Hensey M, Fam N, et al. Transcatheter mitral valve replacement with the transseptal EVOQUE system. *JACC Cardiovasc Interv.* 2020;13:2418–2426.
28. Fam NP, von Bardeleben RS, Hensey M, et al. Transfemoral transcatheter tricuspid valve replacement with the EVOQUE system: a multicenter, observational, first-in-human experience. *JACC Cardiovasc Interv.* 2021;14:501–511.
29. Webb JG, Murdoch DJ, Boone RH, et al. Percutaneous transcatheter mitral valve replacement: first-in-human experience with a new transseptal system. *J Am Coll Cardiol.* 2019;73:1239–1246.
30. Modine T, Vahl TP, Khalique OK, et al. First-in-human implant of the cephea transseptal mitral valve replacement system. *Circ Cardiovasc Interv.* 2019;12:e008003.
31. Vahl TP, Grogan A, Cheng Y, et al. Experimental evaluation of a novel percutaneous transseptal catheter-based mitral valve replacement technology. *Circ Cardiovasc Interv.* 2019;12:e008002.

32. Alperi A, Dagenais F, Del Val D, et al. Early experience with a novel transfemoral mitral valve implantation system in complex degenerative mitral regurgitation. *JACC Cardiovasc Interv*. 2020;13:2427–2437.
33. Alperi A, Del Val D, Ferreira-Neto AN, et al. Device profile of the AltaValve system for transcatheter mitral valve replacement: overview of its safety and efficacy. *Expert Rev Med Dev*. 2020;17:627–636.
34. Anselmi A, Aymami M, Tomasi J, et al. Durability of mitral valve replacement with a 3 rd generation bioprosthesis. *Ann Thorac Surg*. 2021;. Available from: https://doi.org/10.1016/j.athoracsur.2021.03.097. S0003-4975(21)00724-00724.

Chapter 20

Novel Developments in Tricuspid Valve Assessment and Treatments

Jordan Groskurth[1] and Tomasz Timek[2,3]

[1]Cardiothoracic Surgery Fellow, Spectrum Health, Grand Rapids, MI, United States, [2]Division of Cardiothoracic Surgery, Spectrum Health, Grand Rapids, MI, United States, [3]Michigan State University College of Human Medicine, Grand Rapids, MI, United States

Chapter Outline

Background

The tricuspid valve (TV) complex consists of three leaflets (anterior, posterior, and septal), chordae tendineae, the tricuspid annulus, three discrete papillary muscles, and the right ventricular (RV) myocardium (see Fig. 20.1). Proper function of the valve requires coordination of all of these elements to achieve timely and competent valve closure. Dysfunction of the tricuspid valve has been increasingly recognized as contributing to morbidity and mortality, and a recent study has been directed at options for intervention and optimal timing of such interventions.[1] Functional assessment of the tricuspid valve was hampered in the past by suboptimal clinical imaging, but our current understanding of the morphology and pathophysiology of tricuspid regurgitation (TR) has been significantly enhanced due to improvements in three-dimensional echocardiographic techniques and cardiac magnetic resonance imaging (MRI). Recent anatomic reevaluation of the tricuspid valvular complex[2] has focused on elucidating surgical anatomy to facilitate more durable repair techniques. The "forgotten" valve is increasingly recognized as an important contributor to cardiac pathophysiology as the incidence of trivalvular dysfunction is accelerating in Europe[3] and the United States.[4]

Tricuspid stenosis

Tricuspid stenosis is most commonly rheumatic and has anatomic features similar to rheumatic mitral stenosis with thickened leaflets and shortening of chordae. In advanced disease, the leaflets edges can be fused, and calcifications can be present. Even a small diastolic gradient is significant, and as right atrial pressure increases, venous congestion leads to jugular distention, ascites, pleural effusions, and liver dysfunction. Clinical signs and symptoms are consistent with low right-heart output including fatigue, ascites, and peripheral edema. Severe obstruction can sometimes mask features of mitral stenosis such as pulmonary congestion due to the reduced blood flow to the left heart. Echocardiographic signs of tricuspid stenosis include thickened tricuspid leaflets with reduced mobility and doming during diastole. The valve orifice is reduced and the Doppler flow pattern has an elongated slope during antegrade flow. Tricuspid valve stenosis

Advances in Cardiovascular Technology. DOI: https://doi.org/10.1016/B978-0-12-816861-5.00035-6

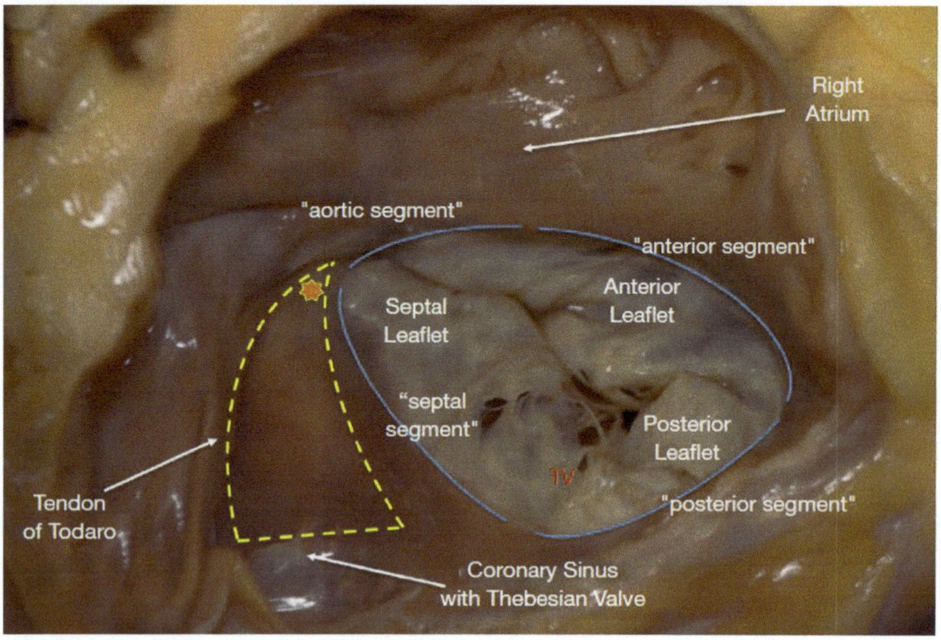

FIGURE 20.1 Anatomy of the tricuspid valve and surrounding structures. Dotted line depicts the triangle of Koch, delineated by the septal leaflet, then tendon of Todaro, and the coronary sinus. The star marks the location of the atrioventricular node. *Reproduced with permission from Taramasso M, Pozzoli A, Basso C, et al. Compare and contrast tricuspid and mitral valve anatomy: interventional perspectives for transcatheter tricuspid valve therapies.* EuroIntervention. *2018;13:1889–1898. https://doi.org/10.4244/eij-d-17-00704.*[5]

is a lesion rarely seen in the United States but has been associated with carcinoid disease, radiation treatment, and endocarditis.[6]

Tricuspid regurgitation

TR can be classified based on the underlying mechanism as primary, secondary, or isolated. Primary TR results from any abnormality of the tricuspid valve apparatus including the annulus, leaflets, chordae, or papillary muscles and can be due to acquired or congenital causes. The most common congenital cause is Ebstein's anomaly in which there is apical displacement of the leaflets with no identifiable chordae.[7] Acquired primary TR can result from rheumatic disease, endocarditis, radiation, iatrogenic injury.[8] An increasingly common cause of primary TR is transvenous pacemaker or defibrillator leads.[9] Secondary or functional tricuspid regurgitation (FTR) is caused by RV dilation and subsequent tethering of the leaflets and annular dilation resulting in poor coaptation.[10] The most common causes of secondary TR include left-sided valvular disease and myocardial disease with increased left atrial pressures and RV afterload. RV dilation in the setting of ischemia and inferior myocardial infarction, pulmonary hypertension, and cardiomyopathy can also lead to TR.[11,12] Thirty to fifty percent of patients with severe mitral regurgitation have concomitant significant TR (moderate to severe).[13] Moreover, progression or late onset TR can occur after surgical or transcatheter treatment of mitral regurgitation, contrary to previous belief.[14–16] Isolated TR occurs in the absence of pulmonary hypertension or left-sided heart disease due to right atrial enlargement and concurrent dilation of the tricuspid annulus.[17,18] Chronic atrial fibrillation is a leading cause of this pathophysiology.

Functional tricuspid regurgitation

FTR is thought to result from altered tricuspid annular (TA) size and RV geometry in the setting of anatomically normal leaflet and chordal structure.[10,19] Uncorrected moderate or severe TR during left heart surgery is associated with increased morbidity and mortality.[20] FTR is a marker of worse long-term survival regardless of left ventricular ejection fraction or pulmonary artery pressure[21] and is a predictor of mortality in degenerative mitral valve disease,[22] congestive heart failure,[23] and left ventricular assist device[24] patients. The increasing incidence of TR in the United States has recently been recognized as a "public health crisis."[4]

In functional dilation, the septal portion of the annulus is typically spared, and the annulus dilates along the anterior and posterior leaflet attachments which causes the annulus to flatten and become more circular.[25] Degree of TR is related to preload, afterload, and RV function which can alter the observed severity of regurgitation on echocardiography. As such, significant TR may not be detected on trans esophageal echocardiogram (TEE) despite substantial dilation of the tricuspid valve annulus. The practice of assessing severity of TR and treating the primary lesion (mitral valve) alone only reduces RV afterload and does not address TA dilation, RV preload, or function. Once the tricuspid annulus has enlarged, it can continue to dilate instead of returning to normal size. In one meta-analysis of patients with mild to moderate TR undergoing mitral valve surgery, 14%–49% of patients had progression to significant TR.[26] Given that TA dilation is a primary cause of FTR, Dreyfus and colleagues hypothesized that TA size may be a better predictor of outcomes than grade of TR. Additionally, addressing TA dilation may be necessary even when regurgitation is mild. These authors performed tricuspid annuloplasty in patients undergoing concomitant mitral valve surgery with tricuspid annulus 70 mm or greater (measured by ruler intraoperatively), regardless of the grade of TR.[27] In patients that underwent mitral valve replacement alone, 48% had TR that increased by 2 or more grades postoperatively compared to 2% of patients that underwent tricuspid valve annuloplasty (TVA). Interestingly, 88% of patients with TA dilation of 70 mm or more had only grade 0 to 1 TR on echocardiogram during preoperative assessment. The authors concluded that there is little or no correlation between tricuspid dilation and regurgitation and that tricuspid dilation is more reliable than TR when assessing functional tricuspid valve disease. Recent American Heart Association and European Society of Cardiology (AHA/ACC and ESC/EACTS) guidelines recommend concomitant TVA in patients with tricuspid annulus end-diastolic diameter of 40 mm or greater on TEE, even if the TR is mild or less (class IIa recommendation).[28,29]

Surgical techniques

Ring annuloplasty

The main surgical approach to the treatment of FTR is reduction of the annulus with rigid or flexible annuloplasty prostheses (Fig. 20.2). There are multiple devices available on the market with differential shape, three-dimensional geometry, and material properties.[30] This reduction allows appropriate coaptation of the leaflets and improvement in TR grade. The length of the base of the septal leaflet determines the size of the ring or band. Mattress sutures are placed circumferentially, avoiding sutures in the region of the atrioventricular (AV) node at the apex of the triangle of Koch. The sutures are placed with wider bites through the annulus and smaller corresponding bites through the ring, which results in annular plication and improved leaflet coaptation.

Although prosthetic ring tricuspid annuloplasty represents the contemporary surgical approach for correcting FTR, suboptimal results have been reported in patients with severe TR.[31] Mild residual FTR is present in more than 40% of patients immediately after ring annuloplasty.[32] A durability study by McCarthy and colleagues found 14% of patients undergoing TVA had 3 or 4 + TVR 1 week after surgery[33] while residual moderate or greater FTR was reported in 24% of patients after undergoing ring repair of severe TR by Calafiore and colleagues.[31] Furthermore, experimental

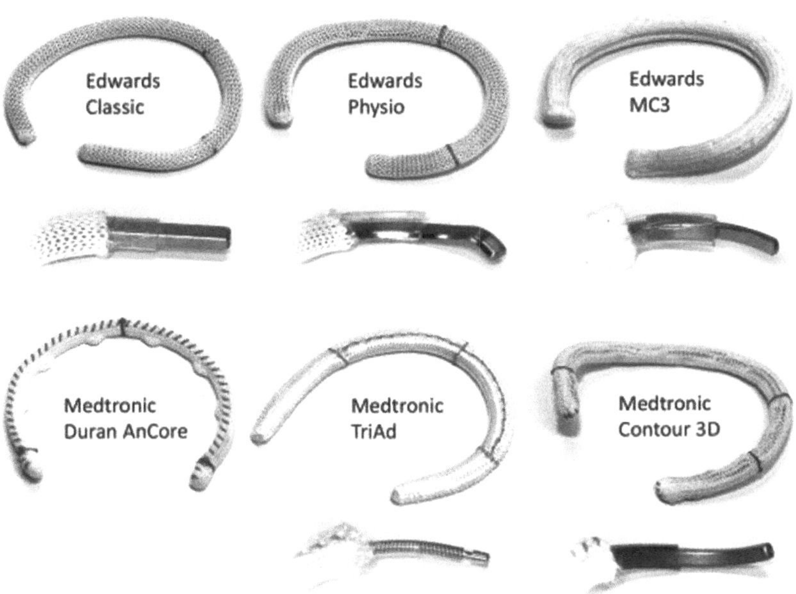

FIGURE 20.2 Annuloplasty rings with various geometry, shape, and stiffness. Note the gap to avoid suture placement into the AV node. *Used with permission from Mathur M, Malinowski M, Timek, TA, Rausch MK. Tricuspid annuloplasty rings: a quantitative comparison of size, nonplanar shape, and stiffness.* Ann Thorac Surg. *2020; 110:1605–1614.*[30]

evidence[34] suggests that aggressive annular undersizing to treat FTR may have deleterious effects on regional RV function. Recurrent FTR rates reach approximately 30% beyond 1 year after surgery.[35,36] A recent study from the Mayo Clinic of 767 patients with predominantly severe FTR confirmed similar 1-year recurrence rates reaching almost 50% recurrent moderate or greater FTR at 6 years.[37] Reoperative surgery for severe FTR is fraught with a perioperative mortality of between 11% and 50%,[38,39] with the majority of patients not being offered surgical intervention owing to a prohibitively high risk profile. Multiple studies have shown that the presence of leaflet tethering is a predictor of residual or recurrent TR.[10,40,41] Other risk factors include atrial fibrillation and excessive annular dilation with malcoaptation of the leaflets.[42,43] In cases of leaflet tethering, the anterior leaflet is restricted from coaptation due to dilation of the free wall of the RV and anterior papillary muscle displacement. This deviation of the papillary muscle away from the annulus causes the portion of the leaflet that is participating in coaptation to shift from the body in a normal functioning valve toward the edge in a tethered valve, eventually there is no coaptation at all. Based on these data, Dreyfus and colleagues have proposed an alternative method of evaluating and approaching FTR (Fig. 20.3) in which the severity of regurgitation, annular diameter, and mode of leaflet coaptation are considered in determining treatment.[10]

The disease is divided into three stages. In stage 1, mild or less regurgitation and annular diameter 40 mm or less with normal coaptation is present. Stage 2 represents mild to moderate regurgitation with dilated annulus 40 mm or more with normal or near normal coaptation. In these patients, TVA should be performed. In contrast to clinical outcomes for late-stage severe TR, early intervention has demonstrated to be a good strategy with durable results. In one study, 86% of patients undergoing ring annuloplasty had less than moderate TR after 7 years.[44] Another study revealed that 90% of such patients undergoing annuloplasty with 3D ring had less than moderate TR after 4 years.[45] In stage 3, there is severe regurgitation and dilated annulus in addition to absent coaptation with possible leaflet tethering. As isolated ring annuloplasty in such patients has not been shown to be predictably effective, surgical adjunct should be considered to correct leaflet tethering in addition to TVA. This algorithm echoes the more aggressive and early treatment of FTR which has been advocated by many.[43]

Adjunctive surgical repair techniques

Several repair options have been proposed as adjuncts to TVA in patients with significant leaflet tethering including augmentation of the anterior leaflet, approximation of the papillary muscles, and RV volume reduction. In augmentation

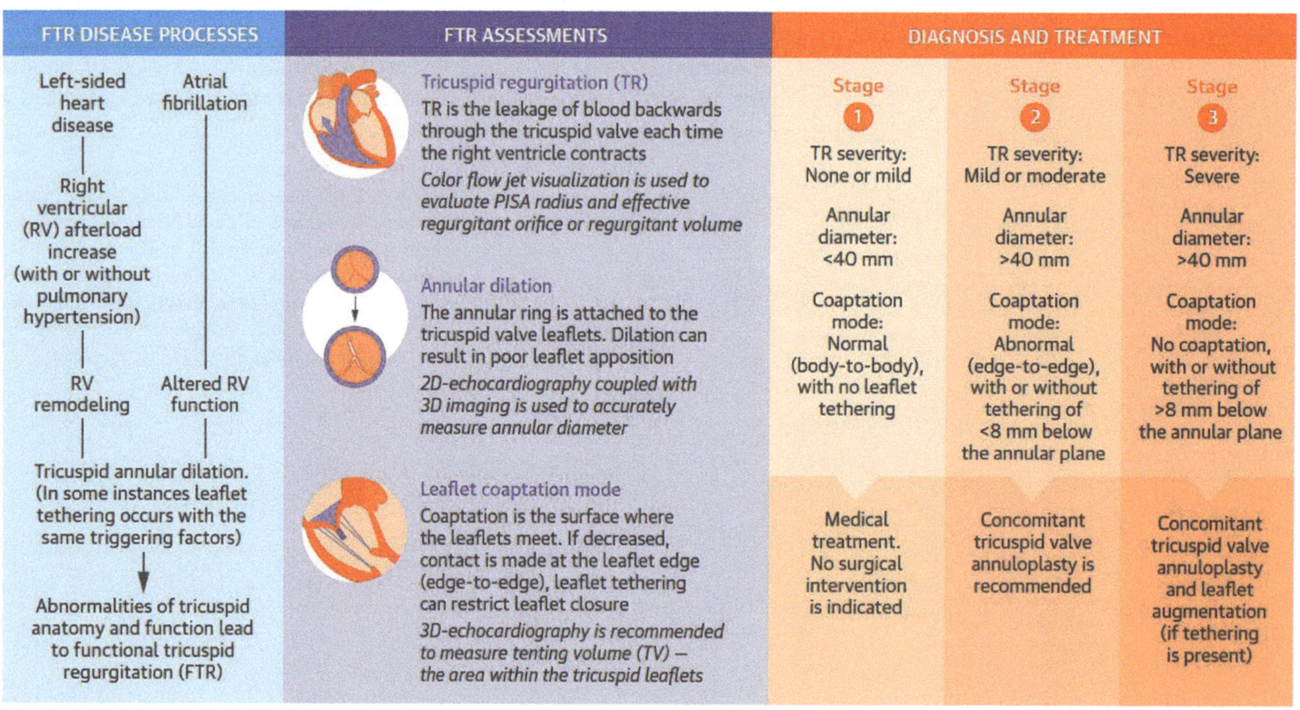

FIGURE 20.3 Functional tricuspid regurgitation development, assessment, diagnosis, and treatment. *Proposed new staging scheme by Dreyfus GD, Martin RP, Chan KMJ, Dulguerov F, Alexandrescu C. Functional tricuspid regurgitation: a need to revise our understanding.* J Am Coll Cardiol. *2015;65:2331–2336.*[10]

of the anterior leaflet (Fig. 20.4), the anterior leaflet is detached along its entire length from anteroseptal to the anteroposterior commissure. A patch of autologous pericardium is excised in the shape of an oval with width the distance between the commissures and height the greatest distance between the detached leaflet and the annulus. The patch is sutured into position with one side to the annulus and the other side to the detached free edge of the leaflet. This allows the leaflet to act as the coaptation surface while the patch functions as the body of the leaflet. TVA with a band is then completed.[46]

This technique maintains continuity of the chordae to the leaflet and permits the anterior leaflet to accommodate the dilated RV geometry by allowing coaptation to occur at a new normal deeper point below the annulus. Two studies of leaflet augmentation showed durable control of TR at 2 years with a freedom of TR ≥ 2 of 84%[47] and 77%.[48] These patients had an average leaflet tethering height of 1 cm which is an identified predictor of recurrent FTR with ring annuloplasty alone.

Other techniques involve repair at the subvalvular level including approximating the anterior and posterior papillary muscles with suture or PTFE sling.[49,50]

These subvalvular approaches aim to return the anterior papillary muscle to a more normal position, allowing for coaptation closer to the annular plane. The result is mild or less TR postoperatively with significant improvement in leaflet motion and reduced tethering on postoperative echo. Multiple surgical techniques to approximate the papillary muscles have been described including interrupted subendocardial sutures to plicate the RV wall at the level of the papillary muscles (Fig. 20.5),[49] reduction sutures placed outside the lateral RV wall,[51] and using a 4 mm synthetic graft to approximate the papillary muscles.[50] Each of these techniques led to improvement of TR from 4+ to 2 or less as well RV size reduction after surgery that persisted at late follow-up.[49–52] In severe cases of leaflet tethering, tricuspid valve replacement may be necessary.

Tricuspid replacement

When the tricuspid valvular complex is altered beyond repair, valve replacement should be considered. In tricuspid valve replacement, pledgeted mattress sutures are used to fixate the prosthetic valve using an everting technique in mechanical valves and supra-annular or intra-annular for biological protheses. The tricuspid septal and posterior leaflets are left in place which helps to avoid injury to the conduction system. The central portion of the anterior leaflet can be excised while preserving chordal attachments in order to prevent obstruction of the RV outflow tract. The patient's age, contraindications to anticoagulation, and social issues should all be considered when choosing the type of valvular prosthesis. Historically, mechanical valves suffered from a high rate of valve thrombosis, but the St Jude bileaflet valve has demonstrated improved clinical outcomes.[53,54] Compared to mitral replacement, bioprostheses in the tricuspid position have better durability with longer freedom from structural degeneration.[55] However, chronic pannus and thrombus formation is thought to be more frequent in the tricuspid position compared to the mitral.[55] Recent summary of the available data suggests equivalent clinical outcomes with both mechanical and bioprosthetic tricuspid valve replacement with the choice of prosthesis tailored to patient risk profile.[56]

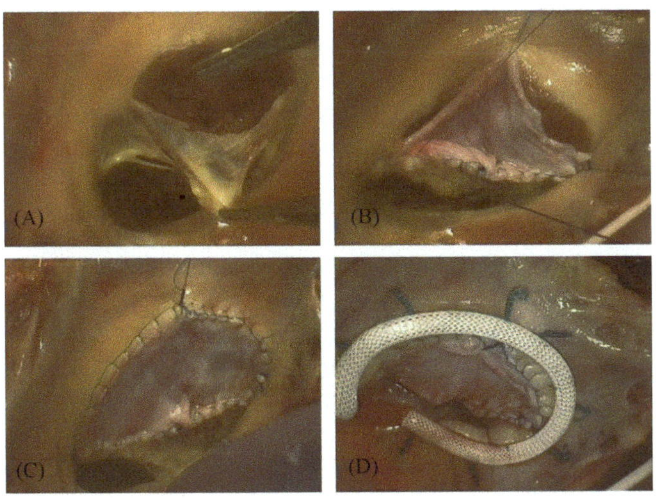

FIGURE 20.4 Operative pictures demonstrating tricuspid leaflet augmentation: (A) detaching the anterior leaflet from the tricuspid annulus, (B) sewing the pericardial patch onto the detached anterior leaflet, (C) the pericardial patch sewn onto the detached anterior leaflet and the annulus, (D) the completed repair with ring annuloplasty. *Used with permission from Dreyfus GD, Raja SG, John Chan KM. Tricuspid leaflet augmentation to address severe tethering in functional tricuspid regurgitation.* Eur J Cardiothorac Surg. *2008;34:908–910.*[46]

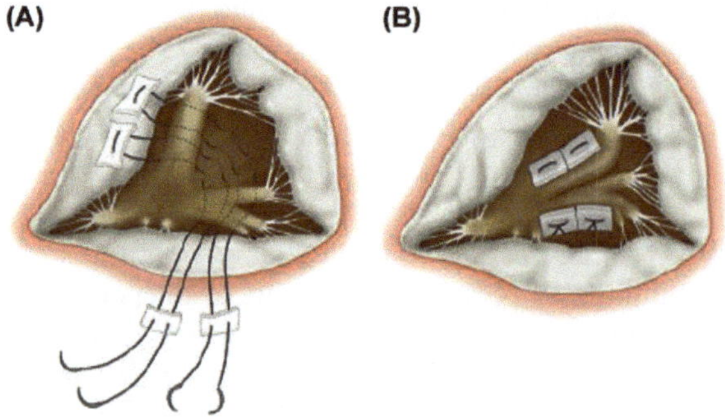

FIGURE 20.5 Schema of right ventricular papillary muscle approximation showing (A) semi circumferentially placed horizontal mattress sutures with multiple bites on the subendocardium of the right ventricle and (B) the position of papillary muscles after the approximation. *Used with permission from Matsumiya G, Kohno H, Matsuura K, et al. Right ventricular papillary muscle approximation for functional tricuspid regurgitation associated with severe leaflet tethering.* Interact Cardiovasc Thorac Surg. *2018;26:700–702.*[49]

In addition to standard prosthetic annular reduction, mitral valve repair techniques including pericardial patching of perforations, partial leaflet resections, chordal transfer, and artificial chordae using PTFE can also be harnessed in tricuspid valve repair to achieve valvular competence and avoid replacement. In one retrospective study, compared to repair, tricuspid valve replacement was associated with a higher rate of severe RV dysfunction, severe TR, and congestive heart failure.[37] The replacement cohort also had worse distant survival.

Another consideration in the surgical approach to the tricuspid valve is cardioplegic arrest versus beating heart surgery. Beating heart technique was demonstrated to be associated with lower risk of renal failure and stroke in one retrospective study,[57] but mortality, both per-operative and long-term, was similar to cardioplegic arrest.[57] When a composite outcome of cardiac death and reoperation was considered, beating heart technique was associated with improved long-term outcomes. In another study, beating heart technique had similar short- and long-term outcomes to arrested heart surgery despite a higher risk profile patient group.[58]

Transcatheter technologies

Treatment options for patients with multiple comorbidities and high surgical risk have been limited until recent development of percutaneous treatments aimed at addressing FTR. Current percutaneous strategies under development include heterotopic caval valve implantation (CAVI), percutaneous tricuspid annuloplasty, leaflet edge-to-edge repair, and atrioventricular valved stents.[59] Transcatheter tricuspid valve interventions (TTVI) show promise in early experience with one registry of multiple TTVI technologies showing reduction of TR to $\leq 2+$ in 72.8% of treated cases. In patients who had successful improvement in TR, actuarial survival was improved at one year to 90% versus 70% for patients with failed interventions.[60] This registry illustrated the importance of patient selection and timing of intervention in patients with functional TR. Successful reduction of TR to 2 + or less was an independent predictor of midterm survival. Factors that predicted treatment failure included advanced leaflet tethering and torrential TR. Thus, survival could potentially be improved with early intervention in patients with TR before it is too clinically advanced. These implications mirror the trends in the surgical literature. Further study of optimal timing for intervention and factors predicting decompensated RV failure is imperative to our understanding of this complex disease pathophysiology and tailoring of appropriate therapies.

There are, however, multiple challenges facing percutaneous treatment of the tricuspid valve. The nonplanar geometry of the tricuspid annulus, significant annular dilatation, and lack of calcium compared to other cardiac valves present significant hindrances to effective percutaneous TV replacement. Potential stented valves may need to be up to twice the size of a transcatheter aortic valve, and as such, the introducer sheath and delivery system need to accommodate much larger prostheses. Additionally, the angulation between the superior and inferior vena cava (IVC), the TV, and the trabeculated RV poses an anatomical challenge. The thin ventricular apex and multiple chordae pose difficulties for a transapical approach,[61] and the proximity of the AV node, aortic valve, right ventricular outflow tract, right coronary artery, and coronary sinus all need to be accounted for in considering size and placement of any transcatheter device. Lastly, the presence of the pacemaker or defibrillator lead limits the feasibility of transcatheter treatment and is currently considered a contraindication for many percutaneous TV therapies. In addition to the above anatomic and physiologic concerns, intraprocedural imaging of the tricuspid valve requires significant expertise due to the difficulty of obtaining appropriate views on TEE. As the TV is not coaxial to the esophagus, TEE images are off-axis and imaging of all three leaflets is often challenging.

Caval valve implantation

CAVI has been suggested as a potential treatment for high-risk patients with severe TR. First described in 2011 by Lauten et al.,[62] self-expanding or balloon expandable stented valves are placed in the IVC and sometimes superior vena cava (SVC) with the goal of eliminating regurgitation of blood into the caval veins in patients with severe, chronic TR.[63] The application of this strategy is thus limited to patients with caval dilation and severe regurgitation which is a late finding in the pathogenesis of FTR. One challenge facing CAVI is the variation in caval dilation and diameter and the proximity to the right atrium and hepatic veins. In addition, there is concern about the effect of ventricularization of the right atrium including atrial fibrillation and RV volume overload. Currently the efficacy, safety, and benefits of this technology are being studied with long-term follow up in the HOVER trial in the United States.[64]

Percutaneous annuloplasty

Several percutaneous treatments of addressing FTR under investigation are designed to mimic surgical annuloplasty in reducing the degree of valvular insufficiency. One such device, the Cardioband (Edwards Lifesciences LLC; Irvine, CA) is under study for feasibility in the United States currently (Fig. 20.6).[65] Nickenig and colleagues have reported 6-month outcomes with this device in 30 patients enrolled in a prospective, single-arm, multicenter study. The TRI-REPAIR study (Tricuspid REgurgitations RePAir with Cardioband Transcatheter System) included patients with symptomatic moderate to severe TR in the absence of left-sided disease who were deemed inoperable. Mean patient age was 75 years, 36.6% had previous cardiac surgery, 52% had massive or torrential TR, and 83% were in New York Heart Association (NYHA) class III/IV at baseline. The device was implanted successfully in 100% of the patients, and perioperatively, two patients died, four had bleeding, three experienced coronary complications, and one had conduction disturbance. At 6 months, the average annular septolateral diameter was reduced from 42 to 38 mm and the effective regurgitant orifice area (EROA) decreased by 50%. In 76% of the patients, improvement in at least 1 NYHA class was observed. Despite only a modest reduction in TR, the authors concluded that Cardioband is feasible, safe, and effective.[65] The authors did report challenges in screening patients with regard to the proximity of potential device implantation site to the right coronary artery. Reliable intraprocedural imaging and the need for multiple anchors, increasing the risk of damage to adjacent structures and bleeding complications, remain the current key limitations in more widespread application of the device.

Other devices which act similar to band annuloplasty include MIA by Micro Interventional devices (Micro Interventional Devices Inc.; Newtown, PA) and Millipede IRIS by Boston Scientifics (Boston Scientific; Marlborough, MA) are currently under investigation and have limited experience in humans.[66]

Two devices that mimic annular reduction with suture annuloplasty are also under development. One, the TriaAlign, by Mitralign (Mitralign Inc; Boston, MA), functionally obliterates the posterior leaflet resulting in a bicuspidization of the tricuspid valve similar to the Kay suture annuloplasty technique.[67] In a feasibility study of 15 patients, 80%

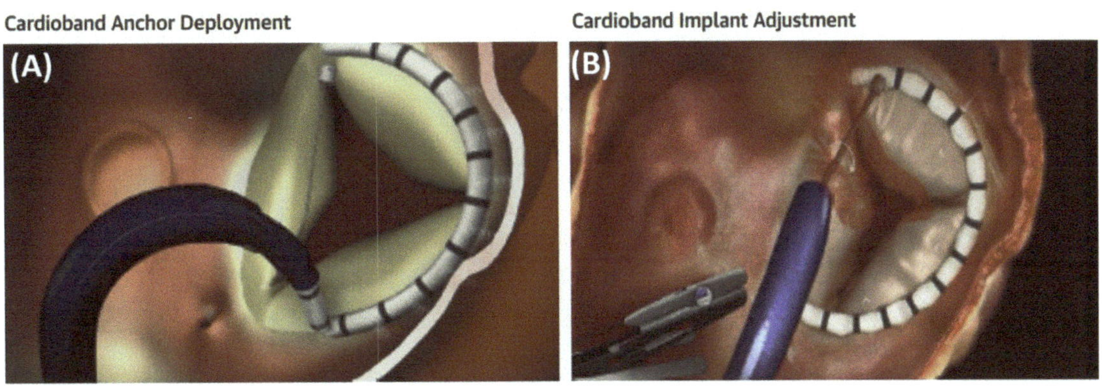

FIGURE 20.6 (A) Cardioband anchor deployment through implant to the tissue. (B) Adjustment of the implant. *Used with permission from Nickenig G, Weber M, Schueler R, et al. 6-Month outcomes of tricuspid valve reconstruction for patients with severe tricuspid regurgitation.* J Am Coll Cardiol. *2019;73:1905–1915.*[65]

experienced technical success at 30 days with three patients having single pledget annular detachment without need for reintervention. In the other 12 patients, there was significant reduction in TA area (12.3−11.3 cm^2) and EROA (0.51 vs 32 cm^2). When measured by intention-to-treat, NYHA and 6-minute walk test were significantly improved.[68] The other suture annuloplasty device, TriCinch by 4Tech (4TECH; Galway, Ireland), mimics De Vega suture annuloplasty by applying tension along the anterior and posterior portions of the annulus (Fig. 20.7).[67,69]

In the Percutaneous Treatment of Tricuspid Valve Regurgitation with the TriCinch System (PREVENT) trial, 24 patients were treated with successful implantation in 18 cases. A significant (≥1 grade) reduction of TR was achieved in 94% of patients. Of the six failed implantations, two patients experienced hemopericardium after the procedure and late detachment of the TA anchor occurred in four. Follow-up at 6 months showed significantly improved 6-minute walk distance and quality of life, with 75% of the patients being in NYHA class I or II.[70]

Coaptation modifiers

Two devices that extend the mitral edge-to-edge technique to tricuspid valve repair are currently being explored: the Triclip by Abbott Structural Heart (Abbott Laboratories; Chicago, IL) and PASCAL by Edwards Lifesciences (Edwards Lifesciences; Irvine, CA). The Triclip is currently the most common off-label procedure to treat severe TR (Fig. 20.8).[60] It was originally designed to repair mitral regurgitation but has been adapted to approximate TV leaflets, improve coaptation, and reduce TR. More than 1000 cases have been performed worldwide and several feasibility studies and registries are forthcoming.[71−73] Several difficulties have become apparent in adapting this technology to the tricuspid valve. Imaging the tricuspid leaflet coaptation gap is critical and more difficult to visualize compared to the mitral valve, the large TV leaflet gaps compared to mitral pathology are more difficult to bridge, the fragile tissue of the tricuspid leaflets may predispose to poor anchorage of the device, and the delivery system was not designed specifically for the tricuspid valve.[74] One registry of 249 patients with 1-year follow-up data (TRIVALVE Registry) demonstrated that 72% of patients had durable reduction of TR from 3−4+ to 2+ or less.[73] The TriClip, a modified Mitraclip delivery system by Abbott (Abbott Laboratories; Chicago, IL), is under investigation in the TRILUMINATE study and has shown promising results after 6 months with 87% of patients having reduction of TR by one or more grades and an adverse event rate of only 3.7%.[59] Current clinical experience and ex vivo models suggest that clipping the anterior or posterior leaflet to the septal leaflet results in better control of TR.[75] If the anterior and posterior leaflets are approximated, regurgitation may actually increase due to distortion of the coaptation zone. Similar to surgical coaptation approaches, large coaptation gaps and leaflet tethering have been shown to be risk factors for treatment failure and recurrent TR.

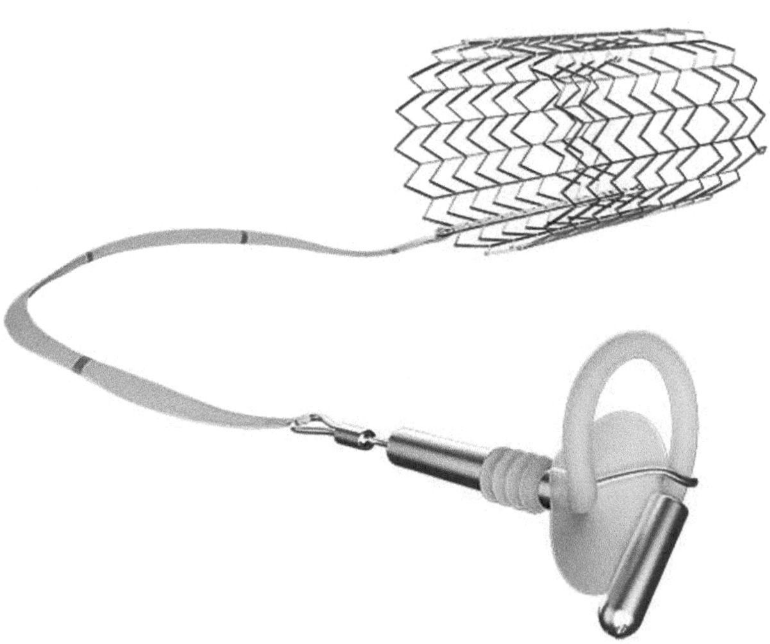

FIGURE 20.7 TriCinch device. The anchor portion is placed through the anterior portion of the tricuspid annulus and the stent is placed in the IVC, placing tension and reducing the diameter of the TA. *Used with permission from https://cardiovascularnews.com/first-us-implantations-of-tricinch-coil-tricuspid-valve-repair-system-performed/.*

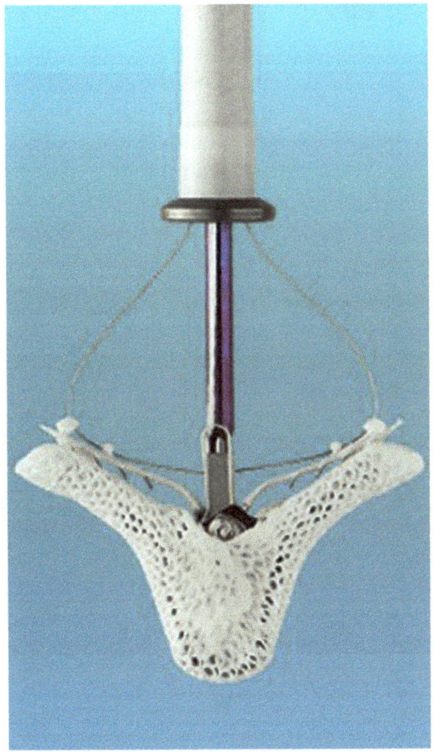

FIGURE 20.8 Trivalve device on application catheter. *Used with permission from https://www.abbott.com/corpnewsroom/products-and-innovation/mitraclip-now-helps-many-more-with-smr.html.*

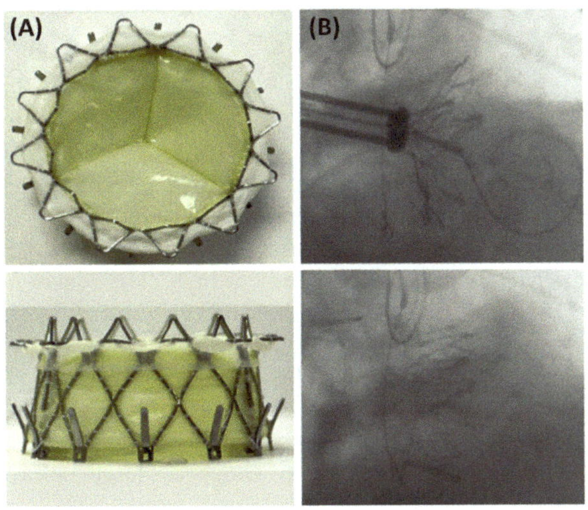

FIGURE 20.9 (A) Navigate valve. Inflow and lateral views. (B) Fluoroscopic views of Navigate valve. *Used with permission from Asmarats L, Puri R, Latib A, Navia JL, Rodés-Cabau J. Transcatheter tricuspid valve interventions: landscape, challenges, and future directions.* J Am Coll Cardiol. *2018;71:2935–2956.*[66]

Transcatheter valve replacement

Transcatheter tricuspid valve replacement (TTVR) has been under investigation for several years with numerous devices currently in early development, including the Navigate (Navigate Cardiac Structures, Inc; Lake Forest, CA) and Evoque systems (Medtronic PLC; Dublin, Ireland).[76,77] The Gate valve is a trileaflet biologic tissue valve fixed on a nitinol stent frame with ventricular anchors (Fig. 20.9). Available diameters range from 36 to 52 mm with an introducer sheath that is 42 Fr and utilizes a transjugular approach. Direct surgical implantation through a transatrial approach has also

FIGURE 20.10 (A) EVOQUE transcatheter valve: Nitinol frame with bovine pericardial leaflets, intra-annular sealing skirt, and ventricular anchors. (B) The 28 Fr EVOQUE tricuspid delivery system. *Used with permission from Fam NP, von Bardeleben RS, Hensey M, et al. Transfemoral transcatheter tricuspid valve replacement with the EVOQUE system: a multicenter, observational, first-in-human experience.* JACC Cardiovasc Interv. *2021;14:501–511.*[76]

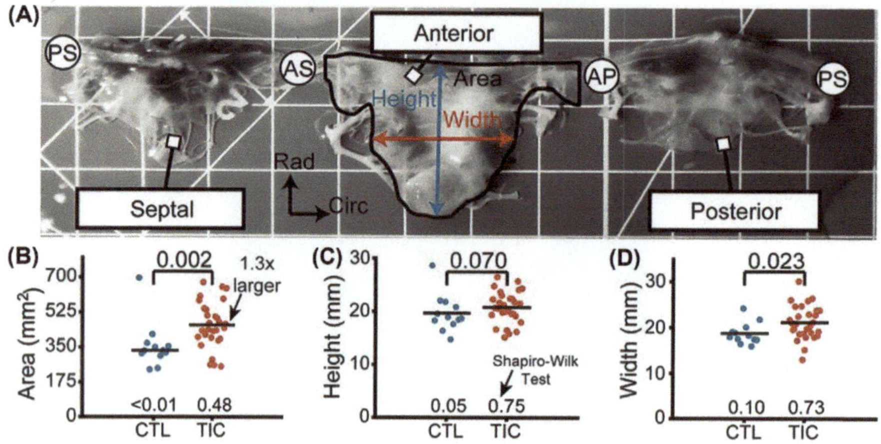

FIGURE 20.11 Tricuspid valve anterior leaflet area increases in sheep with tachycardia-induced cardiomyopathy (TIC) and FTR. (A) An ovine tricuspid valve separated at the posterior-septal commissure (PS). Located between the antero–septal (AS) and antero–posterior (AP) commissures, the anterior leaflet and its measured area (*black*), height (*blue*), and width (*red*) are shown. Grid scale = 1 cm. (B–D) Comparisons between control (CTL, *blue*, $n = 12$) and TIC (*red*, $n = 29$) anterior leaflet (B) area, (C) height, and (D) width. Black bars represent data mean if normal, and data median if nonnormal as determined by Shapiro-Wilk test *P*-value below data. *Values above data represent* P*-values from Student's* t*-test or Wilcoxon Rank Sum test, as appropriate.*[81]

been performed via a minimally invasive right thoracotomy.[78] There have been two small series showing feasibility of human implantation, and further developments in reduction of sheath size and delivery mobility could contribute to increasing clinical potential.[78,79]

The EVOQUE valve is a self-expanding nitinol frame with bovine pericardial leaflets (Fig. 20.10) available in 44 and 48 mm sizes and utilizes a 28 Fr transfemoral sheath system. Early feasibility study in 25 patients showed successful implantation in 92% of study cohort. Over 90% of patients with successful implantation had improvement in TR grade to $\leq 1+$, and 19/25 patients had significant improvement in NYHA functional status at 30 days. Interestingly, 36% of study patients had preexisting pacemaker leads and all had successful treatment with no or mild paravalvular TR and no change in pacemaker function.[76] This possible application represents a frequently encountered clinic challenge as patients with lead-induced TR are usually not candidates for transcatheter leaflet repair or annuloplasty.

Both Gate and EVOQUE valves have anchors on the ventricular as well as atrial side of the valve to stabilize the prosthesis in the annulus and prevent movement during systole. These prostheses have shown early feasibility with good control of TR.[76,80] One drawback of the NaviGate system is its larger sheath size as well as the transjugular approach versus the femoral approach of the EVOQUE system.

Noninterventional therapies

Traditionally, FTR has been thought to result from altered TA and RV geometry in the setting of anatomically "normal" leaflet and chordal structure. However recent experimental data challenge that view and suggest valve tissue maladaptation and remodeling may contribute to pathogenesis of FTR[81] (Fig. 20.11).

This is a significant development as it opens the possibility of treating functional TR by preventing valve remodeling possibly through pharmacological means, rather than by surgical intervention. Increasingly, experimental and clinical studies bring into question a solely "functional" etiology of functional atrioventricular valve insufficiency. Prior ovine experiments on functional mitral regurgitation revealed that although valve tissue appears grossly normal, mitral leaflets remodel significantly during the disease process[82] with changes in leaflet composition, structure, and valve cell phenotype.[83] Furthermore, ex vivo mechanical tests conducted on human mitral leaflets and chordae from congestive heart disease showed significant alterations and fibrosis that could contribute to insufficient leaflet coverage of the valve orifice.[84] Indeed, clinical studies have shown that tricuspid leaflet area increased by 49% in patients with pulmonary hypertension with those developing severe FTR demonstrating inadequate leaflet enlargement relative to annular and ventricular dilation.[84,85] Ovine experiments have revealed that anterior leaflet strains are altered in FTR in vivo further suggesting pathologic remodeling within the leaflet tissue. Although tricuspid valve leaflets have been viewed as inert flaps at the mercy of the hemodynamic milieu of the right ventricle, it is possible that leaflet remodeling may in the future offer an adjunctive therapeutic target. Innovative studies of postinfarction ovine mitral valves have shown this to be a possibility as Losartan treatment was found to modulate fibrotic changes in the valve while maintaining adaptive growth through inhibition of transforming growth factor-beta (TGF-B).[84–86] Such biochemical pathways remain to be studied in functional TR but hold possible promise of an alternative therapeutic approach.

Conclusion

Functional tricuspid valve regurgitation associated with left sided valvular lesions represents the majority of surgical tricuspid pathophysiology seen in modern cardiothoracic practice. Standard prosthetic ring annuloplasty offers durable results in the setting of an aggressive treatment algorithm focused on early intervention before evolution of RV dysfunction or severe regurgitation. However, with advanced ventricular remodeling and severe valvular insufficiency, isolated reductive annuloplasty offers suboptimal clinical outcomes and adjunctive surgical procedures may be needed to improve results. Late-stage isolated TV surgery continues to be associated with one of the highest risk profiles in current surgical practice, and given the poor results of medical therapy, there is a significant clinical need for development of surgical and minimally invasive treatments for these very challenging patients. Multiple studies have now demonstrated feasibility of transcatheter treatments with resultant functional and quality of life improvements. TTVR holds the potential promise of effectively abolishing FTR with minimal or no insufficiency in most cases, yet many procedural and technical difficulties remain to be solved. Maturation of imaging technique, further development and innovation of devices, and continued clinical experience should improve procedural and clinical outcomes. Future trials are needed to address optimal procedural timing, clinical outcomes, and device performance and durability.

References

1. Kim H-K, Lee S-P, Kim Y-J, Sohn D-W. Tricuspid regurgitation: clinical importance and its optimal surgical timing. *J Cardiovasc Ultrasound.* 2013;21:1–9.
2. Kawada N, Naganuma H, Muramatsu K, Ishibashi-Ueda H, Bando K, Hashimoto K. Redefinition of tricuspid valve structures for successful ring annuloplasty. *J Thorac Cardiovasc Surg*. 2018;155:1511–1519. e1.
3. Vieitez JM, Monteagudo JM, Mahia P, et al. New insights of tricuspid regurgitation: a large-scale prospective cohort study. *Eur Heart J Cardiovasc Imaging*. 2021;22:196–202.
4. Enriquez-Sarano M, Messika-Zeitoun D, Topilsky Y, Tribouilloy C, Benfari G, Michelena H. Tricuspid regurgitation is a public health crisis. *Prog Cardiovasc Dis*. 2019;62:447–451.
5. Taramasso M, Pozzoli A, Basso C, et al. Compare and contrast tricuspid and mitral valve anatomy: interventional perspectives for transcatheter tricuspid valve therapies. *EuroIntervention*. 2018;13:1889–1898. Available from: https://doi.org/10.4244/eij-d-17-00704.
6. Cevasco M, Shekar PS. Surgical management of tricuspid stenosis. *Ann Cardiothorac Surg*. 2017;6:275–282.
7. Paranon S, Acar P. Ebstein's anomaly of the tricuspid valve: from fetus to adult. *Heart*. 2008;94:237–243. Available from: https://doi.org/10.1136/hrt.2006.105262.
8. Adler DS. Non-functional tricuspid valve disease. *Ann Cardiothorac Surg*. 2017;6:204–213.
9. Höke U, Auger D, Thijssen J, et al. Significant lead-induced tricuspid regurgitation is associated with poor prognosis at long-term follow-up. *Heart*. 2014;100:960–968. Available from: https://doi.org/10.1136/heartjnl-2013-304673.
10. Dreyfus GD, Martin RP, Chan KMJ, Dulguerov F, Alexandrescu C. Functional tricuspid regurgitation: a need to revise our understanding. *J Am Coll Cardiol*. 2015;65:2331–2336.
11. Hung J, Koelling T, Semigran MJ, William Dec G, Levine RA, Di Salvo TG. Usefulness of echocardiographic determined tricuspid regurgitation in predicting event-free survival in severe heart failure secondary to idiopathic-dilated cardiomyopathy or to ischemic cardiomyopathy. *Am J Cardiol*. 1998;82:1301–1303. Available from: https://doi.org/10.1016/s0002-9149(98)00624-9.
12 Vatterott PJ, Nishimura RA, Gersh BJ, Smith HC. Severe isolated tricuspid insufficiency in coronary artery disease. *Int J Cardiol*. 1987;14:295–301.
13. Shiran A, Sagie A. Tricuspid regurgitation in mitral valve disease incidence, prognostic implications, mechanism, and management. *J Am Coll Cardiol*. 2009;53:401–408.

14. De Bonis M, Lapenna E, Sorrentino F, et al. Evolution of tricuspid regurgitation after mitral valve repair for functional mitral regurgitation in dilated cardiomyopathy. *Eur J Cardiothorac Surg*. 2008;33:600–606.
15. Goldstone AB, Howard JL, Cohen JE, et al. Natural history of coexistent tricuspid regurgitation in patients with degenerative mitral valve disease: implications for future guidelines. *J Thorac Cardiovasc Surg*. 2014;148:2802–2809.
16. Benedetto U, Melina G, Angeloni E, et al. Prophylactic tricuspid annuloplasty in patients with dilated tricuspid annulus undergoing mitral valve surgery. *J Thorac Cardiovasc Surg*. 2012;143:632–638.
17. Park J-H, Shin S-H, Lee M-J, et al. Clinical and echocardiographic factors affecting tricuspid regurgitation severity in the patients with lone atrial fibrillation. *J Cardiovasc Ultrasound*. 2015;23:136–142.
18. Mutlak D, Lessick J, Reisner SA, Aronson D, Dabbah S, Agmon Y. Echocardiography-based spectrum of severe tricuspid regurgitation: the frequency of apparently idiopathic tricuspid regurgitation. *J Am Soc Echocardiogr*. 2007;20:405–408.
19. Fukuda S, Marc Gillinov A, Song J-M, et al. Echocardiographic insights into atrial and ventricular mechanisms of functional tricuspid regurgitation. *Am Heart J*. 2006;152:1208–1214. Available from: https://doi.org/10.1016/j.ahj.2006.07.027.
20. Kwak J-J, Kim Y-J, Kim M-K, et al. Development of tricuspid regurgitation late after left-sided valve surgery: a single-center experience with long-term echocardiographic examinations. *Am Heart J*. 2008;155:732–737.
21. Nath J, Foster E, Heidenreich PA. Impact of tricuspid regurgitation on long-term survival. *J Am Coll Cardiol*. 2004;43:405–409.
22. Essayagh B, Antoine C, Benfari G, et al. Functional tricuspid regurgitation of degenerative mitral valve disease: a crucial determinant of survival. *Eur Heart J*. 2020;41:1918–1929.
23. Messika-Zeitoun D, Verta P, Gregson J, et al. Impact of tricuspid regurgitation on survival in patients with heart failure: a large electronic health record patient-level database analysis. *Eur J Heart Fail*. 2020;22:1803–1813. Available from: https://doi.org/10.1002/ejhf.1830.
24. Veen KM, Mokhles MM, Soliman O, et al. EUROMACS Investigators, Clinical impact and "natural" course of uncorrected tricuspid regurgitation after implantation of a left ventricular assist device: an analysis of the European Registry for Patients with Mechanical Circulatory Support (EUROMACS). *Eur J Cardiothorac Surg*. 2021;59:207–216.
25. Mahmood F, Kim H, Chaudary B, et al. Tricuspid annular geometry: a three-dimensional transesophageal echocardiographic study. *J Cardiothorac Vasc Anesth*. 2013;27:639–646.
26. Kara I, Koksal C, Erkin A, et al. Outcomes of mild to moderate functional tricuspid regurgitation in patients undergoing mitral valve operations: a meta-analysis of 2,488 patients. *Ann Thorac Surg*. 2015;100:2398–2407.
27. Dreyfus GD, Corbi PJ, John Chan KM, Bahrami T. Secondary tricuspid regurgitation or dilatation: which should be the criteria for surgical repair? *Ann Thorac Surg*. 2005;79:127–132. Available from: https://doi.org/10.1016/j.athoracsur.2004.06.057.
28. Otto CM, Nishimura RA, Bonow RO, et al. ACC/AHA guideline for the management of patients with valvular heart disease: a report of the American College of Cardiology/American Heart Association Joint Committee on Clinical Practice Guidelines. *Circulation*. 2020;143(2021): e72–e227.
29. Baumgartner H, Falk V, Bax JJ, et al.ESC Scientific Document Group ESC/EACTS Guidelines for the management of valvular heart disease. *Eur Heart J*. 2017;38(2017):2739–2791.
30. Mathur M, Malinowski M, Timek TA, Rausch MK. Tricuspid annuloplasty rings: a quantitative comparison of size, nonplanar shape, and stiffness. *Ann Thorac Surg*. 2020;110:1605–1614.
31. Calafiore AM, Foschi M, Kheirallah H, et al. Early failure of tricuspid annuloplasty. Should we repair the tricuspid valve at an earlier stage? The role of right ventricle and tricuspid apparatus. *J Card Surg*. 2019;34:404–411.
32. Pfannmüller B, Doenst T, Eberhardt K, Seeburger J, Borger MA, Mohr FW. Increased risk of dehiscence after tricuspid valve repair with rigid annuloplasty rings. *J Thorac Cardiovasc Surg*. 2012;143:1050–1055. Available from: https://doi.org/10.1016/j.jtcvs.2011.06.019.
33. McCarthy PM, Bhudia SK, Rajeswaran J, et al. Tricuspid valve repair: durability and risk factors for failure. *J Thorac Cardiovasc Surg*. 2004;127:674–685.
34. Jazwiec T, Malinowski M, Ferguson H, et al. Effect of variable annular reduction on functional tricuspid regurgitation and right ventricular dynamics in an ovine model of tachycardia-induced cardiomyopathy. *J Thorac Cardiovasc Surg*. 2021;161:e277–e286.
35. Onoda K, Yasuda F, Takao M, et al. Long-term follow-up after carpentier-Edwards ring annuloplasty for tricuspid regurgitation. *Ann Thorac Surg*. 2000;70:796–799. Available from: https://doi.org/10.1016/s0003-4975(00)01681-7.
36. Ton-Nu T-T, Levine RA, Handschumacher MD, et al. Geometric determinants of functional tricuspid regurgitation. *Circulation*. 2006;114:143–149. Available from: https://doi.org/10.1161/circulationaha.106.611889.
37. Pahwa S, Saran N, Pochettino A, et al. Outcomes of tricuspid valve surgery in patients with functional tricuspid regurgitation. *Eur J Cardiothorac Surg*. 2021;59:577–585.
38. Kwon D-A, Park J-S, Chang H-J, et al. Prediction of outcome in patients undergoing surgery for severe tricuspid regurgitation following mitral valve surgery and role of tricuspid annular systolic velocity. *Am J Cardiol*. 2006;98:659–661. Available from: https://doi.org/10.1016/j.amjcard.2006.03.047.
39. Hornick P, Harris PA, Taylor KM. Tricuspid valve replacement subsequent to previous open-heart surgery. *J Heart Valve Dis*. 1996;5:20–25.
40. Fukuda S, Gillinov AM, McCarthy PM, et al. Determinants of recurrent or residual functional tricuspid regurgitation after tricuspid annuloplasty. *Circulation*. 2006;114:I582–I587.
41. Min S-Y, Song J-M, Kim J-H, et al. Geometric changes after tricuspid annuloplasty and predictors of residual tricuspid regurgitation: a real-time three-dimensional echocardiography study. *Eur Heart J*. 2010;31:2871–2880.

42. Xu J, Han J, Zhang H, et al. Risk factors for postoperative recurrent tricuspid regurgitation after concomitant tricuspid annuloplasty during left heart surgery and the association between tricuspid annular circumference and secondary tricuspid regurgitation. *BMC Cardiovasc Disord.* 2021;21:1–9.
43. Calafiore AM, Lorusso R, Kheirallah H, et al. Late tricuspid regurgitation and right ventricular remodeling after tricuspid annuloplasty. *J Card Surg.* 2020;35:1891–1900.
44. Carino D, Zancanaro E, Lapenna E, et al. Long-term results of tricuspid annuloplasty with 3-dimensional-shaped rings: effective and durable!. *Eur J Cardiothorac Surg.* 2021;. Available from: https://doi.org/10.1093/ejcts/ezab111.
45. Hata H, Fujita T, Shimahara Y, Sato S, Kobayashi J. Early and mid-term outcomes of aggressive tricuspid annuloplasty with the MC3 ring. *J Heart Valve Dis.* 2014;23:601–608.
46. Dreyfus GD, Raja SG, John Chan KM. Tricuspid leaflet augmentation to address severe tethering in functional tricuspid regurgitation. *Eur J Cardiothorac Surg.* 2008;34:908–910.
47. Pettinari M, Bertrand P, Van Kerrebroeck C, Vandervoort P, Gutermann H, Dion R. Mid-term results of leaflet augmentation in severe tricuspid functional tethering. *Eur J Cardiothorac Surg.* 2016;50:504–508.
48. Rescigno G, Hothi S, Bond C, Uddin M, Bhatti V, Billing JS. CorMatrix anterior leaflet augmentation of the tricuspid valve: midterm results. *Heart Surg Forum.* 2021;24:E261–E266.
49. Matsumiya G, Kohno H, Matsuura K, et al. Right ventricular papillary muscle approximation for functional tricuspid regurgitation associated with severe leaflet tethering. *Interact Cardiovasc Thorac Surg.* 2018;26:700–702.
50. Al-Attar N, Hvass U. Right papillary muscle sling: proof of concept and pilot clinical experience. *Eur J Cardiothorac Surg.* 2013;43: e187–e189.
51. Ouda A, Matschke K, Ghazy T, et al. Right ventricular reduction for repair of functional tricuspid valve regurgitation: one-year follow up. *J Heart Valve Dis.* 2013;22:754–761.
52. Lohchab SS, Chahal AK, Agrawal N. Papillary muscle approximation to septum for functional tricuspid regurgitation. *Asian Cardiovasc Thorac Ann.* 2015;23:747–750.
53. Kaplan M, Kut MS, Demirtas MM, Cimen S, Ozler A. Prosthetic replacement of tricuspid valve: bioprosthetic or mechanical. *Ann Thorac Surg.* 2002;73:467–473.
54. Scully HE, Armstrong CS. Tricuspid valve replacement. Fifteen years of experience with mechanical prostheses and bioprostheses. *J Thorac Cardiovasc Surg.* 1995;109:1035–1041.
55. Cohen SR, Silver MA, McIntosh CL, Roberts WC. Comparison of late (62 to 140 months) degenerative changes in simultaneously implanted and explanted porcine (Hancock) bioprostheses in the tricuspid and mitral valve positions in six patients. *Am J Cardiol.* 1984;53:1599–1602.
56. Negm S, Arafat AA, Elatafy EE, Fawzy HF. Mechanical vs. bioprosthetic valve replacement in the tricuspid valve position: a systematic review and meta-analysis. *Heart Lung Circ.* 2021;30:362–371.
57. Russo M, Di Mauro M, Saitto G, et al. Beating versus arrested heart isolated tricuspid valve surgery: long-term outcomes. *Ann Thorac Surg.* 2021;. Available from: https://doi.org/10.1016/j.athoracsur.2021.03.070.
58. Flagiello M, Grinberg D, Connock M, et al. Beating versus arrested heart isolated tricuspid valve surgery: an 11-year experience in the current era. *J Card Surg.* 2021;36:1020–1027.
59. Santaló-Corcoy M, Asmarats L, Li C-H, Arzamendi D. Catheter-based treatment of tricuspid regurgitation: state of the art. *Ann Transl Med.* 2020;8:964.
60. Taramasso M, Alessandrini H, Latib A, et al. Outcomes after current transcatheter tricuspid valve intervention: mid-term results from the International TriValve Registry. *JACC Cardiovasc Interv.* 2019;12:155–165.
61 Taramasso M, Vanermen H, Maisano F, Guidotti A, La Canna G, Alfieri O. The growing clinical importance of secondary tricuspid regurgitation. *J Am Coll Cardiol.* 2012;59:703–710.
62. Lauten A, Ferrari M, Hekmat K, et al. Heterotopic transcatheter tricuspid valve implantation: first-in-man application of a novel approach to tricuspid regurgitation. *Eur Heart J.* 2011;32:1207–1213.
63. O'Neill BP, Negrotto S, Yu D, et al. Caval valve implantation for tricuspid regurgitation: insights from the United States Caval Valve Registry. *J Invasive Cardiol.* 2020;32:470–475.
64. Rodés-Cabau J, Hahn RT, Latib A, et al. Transcatheter therapies for treating tricuspid regurgitation. *J Am Coll Cardiol.* 2016;67:1829–1845. Available from: https://doi.org/10.1016/j.jacc.2016.01.063.
65. Nickenig G, Weber M, Schueler R, et al. 6-Month outcomes of tricuspid valve reconstruction for patients with severe tricuspid regurgitation. *J Am Coll Cardiol.* 2019;73:1905–1915.
66. Asmarats L, Puri R, Latib A, Navia JL, Rodés-Cabau J. Transcatheter tricuspid valve interventions: landscape, challenges, and future directions. *J Am Coll Cardiol.* 2018;71:2935–2956.
67. Besler C, Meduri CU, Lurz P. Transcatheter treatment of functional tricuspid regurgitation using the trialign device. *Interv Cardiol Rev.* 2018;13:8.
68. Hahn RT, Meduri CU, Davidson CJ, et al. Early feasibility study of a transcatheter tricuspid valve annuloplasty: SCOUT trial 30-day results. *J Am Coll Cardiol.* 2017;69:1795–1806.
69. Rosser BA, Taramasso M, Maisano F. Transcatheter interventions for tricuspid regurgitation: TriCinch (4Tech). *EuroIntervention.* 2016;12: Y110–Y112.

70. Update on the current landscape of transcatheter options for tricuspid regurgitation treatment, (n.d.). https://doi.org/10.15420/icr.2019.5.1. Accessed June 7, 2021.
71. Muntané-Carol G, Alperi A, Faroux L, Bédard E, Philippon F, Rodés-Cabau J. Transcatheter tricuspid valve intervention: coaptation devices. *Front Cardiovasc Med.* 2020;7. Available from: https://doi.org/10.3389/fcvm.2020.00139.
72. Nickenig G, Weber M, Lurz P, et al. Transcatheter edge-to-edge repair for reduction of tricuspid regurgitation: 6-month outcomes of the TRILUMINATE single-arm study. *Lancet.* 2019;394:2002–2011. Available from: https://doi.org/10.1016/s0140-6736(19)32600-5.
73. Mehr M, Taramasso M, Besler C, et al. 1-Year outcomes after edge-to-edge valve repair for symptomatic tricuspid regurgitation: results from the TriValve Registry. *JACC Cardiovasc Interv.* 2019;12:1451–1461.
74 Singh S. MitraClip in the tricuspid position: slowly, but surely, the march continues. *JACC Cardiovasc Interv.* 2019;12:1462–1464.
75. Braun D, Orban M, Orban M, et al. Transcatheter edge-to-edge repair for severe tricuspid regurgitation using the triple-orifice technique versus the bicuspidalization technique. *JACC Cardiovasc Interv.* 2018;11:1790–1792.
76. Fam NP, von Bardeleben RS, Hensey M, et al. Transfemoral transcatheter tricuspid valve replacement with the EVOQUE system: a multicenter, observational, first-in-human experience. *JACC Cardiovasc Interv.* 2021;14:501–511.
77. Asmarats L, Dagenais F, Bédard E, et al. Transcatheter tricuspid valve replacement for treating severe tricuspid regurgitation: initial experience with the NaviGate bioprosthesis. *Can J Cardiol.* 2018;34:1370.e5–1370.e7.
78. Hahn RT, George I, Kodali SK, et al. Early single-site experience with transcatheter tricuspid valve replacement. *JACC: Cardiovasc Imaging.* 2019;12:416–429. Available from: https://doi.org/10.1016/j.jcmg.2018.08.034.
79. Navia JL, Kapadia S, Elgharably H, et al. First-in-human implantations of the NaviGate bioprosthesis in a severely dilated tricuspid annulus and in a failed tricuspid annuloplasty ring. *Circulation: Cardiovasc Interv.* 2017;10. Available from: https://doi.org/10.1161/circinterventions.117.005840.
80. Elgharably H, Harb SC, Kapadia S, Svensson LG, Navia JL. Transcatheter innovations in tricuspid regurgitation: navigate. *Prog Cardiovasc Dis.* 2019;62:493–495.
81. Meador WD, Mathur M, Sugerman GP, et al. The tricuspid valve also maladapts as shown in sheep with biventricular heart failure. *Elife.* 2020;9. Available from: https://doi.org/10.7554/eLife.63855.
82. Timek TA. Mitral leaflet remodeling in dilated cardiomyopathy. *Circulation.* 2006;114:I–518. Available from: https://doi.org/10.1161/circulationaha.105.000554.
83. Stephens EH, Timek TA, Daughters GT, et al. Significant changes in mitral valve leaflet matrix composition and turnover with tachycardia-induced cardiomyopathy. *Circulation.* 2009;120:S112–S119. Available from: https://doi.org/10.1161/circulationaha.108.844159.
84. Grande-Allen KJ, Jane Grande-Allen K, Edward Barber J, et al. Mitral valve stiffening in end-stage heart failure: evidence of an organic contribution to functional mitral regurgitation. *J Thorac Cardiovasc Surg.* 2005;130:783–790. Available from: https://doi.org/10.1016/j.jtcvs.2005.04.019.
85. Afilalo J, Grapsa J, Nihoyannopoulos P, et al. Leaflet area as a determinant of tricuspid regurgitation severity in patients with pulmonary hypertension. *Circ Cardiovasc Imaging.* 2015;8. Available from: https://doi.org/10.1161/CIRCIMAGING.114.002714.
86. Bartko PE, Dal-Bianco JP, Guerrero JL, et al. Leducq transatlantic mitral network, effect of losartan on mitral valve changes after myocardial infarction. *J Am Coll Cardiol.* 2017;70:1232–1244.

Chapter 21

Innovative Surgical Techniques for Complex Aortic Repair

Ruggero De Paulis[1,2], Raffaele Scaffa[1], Giulio Folino[1], Luca Weltert[1,2] and Andrea Salica[1]
[1]*Cardiac Surgery Department, European Hospital, Rome, Italy,* [2]*Unicamillus University Rome, Italy*

Chapter Outline

Simplicity is the ultimate sophistication.

Leonardo Da Vinci

Introduction

In the last 20 years, the aortic root has received increasing attention in an effort to preserve the native aortic valve. The complex interaction between the aortic valve leaflets and the skeleton of the aortic root—annulus and sinotubular junction (ST (Sino-Tubular) junction)—is the basis for a normally functioning aortic valve.

Computed tomography, cardiac magnetic resonance imaging, and three-dimensional echocardiography have concurred to illustrate anatomical details of the aortic root with a quality unimaginable until a few years ago. The potential of this multimodal imaging opens new boundless horizons toward an increasingly patient-tailored surgical planning of the aortic repair.

Even though the aortic valve repair represents one of the most important advances in cardiac surgery, its complexity and relatively long learning curve have prevented a more rapid diffusion of the techniques, and have slowed its adoption as a routine surgery.

To attempt aortic valve repair, insights in anatomy and pathophysiology of the aortic root are paramount. To achieve a durable repair, every component of the aortic root must be recognized and addressed, particularly the annulus, ST junction, and valve leaflets. To simplify this complex surgery, standardized techniques and new devices are currently used worldwide among aortic surgical teams.

Aortic annuloplasty (techniques and devices)

Accurate measurement of the aortic annulus is important when approaching an aortic valve repair. Many clinical studies have shown that the dimension of the annulus is one of the most important parameters for recurrence of aortic regurgitation (AR) and for valve reoperation at mid- and long-term follow-up.[1,2] A properly reduced and supported aortic annulus is important in obtaining and maintaining a large leaflet coaptation and preventing cusp stress.[3–5]

Even though the annuloplasty was considered essential in aortic valve repair, the most appropriate technique is still highly controversial. Ideal aortic annuloplasty should be easy to perform without interference with the coronary arteries, conduction system, and leaflets mobility, while ensuring good valve function and long-term stability of annular diameter reduction. Different annuloplasty techniques and devices are available and are based, substantially, on an external or

Advances in Cardiovascular Technology. DOI: https://doi.org/10.1016/B978-0-12-816861-5.00014-9

internal approach. In the great majority of cases of a regurgitant tricuspid valve configuration, the main cause is the whole dilatation of the aortic root (i.e., root aneurysm), often linked to connective tissue disorders or to a long-standing hypertensive condition, among other, rarer, factors. On the other hand, when considering a bicuspid configuration, the valve regurgitation can occur not only in association with a root aneurysm, but also in the presence of a root of normal diameter.

Among the aortic valve-sparing (AVS) procedures, the reimplantation procedure described by David[6] does provide an annuloplasty with its proximal suture line and reshapes the annulus by enclosing it into a Dacron graft placed "externally" to the entire aortic root skeleton. After removing the Valsalva's sinuses, a series of sutures is placed under the annulus at the same horizontal level, passing for the nadirs of each cusp without modifying the valve geometry. Then, these sutures are passed through the base of the Dacron prosthesis and tied to secure annulus inclusion and reduction.

This procedure is also fundamental in reestablishing the proper proportion between the annulus and the ST junction, and therefore a good geometrical condition for the aortic valve. Although reimplantation can be performed using completely straight tubular grafts, the use of a specifically designed grafts[7,8] with the ability to reproduce the normal shape and function of sinuses of Valsalva have allowed an optimal reproduction of normal aortic root anatomy, and consequently, improved physiologic results. The Valsalva graft (Terumo Aortic Renfrewshire, UK) has been available since the year 2000, and is the most widely used graft for this type of surgery (Fig. 21.1). In any case, the reimplantation procedure, with or without its sophisticated modifications to reproduce the sinuses of Valsalva, allows aortic root replacement and aortic annular stabilization at the same time, with excellent long-term results.[9–12]

In the original remodeling or Yacoub procedure,[13] the portion of the Dacron graft is tailored in the form of three tongues (by excising three "triangles" to match the interleaflet triangles) and trimmed to match the shape of the sinuses. Unfortunately, when the pathology affects the annulus (the cutoff value varies, but it seems reasonable to consider an annular dimension >25 mm), the remodeling alone fails to restore the normal annulus diameter and to prevent future dilatation.[1] Theoretically, in AVS procedures, patients with a large preoperative aortic annulus diameter are best treated with the reimplantation technique, whereas those with a smaller diameter can also be treated with the remodeling technique. However, the procedure selection should be based upon surgeon preference and experience and patient factors, without consideration of the other two AVS procedures.[14]

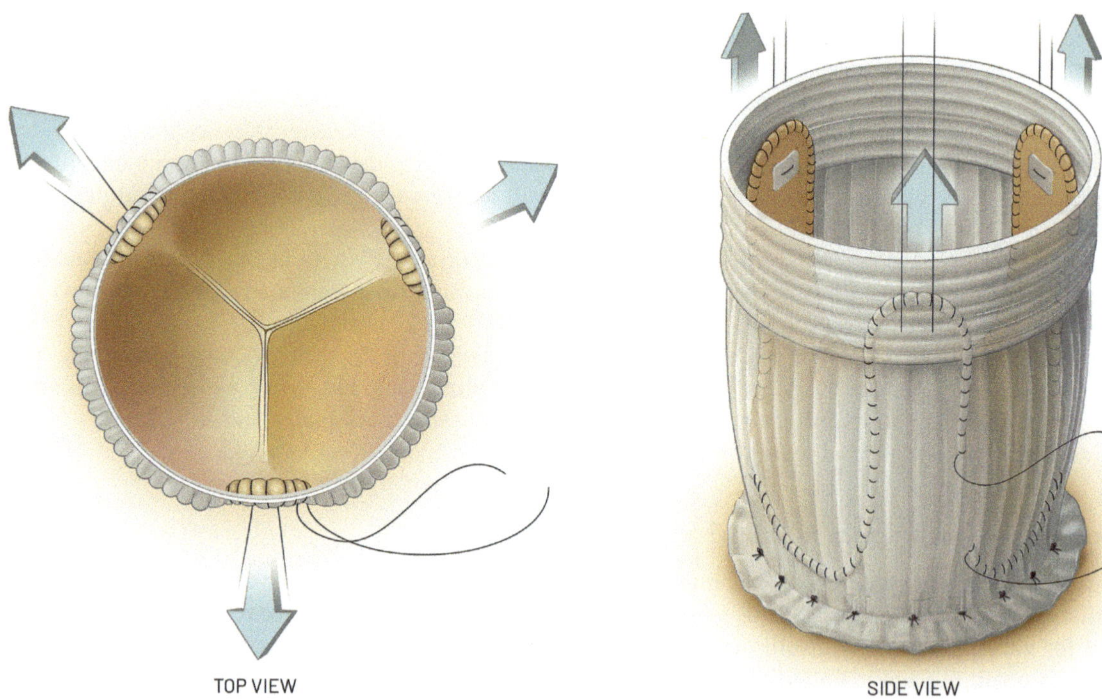

FIGURE 21.1 Schematic view of the Valsalva graft (Terumo Aortic Renfrewshire, UK) in a reimplantation procedure. The skeleton of the aortic root is reconstructed, enclosing it into the "external" Valsalva graft. This graft provides the trilobate conformation similar to the native sinuses of Valsalva. *Published with permission from Scaffa R, Weltert L, Salica A, De Paulis R. The skirt and collar of the valsalva graft: one dress fits all.* Surg Technol Int. *2021;38:sti38/1390.*

In any case, to overcome the drawback of the remodeling technique, and to address the issue of annular dilatation, several types of annuloplasty have been devised. The first technique utilized to reduce the aortic annulus is the "subcommissural annuloplasty" (or Cabrol stitch, for the surgeon who first described it) that was used to decrease the annulus diameter by virtually abolishing the interleaflet triangle by means of three U-stitches supported by pledgets. However, in practice, this technique was unable to prevent annular redilatation.[15] Furthermore, it alters the normal anatomy and physiology of the interleaflet triangles. In fact, a subcommissural annuloplasty eliminates or dampens the fundamental role of the interleaflet triangles (i.e., the ability to transmit the ventricular pressure up to the top of the commissures).

The desire to find a better solution was interpreted in two ways. One proposal was an external approach[16] by means of a flexible ring (Fig. 21.2) positioned around the annulus and fixed by a series of subvalvular sutures (in a manner similar to that used for the reimplantation technique). This ring (Extra-Aortic, Coroneo Inc., QC, Canada) has been commercially available since 2007 and was first implanted in a prospective, multicentric clinical trial[17] in Europe. For the implant, a deep external dissection like the one needed for the reimplantation procedure is an important prerequisite. However, despite a complete dissection of the root to fix a ring or a Dacron conduit to reduce and stabilize the aortic annulus, a slight asymmetry of the ring proximal seating could still exist.

Anatomical studies have pointed out that, in the region from the left/right commissure to the right/noncoronary commissure, it is more difficult to reach the annulus from outside without incorporating a thick portion of these muscular heart structures (Fig. 21.3).[18] This variability might influence the choice of the ring or conduit size and might prevent a correct positioning of the ring that should ideally be positioned for a plane passing through the nadir of the three cusps. However, our recent study showed that annular stabilization in the setting of the reimplantation procedure is unaffected, even though the Dacron graft is positioned slightly higher than the aortic annulus virtual basal ring at the level of the right sinus and left/right commissure.[19] In the experience of Lansac and colleagues, who have standardized the approach to the surgical management of aortic root aneurysms by exclusively relying on the remodeling technique, the introduction of a calibrated expansile annuloplasty ring has improved the results of their repairs. The long-term outcomes were similar for patients with bicuspid aortic valve (BAV) and tricuspid aortic valve (TAV) repair.[20] In isolated AR with a root of nonsurgical diameter, external aortic ring annuloplasty has proven equally effective for aortic valve repair, with satisfactory long-term results.[21]

A different external annuloplasty technique was introduced in 2009 by Schäfers and applied during aortic root replacement or isolate aortic valve repair. Whenever the aortic annulus exceeds 26–27 mm (usually measured by a

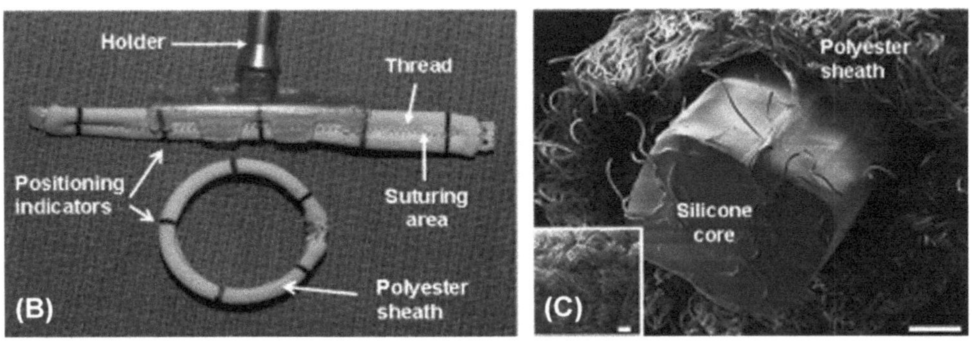

FIGURE 21.2 Extraaortic ring (Coroneo Inc., QC, Canada) before implantation. The ring is available in five different sizes (23–31 mm) for calibrated annuloplasty. (A) Cut-away schematic diagram of the device composed of two silicone elastomer cores covered with polyester textile components. (B) (top): Before use, the ring is opened and mounted on a holder that is attached to the handle. Black indicator marks on the ring and suture windows in the ring holder are incorporated to facilitate optimal suture placement between the two cores. (B) (bottom): After implantation, the ends of the ring are fastened. (C) Scanning electronic microscope image showing a cross-section of the ring with its rectangular silicone core covered by a polyester sheath (insert). Scale bars = 200 μm. *Published with permission from Lansac E, Di Centa I, Raoux F, et al. An expansible aortic ring for a physiological approach to conservative aortic valve surgery.* J Thorac Cardiovasc Surg. *2009;138(3):718–724.*

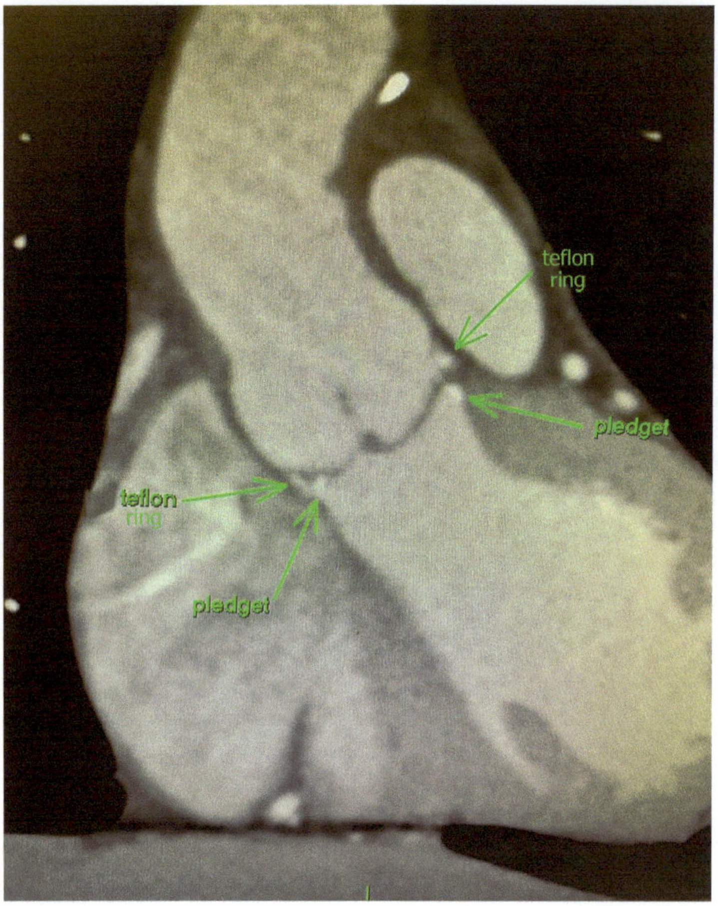

FIGURE 21.3 CT scan after an aortic valve repair for isolated AR. There are clearly visible pledgets positioned at the nadir of the cusps and at the external annuloplasty with a Teflon strip. It is evident that in correspondence of the noncoronary cusp, the Teflon and the pledgets are on the same horizontal level. On the other hand, there is an increased distance and a slight asymmetry in correspondence of the right cusp. Nonetheless, the overall geometry of the root is close to normal.

Hegar dilator or similar), a circumferential suture annuloplasty using polytetrafluoroethylene Gore-Tex 0 is added.[22] This suture is placed from outside the aortic wall, under the coronaries, at the level of the nadirs of the aortic leaflets in a circumferential fashion and tightened around a Hegar dilator of an appropriate size (frequently 23–25 mm). This type of annuloplasty requires limited root dissection and less time, and is associated with satisfactory results and freedom from reoperation.[23] However, the potential risks of suture dehiscence, along with a less accurate annular remodeling, are common arguments against its wider applicability.

The "internal" approach includes the use of strip of Dacron[24] or a rigid Dacron-covered titanium ring (Fig. 21.4). The latter is an internal geometric annuloplasty ring (HAART, BioStable Science and Engineering, Austin, TX, USA), proposed by Rankin in 2011, that restores and reshapes the aortic annulus, potentially facilitating the leaflet plasty. The HAART ring was developed from scans made of healthy humans, translating the consistent geometric pattern into two different types of devices that are implanted below the aortic valve. The HAART rings are available as HAART 300 (with an elliptical shape) for TAV repair, and HAART 200 (with a circular shape) for BAV. The different annular shapes in BAV and TAV were also supported by our recent clinical study that showed a progressive increase in the annular ellipticity when comparing various phenotypes of BAV to a standard tricuspid valve.

Moreover, the study also verified a linear correlation between the commissural orientation and the annular shape along the spectrum of BAVs.[25] A 2-year clinical trial demonstrated that HAART devices for annuloplasty seemed safe and effective and could help standardize aortic valve repair procedures.[26,27] While proponents of the internal rigid rings claim a consistent and predictable reshaping of the annulus to facilitate leaflet plasty and to create a more durable condition, there are still concerns regarding the potential inflammatory reaction, the risk of leaflet restriction, contact and abrasion, and hemolysis or thromboembolism.

As stated before, in the case of isolated AR without aortic root dilation, the annuloplasty is still needed to stabilize the annulus. This is more frequently the case with BAVs, and significantly less frequent in TAVs. Considering the good long-term results and the experience with reimplantation technique, our group prefers in cases of isolate aortic valve

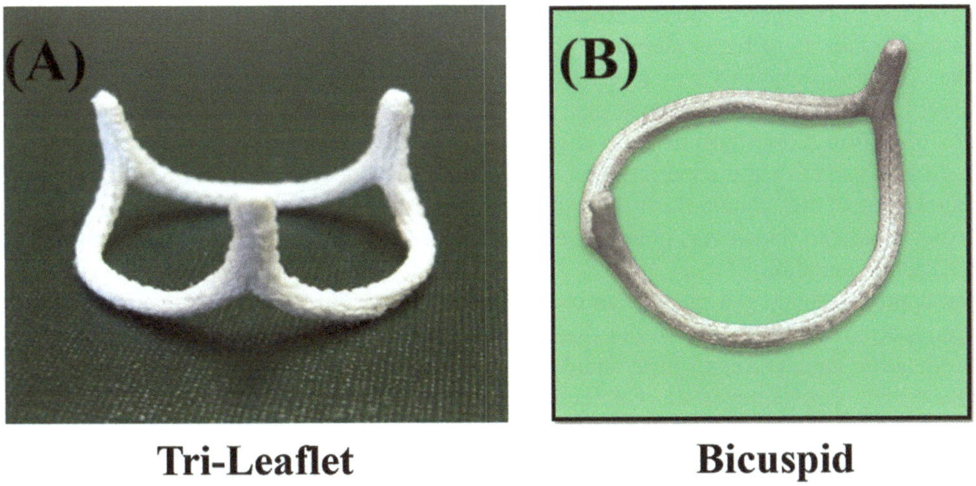

FIGURE 21.4 (A) The aortic tricuspid ring (HAART 300), machined from a single block of titanium and covered with a thin layer of Dacron to promote endothelialization. The shape reflects normal elliptical geometry (2:3 minor-to-major axis diameter ratio). (B) The aortic bicuspid ring (HAART 200) has a circular shape with the two commissures oriented at 180 degrees. The construction of the HAART rings is based on mathematical calculations obtained by analyzing healthy aortic valves with contrast-enhanced computed tomography studies. *Published with permission from Fattouch K, Castrovinci S, Murana G, et al. Functional annulus remodelling using a prosthetic ring in tricuspid aortic valve repair: mid-term results.* Interact Cardiovasc Thorac Surg. *2014;18(1):49–54; discussion 54-5.*[24]

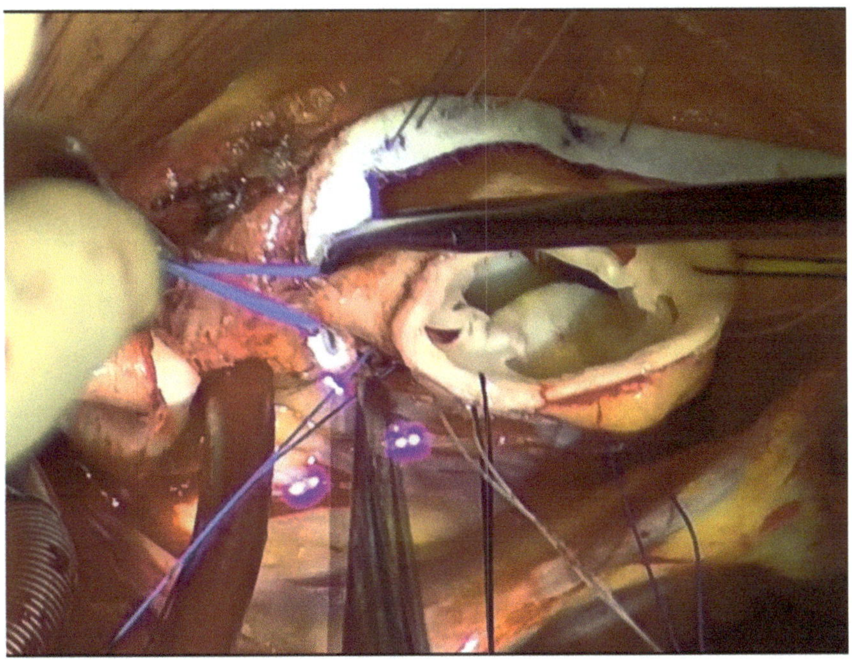

FIGURE 21.5 Intraoperative view of the external ring annuloplasty, with a Teflon strip placed circumferentially below the coronary arteries around the aortic annulus.

repair (usually BAV) to reduce and stabilize the aortic annulus in the external fashion with the placement of a Teflon or Dacron strip.

Briefly, after an extensive root dissection, a series (usually between 6 and 9) of pledgeted sutures is placed circumferentially and horizontally from inside out at the level of 1–2 mm below the annulus. Then the external ring is made by using a 5–7-mm large Teflon strip with a length tailored to reduce the annulus to a diameter between 21 and 23 mm, tightening the suture on Hegar's dilator to reestablish a 1-to-1.3 ratio with the ST junction diameter. This approach seems to ensure an efficient aortic annuloplasty for BAV with anatomical and functional results similar to those of the reimplantation procedure (Fig. 21.5).[28] The main difference is that the two coronary arteries don't need to be excised and reimplanted. There is an ongoing debate as to whether a root replacement is also advisable in patients with regurgitant BAV but who also have a normal or slightly enlarged aortic root. The reasons to be more aggressive and replace the root are based on the possibility of better stabilizing the whole root from the annulus to the ST junction, and possibly reorienting the valve commissure toward a more efficient 180 degrees. The reasons to be more conservative are mostly based on equivalent mid-term results, evidence that root diameter is unlikely to increase in the follow-up, and simplification of the index operation. Long-term data are needed to solve this controversy.

We lack sufficient scientific data to establish the superiority of one technique or device over another. An internal approach might guarantee the placement of the annuloplasty at a proper level and a more accurate reconstruction of annular geometry when a rigid ring is utilized. However, internal devices or internally placed sutures work against the centrifugal forces that tend to redilate the annulus.

On the other hand, an external approach might not provide an optimal annular reshaping but might guarantee a better annular stabilization. In a recent porcine model study on the external approach for annuloplasty, a Dacron ring was found to better reproduce the native aortic annulus than suture annuloplasty, offering more physiological support.[29] The optimal material for an annuloplasty ring is still uncertain. Choice of material and future improvements of dedicated material for the annuloplasty procedures remain a challenge for future studies.

Sinotubular junction plasty (techniques and devices)

The ST junction, with its specific relationship with the other components of the aortic root, plays a fundamental role in the function of the aortic valve.[30] ST junction dilatation moves the commissures away from each other, reducing the coaptation length of the aortic leaflets. In normal healthy hearts, echocardiography shows that the diameter of the ST junction is about 75% of the maximal sinus diameter, which is larger than the aortic annulus, with a mean normal ratio of 1.3.[31] A mismatch in the ratio between the ST junction and the annulus (mostly 1.6 or more) is a frequent cause of secondary AR with normal aortic cusps. However, in a finite element study, dilatation of ST junction was shown to slightly affect leaflet coaptation, with less damage on the leaflet free margins than annular dilatation.[5]

In proximal aortic pathology, isolated dilation of the ST junction is often encountered. Simply replacing the ascending aorta with a Dacron conduit of normal diameter can reshape the sinotubular junction and correct the secondary AR. In some rare cases, the mismatch between the annulus and the ST junction should be corrected even when the aortic dimensions are below the criteria for replacement. Some authors suggest that severe AR may be improved by stabilizing both the aortic annulus and the ST junction using a double annuloplasty ring.[21,32] However, in such cases, a complete root replacement using a reimplantation procedure can probably address both points in a more definite and standardized fashion.

Leaflets plasty (the caliper)

Cusp prolapse (type 2 of the El Khoury classification) is the main mechanism of valve dysfunction in isolated AR[30] and is common in the congenitally fused cusp of BAV. Cusp prolapse causing AR is associated with unique echocardiographic, clinical, and surgical features. Cusp prolapse is defined as a free margin level that is lower than the adjacent level, or an "effective height" that is below the normal value. Similarly, the coaptation length, a measure of the amount of coaptation between two leaflets, is reduced or absent. In addition, cusp prolapse can be the main cause of AR that requires surgery, but can also be induced by the simple reduction of the root diameter after an AVS procedure. In some cases, it can also be induced by imperfect geometric orientation and positioning of the commissures.

Correcting a prolapse and reestablishing a normal effective height of the leaflets, as well as a correct coaptation length (Fig. 21.6), is probably the most critical point for the long-term success of aortic valve repair. Failing to correct a prolapse, whether preexisting or induced after an AVS procedure, will jeopardize the whole procedure and potentially subject the patient to further surgery within a few years. Initially, cusp prolapse was visually assessed intraoperatively by comparison of the cusp-free margins. Differently from the mitral valve, where it is possible to pressurize the ventricle and observe the valve repair from the blood inflow, an aortic valve repair is difficult to assess intraoperatively. Even though it is theoretically possible to pressurize the aortic root, it is virtually impossible to assess the valve competence by direct observation from its inflow. Various attempts with the use of angioscopy have been attempted, without convincing evidence of success.[33]

Since 2004, aortic valve assessment has been revolutionized by direct measurement of specific anatomical and geometrical parameters—namely the geometric height (gH) and the effective height (eH). The gH, also called the cusp height, is defined as the distance between the cusp nadir and the middle of the free margin. Intraoperatively, the gH is measured with a straight ruler along the aortic side of the cusp by applying gentle traction on the free margin to

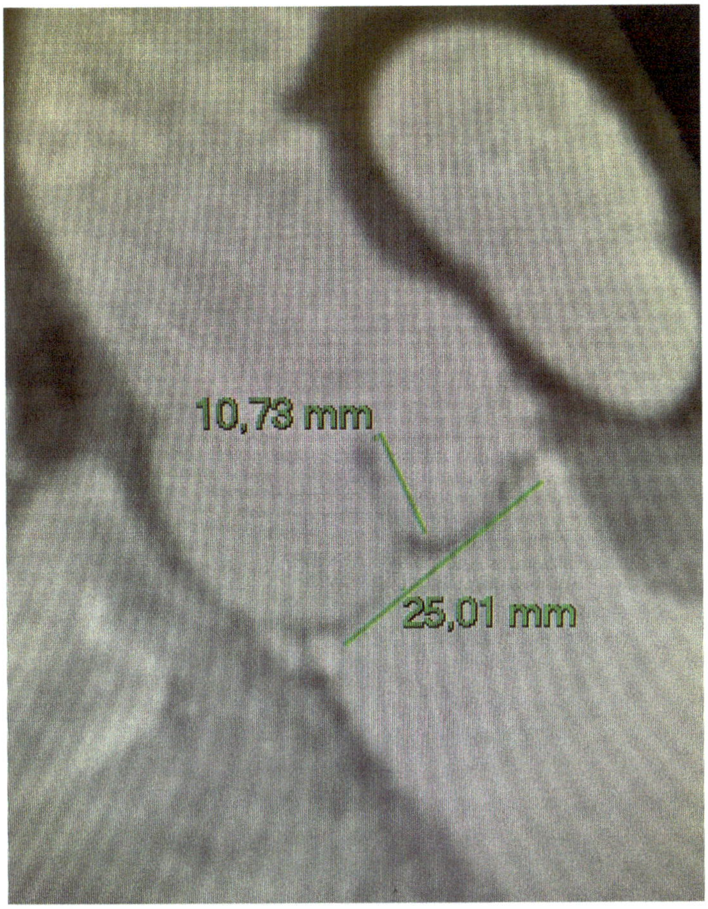

FIGURE 21.6 A good coaptation length and a normal annular diameter are good prognostic factors for the long-term durability of the repair.

straighten the cusp tissue along the ruler. In adults, the cusp is considered retracted when the gH is 16 mm or less in tri-leaflet aortic valves and 19 mm or less in the bicuspid nonfused aortic cusp.[34,35] This parameter is important for the surgeon in assessing the feasibility of the repair. In fact, an insufficient gH indicates the lack of leaflet tissue available for repair and the potential need for pericardial patches. However, the addition of pericardial patch has been found to be a negative prognostic factor for the longevity of the repair. This is an example of how a single anatomical parameter can indicate to the surgeon not only the increased difficulty of the valve repair, but can also serve as an important prognostic indicator.

The eH is the orthogonal distance from the annulus to the middle of the free margin of the cusp, and is an important quantitative parameter of cusp configuration; its normal value is 7–13 mm, or 45% of the gH in the TAV.[4,36] The eH can be determined by echocardiography, as well as intraoperatively. Schäfers has designed a caliper (MSS-1, Fehling Instruments, Karlstein, Germany) that facilitates intraoperative measurement of the eH (Fig. 21.7). To standardize the assessment of the leaflet's prolapse and cusp configuration, the use of this caliper is recommended using an eH of 9–10 mm as a reference. This is important, since precisely assessing effective cusp height is a prerequisite for avoiding small eccentric jets that invariably lead to progressive increase of valve regurgitation and aortic valve repair failure.[37]

The durability of the cusp repair depends mainly on two factors: (1) quality of the valvular tissues, and (2) the technique used for repair. The repair techniques for cusp prolapse are central plication, triangular resection, or free margin shortening by Gore-Tex suture. Central plication appears to be the easiest and most effective technique, although Schäfers et al. reported comparable midterm results among the techniques, with central plication providing slightly greater freedom from AR.[38] The plication of the central portion of the free margin is performed with single 5/0 or 6/0 Prolene sutures; depending on the extent of prolapse, more stitches are placed until an adequate cusp configuration is achieved. In the presence of fibrosis of the free margin (or raphe in case of a BAV), a gentle "shaving" of the cusps is performed to ensure a greater pliability and a normal movement of the leaflets. Use of a pericardial patch is related to suboptimal durability in cases of cusp augmentation, central cusp replacement, and commissural reconstruction.[39]

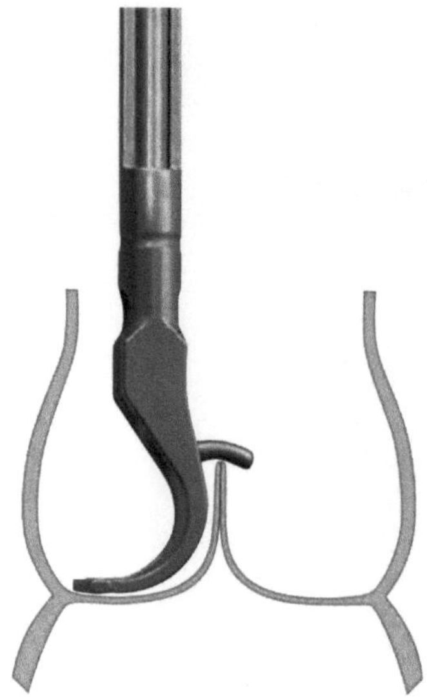

FIGURE 21.7 Effective height is measured by a special caliper (Fehling Instruments, Karlstein, Germany). The base of the caliper is placed at the aortic insertion in the nadir of the sinus, the horizontal bar on the central-free margin. In order to obtain reproducible measurements, it is important to have sufficient and equal tension on the commissural stay sutures and keep the axis of the caliper in line with the axis of the left ventricular outflow tract. *Published with permission from Schneider U, Schäfers HJ. Repair of the bicuspid aortic valve.* Oper Tech Thorac Cardiovasc Surg. *22:91–109 © 2018 Elsevier Inc.*

In the context of reconstructed root with a new reestablished geometry (e.g., remodeling or reimplantantion), the use of a caliper is an important step for intraoperative evaluation of leaflet geometry (often resulting in suboptimal coaptation and/or induced prolapse). The direct and immediate effect of aortic root reconstruction is the reduction of the whole root diameter. This will lead to an excess of tissue in the leaflet-free margin that would consequently tend to prolapse in most of the cases.

As a general rule in a TAV configuration, the right cusp is the one most likely to prolapse, while the fused cusp is almost invariably affected in the BAV configuration. Indeed, with growing experience, we have learned that AVS procedures and leaflet repair are intimately linked, and they always need to be part of the same procedure. Before the introduction of the caliper, small degrees of leaflet sagging or prolapsing after the AVS procedure were barely recognized by the naked eye and were left untreated. Even when the valve looks normal after the resuspension in a Dacron graft and does not present residual regurgitation on echocardiography, it is very unlikely that all the leaflets are at exactly the same level. A competent aortic valve indicates that the coaptation length (the apposition of the leaflets) is sufficient, but fails to indicate that the eH is adequate for each individual leaflet. This stresses once again the importance of measuring the eH for each cusp at the end of the repair. AVS surgery associated with aortic cusp repair has certainly expanded the use of AVS procedures in patients with complex aortic valve regurgitation.

Conclusion

At the start of the 21st century, most patients with AR who were undergoing cardiac surgery received aortic valve replacement rather than aortic valve repair. With the gathering of data from leading centers, aortic valve repair is evolving from its initial experimental stage to a common and well-established procedure. Founded in 2013, the AVIATOR registry, which includes patients with AR and/or aortic aneurysm, analyzes the results of aortic repair vs. replacement and provides a solid "benchmark" to evaluate the role of aortic repair based on long-term patient outcomes.[40]

Today we know that the techniques of aortic valve repair give satisfactory results. We are also learning that some phenotypes of BAV (e.g., 2-sinus BAV, straight commissural orientation) are easy to repair and guarantee better long-term durability.[41] We are also learning that root replacement might also be an option in some cases to facilitate aortic valve repair.

Aortic valve repair remains a complex but increasingly codified surgical strategy. The "modern" surgeon must learn to consider the aortic root as a complex unit, with the aortic cusp one component of this complex mechanism. The importance of annuloplasty in all cases of aortic valve repair, even in the absence of a dilated root, is now clear. Annuloplasty reconstructs geometry, restores the eH, and increases the coaptation length. Essentially, the goals for aortic valve repair (less or more complex) concern the treatment of the dilated diameters (especially annulus and ST junction) combined, if possible, to the preservation of the root dynamics and the restoration of the eH of the cusps.

The currently available techniques and devices guarantee a tailored approach for various anatomical situations, making aortic repair available to a larger number of patients with more advanced or complex pathologies of the aortic root.

References

1. Hanke T, Charitos EI, Stierle U, et al. Factors associated with the development of aortic valve regurgitation over time after two different techniques of valve-sparing aortic root surgery. *J Thorac Cardiovasc Surg*. 2009;137(2):314–319.
2. de Kerchove L, Mastrobuoni S, Boodhwani M, et al. The role of annular dimension and annuloplasty in tricuspid aortic valve repair. *Eur J Cardiothorac Surg*. 2016;49(2):428–437. discussion 437-8.
3. Marom G, Haj-Ali R, Rosenfeld M, Schäfers HJ, Raanani E. Aortic root numeric model: annulus diameter prediction of effective height and coaptation in post-aortic valve repair. *J Thorac Cardiovasc Surg*. 2013;145(2):406–411.
4. Bierbach BO, Aicher D, Issa OA, et al. Aortic root and cusp configuration determine aortic valve function. *Eur J Cardiothorac Surg*. 2010;38(4):400–406.
5. Weltert L, de Tullio MD, Afferrante L, et al. Annular dilatation and loss of sino-tubular junction in aneurysmatic aorta: implications on leaflet quality at the time of surgery. A finite element study. *Interact Cardiovasc Thorac Surg*. 2013;17(1):8–12.
6. David TE, Feindel CM. An aortic valve-sparing operation for patients with aortic incompetence and aneurysm of the ascending aorta. *J Thorac Cardiovasc Surg*. 1992;103:617–621.
7. De Paulis R, De Matteis GM, Nardi P, Scaffa R, Colella DF, Chiarello L. A new aortic Dacron conduit for surgical treatment of aortic root pathology. *Ital Heart J*. 2000;1(7):457–463.
8. Tasca G, Lindner J, Barandon L, et al. Aortic root surgery with the CARDIOROOT vascular graft: results of a prospective multicenter post-market surveillance study. *J Cardiothorac Surg*. 2019;14(1):94.
9. De Paulis R, Chirichilli I, Scaffa R, et al. Long-term results of the valve reimplantation technique using a graft with sinuses. *J Thorac Cardiovasc Surg*. 2016;151(1):112–119.
10. David TE, Feindel CM, David CM, Manlhiot C. A quarter of century of experience with aortic valve-sparing operations. *J Thorac Cardiovasc Surg*. 2014;148(3):872–879.
11. Mastrobuoni S, de Kerchove L, Navarra E, et al. Long-term experience with valve-sparing reimplantation technique for the treatment of aortic aneurysm and aortic regurgitation. *J Thorac Cardiovasc Surg*. 2019;158(1):14–23.
12. Shrestha ML, Beckmann E, Abd Alhadi F, et al. Elective David I procedure has excellent long-term results: 20-year single-center experience. *Ann Thorac Surg*. 2018;105(3):731–738.
13. Sarsam MA, Yacoub M. Remodeling of the aortic valve annulus. *J Thorac Cardiovasc Surg*. 1993;105:435–438.
14. Maskell P, Brimfield M, Ahmed A, Harky A. In patients undergoing valve-sparing aortic root replacement, is reimplantation superior to remodelling? *Interact Cardiovasc Thorac Surg*. 2021;32(3):441–446.
15. de Kerchove L, Vismara R, Mangini A, et al. In vitro comparison of three techniques for ventriculo-aortic junction annuloplasty. *Eur J Cardiothorac Surg*. 2012;41(5):1117–1123. discussion 1123-4.
16. Youssefi P, Di Centa I, Khelil N, Debauchez M, Lansac E. Valve sparing root replacement: remodeling root repair with aortic ring annuloplasty. *Ann Cardiothorac Surg*. 2019;8(3):411–414.
17. Lansac E, Di Centa I, Sleilaty G, et al. An aortic ring to standardise aortic valve repair: preliminary results of a prospective multicentric cohort of 144 patients. *Eur J Cardiothorac Surg*. 2010;38(2):147–154.
18. De Paulis R, Salica A. Surgical anatomy of the aortic valve and root-implications for valve repair. *Ann Cardiothorac Surg*. 2019;8(3):313–321.
19. Irace FG, Chirichilli I, Salica A, et al. Aortic root anatomy after aortic valve reimplantation. *J Thorac Cardiovasc Surg*. 2021;. S0022-5223(21)00695-4.
20. Lansac E, Di Centa I, Sleilaty G, et al. Remodeling root repair with an external aortic ring annuloplasty. *J Thorac Cardiovasc Surg*. 2017;153(5):1033–1042.
21. Lansac E, Di Centa I, Sleilaty G, et al. Long-term results of external aortic ring annuloplasty for aortic valve repair. *Eur J Cardiothorac Surg*. 2016;50:350–360.
22. Schneider U, Aicher D, Miura Y, Schäfers HJ. Suture annuloplasty in aortic valve repair. *Ann Thorac Surg*. 2016;101(2):783–785.
23. Schneider U, Miyahara S, Giebels C, Karliova I, Schäfers HJ. Isolated aortic valve repair-how to do it and long-term results: suture annuloplasty. *Ann Cardiothorac Surg*. 2019;8(3):422–425.
24. Fattouch K, Castrovinci S, Murana G, et al. Functional annulus remodelling using a prosthetic ring in tricuspid aortic valve repair: mid-term results. *Interact Cardiovasc Thorac Surg*. 2014;18(1):49–54. discussion 54-5.
25. Chirichilli I, Irace FG, Weltert LP, et al. A direct correlation between commissural orientation and annular shape in bicuspid aortic valves: a new anatomical and computed tomography classification. *Interact Cardiovasc Thorac Surg*. 2020;30(5):666–670.
26. Rankin JS, Mazzitelli D, Fischlein T, et al. Geometric ring annuloplasty for aortic valve repair during aortic aneurysm surgery: two-year clinical trial results. *Innovations (Phila)*. 2018;13(4):248–253.
27. Rankin JS, Wei ML, Downey RS, et al. Aortic valve repair using geometric ring annuloplasty. Operative techniques in thoracic and cardiovasculary surgeryPublished online *Journal Pre-Proof*. 2021;.

28. Chirichilli I, Irace FG, Salica A, et al. Root reimplantation and aortic annuloplasty with external ring in bicuspid aortic valve: an anatomical comparison. *Semin Thorac Cardiovasc Surg*. 2021;. S1043-0679(21)00307-5.
29. Benhassen LL, Ropcke DM, Sharghbin M, et al. Comparison of Dacron ring and suture annuloplasty for aortic valve repair-a porcine study. *Ann Cardiothorac Surg*. 2019;8(3):342–350.
30. El Khoury G, Glineur D, Rubay J, et al. Functional classification of aortic root/valve abnormalities and their correlation with etiologies and surgical procedures. *Curr Opin Cardiol*. 2005;20(2):115–121.
31. Tamás E, Nylander E. Echocardiographic description of the anatomic relations within the normal aortic root. *J Heart Valve Dis*. 2007;16 (3):240–246.
32. Benhassen LL, Hedensted JH, Sharghbin M, et al. Altered stresses and dynamics after single and double annuloplasty ring for aortic valve repair. *Eur J Cardiothorac Surg*. 2020;57(6):1210–1217.
33. Mourad F, Shehada SE, Benedik J, et al. Endoscopy in aortic valve repair: does it worth it? *J Thorac Dis*. 2018;10(6):3436–3443.
34. El Khoury G, Vanoverschelde JL, Glineur D, et al. Repair of aortic valve prolapse: experience with 44 patients. *Eur J Cardiothorac Surg*. 2004;26(3):628–633.
35. Schäfers HJ, Bierbach B, Aicher D. A new approach to the assessment of aortic cusp geometry. *J Thorac Cardiovasc Surg*. 2006;132:436–438.
36. De Kerchove L, Momeni M, Aphram G, et al. Free margin length and coaptation surface area in normal tricuspid aortic valve: an anatomical study. *Eur J Cardiothorac Surg*. 2018;53(5):1040–1048.
37. Kunihara T, Aicher D, Rodionycheva S, et al. Preoperative aortic root geometry and postoperative cusp configuration primarily determine long-term outcome after valve-preserving aortic root repair. *J Thorac Cardiovasc Surg*. 2012;143(6):1389–1395.
38. Aicher D, Langer F, Adam O, Tscholl D, Lausberg H, Schäfers HJ. Cusp repair in aortic valve reconstruction: does the technique affect stability? *J Thorac Cardiovasc Surg*. 2007;134(6):1533–1538. discussion 1538-9.
39. Karliova I, Schneider U, Ehrlich T, Schäfers HJ. Results of pericardial patches in tricuspid and bicuspid aortic cusp repair. *Ann Thorac Surg*. 2020;109(3):728–735.
40. de Heer F, Kluin J, Elkhoury G, et al. Aortic Valve Repair Research Network Investigators. AVIATOR: an open international registry to evaluate medical and surgical outcomes of aortic valve insufficiency and ascending aorta aneurysm. *J Thorac Cardiovasc Surg*. 2019;157 (6):2202–2211. e7.
41. Michelena HI, Della Corte A, Evangelista A, et al. Summary: international consensus statement on nomenclature and classification of the congenital bicuspid aortic valve and its aortopathy, for clinical, surgical, interventional and research purposes. *Eur J Cardiothorac Surg*. 2021;. ezab039.

Chapter 22

Innovative Technologies for Hybrid Cardiovascular Repair

Michael I. Brener[1] and Isaac George[2]

[1]Division of Cardiology, Columbia University Medical Center, New York, NY, United States, [2]Division of Cardiac, Thoracic and Vascular Surgery, Columbia University Medical Center and the Columbia College of Physicians and Surgeons, New York, NY, United States

Chapter Outline

Introduction

Smaller devices and shrinking delivery platforms have significantly expanded the armamentarium for percutaneous treatment of various cardiovascular conditions ranging from coronary artery disease (CAD) to structural heart disease. However, the feasibility of catheter-based therapies is dictated by the anatomy of the vasculature, which imposes inherent limitations, and the long-term durability of transcatheter therapies remains a subject of ongoing investigation. Hybrid treatment strategies address some of these shortcomings and have traditionally combined percutaneous treatment modalities with conventional surgical techniques. They leverage the lower risk profile of percutaneous procedures with the enhanced technical feasibility and durability of traditional cardiovascular surgery. Over the last decade, novel treatment strategies have used creative combinations of percutaneous and open surgical procedures to substantially expand the scope of hybrid cardiovascular surgery. Fig. 22.1 highlights the myriad conditions where hybrid cardiovascular interventions have developed. Furthermore, hybrid techniques have become mainstays of the "heart team," which brings together interventional cardiologists and cardiovascular surgeons, and is recognized by multiple guidelines and clinical societies as the optimal venue for making treatment decisions for complex patients.[1–5] The goal of this chapter is to review the existing experience and data for hybrid cardiovascular procedures.

Coronary revascularization

The paradigm for hybrid cardiovascular surgery began with coronary revascularization procedures when percutaneous and surgical revascularization were combined to treat patients with multivessel CAD. Given the longevity and survival advantage associated with left internal mammary grafts (LIMA), hybrid procedures combined a LIMA graft to the left anterior descending (LAD) artery with percutaneous intervention for other, non-LAD lesions. The rationale for percutaneous coronary intervention (PCI), particularly with second-generation or newer drug eluting stents (DES), emerged from data showing the poor longevity of saphenous vein grafts (SVG). Vein graft failure occurs in approximately 45% of patients at 12 months postprocedure, yet SVGs remain the default conduit for non-LAD lesions despite guideline recommendations to use arterial conduits.[6] Lesion patency following PCI in the same time frame, in contrast, exceeds 95% but necessitates the use of dual-antiplatelet therapy.[7,8] Therefore, hybrid coronary revascularization merges the

Advances in Cardiovascular Technology. DOI: https://doi.org/10.1016/B978-0-12-816861-5.00017-4

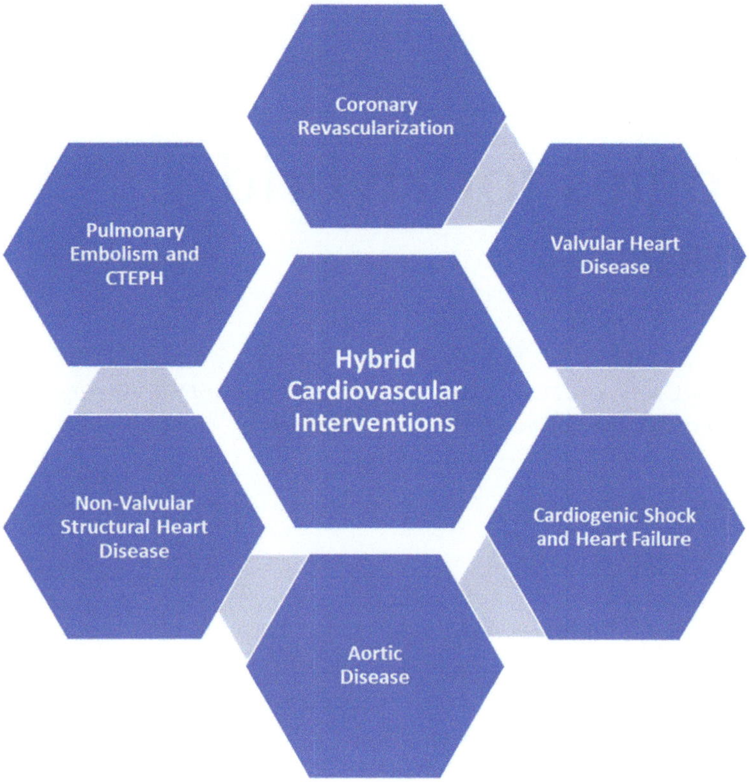

FIGURE 22.1 The scope of hybrid cardiovascular intervention.

optimal revascularization strategy for left main coronary artery and LAD lesions with the best techniques for non-LAD revascularization.

A number of important advances to the surgical aspects of hybrid intervention have refined the procedure further. First, minimally invasive direct coronary artery bypass (MIDCAB), where coronary artery bypass surgery (CABG) is performed on the beating heart and the LIMA is accessed via a lateral thoracotomy or limited sternal incision has substantially hastened postprocedural recovery and improved patient comfort by avoiding a traditional sternotomy (Fig. 22.2).[9] While the procedure is more technically challenging than traditional, open CABG, the reported rates of procedural success and long-term graft patency are very high. Subsequently, the expansion of robotically assisted endoscopic CABG has created a new frontier for hybrid procedures.[10,11] This technique uses three ports placed in the left pleural cavity in lieu of a sternotomy (Fig. 22.3). The LIMA is harvested as a skeletonized vessel and anastomosed to the LAD while off-pump. Overall, this technique significantly reduces hospital and intensive care unit length of stay and completely removes the risk of sternal wound infection.[12,13]

While hybrid coronary revascularization procedures have many advantages over CABG or multivessel PCI, coordinating the timing of two distinct procedures introduces a new layer of complexity. In principle, there are two strategies vis-à-vis procedural timing: "one-step" hybrid revascularization, where PCI and CABG are performed simultaneously; or "two-step" hybrid revascularization, where the procedures are performed separately. One-step hybrid revascularization necessitates a hybrid suite equipped with fluoroscopy where PCI and CABG can be performed simultaneously. Once rare, this layout is becoming increasingly common. Furthermore, dual antiplatelet therapy must be administered in the immediate postoperative period because patients undergo PCI. This increases the risk of bleeding and may complicate recovery. These concerns notwithstanding, one-step hybrid revascularization is associated with excellent short-term outcomes and higher patient satisfaction than off-pump CABG via open sternotomy, while not incurring the risk associated with multiple procedures.[14] Two-step hybrid revascularization has two forms which vary based on which procedure is performed first. Standard two-step hybrid revascularization begins with CABG and PCI follows, typically 4–6 weeks postprocedure. "Reverse" two-step hybrid revascularization begins with PCI and CABG follows afterwards. This strategy is reserved for patients with non-LAD acute coronary syndromes requiring urgent revascularization, but the optimal timing for CABG in reverse two-step hybrid revascularization has not been assessed systematically. Table 22.1 highlights the advantages and disadvantages associated with each strategy.[15] The benefits and risks

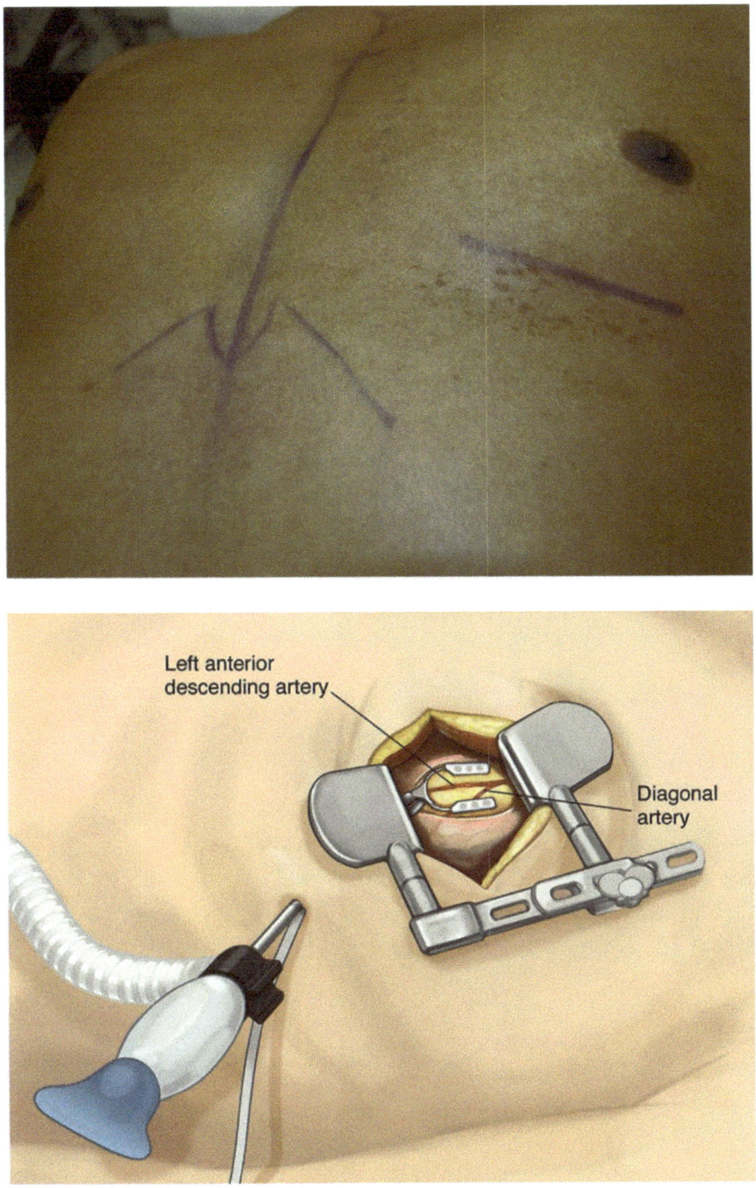

FIGURE 22.2 The minimally invasive direct coronary artery bypass procedure. The top panel illustrates the anterior thoracotomy used in the minimally invasive direct coronary artery bypass (MIDCAB) procedure. Once the appropriate anatomy is exposed, the anastomosis site of the left anterior descending artery is identified, as seen in the bottom panel. *Reproduced with permission from Reddy RC. Minimally invasive direct coronary artery bypass: technical considerations.* Semin Thorac Cardiovasc Surg. *2011;23(3):216–219.*

associated with hybrid revascularization may also depend on the severity and extent of coronary disease. Shen and colleagues stratified a prospective observational study of patients with multivessel CAD and showed that hybrid revascularization performed comparably with both high-risk PCI and CABG for patients with low and intermediate SYNTAX scores. However, in the subset of patients with high SYNTAX scores, hybrid revascularization appeared to be superior to PCI (Fig. 22.4).[16] An additional consideration that has further affected decision-making is the financial implications of a single versus staged procedure, with a staged procedure clearly more attractive from a reimbursement standpoint.

In summary, hybrid revascularization takes advantage of the benefits associated with CABG and PCI while minimizing the risks and neutralizing the limitations associated with each procedure. The evidence supports this theoretical construct, and hybrid revascularization appears to be equivalent to surgical revascularization strategies (i.e., off-pump coronary bypass surgery) in terms of technical procedural success and graft patency while eliminating the need for SVGs.[17] Relative to multivessel PCI, hybrid revascularization may be associated with improved long-term outcomes.[18] The National Heart, Lung, and Brain Institute sponsored Hybrid Trial (NCT01121263) was intended to be the definitive analysis of the hybrid revascularization strategy but was halted due to slow enrollment, and preliminary data are yet to be published. Until then, hybrid revascularization is a viable option to provide complete revascularization in patients

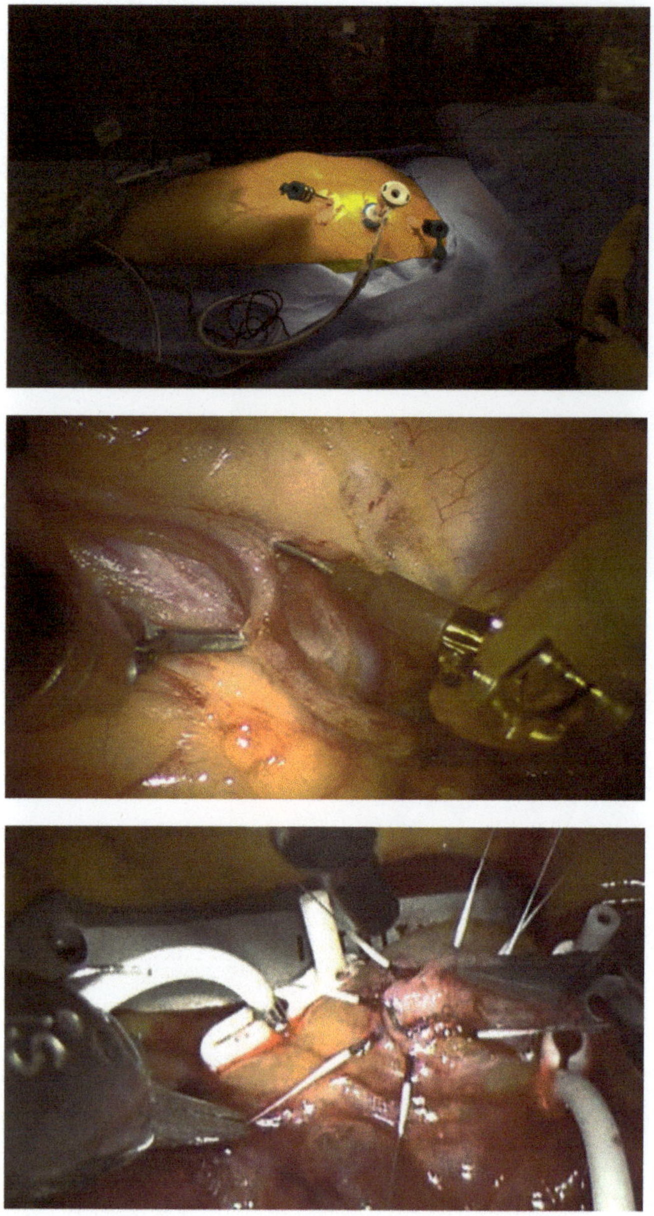

FIGURE 22.3 Technique for robotic coronary bypass surgery. The robotic surgical system is docked via three ports placed in the left pleural cavity (top panel). The middle panel shows how the left internal mammary artery (LIMA) is harvested using the robotic surgical system. Finally, the bottom panel shows the LIMA anastomosis with the left anterior descending artery. *Reproduced with permission from Hemli JM, Patel NC. Robotic cardiac surgery.* Surg Clin North Am. *2020;100(2):219–236; and Yang M, Wu Y, Wang G, Xiao C, Zhang H, Gao C. Robotic total arterial off-pump coronary artery bypass grafting: seven-year single-center experience and long-term follow-up of graft patency.* Ann Thorac Surg. *2015;100(4):1367–1373.*

with multivessel CAD. As with many of the techniques described in this chapter, a learning curve exists for operators and the success of hybrid revascularization hinges on the strength of the heart team.

Structural heart disease

Valvular heart disease, and more broadly, structural heart disease, has become a major forefront for innovative hybrid cardiovascular procedures. Our group previously reported on the combination of PCI and surgical valve replacement, demonstrating the feasibility and safety of this technique in hybrid operating suites.[19] PCI combined with aortic, mitral,

TABLE 22.1 Timing of hybrid coronary revascularization procedures.

	One-step hybrid revascularization	Two-step hybrid revascularization	
		Standard (CABG then PCI)	Reverse (PCI then CABG)
Advantages	Complete revascularization in one procedure, ability to assess anastomosis, shorter hospital stay.	Dual antiplatelet therapy can be started after surgical recovery, ability to assess anastomosis, PCI may be safer after partial revascularization.	Allows for complete hybrid revascularization in patients with any acute coronary syndrome when one-step approach cannot be performed.
Disadvantages	Hybrid suite needed, increased risk of bleeding.	Incomplete revascularization after first procedure.	May require CABG with dual antiplatelet therapy use.

Source: Adapted with permission from Moreno PR, Stone GW, Gonzalez-Lengua CA, Puskas JD. The hybrid coronary approach for optimal revascularization: JACC review topic of the week. *J Am College Cardiol*. 2020;76(3):321–333.

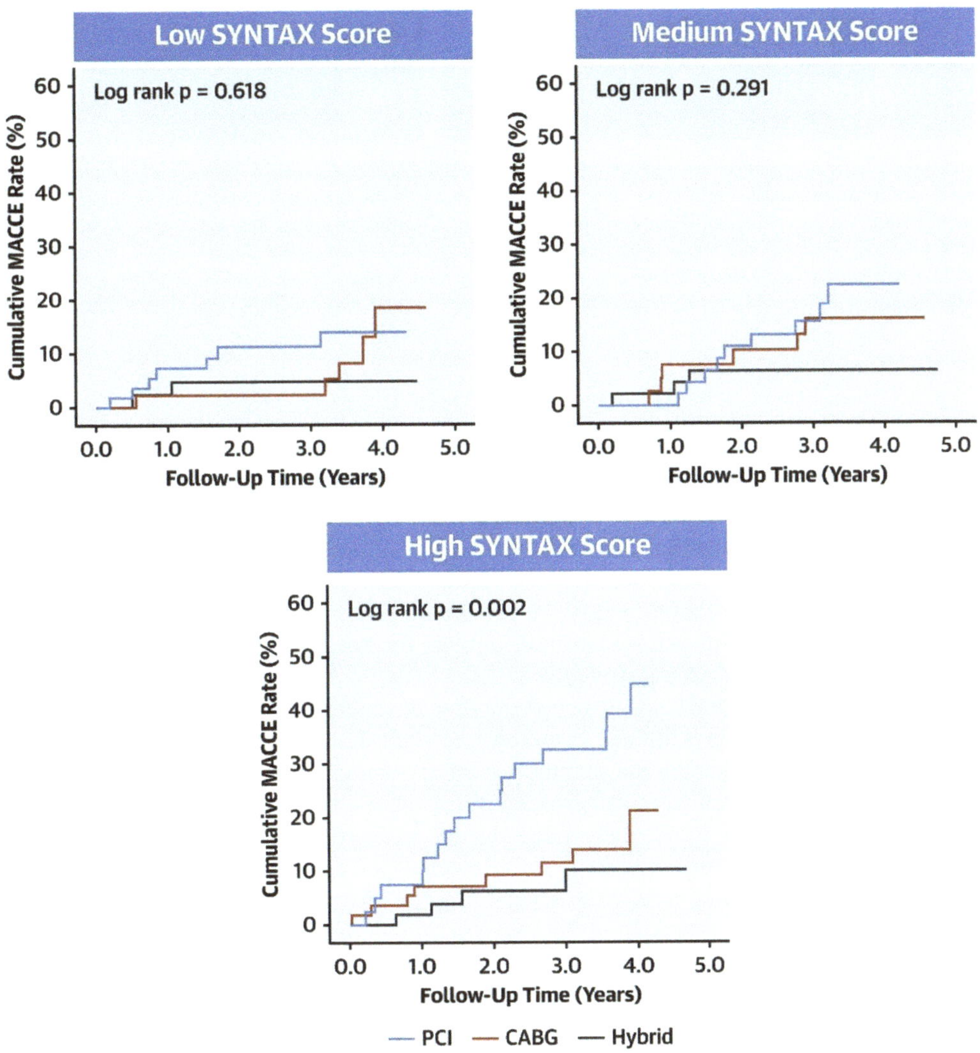

FIGURE 22.4 Hybrid revascularization outcomes as a function of SYNTAX score. Patients without complex CAD (i.e., a low or intermediate SYNTAX score) had similar frequencies of major adverse cardiac and cerebrovascular events, independent of revascularization strategy. In contrast, patients with the highest SYNTAX scores benefited from surgical revascularization, and outcomes were marginally better with hybrid revascularization relative to traditional coronary artery bypass grafting. *Adapted with permission from Moreno PR, Stone GW, Gonzalez-Lengua CA, Puskas JD. The hybrid coronary approach for optimal revascularization: JACC review topic of the week.* J Am College Cardiol. *2020;76(3):321–333.*

or tricuspid surgery in patients with obstructive, non-LAD lesions reduces the overall surgical time relative to CABG plus valvular surgery. More recently, however, with the explosion of transcatheter valve implantation, the field of hybrid cardiac surgery in the valve space has shifted to semisurgical insertion of transcatheter valves, as detailed below.

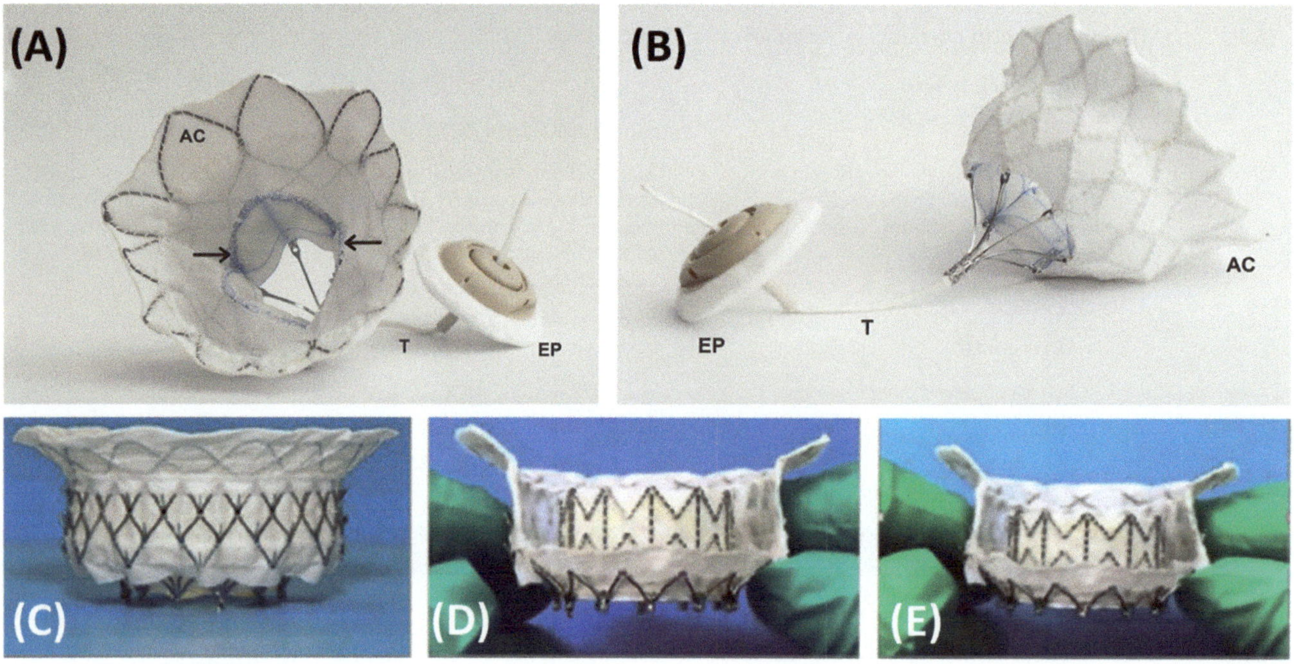

FIGURE 22.5 The Tendyne and Intrepid transcatheter mitral valve replacement device. The Tendyne device, as viewed from the left atrium (A) and the long axis of the left ventricle (B). The device features an atrial cuff (AC) to provide a secure seal with the annulus, as well as tether (T) and epicardial pad (EP) to prevent atrial embolization. Below it is a longitudinal view of the Intrepid device (C), which is a dual stent prosthesis, as seen in (D), that can mold to the shape of the mitral annulus (E). *Adapted with permission from (A–B) Muller DWM, Farivar RS, Jansz P, et al. Transcatheter mitral valve replacement for patients with symptomatic mitral regurgitation: a global feasibility trial.* J Am College Cardiol. *2017;69(4):381–391, and (C–E) McCarthy PM, Kislitsina ON, Malaisrie SC, Davidson CJ. Transcatheter mitral valve replacement with intrepid.* Interv Cardiol Clin. *2019;8(3):287–294.*

Aortic valve disease

The first transcatheter aortic valve replacement (TAVR) procedure was performed in 2002 via an antegrade transseptal approach. TAVR access evolved from a transseptal approach to both a transfemoral retrograde and antegrade transapical approach over time. While transapical access accounted for about 30% of TAVR procedures in the early TAVR experience for patients without femoral access, it has since become obsolete.[20] Alternative access strategies have replaced transapical access with a variety of surgical (i.e., transcarotid, suprasternal, direct aortic) and percutaneous techniques (i.e., transcaval, transaxillary). Even in the contemporary era of shrinking transcatheter aortic valve prostheses and smaller delivery sheaths, surgical access strategies are important options for heart teams given the cost of and complications associated with percutaneous alternative access strategies.

Suprasternal access for TAVR evolved from a direct aortic approach first described by Bapat and colleagues a decade ago, where the ascending aorta is cannulated via a limited sternotomy or right thoracotomy.[21] The approach was refined and evolved into the suprasternal approach, whereby the innominate artery—an unbranched vessel usually measuring ~18 mm in diameter and typically free of significant atherosclerotic disease—is directly accessed through a small incision above the sternal notch, akin to what is routinely done during mediastinoscopy.[22] Subsequently, the superficial structures (i.e., platysma and sternothyroid muscles) are dissected to expose the innominate artery (Fig. 22.6). TAVR proceeds in the standard fashion and the arteriotomy is closed with purse-string sutures. Preprocedural planning is essential for the success of suprasternal TAVR. The innominate artery's suitability is determined by the size of the vessel (minimum lumen diameter ≥7 mm) as well as the burden and location of calcification from cross-sectional imaging. In addition, patients are excluded from suprasternal TAVR if the distance between the sternal notch and the innominate arteriotomy site exceeds 5–6 cm.

Suprasternal TAVR has since expanded beyond Kiser's single-site and single-surgeon experience to be used in an array of patient populations and clinical scenarios, including patients with morbid obesity, advanced peripheral vascular disease, cardiogenic shock, and even a pediatric patient with congenital aortic valvulopathy.[23–26]

Transcarotid and transaxillary TAVR are both alternative access techniques that combine percutaneous and surgical intervention. In the early experience with transcarotid TAVR, the right carotid artery was exposed and a bypass conduit

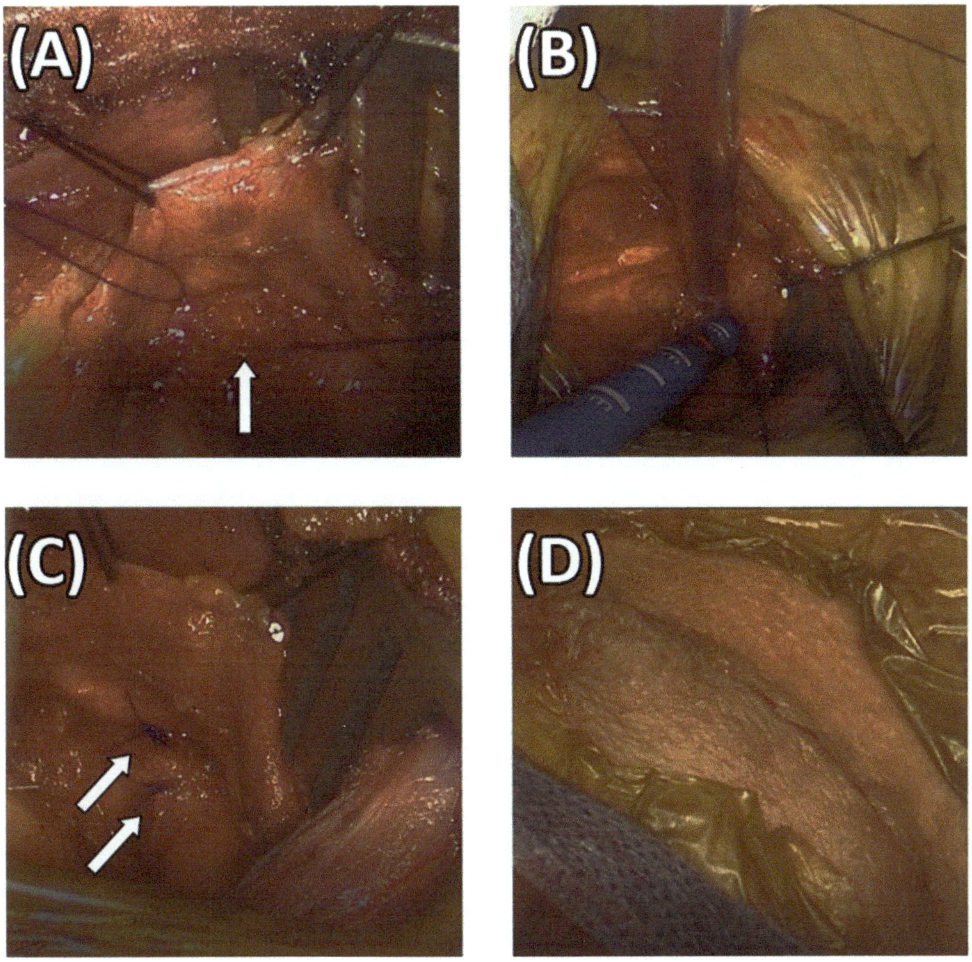

FIGURE 22.6 Technique for suprasternal transcatheter aortic valve replacement. Suprasternal transcatheter aortic valve replacement involves: (A) dissection to the level of the innominate artery, with the planned arteriotomy site identified by the arrow, (B) the Edwards Commander Delivery System positioned in the innominate artery prior to valve deployment, (C) the arteriotomy site sutured after valve deployment, and (D) a representative image of the surgical wound after suturing. *Brener, MI, Olds, A, Nemeth, S, Kurlansky, P, Nazif, TM, Vahl, TP, ... George, I. Suprasternal Versus Transfemoral Access for Transcatheter Aortic Valve Replacement: Insights From a Propensity Score Matched Analysis, Observational Study. J Am Heart Assoc. 2021;10(16): e020491. https://pubmed.ncbi.nlm.nih.gov/34376060/.*

created to perfuse the distal carotid artery from the common femoral artery.[27] With growing experience with transcarotid TAVR, this step is now often omitted. Like transaxillary TAVR, the surgical components of the procedure are now largely limited to the cut-down required to facilitate placement of the delivery sheath. Both access strategies have demonstrated good outcomes, although there appears to be an increased hazard of stroke with transaxillary access.[28,29] Like the transsubclavian alternative access technique, these alternative access strategies may become fully percutaneous and hybrid procedures in the aortic space may be limited to the suprasternal, direct aortic, or transapical techniques.

Mitral valve disease

Minimally invasive mitral valve replacement has also emerged as a key frontier for hybrid cardiovascular intervention.

Patients with severe mitral annular calcification represent an especially high-risk group for traditional mitral valve procedures. Such patients often have multiple comorbidities and are poor surgical candidates. Accordingly, outcomes with traditional valve replacement, with or without calcium debulking, are suboptimal and the procedure is associated with the potentially fatal complication of posterior annular rupture.[30,31] Implantation of a transcatheter valve via an open atriotomy is an example of a hybrid intervention in the mitral position which may minimize the risks of traditional mitral valve repair or replacement.

The technique for transatrial transcatheter mitral valve replacement has been described previously.[32] Briefly, the procedure begins with a median sternotomy or right thoracotomy, followed by aortic and bicaval cannulation to facilitate cardiopulmonary bypass. The left atrium is subsequently exposed through an opening in Waterson's groove and the anterior mitral valve leaflet is resected along with the supporting chordal structures. The posterior leaflet is typically left intact. If the risk of left ventricular outflow tract (LVOT) obstruction is high, a septal myectomy can be performed. Once these preparatory steps are complete, the atrial surface of the mitral valve annulus circumference is lined with

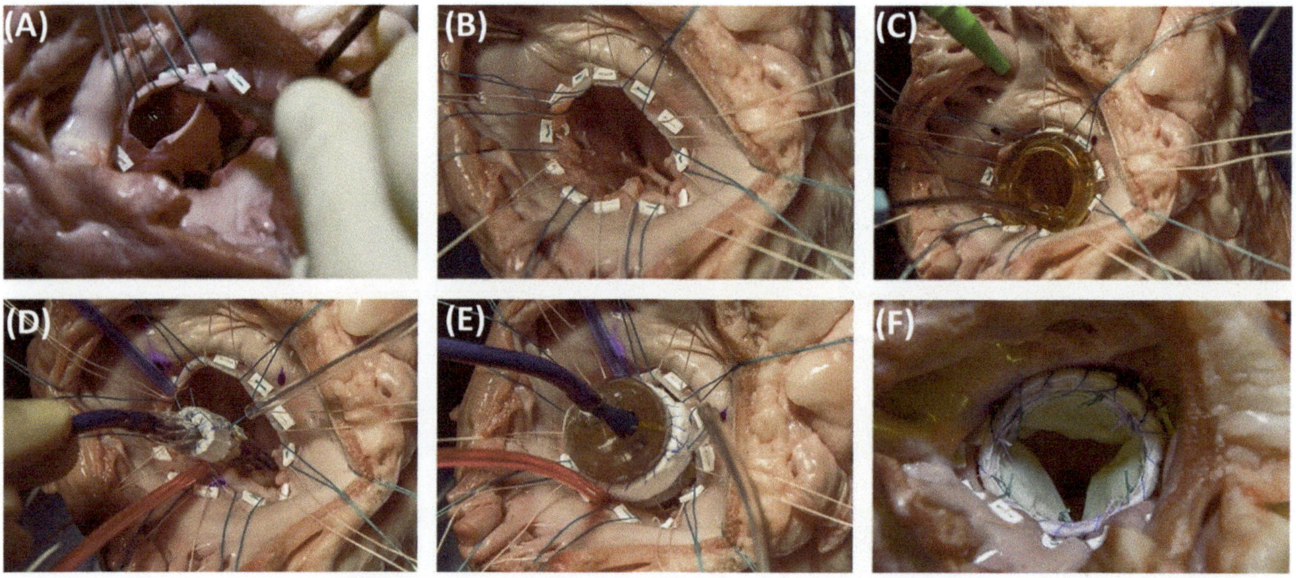

FIGURE 22.7 Technique for transatrial transcatheter mitral valve replacement. (A) Anterior leaflet is resected leaving a rim of atrial tissue and pledgeted sutures are placed through this rim of tissue. (B) After resection of the anterior leaflet, pledgeted valve sutures are placed through aortomitral curtain anteriorly and around the posterior annulus using a rim of left atrial tissue. (C) Using a standard valve sizer with 120 degrees markings, mark where three guiding sutures will be placed. (D) The balloon delivery system and valve are brought up to the table with a standard 0.035-inch J wire placed through the delivery system into the left ventricle. The guiding sutures are placed through the annulus at the previously marked positions and then snared. (E) The balloon is inflated as the three guiding sutures are gently snared. (F) Three guiding sutures are secured, Pledgeted sutures encircling the base of the mitral leaflets are passed through the felt strip and valve frame, and then secured. All photos were taken from simulation laboratory using bovine calf hearts. *LVOT*, Left ventricular outflow tract. *Reproduced with permission from Russell HM, Guerrero ME, Salinger MH, et al. Open atrial transcatheter mitral valve replacement in patients with mitral annular calcification.* J Am College Cardiol. *2018;72(13):1437–1448.*

pledgeted sutures (Fig. 22.7). A transcatheter valve, typically a balloon-expandable Sapien prosthesis (Edwards Lifesciences, Irvine, CA), is appropriately sized and modified with a circumferential felt strip. The guiding sutures are passed through the transcatheter valve frame, with care to avoid damaging the leaflet tissue. Using a two-person technique, the delivery catheter balloon is slowly inflated and orthogonal alignment between the transcatheter valve and the native mitral annulus are maintained. Once appropriate positioning is confirmed, the guiding sutures are subsequently secured. Saline irrigation and transesophageal echocardiography are used to ensure the valve leaflets appropriately coapt and that there is minimal paravalvular leak and no LVOT obstruction.

A report of 26 patients undergoing the aforementioned procedure described 100% technical success but considerable 30-day mortality (27%).[33] Considering the high rates of procedural mortality and adverse outcomes in the recovery period, these data provide encouraging early results to position this hybrid technique as a valuable alternative to traditional surgical techniques. Results from a multicenter observational study are expected in 2021.

A number of mitral valve repair and replacement strategies access the mitral valve via an open atrial incision or other small incisions to treat mitral regurgitation. Among the most vetted is the Tendyne device (Abbott Structural, Abbott Park, IL), which is a nitinol self-expanding trileaflet porcine bioprosthetic valve with an anchor that implants into the left ventricular apex.[34] The valve is mounted onto a double-framed device such that the outer frame forms a seal with the mitral annulus and the left atrium, while the inner frame supports the bioprosthetic valve (Fig. 22.8). The valve comes in many sizes, is inserted via a 34 Fr delivery system through an atriotomy accessed via a left minithoracotomy, and is fully retrievable. In the early feasibility study, Tendyne was associated with technical success in all patients, no acute deaths related to the procedure, and $\leq 1+$ residual mitral regurgitation at 30-days. The Intrepid prosthesis (Medtronic, Dublin, Ireland) is similar to the Tendyne device, but lacks the apical tether. A 27 mm trileaflet bovine pericardial valve is anchored in an internal stent, while an outer stent that is stiffer on the ventricular side and more flexible on the atrial side conforms to the mitral annulus. The device can be implanted transapically or via transseptal access, with the latter slowly phasing out the need for transapical access.[35]

The Millipede IRIS (Boston Scientific, Marlborough, MA) is a semirigid annuloplasty ring that can be placed either through a transatrial incision or percutaneously via a 24 Fr delivery system through a transseptal puncture (Fig. 22.9). The device recreates a traditional complete surgical annuloplasty ring and reduces the septal-lateral length of the

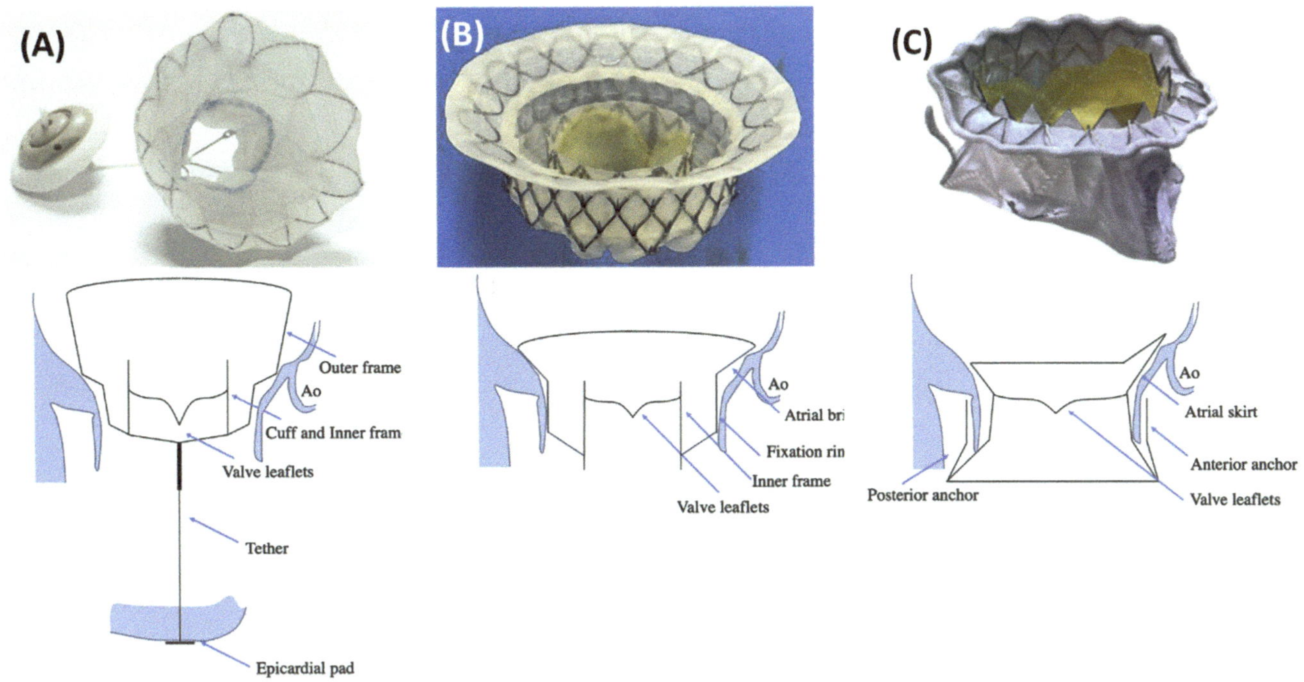

FIGURE 22.8 Transapical transcatheter mitral valve replacement prostheses. Images of the three dominant transcatheter mitral valve replacement options that are placed via transapical access (A, Tendyne; B, Intrepid; C, Tiara) with schematics below, illustrating the prosthesis position with respect to the mitral valve annulus and valve apparatus. *Adapted with permission from Enta Y, Nakamura M. Transcatheter mitral valve replacement.* J Cardiol. *2020.*

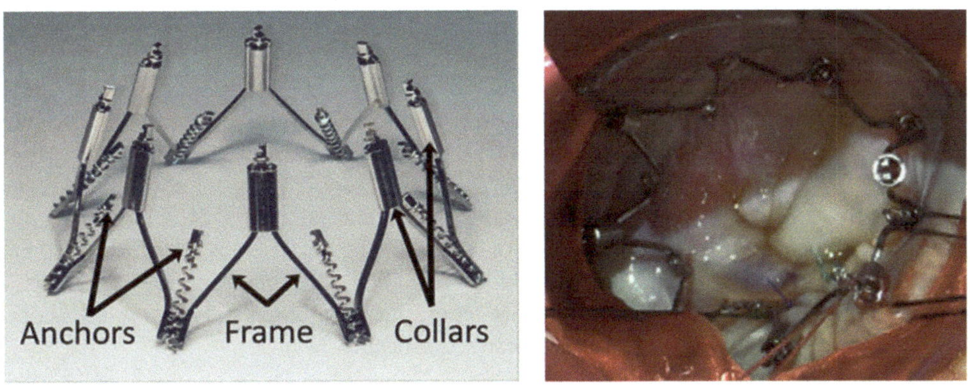

FIGURE 22.9 The Millipede IRIS device. The Millipede IRIS is a nitinol device that can be deliver percutaneously (left panel) or surgically (right panel) and consists of eight anchors that affix to the mitral valve annulus and eight corresponding collars that can be tightened (as seen in the right panel) to reduce the mitral annular dimension. *Reproduced with permission from Rogers JH, Boyd WD, Smith TW, Bolling SF. Transcatheter mitral valve direct annuloplasty with the millipede IRIS ring.* Interv Cardiol Clin. *2019;8(3):261–267.*

annulus to reduce mitral regurgitation. Eight helical anchors attach to the mitral annulus and can be cinched together by collars extending from the atrial side of the device.[36] Initial experience with the device in seven patients (four with transatrial implantation) was positive—technical success and reduction to less than moderate mitral regurgitation was achieved in all cases.[37] An ongoing pilot study is underway (NCT02607527) to evaluate the benefits of the Millipede IRIS in a larger cohort.

The Valcare Amend system (Valcare Medical, Herzliya, Israel) is another catheter-based technique for mitral annuloplasty which can be delivered by transapical access. The semirigid D-shaped annuloplasty ring preserves the mitral valve apparatus, and therefore, allows Valcare Amend system implantation to be combined with other procedures like transcatheter edge-to-edge leaflet approximation. The AMENDTR trial (NCT02602613) is actively recruiting patients

TABLE 22.2 Hybrid transcatheter mitral valve replacement options.

Device	Tendyne	Intrepid	Tiara
Manufacturer	Abbott	Medtronic	Neovasc
Valve shaper	Inner: Circular, Outer: D-shape	Circular	D shape
Frame	Self-expanding	Self-expanding	Self-expanding
Material	Nitinol	Nitinol	Nitinol
Anchoring Mechanism	Apical tether	Radial force and cleats on outer stent frame.	Fibrous trigone capture with native leaflet engagement.
Leaflet number	3	3	3
Leaflet material	Porcine	Bovine	Bovine
Valve position	Intraannular	Intraannular	Intraannular
Delivery catheter size	34 Fr	35 Fr	32/36 Fr
Recapture ability	Yes	No	No
Valve size	30–43 mm	Inner: 27 mm, Outer: 43, 46, and 50 mm	35 and 40 mm

Source: Adapted with permission from Enta Y, Nakamura M. Transcatheter mitral valve replacement. *J Cardiol*. 2020.

and comes in the wake of a single-center pilot investigation showing technical feasibility and significant reduction in annular dimensions and mitral regurgitation severity.[38]

In addition to traditional mitral valve repair strategies, a number of chordal repair devices have further expanded the panoply of hybrid interventions for mitral valve disease. The HARPOON beating heart mitral valve system (Edwards Lifesciences, Irvine, CA) uses expanded polytetrafluoroethylene cords implanted from the ventricular apex to anchor into prolapsing mitral valve leaflets as close as possible to the commissure to reduce mitral regurgitation. A minimum of three cords are implanted (each through their own delivery systems). One-year outcomes with this device showed two patients died and eight (13%) ultimately required reoperation.[39] Another device—the Neochord (Neochord, Minneapolis, MN)—uses a similar technique, but is in earlier stages of development.[40] The Mitral Touch (Mitre Medical, Morgan Hills, CA) is a unique epicardial device that is placed off-pump to surround the mitral annulus and promote positive remodeling and reduce annular dimensions. Technical success was achieved in all five patients in which the device was implanted, and the device represents a simple (implantation achieved in <30 min among two of the five study participants) way to perform a minimally invasive subvalvular annuloplasty in patients already undergoing cardiac surgery.[41] A comprehensive review of all these emerging technologies is outside the scope of this chapter, but the key features of some the main platforms are summarized in Table 22.2.[42]

Tricuspid valve disease

A burgeoning market for tricuspid valve repair and replacement devices has emerged to complement the innovation in the mitral space. The LuX-Valve (Ningbo Jenscare Biotechnology, Ningbo, China) is an orthotopic prosthetic valve delivered via a 32 Fr system inserted directly into the right atrium from a small right thoracotomy. The device consists of four distinct parts (Fig. 22.10): (1) valve leaflets made form bovine pericardium, (2) a self-expandable nitinol stent with an atrial disk, (3) an interventricular septal anchor, and (4) polytetrafluoroethylene-covered "graspers" that attach to the ventricular-side of the tricuspid valve apparatus.[43] Proper device positioning does not rely upon radial force. To date, the LuX-Valve has been implanted in 12 patients on a compassionate use basis; technical success was achieved in all cases and there was evidence of symptom improvement by 30 days postprocedure. Like the LuX-Valve, the NaviGate prosthesis (NaviGate Cardiac Structures, Lake Forest, CA) is a tricuspid valve prosthesis that can be introduced via a right anterolateral minithoracotomy. The delivery system is a 42 Fr catheter and the prosthesis has five sizes, ranging from 36 to 52 mm in diameter. NaviGate is constructed from nitinol and is a self-expanding, cone-shaped stent with a trileaflet valve made from equine pericardium (Fig. 22.11). Since the first-in-human implantation in 2017,

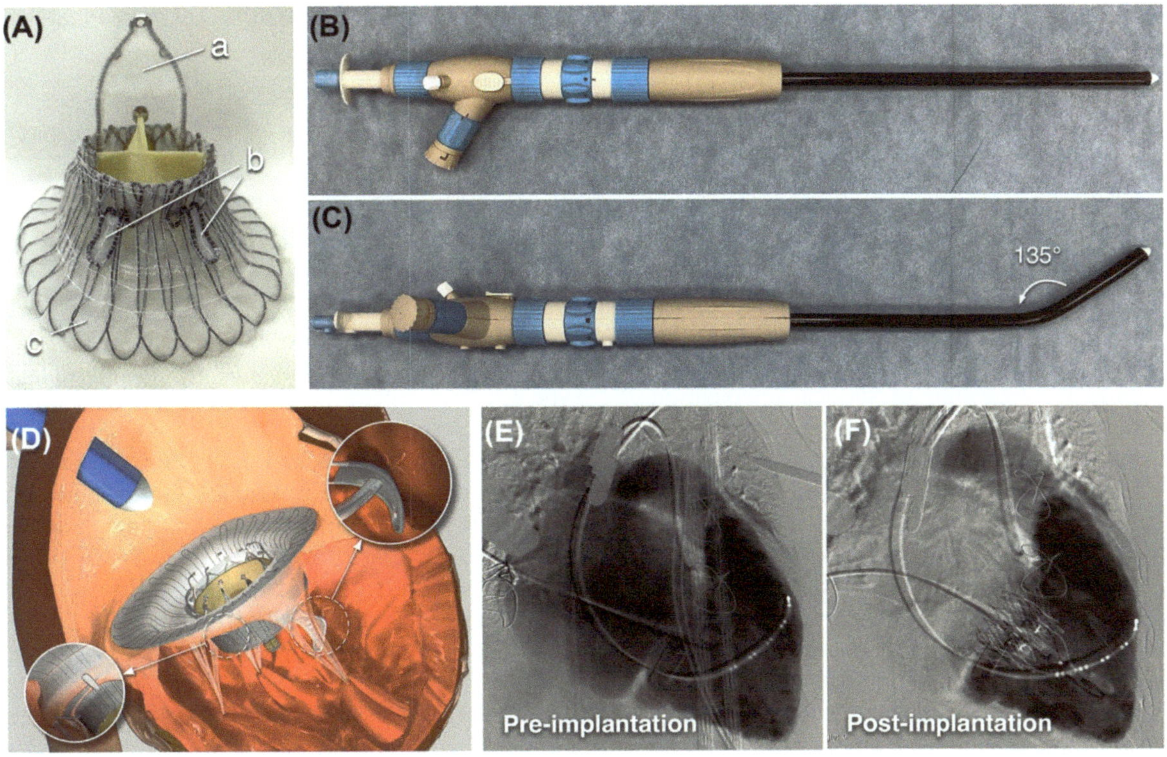

FIGURE 22.10 The LuX transcatheter tricuspid valve replacement platform. (A) The LuX valve consists of one interventricular septal anchor (a), two tricuspid valve graspers (b), and a valve disk (c). (B and C) Orthogonal views of the delivery catheter. (D) Schematic depiction of the implanted LuX-Valve. Insets indicate the interventricular septal anchor (right) and the two valve graspers (left). (E and F) LuX valve implantation significantly reduces tricuspid regurgitation as illustrated by before and after valve implantation right ventriculography. *Adapted with permission from Lu F-L, Ma Y, An Z, et al. First-in-man experience of transcatheter tricuspid valve replacement with LuX-Valve in high-risk tricuspid regurgitation patients.* JACC: Cardiovasc Interv. *2020;13(13):1614–1616.*

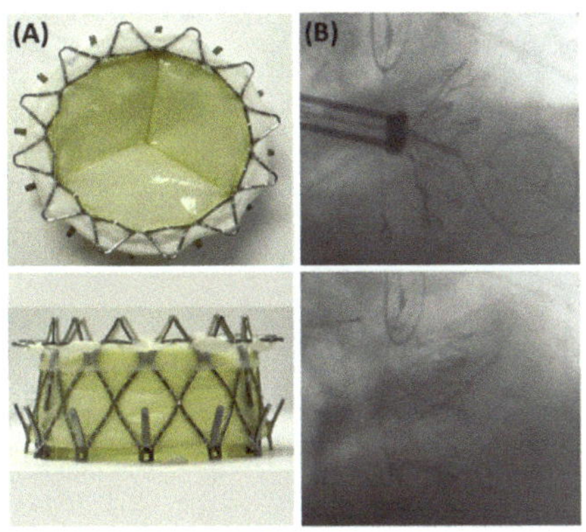

FIGURE 22.11 The NaviGate transcatheter tricuspid valve replacement system. (A) The atrial (top) and lateral (bottom) profiles of the NaviGate valve. (B) Fluoroscopic visualization of the valve during (top) and after (bottom) deployment. *Adapted with permission from Asmarats L, Puri R, Latib A, Navia JL, Rodés-Cabau J. Transcatheter tricuspid valve interventions: landscape, challenges, and future directions.* J Am College Cardiol. *2018;71(25):2935–2956.*

NaviGate was systematically evaluated in a multicenter cohort of 30 patients.[44] Implantation was feasible in 26 out of 30 individuals (of whom 25 had the device placed transatrially), and among those with the device successfully implanted, tricuspid regurgitation was reduced to mild severity or less in 76%. No events related to device implantation were documented at a mean follow-up of 127 days. With these two novel transcatheter tricuspid valve replacement devices, the once "forgotten" valve is now becoming a key area of interest for hybrid cardiac surgery.[45] These valvular devices are already in next-generation design for all transfemoral, transvenous access, and it is expected that the transatrial approach will be abandoned for a fully percutaneous approach in the future.

Aortic disease

Diseases of the aorta (i.e., aneurysm, dissection, intramural hematoma) have been at the forefront of hybrid cardiac surgery since the first endovascular treatment of a thoracic aortic aneurysm in 2000. Endovascular therapies are now part of the routine management of descending aortic disease, so we will focus here on the developing field of endovascular therapy for ascending aortic and aortic arch disease. Ascending aortic and aortic arch pathology have traditionally been treated with open repair, but new techniques have opened the door for endovascular therapy.

Hybrid aortic arch repair has three predominant variations (Fig. 22.12), all of which reduce the total bypass time and minimize the total time of surgical intervention.[46,47] Type I hybrid repairs involve a four-pronged Dacron graft sewn into the native aorta just above the sinotubular junction to perfuse the aortic arch's major branch vessels. Subsequently, a thoracic endovascular repair (TEVAR) graft is used to correct the arch and seal off the major branch vessels. Type II hybrid repair is performed in patients with ascending aortic disease in addition to aortic arch pathology such that there is an inadequate proximal landing zone for the TEVAR graft. In these cases, an ascending aortic graft is placed and the four-pronged Dacron graft is sewn into the ascending aortic graft. TEVAR is subsequently performed in the usual fashion. Finally, Type III hybrid repairs (which are also occasionally grouped with Type II repair techniques) involve a total arch repair combined with a TEVAR. Reoperation rates for hybrid aortic arch repair are low, and outcomes in the short and intermediate term are favorable, especially considering the high-risk profile of patients requiring such interventions.

Ascending aortic pathology is also amenable to endovascular repair under certain circumstances.[48] A satisfactory landing zone (typically >2 cm) that is sufficiently removed from the coronary ostia (> 2 cm) is necessary. Grafts cannot accommodate an ascending aorta >4 cm and the femoral vasculature must be large enough (> 7 mm) and nontortuous to accommodate the graft. Furthermore, individuals should not have severe aortic regurgitation as this may destabilize graft positioning. A variety of TEVAR grafts may be used in the ascending aorta, although the Zenith Ascend (Cook Medical, Bloomington, IN) and Valient (Medtronic, Dublin, Ireland) endovascular grafts are the only ones expressly designed for the ascending aorta. Short-term outcomes are good with ascending aorta TEVAR and it represents a viable option for the nearly one-in-five patients with either acute dissection of the ascending aorta who are prohibitively high risk for surgery. Pseudoaneurysm, while rare, also represents a challenging condition to treat surgically given the risk of hematoma embolization. A hybrid strategy minimizes this risk and has been shown to be a feasible treatment strategy.[49,50]

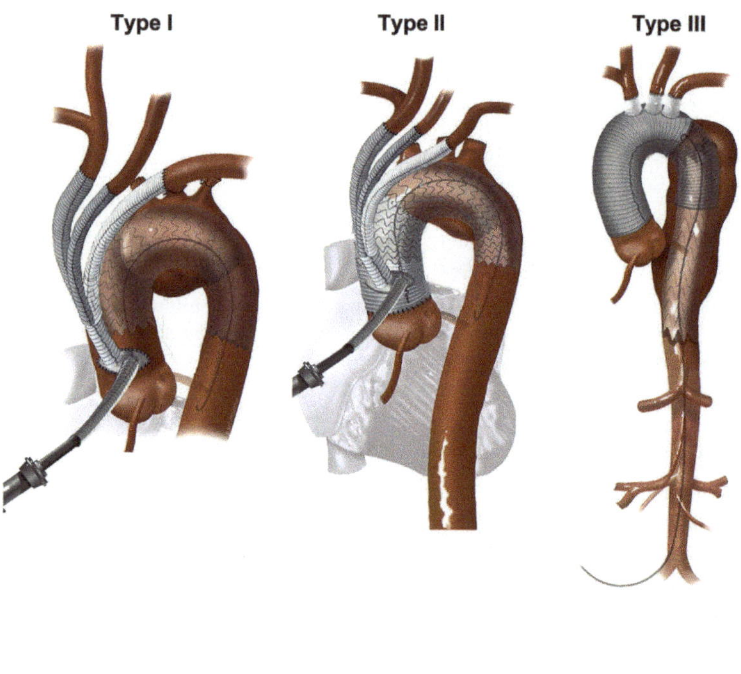

FIGURE 22.12 Hybrid aortic arch repair. Variations of hybrid aortic arch repair involve thoracic endovascular repair combined with different forms of arch intervention: Type I involves sewing branch vessel conduits directly onto the ascending aorta; Type II involves ascending aorta grafting, with the branch vessel conduits sewn into the graft; Type III involves total aortic arch repair combined with TEVAR. *Reproduced with permission from Bavaria J, Vallabhajosyula P, Moeller P, Szeto W, Desai N, Pochettino A. Hybrid approaches in the treatment of aortic arch aneurysms: postoperative and midterm outcomes.* J Thoracic Cardiovasc Surg. *2013;145(3 Suppl.):S85–S90.*

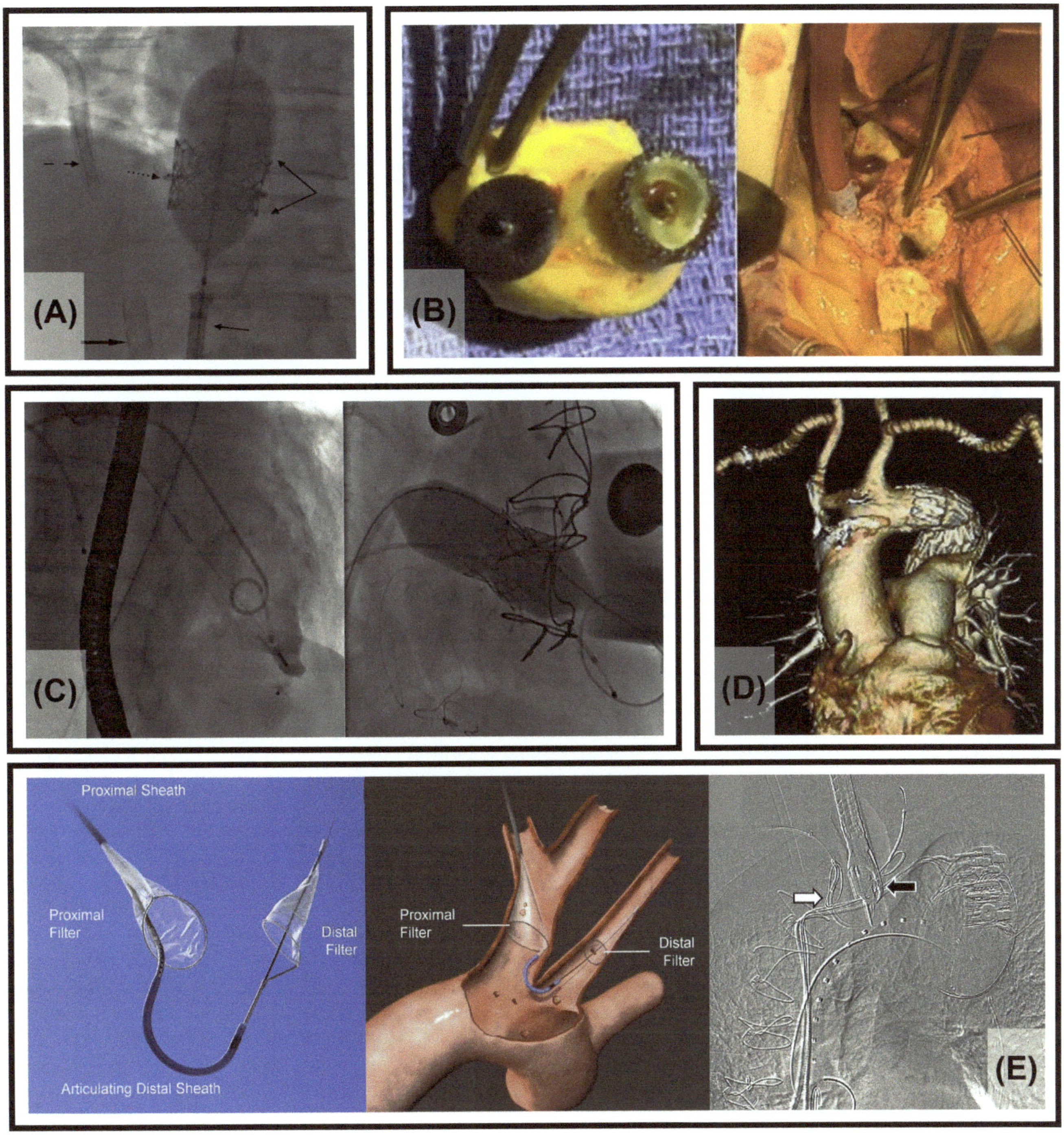

FIGURE 22.13 Forefronts for hybrid cardiovascular surgery (next page). (A) Transcatheter heart valve implantation in the inferior vena cava to limit propagation of the regurgitant jet in a patients with severe tricuspid regurgitation. (B) Close-up and intraoperative images from extraction of an Amplatzer plug device. (C) Illustrations of transcatheter closure of a left ventricular pseudoaneurysm prior to transcatheter aortic valve replacement. (D) Hybrid endovascular repair of a Kommerall diverticulum. (E) Use of a cerebral protection device during thoracic endovascular repair. *Adapted with permission from (A) Mulaikal TA, Pospishil L, George I, Bapat VN, Hahn RT. Surgically assisted transcatheter balloon-expandable valve in inferior vena cava for torrential tricuspid regurgitation. 2018;2(5):174–180; (B) Song J, Ascherman B, Eudailey KW, et al. Long-term failure of Amplatzer plugs in the treatment of aortic pathology.* J Card Surg. *2017;32(7):426–429; (C) Dershowitz L, Wolbinski M, Bapat V, George I, Kodali S. Left ventricular injury: beware the wire.* JTCVS Tech. *2020;3:126–129; (D) Najjar M, Mohar M, Stewart A, George I. Hybrid endovascular repair of Kommerell diverticulum and aberrant right subclavian artery in a patient with repaired coarctation of the aorta.* Heart Lung Vessel. *2015;7 (1):83–55; (E) Kossar AP, Takayama H, Patel V, George I. Novel implementation of a cerebral protection system during ascending thoracic endovascular aortic repair (TEVAR).* Semin Thorac Cardiovasc Surg. *2019;31(2):218–221.*

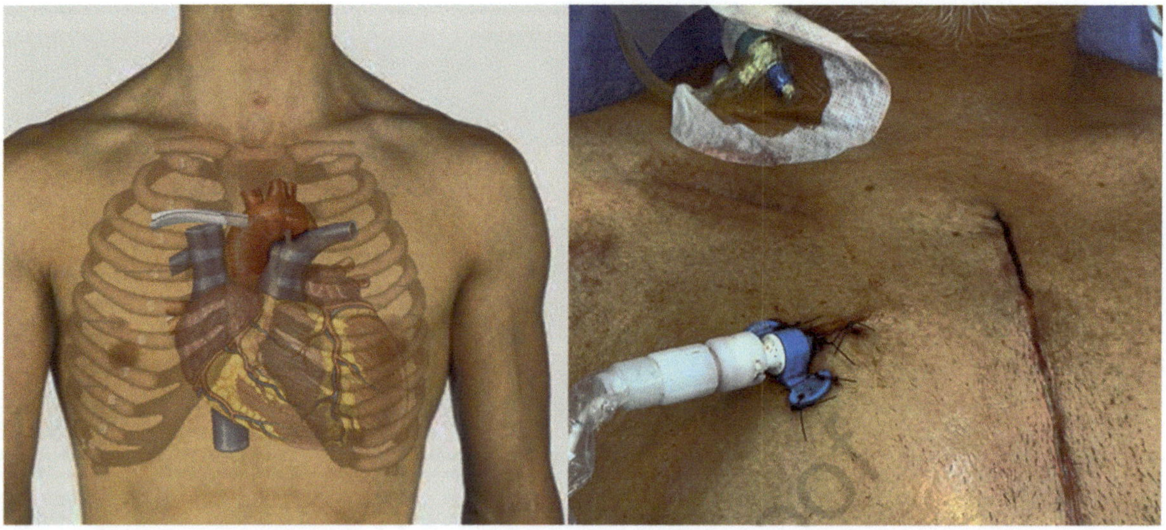

FIGURE 22.14 Impella 5.5 insertion technique. Transthoracic insertion of the Impella 5.5 involves sewing a Dacron graft onto the ascending aorta and placing it through the second and third intercostal spaces. *Reproduced with permission form Anderson M, Smith D, Kane P, Lee R, Khalpey Z, Williams J. Impella 5.5 direct aortic implant and explant techniques.* Ann Thorac Surg. *2020.*

Miscellaneous

The paradigm of hybrid intervention continues to encroach on a number of other fields in cardiovascular medicine in addition to the more established fields described above. Examples are highlighted in Fig. 22.13.[51–55] Acute pulmonary embolism (PE) and chronic thromboembolic pulmonary hypertension (CTEPH) represent one such frontier for hybrid intervention, particularly as "pulmonary embolism response teams" have brought interventional cardiologists and cardiac surgeons together in much the same way that heart teams have for other cardiovascular conditions. Hybrid intervention in this space has emerged in two main settings: first, for patients with acute on chronic thromboembolic disease who undergo an initial percutaneous therapy for acute PE and a surgical endarterectomy for CTEPH, and second, for patients with CTEPH who undergo endarterectomy for proximal pulmonary arterial disease and percutaneous balloon angioplasty to treat distal pulmonary arterial narrowing. Yanaka and colleagues reported symptom improvement in 20 patients who had residual symptoms after pulmonary endarterectomy and subsequently underwent balloon angioplasty, establishing this two-step technique as a viable hybrid option for patients with CTEPH.[56]

Cardiogenic shock and advanced heart failure represent another nascent field for hybrid intervention, especially as device-based therapies expand and "shock teams" increasingly dictate treatment decisions. Temporary ventricular assist devices, like the Impella 5.5 (Abiomed, Danvers, MA), for example, are inserted into the left ventricle through Dacron grafts that connect to the ascending aorta. The graft can be inserted into the aorta either via a median sternotomy or through a transthoracic technique where the graft extends between the second and third intercostal spaces (Fig. 22.14).[57] Hybrid interventions combine devices like the Impella 5.5, which provides full left ventricular support, with percutaneous right ventricular support devices such as the Impella RP (i.e., Bipella) or Protek Duo (LivaNova, London, United Kingdom) to provide biventricular support. For patients with stable, albeit advanced heart failure, a number of other procedures, which recreate traditional surgical ventricular reconstruction through minimally invasive means or percutaneous techniques further illustrate the role of hybrid intervention in patients with advanced heart failure. A comprehensive discussion of these devices and procedures is outside the scope of this chapter, and has been described elsewhere.[58]

Conclusion

Hybrid cardiovascular interventions are becoming increasingly popular treatment options for patients with a range of cardiovascular conditions and epitomize the multidisciplinary care that is the cornerstone of the heart team. Herein we reviewed the many different frontiers of hybrid cardiovascular intervention—from coronary revascularization to structural heart interventions and aortic procedures to therapies for advanced heart failure—and illustrate how these creative therapies leverage the benefits of both traditional, open surgical and percutaneous techniques. As devices and delivery platforms continue to shrink and operators continue to push the boundaries for what can be accomplished through minimally invasive techniques, we can expect hybrid procedural volume to grow and quality to increase.

References

1. Thourani VH, Edelman JJ, Satler LF, Weintraub WS. Surgical aortic valve replacement in the transcatheter aortic valve replacement era: implications for the heart team*. *JACC: Cardiovasc Interv*. 2018;11(21):2157–2159. Available from: https://doi.org/10.1016/j.jcin.2018.07.057.
2. Kolte D, Parikh SA, Piazza G, et al. Vascular teams in peripheral vascular disease. *J Am Coll Cardiol*. 2019;73(19):2477–2486. Available from: https://doi.org/10.1016/j.jacc.2019.03.463.
3. Holmes DR, Rich JB, Zoghbi WA, Mack MJ. The heart team of cardiovascular care. *J Am Coll Cardiol*. 2013;61(9):903–907. Available from: https://doi.org/10.1016/j.jacc.2012.08.1034.
4. Rab T, Ratanapo S, Kern KB, et al. Cardiac shock care centers: JACC review topic of the week. *J Am Coll Cardiol*. 2018;72(16):1972–1980. Available from: https://doi.org/10.1016/j.jacc.2018.07.074.
5. Kabrhel C, Jaff MR, Channick RN, Baker JN, Rosenfield K. A multidisciplinary pulmonary embolism response team. *Chest*. 2013;144(5):1738–1739. Available from: https://doi.org/10.1378/chest.13-1562.
6. Alexander JH, Hafley G, Harrington RA, et al. Efficacy and safety of edifoligide, an E2F transcription factor decoy, for prevention of vein graft failure following coronary artery bypass graft surgery: PREVENT IV: a randomized controlled trial. *JAMA*. 2005;294(19):2446–2454. Available from: https://doi.org/10.1001/jama.294.19.2446.
7. Stone GW, Rizvi A, Newman W, et al. Everolimus-eluting vs paclitaxel-eluting stents in coronary artery disease. *N Engl J Med*. 2010;362(18):1663–1674. Available from: https://doi.org/10.1056/NEJMoa0910496.
8. Bangalore S, Toklu B, Patel N, Feit F, Stone GW. Newer-generation ultrathin strut drug-eluting stents versus older second-generation thicker strut drug-eluting stents for coronary artery disease. *Circulation*. 2018;138(20):2216–2226. Available from: https://doi.org/10.1161/CIRCULATIONAHA.118.034456.
9. Reddy RC. Minimally invasive direct coronary artery bypass: technical considerations. *Semin Thorac Cardiovasc Surg Autumn*. 2011;23(3):216–219. Available from: https://doi.org/10.1053/j.semtcvs.2011.08.011.
10. Bonatti J, Schachner T, Bonaros N, Lehr EJ, Zimrin D, Griffith B. Robotically assisted totally endoscopic coronary bypass surgery. *Circulation*. 2011;124(2):236–244. Available from: https://doi.org/10.1161/circulationaha.110.985267.
11. Halkos ME, Liberman HA, Devireddy C, et al. Early clinical and angiographic outcomes after robotic-assisted coronary artery bypass surgery. *J Thorac Cardiovasc Surg*. 2014;147(1):179–185. Available from: https://doi.org/10.1016/j.jtcvs0.2013.09.010.
12. Hemli JM, Patel NC. Robotic cardiac surgery. *Surg Clin North Am*. 2020;100(2):219–236. Available from: https://doi.org/10.1016/j.suc.2019.12.005.
13. Yang M, Wu Y, Wang G, Xiao C, Zhang H, Gao C. Robotic total arterial off-pump coronary artery bypass grafting: seven-year single-center experience and long-term follow-up of graft patency. *Ann Thorac Surg*. 2015;100(4):1367–1373. Available from: https://doi.org/10.1016/j.athoracsur.2015.04.054.
14. Kon ZN, Brown EN, Tran R, et al. Simultaneous hybrid coronary revascularization reduces postoperative morbidity compared with results from conventional off-pump coronary artery bypass. *J Thorac Cardiovasc Surg*. 2008;135(2):367–375. Available from: https://doi.org/10.1016/j.jtcvs.2007.09.025.
15. Moreno PR, Stone GW, Gonzalez-Lengua CA, Puskas JD. The hybrid coronary approach for optimal revascularization: JACC review topic of the week. *J Am Coll Cardiol*. 2020;76(3):321–333. Available from: https://doi.org/10.1016/j.jacc.2020.04.078.
16. Shen L, Hu S, Wang H, et al. One-stop hybrid coronary revascularization versus coronary artery bypass grafting and percutaneous coronary intervention for the treatment of multivessel coronary artery disease: 3-year follow-up results from a single institution. *J Am Coll Cardiol*. 2013;61(25):2525–2533. Available from: https://doi.org/10.1016/j.jacc.2013.04.007.
17. Sardar P, Kundu A, Bischoff M, et al. Hybrid coronary revascularization versus coronary artery bypass grafting in patients with multivessel coronary artery disease: a *meta*-analysis. *Catheter Cardiovasc Interv*. 2018;91(2):203–212. Available from: https://doi.org/10.1002/ccd.27098.
18. Puskas JD, Halkos ME, DeRose JJ, et al. Hybrid coronary revascularization for the treatment of multivessel coronary artery disease: a multicenter observational study. *J Am Coll Cardiol*. 2016;68(4):356–365. Available from: https://doi.org/10.1016/j.jacc.2016.05.032.
19. George I, Nazif TM, Kalesan B, et al. Feasibility and early safety of single-stage hybrid coronary intervention and valvular cardiac surgery. *Ann Thorac Surg*. 2015;99(6):2032–2037. Available from: https://doi.org/10.1016/j.athoracsur.2015.01.028.
20. Leon MB, Smith CR, Mack M, et al. Transcatheter aortic-valve implantation for aortic stenosis in patients who cannot undergo surgery. *N Engl J Med*. 2010;363(17):1597–1607. Available from: https://doi.org/10.1056/NEJMoa1008232.
21. Bapat V, Thomas M, Hancock J, Wilson K. First successful trans-catheter aortic valve implantation through ascending aorta using Edwards SAPIEN THV system. *Eur J Cardiothorac Surg*. 2010;38(6):811–813. Available from: https://doi.org/10.1016/j.ejcts.2010.03.044.
22. Shin IY, Chung YG, Shin WH, Im SB, Hwang SC, Kim BT. A morphometric study on cadaveric aortic arch and its major branches in 25 korean adults: the perspective of endovascular surgery. *J Korean Neurosurg Soc*. 2008;44(2):78–83. Available from: https://doi.org/10.3340/jkns.2008.44.2.78.
23. Eudailey KW, Olds A, Lewis CT, et al. Contemporary suprasternal transcatheter aortic valve replacement: a multicenter experience using a simple, reliable alternative access approach. *Catheter Cardiovasc Interv*. 2019. Available from: https://doi.org/10.1002/ccd.28460.

24. Elrod J, Prejean S, Daya HA, et al. Suprasternal approach to transcatheter aortic valve replacement in a complex congenital pediatric patient presenting with cardiogenic shock. *Cardiovasc Revasc Med.* 2020. Available from: https://doi.org/10.1016/j.carrev.2020.04.032.
25. Codner P, Pugliese D, Kouz R, et al. Transcatheter aortic valve replacement by a novel suprasternal approach. *Ann Thorac Surg.* 2018;105 (4):1215–1222. Available from: https://doi.org/10.1016/j.athoracsur.2017.10.055.
26. Olds A, Eudailey K, Nazif T, et al. Suprasternal and left axillary transcatheter aortic valve replacement in morbidly obese patients. *Ann Thorac Surg.* 2018;106(6):e325–e327. Available from: https://doi.org/10.1016/j.athoracsur.2018.05.095.
27. Murdock JE, Jensen HA, Thourani VH. Nontransfemoral approaches to transcatheter aortic valve replacement. *Interv Cardiol Clin.* 2015;4 (1):95–105. Available from: https://doi.org/10.1016/j.iccl.2014.09.006.
28. Zhan Y, Saadat S, Soin A, Kawabori M, Chen FY. A meta-analysis comparing transaxillary and transfemoral transcatheter aortic valve replacement. *J Thorac Dis.* 2019;11(12):5140–5151. Available from: https://doi.org/10.21037/jtd.2019.12.07.
29. Bob-Manuel T, Almusawi H, Rezan T, et al. Efficacy and safety of transcarotid transcatheter aortic valve replacement: a systematic review. *Cardiovasc Revasc Med.* 2020;21(7):917–926. Available from: https://doi.org/10.1016/j.carrev.2019.12.012.
30. Ben-Avi R, Orlov B, Sternik L, et al. Short- and long-term results after prosthetic mitral valve implantation in patients with severe mitral annulus calcificationdagger. *Interact Cardiovasc Thorac Surg.* 2017;24(6):876–881. Available from: https://doi.org/10.1093/icvts/ivx043.
31. Casarotto D, Bortolotti U, Thiene G, Gallucci V, Cévese PG. Rupture of the posterior wall of the left ventricle after replacement of the mitral valve: a description of 8 cases. *Artic G Italiano di Cardiologia.* 1977;7(4):387–394.
32. Russell HM, Guerrero ME, Salinger MH, et al. Open atrial transcatheter mitral valve replacement in patients with mitral annular calcification. *J Am College of Cardiol.* 2018;72(13):1437–1448. Available from: https://doi.org/10.1016/j.jacc.2018.07.033.
33. Praz F, Khalique OK, Lee R, et al. Transatrial implantation of a transcatheter heart valve for severe mitral annular calcification. *J Thorac Cardiovasc Surg.* 2018;156(1):132–142. Available from: https://doi.org/10.1016/j.jtcvs.2018.03.016.
34. Muller DWM, Farivar RS, Jansz P, et al. Transcatheter mitral valve replacement for patients with symptomatic mitral regurgitation: a global feasibility trial. *J Am Coll Cardiol.* 2017;69(4):381–391. Available from: https://doi.org/10.1016/j.jacc.2016.10.068.
35. McCarthy PM, Kislitsina ON, Malaisrie SC, Davidson CJ. Transcatheter mitral valve replacement with intrepid. *Interv Cardiol Clin.* 2019;8 (3):287–294. Available from: https://doi.org/10.1016/j.iccl.2019.02.002.
36. Rogers JH, Boyd WD, Smith TW, Bolling SF. Transcatheter mitral valve direct annuloplasty with the millipede IRIS ring. *Interv Cardiol Clin.* 2019;8(3):261–267. Available from: https://doi.org/10.1016/j.iccl.2019.02.001.
37. Rogers JH, Boyd WD, Smith TW, Bolling SF. Early experience with millipede IRIS transcatheter mitral annuloplasty. *Ann Cardiothorac Surg.* 2018;7(6):780–786. Available from: https://doi.org/10.21037/acs.2018.10.05.
38. Cerny S, Benesova M, Skalsky I, Patel NC, Subramanian VA. Persistent reduction of mitral regurgitation by implantation of a transannular mitral bridge: durability and effectiveness of the repair at 2 years—results of a prospective trial. *Eur J Cardio-Thoracic Surg.* 2018;55 (5):867–873. Available from: https://doi.org/10.1093/ejcts/ezy423.
39. Gammie JS, Bartus K, Gackowski A, et al. Safety and performance of a novel transventricular beating heart mitral valve repair system: 1-year outcomes. *Eur J Cardio-Thoracic Surg.* 2020;59(1):199–206. Available from: https://doi.org/10.1093/ejcts/ezaa256.
40. Fatehi Hassanabad A, Kent WDT, Adams C. Review of transapical off-pump mitral valve intervention with NeoChord implantation. *Curr Opin Cardiol.* 2020. Publish Ahead of Print Available from https://doi.org/10.1097/hco.0000000000000825.
41. Thourani VH, George I, Rucinskas K, et al. First in human experience with an epicardial beating heart device for secondary mitral regurgitation. *J Thorac Cardiovasc Surg.* 2020. Available from: https://doi.org/10.1016/j.jtcvs.2020.11.169.
42. Enta Y, Nakamura M. Transcatheter mitral valve replacement. *J Cardiol.* 2020. Available from: https://doi.org/10.1016/j.jjcc.2020.10.020.
43. Lu F-L, Ma Y, An Z, et al. First-in-man experience of transcatheter tricuspid valve replacement with LuX-Valve in high-risk tricuspid regurgitation patients. *JACC: Cardiovasc Interv.* 2020;13(13):1614–1616. Available from: https://doi.org/10.1016/j.jcin.2020.03.026.
44. Hahn RT, Kodali S, Fam N, et al. Early multinational experience of transcatheter tricuspid valve replacement for treating severe tricuspid regurgitation. *JACC Cardiovasc Interv.* 2020;13(21):2482–2493. Available from: https://doi.org/10.1016/j.jcin.2020.07.008.
45. Asmarats L, Puri R, Latib A, Navia JL, Rodés-Cabau J. Transcatheter tricuspid valve interventions: landscape, challenges, and future directions. *J Am Coll Cardiol.* 2018;71(25):2935–2956. Available from: https://doi.org/10.1016/j.jacc.2018.04.031.
46. Vallabhajosyula P, Szeto W, Desai N, Bavaria JE. Type I and Type II hybrid aortic arch replacement: postoperative and mid-term outcome analysis. *Ann Cardiothorac Surg.* 2013;2(3):280–287. Available from: https://doi.org/10.3978/j.issn.2225-31927.2013.05.09.
47. Bavaria J, Vallabhajosyula P, Moeller P, Szeto W, Desai N, Pochettino A. Hybrid approaches in the treatment of aortic arch aneurysms: postoperative and midterm outcomes. *J Thorac Cardiovasc Surg.* 2013;145(3 Suppl.):S85–S90. Available from: https://doi.org/10.1016/j.jtcvs.2012.11.044.
48. Saadi EK, Tagliari AP, Almeida RMS. Endovascular treatment of the ascending aorta: is this the last Frontier in aortic surgery? *Braz J Cardiovasc Surg.* 2019;34:759–764.
49. Blitzer D, Nguyen S, Patel V, George I, Takayama H. *Hybrid Repair of Ascending Aortic Pseudoaneurysm.* Authorea Preprints; 2020.
50. Miller WH, Hsi D, Koulova A, et al. Pseudoaneurysm as a late complication of chronic stanford type A intramural hematoma requiring endovascular repair. *JACC: Case Rep.* 2020;2(15):2470–2475. Available from: https://doi.org/10.1016/j.jaccas.2020.10.017.
51. Mulaikal TA, Pospishil L, George I, Bapat VN, Hahn RT. Surgically assisted transcatheter balloon-expandable valve in inferior vena cava for torrential tricuspid regurgitation. *CASE.* 2018;2(5):174–180. Available from: https://doi.org/10.1016/j.case.2018.04.009.
52. Song J, Ascherman B, Eudailey KW, et al. Long-term failure of Amplatzer plugs in the treatment of aortic pathology. *J Card Surg.* 2017;32 (7):426–429. Available from: https://doi.org/10.1111/jocs.13166.

53. Najjar M, Mohar M, Stewart A, George I. Hybrid endovascular repair of Kommerell diverticulum and aberrant right subclavian artery in a patient with repaired coarctation of the aorta. *Heart Lung Vessel*. 2015;7(1):83–85.
54. Kossar AP, Takayama H, Patel V, George I. Novel implementation of a cerebral protection system during ascending thoracic endovascular aortic repair (TEVAR). *Semin Thorac Cardiovasc Surg Summer*. 2019;31(2):218–221. Available from: https://doi.org/10.1053/j.semtcvs.2018.09.026.
55. Dershowitz L, Wolbinski M, Bapat V, George I, Kodali S. Left ventricular injury: beware the wire. *JTCVS Tech*. 2020;3:126–129. Available from: https://doi.org/10.1016/j.xjtc.2020.02.019.
56. Yanaka K, Nakayama K, Shinke T, et al. Sequential hybrid therapy with pulmonary endarterectomy and additional balloon pulmonary angioplasty for chronic thromboembolic pulmonary hypertension. *J Am Heart Assoc*. 2018;7(13). Available from: https://doi.org/10.1161/jaha.118.008838.
57. Anderson M, Smith D, Kane P, Lee R, Khalpey Z, Williams J. Impella 5.5 direct aortic implant and explant techniques. *Ann Thorac Surg*. 2020. Available from: https://doi.org/10.1016/j.athoracsur.2020.09.069.
58. Brener MI, Uriel N, Burkhoff D. Left ventricular volume reduction and reshaping as a treatment option for heart failure. *Struct Heart*. 2020;4(4):264–283. Available from: https://doi.org/10.1080/24748706.2020.1777359.

Chapter 23

Intraoperative Cardiac Dearing: New Concept and Technology

Jamshid H. Karimov[1,2], Marc Gillinov[3], Kiyotaka Fukamachi[1,2] and Mark S. Lobosky[1]

[1]Lerner Research Institute, Cleveland Clinic, Cleveland, OH, United States, [2]Cleveland Clinic Lerner College of Medicine of Case Western Reserve University, Cleveland, OH, United States, [3]Heart, Thoracic, and Vascular Institute, Cleveland Clinic, Cleveland, OH, United States

Chapter Outline

Background

More than 490,000 cardiac surgery procedures are performed each year in the United States alone, and the number of invasive surgical cases is increasing. In 2020, the US cardiac surgery market was valued at $5.8 billion; despite COVID-19, the market size is expected to increase at a rate of 7.7% to $12.1 billion in 2026, according to report by iData Research.[1] Less invasive approaches have been introduced into practice over the last decade. As a result, the numbers of patients opting for minimally invasive or robotic cardiac surgery, and the numbers of hospitals offering this type of care, are increasing, with continuously improving postoperative outcomes and quality of life. Despite tremendous success and technological breakthroughs in surgical instrumentation and operative platforms, there is still a lack of effective solutions for issues related to alteration to tissue integrity during incisions for open heart access, cardiopulmonary bypass (CPB) perfusion, and intraoperative air embolism.

Since cardiac surgery was first practiced, retained intracardiac air has been an important problem. The effects of air embolism during heart operations and methods intended to prevent it have long been studied.[2–4] Although the introduction of ultrasound visualization (transesophageal echocardiography [TEE]) has allowed for detection of air and guided dearing procedures, adequate air removal is not always attained due to the individual preferences of each surgeon. Air removal procedures appear to vary a great deal among institutions, and are not necessarily based on firm scientific evidence.[5,6]

In all open-heart surgeries, air inevitably gets introduced into the heart during repair and remains accumulated within cardiac chambers. Once cardiac repair is completed, air retained inside the heart must be removed to preclude embolic complications, for example, stroke or death. Any air migrating into the cerebral circulation can be fatal, and air emboli in coronary arteries can cause cardiac arrest, which substantially worsens the surgical outcomes, affects patient survival, delays recovery, and elevates the overall costs of care.[7]

The standard air-removal techniques are based on conventional techniques that involve manual handling of the heart for gentle massage/compression by the surgeon to mobilize air deposited on cardiac walls and between chambers. The surgeon uses TEE to visualize air bubbles as they dislodge and dissipate. However, for minimally invasive cardiac surgeries (MICS) based on operation through small incisions (ranging from 4–8 cm in length), or totally endoscopic procedures, cardiac manipulation is not feasible by any existing means to access the heart. The use of CO_2 in MICS is probably more important compared with open cardiac surgery, as minimal invasive cardiac surgery does not permit normal dearing maneuvers and the minimal invasive surgical wound cavity is much smaller.[8]

Advances in Cardiovascular Technology. DOI: https://doi.org/10.1016/B978-0-12-816861-5.00040-X

Therefore application of known deairing maneuvers in minimally invasive and robotic surgery, given the inherent restrictions in those types of operations, further limits surgical practitioners in addressing known complications, and prevents effective removal of retained air through limited surgical access. This in turn renders altered the less invasive aspect of the procedure due to reduced safety of the deairing component of the surgical procedure.

To address this important unmet clinical need, we have developed the cardiac deairing system (CDS). This innovative device, temporarily introduced under the heart between the inferior cardiac surface and pericardium, provides effective deairing in MICS via motion, by "wiggling" the heart at a required frequency, thereby dislodging any entrapped intracardiac air. The dislodged air is then captured by a dedicated suction vent.

An air embolism can lead to life-threatening complications such as cardiac arrest or stroke, and can prolong operative time. These can compromise patient survival, delay patients' recovery, and adversely affect outcomes. The occurrence of air microemboli in open-heart surgery has been correlated with the degree of postoperative neuropsychological disorders.[9] A recent retrospective analysis based on electronic medical records over a 25-year period showed that air embolism was seen in 94% of in-hospital cases; of these, 77.8% occurred during an operation/invasive procedure and were mainly iatrogenic.[10,11]

The overall mortality rate among stroke patients was 18%, compared with 2% in patients without a complication of stroke, which also nearly doubled the total cost of hospitalization for those who have a postoperative stroke.[12]

The entry of gas into vascular blood flow is a risk in virtually all areas of clinical care. Venous emboli may lead to cardiovascular collapse or to paradoxical arterial emboli. Arterial emboli may occlude end arteries throughout the body, and may cause serious morbidity or death if they occlude cardiac or cerebral vessels.[4] The incidence of iatrogenic air embolisms can only be estimated, since many accidents are not recognized. Clinical manifestations, essentially neurological or cardiovascular disorders, vary greatly.

An air embolism may occur during coronary or cerebral arteriography, CPB, venous catheterization, various types of surgery, or blood transfusion, among other situations. Once air has entered the arterial circulation, the bubble of gas follows the blood flow until it is blocked by a smaller caliber vessel. The progressive diffusion of the air reduces the size of the embolus, which then migrates to smaller and smaller vessels. Subsequent pathological manifestations of air embolism result from mechanical obstruction, leading to ischemia, and inflammatory reactions to air acting as a foreign body.[13]

To address this clinical need for all MICS procedures, we are developing the CDS, which can be introduced easily not only through small incisions, but also through keyhole (stab) incisions and ports placed through the chest. The CDS will use the regular air and vacuum supplies that are standard in the operating room to be activated.

In this chapter, we describe the development of the CDS and report our in vitro research and testing and in vivo feasibility study to test the design features and characteristics of this technology.

Current practice and clinical rationale

All cardiac surgery procedures consist of establishing access to the heart and its internal structures to perform repair. The heart's chambers are usually filled with blood; once accessed from outside, some amount of air enters the heart, lodging inside the chambers, and can cause catastrophic complications, such as cardiac arrest and stroke.[14] Postoperative stroke remains as one of the most devastating complications after cardiac surgery, with air embolism among the primary etiologic contributors for intraoperative stroke.[12]

A recent analysis on adverse events following cardiac surgical procedures in the United States included air embolism in a list of "never events."[14,15] According to the Leapfrog Group (Washington, D.C.), "never events" are defined as "adverse events that are serious, largely preventable, and of concern to both the public and health care providers for the purpose of public accountability."[16] The analysis revealed increased hospital length of stay, hospital charges, and mortality in patients who suffered at least one "never event" compared with those who did not.

Minimally invasive cardiac operative techniques have emerged and progressed during the last decade with many advantages, such as less postoperative pain, lower morbidity and mortality, faster recovery, and shorter duration of operation and hospital stay, associated with lower costs. Modifications of less invasive operations have been developed to offer specific patients and clinical presentations the most suitable minimally invasive approach. The smaller incision per se and the positioning of the surgical tools and catheter through that incision render valve operations technically more complex compared to the standard (full-size incision) approach. Therefore routine techniques, such as institution of the desired venting cannulation, and adequate assessment of retained intracardiac air and its removal techniques, require dedicated skills and experience from surgeons, anesthetists, and perfusionists for a successful result.[17–21]

Traditional air-removal techniques are based on conventional techniques that usually involve manual handling of the heart, that is, gentle massage/compression by the surgeon to mobilize air deposited on cardiac walls and between chambers. Other maneuvers to manage air entrapment inside the cardiac chamber consist of shaking the patient's body, CO_2 insufflation,[8] and surgical position change (Trendelenburg).[22–24] However, there is no evidence of sustained reduction of embolism with CO_2, and Trendelenburg positions do not decrease the cerebral microembolic load compared with the horizontal head position.[22,23,25]

The CO_2 insufflation method has been used with limited success since the 1950s.[8] In a multicenter, prospective, double-blind, randomized, controlled trial, Chaudry et al. compared neurocognitive outcomes in patients ($n = 125$) undergoing open-chamber (left-sided) cardiac surgery who were assigned CO_2 insufflation (through a gas diffuser) or placebo in addition to standardized mechanical deairing maneuvers. The authors reported no difference in neurocognitive outcomes at 6 weeks with CO_2 versus standard mechanical deairing maneuvers alone during open-chamber cardiac surgery.[26] CO_2 use did lead to fewer gaseous emboli in the cardiac chambers during weaning from CPB. This still did not translate into improved neurocognitive outcomes at 6 weeks, however, suggesting that sufficient deairing can be achieved through mechanical deairing maneuvers alone.

In meta-analysis of randomized clinical trials (8 studies), with 668 patients randomized to CO_2 insufflation ($n = 332$) versus standard deairing maneuvers ($n = 336$), the neurocognitive deterioration rate was 12% versus 21% in the CO_2 and control groups, respectively; however, this difference (not statistically significant) did not support any significant protective effect from CO_2 use compared with manual deairing maneuvers in terms of clinical outcomes, including postoperative neurocognitive decline.[27] Due to the lack of powerful evidence, the cerebral and cardiac protection efficacy of CO_2 insufflation during open heart surgery needs to be verified by more high-quality trials.[25,28]

The "deairing time before cardiac ejection" reflects the time required to evacuate air from the left atrium (LA), left ventricle (LV), and aorta (AO). Deairing from these chambers may effectively be achieved by a vent placed in the LV, or aortic root, or both. The "deairing time after cardiac ejection," on the other hand, represents a more complex part of the deairing procedure—namely, the deairing of residual air trapped in the pulmonary veins.[29] However, the limitation of the standard venting approach does not lie in the infectiveness of air emboli evacuation, but in the presence of air pools and entrapped air deposition not susceptible to detachment from cardiac walls and dislodgement (their effective propagation) toward the positioned vents.

Importantly, existing methods focus on air-suction techniques through vents placed in the aorta or ventricle. These techniques are primitive and extremely time-consuming. No existing method addresses air dislodgement inside the heart, which seems to be a primary reason why air, once entrapped within the heart, does not move easily and creates "pools" of air, which are even harder to manage.[24] The existing techniques are even less feasible in MICS than in the standard setting due to limitations in incision size (4- to 7-cm incision length), which limits any manual access to the heart. Because intracardiac air rapidly changes its location and appearance, timely intervention to remove air is critical, especially before weaning from bypass. For this reason, it is important to develop ways to protect against these complications, which can ultimately lead to poorer outcomes and a higher risk of mortality.

Retained intracardiac air remains a hazard for surgical outcome and successful recovery after open heart procedures.[30] Failure to remove entrapped air from the heart results in an increased incremental cost of recovery.[6,7] Data suggest that the detriment in outcomes associated with postoperative air embolism is mostly due to retained intracardiac air, and is detectable in $\sim 100\%$ of cases.[2] Pooled air has been found in different locations and was observable on TEE.[6] Because trapped intracardiac air rapidly changes locations and appearance, continuous TEE monitoring is used for air removal. It is difficult to prove that gaseous microemboli cause organic brain lesions, however, because of the inability of various detection modalities to distinguish reliably between solid and gaseous emboli.[26]

Tingleff et al. reported an extensive TEE evaluation of intraoperative air embolism in open heart surgical cases.[5] The study was divided into three periods: Period I (beginning of deairing to release of the aortic cross-clamp: 2.5–12 minute duration); Period 2 (from release of the aortic cross-clamp to termination of CPB: 15–46 minute duration); and Period 3 (from the end of CPB to the end of the operation: 18–63 minute duration). The substantial presence of retained intracardiac air (from various locations) can further prolong the operating times (aortic cross-clamp time, CPB time, and thus total operating time and anesthesia duration) and negatively affect the postoperative outcomes.

Unrecognized, poorly managed, or unaddressed intracardiac air entrapment leads to adverse patient outcomes and increases resource utilization caused by never events. This situation emphasizes the pressing need for quality improvement surrounding this issue. Minimization of air embolisms, along with effective mobilization of air, could reduce brain embolisms and improve postoperative outcomes and intellectual function in cardiac surgical patients.[5]

Cardiac dealring system

To make the new dealring approach compatible with all MICS procedures, we have designed the CDS (Fig. 23.1). The system consists of a catheter and inflatable device (bladder) coupled to the catheter. In an active state, the inflatable device is sized and configured to transmit motion to the patient's heart sufficient to change its shape.

The CDS operates by means of active inflation (air) and deflation (vacuum) applied to the bladder. The system also includes a remote controller comprising pneumatic valves (4), solid state relays (4), a DC power supply, I/O modules (2), and a self-relieving pressure regulator. It is controlled by a custom logic program written in National Instruments LabVIEW software (LabVIEW, National Instruments, Austin, TX). The system controls the pneumatic valves to inflate and deflate the inflatable device. The logic program provides a user interface to adjust the automatic inflation/deflation parameters to maintain a specific frequency pattern that achieves an adequate displacement of the heart to dislodge air bubbles. The user interface provides additional feedback to the operator on activation/deactivation or operation failure.

When the CDS transmits motion to the heart at a required frequency, cardiac motion dislodges entrapped intracardiac air, which is eventually evacuated through a left ventricular or aortic vent. The intrathoracic portion of the CDS (the bladder and a catheter) has been enabled with a malleability feature, which allows the operator to preshape the catheter before insertion and provide controlled firmness for proper insertion. The "bendiness" of the metallic insert of the CDS allows the operator to feel when the catheter tip abuts the pericardial wall. This is an important requirement to avoid injury to surrounding tissues during insertion. The malleable insert is removed after insertion, and the device will be immediately ready for activation and use.

The primary idea in designing the CDS tool was to enable its application in various types of MICS, including robotic-assisted cases (Da Vinci, Intuitive Surgical, Sunnyvale, CA) and other cardiac surgical procedures performed through smaller incisions. The development of a new device system was the result of extensive bench testing using various early and advanced stage prototypes, extensive material analysis, and subsequent animal experimentation at Cleveland Clinic that provided the necessary feasibility data. The device parameters were adjusted for the first in-human study (currently pending) with design configuration and parameters that were optimized to fit the chest of cardiac surgical patients in real intraoperative settings.

Design and development

The following requirements have been identified during the early CDS design phase. The system had to be: (1) deployable through a catheter-based external (outer) sheath; (2) malleable via a bendable component (guidewire or insert) that would enable preinsertion shaping of the catheter to ensure correct navigation inside the patient's chest and correct area of CDS delivery, to allow easy and secure device placement and avoid cardiac laceration or penetration by a catheter tip prior to activation; (3) retrievable and repositionable; and (4) enabled for proper operation in wet and naturally-lubricated thoracic environment via an optimized bladder surface.

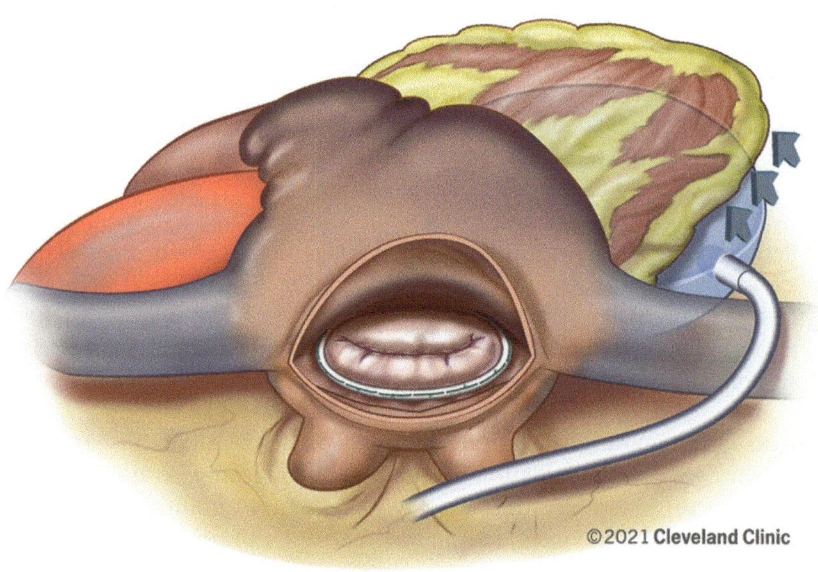

FIGURE 23.1 The cardiac dealring system.

The improved CDS would be constructed to obtain these measurements: (1) bladder volume and activation parameters (inflation, deflation, and "dwell" times); (2) an ergonomic handle that allows tactile feeling by the surgeon during insertion; (3) controller box feedback on activation, deactivation, troubleshooting, and potential failure (low/high air pressure, low/high vacuum, pressure release valves, and light indicators for CDS activation); and (4) controller software optimization for smooth, seamless device operation (to allow an uninterrupted change from one functional mode to another without device deactivation in between, as in the current prototype's design). The control box of the CDS has been shielded and certified against electrical interference with other medical equipment and safety.

Additionally, thorough cost analysis, safety evaluation, troubleshooting, and detailed end-users' feedback has been obtained at this phase to create an ergonomic device design ready for clinical evaluation. (This will include analysis on device handling, operation, and compatibility with standard surgical techniques.)

The initial CDS prototypes were fabricated and tested in a constructed bench-test model. The bladder portion was made of different sizes and shapes so that the device performance and orientation inside the chest could be tested (Fig. 23.2). For the earlier prototype, a PC-based (LabVIEW, National Instruments, Austin, TX) device controller was used to regulate the device (air, 40–60 psi; vacuum, 17–20 mm Hg) through two solenoid valves (NUMATICS Pneumatic Valves, ASCO, Florham Park, NJ) and a pressure-release valve (Qosina, Ronkonkoma, NY) (Fig. 23.3).

The catheter engineering has involved prototyping and optimization of the malleable catheter insert, outer sheath and handle for fully collapsible and retrievable bladder portion and bench-testing it to validate functional efficacy. The bladder (inflatable–deflatable unit) was introduced under the heart during the procedure through the thoracic port, directly through the surgical incision, or through the outer catheter-based sheath. The unit was then connected to an external controller that can regulate (inflate/deflate) the bladder unit. Inflation/deflation occurred by means of injection (air) and deflation (vacuum) through the controller connected to the external air and vacuum source available in the operating room.

The bladder portion of the CDS needed to fit into the 5–12-mm catheter sheath to be suitable for a range of procedures that use working ports placed through intercostal space. The ultimate controller design needed to provide additional feedback to the end-user on CDS activation, deactivation, or activation or operation failure. This should include implementation of a slider or rotary control into the operator's graphical user interface (GUI), using the existing LabVIEW code, a vacuum switch to detect air pressure and optionally show status of the system through controller interface, indicator light(s) to show the status of the system, and pressure release valves with pressure switches to detect air pressures (Fig. 23.4).

In order to reduce the size of the controller unit, several handheld design features and device embodiments were tested and evaluated. The advanced control unit design approach was to create a fully contained battery-operated unit, rendered as an easy-to-use, ergonomic solution that will be operable by a single hand, and be enabled of all controller features such as frequency of bladder inflation, dwell times, and adjustment control buttons. The most ergonomic design

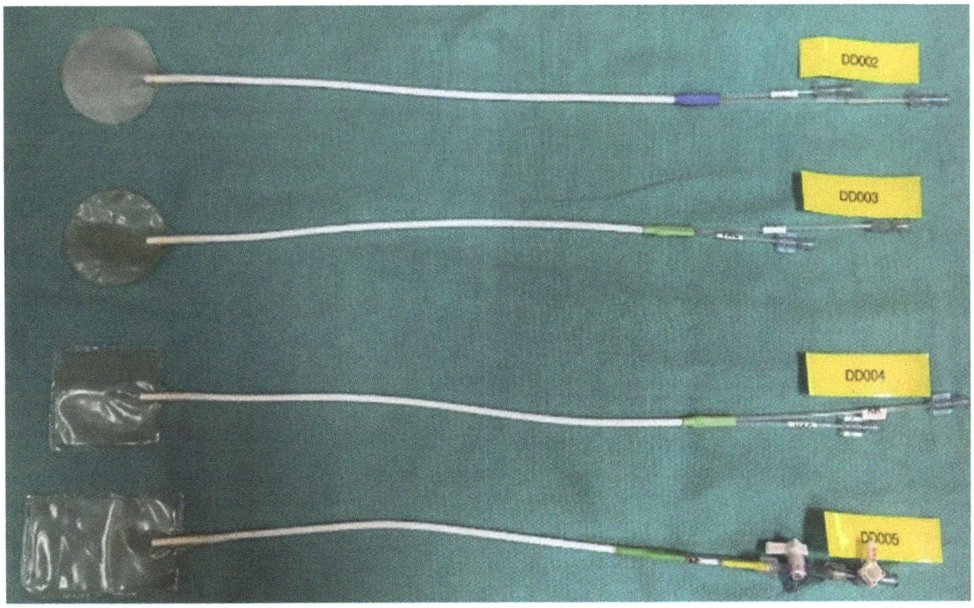

FIGURE 23.2 Cardiac dearing system with double-lumen extruded catheters with Luer connectors. Various size and shape bladders are shown.

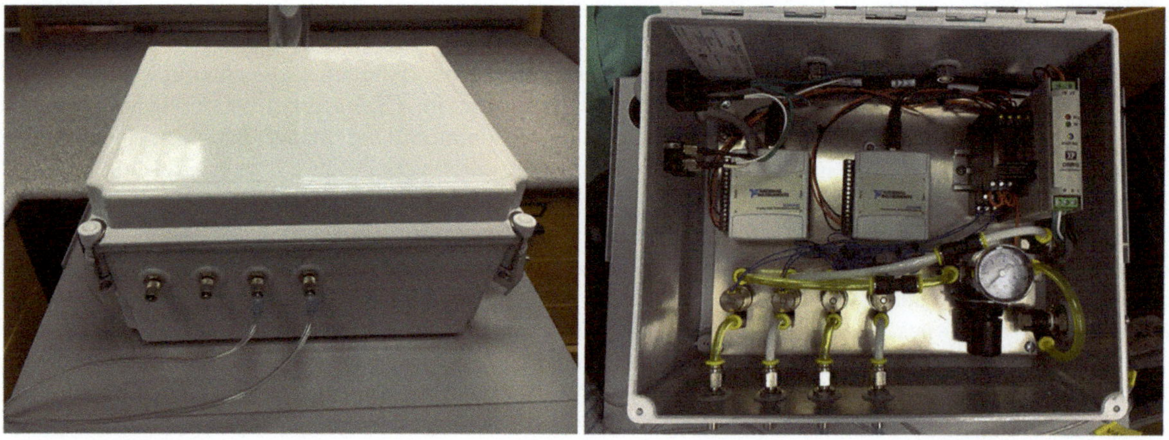

FIGURE 23.3 Early prototype of the cardiac dealring system controller unit (left). Two sets of metal Luer connectors are provided for inflation (air) and deflation (vacuum), internal power source, pressure manometer with regulators (right).

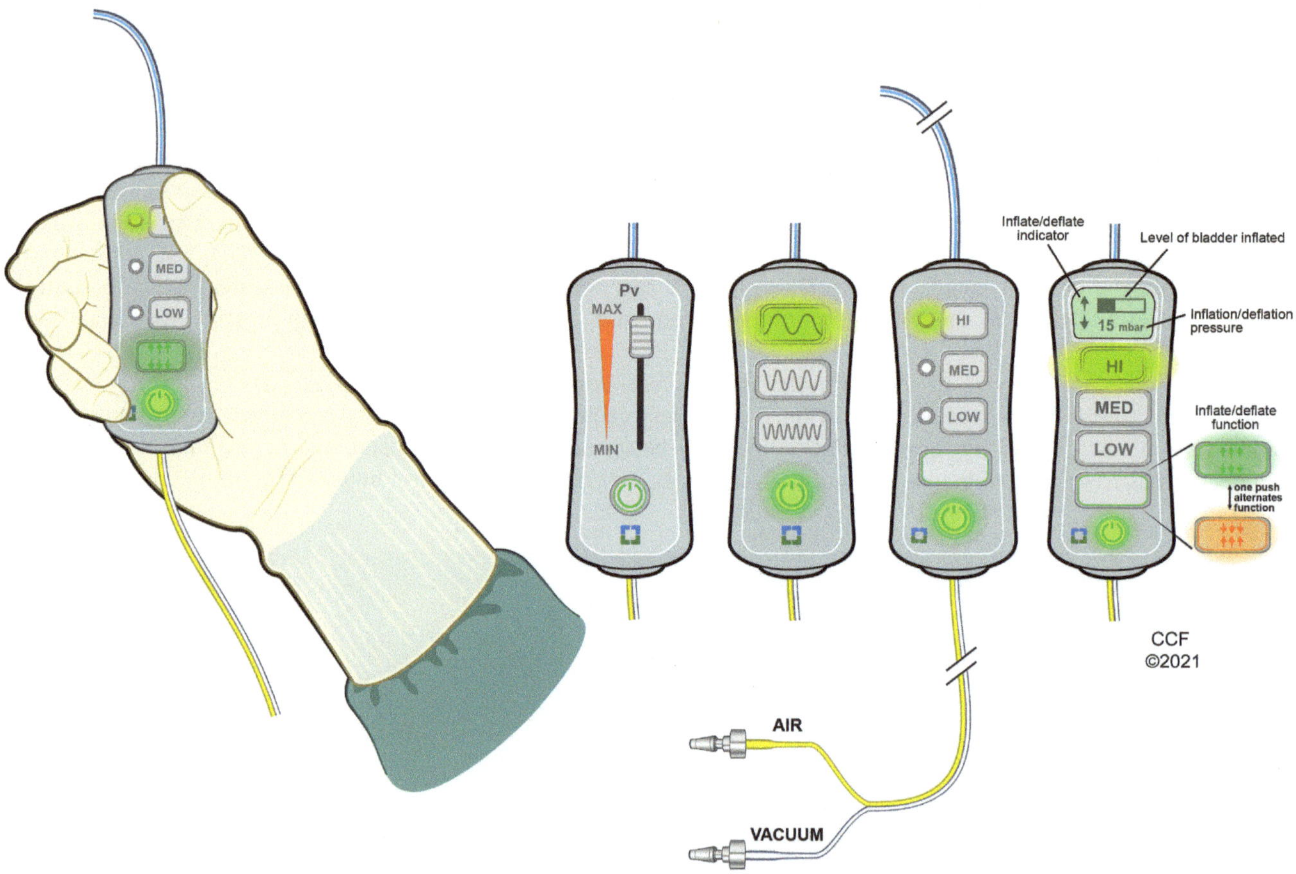

FIGURE 23.4 Schematic illustration of selected design requirements and features for battery-operated handheld cardiac deairing system controller unit.

of the new prototype has been selected: the device has incorporated two stainless steel solenoid valves, power supply (off-the-shelf 9 V internal battery), and an internal electric architecture built on an open source circuit board platform (Arduino LLC, Boston, MA) that enabled unit programmability and parameter adjustments (Fig. 23.5). The functional range from 0 to 50 Hz has been established; however, higher ranges with the advanced CDS prototype have also been tested, as bladder size may cause some changes in effectiveness of the stroke-motion that it generates during activation.

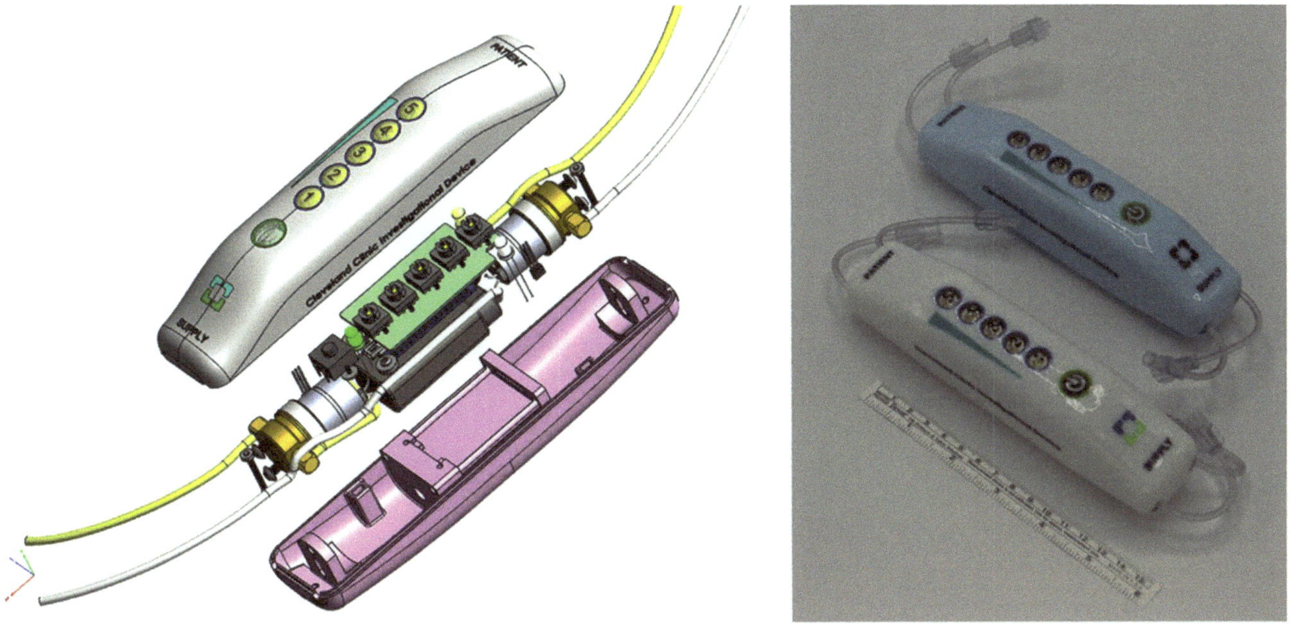

FIGURE 23.5 Advanced prototype of handheld cardiac dearing system controller unit. Membrane switches (programmable within intended operational range 0–12 Hz) in the mid-portion of the front housing; "switch" levels lowered, and placed "into" the housing; smaller, ergonomic location for buttons; LED assigned to each preset work mode; eliminated sharp angles for improved intraoperative device handlings; air and vacuum lines exteriorized with different Luer ends (male and female) on both device ends to prevent errors during connections; miniature internal circuit board for parameter adjustments and device control.

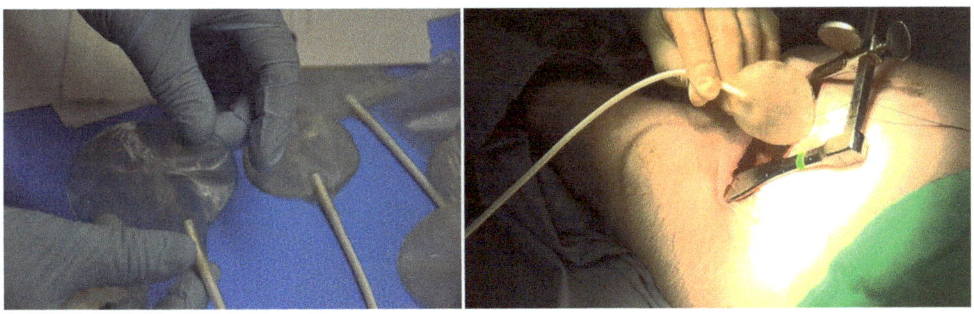

FIGURE 23.6 Intraoperative views of thoracotomy, testing inflatable chambers (bladders) and device insertion through incisions.

The in vivo study has been approved by Cleveland Clinic's Institutional Animal Care and Use Committee. Six healthy Yorkshire pigs (70–100 kg) were used to evaluate the efficacy of the CDS in an acute model. The acute in vivo procedures were performed without a CPB, since performing any steps on intracardiac structures was outside the scope of the study and opening the heart was not planned.

Right thoracotomy access was constructed for device insertion in the 5th–7th intercostal space (Fig. 23.6). Stay sutures were placed to retract the pericardial tissue for better visualization. Once the atria were exposed, the inflatable portion of the system was advanced bluntly through the anatomical structures and positioned under the heart. The optimized CDS was constructed of medical-grade, ethylene oxide sterilizable components, fully sealed, intended for a single use (Fig. 23.7).

The device was delivered through (1) an intercostal port, or directly (2) through a surgical incision, navigated and positioned properly under the heart. The delivery through ports was accomplished by bladder crimping and feeding through the working port (a), or, using a catheter-based approach (b); the bladder was packed into a delivery sheath (Fig. 23.8), delivered, and deployed once the delivery sheath tip was positioned within anatomical structure, beneath the heart (Fig. 23.9).

The outer sheath of the delivery catheter was withdrawn and the bladder portion was deployed. The catheter was then connected to the external controller (outside the chest), which regulated the device's inflation/deflation frequency

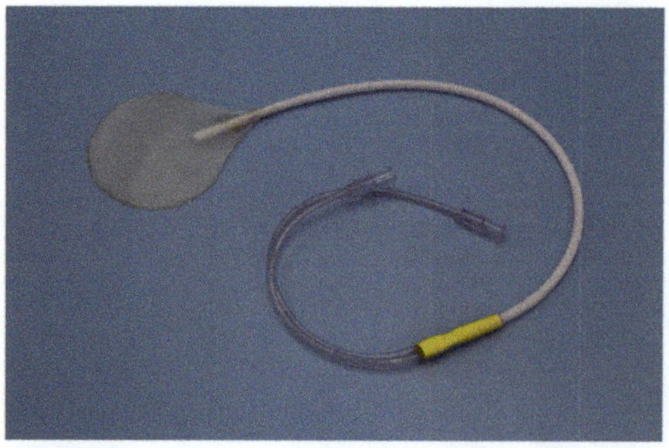

FIGURE 23.7 The optimized inflatable unit (bladder). Double-lumen extrusion tubing (36 cm, measured from the distal edge of the bladder until the Y-connector); small size catheters (OD 8.2 cm; from that up to ~0.2 mm for sealing area); larger catheters (OD 10.2 cm; from that up to ~0.2 mm for sealing area); even distribution of sealed areas at the edge.

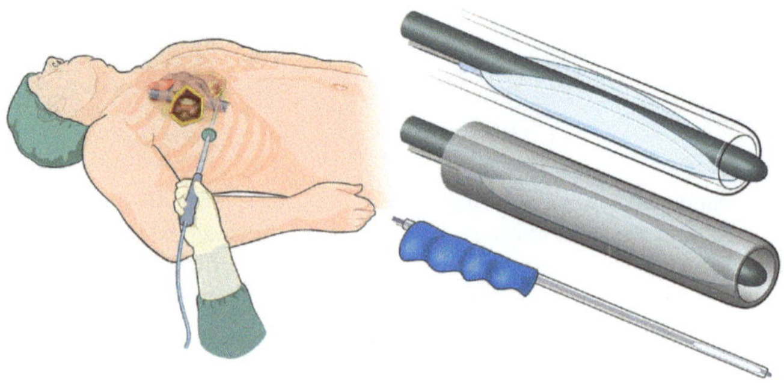

FIGURE 23.8 Transcatheter cardiac deairing system device concept.

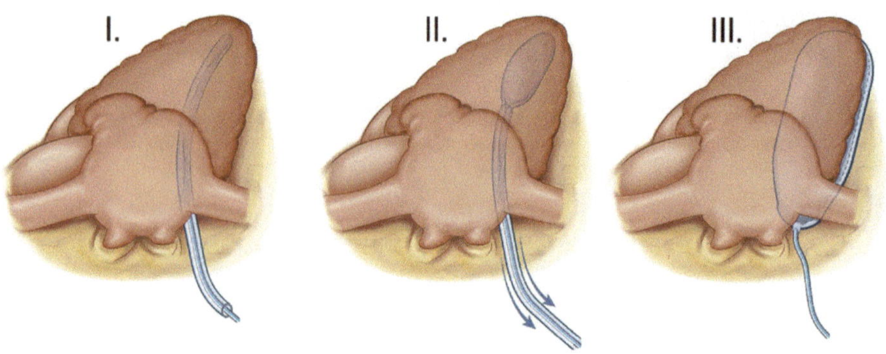

FIGURE 23.9 Step-by-step insertion of cardiac deairing system bladder through the delivery sheath. The option of device insertion under the inferior vena cava is shown.

and volume. The device was activated and tested at different working modes, which consisted of different device inflation/deflation patterns and frequencies. CDS activation data in the 2–33-Hz range were recorded.

The CDS controller software enabled frequency adjustments at any range, which helped us to narrow down the most effective CDS working modes. CDS motion was successfully transmitted to the heart in all cases ($n = 6$). TEE and epicardial echocardiography imaging techniques were feasible in all cases, and cardiac motion was recorded (Fig. 23.10). In some cases, air was introduced through a long needle (16 G) placed into the left ventricle (LV); successful air dislodgement from cardiac walls and septum during CDS activation was observable on TEE. The pooled air remained within the LV when the CDS was off (Fig. 23.11). A 10-mm endoscope (Karl Storz, El Segundo, CA) was used to obtain additional device-related data during the in vivo studies. The endoscope visualized the device at work, where the unique interaction between the inferior cardiac surface and pericardium was recorded on-site during device activation (Fig. 23.12) in two cases 30 degree angle endoscope. Once the study was completed, the device was effortlessly removed from the chest.

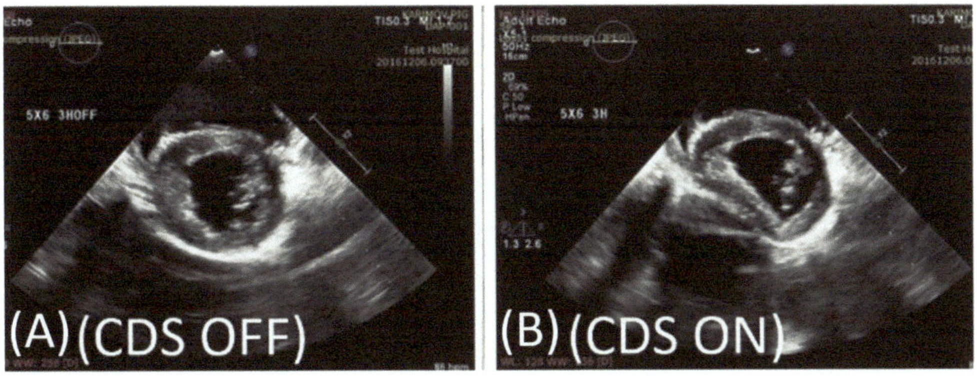

FIGURE 23.10 Transesophageal echocardiography during cardiac dearing system testing (A) normal round heart; (B) cardiac shape during device activation, as intended.

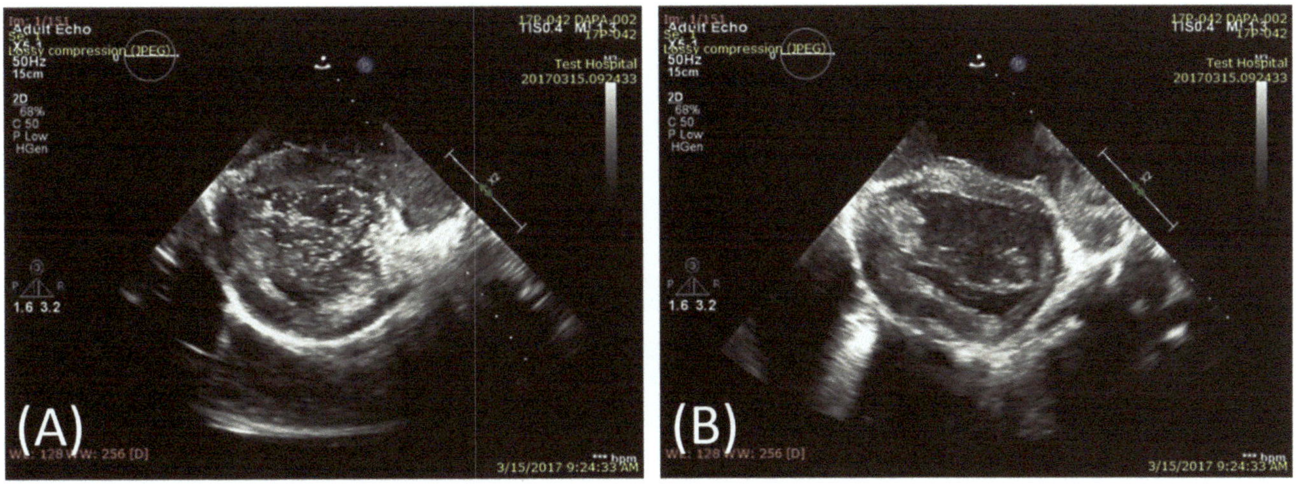

FIGURE 23.11 Transesophageal echocardiography during cardiac dearing system testing. (A) Device OFF, myocardial standstill, air present in the left ventricle (note absence of noticeable movement inside cardiac chamber); (B) Device ON, air depositions successfully dislodged (note movement of air bubbles introduced into the left ventricle).

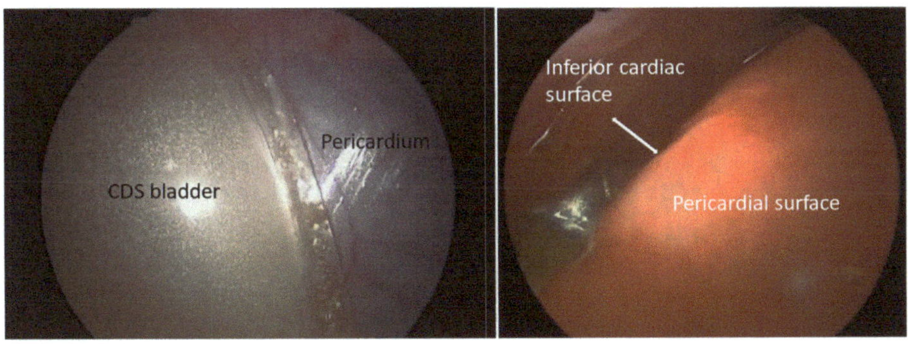

FIGURE 23.12 Endoscopic view: (A) cardiac dearing system properly located over pericardium; (B) side view of cardiac dearing system expansion between heart and pericardial surface.

The in vivo studies showed the optimal CDS activation at 1.5–33 Hz, with fixed pressure parameters. The motion transmitted to the cardiac wall by the bladder effectively dislodged the entrapped air and assisted its detachment from the internal cardiac surface in all cases. The device expansion at the intended range was achieved and shown to be adjustable and fully operable at preset functional modes. No tissue dissection was required for device placement during the acute setting.

An important finding from our in vivo studies is the ability of the CDS to be easily introduced under the heart and to keep its position, without migration, within the partially intact pericardial sac. These results suggest that the CDS

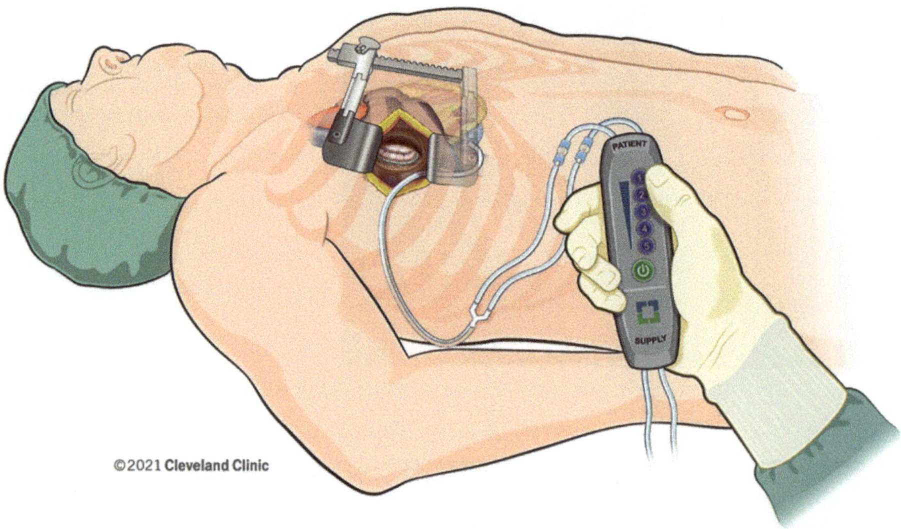

FIGURE 23.13 Advanced iteration of the single-use cardiac dealring system device. The device is surgeon-controlled, suitable for through-port or direct insertion, and activated pressure and vacuum sources in the operating room.

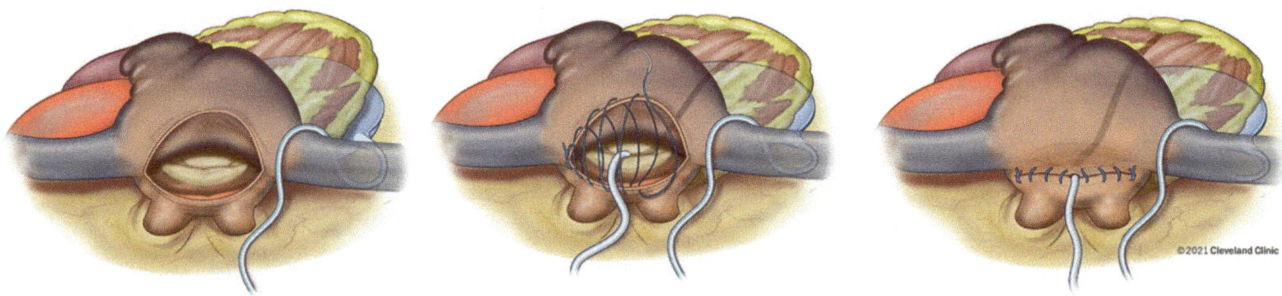

FIGURE 23.14 Cardiac dealring system setup incorporated with standard surgical procedure.

will be easy to insert and use, and that the current prototype models are effective in dislodging intracardiac air. This could have a significant preventive effect on surgical outcomes through reduction of operating room time, as the air-removal technique has been demonstrated to be a fully controlled and measurable technique.

Based on prior research and development, the advanced system was designed for easy introduction not only through small incisions but also via small keyhole (stab) incisions and working ports placed through the chest (Fig. 23.13). Without alterations to existing deairing protocols that are based primarily on vent placement in the LV, the system would seamlessly be incorporated in the surgical workflow (Fig. 23.14). Its applications can potentially be expanded to other cardiac surgical procedures operationalized through smaller scale incisions (Fig. 23.15), where any existing technique addressing retained intracardiac air removal will not be considered an option due to limitations of manual deairing and restrained operative space.

Discussion

In the July 1914 edition of *Annals of Surgery*, Alexis Carrel wrote on his milestone work on a course of experiments conducted in 1913 and 1914 to develop techniques in which plastic operations on the pulmonary and the aortic orifices of the heart could be performed. He stated that the opening of the ventricles or of the pulmonary artery and the aorta is always followed by entrance of air into the heart.[2] Since the early conception of open cardiac surgery techniques, the field has undergone a myriad of changes and innovation; however, the issue of retained cardiac air still persists and represents an important complication of modern surgical practice.

The great risks that the presence of air microemboli in open-heart surgery implies have recently been emphasized by reports that their number correlates with the degree of postoperative neuropsychological disorder.[28] The deairing procedures are rather similar among institutions but not fully validated scientifically.[7] The notoriously ineffective

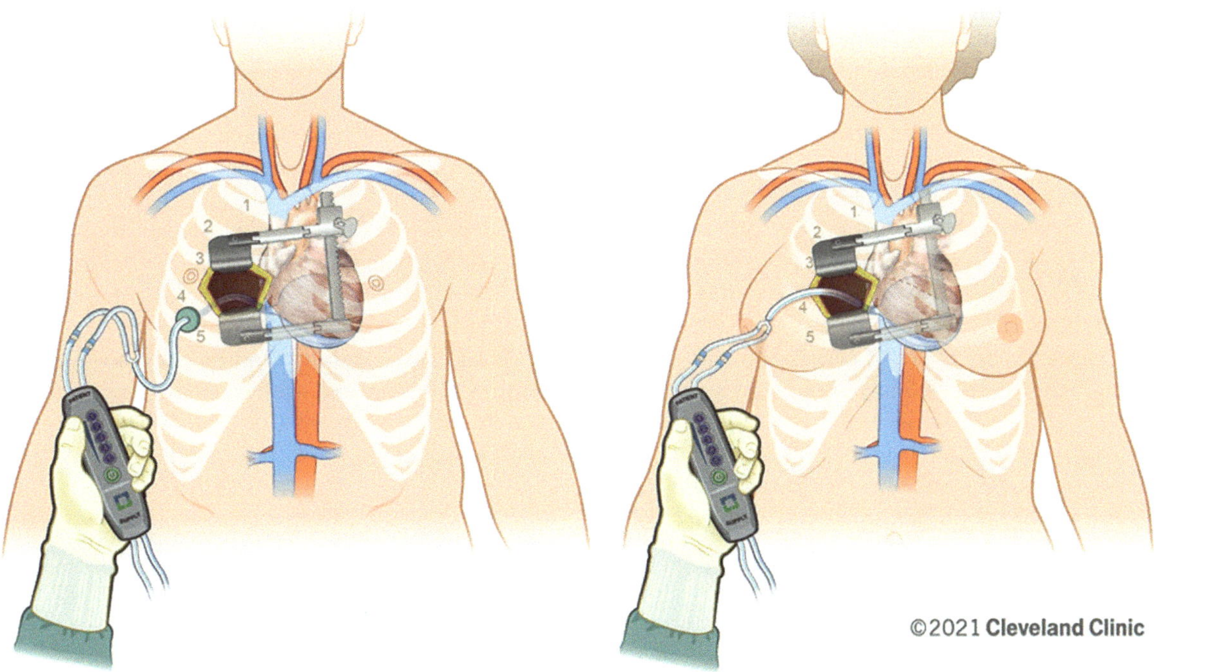

FIGURE 23.15 The cardiac dearing system could be placed through the working port in the intercostal space (left) or directly via surgical incision (right).

manual dearing techniques have proven unable to eliminate the microemboli.[4,10] Even if the techniques are meticulously adhered to, large numbers of microemboli still occur.[11,12] Therefore there is an urgent need to explore other possibilities and develop new, effective solutions.

Air embolisms frequently arise from cardiac surgery and can potentially worsen surgical outcomes such as survival and adequate postoperative recovery.[7,30] Rapid recognition and intervention are critical for reducing morbidity and mortality.[10] Significant progress has been made in intraoperative visualization of cardiac chambers using TEE. However, there is no established technique or dedicated device that could prevent air embolism during cardiovascular surgical procedures and interventions. Effective strategies for retained cardiac air removal will improve outcomes, save patients' lives and operating team time, and reduce costs of hospitalization (being cost-effective or cost-neutral).

The incidence of iatrogenic air embolism can only be estimated, since many accidents are not recognized. As even a small amount of air in the arterial circulation can result in devastating outcomes, timely diagnosis and management are imperative.[31]

In a nationwide survey and literature review, Orihashi et al. reported that 94.9% (262/276) of surgeons and 95.2% (80/84) of anesthesiologists consider dearing an important and frequent event, since they had encountered adverse events of air embolism. According to the anesthesiologists, air-related events were mostly cardiac (97.4%, 76/78) and neurologic (32.1%, 25/78). According to the surgeons, five procedures are routinely performed in the majority of institutions: postural change (87.8%, 244/278); inflation of the lung (98.6%, 274/278); air removal via an LV vent (92.4%, 257/278); root vent (99.3%, 276/278); and direct aspiration of air (33.5%, 93/278).

Several methods for dearing were reported in the replies, including retrograde terminal cardioplegia, aspiration of the right coronary artery with a 27 G needle, and alternative aspiration via the root vent and LV vent, while the cardiac chambers are filled and evacuated repeatedly. CO_2 insufflation was performed in 82.5% of institutions, which have acknowledged their expectations for a dedicated device for dearing.[7]

Clinical manifestations, essentially neurological or cardiovascular disorders, vary greatly.[13] The actual results of dearing procedures or events related to air embolism have not yet been reported on a national basis. The presence of pooled air was significantly associated with the occurrence of cardiac events such as regional wall motion abnormalities, ST elevation, and heart block.[7] Additionally, the neurological sequelae of iatrogenic air embolism have been estimated to reach 19%–50% of the patients.[13] The pathologic features of cerebral air emboli are due to arrested gas emboli in the arterioles, which cause obstruction and decreased blood flow, immediate vasoconstriction followed by prolonged vasodilation, and dysfunction of the blood–brain barrier in association with extravasation of fluid and electrolytes and subsequent cerebral edema.[30]

Although echo visualization has allowed detection of air and guided deairing procedures, adequate air removal is not always attained. Notably, the removal of retained air has been attempted in each surgeon's manner without a solid standard or evidence.[5,24,29,32,33] The presence of residual air bubbles in the cerebral vasculature and of organs throughout the body can decrease blood flow to areas affected by such air embolisms. The resulting deterioration of organ function can adversely affect surgical recovery and postoperative survival.[34,35]

Cerebral and coronary air emboli are still a hazard when the left side of the heart is opened during intracardiac operations, despite increased awareness of the danger and seemingly adequate techniques for evacuating the heart of entrapped air.[36] The air may be introduced into the circulation, traveling to the brain and multiple organs. Such air emboli block blood flow to these areas and are detrimental to long-term survival.[37]

Despite the widespread use of carbon dioxide insufflation (CDI) in cardiac surgery and recent innovations, the present meta-analysis did not find any significant protective effect from CDI compared with manual deairing maneuvers in terms of clinical outcomes, including postoperative neurocognitive decline.[27] Even CO_2 itself may cause significant embolism in the form of pooled type. A potential risk of hypercapnia should also be kept in mind.[38]

Three important elements of air removal remain in force:

1. mobilization of the air (positive chamber filling, stretching of the atrial wall, and ballottement are critical);
2. removal of mobilized air (continuous ascending aorta-venous shunting and nonsuction venting of the left atrium are very important); and
3. proof of elimination of air before CPB is terminated.[6]

In a real clinical scenario, other specific concerns could be related to CO_2 insufflation. This could have effects on arterial blood gases (can be negated by its application toward the end of the procedure, prior to the deairing phase), because it is not needed until the aortic cross-clamp has been removed.[39]

The proposed CDS technology has been developed with a primary application in MICS procedures where retained air is a problem. The system is also envisioned for its application in a greater range and number of surgical procedures. In addition to techniques related directly to enabling intraoperative retained cardiac removal, the CDS is expected to assist with the crucial reduction of operating times (more specifically, the cross-clamp, CPB, and overall operating times), which would contribute an additional value to optimizing postoperative outcomes.

Summary

The design and development of the cardiac deairing system (CDS) provides required engineering data, clinical parameters and important insights on surgeons' experience on intrathoracic (subcardiac) device placement (direct or through a port), device activation, use, and removal. How well the device will fit into the typical operating room setting has been also evaluated. These data would support successful clinical trial efforts and contribute to a straightforward clinical acceptance of this technology.

To date, no reliable method of cardiac deairing during MICS exists to address the potential failure to evacuate intracardiac air, which is known to worsen clinical outcomes. Development of the proposed CDS will address a high-impact clinical issue and improve outcomes at a low cost of development.

The CDS, together with optimized strategies for cardiac air removal, will improve outcomes, favorably affect operating times, and reduce hospitalization costs (thus being cost-effective or cost-neutral), which is in line with the objectives of any clinical intervention and of targeting technologies. CDS can allow surgeons to actively deair the heart and reduce the time the patient spends on bypass, in addition to the unnecessary cost of the passive "watch and wait" approach currently used.

Further, an inflatable device of systems as described here can be introduced under the heart and maintain its position, without migration, within the partially intact pericardial sac. The CDS described here enables cardiothoracic surgeons to perform active deairing of the heart during MICS and traditional cases; reduce surgical time (for total anesthesia, CPB, and/or cross-clamping), thereby improving associated clinical outcomes; and provide a fully measurable, preventive technique of air removal, controlled directly by the surgeon.

Patient preferences continue to move toward less invasive approaches, which in turn impose technically challenging operations, and impose more limitations, such as working space, field of view, and an overly busy operative field. Also, an increasing number of patients undergoing cardiac operations are elderly, with more comorbidities and complicated treatment histories. The minimally invasive surgical approach requires more meticulous preoperative planning and availability of advanced infrastructure, dedicated instrumentation that could assist with bettering of the postoperative prognosis, and achievement of outstanding results.

The physiologic consequences of retained cardiac air are complex, however, and future improvements in MICS will demand close clinical supervision and specifically tailored device solutions rather than just a simple protocol. Effective measures against retained cardiac air removal have to be addressed intraoperatively; however, the consequences of air embolism are usually suspected after the procedure, when neurological deficits are discovered during patient recovery. Therefore timely acknowledgment of the problem and availability of effective tools and techniques are absolutely critical to avoid the deleterious effects of open heart surgery and thus of an air entrapment into cardiac chambers.

Reassessment and standardization of existing deairing maneuvers, scientific verification of intracardiac air, and application of revised protocols for cardiac surgical procedure, particularly minimally invasive and robotic surgeries, will become important steps in establishing the most beneficial tools and techniques suitable for each clinical scenario. High surgical suspicion, better collaboration within surgical and cardiac anesthesia teams, and early acknowledgment of retained air with effective intervention will remain critical for patient survival and warrant superior procedural outcomes.

References

1. Cardiac Surgery Market Analysis, Size, Trends I United States I 2020–2026 I COVID19 I MedSuite: <https://idataresearch.com/product/cardiac-surgery-market-united-states/>.
2. Carrel A. Experimental operations on the orifices of the heart. *Ann Surg*. 1914;60:1–6.
3. Spencer FC, Rossi NP, Yu SC, Koepke JA. The significance of air embolism during cardiopulmonary bypass. *J Thorac Cardiovasc Surg*. 1965;49:615–634.
4. Muth CM, Shank ES. Gas embolism. *N Engl J Med*. 2000;342:476–482.
5. Tingleff J, Joyce FS, Pettersson G. Intraoperative echocardiographic study of air embolism during cardiac operations. *Ann Thorac Surg*. 1995;60:673–677.
6. Oka Y, Inoue T, Hong Y, Sisto DA, Strom JA, Frater RW. Retained intracardiac air. Transesophageal echocardiography for definition of incidence and monitoring removal by improved techniques. *J Thorac Cardiovasc Surg*. 1986;91:329–338.
7. Orihashi K, Ueda T. "De-airing" in open heart surgery: report from the CVSAP nation-wide survey and literature review. *Gen Thorac Cardiovasc Surg*. 2019;67:823–834.
8. Nyman J, Svenarud P, van der Linden J. Carbon dioxide de-airing in minimal invasive cardiac surgery, a new effective device. *J Cardiothorac Surg*. 2019;14:12.
9. Svenarud P, Persson M, van der Linden J. Effect of CO2 insufflation on the number and behavior of air microemboli in open-heart surgery: a randomized clinical trial. *Circulation*. 2004;109:1127–1132.
10. McCarthy CJ, Behravesh S, Naidu SG, Oklu R. Air embolism: diagnosis, clinical management and outcomes. *Diagnostics (Basel, Switz)*. 2017;7.
11. Shah J, Jiwa N, Mamdani N, Hill D. Venous and arterial air embolism: a rare phenomenon with fatal consequences. *BMJ Case Rep*. 2016;2016.
12. LaPar DJ, Quader M, Rich JB, et al. Institutional variation in mortality after stroke after cardiac surgery: an opportunity for improvement. *Ann Thorac Surg*. 2015;100:1276–1282. discussion 1282-1273.
13. Bacha S, Annane D, Gajdos P. [Iatrogenic air embolism]. *Presse Medicale (Paris, France: 1983)*. 1996;25:1466–1472.
14. Rahman ZU, Murtaza G, Pourmorteza M, et al. Cardiac arrest as a consequence of air embolism: a case report and literature review. *Case Rep Med*. 2016;2016:8236845.
15. Robich MP, Krafcik BM, Shah NK, Farber A, Rybin D, Siracuse JJ. Analysis of "never events" following adult cardiac surgical procedures in the United States. *J Cardiovasc Surg*. 2017;58:755–762.
16. Half of US Hospitals Reporting to Leapfrog Say They Won't Bill For a "Never Event." TheLeapFrogGroup: <https://www.leapfroggroup.org/influencing/never-events> 2007.
17. Glauber M, Farneti A, Solinas M, Karimov J. Aortic valve replacement through a right minithoracotomy. *Multimed Man Cardiothorac Surgery: MMCTS*. 2006;2006. mmcts.2005.001826.
18. Karimov JH, Santarelli F, Murzi M, Glauber M. A technique of an upper V-type ministernotomy in the second intercostal space. *Interact Cardiovasc Thorac Surg*. 2009;9:1021–1022.
19. Glauber M, Karimov JH, Farneti PA, et al. Minimally invasive mitral valve surgery via right minithoracotomy. *Multimed Man Cardiothorac surgery: MMCTS*. 2009;2009:mmcts.2008.003350.
20. Karimov JH, Bevilacqua S, Solinas M, Glauber M. Triple heart valve surgery through a right antero-lateral minithoracotomy. *Interact Cardiovasc Thorac Surg*. 2009;9:360–362.
21. Karimov JH, Solinas M, Latsuzbaia K, Murzi M, Cerillo AG, Glauber M. Surgical treatment of double and triple heart valve disease through a limited single-access right minithoracotomy. *Multimed Man Cardiothorac Surg: MMCTS*. 2010;2010:mmcts.2009.004036.
22. Persson M, Svenarud P, van der Linden J. What is the optimal device for carbon dioxide deairing of the cardiothoracic wound and how should it be positioned? *J Cardiothorac Vasc Anesthesia*. 2004;18:180–184.

23. Secretain F, Pollard A, Uddin M, et al. A novel software program for detection of potential air emboli during cardiac surgery. *Cardiovascular Ultrasound*. 2015;13:3.
24. Orihashi K. Efficient removal of retained intracardiac air utilizing buoyancy. *Ann Thorac Surg*. 2016;102:e587–e590.
25. Sun Y, Ji B, Zhu X, Zheng Z. Efficacy of carbon dioxide insufflation for cerebral and cardiac protection during open heart surgery: a systematic review and meta-analysis. *Artif Organs*. 2013;37:439–446.
26. Chaudhuri K, Storey E, Lee GA, et al. Carbon dioxide insufflation in open-chamber cardiac surgery: a double-blind, randomized clinical trial of neurocognitive effects. *J Thorac Cardiovasc Surg*. 2012;144:646–653. e1.
27. Benedetto U, Caputo M, Guida G, et al. Carbon dioxide insufflation during cardiac surgery: a meta-analysis of randomized controlled trials. *SemThorac Cardiovasc Surg*. 2017;29:301–310.
28. Giordano S, Biancari F. Does the use of carbon dioxide field flooding during heart valve surgery prevent postoperative cerebrovascular complications? *Interact Cardiovasc Thorac Surg*. 2009;9:323–326.
29. Landenhed M, Cunha-Goncalves D, Al-Rashidi F, Pierre L, Höglund P, Koul B. Pulmonary collapse alone provides effective de-airing in cardiac surgery: a prospective randomized study. *Perfusion*. 2016;31:320–326.
30. Armon C, Deschamps C, Adkinson C, Fealey RD, Orszulak TA. Hyperbaric treatment of cerebral air embolism sustained during an open-heart surgical procedure. *Mayo Clin Proc*. 1991;66:565–571.
31. Malik N, Claus PL, Illman JE, et al. Air embolism: diagnosis and management. *Future Cardiol*. 2017;13:365–378.
32. Koul BL, Al–Rashidi F, Roijer A, Meurling C. A new technique to reduce residual air emboli in open left cardiac surgery. *J Thorac Cardiovasc Surg*. 2009;138:1438–1439.
33. Al-Rashidi F, Landenhed M, Blomquist S, et al. Comparison of the effectiveness and safety of a new de-airing technique with a standardized carbon dioxide insufflation technique in open left heart surgery: a randomized clinical trial. *J Thorac Cardiovasc Surg*. 2011;141:1128–1133.
34. Karimov JH, Horvath DJ, Byram N, et al. Deairing Techniques for Double-Ended Centrifugal Total Artificial Heart Implantation. *Artif Organs*. 2017;41:568–572.
35. Karimov JH, Gao S, Fukamachi K, Grady P. Continuous-flow total artificial heart port-to-port connection technique using dedicated de-airing sleeve. *Perfusion*. 2020;35:861–864.
36. Fishman NH, Carlsson E, Roe BB. The importance of the pulmonary veins in systemic air embolism following open-heart surgery. *Surgery*. 1969;66:655–662.
37. Karimov JH, Gao S, Dessoffy R, et al. Novel technique for airless connection of artificial heart to vascular conduits. *J Artif Organs: Off J Japanese Soc Artif Organs*. 2017;20:386–389.
38. Orihashi K. Reply. *Ann Thorac Surg*. 2018;105:330.
39. Poullis M. New deairing technique: questions and caveats. *J Thorac Cardiovasc Surg*. 2011;142:1285.

PART V

Treatment of Cardiac Rhythm Disturbances

Chapter 24

Cutting-Edge Technologies for Cardiac Rhythm Treatment: Recent Progress of Catheter Ablation Using Radiofrequency Current and Pulsed Electric Field

Hiroshi Nakagawa[1], Atsushi Ikeda[2], Yoshimori An[1], Shunsuke Kuroda[1], Katsuaki Yokoyama[2], Oussama M. Wazni[1] and Ayman A. Hussein[1]

[1]*Department of Cardiovascular Medicine, Cleveland Clinic, Cleveland, OH, United States,* [2]*Department of Medicine, Nihon University School of Medicine, Tokyo, Japan*

Chapter Outline

Introduction

Various techniques and energy sources have been developed for catheter ablation of tachyarrhythmias, including atrial fibrillation (AF) and ventricular tachycardia (VT).[1–3] Radiofrequency (RF) current has been the most commonly utilized energy source for ablation with very good safety and efficacy. There are, however, several limitations to RF ablation: (1) relatively long procedural times and high recurrence rate, especially for ablation of AF and VT[2,3]; (2) excessive tissue heating potentially leading to serious complications, including steam pop (resulting in cardiac perforation), pulmonary vein (PV) stenosis, coronary arterial injury, and thromboembolism[1–6]; and (3) tissue cooling by arterial blood flow (providing greater heat sink) may limit RF lesion formation, leading to gaps in ablation lines, increasing as such the risk of reentrant tachycardias.[7,8]

In order to overcome these limitations of RF ablation, several new technologies have been developed.

Recent progress of radiofrequency ablation systems

Contact force sensing catheter

During RF ablation, electrode–tissue contact force is one of the major determinants of lesion size.[9–15] No effective lesion is formed without electrode–tissue contact. In contrast, excessive contact force may lead to excessive tissue heating and an increased risk of deep steam pop (perforation) and collateral damage, such as esophageal and phrenic nerve (PN) injury.[9,10,15]

Many ablation catheters have been introduced to clinical practice and are designed to allow measurement in real-time electrode–tissue contact force during mapping and RF ablation. One system uses three optical fibers to measure the microdeflection of a deformable body in the catheter tip (TactiCath, Endosense SA/Abbott, Inc.), which correlates with catheter-tip force.[16–23] The second system uses a small spring between the ablation tip-electrode and catheter

Advances in Cardiovascular Technology. DOI: https://doi.org/10.1016/B978-0-12-816861-5.00032-0

shaft, with a small magnetic transmitter embedded in the tip and magnetic location sensors proximal to the tip to measure microdeflection of the spring (ThermoCool SmartTouch, Biosense Webster, Inc.), corresponding to catheter-tip force.[24–32] Both systems have high resolution (<1 g) in bench testing and accurately display the direction of contact force.[14,21] These two catheters, equipped with saline irrigated tip electrodes, underwent extensive preclinical and clinical studies, confirming the improved safety and efficacy of RF ablation.[16–32]

Preclinical studies to validate contact force sensing ablation catheter

The initial preclinical studies were performed to test the relationship between contact force and RF lesion size in a controlled environment, using a canine thigh muscle preparation.[16] The ablation catheter containing the optical fiber contact sensor (TactiCath) was held perpendicular to the thigh muscle at contact force of 2, 10, 20, 30, and 40 g. RF energy was delivered at constant power (30 or 50 W) for 60 s. Increasing contact force was associated with a progressive increase in lesion depth from 6.2 to 9.9 mm for 30 W and from 7.1 to 11.2 mm for 50 W (Fig. 24.1A). Importantly, lesion depth was significantly greater at 30 W and moderately high contact force of 40 g than at 50 W and lower contact force of 10 g (median depth 9.9 mm vs 8.5 mm depth, $P < .01$).[16] Lesion diameter also increased significantly with increasing contact force from 8.6 to 13.7 mm for 30 W and from 10.2 to 15.2 mm for 50 W (Fig. 24.1B). These data indicate that contact force had as much influence on lesion size as RF power. In addition, excessive contact force also significantly increased the incidence of steam pop and thrombus formation (Fig. 24.2A and 24.2B).[16,25]

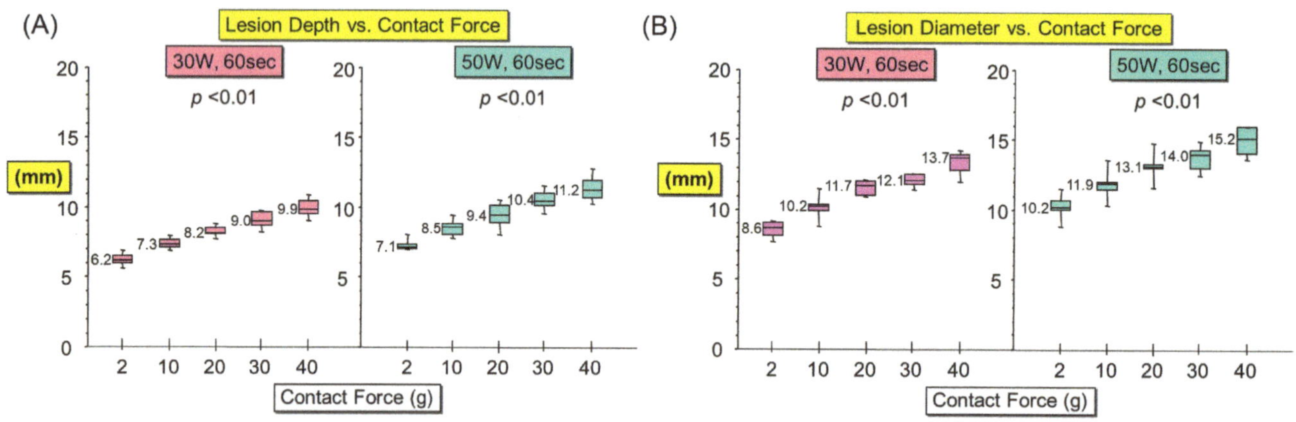

FIGURE 24.1 Relationship between contact force and radiofrequency (RF) lesion size in a canine thigh muscle preparation. Lesion depth (A) and diameter (B) significantly increased by increasing contact force (2, 10, 20, 30, and 40 g) for both 30 W (left panel) and 50 W (right panel) RF applications ($P < .01$). *Modified with permission from Yokoyama K, Nakagawa H, Shah DC, et al. Novel contact force sensor incorporated in irrigated radiofrequency ablation catheter predicts lesion size and incidence of steam pop and thrombus.* Circ Arrhythm Electrophysiol. *2008;1:354–362.*[16]

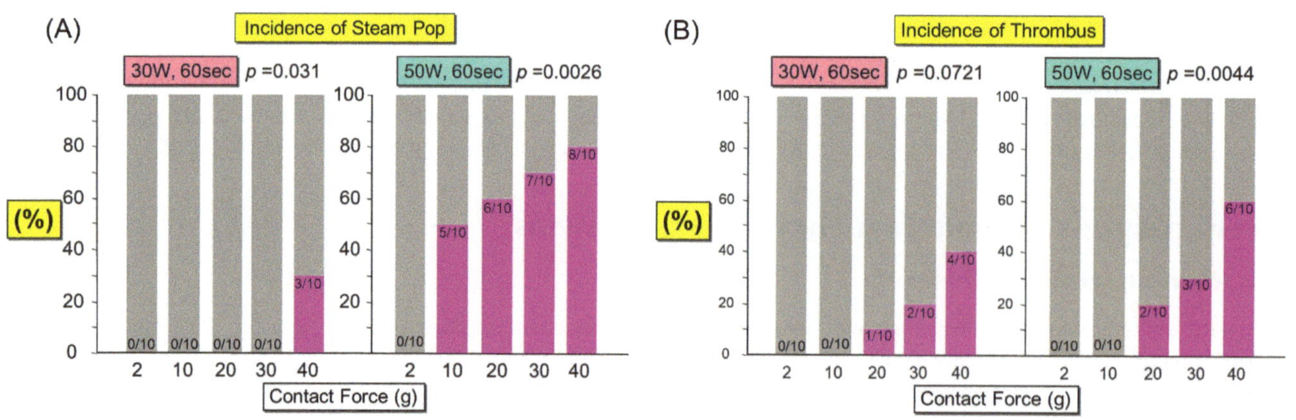

FIGURE 24.2 Relationship between contact force and the incidence of steam pop and thrombus in a canine thigh muscle preparation. (A) The incidence of steam pop increased significantly by increasing contact force at 30 W (left panel) and 50 W (right panel) RF applications ($P = .031$ and $P = .0026$, respectively). (B) The incidence of thrombus formation also increased significantly by increasing contact force at 50 W applications (right panel, $P = .0044$). At 30 W applications (left panel), there was a trend between the incidence of thrombus and contact force ($P = .0721$).

In a study using the same catheter, but adding dynamic contact (simulating the beating heat), RF energy (20 and 40 W for 60 s) was delivered to bovine skeletal muscles with a 20 peak (systolic) and 10 g nadir (diastolic) at 50 and 100 catheter movements/min, or intermittent contact with a 20 g peak and 0 g nadir (loss of contact). Lesion depth and volume correlated linearly with the area under the contact force curve, described as the Force-Time Integral (FTI).[17]

Using the ablation catheter containing a small spring and magnetic location sensors (ThermoCool SmartTouch), preclinical animal research was performed to determine whether an on-line combination of the measurements of contact force, RF power, and application time predicted RF lesion size. Based on the initial study comparing lesion depth to various combinations of contact force, power, and RF time,[24,25,31,32] we developed a logarithmic formula (Force-Power-Time Index (FPTI), Ablation Index (AI)) to predict lesion depth during RF ablation. This formula was tested prospectively to predict lesion depth in the beating canine heart. The formula was able to indicate when to terminate the RF application to obtain lesions of 3, 5, 7, and 9 mm depth within a ± 1 mm accuracy in the beating canine right and left ventricles.[25,31,32] The incidence of steam pop increased significantly by increasing the target lesion depth (no steam pop for 3 and 5 mm target depths, 7/40 RF applications [18%] for 7 mm target depth, and 15/40 RF applications [38%] for 9 mm target depth).

Clinical studies in patients with atrial fibrillation

Using the ablation catheter containing the optical sensor (TactiCath), the EFFICAS I trial was performed to determine the correlation between contact force during PV isolation and the incidence of conduction gaps in PV isolation lines (PV reconnection) at 3 months following ablation.[21–23] The operators were blinded to contact force during PV isolation. RF power ranged between 10 and 40 W and mean contact force during ablation was 19.4 ± 16.2 g. The average FTI for each ablation site was 730 ± 773 gramseconds (gs). All PVs were successfully isolated. Forty of the 46 patients underwent repeat left atrial (LA)/PV mapping at 3 months after ablation. At least one gap was present in 26/40 (65%) patients. Comparing continuously blocked segments (no gap) versus segments with a gap, the minimum contact force applied was 8.1 g versus only 3.6 g ($P < .0001$) and the minimum FTI was 232 gs versus 118 gs ($P < .001$). Segments with a minimum FTI >400 gs had a 95% chance of remaining electrically blocked, while those with a minimum FTI <400 gs had a 79% chance of remaining electrically blocked ($P < .0001$).[21–23]

Another clinical study was performed using the magnetic contact force sensor catheter (ThermoCool SmartTouch) and AI.[31,32] A total of 130 consecutive patients with paroxysmal AF underwent PV isolation.[29] RF current was delivered with a target AI ≥ 400 (target lesion depth of 4 mm) at the posterior/inferior LA wall and ≥ 550 (target lesion depth of 5.5 mm) at the anterior/superior LA wall with an interlesion distance ≤ 6 mm (Fig. 24.3). The first encirclement around the ipsilateral PVs resulted in complete PV isolation in 256/260 (98%) ipsilateral PVs with a relatively short total RF time of 17 ± 5 min and total procedure time of 155 ± 28 min. At 12-month follow-up, single-procedure freedom from recurrent atrial tachyarrythmias was 91.3% without antiarrhythmic drug therapy and 96.2% with antiarrhythmic drug therapy.[31,32]

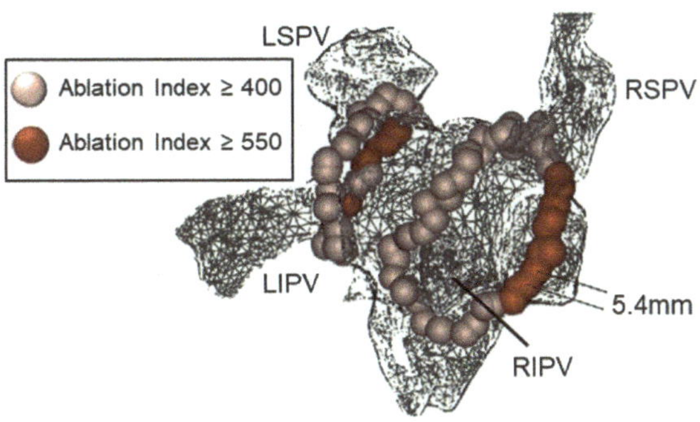

FIGURE 24.3 Ipsilateral pulmonary vein (PV) isolation using force-power-time index (Ablation Index (AI)) in patients with paroxysmal atrial fibrillation (AF).
Point-by-point RF delivery was performed during sinus rhythm aiming for a contiguous circumferential linear ablation around the ipsilateral PV. RF was delivered in power-controlled mode with 25–35 W (irrigation flow up to 30 mL/min) with target AI values of 400 (pale pink tags) at the posterior/inferior wall and 550 (brawn tags) at the anterior/superior wall and an interlesion distance of ≤ 6 mm. *LSPV*, Left superior pulmonary vein; *LIPV*, left inferior PV; *RSPV*, right superior PV; *RIPV*, right inferior PV. *Permission from Taghji P, El Haddad M, Phlips T, et al. Evaluation of a strategy aiming to enclose the pulmonary veins with contiguous and optimized radiofrequency lesions in paroxysmal atrial fibrillation: a pilot study.* JACC Clin Electrophysiol. *2018;4(1):99–108. doi: 10.1016/j.jacep.2017.06.023. Epub 2017 Sep 27.*[32]

These data indicate that the use of contact force sensing catheter and FPTI facilitates acute PV isolation with a relatively short procedure time while reducing the incidence of tissue overheating (steam pop and collateral injury), with a high single-procedure arrhythmias free survival.

Needle RF ablation system to produce deep lesions

Catheter ablation of VT becomes challenging when critical components of the arrhythmogenic substrates are located deep within the myocardium. During RF ablation, active electrode cooling by saline irrigation allows the use of higher RF power and longer application time to produce deeper lesions, while reducing the incidence of thrombus formation and an impedance rise.[33,34] However, RF current density falls rapidly with the fourth power of distance from the ablation electrode, resulting in the tissue only adjacent to the electrode to be rapidly heated ("resistive heating"), while the tissue located at further distance is heated primarily by passive conduction of heat ("conductive heating").[34] Conductive tissue heating is a relatively slow process, resulting in uneven distribution of heat within the tissue. When the tissue temperature underneath the ablation electrode becomes excessive ($>95°C-100°C$), the tissue would desiccate, which is associated with an increase of tissue resistivity (impedance rise), limiting effective RF current delivery, and lesion formation.[33,34] Tissue overheating near the ablation electrode is also associated with steam formation within the tissue, followed by steam expulsion/tissue disruption ("steam pop"), which may result in tissue perforation.[33,34] These fundamental limitations of RF ablation prevent adequate lesion formation deep within the myocardium.

Retractable needle ablation electrodes have been developed to deliver RF current deep within the myocardium to produce deep intramural lesions. A needle ablation electrode is inserted into the tissue and room temperature saline is injected into the tissue through the tip of the needle (1–2 mL/min injection). Cool saline injection increases the electrical conductivity of the tissue and prevents excessive heating at the electrode–tissue interface, enhancing thermal transport deeper into the myocardium (Fig. 24.4).[35,36]

The preclinical study was performed to compare RF lesions produced by a needle ablation catheter to a conventional open irrigated ablation electrode. The needle ablation catheter (RF: 25–35 W for 60 s with saline injection through the needle: 2 mL/min) produced significantly deeper (depth: 9.9 ± 2.7 mm) and larger (width: 8.5 ± 2.5 mm and volume: 1030 ± 362 mm^3, $P < .001$) RF lesions, compared to the lesions (depth: 5.2 ± 2.4 mm, width: 5.2 ± 1.7 mm and volume: 488 ± 384 mm^3) produced by the conventional ablation catheter (RF: 30–35 W for 60 s with saline irrigation of 30 mL/min). Transmural lesions were observed more frequently with the needle ablation (62.5% vs 17%, $P < .01$). Importantly, no steam pop occurred in any of 60 RF applications using the needle ablation catheter, while steam pops were observed in three of 60 RF applications using the conventional catheter.[36]

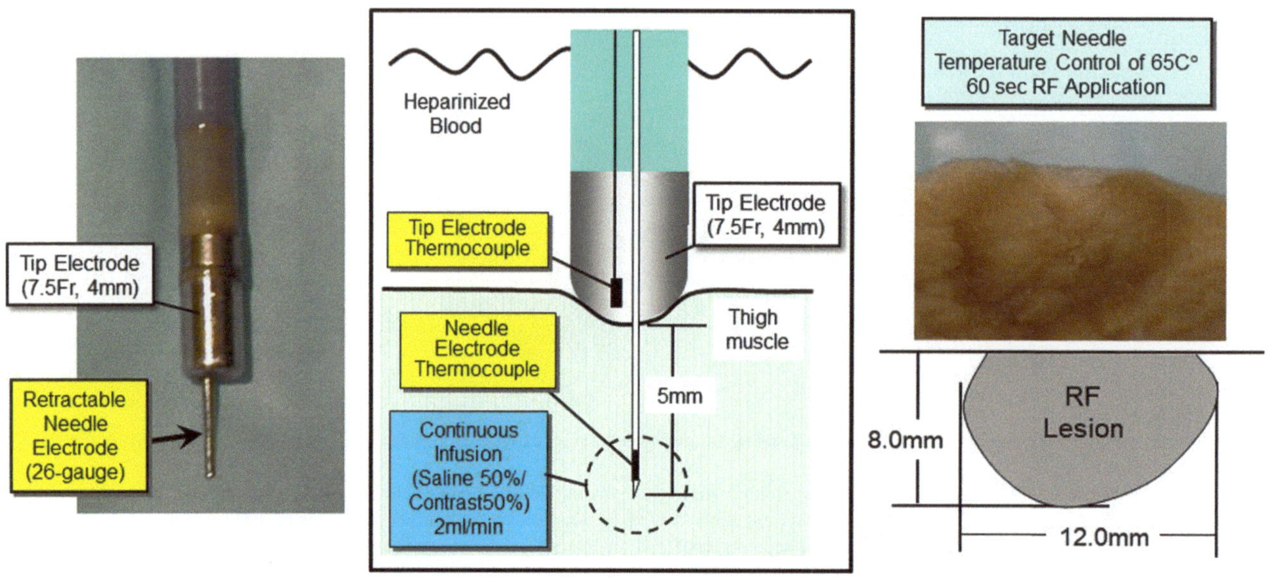

FIGURE 24.4 A deep RF lesion (8 mm in depth) produced by RF needle ablation catheter (temperature control of 65°C for 60 s) in the thigh muscle preparation without steam pop or thrombus formation.

The initial clinical study was performed using the needle ablation catheter in eight patients (nonischemic cardiomyopathy in six and ischemic heart disease in two patients) with recurrent VT despite antiarrhythmic medications and 1–4 prior failed ablation procedures (including a failed epicardial procedure).[37] After the needle electrode was extended 7–9 mm into the myocardium, a median of 19 (range 3–48) RF applications were delivered per patient over a median of 1316 s (131–2686) with a median average power of 8–20 W during heparinized saline infusion at 2 mL/min. All eight patients had at least 1 VT termination or rendered noninducible. During a median of 12 (10–25) months follow-up, four patients were free of recurrent VT, and three patients were improved but had new VTs occur at some point during follow-up. Two patients died of the progression of preexisting heart failure without recurrent VT. Complications included tamponade (resolved with percutaneous pericardial drainage) in one patient and anticipated heart block resulting from RF delivery at the left ventricular basal septum in two patients.[37] These data suggest that intramyocardial infusion-needle catheter ablation is feasible and allows successful ablation of some VTs that have been otherwise refractory to conventional catheter ablation therapy.

It is noteworthy that injected cool (room temperature) saline has the potential downside of acting as a heat sink for RF energy delivery which may result in suboptimal thermal conduction, limiting the size of the ablation lesion being created. Recently, RF ablation technology using a needle electrode combined with heated saline injection has been developed.[38] Normal saline is warmed (60°C) by a saline heater integrated within the proximal portion of the needle electrode. Heated saline is injected into the myocardium via multiple radial outlets in the needle, while RF energy is delivered between the needle and the distal catheter electrode, producing significant "thermal convection" with the creation of full-thickness lesions in the left ventricular myocardium (Fig. 24.5).[38]

Pulsed electric field ablation

The accidental occurrence of complete atrioventricular (AV) conduction block caused by external cardioversion during an electrophysiological study was reported in 1979.[39] It was assumed that the high-energy direct current (DC) was delivered to the catheter positioned over the His bundle region. Shortly afterwards, several animal studies demonstrated that complete AV block can be produced with high-energy DC delivery through a conventional diagnostic catheter positioned at the His bundle region.[40–42] Encouraged by these initial results, closed-chest catheter ablation to produce AV block was developed for the treatment of drug-resistant supraventricular tachycardias.[43] This DC ablation technique was explored for ablation of accessory AV pathways, atrial tachycardias, and life-threatening VT.[44,45]

However, it was abandoned shortly after RF ablation became available in the early 1990s, because DC ablation was associated with serious complications (including pericardial tamponade, lethal ventricular arrhythmias, and sudden

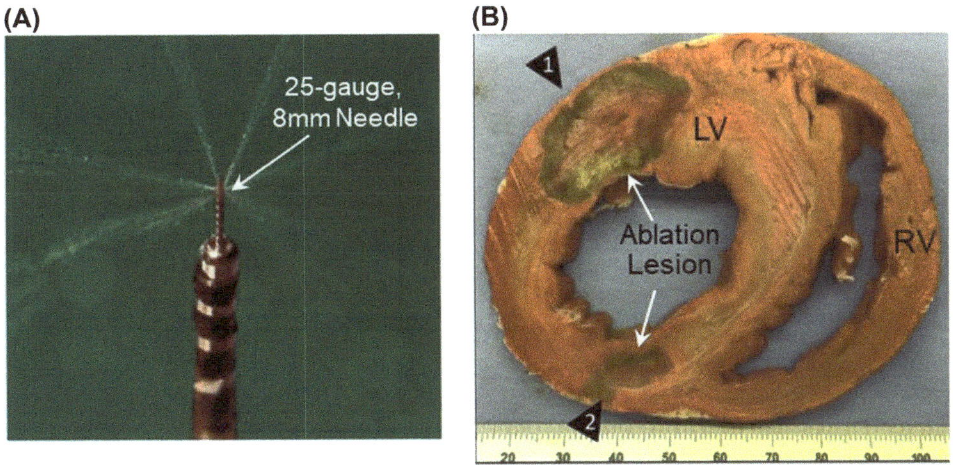

FIGURE 24.5 RF lesion produced by saline-enhanced needle electrode ablation system. (A) An 8 F unidirectional, quadripolar ablation catheter has a 4 mm-tip electrode with a 25-gauge stainless steel needle, which is deployable up to 8 mm from the tip electrode. The closed-end needle has 24 radial side holes for heated saline injection during ablation. (B) Left ventricular lesions (arrows) are created using RF power of 15–45 W for 25–75 s during heated (60°C) saline injection through the needle (10 mL/min). *Permission from Suzuki A, Lehmann HI, Wang S, et al. Impact of myocardial fiber orientation on lesions created by a novel heated saline enhanced radiofrequency needle-tip catheter: an MRI lesion validation study.* Heart Rhythm. *2021;18:443–452.*[38]

cardiac death), which were related to barotrauma resulting from the use of high-energy DC and a small ablation electrode (such a 6Fr, 2 mm electrode).[46,47] Barotrauma is caused by a high-pressure shock wave resulting from an electrically isolating vapor globe, leading to arcing (spark) and explosion.[46] At the end of the DC ablation era, low-energy DC ablation without arcing was developed.[48] Several studies demonstrated that a lower energy level without a high-pressure shock wave is sufficient to create effective myocardial lesions for ablation, such as ablation of accessory AV pathways.[48,49] The main mechanism of lesion formation using low-energy DC ablation is thought to be the formation of pores in cell membranes produced by an application of short high-voltage electric field ("pulsed electric field") and free ion exchange through these pores, resulting in apoptosis of the cell (i.e., "irreversible electroporation").[50,51] The size of the lesions produced by pulsed electric field (PEF) ablation depends on the applied electric field intensity, pulse duration, and the electric resistivity of targeted tissue.[50,51]

PEF is a unique energy source for ablation because lesion formation does not result from tissue heating (such as RF, ultrasound, and laser ablation) or cooling (cryothermia). In contrast, a major limitation of thermal ablation is the nonspecific damage of both cells and extracellular structures through protein denaturation, leading to the risk of complications such as perforation and tamponade, PV stenosis, PN palsy, atrioesophageal fistula, and stroke due to thrombus formation.[1–6] Furthermore, relatively large myocardial lesions may be created in a controlled manner with PEF delivery given that nonthermal electroporation is not affected by local blood flow. Irreversible disruption of cell membranes (necrosis and apoptosis) without thermal injury may also reduce the risk of collateral damage.

Preclinical and clinical studies of pulsed electric field ablation systems

Recent preclinical studies suggest that PEF ablation/irreversible electroporation (IRE) might be effective and safe for the treatment of cardiac arrhythmias.[52–60] However, these studies were performed by several groups using different electric field intensity, pulse duration and number, and electrode configurations.[52–60] Preclinical data on optimal PEF/IRE protocols for clinical use of PEF are still lacking, especially the use of optimized monophasic/biphasic waveforms which remain nondisclosed by the industry.

Using a monophasic external defibrillator (Lifepak 9, Physio-Control, Inc.; Redmond, WA) and a 7Fr, 20 mm circular catheter with 10 ablation electrodes (2 mm electrode length each), Wittkampf et al. performed left ventricular epicardial pulsed field ablation (PFA) in an in vivo porcine model (Fig. 24.6).[53,60] A single circular 200 J application (between an electrode and a large skin patch, 20 J per electrode without vapor globe formation, arcing and a pressure wave), resulted in continuous circumferential lesions. There was a significant relationship between applied current and lesion depth with a slope of 0.17 mm/A. Importantly, thermal injury was not observed directly after ablation.[53]

A clinical feasibility study was then performed in 10 patients with paroxysmal AF using the same monophasic external defibrillator (Lifepak 9) and a custom 8Fr, multipolar circular ablation catheter (2.5 mm electrode length and

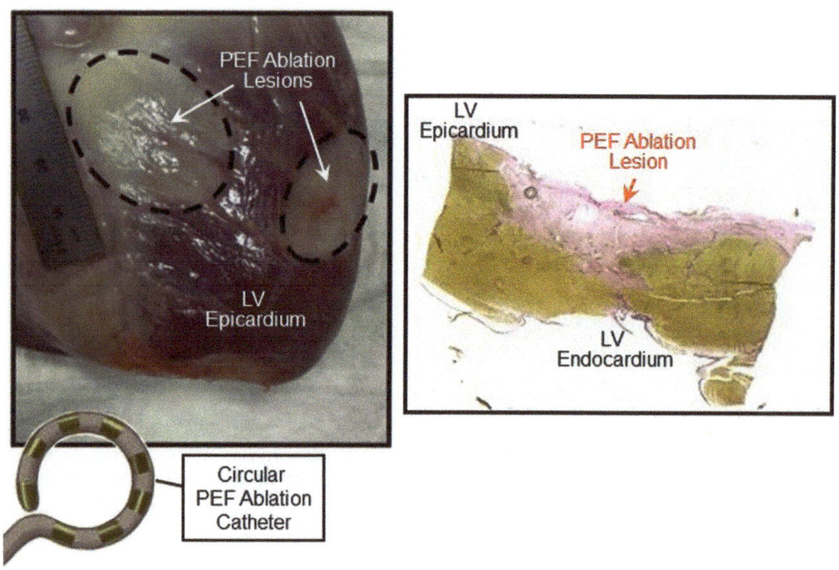

FIGURE 24.6 The custom-made circular electrode catheter used for epicardial PEF ablation and lesions. Left panel: The custom-made circular octopolar device (12 mm in diameter) with 2-mm-long electrodes is placed in the pericardial space, and the nonarcing, nonbarotraumatic DC application (200 J) is delivered. The epicardial lesions in 3 weeks after ablation demonstrate pale regions (30 mm in diameter, delineated by black dashed lines). Right panel: Histological section (elastic van Gieson staining) through the center of the lesion demonstrates a transmural, continuous circumferential lesion (stained in red) with the irregular margin. *Permission from Wittkampf FHM, Van Es R and Neven K. Electroporation and its relevance for cardiac catheter ablation.* J Am Coll Cardiol EP. *2018;4:977–986.*[60]

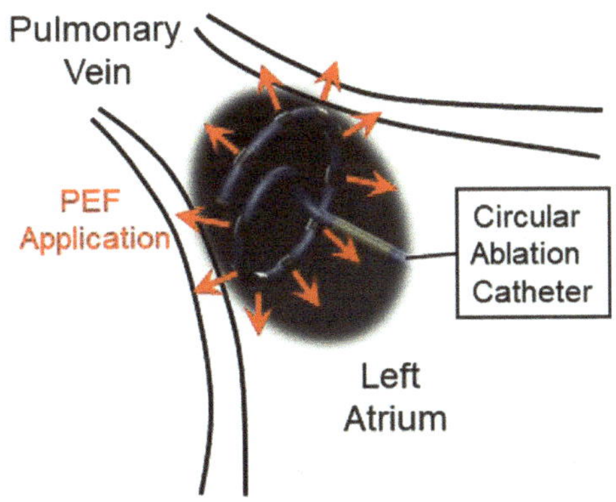

FIGURE 24.7 Schematic representation of pulmonary vein isolation using a circular PEF ablation catheter. A circular ablation catheter is positioned at the ostium of the pulmonary vein and PEF current is applied through all ablation electrodes.

3.5 mm interelectrode spacing, Fig. 24.7).[58] PV isolation was performed delivering nonarcing, nonbarotraumatic 6 Ms, 200 J DC. All 40 PVs were successfully isolated with a mean of 2.4 ± 0.4 PEF applications per PV (mean delivered peak voltage of 2154 ± 59 V and peak current of 33.9 ± 1.6 A) without periprocedural complications.[58]

Evaluation of risk of collateral injury using PEF/IRE

The risk of PN injury during PEF/IRE was evaluated in a porcine model.[55] After the course of PN was identified by pace-mapping in the superior vena cava (SVC), a single 200 J circular PEF was delivered using an octopolar (4 mm ring electrode each) circular ablation catheter positioned close to the PN in firm electrode contact within the SVC. Immediately after ablation, the PN could still be electrically captured by pacing in 17/19 animals and after 30 min in all 19 animals. These data indicate that PEF ablation using a circular ablation catheter at an energy level that creates deep myocardial lesions may spare the PN, even when the PEF current is applied within a distance of 1–2 mm, suggesting significantly lower vulnerability of nerve tissue to PEF.

Esophageal ulceration and atrioesophageal fistula are complications of PV isolation using thermal ablation systems (heating: RF, laser and ultrasound; and cooling: cryothermia), while PEF/IRE is a nonthermal ablation modality for PV isolation. In an in vivo porcine model, the suprasternal esophagus was surgically exposed and a custom linear suction device with a single 35-mm-long and 6-mm-wide protruding linear electrode was used for PFA ablation. Single, nonarcing, nonbarotraumatic, cathodal 100 and 200 J applications were delivered directly at two separate sites on the anterior esophageal adventitia, purposely targeting the esophagus for PEF/IRE using an energy level capable of creating approximately 7-mm-deep lesions in myocardium.[56] Esophageal endoscopy at day 2 and 7 did not show any epithelial erythema, erosions, or ulcerations. After euthanasia at day 60, there were no macroscopically visible lesions on the adventitia or epithelium. Histology showed a small scar at the outer part of the muscular layer, whereas the mucosa and submucosa were intact, suggesting that PEF/IRE appears to be a safe modality for catheter ablation near the esophagus.[56]

Although these preliminary data suggest that PEF/IRE might more selectively result in myocardial tissue (sparing PN and esophagus), this remains a topic of ongoing investigation and more data are required.

Comparison between monophasic and biphasic vaveforms

A preclinical porcine study was performed to compare PEF/IRE using monophasic and biphasic waveforms.[57] PEF ablation was performed using a custom, programmable PEF generator and a 12 Fr, 20-pole ablation catheter (five splines, each containing four electrodes, Farapulse, Inc., Menlo Park, CA).[57] All monophasic and biphasic PEF applications were delivered in bipolar fashion between pairs of electrodes and no ground pad was used. The PEF waveforms consisted of a sequence of microsecond-scale pulses at amplitudes from 800 V (monophasic, seven swine) to 1800 V (biphasic, seven swine). Both waveforms were delivered over 4 or 10 heart beats, respectively, with synchronization of pacing from the coronary sinus catheter and the right ventricular catheter.

Monophasic PEF

All targeted 13 PVs and 7 SVCs were successfully isolated at first pass using monophasic PEF. Monophasic wavefront deliveries were associated with skeletal muscle and diaphragmatic stimulation. Immediately post-PEF ablation, transient ST-T elevation was frequently observed, but coronary angiography showed no coronary spasm. After PEF, the PN was preserved (confirmed by pacing) in all animals. After a mean survival of 63 days, durable isolation was observed in 9/11 (100%) PVs and 1/7 (14%) SVCs without PV and SVC narrowing. Histological examination showed that the majority of lesions was transmural in 138/182 (75.8%) histological sections.

Biphasic PEF

All targeted 14 PVs and 6 SVCs were also successfully isolated at first pass using biphasic PEF. Compared to monophasic applications, muscular stimulation was significantly reduced, not requiring paralytic agents. Significant ST-T elevation was not observed during biphasic delivery and the PN was preserved. Durable isolation was present in 12/12 (100%) PVs and 6/6 (100%) SVCs without PV and SVC narrowing. Histological examination also showed that the majority of lesions was transmural in 138/152 (90.8%).[57] These data suggest that PEF-based vein isolation was feasible without significant collateral injury. Biphasic PEF may provide more durable lesions than monophasic PEF for PV and SVC isolation.

A clinical feasibility study was then performed in 25 patients with persistent AF to assess the safety and lesion durability of PEF ablation for both PV isolation and ablation of the left atrial posterior wall (LAPW).[59] Biphasic, bipolar PEF (1600–2000 V per application) using a multispline catheter (Fig. 24.8) achieved acute isolation of all 96 PVs targeted (median ablation time: 22 min) and complete LAPW ablation in all 24 patients tested (median ablation time: 10 min). Esophageal endoscopy and cardiac computed tomography revealed no mucosal lesions or PV narrowing, respectively. Remapping was performed in 22 of 25 (88%) patients at median of 82 days of follow-up, demonstrating durable PV isolation in 82 of 85 (96%) PVs and 19 of 22 patients (86%).[59]

Using the focal PEF catheter, cavotricuspid isthmus (CTI) ablation was also performed in 13 patients. Acute complete conduction block across the CTI was achieved in all 13 patients (median ablation time: 9 min). Durability of CTI conduction block was reassessed in 12 patients at a median of 87 days of follow-up. In four patients receiving PEF output of 1600 V, only 1 (25%) patient exhibited durable block, while all eight patients receiving 1800 V demonstrated durable bidirectional CTI block.[59]

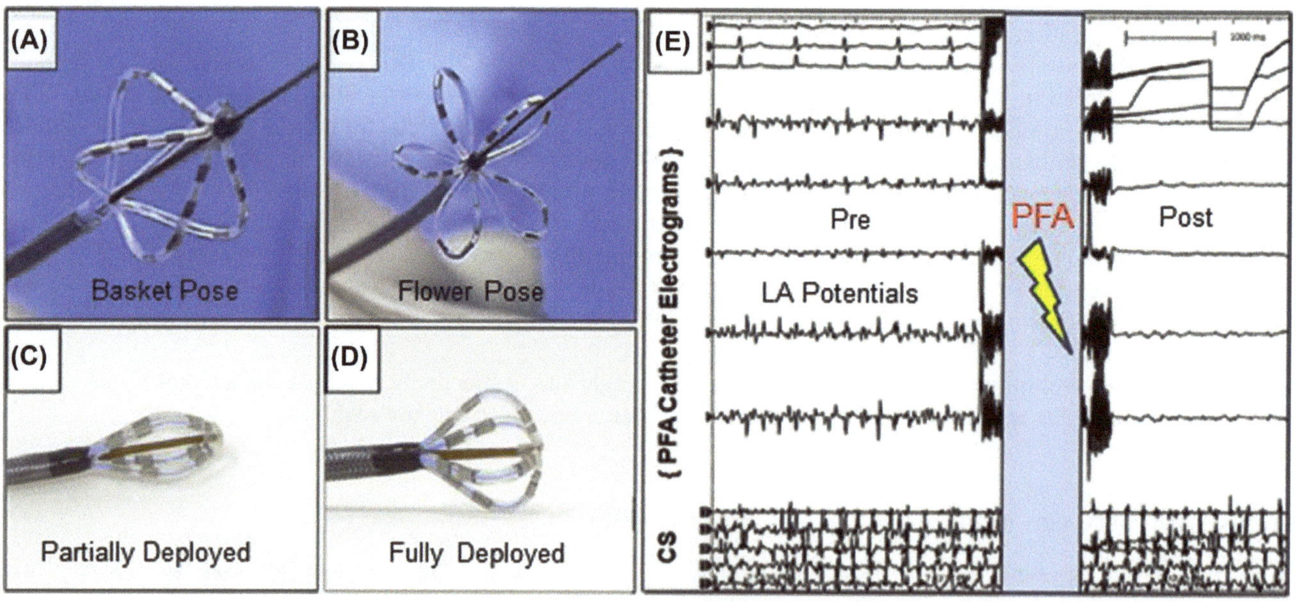

FIGURE 24.8 A multispline pulsed field ablation catheter, partial deployment (panel A) and full deployment (31 mm in diameter, panel B), is used for PV isolation. A focal pulsed field ablation catheter, partial deployment (panel C) and full deployment (17 mm in diameter, panel D), is used for ablation for PV isolation, left atrial posterior wall, and CTI. (Panel E) Bipolar electrograms recorded from the PFA catheter positioned over the left atrial posterior wall demonstrating atrial potentials during AF before (Pre) and after (Post) 3 s pulsed field ablation, showing significant attenuation of left atrial potentials after ablation. *Permission from Reddy VY, Anic A, Koruth J, et al. Pulsed field ablation in patients with persistent atrial fibrillation.* J Am Coll Cardiol. *2020;76:1068–1080.*[59]

These data suggest that, using biphasic, bipolar PEF ablation with a multispline catheter, acute PV isolation and ablation of the LAPW and CTI can be achieved with relatively short ablation times, while preserving the PN and esophagus.

Conclusion

Catheter ablation is an established and important pillar in the treatment of patients with cardiac arrhythmias. Current approaches of cardiac ablation rely on thermal energy to create lesions (heating: RF, laser and ultrasound; and cooling: cryothermia). While thermal ablation has been proven to be effective, there are several limitations, including the risk of collateral damage (injury to PN and esophagus), tissue overheating (steam pop/perforation) and narrowing/occlusion of blood vessels, as well as thrombus formation associated with thermal energy delivery. PEF/IRE offers a unique nonthermal ablation strategy which has the potential to overcome these limitations. However, further preclinical and clinical studies are required to optimize PEF/IRE ablation strategies including PEF generator and ablation catheter configurations.

References

1. Cappato R, Calkins H, Chen SA, et al. Updated worldwide survey on the methods, efficacy, and safety of catheter ablation for human atrial fibrillation. *Circ Arrhythm Electrophysiol*. 2010;3:32–38.
2. Calkins H, Kuck KH, Cappato R, et al. HRS/EHRA/ECAS expert consensus statement on catheter and surgical ablation of atrial fibrillation: recommendations for patient selection, procedural techniques, patient management and follow-up, definitions, endpoints, and research trial design: a report of the Heart Rhythm Society (HRS) Task Force on Catheter and Surgical Ablation of Atrial Fibrillation. Developed in partnership with the European Heart Rhythm Association (EHRA), a registered branch of the European Society of Cardiology (ESC) and the European Cardiac Arrhythmia Society (ECAS); and in collaboration with the American College of Cardiology (ACC), American Heart Association (AHA), the Asia Pacific Heart Rhythm Society (APHRS), and the Society of Thoracic Surgeons (STS). Endorsed by the governing bodies of the American College of Cardiology Foundation, the American Heart Association, the European Cardiac Arrhythmia Society, the European Heart Rhythm Association, the Society of Thoracic Surgeons, the Asia Pacific Heart Rhythm Society, and the Heart Rhythm SocietyHeart Rhythm Society Task Force on Catheter and Surgical Ablation of Atrial Fibrillation. 2012 *Heart Rhythm*. 2012;9:632–696.
3. Stevenson WG, Wilber DJ, Natale A, et al. Ablation Trial Investigators. Irrigated radiofrequency catheter ablation guided by electroanatomic mapping for recurrent ventricular tachycardia after myocardial infarction: the multicenter thermocool ventricular tachycardia ablation trial. *Circulation*. 2008;118:2773–2782.
4. Roberts-Thomson KC, Steven D, Seiler J, et al. Coronary artery injury due to catheter ablation in adults: presentations and outcomes. *Circulation*. 2009;120:1465–1473.
5. Aoyama H, Nakagawa H, Pitha JV, et al. Comparison of cryothermia and radiofrequency current in safety and efficacy of catheter ablation within the canine coronary sinus close to the left circumflex coronary artery. *J Cardiovasc Electrophysiol*. 2005;16:1218–1226.
6. Stavrakis S, Jackman WM, Nakagawa H, et al. Risk of coronary artery injury with radiofrequency ablation and cryoablation of epicardial posteroseptal accessory pathways within the coronary venous system. *Circ Arrhythm Electrophysiol*. 2014;7:113–119.
7. Fuller IA, Wood MA. Intramural coronary vasculature prevents transmural radiofrequency lesion formation: implications for linear ablation. *Circulation*. 2003;107:1797–1803.
8. Wittkampf FHM, van Oosterhout MF, Loh P, et al. Where to draw the mitral isthmus line in catheter ablation of atrial fibrillation: histological analysis. *Eur Heart J*. 2005;26:689–695.
9. Avitall B, Mughal K, Hare J, Helms R, Krum D. The effects of electrode-tissue contact on radiofrequency lesion generation. *PACE*. 1997;20:2899–2910.
10. Haines DE. Determinants of lesion size during radiofrequency catheter ablation: the role of electrode-tissue contact force and duration of energy delivery. *J Cardiovasc Electrophysiol*. 1991;2:509–515.
11. Strickberger SA, Vorperian VR, Man KC, Williamson BD, Kalbfleisch SJ, Hasse C, Morady F, Langberg JJ. Relation between impedance and endocardial contact during radiofrequency catheter ablation. *Am Heart J*. 1994;128:226–229.
12. Zheng X, Walcott GP, Hall JA, et al. Electrode impedance: an indicator of electrode-tissue contact and lesion dimensions during linear ablation. *J Interv Card Electrophysiol*. 2000;4:645–654.
13. Nath S, DiMacro JP, Gallop RG, McRury ID, Haines DE. Effects of dispersive electrode position and surface area on electrical parameters and temperature during radiofrequency catheter ablation. *Am J Cardiol*. 1996;77:765–767.
14. Eick OJ, Wittkampf FH, Bronneberg T, Schumacher B. The LEFTR-principle: a novel method to assess electrode-tissue contact in radiofrequency ablation. *J Cardiovasc Electrophysiol*. 1998;9:1180–1185.
15. Biase LD, Natale A, Barrerr C, et al. Relationship between contact forces, lesion characteristics, "popping," and char formation: experience with robotic navigation system. *J Cardiovasc Electrophysiol*. 2009;20:436–440.
16. Yokoyama K, Nakagawa H, Shah DC, et al. Novel contact force sensor incorporated in irrigated radiofrequency ablation catheter predicts lesion size and incidence of steam pop and thrombus. *Circ Arrhythm Electrophysiol*. 2008;1:354–362.

17. Shah DC, Lambert H, Nakagawa H, Langenkamp A, Aeby N, Leo G. Area under the real-time contact force curve (force-time integral) predicts radiofrequency lesion size in an in vitro contractile model. *J Cardiovasc Electrophysiol.* 2010;21:1038–1043.
18. Thiagalingam A, d'Avila A, Foley L, et al. Importance of catheter contact force during irrigated radiofrequency ablation: evaluation in a porcine ex vivo model using a force-sensing catheter. *J Cardiovasc Electrophysiol.* 2010;21:806–811.
19. Shah DS, Lambert H, Langenkamp A, et al. Catheter tip force required for mechanical perforation of porcine cardiac chambers. *Europcae.* 2010;13:277–283.
20. Ikeda A, Nakagawa H, Lambert H, et al. Relationship between catheter contact force and radiofrequency lesion size and incidence of steam pop in the beating canine heart: electrogram amplitude, impedance, and electrode temperature are poor predictors of electrode-tissue contact force and lesion size. *Circ Arrhythm Electrophysiol.* 2014;7:1174–1180.
21. Kuck KH, Reddy VY, Schmidt B, et al. A novel radiofrequency ablation catheter using contact force sensing: toccata study. *Heart Rhythm.* 2012;9:18–23.
22. Reddy VY, Shah D, Kautzner J, et al. The relationsip between contact force and clinical outcome during radiofrequency catheter ablation of atrial fibrillation in the TOCCATA study. *Heart Rhythm.* 2012;9:1789–1795.
23. Neuzil P, Reddy VY, Kautzner J, et al. Electrical reconnection after pulmonary vein isolation is contingent on contact force during initial treatment: results from the EFFICAS I study. *Circ Arrhythm Electrophysiol.* 2013;6:327–333.
24. Nakagawa H, Kautzner J, Natale A, et al. Locations of high contact force during left atrial mapping in atrial fibrillation patients: electrogram amplitude and impedance are poor predictors of electrode-tissue contact force for ablation of atrial fibrillation. *Circ Arrhythm Electrophysiol.* 2013;6:746–753.
25. Nakagawa H, Jackman WM. The role of contact force in atrial fibrillation ablation. *J Atr Fibrillation.* 2014;7(1):1027. Available from: https://doi.org/10.4022/jafib.1027. eCollection 2014 Jun-Jul.
26. Perna F, Heist EK, Danik SB, Barrett CD, Ruskin JN, Mansour M. Assessment of catheter tip contact force resulting in cardiac perforation in swine atria using force sensing technology. *Circ Arrhythm Electrophysiol.* 2011;4:218–224.
27. Martinek M, Lemes C, Sigmund E, et al. Clinical impact of an open-irrigated radiofrequency catheter with direct force measurement on atrial fibrillation ablation. *Pacing Clin Electrophysiol.* 2012;35:1312–1318.
28. Kumar S, Haqqani HM, Chan M, et al. Predictive value of impedance changes for real-time contact force measurements during catheter ablation of atrial arrhythmias in human. *Heart Rhythm.* 2013;10:962–969.
29. Natale A, Reddy VY, Monir G, et al. Catheter ablation with a contact force sensing catheter. *J Am Coll Cardiol.* 2014;64(7):647–656.
30. Jarman JWE, Panikker S, Das M, et al. Relationship between contact force sensing technology and medium-term outcome of atrial fibrillation ablation: a multicenter study of 600 patients. *J Cardiovasc Electrophysiol.* 2015;26:378–384.
31. El Haddad M, Taghji P, Phlips T, et al. Determinants of acute and late pulmonary vein reconnection in contact force - guided pulmonary vein isolation. Identifying the weakest link in the ablation chain. *Circ Arrhyth Electrophysiol.* 2017;10:e004867. Available from: https://doi.org/10.1161/CIRCEP.116.004867.
32. Taghji P, El Haddad M, Phlips T, et al. Evaluation of a strategy aiming to enclose the pulmonary veins with contiguous and optimized radiofrequency lesions in paroxysmal atrial fibrillation: a pilot study. *JACC Clin Electrophysiol.* 2018;4(1):99–108. Available from: https://doi.org/10.1016/j.jacep.2017.06.023. Epub 2017 Sep 27.
33. Nakagawa H, Yamanashi WS, Pitha JV, et al. Comparison of in vivo tissue temperature profile and lesion geometry for radiofrequency ablation with a saline-irrigated electrode vs temperature control in a canine thigh muscle preparation. *Circulation.* 1995;91:2264–2273.
34. Wittkampf FH, Nakagawa H. RF catheter ablation: lessons on lesions. *Pacing Clin Electrophysiol.* 2006;29:1285–1297.
35. Sapp JL, Cooper JM, Zei P, et al. Large radiofrequency ablation lesions can be created with a retractable infusion-needle catheter. *J Cardiovasc Electrophysiol.* 2006;17:657–661.
36. Berte B, Cochet H, Magat J, et al. Irrigated needle ablation creates larger and more transmural ventricular lesions compared with standard unipolar ablation in an ovine model. *Circ Arrhythm Electrophysiol.* 2015;8:1498–1506.
37. Sapp JL, Beeckler C, Pike R, et al. Initial human feasibility of infusion needle catheter ablation for refractory ventricular tachycardia. *Circulation.* 2013;128:2289–2295.
38. Suzuki A, Lehmann HI, Wang S, et al. Impact of myocardial fiber orientation on lesions created by a novel heated saline enhanced radiofrequency needle-tip catheter: an MRI lesion validation study. *Heart Rhythm.* 2021;18:443–452.
39. Vedel J, Frank R, Fontaine G, Fournial JF, Grosgogeat Y. [Permanent intra-hisian atrioventricular block induced during right intraventricular exploration]. *Archives des maladies du coeur et des vaisseaux.* 1979;72:107–112.
40. Gonzalez R, Scheinman M, Margaretten W, Rubinstein M. Closed-chest electrode-catheter technique for his bundle ablation in dogs. *Am J Physiol.* 1981;241:H283–H287.
41. Bardy GH, Ideker RE, Kasell J, et al. Transvenous ablation of the atrioventricular conduction system in dogs: electrophysiologic and histologic observations. *Am J Cardiol.* 1983;51:1775–1782.
42. Gallagher JJ, Svenson RH, Kasell JH, et al. Catheter technique for closed-chest ablation of the atrioventricular conduction system. *N Engl J Med.* 1982;306:194–200.
43. Scheinman MM, Evans-Bell T. Catheter ablation of the atrioventricular junction: a report of the percutaneous mapping and ablation registry. *Circulation.* 1984;70:1024–1029.
44. Morady F, Scheinman MM, Winston SA, et al. Efficacy and safety of transcatheter ablation of posteroseptal accessory pathways. *Circulation.* 1985;72:170–177.

45. Fontaine G, Tonet JL, Frank R, et al. [Emergency treatment of chronic ventricular tachycardia after myocardial infarction by endocavitary fulguration]. *Archives des maladies du coeur et des vaisseaux*. 1985;78:1037–1043.
46. Bardy GH, Coltorti F, Stewart RB, Greene HL, Ivey TD. Catheter-mediated electrical ablation: the relation between current and pulse width on voltage breakdown and shock-wave generation. *Circ Res*. 1988;63:409–414.
47. Hauer RN, Robles de Medina EO, Borst C. Proarrhythmic effects of ventricular electrical catheter ablation in dogs. *J Am Coll Cardiol*. 1987;10:1350–1356.
48. Ahsan AJ, Cunningham D, Rowland E, Rickards AF. Catheter ablation without fulguration: design and performance of a new system. *Pacing Clin Electrophysiol*. 1989;12:1557–1561.
49. Lemery R, Talajic M, Roy D, et al. Success, safety, and late electrophysiological outcome of low-energy direct-current ablation in patients with the Wolff-Parkinson-White syndrome. *Circulation*. 1992;85:957–962.
50. Tovar O, Tung L. Electroporation of cardiac cell membranes wit monophasic or biphasic rectangular pulses. *Pacing Clin Electrophysiol*. 1991;14:1887–1892.
51. Lee RC, Zhang D, Hannig J. Biophysical injury mechanisms in electrical shock trauma. *Annu Rev Biomed Eng*. 2000;2:47–509.
52. Wittkampf FH, van Driel VJ, van Wessel H, et al. Feasibility of electroporation for the creation of pulmonary vein ostial lesions. *J Cardiovasc Electrophysiol*. 2011;22:302–309. Available from: https://doi.org/10.1111/j.1540-8167.2010.01863.
53. Wittkampf FHM, Van Driel V, Van Wessel H, et al. Myocardial lesion depth with circular electroporation ablation. *Circ Arrhythm Electrophysiol*. 2012;5:581–586.
54. Van Driel VJHM, Neven KGEJ, Van Wessel H, et al. Pulmonary vein stenosis after catheter ablation electroporation vs radiofrequency. *Circ Arrhythm Electrophysiol*. 2014;7:734–738.
55. van Driel VJ, Neven K, van Wessel H, Vink A, Doevendans PA, Wittkampf FH. Low vulnerability of the right phrenic nerve to electroporation ablation. *Heart Rhythm*. 2015;12:1838–1844. Available from: https://doi.org/10.1016/j.hrthm.2015.05.012.
56. Neven K, Van Es R, Van Driel V, et al. Acute and long-term effects of full-power electroporation ablation directly on the porcine esophagus. *Circ Arrhythm Electrophysiol*. 2017;10:e004672.
57. Koruth J, Kuroki K, Iwasawa J, et al. Preclinical evaluation of pulsed!eld ablation: electrophysiological and histological assessment of thoracic vein isolation. *Circ Arrhythm Electrophysiol*. 2019;12:e007781.
58. Loh P, Van Es R, Groen MHA, et al. Pulmonary vein isolation with single pulse irreversible electroporation. *Circ Arrhythm Electrophysiol*. 2020;13:e008192. Available from: https://doi.org/10.1161/CIRCEP.119.008192.
59. Reddy VY, Anic A, Koruth J, et al. Pulsed field ablation in patients with persistent atrial fibrillation. *J Am Coll Cardiol*. 2020;76:1068–1080.
60. Wittkampf FHM, Van Es R, Neven K. Electroporation and its relevance for cardiac catheter ablation. *J Am Coll Cardiol EP*. 2018;4:977–986.

Chapter 25

Modern Technologies for Cardiac Resynchronization Therapy

Niraj Varma
Heart and Vascular Institute, Cleveland Clinic, Cleveland, OH, United States

Chapter Outline

Abbreviations

ECGI electrocardiographic Imaging
RV right ventricle
LV left ventricle
qLV interval from QRS onset to latest LV electrode activation (qLV; a measure of LV activation delay at LV lead site).
LVp-RVs conduction time from LV pacing artifact to the sensed RV iEGM
RVp-LVs conduction time from the RV pacing artifact to the sensed LV iEGM
QRSd QRS duration
LVEF left ventricular ejection fraction
LBBB left bundle branch block
RBB right bundle branch
BiV biventricular pacing
MPP multipoint left ventricular pacing

Introduction

Cardiac resynchronization therapy (CRT) is an electrical therapy designed with the objective of correcting delayed left ventricular (LV) activation among heart failure (HF) patients with dyssynchrony.[1] This has been shown to suppress HF decompensation and improve patient survival, that is, important clinical outcomes. However, CRT benefit is not uniform and patient responses vary.[2,3] This may be due to incomplete electrical resynchronization. Methods to improve and maintain the quality of delivered CRT are needed. There is recent awareness that therapy has to be patient individualized. This requires attention to both electrical substrate (i.e., the disease being treated) and prescription of appropriate pacing therapies. Novel technologies may offer solutions.

The target condition is the electrical substrate created by left bundle branch block (LBBB; Fig. 25.1).[4] Important features to note are intact right bundle branch (RBB), transseptal conduction delay, but normal (i.e., rapid) LV free wall activation. Ventricular resynchronization is traditionally restored with atriobiventricular pacing, that is, atrial, right ventricular (RV), and LV leads (Fig. 25.2). These components merit individual discussion.

Advances in Cardiovascular Technology. DOI: https://doi.org/10.1016/B978-0-12-816861-5.00041-1

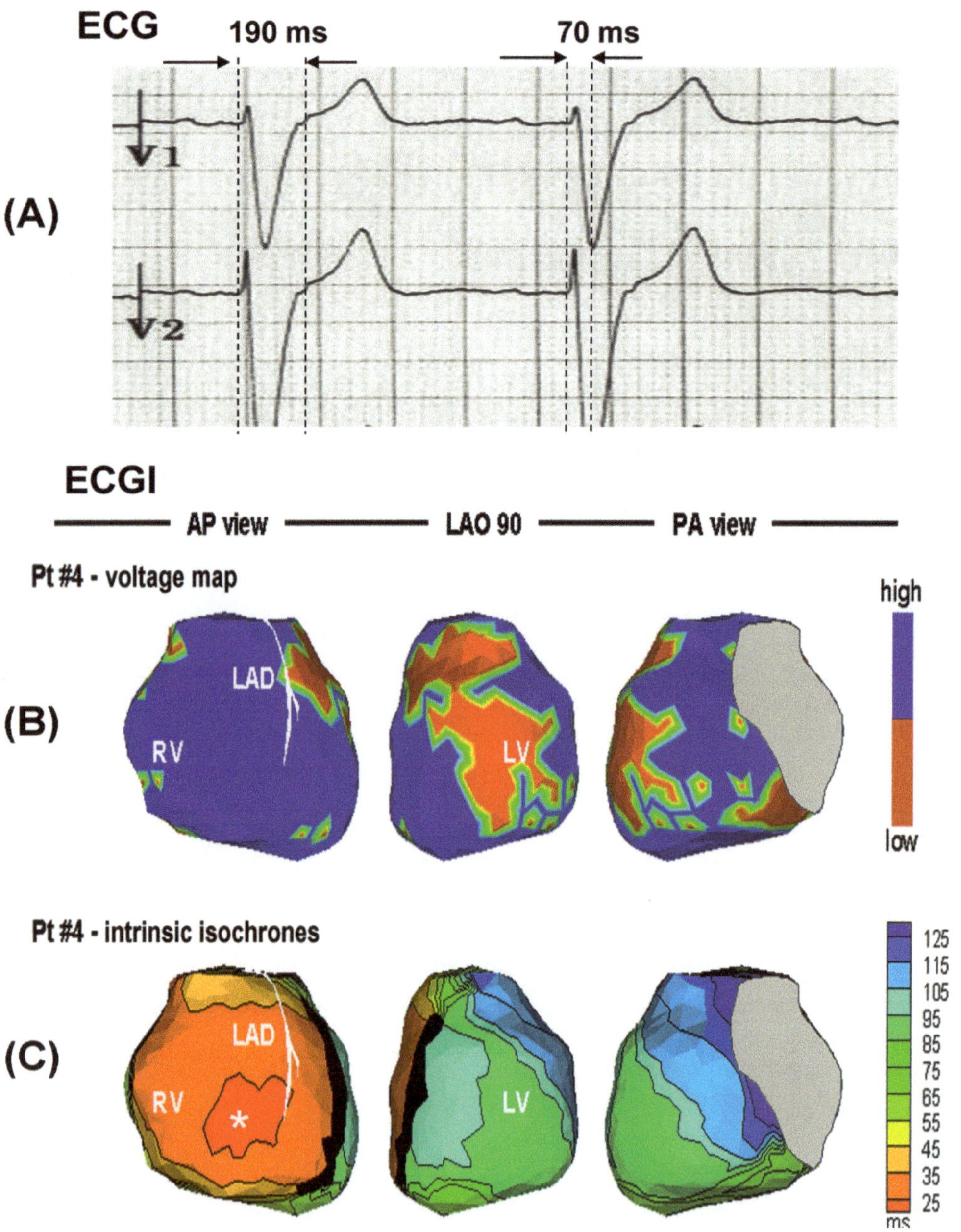

FIGURE 25.1 Ventricular activation during left bundle branch block.

Right ventricular pacing in cardiac resynchronization therapy

RV pacing alone was deleterious among HF patients.[5] This has led to debate about the role of RV pacing in CRT. A result was the production of a device-based pacing algorithm that sought to minimize RV pacing during CRT (AdaptivCRT; Table 25.1).[6] When tested in a trial, it appeared to be noninferior to traditional biventricular pacing, thus seeming to indicate the RV pacing was unnecessary. Although this may apply to small cohort studies with restricted selection criteria (see below), some individuals may benefit from the inclusion of RV pacing (i.e., biventricular pacing). This is because RV stimulation exerts variable effects in HF patients with LBBB.[7,8] Although committing LV activation totally to RV stimulation likely underlay detrimental effects seen among patients with LBBB and LV dysfunction, some initial components of RV stimulation—specifically septal activation—may be useful to CRT (see below).

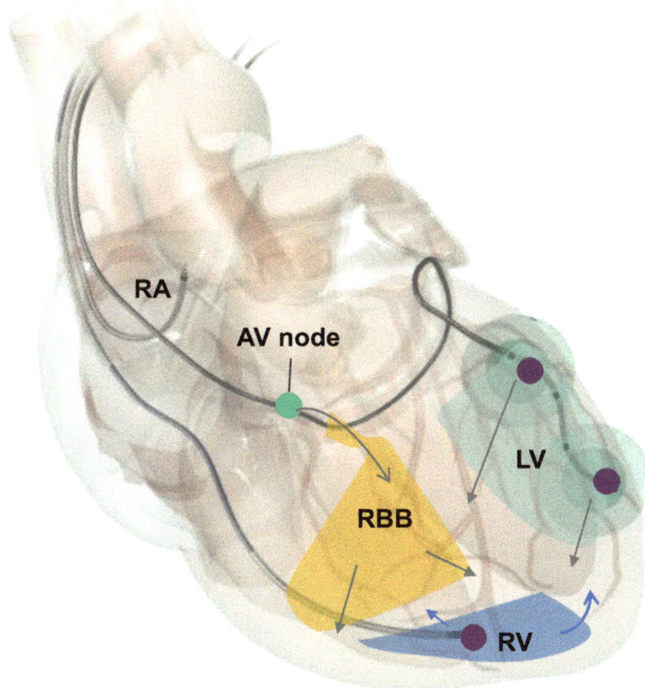

FIGURE 25.2 Atriobiventricular pacing.

Left ventricular pacing in cardiac resynchronization therapy

LV pacing is the core component of CRT. Its purpose is to preexcite the LV, timed to RV activation to restore biventricular synchrony. Although conceptualized to restore confluent LV depolarization, LV paced effects are highly variable. LV epicardial stimulation does not replicate rapid intrinsic conduction even in normal ventricles.[9] In diseased hearts, wavefront propagation may be further retarded by scar and LV enlargement.[10,11] Paced responses to LV stimulation vary from patient to patient and can vary dramatically in an individual patient with a minor change in the position of LV stimulation[12–15] (Fig. 25.3). This unpredictability may limit the ability to pace effectively, let alone restore biventricular resynchronization, and needs to be accounted for during device programming.

The importance of paced effect was shown by reduction in CRT efficacy when LV pacing resulted in slow LV activation (e.g., from scar) manifesting with wide paced QRS (> 200 Ms) or prolongation of transventricular activation time (measured by LV paced to RV sensed interval).[17,18] These challenges may be mitigated by multipolar LV leads (Fig. 25.4).[16] Not only do these improve the success of achieving acceptable pacing thresholds and avoiding PNS,[16] but also permit electrical optimization by selection of the most effective pacing vector).[19] Further, pacing from two electrodes simultaneously (multipoint pacing (see below)) may "jump" conduction barriers or increase the span of ventricular stimulation.[20] This may be useful in LV enlargement[21] (Fig. 25.5).

Atrioventricular interval in cardiac resynchronization therapy

The role of the atrial lead is traditionally to maintain atrioventricular (AV) synchrony and atrial support pacing similarly to its role during DDD pacing for pacemakers. However, control of the AV interval fulfills another important role in CRT when the RBB is intact by determining the degree of RBB-mediated ventricular depolarization (Fig. 25.1).[22] This is illustrated by demonstration of the "fusion band." During biventricular (or LV) pacing, the AV interval is set to minimal (typically 20–30 Ms) then progressively elongated until intrinsic conduction supervenes.[23] Typically the QRS duration narrows then prolongs indicating various levels of fusion of biventricular (or LV) paced wavefronts with RBB-mediated activation. The AV interval providing the narrowest QRS may be considered the optimized AV interval. This highlights another challenge. The intrinsic AV conduction normally is subject to ambulatory variations in response to natural changes in physiologic and autonomic conditions.[24] These variations may

TABLE 25.1 Summary of automatic device based algorithms for CRT.

	QuickOpt (SJM)	SmartDelay (BSC)	AdaptivCRT (MDT)	SonR (Sorin)	SyncAV (Abbott)
Basis	IEGM	IEGM	IEGM	Hemodynamic	IEGM
Programmability (AV/VV)	√	X	X	X	√
Dynamic (Ambulatory)	X	X	√ (per minute)	(√) (Weekly)	√ (every 256 beats)
Trials	**FREEDOM**	**SMART-AV**	**AdaptivCRT**	**RESPOND**	***SyncAV***
Safety	√	√	√	√	*Pending*
Efficacy	**NONINFERIOR** vs Empiric	**EQUIVALENT** vs ECHO-guided or Empiric	**NONINFERIOR** vs Echo BiV	**NONINFERIOR**	*Pending*

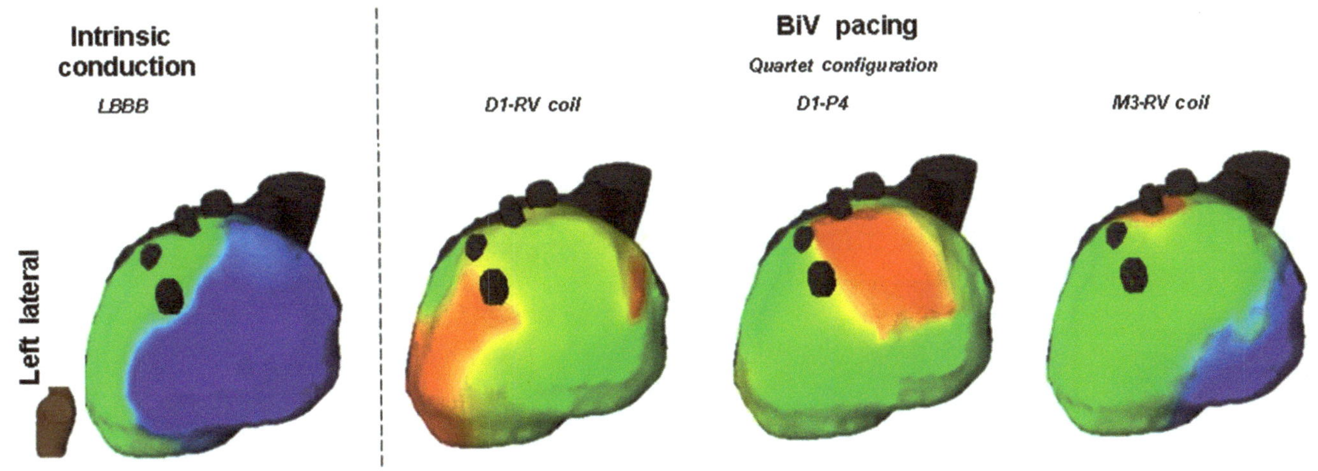

FIGURE 25.3 LV paced effects contrasted among different paced vectors with a quadripolar lead.

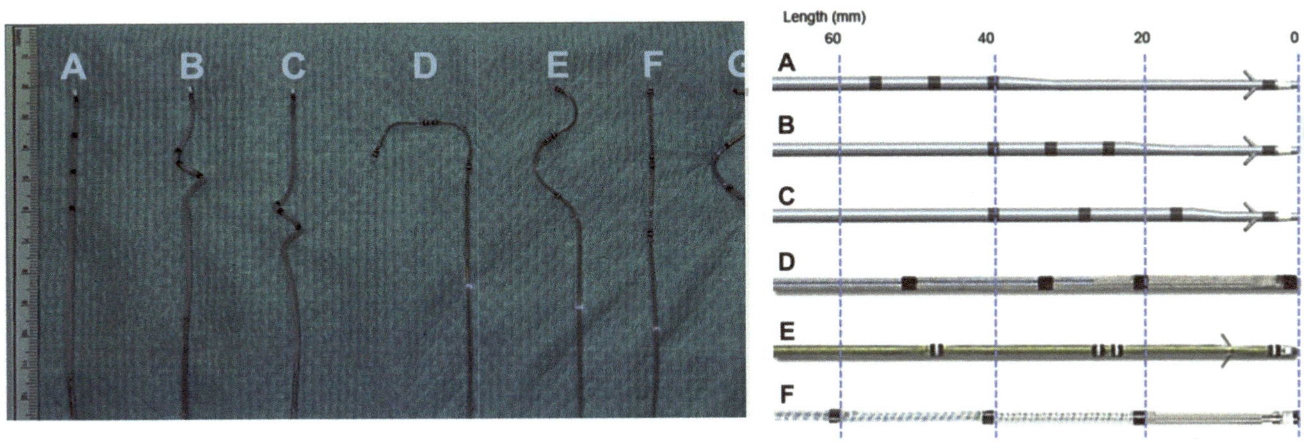

FIGURE 25.4 Multipolar leads. (A) curvature and (B) spacing of currently available quadripolar leads. *Replicated with permission from van Everdingen W.M., Cramer M.J., Doevendans P.A. and Meine M. Quadripolar Leads in Cardiac Resynchronization Therapy. JACC Clin Electrophysiol 2015;1:225–237.*

upset optimized settings that have been so carefully selected and programmed at rest. Accommodating these and preserving optimized settings requires a device-based solution. Automated dynamic control dramatically improves the "quality" of delivered CRT.[25,26]

Cardiac resynchronization therapy

Coordinating these multiple and variegate effects to produce an optimized CRT prescription is complex, particularly since it has to be tailored to each individual (Fig. 25.3). Hence, devices commonly are left at nominal settings, irrespective of intrinsic AV and/or interventricular (VV) intervals.[27] However, there is advantage in achieving postimplant electrical optimization. Studies deliberately targeting electrical resynchronization (e.g., QRS narrowing) as a goal showed changes in acute hemodynamics or chronic ventricular reverse remodeling superior to echocardiographic optimization or nominal device settings.[22,23,28–30] Nevertheless, this still requires significant operator input. An automatic device-based algorithm to manage this entire range of variations in substrate and paced effects is appealing. However, most incorporate some but not all the required features and apply a "one size fits all" solution. Unsurprisingly they have yielded neutral results (Table 25.1). These are briefly reviewed below.

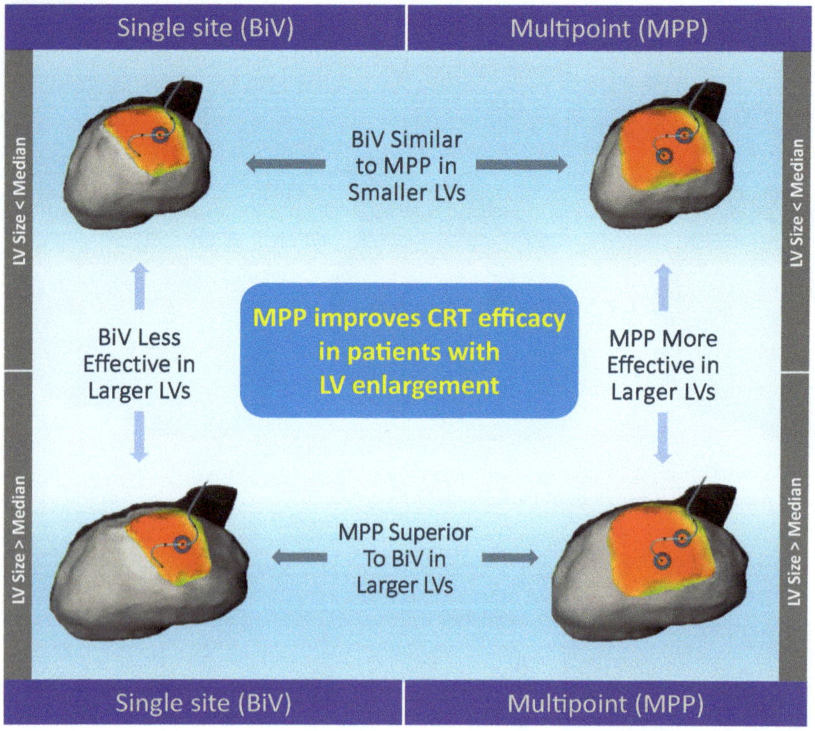

FIGURE 25.5 Multipoint compared to single site left ventricular pacing among patients with left ventricular enlargement during cardiac resynchronization therapy.

Device-based algorithms

SMART delay integrates intrinsic AV intervals, QRS duration, and LV lead location.[31] Correlation between maximal achievable hemodynamic response (% change in LV dP/dt_{max}) and the AV delay predicted by the algorithm was excellent in acute studies. However, the algorithm did not result in chronically improved outcomes when tested in a randomized trial.[32] The reason may be that despite promoting electrical resynchronization, SMART Delay exerts no control over changing AV intervals which can lead to variable LV capture.[25]

The AdaptivCRT algorithm seeks to promote fusion of LV activation during LV pacing with RBB activation of the RV. It aims to achieve this by pacing at a predetermined fraction (approx. 70%) of the intrinsic AV interval, and maintaining this by recalibrating the AV interval every 56 beats after reevaluating intrinsic conduction.[6] This dynamic control was successful in maintaining effective LV capture.[25] However, the algorithm was noninferior to standard biventricular pacing in a randomized trial.[6] Reasons may be multiple. The algorithm is restricted to patients with LBBB and PR interval <200 Ms (i.e., a CRT group already with better prognosis). It may have no advantage when QRS duration is >150 Ms.[33] This is because the main source of LV delay in LBBB is septally based and maintained (and not resolved) by RBB activation. On the other hand, this barrier may be mitigated by the initial effects of RV pacing, i.e., biventricular pacing is more effective (Fig. 25.6).[34] Results of fusion optimization have shown that LV only fusion pacing is best in only a minority and that the AV interval that best delivers this varies among individuals with LBBB.[23] Thus the main deficits of this algorithm are that it does not permit individual programming of AV interval, nor inclusion of RV pacing with LV pacing (or interventricular timings), and thus fails to deliver individualized electrical optimization.

SyncAV

Since past data indicate that a universal programmed strategy (e.g., nominal, LV fusion, etc.[6]) may have limited value, the SyncAV algorithm design was initiated by an investigator at the Cleveland Clinic in collaboration with Abbott. SyncAV enables both electrical optimization and dynamic AV interval control. Delivery of electrical optimization can be patient individualized by permitting BiV or LV pacing as required, in patients with PR intervals up to 300 Ms. At the same time, these optimized settings may be preserved with the automatic dynamic AV control feature. Briefly,

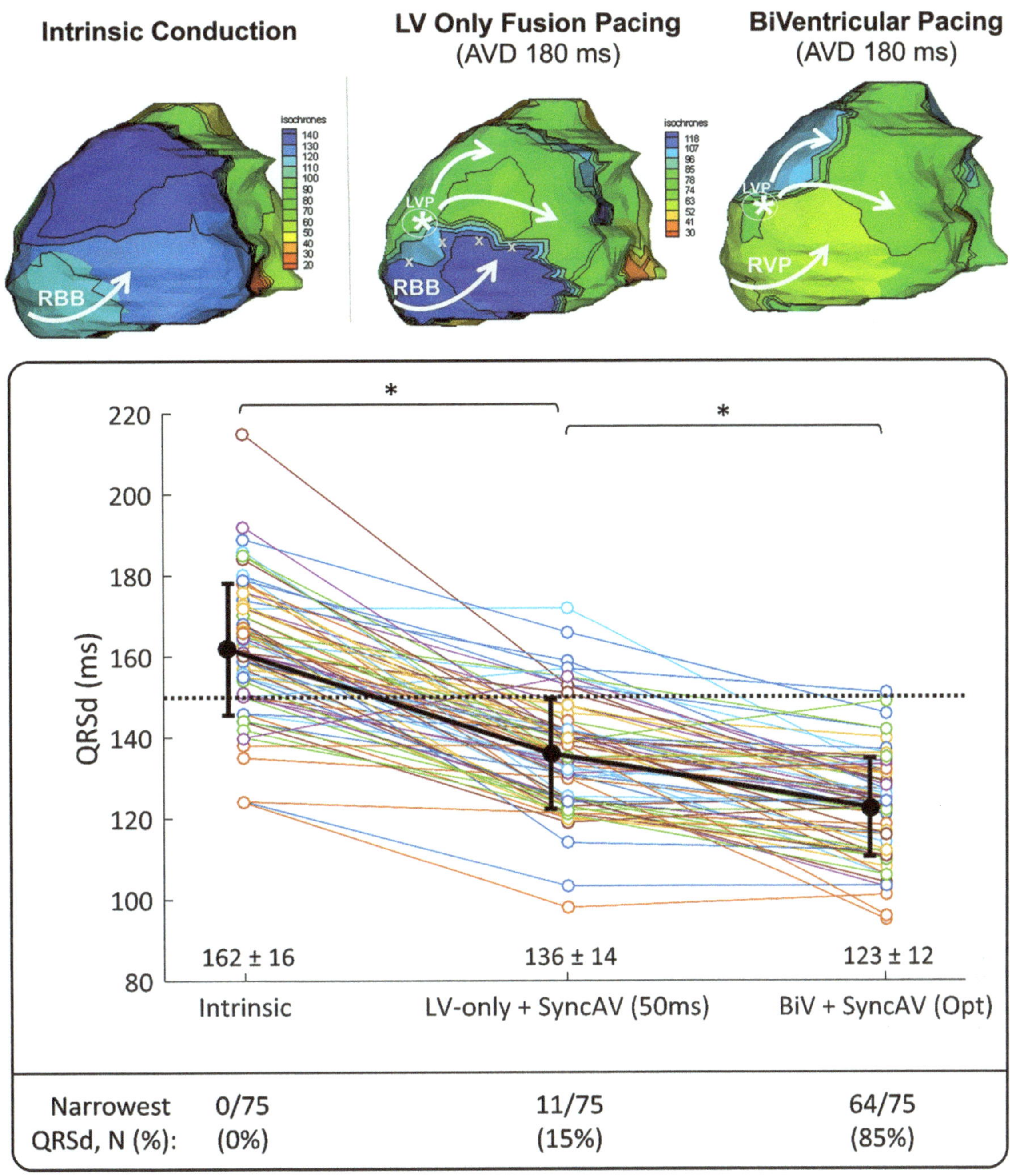

FIGURE 25.6 Left ventricular fusion pacing versus Biventricular fusion pacing.

SyncAV functions as follows. Every 256 beats, the algorithm automatically extends the paced and sensed AV delay for three beats, during which it measures the intrinsic AV interval. With the default SyncAV offset, 50 Ms is subtracted from the measured intrinsic AV interval, and the result applied as the paced AV interval for the following 255 beats (i.e., SyncAV paced AVD = Intrinsic AV interval—50 Ms). The cycle repeats every 256 beats, thus permitting dynamic adjustment of the paced AV interval. The offset value may be reprogrammed across a wide range of offsets (10–120 Ms). Additionally, ventricular pacing configuration may be changed from simultaneous biventricular ("BiV," default) to introduce interventricular timing delays or commit to LV only. Thus any selected set (AV and VV) optimized settings in any individual may be preserved during ambulatory variations in AV interval resulting from changes in heart rate or autonomic tone.

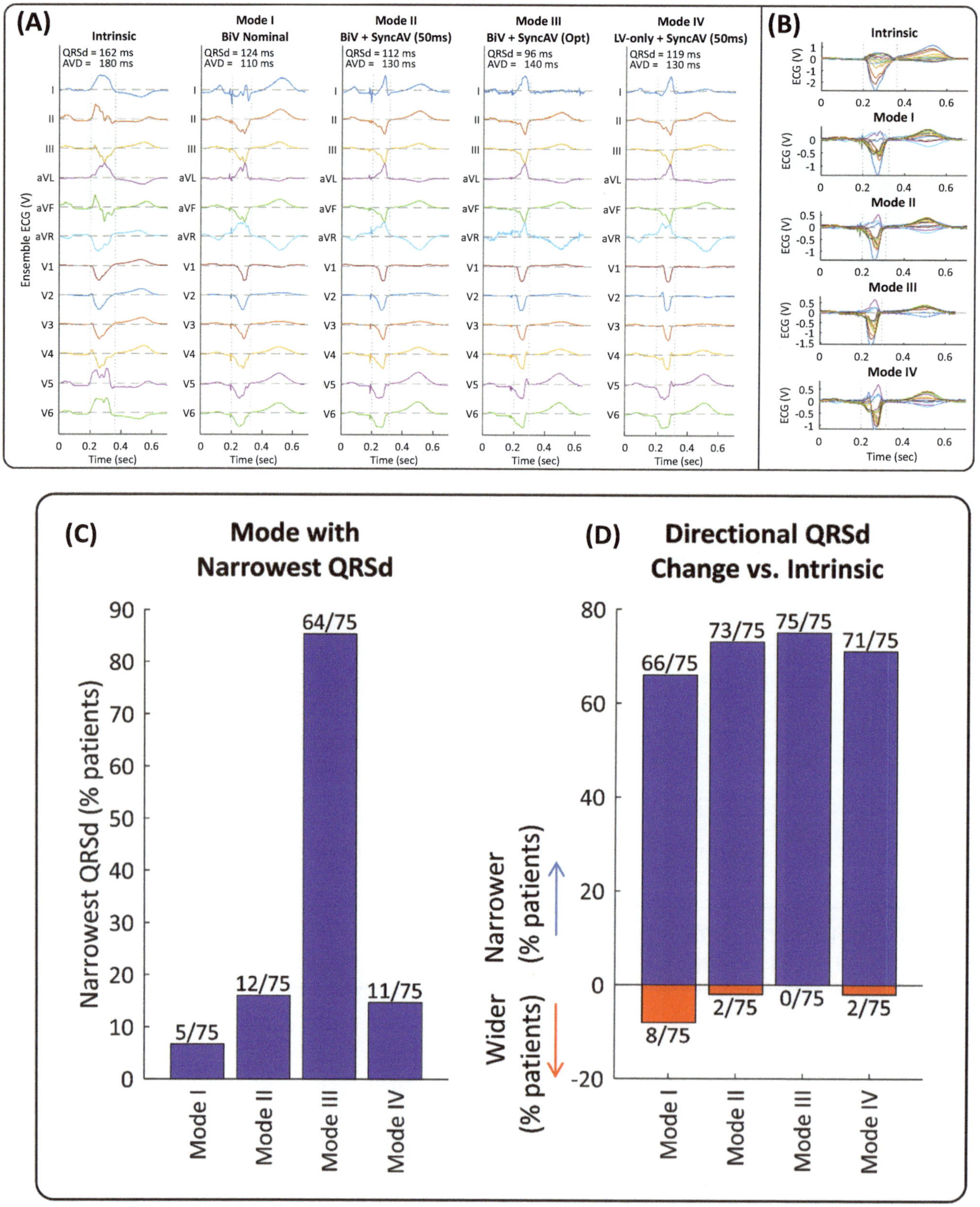

FIGURE 25.7 Electrical resynchronization with different pacing modes.

Clinical studies with SyncAV

The importance of dynamic (but programmable) AV interval optimization was illustrated in an acute electrical optimization study. Intrinsic PR intervals varied widely (range 117–300 Ms) among CRT candidates with LBBB and optimized LV lead placement. Out-of-the-box static settings (which do not account for intrinsic PR) narrowed the QRS duration overall, though only by modest margins, and widened in some candidates, that is, generated further electrical dyssynchrony (Fig. 25.6).[35] Default SyncAV which adapts to intrinsic PR interval by pacing at approximately 80% of its measured value ("50 Ms offset" Mode II) produces longer AV paced intervals which more consistently abbreviated QRS duration, and to a greater degree compared to static settings. When individually optimized, SyncAV narrowed QRSd in *every* patient and resulted in maximal decrement in the whole cohort. Notably, optimized settings varied, that is, the SyncAV offsets most frequently associated with the narrowest QRSd (i.e., offsets used for Mode III) were between 30 and 50 Ms, illustrating the value of AV interval control. LV only fusion pacing was best in only a minority (<15%) confirming observations by others.[22,23] In contrast, by permitting LV and RV timing as well as AV control, adjustments can promote "triple fusion."[23,34,36] These results indicate that postimplant electrical optimization could be significantly improved in otherwise electrically well-selected patients (LBBB and LV lead position) with patient-tailored programming facilitated with the SyncAV dynamic algorithm.

Clinical outcome

These observations have clinical value. Electrical optimization with the goal of QRS narrowing may be useful and intuitive since CRT aims to correct an electrical disorder and systematic reviews support its predictive value.[37] A shorter LV paced QRS duration (index of rapid LV paced effect) was shown to be a strong predictor of clinical response following logistic regression analysis (AUC = 0.74).[17] In one recent RCT, patient-individualized electrical resynchronization improved outcomes measured by structural remodeling compared to nominal settings.[23] Interestingly, electrical optimization was largely dependent on the AV interval and usually required BiV (i.e., RV and LV) rather than LV only pacing to deliver "triple fusion" (BiV pacing at longer AV intervals permits contribution of native right bundle branch conduction[34,36]). One large, real-world investigation of CRT patients reported that application of SyncAV generated a significant reduction in the rate of postimplant HFH, additive to the effect of CRT alone.[38] Furthermore, SyncAV was associated with a reduction in the number of patients with multiple HF events and with 30-day HF rehospitalizations. SyncAV is currently being evaluated in a large international prospective multicenter randomized clinical trial (ClinicalTrials.gov Identifier: NCT04100148).

Future

Opportunities for enhanced electrical resynchronization are being explored. Intracardiac paced intervals, e.g., RV paced to LV sense, and its reciprocal, may direct best programming strategies.[13,18] Use of multipoint pacing may improve LV activation [especially important when the LV is enlarged or scarred (Fig. 25.5)].[39] This may be best visualized with ECGI mapping which show that patterns of wavefront propagation during LV pacing differ from those during intrinsic conduction[10,12,40] and may be modulated by conduction barriers which affect CRT efficacy. Functional conduction barriers to LV-paced wavefronts may develop without underlying scar and limit electrical resynchronization unless there is careful programming.[10,12,34] The SyncAV algorithm, though designed for traditional CRT using LV leads delivered via CS tributaries which remains the only approved therapy for patients with HF and a wide QRS duration, may be adapted to other delivery platforms under current investigation, for example, endocardial LV;[41,42] or His/ LBB area pacing.[43]

Summary

Current expert consensus documents offer little guidance for postimplant management of CRT recipients beyond maximizing %BiV pacing because past CRT optimization trials have yielded neutral results.[44,45] Attempts to improve the degree of CRT response have concentrated on optimization of LV lead position, either anatomically (posterolateral) or electrically (qLV). However, in several studies, CRT efficacy did not differ among inferior, lateral, and anterior sites. The AUC of qLV >95 Ms was modest for predicting CRT reverse remodeling, and siting leads at the point of latest LV activation did not improve results in the prospective ENHANCE CRT trial.[46,47] Ultimately qLV reports substrate for CRT but not effects of therapy (Fig. 25.5). Recent thinking advances the notion that, since the disease being treated is electrical dyssynchrony, it is the CRT paced effect that should be optimized to deliver best electrical resynchronization. This needs to accommodate individual variations in myocardial conduction characteristics in order to gain maximal electrical resynchronization, since nominal programmed settings may be suboptimal or even exacerbate intrinsic electrical dyssynchrony. This condition may occur even when candidates have been selected for LBBB and LV leads sited posterolaterally (Fig. 25.7). However, the wide heterogeneity of paced effects of the components of CRT (atrial, RV and LV leads) among treated individuals indicates that optimization, if performed, should be individualized. This mandates a technical solution that adapts to changing conditions (AV interval, heart remodeling, etc.). SyncAV is unique because although automatic it also enables individualized optimization with dynamic control across a wide range of AV intervals and QRS durations, enabling "triple fusion" shown to have the best acute hemodynamic effects and chronic structural remodeling.[23,34,36] Thus SyncAV promotes delivery of properly timed and maintained CRT pacing.

Representative ECG and ECGI maps during intrinsic conduction in a HF patient with ischemic cardiomyopathy (LVEF 35%) and preserved AV conduction with LBBB.

1. *ECG*. The QRS duration is 190 Ms. In V1 and V2, the rS wave duration of 70 Ms indicates rapid intrinsic forces, that is, the rapid development of myocardial activation wavefront via intact right bundle branch conduction.

   In B and C, ECGI maps of epicardial surfaces of both ventricles are displayed in three views: anterior (AP) which depicts the free wall of the RV, left lateral (LAO 90), and posterior (PA) projections. There is overlap between adjacent views. The left anterior descending (LAD) coronary artery is marked for orientation.
2. *Epicardial voltage map*. Preserved voltage is indicated by blue ("high") and diminished voltage by red zones ("low"). RV epicardial voltage is normal (blue anteriorly) despite extensive LV disease (red zones laterally and posteriorly).
3. *Ventricular activation map* (scale: isochrones marked from QRS onset). RV epicardial free wall breakthrough (*) occurs laterally within 25 Ms of QRS onset. Following this, RV activation is even and rapid (indicated by widely spaced isochrones) and occurs in radial fashion. Free wall depolarization is largely completed within 45 Ms of QRS onset. Following RV activation, the left ventricle is depolarized from apex to base. Thick black markings indicate line/region of conduction slowing in the septum. (Replicated with permission from[4]).

Schematic showing the three pacing components of CRT. The atrial lead permits timing with ventricular pacing and thereby contribution of RBB-mediated depolarization. RV pacing may mitigate a transseptal conduction barrier created by LBBB. The LV lead is used to preexcite delayed LV activation and its paced effects may be promoted by using multipolar leads (a quadripolar lead is depicted here). The term "triple fusion" refers to incorporation of all three elements together during CRT delivery.

ECGI mapping (*red*: earliest; and *blue*: latest activation). *Black* dots represent the electrode positions of a quadripolar lead (interelectode spacing (mm) D1-M2 20; M2-M3 10; and M3-P4 17 mm).

1. Left images show intrinsic conduction in LBBB with early RV and delayed LV activation. The large blue area indicates a large area of terminal activation. Activation (or "qLV") is similar at each electrode site.
2. Posterior view showing LV free wall. LV depolarization (area of earliest activation and pattern of subsequent LV depolarization) differs significantly among paced vectors. D1-P4 pacing activates the LV midsegment early. Stimulation from the anterobasal position (M3-RV coil) produces a small area of early activation locally at electrode site but also generation of an area of slower conductor inferolaterally to generate intra-LV dyssynchrony. Notably these large variations were elicited despite little or no change in qLV along the quadripolar electrodes.

Thus slight positional changes in LV pacing site can generate large variations in ventricular depolarization, which are unrelated to baseline activation. Selection of vector eliciting most rapid and confluent LV activation may improve CRT efficacy. (Replicated with permission from[15]).

3. LV epicardial pacing elicits slowly propagating wavefronts which reduces ability to rapidly activate a LV that is grossly enlarged in HF. Multipoint pacing may overcome these limitations. (Replicated with permission from[21]).

Top: ECGI imaging of a male with ischemic cardiomyopathy (LVEF 10%), typical LBBB (QRS 184 Ms), and PR 200 Ms treated with CRT and laterally placed LV lead was (qLV 160 Ms). LV pacing fails to generate confluent rapid LV activation (*Figure middle*). As a result, LV fusion pacing (at 180 Ms AVD) was insufficient in restoring electrical synchrony, which was accomplished by adding RV pacing (BiV at the same AVD). Transseptal delay- the source of LV delay in LBBB - may have a functional basis that can be mitigated by RV pacing.[12,48] Resolution of conduction delay permits recruitment of a greater proportion of the LV and diminished total activation time, effects which improve hemodynamic function.[20] (Reproduced with permission from[34]).

Bottom:Among 75 patients with LBBB and LV lead location at terminally activated LV regions, LV−only fusion pacing reduced QRS duration (QRSd) by 15.6% relative to intrinsic, but individualized biventricular pacing (BiV) fusion pacing produced a greater effect (23.9% reduction, $P < 0.001$ vs LV fusion pacing). *$P < 0.001$. (Reproduced with permission from[35]).

Thus biventricular stimulation at longer AV intervals introduces the initial effects of RV pacing as well as capturing intrinsic RBB conduction to deliver "triple fusion."[36]

Mode I: Static defaults settings; Mode II: Default SyncAV which adapts to intrinsic conduction by pacing at approximately 80% of the PR interval ("50 Ms offset"); Mode III: optimized SyncAV; Mode IV: LV only SyncAV.

Top: Representative 12-lead ECG for all pacing modes. Simultaneous recordings of ECG leads in a single patient (A) displayed separately, for each Mode (left to right), and (B) depicted as a composite, for each pacing Mode (top to bottom). AVD indicates AV delay; BiV, biventricular pacing; QRSd, QRS duration. AVD indicates AV delay; BiV, biventricular pacing (simultaneous LV-right ventricular pacing; LV, left ventricular. Mode IV results in "LV Fusion") pacing in patients with LBBB.

Bottom: Distribution of pacing Modes associated with the narrowest QRS duration (QRSd) among 75 patients with LBBB and LV lead location at terminally activated LV regions.

(A) Distribution of pacing modes associated with the narrowest QRSd. Note: Total percentage exceeds 100% because the narrowest QRSd can be achieved by multiple pacing modes.

(B) Distribution in directional QRSd change among patients, relative to intrinsic conduction (narrower in blue, wider in red).

LV-only fusion pacing (Mode IV) reduced QRSd by 15.6% relative to intrinsic, but individualized BiV fusion pacing (Mode III) produced a greater effect (23.9% reduction, $P < 0.001$ vs LV-only fusion).

References

1. Wells G, Parkash R, Healey JS, et al. Cardiac resynchronization therapy: a *meta*-analysis of randomized controlled trials. *CMAJ*. 2011;183:421–429.
2. Daubert C, Behar N, Martins RP, Mabo P, Leclercq C. Avoiding non-responders to cardiac resynchronization therapy: a practical guide. *Eur Heart J*. 2017;38:1463–1472.
3. Varma N, Boehmer J, Bhargava K, et al. Evaluation, management, and outcomes of patients poorly responsive to cardiac resynchronization device therapy. *J Am Coll Cardiol*. 2019;74:2588–2603.
4. Varma N, Jia P, Ramanathan C, Rudy Y. RV electrical activation in heart failure during right, left, and biventricular pacing. *JACC Cardiovasc Imaging*. 2010;3:567–575.
5. Wilkoff BL, Cook JR, Epstein AE, et al. Dual-chamber pacing or ventricular backup pacing in patients with an implantable defibrillator: the Dual Chamber and VVI Implantable Defibrillator (DAVID) Trial. *JAMA*. 2002;288:3115–3123.
6. Martin DO, Lemke B, Birnie D, et al. Investigation of a novel algorithm for synchronized left-ventricular pacing and ambulatory optimization of cardiac resynchronization therapy: results of the adaptive CRT trial. *Heart Rhythm*. 2012;9:1807–1814.
7. Varma N. Left ventricular conduction delays in response to right ventricular apical pacing. Influence of LV dysfunction and bundle branch block. *J Cardiovasc Electrophsyiol*. 2008;19:114–122.
8. Varma N. Left ventricular electrical activation during right ventricular pacing in heart failure patients with LBBB: visualization by electrocardiographic imaging and implications for cardiac resynchronization therapy. *J Electrocardiol*. 2015;48:53–61.
9. Wiggers CJ. The muscular reactions of the mammalian ventricles to artificial surface stimuli. *Am J Physiol*. 1925;73:346–378.
10. Jia P, Ramanathan C, Ghanem RN, Ryu K, Varma N, Rudy Y. Electrocardiographic imaging of cardiac resynchronization therapy in heart failure: observation of variable electrophysiologic responses. *Heart Rhythm*. 2006;3:296–310.
11. Leyva F, Foley PW, Chalil S, et al. Cardiac resynchronization therapy guided by late gadolinium-enhancement cardiovascular magnetic resonance. *J Cardiovasc Magn Reson*. 2011;13:29.
12. Varma N, Ploux S, Ritter P, Wilkoff B, Eschalier R, Bordachar P. Noninvasive mapping of electrical dyssynchrony in heart failure and cardiac resynchronization therapy. *Card Electrophysiol Clin*. 2015;7:125–134.
13. Wisnoskey B, Varma N. Left Ventricular paced activation in CRT patients with LBBB and Relationship to its Electrical Substrate. *Heart Rhythm O2*. 2020;1:85–95.
14. van Everdingen WM, Zweerink A, Salden OAE, et al. Pressure-volume loop analysis of multipoint pacing with a quadripolar left ventricular lead in cardiac resynchronization therapy. *JACC Clin Electrophysiol*. 2018;4:881–889.
15. Varma N. Multipolar LV leads to optimize CRT-qLV or LV paced effect? *J Cardiovasc Electrophysiol*. 2020;31:238–239.
16. van Everdingen WM, Cramer MJ, Doevendans PA, Meine M. Quadripolar leads in cardiac resynchronization therapy. *JACC Clin Electrophysiol*. 2015;1:225–237.
17. Kobe J, Dechering DG, Rath B, et al. Prospective evaluation of electrocardiographic parameters in cardiac resynchronization therapy: detecting nonresponders by left ventricular pacing. *Heart Rhythm*. 2012;9:499–504.
18. Ueda N, Noda T, Nakajima I, et al. Clinical impact of left ventricular paced conduction disturbance in cardiac resynchronization therapy. *Heart Rhythm*. 2020;17:1870–1877.
19. Varma N. Variegated left ventricular electrical activation in response to a novel quadripolar electrode: visualization by non-invasive electrocardiographic imaging. *J Electrocardiol*. 2014;47:66–74.

20. Menardi E, Ballari GP, Goletto C, Rossetti G, Vado A. Characterization of ventricular activation pattern and acute hemodynamics during multipoint left ventricular pacing. *Heart Rhythm.* 2015;12:1762–1769.
21. Varma N, Baker 2nd J, Tomassoni G, et al. Left ventricular enlargement, cardiac resynchronization therapy efficacy, and impact of MultiPoint Pacing. *Circ Arrhythm Electrophysiol.* 2020;13:e008680.
22. Arbelo E, Tolosana JM, Trucco E, et al. Fusion-optimized intervals (FOI): a new method to achieve the narrowest QRS for optimization of the AV and VV intervals in patients undergoing cardiac resynchronization therapy. *J Cardiovasc Electrophysiol.* 2014;25:283–292.
23. Trucco E, Tolosana J, Arbelo E, et al. Improvement of reverse remodeling using ECG Fusion-Optimized Intervals in cardiac resynchronization therapy: A randomized study. *JACC: Clin Electrophysiol.* 2018;4:181–189.
24. Cheng A, Landman SR, Stadler RW. Reasons for loss of cardiac resynchronization therapy pacing: insights from 32 844 patients. *Circ Arrhythm Electrophysiol.* 2012;5:884–888.
25. Varma N, Stadler RW, Ghosh S, Kloppe A. Influence of automatic frequent pace-timing adjustments on effective left ventricular pacing during cardiac resynchronization therapy. *Europace.* 2017;19:831–837.
26. Brugada J, Delnoy PP, Brachmann J, et al. Contractility sensor-guided optimization of cardiac resynchronization therapy: results from the RESPOND-CRT trial. *Eur Heart J.* 2017;38:730–738.
27. Lunati M, Magenta G, Cattafi G, et al. Clinical Relevance Of Systematic CRT Device Optimization. *J Atr Fibrillation.* 2014;7:1077.
28. Tamborero D, Vidal B, Tolosana JM, et al. Electrocardiographic vs. echocardiographic optimization of the interventricular pacing delay in patients undergoing cardiac resynchronization therapy. *J Cardiovasc Electrophysiol.* 2011;22:1129–1134.
29. Vatasescu R, Berruezo A, Mont L, et al. Midterm "super-response" to cardiac resynchronization therapy by biventricular pacing with fusion: insights from electro-anatomical mapping. *Europace.* 2009;11:1675–1682.
30. Lecoq G, Leclercq C, Leray E, et al. Clinical and electrocardiographic predictors of a positive response to cardiac resynchronization therapy in advanced heart failure. *Eur Heart J.* 2005;26:1094–1100.
31. Gold MR, Niazi I, Giudici M, et al. A prospective comparison of AV delay programming methods for hemodynamic optimization during cardiac resynchronization therapy. *J Cardiovasc Electrophysiol.* 2007;18:490–496.
32. Ellenbogen KA, Gold MR, Meyer TE, et al. Primary results from the SmartDelay determined AV optimization: a comparison to other AV delay methods used in cardiac resynchronization therapy (SMART-AV) trial: a randomized trial comparing empirical, echocardiography-guided, and algorithmic atrioventricular delay programming in cardiac resynchronization therapy. *Circulation.* 2011;122:2660–2668.
33. Yamasaki H, Lustgarten D, Cerkvenik J, et al. Adaptive CRT in patients with normal AV conduction and left bundle branch block: Does QRS duration matter? *Int J Cardiol.* 2017;.
34. Varma N. Therapy for cardiac resynchronization: When left ventricular–only "fusion" pacing is not enough. *Heart Rhythm Case Rep.* 2020;.
35. Varma N, O'Donnell D, Bassiouny M, et al. Programming cardiac resynchronization therapy for electrical synchrony: reaching beyond left bundle branch block and left ventricular activation delay. *J Am Heart Assoc.* 2018;7:e007489.
36. Ter Horst IAH, Bogaard MD, Tuinenburg AE, et al. The concept of triple wavefront fusion during biventricular pacing: using the EGM to produce the best acute hemodynamic improvement in CRT. *Pacing Clin Electrophysiol.* 2017;40:873–882.
37. Bazoukis G, Naka KK, Alsheikh-Ali A, et al. Association of QRS narrowing with response to cardiac resynchronization therapy-a systematic review and *meta*-analysis of observational studies. *Heart Fail Rev.* 2020;25:745–756.
38. Varma N. Gain in CRT Efficacy with Dynamic Electrical Optimization: Real World Effect of SyncAV on Heart Failure Hospitalizations. Paper presented at: EHRA; 2020; Vienna.
39. O'Donnell D, Wisnoskey B, Badie N, et al. Electrical synchronization achieved by multipoint pacing combined with dynamic atrioventricular delay. *J Interv Card Electrophysiol.* 2020;.
40. Ginks MR, Shetty AK, Lambiase PD, et al. Benefits of endocardial and multisite pacing are dependent on the type of left ventricular electric activation pattern and presence of ischemic heart disease: insights from electroanatomic mapping. *Circ Arrhythm Electrophysiol.* 2012;5:889–897.
41. Reddy VY, Miller MA, Neuzil P, et al. Cardiac resynchronization therapy with wireless left ventricular endocardial pacing: the SELECT-LV study. *J Am Coll Cardiol.* 2017;69:2119–2129.
42. Morgan JM, Biffi M, Geller L, et al. ALternate Site Cardiac ResYNChronization (ALSYNC): a prospective and multicentre study of left ventricular endocardial pacing for cardiac resynchronization therapy. *Eur Heart J.* 2016;37:2118–2127.
43. Huang W, Wu S, Vijayaraman P, et al. Cardiac resynchronization therapy in patients with nonischemic cardiomyopathy using left bundle branch pacing. *JACC Clin Electrophysiol.* 2020;6:849–858.
44. Brignole M, Auricchio A, Baron-Esquivias G, et al. ESC guidelines on cardiac pacing and cardiac resynchronization therapy: the task force on cardiac pacing and resynchronization therapy of the European Society of Cardiology (ESC). Developed in collaboration with the European Heart Rhythm Association (EHRA). *Europace.* 2013;2013(15):1070–1118.
45. Daubert JC, Saxon L, Adamson PB, et al. EHRA/HRS expert consensus statement on cardiac resynchronization therapy in heart failure: implant and follow-up recommendations and management. *Europace.* 2012;2012(14):1236–1286.
46. Gold MR, Birgersdotter-Green U, Singh JP, et al. The relationship between ventricular electrical delay and left ventricular remodelling with cardiac resynchronization therapy. *Eur Heart J.* 2011;32:2516–2524.
47. Singh JP, Berger RD, Doshi RN, et al. Targeted left ventricular lead implantation strategy for non-left bundle branch block patients: the ENHANCE CRT study. *JACC Clin Electrophysiol.* 2020;6:1171–1181.
48. Derval N, Duchateau J, Mahida S, et al. Distinctive left ventricular activations associated with ECG pattern in heart failure patients. *Circ Arrhythm Electrophysiol.* 2017;10.

Chapter 26

Innovative Approaches and Technology Platforms for Pacemaker Lead Extraction

Thomas D. Callahan and Bruce L. Wilkoff
Extraction Center, Cardiac Electrophysiology, Cleveland Clinic, Cleveland, OH, United States

Chapter Outline

Introduction

Permanent pacemaker (PPM) and implantable cardiac defibrillator (ICD) transvenous lead extraction (TLE) is usually a safe procedure with a major complication rate of less than 2% in most studies.[1,2] However, when complications occur, they can be catastrophic, including myocardial avulsion and laceration of the innominate vein or superior vena cava (SVC). Most leads can be removed, and clinical success rates are widely accepted to be around 98%–99%.[2] However, dense fibrous adhesions to the surrounding structures sometimes thwart the best attempts at extraction with existing tools and techniques. Recent advances in TLE have made strides in safety and improved our ability to extract ever more challenging leads. Continued advances in technique and in the tools and technology used to remove leads are critical to the goal of absolute safety and efficacy.

Preoperative phase

Patient selection and risk stratification

Not all pacing and defibrillator devices and leads need to be removed. The risks of TLE or surgical removal must be weighed against the risks of simply abandoning or continuing to use a lead with suboptimal function.[3] In the setting of a cardiac implantable electronic device (CIED) infection, whether localized to the device pocket or an endovascular infection with bacteremia or endocarditis, current guidelines recommend complete removal of the device and all endovascular leads with a class 1 indication[4] (Table 26.1). However, some patients may be so frail that the risks of TLE outweigh the likely benefit. In patients with class 2a and 2b indications for TLE such as leads under advisory, leads that are no longer needed or no longer functioning appropriately, a careful assessment of patient and device factors is required to inform a recommendation for or against extraction and a thorough informed decision-making conversation with the patient is imperative to ensure a clinical decision which meets the patient's individual goals.[3] It is crucial to understand that the difficulty of extraction does not necessarily equate with the acute procedural risk when performed by an experienced operator with appropriate tools. Furthermore, acute risk is largely absent when there is no fibrosis on the leads.

Advances in Cardiovascular Technology. DOI: https://doi.org/10.1016/B978-0-12-816861-5.00018-6

TABLE 26.1 Indications for transvenous lead extraction with level of evidence.

COR	LOE	Recommendation
Infection		
I	B-NR	Complete device and lead removal is recommended for all patients with definite CIED system infection
I	C-EO	Complete removal of epicardial leads and patches is recommended for all patients with confirmed infected fluid (purulence) surrounding the intrathoracic portion of the lead
I	B-NR	Complete device and lead removal is recommended for all patients with valvular endocarditis without definite involvement of the lead(s) and/or device
I	B-NR	Complete device and lead removal is recommended for patients with persistent or recurrent bacteremia or fungemia, despite appropriate antibiotic therapy and no other identifiable source for relapse or continued infection
I	C-EO	Careful consideration of the implications of other implanted devices and hardware is recommended when deciding on the appropriateness of CIED removal and for planning treatment strategy and goals
Chronic Pain		
IIa	C-EO	Device and/or lead removal can be useful for patients with severe chronic pain at the device or lead insertion site or believed to be secondary to the device, which causes significant patient discomfort, is not manageable by medical or surgical techniques, and for which there is no acceptable alternative
Thrombosis/Vascular		
I	C-EO	Lead removal is recommended for patients with clinically significant thromboembolic events attributable to thrombus on a lead or a lead fragment that cannot be treated by other means
I	C-EO	Lead removal is recommended for patients with SVC stenosis or occlusion that prevents implantation of a necessary lead
I	C-EO	Lead removal is recommended for patients with planned stent deployment in a vein already containing a transvenous lead, to avoid entrapment of the lead
I	C-EO	Lead removal as part of a comprehensive plan for maintaining patency is recommended for patients with SVC stenosis or occlusion with limiting symptoms
IIa	C-LD	Lead removal can be useful for patients with ipsilateral venous occlusion preventing access to the venous circulation for required placement of an additional lead
Other		
I	C-EO	Lead removal is recommended for patients with life-threatening arrhythmias secondary to retained leads
IIa	C-EO	Lead removal can be useful for patients with a CIED location that interferes with the treatment of a malignancy
IIa	C-LD	Lead removal can be useful for patients if a CIED implantation would require more than four leads on one side or more than five leads through the SVC
IIa	C-EO	Lead removal can be useful for patients with an abandoned lead that interferes with the operation of a CIED system
IIb	C-LD	Lead removal may be considered for patients with leads that due to their design or their failure pose a potential future threat to the patient if left in place
IIb	C-EO	Lead removal may be considered for patients to facilitate access to MRI. Removal of leads to prevent their abandonment, removal of broken or abandoned leads, or removal of leads to allow implantation of an MRI conditional system
IIb	C-EO	Lead removal may be considered in the setting of normally functioning nonrecalled pacing or defibrillation leads for selected patients after a shared decision-making process

Derived from 2017 HRS expert consensus statement on cardiovascular implantable electronic device lead management and extraction.[3]

Risk stratification is typically done on a case-by-case basis using clinical judgment and risk stratification tools. Notably, prognosis is largely determined by comorbidities rather than the acute procedural risk. The Cleveland Clinic nomogram estimates mortality within 30 days of TLE and is estimated with clinical factors such as age, body mass index, hemoglobin, presence of end-stage renal disease, left ventricular ejection fraction, New York Heart Association functional class, infection as indication for extraction, operator experience, and planned extraction of a dual-coil defibrillator lead to stratify risk.[5] Similarly, the IKAR (Infective indications, Kidney dysfunction, Age >56 yrs, Removal of a high voltage lead) score, assigns points for infective indications, kidney dysfunction, age greater than 56 years, and removal of a high voltage lead. Patients with scores of 2 and below had a 1-year mortality of 16%, compared to 79% for those with a score of 3 or greater.[6] Combined with careful assessment of each patient's unique clinical scenario, these tools help predict risk and aid in patient selection for TLE. However note that these risks are not procedural risk but 1- and 12-month mortality. The acute procedural risk is less than 2% for almost all patients.

Patient populations

Advances in cardiac electrophysiology have resulted in ever-expanding available therapies and indications. As a result, patients undergoing CIED implantation have grown more complex. Simultaneously, the number of patients with CIEDs and with older leads has progressively grown. Patient age, comorbidities, and age of the implanted device all add to the difficulty and complexity of TLE. Studies reporting outcomes in specific populations may provide a deeper understanding of risk and aid in patient selection and operative planning. Efficacy of TLE has previously been reported in patients with congenital heart disease, chronic renal disease, and prosthetic valves. Fender et al. reported complete extraction in 94% of patients with congenital heart disease undergoing TLE with complication rates comparable to controls with normal hearts.[7] While a history of chronic renal insufficiency is known to increase the risk of mortality after TLE, Barakat et al. found no increase in complexity of TLE in this population.[8] Huang et al. reported a 97% success rate in patients with prosthetic valves undergoing TLE for infectious indications.[9] While these studies reported favorable TLE outcomes for patients with congenital heart disease and for patients with prosthetic valves, outcomes data for many populations remain limited.

Recent advances

More recent studies have provided a greater understanding of outcomes in specific patient populations and new methods of risk assessment. The SAFeTY TLE score, described by Jachec' et al., helps predict the risk of major complications during TLE rather than 30-day or 1-year mortality. Patients are assigned points for the sum of lead dwell times, anemia, female gender, number of prior procedures, and age under 30[10] (Fig. 26.1). A score of 10 or greater is associated with a significant increase in risk of major complications. Combined with earlier prediction tools and clinical judgment, it is hoped this tool will guide selection and planning and decrease overall complication rates.

Recent work on specific populations has shed light on TLE outcomes and safety in a broader range of patients. In an analysis of the ELECTRa Registry, female gender was found to be associated with lower rates of radiological and clinical success compared to men.[2] An aging population with CIEDs means that older patients are presenting with CIED malfunction or infection for TLE. Recent research confirms that TLE can be performed in elderly patients. One study examining TLE in octo- and nonagenarians reported similar procedural effectiveness in these older patients compared to younger patients undergoing TLE. The rates of major complications did not differ between the older and younger populations supporting efficacy and safety of TLE in the elderly.[11]

Increasing lead dwell time adds to the difficulty of TLE as leads become more adherent to the vasculature with fibrotic scar and with progressive calcification of that scar. Lead dwell times of more than 10 years have been associated with extraction failure.[12] Continued advances have allowed the extraction of progressively older leads with favorable success rates and safety. Pecha et al. reported outcomes of extraction of leads older than 10 years with clinical success rates of 96.8%. Additional tools including mechanical rotational sheaths and femoral snares were required in almost 17% of patients, but major complications were infrequent at just 3.3%.[13]

Recent studies also shed light on narrower patient groups. Safety and efficacy of TLE has been demonstrated in patients with prior TLE, and in those with arrhythmogenic right ventricular cardiomyopathy.[14,15] High levels of clinical success have also been reported in patients with left ventricular assist devices and cardiac transplantation undergoing TLE.[16] SVC syndrome is a potential complication of transvenous leads, and in some cases, is an indication for extraction. Management of this syndrome is complicated by the presence of the leads, which could be damaged and entrapped if the SVC is stented without extraction, and by the need to reimplant, as newly implanted leads could cause restenosis of a percutaneously or surgically treated vessel (Fig. 26.2). Recent work supports the safety of TLE prior to SVC

(A)

SAFeTY TLE Score

S	Sum of lead dwell times	6.095 points
A	Anemia	2.291 points
Fe	Female gender	2.740 points
T	Treatment (previous procedures)	1.364 points for each procedure
Y	Young patient (first implantation under the age of 30)	2.174 points
TLE	Transvenous Lead Extraction	

The sum of the risk points correlated with the probability of developing major complications during a TLE and the relationship was expressed as the logistic function in the following equation: risk of major complications (%) = 100/(1 + 644/(1.3213x)), where "x" is the number of points obtained

(B)

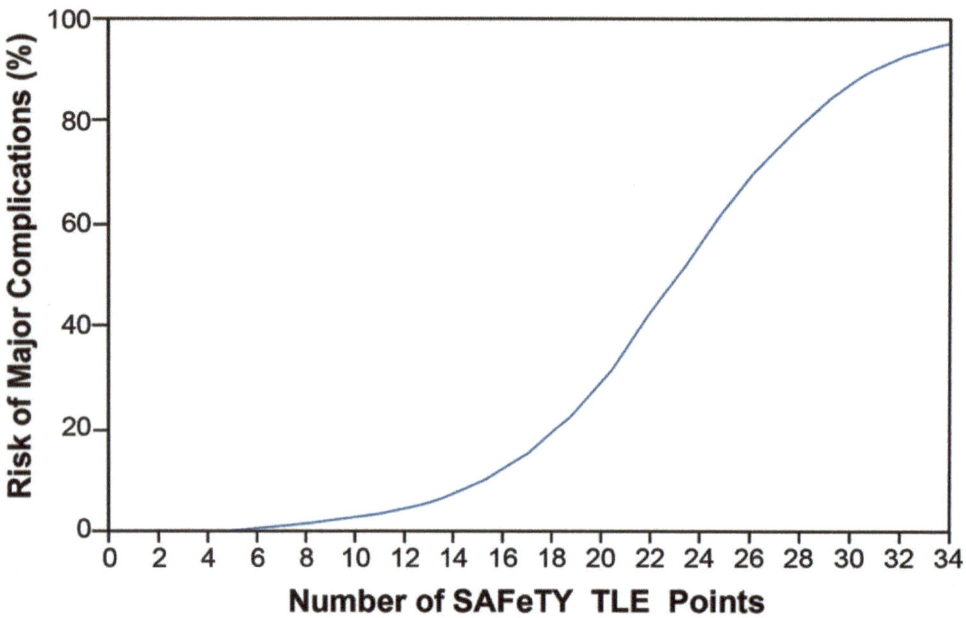

FIGURE 26.1 SAFeTY TLE score. Risk factors are assigned specific point values to create the score (A). Risk of major complications during TLE increases with the number of points (B).[10]

angioplasty with or without stenting.[17] Other treatment strategies including venoplasty without TLE and surgical SVC reconstruction are also described with good short-term results.[18]

Preoperative evaluation and planning

Preoperative evaluation and planning are critical to optimizing safety and the likelihood of success. Preoperative assessment by the involved services such as electrophysiology, cardiac surgery, and anesthesia should be performed in addition to routine preoperative laboratory work. A history should include inquiry into anticoagulant use, prior procedures, and past complications, while a physical examination should investigate for signs of local infection or collateral vessels. It goes without saying that the team must understand which devices and leads are present, which are to be removed, and what will be reimplanted. Recent device reports should be obtained to document the presence and age of leads as

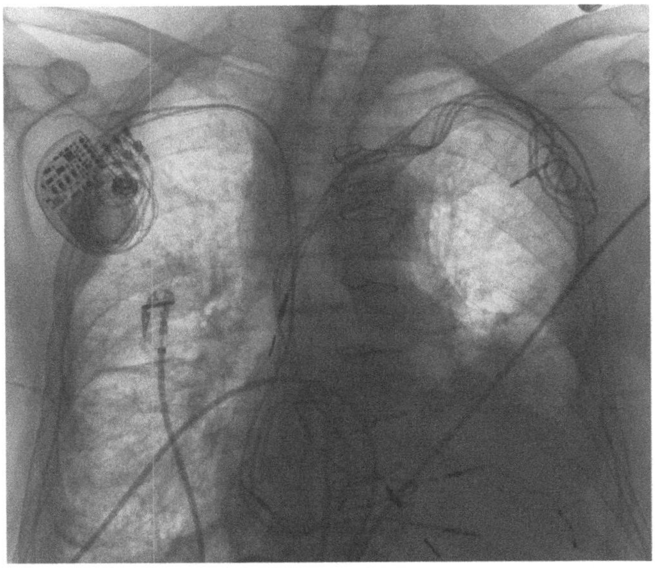

FIGURE 26.2 CIED leads entrapped between the vessel wall and a left subclavian vein stent.

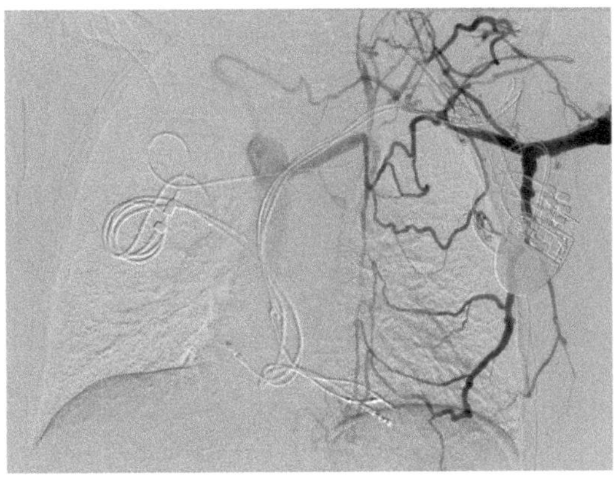

FIGURE 26.3 Left upper extremity venogram demonstrates occlusion of the left subclavian vein with extensive collaterals.

well as the functional status and trends of each. Prior operative reports provide crucial information regarding anatomic variations and challenges that may have arisen during implantation. They can also alert the team to the presence of leads or other devices not listed in the device interrogation. A current device interrogation should be performed to assess device dependency and carefully assess the function of all active leads.

Multiple imaging studies including chest X-ray, transthoracic echocardiogram, transesophageal echocardiogram (TEE), cardiac CT (computed tomography) scan, and a venogram are typically required for surgical planning. A PA (posterior-anterior) and lateral chest X-ray can uncover previously undocumented hardware and provide an important understanding of the position and course of leads. A transthoracic echocardiogram illustrates underlying cardiac structure and function and can further clarify the position and course of leads and devices, and a transesophageal echo can greatly improve upon that information. A TEE is often critical to assess the size and position of vegetations. A venogram performed preoperatively or intraoperatively may be required (Fig. 26.3). A chest CT aids surgical planning for cardiothoracic surgery in the event a sternotomy is required and may clarify anatomy as well as device and lead position and course for the extracting physician.

While a lengthy discussion of the open approach to lead extraction is beyond the scope of this chapter, careful consideration of the optimal extraction approach is imperative. Factors such as the need for open heart surgery to address other issues and the size of lead vegetations or presence of a large PFO (patent foramen ovale) may influence the approach.

TLE should be performed only in centers which have developed and trained all the essential team members. In addition to an electrophysiologist or surgeon with appropriate training and experience in TLE, electrophysiology and

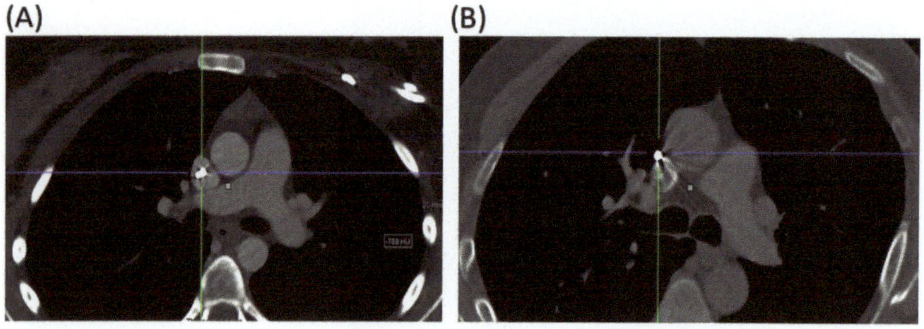

FIGURE 26.4 CT scans examining lead adhesion to the SVC. (A) shows an example of a CIED lead which is central in the SVC with minimal to no adhesion. (B) shows a CIED lead outside the contrast lumen.[19]

surgical OR nurses should be trained for TLE. Anesthesia, perfusionists, vascular surgery, and imaging cardiology play critical roles and should understand and be prepared for the specific demands of TLE.

Recent advances

Patel et al. demonstrated that a preoperative CT could help determine the degree of lead adherence to the vasculature. CT scans were divided into three groups, those with <1 cm of adherence, those with at least one tip adherent >1 cm, and those with at least one lead outside the vein contour. Patients with at least one lead outside the lead contour required significantly longer procedural times and fluoroscopy times compared to patients in the other two groups[19] (Fig. 26.4). In another study ECG (electrocardiogram)-triggered CT with 3D reconstruction was used to help visualize adhesions, perforations, and venous occlusions prior to TLE. Operators used this data to help choose optimal TLE tools and the study reported 100% clinical success and no intraoperative mortalities.[20]

Building and training a team for lead extraction is imperative to the outcomes and safety of the procedure. Experience has been consistently demonstrated to improve outcomes. In a meta-analysis, Di Monaco et al. demonstrated that higher-volume TLE centers have a lower probability of death and minor complications.[21] Similarly, analyses of the ELECTRa registry revealed that high volume centers had higher rates of procedural success and lower rates of procedure-related deaths, major complications, and in-hospital death compared to low-volume centers.[12,22] Efforts to improve operator experience include the development of conventional and virtual reality simulators. In a small study of eight fellows, those randomized to virtual reality simulation were less likely to experience a major simulator complication although the impact of simulator training on real-world outcomes is less certain.[23]

Operative phase

Transvenous lead extraction

TLE is typically performed under general anesthesia with an active type and cross for 2–4 units of packed red blood cells and with cardiac monitoring and defibrillation patches in place. Proper sterile preparation can help minimize infection and complications and facilitate rescue should complications occur. Administration of preoperative antibiotics reduces the risk of infection and is indicated in most cases of TLE. Patient preparation should accommodate bilateral femoral access and allow for immediate subxyphoid epicardial access and sternotomy. Ample venous access should be placed to facilitate administration of pressors as well as rapid volume and blood product administration. An arterial line should be placed for continuous blood pressure monitoring, and venous access to facilitate a venogram should be considered. Additional arterial and venous access to allow prompt transition to cardiopulmonary bypass should be considered in select cases. A temporary transvenous pacing catheter should be placed if prolonged temporary pacing is anticipated. In select cases, a guidewire may be placed from a femoral venous sheath to the SVC or subclavian vein to facilitate contrast venography or temporizing measures in the event of an SVC tear. Intraprocedural imaging by TEE or intracardiac echocardiogram (ICE) can aid in early identification of complications such as tamponade and injury to the tricuspid valve.[24]

The incision over the device should be planned to provide exposure to the device as well as the leads at or near the vein entry. Atypical implantation location, device migration, and skin erosion sometimes necessitate multiple incisions for optimal exposure and access to all the CIED components. The capsule surrounding the CIED may require modification to accommodate a new device or complete debridement in the case of CIED infection.

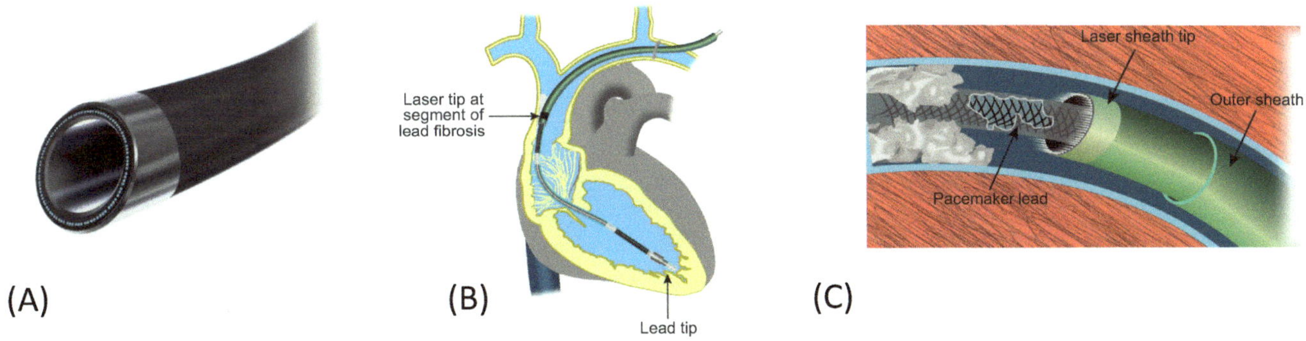

FIGURE 26.5 (A) GuideLight laser extraction sheath. (B) Laser extraction sheath being advanced over pacemaker lead. (C) Magnified view of laser sheath and outer sheath being advanced over pacemaker lead. *(A) Courtesy Philips.*

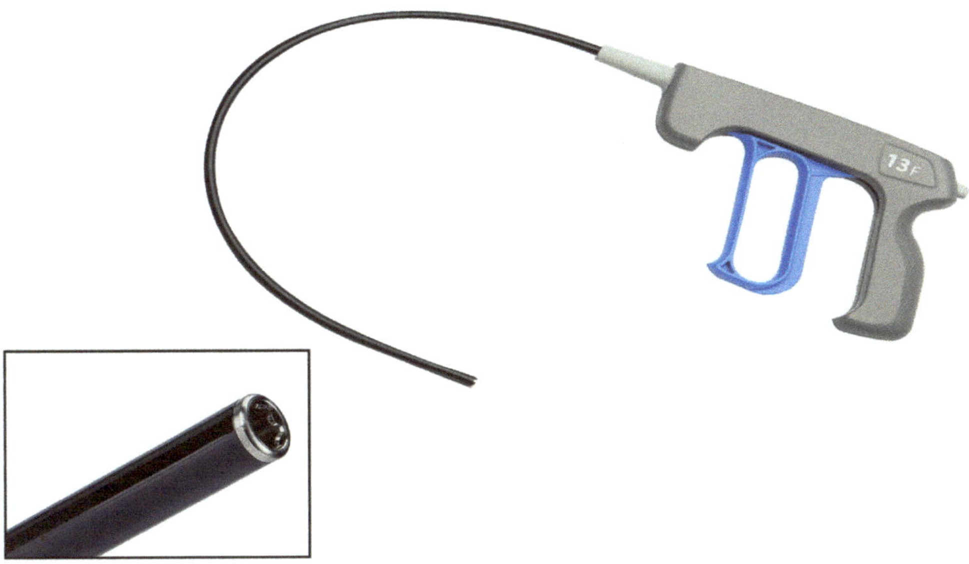

FIGURE 26.6 TightRail mechanical rotational extraction sheath. *Courtesy Royal Philips.*

TLE is performed using a range of tools including telescoping sheaths, mechanical cutting and laser sheaths, locking stylets, and snares. Gaining control over as much of the lead as possible is key to successful extraction. Placement of a locking stylet provides control over the conductor and the distal portion of the lead. Suture has traditionally been used to provide control over the proximal lead, insulation and, in the case of ICD leads, the high-voltage cables.

After securing control of the lead, if the lead cannot be extracted with simple traction, extraction tools are used to disrupt fibrous adhesions. Telescoping sheaths were some of the first tools used to disrupt these adhesions but may fail when dense adhesions are encountered. Laser sheaths may more effectively disrupt dense adhesions but can fail in the setting of very dense or calcified adhesions[25,26] (Fig. 26.5). Mechanical rotational sheaths may be more successful at disrupting calcified adhesions compared to telescoping and laser powered sheaths (Fig. 26.6). Femoral snares allow the capture and extraction of lead fragments and, in some cases, improve success simply by allowing disruption of adhesions from a different approach. They can also be used to secure the distal portion of a lead while extraction tools are advanced beyond an area of stenosis or occlusion (Fig. 26.7). Typically, the extraction tool is advanced along the length of the lead, disrupting adhesions, to a point just short of the lead tip. The lead tip is then pulled back toward the telescoping laser or rotational sheath, thus allowing the myocardium to fall away from the extraction tool as the tip is plucked out of the myocardium.

Incision closure technique varies based on extraction indication, operative findings, and operator preference, and may vary from closure by secondary intention and use of nonabsorbable suture in cases of infection to layers of absorbable suture and dermal adhesive in noninfected cases and reimplantation.

Temporary pacing is sometimes required after TLE. In cases of endovascular infection, reimplantation of a permanent CIED is often delayed until after treatment with a course of antibiotics to minimize the chance of reinfection. To accommodate the need for pacing in dependent patients, externalized active fixation leads may be placed via the right

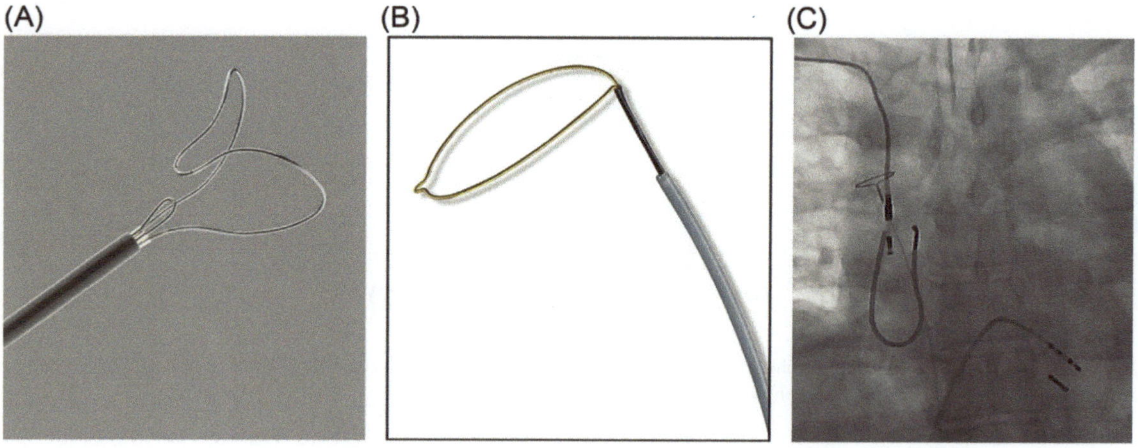

FIGURE 26.7 Femoral snares. (A) Needle Eye snare. (B) Gooseneck snare. (C) CIED lead captured by a femoral snare advanced via the femoral vein and IVC. *(A) Permission for use granted by Cook Medical, Bloomington, Indiana. (B) Reproduced with permission of Medtronic.*

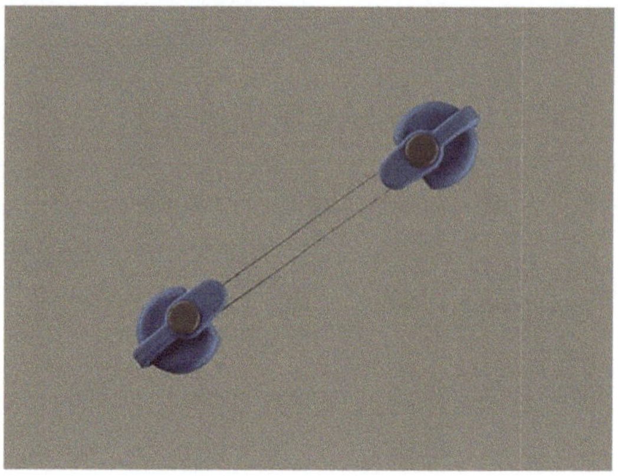

FIGURE 26.8 OneTie compression coil. *Permission for use granted by Cook Medical, Bloomington, Indiana.*

or left internal jugular vein.[27] Similarly, wearable defibrillators provided prior to discharge may provide protection from sudden cardiac death for high-risk patients such as those with secondary prevention indications until the time of staged reimplantation.

Recent advances

Recent advances in the operative phase of TLE include varied intraoperative imaging modalities and a range of new tools to improve efficacy and safety. Several recent studies have closely examined the utility of TEE. In a retrospective study of 3185 patients undergoing TLE, Nowosielecka et al. reported improved clinical success (97.7% vs 95.4%) in patients with TEE guided TLE compared to those without TEE guidance. The rate of serious complications was also lower than expected in the TEE-guided group. They also reported that TEE helped identify looped leads, cardiac perforation, lead-to-lead adhesion, lead-to-myocardium adhesion, and vegetations as well as pulling on cardiac structures.[28] Using intracardiac echocardiography, Sadek at al. were able to identify intracardiac and intravascular lead binding sites and lead-adherent echodensities.[29] In a similar study, Beaser et al. reported that intravascular ultrasound could be used to quantify intravascular lead adherence and that the degree of adherence correlated with extraction difficulty as measured by energy delivery and time required for extraction.[30]

Introduced around 2015, the compression coil (One-Tie, Cook Medical) provides a means of the lead components and locking stylet together to help provide control of the entire length of the lead (Fig. 26.8). Starck et al. performed traction force testing on leads secured with a locking stylet alone, a locking stylet with suture, and a locking stylet with the compression coil. Application of the compression coil resulted in a higher maximal force compared to leads secured with the locking stylet alone and less locking stylet retraction compared to the group secured with a locking stylet and suture.[31]

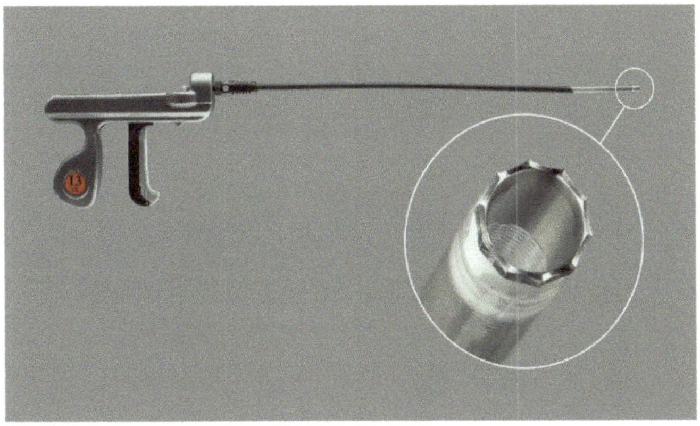

FIGURE 26.9 Evolution RL mechanical rotational extraction sheath. *Permission for use granted by Cook Medical, Bloomington, Indiana.*

Laser-powered and mechanical rotational sheaths have also seen advances. An 80-Hz pulse frequency laser was approved by the Food and Drug Administration in 2012 and was shown to be associated with shorter laser times compared to the prior 40 Hz model.[32] The first iteration of the rotational sheath from Cook Medical (Evolution, Cook Medical) utilized a tip which rotated in a single direction. In 2013, a bidirectional (Evolution RL, Cook Medical) was introduced (Fig. 26.9). In a study comparing these models, clinical success and major complications were similar between the two models, but complete success rates were higher with the newer, RL sheath.[33] The recently reported PROMET study retrospectively examined 2205 patients undergoing TLE utilizing the Cook Evolution rotational sheath and reported complete success in 95% of leads with a procedural mortality of 0.4% and no acute injuries to the extrapericardial portion of the SVC.[34] In a study of another bidirectional mechanical rotational sheath (TightRail, Spectranetics), Mazzone et al. reported 100% complete procedural success and no deaths or major complications.[35] While femoral snares have been used for many years in TLE, more recent data support their safety and efficacy. Recent studies examining the use of femoral snares for TLE reported rates of clinical success at around 97% and major complication rates at 0%–2%.[36–38]

The presence of large vegetations may limit the ability to safely perform TLE for patients with CIED infection. Vegetations are typically dislodged in the process of TLE and large vegetations embolizing to the pulmonary arterial system can result in hemodymnamic collapse. While open surgical lead removal is often pursued in such cases, aspiration of large vegetations to allow TLE has recently been reported. Utilizing an extracorporeal circulation system, Starck et al. reported complete procedural success of the aspiration procedure in 89% of patients with no major complications related to the aspiration.[39] More recent reports demonstrated successful debulking of large vegetations in seven out of eight cases using an angiovac system (Penumbra) and in 94% of 101 patients utilizing an AngioVac cannula (Angiodynamics, USA) and a venovenous extracorporeal filtration circuit.[40]

Leadless pacemakers may provide a new temporary or permanent option for ventricular pacing after TLE. Zucchelli et al. described successful leadless pacemaker implantation with no complications in 30 patients who had undergone TLE.[41] Chang et al. reported successful leadless pacemaker implantation during the same procedure as TLE for infection in 17 patients with no complications and no recurrent infections at a median follow-up of 143 days.[42] Gonzales et al. compared patients undergoing leadless pacemaker implantation after TLE to those treated with temporary externalized pacing and found that those receiving leadless pacemakers had shorter hospital stays compared to controls with temporary transvenous pacing[43] (Fig. 26.10).

Advances in CIEDs and pacing leads may demand new tools and techniques for device and lead extraction. Leadless pacemaker implantation is currently limited to those requiring right ventricular pacing. However, multichamber leadless pacing is on the horizon which will likely result in a sharp increase in implantation of these devices. While multiple studies report successful retrieval of unanchored or embolized devices, more studies are needed to help inform best practices for extraction of leadless devices with longer dwell times.[44–46] His bundle pacing and left bundle branch area pacing are both relatively recently described as alternatives to coronary sinus lead placement for cardiac resynchronization therapy.[47,48] Early reports document safe and complete removal, but these leads may present unique challenges with longer dwell times.[49]

Managing complications

While every effort is taken to prevent complications during TLE, they may occur and can be catastrophic. Cardiac avulsion and laceration of the SVC, subclavian vein, or innominate vein are examples of potentially lethal complications

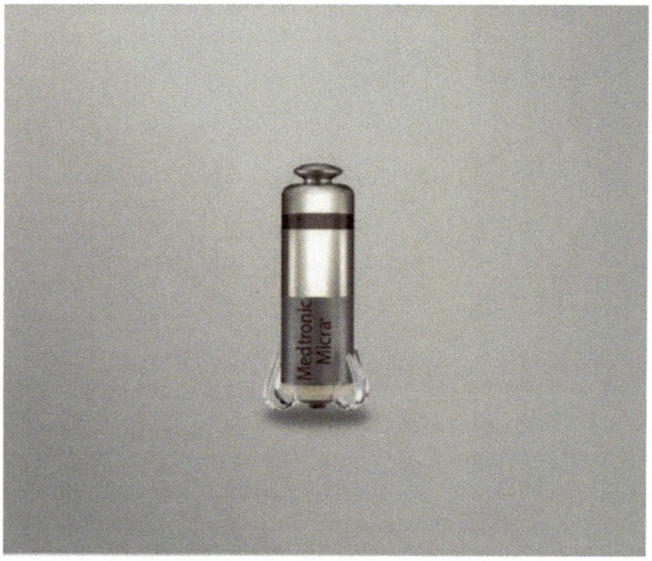

FIGURE 26.10 Micra leadless pacemaker. *Reproduced with permission of Medtronic.*

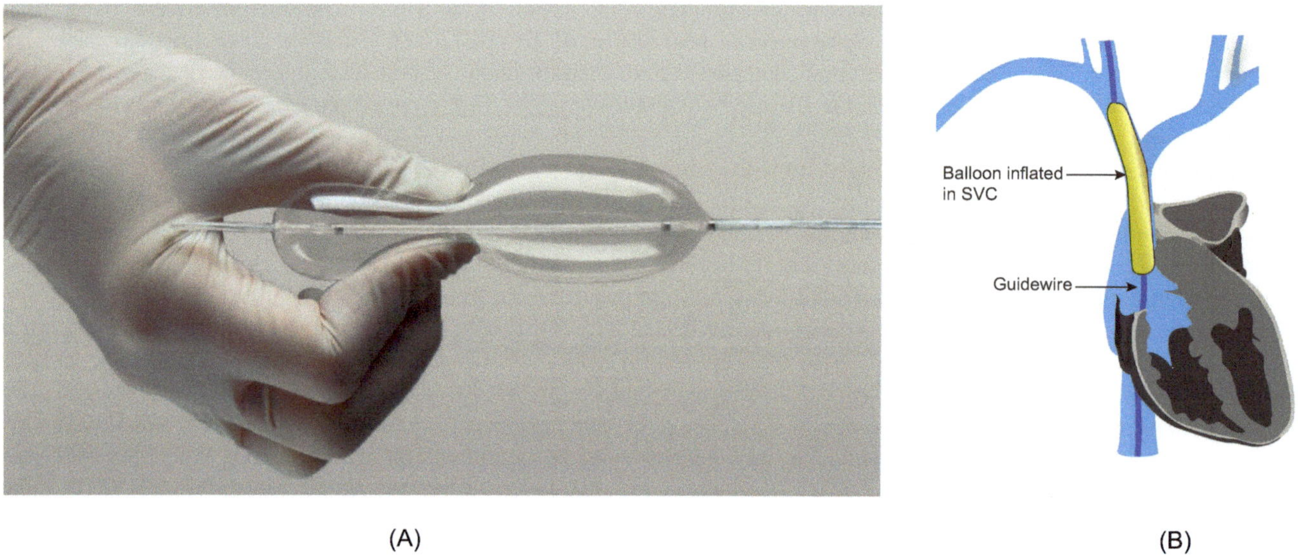

FIGURE 26.11 Bridge balloon. *Courtesy Royal Philips.*

that may occur. In one study, microscopic review revealed segments of vein, most transmural, on over 9% of extracted leads, however, only 1.1% of these cases had a catastrophic complication requiring surgical intervention.[50] Cardiac and vascular complications account for most major complications, in one study representing 84% of major complications with an in-hospital mortality around 30%.[51,52] While training, experience and technique may minimize the chance of these and other major complications, proper planning and preparation are critical to maximizing the chance of rescue. Intraoperative imaging may help identify complications quickly. Early identification and immediate surgical repair are critical to increasing odds of survival with these major cardiac and vascular complications.

Recent advances

One of the most feared complications associated with TLE is SVC laceration which can lead to rapid exsanguination and death. The endovascular occlusion balloon (Bridge Balloon, Spectranetics) was designed to tamponade the SVC,

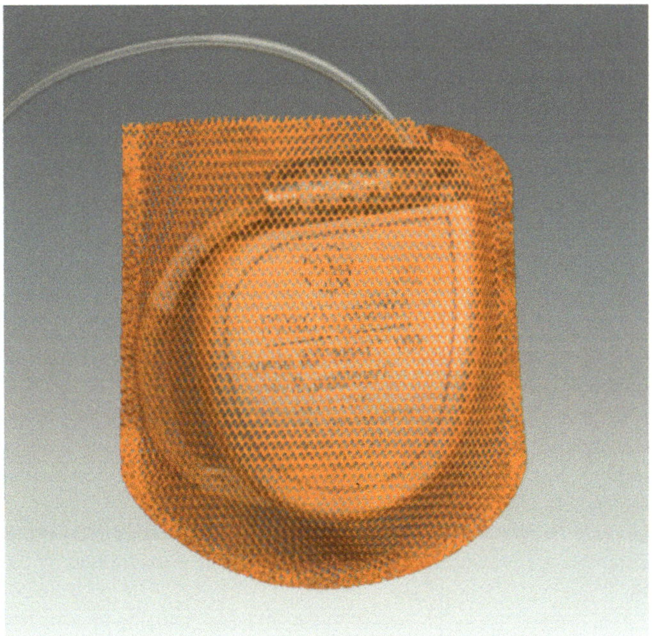

FIGURE 26.12 TYRX antibiotic eluting envelope. *Reproduced with permission of Medtronic.*

stabilizing a patient while a surgeon performs a sternotomy or thoracotomy to repair the injured vessel (Fig. 26.11). Azarrafiy et al. reported survival in 88% of patients with SVC events when the occlusion balloon was properly used compared to just 57% when the balloon was improperly used or not used at al.[53,54]

Hybrid procedural techniques have been described to facilitate immediate identification and treatment of vascular and cardiac injury during TLE. Dai et al. describe a technique utilizing video-assisted throrascopic surgery during TLE.[55] SVC laceration occurred in two out of nine patients. In one case, hemothorax was seen by VATS (video assisted thorocoscopic surgery) prompting immediate sternotomy and repair. In the other, the laceration was below the pericardial reflection resulting in pericardial effusion which was identified by TEE. Sternotomy and repair were successfully performed. In the non-VATS group, one patient had an SVC tear and underwent sternotomy after failed pericardiocentesis. The patient died 7 days later. Major complications occurred in 33% of patients in the VATS group compared to 5% in the non-VATS group.[55] Bontempi et al. reported experience with 12 cases of planned hybrid lead extraction with minithorocotomy in 10 patients and thoracoscopy in two. An SVC laceration was identified and immediately repaired in one case and bleeding from the RA (right atium or right atrial) was controlled with previously placed purse-string sutures in a second. There were no major intraoperative complications associated with thoracoscopy or minithoracotomy and no deaths at 30 days.[56]

The risk of CIED infection is increased in patients with prior infection and in those revision or upgrade of their device. Thus, patients undergoing TLE and subsequent reimplantation of a CIED are at increased risk for infection. The antibiotic eluting envelope (AE) (TYRX, Medtronic) elutes rifampin and minocycline over about 7 days. WRAP-IT, the largest randomized controlled CIED trial to date, demonstrated a 40% reduction in CIED major infections at 12 months for high-risk patients receiving the AE.[57] Use of an AE should be considered in patients undergoing TLE and subsequent CIED reimplantation (Fig. 26.12).

Much work remains to be done to reduce and manage complications of TLE. Advances in hybrid surgical techniques and temporizing tools may reduce the morbidity and mortality of major complications, while advances in extraction tools and techniques seek to reduce the likelihood of major events. Innovations in other specialties may also find application in TLE. For instance, negative pressure wound dressings may decrease wound complications after orthopedic, abdominal, and vascular surgeries.[58–61] Their application in TLE is untested.

Summary

TLE volumes will likely increase in proportion to the growing number of patients with CIEDs and older leads. In experienced centers, these procedures are associated with low rates of major complications. However, when complications do occur, they can be catastrophic. Proper planning and a team approach improve the likelihood of survival in the event of a major complication. Recent work has helped build our understanding regarding outcomes in a wide range of patient populations. Risk stratification tools may provide relevant insights for patient selection and surgical planning. Recent innovations have expanded the available tool set to alert operators to anatomic challenges, improve disruption of adhesions, and rapidly identify and manage complications. Other advances such as antibiotic envelopes hold promise for decreasing late complications. Much work remains in improving the safety and efficacy of TLE. The development of new leads and devices promises to bring new challenges to the field.

References

1. Lensvelt LMH, Egorova AD, Schalij MJ, et al. Mechanical extraction of cardiac implantable electronic devices leads with long dwell time: efficacy and safety of the step up approach. *Pacing Clin Electrophysiol*. 2021;44:120–128.
2. Polewczyk A, Rinaldi CA, Sohal M, group ELi. Transvenous lead extraction procedures in women based on ESC-EHRA EORP European Lead Extraction ConTRolled ELECTRa registry: is female sex a predictor of complications? *Europace*. 2019;21:1890–1899.
3. Kusumoto FM, Schoenfeld MH, Wilkoff BL, et al. HRS expert consensus statement on cardiovascular implantable electronic device lead management and extraction. *Heart Rhythm*. 2017;2017(14):e503–e551.
4. Wilkoff BL, Love CJ, Byrd CL, et al. Heart Rhythm S and American Heart A. Transvenous lead extraction: Heart Rhythm Society expert consensus on facilities, training, indications, and patient management: this document was endorsed by the American Heart Association (AHA). *Heart Rhythm*. 2009;6:1085–1104.
5. Brunner MP, Yu C, Hussein AA, et al. Nomogram for predicting 30-day all-cause mortality after transvenous pacemaker and defibrillator lead extraction. *Heart Rhythm*. 2015;12:2381–2386.
6. Oszczygiel E, Kutarski A, Oszczygiel A, et al. Risk score to assess mortality risk in patients undergoing transvenous lead extraction. *Pacing Clin Electrophysiol*. 2017;40:918–923.
7. Fender EA, Killu AM, Cannon BC, et al. Lead extraction outcomes in patients with congenital heart disease. *Europace*. 2017;19:441–446.
8. Barakat AF, Wazni OM, Tarakji KG, et al. Transvenous lead extraction in chronic kidney disease and dialysis patients with infected cardiac devices. *Circ Arrhythm Electrophysiol*. 2018;11:e005706.
9. Huang XM, Fu HX, Zhong L, et al. Outcomes of transvenous lead extraction for cardiovascular implantable electronic device infections in patients with prosthetic heart valves. *Circ Arrhythm Electrophysiol*. 2016;9.
10. Jachec W, Polewczyk A, Polewczyk M, Tomasik A, Kutarski A. Transvenous lead extraction SAFeTY score for risk stratification and proper patient selection for removal procedures using mechanical tools. *J Clin Med*. 2020;9.
11. Jachec W, Polewczyk A, Segreti L, Bongiorni MG, Kutarski A. Risk factors and long-term survival of octogenarians and nonagenarians undergoing transvenous lead extraction procedures. *Gerontology*. 2021;67:36–48.
12. Bongiorni MG, Kennergren C, Butter C, and Investigators EL. The European Lead Extraction ConTRolled (ELECTRa) study: a European Heart Rhythm Association (EHRA) Registry of Transvenous Lead Extraction Outcomes. *Eur Heart J*. 2017;38:2995–3005.
13. Pecha S, Ziegelhoeffer T, Yildirim Y, et al. Safety and efficacy of transvenous lead extraction of very old leads. *Interact Cardiovasc Thorac Surg*. 2020;.
14. Barakat AF, Zmaili MA, Tarakji KG, et al. Transvenous lead extraction in patients with arrhythmogenic right ventricular cardiomyopathy. *JACC Clin Electrophysiol*. 2019;5:665–670.
15. Hutt E, Diab M, Wazni OM, et al. Transvenous lead extraction in patients with prior extraction procedures: procedural profiles and outcomes. *Heart Rhythm*. 2020;17:1904–1908.
16. Black-Maier E, Piccini JP, Bishawi M, et al. Lead extraction for cardiovascular implantable electronic device infection in patients with left ventricular assist devices. *JACC Clin Electrophysiol*. 2020;6:672–680.
17. Gabriels J, Chang D, Maytin M, et al. Percutaneous management of superior vena cava syndrome in patients with cardiovascular implantable electronic devices. *Heart Rhythm*. 2020;.
18. Arora Y, Carrillo RG. Lead-related superior vena cava syndrome: management and outcomes. *Heart Rhythm*. 2021;18:207–214.
19. Patel D, Sripariwuth A, Abozeed M, et al. Lead location as assessed on cardiac computed tomography and difficulty of percutaneous transvenous extraction. *JACC Clin Electrophysiol*. 2019;5:1432–1438.
20. Vogler J, Pecha S, Azarrafiy R, et al. Navigation of lead extraction-is it possible? Impact of preprocedural electrocardiogram-triggered computed tomography on navigation of lead extraction. *Eur J Cardiothorac Surg*. 2018;54:745–751.
21. Di Monaco A, Pelargonio G, Narducci ML, et al. Safety of transvenous lead extraction according to centre volume: a systematic review and *meta*-analysis. *Europace*. 2014;16:1496–1507.
22. Sidhu BS, Gould J, Bunce C, et al. The effect of centre volume and procedure location on major complications and mortality from transvenous lead extraction: an ESC EHRA EORP European Lead Extraction ConTRolled ELECTRa registry subanalysis. *Europace*. 2020;22:1718–1728.
23. Maytin M, Daily TP, Carillo RG. Virtual reality lead extraction as a method for training new physicians: a pilot study. *Pacing Clin Electrophysiol*. 2015;38:319–325.
24. Oestreich BA, Ahlgren B, Seres T, et al. Use of transesophageal echocardiography to improve the safety of transvenous lead extraction. *JACC Clin Electrophysiol*. 2015;1:442–448.

25. Wazni O, Epstein LM, Carrillo RG, et al. Lead extraction in the contemporary setting: the LExICon study: an observational retrospective study of consecutive laser lead extractions. *J Am Coll Cardiol*. 2010;55:579–586.
26. Wilkoff BL, Byrd CL, Love CJ, et al. Pacemaker lead extraction with the laser sheath: results of the pacing lead extraction with the excimer sheath (PLEXES) trial. *J Am Coll Cardiol*. 1999;33:1671–1676.
27. Kawata H, Pretorius V, Phan H, et al. Utility and safety of temporary pacing using active fixation leads and externalized re-usable permanent pacemakers after lead extraction. *Europace*. 2013;15:1287–1291.
28. Nowosielecka D, Jachec W, Polewczyk A, et al. Transesophageal echocardiography as a monitoring tool during transvenous lead extraction-does it improve procedure effectiveness? *J Clin Med*. 2020;9.
29. Sadek MM, Cooper JM, Frankel DS, et al. Utility of intracardiac echocardiography during transvenous lead extraction. *Heart Rhythm*. 2017;14:1779–1785.
30. Beaser AD, Aziz Z, Besser SA, et al. Characterization of lead adherence using intravascular ultrasound to assess difficulty of transvenous lead extraction. *Circ Arrhythm Electrophysiol*. 2020;13:e007726.
31. Starck CT, Stepuk A, Holubec T, Steffel J, Stark JW, Falk V. Compression coil provides increased lead control in extraction procedures. *Europace*. 2015;17:499–503.
32. Tanawuttiwat T, Gallego D, Carrillo RG. Lead extraction experience with high frequency excimer laser. *Pacing Clin Electrophysiol*. 2014;37:1120–1128.
33. Witte OA, Adiyaman A, Smit JJJ, et al. Success and complication rates of lead extraction with the first- vs. the second-generation evolution mechanical sheath. *Europace*. 2017;19:1717–1722.
34. Starck CT, Gonzalez E, Al-Razzo O, et al. Results of the Patient-Related Outcomes of Mechanical lead Extraction Techniques (PROMET) study: a multicentre retrospective study on advanced mechanical lead extraction techniques. *Europace*. 2020;22:1103–1110.
35. Mazzone P, Melillo F, Radinovic A, et al. Use of the new rotating dilator sheath TightRail for lead extraction: a bicentric experience. *J Arrhythm*. 2020;36:343–350.
36. Uslu A, Kup A, Kanar BG, et al. Transvenous extraction of pacemaker leads via femoral approach using a gooseneck snare. *Herz*. 2021;46:82–88.
37. Yap SC, Bhagwandien RE, Theuns D, et al. Efficacy and safety of transvenous lead extraction using a liberal combined superior and femoral approach. *J Interv Card Electrophysiol*. 2020;.
38. Zhou X, Ze F, Li D, et al. Transfemoral extraction of pacemaker and implantable cardioverter defibrillator leads using Needle's Eye Snare: a single-center experience of more than 900 leads. *Heart Vessel*. 2020;35:825–834.
39. Starck CT, Eulert-Grehn J, Kukucka M, et al. Managing large lead vegetations in transvenous lead extractions using a percutaneous aspiration technique. *Expert Rev Med Dev*. 2018;15:757–761.
40. Starck CT, Schaerf RHM, Breitenstein A, et al. Transcatheter aspiration of large pacemaker and implantable cardioverter-defibrillator lead vegetations facilitating safe transvenous lead extraction. *Europace*. 2020;22:133–138.
41. Zucchelli G, Tolve S, Barletta V, et al. Comparison between leadless and transvenous single-chamber pacemaker therapy in a referral centre for lead extraction. *J Interv Card Electrophysiol*. 2020;.
42. Chang D, Gabriels JK, Soo Kim B, et al. Concomitant leadless pacemaker implantation and lead extraction during an active infection. *J Cardiovasc Electrophysiol*. 2020;31:860–867.
43. Gonzales H, Richardson TD, Montgomery JA, Crossley GH, Ellis CR. Comparison of leadless pacing and temporary externalized pacing following cardiac implanted device extraction. *J Innov Card Rhythm Manag*. 2019;10:3930–3936.
44. Gupta S, Cho K, Papagiannis J, Tisma-Dupanovic S, Borsa J. A novel technique for extraction of a leadless pacemaker that embolized to the pulmonary artery in a young patient: a case report. *HeartRhythm Case Rep*. 2020;6:724–728.
45. Okabe T, Augostini RS, Essandoh MK, Hummel JD. Percutaneous retrieval of an unanchored WiSE-CRT system left ventricular receiver electrode. *HeartRhythm Case Rep*. 2020;6:579–582.
46. Romeo E, D'Alto M, Cappelli M, et al. Retrieval of a leadless transcatheter pacemaker from the right pulmonary artery: a case report. *Pacing Clin Electrophysiol*. 2020;.
47. Cheng L, Zhang J, Wang Z, et al. Efficacy and safety of left bundle branch area pacing vs biventricular pacing in heart failure patients with left bundle branch block: study protocol for a randomised controlled trial. *BMJ Open*. 2020;10:e036972.
48. Strocchi M, Lee AWC, Neic A, et al. His-bundle and left bundle pacing with optimized atrioventricular delay achieve superior electrical synchrony over endocardial and epicardial pacing in left bundle branch block patients. *Heart Rhythm*. 2020;17:1922–1929.
49. Vijayaraman P. Extraction of left bundle branch pacing lead. *JACC Clin Electrophysiol*. 2020;6:903–904.
50. Tarakji KG, Saliba W, Markabawi D, et al. Unrecognized venous injuries after cardiac implantable electronic device transvenous lead extraction. *Heart Rhythm*. 2018;15:318–325.
51. Brunner MP, Cronin EM, Wazni O, et al. Outcomes of patients requiring emergent surgical or endovascular intervention for catastrophic complications during transvenous lead extraction. *Heart Rhythm*. 2014;11:419–425.
52. Zucchelli G, Di Cori A, Segreti L, and Investigators EL. Major cardiac and vascular complications after transvenous lead extraction: acute outcome and predictive factors from the ESC-EHRA ELECTRa (European Lead Extraction ConTRolled) registry. *Europace*. 2019;21:771–780.
53. Azarrafiy R, Tsang DC, Boyle TA, Wilkoff BL, Carrillo RG. Compliant endovascular balloon reduces the lethality of superior vena cava tears during transvenous lead extractions. *Heart Rhythm*. 2017;14:1400–1404.

54. Azarrafiy R, Tsang DC, Wilkoff BL, Carrillo RG. Endovascular occlusion balloon for treatment of superior vena cava tears during transvenous lead extraction: a multiyear analysis and an update to best practice protocol. *Circ Arrhythm Electrophysiol*. 2019;12:e007266.
55. Dai M, Joyce DL, Blackmon S, et al. Outcomes of video-assisted thoracoscopic surgery for transvenous lead extraction. *J Cardiovasc Electrophysiol*. 2018;29:1032–1037.
56. Bontempi L, Vassanelli F, Cerini M, et al. Hybrid minimally invasive approach for transvenous lead extraction: a feasible technique in high-risk patients. *J Cardiovasc Electrophysiol*. 2017;28:466–473.
57. Tarakji KG, Mittal S, Kennergren C, and Investigators W-I. Antibacterial envelope to prevent cardiac implantable device infection. *N Engl J Med*. 2019;380:1895–1905.
58. Bertges DJ, Smith L, Scully RE, et al. A multicenter, prospective randomized trial of negative pressure wound therapy for infrainguinal revascularization with a groin incision. *J Vasc Surg*. 2021;.
59. Chang H, Maldonado TS, Rockman CB, et al. Closed incision negative pressure wound therapy may decrease wound complications in major lower extremity amputations. *J Vasc Surg*. 2020;.
60. Meyer J, Roos E, Abbassi Z, Buchs NC, Ris F, Toso C. Prophylactic negative-pressure wound therapy prevents surgical site infection in abdominal surgery: an updated systematic review and *meta*-analysis of randomized controlled trials and observational studies. *Clin Infect Dis*. 2020;.
61. Tyagi V, Kahan J, Huang P, Li D, Gibson D. Negative pressure incisional therapy and postoperative infection after posterior approach primary total hip arthroplasty. *Cureus*. 2020;12:e7394.

Chapter 27

Surgical Devices for Cardiac Radio-frequency Ablation

Matthew V. Monti, Adam W. Cates and Frank M. Fago
AtriCure, Inc., Mason, OH, United States

Chapter Outline

Introduction

Thermal ablation by heating tissue with various energy modalities has been employed to address a number of disease states offering simpler surgical technique with targeted and precise treatment. Radiofrequency (RF) for thermal ablation, specifically, has emerged as a standard of care energy modality for cardiac arrythmia therapy in both the surgical and endovascular arenas. RF ablation is a well-established technology with a long history of development and clinical application. The French physicist Jacques Arsene d'Arsonval determined the alternating current (AC) frequency to deliver energy without causing muscle stimulation in the late 19th century and subsequent advances in vacuum tube technology enabled sustained high-frequency AC leading to the first clinical application of RF energy by Dr. Harvey Cushing using William T. Bovie's "Bovie" in 1926.[1,2] The fundamental physics and biophysics have remained unchanged leading to nearly a century of refinement by clinical research and engineering development providing the RF technologies available in the clinic today.

Ablation of cardiac tissue to address cardiac arrythmias is a unique challenge. The cardiac system entails nearly every discipline of science and engineering; for example, dynamic mechanics, electricity, fluid dynamics, heat transfer, and biochemistry. Biologically the cardiac system is vastly complex with anatomy including a multitude of tissue types and physiologic processes driving electrical impulses and conduction pathways, cellular coordination, and dynamic healing and regeneration processes, all of which vary from patient to patient and in the presence of injury or disease. Furthermore, precise energy delivery is critical as the cardiac system is intimately adjacent to other anatomy in the thoracic cavity. Controlled destruction of cardiac tissue by means of thermal ablation must recognize all of these challenges and disciplines working in coordination. A multidisciplinary partnership of scientist, engineer, and clinician has been critical for a true understanding of these fundamentals to allow the development of safe and effective cardiac ablation tools and procedures.

Transfer from clinical need to technology—problem definition

Surgical cardiac ablation is rooted in the Cox maze procedure introduced by Dr. James Cox at Barnes-Jewish Hospital in St. Louis in 1987. The procedure was based on creating a number of incisions across both the right and left atrium, which with time would heal to form lines of scar that block electrical wave fronts. Although highly effective, the procedure was challenging and therefore not widely adopted. The Cox Maze is a gold standard in many ways, particularly in

Advances in Cardiovascular Technology. DOI: https://doi.org/10.1016/B978-0-12-816861-5.00028-9

the absolute transmurality of the lines of scar from epicardium to endocardium and their uninterrupted continuity. Absolute, uninterrupted transmurality remains a key clinical driver for any cardiac ablation technology as studies have demonstrated sinus rhythm and atrial fibrillation (AF) can conduct through thin rims of viable tissue.[3]

To successfully deliver absolute transmural lesions a cardiac ablation system must account for inherent variations of the target tissue, procedural constraints, and interoperator variability. The system must also balance efficacy with safe delivery of energy without damaging the target tissue beyond the warranted healing response, not damaging adjacent tissues, or causing emboli.

Tissue variation can be broadly categorized as either geometric or by its makeup. Atrial tissue thickness varies widely and has been reported as thin as 2.3 ± 1.0 mm at the left atrium between the pulmonary veins in normal individuals to 6.5 ± 2.5 mm at the left atrium near the coronary sinus and upwards of 7 mm at the crista terminalis.[4] Additionally, each target lesion may have different lengths between anatomical anchors which vary by patient and course across different thicknesses of cardiac anatomy. The cellular makeup of cardiac tissue varies significantly from patient to patient and across the cardiac anatomy, changing significantly within the left atrium alone. Epicardial and interstitial fat and interstitial fibrosis are commonly encountered, particularly in diseased, older patients and are challenging to predict preoperatively. The summation of geometric and tissue makeup variation results in widely different amounts of energy required to reliably achieve uninterrupted, transmural ablation lines.[5]

Procedural constraints must be considered for any cardiac ablation system as ablation is often concomitant to other procedures. Engagement of the target tissue is influenced by already established access (endovascular, port access, thoracotomy, sternotomy), the patient presentation (body mass index (BMI), overall build), other concomitant procedures (adjacent grafts, valve repair, left atrial appendage (LAA) occlusion/exclusion, previous surgery), the motion of a beating heart versus an arrested heart, cardioplegic temperatures, and of critical importance adjacent structures that must not be unintentionally injured (phrenic nerve, esophagus, valve leaflets).

Interoperator variability can also impact cardiac ablation performance and outcomes. Specifically, intimate, stable contact with the tissue is necessary to couple RF energy into tissue, and force applied to the tissue from the electrode is known to impact the depth of RF ablations.[6] Human factors elements such as grip force, perceived applied force, and dexterity must also be considered as inputs influencing a design.

RF ablation principles

As previously discussed, the basic physics of RF treatment have remained unchanged since d'Arsonval first observed warming by applying an AC circuit. Electrical current (i) is analogous to flowrate, voltage (V) is analogous to pressure, and impedance [or resistance in direct current (DC)] is the resistance to flow. The relationship is governed by Ohm's law which is simply $V = i \times R$. Impedance is a native physical property of tissue and a function of thickness and cellular makeup, with fat and fibrotic tissue having higher impedances than myocardium.

RF ablation is accomplished by applying an oscillating voltage difference between two electrodes in contact with tissue. The frequency for RF tissue ablation is from approximately 50 to 500 kHz and is of high enough frequency to not cause muscle stimulation but low enough to mimic DC resistive heating. Work is required to drive current through a material that resists flow which in turn generates heat from within the material by resistive heating, analogous to thin-wire resistive heating.[8] In the case of RF ablation, tissue provides the resisting pathway causing resistive heating where the current density is highest and conductive heating deeper into the tissue. As tissue temperature reaches a minimum of 50°C necrosis is achieved, resulting in chronic muscle fibrosis.[7]

The pattern of resistive heating within tissue is three-dimensional and extends beyond the physical geometry of the ablation electrodes with the flow of current through tissue dependent on electrode size, shape, and arrangement. Two electrode arrangements are employed clinically, unipolar and bipolar. A unipolar arrangement has one electrode much larger than the other and the electrodes are separated by a large distance (such as cardiac tissue to the patient's lower back) as can be seen in Fig. 27.1. Because current is spread over a broad area at the larger electrode, limited or no heating occurs there. All of the unipolar heating occurs in the vicinity of the smaller electrode (at the target tissue), spreading radially in all directions seeking the path of least resistance to the grounding pad. Any RF device that utilizes a ground pad is a unipolar system, such as a bovie.

A bipolar electrode configuration uses electrodes of similar sizes in close proximity with current flow limited to the tissue between the adjacent bipolar electrodes. Heating occurs equally in the vicinity of each electrode and in the limited tissue volume between the electrodes.[8] As demonstrated in Fig. 27.2 bipolar configurations can be either unidirectional such as an ablation pen in which both electrodes are in contact with either the epicardium or endocardium, or bidirectional such as an ablation clamp in which electrodes are on opposite sides of the target tissue.

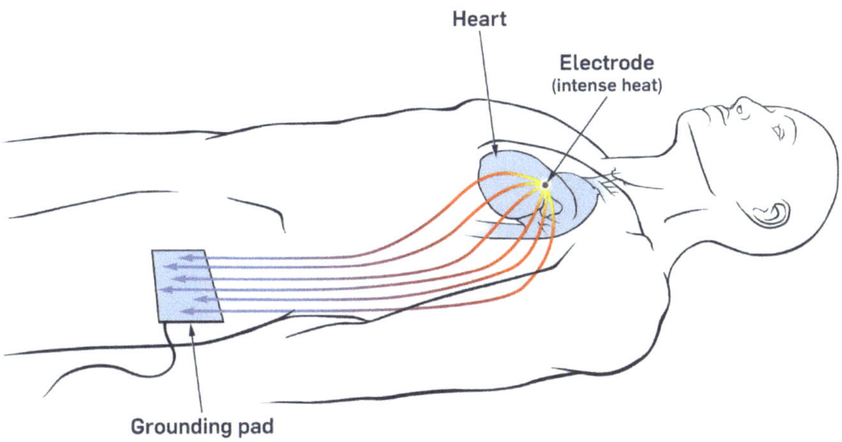

FIGURE 27.1 Unipolar electrode configuration. Current is focused at the point of therapy and flows along the variable path of least resistance to the ground pad. Current density is indicated by the color of flow lines decreasing from yellow to orange, red, purple, and blue.

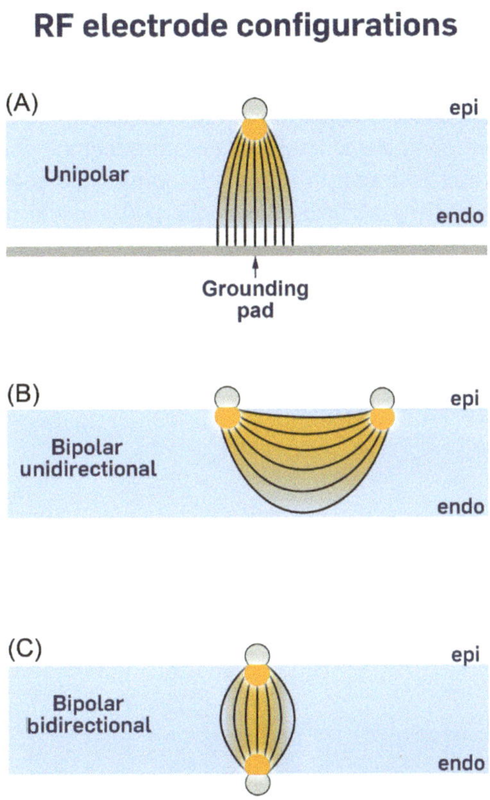

FIGURE 27.2 RF electrode configuration. Blue rectangles represent tissue, gray circles represent electrodes, black lines represent RF field lines. (A) Unipolar. One electrode at the target tissue with current flowing to a remote ground pad typically placed on the patient's back. (B) Bipolar, unidirectional. Both electrodes are placed adjacent on the target tissue and current flows only between the two electrodes. (C) Bipolar, bidirectional. The entire tissue thickness is captured between the electrodes with current flowing only between the two electrodes.

Design solutions

A well-designed system accounts for anticipated variation inherently by its design, delivering the same endpoint regardless of the environment, while relying as little as possible on the operator's judgement and input. Tissue variation, procedural constraints, and interoperator variability each present opportunities for an RF ablation system to address by fundamental design.

Cardiac tissue variation as the summation of both geometric and cellular makeup variation presents the first and most substantial opportunity for an RF ablation system to address by design. The problem can be subdivided to address the geometric (thickness) differences in the target tissue and the cellular makeup.

System designs to address tissue geometry variation

The target tissue must first be captured and stabilized for delivery of RF energy, a challenge that may include compensating for both cardiac and respiratory motion. Ideally the thickness of the tissue is then normalized as much as reasonably possible within the given target ablation site and from site to site, patient to patient.

1. *Clamping the target tissue* is the ideal solution to ensure complete tissue stabilization and uniform mechanical compression, when the entire wall thickness(es) can be anatomically isolated. The design of an RF ablation clamp must ensure:
   a. Uniform force is applied to the tissue along the length of the clamp jaws. This is accomplished by parallel closure from tip to heal of the clamp jaws and enabled by a design with adequate mechanical substructure. Parallel closure accommodates varying tissue thicknesses without creating zones of higher and lower pressure along the length of the jaws which has been shown to directly influence lesion transmurality of clamp ablations.[9] Fig. 27.3 demonstrates the parallel closure of the Isolator Synergy clamp applying uniform tissue pressure regardless of where the tissue is captured with the jaws.
   b. Properly balanced clamping pressure (clamping force ÷ contact area of clamp jaws) such that the tissue is compressed to normalize the thickness and prepare the tissue for energy delivery by expressing interstitial fluid. Clamping pressure must be tuned with the energy delivery algorithm to prevent intramural steam formation potentially causing perforations, but not so much force as to damage the tissue by cutting.
2. *Vacuum stabilization* achieves stable tissue contact, however provides no compression of the tissue. Vacuum stabilization requires careful management of vacuum levels to ensure adequate stabilization without causing mechanical tissue damage but also ensuring tissue is drawn to intimately contact the ablating electrode(s). Appropriate vacuum level is a function of the tissue contacting surface area of the pod and the required force to maintain latch, where vacuum pressure (P) = Latch Force (F)/Pod contact Area (A). $P = F/A$.

   The design of the vacuum pod entails three critical elements in balance: the overall flexibility of the pod to meet the curvature of the tissue, the leading edge of the vacuum pod making the seal on the beating/contracting tissue, and the negative space of the pod that dictates how the tissue is drawn into the pod. The AtriCure EPi-Sense demonstrates this design, as seen in Fig. 27.4.

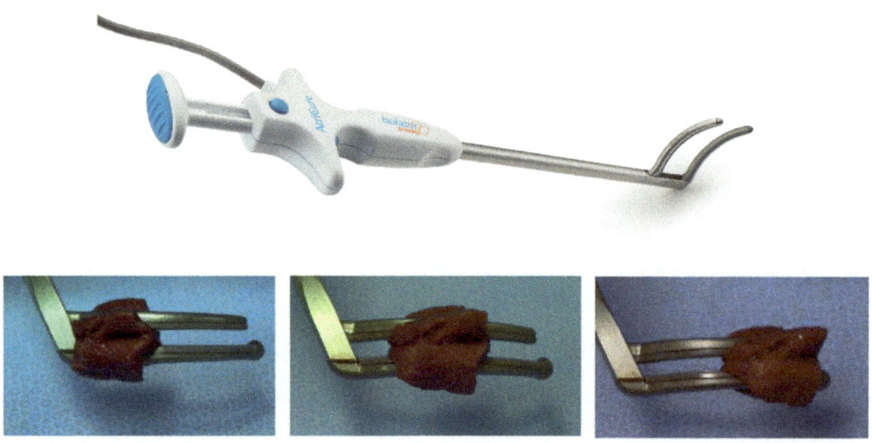

FIGURE 27.3 Top panel: Isolator Synergy clamp (bipolar, bidirectional). Bottom panels demonstrate the parallel closure of the Isolator Synergy clamp regardless of where the tissue is placed along the length of the jaws.

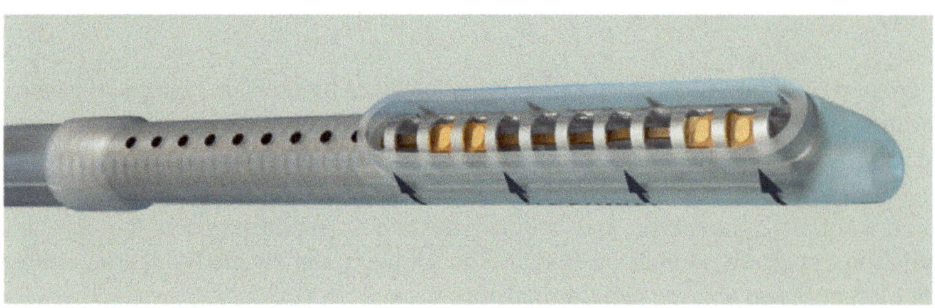

FIGURE 27.4 EPi-Sense Coagulation Device (unipolar, unidirectional). RF ablation device that utilizes vacuum to establish stable tissue contact with a helical ablation electrode.

3. *User applied force* relies on the clinician to carefully stabilize the ablating end effector on the tissue while also applying appropriate force for tissue compression. Endovascularly this challenge is mitigated with contact force sensing technology. Surgically this is enabled with direct vision and tactile feedback via rigid shafted instruments. The design of surgical pens must balance shaft malleability and/or articulation such that the surgeon maintains adequate tactile feedback from the interface of the end effector on the target tissue.

System designs to address tissue cellular makeup variation

Cardiac tissue comprises myocardium with varying degrees of fat and fibrosis significantly impacting the physical properties of the medium to be thermally ablated. RF energy delivery for thermal ablation is a balance. Underdelivery of energy may spare tissue and result in incomplete lesions. Overdelivery of energy either too quickly (too high current or voltage) causing rapid heating or for too long will cause injury beyond the desired insult and cause lateral thermal spread. Additionally rapid heating will result in inadequate lesion depth by creating a self-limiting high impedance surface lesion that RF current cannot penetrate. Safe and efficacious thermal ablation with RF energy requires careful monitoring of the tissue. RF ablation system designs should maximize precise tissue measurement for feedback control of energy delivery and in coordination properly manage heat generated at the tissue–electrode interface for both safety and efficacy reasons. The approach can be subdivided into (1) tissue measurement as feedback into (2) an energy delivery algorithm with (3) electrode design to maximize resistive heating within the tissue and minimize conductive surface heating.

1. *Tissue impedance measurement principles.* Measurement of tissue impedance is accomplished by RF generator control and measurement of voltage and current, in turn allowing simple calculation of tissue impedance by Ohm's law. The engineering design challenge is collecting a precise tissue impedance measurement as feedback into an energy control algorithm. The overall configuration of the system (bipolar vs unipolar) and the electrode geometry primarily influence the measurement. A bipolar, bidirectional configuration allows for the most precise measurement of the entire thickness of the captured tissue. A bipolar, unidirectional configuration allows precise measurement of the tissue pathway between the electrodes, but is physically limited to one side of the tissue and therefore cannot provide a full wall thickness measurement. Unidirectional, unipolar devices also allow tissue impedance monitoring, however the feedback must be taken in terms of relative changes as the current path through the body to the return electrode is variable and uncontrolled. A fixed electrode geometry intimately in contact with the tissue provides the most precise impedance measurement, as measured impedance is inversely proportional to electrode–tissue contact surface area. Fig. 27.5 demonstrates the fixed electrode geometry of the Isolator Synergy clamp. Any means of variable electrical coupling to the tissue such as saline irrigation confounds the impedance measurement and should be interpreted in terms of relative changes.
2. *RF energy delivery algorithms.* Differences in cardiac tissue cellular makeup result in widely different energy requirements to achieve transmurality.[5] Furthermore tissue impedance is dynamic and changes as it is ablated, as

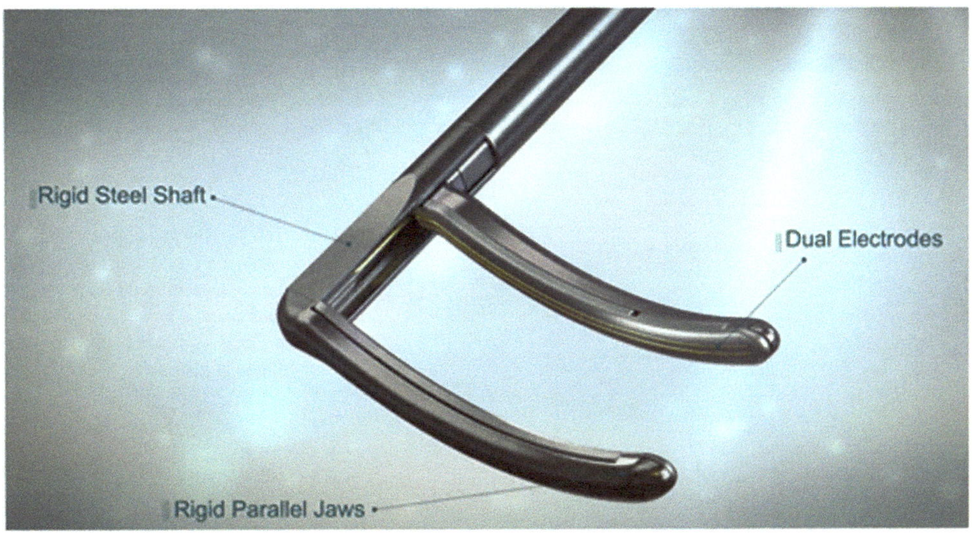

FIGURE 27.5 Isolator Synergy Clamp end effector. Dual electrode pairs are embedded within fixed insulators in each jaw to enable intimate, stable contact with the tissue for precise tissue impedance measurement.

demonstrated in Fig. 27.6, a screen capture of the measured tissue conductance during an RF ablation with the Isolator Synergy clamp. Native tissue impedance is a function of the tissue thickness and cellular makeup (fat, fibrosis, lean cardiac muscle) and varies from patient to patient and with age and disease. Tissue impedance changes as the tissue is heated, first decreasing as the tissue is warmed by RF energy then increasing due to local desiccation and liquid to vapor phase change. Ultimately a stable high impedance between the electrodes can be achieved, which has been strongly correlated to durable transmurality in the case of bipolar, bidirectional ablation clamps.[10] A well-designed RF ablation system should monitor energy delivery as a function of the tissue and in real time adjust accordingly to limit the rate of energy delivery during the course of the ablation and the total energy delivered to safely achieve transmurality.

The rate of energy delivery as Joules per second (Power) is controlled by changing the current and/or voltage delivered to the tissue, where Power (P) = Current (i) × Voltage (V). P = iV. There are three primary power delivery strategies in the clinic today:

a. *Dynamic power by continuous tissue impedance feedback*. Employed by the AtriCure Isolator Synergy ablation clamp this strategy requires accurate, precise, and real-time tissue impedance measurement and therefore can only be enabled in a bipolar configuration with a fixed (dry) electrode geometry.

   As depicted in Fig. 27.7 the system first establishes a voltage potential during power ramp-up, which enables calculation of tissue impedance by Ohm's law. The appropriate power level is then calculated per the ablation algorithm. The power is changed and delivered to the tissue causing a tissue effect, which is again measured. The system runs at 50 Hz, looping through this sequence every 20 ms. Upon achieving a stable, empirically determined high impedance the system triggers transmurality. To ensure the safe delivery of energy, current and voltage limits have been established and are continuously balanced to achieve the target power output throughout the cycle. It should be noted that the course of an ablation requires varying power for optimum efficiency and

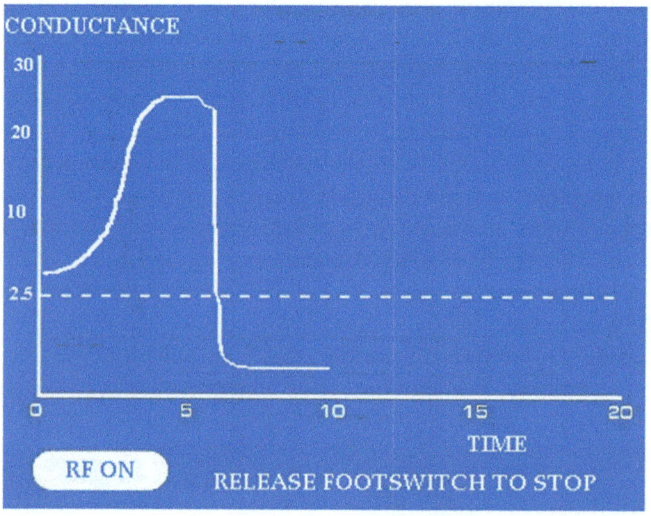

FIGURE 27.6 Screenshot of an ablation with the Isolator Synergy Clamp and Ablation and Sensing Unit (RF Generator). Tissue conductance is displayed on the y-axis, time on the x-axis. Initially conductance rises as the tissue is warmed eventually followed by a sharp drop to a stable, low conductance indicating a transmural effect.

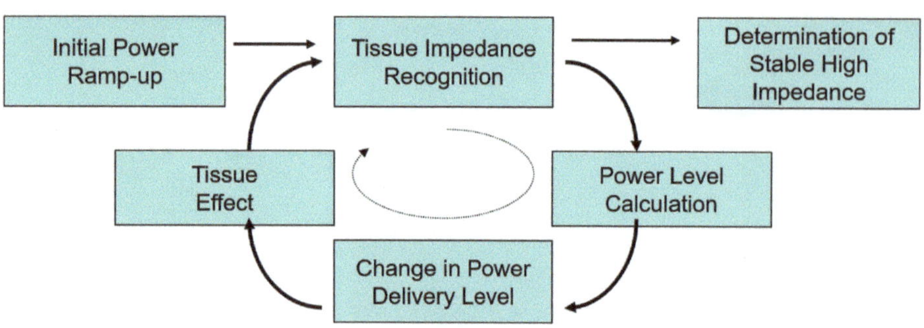

FIGURE 27.7 Block diagram of the bipolar, impedance based Isolator Synergy ablation clamp algorithm. Tissue impedance is measured, and power is adjusted at 50 Hz. The cycle continues until a stable high impedance is achieved.

safe delivery of energy. Initially tissue accepts energy at a higher rate (power) as the tissue becomes more conductive; however, the subsequent rate of energy delivery (power) should be limited to achieve a full thickness tissue effect and avoid causing a self-limiting, high impedance surface effect.

b. *Constant power limited by continuous impedance monitoring*. Employed by the AtriCure bipolar pens applied to one side of the tissue wall, this strategy also requires an accurate and precise, real-time tissue impedance measurement between the ablation electrodes, as enabled by a bipolar configuration with fixed (dry) electrode geometry. Fig. 27.8 demonstrates this electrode configuration. A voltage potential is established between the adjacent bipolar electrodes which enables measurement of the tissue's impedance by Ohm's law as described earlier. RF energy flows from one electrode taking the path of least resistance to the adjacent electrode. Resistive heating occurs within the tissue and in combination with conductive heating the isotherm for cell death is driven into the tissue thickness. To prevent overdelivery of energy the algorithm monitors impedance at 50 Hz and tapers power at predetermined voltage and current limits.
c. *Constant power limited by monitoring the rate of impedance change*. Employed by the AtriCure EPi-Sense system, impedance is continuously monitored between the saline irrigated ablation electrode and the ground pad electrode on the patient's back. Nominal power is delivered within a predetermined impedance range to ensure proper electrode and ground pad contact and that the tissue is within the anticipated range for safe energy delivery. When a predetermined rate of impedance increase (dZ/dt) is exceeded power is reduced and slowly ramped back to nominal. This approach limits tissue impedance spiking (tissue desiccation by phase change of interstitial fluid) to prolong energy delivery and limit surface heating. The algorithm terminates energy delivery at a preset, empirically determined fixed time.

3. *Electrode design*. A properly designed RF ablation end effector does not generate heat itself. Resistive heating within the tissue conducts heat to its surrounding environment which includes the target tissue, surrounding anatomy, and the ablation tool. RF energy generates an electric field within the tissue causing maximal resistive heating where the field is most concentrated, at the electrode tissue interface. This inherently biases the ablation to the electrode/tissue interface. To maximize ablation depth, minimize lateral thermal spread, and minimize heat transferred from the tissue to the ablation end effector the design of the ablation tool must incorporate a means to limit heating at the electrode/tissue interface. To accomplish this four electrode design strategies are currently employed: (a) Multielectrode pulsed RF, (b) electrode thermal mass, (c) internal liquid cooling of electrodes, and (d) electrode irrigation.
   a. *Multielectrode, pulsed RF*. Employed by the AtriCure Synergy clamp. In the case of a bipolar clamp with a single pair of electrodes, lesion formation commences at the tissue–electrode interface and bidirectionally grows inward forming an hour glass-shaped lesion as depicted in Fig. 27.9. This pattern forms as RF field lines are most concentrated at the tissue–electrode interface. The resulting lesion has potential for the breakdown of transmurality within the tissue.

      Pulsing RF energy, by momentarily pausing energy delivery, lowers the electrode–tissue interface temperature which in turn maintains tissue conductivity to maximize resistive heating. Considerations to pulsing RF Energy for thermal effect include longer ablation times and careful balance of the on/off duty cycle so that tissue heating is accomplished. Electrode arrays can be used to manipulate RF field current density and can be designed such that the RF fields are focused or overlapped within the tissue to create focused ablation volumes.

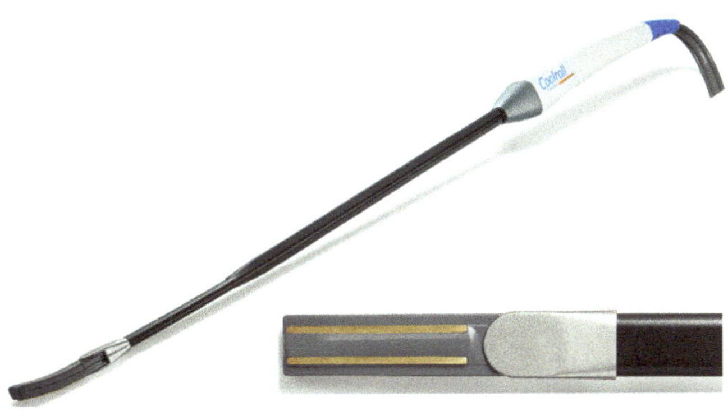

FIGURE 27.8 Coolrail Liner Pen. Bipolar, unidirectional RF ablation pen with internal electrode fluid cooling delivering a fixed constant power limited by continuous tissue impedance monitoring.

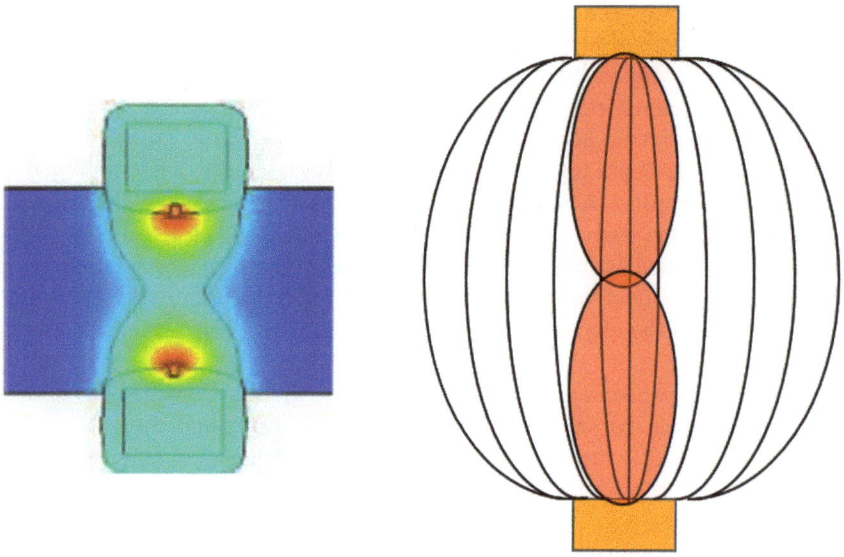

FIGURE 27.9 Left panel: Computational simulation of the first generation Isolator bipolar clamp with a single electrode pair demonstrating zone of maximal heating is centered near the electrode tissue interface. Right panel: Line drawing representing RF field lines (black) resulting in an hour glass shape (pink) lesion geometry with a single electrode bipolar ablation clamp (gold blocks represent electrodes).

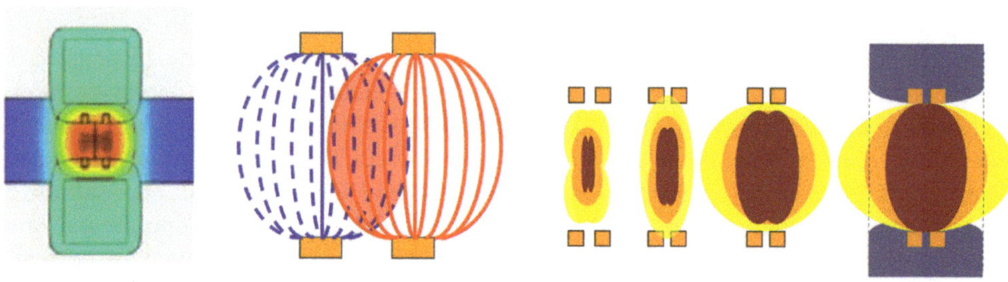

FIGURE 27.10 Left panel: Computational simulation of the Isolator Synergy bipolar ablation clamp with a pulsed on/off dual electrode pair demonstrating zone of maximal heating is centered within the target tissue (blue block represents tissue, green blocks clamp jaws, gold blocks electrodes). Center panel: line drawing representing RF field lines of the dual electrode pair (blue, red) with zone of maximum density the overlapping region within the tissue (as viewed in cross section, gold blocks represent electrodes). Right panel: Resulting temperature within tissue as a function of time moving right (gold blocks represent electrodes, deep red is highest temperature). The resulting lesion is barrel shaped in profile.

Electrode arrays are often employed in solid organ tumor ablation where a fixed ablation depth is known. The combination of pulsed RF with multiple electrode pairs offers a unique solution for bipolar RF clamp ablation of cardiac tissue. In the case of the AtriCure Synergy clamp two electrodes are situated adjacent in both jaws, coupled to a pair of opposing electrodes in the opposite (distal/proximal) jaw of the clamp. Adjacent electrode separation is chosen to create an overlapping RF field within the clamped tissue. The electrode pairs are also pulsed at a 50% on/off duty cycle reducing the electrode–tissue interface temperature. The combination of pulsed, dual pairs creates a focused zone within the tissue, as depicted in Fig. 27.10, that sees constant RF energy, causing heating within the tissue and an ablation that forms outward toward the electrodes with little to no lateral thermal spread.

b. *Electrode thermal mass and material selection* is a significant design variable for any RF ablation system, whether bipolar or unipolar, bidirectional or unidirectional. Electrode mass in conjunction with the material properties of the electrodes impacts the rate of electrode–tissue interface heating and can be tuned to limit conductive heating by the electrodes. Two physical properties should be considered when designing RF electrodes for heat transfer: 1) the electrode material's ability to absorb and retain heat, in thermodynamic terms the specific heat capacity (C_p) of the material, which is defined as the energy (J) to raise a unit mass of material (g) by one degree (C). Cp = J/(g C), and 2) the rate at which heat can conduct through the electrode material, known as thermal conductivity (k), where k = Watt/(meter C). In the case of the AtriCure Isolator Linear Pen the electrodes are gold-plated copper having high thermal conductivity to conduct heat away from the electrode–tissue interface with electrode mass maximized within the end effector to absorb the heat.

c. *Internal liquid electrode cooling*. Continuous, repeated use of an RF ablation tool with fixed electrode mass will result in eventual end effector heat soak requiring either rest or quenching for the electrodes to cool.

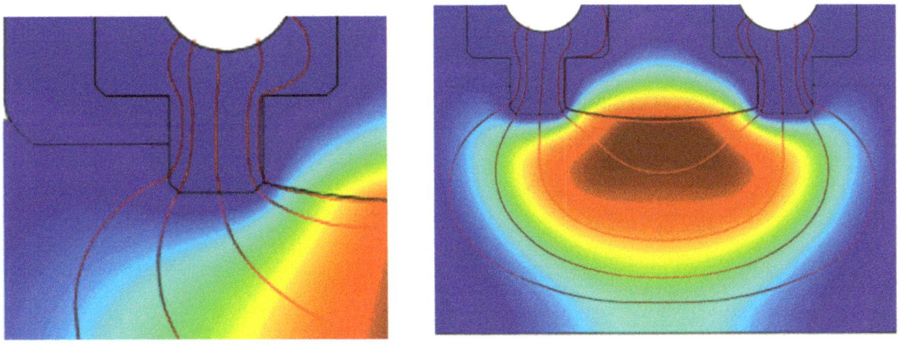

FIGURE 27.11 Computational simulation of the Coolrail Linear Pen with internally fluid cooled electrodes demonstrating low electrode–tissue interface temperature. Cross-sectional view, red lines represent RF field, deep red is highest temperature.

To overcome this, fluid can be pumped through lumens inside the electrodes in a closed circuit without draining fluid into the surgical field. The fluid and flow rate must be selected to cool the electrodes but not prevent the desired tissue heating effect, and the fluid flow path must be designed to cool the warmed fluid in preparation for eventual return to the electrodes. If properly designed, internal liquid electrode cooling allows for maximal heating within the tissue, as shown in Fig. 27.11, and can be run indefinitely without the need to further cool the end effector. This cooling technique is employed by the AtriCure Coolrail Linear Pen.

d. *Irrigated electrodes*. Liquid cooling of the electrode–tissue interface can also be accomplished by expressing fluid from the end effector directly onto the surface of the tissue, commonly known as irrigation. Irrigation by default implies a "wet" electrode design, in that the electrically conductive irrigating fluid expressed from the end effector becomes a virtual, dynamic electrode by coupling between the physical electrode and the tissue surface. Irrigated RF ablation systems are therefore unable to provide precise tissue impedance measurements and are limited to relative impedance monitoring. Two irrigated designs are currently employed: open irrigation, which drains expressed fluid into the body, and closed irrigation which contains and recaptures the fluid within a vacuum pod as employed by the AtriCure EPi-Sense device. Both design strategies allow for fluid contact of the electrode and tissue surface.

Future advancements in cardiac ablation

Significant research continues in the field of cardiac ablation with a focus on procedural approaches and techniques to deliver the ablative therapy, new applications for RF ablation, as well as a novel energy modality and cell death mechanisms.

Advancements in procedural approaches have the potential to significantly further the field of cardiac ablation. Device design solutions to simplify surgical procedures by decreasing procedure time, decreasing the number of device placements, and enabling simpler placement of current ablation tools at the target structures can increase adoption. Specific to AF ablation therapy, lesions are created on the posterior left atrium and limiting or eliminating the careful dissection of pericardial reflections would decrease the technical nature of the procedure. Redesigning existing ablation tools to follow specific concomitant approaches and procedures is a near-term opportunity.

Clinicians are also making advancements in multidisciplinary approaches where the cardiac surgeon and electrophysiologist work in tandem bringing together the strengths of both skillsets and the ability to work both epicardially and endocardially simultaneously. This collaboration is most recently demonstrated by advancements in the treatment of Inappropriate Sinus Tachycardia (IST). Historically cardiac ablation therapy for IST has been limited with often dismal results and serious complications reported, depending on the approach, resulting in a Heart Rhythm Society Expert Consensus class III recommendation to not treat with surgical ablation.[11] By pulling from the strengths of both surgeon and electrophysiologist a better understanding of IST is forming, and a procedure has been developed that leverages both the endocardial access and targeted mapping techniques of the Electrophysiologist with the ablation tools and epicardial access of the surgeon. Initial results of this real-time, collaborative approach are promising.[12] New clinical approaches in turn provide opportunities for new technologies, in this instance tools specifically to enable a novel procedure.

Pulsed field ablation (PFA) has gained significant interest at the time of writing with initial clinical results very promising for the treatment of AF from a percutaneous, endocardial approach. PFA delivers short DC bursts (nanoseconds or microseconds) at high voltages to irreversibly electroporate cells, resulting in cellular death. PFA may inherently address a number of challenges with RF ablation. Properly implemented PFA may be both nonthermal and tissue selective as different tissue types have different electric field (V/cm) thresholds for irreversible poration with cardiac tissue requiring a lower threshold for cell death than skeletal and smooth muscle. Regardless of the approach, PFA has the potential to mitigate the risk of unintended damage to nontarget structures (particularly approaches where energy is directed outward from the cardiac wall towards the esophagus and phrenic nerves), reduce operative time, and increase ablation effectiveness.[13–15]

References

1. Reif-Acherman S. Jacques Arsene d'Arsonval: his life and contributions to electrical instrumentation in physics and medicine. Part III: High-frequency experiences and the beginnings of diathermy [scanning our past]. *Proc IEEE*. 2017;105(2):394–404. Available from: https://doi.org/10.1109/JPROC.2016.2640638.
2. Carter PL. The life and legacy of William T. Bovie. *Am J Surg*. 2013;205(5):488–491. Available from: https://doi.org/10.1016/j.amjsurg.2012.12.005.
3. Melby SJ, Lee AM, Zierer A, et al. Atrial fibrillation propagates through gaps in ablation lines: implications for ablative treatment of atrial fibrillation. *Heart Rhythm*. 2008;5(9):1296–1301. Available from: https://doi.org/10.1016/j.hrthm.2008.06.009.
4. Schuessler RB, Lee AM, Melby SJ, et al. Animal studies of epicardial atrial ablation. *Heart Rhythm*. 2009;6(12 Suppl.):S41–S45. Available from: https://doi.org/10.1016/j.hrthm.2009.07.028.
5. Melby SJ, Zierer A, Voeller RK, et al. Wide variations in energy delivery using an impedance-controlled algorithm in bipolar radiofrequency ablation: evidence against fixed time ablation. *Innovations (Phila)*. 2007;2(2):67–72.
6. Wood MA, Ellenbogen AL, Pathak V, Ellenbogen KA, Kasarajan V. Efficacy of a cooled bipolar epicardial radiofrequency ablation probe for creating transmural myocardial lesions. *J Thorac Cardiovasc Surg*. 2010;139(2):453–458. Available from: https://doi.org/10.1016/j.jtcvs.2009.06.028.
7. Dewhirst M, Viglianti BL, Lora-Michiels M, Hoopes PJ, Hanson MA. Thermal dose requirement for tissue effect: experimental and clinical findings. In: Ryan TP, (ed.); 2003:37. Available from: https://doi.org/10.1117/12.476637.
8. Difrancesco M, Park C, Martin KE, Glithero J, Privitera S, Cassidy BP. Technological advances in minimally invasive radiofrequency ablation of cardiac tissues. *Innovations (Phila)*. 2010;5(2):e138–e146. Available from: https://doi.org/10.1097/IMI.0b013e3181d76900.
9. Varzaly JA, Chapman D, Lau DH, et al. Contact force and ablation assessment of surgical bipolar radiofrequency clamps in the treatment of atrial fibrillation. *Interact Cardiovasc Thorac Surg*. 2019;28(1):85–93. Available from: https://doi.org/10.1093/icvts/ivy191.
10. Gaynor SL, Ishii Y, Diodato MD, et al. Successful performance of Cox-Maze procedure on beating heart using bipolar radiofrequency ablation: a feasibility study in animals. *Ann Thorac Surg*. 2004;78(5):1671–1677. Available from: https://doi.org/10.1016/j.athoracsur.2004.04.058.
11. Sheldon RS, Grubb BP, Olshansky B, et al. 2015 heart rhythm society expert consensus statement on the diagnosis and treatment of postural tachycardia syndrome, inappropriate sinus tachycardia, and vasovagal syncope. *Heart Rhythm*. 2015;12(6):e41–e63. Available from: https://doi.org/10.1016/j.hrthm.2015.03.029.
12. de Asmundis C, Chierchia G-B, Sieira J, et al. Sinus node sparing novel hybrid approach for treatment of inappropriate sinus tachycardia/postural orthostatic sinus tachycardia with new electrophysiological finding. *Am J Cardiol*. 2019;124(2):224–232. Available from: https://doi.org/10.1016/j.amjcard.2019.04.019.
13. Reddy VY, Anter E, Rackauskas G, et al. Lattice-tip focal ablation catheter that toggles between radiofrequency and pulsed field energy to treat atrial fibrillation: a first-in-human trial. *Circulation: Arrhythmia and Electrophysiology*. 2020;13(6):e008718.
14. Reddy VY, Neuzil P, Koruth JS, et al. Pulsed field ablation for pulmonary vein isolation in atrial fibrillation. *J Am Coll Cardiol*. 2019;74(3):315–326. Available from: https://doi.org/10.1016/j.jacc.2019.04.021.
15. Reddy VY, Koruth J, Jais P, et al. Ablation of atrial fibrillation with pulsed electric fields. *JACC Clin Electrophysiol*. 2018;4(8):987–995. Available from: https://doi.org/10.1016/j.jacep.2018.04.005.

Chapter 28

Novel Devices for Epicardial Left Atrial Appendage Management

Frank M. Fago, Adam W. Cates and Matthew V. Monti
AtriCure, Inc., Mason, OH, United States

Chapter Outline

Introduction

Atrial fibrillation (AF), the most common arrhythmia worldwide, is a serious disease with severe morbidity and mortality. Patients with AF are at a risk of stroke that is five times greater than patients without AF, and AF accounts for 20% of all strokes.[1] Overall, 90% of strokes are ischemic, resulting from thromboembolic events, with AF accounting for 25% of the ischemic strokes. In addition to treatments for stopping the arrhythmia, AF treatments have focused on reducing stroke risk by eliminating the source of thrombi. Since the mid-1900s, the left atrial appendage (LAA), a small pouch-like extension of the left atrium, has been implicated as a source of thrombi in AF, and its obliteration has been recommended to reduce stroke risk. Indeed, studies have found the LAA is responsible for more than 90% of thrombi in patients with nonvalvular AF and up to 38% of thrombi in patients with cardiomyopathy.[2] Surgical treatment strategies to manage the LAA initially consisted only of endocardial or epicardial suturing, epicardial stapling, or amputation and suturing while on cardiopulmonary bypass, but technological advances in epicardial exclusion devices have created a growing option for managing the LAA. These devices have been designed not only to prevent formation and dislodgement of clots but also to exclude the LAA entirely from the left atrium, resulting in atrophy of the appendage tissue and thereby eliminating its potential contribution to stroke risk. Furthermore, these epicardial devices provide electrical isolation, an important feature given that the LAA has also been implicated in perpetuating atrial arrhythmia via electrical triggers.[3,4]

Transfer from clinical need to technology

Left atrial appendage anatomy

The LAA is a remnant of the primitive atrium formation during the third or fourth week of human embryonic development. The structures in the left atrium are formed from the primitive pulmonary veins. At roughly 30 days, development of the left atrium begins. At approximately day 50 of embryologic development, the heart and left atrium are completely formed and the LAA remains as a relic of the developmental process (Fig. 28.1).

The LAA is a dead-end tubular structure arising from the left atrium. It connects with the left atrium via an ostium, which is typically oval-shaped and varies in size but is small relative to the depth of the appendage.[6] The shape of the ostium and its location out of the way of the main flow of blood through the left atrium, make the LAA particularly

Advances in Cardiovascular Technology. DOI: https://doi.org/10.1016/B978-0-12-816861-5.00027-7

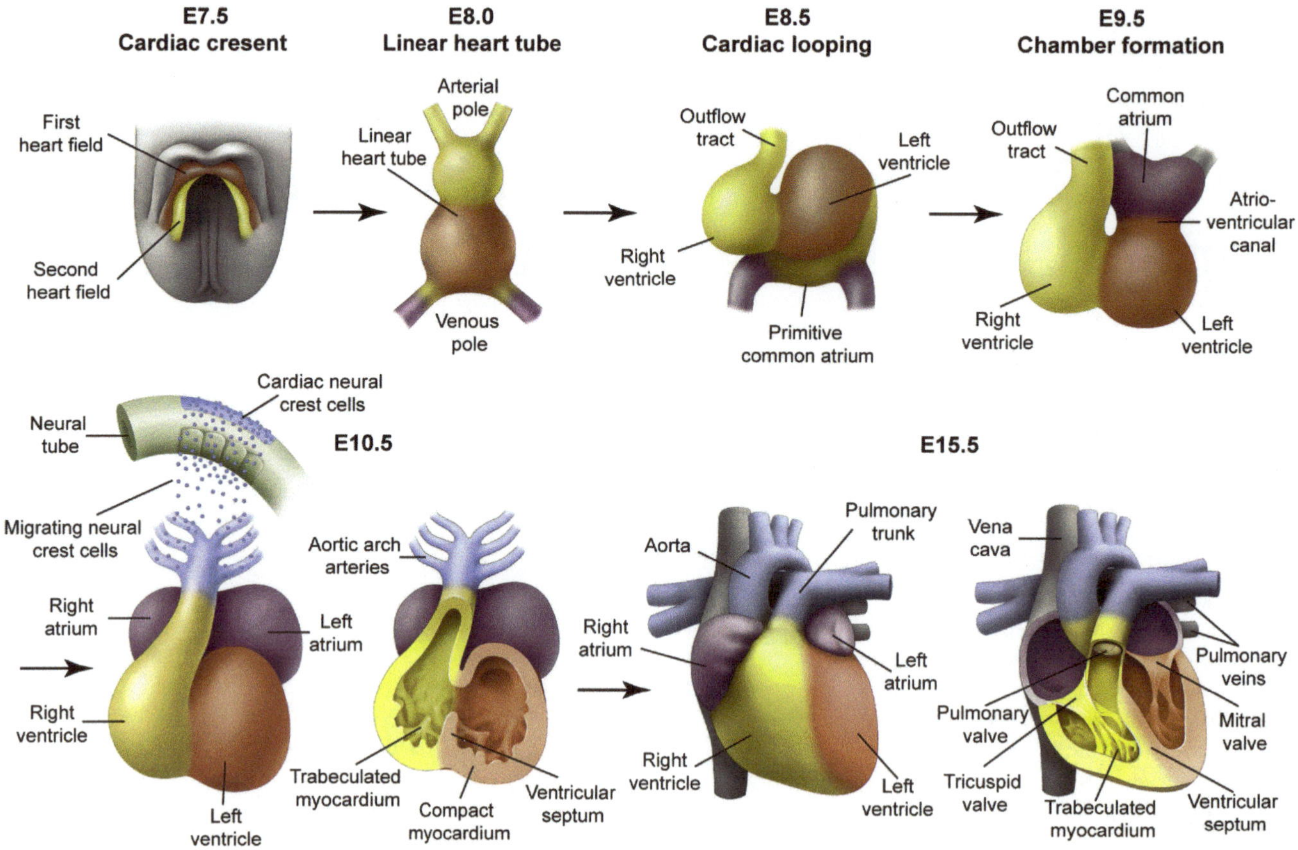

FIGURE 28.1 Progression of embryotic development of the heart showing the formation of the left atrial appendage. The heart originates from mesodermal cells in the primitive streak. During gastrulation, cardiac progenitors migrate to the splanchnic mesoderm to form the cardiac crescent. At E7.5 in the mouse, the cardiac crescent can be divided into two heart field lineages based on differential gene expression and their respective contribution to heart, a first heart field (*red*) and a second heart field (*yellow*), which is located posteriorly and medially to the first heart field. At E8.0, the linear heart tube is present. At E8.5, the looping is associated with uneven growth of cardiac chambers. The outflow tract is at the arterial pole and the inflow tract and primitive atria are at the venous pole. By E9.5, the common atrium has moved superior to the ventricles and is separated by a distinct atrioventricular canal. By E10.5, cardiac neural crest cells from the dorsal neural tube migrate via the pharyngeal arches to the cardiac outflow tract. Further cardiac development involves a series of septation events and myocardial trabeculation that result in a mature four-chambered heart integrated with the circulatory system depicted at E15.5.[5]

susceptible to stagnant flow that can result in thrombi, especially when the heart functionality is compromised with conditions such as AF. While the endocardium near the ostium is smooth, like the left atrium, the distal portion of the LAA is trabeculated, further creating the propensity for stagnant blood flow and thrombi.[7] For adequate exclusion, the appendage must be sealed as close as possible to its base, near the ostium. Specifically, remnants of the LAA should be less than 1 cm to avoid leaving a pouch and/or trabeculations that would promote clot formation.[6,8–10] Because of the importance of location, the devices should also be easily repositioned to assure adequate placement.

The morphology of the LAA varies considerably but there are four categories of shapes commonly referenced: (1) cauliflower, (2) chicken wing, (3) cactus, and (4) windsock.[11,12] The chicken wing is the most common morphology, and while it was shown in at least one study to be the least susceptible to thrombus, other studies have found no difference in stroke risk based on LAA morphology (Fig. 28.2).[13] Accordingly, exclusion devices must account for all variations in morphology and size.

Eliminating the LAA has advantages other than reducing the risk of stroke. A secondary benefit is that it reduces the possibility of AF recurrence. The LAA is a known site of electrical impulse firing that causes the heart to beat out of rhythm. In a study of 987 patients undergoing redo ablation, 266 patients (26%) had electrical impulses firing from the LAA that caused premature atrial contractions and 86 patients (8.7%) had electrical impulses firing from the LAA that were the only source of arrhythmia.[3] Managing the LAA should therefore provide electrical isolation in addition to mechanical exclusion.

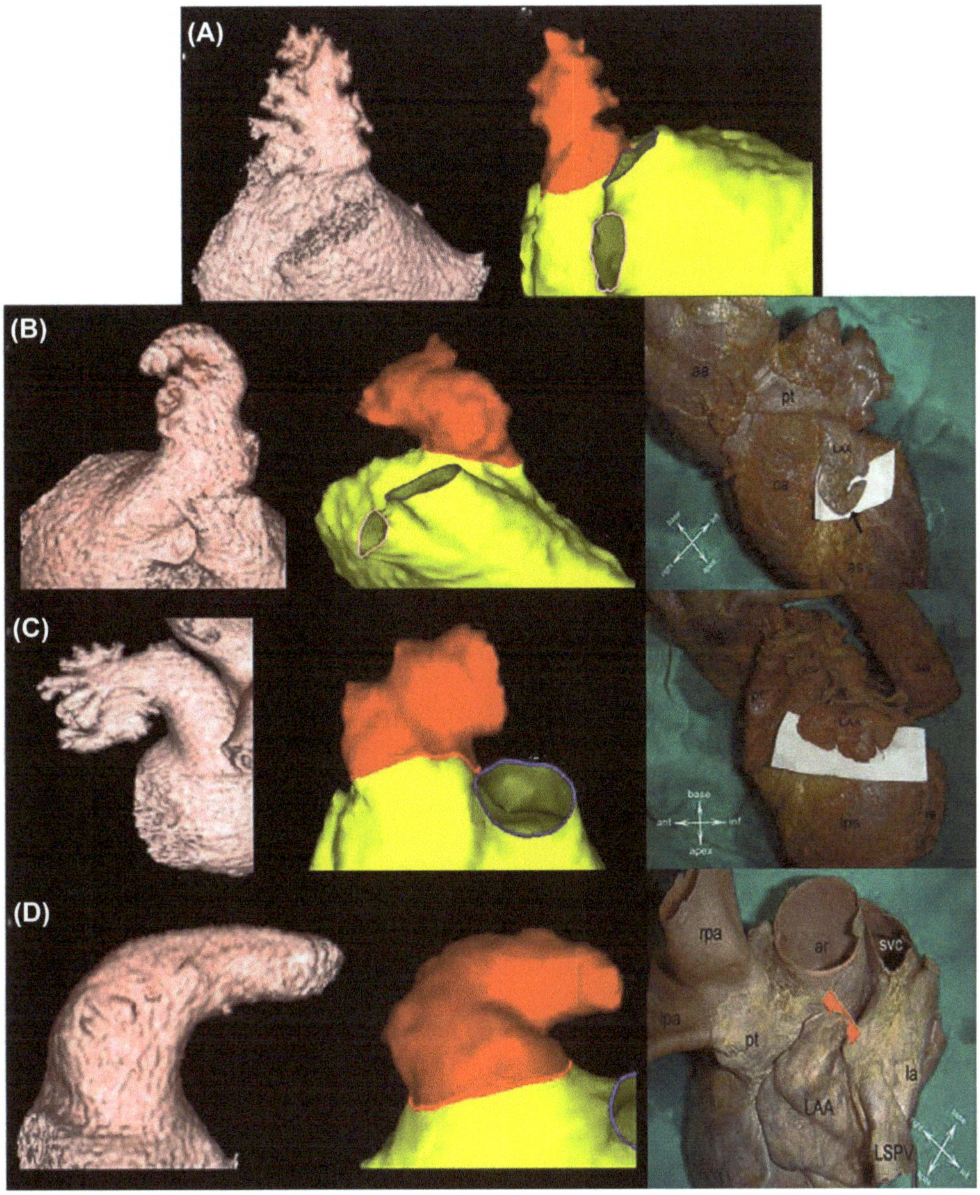

FIGURE 28.2 Left atrial appendage (LAA) morphologies. The four most common LAA morphologies are illustrated on the left by cardiac computed tomography, in the middle by cardiac magnetic resonance imaging, and on the right by cardiac autopsies. (A) Cactus LAA morphology is composed of a dominant central lobe with secondary lobes extending from the central lobe in both the superior and inferior directions. (B) Windsock LAA morphology has one dominant lobe as the primary structure, which is larger than the second or distal portions of the LAA. Variations of this LAA type arise with the location and number of secondary or even tertiary lobes arising from the dominant lobe. (C) Cauliflower LAA morphology presents with a more complex internal characteristics than the chicken wing or windsock morphologies. This LAA type has an irregular shape of the LAA ostium (oval vs round) and a variable number of lobes with no dominant lobe. (D) Chicken wing LAA morphology presents an obvious bend in the proximal or middle part of the dominant lobe or folding back of the LAA anatomy on itself at some distance from the perceived LAA ostium. This type of LAA can have secondary lobes or twigs. AR, aorta; CA, coronary artery; LPA, left pulmonary artery; PT, pulmonary trunk; RPA, right pulmonary artery.[11]

Clinical use environment

The primary treatment for AF patients to reduce risk of stroke is oral anticoagulation. However, anticoagulation requires careful tritration of dosing and has an inherent risk of leading to hemorrhagic stroke. Warfarin, in particular, requires regular visits to the clinic and is not tolerated by all patients. Patient compliance to the anticoagulation medication regime can also be problematic. Device-based management of the LAA is therefore more attractive if it can obviate the need for anticoagulation therapy.[8]

There are various procedural approaches for addressing the LAA, each with different risk-benefit relationships. Endocardial devices inserted into the LAA via vascular access provide a less invasive procedure for managing the LAA than surgical procedures but do not provide electrical isolation and leave an implanted device within the bloodstream. Endocardial devices have more anatomical exclusions than epicardial devices. Currently, an anticoagulation regimen is required for at least a limited time post therapy of endocardial procedures.[8]

Epicardial approaches require a more invasive procedure than endocardial approaches but can provide direct visualization and do not leave a device in the bloodstream. Epicardial LAA management has historically been done concomitantly with other cardiac procedures, such as Coronary artery bypass graft surgery (CABG), mitral valve surgery, or AF ablation, but may also be performed as a stand-alone procedure. Surgical access may include a median sternotomy, mini sternotomy, thoracotomy or thoracoscopy, and may involve an arrested heart on cardiopulmonary bypass or on a beating heart. A median sternotomy is most often used and results in low failure rates and excellent proven long-term outcomes.[14] A median sternotomy provides the ample space and direct visualization for placing the epicardial LAA device. Procedural and device solutions used during median sternotomies include clips, staplers, endo loops and the cut-and-sew technique. The cut-and-sew technique has been widely accepted as the gold standard for treating the LAA.[15] In this technique, the appendage is amputated so the remaining incision must be sewn closed. Although there is no device cost for the cut-and-sew method, other considerations for safety and time must be considered. The cut-and-sew technique must be performed on-pump on an arrested heart, and it is important for the patient's safety to minimize the time that the heart is arrested during a procedure. However, the technique can vary considerably depending on the surgeon and requires careful, thorough suturing to avoid intra-operative and postoperative bleeding complications. Some appendages are friable so the sutures can tear and significantly complicate the procedure. Epicardial LAA exclusion devices are designed to reduce the additional time and risks of the cut and sew procedure. This includes eliminating leaks, minimizing manipulation of the friable LAA tissue, and reducing the interoperator variability.

Minimally invasive thoracoscopic techniques are available to manage the LAA. Minimally invasive surgical access approaches can be either a thoracotomy or a port known as a trocar. A thoracotomy is a surgical procedure in which an incision is made between the ribs to visualize and gain access to the heart. A trocar is a medical device that is made up of an awl (which may be a metal or plastic sharpened or nonbladed tip), a cannula (essentially a hollow tube), and a seal. The seal facilitates insufflation to enable the thoracic cavity to be pressurized with CO_2 to provide space for visualization and manipulation of surgical instruments. Trocars are placed through the thoracic cavity during laparoscopic surgery in order to gain access for surgical tools. Trocars minimize recovery time since the incision injury is small and, in some cases, do not require stitches to close. Minimally invasive techniques are performed concomitantly with other cardiac procedures so they can utilize some of the ports already being used. Three ports are normally utilized for a clip procedure. One port is for a camera necessary for visualization of the heart and the appendage. The port for the camera can be a 5-mm diameter depending on the size of the camera. Another port is used for a manipulation device to assist positioning the appendage into the clip. Devices such as endoscopic kittners and graspers are used to manipulate the appendage. These devices commonly fit down a 5-mm trocar. The third is 12-millimeters to gain access for the clip applier generally placed as the inferior (caudal) port at about the sixth intercostal space. Other approaches such as subxyphoid and right-sided approaches through the transverse sinus have been utilized but those approaches are more difficult and require special skills.

Mechanisms and measurements of exclusion

Optimal LAA exclusion results in a smooth endothelial wall where the ostium is closed as well as atrophy of the LAA tissue distal to the closure device. Closure of the LAA ostium needs not only to discontinue blood flow between the LAA and LA chambers but should also exclude any myocardial tissue from communicating electrically between the LA and the residual LAA stump. Mechanical exclusion occurs by approximating the walls of the ostium with sufficient force to cause ischemic necrosis of the LAA myocardial tissue. The ischemic tissue distal to the ligature cannot electrically conduct action potentials. Chronically, the LAA ostium is sealed with smooth fibrous tissue and subsequently endothelializes. The persistence of adequate force during this healing process is important to assure no leaks develop as the tissue remodels.

Epicardial devices routinely use the assistance of transesophageal echocardiography (TEE) to ensure the device is placed correctly and that blood flow between the LAA and the left atrium is blocked. Because the esophagus is so close to the atria, TEE provides very clear images of the LAA. TEE is a useful tool not only for verifying an absence of clots prior to exclusion but also for chronically monitoring for device migration and the development of leaks.[16] LAA closure success is defined as no residual flow and less than a 1-cm residual stump. Minimizing the length of the residual stump facilitates uniform blood flow to prevent clots from forming.

AtriCure markets two AtriClip devices that are specifically designed for safety and efficacy for excluding the LAA. Both versions of the AtriClip are identical from a force and pressure profile, however they provide different geometry to allow for varying surgical approaches. The AtriClip does not penetrate the LAA to mitigate any bleeding concerns. The AtriClip relies on spring force to continually apply pressure to the LAA to prevent blood flow and electrically isolate the appendage. Since the clip does not penetrate the LAA, it can easily and safely be repositioned if the initial placement is not optimal. Since the blood flow to the LAA is eliminated, the appendage eventually atrophies while the continual pressure facilitates ischemic necrosis of the LAA myocardial tissue. The clip mechanism is covered with a knit braded polyester that promotes tissue ingrowth to anchor the clip in place and prevent migration as the structures remodel. The clip provides a smooth endocardial surface to facilitate a smooth remodeling compared to the cut-and-sew technique that can leave small pockets.

Development path to state of art technologies

The development of appropriate test beds and models has been instrumental throughout the development of technological solutions to exclude the LAA. Because no one model has been completely capable of fulfilling all the needs, the development process has used a combination of various test methods and models. From an anatomy and access approach, the cadaver model is the most realistic. A shortcoming of the cadaver model however is the heart is deflated and not beating. A system has been developed by AtriCure to position balloons in the chambers of the heart and pulse them to simulate a beating heart. This is a good representation of a living patient that simulates the anatomy as well as the challenges associated with motion.

Various bench test models are used whenever possible. Simulated polymer thoracic anatomy models are used to develop initial device geometry and functionality. They provide the locations of the various areas of interest; however, they lack the texture and flexibility of a living model. Other mechanical bench models are used that target specific functional aspects of the clip. One such aspect is the impact of continuous cycling and movement due to the heartbeat. A test bed was fabricated that simulates the appendage size and heartbeat forces to make sure the clip is not susceptible to cyclic fatigue or any other unknown issues related to a dynamic heartbeat for an extended period. This testbed can cycle several clips concurrently to simulate millions of heartbeats to evaluate the clip design as the appendage remodels.

While bench testing is the first choice, animal models are also used. The porcine model is useful since the heart is similar in size to a human heart although the anatomy differs significantly. Since the porcine thoracic cavity does not closely represent human anatomy, it is rarely used to develop surgical procedures. The porcine right atrial appendage is a tubular structure and suitable for *in vivo* tissue response studies, however the porcine LAA is not a tubular structure and therefore not a suitable model. The canine model is useful for in vivo survival studies.[17,18] The canine heart is similar to the human anatomy and the healing process is comparable, so it is a good choice for survival studies. Tissue ingrowth usually starts within weeks of implantation but takes about one to three months for the clip to be totally encapsulated.[17,19]

Staplers have also been used to close the LAA. Staplers are general purpose surgical tools and are not specifically designed for LAA closure. Staplers close the appendage in a linear fashion But complete closure is not always achieved.[16] Staples are designed to close the tissue but are not designed to apply enough pressure to stop the vascular blood flow. This permits the LAA to survive and therefore it can continue to generate electrical impulses that contribute to AF. Staples also penetrate the heart tissue, creating an inherent safety concern related to bleeding.

State of the art

The AtriClip is a ligating device specifically designed to exclude the LAA. The design utilizes linear, dynamic closure that does not penetrate the appendage in order to mitigate safety concerns characteristic with other solutions. The AtriClip closes the LAA linearly and with enough force that erroneous electrical impulses cannot be transmitted from the LAA to the heart.[20] The AtriClip consists of two rigid beams that close in a parallel manner to provide consistent force applied along the beams. The parallel beams close the appendage in a linear fashion which is consistent to the natural shape of the base of the appendage. With a linear closing method, there is no bunching of tissue and therefore no path for blood flow compared to circular closing methods.

Dynamic closure for the beams is applied with springs. The springs are designed to continuously apply force to the tissue and retain the pressure as the appendage atrophies and remodels. The distal appendage atrophies over time and is completely gone within about 90-days of implantation of the clip (Fig. 28.3).

The clip is attached to a rigid applier that allows the surgeon to apply necessary force to ensure clip placement at the base of the LAA. Once the clip is applied to the appendage, its position can be checked using TEE. If the positioning is unsatisfactory, it can be re-opened and moved as many times as desired. When the physician is satisfied with the positioning of the clip, it is released from the applier. The clip relies on compression and does not pierce the LAA so the risk of bleeding related issues is mitigated. Since the clip does not require a time-consuming, tedious process like the cut-and-sew technique, it is much faster to position and deploy, minimizing procedure time. The inherent ease of use limits interoperator variability (Figs. 28.4–28.7).

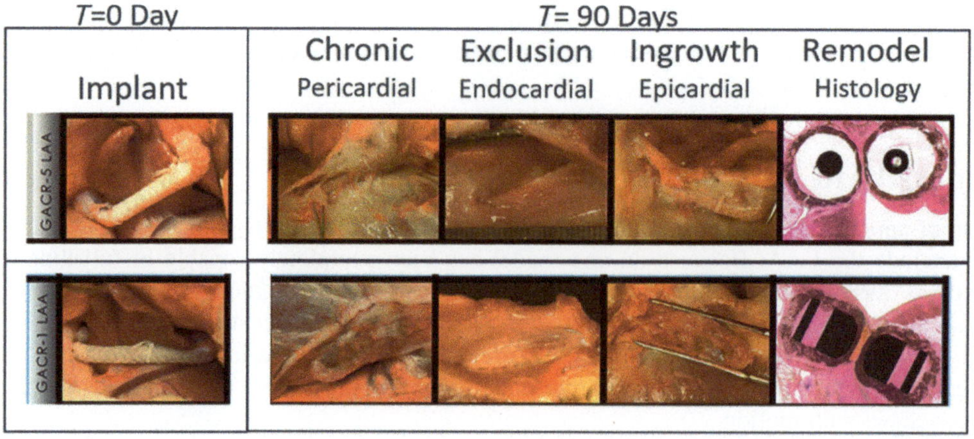

FIGURE 28.3 The top row are images of the AtriClip and the second row is the AtriClip-V. The first column is images of the clip at the time of implantation. The next set of images were taken 90 days after implantation. The Chronic Pericardial images show the epicardial ingrowth of the clip through the pericardium. The Exclusion Endocardial column shows the endocardial remodeling of the appendage exhibiting a smooth endocardial surface. The Epicardial Ingrowth is an image of the clip with the pericardium removed showing the effectiveness of the knit braded polyester covering for promoting ingrowth. In the last column shows the Remodel Histology indicating the appendage in both cases have completely atrophied.

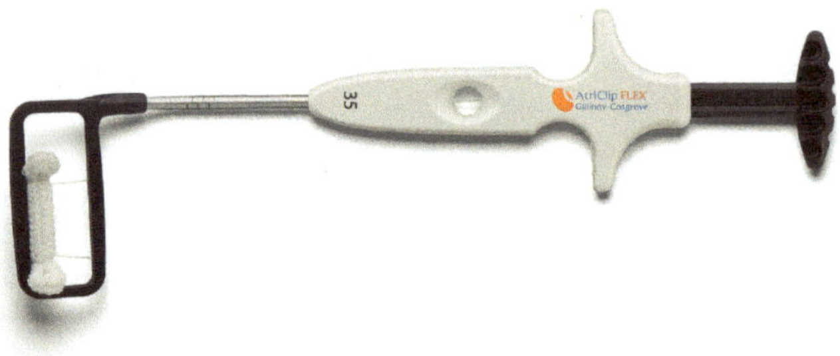

FIGURE 28.4 AtriClip FLEX device used in open procedures utilizing the AtriClip.

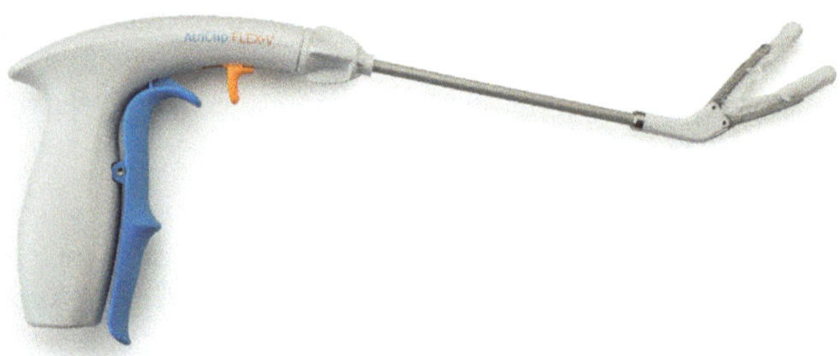

FIGURE 28.5 AtriClip FLEX-V device used in open procedures utilizing the AtriClip-V.

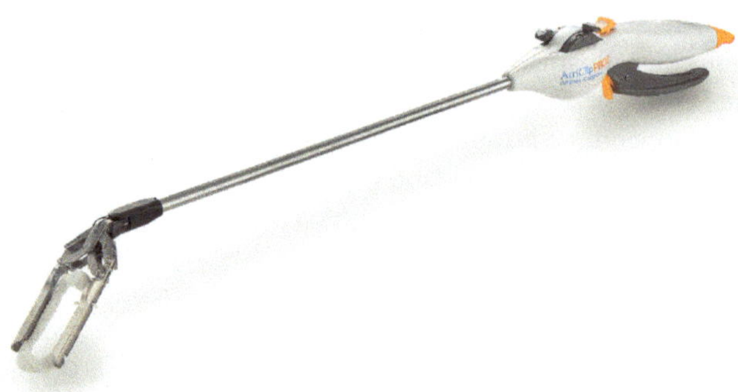

FIGURE 28.6 AtriClip PRO2 device used in minimally invasive procedures utilizing the AtriClip.

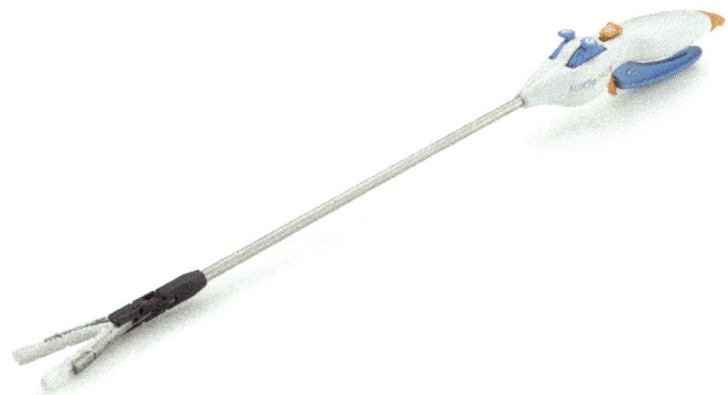

FIGURE 28.7 AtriClip PRO · V device used in minimally invasive procedures utilizing the AtriClip-V.

The clip is enclosed in a knit braided polyester fabric to promote tissue ingrowth as well as provide friction to keep the clip from moving. The spring force and friction of the clip ensures it is secure at the time of placement and the fabric promotes tissue ingrowth for robust long-term security. Tissue ingrowth is also an important consideration to prevent erosion of surrounding structures over time (Figs. 28.8 and 28.9).

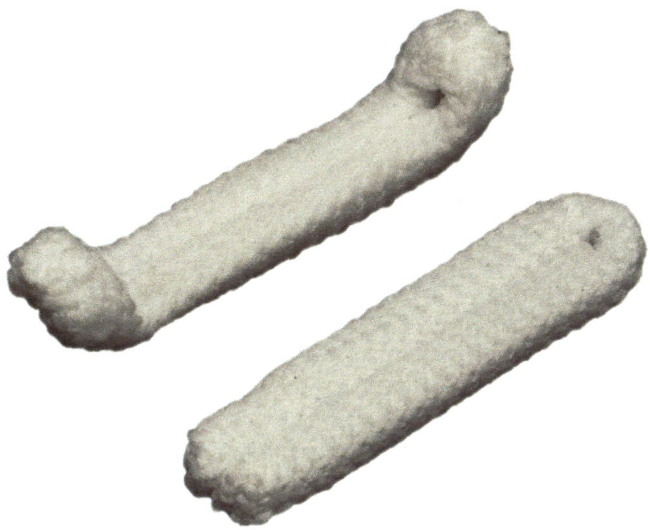

FIGURE 28.8 The top image is a closed AtriClip. The bottom image is an AtriClip-V.

FIGURE 28.9 Images of the AtriClip with the fabric removed to show the mechanism. The top image is an AtriClip. The bottom image is an AtriClip-V.

The force required to resist the pressure in the left atrium is relatively low. The blood entering the left atrium is fed from the capillary exchange in the lungs that oxygenate the blood, so the pressure is lower than typical atrial pressures. The left atrium normally has a peak systolic pressure of 10–15 mm-Hg. Abnormally high peak systolic left atrium pressure in a diseased heart may be as high as 25 mm-Hg or 0.48 psi.[21] The AtriClip applies force and safety factor to withstand the internal pressure to ensure the appendage is completely closed.

AtriCure has two variations for the LAA clips. One fully enclosed version is known as the AtriClip and is configured with nitinol springs positioned at both sides of the beams. This clip requires an approach over the top of the appendage. The springs located at each end of the beam provide a visual indication that the appendage is completely encapsulated and centered within the clip. Since the distal portion of the appendage can sometimes be larger than the base, some manipulation may be required to fit the clip over the appendage. The various appendage morphologies have very little impact on the usability of the clip.

Another version of the clip, known as the AtriClip-V, is configured in a V-shape. The force and tissue contacting geometry of both clips are identical. The AtriClip-V can be slid on to the base of the appendage from the side so the morphology of the distal portion of the appendage is of little concern. The side approached in minimally invasive procedures can be advantageous depending on port placement. The AtriClip-V has less concern for dislodging clots that may already exist in the LAA since it does not disturb the distal end of the appendage. The AtriClip-V is made from a single piece of titanium and it does not utilize Nitinol springs in the design as does the AtriClip. Nitinol is a nickel-titanium alloy, introducing a potential concern for patients with a sensitivity to nickel, although there have not been any reported complaints linked to a nickel sensitivity. The AtriClip-V is an option if there is still a concern since it is completely nickel free.

Since the release of the AtriClip in 2010, the safety and efficacy profile has been exceptional. Approximately 300,000 clips have been placed over the last 10 years. A meta-analysis of more than 11 peer reviewed publications on the AtriClip devices reported a 97% successful LAA closure rate on average with no device migrations or related complications.[22] This success rate indicates little variability in performance across operators.

Future directions

Future direction for LAA management is potentially with a stand-alone procedures and prophylactic applications for patients that do not currently present with A-fib. Reducing the size of epicardial exclusion devices may facilitate less invasive surgical procedures, leading to an increase in stand-alone procedures. Evidence suggests there may be benefit in broadening the adoption of LAA exclusion devices to prophylactic applications in patients undergoing cardiac surgery and at risk of postoperative AF.[23] The ATLAS clinical trial will compare impact of postoperative atrial fibrillation (POAF) among two randomized treatment arms; patients with POAF and surgical LAA closure (using AtriClip Gillinov-Cosgrove LAA Exclusion Systems) versus patients with POAF and no surgical LAA closure; evaluate healthcare resource utilization (i.e., hospital length of stay (LOS), emergency room and/or hospital re-admissions, and costs associated with specific adverse events that may be related to AF). At 3 months, prophylactic LAA exclusion in cardiac surgery patients with elevated CHADS2VASc and HASBLED resulted in significantly lower incidence of CVA (stroke and TIA) without increase in adverse events. Further trials to evaluate the long-term effects of LAAE during cardiac surgery are needed.[24]

In addition to preventing stroke and arrhythmia, LAA exclusion may also be beneficial for treating hypertension. Studies have indicated that excluding the LAA can reduce blood pressure in patients with AF and hypertension. In a study of 247 patients undergoing LAA exclusion,[25] systolic blood pressure (SBP) was reduced 7.4 mm-Hg after 3 months and 8.9 mm-Hg at 1 year. Diastolic Blood Pressure (DBP) was reduced 1.8 mm-Hg at 1 year. This decrease could reduce or potentially eliminate the need for blood pressure medication post surgery.

References

1. Gage BF, Waterman AD, Shannon W, Boechler M, Rich MW, Radford MJ. Validation of clinical classification schemes for predicting stroke: results from the National Registry of Atrial Fibrillation. *JAMA*. 2001;285(22):2864–2870. Available from: https://doi.org/10.1001/jama.285.22.2864.
2. Blackshear JL, Odell JA. Appendage obliteration to reduce stroke in cardiac surgical patients with atrial fibrillation. *Ann Thorac Surg*. 1996;61(2):755–759. Available from: https://doi.org/10.1016/0003-4975(95)00887-X.
3. Di Biase L, Burkhardt JD, Mohanty P, et al. Left atrial appendage: an underrecognized trigger site of atrial fibrillation. *Circulation*. 2010;122(2):109–118. Available from: https://doi.org/10.1161/CIRCULATIONAHA.109.928903.

4. Hocini M, Shah AJ, Nault I, et al. Localized reentry within the left atrial appendage: arrhythmogenic role in patients undergoing ablation of persistent atrial fibrillation. *Heart Rhythm*. 2011;8(12):1853–1861. Available from: https://doi.org/10.1016/j.hrthm.2011.07.013.
5 Epstein JA, Aghajanian H, Singh MK. Semaphorin signaling in cardiovascular development. *Cell Metabol*. 2015;. Available from: https://doi.org/10.1016/j.cmet.2014.12.015.
6. Cox JL. Mechanical closure of the left atrial appendage: is it time to be more aggressive? *J Thorac Cardiovasc Surg*. 2013;146(5):1018–1027.e2. Available from: https://doi.org/10.1016/j.jtcvs.2013.07.011.
7. Sakellaridis T, Argiriou M, Charitos C, et al. Left atrial appendage exclusion-Where do we stand? *J Thorac Dis*. 2014;6(Suppl 1):S70–S77. Available from: https://doi.org/10.3978/j.issn.2072-1439.2013.10.24.
8. Bedeir K, Holmes DR, Cox JL, Ramlawi B. Left atrial appendage exclusion: An alternative to anticoagulation in nonvalvular atrial fibrillation. *J Thorac Cardiovasc Surg*. 2017;153(5):1097–1105. Available from: https://doi.org/10.1016/j.jtcvs.2016.12.040.
9. Ailawadi G, Gerdisch MW, Harvey RL, et al. Exclusion of the left atrial appendage with a novel device: early results of a multicenter trial. *J Thorac Cardiovasc Surg*. 2011;142(5):1002–1009.e1. Available from: https://doi.org/10.1016/j.jtcvs.2011.07.052.
10. Beaver TM, Hedna VS, Khanna AY, et al. Thoracoscopic ablation with appendage ligation vs. medical therapy for stroke prevention a proof-of-concept randomized trial. *Innov Technol Techn Cardiothor Vasc Surg*. 2016;11(2):99–105. Available from: https://doi.org/10.1097/imi.0000000000000226.
11. Di Biase L, Santangeli P, Anselmino M, et al. Does the left atrial appendage morphology correlate with the risk of stroke in patients with atrial fibrillation? Results from a multicenter study. *J Am Coll Cardiol*. 2012;60(6):531–538. Available from: https://doi.org/10.1016/j.jacc.2012.04.032.
12. Wang Y, Di Biase L, Horton RP, Nguyen T, Morhanty P, Natale A. Left atrial appendage studied by computed tomography to help planning for appendage closure device placement. *J Cardiovasc Electrophysiol*. 2010;21(9):973–982. Available from: https://doi.org/10.1111/j.1540-8167.2010.01814.x.
13. Di Biase L, Natale A. Romero J. thrombogenic and arrhythmogenic roles of the left atrial appendage in atrial fibrillation. *Circulation*. 2018;138(18):2036–2050. Available from: https://doi.org/10.1161/CIRCULATIONAHA.118.034187.
14. Julian OC, Lopez-Belio M, Dye WS, Javid H, Grove WJ. The median sternal incision in intracardiac surgery with extracorporeal circulation; a general evaluation of its use in heart surgery. *Surgery*. 1957;42(4):753–761.
15. Stulak JM, Dearani JA, Sundt TM, et al. Superiority of cut-and-sew technique for the Cox maze procedure: comparison with radiofrequency ablation. *J Thorac Cardiovasc Surg*. 2007;133(4):1022–1027. Available from: https://doi.org/10.1016/j.jtcvs.2006.09.115.
16. Kanderian AS, Gillinov AM, Pettersson GB, Blackstone E, Klein AL. Success of surgical left atrial appendage closure: assessment by transesophageal echocardiography. *J Am Coll Cardiol*. 2008;52(11):924–929. Available from: https://doi.org/10.1016/j.jacc.2008.03.067.
17. Sunagawa G, Karimov JH, Breitbach M, Robinson NA, Fukamachi K. Impact of a refined advanced design for left atrial appendage exclusion. *Eur J Cardiothorac Surg*. 2017;52(6):1098–1103. Available from: https://doi.org/10.1093/ejcts/ezx188.
18. Kamohara K, Fukamachi K, Ootaki Y, et al. A novel device for left atrial appendage exclusion. *J Thorac Cardiovasc Surg*. 2005;130(6):1639–1644. Available from: https://doi.org/10.1016/j.jtcvs.2005.08.019.
19. Kamohara K, Fukamachi K, Ootaki Y, et al. Evaluation of a novel device for left atrial appendage exclusion: the second-generation atrial exclusion device. *J Thorac Cardiovasc Surg*. 2006;132(2):340–346. Available from: https://doi.org/10.1016/j.jtcvs.2006.04.021.
20. Starck CT, Steffel J, Emmert MY, et al. Epicardial left atrial appendage clip occlusion also provides the electrical isolation of the left atrial appendage. *Interact Cardiovasc Thorac Surg*. 2012;15(3):416–418. Available from: https://doi.org/10.1093/icvts/ivs136.
21. Geske Jeffrey B, Sorajja P, Nishimura Rick A, Ommen Steve R. Evaluation of left ventricular filling pressures by Doppler echocardiography in patients with hypertrophic cardiomyopathy. *Circulation*. 2007;116(23):2702–2708. Available from: https://doi.org/10.1161/CIRCULATIONAHA.107.698985.
22. Toale C, Fitzmaurice GJ, Eaton D, Lyne J, Redmond KC. Outcomes of left atrial appendage occlusion using the AtriClip device: a systematic review. *Interact CardioVasc Thorac Surg*. 2019;29(5):655–662. Available from: https://doi.org/10.1093/icvts/ivz156.
23. Squiers JJ, Edgerton JR. Surgical Closure of the Left Atrial Appendage: The Past, The Present, The Future. *J Atr Fibrillation*. 2018;10(5):1642. Available from: https://doi.org/10.4022/jafib.1642.
24. ICC - Prophylactic Left Atrial Appendage Exclusion Reduces Stroke Incidence In High Risk Cardiac Surgery Patients: Results From The Atlas Randomized Control Trial. Accessed March 1, 2021. <https://internationalcoronarycongress.com/program/2019/135.cgi>.
25. Turagam MK, Vuddanda V, Verberkmoes N, et al. Epicardial left atrial appendage exclusion reduces blood pressure in patients with atrial fibrillation and hypertension. *J Am Coll Cardiol*. 2018;72(12):1346–1353. Available from: https://doi.org/10.1016/j.jacc.2018.06.066.

Chapter 29

Technology for Surgical Cardiac Cryoablation Procedures

Adam W. Cates, Frank M. Fago and Matthew V. Monti
AtriCure, Inc., Mason, OH, United States

Chapter Outline

Introduction

Cryoablation procedures use extreme cold to destroy specific tissues for therapeutic benefit. Surgeons began using cryoablation for cardiac arrhythmia treatments in the 1970s, taking advantage of closed system cryoprobes that cooled with expanding nitrous oxide. These early probes were designed for spot ablation and were used throughout the 1980s to treat arrhythmias such as supraventricular tachycardia, Wolf-Parkinson-White, atrioventricular nodal reentrant tachycardia, ventricular tachycardia, atrial tachycardia, and atrial fibrillation (AF).[1,2] AF treatment has become the most common use of cardiac cryoablation due to the development of a set of scar lines throughout the atria, developed by Dr. James Cox, that demonstrated remarkable success at freedom from AF.[3,4] The lesion set, the Cox Maze, would be refined by replacing the surgical scars with ablation lesions to expedite the procedure, and it was this lesion set that led to the need for linear cryoprobes. State-of-the-art technologies for cardiac cryoablation have been specifically designed to meet the challenges and needs associated with creating transmural lesions as part of the Cox Maze IV (CMIV) lesion set during open concomitant procedures as well as subsets of the CMIV lesion sets with minimally invasive procedures.

Translating clinical drivers to technology needs

Clinical use environment

The success of the CMIV lesion set for treating AF led to a recommendation in the guidelines from the major surgical societies[5,6] and experts[7] for performing surgical ablation of AF during concomitant surgery. The full CMIV lesion set consists of eight lesions on the left atria and four lesions on the right atria.[8] Radiofrequency (RF) ablation is primarily used, with cryoablation complementing RF, specifically for the tricuspid valve line, mitral valve isthmus line, and the coronary sinus line because cryoablation does not damage the extracellular matrix and can safely create these lines without damaging the vein or valves.[9] However, cryoablation is not limited to the valve lines but has been used to perform all the ablations.[10] The tissue variation among these lesions is considerable, requiring versatility in shape and performance to match the angles and the curves of these lines. Furthermore, atrial myocardial thickness can range from 1 to 17 mm, and is especially thick near the atrioventricular (AV) groove, where the mitral valve isthmus line is placed endocardially and the coronary sinus line is placed epicardially. There may also be extensive fat epicardially, which given that fat has one third the thermal conductivity of myocardium presents a challenge to cryoablation.[11]

Advances in Cardiovascular Technology. DOI: https://doi.org/10.1016/B978-0-12-816861-5.00029-0

Cryoablation during open concomitant surgeries are performed with the heart on cardiopulmonary bypass and cardioplegiacally arrested. A median sternotomy or right mini-thoracotomy allows the surgeon to gain access via atriotomy to the endocardium without a pool of blood.[9] Epicardial cryoablations are performed across the coronary sinus and often encounter fat common along the AV groove. While performing cryoablation with the heart on cardiopulmonary bypass provides the most favorable conditions for transmurality, surgeons must also consider minimizing cardiac ischemic time while on bypass. The move toward minimally invasive thoracoscopic procedures has presented more difficult challenges for cryoablations, such as restricted access size as well as beating hearts without cardiopulmonary bypass. With beating hearts, the endocardial blood flow provides a convective heat source that cryoablation technologies must overcome.[12]

Tissue contact is critically important because the effect of cryoablation is driven by conductive heat transfer. A stiff rigid probe allows the surgeon to distribute a lot of force to the tissue, assuring solid contact at the probe tissue interface for smooth, flat tissue and straight lines. However, the anatomy often dictates that lesions curve (Fig. 29.1), which requires a flexible probe that the surgeon can shape to match the tissue and create consistent contact along the length of the probe. Consequently, modern cryoprobes are designed with the tradeoff to be malleable enough for intraoperative manipulation to match the anatomy while maintaining sufficient rigidity to impart consistent force along the probe.

Mechanisms of tissue destruction with cryoablation

Cryoablation therapy is a thermal assault that subjects cells to a cascade of destructive forces during both the tissue freezing and thawing phases. During rapid freezing, the ice crystal formation due to water freezing inside the cells causes direct mechanical trauma to the cell membrane and organelles.[13] Extracellular ice formation produces a hyperosmotic solution outside the cells that pulls water from inside the cell, leading to cell dehydration. Furthermore, the extracellular ice crystals can disrupt cell membranes, especially in dense tissues. Damage from both intracellular and extracellular ice formation comprises the acute necrotic region from cryoablation.[14]

Cold temperatures that freeze blood in the microcirculation cause capillaries and small vessels to burst leading to microvascular stasis. The ruptured vessels create edema and clotting around the cells and prolong the ischemic insult even as the tissue is warmed. Lastly, cells subjected to freezing temperatures that do not produce intracellular or extracellular ice formation may still be irreparably damaged due to impairment of metabolism and other cellular function during the freeze and thaw cycle. These cells may survive the initial freeze insult but then undergo apoptosis, or programmed cell death, in the hours to days afterwards.[14]

The combination of these mechanisms leads to the formation of a cryolesion, and their effects are directly related to the thermal profile established during the cryoablation. An active cryoprobe on the surface of tissue establishes a

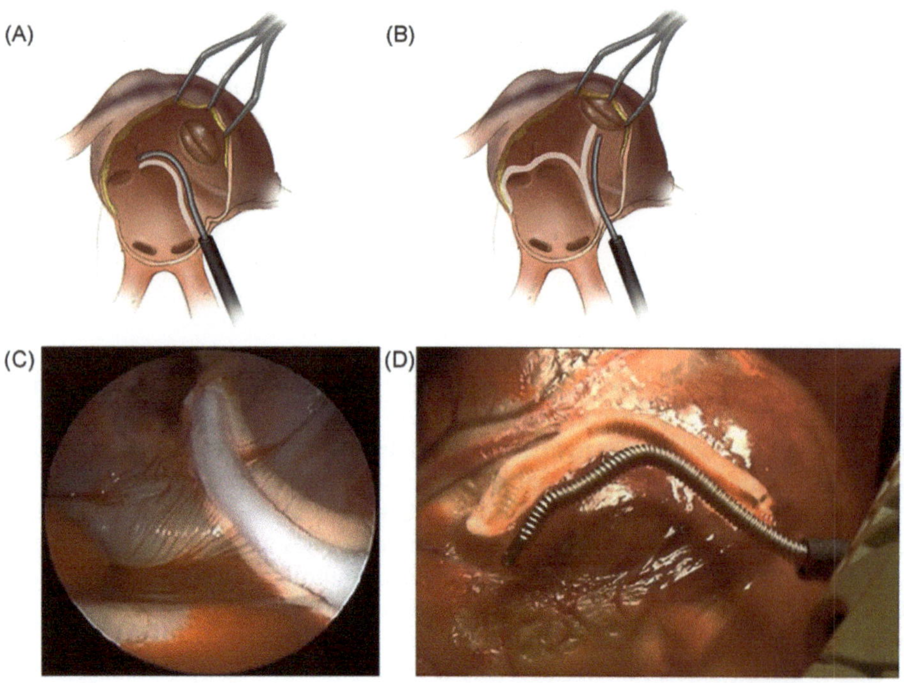

FIGURE 29.1 Examples of various shapes required for cryoablation probes. (A) Schematic of an inferior pulmonary vein line. (B) Schematic of a mitral isthmus line. (C and D) Photographs of preclinical lesions performed with the probes configured in different shapes.

thermal gradient in which the coldest temperatures occur nearest the probe and temperatures warm with distance away from the probe. The 0°C isotherm approximates the edge of the frozen volume of tissue (i.e., "ice ball"). All tissue destruction occurs within this frozen volume of tissue, with the most certain and permanent cellular destruction occurring in the colder isotherms and the least reliable or transient damage occurring in the warmer isotherms. The resulting chronic lesion is therefore slightly smaller than the size of the frozen volume of tissue.[14,15]

Specific temperatures and durations that result in cell death vary by cell type. A threshold of −40°C to −50°C for immediate cell death has historically been referenced, but this threshold is based off the work on cancer cells.[13] Cardiac cells may be susceptible at temperatures warmer than −30°C.[2,16,17]

The thermal gradients established by a cryoprobe vary with time, with the volume of frozen tissue expanding until a thermal equilibrium is met with the surrounding tissue. The time it takes to reach steady state varies based on the initial temperature of the tissue, the thermal conductive properties of the tissue and the presence of any nearby heat sources that counteract the heat extraction, such as volumes of circulating blood, coronary arteries or tissue perfusion. The duration of time the cryoprobe is active therefore influences the extent of tissue destruction not only by increasing the time the target tissues are exposed but also by increasing the volume of tissue exposed to colder temperatures.[18]

Freezing is more effective when it is fast, as lethal intracellular ice is formed before extracellular ice formation dehydrates the cell. Thawing, on the other hand, should be as slow as possible to prolong the ischemic assault. Furthermore, recrystallization, which is a phenomenon that occurs during warming in which the ice crystals rearrange into larger crystals, shears apart cell membranes and structures and is more destructive when warming is slow. Therefore, the thawing rate should be as slow as possible, ideally warmed only by the heat generated by the tissue and its surrounding environment, to assure maximal destruction.[13]

Cryoablation aims to maximize lesion size within the targeted tissue but, due to its unipolar nature, the thermal expansion is difficult to contain, potentially exposing surrounding tissues and structures to damage. Fortunately, the volume of tissue affected is easily identified by the frozen volume of tissue, so avoiding unwanted tissues from being enveloped in the frozen tissue will prevent unintentional injury. Blood, blood vessels, the esophagus, nerves, and lungs are all within the vicinity of a cardiac cryoablation and should not be included in the frozen volume of tissue. Concerns about blood freezing include the potential for embolism and subsequent infarct or stroke and the lysing of red blood cells, but these concerns apply only to beating heart applications and are less likely than with RF ablation.[19–21] The effect of freezing blood vessels is not clearly established in the literature, although it is generally accepted that blood vessels, in particular veins, are very resilient to freezing. In larger coronary arteries, the muscle layer and endothelium can be damaged, potentially leading to infarct, intimal hyperplasia or stenosis.[22] The esophagus is located very close to the posterior wall of the left atrium and may be susceptible to thermal damage. Although the outer esophagus is resilient to freeze, the muscle layers and endothelium can be damaged, leading to dysphagia. Clinically, atrioesophageal fistula formation is less common with cryoablation than with RF ablation.[22] The phrenic nerve is also often in the vicinity of atrial lesions and should be avoided, because if phrenic nerve temperatures reach < -15°C, it may take weeks to heal. Respiratory structures should also be avoided during cryoablation. Cryoablation within the pulmonary veins has extended into the lungs and resulted in a patient coughing up dried blood. Additionally, trachea respond to freezing in a manner similar to the esophagus.[22]

Several indicators are used to gauge the completeness of an ablation. Acutely, probe temperature is monitored by internal thermocouples to assure the probe is functioning adequately and reaching desired temperatures. However, these temperatures represent the probe and not the tissue temperatures that determine lesion success. Ablation duration is monitored and follows prescribed settings based on conventional standards but there is considerable variation among surgeons.[8] Two-minute freezes have been the standard and have shown success preclinically both in vivo[23] and with bench tests[24] but are often misinterpreted.[2] Ideally, the ablation should persist for 2 min after the ice ball has encompassed the entire length of tissue segment targeted all the way through the wall.[2] Assessing the frozen volume of tissue, therefore, can provide some visual feedback of the extent of tissue injury and can be used to guide lesion duration, but it represents a maximum potential effect and is potentially more useful for assuring that adjacent unintended tissues are not affected. Pacing and sensing tests for entrance block and exit block theoretically can be performed, but these will only demonstrate incomplete lines and may provide false positives of complete lines, as cool tissue will fail to propagate action potentials temporarily even when not permanently destroyed. The ultimate measure of success is histological evidence of transmurality along the length of the lesion, which can be measured in preclinical studies but not clinically. Cryolesions have clear, well-demarcated boundaries and appear homogenous throughout compared to RF lesions.[25] Clinically, chronic endpoints include freedom from arrhythmia, preservation of valve tissue and blood vessels, and absence of stroke.[8]

Development path to state-of-the-art technologies

Earliest medical cryoablation technology delivered liquid nitrogen to an exposed metal tip, and this technology still works well today for certain applications, such as dermatological ablations.[1] Liquid nitrogen has limitations, however, especially when incorporated into long narrow probes, where the evaporation from liquid to gaseous state can result in an enormous volume expansion, creating a vapor lock that restricts further flow of liquid.[26] The advent of Joule–Thomson probes, which operate on the fundamental concept that a high pressure gas flowing through a small orifice will rapidly expand and cool, coincided with the earliest preclinical myocardial ablation studies and clinical cardiac arrhythmia uses. Initial probes were reusable with various shaped tips that facilitated spot ablations, but the need for linear ablations to perform the lesion set for AF treatment led to the development of the current rigid linear probe designs. In addition to shapes, the new probes were designed to be malleable to allow surgeons to shape them to match the target anatomy yet rigid enough to impart adequate force for solid contact. The potential damage to the fine tubes and orifices in the probes due to manipulation led to the replacement of reusable probes with single-use handpieces to assure consistent performance from procedure to procedure.

To enable technological improvements in probe design, various test methodology was developed to provide incremental understanding of the performance in a variety of increasingly aggressive challenges. The foundational test is an ablation in air, which allows engineers to monitor temperatures achieved along the probe with minimal heat load. This test proves useful for refining the physics of the Joule–Thomson effect. These tests do not depict how the system performs when the surrounding medium freezes. Water bath tests (Fig. 29.2) are useful for determining the thermal gradient creation in a steady state, homogeneous environment with a surrounding medium that freezes. In these tests, the cryoprobe is partially submerged horizontally in a bath of room temperature water, and the size of the resulting ice ball is used to gauge the cooling capacity as a function of time.[24] However, these tests do not determine how the system will perform in a scenario with constant heat sources counteracting the cooling effect of the probe and therefore do not accurately capture the cooling power of the system. Calorimetric tests evaluate the power of the system to withdraw heat against a steady supply of heat, forming the overstress challenge of a system that will not freeze. In these tests, a probe is placed vertically in an insulated chamber of water that is constantly heated and stirred to prevent freezing. The resulting drop in temperature of the water is then calculated to assess the power of the system.[27]

Although very useful for characterizing the physics of the cryoablation systems, none of these bench tests directly addresses the fundamental goal of freezing tissue. Freezing harvested bovine myocardium, a typical bench tissue model for RF ablation, presents problems for measuring cryolesions because the lesions are not visible once the tissue warms and do not appear with standard staining techniques like triphenyl tetrazolium chloride, so the measurements of lesion depth, in particular, are difficult. Placing thermocouples on the underside of harvested myocardium (Fig. 29.3) provides some insight to lesion transmurality[24] but relies on arbitrarily defining thermal thresholds for tissue destruction and is very sensitive to thermocouple placement in the coldest regions. Another method for assessing lesion depth with myocardium consists of placing harvested myocardium on floral foam in a saline bath. The floral foam has similar thermal conductivities to myocardium and freezes as an approximation of lesion depth and does not thaw as fast as tissue, providing a more consistent method of determining freeze depth.[27] However, based on the mechanisms of cryoablation, freezing does not equal ablation, so the frozen volume is only an estimate of the actual depth of tissue destruction.

The ultimate test for cryoablation systems is in vivo testing. Large animals, in particular sheep and pigs, have proved to be useful models, although acute studies are limited because they do not capture the full extent of tissue destruction that includes cellular apoptosis. Chronic animal models are the gold standard for performance even though

FIGURE 29.2 Water bath test method. (A) Side-view schematic of probe positioned horizontally and submerged halfway in water. (B) Top view photograph of side-by-side probe freezes in water bath. (C) Diagram illustrating measurements of ice ball size.

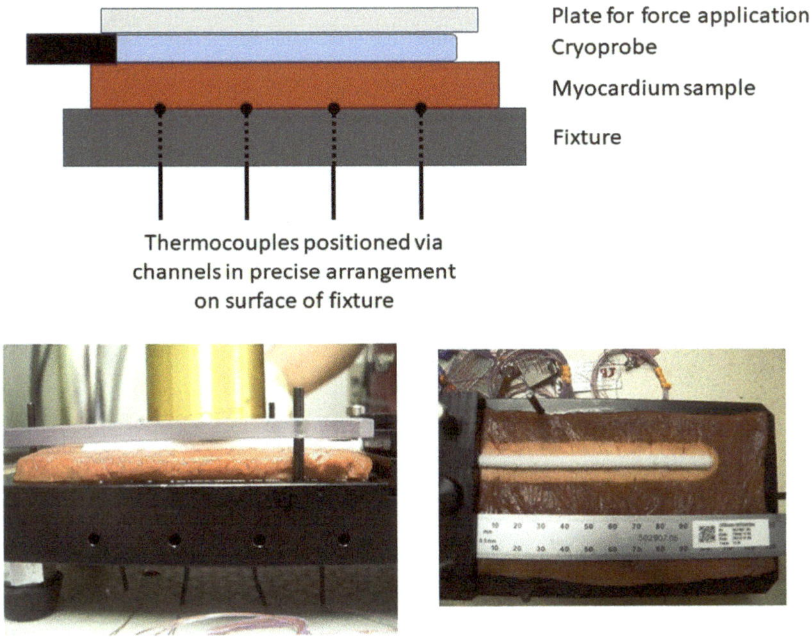

FIGURE 29.3 Bench model using harvested bovine myocardium to measure temperatures achieved at tissue depth.

a consistent survival period has not been established. 24-h studies seem to be the bare minimum, while 1 week minimum are an acceptable alternative. Fully formed lesions are not appreciated until 3 months or longer.

In addition to cardiac ablations in large animals, thigh muscle ablations have proved useful for assessing cryolesions, with at least a 24-h survival. While not directly applicable to cardiac tissue, these offer a reasonable alternative for higher sample sizes, as long as it is not crucial to understand the convective heat losses due to blood flow through the cardiac chambers.

State-of-the-art technologies

The goal for cryoablation systems is to extract as much heat as possible from the target tissue to lower its temperatures below critical thresholds for destruction. Simplistically, these systems must have adequate work capacity to remove heat via a cold probe that offers as little resistance to thermal conduction as possible. To cool the probes, current cardiac cryoablation systems take advantage of the Joule–Thomson effect, in which a rapidly expanding gas through a small orifice cools dramatically. Compressed gas, stored in a cylinder, is delivered via a tube to the tip of the probe where a small orifice throttles the gas flow, resulting in rapid decompression to essentially atmospheric pressure and a reduction in temperature. The cooled gas then flows along the external shell of the metal probe and back out of the device where it is vented to atmosphere (Fig. 29.4A). For long linear probes, multiple throttles may exist at various locations along the length of the probe to distribute the cold temperature (Fig. 29.4B). Maintaining the temperature along the length of the probe is a critical design factor, as uniform cooling is desired, but the cryogen warms as it flows away from the expansion nozzle. The gas temperature in the probe achieved from the Joule–Thomson effect is an intrinsic property of the gas and is proportional to the pressure drop. The minimum temperature is theoretically limited by the boiling temperature (condensation temperature) of the gas. These systems gain an even bigger advantage for removing heat when the expanding gas condenses and a liquid-gas mixture forms. This point is achieved faster by designing probes so that the cooled, venting gas flows alongside the incoming compressed gas, effectively precooling the inlet flow. Because these Joule–Thomson probes rely on pressurized gas escaping through small orifices, these systems require gas with very few impurities to prevent clogging of the orifices.

The transfer of heat from the tissue to the cooled probe can theoretically be described by the heat transfer equation:

$$\dot{Q} = \dot{m} \cdot c_p \cdot \Delta T$$

The rate that heat is removed from the tissue, $\dot{Q}$, is a function of the mass flow rate ($\dot{m}$) of the fluid, in this case the cooled cryogen gas/liquid mix, the specific heat of the cryogen c_p, and the difference in temperature (ΔT) between the cryoprobe and the tissue. The mass flow rate is a function of the pressure of the cryogen, the density of the cryogen,

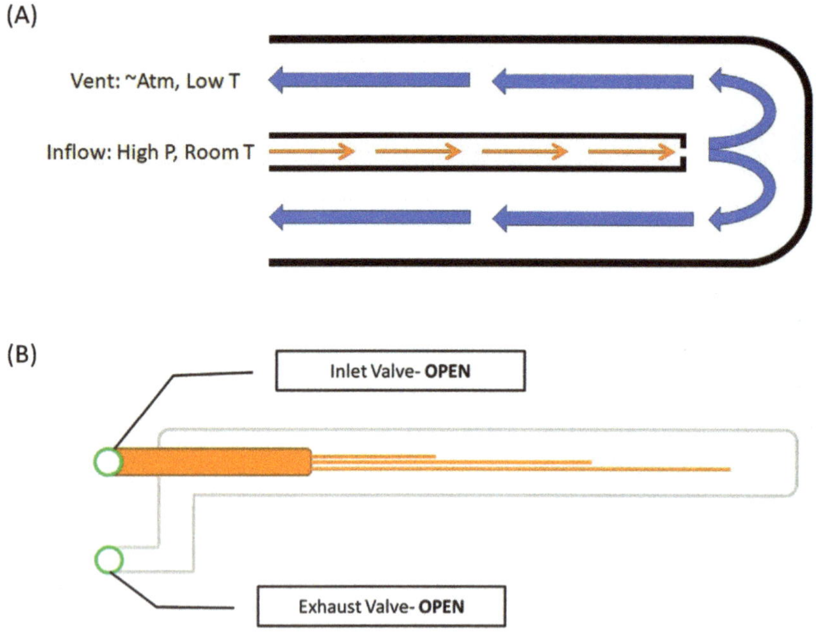

FIGURE 29.4 Schematic of Joule–Thomson probe (A) and example of the use of multiple Joule–Thomson orifices within a single probe (B) to produce more uniform cooling. *T*, Temperature; *P*, pressure; *Atm*, standard atmosphere.

and the resistance of the tubing, including the throttle size. Most systems include a pressure regulator that assures a consistent mass flow rate. The specific heat is an intrinsic property of the cryogen that affects the amount of heat that can be absorbed by a given mass of the cryogen. The higher the specific heat, the more heat the cryogen can extract for a given mass.

In addition to cooling the probe, these systems strive to maintain intimate contact between the tissue and the cryogen, accomplished by minimizing the resistance to heat conduction through the probe. Therefore, the thermal conduction through the probe is an important characteristic and is defined by the following equation:

$$Q = \frac{k \cdot \mathrm{A} \cdot \Delta\mathrm{T}}{t}$$

The transfer of heat through the probe walls (Q), which is directly proportional to the temperature difference (ΔT), is facilitated by the thermal conductivity, k, of the probe metal and the surface area (A) of the probe in contact with tissue and hindered by the thickness of the probe, t. This design constraint for minimizing thermal conduction must be balanced with the clinical needs of tissue contact, malleability, flexibility, force transfer, length of application, biocompatibility and hoop strength for high pressure conditions such as during defrost. The two metals currently used are aluminum and stainless steel. Aluminum offers a much higher thermal conductivity (222 W/m-K) than stainless steel (16 W/m-K) but stainless steel, due to its higher strength and resistance to fatigue, provides a more suitable material for the thin-walled corrugated designs designed for increased flexibility.

Commercially available surgical cardiac cryoablation systems use either nitrous oxide or argon as the cryogen. Each cryogen offers advantages and disadvantages from the design standpoint. Nitrous oxide has a warmer boiling point (−89°C vs −185°C) but a higher specific heat capacity than argon (0.890 kJ/kg vs 0.541 kJ/kg). Nitrous oxide probes typically run at −60°C to −70°C and use gas pressurized to around 700–800 psi. Argon probes typically run at −160°C and use pressures around 3500–6000 psi.

Nitrous oxide surgical cryoablation system

The nitrous oxide (N_2O) system available for surgical cardiac cryoablation (AtriCure Cryosurgical System, Fig. 29.5) consists of a generator (AtriCure CryoModule, ACM), and 3 single-use cryoprobes (CryoICE Ablation probes CRYO2 and CRYO3 and CryoICE CryoFORM). The generator delivers nitrous oxide to the handpieces and consists of a compressed N_2O gas cylinder, an N_2O gas line hose, N_2O exhaust hose, and a footswitch to control the ablation time. A gas cylinder heater band increases the temperature and therefore the pressure of the N_2O gas to be delivered to the handpieces. Standard ablation times are 120 s after the probe has initially cooled to −40°C, although the surgeon can adjust the duration time.

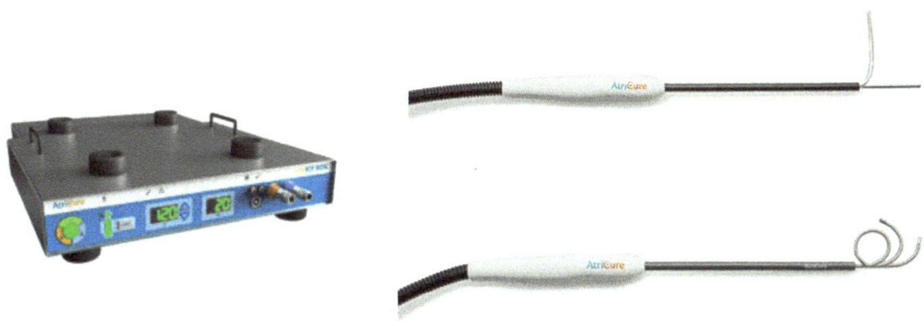

FIGURE 29.5 Surgical nitrous oxide cryoablation system with the generator, an aluminum shell handpiece in straight and bent configuration, and a corrugated stainless-steel shell in various bent configurations.

Each cryoprobe handpiece consists of a cryotip end effector, shaft, handle, thermocouple, inlet tube and exhaust tube. Two of the handpieces, the CRYO2 and CRYO3, incorporate a 10 cm smooth aluminum probe, 4 mm in diameter, with three internal Joule–Thomson orifices to provide uniform cooling. Both the CRYO2 and CRYO3 include a tool that allows the surgeon to form the shape of the probe to match the desired lesion shape, and the CRYO3 uses a different grade of aluminum with slightly lower yield strength than CRYO2 to facilitate forming of the probe. The CryoFORM probe is a corrugated stainless-steel probe that provides substantial flexibility for minimally invasive procedures and robotic manipulation of the probe. Each handpiece includes a rigid shaft with an extendable insulator that allows the surgeon to vary the exposed portion of the probe from 1 to 10 cm to control lesion length while protecting surrounding structures from freezing. All probes include a thermocouple attached to the proximal external surface of the probe to indicate the real-time temperatures at the warmest probe-tissue interface point.

The nitrous oxide system offers a unique defrost feature. By closing the exhaust line upon completion of the timed ablation, pressure in the probe builds and temperature rises rapidly, allowing the probe to defrost and disengage the tissue. This feature not only maximizes thaw time but also improves ease of use as the surgeons are able to move from lesion to lesion more quickly.

Argon surgical cryoablation system

The argon system available for surgical cardiac cryoablation (Medtronic Cardioblate CryoFlex system) consists of a console and 3 single use disposable cryoprobes (CardioFlex 7 cm and 10 cm, CardioFlex 10-S 10 cm). The console delivers argon gas to the cryoprobes and consists of a control panel, a cylinder of compressed argon, a cylinder holder, and a pressure regulator. All the cryoprobes incorporate a corrugated stainless-steel shell to provide probe flexibility. The CardioFlex 10-S uses a higher strength stainless steel in its shell compared to the CardioFlex 10 to increase rigidity of the probe and improve tissue contact. Additionally, the probes can be integrated into a clamp device that helps to improve contact force. Because there is no defrost capability with the argon system, surgeons may apply saline to expedite disconnecting the probe from the tissue. Probes include a retractable insulation sheath to allow the surgeon to expose the desired length of probe for each lesion. Each probe has a thermocouple located near the gas expansion nozzle where temperatures are coldest.

Although the systems have different engineering solutions, both technologies have demonstrated clinical success treating AF, especially for endocardial ablation during concomitant procedures, with rates of 1-year conversion to sinus rhythm equal to 60%–84% in AF patients.[28] A retrospective study by Goette et al.[29] directly compared the use of argon and nitrous oxide in a series of 120 patients with AF undergoing minimally invasive mitral valve surgery with concomitant AF ablation. In this cohort, freedom from AF was higher in the patients treated with the nitrous oxide system than the argon system (87% vs 71%). Despite their success, these systems still have limitations which impact their broader use, particularly for epicardial beating heart ablations. While some preclinical evidence shows potential for cryoablation systems for epicardial beating heart applications,[21,30,31] this practice has not gained widespread clinical use due to concerns about transmurality with the convective heat load of the endocardial blood volume that the cryoablation systems must overcome. Cryoablations also take longer than most RF ablations, with standard applications requiring at least 2 min per lesions. Also, because cryoablation relies solely on conductive heating from one side of the tissue, the depth of effect may be too limited for use in treating ventricular arrhythmias, where the ventricular walls are much thicker than atrial walls.

Future directions

Cardiac cryoablation has been successfully used for over 30 years and has been established as a mainstay for complementary therapy for surgical AF ablation. However, there remain several areas of procedural and technological advancement for this therapy which aim to expand its use and improve its benefit to patients.

Because of cryoablation's benefit for direct visualization of impacted tissue and safe treatment relative to connective tissues, a prospective study is underway to explore the impact of using only cryoablation for the entire Cox MAZE lesion set [clinicaltrials.gov: NCT03732794]. Due to the ability to shape the cryoprobes, multiple lesions from the CMIV lesion set can be accomplished with single ablations, thereby reducing the time required for cryoablation only procedures. Retrospective analyses of procedures performed with cryoablation only compared to the conventional combination of radiofrequency and cryoablation lesions reported similar survival outcomes with improved freedom from AF and reduced stroke risk with cryoablation only procedures.

Additionally, cryoablation is being considered for treatments other than AF. Surgical cardiac cryoablation has been used for ventricular tachycardia[32,33] especially during left ventricular assist device implantation procedures,[34] and its use for conditions like inappropriate sinus tachycardia[35,36] may still be under consideration. Beyond cardiac arrhythmia, there has been renewed interest for using cryoablation to treat nerves for postthoracotomy pain relief to reduce the need for opioids.[37]

There are also ongoing efforts to introduce more potent cryogens that provide greater cooling power at lower temperatures than either nitrous oxide or argon. Both near-critical nitrogen[38] and supercritical nitrogen[27] technologies have shown promise for eliminating the need for high-pressure cryogen tanks while delivering lower temperatures and greater specific heat capacities. These technologies may be able to address beating heart epicardial cryoablation and/or reduce the time for each cryolesion performed but are still in the experimental phase.

Last, other forms of ablation threaten to replace cryoablation for cardiac arrhythmia treatment. In particular, pulsed field ablation, an electrical ablation technique that uses high intensity, short duration bursts of direct current voltage to irreversibly electroporate cell membranes, offers the same benefit of preserving connective tissue with a much shorter duration than cryoablation requires. While this technique is still in early human feasibility evaluation, if proven effective, it could alter the use of cryoablation for cardiac arrhythmia.

References

1. Lustgarten DL. *History of cardiac cryosurgery and cryoablation. Cryoablation of Cardiac Arrhythmias*. Elsevier; 2011:3–12.
2. Cox JL, Malaisrie SC, Churyla A, et al. Cryosurgery for atrial fibrillation: physiologic basis for creating optimal cryolesions. *Ann Thorac Surg*. 2020;112(2):P354–P362. Available from: https://doi.org/10.1016/j.athoracsur.2020.08.114. Published online December 3.
3. Cox JL, Schuessler RB, D'Agostino HJ, et al. The surgical treatment of atrial fibrillation. III. Development of a definitive surgical procedure. *J Thorac Cardiovasc Surg*. 1991;101(4):569–583.
4. Sundt TM, Camillo CJ, Cox JL. The maze procedure for cure of atrial fibrillation. *Cardiol Clin*. 1997;15(4):739–748. Available from: https://doi.org/10.1016/s0733-8651(05)70372-6.
5. Badhwar V, Rankin JS, Damiano RJ, et al. The society of thoracic surgeons 2017 clinical practice guidelines for the surgical treatment of atrial fibrillation. *Ann Thorac Surg*. 2017;103(1):329–341. Available from: https://doi.org/10.1016/j.athoracsur.2016.10.076.
6. Calkins H, Hindricks G, Cappato R, et al. HRS/EHRA/ECAS/APHRS/SOLAECE expert consensus statement on catheter and surgical ablation of atrial fibrillation. *Heart Rhythm*. 2017;14(10):e275–e444. Available from: https://doi.org/10.1016/j.hrthm.2017.05.012.
7. Ad N, Damiano RJ, Badhwar V, et al. Expert consensus guidelines: examining surgical ablation for atrial fibrillation. *J Thorac Cardiovasc Surg*. 2017;153(6):1330–1354. Available from: https://doi.org/10.1016/j.jtcvs.2017.02.027. e1.
8. Philpott J, Zemlin C, Damiano RJ. *Surgical Treatment of Atrial Fibrillation: A Comprehensive Guide to Performing the Cox Maze IV Procedure*. Academic Press; 2017.
9. Lawrance CP, Henn MC, Damiano RJ. Surgical ablation for atrial fibrillation: techniques, indications, and results. *Curr Opin Cardiol*. 2015;30(1):58–64. Available from: https://doi.org/10.1097/HCO.0000000000000125.
10. Mack CA, Milla F, Ko W, et al. Surgical treatment of atrial fibrillation using argon-based cryoablation during concomitant cardiac procedures. *Circulation*. 2005;112(9 Suppl.):I1–I6. Available from: https://doi.org/10.1161/CIRCULATIONAHA.104.524363.
11. Schuessler RB, Lee AM, Melby SJ, et al. Animal studies of epicardial atrial ablation. *Heart Rhythm*. 2009;6(12 Suppl.):S41–S45. Available from: https://doi.org/10.1016/j.hrthm.2009.07.028.
12. Xu J, Luc JGY, Phan K. Atrial fibrillation: review of current treatment strategies. *J Thorac Dis*. 2016;8(9):E886–E900. Available from: https://doi.org/10.21037/jtd.2016.09.13.
13. Gage AA, Baust J. Mechanisms of tissue injury in cryosurgery. *Cryobiology*. 1998;37(3):171–186. Available from: https://doi.org/10.1006/cryo.1998.2115.
14. Snyder KK, Baust JG, Baust JM, Gage AA. *Mechanisms of cryoablation. Cryoablation of Cardiac Arrhythmias*. Elsevier; 2011:13–21.

15. Markovitz LJ, Frame LH, Josephson ME, Hargrove WC. Cardiac cryolesions: factors affecting their size and a means of monitoring their formation. *Ann Thorac Surg*. 1988;46(5):531–535.
16. Snyder K, Baust J, Cox J, Buskirk R, Baust J. An in vitro investigation of cardiomyocyte response to cryoablation. *Cryobiology*. 2010;61:363–364. Available from: https://doi.org/10.1016/j.cryobiol.2010.10.012.
17. Goff RP, Quallich SG, Buechler RA, Bischof JC, Iaizzo PA. Determination of cryothermal injury thresholds in tissues impacted by cardiac cryoablation. *Cryobiology*. 2017;75:125–133. Available from: https://doi.org/10.1016/j.cryobiol.2017.01.002.
18. Wood MA, Parvez B, Ellenbogen AL, et al. Determinants of lesion sizes and tissue temperatures during catheter cryoablation. *Pacing Clin Electrophysiol*. 2007;30(5):644–654. Available from: https://doi.org/10.1111/j.1540-8159.2007.00726.x.
19. Lustgarten DL, Bell S, Hardin N, Calame J, Spector PS. Safety and efficacy of epicardial cryoablation in a canine model. *Heart Rhythm*. 2005;2 (1):82–90. Available from: https://doi.org/10.1016/j.hrthm.2004.09.022.
20. Melby SJ, Schuessler RB, Damiano RJ. Ablation technology for the surgical treatment of atrial fibrillation. *ASAIO J*. 2013;59(5):461–468. Available from: https://doi.org/10.1097/MAT.0b013e3182a394f1.
21. Masroor S, Jahnke M-E, Carlisle A, et al. Endocardial hypothermia and pulmonary vein isolation with epicardial cryoablation in a porcine beating-heart model. *J Thorac Cardiovasc Surg*. 2008;135(6):1327–1333. Available from: https://doi.org/10.1016/j.jtcvs.2007.12.038.
22. Avitall B, Kalinski A. Cardiac ablative technologies. In: Iaizzo PA, ed. *Handbook of Cardiac Anatomy, Physiology, and Devices*. Springer International Publishing; 2015:521–542. Available from: http://doi.org/10.1007/978-3-319-19464-6_29.
23. Weimar T, Lee AM, Ray S, Schuessler RB, Damiano RJ. Evaluation of a novel cryoablation system: in vivo testing in a chronic porcine model. *Innovations (Phila)*. 2012;7(6):410–416. Available from: https://doi.org/10.1097/IMI.0b013e31828534e5.
24. Weimar T, Lee AM, Ray S, Schuessler RB, Damiano RJ. Evaluation of a novel cryoablation system: in vitro testing of heat capacity and freezing temperatures. *Innovations (Phila)*. 2012;7(6):403–409. Available from: https://doi.org/10.1097/IMI.0b013e3182853e74.
25. Khairy P, Chauvet P, Lehmann J, et al. Lower incidence of thrombus formation with cryoenergy versus radiofrequency catheter ablation. *Circulation*. 2003;107(15):2045–2050. Available from: https://doi.org/10.1161/01.CIR.0000058706.82623.A1.
26. Littrup PJ, Babkin A. Evolving concepts: near-critical cooling-based technologies. *Cryoablation Card Arrhythm*. 2011;107–115. Available from: https://doi.org/10.1016/B978-1-4377-1615-3.00009-5. Published online January 1.
27. Baust JM, Robilotto A, Guerra P, et al. Assessment of a novel cryoablation device for the endovascular treatment of cardiac tachyarrhythmias. *SAGE Open Med*. 2018;6. Available from: https://doi.org/10.1177/2050312118769797.
28. Camm CF, Nagendran M, Xiu PY, Maruthappu M. How effective is cryoablation for atrial fibrillation during concomitant cardiac surgery? *Interact Cardiovasc Thorac Surg*. 2011;13(4):410–414. Available from: https://doi.org/10.1510/icvts.2011.271676.
29. Goette J, Weimar T, Vosseler M, et al. Freezing equals freezing? Performance of two cryoablation devices in concomitant mitral valve repair. *Thorac Cardiovasc Surg*. 2016;64(8):672–678. Available from: https://doi.org/10.1055/s-0035-1563537.
30. Doll N, Kornherr P, Aupperle H, et al. Epicardial treatment of atrial fibrillation using cryoablation in an acute off-pump sheep model. *Thorac Cardiovasc Surg*. 2003;51(5):267–273. Available from: https://doi.org/10.1055/s-2003-43086.
31. Gallegos RP, Rivard AL, Rajab TK, et al. Transmural atrial fibrosis after epicardial and endocardial argon-powered CryoMaze ablation. *J Card Surg*. 2011;26(2):240–243. Available from: https://doi.org/10.1111/j.1540-8191.2011.01214.x.
32. Mulpuru SK, Feld GK, Madani M, Sawhney NS. A novel, minimally-invasive surgical approach for ablation of ventricular tachycardia originating near the proximal left anterior descending coronary artery. *Circ Arrhythm Electrophysiol*. 2012;5(5):e95–e97. Available from: https://doi.org/10.1161/CIRCEP.112.975284.
33. Berte B, Sacher F, Wielandts J-Y, et al. A new cryoenergy for ventricular tachycardia ablation: a proof-of-concept study. *Europace*. 2017;19 (8):1401–1407. Available from: https://doi.org/10.1093/europace/euw217.
34. Mulloy DP, Bhamidipati CM, Stone ML, et al. Cryoablation during left ventricular assist device implantation reduces postoperative ventricular tachyarrhythmias. *J Thorac Cardiovasc Surg*. 2013;145(5):1207–1213. Available from: https://doi.org/10.1016/j.jtcvs.2012.03.061.
35. Friedman PL, Bronson J, Macveigh TJ, Montgomery S. P4-111: Sinus node modification using cryoenergy. *Heart Rhythm*. 2006;3(5): S255–S256. Available from: https://doi.org/10.1016/j.hrthm.2006.02.766.
36. Roshan J, Mears JA, Lachman N, Christensen K, Ladewig DJ, Asirvatham SJ. Role of cryoablation for superior vena cava isolation and ablation of inappropriate sinus tachycardia. *Cryoablation Card Arrhythm*. 2011;191–203. Available from: https://doi.org/10.1016/B978-1-4377-1615-3.00017-4. Published online December 1.
37. Graves CE, Moyer J, Zobel MJ, et al. Intraoperative intercostal nerve cryoablation during the Nuss procedure reduces length of stay and opioid requirement: a randomized clinical trial. *J Pediatr Surg*. 2019;54(11):2250–2256. Available from: https://doi.org/10.1016/j.jpedsurg.2019.02.057.
38. Bourier F, Takigawa M, Lam A, et al. Ultralow temperature cryoablation: safety and efficacy of preclinical atrial and ventricular lesions. *J Cardiovasc Electrophysiol*. 2021;32(3):570–577. Available from: https://doi.org/10.1111/jce.14907. Published online January 21.

PART VI

Hemodynamic Patient Monitoring

Chapter 30

Basic Hemodynamic Parameters

Anna Corsini[1], Laura Cercenelli[2], Margherita Zecchi[3], Emanuela Marcelli[2] and Ivan Corazza[3]

[1]*Cardiology Unit, Department of Experimental Diagnostic and Specialty Medicine (DIMES), Alma Mater Studiorum—University of Bologna, Bologna, Italy,* [2]*eDIMES Lab, Laboratory of Bioengineering, Department of Experimental Diagnostic and Specialty Medicine (DIMES), Alma Mater Studiorum—University of Bologna, Bologna, Italy,* [3]*Medical Physics Coordination Center, Department of Experimental Diagnostic and Specialty Medicine (DIMES), Alma Mater Studiorum—University of Bologna, Bologna, Italy*

Chapter Outline

Introduction to the cardiovascular system

The main functions of the cardiovascular system are to carry oxygen, nutrients, and other essential molecules to peripheral tissues, and to extract carbon dioxide and waste metabolic products from peripheral tissues to organs (lungs, liver, kidneys), which contribute to their elimination. To provide these functions a system consisting of two circulating circuits in series, that is, the systemic and the pulmonary circulation, has evolved.

Systemic circulation consists of different vascular beds, with relatively high resistance, specialized in carrying the oxygen and nutrients to different tissues, and in extracting the carbon dioxide and waste metabolic products from peripheral tissues.

Pulmonary circulation is a vascular bed with low resistance, designed for bidirectional blood gas exchange with the outside, introducing the oxygen in the bloodstream and eliminating the carbon dioxide by diffusion in the alveoli.[1]

The blood flows sequentially through both systemic and pulmonary circulatory systems, being propelled by two pumps in series, which constitute the heart.

Each half of the heart consists of two chambers: (1) the atrium with a thin wall, which receives the blood from the venous system of its respective circulation, and provides an auxiliary pump function in propelling the blood into the ventricle through the atrioventricular valve during diastole; and (2) the ventricle with thick muscular wall, which pumps the blood in the arterial system of its respective circulation through the semilunar valve during systole (Fig. 30.1).

Each vascular bed (systemic and pulmonary) is composed of two systems: (1) the arterial system, characterized by high elastance (elastance $E = dP/dV$; P = pressure, V = volume), in which a little variation in volume is needed to produce a great variation in pressure; and (2) the venous system, characterized by high compliance (compliance $C = dV/dP = 1/E$), in which a great variation in volume is needed to produce a little variation in pressure.[2]

The cardiac contractility is the intrinsic ability of the cardiac muscle to contract, ejecting a stroke volume (SV) at a given afterload (arterial pressure) and preload (end-diastolic volume). The pump activity of both ventricles is considered normal when an adequate quantity of blood perfuses the respective vascular bed with normal heart filling pressures, that is, mean right and left atrial pressures (LAP), which correspond, in the absence of atrioventricular valve stenosis, to end-diastolic right and left ventricular pressures respectively (Fig. 30.2).

Advances in Cardiovascular Technology. DOI: https://doi.org/10.1016/B978-0-12-816861-5.00021-6

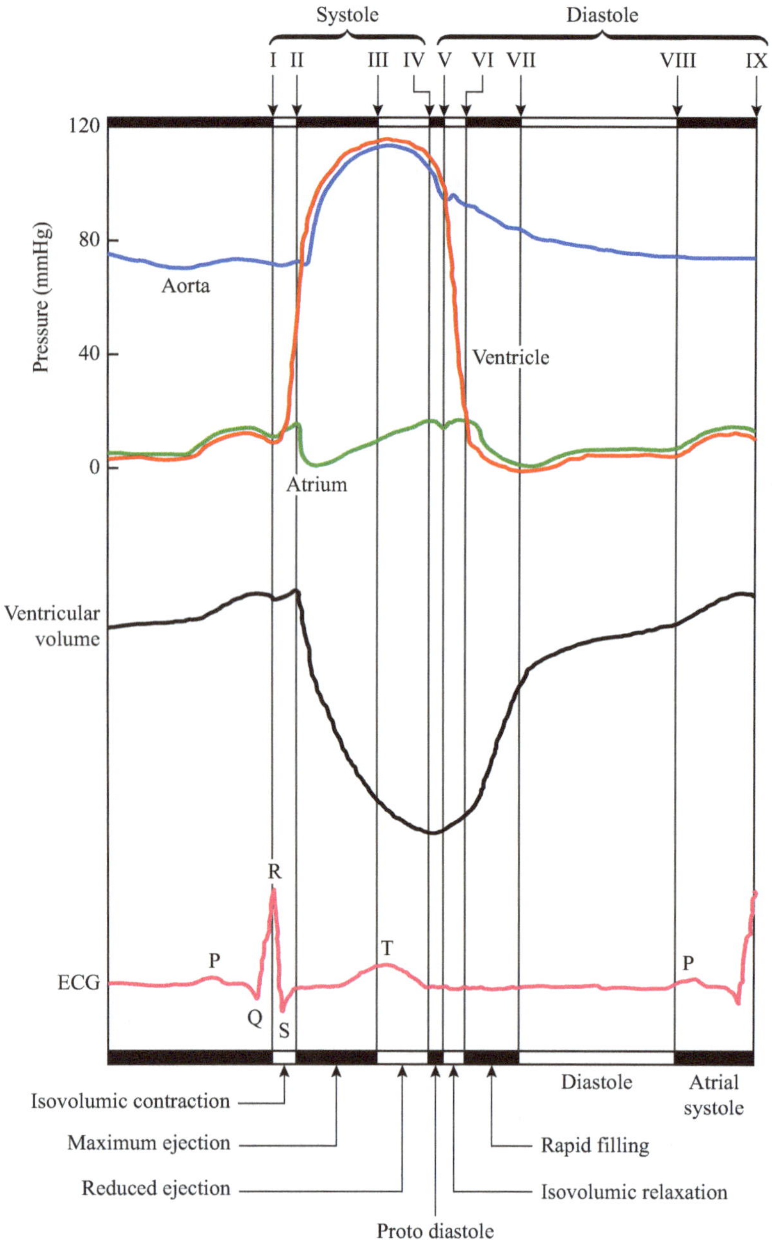

FIGURE 30.1 Arterial, ventricular, and atrial pressures according to electrical and mechanical contraction phases. I: T1, the first heart sound results from the closing of the mitral and the tricuspid valve (atrioventricular valves) and the beginning of the ventricular contraction and ejection phase (ventricular systole) during which the blood is propelled into the aorta (as shown in Fig. 30.1) and into the pulmonary artery; IV: T2, the second heart sound represents closure of the aortic and pulmonary valve (semilunar valves) and the beginning of the ventricular relaxation phase (ventricular diastole) during which the blood flows from the atria into the ventricles; VIII: atrial systole, the active part of the ventricular diastole, during which the blood is actively propelled into the ventricle by atrial contraction.

A correct evaluation of the hemodynamic performances of the cardiovascular system should consider the pump function (heart) and the system of transport, distribution, and collection of blood (arteries and veins), both for the systemic (left heart) and pulmonary (right heart) circulation. Because blood flow changes accordingly to the district (i.e., pulsatile, constant, or mixed), the precise description of hemodynamics is quite difficult. The variable and complex elastic behavior of the structures affects both the energy transfers and the pressure transmissions from the center to the periphery, and vice versa. Moreover, hemodynamic is related to blood's structure too, since an increase in viscosity can lead to a reduction in cardiac output (CO) and can affect the hemodynamic of all the cardiovascular system.[3] It is a complex discussion that would require an in-depth dissertation and that we will necessarily try to simplify in this chapter.

Left heart and systemic circulation

During the cardiac cycle, the pump (ventricle) actively modifies its elastance ($E = dP/dV$, where dP is the pressure increase caused by the volume increase dV), passing from very low values in the filling phase (diastole) to very high

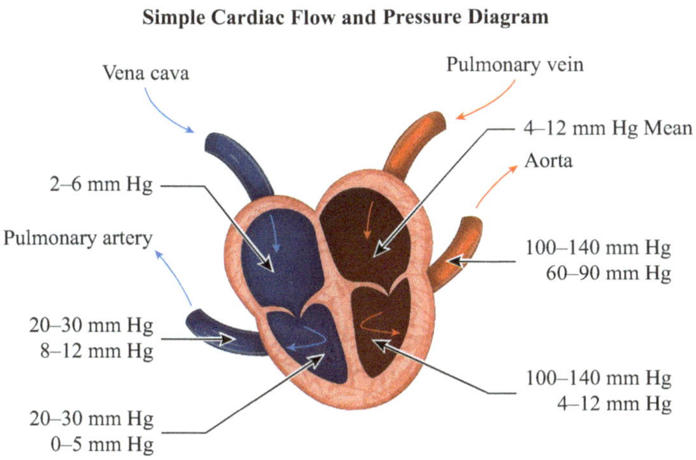

FIGURE 30.2 Normal pressure values in the aorta during systole (100–140 mmHg) and during diastole (60–90 mmHg); normal pressure values in the left ventricle during systole (100–140 mmHg) and during diastole (4–12 mmHg); normal mean pressure value in the left atrium-pulmonary veins (4–12 mmHg); normal pressure values in the pulmonary artery during systole (20–30 mmHg) and during diastole (8–12 mmHg); normal pressure values in the right ventricle during systole (20–30 mmHg) and during diastole (0–5 mmHg); normal mean pressure values in the right atrium–vena cava (2–6 mmHg).

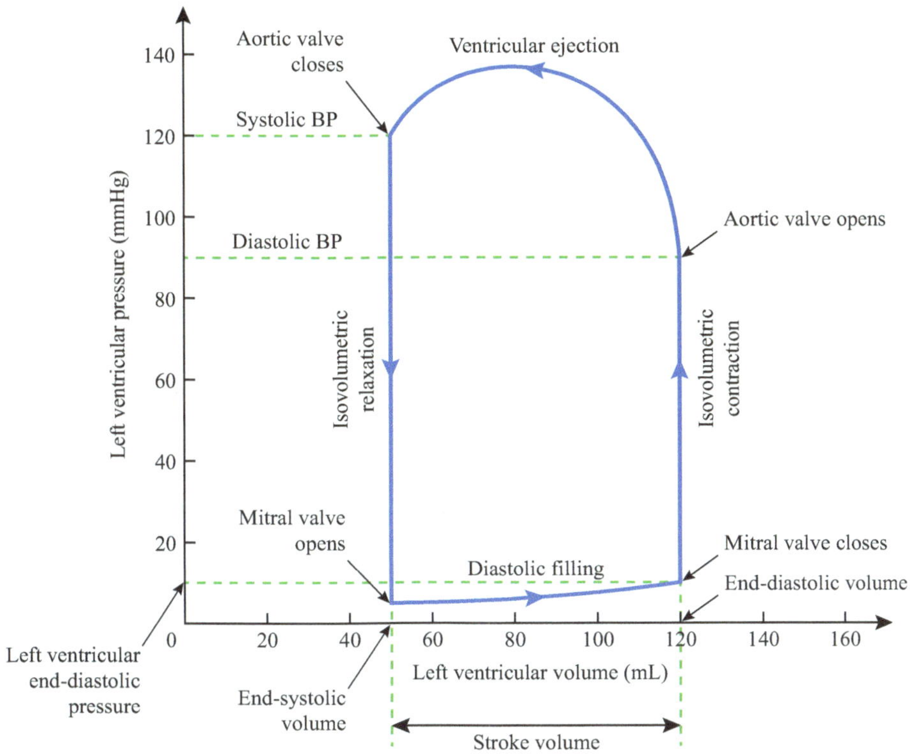

FIGURE 30.3 Cardiac pressure–volume loops (*PV*-loops).

values in the contraction phase (systole). This determines both a rapid increase in intracavitary pressure, and a partial emptying of the blood contained. When the ventricular pressure exceeds the pressure in the aorta (afterload), the aortic valve opens and the blood flows into the arterial vessel with a flow (mL/s) that follows the dynamics of the ventricular walls. Entering the ascending aorta, the blood flow stretches the elastic wall generating a pressure curve, equal in the ventricle and in the aorta, which depends on both the ventricular flow and the arterial compliance ($C = dV/dP$, the reciprocal of elastance).

The correct evaluation of the energetic behavior of the ventricles is commonly represented by the *PV*-loops,[4,5] which instantly reflect the relationship between ventricular pressure and volume during the cardiac cycle (Fig. 30.3).

The ventricular performance, expressed by the area of the *PV*-loop curve, depends on the chamber's filling (Starling's Law), which also involves atrial pressure and contraction (contributing to the preload).[6,7] The ventricle is then energetically coupled afterwards with the aorta, within which the blood circulates and moves toward the periphery, and upstream with the atrium which modulates its filling, and consequently the contractile efficiency, beat by beat. Arteries also have a variable elastance, which affects their dilation and internal pressure in the filling phase (systole), as well as their narrowing and maintenance of an adequate pressure during diastole. Moreover, the arterial elastance also affects the way by which arterial pressure deforms arterial walls, generating a pulse wave (PW), which travels on arterial elastic walls with a certain speed called pulse wave velocity (PWV). Whereas the blood flows in the aorta in a pulsatile way and with a speed of less than 1 m/s, the PW transmitted along the elastic arterial walls has a speed (PWV) between 8 and 20 m/s, depending on arterial elastance and stiffness. The blood speed depends on the CO, the PWV depends on the elastic properties of the vascular wall. The PW propagation to the periphery, meeting constrictions and bifurcations, generates phenomena of reflection from the periphery to the center, and consequently part of the PW energy is reflected back toward the ventricle. This reflex pulse wave can have an amplitude of up to 40% of the incident one. It returns to the aorta and it is added to the incident pulse wave, carrying positive or negative effects. If the elastance is low, the PWV is low; thus the reflex pulse reaches the aorta at the end of the ventricular systole and helps to keep the pressure high in the diastolic phase, favoring coronary perfusion. On the other hand, if the pulse travels quickly (high elastance), the reflex pulse returns to the ascending aorta at the beginning of the ventricular systole adding to the incident PW, thus increasing afterload and ventricular contractile tension, which is not advantageous in terms of energy efficiency. In fact, since the left ventricle and the aorta are both elastic structures, they should have instantly a quite similar elastance, this being the condition for optimal energy transfer. There is no doubt that the ventricular elastance and the arterial elastance, which vary continuously during the cardiac cycle, are closely related to the hemodynamic phenomena, and represent two parameters of paramount importance to characterize the cardiocirculatory system from a mechanical point of view. When, for physiological reasons (e.g., age), or pathological events (e.g., arterial hypertension, arteriosclerosis, coronary artery disease, heart failure), or both, the left ventricle or the aorta modify their elastic characteristics, the energetic coupling also changes, and consequently the overall performance of the cardiovascular system.

As the arterial tree branches and thins, the total section of the arterioles and mostly of the capillaries increases, allowing the blood distribution to all tissues since, with the same CO, a reduction in local speed allows the gas exchange between blood and tissues at the capillary level.

Despite the great changes that can occur in local blood perfusion, arterial pressure is kept constant both at central and peripheral levels. This is possible thanks to the regulation of arterial caliber: in the medium-size arteries (1–10 mm) the wall stress is maintained stable in conditions of different flow; moreover, vascular resistance and tissue perfusion are modified by regulatory mechanisms involving precapillary arterioles.

Once the oxygen and nutrients exchange has happened, the venous system returns the blood to the heart. At the venous level there is no pump that pushes the blood from the periphery to the center, and the blood motion is obtained through a system of valves. When the venous sector between two successive valves is full, the venous wall contracts, generating enough pressure to open the following valve, and to move blood to the following sector. The venous vessels are characterized by a large caliber, consequently the hydraulic resistances are minimal, and very small pressure differences are sufficient to determine the blood return from the periphery toward the heart. Due to this particular behavior, we describe the venous system as an apparatus with very high compliance, where blood flow is not driven by variations of pressure or hydraulic resistance, but by variations of vascular volume. We no longer speak of elastance or hydraulic resistance, but of venous tone, which indicates how the blood volume of a venous district changes along with the pressure at that level. The venous tone is strongly influenced by the autonomic nervous system, and a small change in sympathetic or vagal driving is enough to determine large variations in volume. This affects the blood return to the center, thus influencing, through the Starling mechanism, the ventricular contraction. Another mechanism that could increase the blood return to the center through the "muscular pump" is physical exercise. If these mechanisms were not so efficient, it would be difficult to explain how the heart can greatly increase the CO in a very short period of time.

The elastic and structural characteristics of the vessels are also closely related to the way by which the blood flows inside them. Since the blood is an incompressible and viscous fluid, depending on the size of the vessel in which it moves, its velocity profile changes: from an almost flat profile, with almost equal speeds in the center and at the edges, in large diameter and short vessels (for example abdominal tract of the aorta) to an accentuated parabolic profile, with much greater speed in the center than at the edges, in small caliber's vessels. In any case, under normal conditions, the flow is laminar and the fluid lines, while traveling at different speeds, move parallelly. Alterations in the vessel structure, both physiological and pathological, can lead to alteration of the velocity profiles, exceeding critical thresholds and triggering turbulent flows, which originate the "vascular murmurs," which can also be perceived from the outside. A murmur is always an expression of abnormality, and reflects chaotic alterations of the wall shear stress (WSS), which could determine paradoxical effects (e.g., further structural alterations, such as poststenotic aneurysms).

Right heart and pulmonary circulation

While the left side of the heart is responsible for peripheral perfusion, linked to muscular, visceral, and cerebral activity, the right side of the heart, which is functionally very similar to the left side and in series with it, is aimed at bidirectional blood gas exchanges in the lungs. Here two systems overlap, the vascular one, which controls blood circulation in the lungs, and the respiratory one, which manages the supply of oxygen and the removal of carbon dioxide at the alveolar level. It is clear that problems can arise due to the malfunction of one of the two systems or both. We will limit our analysis to the vascular system, but we cannot forget that in any case the result is "global," and that respiratory problems can nullify vascular interventions.

Pulmonary and systemic circulation have different hydraulic resistance which leads, with the same CO, to much lower right pressure regime compared to the left one: the right pressures are about one fifth of the left ones. What is of interest from the hemodynamic point of view are two important aspects, which could be responsible of an increase in pulmonary arterial pressure (PAP): (1) a "primitive" increase in pulmonary arterial resistance (>3 Wood units), which determines a direct increase of arterial pressure and consequently in right ventricular pressure ("primitive" pulmonary hypertension); (2) a "reactive" increase in PAP, which can lead to an increase in pulmonary arterial resistance, due to an increase in left atrial pressure reflecting a left-side heart dysfunction ("reactive" pulmonary hypertension). In both cases the effect on the right ventricle is the same: it is forced to generate a pressure much higher than normal to ensure the pulmonary perfusion; this generates right ventricular hypertrophy first, than results in right ventricular dilation and dysfunction.[8] To differentiate precapillary ("primitive") to postcapillary ("reactive") pulmonary hypertension it is sufficient to measure the left atrial pressure, and since this cannot be done directly unless by crossing the interatrial septum or by left heart catheterization, a catheter can be inserted via the pulmonary artery until it is wedged into a peripheral artery to block its flow. In this condition of blocked blood flow there is no longer the pressure drop of the pulmonary arterioles, and the catheter will register, with a little time delay, the left atrial pressure. If the left atrial pressure is within normal ranges, pulmonary hypertension has a "primitive" origin (for example pulmonary arterial hypertension, pulmonary hypertension due to lung disease and/or hypoxia, chronic thromboembolic pulmonary hypertension or other pulmonary artery obstructions); if, on the other hand, left atrial pressure is increased, pulmonary hypertension is "reactive," secondary to left heart disease, and could be resolved by intervening on the left side. Long-standing "reactive" pulmonary hypertension can lead to an increase in pulmonary vascular resistance (>3 Wood units), a condition named "combined postcapillary and precapillary pulmonary hypertension."[9]

It is clear that the cardiovascular system is complex and interconnected, and small deviations from normality in a specific sector can cause alterations to the entire system by modifying pressures and flows. The correct evaluation of hemodynamics, therefore should not be separated from an evaluation of the biomechanics of the system. For a correct functional characterization, it would be important to evaluate not only the cardiac, arterial, and venous pressures and flows, but also the biomechanical behavior of the structures involved, as well as all those quantities that directly or indirectly affect the way the blood flows inside the vessels. For example:

1. the heart rate (HR) and how it changes to modify the CO in a compensatory way, sometimes leading to arrhythmias that introduce alterations in cardiac contractility;
2. the CO and district perfusion changes;
3. the oxygen consumption, therefore the ability that our cardiovascular system has to exchange gas with the tissues, both upstream (lungs) and downstream (capillaries); and
4. the endovascular shear stress.[10]

Hemodynamic parameters of the systemic circulation

Systemic arterial pressure

The arterial pressure, which is defined as the force that is exerted by the blood on the arterial wall, depends on CO and peripheral vascular resistance, and produces organ perfusion. The left ventricular systolic pressure, in the absence of obstruction to the outflow (i.e., aortic stenosis), corresponds to the central aortic systolic pressure. Central aortic pressure (CAP) and brachial/peripheral arterial pressure are different, due to wave reflection: the systolic blood pressure (SBP) increases in the peripheral arteries (femoral, radial) as a result of the PW that is reflected from arterial bifurcation points and small peripheral vessels. Moreover, the arterial waveform base narrows in the peripheral arteries, and the diastolic blood pressure (DBP) decreases 1–2 mmHg toward the peripheral arteries (Fig. 30.4). Consequently, the pulse pressure (PP), which is the difference between systolic and diastolic arterial blood pressure, increases, while the mean

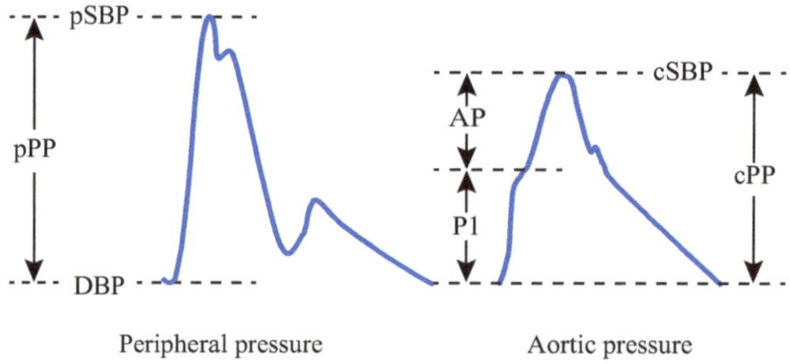

FIGURE 30.4 Peripheral (radial) and central (aortic) pressure waveforms. P1 is determined by interaction of left ventricular ejection with the impedance of the circulation and is related to aortic stiffness. AP is thought to be determined by pressure wave reflection from the periphery of the circulation and/or by the functional compliance of the aorta. AP, augmentation pressure; cPP, central pulse pressure; cSBP, central systolic blood pressure; DBP, diastolic blood pressure; P1, pressure of the incident wave; pPP, peripheral pulse pressure; pSBP, peripheral systolic blood pressure.[11]

arterial pressure is unchanged.[11] The mean arterial pressure may be approximated by (2 DBP + SBP)/3. This formula is applicable at lower HR ($\sim$60 bpm), and becomes less accurate with tachycardia, wherein the systole occupies a larger part of the cardiac cycle, therefore SBP becomes a larger contributor to mean arterial pressure.

Normal values for SBP are 100–140 mmHg; normal values for DPB are 60–90 mmHg; and normal values for mean arterial pressure are 70–105 mmHg.

Systolic, diastolic, and pulse aortic pressures are affected by many factors, as reported below:

1. Systolic blood pressure (SBP) is influenced by: (1) SV; (2) arterial compliance (inverse correlation); (3) number of peripheral waves that are reflected back to the aorta in systole → high SV and low arterial compliance increase SBP;
2. Diastolic blood pressure (DBP) is influenced by: (1) peripheral resistance; (2) arterial compliance; (3) HR → low peripheral resistance, low arterial compliance and bradycardia (low HR) decrease DBP;
3. PP is influenced by: (1) SV; (2) arterial compliance; (3) aortic insufficiency → low SV decreases SBP and increases DBP (vasoconstriction), thus decreasing PP, along with a narrow base of the CAP; low arterial compliance, as well as aortic insufficiency (AI), increases SBP and decreases DBP, thus raising PP.

Several tracing patterns of CAP are suggestive of disease (Fig. 30.5)[12]:

1. Sharp downslope of the DBP with a widened PP, larger than half of SBP or larger than 60 mmHg, suggests AI. In addition, an attenuated or absent dicrotic notch suggests AI. In AI, the PP is wide and DBP is low, and a bifid pulse may be seen (pulsus bisferiens). The first peak results from LV ejection, whereas the second peak arises from the periphery, as a reflected wave (the large SV in chronic AI explains a large systolic pressure peak leading to more peripheral reflection).
2. Late peaking suggests aortic stenosis (AS). In critical AS, a prominent anacrotic notch may be seen on the systolic upstroke; the aortic valve starts to open but stops "short," which impedes trans-aortic flow (pulsus parvus and tardus, or anacrotic pulse). The aortic upstroke starts normally, then is sharply impeded after the valve stops opening, creating this bend called anacrotic notch. In addition, the PP may be reduced because of reduced SV. In elderly patients with reduced arterial compliance, the PP may not be reduced, and the anacrotic notch may be absent.
3. Double peaking in systole suggests AI (pulsus bisferiens) or *hypertrophic obstructive cardiomyopathy* (HOCM), with "spike and dome" appearance in case of HOCM in which the late dynamic obstruction explains the late "doming" of SBP.
4. The loss of dicrotic notch suggests severe AI, poor arterial compliance (elderly, hypertensive, diabetic patients or patients with diffuse peripheral atherosclerosis), or very low peripheral vascular resistance. The dicrotic notch corresponds to the rebound in the arterial pressure from continuous elastic recoil of the arterial tree, and the muscular tone of small vessels. The systemic arteries make an "effort" to maintain the systemic pressure after the ventricle relaxes. This effort is absent if systemic arteries are not compliant or if the peripheral vascular resistance is very low, and it fails in case of a wide-open AI (in those patients, the aortic pressure continues to drop in diastole without a dicrotic notch). Conversely, the dicrotic notch is accentuated in case of vasoconstriction or good arterial compliance.
5. Gradual reduction of the SBP by more than 10 mmHg during normal quiet inspiration suggests pulsus paradoxus, which is seen in cardiac tamponade. Furthermore, in tamponade, there is not only fluctuation of pressure, but also a narrow PP and a narrow-base aortic tracing secondary to the low CO.

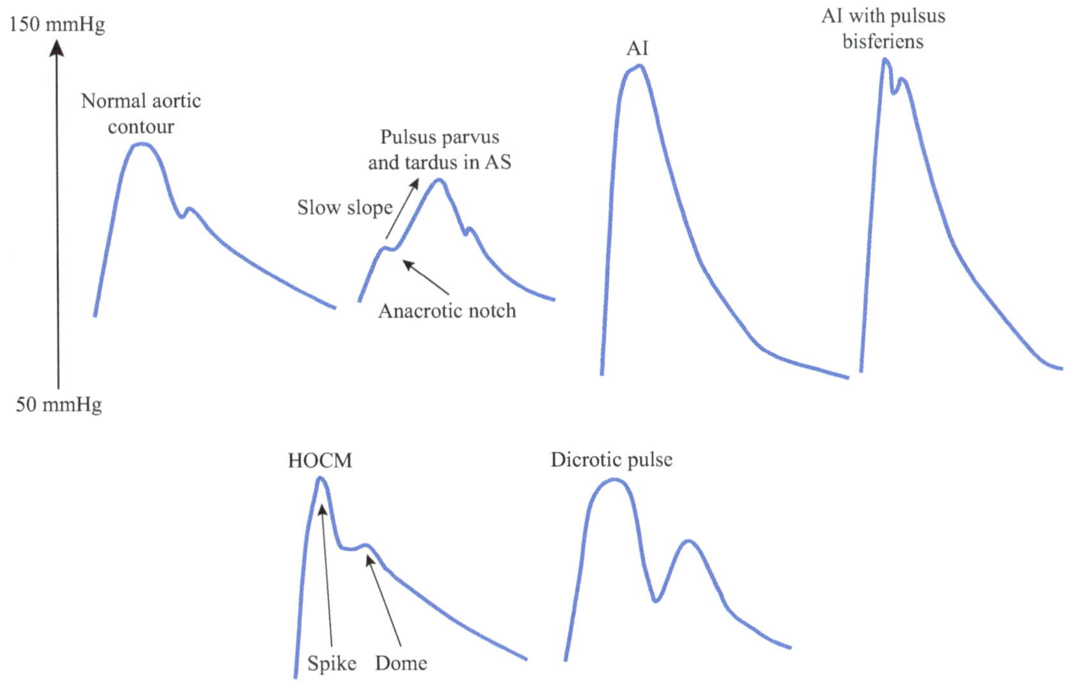

FIGURE 30.5 Aortic pressure contours in various disease states. AI, aortic insufficiency; AS, aortic stenosis; HOCM, hypertrophic obstructive cardiomyopathy.

6. Narrow PP (<30% of systolic arterial pressure) with a narrow-base arterial pressure, plateaued diastole, and prominent dicrotic wave (dicrotic pulse) strongly indicate low SV, and may be seen in heart failure (HF) or cardiac tamponade, as well as in hypovolemia. The dicrotic pulse is characterized by two peaks, one in systole, and one in diastole (differently from pulsus bisferiens, where two aortic peaks occur in systole). It is seen in cases of low SV, such as in left and/or right HF, cardiac tamponade, hypovolemia, where vasoconstriction raises the diastolic pressure and accentuates the dicrotic wave.
7. Every-other-beat alternation in SBP suggests pulsus alternans, which reflects a severe reduction in CO, as in severe left ventricular dysfunction or severe AS, and carries a poor prognosis. Pulsus alternans is often initiated by a premature beat. The strong beat leaves a small residual end-systolic volume that reduces inotropism of the next beat. The end-systolic volume is increased after the weak beat (because of decreased ejection), which leads to more force development in the next beat. In addition to the arterial pressure tracing (CAP or peripheral arterial pressure tracing), pulsus paradoxus and pulsus alternans are seen on the ventricular pressure tracing, and may be seen on the right or left side.

Peripheral systolic amplification can be as high as 20–50 mmHg, and is particularly marked in case of AI, poorly compliant arteries (elderly), peripheral arterial diseases, aortoiliac or femoral stenoses, tachycardia, or hyperdynamic circulation. To note, after the age of 60 years, the reflected waves not only amplify the peripheral arterial pressure, but also return to the central aorta quickly as a result of heightened arterial stiffness which produces increased PWV. Thus these waves arrive to the aortic root in late systole, which leads to augmentation of the CAP in conjunction with the peripheral pressure. This explains why the systolic aortic pressure is closer to the peripheral systolic pressure in the elderly than in the young patient. In fact, in the elderly patient, the systolic and pulse aortic pressure are increased because of the ejection of SV in a stiff aorta and increased PWV that allows the reflected waves to return to the aorta in systole. In contrast, in the young patient, the reflected waves return to the central aorta in early diastole, which increases the diastolic pressure and potentially the coronary filling.

While CAP measurement is limited to left heart catheterization, peripheral arterial pressure monitoring is important in the care of any hospitalized patient. Frequent monitoring is important in critically ill patients and surgical patients with an increased risk of morbidity and mortality. This can be achieved through intermittent monitoring, which is noninvasive, but only provides snapshots in time, or through continuous invasive monitoring. The most common way to do this is the arterial pressure monitoring via the cannulation of a peripheral artery. The most common site of arterial cannulation is the radial artery due to ease of accessibility. Other sites are the brachial, femoral, and dorsalis pedis artery.

For the following patient care scenarios, an arterial line would be indicated[13]:

1. Critically ill patients in the Intensive Care Unit (ICU), who require close monitoring of hemodynamic status, especially those treated with vasoactive medication or patients for whom blood pressure control is of paramount importance (i.e., patients affected by acute aortic syndromes).
2. Surgical patients who are at increased risk of morbidity or mortality, either because of preexisting comorbidities (cardiac, pulmonary, anemia, etc.) or because of more complicated procedures (these include but are not limited to neurosurgical procedures, cardiopulmonary procedures, and procedures in which a large volume of blood loss is anticipated).
3. Patients who need frequent laboratory tests. These include patients on prolonged mechanical ventilation, which require an analysis of arterial blood gas for titration of ventilator settings. Arterial line minimizes patient discomfort, as well as decreases the infection risk as the integrity of the skin does not need to be violated with each lab draw.

For certain patients, the cannulation of an artery is contraindicated. These include infection at the site of insertion, an anatomic variant in which collateral circulation is absent or compromised, presence of peripheral arterial vascular insufficiency, and peripheral arterial vascular diseases, such as small to medium vessel arteritis. Additionally, while not absolute contraindications, careful consideration should be made in patients who have coagulopathies or who are taking medications that prevent normal coagulation.

Pulmonary capillary wedge pressure and mean left atrial pressure

Pulmonary capillary wedge pressure (PCWP) corresponds to mean left atrial pressure (LAP) in absence of pulmonary vein obstruction. Mean LAP corresponds to end-diastolic left ventricular pressure in absence of mitral stenosis/altered mitral orifice and represents an estimation of left ventricular preload. In the clinical practice, monitoring PCWP is important to evaluate the left ventricular filling pressure, thus to quantify the severity of left ventricular failure (which determines PCWP increase according to its severity), to guide therapeutic approaches and quantify therapeutic efficacy. PCWP is also important to discriminate precapillary from postcapillary pulmonary hypertension: in the first case, only pulmonary artery systolic pressure is increased; in the second case, both pulmonary artery systolic pressure and PCWP are increased. Moreover, PCWP is also useful in evaluating blood volume status when fluids are administered during hypotensive shock. Normal values for PCWP/mean LAP are 4–12 mmHg.

Echocardiographic evaluation of transmitral flow pattern, combined with Tissue Doppler Imaging (TDI) and LA volume, are indirect elements for mean LAP estimation. PCWP, measured invasively by the Swan-Ganz catheter (i.e., pulmonary artery catheter-PAC), provides a precise estimate of mean LAP/end-diastolic LV pressure.

Swan-Ganz monitoring provides a comprehensive evaluation of patient's hemodynamic status. In particular, it provides information about:

1. Mixed venous oxygen saturation
2. Right atrial pressure
3. Right ventricular function
4. Pulmonary artery systolic pressure
5. Pulmonary artery diastolic pressure
6. PCWP
7. SV and CO
8. Systemic and pulmonary vascular resistance

The clinical impact of Swan-Ganz catheter use in critically ill patients has been investigated by several large randomized controlled trials comparing survival and clinical outcomes of critically ill patients treated with the use of a PAC, to those treated without the use of a PAC.[14–16] The results of these trials indicate no clear evidence of benefit or harm by managing critically ill patients with a PAC, resulting in declining in its use. Currently, recommendations state that invasive monitoring must be restricted for the most severe patients and complex clinical situations, such as patients with refractory shock with or without acute respiratory distress syndrome. In these settings, according to the most recent recommendations of the European Society of Intensive Care Medicine,[17] it has been suggested to use an invasive monitoring device, such as a Swan-Ganz catheter or transpulmonary thermodilution systems.

Hemodynamic parameters of the pulmonary circulation

Pulmonary arterial pressure

Right ventricular systolic pressure, in absence of obstruction to the outflow (i.e., pulmonary stenosis), corresponds to systolic PAP (PAPs). Monitoring of pulmonary arterial pressure (PAP) in the clinical practice gives information about the right ventricular function, and the presence of some degree of pulmonary hypertension (precapillary, postcapillary or both). Normal values for PAPs are 13–31 (mean 19) mmHg, for diastolic PAP (PAPd) are 6–15 (mean 10) mmHg, for mean PAP (PAPm) are 8–24 (mean 13) mmHg.

In presence of tricuspid regurgitation, PAPs can be estimated noninvasively by echocardiogram, that is, measuring the right ventricle–right atrium gradient from peak velocity of the tricuspid valve regurgitation, using the simplified Bernoulli equation and adding this value to an estimate of the mean right atrial pressure (RAP): $PAPs = 4\ (V\ TR)^2 +$ mean RAP. Similarly, the PAPd can be estimated from the end-diastolic pulmonary regurgitation velocity, and the PAPm from the early pulmonary regurgitation velocity using the simplified Bernoulli equation, always adding the mean RAP. The Mahan's equation is another practical method for estimating PAPm, by using the pulmonary artery acceleration time, measured by sampling the right ventricular outflow with the pulsed Doppler: $PAPm = 79 - [0.45 \times$ pulmonary artery acceleration time (Ms)$]$.[18] The PAPm can also be approximated by $(2\ PAPd + PAPs)\ /\ 3$.

PAP can also be measured invasively using the Swan–Ganz catheter, in the intensive care setting.

Central venous pressure and mean right atrial pressure

Central venous pressure (CVP), which is a measure of the pressure in the vena cava, corresponds to mean RAP in the absence of cava vein obstruction. Mean RAP corresponds to end-diastolic right ventricular pressure in absence of tricuspid stenosis/altered tricuspid orifice and can be used as an estimation of right ventricular preload. The CVP/mean RAP target in healthy subjects is between 0 to 8 (mean 5) mmHg. This value is altered by the total blood volume and/or compliance of the central venous compartment.[19]

In the clinical practice, low or elevated CVP values can be the spy of several different conditions:

1. Low CVP: hypovolemia or venodilation can decrease CVP by decreasing venous return; moreover, a decrement in intrathoracic pressure induced by forced inspiration causes the vena cava to collapse, thus shrinking the venous return and, in turn, decreasing the CVP.
2. Elevated CVP: elevated CVP can occur in Heart Failure (HF) due to decreased contractility, valve abnormalities, and arrhythmias. Any patient on ventilator assistance, who has excessive positive end-expiratory pressure, would have an upsurge in pulmonary arterial resistance, which causes an increment in CVP. However, a risen CVP, due to increased pulmonary arterial resistance, can also be affected by: a decrease in the fraction of inspired oxygen; an increase in ventilation/perfusion abnormalities in the lung due to acute pulmonary emboli or acute respiratory distress syndrome; an increase in intra-abdominal pressure, which would raise the thoracic pressure. Increased juxtacardiac pressure (i.e., tension pneumothorax, pericardial tamponade, right ventricular outflow obstruction) represents an impediment to venous return, resulting in elevated CVP.[20]

CVP and mean RAP can be estimated by clinical examination by the jugular venous pressure evaluation, by ultrasound, and, invasively, by a central venous catheter.

It has been taught that the best method for evaluating the CVP is to position the patient supine in bed, elevate the patient's head to approximately 30–45 degrees, and measure or estimate the vertical height of the meniscus of the right internal or external jugular vein above the sternal angle (angle of Louis), which is approximately 5 cm above the mid right atrium. If the vein is just visible above the clavicle, it is assumed that the venous pressure approximates 7–9 cm H_2O, with 9 cm H_2O assumed to be the upper limit of healthy subjects. To measure higher pressures placing a ruler vertically on the sternal angle and measuring the height of the vein above this reference point has been felt to add a degree of semi-quantitative accuracy to the clinical assessment.[21]

Ultrasound measurement of the inferior vena cava (IVC) diameter and its variations with respiration reliably predicts CVP and mean RAP during spontaneous (negative pressure) ventilation[18]:

1. IVC diameter $\leq$2.1 cm with $>$50% respiratory changes → RA pressure 0–5 (3) mmHg;
2. IVC diameter $\leq$2.1 cm with $<$50% respiratory changes → RA pressure 5–10 (8) mmHg;
3. IVC diameter $>$2.1 cm with $>$50% respiratory changes → RA pressure 5–10 (8) mmHg;
4. IVC diameter $>$2.1 cm with $<$50% respiratory changes → RA pressure 15 mmHg.

CVP and mean RAP monitoring is often used as an assessment of the patient hemodynamic status, particularly in the ICU. A systematic review from 2008 demonstrated a very poor relationship between CVP and blood volume, as well as the inability of CVP/ΔCVP to predict the hemodynamic response to a fluid challenge.[22] As a matter of fact, CVP should not be used as a single parameter to make clinical decisions regarding the fluid management, because in certain patient populations (i.e., patients with right ventricular dysfunction, right-sided valve disease, acute pulmonary embolism) the CVP reflects the right ventricular function and the CO, more than the volume status; moreover, CVP is also influenced by several factors including orthostasis (changing from a standing position to supine), venous dilation, and preload (which may be increased by abdominal muscle or limb contraction, as well as renal failure resulting in fluid retention).

Cardiac output

CO represents the volume of blood being pumped by the left and the right ventricle per time unit. CO is the product of the HR (i.e., the number of heart beats per minute-bpm), and the SV (i.e., the volume of blood pumped from the ventricle per beat); thus CO = HR × SV (L/min). Normal CO values ranges from 4 to 8 L/min; HR from 60 to 100 bpm; SV from 60 to 100 mL/beat.

CO can be calculated noninvasively by imaging techniques (e.g., echocardiography, cardiovascular magnetic resonance imaging), or invasively through right heart catheterization using the Fick or the thermodilution system.

CO measure is important in diagnosis and quantification of HF and of cardiac shunts. CO monitoring is a common practice in anesthesia and critical care by Swan-Ganz catheter,[23] where it is used to guide treatments, in particular fluid resuscitation, and the use of vasoactive and inotropic drugs.

Secondary parameters: vascular compliance, elastance, resistance

A correct evaluation of hemodynamic performance of the cardiovascular system should consider, along with pressures and flows, other parameters such as compliance, elastance and resistance of the cardiovascular structures.[24] In the clinical practice, the systemic and pulmonary resistances could be evaluated invasively, through cardiac catheterization. Among noninvasive methods, the PWV measurement, an index of aortic stiffness, which increases by age and arteriosclerosis degree, is recommended by guidelines in the assessment of hypertension-mediated organ damage.[25] A carotid–femoral PWV >10 m/s is a predictor of future cardiovascular events, and all-cause mortality.[26]

Oxygen saturation

Oxygen saturation measures how much hemoglobin is currently bound to oxygen. Each gram of hemoglobin is capable of carrying 1.36 mL of oxygen. The solubility coefficient of oxygen in plasma is 0.003. This coefficient represents the volume of oxygen in mL that will dissolve in 100 mL of plasma, for each 1 mmHg increment in the partial pressure of oxygen in arterial blood (PaO_2). A formula then calculates the total oxygen content, so that Oxygen Content = (0.003 × PaO_2) + (1.36 × Hemoglobin × Oxygen Saturation). This formula demonstrates that dissolved oxygen is a sufficiently small fraction of total oxygen in the blood.[27]

The oxygen–hemoglobin dissociation curve has a sigmoidal shape (Fig. 30.6) due to the binding nature of hemoglobin: when oxygen molecule bounds, hemoglobin undergoes a conformational change to allow subsequent oxygens to bind, so that each oxygen that binds to hemoglobin increases its affinity to bind more oxygen. In the lungs, the alveolar partial pressure of oxygen (PAO2) is 100 mmHg. However, due to the high affinity for the fourth oxygen molecule, oxygen saturation will remain high, even at a partial pressure of oxygen (PO_2) of 60 mmHg. As the PO_2 decreases under a critical level, hemoglobin saturation will drop rapidly: at a PO_2 of 40 mmHg hemoglobin is 75% saturated, and at a PO_2 of 25 mmHg hemoglobin is 50% saturated. Both parenchymal and pulmonary vascular disease can alter blood oxygenation. Normal PaO_2 is approximately 75–100 mmHg. Values of PaO_2 under 60 mmHg usually indicate the need for supplemental oxygen. A pulse oximeter, a noninvasive device placed over a patient's finger, can measure the oxygen saturation. The use of pulse oximetry has become a standard of care in medicine. Normal pulse oximeter readings usually range from 95%–100%. Values under 90% are considered low and generally indicate the need for supplemental oxygen. Indications for pulse oximetry include any clinical setting where hypoxemia may occur. These settings include patient monitoring in emergency departments, operating rooms, postoperative recovery areas, endoscopy suites, cardiac catheterization suites, intervention requiring conscious sedation, sleep and exercise laboratories, and labor and delivery wards.[27]

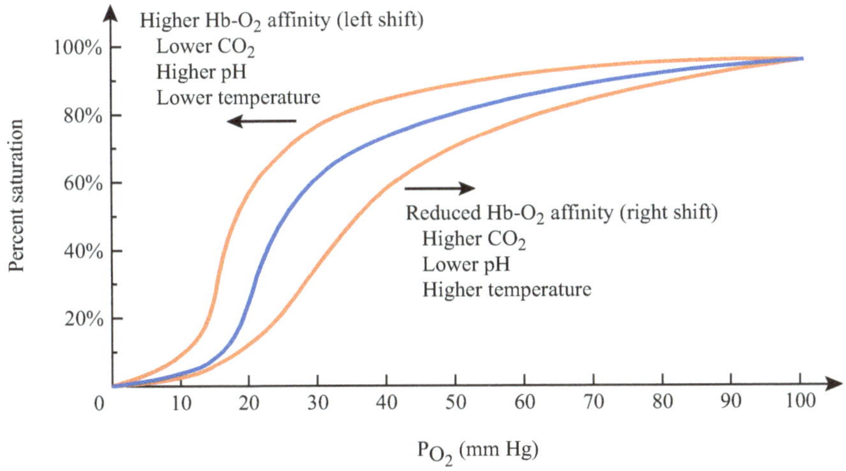

FIGURE 30.6 The oxygen–hemoglobin dissociation curve.

References

1. Braunwald's Heart Disease: A Textbook of Cardiovascular Medicine Single Volume n.d. <https://www.eu.elsevierhealth.com/braunwalds-heart-disease-a-textbook-of-cardiovascular-medicine-single-volume-9780323462990.html> Accessed 09.02.21.
2. O'Rourke MF. Vascular mechanics in the clinic. *J Biomech.* 2003;36:623–630. Available from: https://doi.org/10.1016/S0021-9290(02)00440-2.
3. Somer T, Meiselman HJ. Disorders of blood viscosity. *Ann Med.* 1993;25:31–39. Available from: https://doi.org/10.3109/07853899309147854.
4. Elzinga G, Westerhof N. How to quantify pump function of the heart. The value of variables derived from measurements on isolated muscle. *Circ Res.* 1979;44:303–308. Available from: https://doi.org/10.1161/01.res.44.3.303.
5. Zannoli R, Corazza I, Branzi A, Rossi PL. Aortoventricular mechanical matching: Simulation of normal and pathological conditions. *Journal of Mechanics in Medicine and Biology.* 2008;8:109–120. Available from: https://doi.org/10.1142/S0219519408002498.
6. Schwinger RH, Böhm M, Koch A, et al. The failing human heart is unable to use the Frank-Starling mechanism. *Circ Res.* 1994;74:959–969.
7. Sequeira V, van der Velden J. The Frank-Starling Law: a jigsaw of titin proportions. *Biophys Rev.* 2017;9:259–267. Available from: https://doi.org/10.1007/s12551-017-0272-8.
8. Corazza I, Pinardi C, Manco L, et al. Biomechanical approach to the clinical treatment of pulmonary arterial hypertension. *J Mech Med Biol.* 2013;13, UNSP 1340005. Available from: https://doi.org/10.1142/S0219519413400058.
9. Galiè N, Humbert M, Vachiery J-L, et al. ESC/ERS guidelines for the diagnosis and treatment of pulmonary hypertension: the joint task force for the diagnosis and treatment of pulmonary hypertension of the European Society of Cardiology (ESC) and the European Respiratory Society (ERS): Endorsed by: Association for European Paediatric and Congenital Cardiology (AEPC), International Society for Heart and Lung Transplantation (ISHLT). *Eur Heart J.* 2015;37:67–119, 2016. Available from: https://doi.org/10.1093/eurheartj/ehv317.
10. Pyke KE, Tschakovsky ME. The relationship between shear stress and flow-mediated dilatation: implications for the assessment of endothelial function. *J Physiol.* 2005;568:357–369. Available from: https://doi.org/10.1113/jphysiol.2005.089755.
11. Cecelja M, Jiang B, Spector TD, Chowienczyk P. Progression of central pulse pressure over 1 decade of aging and its reversal by nitroglycerin a twin study. *J Am Coll Cardiol.* 2012;59:475–483. Available from: https://doi.org/10.1016/j.jacc.2011.10.871.
12. Hanna EB, Luke Glancy D. *Practical Cardiovascular Hemodynamics with Self-Assessment Problems.* Demos Medical Publishing; 2012.
13. Nguyen Y, Bora V. *Arterial Pressure Monitoring.* Treasure Island (FL): StatPearls Publishing; 2020.
14. Richard C, Warszawski J, Anguel N, et al. Early use of the pulmonary artery catheter and outcomes in patients with shock and acute respiratory distress syndrome: a randomized controlled trial. *JAMA.* 2003;290:2713–2720. Available from: https://doi.org/10.1001/jama.290.20.2713.
15. Rhodes A, Cusack RJ, Newman PJ, Grounds RM, Bennett ED. A randomised, controlled trial of the pulmonary artery catheter in critically ill patients. *Intensive Care Med.* 2002;28:256–264. Available from: https://doi.org/10.1007/s00134-002-1206-9.
16. Harvey S, Harrison DA, Singer M, et al. Assessment of the clinical effectiveness of pulmonary artery catheters in management of patients in intensive care (PAC-Man): a randomised controlled trial. *Lancet.* 2005;366:472–477. Available from: https://doi.org/10.1016/S0140-6736(05)67061-4.
17. Cecconi M, De Backer D, Antonelli M, et al. Consensus on circulatory shock and hemodynamic monitoring. Task force of the European Society of Intensive Care Medicine. *Intensive Care Med.* 2014;40:1795–1815. Available from: https://doi.org/10.1007/s00134-014-3525-z.
18. Lancellotti P, Zamorano JL, Habib G, Badano L. *The EACVI Textbook of Echocardiography.* 2nd ed. Oxford: OUP; 2016.
19. Shah P, Louis MA. *Physiology, Central Venous Pressure.* Treasure Island,FL: StatPearls Publishing; 2020.
20. Berlin DA, Bakker J. Starling curves and central venous pressure. *Crit Care.* 2015;19:55. Available from: https://doi.org/10.1186/s13054-015-0776-1.
21. Cook DJ, Simel DL. The rational clinical examination. Does this patient have abnormal central venous pressure? *JAMA.* 1996;275:630–634.
22. Marik PE, Baram M, Vahid B. Does central venous pressure predict fluid responsiveness? A systematic review of the literature and the tale of seven mares. *Chest.* 2008;134:172–178. Available from: https://doi.org/10.1378/chest.07-2331.

23. Drummond KE, Murphy E. Minimally invasive cardiac output monitors. Continuing education in anaesthesia. *Critical Care Pain.* 2012; 12:5–10. Available from: https://doi.org/10.1093/bjaceaccp/mkr044.
24. Corazza I, Bianchini D, Marcelli E, Cercenelli L, Zannoli R. Passive aortic counterpulsation: Biomechanical rationale and bench validation. *J Biomech.* 2014;47:1618–1625. Available from: https://doi.org/10.1016/j.jbiomech.2014.03.001.
25. 2018 ESC/ESH Guidelines for the management of arterial hypertension. Rev Esp Cardiol (Engl Ed), 2019;72:160. <https://doi.org/10.1016/j.rec.2018.12.004>.
26. Vlachopoulos C, Aznaouridis K, Stefanadis C. Prediction of cardiovascular events and all-cause mortality with arterial stiffness: a systematic review and *meta*-analysis. *J Am Coll Cardiol.* 2010;55:1318–1327. Available from: https://doi.org/10.1016/j.jacc.2009.10.061.
27. Hafen BB, Sharma S. *Oxygen Saturation.* Treasure Island (FL): StatPearls Publishing; 2020.

Chapter 31

Wearable Technologies for Cardiac Monitoring

Colin K. Drummond[1], Jeffrey J. Hsu[2], Dermot M.J. Phelan[3], Dhruv Seshadri R.[1] and Michael Emery S.[4]

[1]*Department of Biomedical Engineering, Case Western Reserve University, Cleveland, OH, United States,* [2]*Ahmanson-UCLA Cardiomyopathy Center, Division of Cardiology, UCLA Health, Los Angeles, CA, United States,* [3]*Sanger Heart & Vascular Institute—Charlotte, Charlotte, NC, United States,* [4]*Sports Cardiology Center, Department of Cardiovascular Medicine, Section of Clinical Cardiology, Miller Family Heart, Vascular and Thoracic Institute, Cleveland Clinic, Cleveland, OH, United States*

Chapter Outline

Introduction

Over the past two decades individuals have become increasingly more vigilant in taking responsibility for their own health and fitness, empowered in part through wrist-worn personal fitness devices: noninvasive "wearables." Further momentum for the so-called "quantified self" phenomena has grown as consumer electronics have fostered accessibility of wearable sensor data at the same time as the cost of wearables has decreased. Personal fitness training is no longer exclusively the province of gyms and fitness centers—just about anyone can safely and easily enjoy competing with others on, say, inertial measurements of "step counts" at home or in the office. As recreational devices with no medical claim, the wearables market has grown. Manufacturers continue to add features and benefits to wearables that in some ways appear to function similar to medical diagnostic devices, but critical intrinsic constraints limit this strategy. This chapter provides insight on the role wearables are positioned to take—but are currently unrealized—in clinical decision-making for cardiac care as well as in risk management devices for patients at home.

Medical device development has traditionally been dominated by market forces reflecting an investibility litmus test of strong intellectual property, existing reimbursement codes, and regulatory clearance. This "trifecta" for technology commercialization ensures a rational starting point for a well-defined product development path that has significant promise for return-on-investment, but no guarantees. While it has been suggested the evaluation of medical devices has plenty of room for improvement,[1] and balancing patient safety with innovation remains a challenge,[2] the existing product development model is expected to dominate the paradigm for technology commercialization for the foreseeable future.

Growth in the number and variety of wearables is readily evident from the millions of results generally obtained from an internet search of the term "wearables." Refining the search to academic peer-reviewed publications reduces the numbers dramatically. At least one study of the most popular databases[3] indicates that filtering internet searches for content specific to wearables for medical applications still results in hundreds of articles annually. Journal numbers increase year to year and, by any measure, it is a challenge to find and keep up with meaningful reported developments.

Advances in Cardiovascular Technology. DOI: https://doi.org/10.1016/B978-0-12-816861-5.00008-3

Two aspects of this situation are clear: (1) the field of wearables is abundant with perceived value to consumers; and (2) exploration of technologies amenable to suitable wearables for medical practice is vibrant and producing encouraging peer-reviewed studies. In the first case, investments in consumer devices will continue to result in devices that are lower in cost, smaller in size, have longer battery life and increased usability, and generate higher connectedness. In the second case, product reviews, literature reviews, and published data on clinical trials continue to provide guidance and design feedback to device developers.

Two questions are the impetus for this chapter on wearables. First, as the features and benefits of wearable devices have expanded, we begin to wonder: "At what point is a wearable device no longer an unregulated fitness tracker?" Second, as healthcare shifts to more patients requiring care for chronic conditions and increased management drive for cost accountability, a reasonable question is "what will it take for wearables to have a role in clinical decision-making?" These questions approach the same issue from different directions, but in both cases the conventional medical device development process must prevail if a wearable intends to play a role in a clinical setting.[4] Fig. 31.1 suggests there is an opportunity in a "middle ground" for wearable-driven care that sits between consumer driven and clinical care. Indeed, it can be argued that the COVID-19 pandemic prompts a more urgent look at—and has offered a temporary test bed for—the role wearables can play in serving, for instance, isolated patients or cardiac patients where safe monitoring at home is an acceptable alternative to in-patient hospital care.[5]

Our examination of wearables and sensors for cardiac monitoring in this chapter does not duplicate the numerous reviews and surveys of products found elsewhere. The dynamic nature of a consumer-driven market makes a book chapter survey irrelevant quickly. More appropriately, we would rather center on ideas providing a more enduring understanding of the "state-of-the-art" as it is and—if history is an indicator—what is likely to be the situation for some time to come. Paradigm shifts in healthcare do not occur overnight for very good reasons and our objective is to frame a perspective that is moderately nonperishable, actionable, and strategic, not tactical.

We proceed in three parts:

1. The evolution of noninvasive medical instruments from a historical perspective to highlight that product development serving the practice of medicine is not exclusively about the latest technology.
2. A look at two contemporary clinical applications where there is significant potential value to be realized with the right products, but that we have not yet reached the finish line.
3. The "grand challenges" we believe that need to be addressed for technology to be clinically accepted but that the consumer market may not immediately embrace.

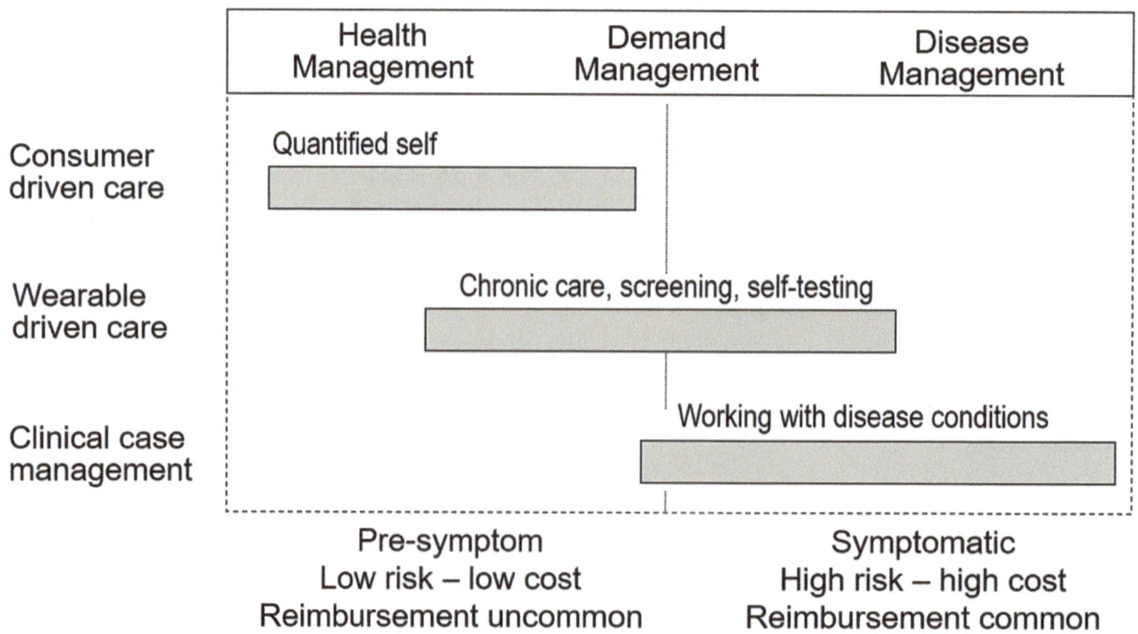

FIGURE 31.1 Landscape of care reflecting the emergence of a consumer-based paradigm of care as presymptom complement to traditional clinical case management.

Historical innovations for noninvasive measurement devices

Technology enables us to create new products and services, but the "go to market" strategy is rarely a linear and rational process. While barriers to technology commercialization are frustrating to impatient newcomers, examining prior noninvasive device developments is enlightening and likely encouraging that the "road ahead" may not be as new or unique as we think. Indeed, we can argue that a retrospective look offers insight on challenges to technology use and adaptation in complex settings. Technology may expand boundaries but that does not mean they are easily crossed. So, what insights can we gain from the history of noninvasive measurements? Let us take a quick look at just four innovations that continue to impact patient care to this day.

The first device mixed creativity with an understanding of clinical need. It is said that during the Victorian era a physician listening to the heart by placing his ear on the patient's chest was not always easy or welcome. Rene Laennec solved the issue with a paper tube placed on the chest—this is believed to be the first noninvasive cardiac device. Not only did this launch the practice of auscultation of the heart in 1819, but upon refinement by other clinical users, the system found new uses eventually leading to a shift from subjective to objective data about the internal body. Despite the technological simplicity and clear evident need, it still took nearly five decades for many contributors to help evolve the device into a form we recognize today. From this we see that being first to market can create value to patient care, but also that open discussion and incremental innovation are a central part of improving patient care with a device eventually aligned with clinical practice.

Our second case illustrates how combining two different technologies adds value each could not fulfill on its own. The sphygmomanometer concept first arose in 1881 but it was not combined with the stethoscope until 1905 when Dr. Nikolai Korotkov identified a repeatable and robust process for blood pressure measurement to take place.[6] By occluding blood flow then releasing in a controlled fashion, the five Korotkov sounds due to turbulent flow were characterized as distinctively different from heart valve vibrations. This unique combination of existing technology required only two decades to achieve clinical utility, but was quite transformative as a new capability to collect objective data about the state of the heart and blood vessels. It is conceivable that this also opened the door to training more than the physician on the collection of vital signs. Linking a technique for noninvasively linking these "new sounds" to health required an understanding of hydrodynamics and different forms of pressure waves established much earlier by Bernoulli's math models in the early 1700s. In this case, a basic understanding of physics enabled devices to be combined in new ways to provide insights into mammalian physiology.

A third system introduced electrical technology to medicine. Understanding the importance of electrical signals originating from the heart took longer to get to market since the clinical significance of the electrical signals was not evident. While the moderately small voltages could be recorded, no immediate clinical implication was evident, and thus electrical tracings were initially marked quite simply with the PQRST method of the mathematician René Descartes (also responsible for the arbitrary use of X as an independent variable). Around 1900 William Einthoven enabled practical application with his work on the string galvanometer and a triangle for ECG lead placement.[7] After three decades of development, by 1930 the electrocardiogram enabled specific heart conditions to be studied in humans with the insight that, for instance, measurable electrical activity of the heart precedes ventricular contraction. In the early 1950s, the ECG emerged as a tool to help differentiate between acute cardiac and noncardiac chest pain.[8] Scientific curiosity preceded clinical application and underscores how basic research is valuable even when the clinical application may not be readily apparent. This also underscores the value of data exchange standards in characterizing data and making the exchange of data between collaborators easier.

As a last example, the measurement of oxygen saturation (SpO_2) originated when the fields of biology and chemistry were blended (decades in advance of the term "biochemistry" coined in 1903 for this new field). Around 1876 it was recognized there was a change in pigment in red blood cells when combined with oxygen; further, red blood cells without oxygen absorb light differently than red blood cells with oxygen. Not only that, but it was also realized that as blood flow increased reflected light changed. Two measurement opportunities were now present. First, different wavelengths of light result in different blood light absorption levels, and this could be correlated with blood oxygen content. Second, reflected light is also a marker of blood flow and its changes. Reflected light data is recorded as a photoplethysmogram (PPG), but the data requires interpretation to be useful. At the risk of oversimplifying details, whether the PPG is used to determine SpO_2 or blood flow rate is dependent on the way in which the output data from the PPG is analyzed. Peak-to-peak waveform data is a measure of heart rate (HR), somewhat like what can be extracted from an ECG. It was not until about 1935 when the earlobe technology of Millikan and Wood was refined by Matthes that a system of clinical significance for this noninvasive technology entered practice.[9] Time to market: five decades. Even then, the complicating effect of the pulsatile nature of blood flow when measuring blood oxygen levels was not resolved until advanced signal processing innovations of the 1970s were in place. The impact of interdisciplinary technology developments is not unique, and advocates need the "long view" for the transition to patient care. More fundamentally, though, we see that signal analysis (analytics) is a powerful tool to tease out meaning from data.

The medical device innovations described above, in an admittedly cursory fashion, bring out several historical lessons that are pertinent today:

1. Technology often took decades to impact standard of care as evidence-based cases emerged.
2. Innovations can be built on one another through sharing of information and open discussions.
3. Multidisciplinary teams brought technology together in unique ways.
4. Adequate signal analysis often lagged sensor technology.

We further suggest that more subtle "lessons learned" include:

1. Clinical need drives problems that are a catalyst for innovation.
2. Team science facilitates the conversion of data into information.
3. Noninvasive clinical devices draw on clinically validated hypotheses.

In the next section we explore wearables at a more contemporary stage and time, then explore the challenges they present in the context of cardiovascular care.

Contemporary applications

Contemporary noninvasive medical devices are similar to historical devices in the intent to characterize a cardiac condition based on objective data measurements. Measurement of vital signs traditionally dominates health assessment, but the term "biomarker" has entered the lexicon as it becomes clear that new technology can expand what we measure and calculate. For clarity, we draw on the consensus statement that a biomarker is "a defined characteristic that is measured as an indicator of normal biological processes, pathogenic processes or responses to an exposure or intervention"[50], and therefore recognize vital signs as a subset of biomarkers. Definitions do not detract from the most common questions central to clinical care of a patient:

- What is the blood pressure?
- What is the HR?
- What is the blood oxygen saturation?
- What is the temperature?

The promise of wearables in cardiac care is to not only to collect data at a point in time, but to inform clinical decision-making through time-histories of these markers of health. Further, the sensitivity and specificity of the devices must be assured, particularly in cases where "measurements," such as HR variability, are derived from measurement and not directly measured. Validated technology offers the promise of enabling this data to be collected at home instead of confining measurements to the clinical setting. Indeed, the commercial opportunity is so great that at the time of writing there are at least 300 devices by hundreds of companies seeking to answer these basic questions.[10] With so much activity and innumerable press releases about developments, it is easy to get lost in the fray. The euphoria simply suggests that achieving success in wearables has not been easy. Fundamentally, there is no avoiding the need to be equivalent to today's "gold standards" for answering the four questions above.

At the beginning of this chapter, we noted a paradigm shift in patient care, and we asked the question: What will it take for wearables to have a role in clinical decision-making? Management of cardiovascular health is broad and complex, reflected by a large and growing number of cardiovascular subspecialties, so to be more specific in our response it is useful to focus the discussion on two specific care scenarios: atrial fibrillation (AF) and COVID-19. Both cases have significant population health implications in which wearables offer promise in health management. Both are contemporary: (1) AF due to the structural shift in healthcare that opens the door for wearables to complement a growing health concern, and (2) COVID-19 due to the impact the pandemic has had on millions of people and the renewed emphasis on precursors of health as well as the challenge of "scale" in providing healthcare. These cases are not intended to exclude the potential role of wearables in other important care situations. Rather, they are presented to illustrate two fundamentally different types of care addressed by wearable technology and to better understand the ability of technology to meet the gold standard for care in these situations.

With hundreds of devices being developed, specific designs will vary widely, but all designs must follow the basic system design diagram shown in Fig. 31.2. In the subsections to follow, we examine more closely the technology and

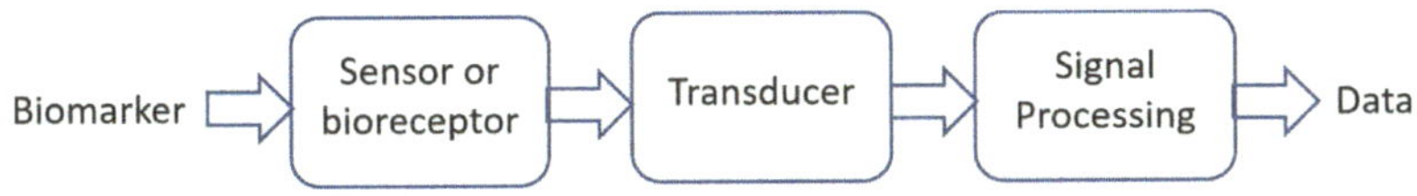

FIGURE 31.2 Basic block diagram for wearable sensor information flow.

issues present in the care scenarios mentioned above. Designs meeting the clinical needs of these two scenarios are most likely to eventually prevail in clinical care, even if the patient is being monitored remotely at home.

Atrial fibrillation

AF is a common type of arrhythmia that occurs when the normal sinus rhythm of the atria is replaced with irregular and often rapid electrical depolarizations. AF can be viewed as "chaotic" electric activity within the atria that interferes with the functioning of the sinoatrial node, thus disrupting the organized transmission of electrical signals to the atrioventricular (AV) node. Subsequent irregularity of AV node signals leads to an irregular pattern of the QRS complex on an ECG recording and a clear departure from normal cardiovascular rhythm. AF is clinically significant since the chaotic electrical impulses in the atria do not allow for normal atrial contraction and, hence, atrial contribution to ventricular filling. Additionally, the chaotic electrical activity from the atria is rapidly "bombarding" the AV node and can result in rapid activation of the ventricles. Persistent, uncontrolled AF is associated with heart failure and the mere presence of AF increases the risk of stroke in many individuals.[11] Some estimates are that AF has grown consistently over the past 50 years with a global prevalence of over 45 million individuals,[12] with over five million people in the United States affected.[13] Additionally, given that some individuals with AF do not experience overt symptoms, it is estimated that as many as 700,000 people in the United States may be undiagnosed.[14] While increases in AF incidence are often associated with age and lifestyle, the etiology is usually multifactorial.[15] The potential to identify and treat asymptomatic AF and thereby reduce the risk of stroke has been a catalyst for technology development. Cases where risk factors suggest the likelihood of AF will occur—but is undiagnosed—present a large vulnerable population to potentially identify. At the level of population health management there is the additional design constraint that technology needs to be deployed in a resource-efficient manner. Wearables present an attractive complement to traditional clinical monitoring particularly given the potential to significantly impact public health.

Progress in identifying modifiable AF risk factors is improving the ability of clinicians to manage health outcomes. For instance, hypertension, obstructive sleep apnea, obesity, and coronary heart disease may modify the electrical properties of the heart and thus increase the probability of AF. Assessing such risk factors enables patient counseling on behavior and risk modification. The pathogenesis of AF is an active area of research that continues to illuminate the triggers and causal effects underpinning susceptibility to AF. Hypotheses for cellular action potential modifications, inflammatory mediators, tissue properties, and electrophysiology alterations enhance our understanding of the pathophysiology of AF, but many questions are unresolved.[16] Recalling the wearable sensor information processing sequence shown in Fig. 31.2, it is unlikely a singular biomarker will adequately represent all potential triggers and risk factors of AF. Subsequently, it must be appreciated that a diagnostic system based on just one parameter is intrinsically limited; that is, the system—at the most fundamental level—is designed with incomplete information despite attempts to compensate by placing special emphasis on signal analysis.

To better understand the way technology can facilitate decision-making, it helps to have characterizations of AF that assist in explaining the diagnosis, care planning, and tracking of risk factors even when the exact etiology remains unknown.[17] Patients may be experiencing AF for the first time or it may be recurrent. If recurrent, AF can be occasional (paroxysmal), persistent, or permanent. Paroxysmal AF may have symptoms that come and go for short periods that may call for only a monitoring strategy, while persistent or permanent AF may indicate the need for a combination of surgical intervention, monitoring, and long-term medication therapy. Comorbidities may confound the situation since AF symptoms such as shortness of breath and weakness are not exclusively due to AF.[18] Technical solutions must therefore selectively differentiate phenomena in a repeatable and reliable fashion as well as provide data output in a form that enhances communication through an interface that is informative, not cryptic and puzzling. Data in isolation does not necessarily take the form of useful information if the context is not clear. AF characterization and/or symptoms of unclear correlation may warrant continuous monitoring outside the clinical setting to collect information for hours, days, or weeks. This is an ideal scenario where wearables can add value to clinical practice. Several studies have been undertaken that highlight this clinical potential, but given current limited clinical trial outcome data, physicians should use caution when attempting to incorporate information from wearable devices into clinical decision-making.[19,20]

Credibility for wearable device applications to medicine appears to have emerged around the time the FDA cleared Kardia Band (KB) accessory in November 2017 through a 510(k) K171816. The KB watchband (now discontinued although the parent company continues with its nonwearable devices) had an integrated electrode that was activated by pressing a button tightly against the skin that would collect and send the data to a smart device. Then, once connected to a smartphone for data analysis, individuals had their own personal single lead ECG system. A clinical trial at the Cleveland Clinic in 2018 affirmed that the KB algorithm could accurately differentiate AF from sinus rhythm.[21] The study found the automated AF detection algorithm had 93% sensitivity, 84% specificity, and a K coefficient of 0.77 when recordings were compared with corresponding 12-lead ECG tracings.

The popularity of the KB was short-lived once Apple Corporation announced in July 2019 the release of an ECG algorithm (an "app") for the Apple Watch (AW), which had a profound effect on the direct-to-consumer market since consumers could collect heart rhythm data (that was often inferred to be a single channel ECG) in a more integrated (and convenient) fashion directly from their watch. Unlike the stainless-steel electrodes of the KB at the time (and the KB product line has since evolved away from the AW designs), the early AW designs used photoplethysmography (PPG) data to feed HR calculations. PPG is an indirect measurement of electrical signal, and it is important to note that FDA clearance was for the AW algorithm; the device itself was not "approved."

The wearables market is not static, and subsequent research and AW product development led to the Apple Watch 4 (AW4) and a 510(k) clearance in October 2020 (K201525). The AW4 features both an optical sensor and an electrode—this expanded sensor suite allowed the collection of a true ECG rhythm strip with the AW, beyond just the PPG pulse rate and regularity. Implications that smartwatches were "coming of age" to serve the clinical market spawned a variety of clinical trials to provide validation studies. Some large-scale studies established foundational work,[22] but care with subject self-selection and enrollment suggest limitations. Success with an algorithm does not mean the device output is equivalent to the 12-lead ECG.

Comparing one wrist-worn device to another is helpful for competitive analyses but fails to persuade that these devices are closer to the gold standard. Fundamentally, wrist-worn devices are likely to be less accurate than chest-worn devices.[19] Of practical importance is whether the wearable device based, *single-lead* ECG patterns correlate with the 12-lead ECG standard, and if they do, what are the differences and how might clinicians distinguish between differences in data collection devices and waveform departures of pathophysiological significance. The single-lead rhythm strip is not a full 12-lead (vector) ECG and hence not capable of producing the full diagnostic potential of the 12-lead gold standard. Device operation steps can also be an issue. For instance, what are the differences between pressing the KB electrode with a specific pressure against the skin and the AW variance in PPG signal produced by watchband tightness? There is no standard by which to make such an assessment.

Other aspects of wearable devices that produce vital signs are emerging. Research from Duke University in 2020 explained why PPG-based HR sensors are inaccurate in individuals with different skin pigmentation.[23] PPG is more challenging in people with higher melanin content because melanin absorbs light. Some wearable companies have overcome this by increasing the light intensity (which may decrease battery life). As we translate wearable devices for patients, inaccuracies are confounded. For instance, the presence of edema increases the resistance of the overall "circuit" thereby affecting the absorption of light and affecting the signal to noise ratio of the sensor leading to inaccuracies in measurement. That type of confounder compounded with different melanin content will magnify the variance in signal between a wrist monitor and, say, a standard of care instrument (e.g., 12-lead ECG). The state of technology today in medical care is that every device should be validated in the situation or multiple situations with which it may be utilized, however, the multiple confounders affecting wearable usage provide a potential hurdle.

AF characterization leads to several scenarios for the early detection and onset of AF that can also expand public awareness of modifiable risk factors. Wearables have been launched that move beyond ambulatory ECG systems, with devices offering continuous and real-time monitoring of HR, blood pressure, and glucose levels, and while remote monitoring of a combination of vital signs is an exciting frontier, the sparse number of feasibility studies thus far illustrate the need for continued development in order to have significant clinical impact.[24]

COVID-19

In 2020, the COVID-19 pandemic changed the landscape of medicine much like the influenza pandemic changed healthcare in the early 1900s.[25–27] This new and highly contagious viral pathogen threatened to overwhelm healthcare systems from a supply chain (e.g., personal protective equipment), staff (physicians, nurses, etc.), and bed availability standpoint. Additionally, in order to reduce potential spread, the public was urged to socially distance and many businesses including medical caregivers temporarily closed their offices. This resulted in a significant reduction in access to medical care providers for not only routine health maintenance but also acute care concerns such as myocardial infarctions.[28,29]

With the prospect that some healthcare systems could become overwhelmed and patients underserved, technologies that had been historically underutilized, such as, for instance, telemedicine, found new importance,[30] and the pandemic motivated temporary expansion of telemedicine reimbursement.[31] This spurred broader discussions such as the ability of our existing healthcare structure to adequately serve population health and wearable technology became a prominent a part of that discussion.[32,33] As an example, Atrium Health (North Carolina, USA) expanded capability with the development and launch of a virtual hospital model that included home monitoring kits allowing "mobile clinicians" remote ECG, vital signs including SpO_2, blood tests, and human touch for isolated patients in their homes. This model provided comprehensive services to nearly two-thirds of the health system's COVID-19 patients.[5] The virtual hospital complemented in-person care during the pandemic, but understanding what is clinically acceptable in the future will take further study.[34] Other healthcare associations and providers are collaborating and advocating the home as a site for healthcare—specifically encouraging telehealth and remote patient monitoring—to "change the way policymakers think about the home as a site of clinical service."[35]

The role of wearables in assisting in the detection and monitoring of COVID-19 quickly became an area of active inquiry during the pandemic. This type of utilization is largely a function of the sensor data that can be collected. Realistically, there are actually very few basic measured parameters from a wearable device; they are: (1) voltages from electrodes, (2) measurements of reflected light, and (3) temperature. The latter, of course, remains a challenge for wrist measurement systems, and for temperature the state-of-the-art seems to be that a second wearable device must be added that communicates with a smartwatch or smartphone. Regardless, for all three parameters it is easy to assume that physiological metrics displayed on, say, a smartphone are accurate measured quantities, but this can be misleading. As a practical matter, Fig. 31.2 underscores that measured quantities and displayed biometrics such as vital signs are very different. Vital signs to inform about health are *derived* quantities from basic measured parameters, that is, algorithms are in the critical path between, say, a skin electrode and the display of HR. Clearly, there is potential for error between the HR metric and the electrode sensing modality. Given that cardiac function is strongly influenced by COVID-19 infection, Table 31.1 (adapted from Seshadri et al.[36]) summarizes sensor modalities for nine physiologic metrics of most interest. Realizing that only a subset of the metrics (HR, HRV, RR, and SpO_2) are readily available with common wearable devices, this is a limitation to wearable devices in the market. What might also limit the potential role of wearables in COVID-19 is understanding exactly how physiologic metrics are computed and what role the virus has on computational algorithms—unfortunately, the absence of access to wearable algorithms limits the ability to fully understand underlying assumptions that may influence interpretation of results. Thus, a dilemma with wearables is that the prevalence and convenience of wearables as a commercial product has intrinsic limitations when thinking of COVID-19 applications.

Current limitations with the ability of wearables to support COVID-19 clinical decision-making in line with gold standards for physiologic metrics does not rule out the potential for an important role in public health. It is clear that COVID-19 precursors are valuable for detection of viral infection, especially so for cardiac patients or those with other preexisting medical conditions.[37,38] Clinical studies suggest that smartwatches and activity trackers could produce data for which HR, variation in resting HR, respiratory rate, and sleep activity could be a useful predictor of the onset of COVID-19.[39–41] For instance, increasing HR, decreasing HR variability, and increasing respiratory rate are signs of viral infection onset. Other parameters for insight can be inferred, for instance, rising resting HR suggesting core body temperature rising, but then temperature may be rising for reasons unrelated to infection. Many inferences are possible, however, and specific hypotheses remain to be validated through clinical trials. Validated algorithms to identify variance in health metrics are a promising step to enabling correlations with disease states, as such, correlations could provide highly informative screening devices in the realm of public health, particularly if a patient predisposition is suspected. Precursors can provide advance warning and the benefit of early intervention. To be robust and informative of meaningful events, it remains that clinical trials for algorithm validation be conducted through the lens of evidence-based medicine and not the commercialization of technology.

Overall, clinical trial results are encouraging and motivate expanded relevance of wearables in patient care, however, these studies underscore the challenges in (1) establishing averaging schemes for large data streams, (2) filtering data for artifacts and low signal-to-noise, (3) compliance with *all* steps in the smartwatch data collection process, and (4) bias in training data sets based on smartwatch ownership among subjects under the age of 65 (i.e., potentially a less vulnerable population).

The multistep data process that starts with the patient collecting data and ultimately presents information for clinical decision-making poses significant quality issues that have to be addressed for the field to advance. The repercussions to

TABLE 31.1 Sensor modalities for monitoring physiological metrics relevant to COVID-19.

	HR	Heart rhythm	HRV	RHR	RR	SpO_2	Skin temp	Core temp*	Sleep
ECG	x	x	x	x	x			x	
PPG	x	x	x	x	x	x		x	
Accelerometer	x				x				x
Temperature							x	x	

HR, heart rate; *HRV*, heart rate variability; *RHR*, resting heart rate; *RR*, respiration rate; *SpO_2*, blood oxygen saturation; *ECG*, electrocardiogram; *PPG*, photoplethysmography; *Temp*, temperature.
Core temperature is measured based on HR and skin temperature and cannot be measured as a stand-alone metric.
Source: From Seshadri DR, Davies EV, Harlow ER, et al. Wearable sensors for COVID-19: a call to action to harness our digital infrastructure for remote patient monitoring and virtual assessments. *Front Digital Health* 2020;2:8. doi:10.3389/fdgth.2020.00008.

safety associated with false negatives and the burden on the healthcare system with widespread false positives are issues that have significant consequences. Standards development and application are on the rise and will be an important part of a sustainable and manageable part of large-scale home health monitoring programs.[42–44] Very clearly, success in wearables impacting the shifting healthcare landscape is not exclusively about the latest consumer wearables technology.

Development challenges

Despite a highly promising—but unrealized—technology readiness level of wearables outlined in the previous discussion, there remain several challenges in the continuing effort to translate the technology into clinical practice. To the extent that wearables are intrinsically unable to meet or exceed current standard of care, the cardiac wearables face three further fundamental challenges:

- Limitations to the crossover product development strategy for wearables.
- Failure of secure interoperability essential to aligning the technology with clinical workflow.
- Substantial equivalence shortcuts to evidence-based medicine for standard of care.

Crossover development strategy limitations

There are two overarching corporate development strategies to bring a medical device to market. One strategy follows a pathway of needs identified within a clinical context and product development vetted through clinical examination and formalized studies to assess hypotheses. This process is aligned with the highly regulated FDA "waterfall" process. The FDA design process for medical products is not elective—it is mandated by law—and has established prevailing design requirements allowing devices to be marketed that are safe and efficacious for patient care. Finger pulse-oximeters for home use are an example of a consumer device with reduced cost and complexity that is derived from a more expensive clinical version.

Another strategy involves investments to repurpose or convert consumer electronic devices into medical grade devices, and this will be referred to as a "crossover strategy." Without the insight provided by open innovation processes, we can only infer specific development strategy, but by all appearances the process of enhancing device features to crossover into healthcare seems common. This takes the form of early versions of a device becoming accepted in a consumer market then features and benefits added with each new release. Eventually, the product takes on the attributes of a medical device. As mentioned earlier, Apple released a software upgrade to the Apple Watch Series 4 in 2018. This added device option leveraged an earlier consumer device platform to achieve FDA Class II "de novo" clearance (but not approval) for its ECG and irregular heart rhythm notification functions.[4] Significant here is that the FDA opened the door in 2018 to the Apple device crossing over into the healthcare space, and five devices by other companies were cleared a year later. These trends will continue to encourage the crossover strategy, yet clinical trial work remains a critical path to success for a device to be accepted in mainstream medical practice.

A fundamental issue with the crossover strategy can be described with the assistance of Fig. 31.3 that expands on the block diagram shown earlier in Fig. 31.2. Consider a hypothetical situation where, for an indirect biomarker measurement, an algorithm requires two sets of electrical signals S_n and T_n, shown in Fig. 31.3, but the wearable device can only produce a subset of the required signals, S_n. The other set of sensor signals T_n ideally needed for the algorithm are unknown or unavailable. In the spirit of "close enough" the subsequent data filtering and sampling rates (that may vary) produce the proxies X_n. The task of the algorithms for integrating and processing the data is to compensate as best as possible for absent information. The output of the algorithm Y_n produces a proxy for the clinical information needed. The logical process of fine-tuning data and algorithms is undertaken to produce, say, an adequate correlation coefficient. Increasingly, the deficiency of a sensor suite is being compensated with added "machine learning" software to provide insight the default data simply does not contain. While the results may be "close enough" for some applications, an argument can be made that the results are not adequately close to gold standards of care. As an example, the 12-lead ECG provides the standard of care in assessing, say, comprehensive ST-elevation for acute coronary syndrome. A substitute system based on just one lead may, indeed, provide some insight on cardiac health, but the practitioner would need significant evidence-based results from a formal clinical trial endorsed by peers to begin to replace the standard of care. This is not a situation where "cleared" vs. "approval" from the FDA is sufficient, since clearance through the 510(k) process only assures one device is substantially (but not exactly) equivalent to another device on the market. Approved devices typically need the credibility derived from a clinical trial. In short, a single-lead system seems

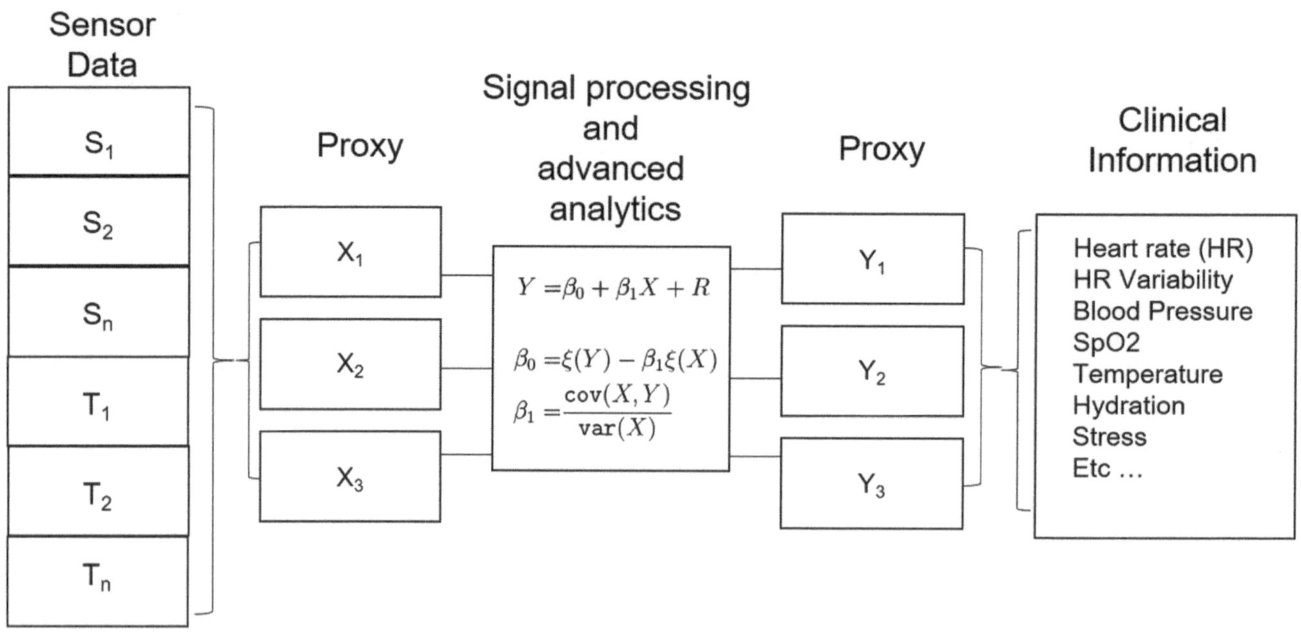

FIGURE 31.3 The role of signal processing to adjudicate incomplete sensor data suites.

unlikely to displace the 12-lead system but could serve quite successfully as a risk management device for ambulatory patients.

Failure of secure interoperability essential to aligning the technology with clinical workflow

Data breaches in the press are hardly news anymore prompting data security to be of increasing concern to consumers. This is quite unlike the environment for care providers who have long understood the legal and ethical ramifications of respecting patient and data privacy, confidentiality, and as a minimum, the 18 measures of protected health information. Clinical information systems that are disperse, wireless, and with multiple interface requirements end up providing a fragmented system of data exchange and a corresponding opportunity for data breaches in various forms.

Sports enthusiasts regularly log into their wearable's cloud as a recreational endeavor to check real-time updates of data for their own information. The clinical scenario is much different. From the clinician's perspective, there are already valid concerns of excessive workload in the use of electronic medical records (EMR) in daily practice compounded by the complications of even simple interoperability issues such as sending discharge planning documents to the next patient care destination. It is challenging to think that hospital administrations would find it appealing for a physician to spend time logging into wearable cloud services to check (on one of many required) patient vital signs or biomarkers if virtual monitoring is capped by insurance at $600/year. The workflow of a clinician is simply not aligned with wearable technology data access and review. Interoperability between systems does not appear to be a priority for wearable systems, and if anything, the economic model is to monetize data access through subscription services designed with the user in mind rather than the clinician.

Commercial entities have no obligation to "open up" their systems for others to inspect and analyze. The free market only requires that the limited monopoly provided by a UTPO patent award document the "teachings" and the "claims" of a patent. Still, as a courtesy, many wearable systems do offer ways to download data in XLS, CSV, or JSON formats, but the burden of proof is on the user or clinician to interpret and attempt to reconcile with clinical outcomes. This is not to say that the technology required is all that difficult to manage, but absent standards, the requisite variety of custom modules leads to a nontrivial effort to gather and understand output—this is an indirect cost not already accounted for in the off-the-shelf cost of the wearable device itself. In contrast, in other fields such as medication management, it is now routine to exchange prescription data between a hospital and a pharmacy. The standardization offered by the process of electronic medication prescriptions has been greatly facilitated by the FDA National Drug Code Directory and naming systems provided by the National Institutes of Health RxNorm; no such equivalent exists for wearables.

The paradox of success arises when extreme success carries blind spots that are ultimately a limitation to greater success. Suppose that, in fact, wearables do achieve clinical relevance and suddenly a practitioner caring for hundreds of patients begins to obtain continuous data on AF warnings, HR, and BP. If processes are absent to convert this data into meaningful clinical information, then what is the value of this data? We are aware of the issues of "big data" in terms of the challenges of volume, velocity, variety, veracity, and variability,[45] but from a liability perspective there must be processes in place to transform data into manageable information. What potential liability issues exist for the provider if the individual has an adverse event (e.g., stroke) and within all of that data was a signal that could have identified the problem (e.g., brief run of AF in 6 months of data). Additionally, there are lessons to be learned from physician burnout due to EMR overload[46] that apply to the development of methods and interfaces for the expected tsunami of wearables data.

To summarize, the context within which wearables must operate offer several unique challenges:

- Most novel and interesting activities are easy to support in the short run. In medical care, long-term value of wearables derives from long-term compliance to processes and procedures. Unfortunately, multiple interfaces for data collection do not align with efficient clinician workflow.
- The lack of interoperability reflects a lack of data exchange standards and readily available open-source modules for data exchange. Widespread wearable usage will produce an expected volume of data that may create the same unintended consequences that medicine is now experiencing with broad adaption of EMR systems.
- System and data security is paramount in medical care and the unintended consequences of poor system security are not always robust or easily addressed through "patching" the software. The system must be designed with data security as an integral part of the process, even at the expense of wearable processing power.

While favorable conditions inspire crossover development activity, we now turn our attention to two grand challenges that must be addressed:

- Inability to make meaningful connection between data and reliable knowledge about health.
- Wearables viewed as the province of industry and not a basic research endeavor.

Substantial equivalence shortcuts to evidence-based medicine for standard of care

A crossover product development strategy presents a challenge to both commercial entities and the medical community they seek to serve. Those new to medical device design might be inclined to emphasize "speed to markets" or a cost-effective strategy of "substantially equivalent" devices. While consideration of cost and speed might make for a competitive product from a commercial perspective, this overlooks the role and value of regulatory agencies in the medical product development and launch process. It is often underappreciated that the creation and subsequent growth in the regulatory power of the FDA throughout its brief history has frequently been a direct result of individuals harmed by mislabeled, untested, adulterated, or toxic products that were well-marketed, but not well-tested—in this regard the FDA can be credited for "...help[ing] pull medicine into the modern era"[51]. Anticipating that most medical devices need to be cleared or approved by the FDA prior to being marketed for sale in the United States, the advantages of low-cost devices are irrelevant if the device cannot be presented to the marketplace due to failed regulatory compliance. Formalized and intentional clinical studies embrace the context of FDA regulation from several perspectives since it was the unanticipated consequences of new technology—and the way the technology was brought to market—that spurred the modern medical device regulatory framework.

The integrity of product development processes permits no shortcuts if the intention is for a device to be as good as or exceed existing standards of care. Many of the processes shown in Fig. 31.4 highlight steps that are essential in the product development process.

There is a compelling role in the medical care of patients for devices that offer "good enough" fidelity under certain care conditions and patient monitoring needs to make them plausible in the public health realm. Not all devices have to be "medical grade" to positively impact health. The more precise the data point needs to be to approximate "medical grade" data, the more the device must mirror that requirement. Acceptable scenarios depends on the question asked. A good example is a HR monitor. When used to provide training zones for an athlete to improve fitness, the "good enough" model works great. When used to provide HR limits within which to exercise for an individual with severe heart disease, the data needs to be much more accurate and "medical grade" may be required.

It can be argued that as a risk management tool, wearables offer the potential value of continuous monitoring and real-time data transfer with some analytics to alert clinical decision support systems. Several systemic limitations have yet to be overcome for wearables to replace several gold standards of care that are an essential part of medical practice.

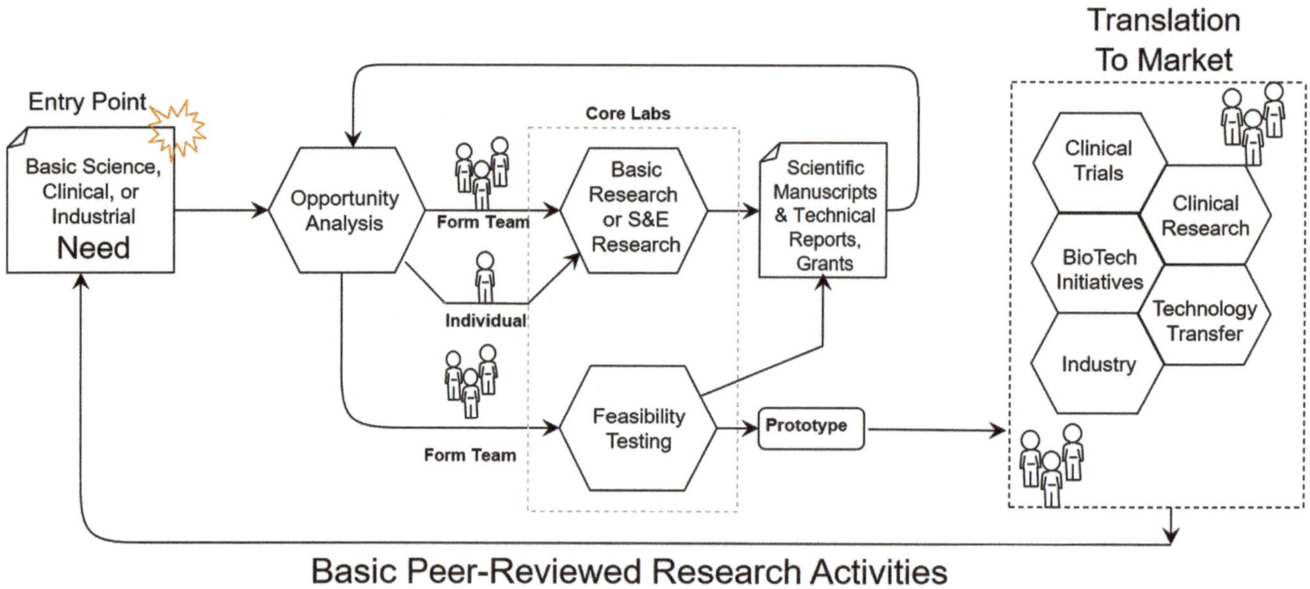

FIGURE 31.4 Conventional peer-reviewed research cycle with clinical trials supporting evidence-based outcomes.

Summary

Wearable technology and sensors for cardiac monitoring present an exciting new chapter in patient care both at home and in clinical settings. The promise of "better, faster, cheaper" is encouraging and the broad use and acceptance of wearables is positioned to have an impact in the long run. Cautious optimism is based on the lessons learned over the past century for which wearables are not exempt:

1. Technology will eventually impact standard of care as evidence-based cases emerged.
2. Sharing of information and open discussions will be the centerpiece of systemic innovation.
3. A multidisciplinary team approach to healthcare ensures systems fit the people they are meant to serve.
4. Adequate signal analysis continues to lag sensor technology.

The structural shift in healthcare driven by wearables for the "quantified self" suggested by Fig. 31.1 has significance beyond personal fitness trends—a new role for wearables to leverage existing digital infrastructure for remote patient monitoring and care looks promising as well.[39,47,48] Surveys of electrophysiology professionals indicate that attitudes have changed favorably about adopting digital health, but concerns remain about guidelines for use, infrastructure, and reimbursement.[49] This more systemic view recognizes that while it is important to understand the device and its impact on patient care, it is equally important that complete understanding of the entire healthcare system be taken into account when advocating the benefits of wearables.

References

1. Kramer DB, Yeh RW. Practical improvements for medical device evaluation. *JAMA*. 2017;318:332–334. Available from: https://doi.org/10.1001/jama.2017.8976.
2. Wizemann, T. (Ed.). *Public Health Effectiveness of the FDA 510(k) Clearance Process: Balancing Patient Safety and Innovation: Workshop Report*. National Academies Press (US), Washington (DC); 2010. https://doi.org/10.17226/12960.
3. Loncar-Turukalo T, Zdravevski E, Silva JM, da Chouvarda I, Trajkovik V. Literature on wearable technology for connected health: scoping review of research trends, advances, and barriers. *J Med Internet Res*. 2019;21:e14017. Available from: https://doi.org/10.2196/14017.
4. Krummel TM. The rise of wearable technology in health care. *JAMA Netw Open*. 2019;2:e187672. Available from: https://doi.org/10.1001/jamanetworkopen.2018.7672.
5. Sitammagari K, Murphy S, Kowalkowski M, et al. Insights from rapid deployment of a "virtual hospital" as standard care during the COVID-19 pandemic. *Ann Intern Med*. 2020;. Available from: https://doi.org/10.7326/M20-4076.
6. Ur A, Gordon M. Origin of Korotkoff sounds. *Am J Physiol Legacy Content*. 1970;. Available from: https://doi.org/10.1152/ajplegacy.1970.218.2.524.
7. Rivera-Ruiz M, Cajavilca C, Varon J. Einthoven's string galvanometer. *Tex Heart Inst J*. 2008;35:174–178.

8. Master AL, Jaffe H, Pordy L. *Cardiac and Non-cardiac Chest Pain*. Annals of Internal Medicine; 1954.
9. Beach J. Pathway to retinal oximetry. *Transl Vis Sci Technol*. 2014;3. Available from: https://doi.org/10.1167/tvst.3.5.2.
10. Vandrico. Wearable Technology Database [WWW Document]. Wearables database. https://vandrico.com/wearables.html; 2021. Accessed 15.01.21.
11. Kannel WB, Benjamin EJ. Current perceptions of the epidemiology of atrial fibrillation. *Cardiol Clin*. 2009;27:13–24. Available from: https://doi.org/10.1016/j.ccl.2008.09.015. vii.
12. Schnabel RB, Yin X, Gona P, et al. 50 year trends in atrial fibrillation prevalence, incidence, risk factors, and mortality in the Framingham Heart Study: a cohort study. *Lancet*. 2015;386:154–162. Available from: https://doi.org/10.1016/S0140-6736(14)61774-8.
13. Benjamin EJ, Muntner P, Alonso A, et al. Heart disease and stroke statistics-2019 update: a report from the American Heart Association. *Circulation*. 2019;139:e56–e528. Available from: https://doi.org/10.1161/CIR.0000000000000659.
14. Turakhia M, Shafrin J, Bognar K, et al. Estimated prevalence of undiagnosed atrial fibrillation in the United States. *PLoS One*. 2018;13: e0195088. Available from: https://doi.org/10.1371/journal.pone.0195088.
15. Lip GYH, Kakar P, Watson T. Atrial fibrillation—the growing epidemic. *Heart*. 2007;93:542–543. Available from: https://doi.org/10.1136/hrt.2006.110791.
16. Staerk L, Sherer JA, Ko D, Benjamin EJ, Helm RH. Atrial fibrillation: epidemiology, pathophysiology, and clinical outcomes. *Circ Res*. 2017;120:1501–1517. Available from: https://doi.org/10.1161/CIRCRESAHA.117.309732.
17. Markides V, Schilling RJ. Atrial fibrillation: classification, pathophysiology, mechanisms and drug treatment. *Heart*. 2003;89:939–943.
18. Kornej J, Börschel CS, Benjamin EJ, Schnabel RB. Epidemiology of atrial fibrillation in the 21st century: novel methods and new insights. *Circ Res*. 2020;127:4–20. Available from: https://doi.org/10.1161/CIRCRESAHA.120.316340.
19. Gillinov S, Etiwy M, Wang R, et al. Variable accuracy of wearable heart rate monitors during aerobic exercise. *Med Sci Sports Exerc*. 2017;49:1697–1703. Available from: https://doi.org/10.1249/MSS.0000000000001284.
20. Seshadri D, Bittel B, Browsky D, et al. Accuracy of apple watch for detection of atrial fibrillation. *Circulation*. 2020;141:702–703. Available from: https://doi.org/10.1161/CIRCULATIONAHA.119.044126.
21. Bumgarner JM, Lambert CT, Hussein AA, et al. Smartwatch algorithm for automated detection of atrial fibrillation. *J Am Coll Cardiol*. 2018;71:2381–2388. Available from: https://doi.org/10.1016/j.jacc.2018.03.003.
22. Perez MV, Mahaffey KW, Hedlin H, et al. Large-scale assessment of a smartwatch to identify atrial fibrillation. *N Engl J Med*. 2019;381:1909–1917. Available from: https://doi.org/10.1056/NEJMoa1901183.
23. Bent B, Goldstein BA, Kibbe WA, Dunn JP. Investigating sources of inaccuracy in wearable optical heart rate sensors. *Npj Digital Med*. 2020;3:1–9. Available from: https://doi.org/10.1038/s41746-020-0226-6.
24. Bekfani T, Fudim M, Cleland JGF, et al. A current and future outlook on upcoming technologies in remote monitoring of patients with heart failure. *Eur J Heart Fail*. 2020;. Available from: https://doi.org/10.1002/ejhf.2033.
25. Cohen E, Cohen MI. COVID-19 will forever change the landscape of telemedicine. *Curr Opin Cardiol*. 2021;36:110–115. Available from: https://doi.org/10.1097/HCO.0000000000000806.
26. Robinson KR. Comparing the Spanish flu and COVID-19 pandemics: lessons to carry forward. *Nurs Forum*. 2020;. Available from: https://doi.org/10.1111/nuf.12534.
27. Abrams EM, Singer AG, Shaker M, Greenhawt M. What the COVID-19 pandemic can teach us about resource stewardship and quality in health care. *J Allergy Clin Immunol Pract*. 2021;9:608–612. Available from: https://doi.org/10.1016/j.jaip.2020.11.033.
28. De Filippo O, D'Ascenzo F, Angelini F, et al. Reduced rate of hospital admissions for ACS during Covid-19 outbreak in Northern Italy. *N Engl J Med*. 2020;383:88–89. Available from: https://doi.org/10.1056/NEJMc2009166.
29. Solomon MD, McNulty EJ, Rana JS, et al. The Covid-19 pandemic and the incidence of acute myocardial infarction. *N Engl J Med*. 2020;383:691–693. Available from: https://doi.org/10.1056/NEJMc2015630.
30. West HJ. Telemedicine in oncology: delivering on an overdue promise in the COVID-19 era. *Front Oncol*. 2020;10. Available from: https://doi.org/10.3389/fonc.2020.578888.
31. Kichloo A, Albosta M, Dettloff K, et al. Telemedicine, the current COVID-19 pandemic and the future: a narrative review and perspectives moving forward in the USA. *Fam Med Community Health*. 2020;8. Available from: https://doi.org/10.1136/fmch-2020-000530.
32. Koller C, Conway P. *Reexamining Medicare Payment Reforms in Light of Covid-19*. NEJM Catalyst Innovations in Care Delivery; 2020.
33. McElroy JA, Day TM, Becevic M. The influence of telehealth for better health across communities. *Prev Chronic Dis*. 2020;17. Available from: https://doi.org/10.5888/pcd17.200254.
34. Gilbert AW, Billany JCT, Adam R, et al. Rapid implementation of virtual clinics due to COVID-19: report and early evaluation of a quality improvement initiative. *BMJ Open Qual*. 2020;9. Available from: https://doi.org/10.1136/bmjoq-2020-000985.
35. PRNewswire. Leading Health Innovators Launch Alliance To Advance Care In The Home [WWW Document]. https://www.prnewswire.com/news-releases/leading-health-innovators-launch-alliance-to-advance-care-in-the-home-301239624.html; 2021. Accessed 03.04.21.
36. Seshadri DR, Davies EV, Harlow ER, et al. Wearable sensors for COVID-19: a call to action to harness our digital infrastructure for remote patient monitoring and virtual assessments. *Front Digital Health*. 2020;2:8. Available from: https://doi.org/10.3389/fdgth.2020.00008.
37. Ferrari R, Di Pasquale G, Rapezzi C. Commentary: what is the relationship between Covid-19 and cardiovascular disease? *Int J Cardiol*. 2020;310:167–168. Available from: https://doi.org/10.1016/j.ijcard.2020.03.074.
38. Guzik TJ, Mohiddin SA, Dimarco A, et al. COVID-19 and the cardiovascular system: implications for risk assessment, diagnosis, and treatment options. *Cardiovasc Res*. 2020;. Available from: https://doi.org/10.1093/cvr/cvaa106.

39. Mishra T, Wang M, Metwally AA, et al. Pre-symptomatic detection of COVID-19 from smartwatch data. *Nat Biomed Eng.* 2020;4:1208–1220. Available from: https://doi.org/10.1038/s41551-020-00640-6.
40. Natarajan A, Su H-W, Heneghan C. Assessment of physiological signs associated with COVID-19 measured using wearable devices. *Npj Digital Med.* 2020;3:1–8. Available from: https://doi.org/10.1038/s41746-020-00363-7.
41. Quer G, Radin JM, Gadaleta M, et al. Wearable sensor data and self-reported symptoms for COVID-19 detection. *Nat Med.* 2021;27:73–77. Available from: https://doi.org/10.1038/s41591-020-1123-x.
42. IEEE. IEEE/ISO 11073-10441-2008 - IEEE/ISO International Standard - Health Informatics - Personal Health Device Communication - Part 10441: Device Specialization - Cardiovascular Fitness and Activity Monitor [WWW Document]. https://standards.ieee.org/standard/11073-10441-2008.html; 2009. Accessed 03.05.21.
43. Abdolkhani R, Gray K, Borda A, DeSouza R. Quality assurance of health wearables data: participatory workshop on barriers, solutions, and expectations. *JMIR mHealth uHealth.* 2020;8. Available from: https://doi.org/10.2196/15329.
44. Abdolkhani R, Gray K, Borda A, DeSouza R. Patient-generated health data management and quality challenges in remote patient monitoring. *JAMIA Open.* 2019;2:471–478. Available from: https://doi.org/10.1093/jamiaopen/ooz036.
45. Ristevski B, Chen M. Big data analytics in medicine and healthcare. *J Integr Bioinform.* 2018;15. Available from: https://doi.org/10.1515/jib-2017-0030.
46. Tajirian T, Stergiopoulos V, Strudwick G, et al. The influence of electronic health record use on physician burnout: cross-sectional survey. *J Med Internet Res.* 2020;22:e19274. Available from: https://doi.org/10.2196/19274.
47. Abbasi J. Using smartwatch data to detect COVID-19 cases early. *JAMA.* 2020;324:2247. Available from: https://doi.org/10.1001/jama.2020.23696.
48. Zhu T, Watkinson P, Clifton DA. Smartwatch data help detect COVID-19. *Nat Biomed Eng.* 2020;4:1125–1127. Available from: https://doi.org/10.1038/s41551-020-00659-9.
49. Han JK, Al-Khatib SM, Albert CM. Changes in the digital health landscape in cardiac electrophysiology: a pre-and peri-pandemic COVID-19 era survey. *Cardiovascular Digital Health J.* 2021;2:55–62. Available from: https://doi.org/10.1016/j.cvdhj.2020.12.001.
50. FDA-NIH Biomarker Working Group. BEST (Biomarkers, EndpointS, and other Tools) Resource. Silver Spring (MD): Food and Drug Administration (US); Bethesda (MD): National Institutes of Health (US), www.ncbi.nlm.nih.gov/books/NBK326791/ (2016, accessed 22 September 2017).
51. Hamburg MA. Innovation, Regulation, and the FDA. *New England Journal of Medicine.* 2010;363:2228–2232. Available from: https://doi.org/10.1056/NEJMsa1007467.

Chapter 32

Intelligent Systems and Smart Devices for the Continuous Monitoring of Cardiac Hemodynamics

Seraina Anne Dual[1] **and Marianne Schmid Daners**[2]

[1]Department of Radiology, Stanford University, Palo Alto, CA, United States, [2]Product Development Group Zurich, Mechanical and Process Engineering, ETH Zurich, Zurich, Switzerland

Chapter Outline

Abbreviations

ECG	electrocardiogram
EDV	end-diastolic volume
LOA	limits of agreement
LV	left ventricle/ventricular
LVAD	left ventricular assist device
μ	mean bias
MEMS	microelectromechanical system
MRI	magnetic resonance imaging
PV	pressure–volume
RV	right ventricle/ventricular

Introduction

The clinical management of patients with cardiovascular disease relies on the availability of real-time and continuous hemodynamic data. The hemodynamics of a patient are specified if data on the cardiovascular system's pressure, volume, and flow are available. Ideally, a clinician would have access to pressure, volume (PV relationship), and flow at any given point in time and at any location within the cardiovascular system. However, due to a lack of invasive methods many fields of cardiovascular medicine have accepted working with the indirect data that is available in order to diagnose, follow-up, and treat patients with cardiovascular disease.

We would like to start this chapter with a couple of illustrative examples on how pressure, volume, and flow data inform clinical practice today. Congestive heart failure has become more and more common[1] and may result in

Advances in Cardiovascular Technology. DOI: https://doi.org/10.1016/B978-0-12-816861-5.00007-1

progressive remodeling of the heart if left untreated. One out of five heart failure patients die within 1 year of diagnosis.[2] The key dynamic of congestive heart failure is volume overload. The disease requires a careful balancing of volume status satisfying patient perfusion needs, while not overloading the damaged heart. Hence, stabilization of patients with heart failure could be achieved using long-term monitoring of the volume of the heart. While in coarctations the pressure difference is important, other diseases require the assessment of overall levels of pressure. For example, pulmonary hypertension is diagnosed by means of the mean pulmonary arterial pressure.[3] Similarly, heart failure with preserved ejection fraction is associated with high left ventricular (LV) filling pressures. In all cases long-term pressure monitoring would help diagnose patients early[4] and allow a timely initiation of therapy.

In the past and currently, hemodynamic assessment has been performed mostly upon consultation at the hospital. At the same time, advanced technologies evidence the potential benefits of remote patient monitoring.[5] But long-term continuous and real-time hemodynamic patient monitoring remains technically challenging. Clinically, we use intermitted magnet resonance imaging (MRI) and ultrasound, as well as continuous invasive pressure measurements for hemodynamic assessment. Due to the bulky size of the apparatus required or the invasiveness of those methods, the application of such technologies usually remains limited to the hospital setting.

Electric signals have been acquired for some time already, by either portable electrocardiogram (ECG) monitors, the apple watch, or medical devices such as pacemakers. In the latter, telemetric recordings of pacemakers have helped to manage these patients based on events that happen in the home setting. However, hemodynamic signals such as volumes, pressures, and flows are innately more difficult to measure than pacemaker signals. Very recently, it was shown that management of patients using the CardioMEMS device (Abbott Laboratories, Atlanta, GA, United States) has helped to reduce hospitalizations by one third.[6] Motivated by the prospect of improving patient care, hemodynamic monitoring is very much an active field of research with applications in numerous fields of cardiovascular medicine.[7]

The aim of this chapter is to introduce the reader to a general framework of what recent advances have been made in the field of intelligent systems and smart devices for real-time and continuous hemodynamic monitoring of pressure, volume, and flow. We intend to show how we can transition from clinical methods to implantable systems, and finally integrate such sensors into smart medical devices (Fig. 32.1). Stand-alone implantable systems serve the purpose to monitor the patient's health status, which gives clinicians the chance to quickly adjust therapy. In research, physiological control of, for example, an LV assist device (LVAD) has been attempted and is challenging because of the lack of the respective control signal needed.[14–18] Hence, multiple efforts have been made to enable continuous measurements from the cannula of an LVAD. We present recent implantable and integrated sensors, discuss their accuracy, need for patient-specific calibration, time resolution, and biocompatibility. Accuracy is reported as mean offset values (μ) and 95% limits of agreement (LOA).

Pressure sensors

Clinical systems

In the clinic as well as at home, arterial pressure is generally measured noninvasively and only intermittently by cuff sphygmomanometry, which is a very old technology.[19] A cuff is wrapped around the upper arm, the wrist, or the finger measuring the peripheral arterial pressure.[20,21] Patients autonomously and manually monitor their blood pressure at home and report regularly to their general practitioner or clinician who manage their therapy. In contrast, invasive cardiac catheterization measures hemodynamic pressure waveforms, which render accurate systolic and diastolic pressures. Compared to hemodynamic pressures, the cuff pressures show comparable systolic pressures, but tend to overestimate diastolic pressures.[22]

During surgery or in the intensive care unit continuous invasive pressure recordings are employed. A transducer connected to the radial or femoral artery continuously measures arterial pressure and displays the recordings. A regular central venous catheter[23] records central venous pressure while a Swan–Ganz catheter[24,25] allows additional pulmonary arterial and left atrial pressure (Wedge pressure) measurements.[26] Central venous and left atrial pressure correspond to right and LV preload. Doppler ultrasound is a noninvasive method to estimate right ventricular (RV) and pulmonary arterial pressure.[27] The RV pressure is calculated as the sum of right atrial pressure, estimated from the diameter of the vena cava, and the pressure gradient across the tricuspid valve, estimated using the Bernoulli equation. Validated against invasive hemodynamics, Doppler echocardiography estimates RV systolic pressures with good accuracy (μ <5.5 mmHg), but wide LOA for experienced readers (15.9 mmHg) and for clinical reports ($\pm$ 33.5 mmHg) suggesting low precision.[27]

A large body of research focuses on alternative ways to noninvasively measure continuous pressure based on a combination of various signals available in clinical practice. For instance, blood pressure may be estimated by

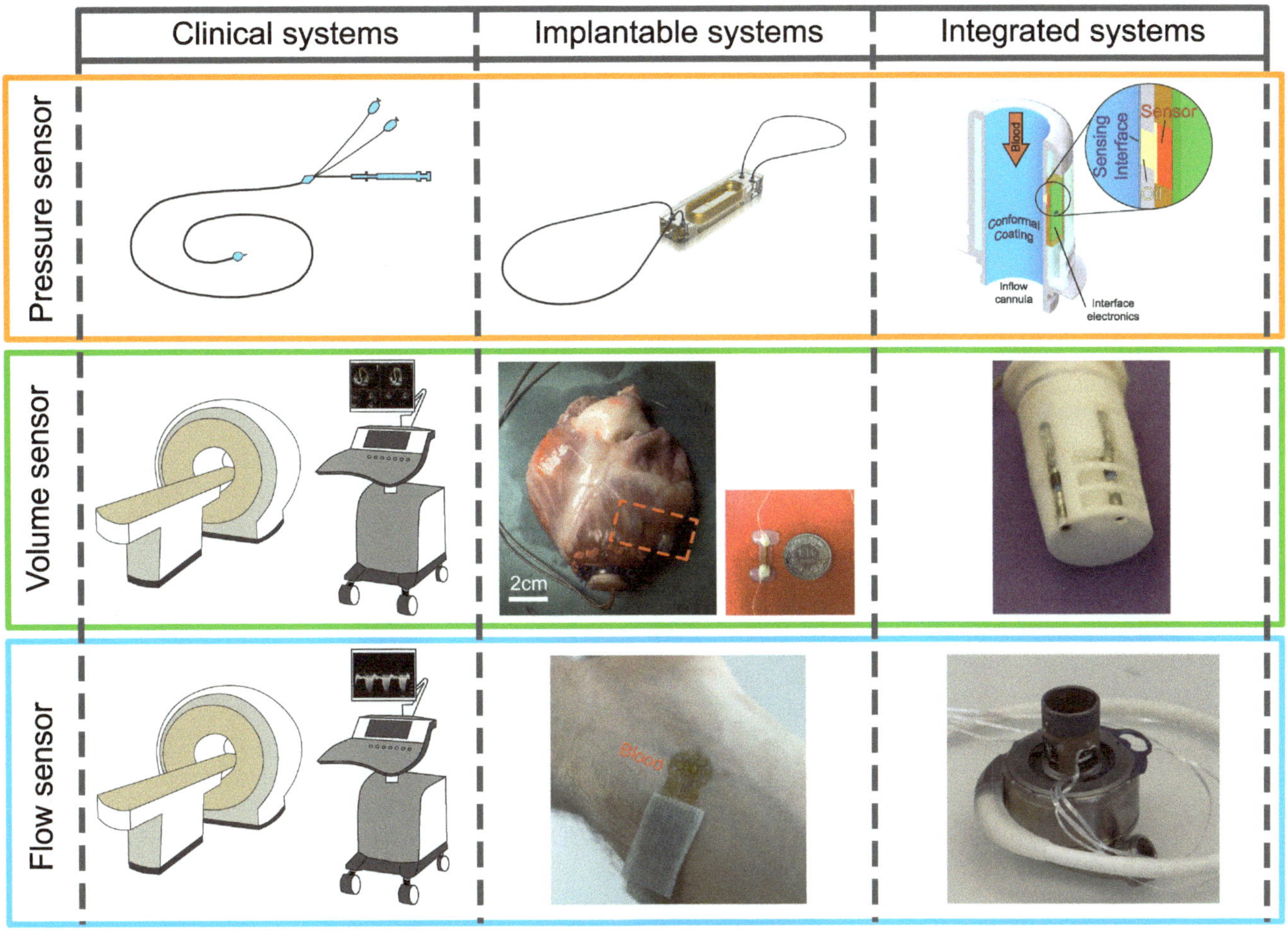

FIGURE 32.1 Clinical systems measure the hemodynamic variables pressure, volume, and flow, but are invasive or bulky. Large progress has been made recently toward continuous hemodynamic monitoring using implantable and medical device integrated technologies.[8–13] *Credit: Top right: Illustration courtesy of Staufert, S. Conformal Parylene-C media separating membranes for pressure sensing applications in ventricular assist devices. (Ph.D. thesis), ETH Zurich, 2018. Middle: Reprinted, with permission, from Dual, S. A. et al. Continuous heart volume monitoring by fully implantable soft strain sensor. Adv Healthc Mater M, 1–11, 2020. Bottom middle: Reprinted, from Webb et al. under CC BY-NC 4.0, 2015, Science Advances.*

machine-learning algorithms applied to the ECG,[28] directly using the R-wave magnitude of lead II of the ECG,[29] or by the pulse transit time assessed from the ECG and plethysmogram, both of which are standard signals available in anesthesia monitors.[21,30]

Implantable stand-alone systems

The CardioMEMS device is implanted in the pulmonary artery and transmits data wirelessly.[31] The FDA approved CardioMEMS as an implantable and biocompatible telemetric system to monitor cardiac hemodynamics[32,33] (Fig. 32.2). The telemetric surveillance system enables outpatient monitoring and may reduce the number of readmissions.[6] It measures the pulmonary arterial pressure and demonstrates that this pulmonary arterial pressure can monitor the volume status changes in heart failure patients more accurately than with traditional methods,[34,35] for example, weight monitoring.[36] After 6 months, comparable results were obtained from simultaneous pressure surveillance and hemodynamic monitoring by the CardioMEMS sensor and an invasive Millar catheter.[32,37]

Various other implantable hemodynamic monitoring systems have been developed to continuously measure RV and pulmonary arterial pressures[38,39] or left atrial pressure by percutaneous transseptal implantation of a pressure sensor[40] and were tested in clinical trials.[41] The Titan telemetric implant is a microelectromechanical pressure sensor system with transcutaneous data readout that has shown promising preliminary clinical results.[42,43] Furthermore, during in-vitro and in-vivo studies, a magnetic sensor mounted at the outside of the artery was shown to measure the aortic pressure by the distension of the vessel.[44]

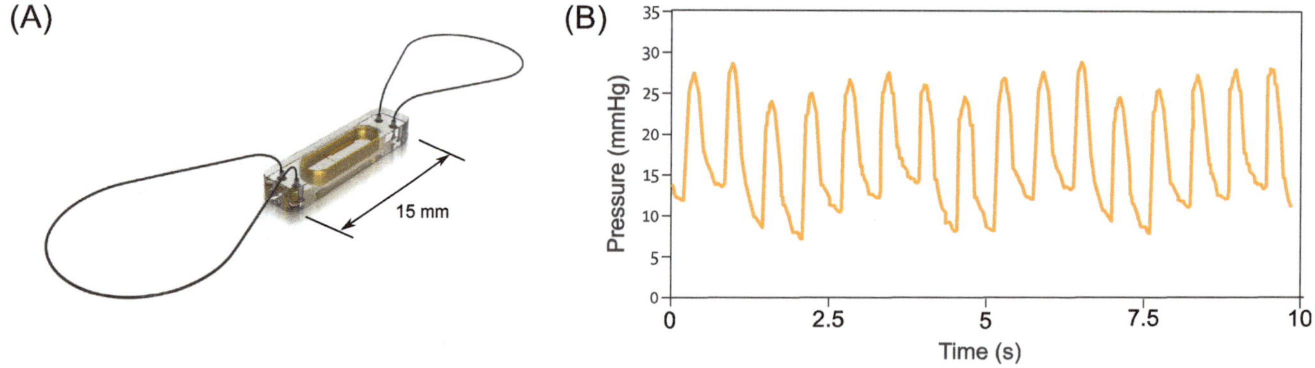

FIGURE 32.2 The CardioMEMS pressure sensor (A) is implanted in the pulmonary artery where it measures invasive pressures continuously and remotely (B). *Credit: Figure reprinted, with permission, from Anker, S. D., Koehler, F. & Abraham, W. T. Telemedicine and remote management of patients with heart failure. Lancet 378, 731–739 (2011).*

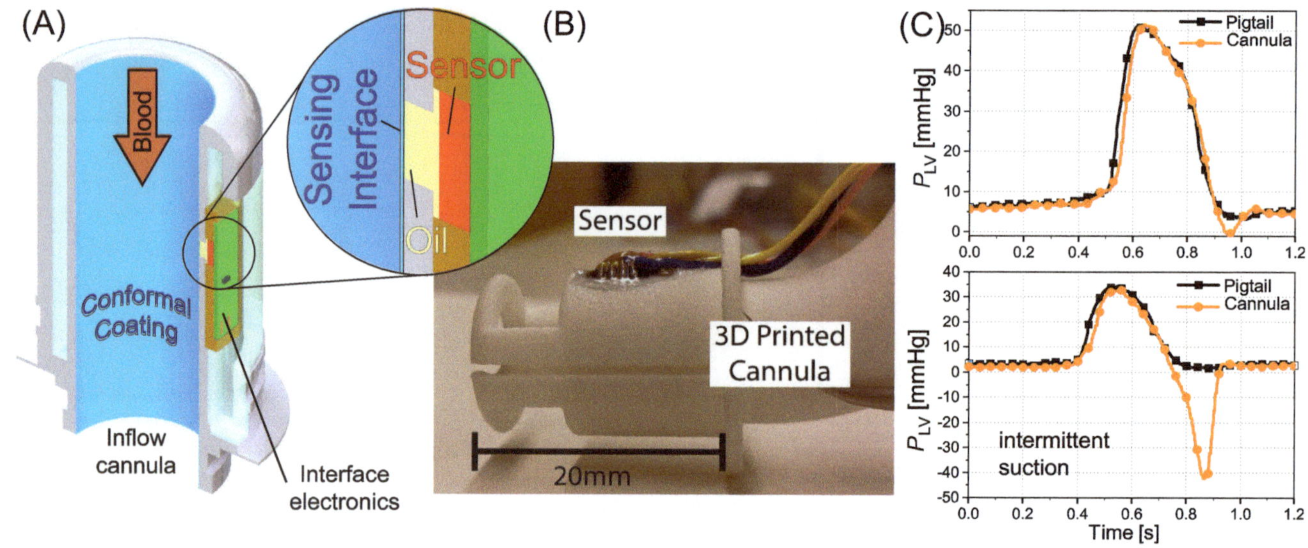

FIGURE 32.3 (A) The pressure sensor is integrated into the cannula of an LVAD using an oil interface and diaphragm.[50] The integrated sensors (B)[14] are sensitive to adverse events such as intermittent suction (C).[8] *Credit: Illustration courtesy of Staufert, S. Conformal Parylene-C media separating membranes for pressure sensing applications in ventricular assist devices. (Ph.D. thesis), ETH Zurich, 2018.*

Integrated systems

Thus far, pressure sensors integrated for instance on the inflow cannula of an LVAD are not available. One problem is the long-term stability required that has not been achieved to date. One pressure sensor was developed that is based on the microelectromechanical system (MEMS) technology to be used for medical applications.[45] It is of particular interest as it consumes little power, is small with a size of 3 mm^2, is biocompatible, and has the potential to be combined with telemetric power and data transmission. Another pressure sensor specifically designed for the LVAD cannula could not overcome the inherent drift, which varied between 180 and −140 mmHg over a 4-week period, but accurately resolved pressures throughout the cardiac cycle.[46] A further design based on the Fabry-Pérot interferometer principle included a Parylene-C coated diaphragm separating the bloodstream from the sensor and showed promising results in vitro.[47] In an alternative approach, the pressure sensor was coated with Parylene-C. That sensor showed a good in vitro stability of less than 0.05 mmHg/day drift after 50 h.[48] By enhancing the intrinsic hydrophobicity of parylene the coated sensor increased the adsorption of plasma proteins and minimized the adhesion of platelets on the surface.[49]

Long-term stable pressure measurements are possible by coating (Fig. 32.3).[50] Measurement results of over 30 days showed no statistically significant drift and the maximum absolute error recorded for a single measurement was at −0.78 mmHg, tested across a typical pressure range. Parylene-C coating has excellent mechanical properties to transmit the blood pressure to the sensor while being biocompatible. However, the long-term stability of Parylene-C is limited by poor adhesion properties to commonly used implant alloys. In conjunction with novel titanium oxide nanostructures the membrane may improve its long-term adhesion stability in permanent contact with fluids.[51] In a recent study, pressure sensors were integrated into a modified tapered inflow cannula of a commercial LVAD.[10] The cannula was 3D printed based on selective laser melting and the sensors were seamlessly integrated by the covering membrane. Tested in vitro, the cannula prototypes enabled continuous pressure measurements from within the cannula with a pressure range of −100–200 mmHg.

Volume sensors

Clinical systems

Volumes of the right and left heart are regularly assessed on a clinical basis. Bedside ultrasound tests are used most regularly to assess ventricular dimensions (2D), biventricular volumes (3D), and volumes of the cardiac cycle (4D). While echocardiography (echo) is considered cheap, it typically requires the patient to be in hospital and a clinician to be present each time a measurement is required. Transthoracic echo is complicated by the high interobserver variability, which results in LOA of 18 or 12 mL using 2D or 3D images, respectively.[52] Similarly, MRI yields time-resolved volumes at increased accuracy, but still requires clinical intervention. Furthermore, some patients may not be eligible for MRI, that is, patients with implants that are not MR compatible.

At home, so-called continuous estimation of volume is effected using weight tracking or cuff blood pressure monitoring performed manually by the patient. However, in either case, the measurement should be considered indirect and is thus a surrogate for volume measurement only. Recent advances in the computational efficiency of smartphones render ultrasonic systems smaller and portable[53] or even enable invasive hemodynamic assessments.[54] Additionally, efforts have gone into indirect measurements of cardiac output and stroke volume via electric impedance tomography[55–57] or plethymography[58] and ballistocardiography,[59] which are integrated into smart weight scales.[60] These multimodality approaches may offer real-time and robust measurement systems and are thus quite promising.

Implantable stand-alone systems

A fully portable or implantable system that measures LV volume continuously is currently unavailable for monitoring patients with cardiovascular disease. An implantable system to measure LV volume was proposed using a flexible epicardial strain sensor.[13] Using sutures, the strain sensor is attached to the epicardial wall close to the apex. Upon filling of the LV, the myocardium stretches, which in turn is picked up by the resistive strain sensor. The system is suggested to be helpful in monitoring patients after cardiac surgery for signs of vasoplegia during their time in the intensive care unit. The sensor is made from biocompatible materials (gold, silicone), has myocardium-like material parameters, and is just 1 cm long (Fig. 32.4).[13] As such, the sensor does not affect the native heart function. The sensor has proved to yield stable readings while immersed in a blood-mimicking solution for a week. The material is durable enough to measure for three days of continuous stretching. The sensor was found to be more accurate than echo

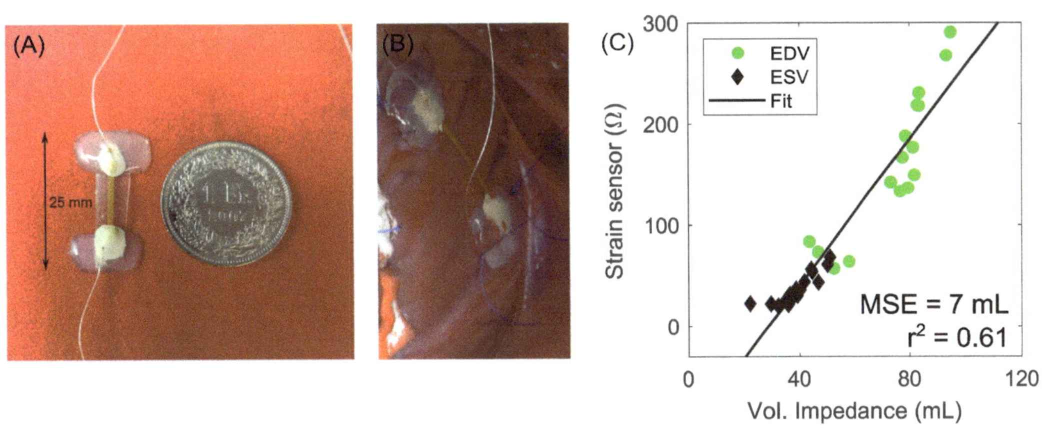

FIGURE 32.4 A soft strain sensor (A) was implanted epicardially in vivo (B). The proof of concept study showed promising accuracy for estimating end-diastolic volume (C). *Credit: Reprinted, with permission, from Dual, S. A. et al. Continuous Heart Volume Monitoring by Fully Implantable Soft Strain Sensor. Adv Healthc Mater M, 1–11 (2020).*

(LOA: 7 mL vs 14 mL) during severe hemodynamic perturbations. Furthermore, the sensor can easily detect sudden acute changes in LV volume mimicked by sudden blood loss in the study. However, the reading requires a subject-specific calibration and is thus indirect. The sensor is small enough to be implanted in other cardiovascular settings, such as to measure strain of the atria or the RV. To date, this has not been attempted.

Similar epicardial strain measurements have been attempted before. A 3D multifunctional membrane was suggested to be used to circumvent the entire heart while recording the local pH, temperature and strains using semiconductor materials which are typically not biocompatible.[61] This system is technically intriguing, but surgically difficult to place around the heart. A liquid-metal-based RV strain sensor was further proposed to monitor RV strains post LVAD implantation.[62] No attempt was made to relate such measurements with the RV volume, and the sensor is rather large.

Integrated systems

The LV end-diastolic volume (EDV) has shown to be a robust control parameter for a physiologic adaptation of LVADs.[14] Preload-based controllers outperformed most alternative control approaches in an extensive in vitro study on LVAD control.[16] Moreover, EDV-based physiologic control has shown to be robust in vitro[63] and feasible in vivo.[14] However, the real-time measurement of the EDV remains technically challenging.

Ultrasound echo is clinically the most established method to measure LV volume. At the same time, significant technological improvements have been made in computational efficiency as well as in miniaturizing ultrasonic systems ($1.7\ cm^3$).[64] Together, this renders ultrasound as a prime candidate for integration into the cannula of an LVAD. Multiple ultrasound transducers on the cannula of an LVAD have been shown to achieve accuracies that are similar to those attained by echo, that is, μ: 10% offset and LOA: ±40%, for volume measurements in anatomically detailed patient-specific heart phantoms.[65] In contrast to transthoracic echo, the placement of the sensor inside the LV makes the measurement more robust against disturbances and additionally allows continuous measurements. A similar setup enabled the prediction of the position of the interventricular septum (IVS) using an ultrasonic sensor. A leftward IVS shift is a condition that should be avoided during LVAD support for preserving mechanical RV integrity.[66] The prediction of the septum position is particularly relevant in small hearts, which overproportionally suffer from right-sided failure after an LVAD implant.[67]

In electric impedance, two electrodes inject an electric current, and voltages between the other electrodes are measured at different levels of the LV. From the voltages measured the impedance is calculated and the volume is assessed. The impedance was measured between left- and right-ventricular leads with a LOA of ±25 mL (9.7%) in 22 patients and with a two-point calibration.[68] Most recently, electric impedance has been integrated in an LVAD cannula, and it correlated reasonably well with the end-diastolic diameter of the LV.[9,69] Currently, the impedance measurement in the cannula is mostly sensitive to the apical LV enlargement. Even though the measurement of impedance is generally a very simple approach, the method requires a patient-specific two-point calibration.

The unipolar measurement of the intracardiac electromyogram (iEMG) offers an alternative, which may capture more of the LV blood pool and does not require an electric current to be injected in the LV. The depolarization amplitude of the iEMG is physically related to the measurements of the LV blood volume via the Brody effect. Recent advancements in unipolar measurement technology[11,70] make the Brody effect practicable for the beat-wise estimation of the EDV.[71] Conventional pacemaker electrodes can be used for the measurement, which makes this approach applicable for integration into the cannula of an LVAD (Fig. 32.5). Compared against electric impedance and

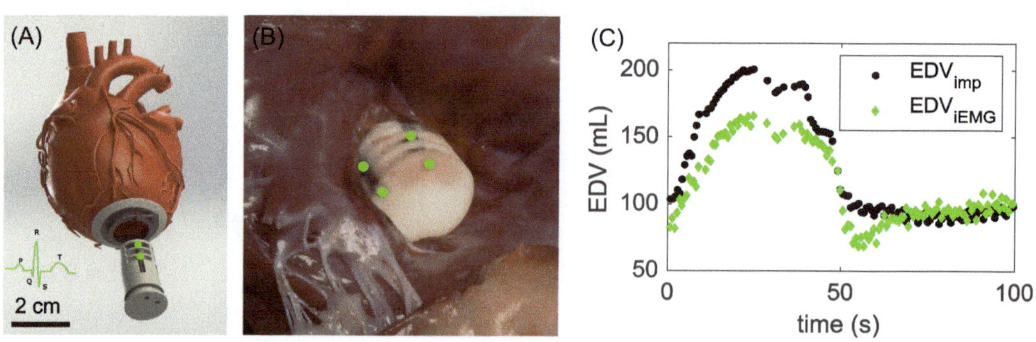

FIGURE 32.5 Four commercial pacemaker electrodes were integrated on the dummy cannula of left ventricular assist device (A). The positioning within the blood pool allowed the results of the real-time estimation of the end-diastolic volume to be similar to those of the catheter electric impedance estimates.[11,71] *Credit: Reprinted, from Schmid Daners, M. et al. Real-Time Ventricular Volume Measured Using the Intracardiac Electromyogram. ASAIO J Publish Ah, 1–9 (2021); under CC BY-NC-ND 4.0.*

transesophageal echo (TEE), the estimates of EDV showed excellent agreement (r-squared = 0.85 and r-squared = 0.83) and resulted in LOA values of ± 15.8 and ± 15.2 mL, respectively. Furthermore, the sensor predicted the ventricular response to an occlusion of the descending aorta very effectively. However, the current results suggest a patient-specific calibration to be necessary.

Typically, systems measure LV volume using one physical quantity alone. However, recent studies focus on the combination of multiple modalities using advanced analytics, such as optimization. A recent study has shown that by combining ultrasonics and electric impedance approaches one can omit patient-specific calibration while preserving the desired accuracy.[72]

Flow sensors

Clinical systems

Clinicians have access to low- and high-resolution 2D, 3D, and 4D imaging, such as computer tomography (CT), MRI, and echo. Besides volume, specific types of imaging systems also resolve the blood flow and velocity distributions in specific compartments of the cardiovascular system. However, all clinical flow recordings remain limited to one-point measurements.

In contrast to larger MRI and CT scanners, echo is more portable and is thus typically employed as a bedside monitoring unit. The 2D echo Doppler examinations acquire colorful maps of velocity data and allow the study of valvular flow patterns such as mitral regurgitation.[73] echo Doppler examinations are most important for the observation of short- and long-term outcomes post aortic valve replacement.[74] Overall, echo Doppler guides the treatment of LVAD patients.[75] In patients with an LVAD, the cardiac output can be assessed by measuring the flow through the LVAD and the aortic valve.[76] Transthoracic echo of the aortic valve guides pump speed settings or is sensitive to detect device-related complications, such as reoccurring heart failure, or aortic regurgitation. Furthermore, continuous monitoring may help to reduce the development of aortic insufficiency.[77] Two limitations of echo are observer variability in the choice of the imaging plane perpendicular to the flow and the rather low image resolution.

The 4D flow MRI measurements[78] are used to record flow dynamics like pulsatile flow patterns in cavities, through heart valves or in large vessels, and are particularly relevant in clinic to treat congenital heart defects. In MRI, the flow information is encoded in the phase of the recorded signals and is therefore often referred to as phase-contrast imaging. Clinically, high-resolution flows are acquired in 2D and, with a lower resolution for the assessment of qualitative flow fields, in 4D.[79] In the clinics, 4D scans typically last 10 min resulting in a 2-mm isotropic spatial resolution. Patients with medical implants such as an LVAD may have limited access as the devices may not be MR-safe. One further limitation is that children need to be sedated for MRI scanning to avoid movement artifacts. Additionally, the pulse transit time measured by flow can be used to derive vessel wall stiffness in vivo,[80] an approach that was recently validated in a synthetic model of the thoracic aorta.[81] In another study, LV blood flow was studied under stress in healthy pigs using 4D flow MRI characterizing representative flow fields for rest and exercise conditions.[82]

CT angiography is used to evaluate non-physiologic flow in arterial stenosis or occlusions for instance to diagnose an acute stroke,[83] aortic diseases, suspected pulmonary embolism or renal artery stenosis.[84] In CT angiography, the measured anatomical vessel diameter is correlated to flow, while in fractional flow reserve-computed tomography the measurement is further enhanced by integrating it with a computational cardiovascular model.[85] For instance, myocardial blood flow can be assessed by contrast enhanced cardiac CT, which supports the diagnosis of patients with a possible ischemic heart disease.[86] One important advantage of CT over MR is that there are no limitations to access a CT scanner due to an implant, however there is exposure to radiation.

Implantable stand-alone systems

Most technologies that are used to measure flow using an implantable system rely on wrapping a sensor system around the blood vessel of interest. They could thus provide chronic monitoring of blood flow in patients in the future. A MEMS-based implantable micro flow sensor is designed to measurearterial blood flow (Fig. 32.6A).[87] The sensor is based on the principle of thermotransfer mass flow and is intended to be mounted at the outside of the arterial wall, for example, the coronaries. During simulation, the temperature changed from initial 311.5 K to 310 and 320 K up- and downstream, respectively. Up to a difference of 10 K that corresponds to a flow rate of 3.5×10^8 $\mu m^3/s$, the sensor is sensitive to flow rate variations before it becomes saturated. Another compact and fully implantable wireless blood flow monitoring system based on Doppler ultrasound with two piezoelectric transducers was tested in vivo.[88] Venous flow in pigs' limbs was measured accurately across its calibrated range with a relative standard error of less than ± 5%

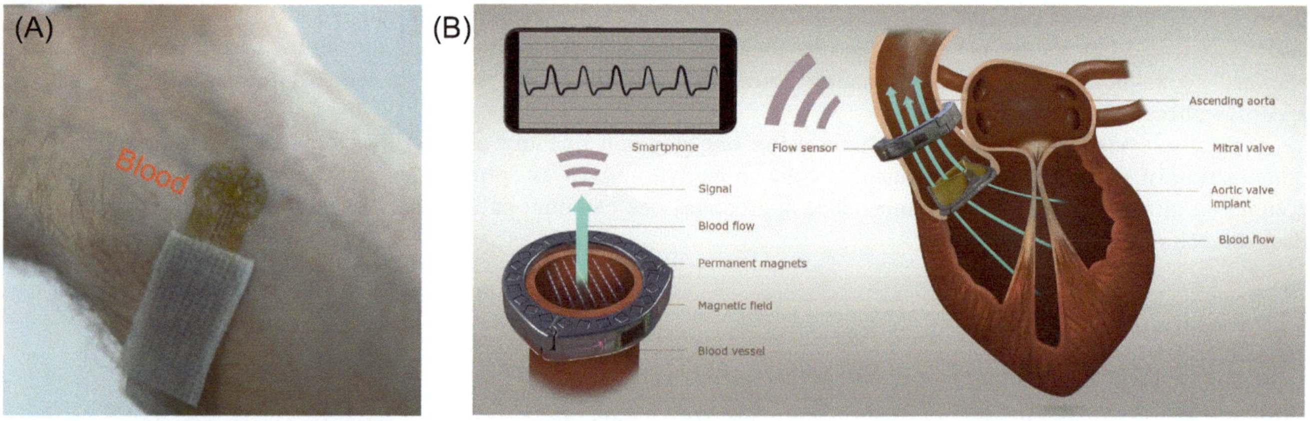

FIGURE 32.6 Examples of implantable systems to monitor and measure flow with wireless readout: (A) peripheral arterial flow sensor,[12] and (B) aortic flow and prosthetic heart valve function.[87,89] *Credit: A Reprinted, from Webb, R. C. et al. Materials Science/Clinical Medi Cine: Epidermal devices for noninvasive, precise, and continuous mapping of macrovascular and microvascular blood flow. Sci Adv 1, 1–13 (2015); under CC BY-NC 4.0. B Reprinted, from Vennemann, B., Obrist, D. & Rösgen, T. A smartphone-enabled wireless and batteryless implantable blood flow sensor for remote monitoring of prosthetic heart valve function. PLoS One 15, e0227372 (2020), under CC BY 4.0.*

above 8 mL/min and between ±0.8% and 1.2% at the largest calibrated flow rate. Furthermore, in silico and in vitro results proved a clamp-on magnetic blood flow sensor exploiting Faraday's law to be most promising (Fig. 32.6 B).[89] The sensor system not only measures ascending aortic blood flow, but also monitors the quality of prosthetic heart valves. The system was miniaturized and optimized for low power consumption and the data was transmitted wirelessly to the patient's smartphone.

A recent study has proposed an epidermal, peripheral device for telemetric blood flow monitoring of patients with cardiovascular disease.[12] Although this epidermal device is currently not an implantable system, the flexible format renders it promising to be implanted or integrated with an implantable device. Presently, it is being investigated attached to the surface of the skin. At this peripheral location, it measures macro- and microvascular flow at a targeted area beneath the skin. Experimental studies on healthy subjects showed accurate results that compare well to conventional flow recordings under physiologic conditions. The readout allows the observation of peripheral blood flow and tissue perfusion based on the thermal distribution applied to a surface vessel.

Integrated systems

Measuring flow through an implant is most valuable for monitoring or control purposes. In the HeartAssist 5 and aVAD (ReliantHeart Inc., Houston, TX, United States) an integrated flow probe is located at the outflow graft, and they feature real-time telemetric surveillance capabilities.[90–92] In a cohort of 11 patients, parameters and alarms, such as pump flow, motor speed, and power consumption, were transmitted successfully via a remote monitoring platform to the hospital staff. Low-flow alarms indicating hypovolemia or suspected pump thrombosis were captured, which facilitated immediate action.[5] More recently, researchers proposed another system to measure flow that is based on two pressure sensors integrated in a tapered cannula of an LVAD (Fig. 32.7).[10] This sensor principle allows pump flow to be measured with a bias of $\mu = -0.14$ and LOA of ±0.08 L/min without requiring a dedicate flow sensor to be in place.

Alternatively to novel sensors, the estimation of blood flow through LVADs can also be estimated based on internal pump signals. For instance, blood flow was estimated with (1) the motor current and rotational speed combined with a mathematical model of a centrifugal pump,[93] (2) the intrinsic pump feature, a model of the human circulatory system assisted by an LVAD and an extended Kalman filter,[94] (3) static and dynamic properties of the brushless direct current model of a centrifugal pump,[95] and (4) the pump speed and pressure difference across the LVAD.[96] Four estimators have been compared in vitro under standardized testing conditions.[97] The mean errors of the overall estimated flow ranged from 0.002 to 0.38 L/min with LOA of 2 L/min.

Outlook

Major advancements have been made in all three fields of pressure, volume, and flow sensor systems. While implantable pressure sensors are already used in clinical practice, their integration into medical devices has been slower.

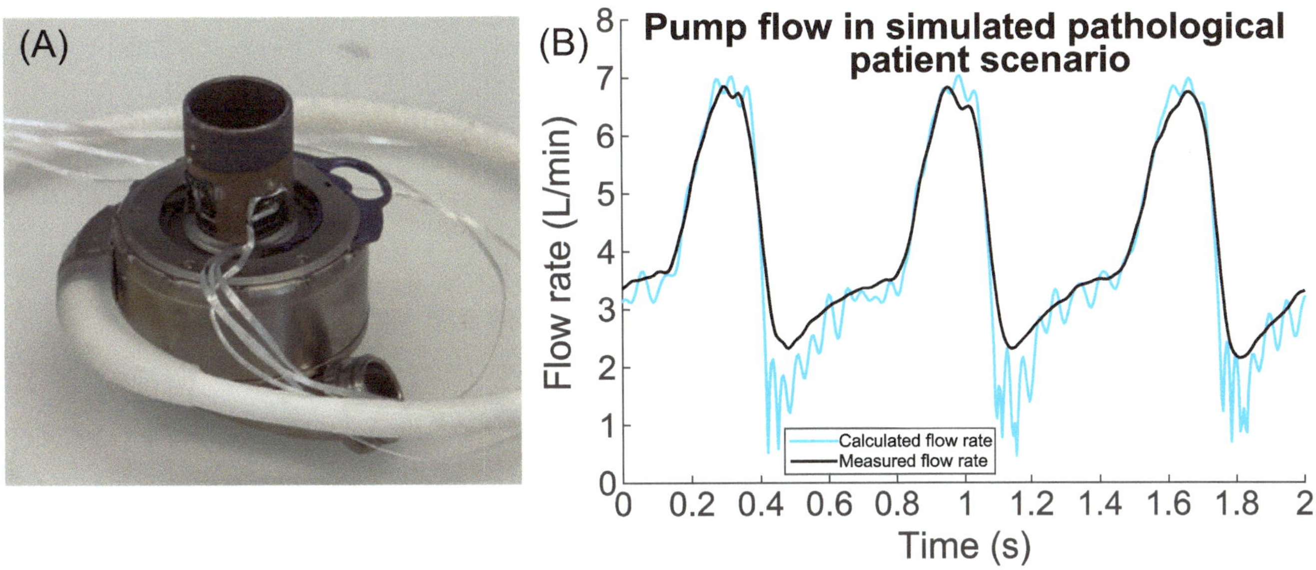

FIGURE 32.7 (A) Commercial LVAD with integrated pressure sensors in tapered inflow cannula and respective flows measured and calculated (B).[10]

Sensors measuring volume continuously have been proposed in research only, and multiple sensor principles have been investigated showing promising results. Continuous flow measurement techniques remain technically most challenging; however, integrated approaches are on the horizon.

Soon, we can expect both volume and pressure data to become available continuously in the clinic. A combination of both modalities is highly desirable, as it allows the calculation of derivative measures including the contractility or compliance of the heart (using for example the PV loop at the approximation by Klotz et al.[98]). Knowledge of contractility and compliance over time will impact our physiologic understanding of the progression of systolic as well as diastolic heart failure. In addition, to the assessment of heart dynamics, stiffness can also be estimated from volume and pressure for vessels, or as done through the skin in the periphery. Additional to flow, implantable devices that use ultrasound may well provide further insights into the time dynamics of valve function and dilation as well as the position of the septum. While continuous flow measurements are further away, they will allow a direct and continuous estimation of cardiac output.

Implantable and/or integrated volume and pressure sensors will not only enable long-term observations of the patient health status and management, but also to autonomously control LVAD flow according to the patient's perfusion demand. Wireless power and data transmission will (1) guarantee energy supply and (2) furnish telemetric data readout, which impacts timely treatment decision making. The immediate detection of alarming changes in health status will enable automatic adaptation of medical devices. At the same time, the data can help to build important data sets that can further increase our knowledge and understanding of individual as well as common health and disease progression.

Today, the potential applications for artificial intelligence algorithms and machine learning are overwhelming. However, these time series analyses require continuous data to be trained, to learn and to apply. Such dynamic data is also extremely rich in physiologic information. However, prior to being able to profit from such sophisticated analysis tools, we need to continue work on the sensor systems.

While the recent technical advancements are notable, we must remember that translation of these technical systems into clinical benefits may take time, owing to economic and regulatory considerations. The latter is a long-term procedure from development to preclinical studies to implants that are certified and approved for implantation (CE market and/or FDA approval). Besides economics and the official authorization of such technologies, the acceptance by clinicians and patients is needed in order for such technologies to be applied and before the management of heart failure patients can remarkably improve in the future.

References

1. Virani SS, et al. Heart disease and stroke statistics-2021 update: a report from the American Heart Association. *Circulation*. 2021;143.
2. Gerber Y, et al. A contemporary appraisal of the heart failure epidemic in Olmsted County, Minnesota, 2000 to 2010. *JAMA Intern Med*. 2015;175:996–1004.

3. Simonneau G, et al. Haemodynamic definitions and updated clinical classification of pulmonary hypertension. *Eur Respir J*. 2019;53.
4. Brown LM, et al. Delay in recognition of pulmonary arterial hypertension: Factors identified from the REVEAL registry. *Chest*. 2011;140:19–26.
5. Hohmann S, et al. Initial experience with telemonitoring in left ventricular assist device patients. *J Thorac Dis*. 2019;11:S853–S863.
6. Kanat N, Nichols M. 6 CardioMEMS for effective management of heart failure: reducing healthcare utilization and 30 Day readmissions. *Hear Lung*. 2017;46:213–214.
7. Hong YJ, Jeong H, Cho KW, Lu N, Kim D. Wearable and implantable devices for cardiovascular healthcare: from monitoring to therapy based on flexible and stretchable electronics. *Adv Funct Mater*. 2019;29:1808247.
8. Staufert, S. Conformal Parylene-C media separating membranes for pressure sensing applications in ventricular assist devices. (*Ph.D. thesis*), ETH Zurich, 2018.
9. Cysyk J, et al. Cannula tip with integrated volume sensor for rotary blood pump control: Early-stage development. *ASAIO J*. 2019;65:318–323.
10. von Petersdorff-Campen K, et al. Pressure and Bernoulli-based flow measurement via a tapered inflow VAD cannula. *IEEE Trans Biomed Eng*. 2021. Available from: https://doi.org/10.1109/TBME.2021.3123983.
11. Dual SA, Jacob D, Meboldt M, Schmid Daners M. Unipolar amplifier enabling measurement of far-field intra-cardiac electromyogram for blood pump control. In *BIODEVICES*, 40–48, SCITEPRESS. *Science and Technology Publications*. 2021.
12. Webb RC, et al. Materials science/clinical medi cine: Epidermal devices for noninvasive, precise, and continuous mapping of macrovascular and microvascular blood flow. *Sci Adv*. 2015;1:1–13.
13. Dual SA, et al. Continuous heart volume monitoring by fully implantable soft strain sensor. *Adv Healthc Mater*. 2020;M:1–11.
14. Ochsner G, et al. In vivo evaluation of physiologic control algorithms for left ventricular assist devices based on left ventricular volume or pressure. *ASAIO J*. 2017;63:568–577.
15. Rüschen D, Opitz S, Korn L, Leonhardt S, Walter M. Robust assistance control of left ventricular assist devices. *EMBEC ¬ NBC*. 2017;65:294–297.
16. Petrou A, et al. Standardized comparison of selected physiological controllers for rotary blood pumps: in vitro study. *Artif Organs*. 2018;42: E29–E42.
17. Giridharan, G. G. G., Pantalos, G., Koenig, S., Gillars, K. & Skliar, M. Achieving physiologic perfusion with ventricular assist devices: comparison of control strategies. Proceedings of the American Control Conference 2005. 3823–3828 (2005). Available from: https://doi.org/10.1109/ACC.2005.1470570.
18. Fetanat M, Stevens M, Hayward C, Lovell NH. A physiological control system for an implantable heart pump that accommodates for interpatient and intrapatient variations. *IEEE Trans Biomed Eng*. 2020;67:1167–1175.
19. Bartels K, Esper SA, Thiele RH. Blood pressure monitoring for the anesthesiologist: A practical review. *Anesth Analg*. 2016;122:1866–1879.
20. Michard F, Sessler DI, Saugel B. Non-invasive arterial pressure monitoring revisited. *Intensive Care Med*. 2018;44:2213–2215.
21 Peter L, Noury N, Cerny M. A review of methods for non-invasive and continuous blood pressure monitoring. *IRBM*. 2014;35(5):271–282. Available from: https://doi.org/10.1016/j.irbm.2014.07.002.
22. Ohte N, et al. Relationship between blood pressure obtained from the upper arm with a cuff-type sphygmomanometer and central blood pressure measured with a catheter-tipped micromanometer. *Heart Vessel*. 2007;22:410–415.
23. Pittman J, Ping J, Mark J. Arterial and central venous catheters. *Int Anesthesiol Clin*. 2004;42:13–30.
24. Putterman C. The Swan-Ganz catheter: a decade of hemodynamic monitoring. *J Crit Care*. 1989;4:127–146.
25. Chatterjee K. The Swan-Ganz catheters: past, present, and future: a viewpoint. *Circulation*. 2009;119:147–152.
26. Verdejo HE, et al. Comparison of a radiofrequency-based wireless pressure sensor to Swan-Ganz Catheter and echocardiography for ambulatory assessment of pulmonary artery pressure in heart failure. *J Am Coll Cardiol*. 2007;50:2375–2382.
27. Amsallem M, et al. Addressing the controversy of estimating pulmonary arterial pressure by echocardiography. *J Am Soc Echocardiogr*. 2016;29:93–102.
28. Simjanoska M, Gjoreski M, Gams M, Bogdanova AM. Non-invasive blood pressure estimation from ECG using machine learning techniques. *Sensors (Switzerland)*. 2018;18.
29. Dual SA, et al. Acute changes in preload and the QRS amplitude in advanced heart failure patients. *Biomed Phys Eng Express*. 2019;5:045015.
30. Fung P, Dumont G, Ries C, Mott C, Ansermino M. Continuous noninvasive blood pressure measurement by pulse transit time. *Annu Int Conf IEEE Eng Med Biol - Proc*. 2004;26(I):738–741.
31 Fonseca MA, Allen MG, Kroh J, White J. Flexible wireless passive pressure sensors for biomedical applicatons. *Tech Dig - Solid-State Sensors, Actuators, Microsyst Work*. 2006;37–42. Available from: https://doi.org/10.31438/trf.hh2006.9.
32. Mangi MA, Nesheiwat Z, Kahloon R, Moukarbel GV. CardioMEMSTM system in the daily management of heart failure: review of current data and technique of implantation. *Expert Rev Med Devices*. 2020;17:637–648.
33. Abraham WT, et al. Wireless pulmonary artery haemodynamic monitoring in chronic heart failure: a randomised controlled trial. *Lancet*. 2011;377:658–666.
34. Anker SD, Koehler F, Abraham WT. Telemedicine and remote management of patients with heart failure. *Lancet*. 2011;378:731–739.
35. Wolfson AM, Fong M, Grazette L, Rahman JE, Shavelle DM. Chronic heart failure management and remote haemodynamic monitoring. *Heart*. 2018;104:1910–1919.
36. Jermyn R, Alam A, Kvasic J, Saeed O, Jorde U. Hemodynamic-guided heart-failure management using a wireless implantable sensor: Infrastructure, methods, and results in a community heart failure disease-management program. *Clin Cardiol*. 2017;40:170–176.
37. Hoppe UC, et al. Chronic monitoring of pulmonary artery pressure in patients with severe heart failure: multicentre experience of the monitoring pulmonary artery pressure by implantable device responding to ultrasonic signal (PAPIRUS) II study. *Heart*. 2009;95:1091–1097.

38. Magalski A, et al. Continuous ambulatory right heart pressure measurements with an implantable hemodynamic monitor: a multicenter, 12-month follow-up study of patients with chronic heart failure. *J Card Fail*. 2002;8:63–70.
39. Lander MM, Aldweib N, Abraham WT. Wireless hemodynamic monitoring in patients with heart failure. *Curr Heart Fail Rep*. 2021;18:12–22.
40. Troughton RW, et al. Direct left atrial pressure monitoring in severe heart failure: long-term sensor performance. *J Cardiovasc Transl Res*. 2011;4:3–13.
41. Potkay JA. Long term, implantable blood pressure monitoring systems. *Biomed Microdevices*. 2008;10:379–392.
42. Tchantchaleishvili V, et al. Clinical implications of physiologic flow adjustment in continuous-flow left ventricular assist devices. *ASAIO J*. 2017;63:241–250.
43. Hubbert L, Baranowski J, Delshad B, Ahn H. First implantation in human of a wireless miniaturized intracardiac pressure sensor in a patient with a HeartMate II™. *J Hear Lung Transpl*. 2014;33:S13.
44. Ruhhammer J, et al. Magnetic sensor for arterial distension and blood pressure monitoring. *Biomed Microdevices*. 2014;16:815–827.
45. Hierold C, et al. Low power integrated pressure sensor system for medical applications. *Sens Actuators, A Phys*. 1999;73:58–67.
46. Fritz B, Cysyk J, Newswanger R, Weiss W, Rosenberg G. Development of an inlet pressure sensor for control in a left ventricular assist device. *ASAIO J*. 2010;56:180–185.
47. Zhou MD, Yang C, Liu Z, Cysyk JP, Zheng SY. An implantable Fabry-Pérot pressure sensor fabricated on left ventricular assist device for heart failure. *Biomed Microdevices*. 2012;14:235–245.
48. Brancato L, Keulemans G, Verbelen T, Meyns B, Puers R. An implantable intravascular pressure sensor for a ventricular assist device. *Micromachines*. 2016;7:135.
49. Brancato L, Keulemans G, Gijsenbergh P, Puers R. Plasma enhanced hydrophobicity of parylene-C surfaces for a blood contacting pressure sensor. *Proc Eng*. 2014;87:336–339.
50. Staufert S, Hierold C. *Novel sensor integration approach for blood pressure sensing in ventricular assist devices. Procedia Engineering*. 168. Elsevier Ltd.; 2016:71–75.
51. Staufert S, Gutzwiller P, Mushtaq F, Hierold C. Surface nanostructuring of Ti6Al4 V surfaces for parylene-C coatings with ultradurable adhesion. *ACS Appl Nano Mater*. 2018;1:1586–1594.
52. Jenkins C, Bricknell K, Hanekom L, Marwick TH. Reproducibility and accuracy of echocardiographic measurements of left ventricular parameters using real-time three-dimensional echocardiography. *J Am Coll Cardiol*. 2004;44:878–886.
53. Spencer KT, et al. Focused cardiac ultrasound: recommendations from the American Society of Echocardiography. *J Am Soc Echocardiogr*. 2013;26:567–581.
54. Begot E, et al. Hemodynamic assessment of ventilated ICU patients with cardiorespiratory failure using a miniaturized multiplane transesophageal echocardiography probe. *Intensive Care Med*. 2015;41:1886–1894.
55. Leonhardt S, Lachmann B. Electrical impedance tomography: the holy grail of ventilation and perfusion monitoring? *Intensive Care Med*. 2012;38:1917–1929.
56. Leonhardt S, Pikkemaat R, Stenqvist O, Lundin S. *Electrical impedance tomography for hemodynamic monitoring. 2012 Annual International Conference of the IEEE Engineering in Medicine and Biology Society*. 2012. IEEE; 2012:122–125.
57. Korn L, et al. In silico and in vitro conductivity models of the left heart ventricle. *J Electr Bioimp*. 2020;11:62–71.
58. Strobeck JE, Silver MA. Beyond the four quadrants: the critical and emerging role of impedance cardiography in heart failure. *Congest Hear Fail*. 2004;10:1–6.
59. Inan OT, et al. Ballistocardiography and Seismocardiography: a review of recent advances. *IEEE J Biomed Heal Inform*. 2015;19:1414–1427.
60. Inan OT, Etemadi M, Paloma A, Giovangrandi L, Kovacs GTA. Non-invasive cardiac output trending during exercise recovery on a bathroom-scale-based ballistocardiograph. *Physiol Meas*. 2009;30:261–274.
61. Xu L, et al. 3D multifunctional integumentary membranes for spatiotemporal cardiac measurements and stimulation across the entire epicardium. *Nat Commun*. 2014;5:3329.
62. Van Story, D. et al. Approaches to real-time ventricular wall strain measurement for the control of soft robotic ventricular assist devices. in *The Ninth Hamlyn Symposium on Medical Robotics* (2017).
63. Ochsner G, et al. A physiological controller for turbodynamic ventricular assist devices based on a measurement of the left ventricular volume. *Artif Organs*. 2014;38:527–538.
64. Rothfuss MA, Unadkat JV, Gimbel ML, Mickle MH, Sejdić E. Totally implantable wireless ultrasonic doppler blood flowmeters: toward accurate miniaturized chronic monitors. *Ultrasound Med Biol*. 2017;43:561–578.
65. Dual SA, et al. Ultrasonic sensor concept to fit a ventricular assist device cannula evaluated using geometrically accurate heart phantoms. *Artif Organs*. 2019;43:467–477.
66. Dual SA, Anthamatten L, Shah P, Meboldt M, Schmid Daners M. Ultrasound-based prediction of interventricular septum positioning during left ventricular support—an experimental study. *J Cardiovasc Transl Res*. 2020;13:1055–1064.
67 Dual SA, et al. Does size matter for female continuous flow LVAD recipients? A translational approach to a decade long question. *ASAIO J*. 2022;68(1):21–27.
68. Haines DE, et al. Validation of a defibrillation lead ventricular volume measurement compared to three-dimensional echocardiography. *Hear Rhythm*. 2017;14:1515–1522.
69. Newswanger R, et al. Development of a cannula tip with integrated volume sensor for control of continuous flow left ventricular assist devices. *ISMCS Abstr Book*. 2017;136.

70. Dual, S. A., Schmid Daners, M., Meboldt, M., Jacob, D. & Pergantis, P. Cardiac device, method and computer program product. 1–38 (2020).
71 Schmid Daners M, et al. Real-time ventricular volume measured using the intracardiac electromyogram. *ASAIO J*. 2021;67(12):1312–1320.
72 Korn L, et al. Dual-modality volume measurement integrated on a ventricular assist device. *IEEE Trans Biomed Eng*. 2021. Available from: https://doi.org/10.1109/TBME.2021.3115019.
73. Otsuji Y, et al. Restricted diastolic opening of the mitral leaflets in patients with left ventricular dysfunction: evidence for increased valve tethering. *J Am Coll Cardiol*. 1998;32:398–404.
74. Bach DS. Echo/Doppler evaluation of hemodynamics after aortic valve replacement. Principles of interrogation and evaluation of high gradients. *JACC: Cardiovascular Imaging*. 2010;3:296–304.
75. Hannawi B, Estep JD. A comprehensive imaging approach to guide the management of patients with continuous-flow left ventricular assist devices. *Cardiol Clin*. 2018;36:583–597.
76. Estep JD, Stainback RF, Little SH, Torre G, Zoghbi WA. The role of echocardiography and other imaging modalities in patients with left ventricular assist devices. *JACC: Cardiovasc Imag*. 2010;3:1049–1064.
77. Cowger JA, Aaronson KD, Romano MA, Haft J, Pagani FD. Consequences of aortic insufficiency during long-term axial continuous-flow left ventricular assist device support. *J Hear Lung Transpl*. 2014;33:1233–1240.
78. Dyverfeldt P, et al. 4D flow cardiovascular magnetic resonance consensus statement. *J Cardiovasc Magn Reson*. 2015;17:72.
79. Markl M, Frydrychowicz A, Kozerke S, Hope M, Wieben O. 4D flow MRI. *J Magn Reson Imaging*. 2012;36:1015–1036.
80. Chirinos JA. Arterial stiffness: Basic concepts and measurement techniques. *J Cardiovasc Transl Res*. 2012;5:243–255.
81. Zimmermann J, et al. On the impact of vessel wall stiffness on quantitative flow dynamics in a synthetic model of the thoracic aorta. *Sci Rep*. 2021;11:6703.
82 Cesarovic N, et al. Left ventricular blood flow patterns at rest and under dobutamine stress in healthy pigs. *NMR Biomed*. 2018;31(1):e4022. Available from: https://doi.org/10.1002/nbm.4022.
83. Shrier DA, et al. CT angiography in the evaluation of acute stroke. *Am J Neuroradiol*. 1997;18.
84. Prokop M. Multislice CT angiography. *Eur J Radiol*. 2000;36:86–96.
85. Douglas PS, et al. Outcomes of anatomical vs functional testing for coronary artery disease. *N Engl J Med*. 2015;372:1291–1300.
86. Alessio AM, et al. Accuracy of myocardial blood flow estimation from dynamic contrast-enhanced cardiac CT compared with PET. *Circ Cardiovasc Imaging*. 2019;12:e008323.
87. Mistry KK, Mahapatra A. Design and simulation of a thermo transfer type MEMS based micro flow sensor for arterial blood flow measurement. *Microsyst Technol*. 2012;18:683–692.
88. Rothfuss MA, et al. A System for simple real-time anastomotic failure detection and wireless blood flow monitoring in the lower limbs. *IEEE J Transl Eng Heal Med*. 2016;4.
89. Vennemann B, Obrist D, Rösgen T. A smartphone-enabled wireless and batteryless implantable blood flow sensor for remote monitoring of prosthetic heart valve function. *PLoS One*. 2020;15:e0227372.
90. DeBakey ME. *A miniature implantable axial flow ventricular assist device. Annals of Thoracic Surgery*. 68. Elsevier; 1999:637–640.
91. Pektok E, et al. Remote monitoring of left ventricular assist device parameters after HeartAssist-5 implantation. *Artif Organs*. 2013;37:820–825.
92. Schmitto JD, et al. First implantation of a novel left ventricular assist device: the ReliantHeart aVAD. *Ann Thorac Surg*. 2017;104:e311–e313.
93. Kitamura T, et al. Physical model-based indirect measurements of blood pressure and flow using a centrifugal pump. *Artif Organs*. 2000;24:589–593.
94. Giridharan GA, Skliar M. Physiological control of blood pumps using intrinsic pump parameters: a computer simulation study. *Artif Organs*. 2006;30:301–307.
95. Granegger M, Moscato F, Casas F, Wieselthaler G, Schima H. Development of a pump flow estimator for rotary blood pumps to enhance monitoring of ventricular function. *Artif Organs*. 2012;36:691–699.
96. Pennings KAMA, et al. Pump flow estimation from pressure head and power uptake for the HeartAssist5, HeartMate II, and HeartWare VADs. *ASAIO J*. 2013;59:420–426.
97. Petrou A, Kuster D, Lee J, Meboldt M, Schmid Daners M. Comparison of flow estimators for rotary blood pumps: an in vitro and in vivo study. *Ann Biomed Eng*. 2018;46:2123–2134.
98. Klotz S, et al. Single-beat estimation of end-diastolic pressure-volume relationship: a novel method with potential for noninvasive application. *Am J Physiol Circ Physiol*. 2006;291:H403–H412.

Chapter 33

Microcirculation and Tissue Perfusion Assessments for Complex Cardiovascular Disease Care

Knut Kvernebo

Institute of Clinical Medicine, University of Oslo, Department of Thoracic Surgery, Oslo University Hospital, Oslo, Norway

Chapter Outline

Abbreviations

CAVM computer-assisted video microscopy
CF circulatory failure
CFV capillary flow velocity
cLab central analyzing platform
DRS diffuse reflectance spectroscopy
ECTT erythrocyte capillary transit time
FCD functional capillary density
HC healthy control
LDPM laser Doppler perfusion measurements
LDPI laser Doppler perfusion imaging
mLab mobile microvascular laboratory
NIRS near infrared spectroscopy
O_2 oxygen molecule
ODIN oxygen delivery index
PVR pulse volume recordings
RBC red blood cells
SaO_2 arterial oxygen saturation
$SmvO_2$ microvascular oxygen saturation
$TcPO_2$ transcutaneous oxygen tension measurement

Advances in Cardiovascular Technology. DOI: https://doi.org/10.1016/B978-0-12-816861-5.00006-X

Introduction

Intensive care unit (ICU) treatment is hampered by the lack of coherence between patient outcomes and routine assessments, that is, clinical assessments of tissue perfusion and the hemodynamic parameters routinely monitored. Central hemodynamics provides reliable information of filling pressures, arterial pressure, and cardiac output, yet the parameters are merely indirect signs of circulatory function.

For example in patients with shock, the value of the sole guideline-recommended biomarker—blood lactate—is questionable.[1] Blood lactate levels can, for example, be normal during an occluding saddle embolus of the aortic bifurcation and severe lower limb ischemia but be critically high after embolectomy and reperfusion because of washout of lactate from ischemic tissue.

Clinical assessments of tissue perfusion can be misleading, for example, for a rare disorder of skin perfusion—erythromelalgia—for which skin can be red and warm yet hypoxic.[2] The explanation is maldistributed perfusion, that is, increased perfusion and arteriovenous shunting in the deep skin thermoregulatory plexus, and inadequate perfusion in nutritive papillary capillaries.

This chapter gives an overview of microvascular anatomy and function, and of diagnostic technologies available for clinical use. An overview of the new diagnostic Oxygen delivery index (ODIN) concept, including technologies, clinical examination protocols and an analyzing platform, as well as proof-of-concept studies will also be given.

Microcirculation—the business side of circulation

Microcirculation is usually defined as the systemic circulation's terminal vascular network, consisting of microvessels—arterioles, capillaries, and postcapillary venules—with diameters <100 μm.

This network is where blood and the body's tissues exchange heat, respiratory gases, nutrients, waste products, water, and hormones.

Arterioles and capillaries

Arterioles are best defined by the structure of the vessel wall, in terms of layers ($n = 1-4$) of smooth muscle cells. Through variation in their vascular smooth muscle tone and thereby their diameters (20−150 mm), arterioles control the distribution of blood flow between and partly within tissues. They provide $\approx$80% of the total resistance to blood flow through the body and reduce blood pressure from arterial down to precapillary values.

Capillaries are the smallest blood vessels, with diameters in the range 5−20 μm (size of a human erythrocyte is $\approx$8 μm) and a wall thickness equal to one endothelial cell. Its wall is supported on the basement membrane—a thin flexible sheet-like type of extracellular matrix.

Many substances are exchanged between capillaries and the surrounding interstitial fluid. The thin capillary wall is especially important for O_2 delivery to cells, since the O_2 molecule, in contrast to CO_2, has a poor diffusion capacity in human tissues.

Microvascular blood flow—regulation and hemorheology

In the microcirculation, blood flow depends on perfusion pressure (PP) and the diameter of the vessel and on hemorheological factors (e.g., blood viscosity). Within vessel walls of arteries and arterioles, smooth muscle cells contract or dilate to create the vascular resistance (R) needed to ensure sufficient blood flow (Q) for a given PP; the relationship is expressed in the equation $Q = PP/R$.

Regulation of perfusion in individual capillaries, has been attributed to precapillary smooth muscle cell sphincters, despite the lack of evidence for the existence of such sphincters.

While smooth muscle cells in arterioles are key to the distribution of blood between organs, pericytes seem to have a similar role for regulation of capillary perfusion within the tissues. Pericytes surround capillaries and contain contractile proteins (i.e., actin and myocin), but are not innervated - (Fig. 33.1).[3]

Pericytes can close off active capillaries and open up/recruit inactive capillaries.

A plausible mechanism is contraction in response to sufficient O_2 supply and dilation in response to acidic extracellular pH and elevated pCO_2.

Blood viscosity is determined by plasma viscosity, hematocrit values, mechanical properties of red blood cells (RBC), and concentration of white blood cells. Healthy erythrocytes are flexible and can change configuration to pass through small capillaries, resulting in decreased blood viscosity as the tube's diameter decreases—the Fåhræus−Lindqvist effect. An opposite effect is seen in patients with sickle cell anemia and thalassemia, who have inflexible red blood cell membranes, that is, increased resistance to microvascular blood flow. Sickle cell patients may develop *sickle cell crisis* with occluded capillaries

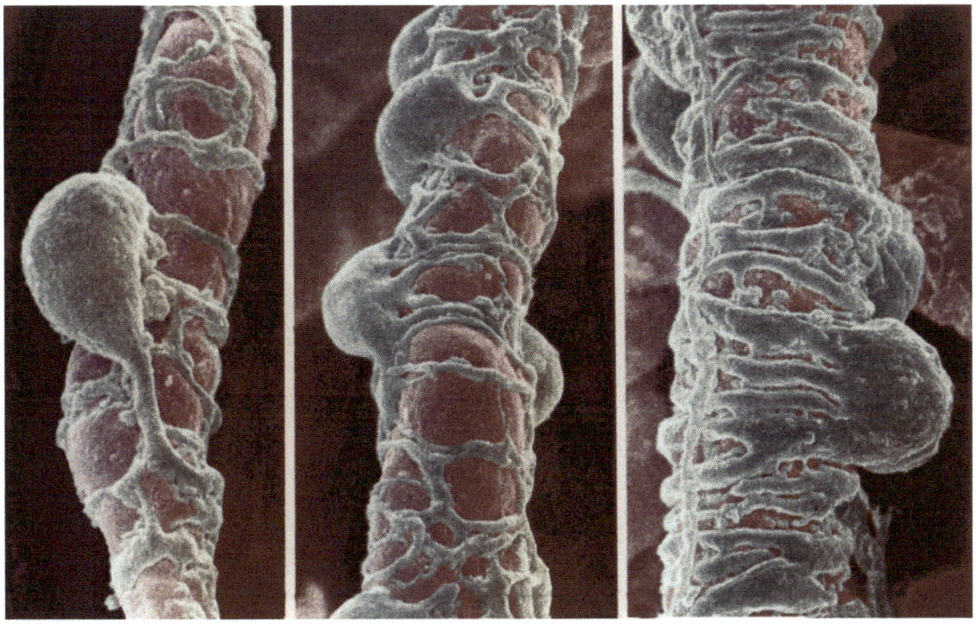

FIGURE 33.1 Pericytes surrounding capillaries.

and tissue hypoxia. Similar symptoms caused by the same mechanism may be seen in patients with increased levels of circulating large, sticky white blood cells (e.g., acute myeloid leukemia): *leukemic crisis.*

Circulatory failure and oxygen delivery

Without sufficient O_2 delivery, human cells die when all physiologically compensatory systems have been exhausted (i.e., temporary anaerobic metabolism with lactate production and hibernation with reduced cell metabolism).

With a healthy lung function ($\approx$100% SaO_2), circulatory failure (CF) can best be defined as the inability of the cardiovascular system to transport sufficient amounts of O_2 to meet the metabolic demands of all cells of the body ($n \approx 10^{14}$ cells). A preserved and well-regulated microcirculatory function is the key to O_2 delivery.

Oxygen diffusion and function capillary density

Gas exchange between the alveoli and the blood in lung capillaries, occurs ~20 times faster for CO_2 than for O_2.[4] The limited O_2 diffusion capacity also applies to O_2 delivery from peripheral capillaries to tissues. August Krogh, the Nobel prize laureate in Physiology or Medicine in 1920, postulated that metabolically active cells must be located within a given distance of a perfused capillary—the Krogh diameter—to get sufficient O_2 supply.[5] He stated that within a virtual tissue cylinder served by one capillary, O_2 availability decreases exponentially at increasing distances from the center of the vessel. Cells outside such a cylinder will suffer from hypoxia, irrespective of the blood flow in the capillary (Fig. 33.2).

Studies of cancer cell growth have shown that the maximal O_2 diffusion distance from a perfused capillary before chronic hypoxia develops, is in the range 40–140 μm.[6] Cell necrosis is seen at distances in the range 80–300 μm from a perfused vessel, depending on the tissue's metabolic rate.

The metabolic needs of tissues are decisive for how many perfused/functional capillaries are active within a given tissue volume. When more O_2 is needed more capillaries are opened (recruited); when less is needed, superfluous capillaries are turned off (become inactive). This fascinating flexibility is in line with Darwin's fundamental law of biological evolution: *In nature, in the fight for survival, energy is not wasted.*

FCD also depends on body temperature since O_2 diffusion is improved at higher temperatures and reduced at lower temperatures, and the metabolic rate increases with 6–10% per degree Centigrade elevation.[7]

Oxygen delivery and capillary flow velocities

In healthy individuals capillary blood flow velocities (CFV) are also regulated within narrow limits to optimize O_2 delivery with the least use of energy. When erythrocytes move slowly through capillaries, the pO_2 in erythrocytes has

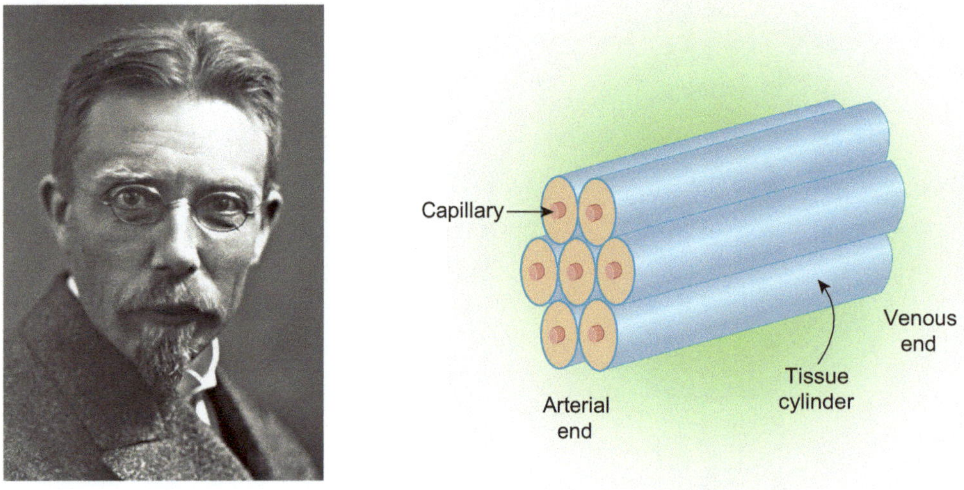

(A) (B)

FIGURE 33.2 (A) August Krogh. (B) Illustration of Krogh cylinder, redrawn from an original illustration used by Krogh.

time to equilibrate with that in the pericapillary tissue (long ECTT); when they move quickly the capillary may function as a physiological shunt with short ECTT (Table 33.1 and Fig. 33.3).

CFV reflects the number of erythrocytes that pass through a capillary—there is a linear relationship. CFV category 3 is the optimal compromise between number of erythrocytes passing through a capillary and sufficient ECTT for unloading O_2.

At low CFV (near 0) and long ECTT, O_2 delivery is low despite maximal O_2 extraction; few erythrocytes pass through capillaries while ECTT is sufficiently long for equilibration of pO_2 across the capillary wall.

At high CFV (category 5) and short ECTT, the capillary serves as a physiological a-v shunt with low O_2 extraction from each erythrocyte, but O_2 delivery may still be maintained, because the pO_2 difference across the capillary wall is maintained because of the large number of erythrocytes passing through the vessel.

In hypoxic tissue, the O_2 delivery increases (upper line).

In cases with a diffusion barrier in the capillary wall (e.g., poorly regulated diabetes), delivery decreases (lower line).

Tissues for noninvasive microvascular examinations

Skin, conjunctiva of the eye, and the mucous membranes of the oral cavity are readily available for noninvasive examination of human microcirculation during systemic CF.

Skin microcirculation

Inspection of color and temperature of skin—the largest organ of the human body—are sometimes misleadingly used as clinical indicators of local and systemic circulatory function. The primary functions are regulation of body temperature and delivery of nutrients and O_2 for cell metabolism. When skin is injured, the microcirculation must also have capacity to increase oxygen delivery needed for the repair process.

Most tissues receive blood flow in proportion to the metabolic rate, but the skin is an exception where the priority is thermal rather than metabolic.

The structure of skin blood vessels and regulation of blood flow are complex and have evolved to serve functional needs (Fig. 33.4).

Arterialized blood enters the skin via small arterioles that traverse the adipose septae of the subcutis and connect to the deep vascular plexus. Ascending arterioles supply the superficial (distributional) vascular plexus made up of anastomosing small-caliber arterioles and venules. Papillary nutritive capillaries branch off to deliver O_2 close to the continuous proliferating epidermal stem cells—located at the epidermal basement membrane—to replace superficial epidermal cells worn off from the skin surface.[8]

The main function of the deep plexus is to regulate body temperature. This plexus is innervated and regulated by autonomous nerves and is the body's radiator.

TABLE 33.1 **Capillary flow velocities.**

Capillary flow velocity categories	Definition
0	*No flow:* visible erythrocytes, no movement
1	*Sluggish flow:* intermittent no flow/slow flow
2	*Slow flow:* continuous slow flow or intermittent "no flow" and rapid flow
3	*Continuous flow*
4	*Rapid flow:* continuous rapid flow or intermittent continuous/brisk flow
5	*Brisk flow:* continuous brisk flow
6	*Uncertain:* velocity impossible to determinate

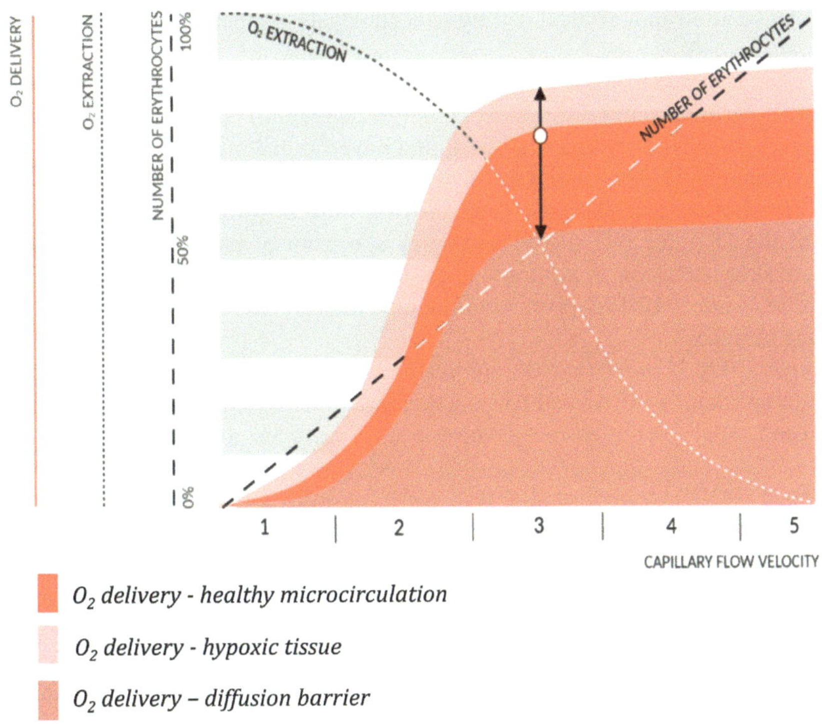

FIGURE 33.3 A theoretical model of O_2 delivery from a single capillary. Capillary flow velocity given in categories.

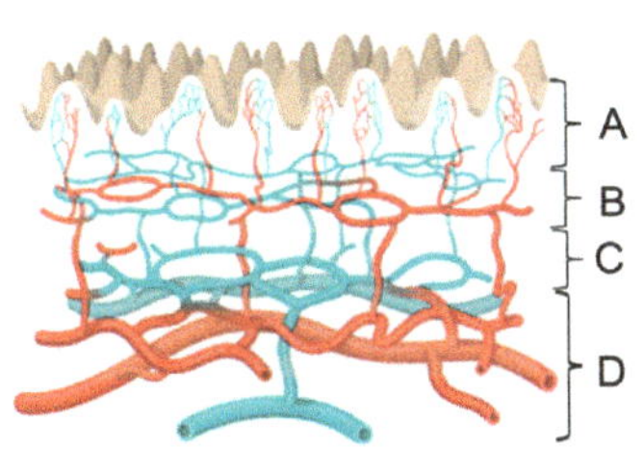

FIGURE 33.4 The structure of skin microcirculation distal to the wrist and the ankle. (A) skin nutritive papillary capillaries; (B) superficial distributional plexus; (C) ascending arterioles and descending venules; (D) deep thermoregulatory plexus.

Distal to the wrist and to the ankle, the plexus contains anatomical arteriovenous innervated shunts. If the body needs to get rid of heat, the shunts open and fills the plexus veins. The skin looks red, and the skin is warm. If the body needs to preserve heat, shunts close, the veins empty—Caucasian skin looks pale. The cause of blushing and becoming pale with fear is also autonomous regulation of blood volume in this plexus.

Papillary capillary perfusion is regulated by the tone of their noninnervated pericytes.

Under thermoneutral conditions >80% of skin perfusion in the hand occurs in the deep plexus where O_2 extraction is low ($\approx$5%). A minor part of perfusion occurs in the papillary nutritive capillaries where O_2 extraction is high ($\approx$50%).[9]

Erythromelalgia, a model of microvascular shunting, is a condition with warm, red, and painful extremities caused by maldistribution of blood flow between the deep plexus and the nutritive capillaries. These patients can develop hypoxic ulcers and even gangrene because of inadequate nutritive perfusion, despite hyperperfusion in the thermoregulatory compartment causing bright red and warm skin.[10]

Technologies for examination of skin circulation

A number of technologies are available for assessing human skin circulation, but no technique for direct assessment of microvascular perfusion is used in routine clinical medicine. Part of the reason is that available techniques are not able to differentiate between blood perfusion in nutritive papillary capillaries and in the deep thermoregulatory plexus. Whether local skin microvascular dysfunction can be used in a clinical setting as a surrogate biomarker for generalized microvascular dysfunction is still debated.[11]

Infrared cameras and thermometers assessing skin temperature are used for screening of individuals with increased body temperature caused by infections (e.g., screening in airports during the Covid-19 pandemic). These techniques are only indirect measure of skin perfusion, they have low sensitivity and precision, and have limited clinical value.

Laser Doppler perfusion measurements and imaging (LDPM and LDPI) send monochromatic laser light into the skin where photons are absorbed or scattered. If photons are scattered from moving objects (i.e., moving red blood cells) the light frequency changes as a result of the Doppler effect. The frequency spectrum of photons emitted back from the tissue are analyzed and used to estimate skin perfusion in arbitrary units (flux = number of particles x average velocity) in the measuring volume in the range of 1 mm^3. LDPM gives continuous dynamic curves in one measuring volume; LDPI gives perfusion maps over a tissue surface.

Since the first report of a LDPM instrument in 1980,[12] about 15,000 publications with the MeSH "laser Doppler" are listed in PubMed up to 2021, of which the large majority is related to skin perfusion measurements. The technique cannot be used to assess skin nutritive perfusion, since the measuring volume also includes the thermoregulatory plexus.

Transcutaneous oxygen tension measurement ($TcPO_2$) is a noninvasive technique previously used in neonatal medicine to monitor the partial pressure of oxygen in papillary nutritive capillaries. Because of the poor diffusion capacity of O_2, the skin surface probe incorporates a heating element to facilitate O_2 diffusion though the epidermis. The technology may be useful for monitoring neonates, but application studies in adult patients have shown the limited sensitivity and value of this technique.[13]

Pulse volume recordings (PVR) noninvasively assess pulse-synchronous volumes changes in a tissue by blood pressure cuffs, an ultrasound probe or an optical sensor. The technologies assess arterial blood flow, and have little value for microvascular assessments.[13]

Pulse oximetry is a noninvasive technology for assessing lung function (a surrogate parameter for SaO_2), but not a test for skin microvascular perfusion.

Near-infrared spectroscopy (NIRS) is used in several clinical settings for estimating an average O_2 saturation in tissue volumes in the range of a few cm^3 (containing arterioles, capillaries and venules) and the method can therefore not assess O_2 saturation of capillary erythrocytes, thus the results have limited value for assessing tissue oxygenation.[14] The technique has gained some popularity for assessing trends of changing tissue oxygenation (e.g., in the brain during open heart surgery and circulatory arrest).

In vivo microscopy

William Harvey in 1628, proposed the existence of a cardiovascular circulatory system with the heart as a pump and arteries and veins as transport vessels, but not until 1661 did Marcello Malphigi describe capillary networks in the lung using a microscope.[15] It was not until the 20th century that the first reports of changes in human microcirculation during diseases appeared (reduced functional capillary density (FCD) during sepsis, and local microvascular pathology in the region around venous leg ulcers).[16,17] White light video capillaroscopy of the highly specialized nail fold capillaries are increasingly used as a diagnostic tool in autoimmune rheumatic diseases.

Orthogonal polarization spectral imaging (OPS imaging) and Sidestream dark field imaging are methods for microvascular imaging in tissues like the mucous membranes of the oral cavity (e.g., sublingually).[18] The techniques give high-resolution microvascular films, and have been used in many clinical studies over the last two decades. The techniques are not usable in adult skin due to the loss of signal through the epidermis.

The oxygen delivery index concept

Whether local skin microvascular dysfunction can be used in a clinical setting as a surrogate biomarker for systemic microvascular dysfunction is for several reasons still a topic for discussion.[19] There has been no clear definition of the term "circulatory failure," and studies have used different methods with different measuring depths and volumes, different data acquisition protocols, and different platforms with a variety of output parameters for analyzes of recorded files. For skin measurements, laser Doppler technologies and NIRS technologies have for example measuring volumes including perfusion in skin papillae, the superficial distributional, and the deep thermoregulatory vascular plexuses, containing arterioles, capillaries, and venules. Different manufacturers have different designs of equipment and there is no consensus on protocols for data acquisition in clinical settings.

Over the last three decades the author and his research group of clinicians with a focus on microvascular hemodynamics at the University of Oslo, Norway, have cooperated with medical technology engineers at Linköping University, Sweden. Gradually the increased knowledge of the complexity of skin microvascular structure and function, and a need for high-resolution technologies with small measuring volumes that only contain locally regulated papillary nutritive capillaries in skin, has emerged.

The patented ODIN concept for assessment of microvascular function—a result of this cooperation—is built on insight into skin microvascular structure and function. Standardized technologies, clinical data acquisition protocols, and an analyzing platform with defined result parameters have been developed.

Technologies for oxygen delivery index examination

The concept uses two noninvasive technologies with measuring volumes of $\approx 0.1\ mm^3$ for data acquisition: computer-assisted video microscopy (CAVM) and diffuse reflectance spectroscopy (DRS).

Computer-assisted video microscopy

A hand-held digital microscope uses white light to penetrate epidermis and visualize subepidermal vessels, see Fig. 33.4. Perfused dermal papillary capillaries can be seen, in contrast to empty nonperfused capillaries and deep thermoregulatory plexus vessels. The microscope has sufficient resolution to visualize individual erythrocytes (size $\approx 8-9$ microns). The field of view is in the range 1×1 mm. CFV in individual capillaries is recorded though video sequences (Fig. 33.5).

Diffuse reflectance spectroscopy

The DRS technique measures O_2 saturation of capillary erythrocytes in the measuring volume.

DRS technique provides noninvasive measurements of oxygen saturation of erythrocytes in tissue volumes of around $0.1\ mm^3$. Visible white light illuminates the skin surface, and spectra of reflected light are analyzed with complex algorithms (Fig. 33.6).

The mobile microvascular laboratory (mLab) and examination protocols

mLab consists of a trolley, the microscope, the spectroscope, and a computer unit with a screen. The screen is used for plotting patient administrative data. When the spectroscope is used, the screen shows the spectra. Twelve spectra are recorded from nearby volumes. When the microscope is used, the screen shows the microscopic field. Five films of 20 s are recorded from nearby volumes. Files are stored in the computer and can be sent via the Cloud for analyses.

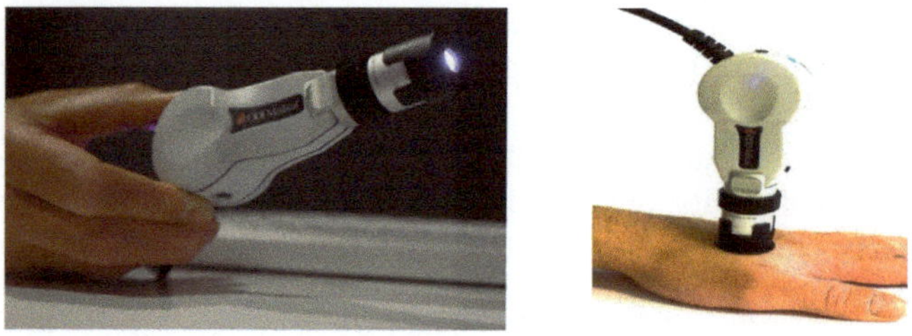

FIGURE 33.5 A handheld digital microscope.

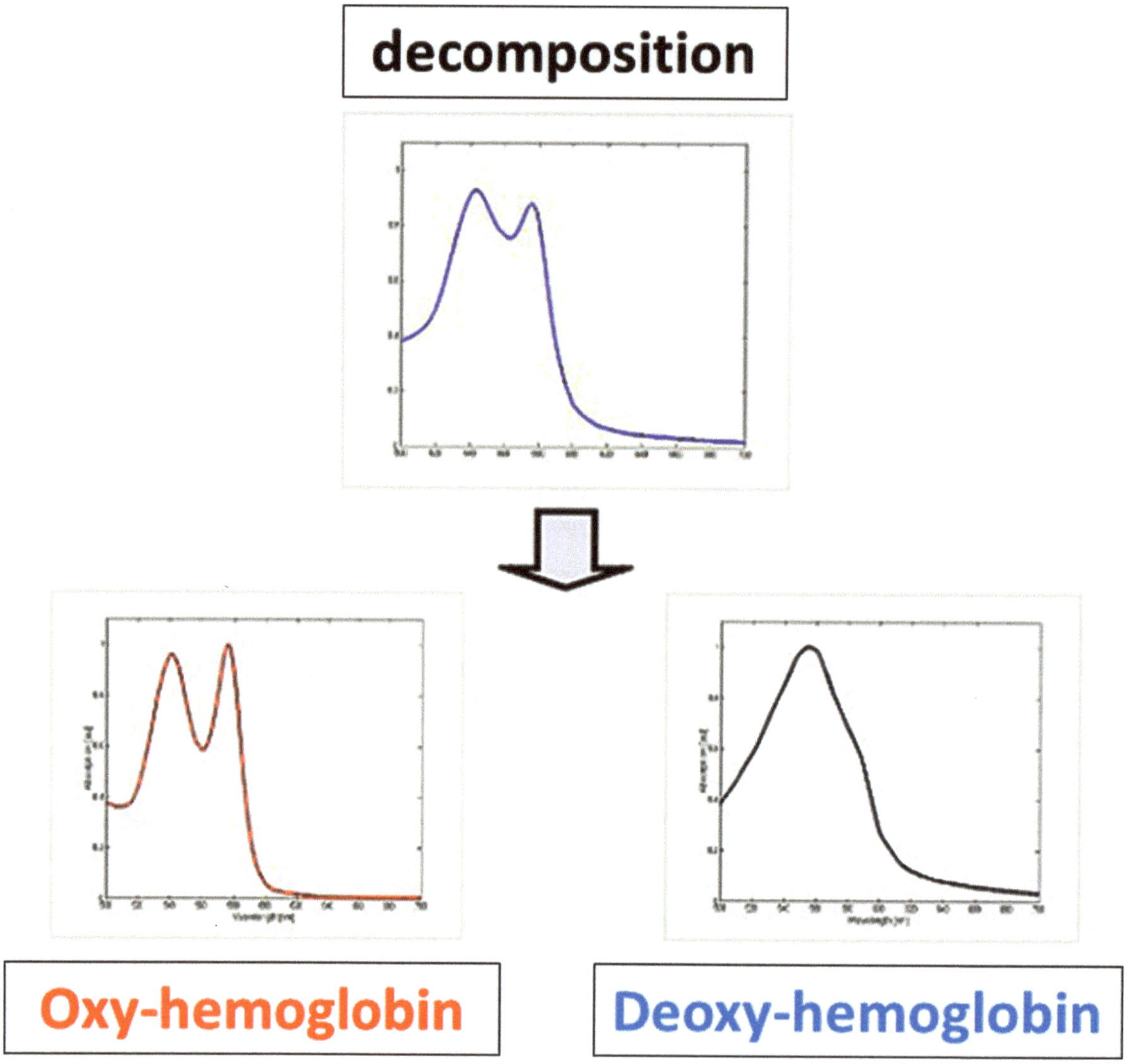

FIGURE 33.6 Diffuse reflectance spectroscopy decomposition algorithm: The absorption spectrum (upper panel) of a diffuse reflectance spectroscopy recording from human blood: x ($SmvO_2$) times the oxyhemoglobin spectrum + y times deoxyhemoglobin spectrum. *Figure kindly supplied by Professor Salerud, Linköping University.*

Centralized analyzing platform (cLab)

Files from mLab are analyzed in cLab using custom-made AI-based algorithms. The cLab output are six parameters describing microvascular function are output from cLab:

1. FCD
2. Heterogeneity of FCD among the recorded films
3. Capillary flow velocities (CFV)

4. Heterogeneity of CFV among individual capillaries
5. Microvascular oxygen saturation ($SmvO_2$)
6. Heterogeneity of $SmvO_2$ among recorded spectra

Functional capillary density

Nutritive papillary capillaries can be arranged in vertical loops (Fig. 33.5), or they can be positioned more parallel with the skin's surface. Fig. 33.7 shows a frame from the dorsum of the hand, and FCD can be expressed as number of loops/mm^2. When capillaries are more horizontal, the density can be expressed as the number of capillaries crossing the grid of lines/mm grid line.[9,20]

Heterogeneity of functional capillary density

Heterogeneity of FCD can be expressed as the coefficient of variation (CoV: SD/Mean) of FCD from five films from neighboring measuring volumes. If the regulatory system (i.e., pericyte function) is overstretched or failing, heterogeneity of FCD may increase even before the mean FCD is changed.

Capillary flow velocities

The importance of well-regulated CFV for O_2 delivery was previously underlined, see Fig. 33.3. Optimal CFV is a compromise between the number of erythrocytes passing through a capillary and sufficient transit time for equilibration of pO_2 inside and outside a capillary. Assessments of CFV can be done visually and expressed as fractions of capillaries in a predefined velocity categorial scale (Table 33.1).

Heterogeneity of capillary flow velocities

Increased variation of CFV between individual capillaries is also a sign of microvascular dysregulation. Increased variability of CFV may, as for parameter 2 (heterogeneity of FCD), be a more sensitive and early warning parameter of disease than average CFV. This parameter may be expressed as a histogram showing fractions of capillaries in each velocity category.

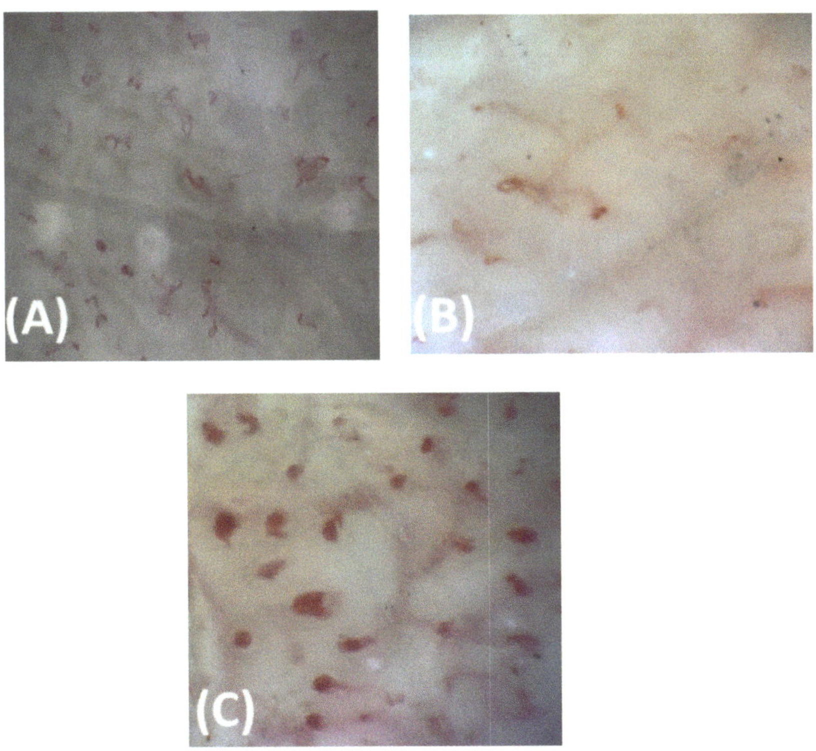

FIGURE 33.7 Microscopic frames of the skin at the dorsum of the hand. (A) Healthy control. The tops of the papillary loops (see Fig. 33.4) are clearly seen. FCD can be expressed as number of loops/mm^2. (B) Patient with severe chronic heart failure. Reduced number of capillary loops branching off from superficial plexus vessels are seen. (C) Patient with severe chronic heart failure. High number of dilated loops branching off from superficial plexus vessels are seen. CFV in these loops were severely reduced (CFV in groups 0 or 1, see Table 33.1).

Microvascular oxygen saturation ($SmvO_2$)

According to Fick's first law, a gas diffuses passively across a membrane from high concentration to low concentration, with a magnitude that is proportional to the concentration gradient. In our case we measure SaO_2 and can estimate pO_2 inside the capillary from the oxyhemoglobin dissociation curve. With healthy lung function, SaO_2 is near to 100%. If $SmvO_2$ is measured at, for example, 55% (typically found in skin of healthy individuals), 45% of O_2 molecules are extracted. If we on the other find $SmvO_2$ of 20% (values we typically have found in Covid-19 patients successfully treated with respirator and O_2 supplements ($SaO_2 > 95\%$), but still dying from the infection), we can deduce that the pO_2 gradient across the capillary wall is increased, and that the tissue is hypoxic because of dysregulated peripheral microcirculation (unpublished data).

Heterogeneity of $SmvO_2$

According to the standardized ODIN data acquisition protocol 12 spectra from nearby measuring volumes are recorded. Increase in heterogeneity of $SmvO_2$ is, similar to FCD and CFV, an early sign of microvascular dysregulation.

Oxygen delivery index proof of concept studies

Safety: No adverse effects of using the ODIN concept in human studies were observed. In a study on healthy neonates, measurements were performed during 15 minutes on the first, second, and third day of life, with parents present and without discomfort for the baby.[20] In another study examinations were performed on critically ill newborns, without reported side effects.[7]

Reproducibility: has been studied in healthy newborns.[20]

Measuring depths in skin of the technologies, corresponding to skin papillary nutritive capillaries, have been documented in studies of the trauma response in skin of healthy subjects.[9]

Sensitivity for detecting changes in systemic metabolism: was studied in asphyxiated newborns where the infant's mean core temperature was reduced to 33.5°C, corresponding to a reduction in metabolic rate and oxygen consumption of approximately 20%–30%. CFV was reduced significantly, while FCD increased (O_2 diffusion is reduced with lower temperature—the Krogh diameter is smaller), and $SmvO_2$ decreased (capillary transit time was increased), as could be expected.[7]

Sensitive for detecting local differences in metabolic rate: In ODIN recordings of conjunctival microcirculation, increased FCD was shown in sleeping piglets in limbus (the ≈ 1 mm circular zone surrounding cornea with a high metabolic rate because of stem cell production of new corneal cells) as compared to the rest of conjunctiva with a low metabolic rate.[21] In healthy humans FCD and oxygen extraction we found it to be higher (lower $SmvO_2$) in limbus compared with bulbar conjunctiva (unpublished data).

Examples of microvascular parameters in health and disease

Healthy individuals

FCD in healthy individuals is in the same range in different tissues of different species, as an evolutionary consequence of the diffusion properties of the O_2 molecule: In the groin skin of sleeping healthy piglets, FCD of 10.4–13.2 grid line crossing/mm line was found[21]; in the skin of healthy newborns FCD of 12.4 ± 1.6 (mean $\pm$ SD) c/mm line was found in the hand and 10.7 ± 1.6 c/mm line in the warmer skin at the chest (O_2 diffusion increases with temperature).[20]

In the conjunctiva of sleeping piglets, densities were higher in the metabolically active limbus (13–15 c/mm line) than in bulbar conjunctiva with a low metabolic rate (6–12 c/mm line).[22] In the conjunctiva of healthy human controls, we have in a pilot study found FCD of 10 individuals to be 11.2 ± 1.5 c/mm in limbus and 5.0 ± 1.2 c/mm in bulbar conjunctiva (unpublished data).

Sepsis

Decreased FCD in sepsis was described already in 1922 by Freedlander and Lenhart at Cleveland Clinic.[17] Similar findings have been confirmed both in animals and humans.[23,24]

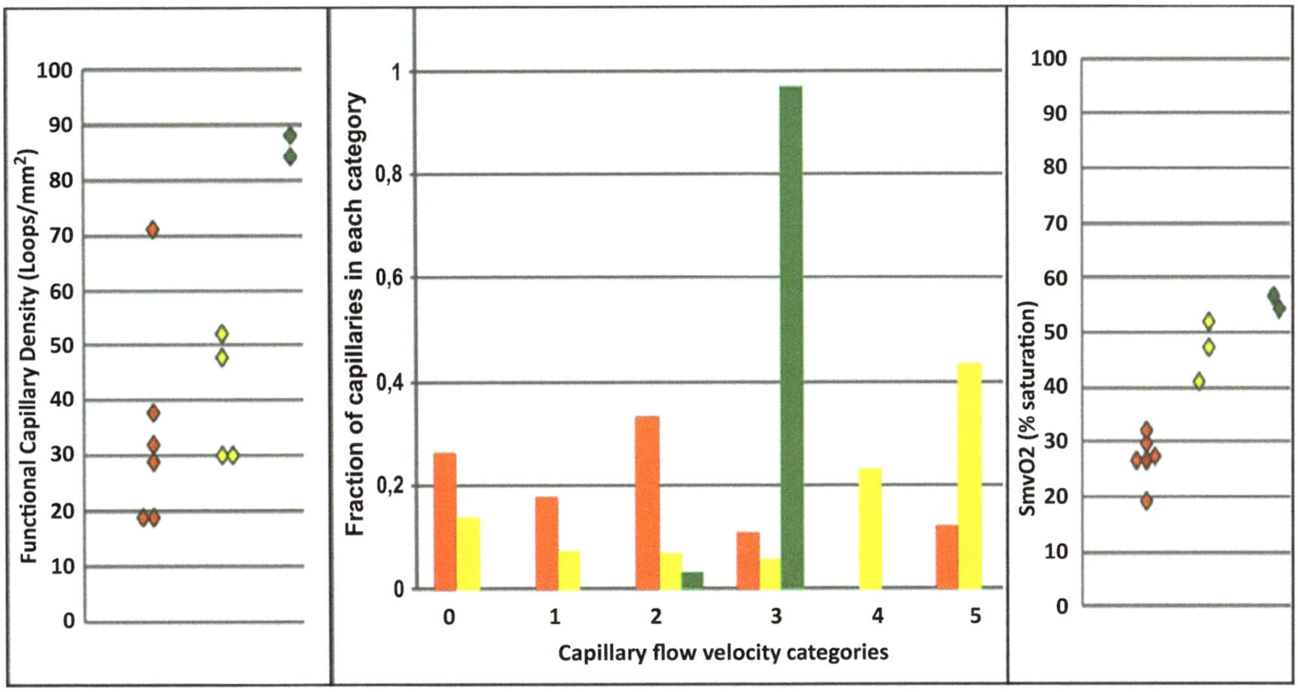

FIGURE 33.8 Oxygen delivery index parameters in patients va-ECMO and controls. Green: two healthy controls (group a). Yellow: 4 patients survived ECMO (group b). Red: 6 patients died on ECMO (group c). Median FCD was lowest in group c. Almost all CFV in group a were in category 3 (optimal velocity, see Fig. 33.3), while all ECMO patients (nonpulsatile pump) had dysregulation of velocities. Group b tended to increase mean velocities, while group c had reduced mean velocities. Group c had lower $SmvO_2$ values as compared with group a and b (unpublished data).

Venoarterial extracorporeal membrane oxygenation

Skin in vivo microscopy of patients treated with va-ECMO for acute heart failure showed that ECMO survivors had FCD and CFV in the same range as healthy controls, while patents dying on ECMO had reduced FCD, lower CFV, and larger heterogeneity of velocities among individual capillaries.[25] In another study sublingual microscopy found that va-ECMO survivors had had significantly higher FCD ("perfused small vessel density") than nonsurvivors.[26]

In pigs with heart arrest, treated with 6 h of pulsatile or nonpulsatile cardiopulmonary bypass, in vivo microscopy of conjunctival capillaries showed that pulsatile flow preserved CFV better than continuous flow.[27]

In a series of 10 patients treated with ECMO for acute heart failure, and two controls, ODIN examinations were performed at the dorsum of the left hand within 24 hours on ECMO (Fig. 33.8). HC had well-regulated CFV, and FCD and $SmvO_2$ were in the same range as other cohorts of HC. All patients on nonpulsatile ECMO had dysregulation of CFV, with a tendency to increased velocities for survivors and reduced velocities for patients dying on ECMO. The $SmvO_2$ separated between HC, ECMO survivors, and patients dying on ECMO. One patient dying on ECMO had FCD of 71 capillary loops/mm^2 but at the same time a majority of capillaries with CFV in category 0 or 1.

Summary

Already in 1922 Freedlander and Lenhart at the Cleveland Clinic stated the importance of understanding microvascular function: "It is evident that the rest of the cardiovascular system exists only to regulate the blood flow through the capillaries, for here takes place the exchange of gases necessary for internal respiration and the exchange of materials necessary for metabolism. Any attempt to measure cardiovascular function is an indirect attempt to measure the efficiency of the capillary circulation. For example, blood pressure determinations are supposed to give some indication of the peripheral blood flow, but as will be seen later, blood pressure is often a poor index of the state of the capillary circulation."[17]

CF can best be defined as insufficient capacity by the respiratory/circulatory systems to deliver sufficient O_2 to all the $\approx 10^{14}$ cells of our body. Clinical examination cannot be used for estimating O_2 delivery, and available laboratory technologies assess only indirectly parameters of local or systemic CF. It has been an open question whether assessments of skin circulation can be used as a surrogate biomarker for systemic microvascular dysfunction. Lack of understanding of function and regulation of skin perfusion, of knowledge of measuring depths of technologies used for assessments of skin perfusion and lack of consensus of clinical protocols for data acquisition, has hindered the development of clinically applicable concepts.

In this chapter we have clarified the prerequisites for using the skin for assessing systemic circulatory function by examining skin nutritive papillary capillary perfusion. These capillaries are regulated by pericyte activity, in the same way as nutritive perfusion in other organs (e.g., the brain). We have presented the ODIN concept, including technologies, clinical protocols, and analyzing parameters. Examples of proof-of-concept studies have also been given. The equipment can be used bedside, and a centralized analysis service ensures the quality control of results.

Clinical examination of heart function using a stethoscope and palpation of peripheral pulse is useful, but cardiology was transformed by the introduction of ultrasound technology. In the same way, assessments of central hemodynamics, blood gases, and lactate are useful, but noninvasive examination capillary perfusion using in vivo microscopy and spectroscopy may be used to give (1) a diagnosis of the function of the circulatory system; (2) to give prognostic information; and (3) to test the effect of therapeutic interventions in individual patients. Microvascular monitoring may help to give the right treatment to the right patient at the right time, resulting in improved outcomes for individual patients and economic savings for health care systems.

References

1. Antonelli M, et al. Hemodynamic monitoring in shock and implications for management. International Consensus Conference, Paris, France, 27-28 April 2006. *Intensive Care Med.* 2007;33:575–590. Available from: https://doi.org/10.1007/s00134-007-0531-4).
2. Mørk C, Kvernebo K. Erythromelalgia – a mysterious condition? *(Editor) Arch Dermatol.* 2000;136(3):406–409.
3. Bergers G, Song S. The role of pericytes in blood-vessel formation and maintenance. *Neuro-Oncology*. 2005;7:452–464. Available from: https://doi.org/10.1215/S1152851705000232 (Posted to Neuro-Oncology [serial online], Doc. 05–023, Septem- ber 1, 2005. URL. Available from: http://neuro-oncology.mc.duke.edu.
4. Santolicandro A, Prediletto R, Fornai E, et al. Mechanisms of hypoxemia and hypocapnia in pulmonary embolism. *Am J Respir Crit Care Med.* 1995;152(1):336–347. Available from: https://doi.org/10.1164/ajrccm.152.1.7599843. PMID: 7599843.
5. Krogh A. The supply of oxygen to the tissues and the regulation of the capillary circulation. *J Physiol.* 1919;52(6):457–474.
6. Forster JC, Harriss-Phillips WM, Douglass MJ, Bezak E. A review of the development of tumor vasculature and its effects on the tumor micro-environment. *Hypoxia (Auckland, NZ).* 2017;5:21–32.
7. Fredly S, Fugelseth D, Nygaard CS, Salerud EG, Stiris T, Kvernebo K. Noninvasive assessments of oxygen delivery from the microcirculation to skin in hypothermia-treated asphyxiated newborn infants. *Pediatric Res.* 2016;79(6):902–906. Available from: https://doi.org/10.1038/pr.2016.16. Epub 2016 Feb 8.
8. Braverman IM. The cutaneous microcirculation: ultrastructure and microanatomical organization. *Microcirculation.* 1997;4(3):329–340. Available from: https://doi.org/10.3109/10739689709146797. PMID: 9329009.
9. Sundheim LK, Sporastøyl AH, Wester T, Salerud G, Kvernebo K. Acute skin trauma induces hyperemia, but superficial papillary nutritive perfusion remains unchanged. *Microcirculation.* 2017. Available from: https://doi.org/10.1111/micc.12389.
10. Kvernebo K. Erythromelalgia. *VASA J Vasc Dis.* 1998;51(Suppl):1–37.
11. Cracowski JL, Roustit M. Human skin microcirculation. *Compr Physiol.* 2020;10(3):1105–1154. Available from: https://doi.org/10.1002/cphy.c190008.
12. Nilsson GE, Tenland T, Oberg PA. A new instrument for continuous measurement of tissue blood flow. *IEEE Trans Biomed Eng.* 1980;12–19. BME-27.
13. Kvernebo K, Megerman J, Hamilton G, Abbott WM. Response of skin photoplethysmography, laser Doppler flowmetry and transcutaneous oxygen tensiometry to stenosis-induced reductions in limb blood flow. *Eur J Vasc Surg.* 1989;3(2):113–120.
14. Kurth CD, Steven JM, Benaron D, Chance B. Near-infrared monitoring of the cerebral circulation. *J Clin Monit.* 1993;9(3):163–170. Available from: https://doi.org/10.1007/BF01617023.
15. Awan ZA, Wester T, Kvernebo K. Human microvascular imaging: a review of skin and tongue videomicroscopy techniques and analysing variables. *Clin Physiol Funct Imaging.* 2010;30(2):79–88.
16. Gilje O. Capillary microscopy in the differential diagnosis of skin diseases. *Acta Derm Venereol.* 1953;33:303–317.
17. Freedlander SO, Lenhart CH. Clinical observations on the capillary circulation. *Arch Intern Med (Chic).* 1922;29(1):12–32. Available from: https://doi.org/10.1001/archinte.1922.00110010017002.
18. De Backer D, Hollenberg S, Boerma C, et al. How to evaluate the microcirculation: report of a round table conference. *Crit Care.* 2007;11:: R101. Available from: https://doi.org/10.1186/cc6118. Available from: http://ccforum.com/content/11/5/R101.
19. Holowatz LA, Thompson-Torgerson CS, W. Larry Kenney WL. The human cutaneous circulation as a model of generalized microvascular function. *J Appl Physiol.* 2008;105:370–372. Available from: https://doi.org/10.1152/japplphysiol.00858.2007.
20. Awan ZA, Häggblad E, Wester T, Kvernebo MS, Halvorsen PS, Kvernebo K. Diffuse reflectance spectroscopy: systemic and microvascular oxygen saturation is linearly correlated and hypoxia leads to increased spatial heterogeneity of microvascular saturation. *Microvasc Res.* 2011;8 (3):245–251.
21. Fredly S, Fugelseth D, Wester T, Häggblad E, Kvernebo K. Skin microcirculation in healthy term newborn infants - assessment of morphology, perfusion and oxygenation. *Clin Hemorheol Microcirc.* 2015;59(4):309–322.

22 Kvernebo AK, Miyamoto T, Drolsum L, et al. Ocular surface microcirculation is better preserved with pulsatile versus continuous flow during cardiopulmonary bypass-An experimental pilot. *Artif Organs*. 2021. Available from: https://doi.org/10.1016/j.exer.2020.108312.

23. Wester T, Haggblad E, Awan ZA, et al. Assessments of skin and tongue microcirculation reveals major changes in porcine sepsis. *Clin Physiol Funct Imaging*. 2011;31(2):151–158.

24. De Backer D, Creteur J, Preiser JC, Dubois MJ, Vincent JL. Microvascular Blood Flow Is Altered in Patients with Sepsis. *Am J Respir Crit Care Med*. 2002;166:98–104. Available from: https://doi.org/10.1164/rccm.200109-016OC.

25. Spanos A, Jhanji S, Vivian-Smith A, Harris T, Pearse RM. Early microvascular changes in sepsis and severe sepsis. *Shock*. 2010;33(4):387–391. Available from: https://doi.org/10.1097/SHK.0b013e3181c6be04.

26. Wester T, Awan ZA, Kvernebo TS, Salerud G, Kvernebo K. Skin microvascular morphology and hemodynamics during treatment with veno-arterial extra-corporeal membrane oxygenation. *Clin Hemorheol Microcirc*. 2014;56(2):119–131.

27. Yeh YC, Lee CT, Wang CH, et al. Investigation of microcirculation in patients with venoarterial extracorporeal membrane oxygenation life support. *Crit Care*. 2018;22:200. Available from: https://doi.org/10.1186/s13054-018-2081-2.

Chapter 34

Technologies for Hemodynamic Measurements: Past, Present and Future

Ivan Corazza[1], Margherita Zecchi[1], Anna Corsini[2], Emanuela Marcelli[3] and Laura Cercenelli[3]

[1]*Medical Physics Coordination Center, Department of Experimental Diagnostic and Specialty Medicine (DIMES), Alma Mater Studiorum—University of Bologna, Bologna, Italy,* [2]*Cardiology Unit, Department of Experimental Diagnostic and Specialty Medicine (DIMES), Alma Mater Studiorum—University of Bologna, Bologna, Italy,* [3]*eDIMES Lab, Laboratory of Bioengineering, Department of Experimental Diagnostic and Specialty Medicine (DIMES), Alma Mater Studiorum—University of Bologna, Bologna, Italy*

Chapter Outline

Introduction

This chapter will describe the most known and commonly used technologies for hemodynamic monitoring. We gave this term a "quite wide" meaning, trying to include all the techniques used to measure and acquire the two most common physical parameters used to describe the dynamics of blood moving inside the vessels: the pressure and the flow. We decided to include techniques used both in specialized units, for example, intensive care units, surgical, and invasive cardiovascular laboratories, and in common clinical practice settings. Most of the described techniques are "old" (e.g., introduced in the past and currently well-established), but their usage has been improved in the last decades together with the technology advancements in the field of medical instrumentation. We did not deepen the described numerical methods and the biomechanical models underlying these approaches, since it would be out of the scope of this chapter, but we made the effort to introduce the bibliography addressing these concepts.

We chose to not include the techniques to measure other important parameters (distance, areas, and volumes), strictly correlated to "dynamics," since it would have involved a thorough discussion of all the technologies and techniques related to medical imaging.

Together with the pressure and flow measurement techniques, we decided to describe some complementary approaches to measure additional parameters directly correlated to hemodynamics (i.e., pulse wave velocity, oxygen saturation, ballistocardiography, cardiac contractility, cardiac wall motion, etc.), since their measuring and monitoring are quite common in different hospital settings.

The chapter is divided into paragraphs related to each single parameter of interest, and each paragraph contains both invasive and noninvasive measurement methods. At the beginning of each paragraph, a table summarizes the techniques considered for each parameter, which are then described in detail.

At the end of the chapter, we included a short appendix describing the basic functioning principles of the most common pressure transducers.

Advances in Cardiovascular Technology. DOI: https://doi.org/10.1016/B978-0-12-816861-5.00022-8

Arterial pressure

The gold standard method for the measurement of endovascular or cardiac pressures is the application of a pressure micro-transducer (micromanometer) inside the cavity of interest. Nowadays, there are different kinds of pressure sensors, commonly used for this purpose and commercially available with different monitoring systems.[1,2] Despite this approach representing the correct way to directly measure the pressure, it is invasive, hence quite difficult and dangerous to use in daily clinical practice. For this reason, during the last century, different methods to noninvasively estimate arterial pressure have been proposed and validated, and they are currently used for ambulatory pressure evaluation (Table 34.1).

In the next paragraphs, we will try to give a schematic overview of the physics principles forming the basis of these measurements systems and approaches, highlighting limits and errors. We will divide the methods based on the target of measurement (e.g., arteries, veins, ventricles, etc.).

Noninvasive measurements

Auscultatory technique (Riva Rocci + Korotkoff): principles

Cardiac contraction generates pulsatile arterial pressure and flow; to estimate the arterial wall tension, the intracavity pressure is measured by occluding the artery through the application of a pressure from the outside, and by detecting the characteristic noises during the deflation phase (Fig. 34.1).

An inflatable band (cuff) connected to a pressure gauge is applied to the arm, at the height of the heart, and a stethoscope is positioned immediately downstream. The cuff is rapidly inflated to a pressure value higher than the maximum pressure existing in the brachial artery, thus producing the occlusion of the artery (disappearance of the radial pulse). The band is then slowly deflated by the operator; when the occlusion pressure falls below the arterial one, the artery reopens and the operator perceives Korotkoff tones; when the arterial pressure drops below the external applied pressure, the artery occludes again. As the deflation continues, the opening intervals will be wider and the tones clearer. It is easy to observe in Fig. 34.1 how the first tone is perceived in correspondence with a band pressure close to the systolic value, and the last tone in correspondence with the diastolic one.[3–6]

The method appears simple and it is used without problems by all healthcare professionals, but it is necessary to consider the underlying physical principles to obtain a reliable measurement, for example, the correct dimension and positioning of the cuff, as well as the right inflating and deflating speeds that are mandatory to make a correct measurement.[3,6–12]

Moreover, the need to detect the pressure in correspondence with the Korotkoff tones makes the described technique difficult to automate and sometimes very subjective.

Auscultatory technique: technology

To apply this method an inflatable band of suitable size is mandatory, connected to a manual pump and a pressure gauge (sphygmomanometer). It is a simple technology and there are many similar products on the market that are commonly used in the clinical setting. Over the years, compared to the first application of the method, the only evolution has been in the use of liquid column manometers since mercury is toxic and forbidden in the clinical setting. Currently, the most widely used sphygmomanometer is the mechanical aneroid one, which uses a Bourdon tube.[1,13]

Sphygmomanometers with digital displays, on the other hand, use different pressure transducers (strain gages, piezoelectric transducers, etc.). For more details on pressure transducers, refer to the Appendix.

Oscillometric: principles

An alternative to measure the arterial pressure is represented by the oscillometric method, which requires the application of a sensor to detect pressure oscillations in the occlusion band (Fig. 34.2). During the deflation phase, as soon as the occlusion pressure falls below the maximum value existing in the artery, the artery, when reopening, transmits an impulse that is detected as a pressure variation in the occlusion band. This pulse gradually becomes wider as the occlusion pressure drops, until it reaches a maximum amplitude that corresponds to the mean pressure (MP). Subsequently, it gradually decreases and, at the minimum pressure value (diastolic pressure, DP), the pulse amplitude trend changes.[14–17]

With this method, MP is measured directly: the systolic and diastolic pressures (SP and DP) are obtained by knowing the mathematical relationship between the amplitudes of the oscillations and the different values of the external pressure (EP). There are two possible approaches: the estimate of SP and DP based on the amplitude of the oscillations, or the estimate based on the slope of the envelope interpolating the oscillations.[18]

TABLE 34.1 Arterial pressure measurements methods.

	Method		Primary measured parameters	Secondary measured parameters	Indirect parameters	Clinical setting	Pitfalls
Noninvasive	Auscultatory technique		Pressure in the arm cuff Korotkoff Sound	Systolic Pressure Diastolic Pressure		Medical Office Intensive care units (ICU)	Difficulty in perceiving Korotkoff sounds Errors due to not correct measuring procedure
	Oscillometric technique		Pressure in the arm cuff Pressure oscillations in the cuff	Mean pressure	Systolic Pressure Diastolic Pressure	Ambulatory ICU	Direct measure only of the mean blood pressure
	Volume clamp		Pressure waves inside the finger cuff Light absorption in the finger	Finger pressure waves	Brachial pressure wavesAortic pressure waves	Medical office ICU Surgery lab Invasive cardiovascular units	Finger pressure waveform is different from the brachial o aortic one; Depends on peripheral vasoconstriction.
	Tonometry		Pressure waves outside a "superficial" artery (carotid, radial)	Pressure waves inside the artery	Brachial pressure waves Aortic pressure waves	Medical office ICU Surgery lab Invasive cardiovascular units	Depends on operator Need always calibration with another methodology
	From "surrogates" pressure waves	Brachial Volume displacement	Pressures oscillations in the brachial inflated cuff	Brachial pressure waves	Aortic pressure waves	Medical office ICU Surgery lab Invasive cardiovascular units	Need a mathematical model to obtain brachial pressure Necessary calibration with other method
		From pletismography	Volumes changes in peripheral artery	Pressure waves	Brachial and/or Aortic pressure waves	Medical office ICU Surgery lab Invasive cardiovascular units	Based on biomechanical models depending on viscoelastic properties of vessels and blood Necessary calibration with other method Motion artifacts Complex mathematical approach
		From velocity profiles	Arterial flow with Doppler	Pulse wave velocity	Radial or aortic pressure waves	Medical office ICU	Depending on biomechanical conditions of the arteria tree and pathologies
	Catheter + external transducer		Systolic and diastolic pressures	Mean pressure		Surgery lab Invasive cardiovascular units	Measurement of systolic, diastolic and mean pressures with an error up to 10%

(*Continued*)

TABLE 34.1 (Continued)

	Method	Primary measured parameters	Secondary measured parameters	Indirect parameters	Clinical setting	Pitfalls
Invasive	Microtransducer	Pressure waves			Surgery lab Invasive cardiovascular units	Cost of the transducers
	Implantable pressure monitors	Pressure waves	Mean, systolic, diastolic Pulmonary artery pressure (PAP) Left atrial pressure (LAP)		Implanted in invasive units but read in ambulatory setting	Possible sensor drift caused by biological processes (fibrous encapsulation after implantation)

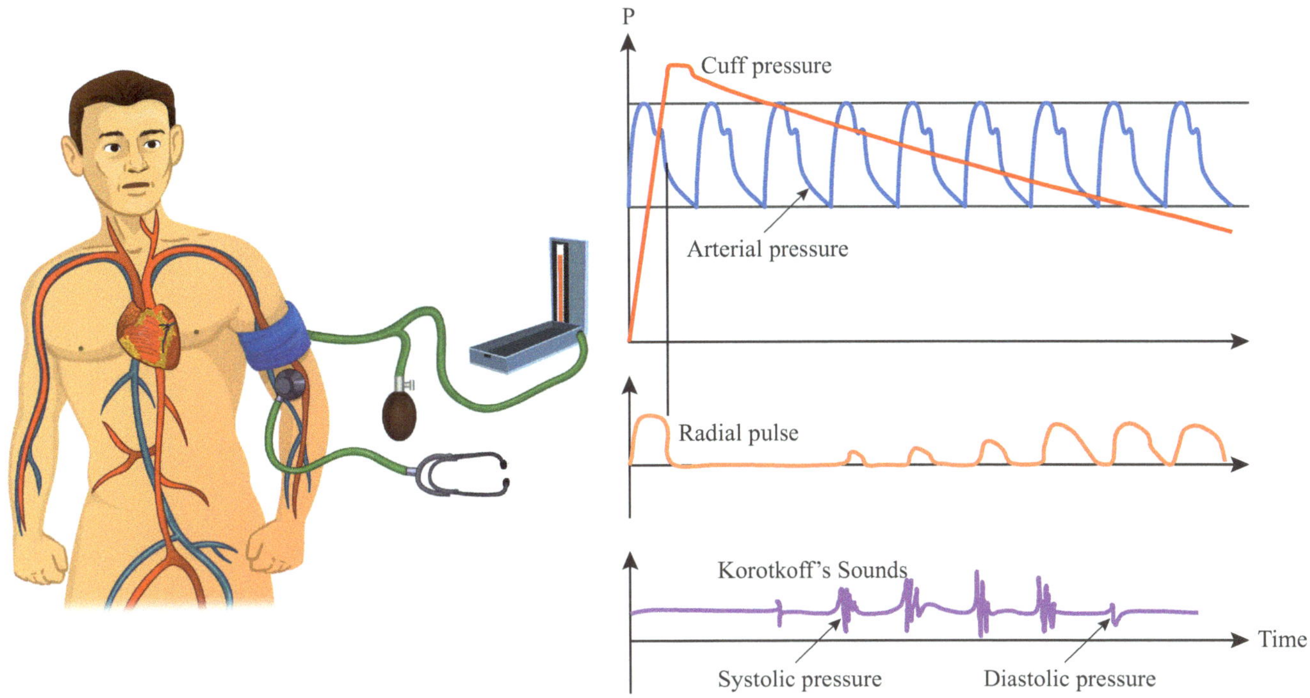

FIGURE 34.1 (Left) Blood pressure measurement with occlusion technique. (Right) graphic representation of the pressure inside the artery (in blue) and in the external cuff (in red). When the artery reopens, the operator perceives the Korotkoff sounds (in purple) and slowly the radial pulse (in orange) reappears.

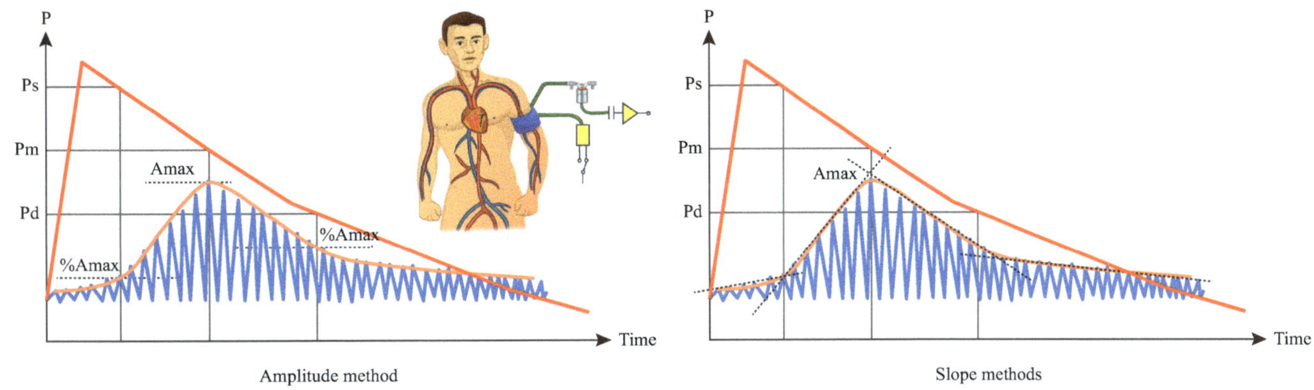

FIGURE 34.2 Pressure measurement with oscillometric technique.

Amplitude method: systole and diastole are fixed percentages of the maximum oscillation amplitude (corresponding to MP). Then, SP and DP are calculated as the cuff pressures corresponding to these percentages in oscillations amplitude. There are no standard percentage values to determine SP and DP: each manufacturer uses its own height-based algorithm (on the basis of their own clinical trials), and fixes its percentages as a function of MP pressure.

Slope method: the oscillations corresponding to SP and DP are calculated based on the slope of the envelope (see Fig. 34.2). As for height-based ones, there is no standard for these algorithms and each instrument uses its own numerical method.

Oscillometric: technology

To apply the oscillometric method, an inflatable band and an automatic compressor for controlled inflation and deflation are necessary together with a system for accurately measuring the band pressure and an electronic device capable of amplifying the pressure signal, extracting the oscillations, and relating them to the band pressure.

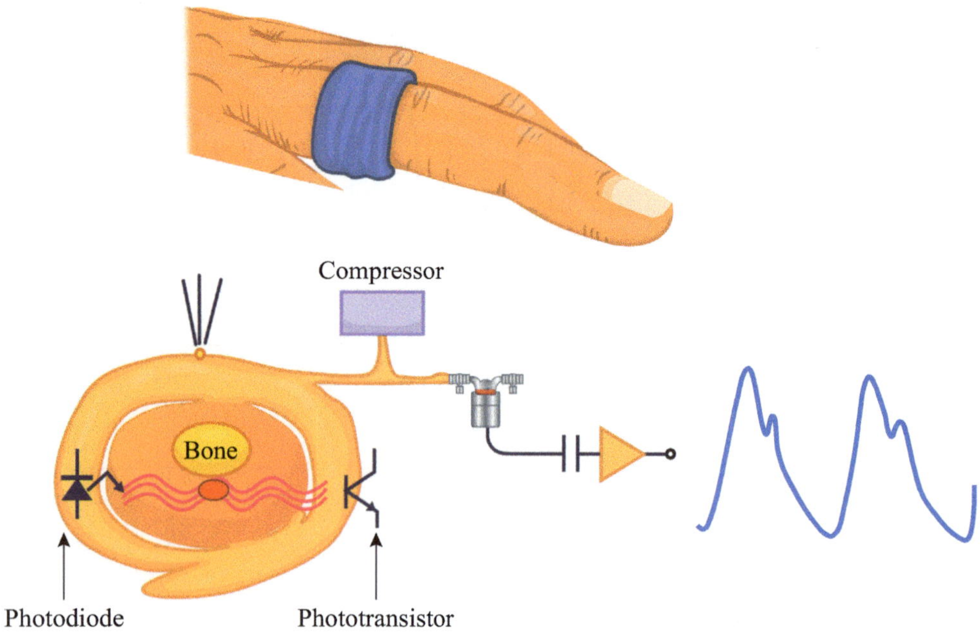

FIGURE 34.3 Schematic view of the volume-clamp method.

The currently used devices directly measure the MP at the maximum amplitude oscillation and then calculate SP and DP by analyzing either the amplitude or the slope of the other curves.

Although easy to use, the tools based on the oscillometric method have many critical issues, due to the nonstandardization of the algorithms to extract the SP and DP values, and to their intrinsic accuracy which could lead, in case of pathology (e.g., hypertension), to incorrectly evaluating the maximum and minimum pressures.[12] Currently, more accurate indicators of the position of SP and DP, based on the study of the morphology of the oscillations, are under investigation, but there are still no validated methods and commercially usable tools.[18–24]

Volume clamp: principles

Blood pressure waves are acquired by means of an inflatable band that is placed at the level of the second or third phalanx of the middle or ring finger. The inflatable cuff is connected to an air compressor for inflation and deflation (Fig. 34.3). At both ends of the band are placed a photodiode (emitting light) and a phototransistor (detecting light). The light emitted by the photodiode passes through the finger section and is absorbed by the blood. Absorbed light is directly related to vessel diameter: the greater the diameter, bigger the absorbed light intensity. Consequently, the light intensity measured by the phototransistor is inversely related to the vessel volume. Since during the cardiac cycle the diameter of the artery varies (it increases in systole and decreases in diastole), by monitoring the transmitted light it is possible to monitor changes in the vessel diameter (plethysmography).[25]

During the measurement phase, the optical system controls the valve at the compressor outlet in order to keep the vessel diameter constant in time, through a feedback drive mechanism. If the diameter tends to increase (the absorption of light increases), the compressor will inflate the band, while, when the diameter tends to decrease (less light is absorbed), the compressor is deflated. If the diameter remains constant, it means that the pressure inside the artery (inside pressure) and the EP are instantly equal (Laplace's law for a cylindrical duct). Therefore, by acquiring via computer the pressure generated instant by instant by the compressor, the result is a pulse wave corresponding to the intravessel pressure.

Volume clamp: technology

Lots of device families, based on Penaz' first method,[25] are available on the market,[26–32] and can be divided into:

1. simple technologies that allow pressure measurement without any correction/calibration;
2. systems with correction for the difference in hydrostatic pressure between the height of the finger and the brachial artery (in case of measurements made in an orthostatic position);

3. systems with finger pressure calibration performed on the diastolic, systolic and average pressure values measured in the arm with an oscillometric technology; and
4. systems with transfer function (TF) to transform the pressure waves on the finger into brachial or aortic waves.

The method, without corrections, provides an evaluation of the pressure at the finger level, therefore the morphology of the obtained curve is different from the brachial or aortic wave (for reflections, stiffness, distances, etc.), and consequently also SP and DP values are different.[27,33–37] Despite this, however, the system works correctly if the pressure variations are seen, during clinical procedures, with respect to a basal trend. In this case, the tool provides correct and reliable DP estimates.

To use these instruments as systolic and diastolic, brachial, or aortic blood pressure gauges, it is necessary that the acquired pressure waves are calibrated (using the methods written above). This involves the use of other methods (e.g., oscillometric or TFs) having their intrinsic uncertainties that are superimposed on the measurement errors due to the method itself.

Arterial tonometry: principles

If the artery is relatively superficial (e.g., radial artery, carotid artery), a device equipped with a pressure transducer applied to the skin can induce the deformation of the external surface of the vessel by the compression of the upper structures (Fig. 34.4). If the deformation is such as to flatten the vascular wall, locally the EP becomes equal to the internal one (Laplace law for cylindrical tube). By measuring this value with the pressure transducer positioned on the tip of the tonometer, it is possible to acquire the pressure waveform inside the artery.[38–41] Since the arterial pressure changes over time and the tonometer is fixed in an external position, the arterial wall is not flat over the complete cardiac cycle and so the measured pressure does not always correspond to the internal one.

This technology assumes that the vessel deforms appropriately, and the deformation is constant over time and independent of the operator. It is therefore difficult to apply this method correctly. For this reason, it is used manually only for the simple detection of the pressure pulse in order to evaluate the presence of the wave [e.g., for pulse wave velocity (PWV) measurement]. This approach presupposes the positioning of the sensor by means of a special mechanical support that always keeps it fixed in the same condition. To calibrate the pressure waveforms, either oscillometric systems are used or the measurements are referred to the brachial or aortic pressure by means of an appropriate TF.[42–46]

Arterial tonometry: technology

Although conceptually simple, tonometry is strongly linked to the operator who uses it, to the correct positioning of the tonometer, and to the methods applied for calibrating the system which themselves have limitations. For this reason, today, tonometry is very useful if you have to verify the presence and temporal position of a pressure wave (e.g., to calculate PWV), but it is not recommended if you want to correctly measure SP and MP values or to analyze the morphology of the curve (e.g., to calculate the augmentation index or cardiac output, CO, by numerical methods). In these cases, a fixed positioning system is required, and a good calibration procedure is mandatory. Calibration is often made using SP, MP, and DP measured with other techniques (e.g., oscillometric, auscultatory).

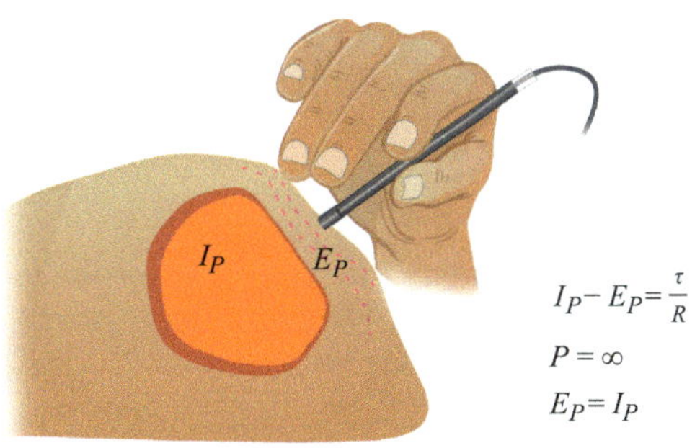

FIGURE 34.4 Example of tonometry application: due to the external pressure applied by the operator, the vascular wall flattens and, locally, the external pressure (EP) becomes equal to the internal one (IP) (R, radius; τ, wall elastic tension).

From radial or finger pressure (by volume clamp or tonometry) to brachial and aortic pressure with numerical methods: principles

To obtain the central aortic (or brachial) pressure wave in a noninvasive way, it is possible to acquire a peripheral non-invasive pressure through volume clamp or tonometry and then, to use an appropriate TF or a biomechanical-based mathematical model. These algorithms are generally integrated in the measurement systems, and they have been previously obtained from other researches by directly measuring the intravessel pressures.[45–54]

These approaches, used when the interest is to obtain an aortic wave morphology (e.g., for the measurement of the augmentation index), have several limitations, due to the fact that the TFs or the biomechanical models vary by age, sex, and pathology, therefore their use must be preceded by a correct evaluation of the patient's condition. Together with this evidence, it is necessary to take into account all the limitations related to the original peripheral pressure waves acquisition system.[55–59]

From pressure "surrogate waves" to pressure waveforms: principles

Several methods are based on the relation between pressure and other biomechanical parameters changing with it (e.g., arterial diameter, blood velocity, volume changes).

From velocity profiles by means of ultrasound

It is possible to calculate the pressure waves by referring to the value of the velocity profiles[60] measured inside a large artery and to the quantification of the PWV by means of an ultrasound (US) method.[61]

From the Moens–Kortweg equation:

$$\mathrm{PWV} = \sqrt{\frac{E \cdot h}{2\rho R}}$$

And the Bramwell–Hill one:

$$\mathrm{PWV} = \sqrt{\frac{A_0 \cdot dP}{\rho \cdot dA}}$$

It is possible to obtain pressure in function of time ($P(t)$):

$$dP = \frac{\mathrm{PWV}^2 \cdot \rho}{A_0^2} \cdot dA$$

$$P(t) - P(0) = \int_{A(0)}^{A(t)} \frac{\mathrm{PWV}^2 \cdot \rho}{A_0^2} \cdot dA$$

where E, Young module of the arteria wall; h, wall thickness; R, arterial radius; A, arterial section area; A_0, end-diastolic area; and ρ, density of the fluid.

By measuring with an US device the arterial section's changes in time (A) and the PWV (with Doppler technique),[61] it is possible to obtain pressure waveforms.

Directly related to this approach, some authors propose to evaluate the blood pressure by flow, estimated with other methods than Doppler, for example, the pulse transit time technology,[62,63] combining different technologies (e.g., ballistocardiography[64]).

From brachial volumetric displacement with subdiastolic oscillometry

Using an oscillometric pressure measurement system with a high-precision pressure sensor, the band is inflated to a pressure lower than the DP: volume variations of the brachial artery generate oscillations in the cuff. These waves reflect volume changes in the artery and are transformed by means of an appropriate TF to either the brachial pressure wave or the central aortic wave.[33,65–69]

From brachial volumetric displacement with sopra-systolic oscillometry

Using an oscillometric pressure measurement system with a high-precision pressure sensor, the band is inflated at a pressure higher than the systolic one (about 30 mmHg) for a fixed time interval (generally 10 s). The brachial artery occludes and acts as a reflection surface for pressure waves.

Pulse wave moving from the aorta reaches the occlusion and then it is reflected; during reflections part of the energy transmits to tissues around the brachial artery and reaches the external inflated cuff. The oscillations generated from these interactions are evaluated by a sensible pressure sensor connected to the cuff. These fluctuations can be directly related to the intraarterial pressure oscillations. By estimating the reflection characteristics at the ends of the tube (occlusion) and the transmission time for pressure wave propagation, the model allows the estimation of the pressure waveform in the aorta[37,70].

From photoplethysmography

Photoplethysmography (PPG), as previously described in the volume clamp method, is the measurement of the light absorption by a vessel during cardiac cycles. Since absorption depends on the diameter of the vessel and diameter varies according to the pressure, PPG can be considered a direct surrogate of pressure measurement inside the artery (Fig. 34.5).

Different methods have been investigated to extract blood pressure from PPG, based on the combination between PPG and ECG or sonography. These methods are based on PPG waveforms analysis and depend on the reliability of the acquired signals. They are strictly dependent on the biomechanical model used to couple flow, volumes, and pressures through the correct evaluation of the viscoelastic characteristics of the vessels and related phenomena (e.g., pressure pulse reflection). Moreover, some of them require external calibration with another pressure measurement method, and the accuracy of these systems is still under evaluation.[72–80]

From pressure surrogates to pressure waveforms: technologies

All these systems, based on mathematical or mechanical assumptions, are patient-dependent, since all the underlying models are based on assumptions regarding the arterial stiffness, reflections, age, presence or absence of pathologies, which make it difficult to give an absolute meaning to the obtained results. Despite that, if the procedure of measure is applied correctly and in a repeatable manner, these approaches can give reliable indications about the evolution of the patient's health during the hospitalization period.[33,81–83]

Pressure from magnetic resonance imaging: principles and technology

With phase contrast magnetic resonance imaging (MRI) it is possible to measure both the cross-sectional area changes of the aorta and the aortic flow. By applying appropriate algorithms, the aortic pressure waves can be derived. This approach requires the segmentation of the aortic circumference, therefore errors can occur during the low flow phase of the cardiac cycle and in the case of poor-contrast MRI images.[33,84–86] The same approach can be used to measure the pulmonary artery (PA) pressure.[87–89]

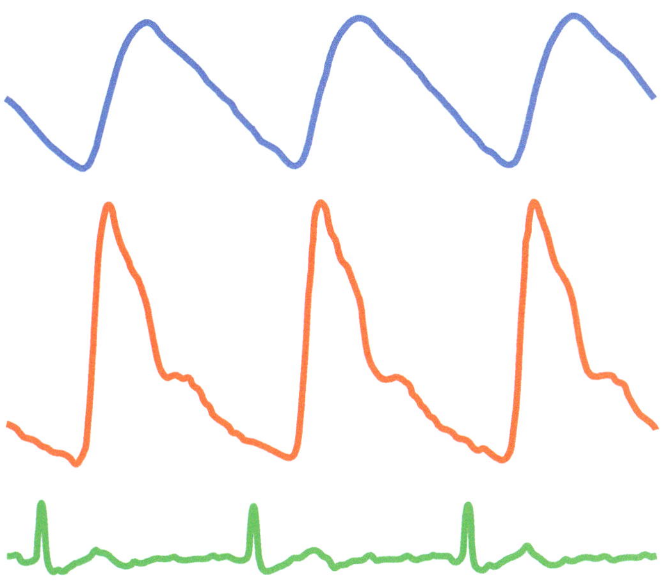

FIGURE 34.5 Example of ECG signal (in green), arterial blood pressure (in red) (volume clamp method), and finger photoplethysmography waveforms (in blue).[71]

Right and left heart chambers pressures by means of US: principles and technology

Measurement of heart or arterial pressures with an echo-Doppler approach are based on numerical estimation starting from US acquisitions of velocity profiles and CO.[90–93] A more detailed approach on ultrasound-based techniques is described in paragraph about noninvasive measurement of venous pressure.

Invasive measurement

Catheter + external transducer: principles

Since it is very difficult to insert the pressure sensor inside the cavity of interest, in most cases it is necessary to make a mechanical connection between the cavity and the sensor, which is located outside the patient (Fig. 34.6). This can be accomplished by using a liquid-filled tube (liquid column) that connects the cavity with the transducer. If the transducer is at the same height as the cavity and considering the law of communicating vessels, the measured pressure should not differ from the intracavitary one. This is certainly true for constant pressures. In the case of rapidly varying pressures, such as in arteries and cardiac cavities, the situation is more complex, presenting the transfer line with elastic characteristics (elasticity of the structures), inertial (mass of the moving liquid), resistive (changes in diameter of the tubes), and damping (viscosity of the liquid). That is, the sampling system behaves like a damped oscillatory system with its own resonant frequency.

Therefore, the system abnormally amplifies the harmonics of frequency close to the resonance one (oscillations appear on the trace), attenuates the high frequency harmonics (the notches from the trace disappear), modifies the local slopes of the curve.

To obtain more accurate measurements, it is possible to optimize the mechanical construction of the transfer line (short connecting pipes made with low elastic material and transducers with rigid membrane) or to use transducers that can be inserted directly into the cavity. There are different types of pressure transducers that can be used to quantify the pressure of the liquid column. We discuss the most common in the following paragraphs.

Catheter + external transducer: technology

The approach is the most used in an interventional cardiology setting, since it is quite easy to perform, and it is not expensive. The approach described above is useful when the absolute value of the pressure is not important, but you

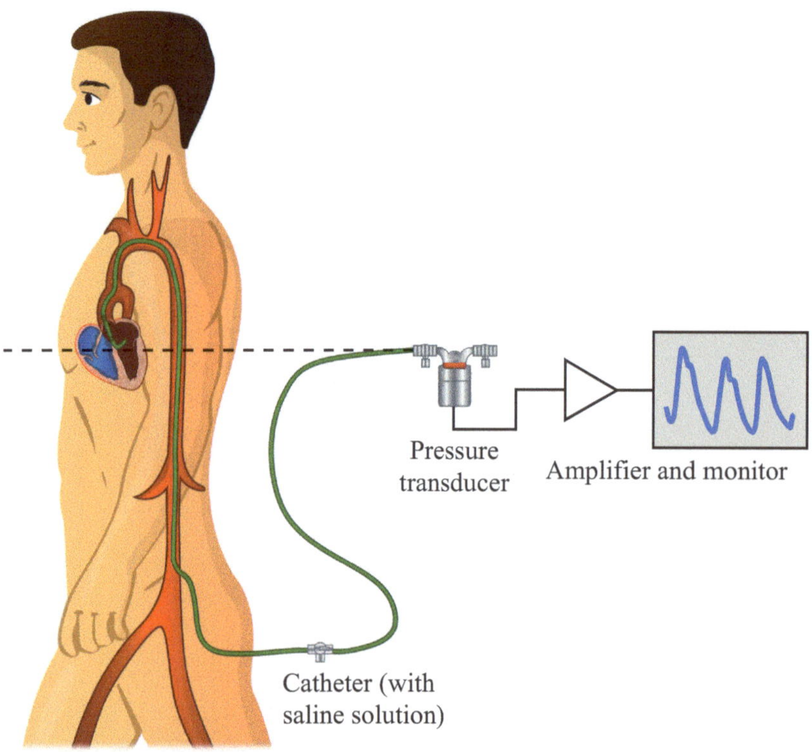

FIGURE 34.6 Schematic representation of invasive pressure measurement with an external transducer.

want to check the trend of the systolic, diastolic, and average pressures inside a vessel or cavity during an interventional procedure (e.g., to check for sudden drops in blood pressure during surgery or invasive cardiovascular procedures).

If it is necessary to measure pressure values with greater accuracy or to study pressure wave trends, it is necessary to use miniaturized pressure transducers that can be inserted directly into the cavity of interest.[94–99]

Microtransducers on catheters tips: principles

To insert a transducer inside a cavity (artery, vein, atrium, or ventricle) it must be miniaturized. In the appendix there is a brief excursus on the main technologies used for pressure transducers.[1,100,101]

This approach still represents the gold standard, although its widespread and routine use is limited by the risk of complications, including infections, catheter-induced thrombosis, and arrhythmias. Moreover, the line insertion requires a certain level of expertise, is time-consuming, and is not practical in prehospital settings.

In clinical practice, the most common accesses for catheter insertion, depending on the target, are:

- femoral or jugular veins to reach the right atrium (RA), the right ventricle (RV), and PA;
- femoral, radial, and brachial (rare) arteries for left atrium (LA), left ventricle (LV), and aorta;
- venous access to the right part of the heart and a trans-septal puncture (interatrial septum) to reach the LA and the left part of the heart.

Implantable pressure monitors: principles

For obtaining miniaturized high-performance implantable pressure sensors, the MEMS (microelectromechanical systems) technology is typically used (see Appendix).

Implantable pressure monitors are particularly demanding because the pressure sensitive part of the sensor must be in direct or indirect physical contact with the medium for which we want to detect the pressure, in contrast to measurement of, for example, acceleration, where direct contact with the surrounding fluid or tissue is not required. For the development of a permanent implantable pressure sensor, drift caused by biological processes is a main concern, since the final fibrous encapsulation after implant might result in sensor failure in the long term. Also corrosion, caused by aggressive body fluids, may alter the sensor characteristics and thereby sensor stability. For this reason packaging is extremely important. The packaging materials must provide for the minimum expected lifetime and have sufficient protection properties toward the sensor, for example, it must not cause any critical degradation of sensor performance and characteristics.

Implantable pressure monitors: technology

Various implantable technological solutions based on pressure sensors have been proposed for long-term monitoring of cardiac dysfunction, such as heart failure.[102,103]

Only one of them is currently FDA-approved and commercially available.[104,105] It consists of a miniaturized wireless capacitive MEMS sensor, which is implanted into a branch of the PA with a minimally invasive procedure. The pressure in the PA deflects the capacitive sensor surface, causing a shift in the resonant frequency of the coupled LC resonant circuit. An external antenna is used for powering the sensor and for reading the changes in the PA pressure (Fig. 34.7).

Another investigational device, currently under clinical trial, is a wireless sensor system, which is implanted permanently in the interatrial septum to measure and transmit the left atrial pressure (LAP), also including recording of the continuous high-resolution LAP waveform.[106] The system is composed of a MEMS pressure transducer associated to a micro application-specific integrated circuit (ASIC) with an antenna, and a braided anchor made of nitinol (Fig. 34.8). The system also includes a light belt to be worn by the patient, which remotely powers and interrogates the implanted sensor via radio frequency bidirectional communication upon activation, and then transmits the collected LAP data to a secure cloud-based database for clinician review.

Venous pressure

Different methods have been proposed and are currently used in clinical practice for the assessment of central venous pressure (CVP); they can be categorized as invasive and noninvasive techniques (see Table 34.2).

In the next paragraphs we will try to give a schematic overview of the physics principles forming the basis of these measurements systems and approaches, highlighting limits and errors.

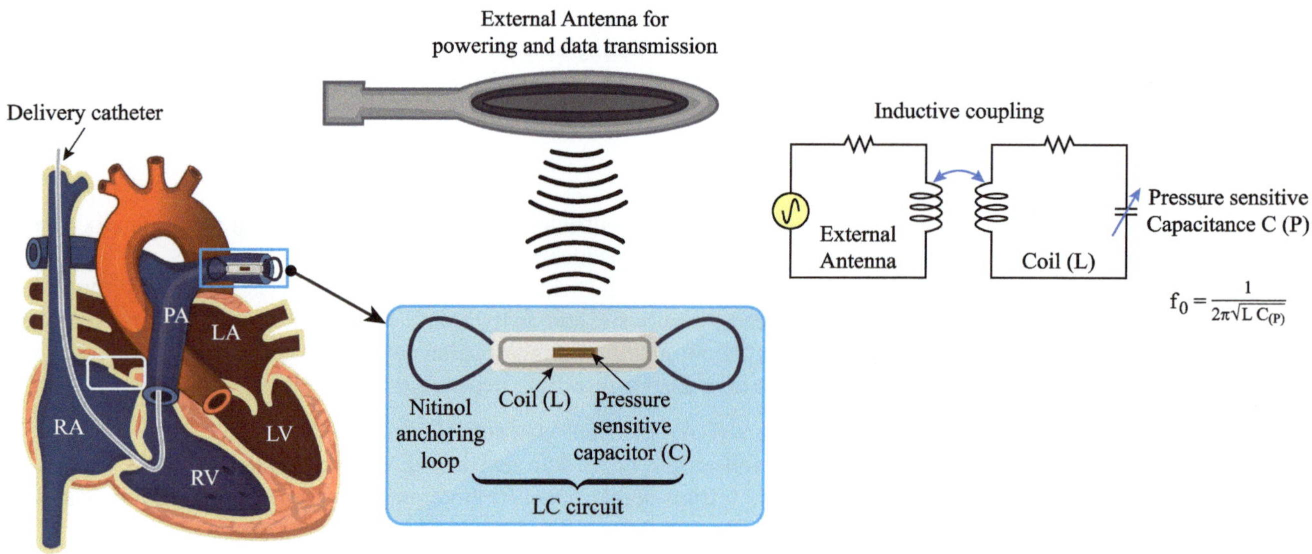

FIGURE 34.7 The implantable wireless pressure sensor to monitor the pulmonary artery (PA) pressure.

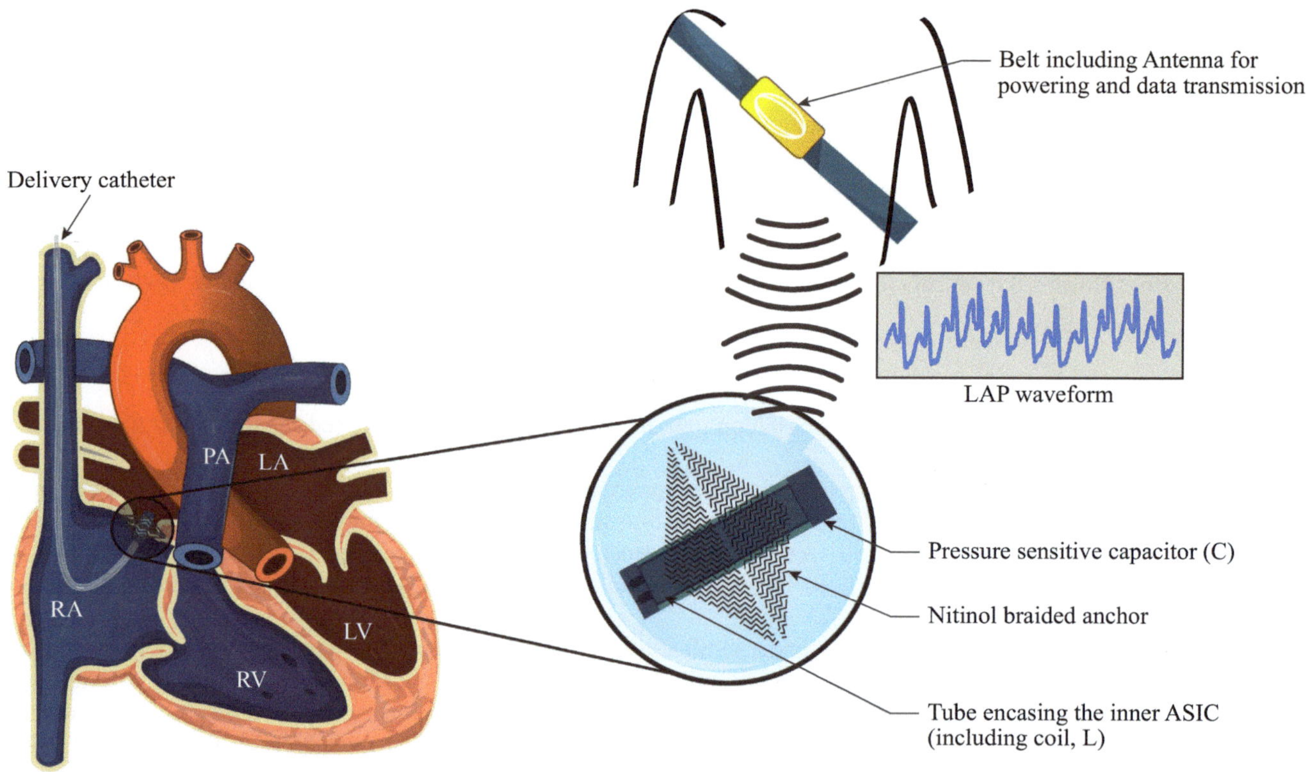

FIGURE 34.8 Scheme of the implantable wireless pressure sensor device to monitor left atrial pressure (LAP).

Noninvasive measurements

Ultrasound-based methods: principles

The inferior vena cava (IVC) is a compliant vessel whose size and shape vary with changes in CVP and intravascular volume. Parameters such as the IVC diameter and the IVC collapsibility index can be assessed by sonographic clinic measurements in order to derive an indirect indicator of CVP.[90,107]

In general, ultrasound-based techniques are well-accepted noninvasive options for CVP assessment, being favorably validated compared to invasive hemodynamic measurements,[108,109] although a recent review does not support

TABLE 34.2 Venous pressure measurements methods.

	Method	Primary measured parameters	Secondary measured parameters	Indirect parameters	Clinical setting	Pitfalls
Noninvasive	Ultrasound (US)-based methods	Inferior vena cava (IVC) diameter Skin pressure to exercise in order to obtain the complete compression of the jugular vein detected by US	IVC collapsibility index (IVCCI)	Central venous pressure (CVP)	Medical office ICU	Operator-dependent Limited accuracy (only wide ranges of CVP values are identified)
	Plethysmographic technique	Pressure of the occlusive cuff corresponding to a volumetric response in the plethysmographic sensor		Peripheral venous pressure (PVP), Central venous pressure (CVP)	Medical office ICU	Limited accuracy
	Optical technique	Blood volume changes in the jugular vein	Jugular venous pulse (JVP)	Central venous pressure (CVP)	Medical office ICU	Impractical for routine use outside the hospital setting
Invasive	Centrally inserted venous catheters	Pressure in a central great vein (central venous pressure, CVP)			Surgery lab Invasive cardiovascular units ICU	Invasiveness, risk of catheter-induced complications
	Peripherally Inserted venous catheters	Pressure in peripheral veins (e.g., arms)		Central venous pressure (CVP)	Surgery lab Invasive cardiovascular units ICU	Invasiveness, risk of catheter-induced complications

the measurement of IVC diameter by ultrasonography as an acceptable surrogate to assess CVP among critical patients.[110]

As a disadvantage, they remain highly operator-dependent, and generally suffer from a limited accuracy, since only wide ranges of CVP (0–5, 5–10, 10–20 mmHg) can be identified.

Ultrasound-based methods: technologies

The IVC parameters for noninvasive CVP estimation can be measured using 2D or M-mode ultrasound imaging. M-mode allows for high–frame rate measurements of IVC size changes throughout the respiratory cycle, and the transthoracic echocardiographic subcostal window can be used to view the IVC in the sagittal plane.[107]

Ultrasound measurement of the IVC diameter and its variations with respiration reliably predict CVP and mean RA pressure during spontaneous (negative pressure) ventilation[111]:

- IVC diameter $\leq$2.1 cm with >50% respiratory changes → RA pressure 0–5 (3) mmHg;
- IVC diameter $\leq$2.1 cm with <50% respiratory changes → RA pressure 5–10 (8) mmHg;
- IVC diameter >2.1 cm with >50% respiratory changes → RA pressure 5–10 (8) mmHg;
- IVC diameter >2.1 cm with <50% respiratory changes → RA pressure 15 mmHg.

In recent years, portable ultrasonography devices have been proposed for noninvasive CVP measurement, and are currently widely used and recommended as an alternative approach to invasive CVP assessment.[112,113]

A technique, named "Compression Ultrasound" is based on a portable ultrasound system and a translucent pressure manometer.[114] In this case, the CVP is estimated via an ultrasound evaluation of the skin pressure to be exercised to obtain the complete compression of a superficial vein at the forearm (preferentially the distal cephalic vein), which is observed by ultrasound imaging. As a major limitation, the technique is characterized by high result variability depending on the expertise and training of the physician performing the echographic examination.[115]

Plethysmographic method: principles

The basic principle of plethysmographic techniques is the venous occlusive plethysmography, for example, the assessment of volume changes in a portion of a limb (e.g., the forearm) after the application of an occlusive pressure in the upstream arm. This physical principle is similar to the one used for noninvasive peripheral flow measurement.

Plethysmographic method: technologies

Examples of plethysmographic devices for CVP assessment were reported by Ward et al.,[116,117] but to date they have not yet achieved a clinical application. In these methods, an occlusive cuff pressure surrounding the upper arm is inflated to a single pressure value (40 mmHg) lower than the diastolic arterial pressure and kept for about 60 s. At this pressure the arterial blood continues to flow into the arm, but venous return is decreased allowing the forearm to swell. During this time the patient's forearm volume increases, and this volume variation is detected by a mercury-in-silastic strain gauge positioned around the forearm or using electrodes for tetrapolar impedance plethysmography. At the end of the inflation hold period, the occlusive cuff is rapidly deflated, and the CVP is noninvasively determined as the cuff pressure noted at the maximum derivative of the volume change during deflation (Fig. 34.9). A similar system has been more recently proposed.

Optical method: principles

The optical method for CVP assessment is based on the principle of near-infrared spectroscopy (NIRS), which has been widely used in various monitoring devices during cardiac surgery, including Swan–Ganz catheter, SpO_2 plethysmograph, and continuous noninvasive arterial pressure monitor.

In this case, infrared sensors of the size of the superficial veins are applied in the neck to detect the blood volume changes in the jugular vein, and therefore to yield a pressure waveform (e.g., JVP, jugular venous pulse).

Optical method: technologies

Commercially available systems based on the NIRS principle use single wavelength LED and photodetectors to detect changes in the jugular vein blood volume (therefore the JVP), and the height of the JVP column relative to the superior vena cava. To obtain an accurate measurement with this technique the patient has to stay at an inclination angle which aligns the JVP pulse in the range of the sensor.[118] Moreover, even if good correlation has been reported between noninvasive CVP assessments and invasive measurements, this optical method remains impractical for routine use outside the hospital environment.

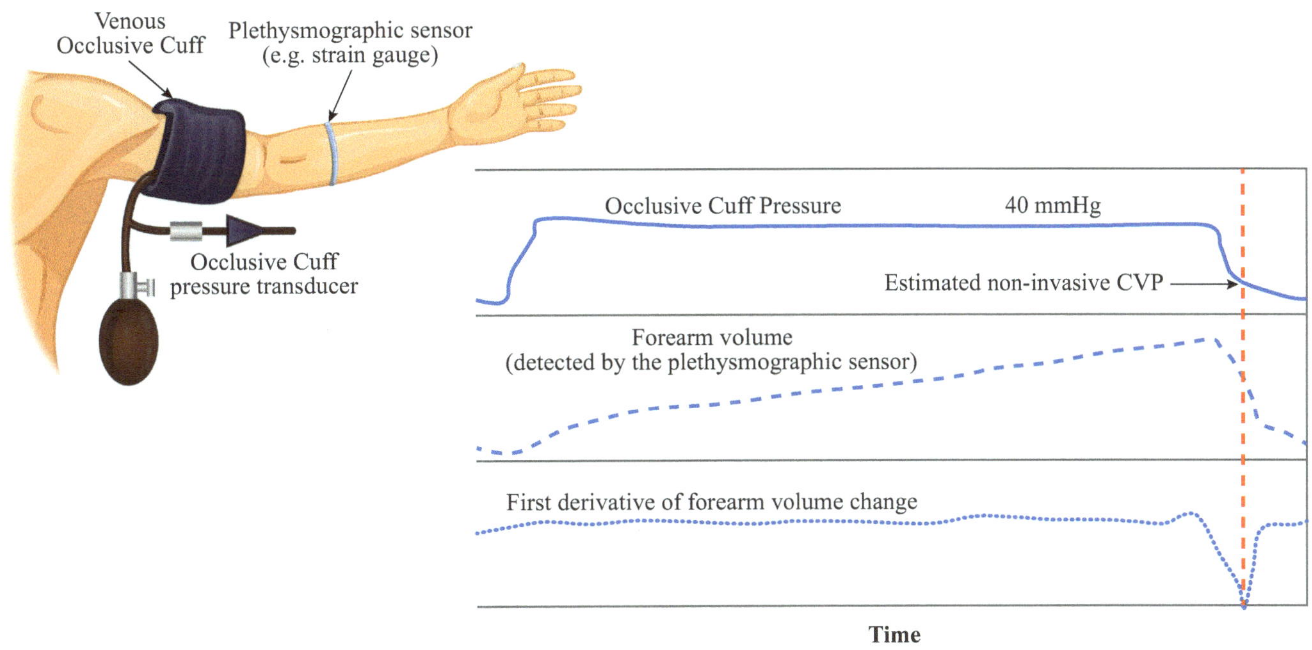

FIGURE 34.9 Scheme of plethysmographic technique for noninvasive assessment of central venous pressure.

Invasive measurements

Venous catheterization: principles

Invasive techniques for CVP assessment require the preparation of a venous (central or peripheral) line for the insertion of a catheter, which is advanced and placed in the superior vena cava. The CVP is measured using a manometer or an electronic transducer connected to the prepared patient's venous line. First, the transducer must be zeroed to atmospheric pressure. Then, the transducer must be aligned to the horizontal plane of the tricuspid valve.

This approach still represents the gold standard in the clinical practice, although its routine use is limited by the risk of complications, including infection, catheter-induced thrombosis, and arrhythmias. Moreover, the central line insertion requires a certain level of expertise, is time-consuming, and is not practical in prehospital settings.[119,120]

Flow measurements

The volumetric blood flow or CO is the volume (*d*Vol) of blood ejected by the heart per unit time (*dt*).

$$\mathrm{CO} = \frac{dVol}{dt}$$

In stationary condition (constant CO), CO can be calculated as the product of the heart rate (HR) and the stroke volume (SV), defined as the quantity of blood ejected in a single beat.

$$\mathrm{CO} = \mathrm{SV} \cdot \mathrm{HR}$$

For an incompressible fluid, considering the continuity equation,

$$\mathrm{CO} = \int \vec{v} \cdot d\vec{S}$$

CO can be approximated by:

$$\mathrm{CO} = \vec{v}_{mean} \cdot \vec{S}$$

where $\vec{v}$ is the instantaneous blood velocity along the internal surface of the vessel; $\vec{v}_{mean}$ is the mean velocity along the internal surface; and $\vec{S}$ is section area of the vessel.

As we did for pressure, we will consider "direct" the methods which give a flow as main result of the measurement (see Table 34.3).

TABLE 34.3 Blood flow measurements methods.

	Method	Primary measured parameters	Secondary measured parameters	Indirect parameters	Clinical setting	Pitfalls
Noninvasive	Doppler CW	Doppler frequency	Blood velocity	Volumetric flow	Medical office ICU	• Superimposition of signals of near vessels • Strictly dependence on US transducer orientation. • Operator dependent
	Doppler PW	Doppler frequency	Blood velocity	Volumetric flow	Medical office ICU	• Impossibility to measure high blood speeds (e.g., for aortic insufficiency diagnosis) • Strictly depends US transducer orientation.
	Strain gauges	Arm circumference increase	Arterial peripheral arm flow		Medical office (primarily for research)	• Very long and complex procedure • Very delicate instruments
	Thoracic bioimpedance	Thoracic impedance	SV CO		Medical office ICU	• Motion and electrical artifacts • Problems with obese patients or pulmonary pathologies (e.g., patients with thoracic fluid overloads) • Dependent on a mathematical model • Errors due to arrhythmias
	Thoracic bioreactance	Voltage phase shift due to changings in reactance	CO SV		Medical office ICU	• Motion and electrical artifacts • Problems with obese patients or pulmonary pathologies • Dependent on a mathematical modeler • Errors due to arrhythmias • Limited accuracy in low flow conditions
	Phase contrast MRI	Velocity area	flow		Medical office	• Expensive and not able to perform a correct sampling of a flow wave

	Numerical methods	Blood pressure waves	Flow waves	CO	Medical office ICU Surgery lab Invasive cardiovascular units	• Limitations of the pressure waves acquisition systems • Mathematical models depending on age, sex and pathology (e.g., arrhythmias, valvular diseases, etc.)
	Fick Method	Oxygen saturation in arterial and venous blood; ventilation and oxygen concentration in breath	CO		ICU Medical office	• Necessity of different apparatuses • Limitations related to each instrument. • Problems with patients with high output $Co(O_2)$ • Difficult to perform in patients with O_2 therapy
Invasive	Electromagnetic flowmeter	Blood velocity	Flow		Surgery labSimulators for research	• Surgery approach necessary • Powerful but fragile technology
	Transit time US flowmeter	Time shift	Blood velocity	Flow	Surgery lab Simulators for research	• Limitations due only to positioning and invasiveness • Not accurate if not calibrated
	Transesophageal Doppler	Doppler frequency	Blood velocity	Volumetric flow of the descending aorta	Medical office ICU	• Operator dependent • Artifacts due to movements • Necessity to correctly estimate aorta diameter
	Continuous thermodilution	Blood temperature	Coronary flow		Surgery lab Invasive cardiology units	
	Bolus injection of green indocyanine	Green indocyanine concentration	CO		Surgery lab Invasive cardiology units	Recirculation dependent
	Bolus Thermodilution	Temperature	CO		Surgery lab Invasive cardiology units	Recirculation dependent

(*Continued*)

TABLE 34.3 (Continued)

	Method	Primary measured parameters	Secondary measured parameters	Indirect parameters	Clinical setting	Pitfalls
Continuous warm thermodilution	Temperature	CO		ICU Surgery lab Invasive cardiology units		
Lithium chloride dilution	Lithium chloride concentration	CO		ICU Surgery lab Invasive cardiology units	Not indicated in patients with lithium therapy Recirculation dependent	
Transpulmonary thermodilution	Blood temperature	CO		Surgery lab Invasive cardiology units	Radial artery not suitable Limited use in surgery lab	
Partial carbon dioxide gas rebreathing method	Carbon dioxide saturation in arterial and venous blood; ventilation and CO_2 concentration in breath	CO		ICU	• Need of intubated patients • Controversial • Difficult to stand by patients	

Noninvasive measurements

US Doppler: principles

Continuous wave

CO (which is a blood flow) depends on the section area of the vessel and the velocity of the blood; if the section is constant, the flow depends only on the velocity. By integrating the flow waveform over a cardiac cycle, it is possible to estimate the blood volume ejected in a particular vessel. If the volume is calculated outside the ventricle, the volume corresponds to the SV, and it is easy to calculate the CO by multiplying for the HR.

In the continuous wave (CW) Doppler flowmeter, a transducer composed of two piezoelectric crystals is placed in contact with the tissue, with the axis oriented in the direction of the vessel to be examined. A crystal continuously emits an ultrasound wave of frequency f_0, which hits the vessel and is diffused by the corpuscles carried by the blood. A second crystal, placed near the first, receives the diffused wave and converts it into an electrical signal. An electronic device continuously measures the difference between the frequency of the emitted signal (f_0) and that of the detected signal (f). This difference is instantly proportional to blood velocity in the vessel (Fig. 34.10).

Since the blood motion can be considered "laminar," the velocity in the center of the section area of the vessel will be different from that detectable near the vessel wall, and the flow curve will have a more or less wide shape, due to this difference.

The main problem of the CW Doppler consists of the impossibility of "localizing" the measurement: if two or more vessels are invested simultaneously by the US beam, the receiving system is not able to separate the signals, and the result is the superposition of each single determination.

Pulsed wave

To refer the velocity measurement to a well-defined point, it is possible to use a Doppler flowmeter with impulsive excitation.

The device has a single transducer which is impulsively excited and then placed in reception. The delays in echo return define the position of the measurement point: by carrying out the harmonic analysis on the corresponding echo signal, it is possible to extract the frequency of the detected signal (Fig. 34.11). This is compared with the emitted frequency, and the difference is calculated. For each pulse emitted, a speed measurement is made at the chosen point. In this way a "sampling" of the Doppler signal is obtained at the pulse repetition frequency (PRF).

According to Shannon's theorem, if the sampling frequency (PRF) is not at least double the frequency of the received Doppler signal, there is an incorrect reconstruction of the signal (aliasing).

The PRF of the system is conditioned by the observation depth, because an impulse cannot be emitted before the echoes generated by the most distant structures have returned to the transducer.

The pulsed wave (PW) Doppler technique allows for the measurement of the velocity in the desired points, with the possibility of simultaneous echographic presentation of the structures (echo-Doppler). It has the limitation of not allowing correct measurements of high speeds for structures far from the transducer (aliasing).

To be able to represent the distribution of velocities in the observation sector, color coding is used. The speed measurement is always in the direction of the transducer; therefore the method provides an evaluation that is only partially objective, since it is up to the operator to identify the optimal scan plane .

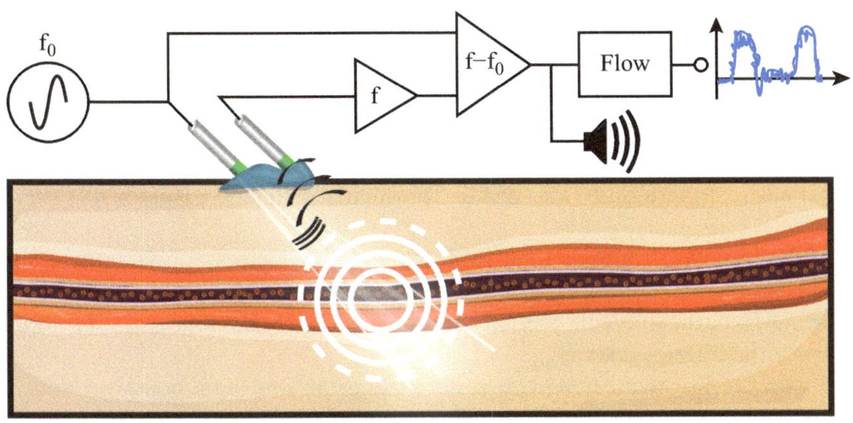

FIGURE 34.10 Continuous Wave Doppler Flowmeter (CW).

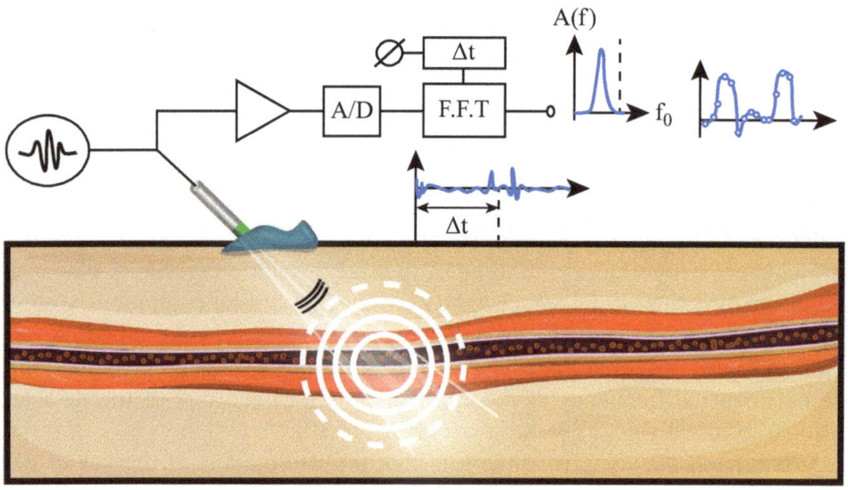

FIGURE 34.11 Pulsed Wave Doppler Flowmeter (PW).

US Doppler: technology

Both CW and PW Doppler technologies are commonly used in clinical practice, depending on the target of the desired measure. For example, to evaluate the blood velocity in case of aortic valve insufficiency, since the speed is high, a CW Doppler system should be used, with all the problems related to the impossibility to exactly locate the measurement site.[121]

Moreover, if the integration of a flow wave is necessary to measure the SV, the accurate choice of the flow profiles is often difficult due to the not clear contour of the pulse wave. Paradoxically, the quality of Doppler signals is worst in more critical conditions when a reliable measurement is mandatory to correctly evaluate flows.

Both approaches present problems related to difficulty in knowing the orientation of the US beam with respect to the vessel axis, so the measurements present an intrinsic error.[122–132]

Cardiac output with echocardiography: principles

The CO, which is the blood volume ejected by a ventricle per minute, may be measured noninvasively by echocardiogram.[133] The SV, for example, the blood volume ejected by a ventricle at a given cardiac cycle (normal value 50–80 mL), could be estimated by the biplane Simpson's method,[134,135] and the 3D full-volume echocardiography. CO can be also derived as the product of the left ventricular outflow tract (LVOT) area and LVOT time velocity integral, which are obtained by PW Doppler of LVOT velocity (Fig. 34.12).

Cardiac output with echocardiography: technology

Despite the mathematical models underlying this approach being quite consolidated, the measurement of SV and then CO by echographic imaging is highly operator-dependent, since the echographic images and the Doppler signal quality are strictly dependent on the instruments and on the skills of the clinician. For example, the measurement of RV dimensions is related to the probe orientation, therefore on the 2D plane investigated by US. Moreover, PW Doppler presents the limitations already cited in the above paragraph, and sometimes it is very difficult to accurately estimate the LVOT. For these reasons, the use of transthoracic echography is suggested for an already stabilized patient's follow-up, but not for evaluation of patients with severe and acute pathologies. In these latter cases, a more accurate and reliable technique is mandatory [e.g., thermodilution, transpulmonary thermodilution (TPTD), Fick method].[136,137]

Peripheral flow: principles

This technique involves rapid venous occlusion upstream of a limb (arm or leg) to which a volume measurement system has been applied.

Following venous occlusion, the volume of the limb will continue to grow due to the blood supply through the arteries; the slope of the volume growth curve expresses the blood flow of muscle tissue.

Venous occlusion can be achieved by placing an inflatable cuff around the limb and applying a pressure of about 50 mmHg, in order to occlude the veins and leave the arteries open (Fig. 34.13).

The volume of the limb can be measured with the impedance technique, immersing the limb in a container with a liquid and detecting the level changes, or by measuring the circumference of the limb with strain gauges, in the assumption of a homogenous volumetric change of structure (proportional enlargement of all sections).

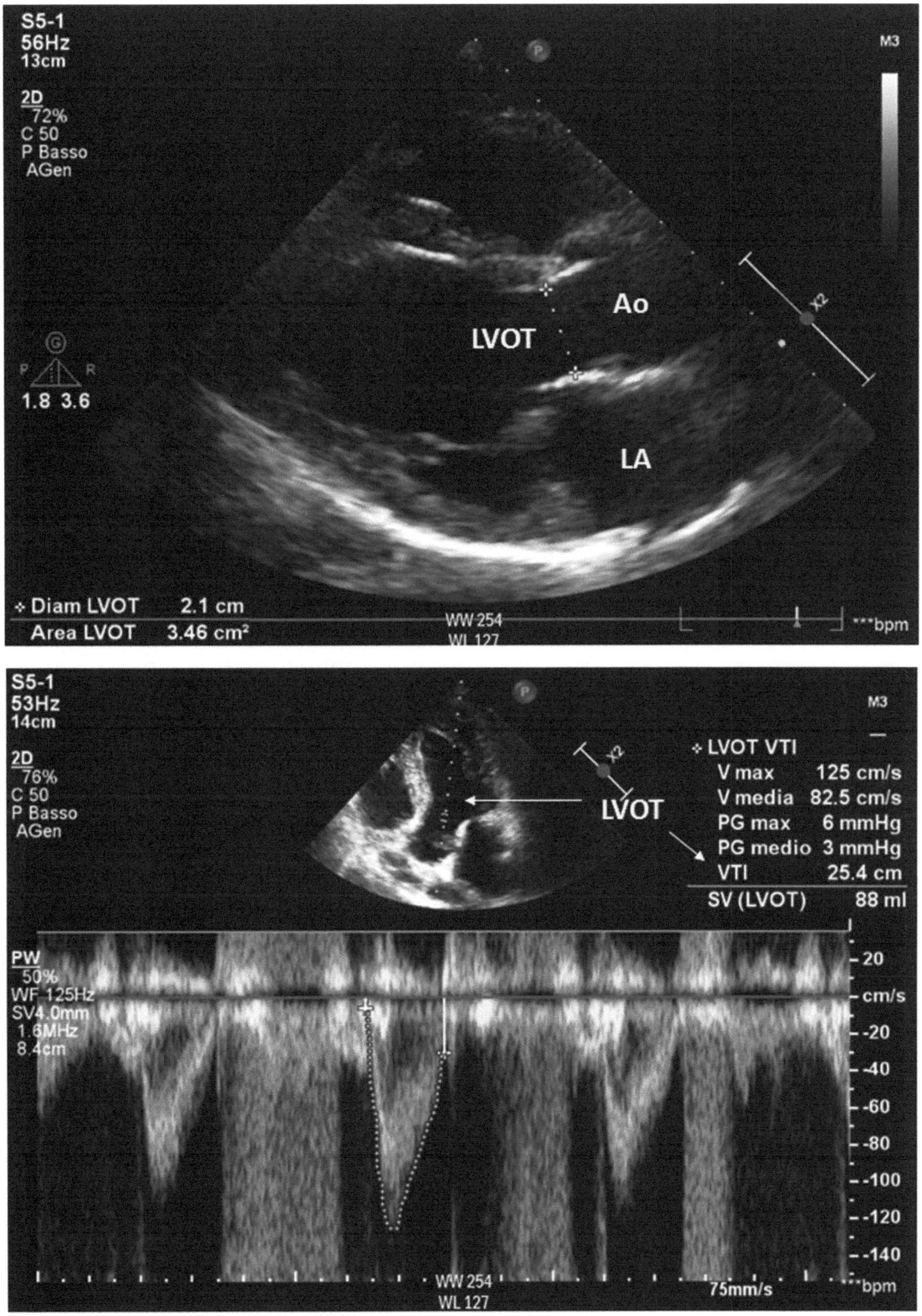

FIGURE 34.12 2D echo measurements needed for calculations of SV.

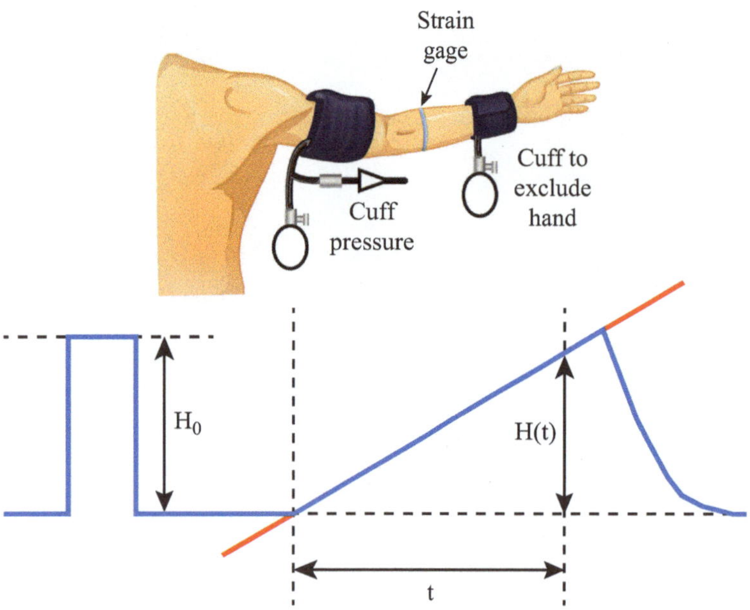

FIGURE 34.13 Muscular flow measurement with strain gauges. A first cuff is inflated at a pressure higher than the venous one. A second cuff is positioned at wrist level and inflated at a pressure higher than the systolic one to exclude hand circulatory system. The arm circumference depends only on the volume variations of the arteries of the arm.

By calculating the dependence of flow (F), initial volume (V_0) of the arm, initial circumference (C_0), and circumference variations (ΔC):

$$\frac{F}{V_0} = \frac{2 \cdot \Delta C}{C_0 \cdot \Delta t}$$

The flow (F), referred to the initial volume (V_0), is directly proportional to the circumference variation (ΔC), referred to the initial circumference (C_0).

Peripheral flow: technology

Peripheral flow is not commonly measured in clinical practice but only for research purposes, since it requires great attention and a long time to perform the whole procedure. Despite that, it is a very sensible method to measure peripheral flow or detect arteries constriction to evaluate changes in peripheral resistances due to autonomic nervous system activation. For this reason, very few commercial instruments based on this principle are available, and lots of research laboratories have built their own systems.[138–142]

Thoracic bioimpedance: principles

Human tissues present resistive and capacitive layers. The total electrical impedance therefore depends on the electrical resistance of the layer (and therefore on their geometry) and on the capacitance introduced by the insulating layers. The response of the tissues to the passage of an alternating electric current depends on the frequency of the applied electric signal.[31,143,144]

Using frequencies of 10–100 kHz, the current is distributed uniformly in the tissues, the capacitive component of the impedance becomes negligible and the impedance is correlated with the volume of the conductor (therefore, section and length). Assuming that the main conductor at the thoracic level is made up of blood flowing into the aorta, the transthoracic impedance instantly depends on the amount of blood and, therefore, on the CO.

A constant-amplitude electric current at high frequency is injected in the chest, applying a voltage difference between two electrodes placed at the extremities (neck and groin). Then, the instantaneous voltage difference between two electrodes placed in the intermediate area is continuously acquired. Since the applied current is constant, the measured voltage depends only on the electrical impedance of the tissues which are placed between the two measuring electrodes (Fig. 34.14).

Since the distance (L) between the measuring electrodes remains constant, the method provides a measurement of the average conduction section which is proportional to the volume of the conductor between the electrodes.

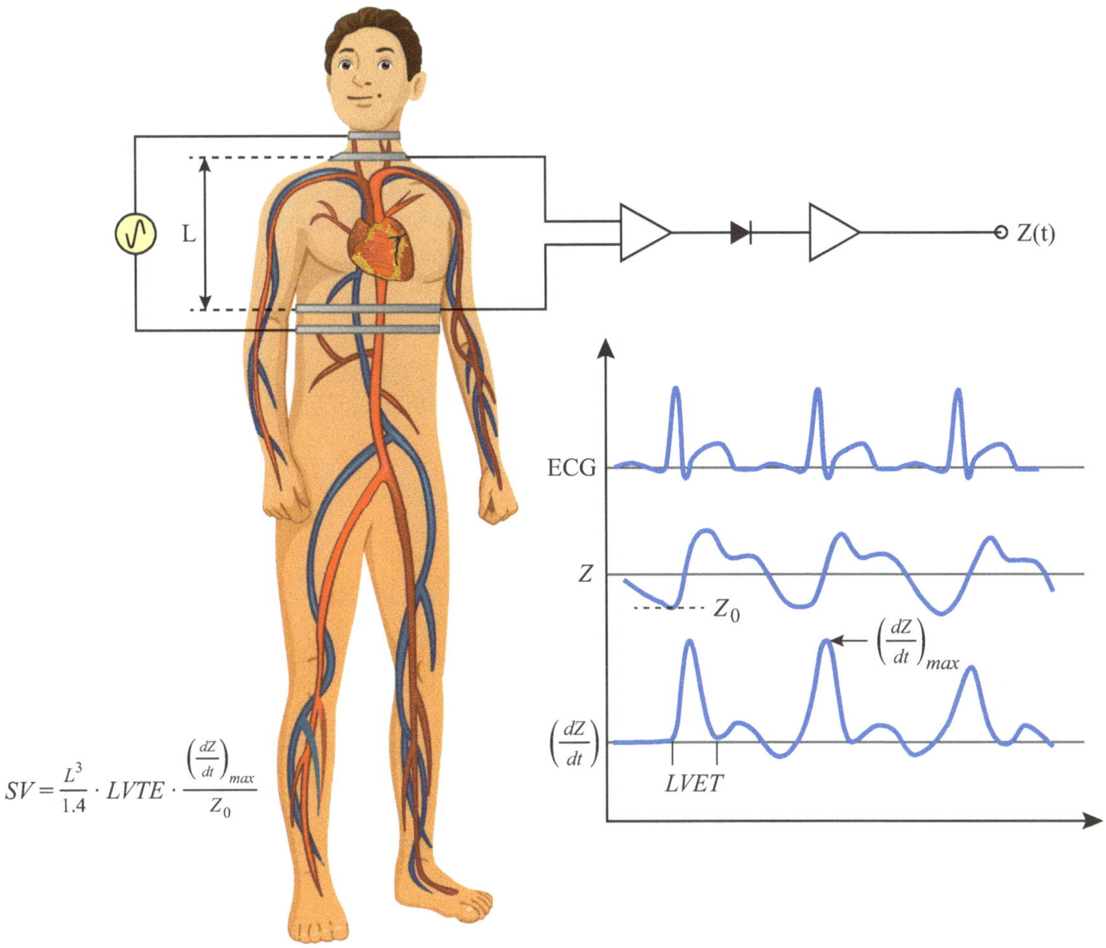

FIGURE 34.14 Impedance ($Z(t)$) measurement for cardiac output (CO) calculation. The graphs represent, in addition to the ECG, the trend over time of the impedance Z and its derivative. For the calculation of the SV one of the most used formulas is that of Sramek.[145] *LVET*, left ventricle ejection time; Z_0, minimal impedance; L, distance between electrodes.

Volume and impedance have an inversely proportional trend, not always linear in the specific operating conditions; moreover, the signal related to the parameter to be measured can be superimposed on others that also determine impedance variations (e.g., air present in the lungs due to respiration, venous return, ejection of the RV, blood redistribution in the lung, etc.).

For this reason, the desired parameter (CO, SV) is calculated using formulae that relate values obtained from the impedance derivative, possibly with the addition of correction factors coming from comparison with results of other techniques.

The impedance-based methods have the advantage of being noninvasive and easily applicable in all operating situations; their limitation consists of not providing absolute results, but being referrable to the particularity of the operative situation, as well as to the patient's anatomy. They are typically associated with the application of a more rigorous method, which is used as a calibration and verification reference. Even in this case, however, the correspondence with other methods is guaranteed only for limited ranges of variation of the parameters, and in particular conditions (e.g., basal conditions).

Thoracic bioimpedance: technology

Several commercial systems applying this approach are available, easy to use, and suitable for noninvasive monitoring of changes in CO of patients in intensive care units or during ambulatory procedures. Despite that, transthoracic bioimpedance reliability depends on the correct application of the electrodes, the mathematical model that links the changes in Z to blood volume changes inside the thoracic arteries, the kind of subject (normal or obese), and on the presence of arrhythmias or pulmonary pathologies.[31,36,143–160]

Thoracic bioreactance: principles

The bioreactance technique is similar to the bioimpedance one, but it is based on the phase shifts between the voltage used to generate the high-frequency alternating current and the voltage measured by the reading electrodes.[143] These phase shifts are proportional to the volume of blood present in the aorta and therefore to the CO (Fig. 34.15).

To apply this technique, four pairs of electrodes are placed on the chest: the outer pairs are used for the injection of the alternating current (75 kHz) and the innermost ones to detect the voltage changes. The voltage measured on the two pairs is summed and the phase shift with respect to the injected voltage is calculated over time. In this way, a signal proportional to the volume of blood present in the aorta, and therefore to the CO, is obtained.

Thoracic bioreactance: technology

These approach presents the same benefits and limitations as the thoracic bioimpedance. Moreover, the systems based on this approach make CO readings averaged over a 60 s period, so they are not able to evaluate rapid CO changes.[143,153,161–171]

Phase-contrast MRI: principles

The motion of magnetic spins in a magnetic field gradient gives the possibility to measure their velocity in a specific direction.[172–174]

Arterial flow is obtained by using a 2D plane through the cross section of the vessel, and then integrating all the speeds over this plane.

Phase-contrast MRI: technology

Standard 2D phase contrast magnetic resonance imaging (PC-MRI) allows for a flow sampling period of about 50 ms. To overcome this limitation, which makes MRI not able to acquire the high-frequency band of the flow wave, some more complex machinery is necessary, making this technology really expensive and not so simple for use in clinical practice, despite its good reliability.

From arterial pressure: principles

Different algorithms were proposed to relate arterial pressure, SV, and CO. The most used and implemented in commercial instruments are pulse contour analysis, pulse power analysis, systolic area, and diastolic decay.[175]

Pulse contour analysis, for example, is based on the statement that the area under the systolic part of the arterial pressure waveform is proportional to the SV.[156,176–182] In this method the area under the pressure wave is measured postdiastole to the end of the ejection phase and then divided by the aortic impedance.

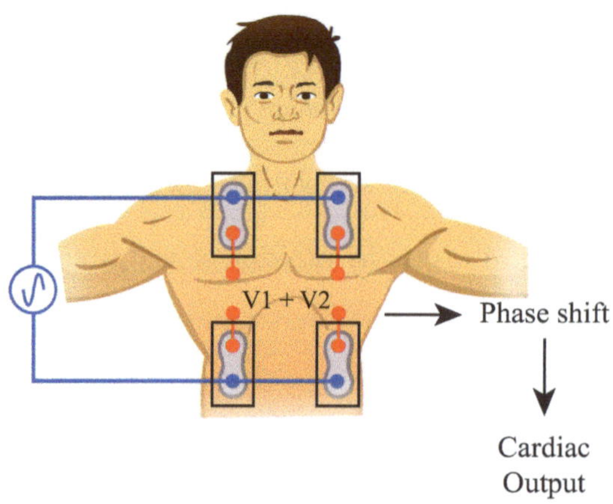

FIGURE 34.15 Bioreactance system for CO calculation: alternating current is applied to the two pairs of blue electrodes and then the voltage is acquired by the internal red electrodes couples.

From arterial pressure: technology

All these approaches need a correct acquisition of the arterial pressure waveforms, and a mathematical model to extract SV and CO from the original data. The first limitation of these methods is related to the systems used to acquire the pressure (generally volume clamp or tonometry, but also PPG) which present themselves some intrinsic errors or uncertainties. Moreover, the mathematical models are strongly dependent on the mechanical properties of the vessels, so the algorithms are related to vascular resistances, compliance of the arterial tree, wave reflection, damping of the transducer and aortic systolic outflow, and all the physiological and pathological factors affecting them.

Some commercial systems based on these kinds of algorithms have been improved by giving the possibility to the operator to calibrate the acquired data with a direct measurement of CO obtained with other methods (i.e., TPTD, lithium dilution, etc.). Despite the power of this approach, these features transform these systems into minimally invasive ones with all the risk connected with these kinds of apparatuses.

These methods can be applied to arterial waveforms, which are invasively acquired; in these cases, although the waveforms are correct and reliable, the uncertainties related to the biomechanical models and the dependency from the pathological conditions of the patients still remain.[183–185]

Fick method: principles

During breathing, there is a continuous dilution of oxygen in the blood: measuring the rate of introduction of oxygen into the blood (oxygen consumption, VO_2) and the oxygen concentrations in the oxygenated blood [arterial $Ca(O_2)$, downstream of the lungs] and in venous blood [venous $Cv(O_2)$, upstream of the lungs], it is possible to calculate the flow rate (CO), by the following formula:

$$CO = \frac{VO_2}{Ca(O_2) - Cv(O_2)}$$

To calculate the amount of oxygen dissolved in the blood, an oximeter can be used. To calculate oxygen consumption, it is necessary to measure the ventilation ($F(t)$) and the amount of oxygen present ($Cair\_O_2(t)$) in the inhaled and exhaled air[168,186–193] (Fig. 34.16).

$$VO_2 = \int F(t) \cdot Cair\_O_2(t) \cdot dt$$

Fick method: technology

To correctly calculate CO with Fick's method it is necessary to measure all the parameters of interest, from oxygen consumption to oxygen saturation in arterial and venous blood by appropriate methods (e.g., breath-by-breath, mixing

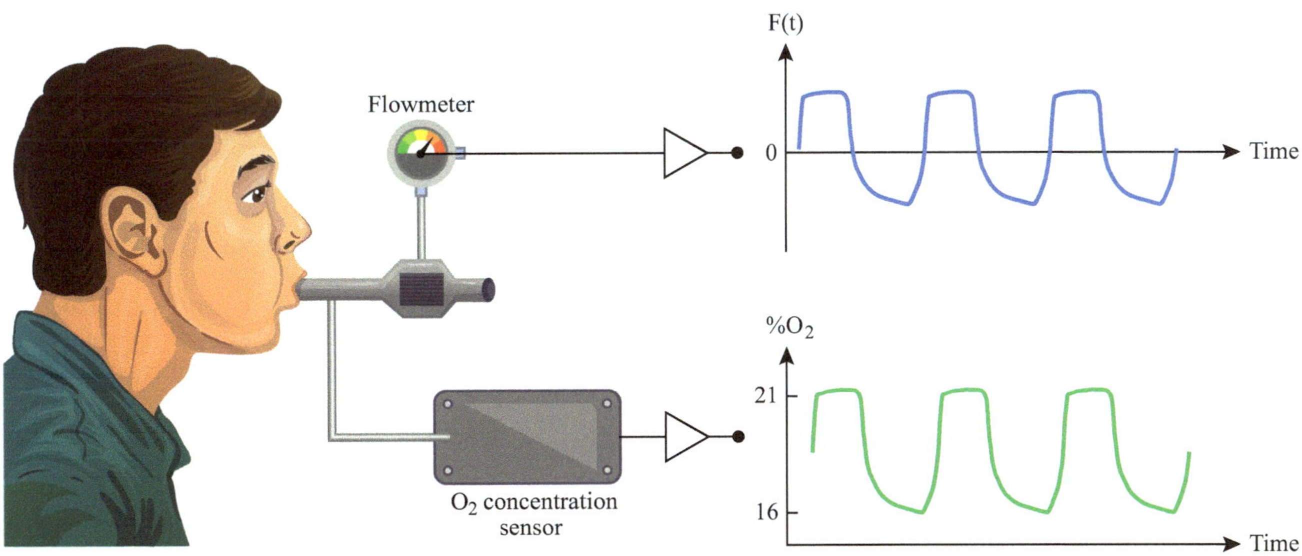

FIGURE 34.16 Example of breath-by-breath method to estimate VO_2.

chamber, collection bag) and reliable measurement instruments (e.g., pneumotachometer, turbine sensor, hot wire sensor for ventilation and paramagnetic, chemical, combustion oximeter, mass spectrometer for oxygen concentration in arterial and blood samples).[194–196]

Moreover, since $Ca(O_2)$ and $Cv(O_2)$ are related to the oxygen saturation in arterial ($Sa(O_2)$, %) and venous blood ($Sv(O_2)$, %) and to the hemoglobin quantity (Hb, g/dL), and considering that 1 g of hemoglobin is capable of carrying 1.36 mL of oxygen, the CO can be calculated as[197]:

$$CO = \frac{VO_2}{1.36 \cdot Hb \cdot (Sa(O_2) - Sv(O_2)) \cdot 10}$$

The Fick method is particularly useful in patients with low CO and a large arteriovenous O_2 difference where instantaneous variability in measurements does not affect CO calculation; conversely, in the presence of high output states, $Sv(O_2)$ is high and ($Sa(O_2)$−$Sv(O_2)$) is reduced so that a small variability in $Sa(O_2)$ or $Sv(O_2)$ leads to a relatively large change in their difference, and thus a large change in the CO calculated with the Fick method. Moreover, in patients requiring O_2 therapy, $Sv(O_2)$ could be disproportionately increased compared to $Sa(O_2)$; this could lead to overestimation of CO.

Sometimes, to reduce complexity, the VO_2 is not measured but calculated as 125 mL/min/m^2 of body surface area. The assumed oxygen consumption could be responsible for the underestimation of CO in young anxious patients, in obese subjects, or for overestimation in patients with heart failure and reduced O_2 consumption.[198,199]

Partial gas rebreathing: principles

It is used in intubated patients under mechanical ventilation. At steady state, the amount of CO_2 entering the lungs via the PA is proportional to the CO and equals the amount exiting the lungs via expiration and pulmonary veins.

During 30 s of rebreathing, the amount entering does not change, but the amount eliminated by expiration decreases, and the endtidal CO_2 increases in proportion to the CO.

CO is calculated according to following formula:

$$CO = \frac{VCO_2}{Cv(CO_2) - Ca(CO_2)}$$

Here VCO_2 is CO_2 consumption, $Ca(CO_2)$ and $Cv(CO_2)$ are arterial and venous CO_2 content, respectively. The diffusion rate of carbon dioxide is 22 times more rapid than that of oxygen, and it is assumed that no difference $Cv(CO_2)$ will occur, whether under normal or rebreathing conditions. A disposable circuit is connected to the ventilator circuit along with infrared CO_2 sensor, pneumotachometer and a rebreathing valve. Partial rebreathing is initiated every 3 min by opening the valve, and the pulmonary blood flow is calculated by difference between normal and rebreathing ratio.

Partial gas rebreathing: technology

Major limitation is that tracheal intubation with fixed ventilator setting is required. Validation studies are controversial, since some researchers found good reliability of the method,[200,201] while others did not find the accuracy of this device with bolus pulmonary thermodilution. Thus, it has limited clinical applicability. Moreover, an arterial catheter is necessary to estimate pulmonary shunting.[183,202]

Invasive measurements

Transesophageal Doppler: principles

The CWD or PWD transducer is mounted on a small probe that can be inserted inside the esophagus. The probe is of the size of an orogastric tube and can be placed for a longer period in intubated patients. These systems measure the flow in the descending aorta (about 70% of total CO) and, if used to extrapolate the total CO, a correction should be introduced to overcome this underestimation.

Transesophageal Doppler: technology

The technology and its related limitations are the same as already discussed in the paragraph above. Moreover, with an invasive approach it is more difficult to control the US beam propagation axis, and therefore the angle between the waves and the blood speed in aorta.[203–216]

Ultrasound transit time flowmeter: principles

An US pulse traveling in the direction of a liquid flow moves faster than a US beam moving against the flow. If we apply this evidence to blood moving inside a vessel, we can calculate its velocity (u) and then its flow.

For example (Fig. 34.17), with two probes, one emitting and the other receiving the US pulse, since the distance (D), the spatial orientation (angle ϑ) of the transducers, the velocity of US in tissue (c) are known, by measuring the time t, it is possible to evaluate the blood velocity (u):

$$t = \frac{D}{v}$$

where t, US pulse transit time to cover a distance D; v, mean ultrasound pulse velocity; u, blood speed; and ϑ, angle between US propagation axis and blood velocity

$$t = \frac{D}{c \pm u \cdot \cos\vartheta}$$

The positive or negative sign depends on the vs of u (so of the blood flow).

Obviously, the blood flow is pulsatile and laminar, so the u velocity is an averaged value along the US beam propagation. To overcome this problem some corrections were proposed to evaluate speed in the case of turbulent flow, too.[217]

US transit time flowmeter: technology

Since the transit time is very short, to improve the measuring system, different configurations have been proposed and the most used is the one reported in Fig. 34.18. This configuration includes two transducers, both emitting and receiving and a metallic reflective surface.

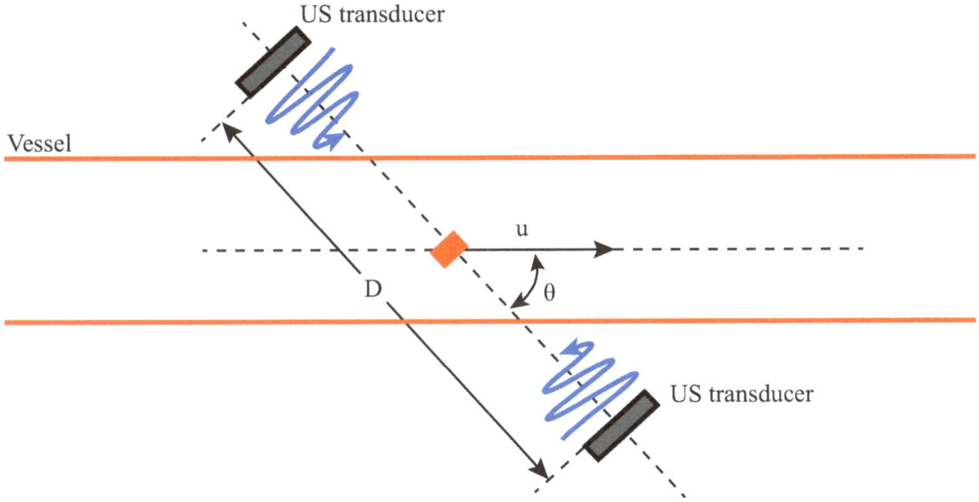

FIGURE 34.17 Transit time measuring system with two opposite US transducers.

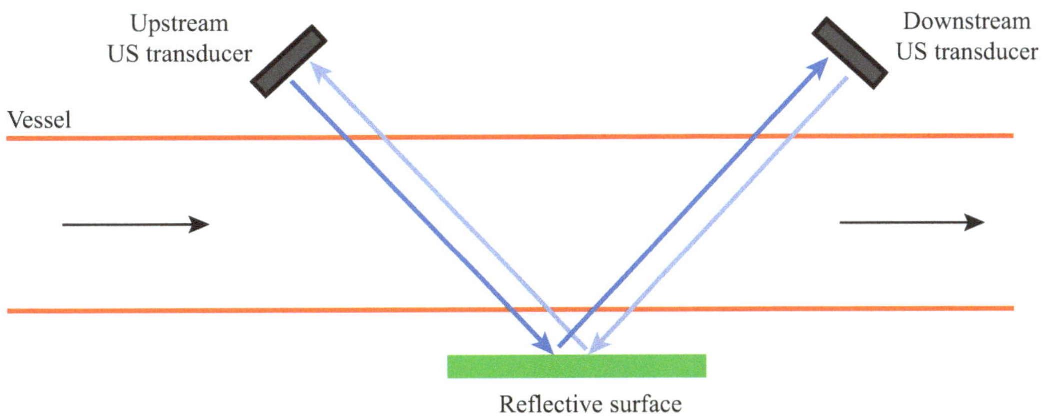

FIGURE 34.18 US Transit time monitoring system with two transducers and a reflective surface.

An ultrasound pulse signal is contemporary emitted from an upstream and a downstream crystal: both the waves reach the other transducer after traveling for a known path (direct or with a reflective surface). Since ultrasound travels faster when moving in the same direction as flow, a small-time difference for the two signals, as expressed in a phase difference, can be determined. The angle between probes and vessel is not critical, since upstream and downstream crystals and reflector (if present) are mounted in fixed positions within the probe. Calibration is suggested to obtain more accurate and reliable measurement.

This approach is commonly used during surgery procedures or for research purposes in mechanical mocks of the cardiovascular system.[183,217–220]

Nowadays some optical sensors based on the same principle have been proposed and validated. The usage of light instead of US allows to miniaturize the sensors making them suitable to be implanted.[221,222]

Electromagnetic flowmeter: principles

Variable magnetic fields produce induced electromotive forces (i.e.m.f.) at the ends of the affected circuits: the variation of the field may be due to the variation of the currents that generate it or to the movement of the circuit. In the case of conductors moving in a uniform magnetic field, an i.e.m.f. that is proportional to the magnetic field (B), the length of the conductor (l), and the speed of movement (v) is revealed:

$$\text{i.e.m.f.} = B \cdot l \cdot v$$

This behavior is used to measure the speed of blood in the arteries, using electromagnetic flowmeters (Fig. 34.19). The electromagnetic flowmeter has the shape of a sleeve which is tightly applied to the external surface of the vessel in which the blood velocity, and therefore the blood flow ($Q = v \cdot S$), is to be measured. A coil is inserted in the sleeve, which generates a uniform magnetic field which crosses the vessel perpendicularly and invests the blood. Blood is a conductor, so when it moves in the magnetic field, an i.e.m.f. proportional to speed is generated.[223]

The i.e.m.f. which is proportional to the speed and therefore to the blood flow, is detected with two electrodes placed inside the sleeve and resting on the vessel wall.

Electromagnetic flowmeter: technology

Electromagnetic flowmeter technology is quite complex[223–228] and now rarely used in standard clinical practice due to the cost of the apparatuses and their frailty. Despite that, they are still used for research purposes with external circulation systems or mechanical mocks of the cardiovascular system, since they are reliable and accurate.[135,229–231]

Continuous thermodilution: principles

Anatomically, the blood that has perfused the coronary arteries flows back into the RA through two veins that drain the two different branches of the coronary circulation and join to form the coronary sinus.

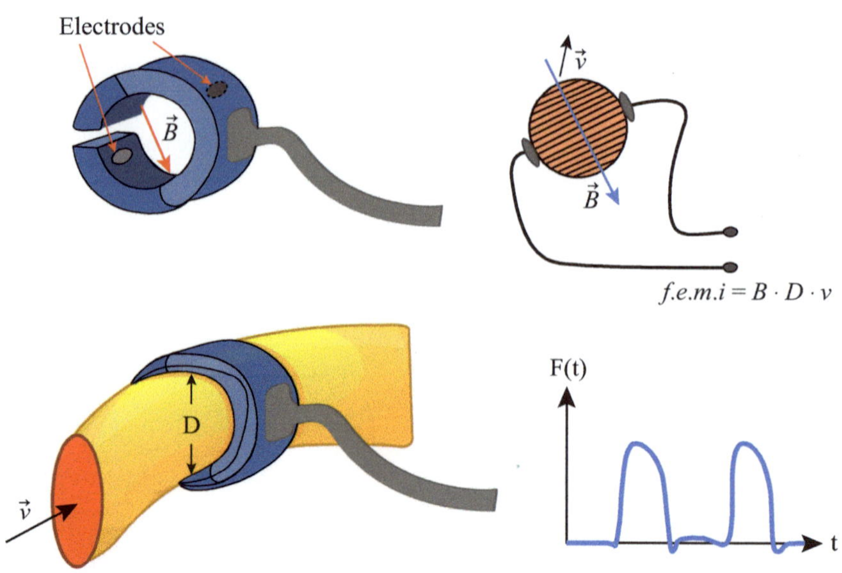

FIGURE 34.19 Electromagnetic flowmeter. The ions cross the magnetic field perpendicularly and generate between the electrodes an induced electromotive force (i.e.m.f.) proportional to their speed.

By introducing a physiological solution at a temperature of 1°C upstream of the coronary sinus, before the joining of the two veins, and monitoring the temperature of the blood downstream of the injection point, it is possible to calculate the coronary flow[232,233] (Fig. 34.20).

Continuous thermodilution: technology

Although continuous thermodilution has been widely used for coronary flow estimation, some researchers proposed the same method to continuously monitor CO in an ICU setting.[232,234,235]

Bolus injection of green indocyanine: principles

A known quantity of green indocyanine is injected into the RA and the concentration in the aorta is continuously measured (by photodensitometer). The concentration curve rises rapidly, then decreases exponentially. After a time, equal to one cardiac cycle, the curve rises due to the effect of recirculation (Fig. 34.21). The underlying area is calculated by excluding this last phase.[236,237]

Bolus injection of green indocyanine: technology

Green indocyanine dye dilution method is considered one of the most accurate and standard methods to measure CO. Despite that, the need for an accurate instrument to measure the green indocyanine concentration in blood makes this approach not commonly used in clinical practice.[238–246]

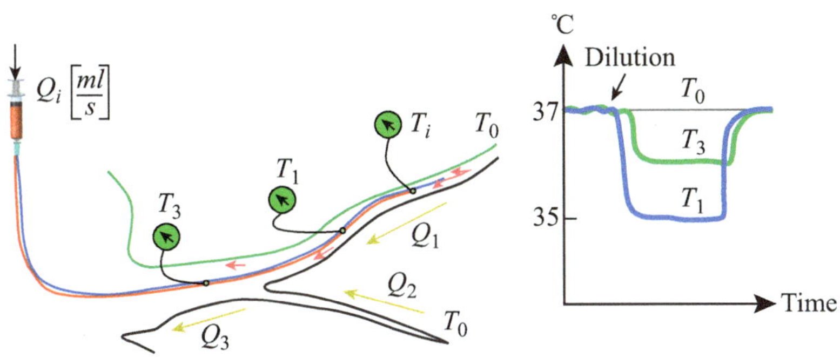

$$Q_1 = \frac{Q_i \cdot (T_1 - T_i)}{T_0 - T_1} \qquad Q_3 = \frac{Q_i \cdot (T_3 - T_i)}{T_0 - T_3} \qquad Q_2 = Q_3 - Q_1$$

FIGURE 34.20 Coronary flow measurement with continuous infusion thermodilution. By knowing the upstream blood temperature (T_0), the infusion flow (Q_i), and the saline infusion temperature (T_i) and by measuring the downstream temperature both before the bifurcation (T_1) and after (T_3), it is possible to calculate the flow in the first branch (Q_1), the flow in the coronary sinus (Q_3), and, by difference, the flow in the second branch of the coronary veins (Q_2).

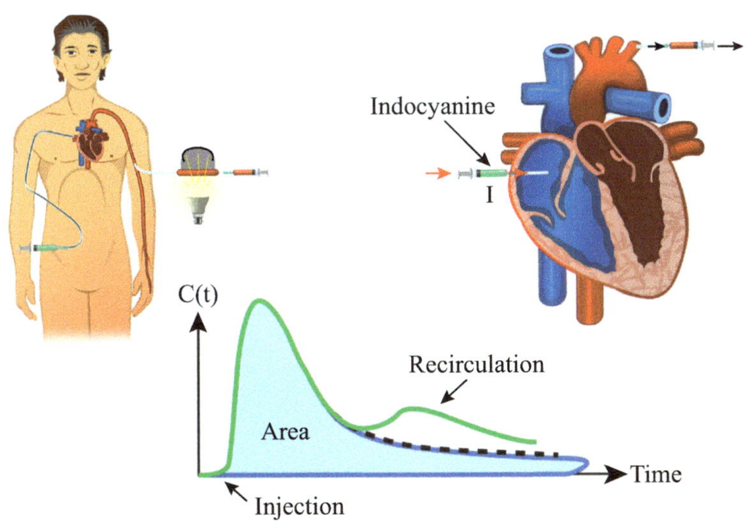

FIGURE 34.21 Bolus injection of green indocyanine in the right atrium.

Bolus thermodilution for CO estimation

A known quantity (I_0) of low-temperature saline (Ti) is rapidly injected into the RA through a pulmonary artery catheter. The pulmonary artery temperature (Tc) is then measured by means of a thermistor placed on top of the catheter. In this case, the flow rate depends on how the temperature varies over time: after a phase in which it rapidly decreases, the temperature rises with an exponential trend[232,236,247] (Fig. 34.22).

Bolus thermodilution for CO estimation: technology

Although Bolus thermodilution represents the gold standard to estimate CO, there are some conditions in which this method is not accurate and can lead to errors:

- Tricuspid regurgitation: part of the injectate goes backward and this produces a prolonged curve; this suggests a falsely lower CO.
- Low CO (<3.5 L/min): the injectate may warm up excessively because it travels slowly up to the PA giving the wrong impression of a slight and brief temperature change at the tip; this suggests a falsely higher CO (using iced saline solution injectate may compensate for this error).
- Right-to-left shunt: part of the injectate is directed to the left side of the heart creating an abbreviated thermodilution curve; this suggests a falsely higher right-sided CO.
- Left-to-right shunt: the shunted blood dilutes the cold injectate and may abbreviate the thermodilution curve; this suggests a falsely lower right-sided flow.[248]

Continuous warm dilution: principles

Continuous warm CO is a modification of standard cold bolus thermodilution: a copper filament in the catheter is positioned in the RV. By means of a high-frequency current, the filament is heated periodically for a fixed time. Consequently, there is an intermittent heating of blood in the right heart, and the temperature is captured by the thermistor near the tip of the catheter. Average value of CO measured over time is displayed on the monitor.[210]

Continuous warm dilution: principles

The main advantage of this method over conventional cold bolus thermodilution is the avoidance of repeated boluses, thus reducing the infection risk and operator errors. With this approach, a catheter should be inserted and left in the right heart during the period of interest.

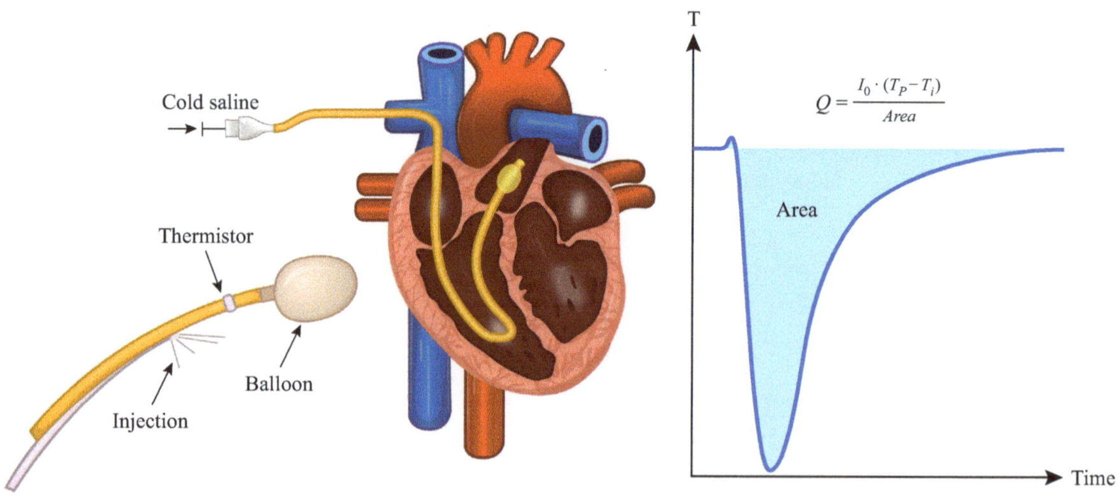

FIGURE 34.22 Bolus thermodilution with Swan–Ganz catheter.

Transpulmonary thermodilution: principles

With this minimally invasive technique, the bolus of cold solution is injected into the superior vena cava and the temperature is measured in the aorta at the level of the iliac bifurcation. With this approach, the total CO is assessed, considering both the right and left sides of the heart.[236,249–251]

Transpulmonary thermodilution: technology

TPTD requires both central venous (femoral or internal jugular) and arterial cannulation (femoral/radial), so it implies all the risks related to invasive procedures.

This kind of approach is used in different commercial systems, and it is typically used to calibrate a different method for CO estimation at the beginning of the patient monitoring.[202,210,252] Recent studies demonstrate the reliability of this method if compared with other traditional and noninvasive techniques in case of severe pathologies (e.g., cardiogenic shock).[136,253,254]

Lithium chloride bolus dilution: principles

It is a minimally invasive technique first described in 1993,[255] and it requires a venous (central or peripheral) line and an arterial catheter. A bolus of lithium chloride is injected into the venous line, and the arterial concentration is measured by withdrawing blood across a disposable lithium sensitive sensor containing an ionophor, which is selectively permeable to lithium. The CO is calculated based on lithium dose and area according to the concentration time circulation.[236]

Lithium chloride bolus dilution: technology

This system cannot be used in patients on lithium therapy and in all cases where lithium chloride can be toxic.[236]

Other technologies

To evaluate speed in microcirculation setting, laser Doppler transducers have been realized and are commonly used to calculate blood velocity in small vessels, both invasively and not (e.g., skin, cerebral, tumor perfusion system).[256–263]

Another method to evaluate the flow is the "pulse wave transit time," which is based on the assumption that the distance between the R-wave in ECG and the pulse wave rise-point assessed by pulse oximetry is inversely related to SV. Although some commercial instruments use this technique to noninvasively measure CO, this approach should need a calibration at the beginning of the measuring procedure. Some data related to calibration factors taking into account age, sex, HR, and other parameters are reported in literature, but further investigation is required.[160]

Additional parameters correlated to hemodynamic monitoring

Noninvasive

Pulse wave velocity: principles

Moens and Korteweg, at the same time and separately discovered the equation that relates to the transmission speed of the pulse (Pulse Wave Velocity, PWV) in a cylindrical elastic tube; the transmission speed depends on the density of the liquid, the radius of the duct, the modulus of elasticity (Young), and the wall thickness, in a nonlinear way:

$$\text{PWV} \cong \frac{1}{\sqrt{\frac{2\rho R}{E\cdot h}}} = \sqrt{\frac{E\cdot h}{2\rho R}}$$

where *PWV*, pulse wave velocity; ρ, density; *R*, duct radius; *E*, wall Young module; and *h*, wall thickness.

The PWV increases by decreasing the radius and increasing the wall thickness. Moving from the aorta to the more peripheral arteries, the speed of the pulses increases significantly.

We must not confuse the PWV, which occurs on the wall and is conditioned by elasticity, with the speed of the blood, which is determined by upstream pressure and vascular resistance.

If we relate the pulse speed to compliance (*C*), we get:

$$C = \frac{1}{\sqrt{\rho}\cdot \text{PWV}}$$

Pulse wave velocity: technology

The arterial pulse wave is traditionally detected and recorded using a pressure probe (tonometry) or Doppler in a proximal artery, for example, the common carotid and in a more distal one, such as the femoral artery.

Since the two vessels are near the body surface, their pulse waves can be easily recorded in a noninvasive way. Furthermore, the pulse wave must travel through most of the aorta and the PWV will therefore be an excellent index of the rigidity of the aortic wall.

With two probes it is possible to directly calculate the delay (Δt) between the rising feet of the two waves (Fig. 34.23).

Using only one probe, it is necessary to acquire an ECG channel as a timing reference.

The distance traveled by the pulse wave is measured on the body surface and PWV is then calculated as a space/time ratio (m/s).[61,69,101,264–270]

Moreover, together with traditional measurement of PWV between carotid and iliac bifurcation, today the PWV can be measured between different sites (e.g., ankle, toe).[270]

Several techniques using MRI have been developed to measure PWV: all of them are based on the transit time of flow or blood velocity waves between different anatomical regions and in the measurement of the distances between the measurement's points.

The measured distance is to be considered an approximation of the real distance traveled by the pulse wave; there seems to be no variations related to the individual's body size, although there are no scientific data to support this and the errors in determining the pathways length between the measuring sites is the major limitation of this kind of approach. The dependence of PWV on HR is controversial: it is therefore advisable to verify that the two waves collected in successive instants of time correspond, however, to conditions of "equal" HR (variability <10%).

To avoid these problems, other ways to measure PWV have been developed. The most important is the area-flow (QA) method: PWV is calculated as the ratio between change in flow and change in cross-sectional area during the reflection-free period of the cardiac cycle.[61] Since the method is related to the correct application of US technique to measure blood flow and vessel area (therefore errors due to angles, high speeds, etc.) this technique should be further investigated to be considered completely reliable.

Other commercial technologies allow the measurement of PWV by an oscillometric-based system: a single pressure measurement is performed and then, after obtaining SP and DP the brachial cuff is inflated to the DP level and held for 10 s to record pulse waves. Using a TF, central pressure curves are obtained and processed using the ARCSolver algorithm (Austrian Institute of Technology, Vienna, Austria) to obtain PWV.[44,271] Another possibility is to extract, from the brachial pressure wave, the two forward and backward components, the first one related to the systolic ejection in aorta and the second one to the reflection from peripheral arteries. The PWV depends on the return time (RT) of the backward wave.[270]

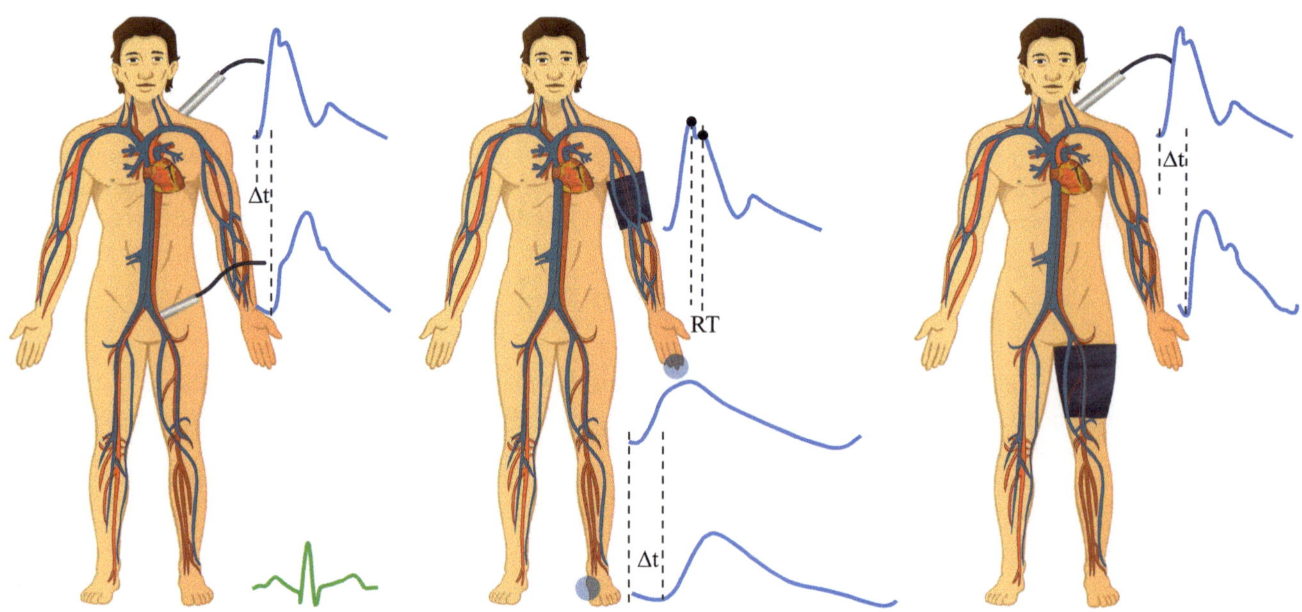

FIGURE 34.23 Current technical approaches to measure PWV.

All these last methodologies are based on biomechanical and mathematical assumptions that are strictly related to the physiological and pathological conditions of the patients.

Oxygen saturation: principles

Light's absorption is not uniform for each frequency, but follows different trends depending on the elements present in the matter and the existing molecular bonds. In the blood, unbound hemoglobin, oxygen-bound hemoglobin, and carbon monoxide-bound hemoglobin have different light absorption curves, as a function of the frequency (Fig. 34.24).

By applying a beam of white light (Fig. 34.25) and interposing transparent filters with well-defined wavelengths (corresponding to the crossing points of the absorption curves), it is possible to introduce the values of emerging light intensity into a system of equations and obtain, by referring to calibration samples, the concentrations of the components present (spectrophotometric method).

Oxygen saturation: techniques

The gold standard method to measure oxygen saturation (SaO_2) is invasive, since a blood sample is mandatory.

The same technique, with some limitations, can be used in vivo, transilluminating a peripheral part of the body (ear, finger) and detecting the intensity variations at the peak of HbO_2 absorption (580 nm). This technique (plethysmography) assumes that all the parameters that affect absorption and therefore the transmission of light (e.g., thickness of the finger, skin color, bone, presence of veins, etc.) remain constant, and that only the percentage of hemoglobin linked to oxygen varies. This approach is quite easy, not invasive, and commonly used in clinical practice, but some limitations remain due to: (1) the possible presence of carboxyhemoglobin (COHb) or methemoglobin (MetHb) for particular pathological states; (2) vasoconstriction and so low light adsorption; (3) improper positioning of the measurement system and motion artifacts; (4) presence of veins superimposed to arteries which can alter the light adsorption; (5) problems with system calibration.

Nowadays some devices measure the light reflected by the blood and not the transmitted one (reflection pulse-oximetry). Sensors (emitter and receiver) are located on the same surface skin, and the device can be applied on any accessible surface (e.g., forehead; internal structures like trachea or esophagus).[272–277]

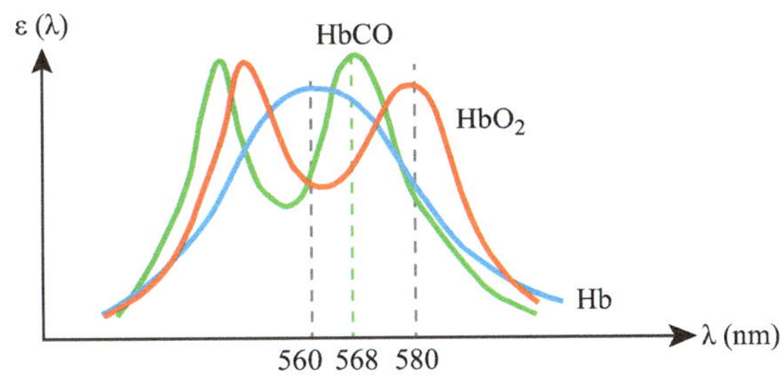

FIGURE 34.24 On the left: Mode of absorption of white light by different types of hemoglobin present in the blood: by carrying out the measurements at well-defined frequencies, it is possible to calculate the relative presence of the various component fractions.

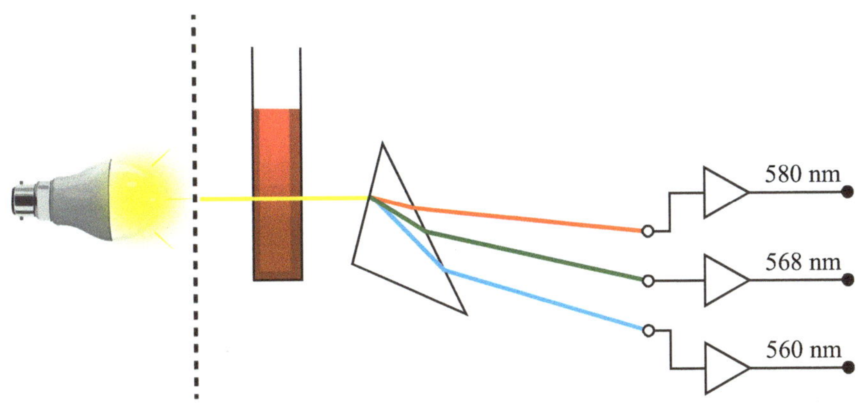

FIGURE 34.25 On the right: the light beam passing through the sample is divided into its spectral components with an optical prism, and the intensity of the desired frequencies is measured with phototubes. The comparisons between the measures allow the calculation of the percentages of the different substances present in the solution.

Ballistocardiography: principles

Ballistocardiogram (BCG) is the measure of forces of the body in reaction to blood ejection from heart to arteries. Since the ventricle applies a force to the blood to pull it into the aorta, for the action-reaction principle of the dynamics, an equal force is applied to the cardiac muscle. If we consider the heart in-built to the whole body and we can measure this force, we can quantify the contraction force (and then pressure) exerted by the heart to the blood (Fig. 34.26).

Ballistocardiography: techniques

Traditional ballistocardiograph systems are bed-based: the patient is positioned on a sensible movable and instrumentated bed: due to force, the bed oscillates and by measuring these movements, velocities, accelerations, and forces can be easily calculated. This approach measures the force component along the body longitudinal axis (head to foot). More recent systems allow to measure accelerations directly on the chest and back surface, to evaluate all the components of the vectors.

Although ballistocardiography is powerful, traditional systems can acquire only one component of the force and the quality of acquired waves depends on the inertial resistance of the systems. Modern accelerometers are easier to apply but acquired BCG waves are affected by noise due to other movements of the body and strictly dependent on the sensor positioning on the body surface.[64,278–284]

Seismocardiography: principles

Seismocardiogram (SCG) represents the local vibrations of the chest surface in response to the heart-induced motion during the heartbeat, which includes linear and rotational displacement, velocity, and acceleration.

SCG signals are caused by cardiac mechanical processes including cardiac muscle contraction, cardiac valve movements and blood flow turbulence. Therefore, the SCG contains useful information on cardiac mechanics, in particular heart sounds and CO. SCG signal was reported to contain a low-frequency wave corresponding to the atrial systole, a high-amplitude wave corresponding to ventricular systole, and some relatively high-frequency waves at the time of the first and second heart sounds.

SCG is typically measured by placing accelerometers or force transducers onto the subject's chest (Fig. 34.27).

Seismocardiography: technology

At the beginning, mainly due to the need for relatively cumbersome instrumentation and considerable time for data interpretation, the clinical applicability of SCG was limited to laboratory tests to collect similar clinical information provided by other systems, such as phonocardiographers and echocardiographers. As soon as the technological advancements (e.g., MEMS technology) provided small and cheap sensors, the SCG earned growing interest.[285,286] As a consequence, many technological solutions, based on the application of MEMS accelerometers in the mid-sternal precordial region, have been proposed for the assessment of cardiovascular indexes, such as myocardial contractility evaluated via a noninvasive measurement of peak endocardial acceleration (PEA),[287,288] and other hemodynamic parameters, for example, SV and cardiac time intervals.[289]

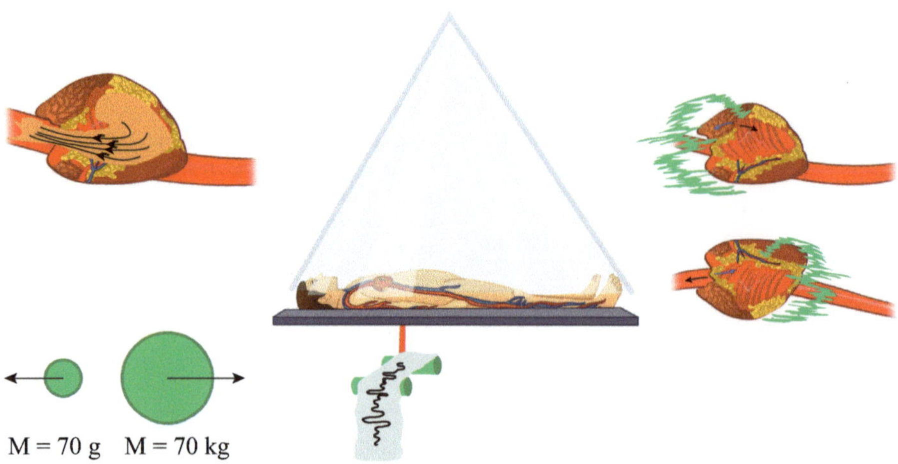

FIGURE 34.26 During contraction, the ventricle exerts a force to the blood (70 g); the blood exerts the same force to the heart in-built to the whole body (70 kg). If the patient is positioned on a movable system, the bed oscillates as a consequence of the contractions and by measuring the amplitudes it is possible to evaluate speeds, accelerations and forces components along the longitudinal axis of the body.

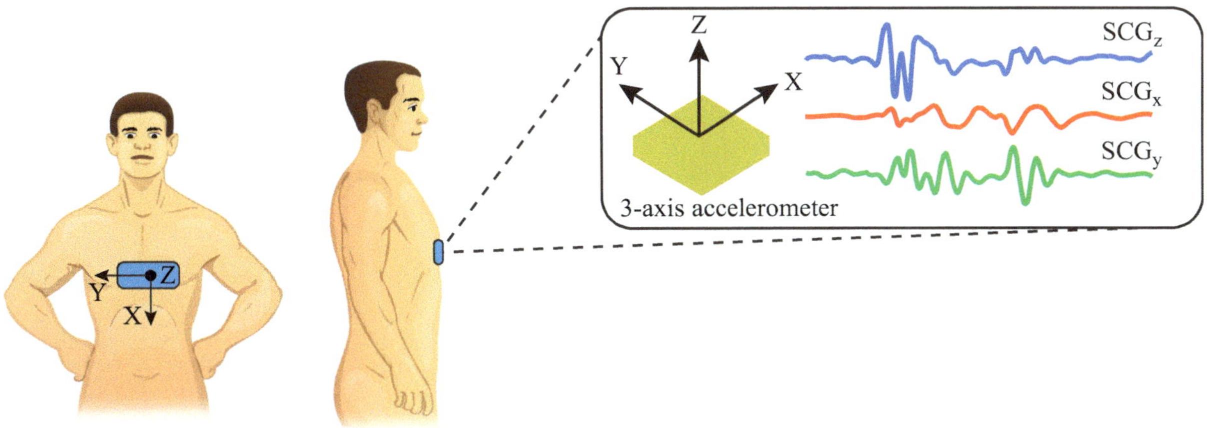

FIGURE 34.27 Scheme of a typical seismocardiography system.

In more recent years, a new class of wearable systems have also been studied and developed for SCG assessment in a wide range of conditions, such as sleep and daily life activities. They include textile-based wearable devices[290] and patch systems[291] that allow for the unobtrusive recording of cardiac-induced mechanical vibrations, together with detection of other vital signs such as ECG and respiration.

A novel SCG approach, named forcecardiography since it is based on the use of force sensors instead of accelerometers, has been recently proposed.[292] In this case, a force-sensitive resistor with a rigid dome fixed on its active area is positioned on a standard chest location in order to measure local cardiac-induced vibrations. This force-based approach may have the advantage of being easily embedded in washable smart garments as opposed to accelerometers that are not waterproof. Moreover, the acquired signal seems to provide new additional information related to the ventricular emptying phase, which cannot be detected by the well-established accelerometric seismocardiography.

Invasive measurements

Implantable sensors for cardiac motion: principles

Implantable sensors for monitoring cardiac-related movements, such as cardiac wall displacement, cardiac rotation, endocardial vibrations, and cardiac valve movements, rely on similar principles adopted for noninvasive seismocardiographic approaches, for example, the use of miniaturized inertial sensors (accelerometers, gyroscopes, miniaturized magnets), that are directly inserted in the cardiac cavities and connected, in a wired or wireless manner, to a powering and data-reading unit.

Implantable sensors for cardiac motion: technology

Implantable endocardial acceleration sensor (for cardiac contractility)

The endocardial acceleration sensor is an implantable micromass accelerometer integrated with dedicated electronics, which is located inside the tip of a conventional RV pacing lead connected to a pacemaker.[293] The sensor detects the mechanical vibrations that are generated by the myocardium during the isovolumetric contraction phase, being responsible, in their audible component, for the first heart sound. Before becoming a commercially available device, the sensor was extensively evaluated,[294–296] showing that changes in the peak of the detected endocardial vibrations (PEA) are closely related to changes in max LV dP/dt, which is a well-established index of cardiac contractility.

Originally, the PEA sensor was implemented only into RV pacemakers (Fig. 34.28). More recently, after the demonstration of its usefulness also in cardiac resynchronization therapy (CRT),[297] the sensor localization was investigated also in an atrial pacing lead, in order to make possible the application of PEA technology also to CRT defibrillators (CRT-D). The resulting commercial device was an endocardial, bipolar pacing lead with an integrated acceleration sensor attached to the atrial wall[298,299]

Implantable magnetic sensors (for cardiac contractility and cardiac kinematics)

To evaluate cardiac contractility, an invasive passive system has been proposed based on the implantation of permanent miniaturized magnets inside atria and ventricles. An external transcutaneous sensor can monitor the magnetic fields variations in time so as to three-dimensionally reconstruct the movements of the magnets. Since the magnets are permanent, no implantable power source is needed.[300]

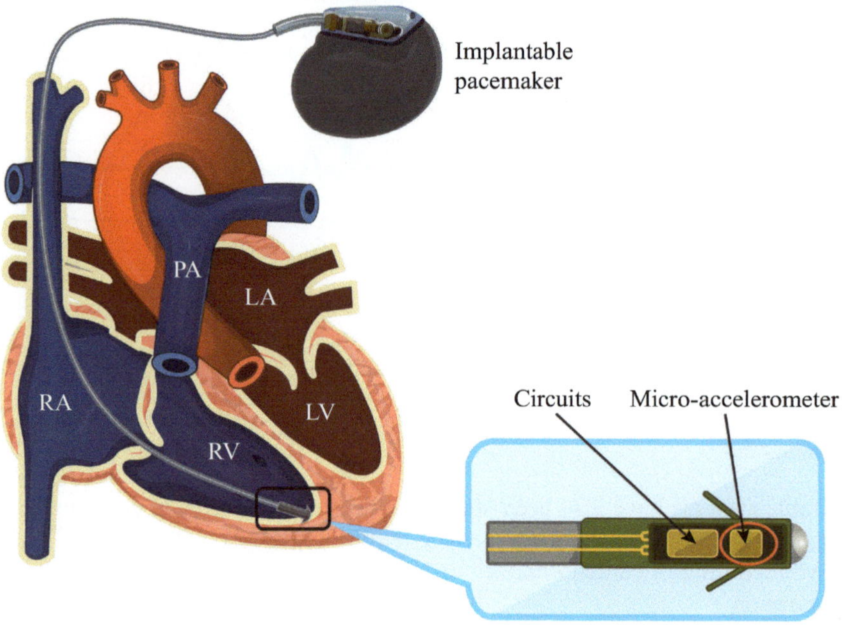

FIGURE 34.28 Scheme of implantable peak endocardial acceleration sensor for monitoring cardiac contractility.

Another similar experience reports the use of a magnetic tracking system to assess the LV kinematics in three-dimensional space during the cardiac cycle. The system consists of a small magnetic receiver sensor epicardially glued on LV apex, a magnetic pulse transmitter, and a control unit. Both the receiver and transmitter contain three orthogonally-placed coils, which are able to generate (transmitter) and detect (sensor) the magnetic pulses in the three orthogonal spatial directions.[301] When the transmitter is located near the sensor, the control unit provides the real-time calculation of x, y, z coordinates of the sensor position in a Cartesian 3D reference frame, by combining the data of magnetic pulses from each coil. These spatial coordinates describe a three-dimensional path of LV apex displacement, that may be used to characterize a normal or altered cardiac condition.

Implantable gyroscopes and accelerometers (for cardiac rotation)

The cardiac rotation represents the twisting movement of the heart during the cardiac cycle. It is mainly due to the obliquely oriented myocardial fibers,[302–304] and it was demonstrated to be a sensitive clinical index in the diagnosis of alterations of myocardial functionality.[305,306]

In recent years, cardiac rotation has been quantified mainly using noninvasive imaging techniques, such as tagged MRI, Doppler tissue imaging, and speckle tracking imaging, that however remain operator-dependent and require periodic imaging examinations.

Implantable alternatives to detect the cardiac rotation and cardiac wall motion have been recently proposed, although they still remain at research and investigational level.

In the first case, the idea is to use a three-axis MEMs gyroscopic sensor, which is inserted in the tip of a standard pacing lead together with a miniaturized printed circuit board. This gyro-based technology has been extensively evaluated in animal experiments during both the normal heart functioning and experimentally induced altered cardiac conditions, starting with uniaxial gyroscopes positioned on the epicardium.[307–309] More recently, an implantable prototype for this technology has been developed in order to assess rotation also in endocardial sites.[310]

In the other case, the use of a three-axis accelerometer, sutured on epicardial surface, is proposed for the continuous real-time assessments of heart wall motion in the longitudinal, circumferential, and radial directions, for possible detection of cardiac alterations such as myocardial ischemia.[311,312]

Impedance-based sensor (applied to prosthetic heart valve implants)

A breakthrough application that can fall into the category of implantable sensors for monitoring cardiac dynamics deals with a novel sensorized prosthetic heart valve (PHV), which is currently being developed as a proof-of-concept prototype.[313] The innovative concept is to apply an impedance-based sensing principle to a PHV. The technology uses miniature electrodes embedded in the supporting structure of the valve to generate a local electric field and to measure the electric field variations caused by the moving valve leaflets during each cardiac cycle. This sensor allows to continuously monitor the valve leaflet motion, and therefore the valve status and indirectly the patient's cardiac hemodynamic status, after implantation of the PHV.

Appendix: Pressure transducers

There are different types of presure transducers on the market, some miniaturized, some not.

Below, the reader can find a quick rundown on the technology used today to build a pressure measurement system.[1,100]

Strain gauges

The physical parameter pressure (force on surface unit placed perpendicular to the direction of the force) is measured by detecting the deformation of a flexible membrane exposed to stress.

For medical applications, the flexible membrane is placed in a transparent dome to allow verification of the correct application. One or more strain gauges are applied to the membrane and are connected to a Wheatstone bridge and powered by constant or alternating current[314,315] (Fig. A.34.1).

Piezoresistive sensors

These sensors have an operating principle similar (but not the same) to that of the resistance strain gauges described above.

They are generally made up of piezoelectric semiconductors for which deformation is mainly associated with a change in resistivity (not in length or section as for strain gauges). Like strain gauges, they are inserted inside a Wheatstone bridge. Being based on semiconductors and resistivity variations, they can be more easily miniaturized.[101]

Piezoelectric sensors

A piezoelectric material, suitably shaped, if subjected to a pressure (force) deforms and generates a voltage difference, which is proportional to the deformation itself (for stresses included in the linearity range of the material)[316,317] (Fig. A.34.2).

Capacitive sensors

Capacitive sensors consist of a capacitor that has a fixed face and another movable one: due to the effect of the pressure acting on the movable plate, the distance d between the plates varies and the electrical capacity of the system changes. The pressure measurement is done through the direct measurement of C, by inserting the capacitor inside an RLC circuit (Fig. A.34.3).

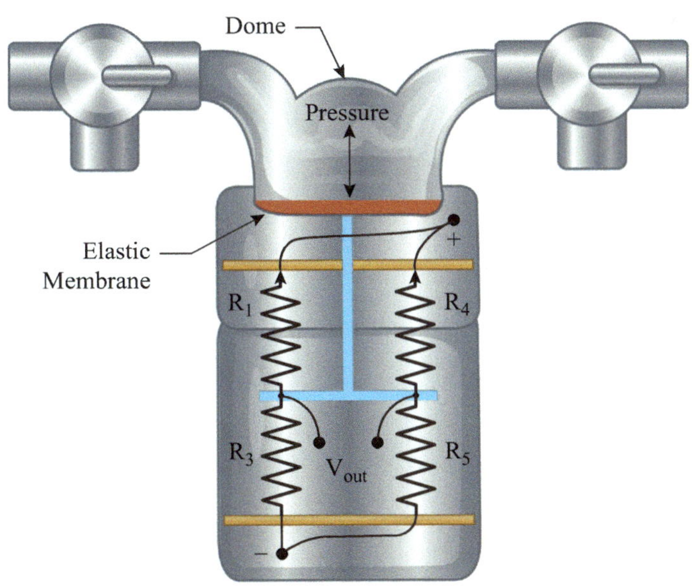

FIGURE A.34.1 Schematic view of a strain gauge transducer. Pressure applied to the elastic membrane causes a deformation of a conductor with a consequent change in resistance.

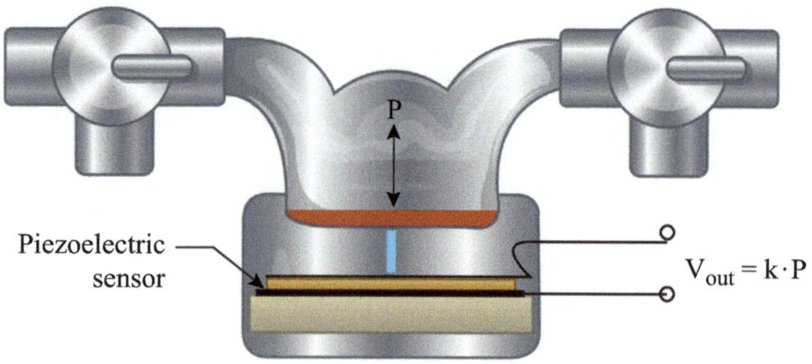

FIGURE A.34.2 Piezoelectric transducer: a pressure over an optimally shaped piezoelectric transducer generates a deformation and so a proportional voltage difference.

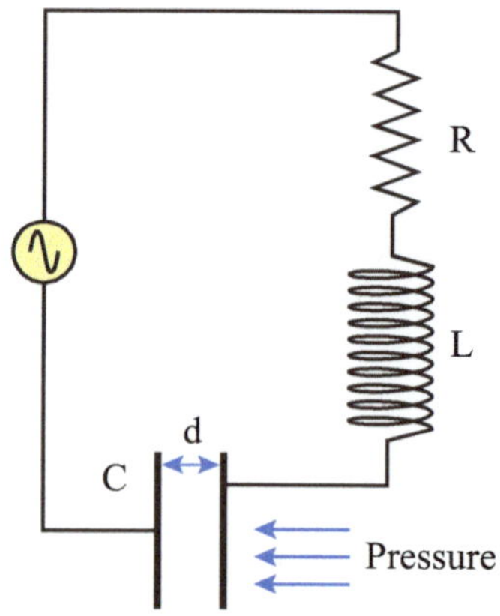

FIGURE A.34.3 Capacitive sensor inside a RLC circuit.

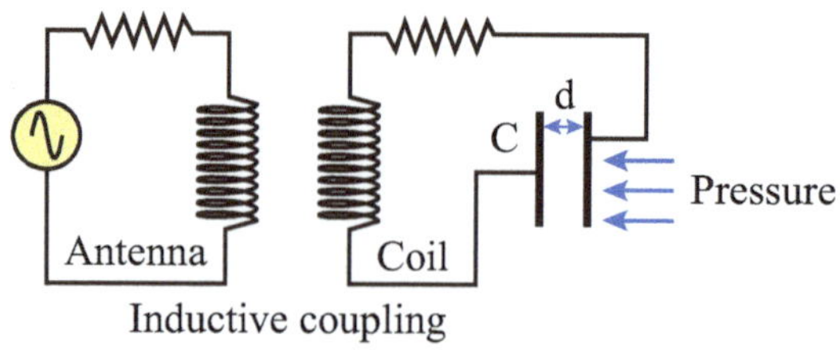

FIGURE A.34.4 An external antenna stimulates the passive transducer which varies its resonant frequency as a function of the variable capacitance. The external device works both as a power supply for the transducer and as a reader of the generated signal.

The change in capacitance can be measured by connecting the sensor in a frequency-dependent circuit such as an oscillator or an LC tank circuit. The resonant frequency of the circuit will change as the capacitance changes with pressure. An oscillator requires some extra electronic components and a power supply, while a resonant LC circuit can be used as a passive sensor, without its own source of power. In the case of passive sensors an external antenna can be used to provide a signal to stimulate the tuned LC circuit (powering by inductive coupling) and so measure the change in resonance frequency (wireless readout). This makes them especially suitable for medical devices that need to be implanted, so to have minimum size and low power consumption (Fig. A.34.4).

Capacitive transducers are suitable for use as passive translators in wireless circuits.[318–321]

The capacitive sensors are mechanically simple, so they can be made rugged with stable output, being suitable for use in harsh environments, and have greater sensitivity and temperature independence over other types of pressure sensors.

On the other hand, they have nonlinear output, although this can be reduced in touch-mode devices, and the electronics for measuring and conditioning the signal need to be placed close to the sensing element to minimize the effect of stray capacitance.

Optical sensors

Optical transducers[322–325] are based on the possibility of mechanically modifying the path of the light emitted by a suitable source, detecting these changes, and correlating them to the external pressure. There are three types of optical transducers depending on the physical principle considered:

1. The pressure acts on a mobile (elastic) device, which obscures part of the light beam emitted by a LED and detected by a photodiode or phototransistor. The greater the obscuration, the lower the intensity of the beam reaching the receiver; the lower the darkening, the greater the light intensity reaching the photodiode. There is an inverse proportionality between the measured light intensity and the EP (Fig. A.34.5).
2. Emitter and receiver are placed side by side: the pressure acts directly on an appropriately shaped reflective surface against which the incident light beam affects. If the distance of the surface varies, part of the light is no longer reflected against the receiver, but is deflected outside the sensor. In this way, the light intensity that is detected is reduced and this reduction is linked to the distance of the reflecting surface, which is influenced by the EP (Fig. A.34.6).
3. In order to obtain measurements with better resolutions, interferometry and a device similar to a Michelson–Morley interferometer must be exploited. Led and photodiode are positioned next to each other: the led sheds a beam of light. When the beam meets the partially reflective surface, a first reflection occurs. Part of the beam, however, continues on its way until it meets the totally reflective surface. The waves reflected by the second surface come back, reemerge from the device, and interfere with those reflected by the first. The interference pattern changes as a function of the difference in the optical path, and therefore of the distance between the two surfaces. If the first (partially reflective) is fixed and the second mobile (because it is subject to EP), the distance d and therefore the interference

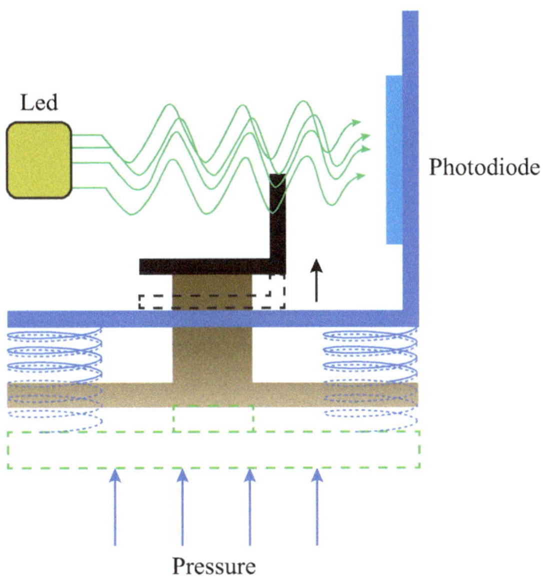

FIGURE A.34.5 Optical transducers: the pressure pushes an elastic device connected to a black screen which modifies the light intensity reaching the phototransistor.

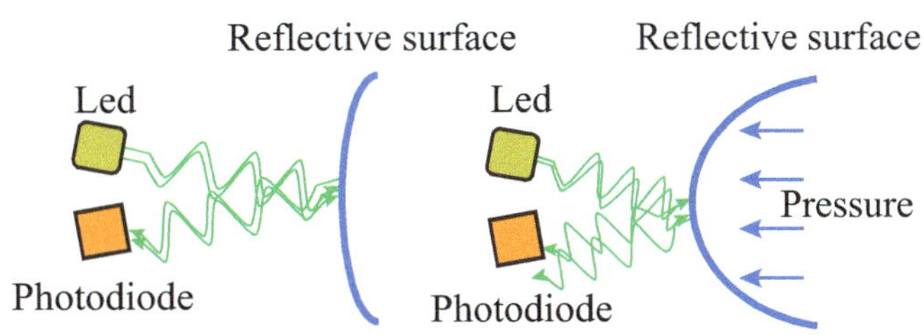

FIGURE A.34.6 Optical transducer: pressure acts modifying the shape of the reflective surface. As a consequence, part of the light beam does not reach the detector.

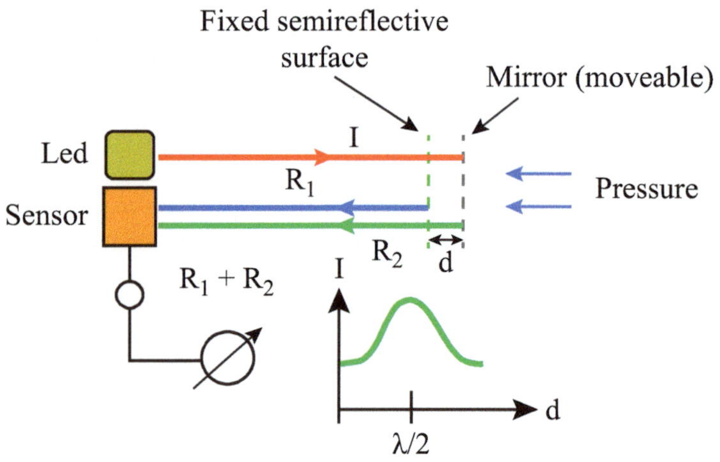

FIGURE A.34.7 The pressure acting on the totally reflective surface modifies the distance and therefore the difference in optical path between the two light rays, which will interfere, generating different light intensities depending on *d*.

pattern depend on the pressure. If *d* is equal to an integer number of wavelengths, we will have a luminous maximum (constructive interference), while if *d* is equal to an odd integer number of wavelengths divided by 4, we will have a shadow zone (totally destructive interference).

By moving the mobile surface, the maximum and minimum interference and intermediate figures will alternate on the detector. Knowing the intensity of the detected light, the difference in the optical path and therefore the distance *d* can be traced (Fig. A.34.7).

Microelectromechanical systems

Modern technology allows the creation of miniaturized pressure transducers containing both mechanical and electronic elements (MEMS). Miniaturization allows the transducer to be inserted into very small cavities (e.g., blood vessels) and to provide a pressure measurement directly at the point of interest (without liquid column).

The most used MEMS can mount capacitive or piezoresistive transduction systems.[101,320,326,327] The use of capacitive transducers allows the creation of passive MEMS, which do not require internal power, but which can be driven "on-demand" from the outside, through an inductive coupling with an appropriate antenna. These kinds of transducers are now implanted inside a cavity (e.g., PA) and provide pressure "on demand," whenever they are stimulated and "read" by a suitable external antenna.

References

1. Morris AS, Langari R. *Pressure Measurement. Measurement and Instrumentation*. Elsevier; 2016:463–491. Available from: https://doi.org/10.1016/B978-0-12-800884-3.00015-0.
2. Geddes LA. *Handbook of Blood Pressure Measurement*. Humana Press; 1991. Available from: https://doi.org/10.1007/978-1-4684-7170-0.
3. Corazza I, Fabbiani L, Marras L, Mariselli M, Marangoni F, Zannoli R. The arterial pressure auscultatory method: new approach for better cardiology teaching. *High Blood Press Cardiovascular Prev*. 2006;13:179–183. Available from: https://doi.org/10.2165/00151642-200613040-00006.
4. Perloff D, Grim C, Flack J, Frohlich ED, Hill M, McDonald M, et al. Human blood pressure determination by sphygmomanometry. *Circulation*. 1993;88:2460–2470. Available from: https://doi.org/10.1161/01.CIR.88.5.2460.
5. Roguin A. Scipione Riva-Rocci and the men behind the mercury sphygmomanometer: the history behind the mercury sphygmomanometer. *Int J Clin Pract*. 2005;60:73–79. Available from: https://doi.org/10.1111/j.1742-1241.2005.00548.x.
6. Strugo V, Glew FJ, Davis J, Opie LH. Update: recommendations for human blood pressure determination by sphygmomanometers. *Hypertension*. 1990;16:594. Available from: https://doi.org/10.1161/01.hyp.16.5.594.
7. Verrij EA, Nieuwenhuizen L, Bos WJW. Raising the arm before cuff inflation increases the loudness of Korotkoff sounds. *Blood Press Monit*. 2009;14:268–273. Available from: https://doi.org/10.1097/MBP.0b013e3283330a67.
8. A'Court C, Stevens R, Sanders S, Ward A, McManus R, Heneghan C. Type and accuracy of sphygmomanometers in primary care: a cross-sectional observational study. *Br J Gen Pract*. 2011;61:e598–e603. Available from: https://doi.org/10.3399/bjgp11X593884.
9. Bos WJW, Van Goudoever J, Van Montfrans GA, Wesseling KH. Influence of short-term blood pressure variability on blood pressure determinations. *Hypertension*. 1992;19:606–609. Available from: https://doi.org/10.1161/01.HYP.19.6.606.
10. Coleman AJ, Steel SD, Ashworth M, Vowler SL, Shennan A. Accuracy of the pressure scale of sphygmomanometers in clinical use within primary care. *Blood Press Monit*. 2005;10:181–188. Available from: https://doi.org/10.1097/01.mbp.0000168398.87167.c2.

11. Lacruz ME, Kluttig A, Kuss O, Tiller D, Medenwald D, Nuding S, et al. Short-term blood pressure variability—variation between arm side, body position and successive measurements: a population-based cohort study. *BMC Cardiovas Disord*. 2017;17. Available from: https://doi.org/10.1186/s12872-017-0468-7.
12. Picone DS, Schultz MG, Hughes AD, Sharman JE. Cuff under pressure for greater accuracy. *Curr Hypertension Rep*. 2020;22. Available from: https://doi.org/10.1007/s11906-020-01103-8.
13. Saxena Y, Saxena V, Gupta R. Clinical validation of aneroid sphygmomanometer. *Indian J Physiol Pharmacol*. 2012;56:255–261.
14. Alpert BS, Quinn D, Gallick D. Oscillometric blood pressure: a review for clinicians. *J Am Soc Hypertens*. 2014;8:930–938. Available from: https://doi.org/10.1016/j.jash.2014.08.014.
15. Bogatu L, Bresch E, Muehlsteff J, Smink J, Woerlee P. Insights into oscillometry: an experimental study for improvement of cuff-based blood pressure measurement technology. *Annu Int Conf IEEE Eng Med Biol Soc*. 2019;2019:7068–7071. Available from: https://doi.org/10.1109/EMBC.2019.8856994.
16. Kafka H. A simple method for blood pressure measurements in the premature and newborn infant. *Pediatrics*. 1967;40:106–108.
17. Wohlfahrt P, Cífková R, Krajčoviechová A, Šulc P, Bruthans J, Linhart A, et al. Comparison of three office blood pressure measurement techniques and their effect on hypertension prevalence in the general population. *J Hypertens*. 2020;38:656–662. Available from: https://doi.org/10.1097/HJH.0000000000002322.
18. Lewis PS, Chapman N, Chowienczyk P, Clark C, Denver E, Lacy P, et al. Oscillometric measurement of blood pressure: a simplified explanation. A technical note on behalf of the British and Irish Hypertension Society. *J Hum Hypertension*. 2019;33:349–351. Available from: https://doi.org/10.1038/s41371-019-0196-9.
19. Celler BG, Argha A, Le PN, Ambikairajah E. Novel methods of testing and calibration of oscillometric blood pressure monitors. *PLoS One*. 2018;13:e0201123. Available from: https://doi.org/10.1371/journal.pone.0201123.
20. Chen Z, Wang X, Wang Z, Zhang L, Hao G, Dong Y, et al. Assessing the validity of oscillometric device for blood pressure measurement in a large population-based epidemiologic study. *J Am Soc Hypertens*. 2017;11:730–736.e4. Available from: https://doi.org/10.1016/j.jash.2017.09.004.
21. Fonseca-Reyes S, Romero-Velarde E, Torres-Gudiño E, Illescas-Zarate D, Forsyth-MacQuarrie AM. Comparison of auscultatory and oscillometric BP measurements in children with obesity and their effect on the diagnosis of arterial hypertension. *Arch Cardiol Mex*. 2018;88:16–24. Available from: https://doi.org/10.1016/j.acmx.2017.01.005.
22. Šelmytė-Besusparė A, Barysienė J, Petrikonytė D, Aidietis A, Marinskis G, Laucevičius A. Auscultatory vs oscillometric blood pressure measurement in patients with atrial fibrillation and arterial hypertension. *BMC Cardiovasc Disord*. 2017;17:87. Available from: https://doi.org/10.1186/s12872-017-0521-6.
23. Liu J, Cheng H-M, Chen C-H, Sung S-H, Hahn J-O, Mukkamala R. Patient-specific oscillometric blood pressure measurement: validation for accuracy and repeatability. *IEEE J Transl Eng Health Med*. 2017;5:1900110. Available from: https://doi.org/10.1109/JTEHM.2016.2639481.
24. Liu J, Cheng H-M, Chen C-H, Sung S-H, Moslehpour M, Hahn J-O, et al. Patient-specific oscillometric blood pressure measurement. *IEEE Trans Biomed Eng*. 2016;63:1220–1228. Available from: https://doi.org/10.1109/TBME.2015.2491270.
25. Penaz J, Voigt A, Teichmann W. Continuous indirect recording of the blood pressure. *Z Fur Die Gesamte Innere Med Und Ihre Grenzgebiete*. 1976;31:1030–1033.
26. Eeftinck Schattenkerk DW, van Lieshout JJ, van den Meiracker AH, Wesseling KR, Blanc S, Wieling W, et al. Nexfin noninvasive continuous blood pressure validated against Riva-Rocci/Korotkoff. *Am J Hypertension*. 2009;22:378–383. Available from: https://doi.org/10.1038/ajh.2008.368.
27. Saugel B, Hoppe P, Nicklas JY, Kouz K, Körner A, Hempel JC, et al. Continuous noninvasive pulse wave analysis using finger cuff technologies for arterial blood pressure and cardiac output monitoring in perioperative and intensive care medicine: a systematic review and meta-analysis. *Br J Anaesth*. 2020;125:25–37. Available from: https://doi.org/10.1016/j.bja.2020.03.013.
28. Wagner JY, Prantner JS, Meidert AS, Hapfelmeier A, Schmid RM, Saugel B. Noninvasive continuous vs intermittent arterial pressure monitoring: evaluation of the vascular unloading technique (CNAP device) in the emergency department. *ScJ Trauma, Resuscitation Emerg Med*. 2014;22:8. Available from: https://doi.org/10.1186/1757-7241-22-8.
29. Silke B, McAuley D. Accuracy and precision of blood pressure determination with the Finapres: an overview using re-sampling statistics. *J Hum Hypertension*. 1998;12:403–409. Available from: https://doi.org/10.1038/sj.jhh.1000600.
30. Berkelmans GFN, Kuipers S, Westerhof BE, Spoelstra-de Man AME, Smulders YM. Comparing volume-clamp method and intra-arterial blood pressure measurements in patients with atrial fibrillation admitted to the intensive or medium care unit. *J Clin Monit Comput*. 2018;32:439–446. Available from: https://doi.org/10.1007/s10877-017-0044-9.
31. van Wijk JJ, Weber F, Stolker RJ, Staals LM. Current state of noninvasive, continuous monitoring modalities in pediatric anesthesiology. *Curr Opin Anaesthesiol*. 2020;33:781–787. Available from: https://doi.org/10.1097/ACO.0000000000000927.
32. O'Brien E, Stergiou G, Palatini P, Asmar R, Ioannidis JP, Kollias A, et al. Validation protocols for blood pressure measuring devices: the impact of the European Society of Hypertension International Protocol and the development of a Universal Standard. *Blood Press Monit*. 2019; 24:163–166. Available from: https://doi.org/10.1097/MBP.0000000000000391.
33. Mynard JP, Kondiboyina A, Kowalski R, Cheung MMH, Smolich JJ. Measurement, analysis and interpretation of pressure/flow waves in blood vessels. *Front Physiol*. 2020;11:1085. Available from: https://doi.org/10.3389/fphys.2020.01085.
34. Zhang Q, Zhou D, Zeng X. Highly wearable cuff-less blood pressure and heart rate monitoring with single-arm electrocardiogram and photoplethysmogram signals. *Biomed Eng Online*. 2017;16:23. Available from: https://doi.org/10.1186/s12938-017-0317-z.
35. Wagner JY, Grond J, Fortin J, Negulescu I, Schöfthaler M, Saugel B. Continuous noninvasive cardiac output determination using the CNAP system: evaluation of a cardiac output algorithm for the analysis of volume clamp method-derived pulse contour. *J Clin Monit Comput*. 2016;30:487–493. Available from: https://doi.org/10.1007/s10877-015-9744-1.

36. Sauder KA, Pokorney PE, McCrea CE, Ulbrecht JS, Kris-Etherton PM, West SG. Noninvasive assessment of hemodynamics: a comparative analysis of fingertip pulse contour analysis and impedance cardiography. *Blood Press Monit.* 2015;20:209–214. Available from: https://doi.org/10.1097/MBP.0000000000000118.
37. Lowe A, Harrison W, El-Aklouk E, Ruygrok P, Al-Jumaily AM. Non-invasive model-based estimation of aortic pulse pressure using suprasystolic brachial pressure waveforms. *J Biomech.* 2009;42:2111–2115. Available from: https://doi.org/10.1016/j.jbiomech.2009.05.029.
38. Salvi P, Safar ME, Parati G. Arterial applanation tonometry: technical aspects relevant for its daily clinical use. *J Hypertens.* 2013;31:469–471. Available from: https://doi.org/10.1097/HJH.0b013e32835e3422.
39. Nelson MR, Stepanek J, Cevette M, Covalciuc M, Hurst RT, Tajik AJ. Noninvasive measurement of central vascular pressures with arterial tonometry: clinical revival of the pulse pressure waveform? *Mayo Clin Proc.* 2010;85:460–472. Available from: https://doi.org/10.4065/mcp.2009.0336.
40. Matthys K, Verdonck P. Development and modelling of arterial applanation tonometry: a review. *Technol Health Care.* 2002;10:65–76. Available from: https://doi.org/10.3233/thc-2002-10107.
41. Meidert AS, Huber W, Müller JN, Schöfthaler M, Hapfelmeier A, Langwieser N, et al. Radial artery applanation tonometry for continuous noninvasive arterial pressure monitoring in intensive care unit patients: comparison with invasively assessed radial arterial pressure. *Br J Anaesth.* 2014;112:521–528. Available from: https://doi.org/10.1093/bja/aet400.
42. Butlin M, Qasem A. Large artery stiffness assessment using sphygmocor technology. *Pulse (Basel).* 2017;4:180–192. Available from: https://doi.org/10.1159/000452448.
43. Chen CH, Ting CT, Nussbacher A, Nevo E, Kass DA, Pak P, et al. Validation of carotid artery tonometry as a means of estimating augmentation index of ascending aortic pressure. *Hypertension.* 1996;27:168–175. Available from: https://doi.org/10.1161/01.hyp.27.2.168.
44. Papaioannou TG, Thymis J, Benas D, Triantafyllidi H, Kostelli G, Pavlidis G, et al. Measurement of central augmentation index by three different methods and techniques: agreement among arteriograph, complior, and mobil-o-graph devices. *J Clin Hypertens (Greenwich).* 2019;21:1386–1392. Available from: https://doi.org/10.1111/jch.13654.
45. Gao M, Rose WC, Fetics B, Kass DA, Chen C-H, Mukkamala R. A simple adaptive transfer function for deriving the central blood pressure waveform from a radial blood pressure waveform. *Sci Rep.* 2016;6. Available from: https://doi.org/10.1038/srep33230.
46. Hope SA, Tay DB, Meredith IT, Cameron JD. Use of arterial transfer functions for the derivation of central aortic waveform characteristics in subjects with type 2 diabetes and cardiovascular disease. *Diabetes Care.* 2004;27:746–751. Available from: https://doi.org/10.2337/diacare.27.3.746.
47. Yao Yang, Wang Lu, Hao Liling, Xu Lisheng, Zhou Shuran, Liu Wenyan. The noninvasive measurement of central aortic blood pressure waveform. *Blood Press Bench Bed.* 2018;. Available from: https://doi.org/10.5772/intechopen.76770.
48. Costello BT, Schultz MG, Black JA, Sharman JE. Evaluation of a brachial cuff and suprasystolic waveform algorithm method to noninvasively derive central blood pressure. *Am J Hypertension.* 2015;28:480–486. Available from: https://doi.org/10.1093/ajh/hpu163.
49. Guelen I, Westerhof BE, Van Der Sar GL, Van Montfrans GA, Kiemeneij F, Wesseling KH, et al. Validation of brachial artery pressure reconstruction from finger arterial pressure. *J Hypertension.* 2008;26:1321–1327. Available from: https://doi.org/10.1097/HJH.0b013e3282fe1d28.
50. Hope SA, Tay DB, Meredith IT, Cameron JD. Use of arterial transfer functions for the derivation of aortic waveform characteristics. *J Hypertension.* 2003;21:1299–1305. Available from: https://doi.org/10.1097/00004872-200307000-00017.
51. Karamanoglu M, Feneley MP. Derivation of the ascending aortic-carotid pressure transfer function with an arterial model. *Am J Physiol Heart Circulatory Physiol.* 1996;271:H2399–H2404. Available from: https://doi.org/10.1152/ajpheart.1996.271.6.h2399.
52. Lehmann ED, Kass DA, Chen C-H, Nevo E, Fetics B, Pak PH, et al. Estimation of central aortic pressure waveform by mathematical transformation of radial tonometry pressure data [1] (multiple letters). *Circulation.* 1998;98:186–187. Available from: https://doi.org/10.1161/01.CIR.98.2.186.
53. Pucci G, Cheriyan J, Hubsch A, Hickson SS, Gajendragadkar PR, Watson T, et al. Evaluation of the Vicorder, a novel cuff-based device for the noninvasive estimation of central blood pressure. *J Hypertension.* 2013;31:77–85. Available from: https://doi.org/10.1097/HJH.0b013e32835a8eb1.
54. Weber T, Wassertheurer S, Rammer M, Maurer E, Hametner B, Mayer CC, et al. Validation of a brachial cuff-based method for estimating central systolic blood pressure. *Hypertension.* 2011;58:825–832. Available from: https://doi.org/10.1161/HYPERTENSIONAHA.111.176313.
55. Karamanoglu M, O'Rourke MF, Avolio AP, Kelly RP. An analysis of the relationship between central aortic and peripheral upper limb pressure waves in man. *Eur Heart J.* 1993;14:160–167. Available from: https://doi.org/10.1093/eurheartj/14.2.160.
56. Chen Chen-Huan, Nevo Erez, Fetics Barry, Pak Peter H, Yin Frank CP, Maughan WLowell, et al. Estimation of central aortic pressure waveform by mathematical transformation of radial tonometry pressure. *Circulation.* 1997;95:1827–1836. Available from: https://doi.org/10.1161/01.CIR.95.7.1827.
57. Karamanoglu M, Feneley MP. On-line synthesis of the human ascending aortic pressure pulse from the finger pulse. *Hypertension.* 1997;30:1416–1424. Available from: https://doi.org/10.1161/01.hyp.30.6.1416.
58. Gallagher D, Adji A, O'Rourke MF. Validation of the transfer function technique for generating central from peripheral upper limb pressure waveform. *Am J Hypertens.* 2004;17:1059–1067. Available from: https://doi.org/10.1016/j.amjhyper.2004.05.027.
59. Stok WJ, Westerhof BE, Guelen I, Karemaker JM. Aortic pressure wave reconstruction during exercise is improved by adaptive filtering: a pilot study. *Med Biol Eng Comput.* 2011;49:909–916. Available from: https://doi.org/10.1007/s11517-011-0795-2.
60. Beulen BWAMM, Bijnens N, Koutsouridis GG, Brands PJ, Rutten MCM, van de Vosse FN. Toward noninvasive blood pressure assessment in arteries by using ultrasound. *Ultrasound Med Biol.* 2011;37:788–797. Available from: https://doi.org/10.1016/j.ultrasmedbio.2011.01.020.

61. Rabben SI, Stergiopulos N, Hellevik LR, Smiseth OA, Slørdahl S, Urheim S, et al. An ultrasound-based method for determining pulse wave velocity in superficial arteries. *J Biomech*. 2004;37:1615–1622. Available from: https://doi.org/10.1016/j.jbiomech.2003.12.031.
62. Vlahandonis A, Biggs SN, Nixon GM, Davey MJ, Walter LM, Horne RSC. Pulse transit time as a surrogate measure of changes in systolic arterial pressure in children during sleep. *J Sleep Res*. 2014;23:406–413. Available from: https://doi.org/10.1111/jsr.12140.
63. Lee J, Sohn J, Park J, Yang S, Lee S, Kim HC. Novel blood pressure and pulse pressure estimation based on pulse transit time and stroke volume approximation. *Biomed Eng Online*. 2018;17:81. Available from: https://doi.org/10.1186/s12938-018-0510-8.
64. He S, Dajani HR, Meade RD, Kenny GP, Bolic M. Continuous tracking of changes in systolic blood pressure using BCG and ECG. *Annu Int Conf IEEE Eng Med Biol Soc*. 2019;2019:6826–6829. Available from: https://doi.org/10.1109/EMBC.2019.8856332.
65. Wassertheurer S, Kropf J, Weber T, Van Der Giet M, Baulmann J, Ammer M, et al. A new oscillometric method for pulse wave analysis: comparison with a common tonometric method. *J Hum Hypertension*. 2010;24:498–504. Available from: https://doi.org/10.1038/jhh.2010.27.
66 Butlin M, Qasem A, Avolio AP. Estimation of central aortic pressure waveform features derived from the brachial cuff volume displacement waveform, 2012;2591–2594. https://doi.org/10.1109/EMBC.2012.6346494.
67. Luzardo L, Lujambio I, Sottolano M, da Rosa A, Thijs L, Noboa O, et al. 24-h ambulatory recording of aortic pulse wave velocity and central systolic augmentation: a feasibility study. *Hypertens Res*. 2012;35:980–987. Available from: https://doi.org/10.1038/hr.2012.78.
68. Shoji T, Nakagomi A, Okada S, Ohno Y, Kobayashi Y. Invasive validation of a novel brachial cuff-based oscillometric device (SphygmoCor XCEL) for measuring central blood pressure. *J Hypertension*. 2017;35:69–75. Available from: https://doi.org/10.1097/HJH.0000000000001135.
69. Hwang M-H, Yoo J-K, Kim H-K, Hwang C-L, Mackay K, Hemstreet O, et al. Validity and reliability of aortic pulse wave velocity and augmentation index determined by the new cuff-based SphygmoCor Xcel. *J Hum Hypertens*. 2014;28:475–481. Available from: https://doi.org/10.1038/jhh.2013.144.
70. Lin ACW, Lowe A, Sidhu K, Harrison W, Ruygrok P, Stewart R. Evaluation of a novel sphygmomanometer, which estimates central aortic blood pressure from analysis of brachial artery suprasystolic pressure waves. *J Hypertension*. 2012;30:1743–1750. Available from: https://doi.org/10.1097/HJH.0b013e3283567b94.
71. Corazza I, Barletta G, Guaraldi P, Cecere A, Calandra-Buonaura G, Altini E, et al. A new integrated instrumental approach to autonomic nervous system assessment. *Computer Methods Prog Biomedicine*. 2014;117:267–276. Available from: https://doi.org/10.1016/j.cmpb.2014.08.002.
72. Chua CP, Heneghan C. Continuous blood pressure monitoring using ECG and finger photoplethysmogram. *Conf Proc IEEE Eng Med Biol Soc*. 2006;2006:5117–5120. Available from: https://doi.org/10.1109/IEMBS.2006.259612.
73. Elgendi M, Fletcher R, Liang Y, Howard N, Lovell NH, Abbott D, et al. The use of photoplethysmography for assessing hypertension. *Npj Digit Med*. 2019;2:1–11. Available from: https://doi.org/10.1038/s41746-019-0136-7.
74. Alian AA, Shelley KH. Photoplethysmography. *Best Pract Res Clin Anaesthesiol*. 2014;28:395–406. Available from: https://doi.org/10.1016/j.bpa.2014.08.006.
75. Allen J. Photoplethysmography and its application in clinical physiological measurement. *Physiol Meas*. 2007;28:R1–R39. Available from: https://doi.org/10.1088/0967-3334/28/3/R01.
76. Castaneda D, Esparza A, Ghamari M, Soltanpur C, Nazeran H. A review on wearable photoplethysmography sensors and their potential future applications in health care. *Int J Biosens Bioelectron*. 2018;4:195–202. Available from: https://doi.org/10.15406/ijbsbe.2018.04.00125.
77. Chua EC-P, Redmond SJ, McDarby G, Heneghan C. Towards using photo-plethysmogram amplitude to measure blood pressure during sleep. *Ann Biomed Eng*. 2010;38:945–954. Available from: https://doi.org/10.1007/s10439-009-9882-z.
78. Martínez G, Howard N, Abbott D, Lim K, Ward R, Elgendi M. Can photoplethysmography replace arterial blood pressure in the assessment of blood pressure? *J Clin Med*. 2018;7. Available from: https://doi.org/10.3390/jcm7100316.
79. Radha M, de Groot K, Rajani N, Wong CCP, Kobold N, Vos V, et al. Estimating blood pressure trends and the nocturnal dip from photoplethysmography. *Physiol Meas*. 2019;40:025006. Available from: https://doi.org/10.1088/1361-6579/ab030e.
80. Xing X, Ma Z, Zhang M, Zhou Y, Dong W, Song M. An unobtrusive and calibration-free blood pressure estimation method using photoplethysmography and biometrics. *Sci Rep*. 2019;9:8611. Available from: https://doi.org/10.1038/s41598-019-45175-2.
81. Kazamias TM, Gander MP, Franklin DL, Ross Jr. J. Blood pressure measurement with Doppler ultrasonic flowmeter. *J Appl Physiol*. 1971;30:585–588. Available from: https://doi.org/10.1152/jappl.1971.30.4.585.
82. Schlöglhofer T, Gilly H, Schima H. Semi-invasive measurement of cardiac output based on pulse contour: a review and analysis. *Can J Anaesth*. 2014;61:452–479. Available from: https://doi.org/10.1007/s12630-014-0135-8.
83. Missant C, Rex S, Wouters PF. Accuracy of cardiac output measurements with pulse contour analysis (PulseCO) and Doppler echocardiography during off-pump coronary artery bypass grafting. *Eur J Anaesthesiol*. 2008;25:243–248. Available from: https://doi.org/10.1017/S0265021507002979.
84. Biglino G, Steeden JA, Baker C, Schievano S, Taylor AM, Parker KH, et al. A non-invasive clinical application of wave intensity analysis based on ultrahigh temporal resolution phase-contrast cardiovascular magnetic resonance. *J Cardiovasc Magn Reson*. 2012;14:57. Available from: https://doi.org/10.1186/1532-429X-14-57.
85. Quail MA, Steeden JA, Knight D, Segers P, Taylor AM, Muthurangu V. Development and validation of a novel method to derive central aortic systolic pressure from the MR aortic distension curve. *J Magn Reson Imaging*. 2014;40:1064–1070. Available from: https://doi.org/10.1002/jmri.24471.
86. Li Y, Hickson SS, McEniery CM, Wilkinson IB, Khir AW. Stiffening and ventricular-arterial interaction in the ascending aorta using MRI: ageing effects in healthy humans. *J Hypertens*. 2019;37:347–355. Available from: https://doi.org/10.1097/HJH.0000000000001886.

87. Swift AJ, Rajaram S, Hurdman J, Hill C, Davies C, Sproson TW, et al. Noninvasive estimation of pa pressure, flow, and resistance with CMR imaging: derivation and prospective validation study from the aspire registry. *JACC: Cardiovas Imaging*. 2013;6:1036–1047. Available from: https://doi.org/10.1016/j.jcmg.2013.01.013.
88. Sieren MM, Berlin C, Oechtering TH, Hunold P, Drömann D, Barkhausen J, et al. Comparison of 4D Flow MRI to 2D Flow MRI in the pulmonary arteries in healthy volunteers and patients with pulmonary hypertension. *PLoS One*. 2019;14. Available from: https://doi.org/10.1371/journal.pone.0224121.
89. Kawakubo M, Akamine H, Yamasaki Y, Takemura A, Abe K, Hosokawa K, et al. Three-dimensional phase contrast magnetic resonance imaging validated to assess pulmonary artery flow in patients with chronic thromboembolic pulmonary hypertension. *Radiol Phys Technol*. 2017;10:249–255. Available from: https://doi.org/10.1007/s12194-016-0383-0.
90. Rudski LG, Lai WW, Afilalo J, Hua L, Handschumacher MD, Chandrasekaran K, et al. Guidelines for the echocardiographic assessment of the right heart in adults: a report from the American Society of Echocardiography. Endorsed by the European Association of Echocardiography, a registered branch of the European Society of Cardiology, and the Canadian Society of Echocardiography. *J Am Soc Echocardiogr*. 2010;23:685–713. Available from: https://doi.org/10.1016/j.echo.2010.05.010.
91. Beigel R, Cercek B, Luo H, Siegel RJ. Noninvasive evaluation of right atrial pressure. *J Am Soc Echocardiogr*. 2013;26:1033–1042. Available from: https://doi.org/10.1016/j.echo.2013.06.004.
92. Watanabe R, Amano H, Saito F, Toyoda S, Sakuma M, Abe S, et al. Echocardiographic surrogates of right atrial pressure in pulmonary hypertension. *Heart Vessel*. 2019;34:477–483. Available from: https://doi.org/10.1007/s00380-018-1264-8.
93. Nagueh SF. Non-invasive assessment of left ventricular filling pressure. *Eur J Heart Fail*. 2018;20:38–48. Available from: https://doi.org/10.1002/ejhf.971.
94. Saugel B, Kouz K, Meidert AS, Schulte-Uentrop L, Romagnoli S. How to measure blood pressure using an arterial catheter: a systematic 5-step approach. *Crit Care*. 2020;24:172. Available from: https://doi.org/10.1186/s13054-020-02859-w.
95. Allan MW, Gray WM, Asbury AJ. Measurement of arterial pressure using catheter-transducer systems. Improvement using the accudynamic. *Br J Anaesth*. 1988;60:413–418. Available from: https://doi.org/10.1093/bja/60.4.413.
96. Boyer TD, Triger DR, Horisawa M, Redeker AG, Reynolds TB. Direct transhepatic measurement of portal vein pressure using a thin needle. *Comparison Wedged Hepatic Vein Press Gastroenterol*. 1977;72:584–589.
97. Groszmann R, Vorobioff JD, Gao H. Measurement of portal pressure: when, how, and why to do it. *Clin Liver Dis*. 2006;10:499–512. Available from: https://doi.org/10.1016/j.cld.2006.08.005. viii.
98. Lambert EH, Wood EH. The use of a resistance wire, strain gauge manometer to measure intraarterial pressure. *Proc Soc Exp Biol Med*. 1947;64:186–190. Available from: https://doi.org/10.3181/00379727-64-15742.
99. Neglen P, Raju S. Ambulatory venous pressure revisited. *J Vasc Surg*. 2000;31:1206–1213. Available from: https://doi.org/10.1067/mva.2000.105669.
100. Al-Qatatsheh A, Morsi Y, Zavabeti A, Zolfagharian A, Salim N, Z Kouzani A, et al. Blood pressure sensors: materials, fabrication methods, performance evaluations and future perspectives. *Sens (Basel)*. 2020;20. Available from: https://doi.org/10.3390/s20164484.
101. Nguyen T-V, Mizuki Y, Tsukagoshi T, Takahata T, Ichiki M, Shimoyama I. MEMS-based pulse wave sensor utilizing a piezoresistive cantilever. *Sens (Basel)*. 2020;20. Available from: https://doi.org/10.3390/s20041052.
102. Afari ME, Syed W, Tsao L. Implantable devices for heart failure monitoring and therapy. *Heart Fail Rev*. 2018;23:935–944. Available from: https://doi.org/10.1007/s10741-018-9687-y.
103. Abraham WT, Perl L. Implantable hemodynamic monitoring for heart failure patients. *J Am Coll Cardiol*. 2017;70:389–398. Available from: https://doi.org/10.1016/j.jacc.2017.05.052.
104. Abraham WT, Adamson PB, Bourge RC, Aaron MF, Costanzo MR, Stevenson LW, et al. Wireless pulmonary artery haemodynamic monitoring in chronic heart failure: a randomised controlled trial. *Lancet*. 2011;377:658–666. Available from: https://doi.org/10.1016/S0140-6736(11)60101-3.
105. Volterrani M, Spoletini I, Angermann C, Rosano G, Coats AJ. Implantable devices for heart failure monitoring: the CardioMEMS™ system. *Eur Heart J Suppl*. 2019;21:M50–M53. Available from: https://doi.org/10.1093/eurheartj/suz265.
106. Perl L, Soifer E, Bartunek J, Erdheim D, Köhler F, Abraham WT, et al. A novel wireless left atrial pressure monitoring system for patients with heart failure, first ex-vivo and animal experience. *J Cardiovasc Trans Res*. 2019;12:290–298. Available from: https://doi.org/10.1007/s12265-018-9856-3.
107. Porter TR, Shillcutt SK, Adams MS, Desjardins G, Glas KE, Olson JJ, et al. Guidelines for the use of echocardiography as a monitor for therapeutic intervention in adults: a report from the American Society of Echocardiography. *J Am Soc Echocardiogr*. 2015;28:40–56. Available from: https://doi.org/10.1016/j.echo.2014.09.009.
108. Dipti A, Soucy Z, Surana A, Chandra S. Role of inferior vena cava diameter in assessment of volume status: a meta-analysis. *Am J Emerg Med*. 2012;30:1414–1419.e1. Available from: https://doi.org/10.1016/j.ajem.2011.10.017.
109. Yildirimturk O, Tayyareci Y, Erdim R, Ozen E, Yurdakul S, Aytekin V, et al. Assessment of right atrial pressure using echocardiography and correlation with catheterization. *J Clin Ultrasound*. 2011;39:337–343. Available from: https://doi.org/10.1002/jcu.20837.
110. Alavi-Moghaddam M, Kabir A, Shojaee M, Manouchehrifar M, Moghimi M. Ultrasonography of inferior vena cava to determine central venous pressure: a meta-analysis and meta-regression. *Acta Radiol*. 2017;58:537–541. Available from: https://doi.org/10.1177/0284185116663045.
111. Lancellotti P, Zamorano JL, Habib G, Badano L. *The EACVI Textbook of Echocardiography*. 2nd ed. Oxford: OUP; 2016.
112. Brennan JM, Ronan A, Goonewardena S, Blair JEA, Hammes M, Shah D, et al. Handcarried ultrasound measurement of the inferior vena cava for assessment of intravascular volume status in the outpatient hemodialysis clinic. *CJASN*. 2006;1:749–753. Available from: https://doi.org/10.2215/CJN.00310106.

113. Rizkallah J, Jack M, Saeed M, Shafer LA, Vo M, Tam J. Non-invasive bedside assessment of central venous pressure: scanning into the future. *PLoS One*. 2014;9:e109215. Available from: https://doi.org/10.1371/journal.pone.0109215.

114. Thalhammer C, Aschwanden M, Odermatt A, Baumann UA, Imfeld S, Bilecen D, et al. Noninvasive central venous pressure measurement by controlled compression sonography at the forearm. *J Am Coll Cardiol*. 2007;50:1584–1589. Available from: https://doi.org/10.1016/j.jacc.2007.07.022.

115. Thalhammer C, Siegemund M, Aschwanden M, Gassmann M, Baumann UA, Jaeger KA, et al. Non-invasive central venous pressure measurement by compression ultrasound–a step into real life. *Resuscitation*. 2009;80:1130–1136. Available from: https://doi.org/10.1016/j.resuscitation.2009.06.027.

116. Ward KR, Tiba MH, Draucker GT, Proffitt EK, Barbee RW, Gunnerson KJ, et al. A novel noninvasive impedance-based technique for central venous pressure measurement. *Shock*. 2010;33:269–273. Available from: https://doi.org/10.1097/SHK.0b013e3181ab9b9b.

117. Ward KR, Tiba MH, Barbee RW, Ivatury RR, Arrowood JA, Spiess BD, et al. A new noninvasive method to determine central venous pressure. *Resuscitation*. 2006;70:238–246. Available from: https://doi.org/10.1016/j.resuscitation.2005.12.013.

118. Sathish N, Singh N, Nagaraja P, Sarala B, Prabhushankar C, Dhananjaya M, et al. Comparison between noninvasive measurement of central venous pressure using near infrared spectroscopy with an invasive central venous pressure monitoring in cardiac surgical Intensive Care Unit. *Ann Card Anaesth*. 2016;19:405. Available from: https://doi.org/10.4103/0971-9784.185520.

119. Teichgräber UK, Gebauer B, Benter T, Wagner HJ. Central venous access catheters: radiological management of complications. *Cardiovasc Intervent Radiol*. 2003;26:321–333. Available from: https://doi.org/10.1007/s00270-003-0112-z.

120. Ruesch S, Walder B, Tramèr MR. Complications of central venous catheters: internal jugular vs subclavian access–a systematic review. *Crit Care Med*. 2002;30:454–460. Available from: https://doi.org/10.1097/00003246-200202000-00031.

121. Corazza Ivan, Zecchi Margherita, Zannoli Romano. Evaluation of low gradient severe aortic stenosis: should we change our outlook in the analysis of clinical data? *Open Heart*. 2021;8(2):e001746. Available from: https://doi.org/10.1136/openhrt-2021-001746.

122. Evans DH, Jensen JA, Nielsen MB. Ultrasonic colour Doppler imaging. *Interface Focus*. 2011;1:490–502. Available from: https://doi.org/10.1098/rsfs.2011.0017.

123. Bashford GR. Ultrasonic measurement of blood flow velocity and applications for cardiovascular assessments. In: Patel VB, Preedy VR, eds. *Biomarkers in Cardiovascular Disease, Dordrecht*. The Netherlands: Springer; 2015:1–31. Available from: https://doi.org/10.1007/978-94-007-7741-5_18-1.

124. Berman W, Alverson DC. Assessment of hemodynamic function with pulsed Doppler ultrasound. *J Am Coll Cardiol*. 1985;5:104S–112S. Available from: https://doi.org/10.1016/s0735-1097(85)80151-0.

125. Coldani G, Ghidetti P, Lombardi R, Corazza I, Zannoli R. An instrument to measure velocity profile by means of ultrasound techniques. *J Mech Med Biol*. 2003;3:21–30. Available from: https://doi.org/10.1142/S0219519403000569.

126. Hoskins PR. Measurement of arterial blood flow by Doppler ultrasound. *Clin Phys Physiol Meas*. 1990;11:1–26. Available from: https://doi.org/10.1088/0143-0815/11/1/001.

127. Kaszczewski P, Elwertowski M, Leszczynski J, Ostrowski T, Galazka Z. Volumetric carotid flow characteristics in Doppler ultrasonography in healthy population over 65 years old. *J Clin Med*. 2020;9. Available from: https://doi.org/10.3390/jcm9051375.

128. Krishnamoorthy VK, Sengupta PP, Gentile F, Khandheria BK. History of echocardiography and its future applications in medicine. *Crit Care Med*. 2007;35:S309–S313. Available from: https://doi.org/10.1097/01.CCM.0000270240.97375.DE.

129. Reneman RS, Hoeks AP. Noninvasive vascular ultrasound: an asset in vascular medicine. *Cardiovasc Res*. 2000;45:27–35. Available from: https://doi.org/10.1016/s0008-6363(99)00282-5.

130. Spampinato RA, Jahnke C, Paetsch I, Hilbert S, Busch F, Schloma V, et al. Quantification of aortic valve regurgitation by pulsed Doppler examination of the left subclavian artery velocity contour: a validation study with cardiovascular magnetic resonance imaging. *J Am Soc Echocardiogr*. 2018;31:42–51. Available from: https://doi.org/10.1016/j.echo.2017.10.004.

131. Veyrat C. Directional Doppler in cardiology: a 50-year journey. *J Am Soc Echocardiogr*. 2018;31:1308–1322. Available from: https://doi.org/10.1016/j.echo.2018.09.014.

132. Yusheng F, Wang Q, Yi J, Song D, Xiang X. A numerical model of blood flow velocity measurement based on finger ring. *J Healthc Eng*. 2018;2018:1–8. Available from: https://doi.org/10.1155/2018/3916481.

133 The EACVI. Textbook of Echocardiography. Oxford University Press; n.d.

134. McGowan JH, Cleland JGF. Reliability of reporting left ventricular systolic function by echocardiography: a systematic review of 3 methods. *Am Heart J*. 2003;146:388–397. Available from: https://doi.org/10.1016/S0002-8703(03)00248-5.

135. Corazza I, Casadei L, Bonafè E, Cercenelli L, Marcelli E, Zannoli R. How to transform a fixed stroke alternating syringe ventricle into an adjustable elastance ventricle. *Rev Sci Instrum*. 2018;89:074301. Available from: https://doi.org/10.1063/1.5030100.

136. Vignon P, Begot E, Mari A, Silva S, Chimot L, Delour P, et al. Hemodynamic assessment of patients with septic shock using transpulmonary thermodilution and critical care echocardiography. *Chest*. 2018;153:55–64. Available from: https://doi.org/10.1016/j.chest.2017.08.022.

137. De Backer D, Bakker J, Cecconi M, Hajjar L, Liu DW, Lobo S, et al. Alternatives to the Swan–Ganz catheter. *Intensive Care Med*. 2018;44:730–741. Available from: https://doi.org/10.1007/s00134-018-5187-8.

138. Feliciani G, Peron C, La Rocca A, Scuppa MF, Malavolta A, Bianchini D, et al. Cold pressor test using strain-gauge plethysmography. *Adv Physiol Educ*. 2016;40:410–417. Available from: https://doi.org/10.1152/advan.00096.2015.

139. Harel F, Denault A, Ngo Q, Dupuis J, Khairy P. Near-infrared spectroscopy to monitor peripheral blood flow perfusion. *J Clin Monit Comput*. 2008;22:37–43. Available from: https://doi.org/10.1007/s10877-007-9105-9.

140. Rezk-Hanna M, Nelson MD, Rader F, Benowitz NL, Rosenberry R, Chang LC, et al. Peripheral blood flow changes to cutaneous and muscular beds in response to acute hookah smoking. *Am J Cardiol.* 2020;125:1725–1731. Available from: https://doi.org/10.1016/j.amjcard.2020.02.042.
141. Stewart JM, Montgomery LD. Regional blood volume and peripheral blood flow in postural tachycardia syndrome. *Am J Physiol-Heart Circul Physiol.* 2004;287:H1319–H1327. Available from: https://doi.org/10.1152/ajpheart.00086.2004.
142. Fehling PC, Arciero PJ, MacPherson CJ, Smith DL. Reproducibility of resting peripheral blood flow using strain gauge plethysmography. *Int J Sports Med.* 1999;20:555–559. Available from: https://doi.org/10.1055/s-1999-8845.
143. Jakovljevic DG, Trenell MI, MacGowan GA. Bioimpedance and bioreactance methods for monitoring cardiac output. *Best Pract Res Clin Anaesthesiol.* 2014;28:381–394. Available from: https://doi.org/10.1016/j.bpa.2014.09.003.
144. Corazza I, Diemberger I, Reggiani E, Zannoli R. A non-invasive biomechanical approach to the clinical follow-up of heart failure patients. *Ser Biomech.* 2015;29:5–13.
145. Fuller HD. Improving the accuracy of impedance cardiac output in the intensive care unit: comparison with thermodilution cardiac output. *Congestive Heart Fail.* 2006;12:271–276. Available from: https://doi.org/10.1111/j.1527-5299.2006.05755.x.
146. Barin E, Haryadi DG, Schookin SI, Westenskow DR, Zubenko VG, Beliaev KR, et al. Evaluation of a thoracic bioimpedance cardiac output monitor during cardiac catheterization. *Crit Care Med.* 2000;28:698–702. Available from: https://doi.org/10.1097/00003246-200003000-00016.
147. Beck R, Milella L, Labellarte C. Continuous non-invasive measurement of stroke volume and cardiac index in infants and children: comparison of Impedance Cardiography NICaS® vs CardioQ® method. *Clin Ter.* 2018;169:e110–e113. Available from: https://doi.org/10.7417/T.2018.2064.
148. Brittain JM, Busk TM, Møller S. Validation of non-invasive haemodynamic methods in patients with liver disease: the Finometer and the Task Force Monitor. *Clin Physiol Funct Imaging.* 2018;38:384–389. Available from: https://doi.org/10.1111/cpf.12425.
149. Dazzani F, Micati M, Caraceni P, Drago GM, Domenicali M, Pacilli P, et al. Transthoracic electrical bioimpedance: a non-invasive technique for the evaluation of the haemodynamic alterations in patients with liver cirrhosis. *Dig Liver Dis.* 2005;37:786–792. Available from: https://doi.org/10.1016/j.dld.2005.05.008.
150. Elwan MH, Hue J, Green SJ, Eltahan SM, Sims MR, Coats TJ. Thoracic electrical bioimpedance vs suprasternal Doppler in emergency care. *Emerg Med Australas.* 2017;29:391–393. Available from: https://doi.org/10.1111/1742-6723.12765.
151. Hafid A, Benouar S, Kedir-Talha M, Abtahi F, Attari M, Seoane F. Full Impedance Cardiography measurement device using Raspberry PI3 and System-on-Chip biomedical Instrumentation Solutions. *IEEE J Biomed Health Inf.* 2018;22:1883–1894. Available from: https://doi.org/10.1109/JBHI.2017.2783949.
152. Harford M, Clark SH, Smythe JF, Gerry S, Villarroel M, Jorge J, et al. Non-invasive stroke volume estimation by transthoracic electrical bioimpedance vs Doppler echocardiography in healthy volunteers. *J Med Eng Technol.* 2019;43:33–37. Available from: https://doi.org/10.1080/03091902.2019.1599074.
153. Jakovljevic DG, Moore S, Hallsworth K, Fattakhova G, Thoma C, Trenell MI. Comparison of cardiac output determined by bioimpedance and bioreactance methods at rest and during exercise. *J Clin Monit Comput.* 2012;26:63–68. Available from: https://doi.org/10.1007/s10877-012-9334-4.
154. Lorne E, Mahjoub Y, Diouf M, Sleghem J, Buchalet C, Guinot P-G, et al. Accuracy of impedance cardiography for evaluating trends in cardiac output: a comparison with oesophageal Doppler. *Br J Anaesth.* 2014;113:596–602. Available from: https://doi.org/10.1093/bja/aeu136.
155. Roth S, Fox H, Fuchs U, Schulz U, Costard-Jäckle A, Gummert JF, et al. Noninvasive pulse contour analysis for determination of cardiac output in patients with chronic heart failure. *ClRes Cardiol.* 2018;107:395–404. Available from: https://doi.org/10.1007/s00392-017-1198-7.
156. Sangkum L, Liu GL, Yu L, Yan H, Kaye AD, Liu H. Minimally invasive or noninvasive cardiac output measurement: an update. *J Anesth.* 2016;30:461–480. Available from: https://doi.org/10.1007/s00540-016-2154-9.
157. Thomas SH. Impedance cardiography using the Sramek-Bernstein method: accuracy and variability at rest and during exercise. *Br J Clin Pharmacol.* 1992;34:467–476.
158. van der Meer BJ, Woltjer HH, Sousman AM, Schreuder WO, Bulder ER, Huybregts MA, et al. Impedance cardiography. *Import Equ Electrode Config Intensive Care Med.* 1996;22:1120–1124.
159. Wassertheurer S, Mayer C, Breitenecker F. Modeling arterial and left ventricular coupling for non-invasive measurements. *Simul Model Pract Theory.* 2008;16:988–997. Available from: https://doi.org/10.1016/j.simpat.2008.04.016.
160. Saugel B, Cecconi M, Wagner JY, Reuter DA. Noninvasive continuous cardiac output monitoring in perioperative and intensive care medicine. *Br J Anaesth.* 2015;114:562–575. Available from: https://doi.org/10.1093/bja/aeu447.
161. Ballestero Y, López-Herce J, Urbano J, Solana MJ, Botrán M, Bellón JM, et al. Measurement of cardiac output in children by bioreactance. *Pediatr Cardiol.* 2011;32:469–472. Available from: https://doi.org/10.1007/s00246-011-9903-1.
162. Ballestero Y, Urbano J, López-Herce J, Solana MJ, Botrán M, Vinciguerra D, et al. Pulmonary arterial thermodilution, femoral arterial thermodilution and bioreactance cardiac output monitoring in a pediatric hemorrhagic hypovolemic shock model. *Resuscitation.* 2012;83:125–129. Available from: https://doi.org/10.1016/j.resuscitation.2011.06.039.
163. Cheung H, Dong Q, Dong R, Yu B. Correlation of cardiac output measured by non-invasive continuous cardiac output monitoring (NICOM) and thermodilution in patients undergoing off-pump coronary artery bypass surgery. *J Anesth.* 2015;29:416–420. Available from: https://doi.org/10.1007/s00540-014-1938-z.
164. Fagnoul D, Vincent J-L, Backer DD. Cardiac output measurements using the bioreactance technique in critically ill patients. *Crit Care.* 2012;16:460. Available from: https://doi.org/10.1186/cc11481.
165. Jones TW, Houghton D, Cassidy S, MacGowan GA, Trenell MI, Jakovljevic DG. Bioreactance is a reliable method for estimating cardiac output at rest and during exercise. *Br J Anaesth.* 2015;115:386–391. Available from: https://doi.org/10.1093/bja/aeu560.

166. Keren H, Burkhoff D, Squara P. Evaluation of a noninvasive continuous cardiac output monitoring system based on thoracic bioreactance. *Am J Physiol Heart Circ Physiol*. 2007;293:H583–H589. Available from: https://doi.org/10.1152/ajpheart.00195.2007.
167. Kossari N, Hufnagel G, Squara P. Bioreactance: a new tool for cardiac output and thoracic fluid content monitoring during hemodialysis. *Hemodial Int*. 2009;13:512–517. Available from: https://doi.org/10.1111/j.1542-4758.2009.00386.x.
168. Rali AS, Buechler T, Van Gotten B, Waters A, Shah Z, Haglund N, et al. Non-invasive cardiac output monitoring in cardiogenic shock: the NICOM study. *J Card Fail*. 2020;26:160–165. Available from: https://doi.org/10.1016/j.cardfail.2019.11.015.
169. Sivakumar S, Lazaridis C. Bioreactance-based noninvasive fluid responsiveness and cardiac output monitoring: a pilot study in patients with aneurysmal subarachnoid hemorrhage and literature review. *Crit Care Res Pract*. 2020;2020:2748181. Available from: https://doi.org/10.1155/2020/2748181.
170. Squara P. Bioreactance: a new method for non-invasive cardiac output monitoring. In: Vincent J-L, ed. *Yearbook of Intensive Care and Emergency Medicine*. Berlin, Heidelberg: Springer; 2008:619–630. Available from: https://doi.org/10.1007/978-3-540-77290-3_58.
171. Weisz DE, Jain A, McNamara PJ, EL-Khuffash A. Non-invasive cardiac output monitoring in neonates using bioreactance: a comparison with echocardiography. *Neonatology*. 2012;102:61–67. Available from: https://doi.org/10.1159/000337295.
172. Pelc NJ, Herfkens RJ, Shimakawa A, Enzmann DR. Phase contrast cine magnetic resonance imaging. *Magn Reson Q*. 1991;7:229–254.
173. Wymer DT, Patel KP, Burke WF, Bhatia VK, Phase-Contrast MRI. Physics, techniques, and clinical applications. *RadioGraphics*. 2020;40:122–140. Available from: https://doi.org/10.1148/rg.2020190039.
174. Chai P, Mohiaddin R. How we perform cardiovascular magnetic resonance flow assessment using phase-contrast velocity mapping. *J Cardiovas Mag Reson*. 2005;7:705–716. Available from: https://doi.org/10.1081/JCMR-200065639.
175. Sun JX, Reisner AT, Saeed M, Heldt T, Mark RG. The cardiac output from blood pressure algorithms trial. *Crit Care Med*. 2009;37:72–80. Available from: https://doi.org/10.1097/CCM.0b013e3181930174.
176. Boisson M, Poignard ME, Pontier B, Mimoz O, Debaene B, Frasca D. Cardiac output monitoring with thermodilution pulse-contour analysis vs. non-invasive pulse-contour analysis. *Anaesthesia*. 2019;74:735–740. Available from: https://doi.org/10.1111/anae.14638.
177. Grensemann J. Cardiac output monitoring by pulse contour analysis, the technical basics of less-invasive techniques. *Front Med*. 2018;5:64. Available from: https://doi.org/10.3389/fmed.2018.00064.
178. Kamoi S, Pretty C, Balmer J, Davidson S, Pironet A, Desaive T, et al. Improved pressure contour analysis for estimating cardiac stroke volume using pulse wave velocity measurement. *Biomed Eng Online*. 2017;16:51. Available from: https://doi.org/10.1186/s12938-017-0341-z.
179. Saugel B, Meidert AS, Langwieser N, Wagner JY, Fassio F, Hapfelmeier A, et al. An autocalibrating algorithm for non-invasive cardiac output determination based on the analysis of an arterial pressure waveform recorded with radial artery applanation tonometry: a proof of concept pilot analysis. *J Clin Monit Comput*. 2014;28:357–362. Available from: https://doi.org/10.1007/s10877-013-9540-8.
180. Truijen J, van Lieshout JJ, Wesselink WA, Westerhof BE. Noninvasive continuous hemodynamic monitoring. *J Clin Monit Comput*. 2012;26:267–278. Available from: https://doi.org/10.1007/s10877-012-9375-8.
181. Westerhof N, Lankhaar J-W, Westerhof BE. The arterial Windkessel. *Med Biol Eng Comput*. 2009;47:131–141. Available from: https://doi.org/10.1007/s11517-008-0359-2.
182. Erlanger J. An experimental study of blood-pressure and of pulse-pressure in man. *Physiol Lab Johns Hopkins Univ*. 1904;.
183. Clement RP, Vos JJ, Scheeren TWL. Minimally invasive cardiac output technologies in the ICU: putting it all together. *Curr Opin Crit Care*. 2017;23:302–309. Available from: https://doi.org/10.1097/MCC.0000000000000417.
184. Zhang J, Critchley LAH, Huang L. Five algorithms that calculate cardiac output from the arterial waveform: a comparison with Doppler ultrasound. *Br J Anaesth*. 2015;115:392–402. Available from: https://doi.org/10.1093/bja/aev254.
185. Huang SJ. Measuring cardiac output at the bedside. *Curr Opin Crit Care*. 2019;25:266–272. Available from: https://doi.org/10.1097/MCC.0000000000000599.
186. Wippermann CF, Huth RG, Schmidt FX, Thul J, Betancor M, Schranz D. Continuous measurement of cardiac output by the Fick principle in infants and children: comparison with the thermodilution method. *Intensive Care Med*. 1996;22:467–471. Available from: https://doi.org/10.1007/BF01712169.
187. Weinbroum AA, Biderman P, Soffer D, Klausner JM, Szold O. Reliability of cardiac output calculation by the fick principle and central venous oxygen saturation in emergency conditions. *J Clin Monit Comput*. 2008;22:361–366. Available from: https://doi.org/10.1007/s10877-008-9143-y.
188. Sun XG, Hansen JE, Ting H, Chuang ML, Stringer WW, Adame D, et al. Comparison of exercise cardiac output by the Fick principle using oxygen and carbon dioxide. *Chest*. 2000;118:631–640. Available from: https://doi.org/10.1378/chest.118.3.631.
189. Selzer A, Sudrann RB. Reliability of the determination of cardiac output in man by means of the Fick principle. *Circ Res*. 1958;6:485–490. Available from: https://doi.org/10.1161/01.res.6.4.485.
190. Fritts HW, Cournand A. The application of the Fick principle to the measurement of pulmonary blood flow. *Proc Natl Acad Sci USA*. 1958;44:1079–1087. Available from: https://doi.org/10.1073/pnas.44.10.1079.
191. Feustel PJ, Perkins RJ, Oppenlander JE, Stratton HH, Cohen IL. Feasibility of continuous oxygen delivery and cardiac output measurement by application of the Fick principle. *Am J Respir Crit Care Med*. 1994;149:751–758. Available from: https://doi.org/10.1164/ajrccm.149.3.8118646.
192. Samet P, Bernstein WH, Kiem I. Reproducibility of cardiac output determination in man by the direct Fick principle. *Dis Chest*. 1964;45:294–300. Available from: https://doi.org/10.1378/chest.45.3.294.
193. Keinänen O, Takala J, Kari A. Continuous measurement of cardiac output by the Fick principle: clinical validation in intensive care. *Crit Care Med*. 1992;20:360–365. Available from: https://doi.org/10.1097/00003246-199203000-00012.

194. Lauber R, Steiner A, Zbinden AM. Accuracy, alarm limits and rise times of 12 oxygen analysers. *Anaesthesia.* 1994;49:1071–1077. Available from: https://doi.org/10.1111/j.1365-2044.1994.tb04359.x.
195. Razvi Y, Ladie DE. *Cardiopulmonary Exercise Testing.* StatPearls, Treasure Island, FL: StatPearls Publishing; 2020.
196. Corazza I, Fabbiani L, Zannoli R. Measurement of oxygen uptake: validation of a "mask-free" method. *Phys Med.* 2007;23:41–47. Available from: https://doi.org/10.1016/j.ejmp.2007.02.002.
197. Hafen BB, Sharma S. *Oxygen Saturation.* StatPearls, Treasure Island, FL: StatPearls Publishing; 2020.
198. Narang N, Thibodeau JT, Levine BD, Gore MO, Ayers CR, Lange RA, et al. Inaccuracy of estimated resting oxygen uptake in the clinical setting. *Circulation.* 2014;129:203–210. Available from: https://doi.org/10.1161/CIRCULATIONAHA.113.003334.
199. Kresoja K-P, Faragli A, Abawi D, Paul O, Pieske B, Post H, et al. Thermodilution vs estimated Fick cardiac output measurement in an elderly cohort of patients: a single-centre experience. *PLoS One.* 2019;14:e0226561. Available from: https://doi.org/10.1371/journal.pone.0226561.
200. Okwose NC, Bouzas-Cruz N, Fernandez OG, Koshy A, Green T, Woods A, et al. Validity of hemodynamic monitoring using inert gas rebreathing method in patients with chronic heart failure and those implanted with a left ventricular assist device. *J Card Fail.* 2020;. Available from: https://doi.org/10.1016/j.cardfail.2020.09.479. S107191642031469X.
201. Marma AK, Opotowsky AR, Fromm BS, Ubeda-Tikkanen A, Porras D, Rhodes J. Noninvasive cardiac output estimation by inert gas rebreathing in pediatric and congenital heart disease. *Am Heart J.* 2016;174:80–88. Available from: https://doi.org/10.1016/j.ahj.2016.01.006.
202. Funk DJ, Moretti EW, Gan TJ. Minimally invasive cardiac output monitoring in the perioperative setting. *Anesth Analg.* 2009;108:887–897. Available from: https://doi.org/10.1213/ane.0b013e31818ffd99.
203. Dark PM, Singer M. The validity of trans-esophageal Doppler ultrasonography as a measure of cardiac output in critically ill adults. *Intensive Care Med.* 2004;30:2060–2066. Available from: https://doi.org/10.1007/s00134-004-2430-2.
204. Feinberg MS, Hopkins WE, Davila-Roman VG, Barzilai B. Multiplane transesophageal echocardiographic doppler imaging accurately determines cardiac output measurements in critically ill patients. *Chest.* 1995;107:769–773. Available from: https://doi.org/10.1378/chest.107.3.769.
205. Gorcsan J, Diana P, Ball BA, Hattler BG. Intraoperative determination of cardiac output by transesophageal continuous wave Doppler. *Am Heart J.* 1992;123:171–176. Available from: https://doi.org/10.1016/0002-8703(92)90762-k.
206. Graeser K, Zemtsovski M, Kofoed KF, Winther-Jensen M, Nilsson JC, Kjaergaard J, et al. Comparing methods for cardiac output: intraoperatively Doppler-derived cardiac output measured with 3-dimensional echocardiography is not interchangeable with cardiac output by pulmonary catheter thermodilution. *Anesth Analg.* 2018;127:399–407. Available from: https://doi.org/10.1213/ANE.0000000000002800.
207. Grigg LE, Wigle ED, Williams WG, Daniel LB, Rakowski H. Transesophageal Doppler echocardiography in obstructive hypertrophic cardiomyopathy: clarification of pathophysiology and importance in intraoperative decision making. *J Am Coll Cardiol.* 1992;20:42–52. Available from: https://doi.org/10.1016/0735-1097(92)90135-a.
208. Kasprzak JD, Drozdz J, Peruga JZ, Rafalska K, Krzemińska-Pakuła M. Definition of flow parameters in proximal nonstenotic coronary arteries using transesophageal Doppler echocardiography. *Echocardiography.* 2000;17:141–150. Available from: https://doi.org/10.1111/j.1540-8175.2000.tb01115.x.
209. Lafanechère A, Albaladejo P, Raux M, Geeraerts T, Bocquet R, Wernet A, et al. Cardiac output measurement during infrarenal aortic surgery: echo-esophageal Doppler vs thermodilution catheter. *J Cardiothorac Vasc Anesth.* 2006;20:26–30. Available from: https://doi.org/10.1053/j.jvca.2005.07.029.
210. Mehta Y, Arora D. Newer methods of cardiac output monitoring. *World J Cardiol.* 2014;6:1022–1029. Available from: https://doi.org/10.4330/wjc.v6.i9.1022.
211. Møller-Sørensen H, Cordtz J, Østergaard M, Nilsson JC, Hansen KL. Transesophageal Doppler reliably tracks changes in cardiac output in comparison with intermittent pulmonary artery thermodilution in cardiac surgery patients. *J Clin Monit Comput.* 2017;31:135–142. Available from: https://doi.org/10.1007/s10877-015-9806-4.
212. Muhiudeen IA, Kuecherer HF, Lee E, Cahalan MK, Schiller NB. Intraoperative estimation of cardiac output by transesophageal pulsed Doppler echocardiography. *Anesthesiology.* 1991;74:9–14. Available from: https://doi.org/10.1097/00000542-199101000-00003.
213. Redberg RF, Sobol Y, Chou TM, Malloy M, Kumar S, Botvinick E, et al. Adenosine-induced coronary vasodilation during transesophageal Doppler echocardiography. Rapid and safe measurement of coronary flow reserve ratio can predict significant left anterior descending coronary stenosis. *Circulation.* 1995;92:190–196. Available from: https://doi.org/10.1161/01.cir.92.2.190.
214. Saran S, Gurjar M, Azim A, Mishra P, Ghosh PS, Baronia AK, et al. Trans-esophageal Doppler assessment of acute hemodynamic changes due to prone positioning in acute respiratory distress syndrome patients. *Shock.* 2019;52:e39–e44. Available from: https://doi.org/10.1097/SHK.0000000000001290.
215. Schober P, Loer SA, Schwarte LA. Transesophageal Doppler devices: a technical review. *J Clin Monit Comput.* 2009;23:391–401. Available from: https://doi.org/10.1007/s10877-009-9204-x.
216. Taams MA, Gussenhoven EJ, Cornel JH, The SH, Roelandt JR, Lancée CT, et al. Detection of left coronary artery stenosis by transoesophageal echocardiography. *Eur Heart J.* 1988;9:1162–1166. Available from: https://doi.org/10.1093/oxfordjournals.eurheartj.a062424.
217. Zhang H, Guo C, Lin J. Effects of velocity profiles on measuring accuracy of transit-time ultrasonic flowmeter. *Appl Sci.* 2019;9:1648. Available from: https://doi.org/10.3390/app9081648.
218. Laustsen J, Pedersen EM, Terp K, Steinbrüchel D, Kure HH, Paulsen PK, et al. Validation of a new transit time ultrasound flowmeter in man. *Eur J Vasc Endovasc Surg.* 1996;12:91–96. Available from: https://doi.org/10.1016/s1078-5884(96)80282-6.
219. Beldi G, Bosshard A, Hess OM, Althaus U, Walpoth BH. Transit time flow measurement: experimental validation and comparison of three different systems. *Ann Thorac Surg.* 2000;70:212–217. Available from: https://doi.org/10.1016/s0003-4975(00)01246-7.

220. Rajita G, Mandal N. *Review on transit time ultrasonic flowmeter. 2nd International Conference on Control, Instrumentation, Energy & Communication (CIEC)*. Kolkata, India: IEEE; 2016; 2016:88–92. Available from: https://doi.org/10.1109/CIEC.2016.7513740.
221. Fiala J, Bingger P, Ruh D, Foerster K, Heilmann C, Beyersdorf F, et al. An implantable optical blood pressure sensor based on pulse transit time. *Biomed Microdev*. 2013;15:73–81. Available from: https://doi.org/10.1007/s10544-012-9689-9.
222. Theodor M, Ruh D, Fiala J, Förster K, Heilmann C, Manoli Y, et al. Subcutaneous blood pressure monitoring with an implantable optical sensor. *Biomed Microdev*. 2013;15:811–820. Available from: https://doi.org/10.1007/s10544-013-9768-6.
223. Kolin A. An electromagnetic flowmeter. Principle of the method and its application to bloodflow measurements. *Proc Soc Exp Biol Med*. 1936;35:53–56. Available from: https://doi.org/10.3181/00379727-35-8854P.
224. Denison AB, Spencer MP, Green HD. A square wave electromagnetic flowmeter for application to intact blood vessels. *Circ Res*. 1955;3:39–46. Available from: https://doi.org/10.1161/01.res.3.1.39.
225. Huber P, Meyer JS, Handa J, Ishikawa S. Electromagnetic flowmeter study of carotid and vertebral blood flow during intracranial hypertension. *Acta Neurochir (Wien)*. 1965;13:37–63. Available from: https://doi.org/10.1007/BF02074644.
226. Matsumura H, Ito Y, Uemura K, Nakai Y, Komatsu Y, Ishikawa E, et al. Prediction of the cerebral hyperperfusion phenomenon after carotid endarterectomy using a transit time flowmeter. *Neurol Med Chir (Tokyo)*. 2020;60:94–100. Available from: https://doi.org/10.2176/nmc.oa.2019-0114.
227. Vennemann B, Obrist D, Rösgen T. A smartphone-enabled wireless and batteryless implantable blood flow sensor for remote monitoring of prosthetic heart valve function. *PLoS One*. 2020;15:e0227372. Available from: https://doi.org/10.1371/journal.pone.0227372.
228. Yang CCH, Yeh ITY, Lai H-Y, Chen H-I, Kuo TBJ. Dynamic effects of respiration on aortic blood flow and its autonomic control in rats. *Clin Exp Pharmacol Physiol*. 2008;35:1294–1300. Available from: https://doi.org/10.1111/j.1440-1681.2008.04998.x.
229. Zannoli R, Corazza I, Cacciafesta P, Branzi A. Evaluating the mechanical behavior of a ventricular simulator. *J Mech Med Biol*. 2005;5:369–373. Available from: https://doi.org/10.1142/S0219519405001503.
230. Zannoli R, Corazza I, Branzi A. Mechanical simulator of the cardiovascular system. *Phys Med*. 2009;25:94–100. Available from: https://doi.org/10.1016/j.ejmp.2008.02.007.
231. Corazza I, Taglieri N, Pirazzini E, Rossi PL, Lombi A, Scalise F, et al. Carbon dioxide coronary angiography: a mechanical feasibility study with a cardiovascular simulator. *AIP Adv*. 2018;8:015225. Available from: https://doi.org/10.1063/1.5016601.
232. Ganz W, Swan HJC. Measurement of blood flow by thermodilution. *Am J Cardiol*. 1972;29:241–246. Available from: https://doi.org/10.1016/0002-9149(72)90635-2.
233. Ganz W, Donoso R, Marcus HS, Forrester JS, Swan HJ. A new technique for measurement of cardiac output by thermodilution in man. *Am J Cardiol*. 1971;27:392–396. Available from: https://doi.org/10.1016/0002-9149(71)90436-x.
234. Yelderman ML, Ramsay MA, Quinn MD, Paulsen AW, McKown RC, Gillman PH. Continuous thermodilution cardiac output measurement in intensive care unit patients. *J Cardiothorac Vasc Anesthesia*. 1992;6:270–274. Available from: https://doi.org/10.1016/1053-0770(92)90137-V.
235. Gutiérrez-Barrios A, Izaga-Torralba E, Rivero Crespo F, Gheorghe L, Cañadas-Pruaño D, Gómez-Lara J, et al. Continuous thermodilution method to assess coronary flow reserve. *Am J Cardiol*. 2020. Available from: https://doi.org/10.1016/j.amjcard.2020.11.011. S0002914920312364.
236. Reuter DA, Huang C, Edrich T, Shernan SK, Eltzschig HK. Cardiac output monitoring using indicator-dilution techniques: basics, limits, and perspectives. *Anesth Analg*. 2010;110:799–811. Available from: https://doi.org/10.1213/ANE.0b013e3181cc885a.
237. Miller DE, Gleason WL, McINTOSH HD. A comparison of the cardiac output determination by the direct Fick method and the dye-dilution method using indocyanine green dye and a cuvette densitometer. *J Lab Clin Med*. 1962;59:345–350.
238. Lund-Johansen P. The dye dilution method for measurement of cardiac output. *Eur Heart J*. 1990;11:6–12. Available from: https://doi.org/10.1093/eurheartj/11.suppl_I.6.
239. De Leeuw PW, Birkenhager WH. Some comments on the usefulness of measuring cardiac output by dye dilution. *Eur Heart J*. 1990;11:13–16. Available from: https://doi.org/10.1093/eurheartj/11.suppl_I.13.
240. Maarek J-MI, Holschneider DP, Harimoto J, Yang J, Scremin OU, Rubinstein EH. Measurement of cardiac output with indocyanine green transcutaneous fluorescence dilution technique. *Anesthesiology*. 2004;100:1476–1483. Available from: https://doi.org/10.1097/00000542-200406000-00020.
241. Alander JT, Kaartinen I, Laakso A, Pätilä T, Spillmann T, Tuchin VV, et al. A review of indocyanine green fluorescent imaging in surgery. *Int J Biomed Imaging*. 2012;2012:1–26. Available from: https://doi.org/10.1155/2012/940585.
242. Maarek JMI, Rubinstein EH, Guo Y, Lane CJ, Campese VM, Holschneider DP. Measurement of cardiac output and blood volume during hemodialysis with fluorescent dye dilution technique. *Ann Biomed Eng*. 2017;45:580–591. Available from: https://doi.org/10.1007/s10439-016-1711-6.
243. Hiyoshi M, Yano K, Nanashima A, Imamura N, Hamada T, Wada T. Reliability of indocyanine green retention and clearance rates at 15 minutes calculated by dye-dilution cardiac output flowmetry in comparison to blood sampling in patients undergoing hepatic resection. *Indian J Gastroenterol*. 2019;38:441–449. Available from: https://doi.org/10.1007/s12664-019-00994-z.
244. Reekers M, Simon MJG, Boer F, Mooren RAG, van Kleef JW, Dahan A, et al. Cardiovascular monitoring by pulse dye densitometry or arterial indocyanine green dilution. *Anesthesia Analgesia*. 2009;109:441–446. Available from: https://doi.org/10.1213/ane.0b013e3181a8d81f.
245. Hofer CK, Buhlmann S, Klaghofer R, Genoni M, Zollinger A. Pulsed dye densitometry with two different sensor types for cardiac output measurement after cardiac surgery: a comparison with the thermodilution technique. *Acta Anaesthesiol Scand*. 2004;48:653–657. Available from: https://doi.org/10.1111/j.1399-6576.2004.00371.x.

246. Louvaris Z, Spetsioti S, Andrianopoulos V, Chynkiamis N, Habazettl H, Wagner H, et al. Cardiac output measurement during exercise in COPD: a comparison of dye dilution and impedance cardiography. *Clin Respir J.* 2019;13:222–231. Available from: https://doi.org/10.1111/crj.13002.
247. Fischer M-O, Joosten A, Desebbe O, Boutros M, Debroczi S, Broch O, et al. Interchangeability of cardiac output measurements between non-invasive photoplethysmography and bolus thermodilution: a systematic review and individual patient data meta-analysis. *Anaesth Crit Care Pain Med.* 2020;39:75–85. Available from: https://doi.org/10.1016/j.accpm.2019.05.007.
248. Hanna Elias B, Luke Glancy D. *Practical Cardiovascular Hemodynamics with Self-Assessment Problems.* Demos Medical Publishing; 2012.
249. Gondos T, Marjanek Z, Kisvarga Z, Halász G. Precision of transpulmonary thermodilution: how many measurements are necessary? *Eur J Anaesthesiol.* 2009;26:508–512. Available from: https://doi.org/10.1097/EJA.0b013e32832a2d23.
250. Monnet X, Teboul J-L. Transpulmonary thermodilution: advantages and limits. *Crit Care.* 2017;21:147. Available from: https://doi.org/10.1186/s13054-017-1739-5.
251. Pauli C, Fakler U, Genz T, Hennig M, Lorenz H-P, Hess J. Cardiac output determination in children: equivalence of the transpulmonary thermodilution method to the direct Fick principle. *Intensive Care Med.* 2002;28:947–952. Available from: https://doi.org/10.1007/s00134-002-1334-2.
252. Litton E, Morgan M. The PiCCO monitor: a review. *Anaesth Intensive Care.* 2012;40:393–409. Available from: https://doi.org/10.1177/0310057X1204000304.
253. Vignon P. Continuous cardiac output assessment or serial echocardiography during septic shock resuscitation? *Ann Transl Med.* 2020;8:797. Available from: https://doi.org/10.21037/atm.2020.04.11. 797.
254. Beurton A, Teboul J-L, Monnet X. Transpulmonary thermodilution techniques in the haemodynamically unstable patient. *Curr OpCrit Care.* 2019;25:273–279. Available from: https://doi.org/10.1097/MCC.0000000000000608.
255. Linton RA, Band DM, Haire KM. A new method of measuring cardiac output in man using lithium dilution. *Br J Anaesth.* 1993;71:262–266. Available from: https://doi.org/10.1093/bja/71.2.262.
256. Low DA, Jones H, Cable NT, Alexander LM, Kenney WL. Historical reviews of the assessment of human cardiovascular function: interrogation and understanding of the control of skin blood flow. *Eur J Appl Physiol.* 2020;120:1–16. Available from: https://doi.org/10.1007/s00421-019-04246-y.
257. Kouadio AA, Jordana F, Koffi NJ, Le Bars P, Soueidan A. The use of laser Doppler flowmetry to evaluate oral soft tissue blood flow in humans: a review. *Arch Oral Biol.* 2018;86:58–71. Available from: https://doi.org/10.1016/j.archoralbio.2017.11.009.
258. Kubasch ML, Kubasch AS, Torres Pacheco J, Buchmann SJ, Illigens BM-W, Barlinn K, et al. Laser Doppler assessment of vasomotor axon reflex responsiveness to evaluate neurovascular function. *Front Neurol.* 2017;8:370. Available from: https://doi.org/10.3389/fneur.2017.00370.
259. Zegarra-Parodi R, Snider EJ, Park PYS, Degenhardt BF. Laser Doppler flowmetry in manual medicine research. *J Am Osteopath Assoc.* 2014;114:908–917. Available from: https://doi.org/10.7556/jaoa.2014.178.
260. Wang CP. Laser Doppler velocimetry. *J Quant Spectrosc Radiat Transf.* 1988;40:309–319. Available from: https://doi.org/10.1016/0022-4073(88)90122-7.
261. Newson TP, Obeid A, Wolton RS, Boggett D, Rolfe P. Laser Doppler velocimetry: the problem of fibre movement artefact. *J Biomed Eng.* 1987;9:169–172. Available from: https://doi.org/10.1016/0141-5425(87)90030-6.
262. Olkkonen H, Simonen P, Immonen T, Jääskeläinen T, Fräki J. Dual diode transducer for laser Doppler skin blood velocimeter. *J Biomed Eng.* 1984;6:75–77. Available from: https://doi.org/10.1016/0141-5425(84)90014-1.
263. Shepherd AP, Öberg PÅ. *Laser-Doppler Blood Flowmetry.* Springer Science & Business Media; 2013.
264. Salvi P, Lio G, Labat C, Ricci E, Pannier B, Benetos A. Validation of a new non-invasive portable tonometer for determining arterial pressure wave and pulse wave velocity: the PulsePen device. *J Hypertens.* 2004;22:2285–2293. Available from: https://doi.org/10.1097/00004872-200412000-00010.
265. Davies JM, Bailey MA, Griffin KJ, Scott DJA. Pulse wave velocity and the non-invasive methods used to assess it: Complior, SphygmoCor, Arteriograph and Vicorder. *Vascular.* 2012;20:342–349. Available from: https://doi.org/10.1258/vasc.2011.ra0054.
266. Grillo A, Parati G, Rovina M, Moretti F, Salvi L, Gao L, et al. Short-term repeatability of noninvasive aortic pulse wave velocity assessment: comparison between methods and devices. *Am J Hypertens.* 2017;31:80–88. Available from: https://doi.org/10.1093/ajh/hpx140.
267. Harbaoui B, Courand P-Y, Cividjian A, Lantelme P. Development of coronary pulse wave velocity: new pathophysiological insight into coronary artery disease. *J Am Heart Assoc.* 2017;6. Available from: https://doi.org/10.1161/JAHA.116.004981.
268. Ma Y, Choi J, Hourlier-Fargette A, Xue Y, Chung HU, Lee JY, et al. Relation between blood pressure and pulse wave velocity for human arteries. *Proc Natl Acad Sci USA.* 2018;115:11144–11149. Available from: https://doi.org/10.1073/pnas.1814392115.
269. Benas D, Kornelakis M, Triantafyllidi H, Kostelli G, Pavlidis G, Varoudi M, et al. Pulse wave analysis using the Mobil-O-Graph, Arteriograph and Complior device: a comparative study. *Blood Press.* 2019;28:107–113. Available from: https://doi.org/10.1080/08037051.2018.1564236.
270. Milan A, Zocaro G, Leone D, Tosello F, Buraioli I, Schiavone D, et al. Current assessment of pulse wave velocity: comprehensive review of validation studies. *J Hypertens.* 2019;37:1547–1557. Available from: https://doi.org/10.1097/HJH.0000000000002081.
271. Weber T, Wassertheurer S, Hametner B, Parragh S, Eber B. Noninvasive methods to assess pulse wave velocity: comparison with the invasive gold standard and relationship with organ damage. *J Hypertens.* 2015;33:1023–1031. Available from: https://doi.org/10.1097/HJH.0000000000000518.
272. Torp KD, Modi P, Simon LV. *Pulse Oximetry.* StatPearls, Treasure Island, FL: StatPearls Publishing; 2020.
273. Van Meter A, Williams U, Zavala A, Kee J, Rebello E, Tsai J, et al. Beat to beat: a measured look at the history of pulse oximetry. *J Anesth Hist.* 2017;3:24–26. Available from: https://doi.org/10.1016/j.janh.2016.12.003.

274. Hess DR. Pulse oximetry: beyond SpO_2. *Respir Care*. 2016;61:1671–1680. Available from: https://doi.org/10.4187/respcare.05208.
275. Nitzan M, Romem A, Koppel R. Pulse oximetry: fundamentals and technology update. *Med Devices (Auckl)*. 2014;7:231–239. Available from: https://doi.org/10.2147/MDER.S47319.
276. Chan ED, Chan MM, Chan MM. Pulse oximetry: understanding its basic principles facilitates appreciation of its limitations. *Respir Med*. 2013;107:789–799. Available from: https://doi.org/10.1016/j.rmed.2013.02.004.
277. Kyriacou PA. Pulse oximetry in the oesophagus. *Physiol Meas*. 2005;27:R1–R35. Available from: https://doi.org/10.1088/0967-3334/27/1/R01.
278. Deuchar DC. Ballistocardiography. *Br Heart J*. 1967;29:285–288. Available from: https://doi.org/10.1136/hrt.29.3.285.
279. Da He D, Winokur ES, Sodini CG. A continuous, wearable, and wireless heart monitor using head ballistocardiogram (BCG) and head electrocardiogram (ECG). *Annu Int Conf IEEE Eng Med Biol Soc*. 2011;2011:4729–4732. Available from: https://doi.org/10.1109/IEMBS.2011.6091171.
280. He DD, Winokur ES, Sodini CG. An ear-worn continuous ballistocardiogram (BCG) sensor for cardiovascular monitoring. *Annu Int Conf IEEE Eng Med Biol Soc*. 2012;2012:5030–5033. Available from: https://doi.org/10.1109/EMBC.2012.6347123.
281. Inan OT. Recent advances in cardiovascular monitoring using ballistocardiography. *Annu Int Conf IEEE Eng Med Biol Soc*. 2012;2012:5038–5041. Available from: https://doi.org/10.1109/EMBC.2012.6347125.
282. Vogt E, MacQuarrie D, Neary JP. Using ballistocardiography to measure cardiac performance: a brief review of its history and future significance. *Clin Physiol Funct Imaging*. 2012;32:415–420. Available from: https://doi.org/10.1111/j.1475-097X.2012.01150.x.
283. Inan OT, Migeotte P-F, Park K-S, Etemadi M, Tavakolian K, Casanella R, et al. Ballistocardiography and seismocardiography: a review of recent advances. *IEEE J Biomed Health Inf*. 2015;19:1414–1427. Available from: https://doi.org/10.1109/JBHI.2014.2361732.
284. Morra S, Gauthey A, Hossein A, Rabineau J, Racape J, Gorlier D, et al. Influence of sympathetic activation on myocardial contractility measured with ballistocardiography and seismocardiography during sustained end-expiratory apnea. *Am J Physiol Regul Integr Comp Physiol*. 2020;319:R497–R506. Available from: https://doi.org/10.1152/ajpregu.00142.2020.
285 Zanetti JM, Tavakolian K. Seismocardiography: past, present and future. In: *2013 35th Annual International Conference of the IEEE Engineering in Medicine and Biology Society (EMBC)*, Osaka, IEEE; 2013, p. 7004–7007. https://doi.org/10.1109/EMBC.2013.6611170.
286. Taebi A, Solar B, Bomar A, Sandler R, Mansy H. Recent advances in seismocardiography. *Vibration*. 2019;2:64–86. Available from: https://doi.org/10.3390/vibration2010005.
287. Bombardini T. Operator independent left ventricular function monitoring during pharmacological stress echo with the new peak transcutaneous acceleration signal. *Heart*. 2001;85:286–289. Available from: https://doi.org/10.1136/heart.85.3.286.
288. Bordachar P, Labrousse L, Ploux S, Thambo J-B, Lafitte S, Reant P, et al. Validation of a new noninvasive device for the monitoring of peak endocardial acceleration in pigs: implications for optimization of pacing site and configuration. *J Cardiovasc Electrophysiol*. 2008;19:725–729. Available from: https://doi.org/10.1111/j.1540-8167.2008.01105.x.
289. Tavakolian K, Blaber AP, Ngai B, Kaminska B. Estimation of hemodynamic parameters from seismocardiogram. *Comput Cardiol*. 2010; 1055–1058.
290 Di Rienzo M, Meriggi P, Rizzo F, Vaini E, Faini A, Merati G, et al. A wearable system for the seismocardiogram assessment in daily life conditions. In: *2011 Annual International Conference of the IEEE Engineering in Medicine and Biology Society*. Boston, MA: IEEE; 2011, p. 4263–4266. https://doi.org/10.1109/IEMBS.2011.6091058.
291. Marcelli E, Capucci A, Minardi G, Cercenelli L. Multi-sense CardioPatch: a wearable patch for remote monitoring of electro-mechanical cardiac activity. *ASAIO J*. 2017;63:73–79. Available from: https://doi.org/10.1097/MAT.0000000000000446.
292. Andreozzi E, Fratini A, Esposito D, Naik G, Polley C, Gargiulo GD, et al. Forcecardiography: a novel technique to measure heart mechanical vibrations onto the chest wall. *Sensors*. 2020;20:3885. Available from: https://doi.org/10.3390/s20143885.
293. Rickards AF, Bombardini T, Corbucci G, Plicchi G, On Behalf of The Multicenter Pea Study Group. An implantable intracardiac accelerometer for monitoring myocardial contractility. *Pacing Clin Electro*. 1996;19:2066–2071. Available from: https://doi.org/10.1111/j.1540-8159.1996.tb03280.x.
294. Plicchi G, Marcelli E, Parlapiano M, Bombardini T. PEA I and PEA II based implantable haemodynamic monitor: pre clinical studies in sheep. *Europace*. 2002;4:49–54. Available from: https://doi.org/10.1053/eupc.2001.0204.
295. Plicchi G, Marcelli E, Cercenelli L. Implantable sensors to assess cardiac function. *J Mech Med Biol*. 2006;06:81–89. Available from: https://doi.org/10.1142/S0219519406001777.
296. Marcelli E, Vanoli E, Mattera GG, Gaggini G, Cercenelli L, Plicchi GAN. Endocardial acceleration sensor for monitoring cardiac function of ischemic hearts. *J Mech Med Biol*. 2006;06:75–80. Available from: https://doi.org/10.1142/S0219519406001753.
297. Delnoy PP, Marcelli E, Oudeluttikhuis H, Nicastia D, Renesto F, Cercenelli L, et al. Validation of a peak endocardial acceleration-based algorithm to optimize cardiac resynchronization: early clinical results. *Europace*. 2008;10:801–808. Available from: https://doi.org/10.1093/europace/eun125.
298. Kaszala K, Ellenbogen KA. Device sensing: sensors and algorithms for pacemakers and implantable cardioverter defibrillators. *Circulation*. 2010;122:1328–1340. Available from: https://doi.org/10.1161/CIRCULATIONAHA.109.919704.
299. Duncker D, Delnoy PP, Nägele H, Mansourati J, Mont L, Anselme F, et al. First clinical evaluation of an atrial haemodynamic sensor lead for automatic optimization of cardiac resynchronization therapy. *Europace*. 2016;18:755–761. Available from: https://doi.org/10.1093/europace/euv114.
300. Corazza I, Diemberger I, Martignani C, Ziacchi M, Rossi PL, Lombi A, et al. Wireless endocardial atrial (and ventricular) sensing with no implanted power source: a proposal. *J Med Syst*. 2019;43:159. Available from: https://doi.org/10.1007/s10916-019-1277-6.

301. Marcelli E, Spolzino S, Cercenelli L, Cappello A, Bagnoli P, Costantino ML, et al. *Assessment of cardiac apex kinematics using a real-time 3D magnetic tracking system. 2008 Computers in Cardiology*. Bologna, Italy: IEEE; 2008:397–400. Available from: https://doi.org/10.1109/CIC.2008.4749062.
302. Arts T, Reneman RS, Veenstra PC. A model of the mechanics of the left ventricle. *Ann Biomed Eng*. 1979;7:299–318. Available from: https://doi.org/10.1007/BF02364118.
303. Guccione JM, Waldman LK, McCulloch AD. Mechanics of active contraction in cardiac muscle: Part II—cylindrical models of the systolic left ventricle. *J Biomech Eng*. 1993;115:82–90. Available from: https://doi.org/10.1115/1.2895474.
304. Bagnoli P, Malagutti N, Gastaldi D, Marcelli E, Lui E, Cercenelli L, et al. Computational finite element model of cardiac torsion. *Int J Artif Organs*. 2011;34:44–53. Available from: https://doi.org/10.5301/IJAO.2011.6313.
305. Sengupta PP, Tajik AJ, Chandrasekaran K, Khandheria BK. Twist mechanics of the left ventricle. *JACC: Cardiovas Imaging*. 2008;1:366–376. Available from: https://doi.org/10.1016/j.jcmg.2008.02.006.
306. Bloechlinger S, Grander W, Bryner J, Dünser MW. Left ventricular rotation: a neglected aspect of the cardiac cycle. *Intensive Care Med*. 2011;37:156–163. Available from: https://doi.org/10.1007/s00134-010-2053-8.
307. Marcelli E, Cercenelli L, Musaico M, Bagnoli P, Costantino ML, Fumero R, et al. Assessment of cardiac rotation by means of gyroscopic sensors. *2008 Comp Cardiol*. 2008;389–392. Available from: https://doi.org/10.1109/CIC.2008.4749060.
308. Cercenelli L, Marcelli E. Cardiac Apex rotation assessed by an implantable gyro sensor: correlation with a Lv pressure-derived myocardial performance index in experimentally induced ischemia. *J Mech Med Biol*. 2015;15:1540013. Available from: https://doi.org/10.1142/S0219519415400138.
309. Marcelli E, Cercenelli L, Parlapiano M, Fumero R, Bagnoli P, Costantino ML, et al. Effect of right ventricular pacing on cardiac Apex rotation assessed by a gyroscopic sensor. *ASAIO J*. 2007;53:304–309. Available from: https://doi.org/10.1097/MAT.0b013e31805370e3.
310. Marcelli E, Cercenelli L. An implantable sensorized lead for continuous monitoring of cardiac Apex rotation. *Sens (Basel)*. 2018;18. Available from: https://doi.org/10.3390/s18124195.
311. Halvorsen PS, Espinoza A, Fleischer LA, Elle OJ, Hoff L, Lundblad R, et al. Feasibility of a three-axis epicardial accelerometer in detecting myocardial ischemia in cardiac surgical patients. *J Thorac Cardiovas Surg*. 2008;136:1496–1502. Available from: https://doi.org/10.1016/j.jtcvs.2008.08.043.
312. Halvorsen PS, Remme EW, Espinoza A, Skulstad H, Lundblad R, Bergsland J, et al. Automatic real-time detection of myocardial ischemia by epicardial accelerometer. *J Thorac Cardiovas Surg*. 2010;139:1026–1032. Available from: https://doi.org/10.1016/j.jtcvs.2009.05.031.
313. Marcelli E, Bortolani B, Corazza I, Cercenelli L. A novel sensorized heart valve prosthesis: preliminary in vitro evaluation. *Sens (Basel)*. 2018;18. Available from: https://doi.org/10.3390/s18113905.
314. Wang Y-J, Chen C-H, Sue C-Y, Lu W-H, Chiou Y-H. Estimation of blood pressure in the radial artery using strain-based pulse wave and photoplethysmography sensors. *Micromachines*. 2018;9:556. Available from: https://doi.org/10.3390/mi9110556.
315. Ortega R, Connor C, Kotova F, Deng W, Lacerra C. Use of pressure transducers. *N Engl J Med*. 2017;376:e26. Available from: https://doi.org/10.1056/NEJMvcm1513613.
316. Park DY, Joe DJ, Kim DH, Park H, Han JH, Jeong CK, et al. Self-powered real-time arterial pulse monitoring using ultrathin epidermal piezoelectric sensors. *Adv Mater*. 2017;29:1702308. Available from: https://doi.org/10.1002/adma.201702308.
317. Gautschi G. Piezoelectric sensors. In: Gautschi G, ed. *Piezoelectric Sensorics: Force Strain Pressure Acceleration and Acoustic Emission Sensors Materials and Amplifiers*. Berlin, Heidelberg: Springer; 2002:73–91. Available from: https://doi.org/10.1007/978-3-662-04732-3_5.
318. Kim J, Chou E, Le J, Wong S, Chu M, Khine M. Soft wearable pressure sensors for beat-to-beat blood pressure monitoring. *Adv Healthc Mater*. 2019;8:1900109. Available from: https://doi.org/10.1002/adhm.201900109.
319. Bingger P, Zens M, Woias P. Highly flexible capacitive strain gauge for continuous long-term blood pressure monitoring. *Biomed Microdev*. 2012;14:573–581. Available from: https://doi.org/10.1007/s10544-012-9636-9.
320. Rao KS, Samyuktha W, Vardhan DV, Naidu BG, Kumar PA, Sravani KG, et al. Design and sensitivity analysis of capacitive MEMS pressure sensor for blood pressure measurement. *Microsyst Technol*. 2020;26:2371–2379. Available from: https://doi.org/10.1007/s00542-020-04777-x.
321. Pruvost M, Smit WJ, Monteux C, Poulin P, Colin A. Polymeric foams for flexible and highly sensitive low-pressure capacitive sensors. *Npj Flex Electron*. 2019;3:1–6. Available from: https://doi.org/10.1038/s41528-019-0052-6.
322. Song P, Ma Z, Ma J, Yang L, Wei J, Zhao Y, et al. Recent progress of miniature MEMS pressure sensors. *Micromachines*. 2020;11. Available from: https://doi.org/10.3390/mi11010056.
323. Haseda Y, Bonefacino J, Tam H-Y, Chino S, Koyama S, Ishizawa H. Measurement of pulse wave signals and blood pressure by a plastic optical fiber FBG sensor. *Sens (Basel)*. 2019;19. Available from: https://doi.org/10.3390/s19235088.
324. Liu Y, Jing Z, Li R, Zhang Y, Liu Q, Li A, et al. Miniature fiber-optic tip pressure sensor assembled by hydroxide catalysis bonding technology. *Opt Exp*. 2020;28:948–958. Available from: https://doi.org/10.1364/OE.380589.
325. Wu T, Shen J, Li Z, Zou T, Xin W, Xing F, et al. Graphene-based ultrasensitive optical microfluidic sensor for the real-time and label-free monitoring of simulated arterial blood flow. *Opt Exp*. 2020;28:16594–16604. Available from: https://doi.org/10.1364/OE.392993.
326. Satake S, Shimura T, Ono T, Shimomura K, Takenoshita S, Kono K. Noninvasive continuous blood pressure monitoring using microelectromechanical system technology. *Blood Press Monit*. 2019;24:155–159. Available from: https://doi.org/10.1097/MBP.0000000000000380.
327. Marcelli Emanuela, Cercenelli Laura, Bortolani Barbara, et al. A Novel Non-Invasive Device for the Assessment of Central Venous Pressure in Hospital, Office and Home. *Medical Devices (Auckl)*. 2021;14:141–154. Available from: https://doi.org/10.2147/mder.s307775.

PART VII

Future Directions of Cardiovascular Technology

Chapter 35

Cardiovascular Device Markets: Where Do We Go Now?

Michael Austriaco[1] and Suzanne Ratzloff[2]

[1]*Cleveland Clinic, Cleveland, OH, United States,* [2]*Health Research International, Lakewood, OH, United States*

Chapter Outline

"But where are we going? And what kind of future will we discover there? Surprisingly, the answers lie in our past."

Dame Judi Dench, Spaceship Earth (EPCOT)

Background

Physicians and other practitioners of modern healthcare are often characterized as being relatively conservative and resistant to change, in part due to the very nature of the scientific method, evidence-based medicine, and the Hippocratic Oath: it takes time to build a sufficiently compelling reason to alter established clinical practice. However, cardiovascular medicine, and its application of medical devices in particular, represents one of the most progressive fields driving the transformation of how patients are diagnosed and treated. This is readily demonstrated via the introduction and rapid adoption of innovative device-based solutions by cardiothoracic surgeons and interventional cardiologists over the past several decades. Such advancements are perhaps not entirely surprising in that cardiovascular disease continues to represent the leading cause of mortality globally, claiming approximately 17.8 million lives worldwide in 2017 and representing the underlying cause of death for nearly 860,000 Americans that same year, accounting for more deaths than all forms of cancer and chronic lower respiratory disease combined.[1] Thus the persisting unmet needs of millions of cardiac patients around the world serve as the mother of invention for countless clinical visionaries, biomedical engineers, and entrepreneurs alike.

In order to understand how cardiovascular device markets might develop in the future, it is prudent to examine their history and evolution to the present. The drive toward less invasive therapy has been prominent in the transformation of cardiovascular device markets, notably in the shift from open surgical approaches to percutaneous, transcatheter devices, and techniques that solve the same problems. Such solutions have not only reduced the overall morbidity of intervention, but have often served to expand the treatable patient population. Two of the most prominent examples of success in the introduction of minimally invasive options for cardiovascular conditions include the emergence of angioplasty and stenting in the treatment of coronary artery disease (CAD), and the more recent introduction of catheter-based aortic valve prostheses used in the management of severe aortic stenosis.

Advances in Cardiovascular Technology. DOI: https://doi.org/10.1016/B978-0-12-816861-5.00026-5

Trends in coronary revascularization

Two of the most significant advances in the modern treatment of cardiovascular disease were pioneered at Cleveland Clinic, including the development of coronary angiography by F. Mason Sones in 1958 and the subsequent development and refinement of coronary artery bypass graft (CABG) surgery, led by Dr. René Favaloro in 1967.[2] These two novel techniques supported the creation of numerous sustainable medical device markets, including those for angiography catheters, heart-lung machines and their related disposables, and more recently, endoscopic vessel harvesting instruments and off-pump, beating heart surgical systems. At the same time, these procedures also paved the way for continued innovation and a transition toward less invasive therapeutic approaches that forged the basis for the rapidly expanding field of interventional cardiology. This evolution was unquestionably buoyed by the introduction of balloon angioplasty through the historic efforts of Dr. Andreas Grüntzig in 1974[3] and his adaptation of the technique to perform the first percutaneous transluminal coronary angioplasty (PTCA) in 1977.[4]

As Fig. 35.1 shows, clinical experience in refining PTCA and its appropriate indications led to rapid adoption and expansion of the treated patient population undergoing myocardial revascularization, and by the mid-1980s, the minimally invasive technique accounted for more than one third of coronary revascularization procedures performed in the United States. By 1993, just 16 years after Dr. Grüntzig's pioneering case, the transcatheter approach had overtaken CABG as the predominant mode of surgical therapy for patients with CAD. Subsequent adoption has firmly established PTCA as the new standard of care, representing roughly three-quarters of the revascularization caseload since the mid-2000s.[5,6]

The PTCA era ushered in a string of technological milestones that served to propel both procedural growth and market value through the sale of coronary stents. While the first clinical use of coronary stenting was reported by Sigwart et al. in 1986,[7] it was not until July 1993 that the US FDA granted approval for the first such device, the Gianturco-Roubin Coronary Flex-Stent (although this honor is frequently attributed to Cordis Corporation's FDA approval of the Palmaz-Schatz stent, which was arguably more widely used, in August 1994). The ability of these tiny implantable scaffolds to reduce the risk of acute restenosis due to vessel recoil and improve clinical outcomes helped

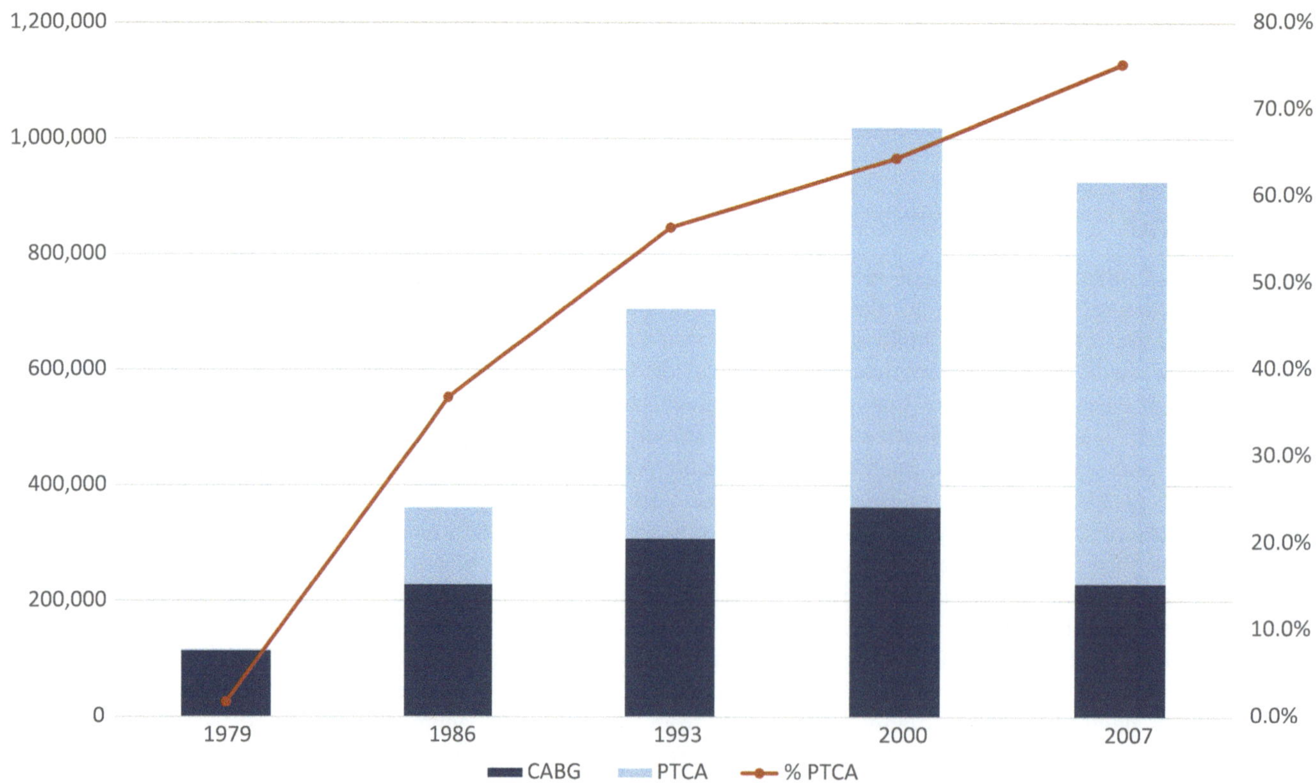

FIGURE 35.1 Trend in US coronary revascularization procedures, 1979–2007. *CABG*, coronary artery bypass graft; *PTCA*, percutaneous transluminal coronary angioplasty. *Data from Gillum RF. et al. Coronary heart disease and related procedures. National Hospital Discharge Survey data. Circulation 1989;79(6 Pt 2):113–118. https://pubmed.ncbi.nlm.nih.gov/2524295/ and HCUP Nationwide Inpatient Sample (NIS). Healthcare Cost and Utilization Project (HCUP). 1993–2007. Agency for Healthcare Research and Quality, Rockville, MD. http://www.hcup-us.ahrq.gov/nisoverview.jsp.*

drive the acceleration of coronary interventions in the cath lab. The adoption of coronary stenting within the PTCA caseload, now more commonly referred to as percutaneous coronary intervention (PCI), was even more rapid than that of PTCA within the revascularization caseload: it took less than 10 years following FDA approval of the first coronary stent to realize market penetration of greater than 90%, and utilization remains ubiquitous to this day.

While acute vessel reclosure was largely solved by coronary stenting, the problem of in-stent restenosis (ISR) secondary to intimal hyperplasia remained persistent with the use of bare metal stents (BMS) in approximately 16%–44% of cases.[8] This unmet clinical need spurred the development and testing of a broad range of interventional therapies targeting the treatment or prevention of ISR, including intravascular brachytherapy, therapeutic ultrasound, cutting balloons, and other plaque modification and debulking techniques, among other approaches. However, drug-eluting stents (DES), first implanted in 1999 by Eduardo Sousa,[9] ultimately became the therapeutic option of choice by reliably reducing restenosis rates to the single digits. With DES, both commercialization timelines and market adoption were even more condensed than with BMS: the first coronary DES, Cordis' Cypher sirolimus-eluting stent, obtained FDA approval in April 2003 just 4 years following first-in-man use. Further, according to data from the HCUP NIS database,[10] DES penetration of the coronary stent caseload rocketed to account for more than three-quarters of the total in the second year of US commercialization. Following the ultimate resolution of safety concerns relating to reports of late-stent thrombosis associated with first-generation devices[11] and temporary dip in penetration, DES utilization climbed back to exceed 80% of coronary stent procedures by 2014, as shown in Fig. 35.2.

These procedural trends were accompanied by significant commercial accomplishments from the companies supporting the investment in research and development of coronary stents, namely Cordis Corporation, then a subsidiary of Johnson & Johnson (now a subsidiary of Cardinal Health and, since 2011, no longer a player in the coronary stent market it effectively revolutionized twice), Boston Scientific, Guidant Corporation (now Abbott Vascular), and Medtronic. These four players dominated the coronary stent market and, by way of effective product bundling, the entire interventional cardiology market, throughout the late 1990s and early 2000s. As illustrated in Fig. 35.3, the global market for coronary stent products increased in value by more than 2.6-fold between the end of the BMS era in 2002 and 2006, which served as a high watermark driven by DES adoption prior to the aforementioned late-stent thrombosis scare.

This rapid growth in market value was achieved through a combination of both swift clinical adoption and, as is the case with many new medical device products, premium product pricing. Whereas BMS prices in the United States around the time of DES market introduction had fallen to less than $1000 per unit, DES products initially sold for well

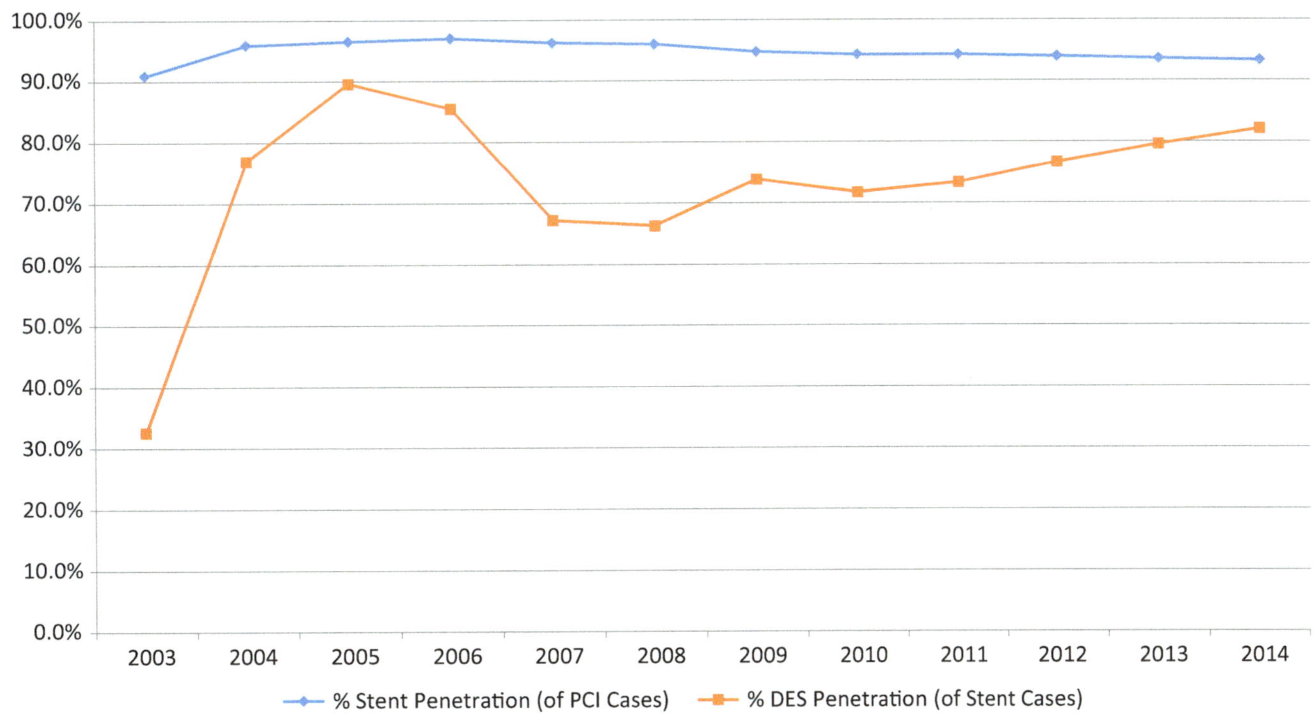

FIGURE 35.2 Trend in US coronary stent and drug-eluting stent market penetration, 2003–2014. *PCI*, percutaneous coronary intervention; *DES*, drug-eluting stent. *Data from HCUP NIS.*

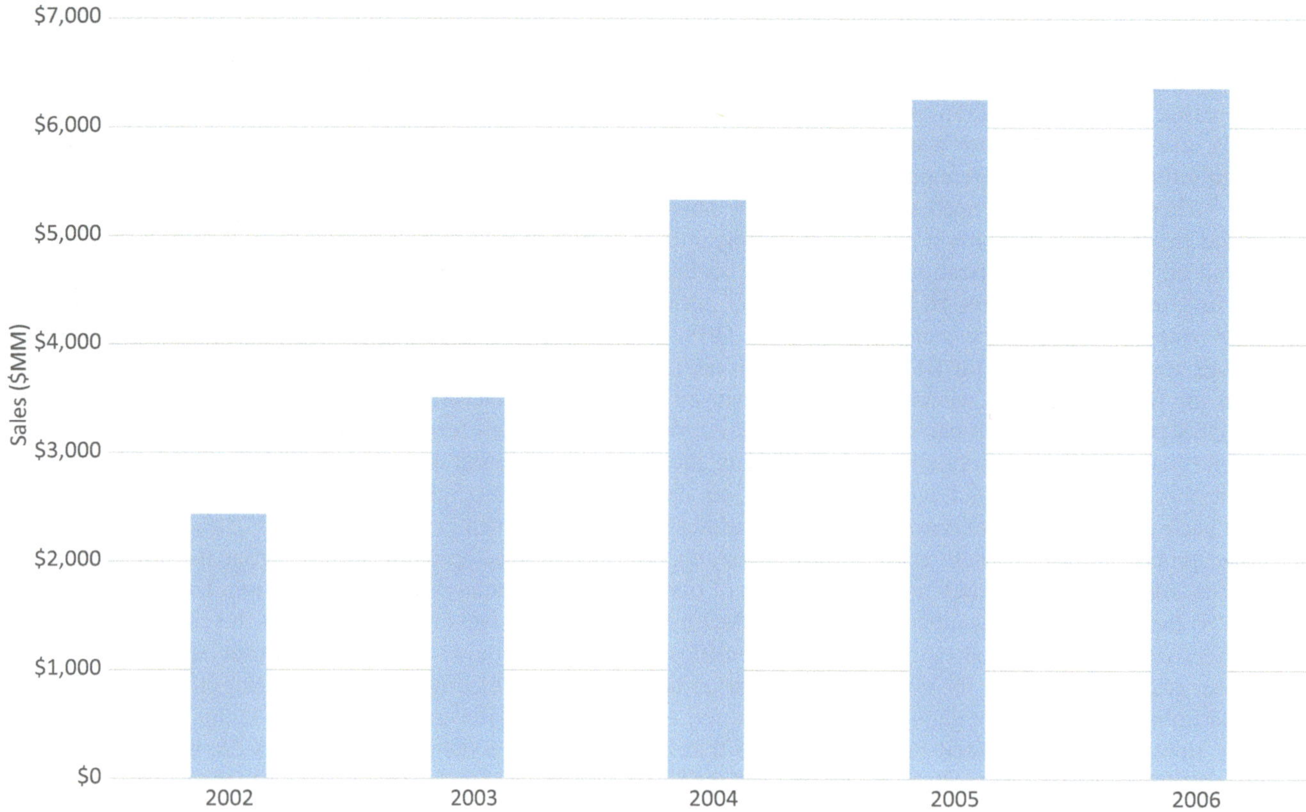

FIGURE 35.3 Global coronary stent market, 2002–2006. *Courtesy Health Research International.*

over $2000. Nevertheless, the clinical benefits brought on by DES, notably a significant reduction in the need for repeat target vessel revascularization, demonstrated cost-effectiveness of the new technology despite the higher index cost associated with its use.[12]

As will be illustrated later in this chapter, it is not uncommon for cardiovascular devices to be sold at market prices that some may view as exorbitant, especially in the case of more "disruptive" technologies that help overcome significant clinical challenges and/or enable the effective treatment of patient populations with otherwise limited medical options. While not every technology is supported with compelling cost-effectiveness and healthcare economic outcomes data, those that are not will ultimately face challenges obtaining adequate reimbursement coverage from public and commercial payors. As providers, including both hospital facilities and physicians specializing in the treatment of cardiovascular disease, are generally (though not always) less likely to pay for and broadly utilize technologies that will not be covered or will result in a financial loss, pricing and reimbursement factors can have a measurable impact on actual clinical adoption. This fact has prompted many investors supporting new medical device development, notably venture capitalists, to demand that startup companies and the entrepreneurs driving them have a solid understanding and clear path to obtaining adequate reimbursement coverage, even in the earliest stages of product development.

A variety of factors helped set the stage for the success of DES, and PCI in general, including but not limited to a large addressable patient population; the introduction of a new and clinically effective solution (which improved significantly through the evolution from plain old balloon angioplasty to BMS and subsequently DES); and strong patient demand for less invasive therapy. The latter point represents a particularly impactful driver in cardiovascular device markets, especially where conventional open surgical options frequently necessitate median sternotomy and cardiopulmonary bypass.

Trends in aortic valve replacement

A more contemporary example of this trend, and perhaps one of the most successful product segments in the history of cardiovascular device markets can be seen with the emergence of transcatheter aortic valve replacement (TAVR), also referred to as transcatheter aortic valve implantation, in the treatment of symptomatic aortic stenosis. Conventional management of this condition has involved open surgical excision of the diseased aortic valve and replacement with a mechanical or tissue-based valve prosthesis, that is, surgical aortic valve replacement (SAVR). While percutaneous balloon aortic valvuloplasty (BAV) has been offered as a less invasive therapy since the 1980s, this option has primarily been reserved for elderly patients who are not surgical candidates due to limited durability and questionable mortality benefit.[13] Similar to the augmentation of PTCA with coronary stent placement, TAVR represents a somewhat natural extension of BAV with the addition of a stented valve prosthesis placed within the native valve annulus. While this approach was first conceived in 1992,[14] it was a full decade later before the first human TAVR procedure was performed in 2002[15] and thereafter, another 5 years before the first TAVR devices were commercialized following the receipt of CE Mark approval.[16,17]

In the United States, FDA approval of the first TAVR device was granted in late 2011, with initial indications limited to inoperable patients. As operator experience grew and continued research and development efforts helped optimize devices, notably with respect to reduced delivery profiles (from 24F initially down to 14F), positive clinical outcomes served to expand FDA-labeled indications across the risk continuum of patients with symptomatic severe aortic stenosis to include those at high surgical risk (2012), intermediate surgical risk (2016), and finally low surgical risk (2019). This expansion propelled growth of TAVR procedures such that by 2019, approximately 8 years after initial FDA approval, the minimally invasive transcatheter caseload overtook SAVR as the predominant mode of therapy for the treatment of symptomatic severe aortic stenosis. This conversion occurred relatively rapidly, especially compared with the transition away from CABG toward PCI, which as mentioned previously took approximately 16 years, or twice the amount of time for TAVR to exceed SAVR. According to registry data from the STS and ACC, the combined performance of surgical and TAVR procedures in the United States grew from approximately 67,000 in 2012 to more than 130,000 in 2019,[18] as shown in Fig. 35.4. This represents compound annual growth of 9.9% over the 7-year period. During that same time, the total population of the United States grew by less than 1% annually, and by 3.3% for the elderly population age 65 years and older,[19] which accounts for the majority of both SAVR and especially TAVR patients. Thus the introduction of TAVR did not just cannibalize the SAVR caseload, but significantly expanded the treated population of patients with symptomatic severe aortic stenosis.

As the clinical adoption of TAVR was brisk, the commercial impact was even more so, owing in part to the average selling price of TAVR devices, which at approximately $32,000 in the United States is well over sixfold that of surgical

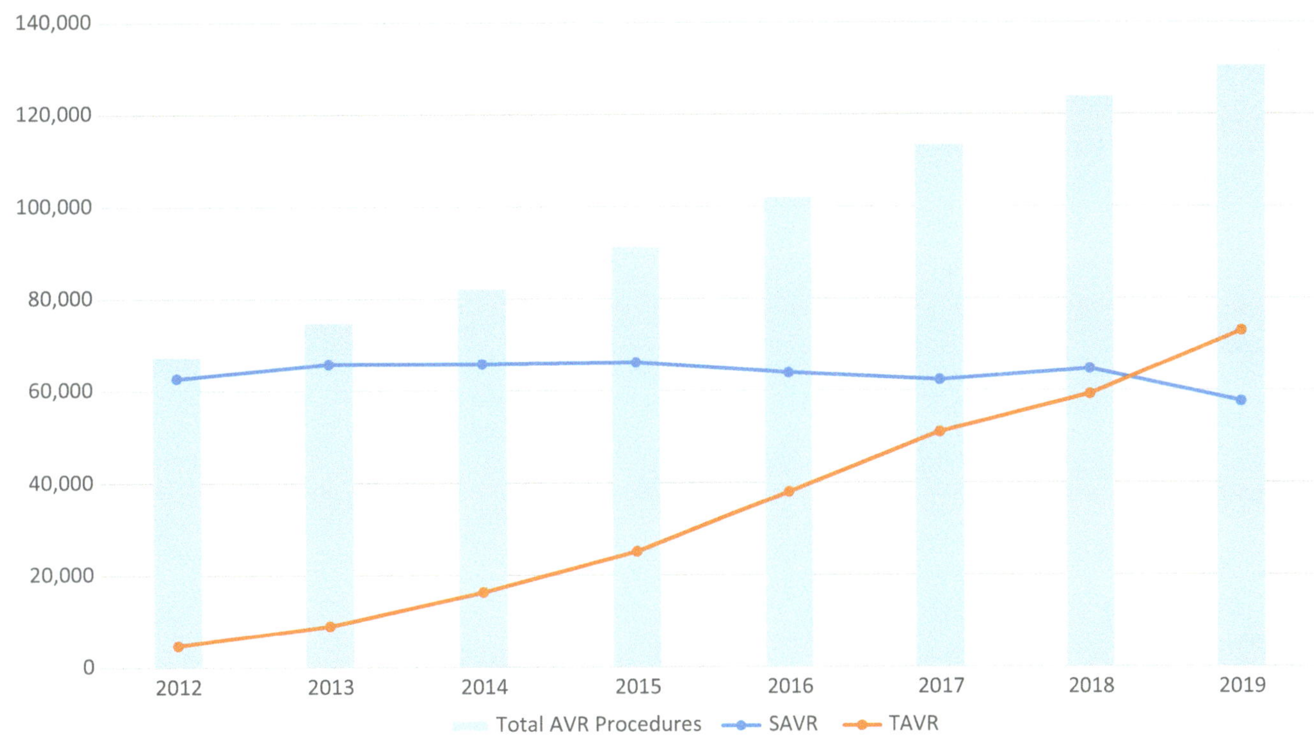

FIGURE 35.4 Trend in US SAVR and TAVR procedures, 2012–2019. *SAVR*, surgical aortic valve replacement; *TAVR*, transcatheter aortic valve replacement. *Data from Carroll et al., STS-ACC TVT registry of transcatheter aortic valve replacement. J Am Coll Cardiol. 2020;76(21):2492–2516. Available from: https://doi.org/10.1016/j.jacc.2020.09.595. https://pubmed.ncbi.nlm.nih.gov/33213729/.*

aortic valve prosthesis (~$5000).[20] While this pricing differential is orders of magnitude greater than the roughly 100% premium that DES carried over BMS upon market introduction, data regarding the cost-effectiveness of TAVR has been reported as "uniformly positive" to date across both high-risk and intermediate-risk populations. Moreover, it would appear that both patient demand and clinical preferences will continue to support and even magnify the dominance of TAVR into the foreseeable future.

Unlike the coronary stent market, where both BMS and DES experienced significant price erosion over time, the leading TAVR manufacturers have been able to maintain their premium price points, especially in the United States due in part to limited competition. Until April 2019, Edwards Lifesciences and Medtronic were the only two companies with FDA approval for TAVR systems. However, the new market entrant, Boston Scientific, initiated a global recall and discontinuation of its LOTUS Edge device in November 2020,[21] leaving the US TAVR market a duopoly for the time being. Even outside the US, Edwards and Medtronic continue to control the market, as their Sapien and CoreValve product platforms, respectively, have benefitted from technological refinements and a far greater evidence base that clinicians have come to rely on. Fig. 35.5 illustrates the growth in global TAVR device sales from 2012 to 2019, during which time revenues are estimated to have grown more than 26% annually to exceed $4 billion.

Lather, rinse, and repeat?

These prominent examples of cardiovascular device market success have certainly been applied elsewhere in the body, although not always with the same outcomes. Although it is arguably true that endovascular techniques have become the standard of care over surgical bypass as a first line treatment of most patients with peripheral arterial disease (PAD), stenting does not enjoy the same level of near universal adoption as seen in the treatment of CAD. Similarly, and perhaps even more poignantly, transcatheter valve replacement technology has been developed for valve positions beyond the aortic, but remains largely investigational in these applications. Both of these examples demonstrate the relative importance of anatomic variability and its potential impact on the translation of technology platforms across similar disease states. For instance, vessel tortuosity and compressive forces from bending of the knee has factored into the relative lack of durable patency following stent placement in the superficial femoral and popliteal arteries. Similarly, whereas familiar retrograde femoral access has provided direct access for the relatively straightforward delivery and placement of TAVR devices, the atrioventricular valves present with considerably more challenging anatomy due to the

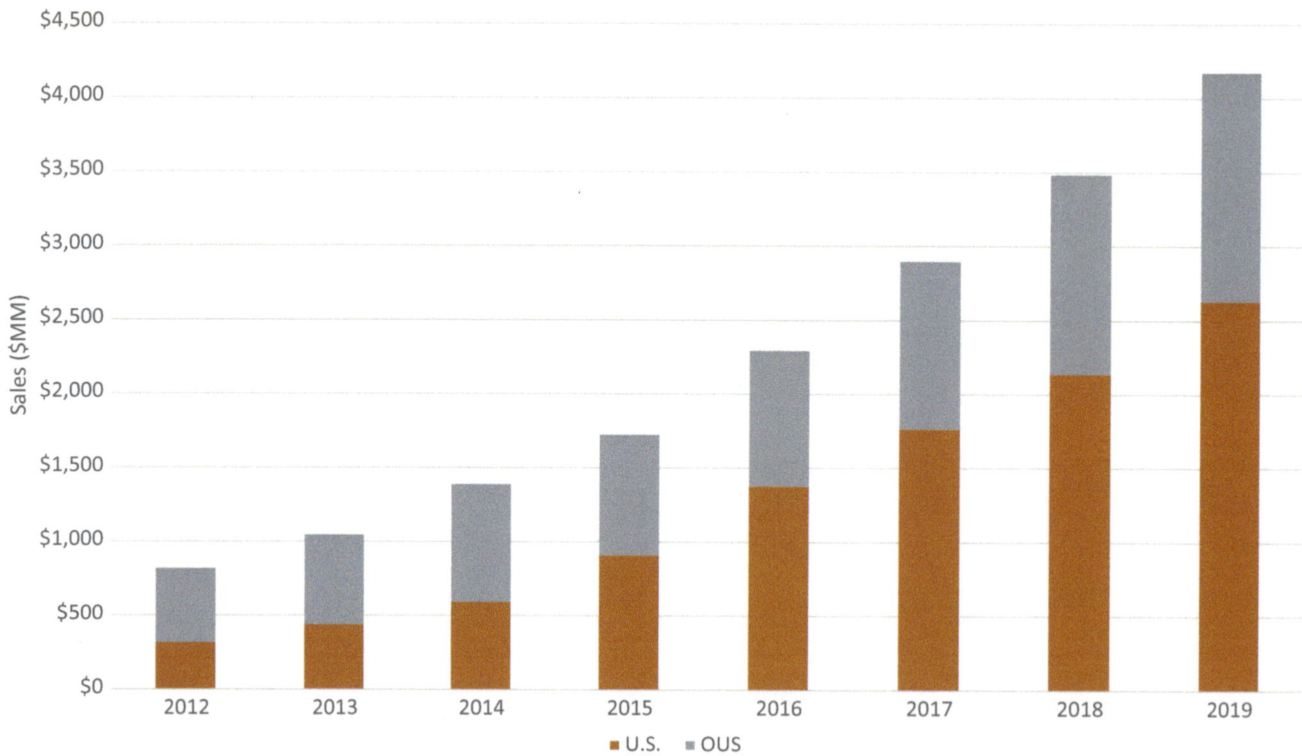

FIGURE 35.5 Global TAVR device market, 2012–2019. *Courtesy Health Research International.*

need for transseptal access (if the familiar transfemoral route is to be possible), the relative lack of adequate "landing zones" for ease of prosthetic securement with simple radial force, negotiating the subvalvular apparatus of both the mitral and tricuspid valves, and in the case of severe tricuspid regurgitation, the need for very large devices to accommodate significant annular dilation. Moreover, valve repair over replacement is the predominant mode of surgical intervention with both atrioventricular valves. Both examples have necessitated novel engineering and therapeutic approaches that continue to be refined.

What is generally true, however, is the fact that cardiovascular device markets will continue to trend toward catheter-based and other minimally invasive approaches. On one hand, there will always be a role and need for open surgery to treat various maladies using well-established and proven techniques that will demand the skills of cardiothoracic and vascular surgeons, which will continue to present opportunities for novel implantable devices and surgical instrumentation. Nevertheless, interventional cardiology and radiology has forever transformed the overall management of cardiovascular disease and will serve as the primary growth driver of related medical device markets into the foreseeable future.

Opportunities in cardiovascular device markets

According to the American Heart Association, heart disease has consistently been the leading cause of death globally. In 2019, it claimed 18.6 million people—representing about one-third of total deaths and a 17% increase over the last decade—accompanied by more than 523 million cases, which represented a nearly 27% increase over 2010. The countries with the highest number of heart disease deaths were, in descending order, China, India, Russia, the United States, and Indonesia.

Looking specifically at the United States, the Centers for Disease Control report that diseases of the heart, including stroke, accounted for 809,046 deaths in 2019, which represented 27% of total deaths and was more than all cancers combined. Moreover, despite advances in diagnosis and treatment of cardiovascular disease, this share is expected to increase gradually through the foreseeable future due to population aging and the rapid growth in obesity, diabetes and, since 2010, hypertension among US adults. Demographics is the major driver of this trend and the one that cannot be modified by lifestyle changes, whereas obesity, diabetes, hypertension, and their sequelae can be modified by lifestyle changes. Unfortunately, efforts in this direction have been largely lacking among the US population to date, as shown in Fig. 35.6.

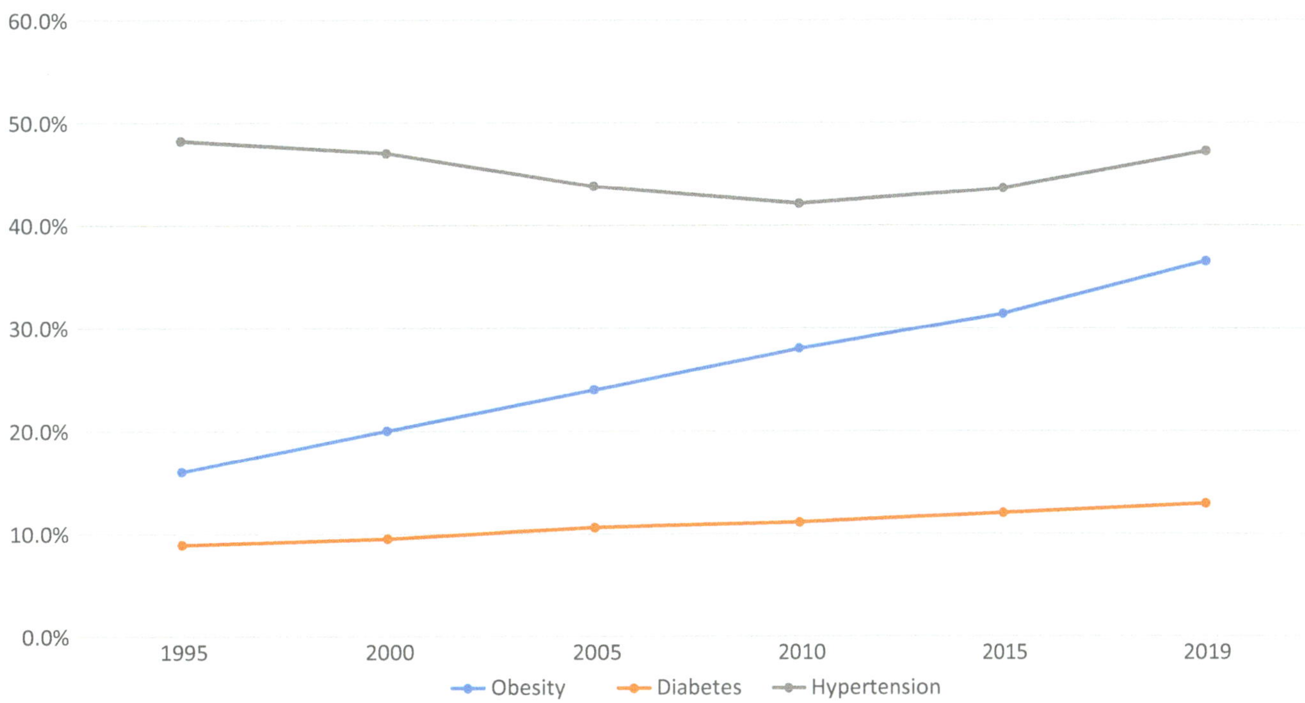

FIGURE 35.6 Trend in share of US adults with obesity, diabetes, and hypertension, 1995–2019. *Data from: Obesity- Centers for Disease Control & Prevention,* Behavior Risk Factor Surveillance System, *2020; Diabetes and Hypertension – National Center for Health Statistics,* National Health and Nutrition Examination Surveys, *1995–2019.*

Contribution of cardiovascular technologies to global medical devices and diagnostics market

Due to the major role heart disease plays in global and US mortality and morbidity, it should not be surprising that cardiovascular devices are one of the three dominant segments within the global medical diagnostics and devices (MD&D) market, accounting for nearly 12% of total revenues in 2019 (see Fig. 35.7). Moreover, adding cardiovascular-focused in vitro diagnostics, imaging systems and agents, energy devices, and other involved technologies to the tally brings total cardiovascular-focused medical devices' contribution to the global MD&D market to 19% in 2019, representing $87.5 billion (see Fig. 35.8). Finally, the United States remains the single largest geographic market for MD&D devices in general (see Fig. 35.9), and typically accounts for a dominant share of product sales across the various segments. This is especially true for cardiovascular devices, where US clinicians tend to be highly receptive to new technologies that improve patient care despite their frequently high cost. What are all of these cardiovascular devices that are being employed to treat heart disease and stroke and where do they fit into the treatment algorithms? For the purposes of this discussion, cardiovascular technologies may be segmented into six major categories based largely on clinical target. These are: (1) cardiac rhythm management (CRM) (pacemakers, implantable cardioverter defibrillators, manual/automatic external defibrillators), primarily addressing cardiac arrhythmias; (2) interventional cardiology (coronary stents, coronary angioplasty and related devices, intravascular ultrasound, fractional flow reserve, optical coherence tomography, transcatheter defect repair and transcatheter valve replacement and repair technologies), designed to treat coronary heart disease and, more recently, valve disease using less invasive transcatheter techniques; (3) peripheral and endovascular (endografts, drug-coated balloons, peripheral stents, peripheral atherectomy and other peripheral/endovascular devices), which target vascular conditions lying outside of the coronary vasculature; (4) cardiac surgery (cardiopulmonary and minimally invasive cardiac surgery, surgical heart valve replacement and repair, and cardiac assist), which represent more traditional, invasive methods of treating and managing heart disease; (5) electrophysiology (mapping and navigation systems, catheters and accessories) employed to treat atrial fibrillation and other cardiac arrhythmias using transcatheter techniques; and (6) neurovascular (catheters, stents, coils, clips, access devices), which enable interventional management of stroke. Fig. 35.10 shows the distribution of the global market among these major technology segments in 2019.

Underserved cardiovascular diseases drive growth opportunities

Clearly, as discussed in the previous section, the treatment of coronary heart disease has been the major beneficiary of cardiovascular product development over the past decades, with interventional cardiology accounting for 29% of total

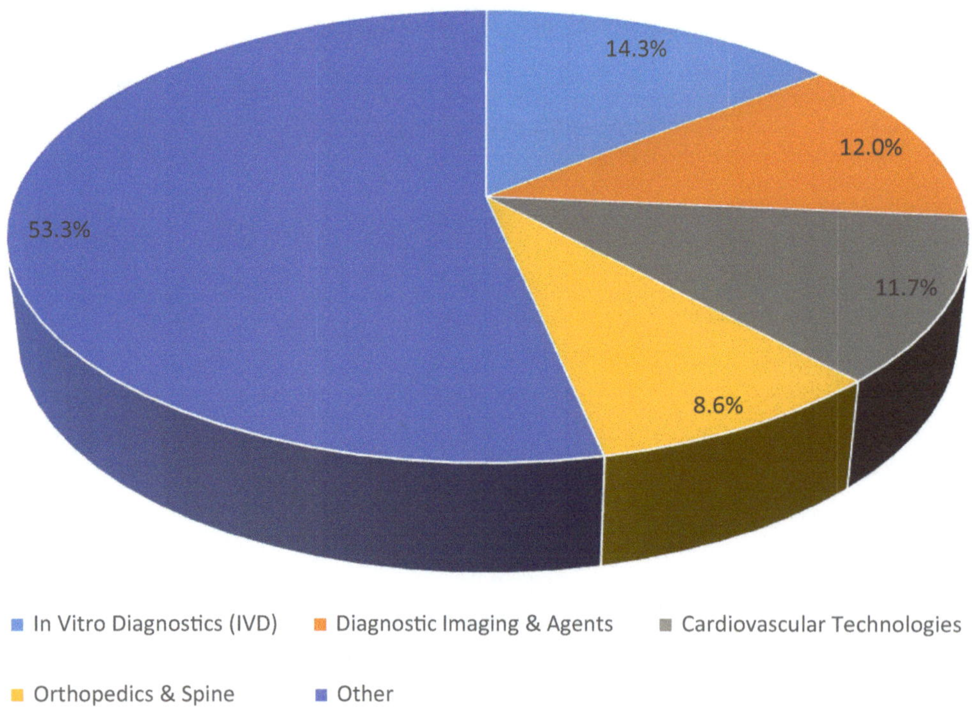

FIGURE 35.7 Global MD&D market by technology, 2019. *Courtesy Health Research International.*

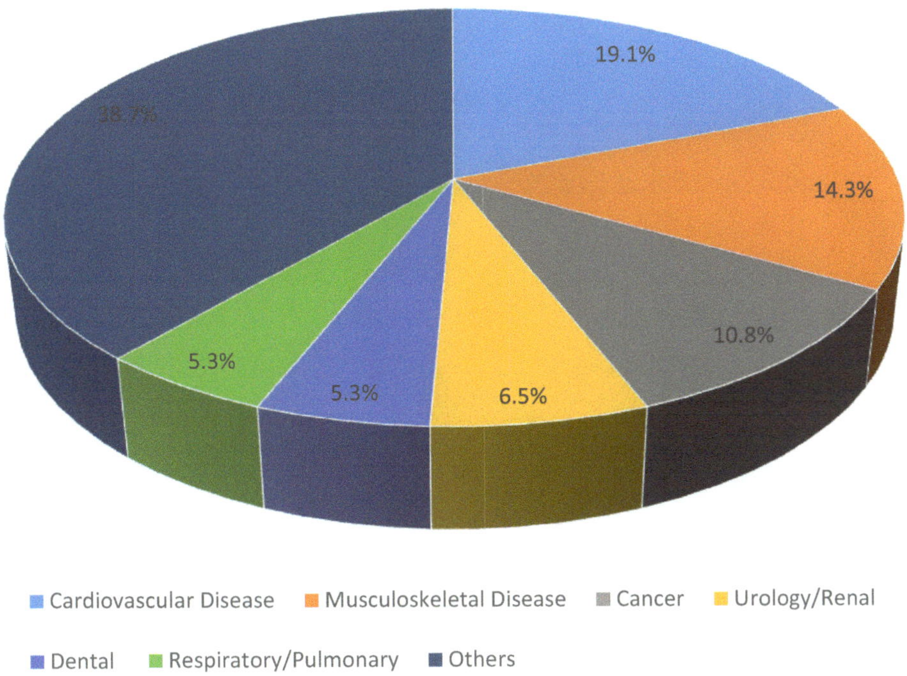

FIGURE 35.8 Global MD&D market by clinical application, 2019. *Courtesy Health Research International.*

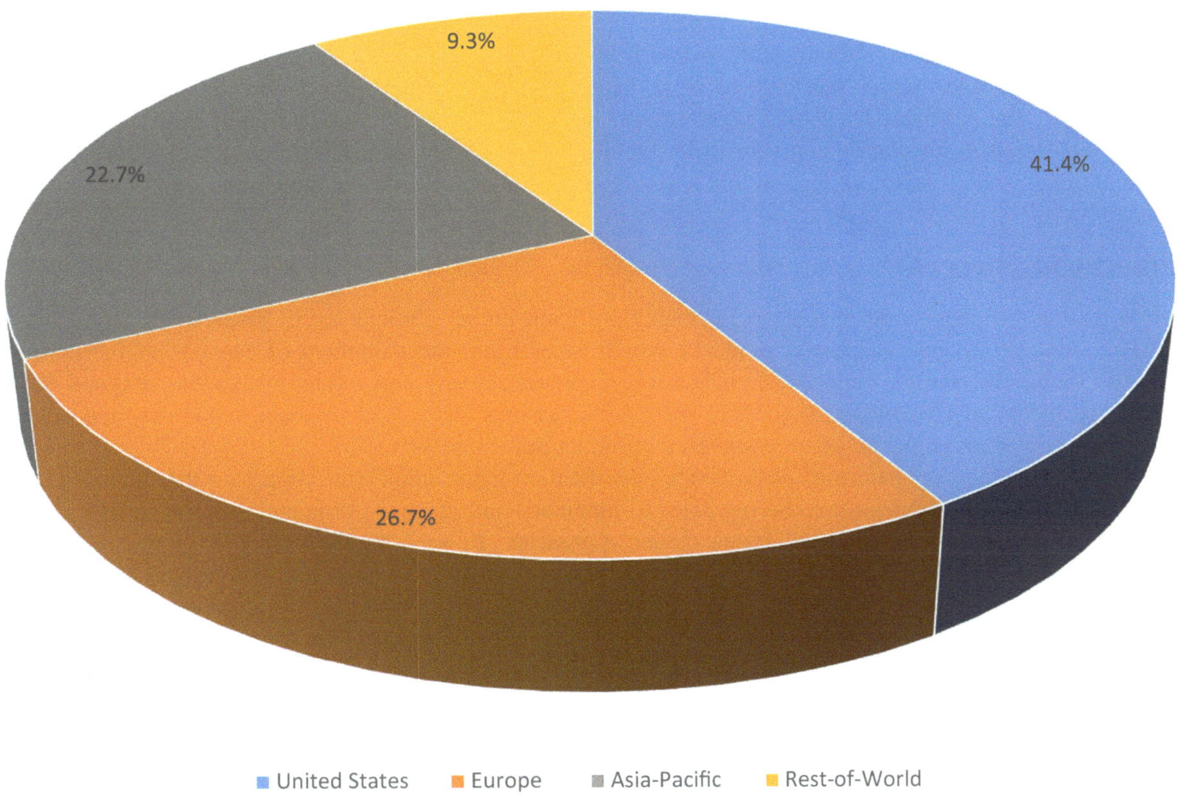

FIGURE 35.9 Global MD&D market by major geography, 2019. *Courtesy Health Research International.*

product revenues in 2019. This, of course, includes the recent, substantial sales of transcatheter aortic valves which, unlike coronary stents and angioplasty devices, have yet to become "commoditized" and so, with transcatheter mitral valve repair devices, are slated for significant continued growth. CRM represents an even more mature market, as does cardiac surgery (with the notable exception of cardiac assist devices targeting heart failure). Other types of highly

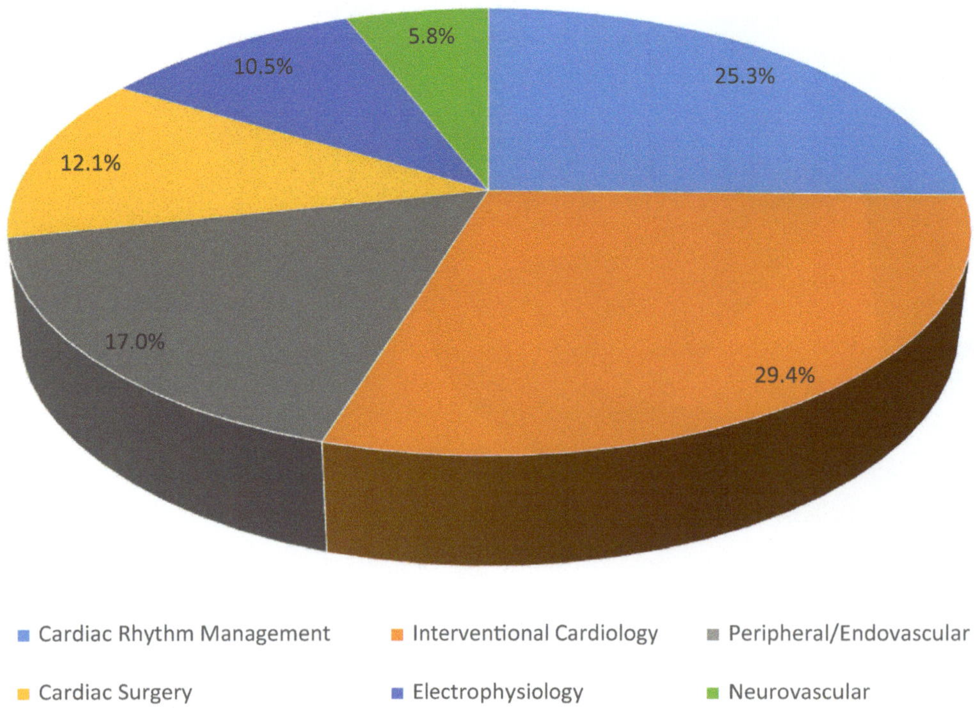

FIGURE 35.10 Global cardiovascular devices market by major segment, 2019. *Courtesy Health Research International.*

prevalent cardiovascular diseases remain underserved from a product standpoint, being heavily dependent on lifestyle changes and drug therapies, which can help to slow disease progression but not address the structural challenges of advanced disease. The list includes PAD, hypertension, heart failure, stroke, cardiac arrhythmias (notably complex arrhythmias such as atrial fibrillation and ventricular tachycardia), and venous disease. Fig. 35.11 presents an overview of these conditions—which represent current and emerging opportunities for device development.

Ingredients for success

In order to realize the opportunity presented by such significant unmet clinical needs, researchers need to be able to translate their innovations into successful products, which necessitates the alignment of capable engineers working in lockstep with thoughtful and experienced clinicians to conceptualize, test, and refine novel ideas. Beyond the obvious intellectual and technical horsepower, successful innovation requires substantial financial resources, and often significant time to advance through the stages of commercial product development.

In the US market, the very nature of FDA device classification has a significant impact on the timelines and investment required to successfully negotiate the regulatory framework and achieve commercialization, which in turn helps direct the clinical problems to be solved with new device approaches. In brief, the FDA classification scheme for medical devices provides for a direct relationship between regulatory control and the perceived risk benefit of the technology. For example, the lowest risk devices (Class I) are generally exempt from premarket notification; medium risk devices (Class II) generally require proving substantial equivalence to a predicate device under the 510(k) clearance process (i.e., Premarket Notification), and the highest risk devices (Class III) generally require Premarket Approval through the PMA pathway. Medical implants, whether or not they are life sustaining, are generally either designated as Class II or Class III, and many of the significant opportunities for cardiovascular devices, both commercial and in the development pipeline, are of the Class III/PMA variety. For instance, vascular stents and stent-grafts, implantable pacemakers and defibrillators, surgical heart valve replacement prostheses, both transcatheter heart valve replacement and repair devices, ventricular assist devices, and ablative devices for the treatment of atrial fibrillation, are all Class III medical devices.

As Fig. 35.12 illustrates, even the most simple Class II devices, that is, those that do not require an Investigational Device Exemption clinical trial, are likely to require an investment of a couple million dollars and up to 2 years to obtain FDA 510(k) clearance. According to an industry survey of more than 200 medical device companies, more complex Class II devices take, on average, more than 6 years and $31 million to advance from concept through FDA clearance. At the far end of the spectrum, Class III devices were reported to require an average of nearly 10 years and

CONDITION	DESCRIPTION	U.S. PREVALENCE 2019	CURRENT TREATMENTS	UNMET NEEDS/DEVELOPMENTS
Peripheral Artery Disease (PAD)	Narrowing or blockage of vessels that carry blood from lungs to lower limbs and other non-cardiac and non-cerebral sites; caused by atherosclerosis; progressive	8-20M	Lifestyle changes, drugs (statins, blood thinners, thrombolytics), surgery (bypass, amputation), interventional (angioplasty, atherectomy, thrombectomy,	Small size of and tortuosity of lower limb vessels can compromise revascularization outcomes; critical limb ischemia results in >160,000 amputations annually
Hypertension	Consistently high pressure in arteries carrying blood from heart to rest of body; contributes to most other cardiovacular diseases	75-120M	Lifestyle changes, drugs (diuretics, variety of blockers and inhibotors); many patients non-compliant or drug refractory	Many patients non-compliant with drug therapy or drug refractory; renal denervation being investigated/revisited as minimally invasive option for refractory hypertension
Heart Failure	Progressive disease in which pumping action of heart muscle is insufficient to meet oxygenation needs of body and lungs; >50% of affected individuals die within 5 years of diagnosis; causes more hospitalizations than all cancers combined	≥6M	Lifestyle changes, drugs (ACE inhibitors, beta blockers, diuretics), surgery to correct underlying contributors (CABG, valve surgery), arrhythmia implants (ICDs, CRT devices) and mechanical circulatory support (VADs, TAHs)	Transcatheter mitral repair and replacement for HF patients developing severe mitral regurgitation, which carries poor prognosis; smaller and more durable cardiac assist devices for less invasive surgery, and transcutaneous energy sources to eliminate
Stroke	Reduction or interruption of blood supply to brain (ischemic stroke) or rupture of cerebral blood vessel (hemorrhagic stroke), causing cell death; medical emergency and major cause of long-term disability in adults; 25% of survivors have another stroke within 5 years	≈800,000 new or recurrent strokes annually, of which ~150,000 die	Ischemic stroke: emergent infusion of tPA to dissolve clot blocking artery and/or removal of clot using stent retriever; Hemorrhagci stroke: drugs or other measures to reduce intracranial pressure, surgical removal of anuerysm or clipping of AVM, endovascular embolization using coils to fill anuerysm and	Need for early, complete resolution of blockage or hemorrhage to prevent long-term sequalae focusing research on neuroprotection and neurorecovery, latter currently using stem calls; transcatheter left atrial appendage occlusion for prevention of recurrent stroke as alternative to long-term oral anticoagulation
Cardiac Arrhythmias	Group of conditions in which disturbance in which normal electrical activity of heart causes it to beat too slowly (bradycardia), too fast (tachycardia) or erratically	>2.3M (atrial fibrillation); >200,000 ventricular tachycardia (incidence)	Drugs (antiarrhythmics), cardioversion, implants (pacemaker, ICD), catheter ablation	Catheter ablation becoming popular as device-based advances are shortening procedure times, increasing precision of ablations, increasing safety and improving outcomes; improved diagnostics via remote monitoring and AI/ML algorithms both in and out of the
Venous Disease	The inability of the venous system to return the entire volume of blood distributed by the arterial system back to the right heart and lungs due to damage to the vein walls or valves	Varicose veins: 15% of adults; deep vein thrombosis: 900,000-2.6 million; chronic venous insufficiency: 3 million annually	Varicose veins and CVI: Sclerotherapy, laser surgery, surgical vein stripping, phlebectomy; DVT: blood thinners, thrombolytic agents, IVC filtesr, mechanical thrombectomy	Catheter-based options are reducing treatment invasiveness; however, development of dedicated venous stents offers greatest opportunity to address venous outflow obstruction

FIGURE 35.11 Unmet needs and opportunities for cardiovascular device development. *Courtesy Health Research International.*

$94 M to obtain FDA approval through the PMA pathway.[22] These extended timelines and substantial development costs, the latter of which are now likely even more inflated given the data from this survey was published over a decade ago, necessitate that market opportunities are sufficiently large to provide the sponsors of innovation with the potential for an acceptable return on investment.

Such barriers to market entry have led some companies to prioritize initial commercialization in markets outside the United States, notably in Europe, where requirements to gain CE Mark approval for high risk devices have been somewhat less stringent, at least historically. For instance, the first TAVR devices were commercially available in Europe approximately four years prior to their FDA approval. This strategy may be less practical for smaller companies without a direct sales channel in local markets, or at the very least is less profitable if distribution partners are a necessity for market access. Further, early commercial uptake of cardiovascular devices has tended to be more limited without the clinical backing of more rigorous safety and efficacy testing associated with pivotal FDA studies. Nevertheless, in recent years, policy reforms in both the European Union (e.g., Medical Devices Regulation) as well as in the United States have provided renewed incentive to refocus development efforts on the generally more lucrative US market, which tends to support both higher pricing and more intensive utilization.

One such mechanism has been FDA's Breakthrough Devices Program (BDP),[23] which was established in 2018 to accelerate development and regulatory review of certain medical devices, in particular those that offer a more effective

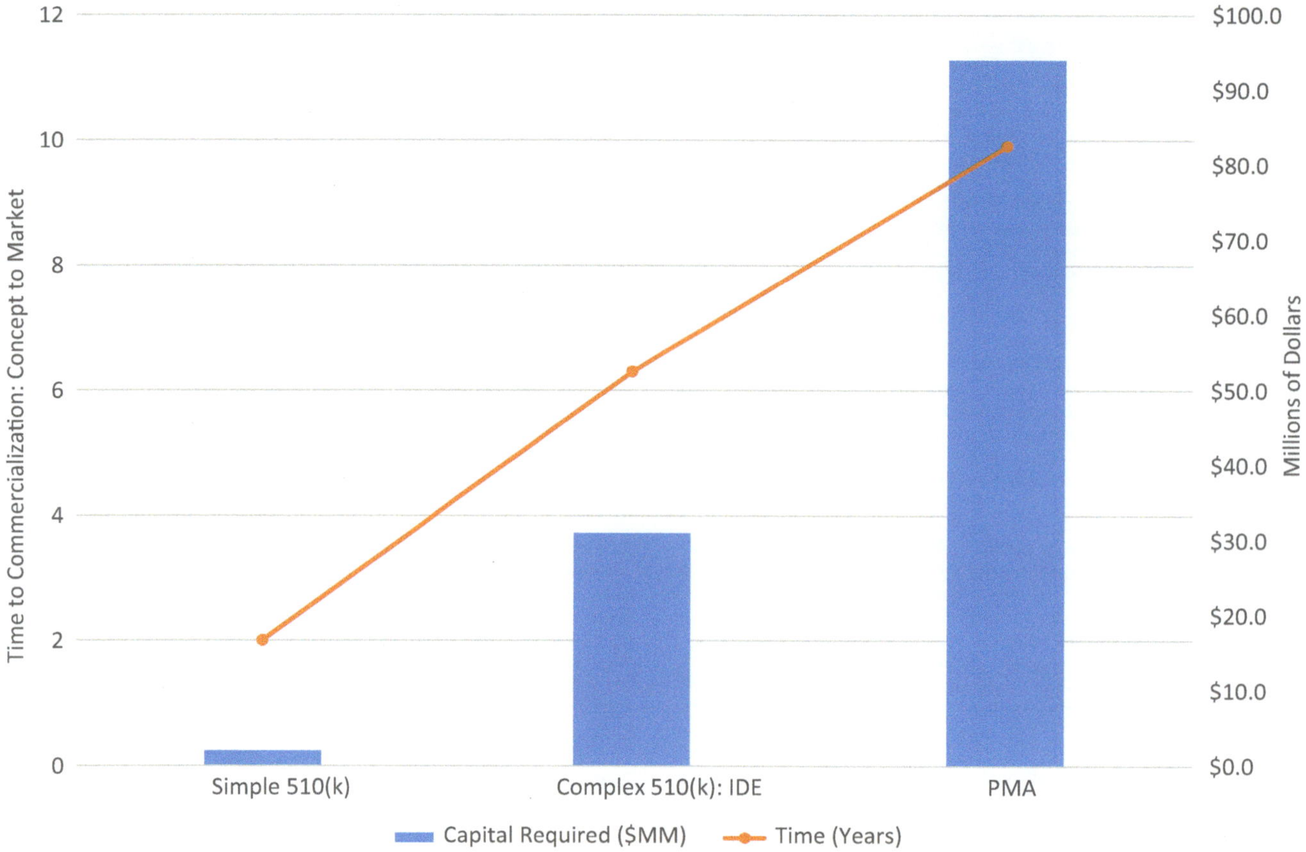

FIGURE 35.12 Impact of regulatory path on US medical device commercialization. *Data from Makower et al.,* FDA Impact on U.S. Medical Technology Innovation. *AdvaMed, with support from the Medical Device Manufacturers Association (MDMA), National Venture Capital Association (NVCA), and multiple state medical industry organizations, November 2010. Retrieved from https://www.advamed.org/sites/default/files/resource/30_10_11_10_2010_Study_CAgenda_makowerreportfinal.pdf and proprietary data.*

treatment or diagnosis for conditions deemed life-threatening or irreversibly debilitating. While the statutory standards for PMA, 510(k) clearance, and De Novo authorization do not change, the BDP provides medical device manufacturers with an opportunity for early and more extensive engagement of FDA experts during the development and premarket stages, a means of obtaining expedited feedback, and the benefit of prioritized review of regulatory submissions to ultimately offer both patients and providers with more timely access to new medical devices. Further, the Centers for Medicare and Medicaid Services (CMS) issued a proposed rule in September 2020 to establish the Medicare Coverage of Innovation Technology (MCIT) pathway, under which national Medicare coverage would be automatically provided for Breakthrough medical devices upon receipt of FDA marketing authorization for up to 4 years.[24] This would provide manufacturers even greater incentive to solve for the unmet needs with creative medical device solutions by reducing additional barriers to adoption often present between regulatory approval and establishment of reimbursement. This proposal complements earlier CMS programs, such as the New Technology Add-On Payment (NTAP) initiative, whereby certain recently FDA-approved medical devices may be eligible to receive temporary incremental payments to help offset their costs when established inpatient reimbursement is insufficient.

The medical device industry has clearly embraced these programs: as of February 2021, more than 400 medical devices have received designations through the BDP, 23 of which have proceeded through PMA approval, 510(k) clearance or De Novo authorization.[25] While the BDP is voluntary and, unlike with premarket approvals and clearances, FDA does not provide a public database of Breakthrough Device designations, a number of medical device manufacturers have leveraged their receipt of such designation as a marketing tool in press releases. A subset analysis 46 Breakthrough Device designations for cardiovascular technologies announced from 2019 to early 2021 clearly shows the prominence of minimally invasive technologies, with nearly three-quarters of the programs involving transcatheter devices. Similarly, this data demonstrates a concentration of device solutions targeting heart failure, peripheral and coronary vascular disease (Fig. 35.13). According to a recent analysis of technologies approved for supplemental

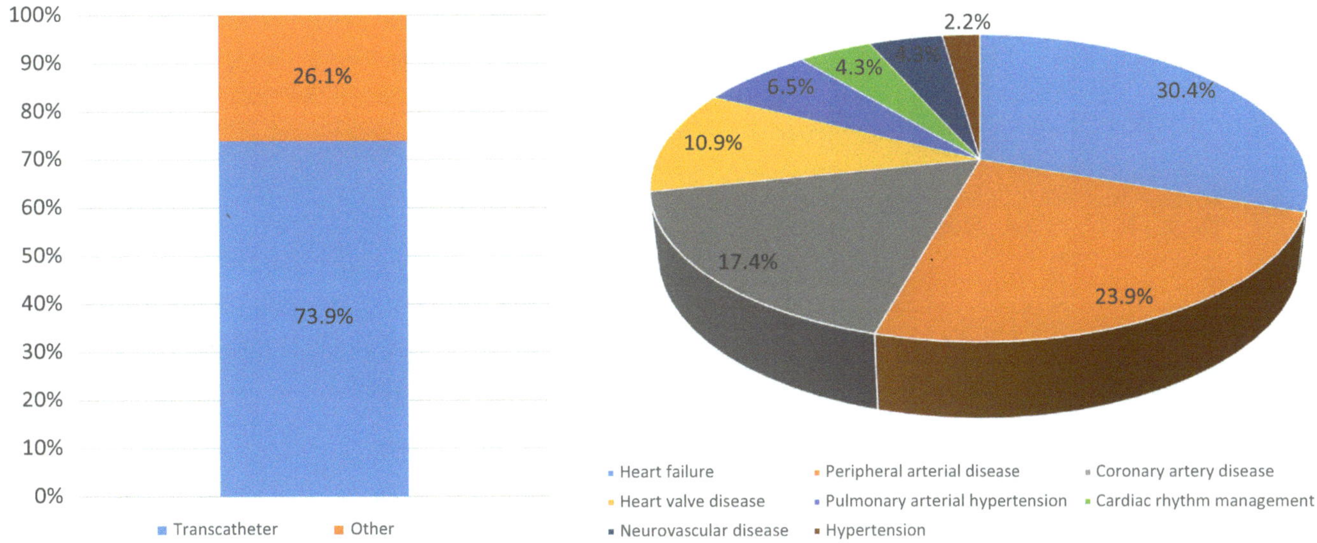

FIGURE 35.13 Cardiovascular breakthrough device designations by technology type and indication. Data set includes several devices with multiple breakthrough designations for different indications. *Data from Company press releases.*

reimbursement through the NTAP program between 2001 and 2016, cardiovascular technologies accounted for more than half of the 17 approved medical devices, with multiple approvals for technologies targeting PAD and heart failure, and a similar concentration of transcatheter technologies (78%) as with the BDP.[26]

Industry trends: dollars and deals

Industry consolidation has been rampant across the medical device market in recent years, and the cardiovascular sector is no exception as one of the largest and fastest growing segments. This consolidation is clearly evident when examining FDA approvals of novel cardiovascular devices. FDA defines novel medical devices as those brought to the US market through the PMA, De Novo, and humanitarian device exemption pathways, as well as certain 510(k) clearances of BDP designated technologies, which encompass those technologies requiring the highest levels of investment for commercialization. As Fig. 35.14 shows, of 459 novel cardiovascular devices authorized by FDA between 1993 (the year the first coronary stent was FDA approved) and 2020, only six companies brought more than a 15 such products to market, and these manufacturers collectively accounted for more than 60% of all authorizations. Medtronic, Abbott, and Boston Scientific alone contributed nearly half of these novel cardiovascular device authorizations. Also noteworthy is the fact that more than half of novel cardiovascular devices have been involved in a strategic acquisition during the analyzed time frame.

Perhaps even more striking are the number and value of merger and acquisition (M&A) transactions that have been executed in the cardiovascular device market over the past two decades. For individual companies, especially larger more mature organizations, M&A transactions provide a means of accelerating growth beyond that which might be attained via conventional, organic research and development efforts. This M&A activity has not been homogenous, but rather characterized by a number of transformational "mega-mergers," so-called "tuck-in" deals whereby larger players extend their product lines with the addition of complementary technologies, as well as more speculative strategic acquisitions of development stage technologies that show promise in unlocking the potential of significant untapped markets. Such significant consolidation has been observed across the entire medical device industry, and cardiovascular devices are certainly no exception.

While not comprehensive, a sampling of select, publicly disclosed M&A activity relating to cardiovascular devices among some of the more prominent industry players over the past two decades encompasses more than 300 individual transactions (see Fig. 35.15). Financial terms were disclosed in nearly 90% of these transactions ($n = 264$), with a corresponding aggregate deal value (including both up-front cash and stock payments, as well as regulatory and commercial milestone payments, which may or may not have been achieved) of more than $130 billion. Sifting through this data, several observations present themselves:

- of 43 unique acquiring companies, more than 40% ($n = 19$) have gone on to be acquired themselves by larger players;

Manufacturer	Novel Device Authorizations	% of Total	Acquired Technologies	% acquired
Medtronic	71	18.4%	27	38.0%
Abbott	63	16.4%	60	95.2%
Boston Scientific	50	13.0%	20	40.0%
Edwards Lifesciences	20	5.2%	12	60.0%
Gore Medical	18	4.7%	1	5.6%
Becton Dickinson	16	4.2%	16	100.0%
All Others (n=65)	147	38.2%	60	40.8%
TOTAL	**385**	**100.0%**	**196**	**50.9%**

FIGURE 35.14 FDA-Authorized Novel cardiovascular devices by manufacturer, 2000–2020. Novel device authorizations including original PMAs, panel track supplement PMAs, De Novos and Humanitarian Device Exemptions overseen by FDA's Cardiovascular Devices Panel; breakthrough 510 (k) clearances are not specified in public reporting and are therefore excluded from this data set; listed manufacturers represent the current parent company for each product; acquired technologies. *Data from Food and Drug Administration.*

	Deal Count w/Value	Disclosed Deal Value ($MM)	Mean Deal Value ($MM)	% of Deals	% of Deal Value
TOTAL	**264**	**$133,367**	**$505**		
Technology Sector					
Transcatheter	199	$72,460	$364	75.4%	54.3%
Surgery	66	$60,907	$923	25.0%	45.7%
*Surgery ex-GDT/STJ**	*64*	*$15,683*	*$245*	24.2%	11.8%
Deal Type					
Mega-Mergers	17	$97,553	$5,738	6.4%	73.1%
Tuck-Ins	159	$22,514	$142	60.2%	16.9%
Development Stage	88	$13,301	$151	33.3%	10.0%
Current Parent Company					
Abbott	32	$40,128	$1,254	12.1%	30.1%
Boston Scientific	37	$39,897	$1,078	14.0%	29.9%
Medtronic	46	$20,660	$449	17.4%	15.5%
Others	149	$32,682	$219	56.4%	24.5%

FIGURE 35.15 Select cardiovascular device merger and acquisition activity. *Excluding Boston' Scientific's acquisition of Guidant and Abbott's acquisition of St. Jude Medical (surgical business unit values only). *Data from Company reports.*

- a small number of companies have invested orders of magnitude more money into these acquisitions: Abbott, Boston Scientific, and Medtronic are the current parent companies of more than 75% of acquired cardiovascular device technologies in this sample as measured by disclosed deal value;
- unsurprisingly, "mega-mergers" of commercial stage businesses valued at $1 billion or more are overrepresented in the sample, accounting for nearly three-quarters of total deal value;
- similar to the regulatory data previously shown, transcatheter technologies dominate among these cardiovascular device acquisitions, accounting for an estimated three-quarters of total transactions; and
- M&A activity relating to technologies directed toward cardiac and vascular surgery, including conventional CRM, appears more frequently associated with divestiture and, excluding two outlier mega-mergers (Boston Scientific's acquisition of Guidant and Abbott's acquisition of St. Jude Medical), smaller average deal values.

While there are surely exceptions to this rule, if one assumes that strategic, development stage transactions serve as a reasonable proxy for startup acquisitions, the above data suggests that a significant amount of innovation in

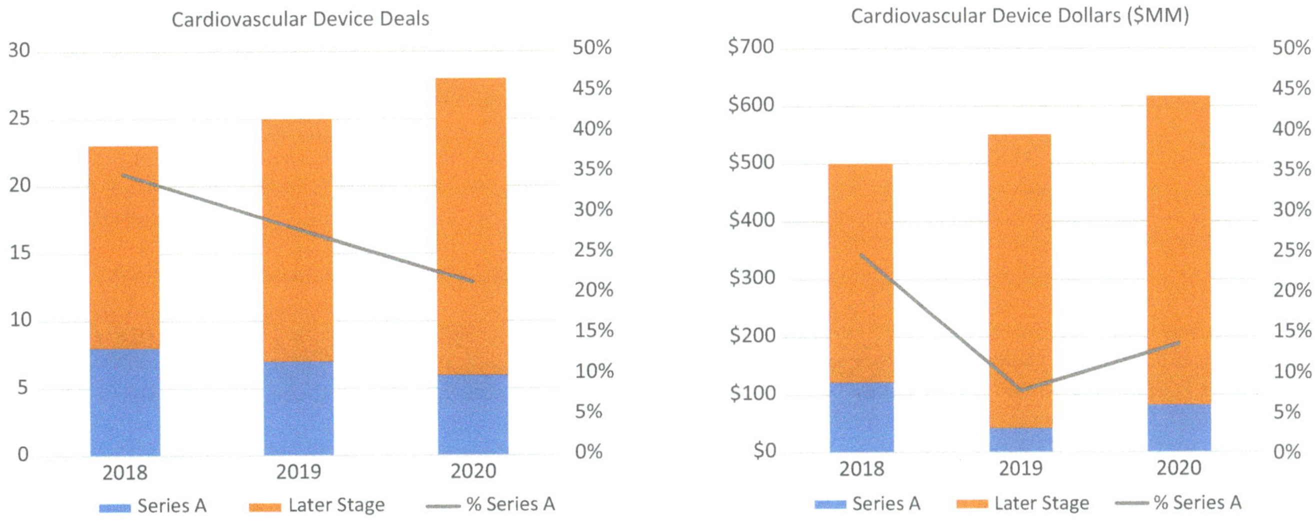

FIGURE 35.16 Series A investments as a share of cardiovascular device deals and dollars, 2018–2020. Series A includes first-time investments from institutional or corporate venture investment in the US and Europe and any first-round investments equal to or greater than $2 million. *Data from Silicon Valley Bank and PitchBook.*

cardiovascular devices originated in the startup community. The data from this sample shows that at least one-third of all cardiovascular device transactions in the past two decades involved the acquisition of startup companies, and is surely undercounted as this figure only encompasses development stage companies and excludes transactions involving startup companies acquired postcommercialization.

According to Silicon Valley Bank research, venture capital investment into cardiovascular devices increased from $500 million in 2018 to $617 million in 2020.[27] However, the percentage of both the total number of deals and dollars invested in those deals contributing to Series A investments fell significantly from 2018 levels, as shown below in Fig. 35.16. This data suggests that investors are increasingly favoring later stage technologies, which serves to extend the amount of money startup companies need to raise and the time horizons for those companies to achieve an exit. Perhaps more importantly, fewer early stage investments may have negative implications for innovation in this segment, as strategic acquirers, have gotten fewer in number through industry consolidation, and have demonstrated an increasing tendency to wait until technologies have been derisked rather than making more speculative acquisitions in the earlier stages of development.

Conclusions

The cardiovascular devices market continues to represent one of the most innovative and highest growth sectors in the medical device industry, driven by persistently large and undertreated patient populations with deadly and debilitating conditions. Although as with conventional investing wisdom, past success provides no guarantee for future results, the combination of continuous improvement to refine existing technologies and, especially the plethora of still significant unmet clinical needs along with unstoppable demographic trends, portend to sustainable growth into the foreseeable future and likely beyond. The advances in more conventional cardiovascular device markets in the surgical and interventional sectors will be elevated by the continued digitization of healthcare via increased utilization and integration of remote monitoring, wearables, and artificial intelligence/machine learning algorithms into comprehensive software as a medical device platforms. This should help facilitate increasing mobility of care, whereby cardiovascular devices can play a greater role in healthcare delivery beyond the operating room and cath lab, into the patient home, while simultaneously inviting new and perhaps some unlikely or nonobvious market entrants to establish a foothold in this expanding sector. At the same time, continued industry consolidation can be anticipated, which also suggests that renewed investment in early stage cardiovascular device companies will be critical to sustain a viable pipeline to feed and ultimately realize the vast opportunities that lie ahead.

References

1. Virani SS, Alonso A, Benjamin EJ, Bittencourt MS, et al. Heart disease and stroke statistics-2020 update: a report from the American Heart Association, On behalf of the American Heart Association Council on Epidemiology and Prevention Statistics Committee and Stroke Statistics

Subcommittee. *Circulation*. 2020;141:e1–e458. Available from: https://www.heart.org/-/media/files/about-us/statistics/2020-heart-disease-and-stroke-ucm_505473.pdf?la = en.

2. Svennson L, et al. The father of coronary artery bypass grafting: Rene Favolaro and the 50th anniversary of coronary artery bypass grafting. *J Thorac Cardiovasc Surg*. 2018;155(6):2324–2328. Available from: https://pubmed.ncbi.nlm.nih.gov/29602424/.
3. Grüntzig A, Hopff H. Percutaneous recanalization after chronic arterial occlusion with a new dilator-catheter (modification of the Dotter technique) (author's transl). *Dtsch Med Wochenschr*. 1974;99(49):2502–2510, 2511. Available from: https://pubmed.ncbi.nlm.nih.gov/4434847/, http://doi.org/10.1055/s-0028-1108161.
4. Gruntzig A. Transluminal dilatation of coronary-artery stenosis. *Lancet*. 1978;1(8058):263. Available from: https://pubmed.ncbi.nlm.nih.gov/74678/, http://doi.org/10.1016/s0140-6736(78)90500-7.
5. Gillum RF, et al. Coronary heart disease and related procedures. National Hospital Discharge Survey data. *Circulation*. 1989;79(6 Pt 2):I13–I18. Available from: https://pubmed.ncbi.nlm.nih.gov/2524295/.
6. HCUP Nationwide Inpatient Sample (NIS). *Healthcare Cost and Utilization Project (HCUP). 1993–2007*. Rockville, MD: Agency for Healthcare Research and Quality. http://www.hcup-us.ahrq.gov/nisoverview.jsp.
7. Sigwart U, et al. Intravascular stents to prevent occlusion and restenosis after transluminal angioplasty. *N Engl J Med*. 1987;316(12):701–706. Available from: http://doi.org/10.1056/NEJM198703193161201, https://pubmed.ncbi.nlm.nih.gov/2950322/.
8. Akin I, Nienaber C. Treatment of coronary in-stent restenosis—evidence for universal recommendation? *J Thorac Dis*. 2015;7(10): 1672–1675. Available from: http://doi.org/10.3978/j.issn.2072-1439.2015.10.16. https://www.ncbi.nlm.nih.gov/pmc/articles/PMC4635288/.
9. Sousa JE, et al. Lack of neointimal proliferation after implantation of sirolimus-coated stents in human coronary arteries: a quantitative coronary angiography and three-dimensional intravascular ultrasound study. *Circulation*. 2001;103(2):192–195. Available from: http://doi.org/10.1161/01.cir.103.2.192, https://pubmed.ncbi.nlm.nih.gov/11208675/.
10. HCUP Nationwide Inpatient Sample (NIS). *Healthcare Cost and Utilization Project (HCUP). 2003–2014*. Rockville, MD: Agency for Healthcare Research and Quality. http://www.hcup-us.ahrq.gov/nisoverview.jsp.
11. Camenzind E, et al. Stent thrombosis late after implantation of first-generation drug-eluting stents: a cause for concern. *Circulation*. 2007;115 (11):1440–1455, Epub 2007 Mar 7. Available from: https://pubmed.ncbi.nlm.nih.gov/17344324/, http://doi.org/10.1161/CIRCULATIONAHA.106.666800.
12. Applegate RJ, et al. Cost-effectiveness of drug-eluting stents vs bare metal stents in clinical practice. *Circ Cardiovasc Qual Outcomes*. 2011;4 (4):408–415, Epub 2011 Jun 21. Available from: https://pubmed.ncbi.nlm.nih.gov/21693724/. http://doi.org/10.1161/CIRCOUTCOMES.110.960187.
13. Cribier A. The development of transcatheter aortic valve replacement (TAVR). *Glob Cardiol Sci Pract*. 2016;2016(4):e201632, Published online 2016 Dec 30. Available from: https://www.ncbi.nlm.nih.gov/pmc/articles/PMC5624190/. http://doi.org/10.21542/gcsp.2016.32.
14. Andersen HR, et al. Transluminal implantation of artificial heart valves. Description of a new expandable aortic valve and initial results with implantation by catheter technique in closed chest pigs. *Eur Heart J*. 1992;13(5):704–708. Available from: http://doi.org/10.1093/oxfordjournals.eurheartj.a060238, https://pubmed.ncbi.nlm.nih.gov/1618213/.
15. Cribier A, et al. Percutaneous transcatheter implantation of an aortic valve prosthesis for calcific aortic stenosis: first human case description. *Circulation*. 2002;106(24):3006–3008. Available from: http://doi.org/10.1161/01.cir.0000047200.36165.b8, https://pubmed.ncbi.nlm.nih.gov/12473543/.
16. CoreValve. *CoreValve Receives CE Mark Approval for its ReValving™ System and Announces Plans to Initiate Expanded Clinical Evaluation* [Press release]; 2007. Retrieved from https://www.businesswire.com/news/home/20070516005698/en/CoreValve-Receives-CE-Mark-Approval-for-its-ReValving-TM-System-and-Announces-Plans-to-Initiate-Expanded-Clinical-Evaluation.
17. Edwards Lifesciences Corporation. *Edwards Lifesciences Receives CE Mark for Edwards SAPIEN Transcatheter Heart Valve* [Press release]; 2007, September 7. Retrieved from https://www.edwards.com/ns20070905.
18. Carroll JD, et al. STS-ACC TVT registry of transcatheter aortic valve replacement. *J Am Coll Cardiol*. 2020;76(21):2492–2516. Available from: http://doi.org/10.1016/j.jacc.2020.09.595, https://pubmed.ncbi.nlm.nih.gov/33213729/.
19. United States Census Bureau (2020). *International Database, Population by Age*. Retrieved from https://www.census.gov/data-tools/demo/idb.
20. Cohen DJ. et al. *Economic Considerations for TAVR Vs. SAVR: Historical Perspective and Future Predictions*. American College of Cardiology website (http://www.acc.org), Latest in Cardiology. https://www.acc.org/latest-in-cardiology/articles/2019/06/18/07/43/economic-considerations-for-tavr-vs-savr; Accessed 04.01.21.
21. Boston Scientific Corporation. *Boston Scientific Announces LOTUS Edge™ Aortic Valve System Voluntary Recall and Product Discontinuation* [Press release]. Retrieved from https://news.bostonscientific.com/2020-11-17-Boston-Scientific-Announces-LOTUS-Edge-TM-Aortic-Valve-System-Voluntary-Recall-and-Product-Discontinuation.
22. Makower J. et al. *FDA Impact on U.S. Medical Technology Innovation*. AdvaMed, with support from the Medical Device Manufacturers Association (MDMA), National Venture Capital Association (NVCA), and multiple state medical industry organizations, November 2010. Retrieved from https://www.advamed.org/sites/default/files/resource/30_10_11_10_2010_Study_CAgenda_makowerreportfinal.pdf.
23. Food and Drug Administration. *Breakthrough Devices Program*. Retrieved from https://www.fda.gov/medical-devices/how-study-and-market-your-device/breakthrough-devices-program.
24. Federal Register. *Medicare Program; Medicare Coverage of Innovative Technology (MCIT) and Definition of "Reasonable and Necessary."* 2020, September 1. Retrieved from https://www.federalregister.gov/documents/2020/09/01/2020-19289/medicare-program-medicare-coverage-of-innovative-technology-mcit-and-definition-of-reasonable-and.

25. Shuren J. *Reflections on a Record Year for Novel Device Innovation Despite COVID-19 Challenges*, 2021, February 16. Retrieved from https://www.fda.gov/news-events/fda-voices/reflections-record-year-novel-device-innovation-despite-covid-19-challenges.
26. Judson J, et al. Evaluation of technologies approved for supplemental payments in the United States. *BMJ*. 2019;365:I2190. Available from: http://doi.org/10.1136/bmj.l2190, https://www.ncbi.nlm.nih.gov/pmc/articles/PMC6890455/.
27. Silicon Valley Bank. Healthcare Investments and Exits: Annual Report 2021. Retrieved from https://www.svb.com/globalassets/library/managedassets/pdfs/healthcare-report-2021-annual.pdf.

Chapter 36

Innovation in Cardiovascular Bioelectronics

Rose T. Yin[1], Yeon Sik Choi[2,3,4], Kedar K. Aras[1], Helen S. Knight[1], Alana N. Miniovich[1] and Igor R. Efimov[1,5]

[1]*Department of Biomedical Engineering, The George Washington University, Washington, DC, United States,* [2]*Center for Bio-Integrated Electronics, Northwestern University, Evanston, IL, United States,* [3]*Querrey Simpson Institute for Bioelectronics, Northwestern University, Evanston, IL, United States,* [4]*Department of Materials Science and Engineering, Northwestern University, Evanston, IL, United States,* [5]*Department of Biomedical Engineering, Northwestern University, Evanston, IL, United States*

Chapter Outline

Introduction

Advances in materials science have enabled new bioelectronics platforms for novel approaches to medicine. Bioelectronics for disease diagnosis and treatment that were once bulky have become miniaturized and lightweight. The rigid geometries that were previously incompatible with tissues and organs are now flexible and stretchable to conform to organ curvatures. Energy sources dependent on batteries can now harvest energy from mechanical motion, static electricity, light, ultrasound, and electromagnetic fields. Materials at the tissue–bioelectronics interface inducing significant foreign body responses have been replaced by materials such as hydrogels and graphene that are much more biocompatible. These innovations have enabled the development of bioelectronics for the treatment of cardiovascular diseases, such as monitors, ablation, pacemaker, and implantable cardioverter defibrillator (ICD) therapy. This portfolio of bioelectronic devices collects high-resolution data across multiple parameters and can deliver the pertinent electrotherapy. The bioelectronic conformal devices serve as the foundation of the medical internet-of-things, which will ultimately improve the accessibility of medicine, the efficiency of the healthcare system, and enhance human health.

Device features

Flexible and conformal electronics

Cardiac electrical mapping is a basic tool for understanding normal conduction and arrhythmia mechanisms. The earliest version of cardiac maps was generated using electrode "socks" developed in the 1980s for epicardial electrical mapping.[1] The devices were handmade designs with up to 247 silver/silver chloride passive sensing electrodes sewn onto a nylon sock or similar fabric.[2] The sock would fit loosely on the heart surface constrained by the dynamic contours of a beating heart. This would result in uneven electrode–tissue surface contact which in turn affected the quality of

Advances in Cardiovascular Technology. DOI: https://doi.org/10.1016/B978-0-12-816861-5.00038-1

recorded signals. While still relevant in the research setting, these devices were limited by lack of scalable manufacturing and the low spatial resolution associated with low density electrode arrays.

Advances in material sciences fabrication techniques and pioneering circuit designs have led to the emergence of novel electronic membranes that can conform to the heterogeneous surface of biological tissues. These devices are built on stretchable contour-fitting membranes customized to the geometry of the specific tissue of the heart. The first generation of these devices are relatively simple, passive systems without integrated electronics.[3] A diverse array of multifunctional sensors and actuators can be placed in custom orientations with the ability to integrate directly across large areas of the epicardium and endocardium in beating hearts. The second-generation devices include an underlying backplane of thin, flexible active electronics that perform local signal amplification and allow for multiplexed addressing.[4] Active electronics enabled scaling to high-density, high-speed measurements far beyond the first-generation devices. However, direct contact between the active electrode and tissue surface can result in leakage of current from the underlying electronics. This current leakage degrades the electronics hardware and could cause cardiac arrhythmias. The third-generation devices have evolved to include hermetic sealing of the underlying electronics such as with an ultrathin, thermally grown layer of silicon dioxide which covers the entire system surface. This layer serves as a dielectric to enable direct capacitive coupling to the silicon nanomembrane transistor arrays. It also acts as a robust, biocompatible barrier to prevent penetration of biofluids.[5] More recent advances have focused on using softer rubbery materials with the eventual goal of achieving mechanical softness and the deforming characteristics of a cardiac tissue.[6]

The combination of exciting new advances in materials and mechanics concepts as well as the recent demonstrations of these novel devices with multiparametric actuators and sensors provides a realistic pathway for utility both in fundamental research and clinical applications. In addition to providing increased levels of data to study pathophysiology of various heart diseases, the devices could eventually be used as transient or chronic implants for cardiac monitoring and patient therapy.

Bioresorbable devices

Bioresorbable devices represent electronic platforms with an ability to maintain their electrical function over a relevant time frame with eventual complete dissolution into the surrounding biofluids and from the body itself by hydrolysis and natural metabolic processes.[7] Recent advances in bioresorbable devices enable pacing of the heart in a battery-free manner. Choi et al. developed a bioresorbable pacemaker that consists of a wireless power harvesting unit, a flexible extension electrode, and a tissue contact pad (Fig. 36.1A). The wireless power harvesting part includes a loop antenna with a bilayer, dual-coil configuration [tungsten-coated magnesium (W/Mg), 700 nm/50 μm], a poly(lactide-*co*-glycolide) (PLGA)-based encapsulation and interlayer (50 μm), and a radiofrequency PIN diode (monocrystalline silicon nanomembrane, Si NM, 320 nm). The extension electrode (W/Mg, 700 nm/50 μm) delivers electrical stimuli from the power harvesting unit to the myocardium. The tissue contact pad consists of the exposed pair of electrodes (W/Mg, 700 nm/50 μm; $2.0 \times 1.4\ mm^2$) with adjacent holes (700 μm diameter) for suturing to the heart with bioresorbable suture (Fig. 36.1A, inset). Optical mapping of an ex vivo Langendorff-perfused mouse heart shows that the pacemaker drives the activation of the myocardium, as evidenced by anisotropic activation of the epicardium that originates from the location of the pacemaker electrode (Fig. 36.1B). The device can be implanted for in vivo pacing with power delivered wirelessly by aligning a transmission antenna (12 mm diameter, 3 turns) onto the chest where the pacemaker is located (Fig. 36.1C). Once the implanted pacemaker is turned on, the electrocardiogram (ECG) signal converts from narrow QRS complexes, consistent with normal rate sinus rhythm, to widened and high-amplitude QRS complexes with shortened R-R intervals, which signifies successful ventricular capture (Fig. 36.1D). To achieve the bioresorbable characteristics of the bioresorbable pacemaker, all parts of the device are made of bioresorbable materials. For example, inorganic materials, such as Mg, W, and Si NM, disappear into nontoxic products, according to ($Mg + 2H_2O \rightarrow Mg(OH)_2 + H_2$), ($2W + 2H_2O + 3O_2 \rightarrow 2H_2WO_4$), and ($Si + 4H_2O \rightarrow Si(OH)_4 + 2H_2$), respectively. Organic constituents, such as PLGA and Candelilla wax, also dissolve by hydrolysis into their monomers and resorb into the body. The constituent materials largely dissolve within 5 weeks, and the remaining residues completely disappear after 7 weeks (Fig. 36.1E). The processes of bioresorption can also be monitored noninvasively and at high resolution using computed tomography (Fig. 36.1F).

Novel energy sources

Due to the limited life span of the battery, implantable device replacement surgery is necessary every 7–10 years for permanent pacemakers[8] or every 3–6 years for ICDs.[9] However, the procedure for the replacement of an implanted

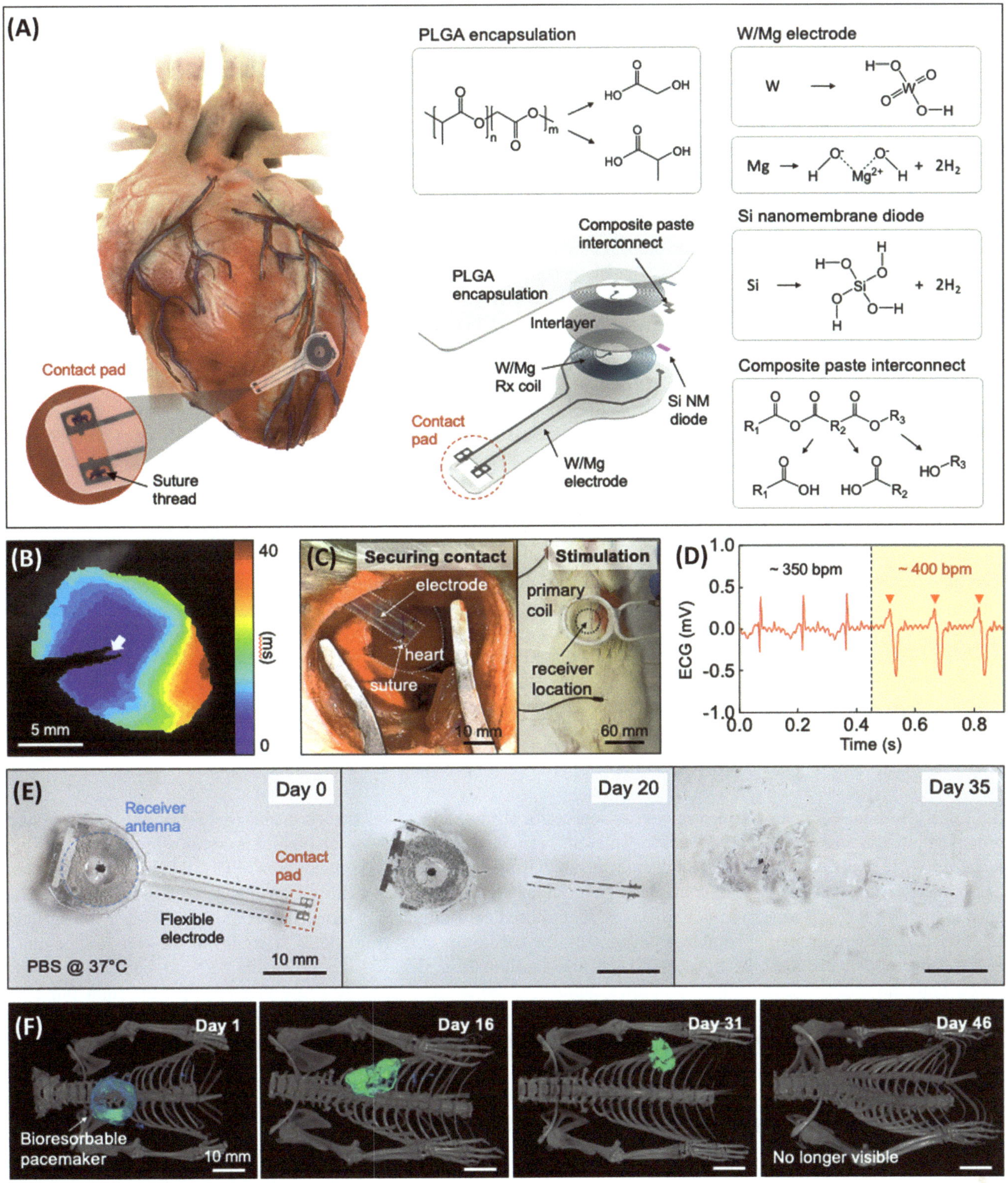

FIGURE 36.1 (A) Schematic illustration of the bioresorbable cardiac pacemaker with their design and constituent materials with their hydrolysis reaction. (B) Activation map of membrane potential for ex vivo demonstration on mouse. The *white* arrow indicates the position of the electrode pad. (C) Photographs of surgical and experimental procedures of bioresorbable pacemaker in a chronic in vivo rat model. (D) ECG signals before (*white* background) and during electrical stimulation (*yellow* background; *red* arrows indicate delivered electrical stimuli). (E) Photographs of the dissolution of a device immersed in PBS (pH = 7.4; 37°C). (F) Three-dimensional-rendered computed tomography images of rats collected over 7 weeks after the implantation of bioresorbable pacemakers. *Reproduced with permission from Choi YS, et al. Fully bioresorbable, leadless, battery-free cardiac pacemaker.* Nat Biotechnol. *2021.*

pacemaker has unavoidable risks, such as tearing of the surrounding blood vessels, perforation of the heart, lodging of blood clots in the lungs, and eventual loss of vascular access.[10] Finding alternative energy sources to increase the lifetime of a conventional pacemaker and to realize a battery-free pacemaker system is an essential challenge.

Mechanical energy

Piezoelectric devices have been suggested as an alternative energy source for cardiac pacemakers.[11] Piezoelectric materials, such as lead zirconatate titanate, single crystalline (1-x) $Pb(Mg_{1/3}Nb_{2/3})O_3 - xPbTiO_3$ (PMN-PT), and barium titanate ($BaTiO_3$), enable us to scavenge inexhaustible mechanical energy since they can convert the mechanical stress to electricity by changing polarization. Hwang et al. developed a self-powered piezoelectric pacemaker has a simple metal–insulator (PMN-PT)–metal (MIM) structure.[12] The piezoelectric harvester directly connects to stimulation electrodes to deliver electrical energy for pacing the heart (Fig. 36.2A (i)). The generated energy (2.7 μJ) from one bending motion of the piezoelectric stimulator is greater than the threshold energy (1.1 μJ) necessary to electrically stimulate the heart (Fig. 36.2A (ii)). When the piezoelectric stimulator cyclically bends and unbends, the ECG shows corresponding pacing peaks (Fig. 36.2A (iii)).

Static electricity

Static electricity is an abundant energy source that can be exploited using triboelectric energy generators.[13] Triboelectric devices consist of a pair of dielectric materials and attached electrodes, and such electrodes are connected via an external circuit (Fig. 36.2B (i)). Contact between two different materials causes a charge transfer between the surfaces of the materials, but there is no electron flow in the external circuit because transferred charges with opposite polarities are fully balanced. As the two surfaces separate, transferred charges on the contact surfaces of both materials induce charge flow on their respective bottom electrodes to reach the equilibrium states. Recently, there have been trials to collect static electricity induced by biomechanical motion, such as the beating heart, muscle contraction and relaxation, and blood circulation, and convert it to an alternative energy source for implantable biomedical devices. An implantable triboelectric nanogenerator (iTENG) converts the static electricity generated from cardiac motion to the electric energy and stores it in the capacitor of the power management unit (PMU). Then, by wireless passive triggering, the PMU turns on the pacemaker to pace the heart (Fig. 36.2B (ii)). Ouyang et al. demonstrated this iTENG-based self-powered pacemaker system in a pig model.[14] The PMU continuously harvests static electricity from normal cardiac motion to fully charge in approximately 200 min. The self-powered system can drive the heart rhythm as evidenced by a captured QRS complex morphology in the ECG (Fig. 36.2B (iii)).

Light

Light can also be used as an alternative energy source for stimulation of cardiac cells and tissues by dopant modulation in the semiconducting materials. Parameswaran et al. fabricated a polymer–semiconductor composite mesh using SU-8 as the polymer support component and coaxial p-type/intrinsic/n-type silicon nanowires (PIN-SiNWs) as the semiconductor modulation component.[15] Neonatal rat ventricular cardiomyocytes cultured on the mesh demonstrate that the mesh is highly biocompatible, and alignment of the cultured cardiomyocytes along the axis of the SU-8 mesh structure suggests that the mesh is a favorable environment for cardiomyocyte (Fig. 36.2C (i)). In an ex vivo Langendorff-perfused rat model, the mesh is placed onto the myocardium (Fig. 36.2C (ii)). When the 532-nm light (19.7 mW) is pulsed onto the mesh at 2 Hz, the ECG shows capture of the heart rhythm at a rate that matches the light pulse frequency (Fig. 36.2C (iii)). Considering that there is no increase in beating frequency by composite mesh or light alone, this result suggests that dopant modulation in Si NW can stimulate the cardiac cells by light.

Magnetic fields

The battery-free operation and control schemes can be achieved by wireless interfaces that exploit magnetic fields. In a magnetic field-powered battery-free pacemaker, the secondary antenna in the implanted pacemaker receives AC radiofrequency power (frequency = 13.56 MHz) from a primary antenna in the external function generator via magnetic resonant coupling. Then, the single wave rectifier, which includes a matching capacitor for tuning the system, rectifies the AC power to DC and provides a stable supply to the microcontroller (Fig. 36.2D (i)). The highly miniaturized harvester and stimulation control unit comprise the receiver of the pacemaker (Fig. 36.2D (ii)). In a magnetic field-powered schematic, the weight of the device can be reduced to just 110 mg. The pacemaker developed by Gutruf et al. harnesses electromagnetic resonant coupling for wireless power transfer to successfully power a highly miniaturized pacemaker that can capture the heart up to postoperative day 6[16] (Fig. 36.2D (iii)).

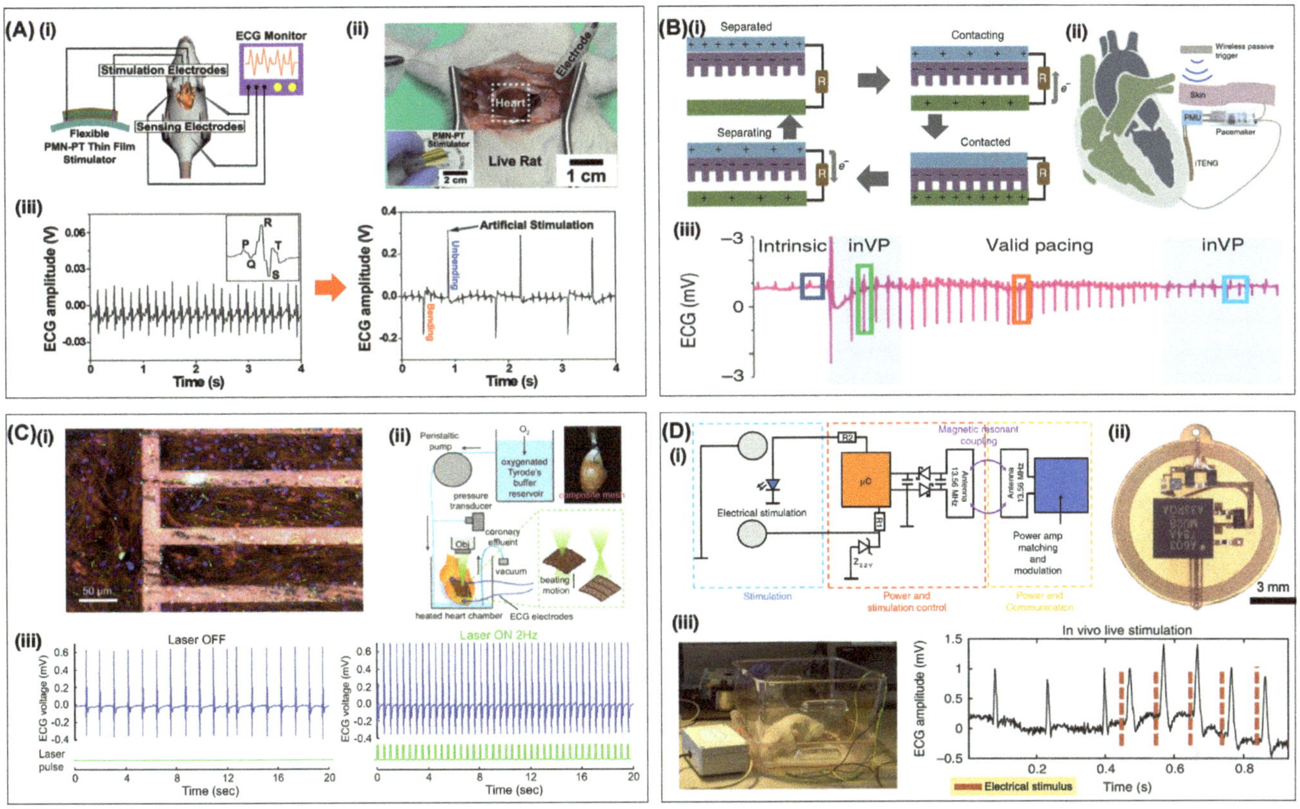

FIGURE 36.2 (A) (i) Schematic illustration of the cardiac pacing setup with PMN-PT-based piezoelectric stimulator. (ii) A photograph of the medical experiment on a living rat for stimulating the heart. The inset is a snapshot of flexible PMN-PT thin film stimulator. (iii) The simultaneously recorded ECG in a normal rat heart before (left) and after (right) subjecting mechanical energy. (B) (i) Schematic diagram of the working principle of triboelectric energy generators. (ii) Schematic illustration of iTENG-based self-powered cardiac pacemaker system. (iii) Changes in ECG during self-powered device operation. InVP, invalid pacing. (C) (i) Confocal fluorescence microscope image of primary neonatal cardiomyocytes cultured partly on a fibronectin-coated composite mesh. (ii) Ex vivo Langendorff optical stimulation setup. (iii) ECG recordings from an adult heart with a composite mesh, before (left) and during (right) optical stimulation. (D) (i) Schematic illustration of the operating principles of magnetic field-powered battery-free pacemaker. (ii) Microscopic image of the receiver and control unit of the device. (iii) Chronic in vivo pacing setup with wireless pacing cage. ECG data shows that rat hearts were captured while the animal was freely moving. *Reproduced with permission from: (A) Hwang GT, et al. Self-powered cardiac pacemaker enabled by flexible single crystalline PMN-PT piezoelectric energy harvester.* Adv Mater. *2014;26:4880–4887. (B) Ouyang H, et al. Symbiotic cardiac pacemaker.* Nat Commun. *2019;10:1821. (C and D) Gutruf P, et al. Wireless, battery-free, fully implantable multimodal and multisite pacemakers for applications in small animal models.* Nat Commun. *2019;10.*

New interface materials

The materials at the interface of the bioelectronic platforms and organ tissues defines the efficiency of the optical, electrical, mechanical, and chemical coupling. In addition, interface materials affect the degree of the tissue's foreign body reaction, which is one of the primary determining factors in the functional lifetime of bioelectronic devices.

Graphene

Graphene is an allotrope of ordinary carbon where carbon atoms are arranged in a two-dimensional lattice structure to exclusively form bonds with itself. Its unique atomic structure gives graphene exceptional conductive, mechanical, optical, and thermal properties, not to mention its excellent biocompatibility. Graphene is a promising material for pacemaker electrode interfaces because it overcomes problems such as rigidity, poor biocompatibility, and limited long-term stability in the body. Cultured cardiomyocytes have been shown to strongly adhere to graphene surfaces.[17] These results suggest that lower amplitude pacing thresholds may be achievable since there is more direct delivery of electrical energy to the cardiac tissue. Graphene has also been used as an electrode material for microelectrode arrays (MEA). Rastogi et al. recorded electrical signals from cultured cardiomyocytes with high temporal resolution.[18] The biocompatibility of cardiomyocytes with the graphene MEA shows that graphene is a suitable material for stable, long-term tissue–electrode interfacing.

Graphene may be able to act as an interface to enable optical energy to depolarize cardiac tissue without exogenous modifications to cells or tissue to potentially realize optical pacemakers. Savtchenko et al. created a graphene interface that can harness light to cause depolarization of cardiac cells[19] (Fig. 36.3A (i)). Photogenerated hot, ballistic electrons from graphene displace cations near the G-biointerface due to capacitive coupling between the cell membrane and the graphene surface, which results in membrane depolarization. To evaluate the efficacy of the graphene interface, a rGO dispersion (0.1 or 0.5 mg/mL) is injected into zebrafish embryos at 3 days postfertilization (Fig. 36.3A (iv)). When the zebrafish heart was illuminated, there was a significant increase in heart rate, whereas optical stimulation of control zebrafish had no effect on their heart contractions (Fig. 36.3A (v)).

Hydrogel adhesives

Hydrogel adhesives can provide a mechanically compliant and electrically conductive interface between the electrode and cardiac tissue. Hydrogel adhesives are typically a mixture of covalently crosslinked bifunctional polyethylene glycol-polylactidediacrylate triblock macromers (PEG-PLA-DA) and ionically crosslinked sodium alginate. A significant advantage of bioadhesive attachment of bioelectronics is that the inherent tissue structure is not disrupted during attachment. In addition, the bioadhesive's low modulus mechanical properties and high levels of elongation minimizes the potential damage at the tissue–bioelectronics interface for both the tissue and the device to further lengthen the functional lifetime of the device. Yang et al. developed a bioadhesive that enables attachment of bioelectronics to the heart.[20] The bioadhesive provides strong adhesion between a miniature pacemaker and the heart across 10 days postsurgery (Fig. 36.3B (ii, middle)). The bioadhesive can successfully pace the heart for up to 8 days postsurgery due to its electrical conductivity and fast electrical response time (Fig. 36.3B (iii)). In comparing suture attachment to bioadhesive attachment, large perforations in the cardiac muscle appear with surrounding fibrotic tissues induced by the suturing process. Perforation can potentially result in compromised mechanical contraction of the myocardium and life-threatening arrhythmias during implantation procedures and after surgery. The bioadhesive attachment approach shows no such effects (Fig. 36.3B (ii, right)).

Disease applications

Ablation

Catheter ablation is one of the primary options to treat atrial fibrillation (AF). In a typical ablation procedure, a catheter threads through the cardiovascular system and into the heart to place tiny scars at specific locations in the tissue in order to restore normal rhythms. However, arrhythmia-free survival after a single catheter ablation procedure of persistent AF has been reported to be as low as 25% at a 7-year follow-up.[21] Moreover, ablation as a therapy involves major risks including stroke, cardiac perfusion, pulmonary vein stenosis, and the rare but often fatal atrioesophageal fistula.[22]

The current gold standard procedure for AF ablation involves a point-by-point ablation via a single-tip catheter, which is usually combined with a three-dimensional (3D) mapping system.[23] The most common energy source for AF ablation is radiofrequency ablation. Other available modalities include cryoablation and, more recently, irreversible electroporation.[24] However, the single-tip catheter technology is limited by tip-to-tissue contact and a steep learning curve of the operators.[25] As a result, achieving contiguous, transmural, permanent lesions necessary to successfully treat AF remains challenging. Optimal lesion formation depends on a number of factors, including the duration of each local ablation attempt, the size of the catheter tip, power setting, contact temperature, and pressure. Current single-tip catheter technologies can only provide limited information on these factors, which often results in suboptimal lesion formation.

The challenges of the single-tip catheter system have led to the advent of balloon-based ablation technologies including the cryoballoon (ArcticFront by Medtronic Inc., Minneapolis, MN, USA), the laser balloon (HeartLight, CardioFocus, Marlborough, MA, USA), and the multielectrode contact mapping and ablation system (Globe, Kardium Inc, Burnaby, BC, Canada). Han et al. have also recently demonstrated a novel multimodal, multilayered balloon catheter technology in animal studies and donor human hearts.[26] The balloon catheter (Fig. 36.4) is capable of conformal tissue contact for simultaneous multimodal mapping of temperature, pressure, electrical sensing and stimulation, radiofrequency ablation, and irreversible electroporation. These features allow physicians to acquire a richer set of physiological data to enable better lesion formation and shorter procedure durations.

These emerging technologies in ablation seek to improve the safety and efficacy of AF ablation therapy, while also reducing procedure time and the associated learning curve. Moreover, the use of these technologies will also enable better understanding of mechanisms underlying AF and consequently lead to improved clinical outcomes for patients undergoing treatment for AF.

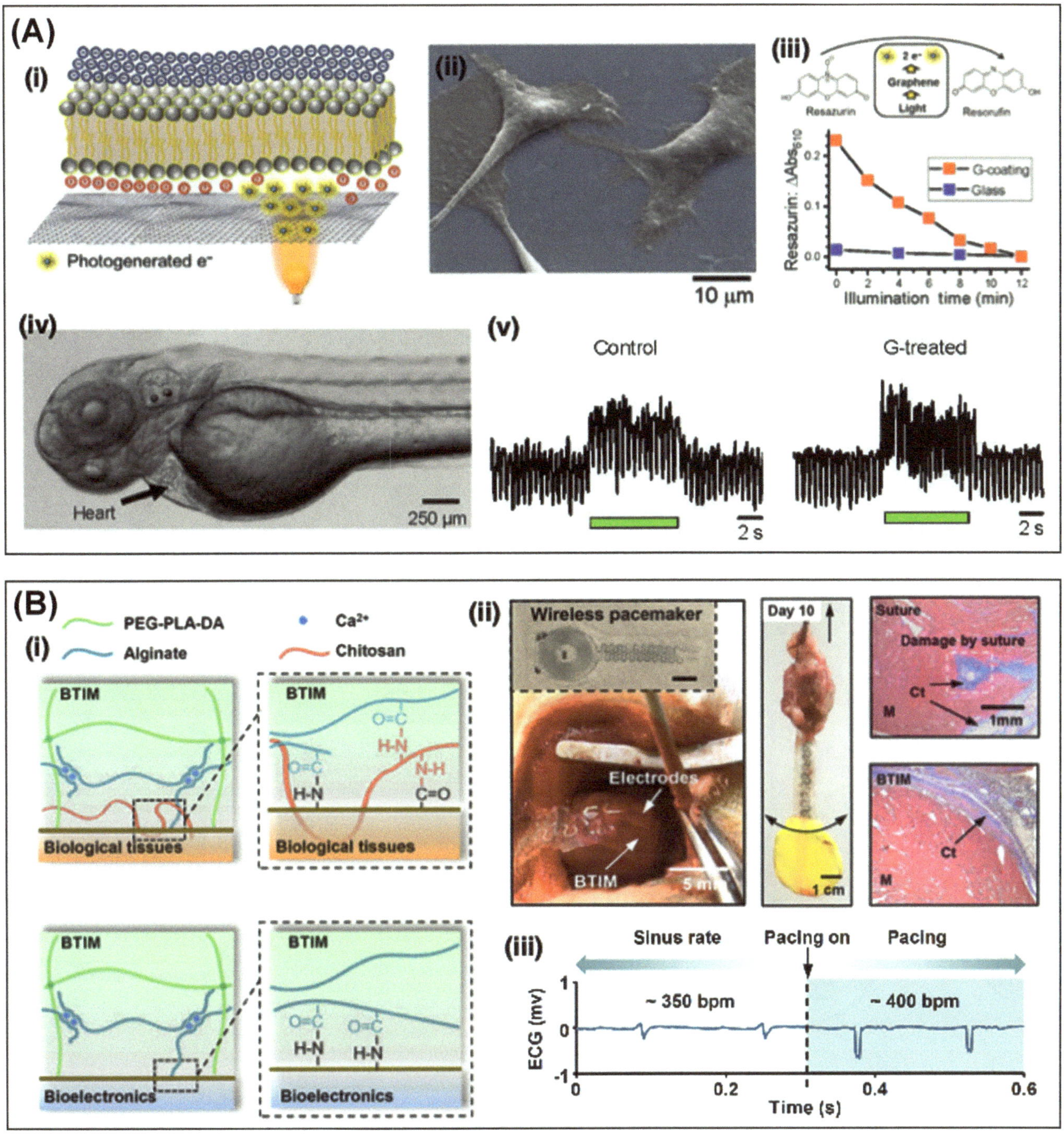

FIGURE 36.3 (A) (i) Illustration of the proposed mechanism of cellular optical stimulation using light-activated G-biointerfaces. (ii) Representative scanning electron microscopy (SEM) image of cardiomyocytes (CMs) on reduced graphene oxide (rGO)-coated coverslips. (iii) Photocatalytic reduction of resazurin to resorufin. (top) Changes in absorption of a 40 mM resazurin solution (at its absorption peak of 610 nM) on G-biointerfaces (red squares) or noncoated glass coverslips (blue circles) as a function of illumination time (465 nm, 1.8 mW/mm^2). (bottom) (iv) Zebrafish embryo 3 dpf. (v) Representative heart contraction traces demonstrating optical stimulation effects (*green* light, 0.8 mW/mm^2) in control (left) and G-treated (0.5 mg/mL; right) zebrafish embryos. (B) (i) Interfacial bonding mechanism of hydrogel adhesive (BTIM) to the surfaces of biological tissues (top) and bioelectronics devices (bottom). (ii) (left) Photograph of the pacemaker and the process of adhering the electrodes to the myocardium in a rat model. (middle) Manual mechanical tests demonstrate the strength of adhesion following removal of the device on day 10 postsurgery. (right) Histological evaluations by Masson's trichrome staining of cross sections of hearts compare implantations secured with sutures (top) and the bioadhesive (bottom). (iii) Representative ECG shows conversion from normal sinus rhythm (*white* background) to pacing rates of 400 beats per minute (*light blue* background). *Reproduced with permission from: (A) Savchenko A, et al. Graphene biointerfaces for optical stimulation of cells.* Sci Adv. *2018. doi:10.1126/sciadv.aat0351. (B) Yang Q, et al. Photocurable, bioresorbable adhesives as functional interfaces between flexible bioelectronic devices and soft biological tissues.* Nat Mater. *2021.*

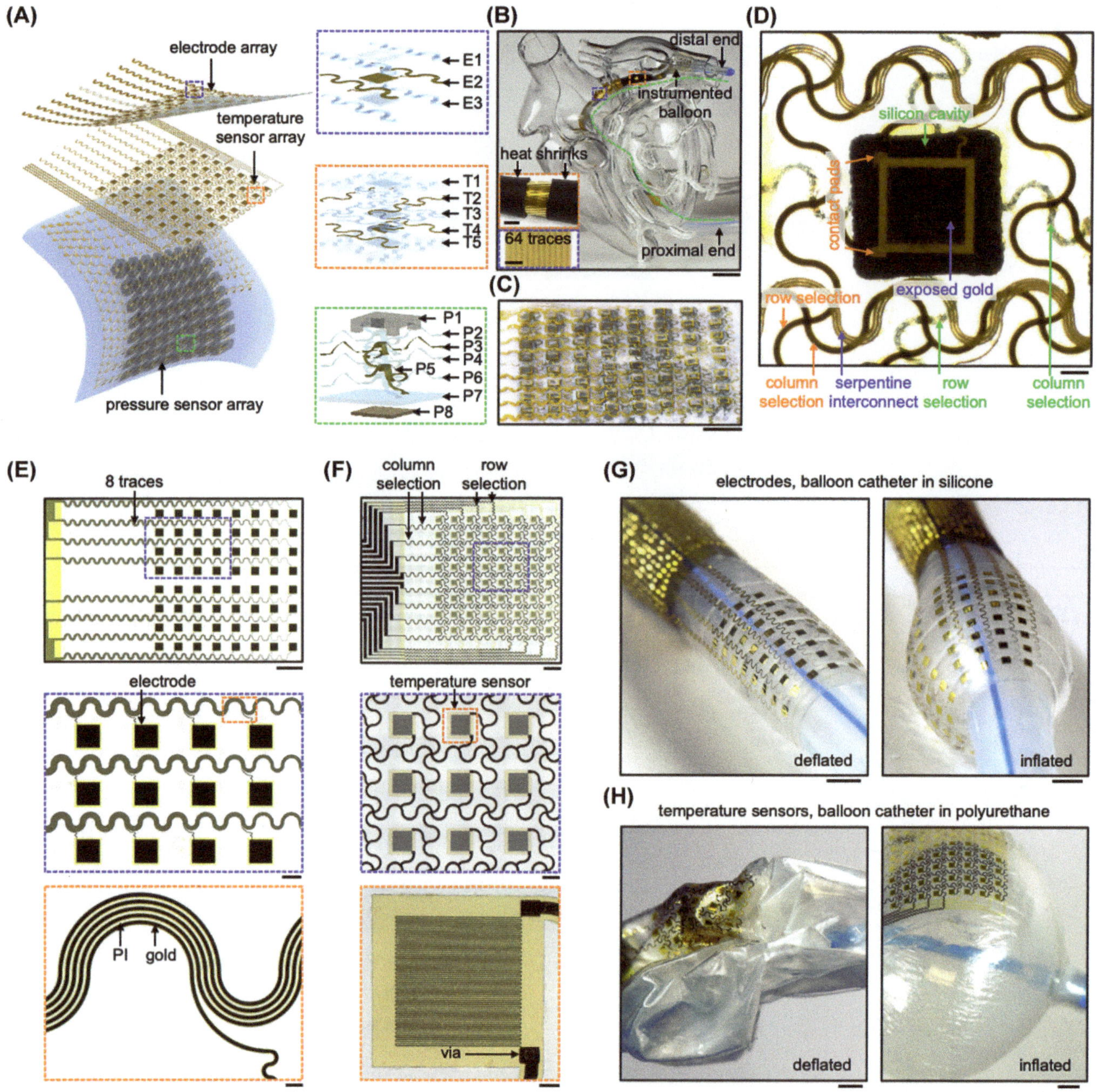

FIGURE 36.4 (A) Schematic illustration of the device arrays in a multilayer format. The top functional system allows electrophysiological mapping and radiofrequency ablation or irreversible electroporation, with layers E1 (the top polymer film with openings), E2 (the array of electrodes and interconnects), and E3 (the bottom polymer film) for the electrode array. The next system enables measurements of temperature distributions, with layers T1 (the top polymer), T2 (the array of temperature sensors and row selection lines), T3 (a polymer interlayer with via openings), T4 (the column selection lines), and T5 (the bottom polymer film). The bottom system supports measurements of pressure distributions, with layers P1 (the array of silicon cavities), P2 (the top polymer film), P3 (the array of pressure sensors and row selection lines), P4 (a polymer interlayer with via openings), P5 (the column selection lines), P6 (the bottom polymer film), P7 (a layer of silicone elastomer), and P8 (an array of rigid islands). (B) Image of an inflated, instrumented balloon catheter inserted into a transparent heart model. The green dashed curve illustrates the path of the catheter. Scale bars, 1 cm (main image); 2 mm (inset). (C) Image of the vertically stacked arrays of electrodes, temperature sensors, and pressure sensors. Scale bar, 2 mm. (D) Image of one unit of the stacked arrays. Arrows and text in blue, orange, and green correspond to layers for electrodes, temperature sensors, and pressure sensors, respectively. Scale bar, 200 μm. Images of the array of electrodes (E) and array of temperature sensors (F) in planar format. (G) Images of the array of electrodes transferred onto a balloon catheter of silicone. (H) Images of the array of temperature sensors transferred on a balloon catheter of polyurethane. Scale bars in e–h, 2 mm for the images in black squares, 500 μm for images in blue outline, 100 μm for images in orange dashed outline. *Reproduced with permission from Han M, et al. Catheter-integrated soft multilayer electronic arrays for multiplexed sensing and actuation during cardiac surgery.* Nat Biomed Eng. *2020;4:997–1009.*

Pacemakers

Pacemakers are implantable devices that treat bradyarrhythmias, heart block, and ventricular dyssynchrony by delivering electrical impulses at a regular rate to maintain a healthy heart rhythm. Over 700,000 pacemakers are implanted each year worldwide.[27] Generally, the modern pacemaker consists of two main components: the generator and the electrodes. The generator is implanted underneath the skin at the chest, and the electrodes' wires thread through the veins and into the heart chamber. The latest advances in pacemaker technology aim to create a more physiological pacing configuration, increase the miniaturization of technology, and eliminate leads that cause complications.

Biventricular pacemakers provide cardiac resynchronization therapy (CRT) to resynchronize ventricular contraction. CRT has been shown to better maintain healthy physiology, such as myocyte survival,[28] sarcomere calcium sensitivity,[29] and adrenergic signaling.[28] Lyu et al. developed a leadless battery-free biventricular pacemaker that eliminates risks associated with lead breakage and minimizes the associated physical bulk and mass of batteries. The pacemaker is flexible and highly miniaturized (11×11 mm) and has two prongs that attach onto the surface of the heart[30] (Fig. 36.6I). The battery-free nature of the device is enabled by wireless inductive power transfer. A transmitting (Tx) signal delivers energy to the receiving (Rx) coil to power the pacemaker in order to deliver energy for pacing. For biventricular pacing, one device is placed onto each ventricle (Fig. 36.6H). Each device operates at different industrial, scientific, and medical (ISM) bands at 13.56 and 40.68 MHz, respectively. Operation at separate ISM bands allows independent control of each device to achieve pacing of each ventricle. The biventricular pacemaker can separately pace each ventricle or provide biventricular pacing (Fig. 36.6J). Importantly, biventricular pacing with zero offset time between leads resulted in a more normal physiological stroke volume.

The multisite pacing strategy has been shown to further improve the efficiency of CRT in reducing dyssynchrony and improving hemodynamics compared to biventricular pacing.[31] Multisite pacing involves placing leads at more than two locations in the heart (posterior wall, anterior wall, lateral wall, apex) to restore globally synchronous contractions in a manner that more closely matches physiological conduction. Different individuals require different optimal site combinations.[31] Conformal mesh arrays for the heart are well-suited to deliver multisite electrical stimulation for pacing at various locations across the heart. These soft, elastic meshes conform to the epicardium to provide multiple contact points for pacing without impeding the contraction and relaxation of the heart. Several groups have created conformal arrays for CRT. Park et al. created an epicardial mesh that mimics the Purkinje fiber system to provide impulses for coordinated contractions of the heart.[32] The mesh is composed of an elastic and highly conductive nanocomposite that can be tailored to specific heart geometries. The mesh also records high-quality electrograms of the heart. In a rat model of myocardial infarction, the mesh restored ventricular activation as evidenced by a normalized QRS duration. Sunwoo et al. devised a stretchable conductive epicardial mesh for ECG monitoring and electrical stimulation.[33] The mesh provides real-time measurements of surface electrograms of the heart to monitor the effect of the electrical stimulations. Biventricular pacing using this mesh demonstrated improved left ventricular fractional shortening in a rat model of myocardial infarction.

Leadless pacemakers are highly miniaturized pacemakers that provide pacing therapy without the need for leads. This leadless design eliminates the complications such as lead fracture. Several groups have created highly miniaturized pacemakers that are small enough to test in mouse models. Laughner et al. developed a wireless battery-free pacemaker that weighs 560 mg with dimensions of 3.3×13.7 mm[34](Fig. 36.5A-C). This miniature device was implanted transvenously to achieve 30 days of pacing (Fig. 36.5D). Gutruf et al. further miniaturized the wireless battery-free pacemaker to make it even more lightweight (110 mg)[16] (Fig. 36.5E-F). In an in vivo rat model, the device achieved 6 days of pacing with capture (Fig. 36.5G). The device is also highly biocompatible, as evidenced by no significant increase in intramyocardial fibrosis. The pacemaker also has a biventricular configuration for simultaneous pacing of both ventricles. Choi et al. built a leadless battery-free pacemaker that can be fully resorbed by the body[35] (Fig. 36.1). The device's constituent materials can all be naturally resorbed by the body through metabolic action and hydrolysis. In circumstances of second degree atrioventricular block, the pacemaker can restore ventricular activation. The implanted device and its resorption over the course of several weeks is highly biocompatible and does not disturb normal physiology or the mechanical function of the heart.

Biventricular pacemakers provide cardiac resynchronization therapy (CRT) to resynchronize ventricular contraction. CRT has been shown to better maintain healthy physiology, such as myocyte survival,[28] sarcomere calcium sensitivity,[29] and adrenergic signaling.[28] Lyu et al. developed a leadless battery-free biventricular pacemaker that eliminates risks associated with lead breakage and minimizes the associated physical bulk and mass of batteries. The pacemaker is flexible and highly miniaturized (11×11 mm) and has two prongs that attach onto the surface of the heart[30] (Fig. 36.5I). The battery-free nature of the device is enabled by wireless inductive power transfer. A transmitting (Tx)

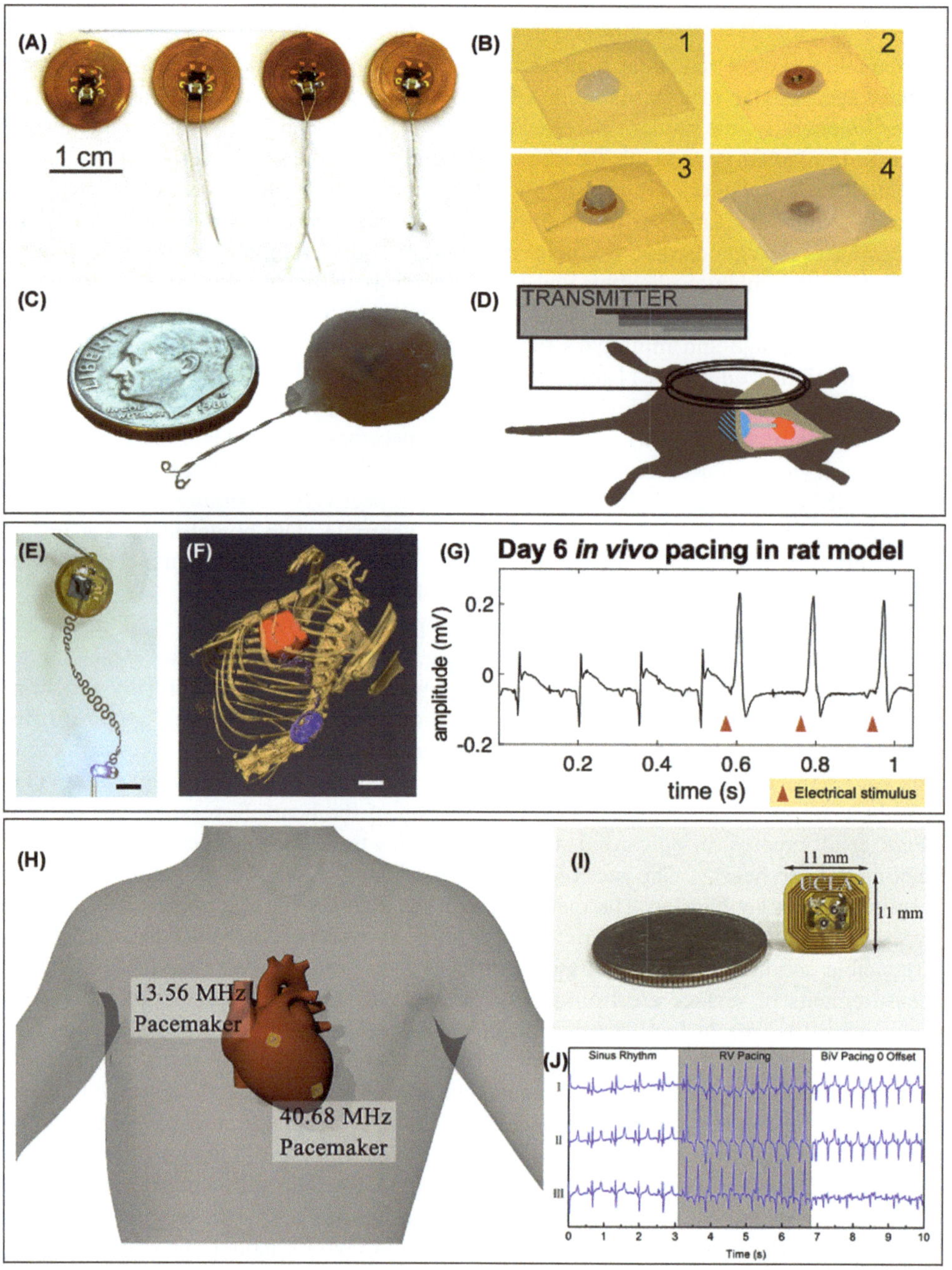

FIGURE 36.5 (A) Teflon-coated silver wire is attached to the circuit board, wound together, and coiled with a 0.5-cc syringe. (B) Process of making pacemaker: (1) a bead of silastic is placed on a piece of parafilm, (2) the device is placed on the bead, (3) coated with an additional layer of Silastic, (4) and topped with a piece of gas permeable film. (C) Final pacemaker compared with a US dime. (D) Artistic rendering of external transmitter interacting with abdominally implanted receiver in mouse. (E) Photographic image with twisted device and active optogenetic stimulator. Scale bar, 5 mm. (F) Three-dimensional segmentation of anatomical positioning of device (blue) with respect to the rat heart (red). Scale bar, 1 cm. (G) Chronic in vivo pacing of rat hearts was achieved by implanted pacemakers for up to 6 days. (H) Illustration of implementation of leadless biventricular pacemakers. (I) Picture of the biventricular pacemaker compared with a US quarter. (J) ECG recordings from an open chest pig study using the leadless biventricular pacemakers showing a transition from right ventricle single-site pacing to biventricular pacing. *Reproduced with permission from: (A–D) Laughner JI, et al. A fully implantable pacemaker for the mouse: from battery to wireless power.* PLoS One *2013;8:1–8. (C–F) Gutruf P, et al. Wireless, battery-free, fully implantable multimodal and multisite pacemakers for applications in small animal models.* Nat Commun. *2019;10. (G–I) Lyu H, et al. Synchronized biventricular heart pacing in a closed-chest porcine model based on wirelessly powered leadless pacemakers.* Sci Rep. *2020;10:1–9. Available from http://doi.org/10.1038/s41598-020-59017-z.*

signal delivers energy to the receiving (Rx) coil to power the pacemaker in order to deliver energy for pacing. For biventricular pacing, one device is placed onto each ventricle (Fig. 36.5H). Each device operates at different industrial, scientific, and medical (ISM) bands at 13.56 and 40.68 MHz, respectively. Operation at separate ISM bands allows independent control of each device to achieve pacing of each ventricle. The biventricular pacemaker can separately pace each ventricle or provide biventricular pacing (Fig. 36.5J). Importantly, biventricular pacing with zero offset time between leads resulted in a more normal physiological stroke volume.

These next generation-pacemakers demonstrate the potential of enhanced electrotherapy for bradyarrhythmias and heart block. These new pacemakers are significantly miniaturized in size and mass through the elimination of batteries by harnessing novel energy sources like wireless inductive power transfer. Conformal arrays provide a wider variation of pacing sites to more closely mimic the native cardiac conductions system. Pacemakers can also be bioresorbable so that the device self-eliminates without the need for surgical removal. The future of cardiac pacemakers will provide pacemakers that more seamlessly integrate into the body for more physiological pacing and self-elimination upon completion of therapy.

Defibrillators

Defibrillator devices treat fibrillation by delivering electrical shocks to the heart to restore normal sinus rhythm. Automated external defibrillators are ubiquitously placed throughout public spaces for emergencies, while patients who have a history of life-threatening fibrillations often have an implantable defibrillator within the body. The conventional ICD consists of a generator, commonly known as the "can," and stimulation coils that extend from the generator to sit at the endocardial apex of the ventricles.

Innovations in defibrillators seek to create a conformal mesh that encompasses the heart for defibrillation. This mesh is outfitted with electrodes at multiple locations around the heart for mapping of cardiac dynamics and spatiotemporal accurate delivery of defibrillating stimuli. This array of electrodes enables feedback-controlled targeted defibrillation which reduces the energy requirements for termination of arrhythmias. These compliant meshes covering the heart have greater coverage of electrically active surfaces for defibrillation so that the native Purkinje fiber system of the ventricles activates in a more physiological manner. In addition, conformal contact minimizes impedance at the tissue–electrode interface to record high-quality signals even during cardiac motion. Park et al. designed a conductive epicardial mesh that conformally integrates to the heart.[32] The mesh is composed of a homogenous nanocomposite of Ag nanowires and styrene–butadiene–styrene rubber to achieve both elasticity and electrical conductivity. The mesh was devised in a serpentine pattern to create softer mechanical properties. The mesh's efficacy as a defibrillator was demonstrated in postinfarct rats whose hearts were integrated with the epicardial mesh. Ventricular fibrillation was induced, and a biphasic shock of 2 J delivered by the mesh successfully terminated the fibrillation. This 2 J shock is significantly lower than the 20–30 J high-energy shocks that are typically delivered by conventional implantable defibrillators.[36]

The designs of the epicardial meshes can be patterned to enable more efficient current transfer from electrode to tissue. Since current is preferentially distributed at the edge of an electrode,[37,38] increasing the ratio of the edge length to the geometric surface area more efficiently transfers current from electrode to tissue. Xu et al. developed a 3D multifunctional integumentary membrane with a fractal mesh geometry so that electrodes maximally cover the surface area of the myocardium without compromising the compliance of the mesh[39] (Fig. 36.6A and B). In an ex vivo Langendorff-perfused rabbit heart model, the stretchable thin membrane induces a virtual electrode pattern by delivering 50 V electrical shocks (Fig. 36.6C and D). The induced virtual electrode pattern is an important feature that determines successful electrotherapy and demonstrates that the membrane is capable of defibrillation. In addition, temperature sensors can be outfitted in the areas between fractal defibrillation patterns, which are useful for characterizing spatial distributions of temperature in order to guide the precision of ablation therapy.

Epicardial membranes and meshes for defibrillation will more seamlessly integrate onto the heart to better engage the native cardiac conduction system and enable lower power defibrillation. Defibrillation electrodes can be outfitted alongside other sensors to more precisely guide therapy.

Cardiac diagnostic monitors

Cardiac diagnostic monitoring devices record hemodynamic and electrocardiogram signals in order to collect data for diagnosis of arrhythmias. These devices can be wearable or implantable, and they provide long-term monitoring of heart

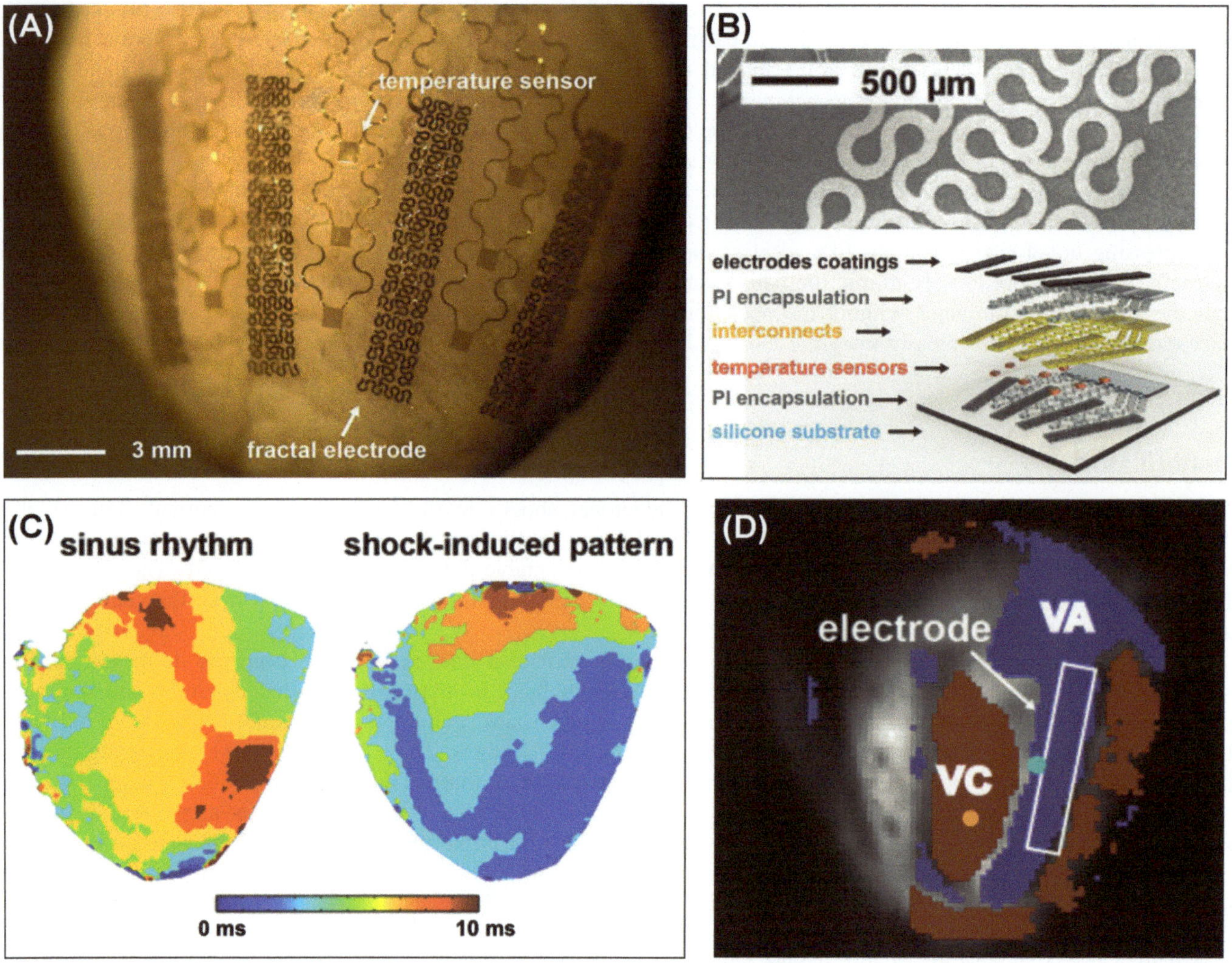

FIGURE 36.6 (A) Image of a representative device integrated on a Langendorff-perfused rabbit heart. The white arrows highlight functional components. (B) SEM image of a fractal electrode (top). Schematic exploded view illustration of device structures and functional components (bottom). (C) Activation maps created from optical signals on one side of the epicardium during a sinus beat (left) and immediately following a 50-V shock (right) from a Langendorff-perfused rabbit heart. (D) Virtual electrode pattern and representative traces from the virtual anode and the virtual cathode illustrating the effect of a 50-V shock applied across the heart from the device. The electrode used is shown in the field of view as highlighted in *white*. SEM, Scanning electron microscopy. *Reproduced with permission from: Xu L, et al. 3D multifunctional integumentary membranes for spatiotemporal cardiac measurements and stimulation across the entire epicardium.* Nat Commun. *2014;5:3329.*

rhythms in order to assist a physician in diagnosis to devise the optimal treatment strategy. The next generation of cardiac monitoring more seamlessly integrates with the body for optimal signal recordings for enhanced diagnosis.

The electrocardiogram is cornerstone to the diagnosis of heart rhythm disorders. To record an electrocardiogram, electrodes are situated at different locations on the body to collect signals. Since it is ideal to collect signals as patients go about their daily life activities, electrodes must be able to maintain contact with the patient in a manner that is comfortable for long-term use.

Monitoring of hemodynamics, or the flow and pressure of blood, gives insight into the mechanical pumping function of the heart. Because electrical excitation and mechanical contraction are closely coupled in the heart, an understanding of a patient's hemodynamics provides insight into the health of the cardiovascular system. These hemodynamic monitors are typically implanted into the body. Son et al. created a bioresorbable stent that monitors blood flow and can be fully resorbed into the body upon completion of diagnosis.[40] This stent is outfitted with range of sensors, such as vessel blood flow monitors and temperature sensors (Fig. 36.7A). Cong et al. devised a real-time blood pressure monitoring system where a cuff is instrumented around a blood vessel. The lightweight device, weighing 130 mg, is battery-free and is wirelessly powered by an external radiofrequency power source. In an in vivo rat model of

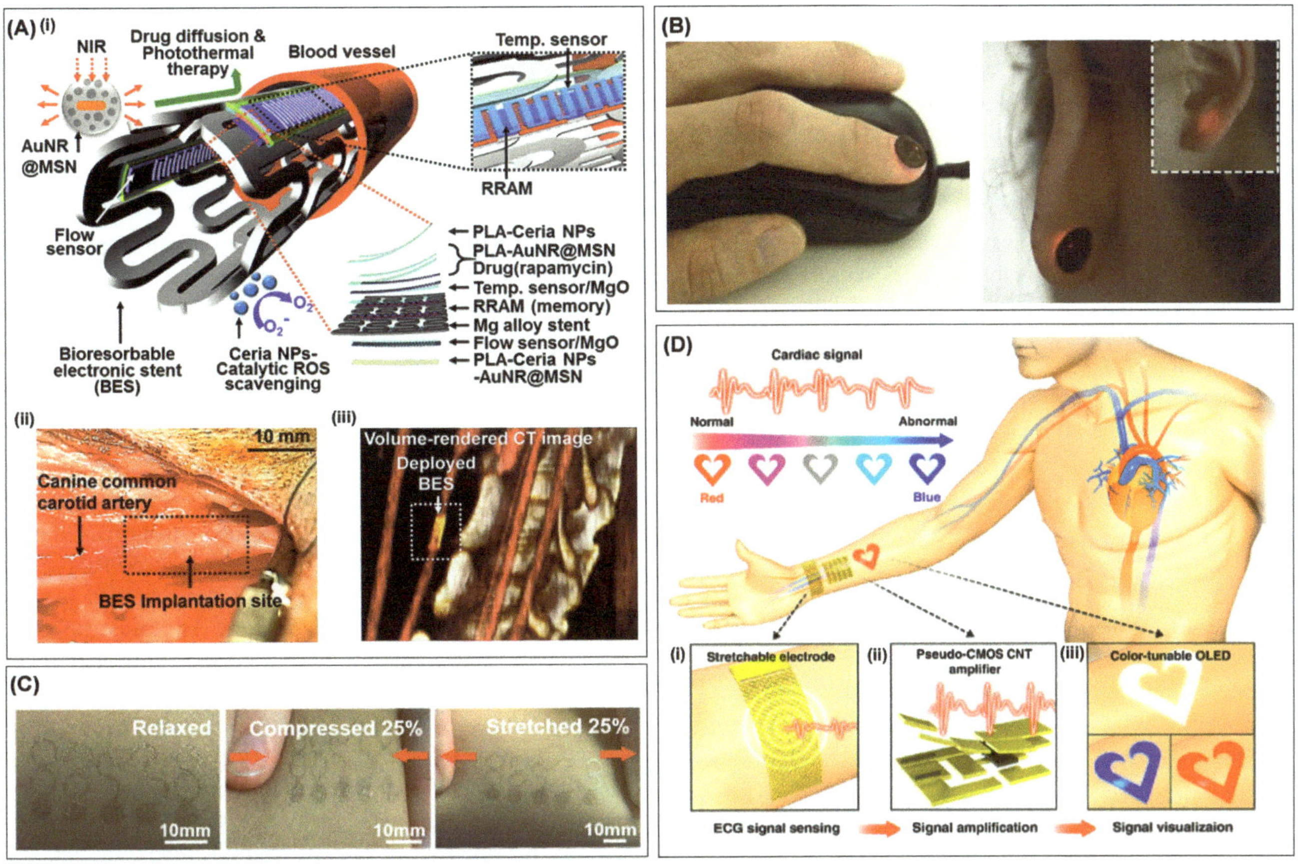

FIGURE 36.7 (A) (i) Schematic illustration of the bioresorbable electronic stent. The bioresorbable electronic stent includes bioresorbable temperature/flow sensors, memory modules, and bioresorbable/bioinert therapeutic nanoparticles. The therapeutic functions are either passive (ROS scavenging) or actively actuated (hyperthermia-based drug release) by near infrared exposure. (ii) Image of the deployed bioresorbable electronic stent in the canine common carotid artery in vivo. (iii) Volume-rendered X-ray computed tomography (CT) image of the deployed bioresorbable electronic stent in the canine common carotid artery. (B) Demonstration of a possible application in which a computer mouse serves as the wireless interface (left). Demonstration of mounting on the back of the earlobe (right). The inset shows the front of the earlobe during the device operation. (C) Graphene electronic tattoo mounted on the skin. The tattoo maintains conformal contact with the skin throughout stretching and compressing of the skin. (D) Schematic illustration of the real-time wearable cardiac-monitoring system. The electrocardiogram (ECG) signals are measured by (i) a stretchable Au electrode and then amplified by (ii) a p-MOS carbon nanotubule inverter for the high signal-to-noise ratio. The retrieved ECG signals are displayed via synchronized color changes of (iii) the wearable color-tunable organic light-emitting diodes. *Reproduced with permission from: (A) Son D, et al. Bioresorbable electronic stent integrated with therapeutic nanoparticles for endovascular diseases.* ACS Nano *2015;9:5937–5946. (B) Kim J, et al. Miniaturized battery-free wireless systems for wearable pulse oximetry.* Adv Funct Mater. *2017;27:1–8. (C) Kabiri Ameri S, et al. Graphene electronic tattoo sensors.* ACS Nano *2017;11:7634–7641. (D) Koo JH, et al. Wearable electrocardiogram monitor using carbon nanotube electronics and color-tunable organic light-emitting diodes.* ACS Nano. *2017;11:10032–10041.*

evaluation, high-quality blood pressure signals can be reliably recorded while animals freely roam around the cage.[41] Small, flexible pulse oximetry devices can be worn on the earlobe or nails in a way that is not irritating or uncomfortable[42] (Fig. 36.7B). The device uses micro-LEDs to optically measure and record heart rate, heart rate variability, and blood oxygen saturation levels.

Wearable electronic devices are a highly noninvasive way to attach electrodes to a patient. One interesting approach to attach these electrodes is by weaving sensors into textiles that subjects wear as clothing. Textiles interwoven with carbon nanotubules or graphene can be employed as wearable electrodes for ECG. These graphene-based textile electrodes are flexible with effective electrical performance and can record ECG signals for a full week with no degradation in signal quality.[43] The textile-based dry, flexible, highly conductive graphene electrodes allow the subject more comfort when wearing the monitoring equipment while recording high-fidelity electrocardiographic signals.[44] Graphene-based e-textiles can be inkjet-printed for wearable electronics that monitor ECG signals, heart rate, and heart rate variability.[45] Users can undergo regular activities, such as taking a shower or running 20 miles, without risking detachment of inkjet-printed electrodes.[46] Graphene can be fashioned into tattoos to deform with the skin to record high-quality signals with superior comfort and mobility[47] (Fig. 36.7C). Fully textile armbands can record high-fidelity signals from a single arm that match ECG signals recorded from conventional gel-based electrodes.[48] Ultrathin wearable textile electrocardiogram monitors can also be tuned to display colors that correspond with the level of abnormality of the heart rhythm to immediately indicate any heart rhythm problems[49] (Fig. 36.7D).

Bioelectronics-enabled medical internet-of-things

Many new bioelectronic systems are available to monitor the human body to collect information about health status. These wearable or implantable devices collect data such as blood pressure, ECG, electroencephalograms, body temperature, and heart rate. Devices wirelessly connect to enable a more interconnected understanding of the body's condition in real time. Information collected by this network of bioelectronic sensors enables early detection, helps make decisions in critical situations, and provides a pathway for more personalized medical diagnosis and treatment.

A network of bioelectronic sensors for human health is the foundation of the medical internet-of-things. The medical internet-of-things is a network of interconnected biomedical devices that collect, process, and identify patterns of the human body to provide information to improve diagnosis and develop enhanced treatment strategies. There are three layers that comprise the medical internet-of-things: perception, network, and application. Perception technologies perceive changes in physiology through electronic sensors that are worn on or implanted into the body to collect data in real time. The collected data are networked together by cloud computing. Data gathered by the sensors are usually very large and is difficult to store and process in the device itself. Cloud computing is useful because it provides a massive data storage infrastructure with greater computing capabilities to process the data. The biomedical data must be stored with security to guarantee the integrity of the data.[50] At the application stage, machine learning (ML) is useful to find patterns in the massive amount of data collected. The goal of the use of ML in the application stage is to improve the physician's diagnostic abilities. ML can automate the annotation and diagnosis of ECG recordings at a performance level similar to that of cardiologists.[51] ECG signals can be collected remotely for automated ML processing to provide patients with high-quality long-distance medical care.[52] ML algorithms may even be able to identify abnormalities in the ECG to diagnose diseases that are difficult to screen, such as left ventricular systolic dysfunction, before its manifestation becomes apparent.[53] Neural networks in ML systems allow for continuous feedback and refinement. ML systems present a more concise and relevant version of the collected information to the physician, who makes the final decision about diagnosis and treatment.

The medical internet-of-things is advantageous because it allows for longer-term monitoring of patient health. For example, long-term monitoring could be useful for patients with arrhythmias that occur infrequently. The patients can be monitored at home for a longer duration of time since incidents of arrhythmias do not always occur during short medical visits. Access to cardiological care can be limited in developing countries or rural areas where medical deserts are common. The medical internet-of-things enables monitoring of health even at long distances where devices can send information directly to physicians who are remote. Greater quantities of high-quality medical data for each individual provide physicians more clues for specific diagnosis and treatment in each case for more personalized care.

Miniaturized conformal cardiac implantable bioelectronics can integrate into the medical internet-of-things networks to provide detailed information about the state of the heart. Currently, there remains a lack of proper cardiac sensors to achieve an internet-of-things for cardiac diseases. These next-generation bioelectronic sensors can measure a more detailed set of parameters about the heart, such as local electrical activity, metabolism, pH, and temperature. Mesh sensor arrays provide higher-resolution monitoring of the heart. Different therapies can be integrated by deploying a combination of devices. For example, heart failure could be treated with a combination of pharmacological, mechanical, and electrical therapy. Data can be wirelessly transmitted from cardiac bioelectronics to enable remote real-time therapy. Bioresorbable bioelectronics can naturally disappear via metabolic action and hydrolysis after diagnosis or therapy is complete to eliminate the need for device extraction surgery, which realizes a more remote therapy process.

References

1. Worley SJ, et al. A new sock electrode for recording epicardial activation from the human heart: one size fits all. *Pacing Clin Electrophysiol.* 1987;10:21–31.
2. Aras K, Burton B, Swenson D, MacLeod R. Spatial organization of acute myocardial ischemia. *J Electrocardiol.* 2016;49:323–336.
3. Xu L, et al. 3D multifunctional integumentary membranes for spatiotemporal cardiac measurements and stimulation across the entire epicardium. *Nat Commun.* 2014;5:3329.

4. Viventi J, et al. Flexible, foldable, actively multiplexed, high-density electrode array for mapping brain activity in vivo. *Nat Neurosci.* 2011;14:1599−1605.
5. Fang H, et al. Capacitively coupled arrays of multiplexed flexible silicon transistors for long-term cardiac electrophysiology. *Nat Biomed Eng.* 2017;1:0038.
6. Sim K, et al. An epicardial bioelectronic patch made from soft rubbery materials and capable of spatiotemporal mapping of electrophysiological activity. *Nat Electron.* 2020;3:775−784.
7. Choi Y, Koo J, Rogers JA. Inorganic materials for transient electronics in biomedical applications. *MRS Bull.* 2020;45:103−112.
8. Mallela VS, Ilankumaran V, Rao NS. Trends in cardiac pacemaker batteries. *Indian Pacing Electrophysiol J.* 2004;4:201−212.
9. Horlbeck FW, Mellert F, Kreuz J, Nickenig G, Schwab JO. Real-world data on the lifespan of implantable cardioverter-defibrillators depending on manufacturers and the amount of ventricular pacing. *J Cardiovasc Electrophysiol.* 2012;23:1336−1342.
10. Buch E, Boyle NG, Belott PH. Pacemaker and defibrillator lead extraction. *Circulation.* 2011;123.
11. Wang ZL. Piezoelectric nanogenerators based on zinc oxide nanowire arrays. *Sci (80-).* 2006;312:242−246.
12. Hwang GT, et al. Self-powered cardiac pacemaker enabled by flexible single crystalline PMN-PT piezoelectric energy harvester. *Adv Mater.* 2014;26:4880−4887.
13. Choi YS, Kim S, Kar-Narayan S. Materials-related strategies for highly efficient triboelectric energy generators. *Adv Energy Mater.* 2021;11:2003802.
14. Ouyang H, et al. Symbiotic cardiac pacemaker. *Nat Commun.* 2019;10:1821.
15. Parameswaran R, et al. Optical stimulation of cardiac cells with a polymer-supported silicon nanowire. *Proc Nat Acad Sci.* 2019;116 (January 2):413–421. Available from https://doi.org/10.1073/pnas.1816428115.
16. Gutruf P, et al. Wireless, battery-free, fully implantable multimodal and multisite pacemakers for applications in small animal models. *Nat Commun.* 2019;10.
17. Kim T, Kahng YH, Lee T, Lee K, Kim DH. Graphene films show stable cell attachment and biocompatibility with electrogenic primary cardiac cells. *Mol Cell.* 2013;36:577−582.
18. Rastogi SK, et al. Graphene microelectrode arrays for electrical and optical measurements of human stem cell-derived cardiomyocytes. *Cell Mol Bioeng.* 2018;11:407−418.
19. Savchenko A, et al. Graphene biointerfaces for optical stimulation of cells. *Sci Adv.* 2018. Available from: https://doi.org/10.1126/sciadv.aat0351.
20. Yang Q, et al. Photocurable, bioresorbable adhesives as functional interfaces between flexible bioelectronic devices and soft biological tissues. *Nat Mater.* 2021.
21. Brooks S, et al. Insights into ablation of persistent atrial fibrillation: lessons from 6-year clinical outcomes. *J Cardiovasc Electrophysiol.* 2018;29:257−263.
22. Cappato R, et al. Updated worldwide survey on the methods, efficacy, and safety of catheter ablation for human atrial fibrillation. *Circ Arrhythmia Electrophysiol.* 2010;3:32−38.
23. Haïssaguerre M, et al. Spontaneous initiation of atrial fibrillation by ectopic beats originating in the pulmonary veins. *N Engl J Med.* 1998;339:659−666.
24. Wittkampf FHM, van Es R, Neven K. Electroporation and its relevance for cardiac catheter ablation. *JACC Clin Electrophysiol.* 2018;4:977−986.
25. Kautzner J, et al. EFFICAS II: optimization of catheter contact force improves outcome of pulmonary vein isolation for paroxysmal atrial fibrillation. *Europace.* 2015;17:1229−1235.
26. Han M, et al. Catheter-integrated soft multilayer electronic arrays for multiplexed sensing and actuation during cardiac surgery. *Nat Biomed Eng.* 2020;4:997−1009.
27. Mond HG, Proclemer A. The 11th world survey of cardiac pacing and implantable cardioverter-defibrillators: calendar year 2009 - a world society of Arrhythmia's project. *PACE - Pacing Clin Electrophysiol.* 2011;34:1013−1027.
28. Chakir K, et al. Mechanisms of enhanced β-adrenergic reserve from cardiac resynchronization therapy. *Circulation.* 2009;119:1231−1240.
29. Kirk JA, et al. Cardiac resynchronization sensitizes the sarcomere to calcium by reactivating GSK-3β. *J Clin Invest.* 2014;124:129−139.
30. Lyu H, et al. Synchronized biventricular heart pacing in a closed-chest porcine model based on wirelessly powered leadless pacemakers. *Sci Rep.* 2020;10:1−9.
31. Qiu Q, et al. Acute effects of multisite biventricular pacing on dyssynchrony and hemodynamics in canines with heart failure. *J Card Fail.* 2017;23:304−311.
32. Park J, et al. Electromechanical cardioplasty using a wrapped elasto-conductive epicardial mesh. *Sci Transl Med.* 2016;8:1−12.
33. Sunwoo SH, et al. Stretchable low-impedance nanocomposite comprised of Ag−Au core−shell nanowires and Pt black for epicardial recording and stimulation. *Adv Mater Technol.* 2020;5:1−8.
34. Laughner JI, et al. A fully implantable pacemaker for the mouse: from battery to wireless power. *PLoS One.* 2013;8:1−8.
35. Choi YS, et al. Fully bioresorbable, leadless, battery-free cardiac pacemaker. *Nat Biotechnol.* 2021;.
36. Niwano S, et al. Measurement of body surface energy leakage of defibrillation shock by an implantable cardioverter defibrillator. *Pacing Clin Electrophysiol.* 2002;25:1212−1218.
37. Ahuja AK, Behrend MR, Whalen JJ, Humayun MS, Weiland JD. The dependence of spectral impedance on disc microelectrode radius. *IEEE Trans Biomed Eng.* 2008;55:1457−1460.
38. Newman J. Resistance for flow of current to a disk. *J Electrochem.* 1966;113:501−502.

39. Xu L, et al. Materials and fractal designs for 3D multifunctional integumentary membranes with capabilities in cardiac electrotherapy. *Adv Mater.* 2015;27:1731–1737.
40. Son D, et al. Bioresorbable electronic stent integrated with therapeutic nanoparticles for endovascular diseases. *ACS Nano.* 2015;9:5937–5946.
41. Cong P, Chaimanonart N, Ko WH, Young DJ. A wireless and batteryless 10-bit implantable blood pressure sensing microsystem with adaptive RF powering for real-time laboratory mice monitoring. *IEEE J Solid-State Circuits.* 2009;44:3631–3644.
42. Kim J, et al. Miniaturized battery-free wireless systems for wearable pulse oximetry. *Adv Funct Mater.* 2017;27:1–8.
43. Lou C, et al. Flexible graphene electrodes for prolonged dynamic ECG monitoring. *Sens (Basel).* 2016;16:1–12.
44. Yapici MK, Alkhidir TE. Intelligent medical garments with graphene-functionalized smart-cloth ECG sensors. *Sens (Switz).* 2017;17:1–12.
45. Karim N, et al. All inkjet-printed graphene-based conductive patterns for wearable e-textile applications. *J Mater Chem C.* 2017;5:11640–11648.
46. Batchelor JC, Casson AJ. Inkjet printed ECG electrodes for long term biosignal monitoring in personalized and ubiquitous healthcare. *Proc Annu Int Conf IEEE Eng Med Biol Soc EMBS* 2015-November, 2015;4013–4016.
47. Kabiri Ameri S, et al. Graphene electronic tattoo sensors. *ACS Nano.* 2017;11:7634–7641.
48. Acar G, Ozturk O, Yapici MK. Wearable graphene nanotextile embedded smart armband for cardiac monitoring. *Proc IEEE Sens.* 2018;1–4.
49. Koo JH, et al. Wearable electrocardiogram monitor using carbon nanotube electronics and color-tunable organic light-emitting diodes. *ACS Nano.* 2017;11:10032–10041.
50. Sun W, et al. Security and privacy in the medical internet of things: a review. *Secur Commun Netw.* 2018;2018.
51. Hannun AY, et al. Cardiologist-level arrhythmia detection and classification in ambulatory electrocardiograms using a deep neural network. *Nat Med.* 2019;25.
52. Yin M, et al. Influence of optimization design based on artificial intelligence and internet of things on the electrocardiogram monitoring system. *J Healthc Eng.* 2020;2020.
53. Attia ZI, et al. Screening for cardiac contractile dysfunction using an artificial intelligence–enabled electrocardiogram. *Nat Med.* 2019; 25:70–74.

Chapter 37

Diagnostics and Prevention: Landscape for Technology Innovation in Precision Cardiovascular Medicine

Joseph Mauch[1], Varun Thachil[2] and W. H. Wilson Tang[3]

[1]*Cleveland Clinic Lerner College of Medicine of Case Western Reserve University, Cleveland, OH, United States,* [2]*Case Western Reserve University, Cleveland, OH, United States,* [3]*Heart, Vascular and Thoracic Institute, Cleveland Clinic, Cleveland, OH, United States*

Chapter Outline

Introduction

Modern advances in cardiovascular care have dramatically changed the outlook for patients with cardiovascular disease (CVD). Like advances in public health and infectious disease preventative measures underlying the significant decline in morbidity/mortality during the early 20th century, the mortality rate of CVD has steadily declined over the past 40–50 years. The age-adjusted annual heart disease mortality per 100,000 fell by 56% from 1950–96.[1] This trend has continued since the turn of the century with coronary heart disease mortality rates downtrending from 168.0 per 100,000 in 2008 to 120.0 in 2017.[2] While most advancements in mortality can be attributed to major strides made in the management of coronary heart disease and stroke risk, care for all CVD has improved dramatically. These successes have come in part due to greater knowledge and interventions aimed at cardiovascular risk factors (smoking, cholesterol, inactivity, diet), in addition to revolutionary treatments including coronary angiography/intervention. Yet, there remains a significant need unmet by current standards. In 2017, there was an estimate age-adjusted prevalence of all types of heart disease of 10.6% and approximately 17.8 million deaths due to CVD globally, representing a 21% increase since 2007.[3]

While this trend can be attributed in part to the challenges of incorporating modern advances of cardiac care globally, there remains significant room for improvement especially in disease prevention and chronic disease management. This gap has been explained through several limitations in care: (1) the vast clinical heterogeneity and interplay of genetic and environmental risk/protective factors in CVD not amenable to a traditional reductionist approach (singular

Advances in Cardiovascular Technology. DOI: https://doi.org/10.1016/B978-0-12-816861-5.00004-6

cause and effect pathways); (2) technological limitations ranging from imaging, devices, patient monitoring, and implants; as well as (3) a lack of large datasets to guide clinical decision-making. As these limitations continue shrink due to technological innovation and robust basic and clinical research initiatives, outcomes for patients will only continue to improve. Cardiovascular medicine has been a leader for decades in embracing the use of technology and innovation through advanced imaging modalities, development of pharmacologic options, and the implementation of robust procedural treatments such as coronary angioplasty in the 1970s and modern-day successes of transcatheter approaches in treating valvular disease (TAVR and MitraClip). In the subsequent chapter, we will discuss current innovations that are changing the care for cardiac patients, both in research and already at the bedside, as well as future directions. The focus will be on: (1) big data and genomic medicine in the prevention and management of CVD; (2) artificial intelligence (AI) and machine learning in imaging modalities; and (3) utilization of personalized digital health tools in cardiology (i.e., wearables and remote patient monitoring).

Genomic medicine in cardiovascular care

A majority of CV morbidity and mortality today now takes on the form of chronic conditions such as coronary heart disease, valvular heart disease, arrhythmic disorders, and heart failure. These chronic conditions often have multiple underlying etiologies unique to each patient rather than one singular causal factor. Given this heterogeneity, using the traditional reductionist approach (breaking the problem into its discrete components) may not be optimal. The shortcomings of this concept of disease are best exemplified by the difficulty of Human Genome Project to deliver on its initial lofty goals to quickly translate genetic information into meaningful changes in clinical practice. The functional interconnectedness that genetics and environment have on disease is difficult to tease apart. This high degree of complexity has been appreciated historically, however, researchers and clinicians did not have the tools to study this. With the inception of computational and systems biology, we are now able to make connections of higher complexity than ever before.

Monogenic cardiovascular diseases

Genetics continues to provide a rich source of innovation inside cardiovascular medicine with the insights gained in "functional knockout studies" of monogenic CVD (cardiomyopathies, arrhythmias, connective tissue disorders, and familial hypercholesterolemia) leading to exciting pharmacological targets. For example, investigation into the low-density lipoprotein receptor and its role in the development of familial hypercholesterolemia led to the first discovered Mendelian CVD.[4] This discovery has led to a more robust understanding of cholesterol metabolism and a subsequent explosion in targeted lipid-lowering therapies. Even where genetics has not yet led to clear pathophysiological insights, the role of genetics on CVD is ubiquitous and is now broadly accessible. In the general population, a positive parental history of premature atherosclerotic CVD increases a patient's risk of disease threefold.[5] While the exact degree of genetic influence throughout the spectrum of CVD is variable, its pervasive nature has made it a research focus for well over 50 years in cardiovascular medicine. Recent studies have revealed that pathogenic or likely pathogenic monogenic genetic variants are more common than otherwise expected in our everyday clinical practice. For example, a recent survey of two large clinical trials in heart failure identified various cardiomyopathy-related genes (especially truncating variants of titin) in up to 3.4% of research participants, an observation that was validated in the large UK BioBank dataset.[6] Interestingly, two thirds of them have a diagnostic label of "ischemic cardiomyopathy," suggesting that our clinical diagnoses are plagued with inaccuracies if genetic data were not readily available. In another dataset of patients referred for cardiac catheterization, 1.7% of the cohort had monogenic CVD, but only 35% were recognized clinically.[7] These findings point to not only the ability to gain better insights into proband disease states, but also the unique opportunity to pursue cascade testing to identify at-risk family members and prevent CVD development.

Genome-wide association studies and polygenic risk scores

One of the first successes that led to the inception of "precision medicine" was the adoption of genome-wide association studies (GWAS). Of note, the first published GWAS study initially published in 2002 was investigating genetic associations to myocardial infarction.[8] This large-scale approach takes a sample of a population of interest and sequences their genome to compare common genetic variations [most often single nucleotide polymorphisms (SNPs)] with their associated phenotype. Due to their extremely large sample size, GWAS studies have the statistical power to detect weak genetic effects. The strength of this study design is that GWAS studies are an unbiased approach that often shed light on: (1) novel variant-trait associations, (2) previously unknown biological mechanisms causing pathology, (3) potential biomarkers for disease states, and (4) new therapeutic targets.[9] Alternatively, limitations of GWAS studies include their limited ability to explain only a modest portion of heritability in CVD and inability to differentiate between causal or associated genes.[9]

To understand the utility of GWAS consider the example of hypertension. As was first described in Page's Mosaic Theory of Hypertension, this disease has classically been understood as a complex interplay between cardiac, neural, endocrine, and vascular biology, each with its own genetic and environmental factors.[10,11] While several monogenic causes of primary hypertension have been identified,[12] almost all patients have secondary hypertension which does not follow the Mendellian single gene mutation as their etiology. Instead, GWAS studies have identified 733 gene locations associated to increased risk for hypertension (Table 37.1). These genes have been implicated in finding novel pathophysiological mechanisms of hypertension such as Uromodulin-NKCC2, endothelin, and natriuretic peptide receptor C.[13] GWAS studies have been conducted throughout the spectrum of CVD including atrial fibrillation,[14] heart failure,[15] and dilated cardiomyopathy.[16] As demonstrated by the NHGRI-EBI catalog of human GWAS studies (Table 37.1), the excitement surrounding this research technique has led to hundreds of studies in focusing on genetic risk of CVD.

As a result of the omnipresence of GWAS studies throughout various human diseases, it has been estimated that 17% of genes are associated with more than one outcome or trait.[17] The concept of pleiotropy, where one gene affects more than one phenotype, has become an interest of study beyond geneticists. An exciting approach utilizing insights from GWAS studies is phenome-wide association studies (PWAS), a systematic study of one or more genetic variants across a broad range of human phenotypes.[18] For example, CDKN2A and CDKN2B variants have been associated with both T2DM[19] and coronary heart disease,[20] suggesting a shared pathophysiological mechanism. In addition to finding shared mechanisms of disease, PWAS studies have the potential to establish relationships between different disease states previously not thought to be associated, to assist in repurposing drugs by identifying suitable alternative diseases, and to identify novel biomarkers.[18,21]

With the robust insights gained from genetic studies as discussed, there is increasing interest in utilizing this knowledge in projecting risk for disease, a concept referred to as a Polygenic Risk Score (PGRS). In broad terms, PGRS aims to estimate patient's cumulative genetic risk for a phenotype using large numbers of genetic variants instead of relying on one SNP, rendering the summarized risk more clinically significant and broadly applicable in complex diseases such as CVD. However, recent attempts to integrate PGRS for hypertension and coronary heart

TABLE 37.1 Genome-wide association studies in cardiovascular disease and associated comorbidities.

Phenotype/disease	Number of genome-wide association studies	Genetic associations
Type II diabetes mellitus	176	4049
Coronary artery (heart) disease	125	1790
Hypertension	90	733
Stroke	65	371
Cardiomyopathies	55	390
Obesity	39	366
Atrial fibrillation	28	545
Myocardial infarction	22	301
Heart failure	16	283
Heart valve disease	7	42
Sudden cardiac arrest	7	54

Obtained from https://www.ebi.ac.uk/gwas/home on July 6, 2021. Buniello A, MacArthur JAL, Cerezo M, Harris LW, Hayhurst J, Malangone C, McMahon A, Morales J, Mountjoy E, Sollis E, Suveges D, Vrousgou O, Whetzel PL, Amode R, Guillen JA, Riat HS, Trevanion SJ, Hall P, Junkins H, Flicek P, Burdett T, Hindorff LA, Cunningham F and Parkinson H. The NHGRI-EBI GWAS Catalog of published genome-wide association studies, targeted arrays and summary statistics 2019. Nucleic Acids Research, 2019, Vol. 47 (Database issue): D1005-D1012.

disease into clinical practice has yielded mixed results regarding their incremental benefit over standard care with existing risk profiling strategies. In some studies (majority being epidemiologic or observational studies), PGRS has demonstrated significant increased risk for the development of coronary heart disease or hypertesnion.[13] Posthoc analyses of PCSK9 clinical trials to assess for utility of PGRS stratification found a strong association between high risk score and subsequent cardiovascular events even after adjusting for traditional risk factors.[22–24] Some research has suggested that primary beneficial effect of PGRS over typical clinical risk scores is in younger populations, giving them a potentially unique role in primary prevention.[25] These exciting results are in contrast to several studies showing a lack of clinical utility of PGRS when added to standard pooled cohort equations.[26–28] One possible explanation may be the fact that genetic predispositions to complex gene–environment interactions are only one of many factors impacting outcomes.

As much as proponents of PGRS have promoted their incremental value, critics have also pointed out that knowledge of increased susceptibility is only clinically useful when the information leads a patient to meaningful change in behavior and risk factors. Current population-based trends suggest that traditional risk factor control is <50%,[29] despite clear benefits to cardiovascular health from tobacco cessation, regular exercise, and healthy diet, suggesting that increased awareness does not always lead to behavior change. To this point, some studies[30] have shown poor rates of behavior and risk factor management has been seen with the integration of genetic risk characteristics, whereas others have attenuation of genetic risk with lifestyle modification.[31,32] Furthermore, existing GWAS datasets have lacked racial diversity, and whether such PGRS predictions can be applicable across various racial and ethnic groups remain largely in question. Taken together, the theoretical benefits of leveraging the known genetic risk have not yet changed the management for either primary or secondary prevention of CVD. Nevertheless, scalable computation of PGRS (especially on multiple chronic diseases beyond CVD, such as cancer and type 2 diabetes that also has robust PGRS applications) are likely here to stay with its unique ability to identify polygenic risk predispositions to improve population health prevention measures.

At this point, genomic medicine has played a larger role in the acceleration and exploration of avenues in basic science study but has failed to translate into clinical practice, despite these aspirational goals. With the excitement surrounding GWAS studies, an important drawback has been the significant lack of diversity in genotypes studied limiting the generalizability. Based on GWAS studies cataloged across human disease, 78% of individuals studied are of European descent whereas African, Hispanic, or Latin American only make up a combined 3%.[33] This has truly been a disadvantage to the scientific community since GWAS is uniquely established to study ethnic variations in genetics.[9] While this significant limitation has begun to shift the conversation about genomic medicine, more work is needed to include overlooked populations to improve disease prediction and risk management.

Proteomics and multiplex cardiovascular biomarkers

In parallel to the advances in gene sequencing, revolutions in the study of other biological systems such as proteins, lipids, and metabolism have exploded our available data to study CVD. The broad term of proteomics, coined over twenty years ago now, denotes the study of proteins in a tissue or blood sample. These laboratory approaches can be used to derive structural information, catalog proteins in a sample, analyze protein interactions/complexes, and quantify protein concentrations in certain disease states or patient subgroups.[34] This field of proteomics has been enhanced with recent technological advances in sample preparation, analysis with mass spectrometry or aptamer-based platforms, robust databases, and bioinformatics techniques. The benefits of proteomics over DNA or RNA approaches are that protein analyses provide insights into protein expression but more importantly posttranslational modification and protein–protein interactions. While this research field is still new, cardiovascular proteomics has the potential to better determine the protein signature of cardiac disease both acute (MI) and chronic (heart failure) disease and develop point-of-care technologies for monitoring of treatment efficacy. They may also work in conjunction with genomic and clinical data (multiomics approach) to gain mechanistic insights in CVD in the clinical setting that lack good animal models (HFpEF).[34] The potential power of this technique can be seen in pilot experiments attempting to characterize the plasma protein expression in different stages of heart failure to better identify possible pathophysiological mediators such as B-type natriuretic peptide, interleukin-18, and complement proteins.[35] Similar revolutions are occurring in the fields of metabolism and lipid biology. The insights gained from each of these fields are useful on their own already. But now with the processing power and innovations in machine learning, the pooling from proteomics, genetics, and electronic medical record in a "multiomics approach" helps to get closer to the goal of precision medicine.

Deep phenotyping with multiomic analysis and network medicine

As already discussed, there is ample room for pleiotropy to exist across subtypes of CVD that does not fit our current model of simplified genotype-phenotype model.[36] In a clinical setting, the desired phenotype of interest is a disease state, the pathophenotype. Over the past twenty years, scientists have a growing appreciation for the fact that the genomic data (genotype) collected does not easily and consistently predict the pathophenotype. An intriguing perspective of biological disease posits that there are numerous discrete biological processes (inflammation, thrombosis, and fibrosis) that play a role in propagating the ultimate pathophenotype (myocardial infarction, for example). Such processes have predictable interactions ("interactomes") that can be mined and predicted, and these biological mechanisms of fibrosis or inflammation could be considered intermediate phenotypes (or endophenotypes) that arise from the interplay of genetic, metabolic, and proteomic profiles.[37,38] In consideration of this model, deep phenotyping aims to broadly characterize a patient through their sequenced genome, proteomic profile, environmental exposures, and clinical data (obtained from electronic health record) and use advanced computing capabilities (network analyses) to better characterize the patient's endophenotype to personalize therapeutic approaches (Fig. 37.1).[37] For example, patient 1 and patient 2 might share the same pathophenotype of a myocardial infarction. But the variation in the underlying biological processes resulting in the pathophenotype are different. These differences can be best detected and characterized by big data sources and complex computing processes. While conceptually appealing, this growing field of "Network Medicine" have yet to be applied clinically as this has been largely applicable in redefining disease categorization or identifying drug targets or repurposing opportunities. Ongoing large-scale efforts are made to capture such multidimensional data for heterogeneous diseases, and interesting use cases to identify individual patient "reticulotypes" (unique network patterns that can render prediction of natural history and disease manifestations) have shown promises in its clinical applications.[39]

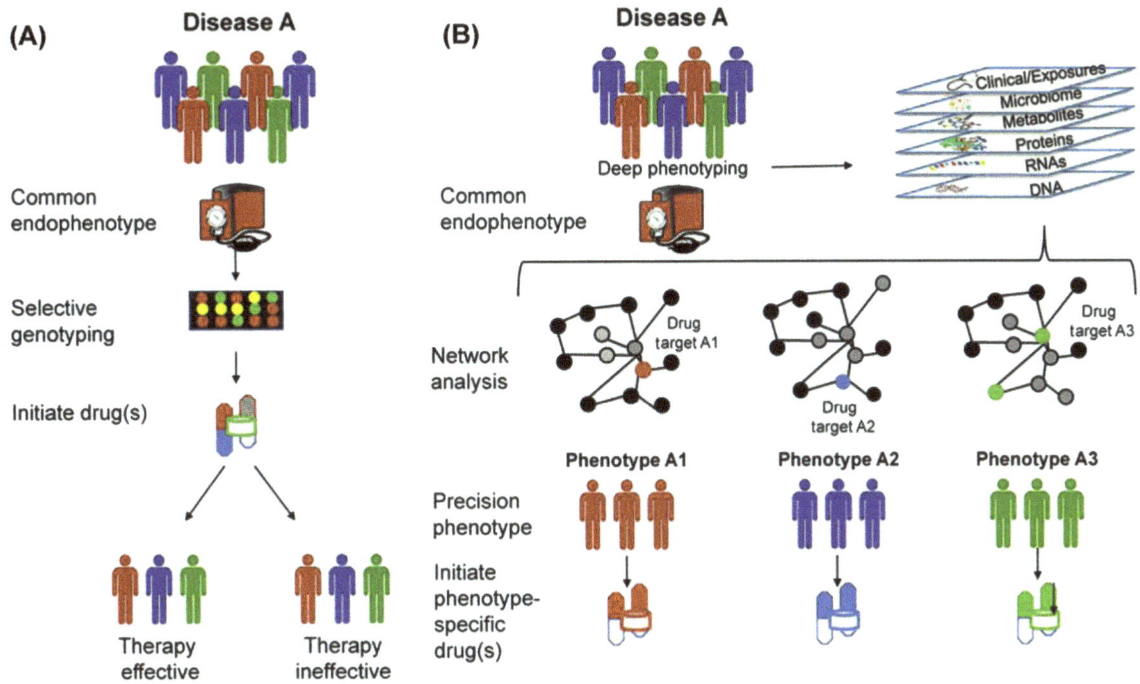

FIGURE 37.1 Network approach to precision phenotype and rational polypharmacy.[37] (A) The conventional reductionistic therapy decision involves identifying patients with a common endophenotype (e.g., obesity) who may otherwise have distinctive underlying biological phenotypes (depicted in *red*, *blue*, or *green*). Some patients may undergo limited genotyping. Drug therapies are initiated in this heterogeneous patient group based on population-based clinical trial results with limited consideration for the individual's underlying biology. (B) The precision medicine approach involves deep phenotyping of each individual with a common endophenotype using multiomic platforms and clinical assessments. These data are integrated using network analysis to determine more precise phenotypes with their key molecular components (*gray nodes*). On the basis of such analysis in the context of the interactome, a drug target(s) (*red*, *blue*, or *green nodes*) can be determined for each phenotype.

Gene editing in cardiovascular medicine

While most CVD is not a result of monogenic genetic conditions, there remains a significant disease burden for conditions such as hereditary amyloidosis, familial hypercholesterolemia (and associated subtypes), inherited arrhythmias, and genetically associated cardiomyopathies. These inherited conditions offer a unique opportunity to reverse the pathogenesis of disease not as readily accessible in most polygenic CVD. With development of new gene editing techniques, the theorized benefit of controlling gene expression therapies has become a real consideration. Gene editing technologies such as zinc finger nucleases (ZFNs) and transcription activator-like effector nucleases (TALENs) were the initial methodologies to be able to insert new genetic material into the genome, pioneered now twenty years ago. The limitations of these techniques including difficulty in designing ZFNs to bind at a desire locus have hampered the ability of these technologies to be rapidly applied across the spectrum of disease.[40]

Yet the introduction of CRISPR/Cas9 (clustered regularly interspaced short palindromic repeats) in 2013, revolutionized the field. This was best exemplified by the presentation of the 2020 Nobel prize in Chemistry to Drs. Emmanuelle Charpentier and Jennifer Doudna for development of the CRISPR/Cas9 technology. While best characterized through *Strep pyogenes*, CRISPR is a widely used molecular process rather ubiquitous among adaptive immunity systems in bacterial species. The benefits provided by CRISPR include use of a 100-nucleotide RNA molecule known as a "guide RNA" which provides specific targeting to a genome which can be altered to vary site specificity quickly. Once a site complementary to the guide RNA is found, Cas9 delivers a blunt-end (no additional nucleotides on either end of DNA strand break) break. Double strand breaks are fixed in vivo through two repair pathways termed nonhomologous end joining (NHEJ) and homology-directed repair (HDR).[41] Since NHEJ does utilize a repair template while reconnecting free ends of the DNA, it is prone to indels (insertions or deletions) that can ultimately cause frameshift mutations causing of loss of function.[40] On the other hand, HDR uses a template, most often sister chromatids, to provide much more efficient, less error prone DNA repair. The goal with CRISPR is to use intrinsic HDR repair pathways with the addition of an exogenous DNA, as the template to be used for repair. While blunt-end double stranded breaks were initially only possible, further advancements in field have allowed for sticky-end double strand breaks or even single strand breaks allowing for broader applications of gene editing with decreased risk of mutagenesis and damage to the genome, respectively.

Another exciting application of genomic medicine in cardiovascular medicine includes RNA therapeutics (Tables 37.2 and 37.3). While similar to the approach taken by gene editing to regulate (either increasing or decreasing) gene expression, RNA therapeutics does not leave a permanent change in the genome as is the case with gene editing. Instead, these genomic medicine technologies, such as antisense oligonucleotides (ASOs) and small interfering RNA (siRNAs), work by inhibiting micro-RNA (miRNA), a subtype of noncoding RNA that functions by influencing protein

TABLE 37.2 Gene editing preclinical studies in cardiovascular disease.

Targeted gene	Description	References
Dystrophin	Function—binds extracellular matrix to the cytoskeleton by F-actin; causal gene for Duchenne muscular dystrophyThree studies in *Science* in 2016 demonstrated in vivo use of CRISPR to recover dystrophin function in DMD.Later demonstrated in canine and pig models.Better recovery in skeletal muscle than cardiac muscle.	Long et al.[46] Nelson et al.[47] Tabebordbar et al.[48] Amoasii et al.[49] Moretti et al.[50]
Salvador (Sav)	Component of Hippo signaling pathway, an inhibitory kinase cascade that represses cardiomyocyte proliferation and renewalTreatment of adeno-associated virus 9 gene therapy with Sav hairpin with local delivery to border zone cardiomyocytes led to improvement in ejection fraction in pig MI model.	Liu et al.[51]
Phospholamban (PLN)	Function—key regulator of Ca2 + reuptake by ATPase in cardiomyocytesZFNs repressing PLN increases cardiac calcium reuptake and contractility in vitro and animal heart failure model	Zhang et al.[52]
PCSK9	Function—targets hepatic LDL-R for degradationCRISPR-Cas9 with adenovirus given in mouse model with reduction of circulating PCSK9 by 90% and reduced cholesterol levels by 40%Further studies done with hybridized mouse model with human hepatocytes with similar results seen.	Ding et al.[53] Wang et al.[54]

TABLE 37.3 Micro RNAs (miRNA) implicated in preclinical studies of cardiovascular disease.

miRNA	Function and/or associated pathophysiology	Reference
miRNA-21	Regulation of ERK-MAP kinase pathway in cardiac fibroblastsUpregulated in heart failure modelsIn vivo silencing of *miR-21* inhibits interstitial fibrosis	Thum et al.[55] Ji et al.[56]
miRNA-29a	By comparison between patients with hypertrophic cardiomyopathy to age and sex-matched controls, miR-29a was found to correlate to hypertrophy and fibrosis.	Roncarati et al.[57]
miRNA-34 family	miR-34a, -34b, and -34c are elevated in heart in response to stressLNA-antimiR-34 in rodent acute stress model reduced cardiac remodeling and atrial enlargement	Bernardo et al.[58]
miRNA-92a	Negative regulator of angiogenesisInhibition of miR-92a in limb ischemia and MI increased blood vessel growth and functional recovery	Bonauer et al.[59]
miRNA-126–5p	Suppresses Dlk1 (Notch1 inhibitor) and subsequently prevents atherosclerosisAdministration of miR-126–5p increased endothelial cell proliferation and limited atherosclerosis in murine model	Schober et al.[60]
miRNA-132	Implicated in capillary rarefaction, interstitial fibrosis, cardiac hypertrophy, and myocardial dysfunctionAdministration of miR-132 in pig and mouse animal models improves systolic function after pressure overload or ischemic insult.[61]Early clinical trial (NCT04045405)[62]—phase Ib trial of ASO, CDR132L in 28 HFrEF patients demonstrated 20% reduction in NT-proBNP compared to standard care	Hinkel et al.[61] Taubel et al.[62]
miRNA-181b	Inhibits activation of NF-kB pathwayReduced levels in apolipoprotein E-deficient miceAdministration of miR-181b inhibited atherosclerotic lesion formation and proinflammatory gene expression	Sun et al.[63]
miRNA-208a	Increases Myh7 expression and cardiac remodeling after pressure overloadTherapeutic inhibition of miR-208a prevents cardiac remodeling in Dahl hypertensive rats	Montgomery et al.[64]
miRNA-712	Mechanosensitive miRNA that is increased by disturbed flow in endothelial cells, inhibits TIMP3 expressionInhibition of miR-712 prevents atherosclerosis in murine models	Son et al.[65]

LNA, locked nucleic acid; *Dlk1*, delta-like 1 homolog; *TIMP3*, tissue inhibitor of metalloproteinase3; *green*, overexpression studies; *red*, inhibition studies.

expression.[42] These miRNA molecules have been implicated in the regulation of disease throughout the cardiovascular field, ranging from angiogenesis, fibrosis, ventricular remodeling, arrhythmias, and cardiac metabolism.[43] While the majority of applications have focused on inhibition of miRNA, RNA therapeutics also includes the theoretical overexpression of beneficial miRNA within a certain tissue or organ. Yet, as discussed below, this augmentation approach is much more difficult than inhibition of miRNA limiting its applications.[42]

Despite the growing interest in using RNA therapeutics, there remain significant obstacles in their application. While a wide pool of evidence supports their usage, most RNA therapeutic research has been done in nonhuman subjects. An important limitation for CVD is the difficulty of targeting heart tissue especially in comparison to other organs such as the kidneys or the liver.[42] In addition, there is significant dosing variation for these drugs, not only across targeted tissues, but also those that target the miRNA system. For example, inhibition of miR-92a in the heart could be achieved with a dose of 8 mg/kg bodyweight and inhibition of miR-34a in the heart required 20 mk/kg.[44,45] In comparison, first application of miRNA targeting in patients for treatment of HCV infection (targeting the liver) required dosages ranging from 3–7 mg/kg bodyweight.[46] An important drawback for overexpression targets in miRNA therapeutics is frequent and often serial high-dose requirements to achieve required effects. Finally, toxicities in RNA therapeutics are important. While on-target (miRNA of interest) binding of the drug can produce expected side effects, off-target binding of similar nucleotide sequences can target other RNA or even DNA and proteins.[42] Additionally, side effects associated with administration not binding include activation of the immune system including Toll-like receptor signaling.[47]

Preliminary clinical applications

RNA therapeutics

The APOLLO trial released its phase 3 results in 2018[48] that quickly led to FDA approval of Onpattro (patisiran), a small interfering RNA, for the treatment of polyneuropathy in hereditary TTR amyloidosis.[49] Although technically not for a cardiovascular indication, this marked the first FDA approval for the use of an RNA therapeutic targeting an exogenous protein. Similarly, Inclisiran, a siRNA targeting PCSK9 synthesis which demonstrated 30%–53% reductions in LDL-C at 6 months in its phase 2 trial,[50] is being studied in an ongoing phase 3 clinical trial (ORION-4, NTC03705234). These two therapies leverage the primary pathophysiological aberration of each disease to selectively reverse the disease course. Both medications aim to change the way we treat CVD soon and will likely be joined by other RNA therapeutics as well.

Gene editing

The incredible breakthroughs in RNA therapeutics were quickly followed by advancements in gene editing. The first clinical proof-of-concept application was conducted in sickle cell disease and beta-thalassemia.[51] Early success with an "exon skipping" strategy has been adopted for the treatment of Duchenne Muscular Dystrophy that can induce production of partially functional dystrophins.[52] Within the field of CVD, preliminary data was released from a phase 1 trial of six patients where CRISPR-Cas9 gene editing resulted in up to 87% reductions of TTR protein in patients with hereditary TTR amyloidosis.[53] While further evidence for the efficacy and safety data will be required prior to approval in clinical use, it is clear that medicine is on the cusp of utilizing gene editing to treat patients. Additionally, gene editing studies conducted on nonhuman primates using a single-strand break variant of CRISPR led to near-complete knockdown of PCSK9 expression with low (<1%) off-target editing suggesting a good safety profile.[54,55] Given the successes of gene editing targeting both PCSK9 and TTR (proteins produced in the liver), near-future applications of gene editing will likely focus on additional liver-synthesized proteins, such as Wilson's disease and hemochromatosis, both of which have cardiovascular manifestations. Yet, continued advances in tissue targeting and delivery routes will continue to broaden the use of gene editing and RNA therapeutics throughout the field of CVD. Alternatively, instead of in vivo gene editing techniques discussed already, a theorized approach to gene editing in CVD would be performing gene editing ex vivo.[40] While currently this remains a distant goal, this technique would circumvent the challenge of targeting adult cardiomyocyte tissue.

Stem cell and tissue engineering in cardiovascular medicine

While the field of stem cell biology traces its roots back 40 years ago in the use of mouse embryo-derived stem cells, the field has proliferated in recent years especially with the breakthrough research of Drs. Takahashi and Yamanaka into induced pluripotent stem cells (iPSCs) in 2006.[56] This ultimately led to the Nobel prize in Physiology in 2012 going to Drs. Yamanaka and Gurdon for their seminal work on stem cells.

In brief, stems cells are undifferentiated cells that retain the capability to differentiate into "adult" cell types as well as self-renew. While there are different types of stem cells varied in a hierarchy of potential for differentiation, pluripotent stem cells are often the focus of most research due to their ability to form cells of all three primary germ layers.[57] iPSCs are actually adult somatic cells that are artificially "reversed" into a state of pluripotency. While initially described as produced through reprogramming, the revolution of genome editing also provides the possibility of making mutant induced pluripotent stem cells. The iPSCs have a distinct advantage of being a potential renewable source of cardiomyocytes as these cells can be expanded. Another type of stem cells commonly described are somatic stem cells which have a restricted range of differentiation and are found throughout the body into adulthood. Examples of somatic stem cells include mesenchymal stem cells (MSC), hematopoietic stem cells, and skin stem cells. While many researchers have investigated the potential for somatic cardiac stem cells, no current research supports their function in clinically relevant regeneration of cardiomyocytes.[58] Of note, MSC have been an early area of focus for regeneration given their antiinflammatory and antifibrotic effects as well as stimulation of endogenous repair pathways.[59] Through a combination of paracrine signaling and exosome release, these stem cells are thought to modulate inflammation, cardiomyocyte death, and reduce oxidative stress. Additional properties of MSCs, such as their relative abundance (found within the bone marrow, adipose tissue, and placenta) and lack of significant host/graft immune reaction (given lack of MHC II expression), make them ideal for clinical use as well. While much excitement surrounds the use of in vitro stem cells, there has also been significant progress in the field of in vivo cardiac reprogramming, where the aim is to generate new cardiomyocytes by converting other cell types to a cardiomyocyte fate without going through an intermediate phase of pluripotency.[60]

Current applications of stem cells in medicine include hematopoietic stem cell transplantation and as a target for pharmacological testing (Fig. 37.2). Future applications relevant to cardiology primarily focus on cell and tissue-based therapies.[61] *Cell-based therapies* refer to the administration of cells that are injected or grafted into a patient for a therapeutic benefit. This technique of cell administration contrasts to gene editing where one seeks to alter the genome of preexisting cells in a patient. One notable advantage in cell-based therapies is the ability to treat diseases of polygenic or even unknown genetic origin, whereas gene editing requires the problem to be monogenic in nature.[59]

In conjunction with cell-based therapies, *tissue-based therapies* focus not on the implementation of stem cells as direct therapy but on the use of stem cells (iPSCs most commonly) to engineer cardiac tissue. Cardiac tissue engineering most broadly requires three components: iPSC-derived cells, a biomaterial scaffold, and mechanical or electrical stimuli for tissue maturation.[61] In recent years differentiation protocols have been developed for iPSC-induced cardiomyocytes, endothelial cells, cardiac fibroblasts, and vascular smooth muscle cells. This variety of cell types helps to mimic the complexity and interplay of cell types in adult cardiac tissue. Most tissue engineering strategies can be categorized into three subtypes: cardiac spheroid/organoids, engineered heart tissue, and 3D bioprinting models. While discussed in depth elsewhere,[61] cardiac spheroids rely on the self-assembly of cells providing 3D microenvironment for cells, whereas with engineered heart tissue researchers can control the structure and assess function more clearly. Most common uses of cardiac tissue engineering focus on drug screening, evaluation of substance's cardiotoxicity, disease modeling, and cardiac repair therapy. While these applications have previously been performed with 2D cultures, tissue engineering provides the obvious benefit of a more physiological model and control over environmental conditions.

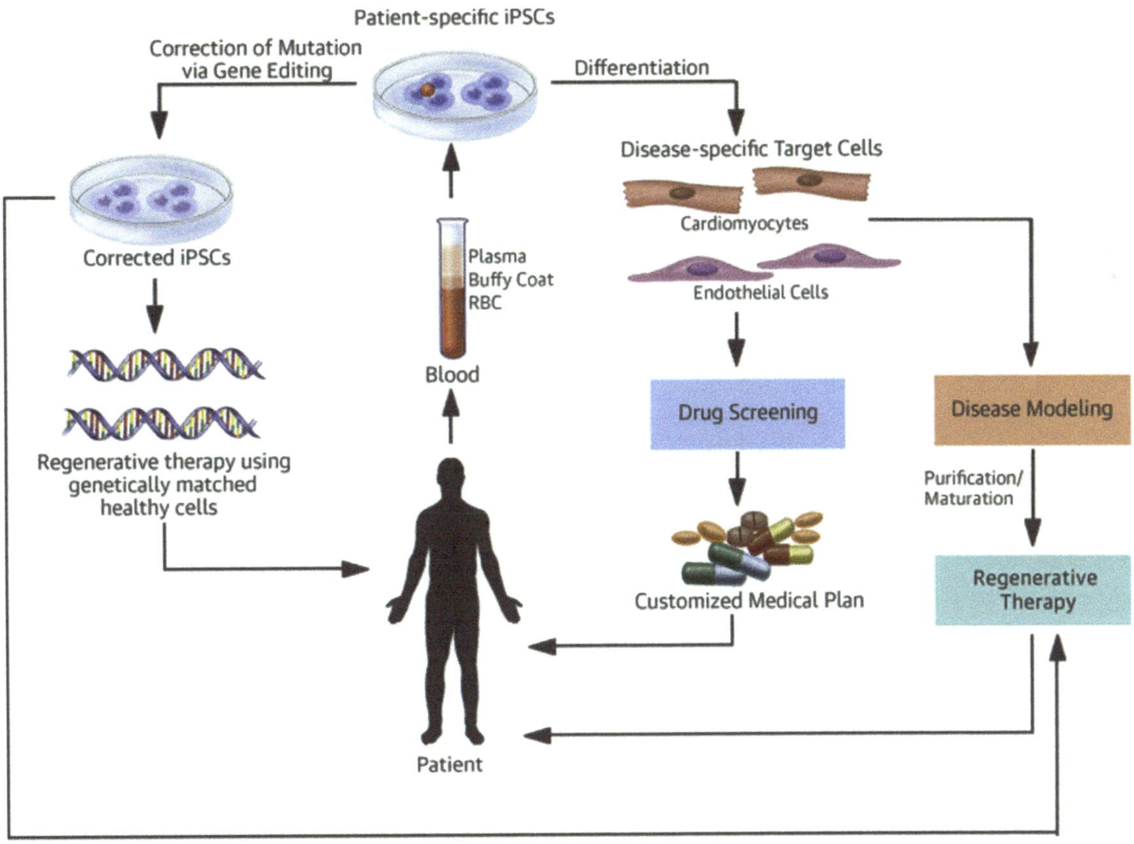

FIGURE 37.2 Potential applications of patient-specific iPSCs.[62] Disease-specific target cells differentiated from patient-specific induced pluripotent stem cells (iPSCs) have three major applications: disease modeling, regenerative therapies, and drug discovery/toxicity studies. iPSCs from patients with genetic mutations could be corrected via genome editing to yield healthy target cells for cell therapy. Both corrected and uncorrected target cells could be used for disease modeling or drug screening. RBC = red blood cells.

Preliminary clinical applications

Cell and tissue-based therapies

With the field of stem cell therapy being so new, there remains a wide variety of ongoing clinical projects to incorporate these exciting findings into patient care. Many initial clinical studies focused on the use of bone marrow mononuclear cells in acute MI which yielded mixed results.[63] Similarly, a late preclinical proof-of-method transplantation of human embryonic stem cell cardiomyocytes in a large animal model displayed modest effects in remuscularization and return of function.[64] Despite mixed initial results, other studies have shown promising results. For example, MSC have shown promising outcomes in early clinical results for cardiac patients in the attenuation of fibrosis in scarred tissues and improvements in EF and endothelial function.[65,66] In addition to mesenchymal stem cell administration, a preliminary case study published in 2015 tested the first intraoperative administration of human embryonic stem cells in pericardial patch to a previously infarcted ventricular wall segment in a patient with severe systolic heart failure.[67] Further follow-up of this case series followed six patients which demonstrated only clinically silent alloimmunization but importantly, increased systolic motion of cell-treated ventricular segments.[68] While iPSCs have not yet made the transition to clinical experimentation, this case series demonstrates the viability of using pluripotent stem cells with a good safety profile from preliminary results.

The variation in results reported from early clinical trials for cell-based therapies is driven in part by huge variation in types of stem cell transplantation, mode of delivery, and even variation in the reported outcomes/results. Better systematic ways to report these findings will be improved as more trials are conducted. In addition, more basic science insights will continue to improve the clinical feasibility and outcomes of these trials. Key limitations being studied include ways to improve functional integration and angiogenesis of grafts through overexpression of genes such as N-cadherin and angiopoietin-1.[69,70] Further development in stem cell collection, isolation, and delivery protocols will help push this field forward into the next decade.

Disease modeling, drug screening, and cardiotoxicity applications

While the use of iPSCs as cell-based therapy has not yet come into the clinical field, iPSCs have transformed the approaches to study and apply knowledge about drug screening, cardiotoxic agents, and pharmacogenomics. The use of iPSC-derived cardiomyocytes offers a theoretical endless supply of cells to study. This technology in tandem with genetic editing techniques and insights gained from GWAS have allowed new insights into what traditionally were variants of unknown significance.[71] By studying cohorts of differentiated iPSCs with diverse genetic variations, researchers can determine genetic expression of nearby genes and resulting functional capacity. Similarly through the combination of gene editing and iPSCs, numerous disease models have come about ranging from cardiomyopathies and metabolic syndromes to arrhythmias and valvular heart disease.[72] Others have used iPSC cells to gain insights into biomarkers associated with chemotherapy-related cardiotoxicity or creating rankings of cardiac safety within drug classes. While these applications are changing medicine now, there is still a theoretical goal of iPSCs being able to provide more evidence for individualized, precision medicine. Since iPSCs contain an individual's unique genome, some have suggested these stem cells could be tested to predict therapeutic and toxic response to medications in a drug class to guide personalized therapy.[71]

Additional clinical applications of genetic insights

Pharmacogenomics

Cardiology over the past 40 years has seen a surge in treatment options ranging from antiarrhythmics and antihypertensives to heart failure medications and anticoagulation/antiplatelet agents. The current approach to drug therapy for each patient most often relies on a combination of clinical guidelines, clinician's experience, and basic trial and error. Pharmacogenomics is a branch of precision medicine that seeks to understand a patient's unique genetic predisposition to a drug's metabolism. By providing personalized information, pharmacogenomics seeks to inform drug therapy decisions to achieve maximal efficacy with less adverse reactions. Much of pharmacogenomics features the study of an individual's metabolism of the drug especially through the CYP450 system in the liver that is known to vary between patients.

Arguably, the largest success story in all of pharmacogenetics comes from within cardiovascular medicine in clopidogrel (P2Y12 inhibitor) and genotypes of liver enzyme CYP2C19.[73] Clopidogrel is a prodrug that undergoes a two-step metabolism with its active metabolites providing inhibition of platelet activation. The 2C19 CYP enzyme is involved in both steps therefore playing a key role in clopidogrel's ability to inhibit platelet function.

With a wide variety of polymorphisms known, this leads to patients being high, intermediate, or low metabolizers. These low metabolizers are at risk of no therapeutic antiplatelet activity. Previous studies demonstrated low metabolizers have increased risk of stent thrombosis leading the FDA to adding a black box warning about reduced effectiveness in 2010.[74] Clinical trials studying the effectiveness of genotype-based therapy have shown most benefit in decreasing a patient's risk for early stent thrombosis but ultimately does not change long-term outcomes.[75] With the addition of other antiplatelet drugs like prasugrel or ticagrelor, there are additional options available to patients, but clopidogrel/CYP testing remains one of the most utilized applications of pharmacogenomics in the clinical setting to date.

Additional clinical scenarios considered for implementation include beta-blockers. Investigations into the polymorphic β1-adrengergic receptor (ADRB) found that polymorphisms, such as Arg389 variant, predispose to heart failure and influence the therapeutic response to beta blockers.[76] Despite these significant findings, the potential for harm in beta-blocker therapy is overshadowed by the robust benefit seen with universal beta-blocker therapy.[77] There is no evidence to date showing that implementation would display improvement in patient outcomes. As a result, β-blocker pharmacogenomics is yet to translate into clinical practice. Similarly, warfarin and statin therapy have been considered for clinical pharmacogenomic approaches but so far have not been widely employed.[78,79]

Direct to consumer genomics

Current AHA guidelines[80] only recommend genetic testing for confirmed or suspected patients with heritable CVD (including those with family members with previously identified pathogenic variant). Continued development of these testing capabilities has yielded profound clinical utility in disease states ranging from cardiomyopathies to heritable arrhythmic, lipid, and connective tissue disorders. Where previously genetic testing was confined to clinical and research environments, it has grown in widespread availability outside of these controlled environments. With the introduction of DTC products such as 23andMe in 2007, there became more widespread interest in the insights to be gained from genetic testing. DTC genetic tests refers to testing in which the results are returned to the consumer without involvement of a clinician or genetic counselor.[81] The medical community has long been hesitant to project genetic risk on an individual level, but commercial products have pushed this issue to the forefront. Initially, these products were often cost prohibitive, limiting their effect. From its initial approval in 2007, the average price for DTC genetic testing decreased from \$1000 to \$99 just 5 years later in 2012.[82] There has been a significant evolution in DTC products offered with FDA approval for carrier screens in 2015 and genetic health risks in 2017. Genetic health risk panels offered include predisposition to cardiac conditions like atrial fibrillation, coronary artery disease, hypertension, type 2 diabetes mellitus, and elevated LDL cholesterol and triglycerides. Programs also offer consumer's carrier status reports for 40 + genetically inherited conditions (sickle cell, inborn errors of metabolism, and cystic fibrosis), as well as pharmacogenomic information about CYP2C19 drug metabolism.[83]

Critics of DTC genetic testing have argued that there is a mismatch between a user's expected utility of the information and what they actually are receiving.[81] One study has estimated that 62% of patients receiving DTC genetic kits were motivated by their own individual health implications.[84] Currently the landscape of third-party genetic testing is either too broad or narrow in focus to provide clinical utility on a large scale.[85] Some narrow tests only look for certain mutations within a gene known to have an established pathological role, whereas others DTC kits provide information on genes and mutations of unknown clinical significance. As federal regulators became involved in DTC genetic testing, companies are required to provide analytical and user comprehension validation as well as separate more clearly the health panels form additional "entertainment" genetic results.[82] It has become clear that these technologies are here to stay as they have begun to become more integrated into the medical system, albeit slowly. The hope is for more collaboration between private sector DTC companies and established medical institutions to help provide valuable health information to patients.

Artificial intelligence in cardiovascular imaging

Much attention has been drawn to ideas surrounding "big data" and AI. With the exponential increase in AI research and clinical tools, it's clear that all clinicians need functional understandings of the technology and its potential implications in clinical care. While the concept of using technology and computing to automate clinical tasks can be traced back as early as the 1950s, removal of previous limitations has made these lofty, theoretical ideas possible. Proof-of-concept of AI in medicine has been performed for well over 20 years now.[86] Yet, recent advances in computing power, along with its ever-increasing availability and reduced cost, as well as a surge in available datasets (integration of electronic health records, advancement in biobank programs, etc.) have pushed AI into the cutting-edge of medical research.

Cardiology has been an early adopter of these research techniques, especially within the data-heavy field of cardiovascular imaging. The application of AI in cardiovascular imaging has the potential to improve the speed and quality of data acquisition, reduce measurement times of cardiac parameters, allow for prompt diagnoses, and provide prognostic risk stratification—all culminating in better patient care. Specifically, AI has demonstrated promise in addressing the labor-intensive challenges associated with traditional approaches to CV imaging, including automating precise measurements of volumes, pressures, and gradients, integration of 2D images for a 4D structure, and functional assessment of a dynamic, moving structure (LV chamber for one). Before we begin to understand the applications of AI, we first must be clear on the terminology surrounding the field.

Definitions and background

Starting off broadly, *AI* is a computational program that performs a task that is classically associated with human cognition (pattern recognition or problem solving).[87] These higher-level functions require not only data collection but also analysis or interpretation of the information. *Machine learning* is a subset of AI that uses statistical methods to enable machines to learn from data sets to make predictions and decisions without being explicitly programmed. In essence, machine learning is a method to achieve the goal of AI. A foundational component to machine learning is an *algorithm*, a process or set of rules to be followed in order to achieve a particular goal.[88] Machine learning can be either unsupervised or supervised.[87] In *unsupervised learning*, an algorithm is trained on unlabeled data with the goal of recognizing general patterns or similarities in the data. In contrast, in *supervised learning*, the raw data is given labels (variables and measurements), and the algorithm is trained to detect correlations between the input data and the labels. Since these labels given to the program (also known as *annotated data*) require manual extraction, there have been significant limitations in the application of structured learning due to inter/intraobserver variability and even feasibility of creating annotated data for complex imaging such as 3D or 4D image acquisition techniques.[89] *Deep learning*, a subset of machine learning, uses combinations of *neural networks* (organizational set of decision points that are processed similarly to human neural network).[87,90] These networks work together as filters in which each layer processes input information and passes the data along to the next network. More broadly, deep learning requires input values and then gives an output of a prediction while the layers/algorithms in which the information travels between in order to obtain the outcomes that are "hidden." Deep learning is one of the fastest growing fields in cardiovascular medicine. The term *big data* simply refers to the idea of analyzing large, complex datasets previously unable to be performed due to technological restraints.

The process of machine learning is best explained in an example. A group of cardiologists are wanting to develop a program to detect cases of dilated cardiomyopathy on cardiac MRI. For the program to "learn" the difference between normal and diseased hearts, the algorithm is given standardized normal hearts and those with dilated CM. Typically, each image would be broken down into its features ranging from pixel density (myocardial enhancement), measurements from the image (LVOT diameter), or even movement or velocity of a structure (apical movement). Then the computer's learned knowledge is tested on another group of images with both diseased and normal hearts.

Applications of artificial intelligence in cardiovascular imaging

Currently, AI is still being studied in the clinical setting with a relatively limited scope with the acquisition of quantifiable measurements or warnings of concerning features.[91] The interest in AI techniques being used for full diagnostic capabilities would still need much more external validation beyond the current models being used.[92] Current and future applications of AI in CV imaging are discussed below.

Phenomapping

A common method in unsupervised learning is clustering, whereby categorization schemes are created based on the variation in the data seen (Fig. 37.3). The benefit of using computing power for this process is twofold: (1) ability to process huge amounts of data that manual methods cannot, and (2) ability to detect slight/subtle differences undetectable to human cognition. This categorization and stratification of patients and/or images can be especially beneficial for the application to heterogenous diseases such as heart failure or hypertension. For example, Katz et al. were able to identify two distinct patient subgroups with hypertension with significant variations in clinical characteristics, cardiac function, and hemodynamics.[93] Better, more delineated unique phenotypes in each heterogenous

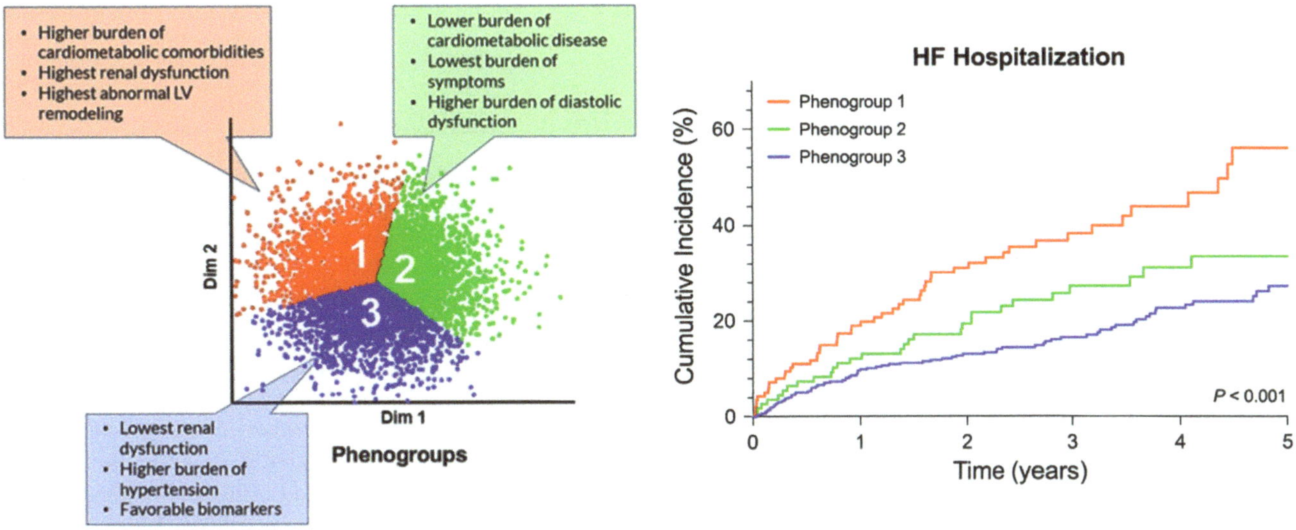

FIGURE 37.3 Phenomapping of heart failure with preserved ejection fraction from the TOPCAT Trial.[98] (Left) Cluster analysis showing three phenogroups with distinct phenotypes in TOPCAT; (Right) cumulative incidence plot for heart failure hospitalization outcomes among participants in the derivation cohort stratified by phenogroups.

disease state are useful for the application to more personalized therapies. The theorized benefit of phenomapping has begun to demonstrate promising results. Further analysis from data from the MADIT-CRT trial (demonstrating cardiac resynchronization benefit) included phenogrouping of the baseline characteristics of participants making four distinct groups. When stratified by group, two groups demonstrated significant improvement with CRT, whereas two others did not.[94] This powerful technique highlights the capabilities of personalizing medicine with the applications of machine learning. Similar approaches are sought in additional cardiac pathologies such as idiopathic nonischemic cardiomyopathy, HFpEF, and paroxysmal atrial fibrillation.[95–97] While this approach aims to seek out more individualized management, clustering can also help to determine the prognosis of an image, flagging concerning images to the appropriate cardiologist.

Image acquisition and processing

Given the challenge of explaining both static and dynamic characteristics of heart structure and function, image processing has become a bottleneck for much of the clinical workflow. AI has been able to provide significant improvement in quality control[99] and definition of imaging[100] modalities like CT and MRI. Since much of the imaging in cardiology relies on the differentiation of different anatomical structures, image segmentation is especially important. *Segmentation* is the extraction of content from an image or sequence of images to create a volume or shape. These can then be layered for modeling of 4D structures. This process of *anatomic modeling* has accelerated with significant improvements with the adoption of AI techniques. For example, 4D modeling of aortic and mitral valve apparatus has been especially sought after with the development of transcatheter approaches to valve replacement. Anatomical modeling has been widely explored even through less invasive modalities such as echocardiogram.

Current limitations and future directions

With the early successes of the application of AI to CV imaging, researchers and clinicians alike are aiming to apply these technologies more broadly into clinical practice. Two current needs in this space include implementation and further research. It is an exciting time for the field of CV imaging with many new applications on the horizon, such as interest in the "digital twin,"[101] meaning a comprehensive tool that integrates clinical data acquired over time.

While the excitement surrounding AI in CV imaging is warranted, further evidence of clinical utility of these technologies must be demonstrated.[87] A common problem seen in the field is *overfitting a model*. This refers to when an algorithm is given too specific training data and then is unable to perform as well on the testing data because the model was "overfit" to the training. Similarly, critics have pointed out the possibility of AI continuing or even worsening established bias between ethnicities and sex with some data suggesting disparate outcomes with use of AI technology.[102] With the goals of broad integration of global databases, training sets for algorithms will be appropriately trained.[103]

Digital health advances in individualized cardiovascular care

The field of cardiology has always relied heavy on objective parameters of physiologic and cardiac functioning to direct management and patient care. From the historical past with revolutionary technology such as the Swan–Ganz catheter for central pressure measurement to today, cardiology has led medicine in developing technology to detect and monitor disease. The advancement of hardware and technological devices as well as the goal of de-centralizing healthcare away from inpatient care have developed in parallel. These combine in the lofty goals/aspirations of ambulatory medicine and remote monitoring. This idea that was just theoretical a decade or two ago has now begun to be standard practice. Early successes of technology such as implantable cardiac defibrillators with an expansive amount of patient data has shown the true capabilities of this field to transform patient care.[104] This idea of remote patient monitoring has been around for decades. But the feasibility (bulky technology, limited recording capacity, invasive) of these interventions made their scope limited to those with severe cardiac disease. As the technology has continued to develop, less invasive and more practical approaches to patient monitoring allow the possibility of more widespread utilization within the general public and with less cost.[105] These technologies have already begun to take hold all across subspecialties in cardiac care, especially electrophysiology and heart failure.[106] As monitoring devices, wearables, smartphone apps, and watches have already begun to become a part of everyday practice, this field is set to expand exponentially within the next decade. Just in 2019 alone there was an estimated $7.9 billion invested in digital health companies.[107]

Basics of ambulatory patient monitoring

In development of patient monitoring there are several components to consider: (1) physiological parameters of interest to collect, (2) sensors and technology required to collect these physiological components, (3) data fusion and algorithms to compile raw data into meaningful insights, and then finally (4) modes of collection from a patient.[105] Physiological parameters collected are the same vital signs that would be collected in clinic (such as heart rate and rhythm, blood pressure, respiratory rate, oxygen saturation, and temperature). There has been a proliferation in technologies available to capture these vitals (see Box 37.1). As these sensors and hardware have dramatically increased in capability and decreased in physical size, the application of these tools has allowed for less invasive approaches such as headbands, watches, eyewear, and clothing.

Types of devices

Monitoring and recording devices have been a vital component of routine cardiovascular care including use of Holter monitors and Zio patches. Prior advancements in pacemakers and defibrillators have helped provide life-saving care with a better safety profile, including less adverse events and infection rates.[109] While initial device technology focused on invasive implantable devices, recent advancements have been made in external device technology. Yet further development of software in devices such as cardiac loop recorders has made some in the field wonder if implantable devices offer better clinical benefit over external ambulatory monitoring. Recent studies have shown that these monitors can be placed safely and provide significant benefit in early detection of disease as well as disease progression. For example, results published from both STROKE-AF[110] and PER DIEM[111] randomized clinical trials demonstrate

BOX 37.1 Physiological parameters collected from implanted or wearable devices

Accelerometer—detects small movements and changes in motion

Balistocardiogram—measures repetitive movement in all three axes to determine movement caused by blood ejection (detect heart rate)

Phenocardiography—acoustic technology that records heart sounds

Seismocardiography—records vibrations made from heart

Photoplethysmography (PPG)—optical measurement with infrared light to measure volumetric variations in blood circulation[108]

implantable monitors having significantly higher rates of AF detection following ischemic stroke as compared to external cardiac monitoring. In addition, innovations made in self-powered devices have broadened the capabilities, feasibility, and safety in implantable devices.[112]

The application of sensors especially through use of balistocardiogram and photoplethysmography has yielded a wide variety of wearable devices being piloted. Different usages include patches, handheld devices, smartphone applications, smartphone cases, and smartphone-connected devices, wristwatch, and head-mounted devices. Even with the huge variety of applications in wearables, most focus on collecting several vitals: blood pressure, temperature, heart rate and arrhythmia detection, and oxygen saturation.

While in recent past there was a clear delineation between medical devices and consumer products, the integration of cardiac monitoring has gone mainstream. The Apple Watch is one of the first consumer products that was published on in medical literature to demonstrate significant findings of clinical benefit. The results from the Apple Heart Study published in 2019 through New England Journal of Medicine demonstrated that 34% of participants with irregular pulse notification ultimately had evidence of atrial fibrillation on subsequent ECG.[113] Additionally, an estimated 84% irregular pulse notifications were shown to be likely AF events on adjudication.[113] These findings have excited the field and began to change the outlook for widespread screening for atrial fibrillation, previously believed to be too costly and difficult to implement.[114] Smartwatch technology has exploded in recent years and there is hope that these devices will provide an additional option for clinicians and patients to use.

Integration of devices and additional clinical data

While these devices that measure a set of vitals can provide clinically useful insights on their own, integration of this data with additional laboratory, imaging, and physical findings can be used for better capture of the patient's true phenotypic presentation (Fig. 37.4).

Device diagnostics

From data in the MultiSENSE study, researchers attempted to integrate multiple features measured in the ambulatory setting (third heart sound, thoracic impedance, heart rate, and patient activity) to predict/alert for heart failure related events. Results demonstrated 70% sensitivity and an unexplained alert rate of 1.47 per patient-year.[115] Furthermore, these exciting results were further improved with the addition of NT-proBNP for risk stratification, adding to the evidence of the power of these devices used in combination with establish risk projection variables.[116] The HeartLogic

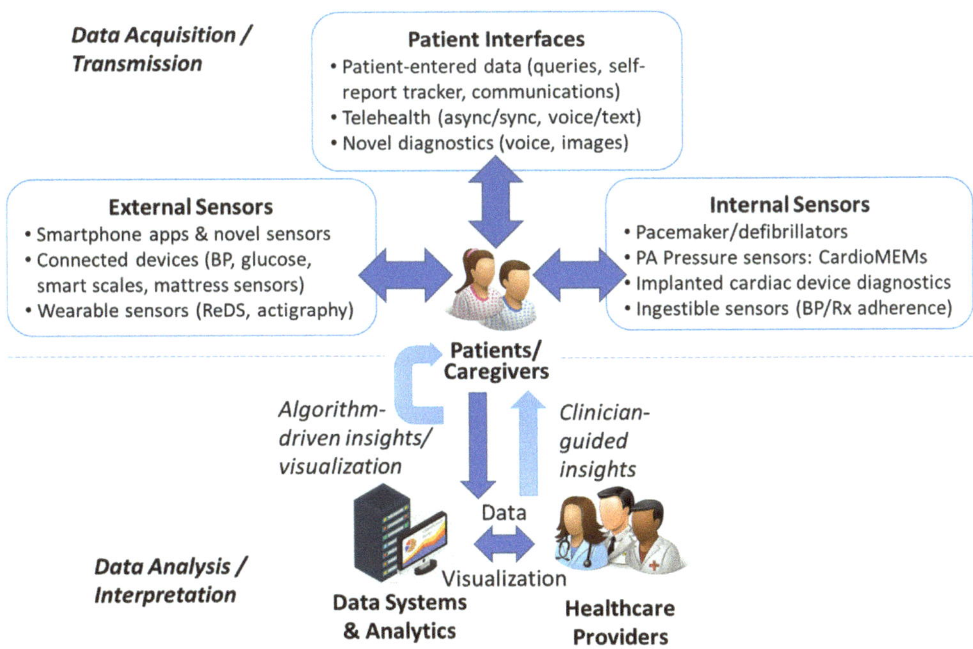

FIGURE 37.4 Considerations for integrating device diagnostics to inform precision cardiovascular care.[118]

algorithm was found to be predictive across a wide spectrum of CVD, including atrial fibrillation, ischemic HF, and reduced LVEF or HFrEF (<35%), and it alone had a higher independent predictive ability than NT-proBNP levels, with event rates of 0.80 versus 0.42/patient-year.[117]

Implantable hemodynamic monitoring

Additional studies have sought to apply wearable technology for monitoring and adjustment of pharmacological treatment. In a unique study looking to titrate GDMT based on ambulatory PA pressure, CHAMPION (Cardio-Microelectromechanical Systems Heart Sensor Allows Monitoring of Pressure to Improve Outcomes in Class III Heart Failure) investigated the ability of CardioMEMS Heart Sensors to analyze pulmonary artery (PA) pressures.[118,119] With initial results first being published in 2011 and long-term follow-up in 2016, the PA pressure group demonstrated reduced occurrence of HFEs by 39% compared to standard care and reduced HF hospitalizations by 28%.[117,118] Clinicians were able to provide treatment for HF as informed by the PA readings before the dysfunction became even worse. Treatment was especially effective for patients with low LVEF (>40%), as the experimental group had 50% lower HFE than the controls.[117]

Remote dielectric sensing (ReDS) has emerged as a leading noninvasive strategy and viable alternative to chest CT. Via a wearable vest with a sensor on the front and back of a patient's thorax, ReDS uses electromagnetic energy to measure the dielectric properties of tissues and quantify lung fluid content.[120] This stands out from other wireless pressure biosensors or implantable devices because the sensors are not affected by unrelated biological factors including skin moisture and electrode position. The technology can function well with acute decompensated HF and nonacute decompensated HF patients alike, and it has specifically been effective in preventing HF hospital readmission—an experiment by Lala et al. found a strong association with the use of ReDS and lower 30-day CV readmission.[121] The Heart Health Program conducted by researchers at Mount Sinai also tested a form of ambulatory remote monitoring. Vital signs like systolic and diastolic BP, weight, and heart rate were measured in real time and relayed via smartphone, similar apps like PatientConnect and HeartMapp. After the program, the 30-day CV readmission rate was 13% lower than Mount Sinai Hospital's rate and 15% lower than the readmission rate nationwide.[122] No specific correlation was found for a specific patient demographic, indicating that their remote monitoring system can be effective across age groups. It should be noted that patient adherence to device usage is a crucial factor in how the results presented.

Another major study called LINK-HF (Multisensor Noninvasive Remote Monitoring for Prediction of Heart Failure Exacerbation) investigated one such device that measured physiological readings including skin temperature, heart rate, sleep, and ECG via a disposable adhesive patch positioned on the chest.[123] Machine learning algorithms were designed to detect further HF dysfunction, and the data gathered from the patch was recorded in a cloud analytics database. Like ReDS, this multisensor was able to reduce rehospitalization rates in the study, comparable to the efficacy of well-established implantable devices.

Wearable technologies

The integration of technology and wearable devices into routine clinical practice is on the cusp of becoming mainstream for the treatment of patients with cardiac disease. With the FDA's recent breakthrough device designation in July 2021 of Biofourmis' BiovitalsHF, this marks the first digital therapeutic receiving this designation for heart failure.[124] This device, similar to other devices previously discussed, integrates continuous physiological data from vital sign monitoring devices and electronic patient reported outcome monitoring of patients to support goal-direct medical therapy.[125] While the results from Biofourmis' randomized clinical trial (ClinicalTrials.gov Identifier: NCT04191330) are not available yet, preliminary data demonstrates that these technologies can increase adherence to guideline medical therapy in routine clinical practice.[126]

While there has been robust application of wearable technologies in the heart failure realm, other fields of cardiology have also been early adopters with exciting results. In the care for patients with atrial fibrillation, the application of wearable technology is aimed to not only detect and screen patients for atrial fibrillation (as discussed in earlier sections), but also take part in the management of existing atrial fibrillation patients. One area of increased focus is to personalize recommendations for long-term anticoagulation based on out-of-clinic monitoring results. A recent pilot study comparing the efficacy and safety of chronic anticoagulation vs. intermittent (if detected on iPhone-based rhythm for >1 hour) has pushed this topic to the forefront.[127] With promising results, still, a large, prospective trial underway will be required to determine differences in bleeding risk and outcomes for stroke.[128]

Additional applications

While wearable applications to patient health are just beginning to be explored, numerous other clinical scenarios have been explored through wearables. Data suggest that remote patient monitoring improves outcomes in patients discharged for acute heart failure exacerbations[129] and acute myocardial infarction,[130] with clear implications for cardiac rehabilitation.

Further research and outlook

Ambulatory patient monitoring holds significant promise as early successes have demonstrated the value of the patient data they provide. While previously physicians were limited to physiological parameters within a controlled healthcare setting, wearables provide granular, personalized detail about a patient's real day-to-day cardiac and physiologic function. These technologies also offer the opportunity to lessen the differences in access and availability of services between rural and urban healthcare environments. As the field proliferates, challenges remain in upscaling production and how to apply these technologies to the public. While preliminary data of wearables are promising, these technologies have not been widely accepted into societal guidelines. This is in part due to the extreme variation in diagnostic capabilities and accuracy between devices. Data from the Apple Heart Study among others are exciting first steps but additional studies assessing for differences in clinical outcomes are required prior to robust implementation. In addition, study of both integration of these technologies alongside more "traditional" approaches and head-to-head comparisons will be required.[131] There remains concern about the technology surrounding wearables as well. For example, PPG, while becoming routine standard of care, has shown wide variations with changes in temperature, skin color, hair, and tattoos.[132] Additionally, patients and clinicians alike have risen concerns surrounding data privacy and usage rights. Patients concerned that the wide variety of data is being shared with insurance companies and data analytic programs is of legitimate concern and current legal protections in the United States do not reflect this ever-growing trove of patient data.[133,134]

While the line will likely blur with time, there remains a contrast between direct-to-consumer products available for the general public (heart rate monitors, health watches, augmented reality) and those approved by the FDA for evaluation of disease states. This dichotomy is seen as untenable given the extent of disease in the population and the

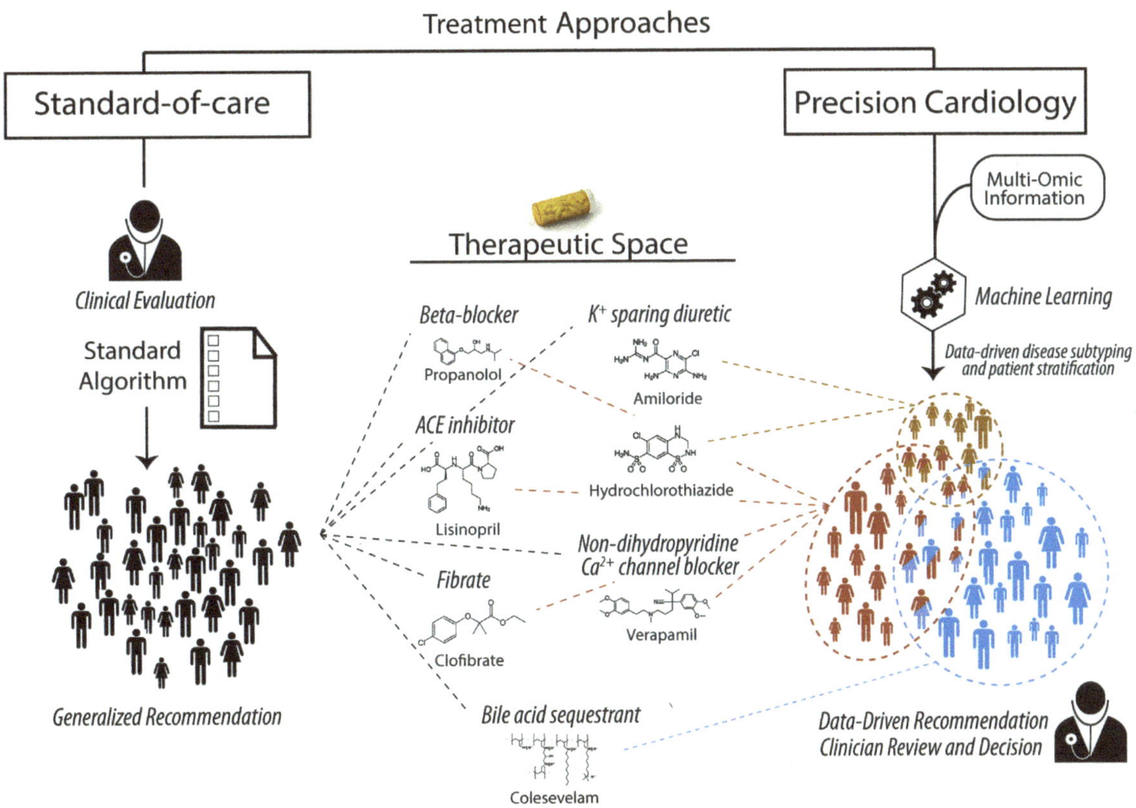

FIGURE 37.5 Conceptualization of a personalized medicine approach to cardiology contrasted with the current standard of care.[135]

widespread use of products. Further advancement in these technologies given to the public will likely require further FDA clearance for validation of accuracy in diagnosis and management of disease.[107]

Summary

Precision cardiovascular medicine is an integrative approach to CVD prevention and treatment that considers an individual's genetics, lifestyle, and exposures as determinants of their cardiovascular health and disease phenotypes. This has the potential of overcoming the simplification of current diagnostic and treatment approaches, which presumes that all patients with the same signs of heart disease share a common pathophenotype and, therefore should be treated similarly based on large-scale clinical trial evidence. While heavily reliant on genomic understanding that has yielded some successes in other medical fields, there is now a growing call to the cardiovascular community for more precise phenotyping to advance precision medicine as a strategy to improve cardiovascular health and prevent disease. In this chapter, we have outlined the progress being made regarding the pursuit of precision cardiology, although incremental benefit of this approach still demands prospective validation (Fig. 37.5).[133]

It is important to highlight four key areas of innovation that are necessary to operationalize precision cardiovascular medicine and move the field forward. First, we need to better gain insights into molecular signatures to redefine pathophysiologic understanding of human CVDs and challenge traditional classifications. Second, we need to uncover extracardiac pathophysiologic pathways as novel therapeutic targets through multiomics evaluation that may impact disease progression. In addition, we need to devise strategies and infrastructure to incorporate, engage, and integrate patient-centric information that can be assessable and interpretable to inform clinical care. Finally, we need to routinely apply novel technologies and analytics to identify and individualize therapeutic targets and refine such strategies to be tested in rigorous clinical trials to establish their incremental values. In summary, the combination of robust infrastructure, increased fundamental knowledge, and policy systems is need to to advance and operationalize precision cardiovascular medicine.

Funding support

Dr. Tang is partially supported by grants from the National Institutes of Health (R01HL126827, R01HL146754).

Disclosure

Dr. Tang is a consultant for Sequana Medical A.G., Owkin Inc., Relypsa Inc., PreCARDIA Inc., Cardiol Therapeutics Inc., Genomics plc, and has received honorarium from Springer Nature for authorship/editorship and American Board of Internal Medicine for exam writing committee participation—all unrelated to the subject and contents of this paper. All other authors have no relationships to disclose.

References

1. Mensah GA, Wei GS, Sorlie PD, et al. Decline in cardiovascular mortality. *Circul Res*. 2017;120:366–380.
2. Rate of coronary heart disease mortality among US adults (18 +); NVSS - United States - Age-Standardized. CDC Chronic Disease and Health Promotion Data & Indicators. 2017.
3. Virani SS, Alonso A, Benjamin Emelia J, et al. Heart disease and stroke statistics—2020 update: a report from the american heart association. *Circulation*. 2020;141:e139–e596.
4. Lehrman MA, Schneider WJ, Südhof TC, Brown MS, Goldstein JL, Russell DW. Mutation in LDL receptor: Alu-Alu recombination deletes exons encoding transmembrane and cytoplasmic domains. *Science*. 1985;227:140–146.
5. Kathiresan S, Srivastava D. Genetics of human cardiovascular disease. *Cell*. 2012;148:1242–1257.
6. Povysil G, Chazara O, Carss KJ, et al. Assessing the role of rare genetic variation in patients with heart failure. *JAMA Cardiol*. 2021;6:379–386.
7. Abdulrahim JW, Kwee LC, Alenezi F, et al. Identification of undetected monogenic cardiovascular disorders. *J Am Coll Cardiol*. 2020;76:797–808.
8. Ozaki K, Ohnishi Y, Iida A, et al. Functional SNPs in the lymphotoxin-alpha gene that are associated with susceptibility to myocardial infarction. *Nat Genet*. 2002;32:650–654.
9. Tam V, Patel N, Turcotte M, Bossé Y, Paré G, Meyre D. Benefits and limitations of genome-wide association studies. *Nat Rev Genet*. 2019;20:467–484.
10. Padmanabhan S, Dominiczak AF. Genomics of hypertension: the road to precision medicine. *Nat Rev Cardiol*. 2021;18:235–250.

11. Harrison DG. The mosaic theory revisited: common molecular mechanisms coordinating diverse organ and cellular events in hypertension. *J Am Soc Hypertens*. 2013;7:68–74.
12. Padmanabhan S, Joe B. Towards precision medicine for hypertension: a review of genomic, epigenomic, and microbiomic effects on blood pressure in experimental rat models and humans. *Physiol Rev*. 2017;97:1469–1528.
13. Evangelou E, Warren HR, Mosen-Ansorena D, et al. Genetic analysis of over 1 million people identifies 535 new loci associated with blood pressure traits. *Nat Genet*. 2018;50:1412–1425.
14. Ellinor PT, Lunetta KL, Albert CM, et al. *Meta*-analysis identifies six new susceptibility loci for atrial fibrillation. *Nat Genet*. 2012;44:670–675.
15. Shah S, Henry A, Roselli C, et al. Genome-wide association and Mendelian randomisation analysis provide insights into the pathogenesis of heart failure. *Nat Commun*. 2020;11:163.
16. Meder B, Rühle F, Weis T, et al. A genome-wide association study identifies 6p21 as novel risk locus for dilated cardiomyopathy. *Eur Heart J*. 2014;35:1069–1077.
17. Sivakumaran S, Agakov F, Theodoratou E, et al. Abundant pleiotropy in human complex diseases and traits. *Am J Hum Genet*. 2011;89:607–618.
18. Bush WS, Oetjens MT, Crawford DC. Unravelling the human genome–phenome relationship using phenome-wide association studies. *Nat Rev Genet*. 2016;17:129–145.
19. Diabetes Genetics Initiative of Broad Institute of H, Mit LU, Novartis Institutes of BioMedical R, et al. *Genome-wide association analysis identifies loci for type 2 diabetes and triglyceride levels. Science (New York, NY)*. 316. 20071331–1336.
20. Samani NJ, Erdmann J, Hall AS, et al. Genomewide association analysis of coronary artery disease. *N Engl J Med*. 2007;357:443–453.
21. Rastegar-Mojarad M, Ye Z, Kolesar JM, Hebbring SJ, Lin SM. Opportunities for drug repositioning from phenome-wide association studies. *Nat Biotechnol*. 2015;33:342–345.
22. Levin MG, Rader DJ. Polygenic risk scores and coronary artery disease. *Circulation*. 2020;141:637–640.
23. Marston NA, Kamanu FK, Nordio F, et al. Predicting benefit from evolocumab therapy in patients with atherosclerotic disease using a genetic risk score. *Circulation*. 2020;141:616–623.
24. Damask A, Steg PG, Schwartz GG, et al. Patients with high genome-wide polygenic risk scores for coronary artery disease may receive greater clinical benefit from alirocumab treatment in the ODYSSEY OUTCOMES trial. *Circulation*. 2020;141:624–636.
25. Mars N, Koskela JT, Ripatti P, et al. Polygenic and clinical risk scores and their impact on age at onset and prediction of cardiometabolic diseases and common cancers. *Nat Med*. 2020;26:549–557.
26. Mosley JD, Gupta DK, Tan J, et al. Predictive accuracy of a polygenic risk score compared with a clinical risk score for incident coronary heart disease. *JAMA*. 2020;323:627.
27. Elliott J, Bodinier B, Bond TA, et al. Predictive accuracy of a polygenic risk score–enhanced prediction model vs a clinical risk score for coronary artery disease. *JAMA*. 2020;323:636.
28. Khan SS, Cooper R, Greenland P. Do polygenic risk scores improve patient selection for prevention of coronary artery disease? *JAMA*. 2020;323:614.
29. Peters SAE, Muntner P, Woodward M. Sex differences in the prevalence of, and trends in, cardiovascular risk factors, treatment, and control in the United States, 2001 to 2016. *Circulation*. 2019;139:1025–1035.
30. Silarova B, Sharp S, Usher-Smith JA, et al. Effect of communicating phenotypic and genetic risk of coronary heart disease alongside web-based lifestyle advice: the INFORM randomised controlled trial. *Heart*. 2019;105:982–989.
31. Khera AV, Emdin CA, Drake I, et al. Genetic risk, adherence to a healthy lifestyle, and coronary disease. *N Engl J Med*. 2016;375:2349–2358.
32. Pazoki R, Dehghan A, Evangelou E, et al. Genetic predisposition to high blood pressure and lifestyle factors: associations with midlife blood pressure levels and cardiovascular events. *Circulation*. 2018;137:653–661.
33. Sirugo G, Williams SM, Tishkoff SA. The missing diversity in human genetic studies. *Cell*. 2019;177:26–31.
34. Lindsey ML, Mayr M, Gomes AV, et al. Transformative impact of proteomics on cardiovascular health and disease. *Circulation*. 2015;132:852–872.
35. Egerstedt A, Berntsson J, Smith ML, et al. Profiling of the plasma proteome across different stages of human heart failure. *Nat Commun*. 2019;10:5830.
36. Webb TR, Erdmann J, Stirrups KE, et al. Systematic evaluation of pleiotropy identifies 6 further loci associated with coronary artery disease. *J Am Coll Cardiol*. 2017;69:823–836.
37. Leopold JA, Maron BA, Loscalzo J. The application of big data to cardiovascular disease: paths to precision medicine. *J Clin Invest*. 2020;130:29–38.
38. Ghiassian SD, Menche J, Chasman DI, et al. Endophenotype network models: common core of complex diseases. *Sci Rep*. 2016;6:27414.
39. Maron BA, Wang R-S, Shevtsov S, et al. Individualized interactomes for network-based precision medicine in hypertrophic cardiomyopathy with implications for other clinical pathophenotypes. *Nat Commun*. 2021;12:873.
40. Strong A, Musunuru K. Genome editing in cardiovascular diseases. *Nat Rev Cardiol*. 2017;14:11–20.
41. Scully R, Panday A, Elango R, Willis NA. DNA double-strand break repair-pathway choice in somatic mammalian cells. *Nat Rev Mol Cell Biol*. 2019;20:698–714.
42. Lucas T, Bonauer A, Dimmeler S. RNA therapeutics in cardiovascular disease. *Circul Res*. 2018;123:205–220.
43. Thum T, Condorelli G. Long noncoding RNAs and microRNAs in cardiovascular pathophysiology. *Circ Res*. 2015;116:751–762.
44. Boon RA, Iekushi K, Lechner S, et al. MicroRNA-34a regulates cardiac ageing and function. *Nature*. 2013;495:107–110.

45. Hinkel R, Penzkofer D, Zühlke S, et al. Inhibition of MicroRNA-92a protects against ischemia/reperfusion injury in a large-animal model. *Circulation*. 2013;128:1066–1075.
46. Janssen HL, Reesink HW, Lawitz EJ, et al. Treatment of HCV infection by targeting microRNA. *N Engl J Med*. 2013;368:1685–1694.
47. Bayraktar R, Bertilaccio MTS, Calin GA. The interaction between two worlds: micrornas and toll-like receptors. *Front Immunol*. 2019;10.
48. Adams D, Gonzalez-Duarte A, O'Riordan WD, et al. Patisiran, an RNAi therapeutic, for hereditary transthyretin amyloidosis. *N Engl J Med*. 2018;379:11–21.
49. Walsh S, Kotz D FDA approves first-of-its kind targeted RNA-based therapy to treat a rare disease. FDA.gov: US Food & Drug Administration; 2018.
50. Ray KK, Landmesser U, Leiter LA, et al. Inclisiran in patients at high cardiovascular risk with elevated LDL cholesterol. *N Engl J Med*. 2017;376:1430–1440.
51. Frangoul H, Altshuler D, Cappellini MD, et al. CRISPR-Cas9 gene editing for sickle cell disease and β-thalassemia. *N Engl J Med*. 2020;384:252–260.
52. Aartsma-Rus A, Straub V, Hemmings R, et al. Development of exon skipping therapies for duchenne muscular dystrophy: a critical review and a perspective on the outstanding issues. *Nucleic Acid Ther*. 2017;27:251–259.
53. Gillmore JD, Gane E, Taubel J, et al. CRISPR-Cas9 in vivo gene editing for transthyretin amyloidosis. *N Engl J Med*. 2021.
54. Musunuru K, Chadwick AC, Mizoguchi T, et al. In vivo CRISPR base editing of PCSK9 durably lowers cholesterol in primates. *Nature*. 2021;593:429–434.
55. Rothgangl T, Dennis MK, Lin PJC, et al. In vivo adenine base editing of PCSK9 in macaques reduces LDL cholesterol levels. *Nat Biotechnol*. 2021.
56. Takahashi K, Yamanaka S. Induction of pluripotent stem cells from mouse embryonic and adult fibroblast cultures by defined factors. *Cell*. 2006;126:663–676.
57. Zakrzewski W, Dobrzyński M, Szymonowicz M, Rybak Z. Stem cells: past, present, and future. *Stem Cell Res Ther*. 2019;10:68.
58. van Berlo JH, Kanisicak O, Maillet M, et al. c-kit + cells minimally contribute cardiomyocytes to the heart. *Nature*. 2014;509:337–341.
59. Bagno L, Hatzistergos KE, Balkan W, Hare JM. Mesenchymal stem cell-based therapy for cardiovascular disease: progress and challenges. *Mol Ther*. 2018;26:1610–1623.
60. Nam YJ. Translational perspectives on cardiac reprogramming. *Semin Cell Dev Biol*. 2021.
61. Cho S, Lee C, Skylar-Scott MA, Heilshorn SC, Wu JC. Reconstructing the heart using iPSCs: Engineering strategies and applications. *J Mol Cell Cardiol*. 2021;157:56–65.
62. Sayed N, Liu C, Wu JC. Translation of human-induced pluripotent stem cells: from clinical trial in a dish to precision medicine. *J Am Coll Cardiol*. 2016;67:2161–2176.
63. Banerjee MN, Bolli R, Hare JM. Clinical studies of cell therapy in cardiovascular medicine. *Circulation Res*. 2018;123:266–287.
64. Chong JJH, Yang X, Don CW, et al. Human embryonic-stem-cell-derived cardiomyocytes regenerate non-human primate hearts. *Nature*. 2014;510:273–277.
65. Hare JM, Fishman JE, Gerstenblith G, et al. Comparison of allogeneic vs autologous bone marrow–derived mesenchymal stem cells delivered by transendocardial injection in patients with ischemic cardiomyopathy: The POSEIDON randomized trial. *JAMA*. 2012;308:2369–2379.
66. Hare JM, DiFede DL, Rieger AC, et al. Randomized comparison of allogeneic vs autologous mesenchymal stem cells for nonischemic dilated cardiomyopathy: POSEIDON-DCM trial. *J Am Coll Cardiology*. 2017;69:526–537.
67. Menasché P, Vanneaux V, Hagège A, et al. Human embryonic stem cell-derived cardiac progenitors for severe heart failure treatment: first clinical case report. *Eur Heart J*. 2015;36:2011–2017.
68. Menasche P, Vanneaux V, Hagege A, et al. Transplantation of human embryonic stem cell-derived cardiovascular progenitors for severe ischemic left ventricular dysfunction. *J Am Coll Cardiol*. 2018;71:429–438.
69. Lou X, Zhao M, Fan C, et al. N-cadherin overexpression enhances the reparative potency of human-induced pluripotent stem cell-derived cardiac myocytes in infarcted mouse hearts. *Cardiovasc Res*. 2020;116:671–685.
70. Tao Z, Loo S, Su L, et al. Angiopoietin-1 enhanced myocyte mitosis, engraftment, and the reparability of hiPSC-CMs for treatment of myocardial infarction. *Cardiovasc Res*. 2021;117:1578–1591.
71. Lam CK, Wu JC. Clinical trial in a dish: using patient-derived induced pluripotent stem cells to identify risks of drug-induced cardiotoxicity. *Arterioscler Thromb Vasc Biol*. 2021;41:1019–1031.
72. Musunuru K, Sheikh F, Gupta RM, et al. Induced pluripotent stem cells for cardiovascular disease modeling and precision medicine: a scientific statement from the american heart association. *Circ Genom Precis Med*. 2018;11:e000043.
73. Duarte JD, Cavallari LH. Pharmacogenetics to guide cardiovascular drug therapy. *Nat Rev Cardiology*. 2021;1–17.
74. Holmes MV, Perel P, Shah T, Hingorani AD, Casas JP. CYP2C19 genotype, clopidogrel metabolism, platelet function, and cardiovascular events: a systematic review and *meta*-analysis. *JAMA*. 2011;306:2704–2714.
75. Pereira NL, Farkouh ME, So D, et al. Effect of genotype-guided oral P2Y12 inhibitor selection vs conventional clopidogrel therapy on ischemic outcomes after percutaneous coronary intervention: the TAILOR-PCI randomized clinical trial. *JAMA*. 2020;324:761.
76. Mialet Perez J, Rathz DA, Petrashevskaya NN, et al. Beta 1-adrenergic receptor polymorphisms confer differential function and predisposition to heart failure. *Nat Med*. 2003;9:1300–1305.
77. Johnson JA, Liggett SB. Cardiovascular pharmacogenomics of adrenergic receptor signaling: clinical implications and future directions. *Clin Pharmacol Ther*. 2011;89:366–378.
78. Cooper GM, Johnson JA, Langaee TY, et al. A genome-wide scan for common genetic variants with a large influence on warfarin maintenance dose. *Blood*. 2008;112:1022–1027.

79. Giacomini KM, Yee SW, Mushiroda T, Weinshilboum RM, Ratain MJ, Kubo M. Genome-wide association studies of drug response and toxicity: an opportunity for genome medicine. *Nat Rev Drug Discov*. 2017;16:1.
80. Musunuru K, Hershberger RE, Day SM, et al. Genetic testing for inherited cardiovascular diseases: a scientific statement from the American Heart Association. *Circul Genomic Precis Med*. 2020;13:e000067.
81. Blell M, Hunter MA. Direct-to-consumer genetic testing's red herring: "genetic ancestry" and personalized medicine. *Front Med*. 2019;6.
82. Allyse MA, Robinson DH, Ferber MJ, Sharp RR. Direct-to-consumer testing 2.0: emerging models of direct-to-consumer genetic testing. *Mayo Clin Proc*. 2018;93:113−120.
83. Health + Ancestry Services. 23andMe. Accessed on 07/07/2021.
84. Wang C, Cahill TJ, Parlato A, et al. Consumer use and response to online third-party raw DNA interpretation services. *Mol Genet Genomic Med*. 2018;6:35−43.
85. Kilbride MK, Bradbury AR. The need to improve the clinical utility of direct-to-consumer genetic tests: either too narrow or too broad. *JAMA*. 2020;323:1443−1444.
86. Ye Q-H, Qin L-X, Forgues M, et al. Predicting hepatitis B virus-positive metastatic hepatocellular carcinomas using gene expression profiling and supervised machine learning. *Nat Med*. 2003;9:416−423.
87. Dey D, Slomka PJ, Leeson P, et al. artificial intelligence in cardiovascular imaging: JACC State-of-the-Art Review. *J Am Coll Cardiol*. 2019;73:1317−1335.
88. A data scientist's guide to data structures & algorithms, part 1. 2018.<https://towardsdatascience.com/a-data-scientists-guide-to-data-structures-algorithms-1176395015a0>.
89. Cheplygina V, de Bruijne M, Pluim JPW. Not-so-supervised: a survey of semi-supervised, multi-instance, and transfer learning in medical image analysis. *Med Image Anal*. 2019;54:280−296.
90. Sermesant M, Delingette H, Cochet H, Jaïs P, Ayache N. Applications of artificial intelligence in cardiovascular imaging. *Nat Rev Cardiol*. 2021;1−10.
91. Stebbing RV, Namburete AIL, Upton R, Leeson P, Noble JA. Data-driven shape parameterization for segmentation of the right ventricle from 3D + t echocardiography. *Med Image Anal*. 2015;21:29−39.
92. Fatima M, Pasha M. Survey of machine learning algorithms for disease diagnostic. *J Intell Learn Syst Appl*. 2017;9:1−16.
93. Katz DH, Deo RC, Aguilar FG, et al. Phenomapping for the identification of hypertensive patients with the myocardial substrate for heart failure with preserved ejection fraction. *J Cardiovasc Transl Res*. 2017;10:275−284.
94. Cikes M, Sanchez-Martinez S, Claggett B, et al. Machine learning-based phenogrouping in heart failure to identify responders to cardiac resynchronization therapy. *Eur J Heart Fail*. 2019;21:74−85.
95. Cohen JB, Schrauben SJ, Zhao L, et al. Clinical phenogroups in heart failure with preserved ejection fraction. *JACC Heart Fail*. 2020;8:172−184.
96. Shah SJ, Katz DH, Selvaraj S, et al. Phenomapping for novel classification of heart failure with preserved ejection fraction. *Circulation*. 2015;131:269−279.
97. Verdonschot JAJ, Merlo M, Dominguez F, et al. Phenotypic clustering of dilated cardiomyopathy patients highlights important pathophysiological differences. *Eur Heart J*. 2020;42:162−174.
98. Segar MW, Patel KV, Ayers C, et al. Phenomapping of patients with heart failure with preserved ejection fraction using machine learning-based unsupervised cluster analysis. *Eur J Heart Fail*. 2020;22:148−158.
99. Tarroni G, Oktay O, Bai W, et al. Learning-Based Quality Control for Cardiac MR Images. *IEEE Trans Med Imaging*. 2019;38:1127−1138.
100. Lossau T, Nickisch H, Wissel T, et al. Motion artifact recognition and quantification in coronary CT angiography using convolutional neural networks. *Med Image Anal*. 2019;52:68−79.
101. Corral-Acero J, Margara F, Marciniak M, et al. The 'Digital Twin' to enable the vision of precision cardiology. *Eur Heart J*. 2020;41:4556−4564.
102. Vinuesa R, Azizpour H, Leite I, et al. The role of artificial intelligence in achieving the sustainable development goals. *Nat Commun*. 2020;11:233.
103. Miller DD, Brown EW. Artificial intelligence in medical practice: the question to the answer? *Am J Med*. 2018;131:129−133.
104. Chatterjee NA, Singh JP. Making sense of remote monitoring studies in heart failure. *Eur Heart J*. 2017;38:2361−2363.
105. Sana F, Isselbacher Eric M, Singh Jagmeet P, Heist EK, Pathik B, Armoundas, Antonis A. Wearable devices for ambulatory cardiac monitoring. *J Am Coll Cardiol*. 2020;75:1582−1592.
106. DeVore AD, Wosik J, Hernandez AF. The future of wearables in heart failure patients. *JACC Heart Fail*. 2019;7:922−932.
107. Tarakji KG, Silva J, Chen LY, et al. Digital health and the care of the patient with arrhythmia. *Circul Arrhythmia Electrophysiol*. 2020;13:e007953.
108. Castaneda D, Esparza A, Ghamari M, Soltanpur C, Nazeran H. A review on wearable photoplethysmography sensors and their potential future applications in health care. *Int J Biosens Bioelectron*. 2018;4:195−202.
109. Cingolani E, Goldhaber JI, Marbán E. Next-generation pacemakers: from small devices to biological pacemakers. *Nat Rev Cardiol*. 2018;15:139−150.
110. Bernstein RA, Kamel H, Granger CB, et al. Effect of long-term continuous cardiac monitoring vs usual care on detection of atrial fibrillation in patients with stroke attributed to large- or small-vessel disease: the STROKE-AF randomized clinical trial. *JAMA*. 2021;325:2169−2177.
111. Buck BH, Hill MD, Quinn FR, et al. Effect of implantable vs prolonged external electrocardiographic monitoring on atrial fibrillation detection in patients with ischemic stroke: the PER DIEM randomized clinical trial. *JAMA*. 2021;325:2160−2168.

112. Zheng Q, Tang Q, Wang ZL, Li Z. Self-powered cardiovascular electronic devices and systems. *Nat Rev Cardiol.* 2021;18:7–21.
113. Perez MV, Mahaffey KW, Hedlin H, et al. Large-scale assessment of a smartwatch to identify atrial fibrillation. *N Engl J Med.* 2019;381:1909–1917.
114. Jones NR, Taylor CJ, Hobbs FDR, Bowman L, Casadei B. Screening for atrial fibrillation: a call for evidence. *Eur Heart J.* 2020;41:1075–1085.
115. Boehmer JP, Hariharan R, Devecchi FG, et al. A Multisensor algorithm predicts heart failure events in patients with implanted devices: results from the MultiSENSE study. *JACC Heart Fail.* 2017;5:216–225.
116. Gardner RS, Singh JP, Stancak B, et al. HeartLogic multisensor algorithm identifies patients during periods of significantly increased risk of heart failure events. *Circul Heart Fail.* 2018;11:e004669.
117. Costanzo MR. The luck of having a cardiac implantable electronic device. *Circulation: Heart Fail.* 2018;11:e004894.
118. Farwati M, Riaz H, Tang WHW. Digital health applications in heart failure: a critical appraisal of literature. *Curr Treat Options Cardiovasc Med.* 2021;23.
119. Abraham WT, Stevenson LW, Bourge RC, Lindenfeld JA, Bauman JG, Adamson PB. Sustained efficacy of pulmonary artery pressure to guide adjustment of chronic heart failure therapy: complete follow-up results from the CHAMPION randomised trial. *Lancet.* 2016;387:453–461.
120. Amir O, Azzam ZS, Gaspar T, et al. Validation of remote dielectric sensing (ReDS™) technology for quantification of lung fluid status: comparison to high resolution chest computed tomography in patients with and without acute heart failure. *Int J Cardiol.* 2016;221:841–846.
121. Lala A, Barghash MH, Giustino G, et al. Early use of remote dielectric sensing after hospitalization to reduce heart failure readmissions. *ESC Heart Fail.* 2021;8:1047–1054.
122. Park C, Otobo E, Ullman J, et al. Impact on readmission reduction among heart failure patients using digital health monitoring: feasibility and adoptability study. *JMIR Med Inf.* 2019;7:e13353.
123. Stehlik J, Schmalfuss C, Bozkurt B, et al. continuous wearable monitoring analytics predict heart failure hospitalization: the LINK-HF multicenter study. *Circul Heart Fail.* 2020;13:e006513.
124. Stultz T. Biofourmis Earns FDA's First-Ever Breakthrough Device Designation for a Novel Digitial Therapeutic for Heart Failure. Biofourmis; <https://www.biofourmis.com/press-releases/biovitalshf-fda-breakthrough-designation-approval/> 2021.
125. ClinicalTrials.gov. Artificil Intelligence Mobile Health Trial Of A Digitial Plaform To Optimize GDMT Using Wearable Sensors (AIM-POWER). 12/9/2019 (ed.) ClinicalTrials.gov: US National Library of Medicine.
126. Desai AS, MacLean T, Bosque-Hamilton J, Dunning J, Fischer C, Fera L. Naigator-Led remote optimization of guideline-directed medical therapy in patients with heart failure with reduced ejection fraction. *Sci Sessions.* 2019;. American Heart Association.
127. Stavrakis S, Stoner JA, Kardokus J, Garabelli PJ, Po SS, Lazzara R. Intermittent vs. continuous anticoagulation therapy in patients with atrial fibrillation (iCARE-AF): a randomized pilot study. *J Interv Card Electrophysiol.* 2017;48:51–60.
128. Bayoumy K, Gaber M, Elshafeey A, et al. Smart wearable devices in cardiovascular care: where we are and how to move forward. *Nat Rev Cardiol.* 2021;1–19.
129. Ong MK, Romano PS, Edgington S, et al. Effectiveness of remote patient monitoring after discharge of hospitalized patients with heart failure: the better effectiveness after transition – heart failure (BEAT-HF) randomized clinical trial. *JAMA Intern Med.* 2016;176:310–318.
130. Spaulding EM, Marvel FA, Lee MA, et al. Corrie health digital platform for self-management in secondary prevention after acute myocardial infarction. *Circ Cardiovasc Qual Outcomes.* 2019;12:e005509.
131. Lopez Perales CR, Van Spall HGC, Maeda S, et al. Mobile health applications for the detection of atrial fibrillation: a systematic review. *Europace.* 2021;23:11–28.
132. Carpenter A, Frontera A. Smart-watches: a potential challenger to the implantable loop recorder? *Europace.* 2016;18:791–793.
133. Banerjee S, Hemphill T, Longstreet P. Wearable devices and healthcare: Data sharing and privacy. *Inf Soc.* 2018;34:49–57.
134. Cohen IG, Mello MM. Big Data, Big Tech, and protecting patient privacy. *JAMA.* 2019;322:1141–1142.
135. Johnson KW, Shameer K, Glicksberg BS, et al. Enabling precision cardiology through multiscale biology and systems medicine. *JACC: Basic Transl Sci.* 2017;2:311–327.

Chapter 38

Devices that Enhance the Biomechanics of the Failing Heart

Luca Rosalia[1,2], Mossab Saeed Y.[3] and Ellen T. Roche[1,4]

[1]Institute for Medical Engineering and Science, Massachusetts Institute of Technology, Cambridge, MA, United States, [2]Harvard-MIT Health Sciences and Technology Program, Massachusetts Institute of Technology, Cambridge, MA, United States, [3]Harvard Medical School, Boston Children's Hospital, Boston, MA, United States, [4]Department of Mechanical Engineering, Massachusetts Institute of Technology, Cambridge, MA, United States

Chapter Outline

Devices for heart failure with reduced ejection fraction

Introduction

Blood-contacting devices like ventricular assist devices (VADs) and total artificial hearts (TAHs) are a viable treatment option for heart failure with reduced ejection fraction (HFrEF), and are reviewed comprehensively in other chapters of this book. One of the shortcomings of these devices is that they can cause thromboembolism and increased risk of stroke,[1] motivating the development of nonblood-contacting extracardiac devices, both passive and active.[2–5] These devices are reviewed in turn in this section and illustrated in Fig. 38.1.

Ventricular restraint devices—reinforcing the ventricles

Mechanical reinforcement or restraint of the ventricular wall has been shown to modulate and attenuate negative remodeling that can lead to dilated ventricles and heart failure (HF). Strategies for ventricular restraint include wraps, meshes, cardiac patches, or fluid-filled bladders, and are comprehensively reviewed elsewhere.[6] Although many devices discussed are of historic interest only, we briefly summarize the prior art below in the context of biventricular and univentricular modulation of biomechanics.

Biventricular restraint sleeves

Biventricular restraint devices reinforce or restrain both ventricles simultaneously to prevent or attenuate adverse remodeling and ventricular dilation, typically postischemic event. Examples are the Acorn CorCap (Acorn Cardiovascular, Inc.) and the ParaCor HeartNet (Paracor Medical). The Acorn CorCap cardiac support device (CSD) is a knitted polyester mesh sleeve that is placed epicardially on both ventricles from the apex to the atrioventricular (AV) groove.[7]

Advances in Cardiovascular Technology. DOI: https://doi.org/10.1016/B978-0-12-816861-5.00015-0

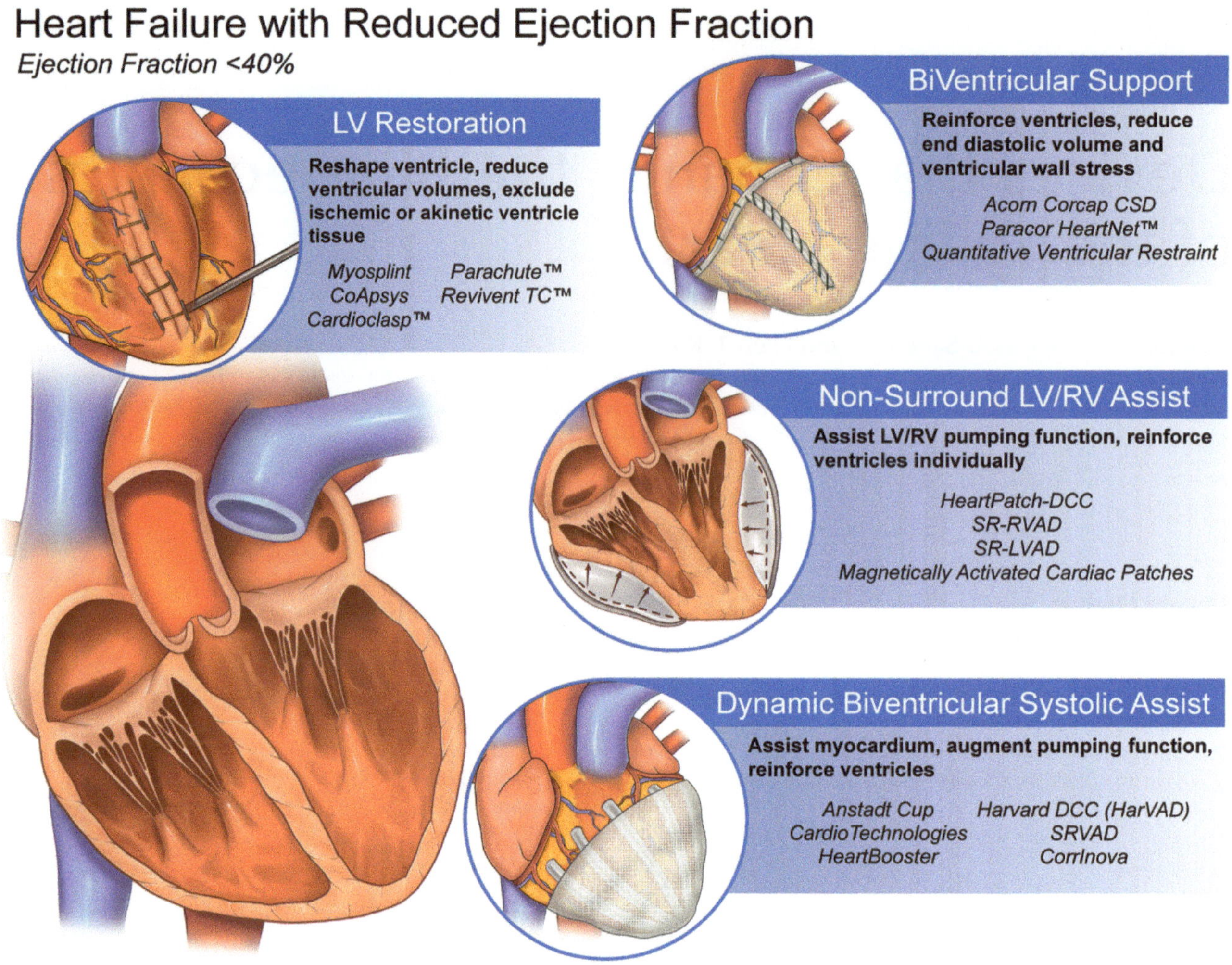

FIGURE 38.1 Summary of the device categories for heart failure with reduced ejection fraction with an illustration of an example of each.

The anisotropic fabric provides a higher level of restraint circumferentially than longitudinally (apex-to-base) to limit ventricular dilation without impeding ventricular filling. Careful adjustment of tension during implantation enables the device to prevent increases in end-diastolic volume over time, and thus reduce the degree of stretching and the magnitude of stress in the ventricular wall.[7] Attenuation of postinfarct dilation and HF progression was demonstrated with ovine animal studies.[8–10] The first-in-human (FIH) safety study reported no device-related mortality and improvements in the New York Heart Association (NYHA) classification of HF.[11,12] The 5-year follow-up in a multicenter randomized controlled trial demonstrated benefits in left ventricular (LV) size and shape in patients with the CSD,[13,14] yet the US Food and Drug Administration (FDA) did not approve the device, potentially due to the fact that no improvement in survival was demonstrated at 1, 3, and 5 years compared to medical management.[13–17]

The HeartNet (Paracor Medical Inc., Sunnyvale, CA) is a nitinol mesh that surrounds the heart and provides ventricular restraint. It is delivered through a minithoracotomy off-pump and can be visualized under fluoroscopy. The device works similarly to the CorCap, and demonstrated benefits in a sheep model of ischemic cardiomyopathy.[18] The feasibility study showed an increase in 6-minute walk distance, and encouraged the company to sponsor a multicenter randomized controlled trial to compare HeartNet to medical therapy for functional capacity in HR patients. The trial was suspended at an interim analysis, and despite benefits in quality of life and echocardiographic measurements of cardiac function, the all-cause mortality at 1 year was similar for the treatment and control groups.[19]

Quantitative ventricular restraint (QVR) was introduced by Ghanta et al.[20–22] The device consists of a semiellipsoidal, flexible, inflatable polyurethane balloon that wraps around both ventricles and can both apply varying levels of restraint—by adjusting the amount of fluid in the balloon through a subcutaneous port—and sense the applied pressure. Studies in sheep found that the level of restraint is important for optimization of acute cardiac function and energetics, and also for attenuating remodeling in the chronic setting.[23] The adjustability and modularity of this platform also enabled studies that concluded that adjustable restraint was superior to the aforementioned nonadjustable restraint devices,[22] and that LV restraint alone was superior to biventricular restraint.[24]

A number of epicardial patches that locally reinforce the infarcted part of the ventricle, providing isotropic[25,26] or anisotropic[27,28] reinforcement to the post-MI heart have been studied and are comprehensively reviewed elsewhere.[6,23]

Left ventricular restoration—reshaping the ventricle

Unlike biventricular support devices, LV restraints or restoration devices focus on reshaping the LV alone and typically have an intracardiac component. Some are of historical interest and no longer in development. For example, the Myosplint (Myocor Inc., Minneapolis, MN) is comprised of two epicardial pads connected by a polyethylene braid in the LV cavity. By tensioning the device and reducing the distance between the pads, the LV was reshaped into a bilobular chamber with a smaller diameter, resulting in decreased wall stress, according to the law of LaPlace.[29] Typically, three Myosplints are implanted longitudinally in the LV, and adjusted until a reduction of 20% of the dilated LV cavity is achieved. Attenuation of ventricular dilation was achieved in a canine model and later in a clinical pilot study of 21 patients, and successfully demonstrated safety and feasibility. However, there were limitations with respect to functional mitral regurgitation as the majority of the pilot study patients ultimately required mitral valve (MV) surgery and cardiopulmonary bypass. The company then focused on the Coapsys device, a similar device that approximates the papillary muscles (PM) by reshaping the ventricle at the level of the subvalvular apparatus with a similar tensioning filament.[30,31] The iCoapsys is a minimally invasive version of this device but unfortunately the company ceased operations in 2008.

Another LV restraint device, the CardioClasp (Cardioclasp Inc., Pine Brook, NJ) reduced LV dilation by reshaping. It consisted of two rigid bars connected through the LV cavity by an adjustable tether.[32] The bars mimic the natural contour of the heart and were secured in place in a longitudinal line on the anterior and posterior LV walls. The tether approximates the bars to achieve a 30% reduction of the initial diastolic dimensions. Multiple studies investigating the acute effects of the CardioClasp in a canine HF model report a decrease in wall stress, increase in fractional area of contraction, enhanced systolic contractility, and positive chamber geometrical changes.[32–34]

The Parachute (Cardiokinetix, Menlo Park, CA, United States) is a ventricular partitioning device which can be delivered minimally invasively in a retrograde manner through the aortic valve. Unlike other devices that restore the LV from the epicardial surface, the Parachute aims to restore the ventricle from within and creates a partition between the normal myocardium and the diseased or aneurysmal myocardium, thus reducing the cavity size.[35] The dynamic part of the ventricle is therefore hemodynamically isolated from the static section (the scarred or aneurysmal portion), thus unloading it. The ventricular volume reduction also reduces the stress on the ventricular wall.[36] The FIH study showed no changes in EF or stroke volume index but a decrease in NYHA class at 1 year.[37] Larger pivotal studies showed improved or maintained functional status and improvements in NYHA class.[38] A cardiac computerized tomographic study showed improvements in LV diastolic mechanics after device implantation in patients with ischemic injury.[39] In the PARACHUTE IV trial, 331 subjects were enrolled, with a primary endpoint of death, hospitalization or worsening HF. The trial was terminated in 2017[40] and investigations on the device may not be continued.[41] Some challenges that were faced with the device include strut fractures, tears in the membrane of the device, and thrombus formation.[38]

The Revivent TC (BioVentrix, San Ramon, CA) is a method for ventricular remodeling in patients with HFrEF post-infarct. A series of microanchor pairs are implanted to exclude the scarred myocardium and to reduce and reshape the LV. The so-called Less Invasive Ventricular Enhancement (LIVE) procedure involves implanting internal anchors on the right ventricular septum through a transcatheter technique and external titanium anchors on the epicardial LV wall and tethering them with a poly-ether -ether ketone (PEEK) intracardiac tether to approximate the anterior septum with the LV free wall. Ongoing trials[42,43] are promising—at the beginning of 2020, the sponsors had recruited 47 patients reported 100% procedural success, no intrahospital mortality, late mortality and a median improvement in NYHA class by one grade.[44]

Dynamic systolic assist devices—augmenting muscle function

Active systolic assist—also known as direct mechanical ventricular assist (DMVA) or direct cardiac compression (DCC) involves implantation of a sleeve, bladder or cup around the heart. The device is pressurized in synchrony with the systolic phase of the cardiac cycle and augments the amount of blood ejected from the ventricles. As this approach is non-blood contacting, it is poised to reduce thromboembolic events and the need for anticoagulant therapy associated with traditional blood-contacting ventricular assist devices (VADs), reviewed elsewhere in this book. DMVA was first introduced for resuscitation of an arrested heart and additional similar technologies have been developed since.[3–5] Typically, these devices are inserted into the pericardial space and fixed in place with negative pressure or sutures. When the device is pressurized or actuated, it compresses the ventricular walls and augments ejection of blood from the ventricles. Depressurization allows the ventricles to fill.

The first example of DCC for cardiopulmonary resuscitation involved pneumatic inflation and evacuation of the pericardial space, but was found to compress the atria.[45] An extracardiac cup that compressed the heart to assist pumping action or mechanically massage the heart was then introduced[46] to reverse cardiac arrest. An elliptically shaped cup covers both ventricles and a semirigid outer layer and inflatable inner diaphragm delivers epicardial compression. Device actuation is not synchronized with the native heartbeat which can lead to rhythm disturbances and myocardial injury.

The Cardiac Assist System (Cardio Technologies Inc, Pine Brook, NJ) is a cup-shaped cardiac assist device with a pneumatically powered inner cuff which inflates to compress the heart. It is designed for short-term support for acutely failing hearts and is removed shortly after recovery. It is customized to the size of the patient's heart, provides adjustable levels of assistance and can be synchronized with the heartbeat.[5] The Heart Booster (AbioMed Inc, Danvers, MA) is a sleeve with longitudinal air bladders that compresses the heart by contracting to a smaller diameter. It was designed for long-term support.[4] Smooth outer surfaces and a proprietary polymer are used to prevent tissue ingrowth during long-term use. The device can be delivered minimally invasively through a minithoracotomy and anchored to the pericardium; however attachment can creep in the long term.[4] The HeartPatch (Heart Assist Technologies Pty Ltd; St Leonards, Australia) is a device composed of two inflatable patches that are placed on each ventricle to allow independent actuation. Patches provide "nonsurround" cardiac assistance to avoid impeding ventricular filling[47] and to enable the ventricles to be actuated independently at varying pressures. Minimally invasive delivery is followed by a period of two weeks before actuating so that biointegration can occur to couple the patches to the heart, which is intended for chronic use. One shortcoming of this device is that it induces tensile peel stresses at the patch edges as the center of the patch inflates and compresses the ventricle, eventually leading to delamination.[48] The device also creates unnatural deformation of the underlying as it inverts the curvature of the ventricular wall, which can lead to impaired valve function.[49] Alternative ways to provide epicardial mechanical assistance to the heart include using cables driven by ultrasonic motors,[50] rotational motors,[51] or by using electroactive polymers and shape memory alloys.[1,52,53] Recently a magnetically activated cardiac patch (MCAP) was described,[54] but as yet it needs a large magnetic actuation system for its operation, and would not be suitable for ambulatory patients.

Our group introduced a biomimetic direct cardiac compression device, which was based on soft robotic technology initially used to simulate the twisting motion of the heart.[55] The silicone sleeve was capable of achieving squeezing and twisting motion, and restoring cardiac output to baseline in an acute model of HF in swine.[56,57] A modular design was then described which had been optimized for coupling to the epicardium and synchronization with the heart rhythm.[58] The sleeve was then designed with a medical textile mesh, which allowed a controlled level of tissue integration for optimal coupling.[59] These devices are being developed further, with a focus on the robustness of the actuators, and minimization of the associated hardware. Corrinova Inc. have introduced a collapsible thin-film soft robotic device with a thin self-deploying frame.[2,60] The focus for this company is on short-term assist for acute HF—for chronic HF patients with acute decompensation as a bridge to decision, or for acute HF syndrome patients as a bridge to recovery. Recently, the company have added viral myocarditis to their indications, partially in response to the COVID-19 pandemic.

Unlike devices for HFrEF where the focus is on restraining the ventricles or actively assisting muscle function, devices to treat heart failure with preserved ejection fraction (HFpEF) aim to expand the ventricles or reduce the preload (unloading the atria), and are discussed in the following section.

Devices for heart failure with preserved ejection fraction

Introduction

The inadequacy of traditional treatment options in the management HFpEF has galvanized the development of device-based solutions, now at various stages of preclinical and clinical testing.[61,163] HFpEF is an umbrella term encompassing cardiac disease of various etiologies and phenotypes, whereby typical symptoms of heart failure occur despite a normal left ventricular ejection fraction.[62–64] In the past few decades, the incidence and prevalence of HFpEF have been growing considerably relatively to HFrEF,[65] leading to over two thirds of current heart failure patients over the age of 65 being affected by HFpEF. This phenomenon has been intensified by the rise in metabolic diseases and other risk factors of HFpEF.[66–68] To date, no treatment has conclusively reduced morbidity and mortality,[69] resulting in a 5-year survival of $<50\%$ and elevated rehospitalization rates.[70,71]

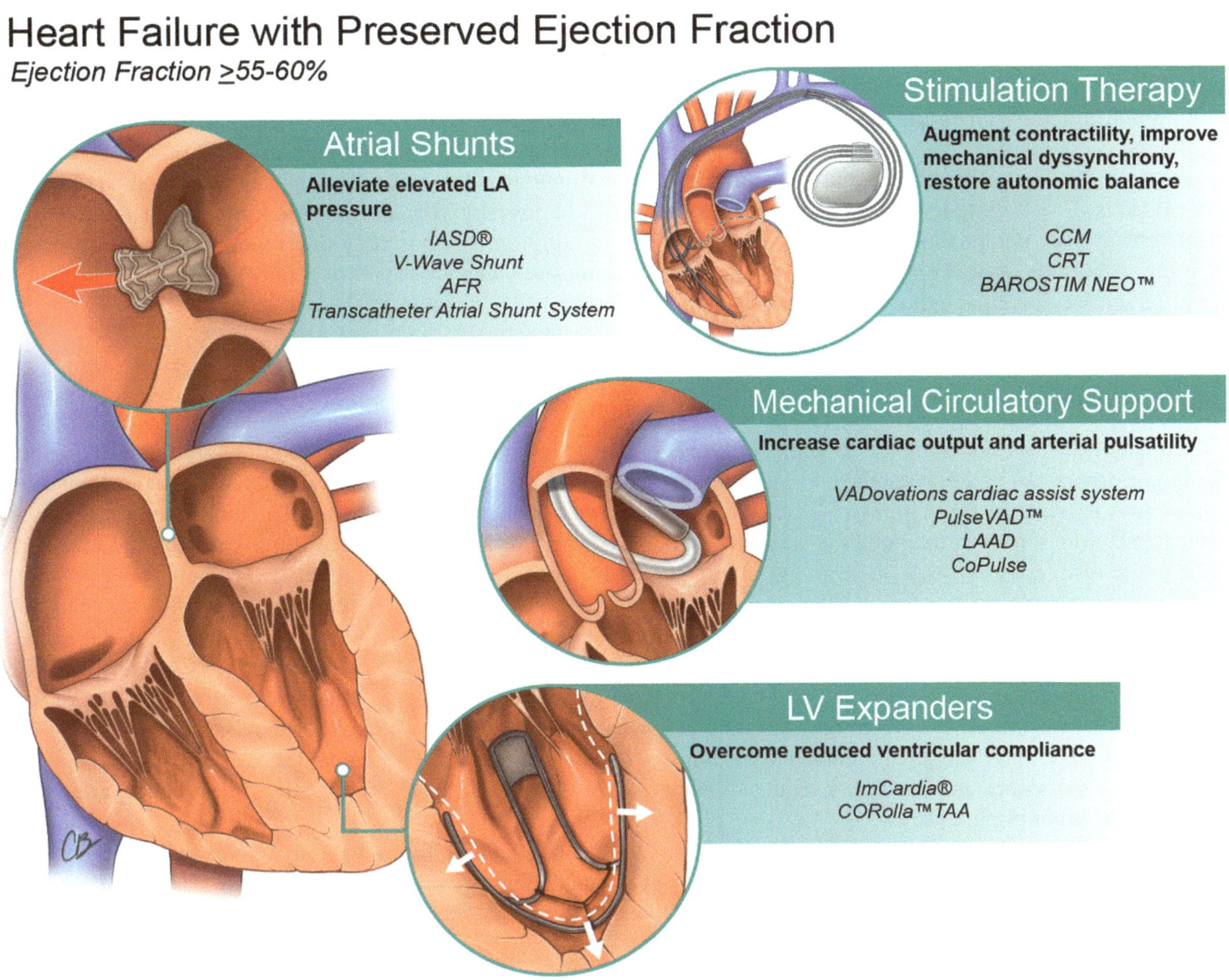

FIGURE 38.2 Summary of the device categories for heart failure with preserved ejection fraction with an illustration of an example of each. HFpEF is defined for ejection fractions ≥ 55% and ≥60% in men and women, respectively.

HFpEF devices can be broadly categorized by the biomechanical or hemodynamic aberrations that they target as follows: (1) atrial shunts to lower left atrial (LA) pressure, (2) LV expanders to augment filling performance, (3) neuromodulation and electrical stimulators to target contractility, mechanical dyssynchrony and autonomic imbalance, and (4) mechanical circulatory support (MCS) devices to increase cardiac output and arterial pulsatility (Fig. 38.2).

Atrial shunting devices

Atrial shunts have drawn attention in the landscape of device-based solutions for HFpEF. Through the creation of a conduit from the LA to other chambers or structures, these devices aim to reduce LA pressure and alleviate pulmonary congestion and associated symptoms, often observed with HFpEF.

Interatrial shunts are the most common device in this category and are currently under evaluation in several large clinical trials.[72–79] The idea of creating a fenestration in the atrial septum derives from the observation that patients affected by the Lutembacher syndrome (i.e., mitral stenosis in the presence of an interatrial septal defect) manifest fewer symptoms than individuals with isolated mitral stenosis.[80] To overcome some of the limitations associated with these devices, including the potential risk of right-to-left shunting and systemic embolization, a novel approach leveraging LA-to-coronary sinus (CS) shunting was recently developed.[81,82]

To date, three interatrial shunt devices exist: the InterAtrial Shunt Device (IASD) (Corvia Medical Inc., Tewksbury, MA, United States)—a bare-metal nitinol frame creating an 8-mm bidirectional shunt; the V-Wave Shunt (W-Wave Ltd., Or Akiva, Israel)—an hourglass-shaped self-expanding nitinol frame covered with porcine pericardial tissue preventing reverse shunting; and the atrial flow regulator (AFR) (Occlutech, Instanbul, Turkey)—a self-expandable double-disk nitinol mesh braided into two flat disks. The Transcatheter Atrial Shunt System (Edwards Life-sciences, Irvine, CA, United States) is the first and only LA-to-CS shunting device currently under investigation. It constitutes a bare-nitinol implant with an internal diameter of 7 mm, which can be delivered via percutaneous atriotomy, a procedure involving access via the right internal jugular vein, CS cannulation, and CS-to-LA puncture.[82]

Whereas the feasibility, safety, and short-term efficacy of interatrial shunting devices has been evaluated in several clinical trials,[78,83–87] long-term use of these devices is yet to be investigated, as concerns remain over induced risks of atrial arrhythmias, paradoxical embolism and right HF.[88] In addition, there is only one report documenting LA-to-CS shunting. Albeit encouraging, the conclusions drawn from this study cannot be yet generalized due to the limited number of patients involved and the short follow-up.[82]

Left ventricular expanders

LV expander devices are designed to apply force to dilate the left ventricle during relaxation—ultimately augmenting filling performance. Diminished relaxation is a key contributor to the HFpEF pathophysiology, and evidence of diastolic dysfunction is required for diagnosis.[62] Loss of ventricular compliance is mediated by structural alterations of various etiologies,[89] which lead to the hemodynamic hallmarks of HFpEF, including reduced filling volumes and elevated diastolic pressures.

LV expanders are passive spring-like devices that store elastic energy during contraction, which they then transfer to the LV wall during relaxation.[90,91] The ImCardia (CorAssist Inc. Haifa, Israel) is an extracardiac device designed for implantation on the epicardial surface, whereas the CORolla TAA (Trans-Apical Approach) (CorAssist Inc. Haifa, Israel) is an intraventricular device that is delivered transapically to adhere to the endocardium. While trials of the former have been suspended as further device optimization was required,[92] the CORolla TAA was shown to be safe and able to reduce LV mass and LA volume, and to improve diastolic function.[93] A clinical trial is currently ongoing to further evaluate the safety and feasibility of the CORolla TAA, as well as its effects on diastolic function and symptom severity on a larger population.[94]

Mechanical augmentation of cardiac relaxation aims to restore adequate filling hemodynamics by directly addressing the diminished ventricular compliance resulting from structural abnormalities or adverse remodeling. However, it is important to investigate the efficacy of these devices over different degrees of hypertrophy and in the context of remodeling as a dynamic process, that may regress due to medical intervention or worsen during the course of the disease. Furthermore, it is uncertain whether the level of support provided can be programmed to provide a patient-specific solution and the effects of expansion forces applied to the ventricular septum on right heart biomechanics and on ventricular interdependence are yet to be studied.

Neuromodulation and electrical therapies

Abnormalities in the autonomic tone and in the heart's electrical activity are major contributors to the pathophysiology and symptomatology of HFpEF. In the context of diminished systolic and diastolic reserve, additional impairments including chronotropic incompetence, reduced heart rate recovery, and attenuated arterial baroreflex sensitivity, may further limit cardiac output and exacerbate symptoms of effort intolerance, particularly in the setting of exercise.[62,63]

Cardiac contractility modulation (CCM) and baroreceptor activation therapy (BAT) target some of these defects through implantable pulse generating systems, currently undergoing clinical evaluation for HFpEF.[95,96] Originally developed for HFrEF, CCM devices aim to augment native cardiac contractility without affecting the cardiac rhythm.[97] In HFpEF, this technique was shown to increase the expression of contractile proteins and augment diastolic function.[98] Conversely, BAT neuromodulation devices such as the BAROSTIM NEO (CVRx Inc., Minneapolis, MN, United States), act to enhance the parasympathetic tone via stimulation of the carotid sinus. Although studies on patients with hypertension suggest that BAT may improve diastolic hemodynamics, and thus potentially address one of the primary treatment goals in HFpEF,[99,100] stronger clinical evidence regarding the efficacy on HFpEF patients is yet to be provided.

Atrial fibrillation and mechanical dyssynchrony are also largely associated with HFpEF. At the AV level, these lead to abnormalities in the hemodynamics of ventricular filling. Studies of the effects of cardiac resynchronization therapy (CRT) have confirmed that bi-atrial CRT,[101,102] biventricular fusion pacing,[103–105] and permanent LA or His Bundle pacing[106] can all improve cardiac hemodynamics and functional class. Nevertheless, it is still unclear whether mechanical dyssynchrony can be effectively corrected in HFpEF, and several clinical trials aiming to evaluate the efficacy of these techniques are currently ongoing.[107–109]

These stimulation therapies are not new in the panorama of cardiovascular disease; their use for the treatment of other conditions is widespread and their efficacy is well documented. However, their impact on cardiac biomechanics in the context of a multifactorial disease such as HFpEF is unknown. While it is unlikely that they act as a panacea for HFpEF, they may revolutionize the management of certain patients who are at increased risk of atrial fibrillation or experience severe autonomic imbalance.

Mechanical circulatory support

A comprehensive review of MCS for HFpEF is provided in a separate chapter of this book and is briefly reviewed here. LVADs, as discussed in the preceding section, can provide hemodynamic support for patients with end-stage HF, as a bridge-to-transplant or long-term destination therapy. Although little evidence exists around the use of LVADs in HFpEF patients, studies suggest that in this population they may pose an increased risk of LA collapse or intraventricular clotting.[110,111] While emphasizing the importance of pump speed control to moderate these risks, computational studies demonstrated that LVADs could offer several hemodynamic benefits to HFpEF patients.[112] Based on these learnings, other LVAD-type devices that pump blood from the LA directly to the arterial system have been modeled in silico, showing that they are associated with a lower risk of LA suction events.[89,164] Examples of analogous devices include the VADovations cardiac assist system (VADovations, Oklahoma City, OK, United States), allowing minimally invasive percutaneous delivery, and the PulseVAD (Northern Development, Strandhaugen, Oslo, Norway), enabling adaptive flow from the LA to the descending aorta. However, scarce documentation exists around these devices and their safety and efficacy are yet to be demonstrated.

LA assist devices (LAADs) are another type of pump designed to provide hemodynamic support in HFpEF patients. They consist of a continuous-flow pump implanted at the MV. Computational and in vitro studies show that this device can lower LA pressure and increase cardiac output[113–115] and suggest that implantation between the LA and the LV overcomes several of the limitations associated with other configurations, including diminished arterial pulsatility and reduced LV end-diastolic volumes. The LAAD is the first and only MCS device for HFpEF undergoing animal testing. Preliminary results in four healthy calves demonstrated improvements in cardiac output and arterial pressure as well as successful decompression of the LA and minimal adverse events.[116] However, more extensive preclinical evaluation of these devices is required, particularly on animals with HFpEF. Further, the risk of elevation of LV end-diastolic pressure, which may in turn affect coronary blood flow and right ventricular function is yet to be investigated.[61]

The CoPulse device is a type of valveless, pulsatile pump designed to be implanted in the apex of the LV. It is composed of an actuation chamber and of a blood compartment, separated by a flexible polyurethane membrane, whose displacement results in a 30 mL or 60 mL stroke volume. Shear stress and associated risk of thrombogenesis are reduced due to its valveless design. In silico and in vitro investigations have demonstrated that the CoPulse pump is highly hemocompatible and able to improve LA hemodynamics and cardiac output,[117,118] although the risk of LA suction and other adverse events is yet to be fully investigated.

Albeit supported by in silico and in vitro data, the use of MCS devices for the treatment of HFpEF requires further validation. Ongoing research is investigating the effects associated with MV removal in LAADs and the durability of the polyurethane membrane of the CoPulse device. A consensus is yet to be reached concerning the most appropriate pump configuration and the expected outcomes associated with each of the various hemodynamic phenotypes seen in HFpEF.

Augmentation of atrioventricular valvular biomechanics

Introduction

The heart has two AV valves interfacing and ensuring unidirectional blood flow from the atria to the ventricles; the MV on the left and the tricuspid valve (TV) on the right. HF and AV valve disease are often intertwined conditions and one is often a sequela of the other. HF can often lead to dysfunction of these valves resulting in mitral/tricuspid regurgitation (MR/TR), and similarly valvular disease can lead to HF. This section reviews emerging technologies that improve native diseased MV or TV function and restore standard functionality without necessitating valve replacement. Here, we identify the anatomical targets of the AV valves that are modulated with these technologies and categorize devices according to their intended geometrical and biomechanical effects. Based on these anatomical targets, we classify

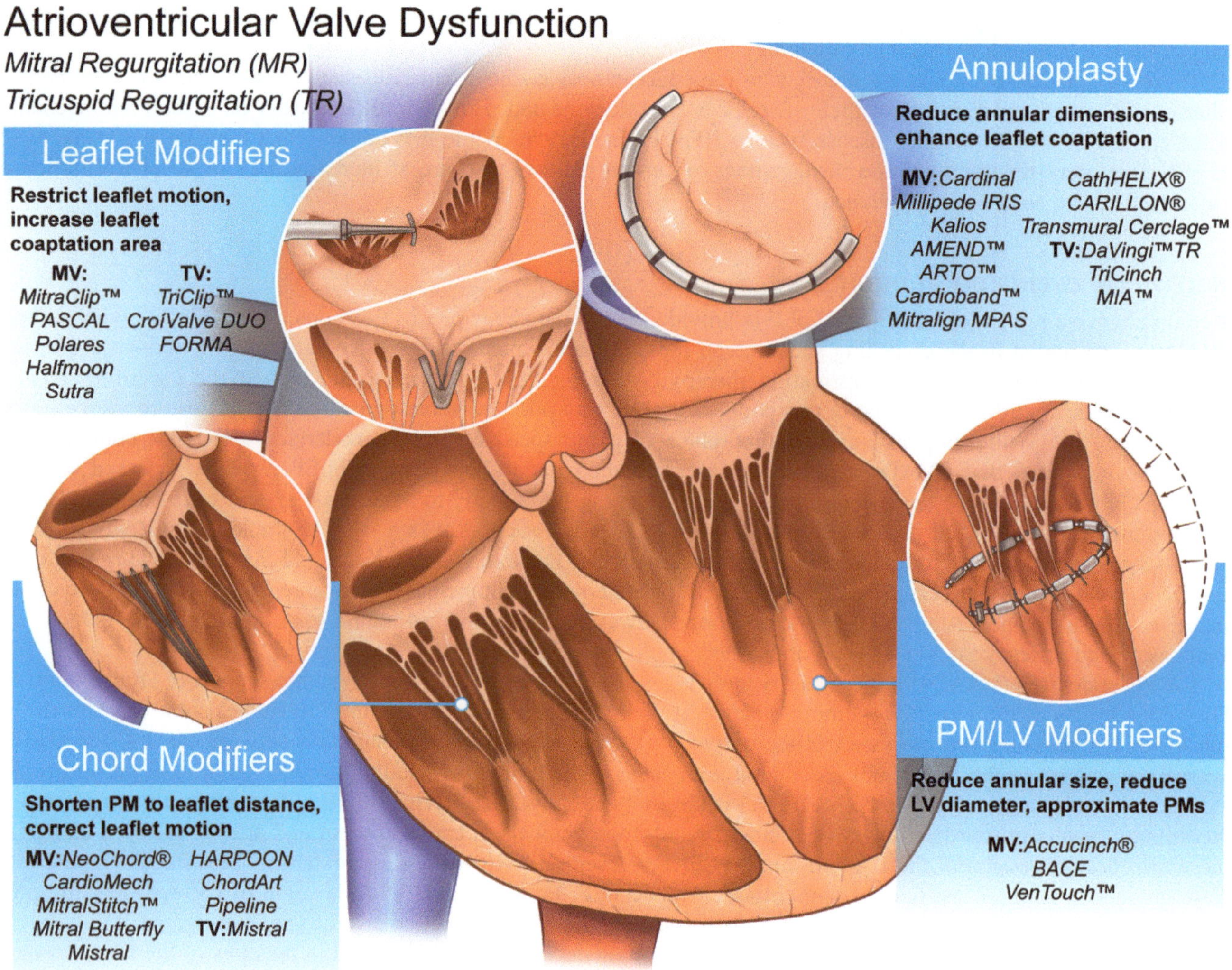

FIGURE 38.3 Summary of the device categories for addressing atrioventricular valve dysfunction with an illustration of an example of each.

devices as follows; devices that modulate (1) the valve annulus, (2) leaflets, (3) chordae tendineae, and (4) PM and/or the ventricular chamber. The classification and examples are illustrated in Fig. 38.3.

Mitral valve

Mitral valve annuloplasty

MV annulus dilation is among the causes of MV insufficiency.[119] Devices typically aim to correct this insufficiency by modulating three geometrical targets; annulus shape, effective orifice area, and the ratio between leaflet and annulus areas. These geometrical adaptations lead to biomechanical effects including variations in annulus motion, valvular flow dynamics, and annulus and leaflet stresses. While lowering MV leaflet stress by reducing the annulus area, annuloplasty may also lead to an increase in annulus stress that is dependent on the degree of annular reduction, saddle height loss, and the stiffness of the material used during intervention. Elevated annulus stress may ultimately lead to device dehiscence from the annulus, with this risk increasing significantly with more than 30% reduction in annulus size.[120–123]

Complete direct annuloplasty devices are implanted on the MV annulus and form a complete ring around the MV anterior and posterior leaflets. They correct MR by changing the annulus 3D saddle shape to a more circular geometry and reducing the dimensions of the annulus and the orifice area. Whilst annuloplasty can successfully reduce MR, unfavorable effects of the intervention include impeding annulus motion, redirecting inflow toward the interventricular septum instead of the apex, and increasing annulus stress.[124,125] Surgically implanted devices include the Cardinal device

(Edwards Lifesciences, Irvine, CA, United States), which can be adjusted after initial fixation during first operation, also, it allows further adjustment any time in the future. The KALIOS device (Affluent medical, Paris, France) utilizes three balloons to adjust the posterior MV annulus any time after surgery under echocardiographic guidance.[126,127] Additional percutaneous implanted devices with adjustability during implantation include the CathHILEX device (Medtentia, Uusimma, Finland)—a single continuous loop that forms an adjustable helix around the annulus; the AMEND device (Valcare Medical, Herzliya Pituah, Israel)—which anchors between the two trigones along the posterior annulus and supports anterior annulus; and the semirigid Millipede IRIS device (Millipede Medical, Santa Rosa, CA, United States).[128,129]

Partial direct annuloplasty devices are percutaneously implanted on a part of the annulus and can correct MR by changing annulus dimensions while partially preserving the 3D saddle shape of the annulus. Compared to complete annuloplasty devices, they can achieve a similar reduction in leaflet stress with a negligible effect on annulus stress.[130] They include Cardioband (Edwards Lifesciences, Irvine, CA, United States) for posterior annuloplasty and postimplant adjustability, and Mitralign (Mitralign, Tewksbury, MA, United States), which is applied to a specific target on the MV annulus from the ventricular side, with a limited effect on annulus 3D shape and motion.[131,132]

Partial indirect annuloplasty devices are percutaneous delivery devices that compress the posterior annulus through the CS. The Carillon mitral contour system (Cardiac Dimensions, Kirkland, WA, United States) affects the mitral annulus on the lateral and anterior–posterior axes and reduces the orifice area. As they are not directly implanted on the annulus, the unfavorable effect on the annulus motion is decreased. The Transmural Cerclage device (Transmural Systems, Andover, MA, United States) also modifies the annulus through the CS and forms a complete ring through the first perforator vein, the right ventricle outlet (RVOT), and the TV. Whilst it can reduce the annulus geometry by 15%, the effect on the TV and RVOT warrants further investigation. The ARTO system (MVRX, San Mateo, CA, United States) modifies the anteroposterior diameter of the MV annulus through a suture that anchors between the CS and interatrial septum above the annular plane.[133–135]

Mitral valve leaflets

Restriction of leaflet motion is a common biomechanical modulation to decrease MR. Notably, the Mitraclip (Abbott, Illinois City, IL, United States) and PASCAL (Edwards Lifesciences, Irvine, CA, United States) devices have demonstrated positive clinical outcomes by shortening the anterior–posterior diameter and creating a double orifice. As intended, local leaflet stress and peak inflow velocities show an inverse and direct proportionality to the number of clips respectively. The resulting MV area may be reduced by 52–57% depending on the type of clip used, the position on the leaflets and the number of clips implanted.[136,137,162]

Increasing the posterior leaflet area addresses MR by adding a new coaptation surface that acts as a new, larger posterior leaflet and restricts native leaflet motion. The Polares (Palo Alto, CA, United States) and Half Moon (Menlo Park, CA, United States) are Polytetrafluoroethylene (PTFE) leaflets. These devices differ in that the Half Moon device is deployed above the annulus while the Polares is anchored to the posterior annulus achieving an annuloplasty-like effect.[138]

Chordae tendineae

LV free wall anchored devices include NeoChord DS 1000 (NeoChord, Inc, St. Louis Park, MN, United States), CardioMech (CardioMech AS, Trondheim, Norway), and Mitral Stitch (Hangzhou DeJin Medtech Co., Ltd, Hangzhou, China)—prosthetic chords that attach to the edge of prolapsed leaflets to correct their motion and restor leaflets coaptation. Conversely, the Harpoon device (Edwards Lifesciences, Irvine, CA, United States) is attached to the belly of the leaflet and the LV free wall.[139–141] Stress at the leaflet insertion point and the entry point to the LV wall is dependent on the end angle of insertion at the leaflet, and the variability in the distance between the different parts of LV wall and the MV annulus (up to 6%–17% variability throughout the cardiac cycle).[142] Leaflet stress is inversely related to the number of artificial chords.[143] Optimal tensioning of chords has important implications for repair result and durability. Undertensioning can increase the stress on the native chords and the leaflet and result in recurrence of MR, while overtensioning can elevate the stress on the artificial chords, potentially leading to leaflet tear or chord rupture.[144,145]

Papillary Muscle (PM) Anchored Chords exhibit lower stresses and length changes compared to those anchored to the LV.[143,144] Typically, PMs are equidistant to the MV annular plane throughout the cardiac cycle—a relationship that plays a crucial role in controlling the geometry and stress in the leaflets and chords. ChordArt (Coremedic GmbH, Tuebingen, Germany) and Pipeline line (W.L. Gore & Associates, Inc. Newark, NJ, United States) are transcatheter devices under investigation that are inserted into the PMs and the edge of the MV leaflet. Stress at the insertion point on the leaflet is inversely related to the number of chords and is dependent on the angle of insertion and optimal tension.[165] The Mitral Butterfly device (AAVie, Vienna, Austria) restricts prolapsed posterior leaflet motion by containing the leaflet between multiple cords that are attached to the ventricle near the MV annulus. Leaflet stress is transferred to an artificial papillary muscle on the ventricular part of the device.[146]

Augmentation of left ventricular and papillary muscle function

Augmenting LV and associated PM function can reduce MR by modifying LV dilation, thereby reducing both the MV annulus dimension and the LV wall stress. The AccuCinch system (Ancora Heart, Santa Clara, CA, United States) is a percutaneous device that anchors directly to the endocardium. The device applies three geometrical changes; it reduces the LV diameter, approximates PMs by correcting the tether muscle position, and decreases MV annulus size. Conversely, the BACE device (Phoenix Cardiac Devices, Cary, NC, United States) and the VenTouch (Mardil Medical, Minneapolis, MN, United States) are sleeves that are surgically applied to the epicardial surface with fluidic chambers that can be selectively pressurized for adjustable modification of LV geometry, approximation of PMs, and reduction of the MV annulus diameter.[147–149]

Tricuspid valve

Tricuspid valve annuloplasty

The TV annulus is dynamic and highly sensitive to the right ventricle (RV) free wall geometry and motion as the larger anterior leaflet is attached to a PM on the RV free wall.[150] TV devices were largely designed primarily to treat MV and act similarly to alter TV geometry and biomechanics. However, proximity of the TV annulus to the right coronary artery (RCA) and the AV node leads to additional complications when implanting these devices. These risks are somewhat mitigated by utilizing advanced integrated imaging modalities and 3D models for enhanced procedural planning.

Complete direct annuloplasty devices have fixation points near the AV node, the RCA, and Koch's triangle—a part of the conduction system of the heart.[151] Reducing the TV annulus diameter causes a reduction in its motion, and may kink or obstruct the RCA.[152] Annuloplasty rings succeed in reducing the RV free wall stress throughout the cardiac cycle.[153] This group includes percutaneous devices that can be adjusted during implantation like the Millipede IRIS device (Millipede Medical, Santa Rosa, CA, United States) and the DaVingi TR System (Cardiac Implants LLC, Tarrytown, NY, United States). The DaVingi TR System requires a second procedure 90 days after implant to allow the device to heal before adjusting the annulus to the target size.

Partial direct annuloplasty devices include the Cardioband (Edwards Lifesciences, Irvine, CA, United States), which aims to achieve partial annuloplasty to avoid interfering with the AV node; the Mitralign Percutaneous Annuloplasty System (MPAS) device (Mitralign, Tewksbury, MA, United States); and the MIA device (Micro Interventional Devices, Inc, Newtown, PA, United States), which comprises multiple PolyCor anchors implanted 10 mm apart around the posterior annulus and pulled together to cinch the annulus and create a bicuspid valve.[154]

Partial indirect annuloplasty devices for the TV include the TriCinch device (4Tech Cardio Ltd, Galway, Ireland), which is delivered percutaneously to reduce the anterio–posterior diameter of the annulus. The device uses a nitinol coil anchor on the epicardial surface of the RV, and a stent of the same material is inserted into the inferior vena cava for stabilization. However, this technique is associated with an elevated risk of bleeding and RCA compression.[155]

Leaflets

Similar to the MV, *restriction of TV leaflet motion* can be achieved with the TriClip device (Abbott, Illinois City, IL, United States), which is the first clip designed precisely for TV. Recently it received a CE mark. The PASCAL device (Edwards Lifesciences, Irvine, CA, United States) is also used for the TV following successful MV use. Both create a double orifice TV, but the induced stress is slightly lower at the clip insertion site compared to the MV, likely due to reduced transvalvular and ventricular pressure on the right side. The resultant bicuspid TV geometry can induce abnormal vortices in the RV which may lead to energy loss and increased RV work.[156]

Enhancement of TV leaflet coaptation surface is another strategy employed to reduce TR, when leaflets do not coapt, leading to a regurgitant jet. Percutaneously delivered devices can provide a supportive surface in the central orifice for TV leaflets to coapt with, effectively reducing any residual orifice or regurgitation. This group includes the FORMA device (Edwards Lifesciences, Irvine, CA, United States)—a balloon that anchors to the RV apex and can be adjusted postimplant with a subcutaneous control port,[157] and the Croí valve Duo (CroíValve, Dublin, Ireland) that is attached through a stent placed in the superior vena cava and contains a centrally located biological valve to compensate for any loss in valve orifice area that may result from device implantation.[158]

Tricuspid valve chordae tendineae

The Mistral device (Mitralix Ltd, Yokne'am Moshava, Israel) is a spiral implant that grasps subvalvular chords together to enhance leaflet approximation and coaptation, mimicking the Clover technique for TV repair.[159] However, in an ex vivo model, this technique showed loss of vortices under the TV leaflets in the RV which may increase the risk of thrombus formation.[158]

Recently, soft robotic actuators with an interventricular septum anchoring device were able to dynamically modify LV and MV geometry during the systolic phase of the cardiac cycle and improve their function. The device approximates the PM and the LV free wall to the septum. Moreover, it reduces the MV anteroposterior dimension and increases leaflet coaptation height.[57,160,161]

Conclusion

Treatment of HFrEF with biomechanical modulators has progressed from historic passive, biventricular restraint devices to adjustable, quantitative, and dynamic VADs. Many of the trials with devices that restore and reshape the LV in post-MI HFrEF patients did not succeed, but trials with the Revivent TC are ongoing. Soft robotic techniques have shown promise in providing adjustability, and tunable augmentation of function, and have transformative potential for the treatment of HFrEF.

In an era where the prevalence of HFpEF is rising and pharmacological therapies for this condition are inadequate, device-based solutions enhancing cardiac mechanics have been gaining momentum, yet they are associated with several challenges. Due to the multifactorial nature of HFpEF and the variety of its etiologies and phenotypes, the development of a device that is capable of addressing all the hemodynamic challenges, as they arise from a multitude of structural and functional abnormalities, is exacting. The plethora of devices engineered to restore adequate hemodynamics, coupled with the abundance of ongoing clinical investigations, suggest that enormous progress has been made in the quest for a treatment of HFpEF and that there is hope that management of these patients may be soon revolutionized.

Finally, devices addressing MV and TV repair were initially designed to passively modulate valve geometry and be surgically implanted in the static heart. However, the advent of minimally invasive, beating heart interventions introduced new challenges. Two core concepts observed in emerging technologies for AV valve repair are miniaturization for less invasive delivery approaches and postimplant adjustability. These approaches have led to multiple innovative devices, as reviewed in this section. Long-term follow-up and advancements in imaging modalities provide a better understanding of patient-specific disease, and engineering perspectives have shed light on the importance of the temporally dynamic biomechanical effects of interventions. Retrospective analysis of the failure points or complications of these devices motivates the use of soft robotic approaches to provide dynamic solutions for augmenting AV valve function. A deep understanding and sophisticated modeling of the biomechanical effects of interventions may improve device development, especially for dynamic organs and anatomical structures like the cardiac valves.

References

1. Shahinpoor M. A review of patents on implantable heart-compression/assist devices and systems. *Recent Patents Biomed Eng*. 2010;3:54–71.
2. Moreno MR, Biswas S, Harrison LD, et al. Development of a non-blood contacting cardiac assist and support device: an in vivo proof of concept study. *J Med Device*. 2011;5. 041007-1-041007–9.
3. A. Alazmani, D.G. Keeling, P.G. Walker, et al., Introducing a hardware-in-the-loop simulation of the cardiovascular system. In: *Proceedings of the Fourth IEEE RAS EMBS International Conference for Biomedical Robotics and Biomechatronics*, 2012, 153–158.
4. Kung RT, Rosenberg M. Heart booster: a pericardial support device. *Ann Thorac Surg*. 1999;68:764–767.
5. Oz MC, Artrip JH, Burkhoff D. Direct cardiac compression devices. *J Heart Lung Transplant*. 2002;21:1049–1055.
6. Varela CE, Fan Y, Roche ET. Optimizing epicardial restraint and reinforcement following myocardial infarction: moving towards localized, biomimetic, and multitherapeutic options. *Biomimetics*. 2019;. Available from: https://doi.org/10.3390/biomimetics4010007.
7. Oz MC, Konertz WF, Kleber FX, et al. Global surgical experience with the Acorn cardiac support device. *J Thorac Cardiovasc Surg*. 2003;. Available from: https://doi.org/10.1016/S0022-5223(03)00049-7.
8. Chaudhry PA, Mishima T, Sharov VG, et al. Passive epicardial containment prevents ventricular remodeling in heart failure. *Ann Thorac Surg*. 2000;. Available from: https://doi.org/10.1016/S0003-4975(00)01755-0.
9. Blom AS, Mukherjee R, Pilla JJ, et al. Cardiac support device modifies left ventricular geometry and myocardial structure after myocardial infarction. *Circulation*. 2005;112:1274–1283.

10. Blom AS, Pilla JJ, Arkles J, et al. Ventricular restraint prevents infarct expansion and improves borderzone function after myocardial infarction: a study using magnetic resonance imaging, three-dimensional surface modeling, and myocardial tagging. *Ann Thorac Surg*. 2007;. Available from: https://doi.org/10.1016/j.athoracsur.2007.06.062.
11. Konertz WF, Shapland JE, Hotz H, et al. Passive containment and reverse remodeling by a novel textile cardiac support device. *Circulation*. 2001;. Available from: https://doi.org/10.1161/hc37t1.094525.
12. Oz MC, Konertz WF, Raman J, Kleber FX. Reverse remodeling of the failing ventricle: surgical intervention with the Acorn cardiac support device. *Congest Hear Fail*. 2004;. Available from: https://doi.org/10.1111/j.1527-5299.2004.00291.x.
13. Speziale G, Nasso G, Piancone F, et al. One-year results after implantation of the CorCap for dilated cardiomyopathy and heart failure. *Ann Thorac Surg*. 2011;. Available from: https://doi.org/10.1016/j.athoracsur.2011.02.006.
14. Acker MA, Jessup M, Bolling SF, et al. Mitral valve repair in heart failure: five-year follow-up from the mitral valve replacement stratum of the Acorn randomized trial. *J Thorac Cardiovasc Surg*. 2011;. Available from: https://doi.org/10.1016/j.jtcvs.2010.10.051.
15. Mann DL, Acker MA, Jessup M, Sabbah HN, Starling RC, Kubo SH. Clinical evaluation of the CorCap cardiac support device in patients with dilated cardiomyopathy. *Ann Thorac Surg*. 2007;84:1226–1235.
16. Starling RC, Jessup M, Oh JK, et al. Sustained benefits of the CorCap cardiac support device on left ventricular remodeling : three year follow-up results from the acorn clinical trial. *Ann Thorac Surg*. 2007;84:1236–1242.
17. Mann DL, Kubo SH, Sabbah HN, et al. Beneficial effects of the CorCap cardiac support device: five-year results from the Acorn Trial. *J Thorac Cardiovasc Surg*. 2012;143:1036–1042.
18. Magovern JA, Teekell-Taylor L, Mankad S, et al. Effect of a flexible ventricular restraint device on cardiac remodeling after acute myocardial infarction. *ASAIO J*. 2006;52:196–200.
19. Costanzo MR, Ivanhoe RJ, Kao A, et al. Prospective evaluation of elastic restraint to lessen the effects of heart failure (PEERLESS-HF) trial. *J Card Fail*. 2012;18:446–458.
20. Ghanta RK, Rangaraj A, Umakanthan R, et al. Adjustable, physiological ventricular restraint improves left ventricular mechanics and reduces dilatation in an ovine model of chronic heart failure. *Circulation*. 2007;115:1201–1210.
21. Ghanta RK, Lee LS, Umakanthan R, et al. Real-time adjustment of ventricular restraint therapy in heart failure. *Eur J Cardio-thOracic Surg*. 2008;34:1136–1140.
22. Lee LS, Ghanta RK, Mokashi SA, et al. Optimized ventricular restraint therapy: adjustable restraint is superior to standard restraint in an ovine model of ischemic cardiomyopathy. *J Thorac Cardiovasc Surg*. 2013;145:824–831.
23. S.A. Clarke, R.K. Ghanta, G. Ailawadi, J.W. Holmes, *Studies in Mechanobiology, Tissue Engineering and Biomaterials*, (Springer-Verlag Berlin Heidelberg, pp: 169–206, 2013.
24. Mokashi SA, Lee LS, Schmitto JD, et al. Restraint to the left ventricle alone is superior to standard restraint. *J Thorac Cardiovasc Surg*. 2013;146:192–197.
25. Contessotto P, Orbanić D, Da Costa M, et al. Elastin-like recombinamers-based hydrogel modulates post-ischemic remodeling in a non-transmural myocardial infarction in sheep. *Sci Transl Med*. 2021;13:eaaz5380.
26. Serpooshan V, Zhao M, Metzler SA, et al. The effect of bioengineered acellular collagen patch on cardiac remodeling and ventricular function post myocardial infarction. *Biomaterials*. 2013;34:9048–9055.
27. Estrada AC, Yoshida K, Clarke SA, Holmes JW. Longitudinal reinforcement of acute myocardial infarcts improves function by transmurally redistributing stretch and stress. *J Biomech Eng*. 2020;. Available from: https://doi.org/10.1115/1.4044030.
28. Olvera D, Sohrabi Molina M, Hendy G, Monaghan MG. Electroconductive melt electrowritten patches matching the mechanical anisotropy of human myocardium. *Adv Funct Mater*. 2020;. Available from: https://doi.org/10.1002/adfm.201909880.
29. McCarthy PM, Takagaki M, Ochiai Y, et al. Device-based change in left ventricular shape: a new concept for the treatment of dilated cardiomyopathy. *J Thorac Cardiovasc Surg*. 2001;. Available from: https://doi.org/10.1067/mtc.2001.115240.
30. Kalogerakos PD, Hassoulas J, Ladopoulos VS. Beyond heart transplantation: Potentials and problems of the shape memory alloy fibers in the treatment of heart failure. *ASAIO J*. 2014;60:263–268.
31. H.N. Sabbah, *Annals of Thoracic Surgery*, 2003.
32. Kashem A, Santamore WP, Hassan S, Crabbe DL, Margulies KB, Melvin DB. CardioClasp: a new passive device to reshape cardiac enlargement. *ASAIO J*. 2002;. Available from: https://doi.org/10.1097/00002480-200205000-00010.
33. Kashem A, Hassan S, Crabbe DL, Melvin DB, Santamore WP, Chitwood WR. Left ventricular reshaping: effects on the pressure-volume relationship. *J Thorac Cardiovasc Surg*. 2003;. Available from: https://doi.org/10.1067/mtc.2003.4.
34. A. Kashem, W.P. Santamore, S. Hassan, et al., *Journal of Cardiac Surgery*, 2003.
35. Sharkey H, Nikolic S, Khairkhahan A, Dae M. Left ventricular apex occluder. Description of a ventricular partitioning device. *EuroIntervention*. 2006;2:125–127.
36. Nikolic SD, Khairkhahan A, Ryu M, Champsaur G, Breznock E, Dae M. Percutaneous implantation of an intraventricular device for the treatment of heart failure: experimental results and proof of concept. *J Card Fail*. 2009;15:790–797.
37. Mazzaferri EL, Gradinac S, Sagic D, et al. Percutaneous left ventricular partitioning in patients with chronic heart failure and a prior anterior myocardial infarction: results of the percutaneous ventricular restoration in chronic heart failure patients trial. *Am Heart J*. 2012;163:812–820. e1.
38. Oliveira GH, Al-Kindi SG, Bezerra HG, Costa MA. Left ventricular restoration devices. *J Cardiovasc Trans. Res*. 2014;7:282–291.
39. Yun C-H, Sun J-Y, Templin B, et al. Improvements in left ventricular diastolic mechanics after parachute device implantation in patients with ischemia heart failure: a cardiac computerized tomographic study. *J Card Fail*. 2017;23:455–463.

40. CardioKinetix, A. Pivotal Trial to Establish the Efficacy and Long-term Safety of the Parachute Implant System (PARACHUTE IV). 2017. Available from: <https://www.clinicaltrials.gov/ct2/show/NCT01614652>
41. Hendriks T, Schurer RAJ, Al Ali L, van Den Heuvel AFM, van der Harst P. Left ventricular restoration devices post myocardial infarction. *Heart Fail Rev*. 2018;. Available from: https://doi.org/10.1007/s10741-018-9711-2.
42. Bioventrix BioVentrix Revivent TC™ System Clinical Study.
43 BioVentrix Randomized Evaluation and Verification of Ventricular Enhancement (REVIVE-HF). 2021.
44. Pillay T, Neves P, Benetti F, Van Bladel K, Wechsler A, Annest L. Minimal access left ventricular reconstruction. *J Card Surg*. 2021; 36:300–306.
45. Bencini A, Parola P. The pneumomassage of the heart. *Surgery*. 1956;39:375–384.
46. Anstadt G, Blakemore W, Baue A. A new instrument for prolonged mechanical massage. *Circulation*. 1965;31/32:43–44.
47. Mau J, Menzie S, Huang Y, Ward M, Hunyor S. Nonsurround, nonuniform, biventricular-capable direct cardiac compression provides Frank-Starling recruitment independent of left ventricular septal damage. *J Thorac Cardiovasc Surg*. 2011;142:209–215.
48. Y. Huang, S. Hunyor, Implantable Direct Cardiac Compression Device and System, US Patent Application. 10/567,520, 2004.
49. D. Melvin, Heart Wall Actuation System for the Natural Heart with Shape Limiting Elements, US Pat. 6,988,982, 2006.
50. Ming Y, Meiling Z, Richardson RC, Levesley MC, Walker PG, Watterson K. Design and evaluation of linear ultrasonic motors for a cardiac compression assist device. *Sens Actuators A Phys*. 2005;119:214–220.
51. D.R. Trumble, Apical torsion device for cardiac assist <http://www.google.com/patents/US8075471>, 2011.
52. M. Shahinpoor, K. Kim, Design, development, and testing of a multifingered heart compression/assist device equipped with IPMC artificial muscles. In: *Proceedings of the SPIE's Eighth Annual International Symposium on Smart Structures and Materials* 4329, 2001, 411–420.
53. M. Shahinpoor, Biomedical Applications of Electroactive Polymer Actuators, (2009), 7: 37-59.
54. Gu H, Bertrand T, Boehler Q, Chautems C, Vasilyev NV, Nelson BJ. Magnetically active cardiac patches as an untethered, non-blood contacting ventricular assist device. *Adv Sci*. 2021;8:2000726.
55. Roche ET, Wohlfarth R, Overvelde JTB, et al. A bioinspired soft actuated material. *Adv Mater*. 2014;1200.
56. E.T. Roche, M.A. Horvath, A. Alazmani, et al., Proceedings of the ASME Design Engineering Technical Conference, (Boston, MA), vol. 5A, 2015.
57. Roche ET, Horvath MA, Wamala I, et al. Soft robotic sleeve supports heart function. *Sci Transl Med*. 2017;9:eaaf3925.
58. Payne CJ, Wamala I, Abah C, et al. An implantable extracardiac soft robotic device for the failing heart: mechanical coupling and synchronization. *Soft Robot*. 2017;. Available from: https://doi.org/10.1089/soro.2016.0076.
59. Horvath MA, Varela CE, Dolan EB, et al. Towards alternative approaches for coupling of a soft robotic sleeve to the heart. *Ann Biomed Eng*. 2018;1–14.
60. Moreno MR, Biswas S, Harrison LD, et al. Assessment of minimally invasive device that provides simultaneous adjustable cardiac support and active synchronous assist in an acute heart failure model. *J Med Device*. 2011;5. 041008-1-041008–9.
61. Miyagi C, Miyamoto T, Karimov JH, Starling RC, Fukamachi K. Device-based treatment options for heart failure with preserved ejection fraction. *Heart Fail Rev*. 2021;. Available from: https://doi.org/10.1007/s10741-020-10067-5.
62. Borlaug BA, Paulus WJ. Heart failure with preserved ejection fraction: pathophysiology, diagnosis, and treatment. *Eur Heart J*. 2011;32:670–679.
63. Borlaug BA. The pathophysiology of heart failure with preserved ejection fraction. *Nat Rev Cardiol*. 2014;11:507.
64. Lam CSP, Donal E, Kraigher-Krainer E, Vasan RS. Epidemiology and clinical course of heart failure with preserved ejection fraction. *Eur J Heart Fail*. 2011;13:18–28.
65. Borlaug BA. Evaluation and management of heart failure with preserved ejection fraction. *Nat Rev Cardiol*. 2020;17:559–573.
66. Dunlay SM, Roger VL, Redfield MM. Epidemiology of heart failure with preserved ejection fraction. *Nat Rev Cardiol*. 2017;14:591–602.
67. Sidney S, Go AS, Jaffe MG, Solomon MD, Ambrosy AP, Rana JS. Association between aging of the US Population and Heart Disease Mortality from 2011 to 2017. *JAMA Cardiol*. 2019;. Available from: https://doi.org/10.1001/jamacardio.2019.4187.
68. Flegal KM, Kruszon-Moran D, Carroll MD, Fryar CD, Ogden CL. Trends in obesity among adults in the United States, 2005 to 2014. *JAMA - J Am Med Assoc*. 2016;. Available from: https://doi.org/10.1001/jama.2016.6458.
69. Ponikowski P, Voors AA, Anker SD, et al. 2016 ESC Guidelines for the diagnosis and treatment of acute and chronic heart failure. *Eur Heart J*. 2016;37:2129–2200. m.
70. Owan TE, Hodge DO, Herges RM, Jacobsen SJ, Roger VL, Redfield MM. Trends in prevalence and outcome of heart failure with preserved ejection fraction. *N Engl J Med*. 2006;355:251–259.
71. Lam CSP, Voors AA, De Boer RA, Solomon SD, Van Veldhuisen DJ. Heart failure with preserved ejection fraction: from mechanisms to therapies. *Eur. Heart J*. 2018;. Available from: https://doi.org/10.1093/eurheartj/ehy301.
72. REDUCE LAP-HF III Corvia Protocol 1701 (REDUCELAPHFIII) ClinicalTrials.gov, 2019.
73. Reduce Lap-Hf Trial II (2017).
74. Reduce Lap-Hf Randomized Trial I ClinicalTrials.gov, 2020.
75. Reducing Right Ventricular Failure in Pulmonary Arterial Hypertension (RELIEVE-PAH). ClinicalTrials.gov, 2020.
76. Reducing Lung CongestIon Symptoms in Advanced Heart Failure (2018).
77. The AFteR Registry - Follow-up Study to Monitor the Efficacy and Safety of the Occlutech AFR in Heart Failure Patients, ClinicalTrials.gov, 2020.

78. The Prelieve Trial - Pilot Study to Assess Safety and Efficacy of a Novel Atrial Flow Regulator (AFR) in Heart Failure Patients (PRELIEVE), 2020.
79. The Prophet Trial - Pilot Study to Assess Safety and Efficacy of a Novel Atrial Flow Regulator (AFR) in Patients With Pulmonary Hypertension. ClinicalTrials.gov, 2020.
80. Kulkarni SS, Sakaria AK, Mahajan SK, Shah KB. Lutembacher's syndrome. *J. Cardiovasc Dis Res*. 2012;. Available from: https://doi.org/10.4103/0975-3583.95381.
81. Early Feasibility Study - Transcatheter Atrial Shunt System (ALt FLOW US), ClinicalTrials.gov, 2020.
82. Simard T, Labinaz M, Zahr F, et al. Percutaneous atriotomy for levoatrial–to–coronary sinus shunting in symptomatic heart failure: first-in-human experience. *JACC Cardiovasc Interv*. 2020;. Available from: https://doi.org/10.1016/j.jcin.2020.02.022.
83. Søndergaard L, Reddy V, Kaye D, et al. Transcatheter treatment of heart failure with preserved or mildly reduced ejection fraction using a novel interatrial implant to lower left atrial pressure. *Eur J Heart Fail*. 2014;16:796–801.
84. Hasenfuß G, Hayward C, Burkhoff D, et al. A transcatheter intracardiac shunt device for heart failure with preserved ejection fraction (REDUCE LAP-HF): a multicentre, open-label, single-arm, phase 1 trial. *Lancet*. 2016;387:1298–1304.
85. Kaye DM, Hasenfuß G, Neuzil P, et al. One-year outcomes after transcatheter insertion of an interatrial shunt device for the management of heart failure with preserved ejection fraction. *Circ Hear Fail*. 2016;. Available from: https://doi.org/10.1161/CIRCHEARTFAILURE.116.003662.
86. Rodés-Cabau J, Bernier M, Amat-Santos IJ, et al. Interatrial shunting for heart failure: early and late results from the first-in-human experience with the V-wave system. *JACC Cardiovasc Interv*. 2018;11:2300–2310.
87. Paitazoglou C, Bergmann MW, Özdemir R, et al. One-year results of the first-in-man study investigating the Atrial-Flow-Regulator for left-atrial shunting in symptomatic heart failure patients: the PRELIEVE study. *Eur J Heart Fail*. 2021;. Available from: https://doi.org/10.1002/ejhf.2119.
88. Gupta A, Bailey SR. Update on devices for diastolic dysfunction: options for a no option condition? *Curr Cardiol Rep*. 2018;20:85.
89. Burkhoff D, Maurer MS, Joseph SM, et al. Left atrial decompression pump for severe heart failure with preserved ejection fraction: theoretical and clinical considerations. *JACC Hear Fail*. 2015;3:275–282.
90. Feld Y, Dubi S, Reisner Y, et al. Energy transfer from systole to diastole: a novel device-based approach for the treatment of diastolic heart failure. *Acute Card Care*. 2011;. Available from: https://doi.org/10.3109/17482941.2011.634012.
91. Feld Y, Dubi S, Reisner Y, Schwammenthal E, Elami A. Future strategies for the treatment of diastolic heart failure. *Acute Card Care*. 2006;. Available from: https://doi.org/10.1080/14628840600548988.
92. ImCardia™ for DHF to Treat Diastolic Heart Failure (DHF) Patient a Pilot Study (ImCardia), 2011.
93. Y. Feld, in PCR innovators day, PCR, 2019.
94. CORolla® TAA for Heart Failure With Preserved Ejection Fraction (HFpEF) and Diastolic Dysfunction (DD), 2020.
95. CCM in Heart Failure With Preserved Ejection Fraction (CCM-HFpEF), 2020.
96. BAROSTIM THERAPY™ In Heart Failure With Preserved Ejection Fraction (HFpEF), 2020.
97. D. Burkhoff Md Phd, I. Shemer, B. Felzen, et al., Heart Failure Reviews, 2001.
98. Tschöpe C, Van Linthout S, Spillmann F, et al. Cardiac contractility modulation signals improve exercise intolerance and maladaptive regulation of cardiac key proteins for systolic and diastolic function in HFpEF. *Int J Cardiol*. 2016;. Available from: https://doi.org/10.1016/j.ijcard.2015.10.208.
99. Kroon A, Schmidli J, Scheffers I, et al. Sustained blood pressure reduction by baroreflex activation therapy with a chronically implanted system: 4-year data of Rheos Debut-HT study in patients with resistant hypertension. *J Hypertens*. 2010;. Available from: https://doi.org/10.1097/01.hjh.0000379526.21089.39.
100. Bisognano JD, de Leeuw P, Bach DS, Lovett EG, Kaufman CL. Improved functional capacity and cardiovascular structure after baroreflex activation therapy™ in resistant hypertension patients with symptomatic heart failure: results from European and United States Trials of the Rheos® system. *J Card Fail*. 2009;. Available from: https://doi.org/10.1016/j.cardfail.2009.06.191.
101. LEft Atrial Pacing in Diastolic Heart Failure (LEAD), ClinicalTrials.gov, 2018.
102. Laurent G, Eicher JC, Mathe A, et al. Permanent left atrial pacing therapy may improve symptoms in heart failure patients with preserved ejection fraction and atrial dyssynchrony: a pilot study prior to a national clinical research programme. *Eur J Heart Fail*. 2013;15:85–93.
103. Pacing for Heart Failure With Preserved Ejection Fraction (HFpEF), ClinicalTrials.gov, 2013.
104. AlTurki A, Lima PY, Vidal A, et al. Fusion pacing in patients with right bundle branch block who undergo cardiac resynchronization therapy. *J Electrocardiol*. 2021;64:66–71.
105. Hsu JC, Birnie D, Stadler RW, Cerkvenik J, Feld GK, Birgersdotter-Green U. Adaptive cardiac resynchronization therapy is associated with decreased risk of incident atrial fibrillation compared to standard biventricular pacing: A real-world analysis of 37,450 patients followed by remote monitoring. *Hear Rhythm*. 2019;16:983–989.
106. Huang W, Su L, Wu S, et al. Benefits of permanent His bundle pacing combined with atrioventricular node ablation in atrial fibrillation patients with heart failure with both preserved and reduced left ventricular ejection fraction. *J Am Heart Assoc*. 2017;6. Available from: https://doi.org/10.1161/JAHA.116.005309.
107. Cardiac Resynchronisation Therapy vs Rate-responsive Pacing in Heart Failure With Preserved Ejection Fraction (PREFECTUS). 2021.
108. Efficacy Study of Pacemakers to Treat Slow Heart Rate in Patients With Heart Failure (RAPID-HF). 2022.
109. Physiologic Accelerated Pacing as a Treatment in Patients With Heart Failure With Preserved Ejection Fraction (PACE HFpEF). 2021.

110. Topilsky Y, Pereira NL, Shah DK, et al. Left ventricular assist device therapy in patients with restrictive and hypertrophic cardiomyopathy. *Circ Hear Fail*. 2011;. Available from: https://doi.org/10.1161/CIRCHEARTFAILURE.110.959288.
111. Muthiah K, Phan J, Robson D, et al. Centrifugal continuous-flow left ventricular assist device in patients with hypertrophic cardiomyopathy: a case series. *ASAIO J*. 2013;. Available from: https://doi.org/10.1097/MAT.0b013e318286018d.
112. Moscato F, Wirrmann C, Granegger M, Eskandary F, Zimpfer D, Schima H. Use of continuous flow ventricular assist devices in patients with heart failure and a normal ejection fraction: a computer-simulation study. *J Thorac Cardiovasc Surg*. 2013;. Available from: https://doi.org/10.1016/j.jtcvs.2012.06.057.
113. Fukamachi K, Horvath DJ, Karimov JH, et al. Left atrial assist device to treat patients with heart failure with preserved ejection fraction: Initial in vitro study. *J Thorac Cardiovasc Surg*. 2020;. Available from: https://doi.org/10.1016/j.jtcvs.2019.12.110.
114. Kado Y, Polakowski AR, Kuban BD, et al. Left atrial assist device function at various heart rates using a mock circulation loop. *Int J Artif Organs*. 2020;. Available from: https://doi.org/10.1177/0391398820977508.
115. Kado Y, Polakowski AR, Kuban BD, et al. The effects of preserving mitral valve function on a left atrial assist device: an: in vitro: mock circulation loop study. *ASAIO J*. 2020;.
116. Miyagi C, Karimov JH, Kuban BD, et al. Development of the left atrial assist device for patients with heart failure with preserved ejection fraction: first in vivo results. *J Hear Lung Transplant*. 2021;40:S176–S177.
117. Granegger M, Dave H, Knirsch W, et al. Pulsatile pump for the treatment of heart failure with preserved ejection fraction: a simulation study. *Cardiovasc Eng Technol*. 2019;10:69–79.
118. Escher A, Choi Y, Callaghan F, et al. A valveless pulsatile pump for heart failure with preserved ejection fraction: hemo- and fluid dynamic feasibility. *Ann Biomed Eng*. 2020;. Available from: https://doi.org/10.1007/s10439-020-02492-2.
119. Scandura S, Mangiafico S, Giaquinta S. Mitral regurgitation: epidemiology, etiology and physiopathology. *Percutaneous Treat Left Side Card Valves*. 2018;49–61.
120. Jensen MO, Jensen H, Smerup M, et al. Saddle-shaped mitral valve annuloplasty rings experience lower forces compared with flat rings. *Circulation*. 2008;118:250–255.
121. Jensen MO, Honge JL, Benediktsson JA, et al. Mitral valve annular downsizing forces: implications for annuloplasty device development. *J Thorac Cardiovasc Surg*. 2014;148:83–89.
122. Gheorghe L, Ielasi A, Rensing BJWM, et al. Complications following percutaneous mitral valve repair. *Front Cardiovasc Med*. 2019;6. Available from: https://doi.org/10.3389/fcvm.2019.00146.
123. Pfannmüller B, Doenst T, Eberhardt K, Seeburger J, Borger MA, Mohr FW. Increased risk of dehiscence after tricuspid valve repair with rigid annuloplasty rings. *J Thorac Cardiovasc Surg*. 2012;143:1050–1055.
124. Witschey WRT, Zhang D, Contijoch F, et al. The influence of mitral annuloplasty on left ventricular flow dynamics. *Ann Thorac Surg*. 2015;100:114–121.
125. Pierce EL, Bloodworth CH, Siefert AW, et al. Mitral annuloplasty ring suture forces: Impact of surgeon, ring, and use conditions. *J Thorac Cardiovasc Surg*. 2018;155:131–139. e3.
126. Maisano F, Falk V, Borger MA, et al. Improving mitral valve coaptation with adjustable rings: outcomes from a European multicentre feasibility study with a new-generation adjustable annuloplasty ring system†. *Eur J Cardio-Thoracic Surg*. 2013;44:913–918.
127. Tozzi P, Hayoz D, Antona C, et al. Mastering mitral leaflets coaptation after valve repair with adjustable mitral annuloplasty ring: proof of concept in mock loop study. *ASAIO J*. 2017;63:168–173.
128. Konerding MA, Simpanen J, Ihlberg L, et al. Comparison of the novel Medtentia double helix mitral annuloplasty system with the Carpentier-Edwards Physio annuloplasty ring: morphological and functional long-term outcome in a mitral valve insufficiency sheep model. *J Cardiothorac Surg*. 2013;8:1–9.
129. Rogers JH, Boyd WD, Smith TW, Bolling SF. Transcatheter mitral valve direct annuloplasty with the millipede IRIS ring. *Interv Cardiol Clin*. 2019;8:261–267.
130. Dagum P, Timek T, Green GR, et al. Three-dimensional geometric comparison of partial and complete flexible mitral annuloplasty rings. *J Thorac Cardiovasc Surg*. 2001;122:665–673.
131. Arsalan M, Agricola E, Alfieri O, et al. Effect of transcatheter mitral annuloplasty with the cardioband device on 3-dimensional geometry of the mitral annulus. *Am J Cardiol*. 2016;118:744–749.
132. Nickenig G, Schueler R, Dager A, et al. Treatment of chronic functional mitral valve regurgitation with a percutaneous annuloplasty system. *J Am Coll Cardiol*. 2016;67:2927–2936.
133. Lipiecki J, Siminiak T, Sievert H, et al. Coronary sinus-based percutaneous annuloplasty as treatment for functional mitral regurgitation: the TITAN II trial. *Open Hear*. 2016;3:e000411.
134. Park YH, Chon MK, Lederman RJ, et al. Mitral loop cerclage annuloplasty for secondary mitral regurgitation: first human results. *JACC Cardiovasc Interv*. 2017;10:597–610.
135. Worthley S, Redwood S, Hildick-Smith D, et al. Transcatheter reshaping of the mitral annulus in patients with functional mitral regurgitation: one-year outcomes of the MAVERIC trial. *EuroIntervention*. 2021;16:1106–1113.
136. Khan F, Winkel M, Ong G, et al. Percutaneous mitral edge-to-edge repair: state of the art and a glimpse to the future. *Front Cardiovasc Med*. 2019;6:1–7.
137. Caballero A, Mao W, McKay R, Hahn RT, Sun W, Comprehensive A. Engineering analysis of left heart dynamics after mitraclip in a functional mitral regurgitation patient. *Front Physiol*. 2020;11:1–17.

138. Mangieri A, Laricchia A, Giannini F, et al. Emerging technologies for percutaneous mitral valve repair. *Front Cardiovasc Med.* 2019;6:1–9.
139. Rucinskas K, Janusauskas V, Zakarkaite D, et al. Off-pump transapical implantation of artificial chordae to correct mitral regurgitation: early results of a single-center experience. *J Thorac Cardiovasc Surg.* 2014;147:95–99.
140. Gammie JS, Bartus K, Gackowski A, et al. Beating-heart mitral valve repair using a novel ePTFE cordal implantation device. *J Am Coll Cardiol.* 2018;71:25–36.
141. Wang S, Meng X, Luo Z, Pan X. Transapical beating-heart mitral valve repair using a novel artificial chordae implantation system. *Ann Thorac Surg.* 2018;106:e265–e267.
142. Weber A, Hurni S, Vandenberghe S, et al. Ideal site for ventricular anchoring of artificial chordae in mitral regurgitation. *J Thorac Cardiovasc Surg.* 2012;143:S78–S81.
143. Sturla F, Votta E, Onorati F, et al. Biomechanical drawbacks of different techniques of mitral neochordal implantation: when an apparently optimal repair can fail. *J Thorac Cardiovasc Surg.* 2015;150:1303–1312. e4.
144. Imbrie-Moore AM, Paulsen MJ, Thakore AD, et al. Ex vivo biomechanical study of apical vs papillary neochord anchoring for mitral regurgitation. *Ann Thorac Surg.* 2019;108:90–97.
145. Caballero A, McKay R, Sun W. Computer simulations of transapical mitral valve repair with neochordae implantation: clinical implications. *JTCVS Open.* 2020;3:27–44.
146. Ticar JM, Gaidulis G, Veith K, Rath C, Jarman J, Mohl W. Mitral butterfly. *JACC Basic Transl Sci.* 2020;5:1002–1014.
147. Reisman M, Wudel J, Martin S, et al. TCT-88 6-month outcomes of an early feasibility study of the AccuCinch left ventricular repair system in patients with heart failure and functional mitral regurgitation. *J Am Coll Cardiol.* 2019;74:B88.
148. Hote M, Seth S, Choudhary S, et al. Report of an external basal annuloplasty device for mitral regurgitation (basal annuloplasty of the cardia externally device implantation). *J Pract Cardiovasc Sci.* 2015;1:200.
149. Taramasso M, Feldman T, Maisano F. Transcatheter mitral valve repair: review of the clinical evidence. *EuroIntervention.* 2018;14: AB91–AB100.
150. Prihadi EA, Delgado V, Leon MB, Enriquez-Sarano M, Topilsky Y, Bax JJ. Morphologic types of tricuspid regurgitation. *JACC Cardiovasc Imaging.* 2019;12:491–499.
151. Asmarats L, Puri R, Latib A, Navia JL, Rodés-Cabau J. Transcatheter tricuspid valve interventions: landscape, challenges, and future directions. *J Am Coll Cardiol.* 2018;71:2935–2956.
152. Malinowski M, Jazwiec T, Quay N, Goehler M, Rausch MK, Timek TA. The influence of tricuspid annuloplasty prostheses on ovine annular geometry and kinematics. *J Thorac Cardiovasc Surg.* 2021;161:e191–e207.
153. Jazwiec T, Malinowski M, Bush J, et al. Right ventricular free wall stress after tricuspid valve annuloplasty in acute ovine right heart failure. *J Thorac Cardiovasc Surg.* 2019;158:759–768.
154. Kohorst K, Pretorius M. Future technology. *Semin Cardiothorac Vasc Anesth.* 2019;23:123–133.
155. Asmarats L, Taramasso M, Rodés-Cabau J. Tricuspid valve disease: diagnosis, prognosis and management of a rapidly evolving field. *Nat Rev Cardiol.* 2019;16:538–554.
156. Nickenig G, Kowalski M, Hausleiter J, et al. Transcatheter treatment of severe tricuspid regurgitation with the edge-to-edge mitraclip technique. *Circulation.* 2017;135:1802–1814.
157. Asmarats L, Perlman G, Praz F, et al. Long-term outcomes of the FORMA transcatheter tricuspid valve repair system for the treatment of severe tricuspid regurgitation: insights from the first-in-human experience. *JACC Cardiovasc Interv.* 2019;12:1438–1447.
158. Noack T, Kiefer P, Besler C, et al. Transcatheter mitral valve repair: review of current techniques. *Indian J Thorac Cardiovasc Surg.* 2020;36:53–63.
159. De Bonis M, Lapenna E, Di Sanzo S, et al. Long-term results (up to 14 years) of the clover technique for the treatment of complex tricuspid valve regurgitation. *Eur. J. Cardio-Thoracic Surg.* 2017;52:125–130.
160. Wamala I, Roche ET, Pigula FA. The use of soft robotics in cardiovascular therapy. *Expert Rev Cardiovasc Ther.* 2017;15:767–774.
161. Saeed MY, Van Story D, Payne CJ, et al. Dynamic augmentation of left ventricle and mitral valve function with an implantable soft robotic device. *JACC Basic Transl Sci.* 2020;5:229–242.
162. Kassar Mohammad, Praz Fabien, Hunziker Lukas, et al. Anatomical and Technical Predictors of Three-Dimensional Mitral Valve Area Reduction After Transcatheter Edge-To-Edge Repair. *J Am Soc Echocardiogr.* 2022;1097-679535(1):96–104. Available from: https://doi.org/10.1016/j.echo.2021.08.021.
163. Rosalia Luca, Ozturk Caglar, Shoar Saeed, et al. Device-Based Solutions to Improve Cardiac Physiology and Hemodynamics in Heart Failure With Preserved Ejection Fraction. *J Am Coll Cardiol Basic Trans Science.* 2021;6(9–10):772–795. Available from: https://doi.org/10.1016/j.jacbts.2021.06.002.
164. Ozturk Caglar, Rosalia Luca, Roche Ellen. A Multi-Domain Simulation Study of a Pulsatile-Flow Pump Device for Heart Failure With Preserved Ejection Fraction. *Front. Physiol.* 2022;13. Available from: https://doi.org/10.3389/fphys.2022.815787.
165. Sturla Francesco, Onorati Francesco, Votta Emiliano, et al. Is it possible to assess the best mitral valve repair in the individual patient? Preliminary results of a finite element study from magnetic resonance imaging data. *J Thorac Cardiovasc Surg.* 2014;1097-685X148 (3):1025–34; discussion 1034. Available from: https://doi.org/10.1016/j.jtcvs.2014.05.071.

Chapter 39

Advancement of Technology and Innovation for Future Cardiovascular Care

James B. Young[1,2]

[1]Kaufman Center for Heart Failure and Recovery, Heart, Vascular and Thoracic Institute, Cleveland Clinic, Cleveland, OH, United States,
[2]Cleveland Clinic Lerner College of Medicine of Case Western Reserve University, Cleveland, OH, United States

Chapter Outline

"It's tough to make predictions,

especially about the future."

"If you don't know where you are going,

you will end up someplace else."

"It's like deja vue all over again"

"The Future ain't what it used to be"

Lawrence Peter "Yogi" Berra (1925–2015)

Major League Baseball Player

Setting the stage

Putting the future into perspective is different than predicting future events. Sometimes this is subtle, but at other times not so much. Sometimes we are successful, oftentimes, not so much. Yogi Berra's pithy and entertaining comments reflect that fact. But what we can do is use our knowledge, research, experience, education, and principles of navigation to help look into the future and drive achievement. To set the stage for predicting the future of cardiovascular disease (CVD) prevention, diagnosis, and treatment we must tell a story. This will help us understand the importance of innovation and advancement of technology. Exactly what the desired achievement is will be determined by further understanding of CVD pathophysiology and an innovative, resilient, persevering spirit. In healthcare the words of Osler summarize our charge nicely; "To prevent disease, to relieve suffering, and to heal the sick—this is what we do." Innovation and technological advancement is essential and must continue ad infinitum to achieve our mission and

Advances in Cardiovascular Technology. DOI: https://doi.org/10.1016/B978-0-12-816861-5.00019-8

prevent backsliding. Lives depend on that. But innovation and technological advancement depend on knowledge about, and understanding of, our phenotypic expression. Improvements based on discovery have germinated countless significant healthcare initiatives over almost 2 centuries. They have been dramatic when measured simply by overall life expectancy (realizing there are still significant differences around the world and in various populations); 78 years predicted in the 2000–50 epoch versus 40 years in the 1850–1900 period.[1] This dramatic increase in life expectancy isn't solely related to better healthcare, but would not have occurred without it.

When contemplating the future state, and navigating from now to then, it is imperative to know where you are and have at least an idea of where it is you wish to go. Knowing where you've been is also important. Those principles of navigation drive our advancement of knowledge, particularly technological knowledge, and allow healthcare innovations generally, and more specifically, those to prevent, diagnose and treat CVD, to be adopted. The many chapters in this book are testament to successes to date remembering that we've a long way to go. Important is understanding that success has been linked to constant paradigm shifts of one sort or another. A paradigm is simply an example or pattern. Thomas Kuhn in his 1962 treatise *The Structure of Scientific Revolutions* coined the phrase "Paradigm Shift" which he described as "extraordinary" or "revolutionary" scientific understanding of events. He pointed out that once a paradigm shift is complete, one cannot, for example, declare "that a miasma causes disease or that ether carries light."[2] And so time after time paradigm shifts driven by discovery, innovation, and technologic advances change the way we approach CVD and disease writ large. Innovation is not an easy task. To change the way we practice healthcare, and the tools we have, requires curiosity, perspective, perseverance, and hard work.

The story of James A. Garfield's assassination shortly after being elected 20th President of the United States is an example of being in caught in the middle of a paradigm shift and dying because of that.[3] Our profession must continue to evolve with study, learning, innovation, and change. And so when we see a counterproductive paradigm we must shift to a different model, and develop interventions that are, hopefully, evidence based in order to force change. Garfield didn't have to die but he was shot before widespread acceptance of innovative antisepsis measures and aseptic surgery became acceptable practice and de rigueur. He was shot in the back July 2, 1881 and died about 3 months later of an eroding abscess which ruptured his splenic artery. He suffered greatly with sepsis caused by repeated barehanded poking and implement prodding of the wound by his physicians. There was no regard for the preaching of Joseph Lister, father of antiseptic surgery, who had just been on a tour of the United States in 1876. He had presented his practices at the 1776 Centenary Celebration in Philadelphia. Lister had been invited by the doyen of surgery at the time, Samuel Gross, to present his findings as a paradigm shift had begun in Great Britain and Europe. Gross, who chaired the scientific portion of the celebration at 70 years of age, had commissioned a portrait of himself operating. Thomas Eakins, arguably the most famous portrait artist of the time, completed the work. It was to be exhibited prominently at the exhibition but instead was relegated to a backyard tent because of its grossness (pun intended). It portrayed the scene of a typical operating theater of the time which was poorly lit and had none of the aseptic maneuvers Lister promoted. It was a disturbing picture. Most, but not all, American surgeons of the time denigrated Lister's theories. Even D. Hayes Agnew of Civil War surgery fame (and Gross's cross town rival) who became a consultant on the Garfield treatment team, didn't put much stock in the concept at first. A paradigm shift had not occurred or even begun. It is interesting that Eakins also painted a portrait of Agnew in his operating theater 14 years later (1889) where dramatic and progressive differences could be seen in the lighting and surgical garb with other evidence of progress including aseptic practices. A paradigm shift had occurred but it was too late to save Garfield.[4] Of course Pasteur and Lister had been diligently working in one fashion or another on the microbial nature of disease for years. Indeed, it was Pasteur's 70th birthday celebration at the Sorbonne in 1882 where those efforts were highlighted a short time after Garfield's demise. The lack of influence Semmelweis (1818–65), Pasteur (1822–95), Lister (1827–1912), and Koch (1843–1910) had on North American surgeons regarding "aseptic surgery," evidenced by a very slow adaptation of that practice which was way behind European practice, has been documented.[5] And so the defense Garfield's assassin gave was something to the effect "I shot Garfield but it was his doctors who killed him." Despite the likely truth of the statement, Charles Guiteau was hung for the crime. It could be argued that Garfield's death was a case of malpractice. The movement in surgery, and medicine generally, to aseptic practices is but one example of a paradigm shift that massively influenced outcomes after surgical interventions. There have been countless other examples of groundbreaking paradigm shifts over the past century.

Why we need to push for paradigm shifts

With respect to the many innovations and advances detailed in this book, the reason for Garfield's story is to remind us to avoid hubris and clinging to the past as new ideas and information come available. Knowing our history can help guide us on our journey to the North Star. The North Star being conquest of CVD with reductions in morbidity and mortality. As we consider where we are going with innovation and technologic advances we should remember Garfield and thousands of other stories emphasizing the evolutionary nature of our field, sometimes the whimsy of progress, and the necessity of championing paradigm shifts to perpetuate our mantra "this is what we do." We will use this springboard to consider the advancement of technology and innovation for future CVD diagnosis, prevention, and care. It will often be speculative, and in many cases simply a dream world. Dreaming is, however, often the first step forward when pursuing innovation. Walt Disney is remembered by the quote "If you can dream it, you can do it. Always remember that this whole thing was started with a dream and a mouse."[6]

Paradigm shifts to new, different, and creative strategies are necessary because the problematic CVD epidemic is devastating. Halting it can only be done by dreaming up new approaches to the dilemma. The American Heart Association (AHA) regularly produces an impressive compendium of heart disease and stroke statistics. This gives us our targets. The latest report summarized in the 2021 Heart Disease and Stroke Statistics Update At-a-Glance details information from the massive document (almost 500 pages).[7] Table 39.1 is a summary of the subjects addressed in the document and serves as a road map for areas requiring innovative technologic intervention today and in the future.

The US data is sobering with CVD causing death of 868,662 individuals in 2017. Coronary heart disease (CHD) remains the number one problem. CVD claims more lives each year than all forms of cancer and lower respiratory diseases combined. Between 2015 and 2018, almost 127 million American adults had CVD of one form or another. The direct and indirect cost of CVD were approaching $365 billion between 2026 and 2017. Minorities are particularly hard hit. In the interval 2015–18, 59% of non-Hispanic Black females and 60% of Black males had some form of CVD. In 2018, CHD was the leading cause (42%) of deaths attributable to CVD in the United States, followed by stroke (17%), high blood pressure (11%), heart failure (10%), arterial disease (3%), and other CVD (17%). Furthermore, the problem

TABLE 39.1 Targets for innovation and progress to prevent and treat CVD.

- Overview
  - Cardiovascular health defined generally
- Health behaviors
  - Smoking/tobacco use
  - Physical inactivity
  - Nutrition
  - Overweight/obesity
- Health risk factors
  - Dyslipidemia
  - Hypertension
  - Diabetes
  - Metabolic syndrome
  - Kidney disease
  - Sleep perturbation
  - Adverse pregnancy outcomes
- Cardiovascular conditions/diseases
  - Coronary heart disease
  - Subclinical ASCVD
  - Peripheral vascular disease
  - Transient cerebral ischemia attacks and stroke
  - Congenital heart and vascular defects
  - Rhythm disturbances (supraventricular)
  - Rhythm disturbances (ventricular)
  - Cardiomyopathy/heart failure
  - Venous thromboembolism (DVT, PE)
  - Chronic venous insufficiency
  - Pulmonary hypertension
- Outcomes of care
  - Quality of care (AHA "Get with the Guidelines")
  - Medical procedures
  - Economics/costs

AHA, American Heart Association; *ASCVD*, atherosclerotic cardiovascular disease; *DVT*, deep vein thrombosis; *PE*, pulmonary embolism.
Source: Modified from Heart Disease and Stroke Statistics—AHA 2021 update. *Circulation*. 2021;143:e254–e793.

is not just an issue in the United States as CVD is the leading cause of death globally, accounting for about 19 million deaths in 2019.

AHA also provides statistics overviewing US cardiovascular health by tracking risk factors for CVD. They have established a goal of improving cardiovascular health of all Americans by focusing on "Life's Simple 7" such that deaths from CVD and stroke will be reduced by 20% over the near future. The Simple 7 include not smoking, increased physical activity, a healthy diet, mesomorphic body habitus, and optimization of lipid levels, blood pressure, and blood sugar. These are particularly robust targets for innovative technology and preventive medicine paradigm shifts. The opportunity to leverage "big data" to create risk factor intervention programs with artificial intelligence and machine learning is but one example.

CVD paradigm shifts: successful innovation and progress

Congestive heart failure (CHF) can be one quick and dramatic example of modern day innovative paradigm shifts in treating CVD. Fifty years ago the syndrome was looked at as only a symptomatic "pump problem" caused largely by cardiomyopathy, valvular, and CHD all with systemic congestion. Medical treatment focused on using the, at the time, 3-year-old diuretic furosemide, and stemming from Withering's 1785 report of Foxglove tea for dropsy, congeners of digitalis. As understanding of the syndrome increased (driven by basic science and clinical investigation which are critical elements of innovation), the heart failure paradigm shifted from simply being a failure of the heart as a pump to heart failure being a complex syndrome rooted in cardiac contractile abnormalities and harmful metabolic and humoral physiologic responses. Understanding this lead to the introduction over the next four decades of angiotensin converting enzyme inhibitors (ACEi), beta blockers (BB), angiotensin receptor blockers (ARB), aldosterone antagonists (AA), angiotensin receptor−neprolysin inhibitors (ARNI), and more recently drugs whose primary purpose was thought to be treating diabetes, sodium-glucose cotransporter-2 inhibitors (SGLT-2 Inhibitors), and Glucagon-like Peptide-1 Agonist (GLP-1 Agonists). It has been dramatic. The development of new diabetes management drugs is an example of basic and clinical science driving innovation by stimulating thinking in nontraditional ways and a bit of dreaming. A glucocentric approach to managing adult type 2 diabetes was trumped by other physiologic benefits of these particular agents. Now the GPL-1 agonist semaglutide has moved into the weight loss arena and the SGLT-2 Inhibitor empagliflozin plays an important role in treating heart failure.[8]

Another dramatic example of paradigm shifts in treating advanced end-stage heart failure (New York Heart Association Class IIIb to IV, and Stage D) is the story of mechanical circulatory support device (MCSD) development. These efforts are ongoing and extensively covered in this book. Briefly however, beginning in the 1930s, John Gibbon, and his wife Mary, ushered in the concept of cardiopulmonary bypass to allow surgical repair of life-threatening cardiac defects which could not be done without circulatory support. The first rudimentary devices ultimately led in the 1950s to machines that allowed "stopped heart" operations where congenital heart disease could be effectively addressed. Progress was slow at first but in the 1960s exploded to include a variety of operations for acquired valvular and CHD, complex congenital cardiac anomalies, as well as devastating large vessel vascular disease. It is important to remember that, at that time, the goal in many cases was total replacement of the heart; orthotopic cardiac transplantation or implanting a total artificial heart (TAH). Indeed, the first successful TAH to have supported a dog for several hours was done in the Cleveland Clinic labs of Wilhelm J. Kolff in 1957.[9] Kolff has been called the "father of artificial organs" because of his work to develop practical hemodialysis and the TAH. In 1954 he became the first President of the American Society for Artificial Internal Organs (ASAIO). And in the 1960s another paradigm shift was beginning in the labs of Michael E. Debakey in Houston, Texas, as well as other venues, where, in addition to working on development of a TAH, a variety of approaches to ventricular assist devices (VAD) were underway.[10] The first successful "Bridge-to-Recovery" VAD implantation was in 1966 by DeBakey's team when a patient could not be weaned from cardiopulmonary bypass (CPB) after valvular heart surgery and had an extracorporeal left ventricular assist device (LVAD) implanted for several days. It was successfully removed after satisfactory cardiac recovery. Also pushing the envelope were the first so-called successful orthotopic heart transplant in December 1967 and first TAH implant in April 1969. Both patients died shortly after surgery and the operations sparked controversy. But paradigm shifts had begun with movement from total permanent replacement of the heart with a TAH (now called "Destination Therapy") to a TAH used as a "Bridge-to-Transplant" and greater attention being placed on MCSD development both as "Destination" and "Bridge" therapies. Contemporary TAH development has not gone smoothly. The first destination therapy TAH was of a Jarvic-7 device developed by Kolff and Jarvik's team at the University of Utah in December 1982. It was a failure.[11] TAH programs progressed haltingly over the next decade, or so, with the TAH almost becoming a bridge-too-far. In June of 2001 Newsweek highlighted a newer edition to the TAH experimental stable with the lead "How technology will heal your heart."[12] That essay attempted to accomplish two decades ago the same thing this chapter is trying to do, predict the future of CVD treatment innovation. In the face of halting successes much more work is needed before we can truly champion consistently the TAH approach.

MCSD VADs on the other hand have made it further along the pathway to success with outcomes in some senses approaching heart transplantation. This has been the result of multiple paradigm shifts. Some notable ones include moving from extracorporeal to intracorporeal pump placement, pulsatile to continuous flow (CF), bulky and noisy gas driven diaphragm pumps to electrically powered pusher plate devices and devices got much smaller with fewer and less complicated moving parts. The debate about pulsatile versus CF devices has been settled with the success of several nonpulsatile CF VADs. Evidence that centrifugal flow designs, particularly one with a fully magnetically levitated driving system, were superior to the impeller driven CF device helped settle the question.[13] Of course, to make, at the least, MCSD technology equal to heart transplant outcomes many remaining challenges must be met including improved transcutaneous transmission of power, wireless device control and monitoring, increased durability of moving parts, and decreased risk of thromboembolism among others. Nonetheless, these two examples of extraordinary innovation and progress in the field of CVD treatment are encouraging and prompt us to consider moving forward even faster and further. We will explore what is on the horizon.

Necessary elements for healthcare innovation and progress

How do we innovate to ensure progress? Reams of manuscripts, numerous textbooks, countless courses, seminars, webinars, and degree granting programs have focused on entrepreneurship and innovation. All address nuances of our challenge to improve healthcare. There are several common themes and principles which should be reviewed and can lay a foundation for advancing technology and innovation. Many things are changing rapidly in our healthcare world, as they should, and we've seen many paradigm shifts over the last several decades, but we need more as evidenced by the AHA statistics.

Table 39.2 is modified from a Brookings report summarizing principles to guide creation of innovation districts.[14] Innovation teams or "districts" can be defined broadly. It might be one individual in a small cubicle, a multitude scattered around the world, or a few occupying a defined space, building, or city district. The Brookings guidelines can be tailored to any organization's size and structure (either small and centralized or large and geographically dispersed). One must realize the diverse nature of organizations embracing innovation and understand the workings of a team-of-teams multidisciplinary approach to solving problems with paradigm shifts. That is how creation of new devices and approaches to CVD prevention, diagnosis, and care will occur. Perhaps an optimal structure is clustering individuals or groups in a manner that stimulates communication and "brain storming." Whether that is done by ZOOM links or in a conference room may not be the most important issue. The COVID-19 pandemic shutdown proved that virtual meetings, when properly done, can be an effective tool to pull teams together. But a diverse team is essential to creativity and driving ideas forward. The example of improvements in MCSD design over time addresses that issue. Change is frightening and frustrating to many but if we are to make a significant impact on preventing, diagnosing, and treating CVD, change we must. And to do that we can't keep thinking along old lines but must bring in new participants who

TABLE 39.2 **Considerations for innovation and progress in prevention/treatment of CVD.**

- An innovation "center" or "district" should be broadly defined
- Cluster together innovation centers and research strengths
- A diverse enterprise is essential
- Diverse thinking and concepts are at the heart of innovation
- Seize a crises such as the COVID-19 pandemic to create targets for innovation
- Ensure participants are diverse with respect to visions, goals and strategies
- Connectivity (and perhaps proximity) necessary
- Vary strategies—immediate and longer range plans important
- Grow skills, strengthen components, build networks
- Social interactions essential for collaboration, learning and inspiration within teams
- Make innovation progress visible to teams and wider audiences
- Be mindful of costs and balancing resources with expectations
- Create innovative multilateral financing for discovery commercialization
- Ensure there is a collaborative team-of-teams approach to governance

Source: Modified from https://www.brookings.edu/blog/metropolitan-revolution/2017/09/08/12-principles-guiding-innovation-districts-2/.

view things differently. Indeed, diverse thinking and new concepts are at the heart of innovation. Particularly important is the concept of carpe diem—seize the day. Meaning, "never waste a crisis." A crisis might be that after countless hours of work building a CF MCSD prototype, during the first mock circulatory circuit run disastrous and massive hemolysis stopped the experiment short. But the crisis might drive a paradigm shift. For example, perhaps moving from an impeller driven mechanism to one with centrifugal design might solve the problem of hemolysis. Or more broadly, look at the examples of paradigm shifts the COVID-19 pandemic and lockdown drove. One striking example being a dramatic increase in the practice of "virtual medicine" which capitalized on telemonitoring and video telehealth platforms. Over time these systems will play a huge, and perhaps even dominating, role in patient management. Having participants able to dream up, see, and understand these epic responses to events in our lives is essential to shifting paradigms. To do that we come back to the importance of diverse teams and thinking. Connectivity of teams and teammates in one way or another is, stated again, necessary to vet ideas and results of experiments. Often this is termed "thinking outside of the box." Though physical distance of the innovation actors might be disruptive, there are ways to get around that challenge. Assessing the need for proximity to drive socialization, however, is still important.

Effective management of innovation requires setting clear-cut objectives, priorities, and timelines, while visioning desired results. Varying strategies and timetables with both short and long term waymarks will facilitate moving ideas downstream. As alluded to, failure should be expected but projects gain from that by being informed. It is also one of the critical elements of successful innovation. Failure drives change. To understand that issue teams need to grow their skills and strengthen their network to support their work. While doing that, and despite the present-day ability to "meet" with a team virtually, one must be mindful of the fact that social interactions are still necessary for collaboration, learning, and inspiration. Again, the COVID-19 pandemic isolation emphasized the challenge with that but there was also demonstration of clever ways to remediate that problem virtually such as using "icebreakers" to open meetings and "virtual cocktail hours" to celebrate successes. The importance of making progress known to both immediate stakeholders as well as wider audiences should be stressed. Obviously, being mindful of the project costs and balancing resources with expectations is a challenge. Offsetting this can be creation of multilateral financing for discovery, creation, and commercialization. Finally, leadership should ensure that there is a collaborative team-of-teams approach to project governance.[15]

Emerging and future innovations

Table 39.3 is a listing of emerging innovations and advances that will be top of the mind as we consider the future of healthcare innovation in the CVD space. There are more that did not make the list but the compendium can serve as a broad overview of some important arenas. Many are paradigm shifts in healthcare generally, such as AI and virtual patient management, while others are quite specific to CVD, such as developments in MCSD technology, pacemakers, defibrillators, heart valves, and other implantable devices. The specifics of several are covered in this book. Several themes emerge and include a different approach to managing CVD prevention, diagnosis and treatment, a different approach to interacting with patients, AI and machine learning applied to massive data sets, newer imaging-based technologies, virtual/augmented reality education platforms, regenerative and molecular/genetic medicine advances and continued innovation with respect to MCSDs, arrhythmia management, percutaneous coronary and peripheral vascular catheter based interventions, and structural implants (valves, vascular stents, medication delivery systems, etc.).

As patient and population-based data is gathered more efficiently, CVD prevention, diagnosis, and treatments evolve. Not new, but impactful and changing over the last 20 years, has been the electronic medical record (EMR). It will continue to evolve. Though often characterized as an annoyance by clinicians, this innovation has dramatically changed the way we interact with our patients. The EMR focuses on data storage that is generally easy to access, manage, and analyze. The technology can be used to improve patient care quality. It can assist with developing value-based care systems. Timely reminders about good guideline-driven care can be programed into systems and patient "chatbots" and healthy living texting pings help with communication. eRx systems improve drug dispensing safety. Some applications can be used to query patient cohorts to gain insight into clinical issues of significance. By doing population health surveys problematic nests of disease can be discovered. One example is the EPIC-based EMR self-service query system, Slicer Dicer, which allows clinicians to generate quick overviews of patient populations. Slicer Dicer data are pulled from the Epic EMR backend data warehouse (Caboodle). Data in Slicer Dicer are current through the prior day. This was the system used to discover the extent of the Flint, Michigan water lead crisis in 2015 for example.[16,17] The EMR can also have other research attributes. They can facilitate clinical trial management and assist with the recruitment of patients for research studies. As cloud and hybrid cloud data storage systems evolve, massive platforms of "big data" can be cobbled together, fed by EMR health insurance portability and accountability act (HIPPA) compliant

TABLE 39.3 Emerging innovation and advances.

- A different approach to managing CVD prevention and treatment data is evolving
  - Patient and population data collection, collation, corroboration, management and evaluation
    - Electronic medical records (EMRs) (a focus on individuals)
    - Population health surveys (a focus on population nests)
    - Public and private databases (CMS and Federal Reserve databases)
    - Cloud and Hybrid Cloud data storage and management
    - Advanced computing applications such as IBM Quantum
- A different approach to interacting with patients is necessary (Value-Based Care)
  - Improved and linked EMRs
    - Blockchain "smart" contracts for EMR data portability
  - "Virtual" telehealth patient interactions enhancing care quality
    - Multidisciplinary disease management "virtual" encounters
    - Telemonitoring
      - Implantable devices (arrhythmia monitoring, pulmonary pressure)
      - Wearable nonimplantable devices (bands, watches, rings)
      - Remote hospital (eHospital) and ICU patient monitoring services
    - Telerehabilitation
      - Robotic
      - Virtual reality
      - Gaming-based devices
    - Mobile platforms and connected health
      - Mobile stroke units
      - Chatbots, automated text and messaging platforms, "smart phones"
- Artificial/augmented intelligence, machine learning, and quantum computing maturing
  - Data analysis
  - CVD prevention
  - Diagnostic probabilities
  - Patient management decisions
- Imaging based technologies
- Virtual/augmented reality education platforms expanding
- Regenerative and molecular genetic medicine advances
- Continued innovation of MCSD, arrhythmia management, percutaneous coronary and peripheral vascular catheter based interventions, structural implants (valves, vascular stents, medication delivery systems, etc.)

deidentified information, and support sophisticated queries (see artificial/augmented intelligence discussion below). Public databases often linked to EMR databases are available such as the Centers for Medicare and Medicaid Services (CMS) and Federal Reserve Bank databases.

EMR portability can be a problem, however. Often, different EMR systems won't interact with one another and this prevents patients from seamlessly moving from one healthcare provider or system to another. Blockchain technology is a decentralized database or ledger that maintains a continuously growing list of data records and transactions across a peer-to-peer network. Participants can use the technology to confirm transactions without the need for a central certifying authority. EMR data can be shared using this approach. A user can request a transaction and a "block" of data representing the request is created. The block is then broadcasted to all of the nodes (computer stations) in the network. All the nodes validate the block and transaction. The block is added to the chain of information. The transaction is verified and executed. It is a secure, trusted, and automated way to transfer massive amounts of information.[18]

As mentioned, "virtual healthcare" is gaining great attention. Virtual visits using a variety of platforms including American Well, ZOOM, Microsoft Teams, Facetime, and others have allowed connectivity that is internet or smartphone-based, allowing for verbal and visual links to a patient, patients' families, and groups of patients (shared medical appointments). In our experience, virtual shared medical appointments have been used in heart failure, diabetes, and obesity management clinics with great success. Surgeons have done routine postoperative follow-up appointments. Urgent care screening visits have been managed this way. When we remember how much healthcare delivery occurred over the telephone during the last century one should not be surprised by this paradigm shift. It is an effective way to obtain a patient history, assess a chief complaint, determine progress over time, review medications, do a limited physical examination, and sometimes gain insight into a patient's surroundings or living conditions and support available (was that a pack of cigarettes I saw on the kitchen counter?). Patients don't have to leave their home, office, or car to travel to a healthcare provider location. They might not have to take time off from work as it is easy to do a visit on a lunch break. Multidisciplinary management can be done bringing together teams of caregivers on a variety of platforms eliminating the need for the patient to bounce from one office or location to another. Some systems have been developed that can allow auscultation to be done, for example, as well as other more traditional physical examination techniques. This is an area ripe for innovative delivery systems to be developed. Challenging is internet connectivity in remote or underserved areas and populations, confidentiality, and security of data and information.

Telemonitoring can be linked to telemedicine visits synchronously or asynchronously. Patients with implantable or wearable arrhythmia monitoring devices, for example, can have information electronically downloaded for clinician review synchronously during virtual visits, asynchronously at prescheduled times, or when symptoms occur. This technology is not new. What is evolving, however, is the magnitude and breadth of data coming available, and the ability of using AI to develop sophisticated prediction and response algorithms. Information systems have broadened to include nonimplantable wearable devices such as monitoring chest or arm bands, watches, and rings.[19–21]

Remote patient monitoring has allowed caregivers to make telemedicine virtual rounding at a distance on hospitalized patients. Links between the tele-expert rounder and the onsite personnel can be made. Conversations with the patient can be established. This has, in particular, been helpful in small, remote hospitals where medical and surgical specialists might not be readily available. eHospital programs can enhance in-hospital care, particularly in intensive care units (ICUs). Having expert eyes on ICU patients and their monitors continuously means that there is better access to intensivists and experienced critical care nurses during night shifts, weekends and holidays. In our healthcare system the eHospital program is located in an operations center on the main campus that is staffed 7 p.m.−7 a.m. 7 days a week. The goal is to facilitate problem identification and swiftly intervene when necessary in our outlying network hospitals. The team can't touch the patient but has access to a great deal of data and can communicate with caregivers on the scene. Algorithms are used to risk stratify patients into higher or lower acuity of care. The system has specific real-time alerts that drive responses to clinical events that assist the local bedside and bunker based caregivers.[22]

Creating telerehabilitation programs can be a way to connect patients remotely and virtually to physical therapy programs. Theoretically a homebound patient could be provided with a robotic device and supervised remotely while it is in use. Virtual reality could be used in creative fashion to stimulate exercise and movement based rehabilitation. Peloton type bikes and treadmills could have virtual video biking and hiking adventures tailored to cardiac rehabilitation patients for supervised and unsupervised rehab programs. These could be remotely monitored with wearable electrocardiographic and blood pressure measurement devices. Gaming devices would provide similar opportunities with an entertainment feature that might enhance compliance.[23]

Mobile stroke response units are examples of an innovative approach that brings an ICU to the patient rather than the patient being brought to a hospital as a first response to an event. Knowing that treatment of certain types of stroke with thrombolytics as quickly as possible preserves potentially brain tissue and saves lives, this therapeutic is important and continues to evolve. It is complicated. To be effective a thrombolytic must be given within only a few hours of stroke symptoms and sophisticated cerebrovascular imaging (computerized tomography; CT scanning) is required to determine the type of event; ischemic (thrombotic) or hemorrhagic. Thrombolytics are contraindicated when hemorrhage is the problem. When called, the dispatched team arrives quickly, scans the brain, and transmits the images wirelessly to a neuroradiologist at home base for immediate interpretation and decision making. If indicated, thrombolytics are quickly given. All of this is a remarkable testament to innovative technology at multiple levels and serves as a base for continued advances in this realm.[24]

Artificial/augmented intelligence and machine learning is driven by data gleaned from many sources including clinical observations and measurements, clinical experiments, wearable and implantable data gathering devices and other relevant biologic determinations. Data acquisition then allows sophisticated electronic storage with formatting, cleaning, normalization, scaling and subsequent analysis. Complex data management technology programs apply complicated regression and other forms of analysis to the data. Decision trees and algorithms can be produced which help create neural networks that drive model development which are able to be constantly updated. Models can then be applied to individuals and populations to help make a diagnoses and create treatment recommendations. As examples, prediction of cardiovascular events such as vascular in-stent stenosis and myocardial infarction have been done utilizing "big data" analysis. The ability of clinicians to make decisions driven by a more accurate and individualized risk determination will change the way we recommend interventions and medications. Advantages will be great value meaning better outcomes at less cost.[25–27]

Chatbots are being developed to facilitate patient visits and can be used both in person and virtually in synchronous and asynchronous modes. The concept is to have a patient do a previsit interview with the chatbot to develop themes for the visit. Practitioners can individualize the chatbot questions to fit their specialty and program the system with follow-up questions based on answers. This can then enhance subsequent conversation or encounters. Real time voice recognition technology continues to advance and it is likely that these systems will replace paper and pen scribbling and keyboard pounding. The AHA has worked with Amazon to create a voice activated program that can work through their personal assistant "Alexa"—"just ask Alexa what are the signs and symptoms of a heart attack or stroke" or "please provide cardiopulmonary resuscitation instructions."[28] Automated text and other messaging platforms with smart phones are also undergoing evaluation to determine if there is value to using them to prompt good health behavior such as walking 10,000 steps in a day (Apple watch) or prompting reminders to take medications at certain times.[29]

A quantitative systematic review of the efficacy of mobile phone interventions to improve medication adherence included 29 studies using texting applications. A majority of the studies found significant improvement in medication adherence. However the investigators suggested "longer term studies characterized by rigorous research methodologies, appropriate statistical and economic analyses and the test of theory-based interventions are needed to determine the efficacy of mobile phones to influence medication adherence." Indeed, in many of the emerging innovations and advances noted in Table 39.3 a common theme is the need for more and better clinical trials and studies designed to determine their value with value defined as the relationship of meaningful outcomes to cost.[30]

Imaging technologies continue to evolve and we can expect dramatic progress in the future. The roentgenographic evolution began in 1895 when a German physicist demonstrated that cathode tube-produced X-rays could pass through flesh and leave an imprint on photographic plates. The limitations of X-rays, however, were many and not just the fact that dense bones got in the way of imaging soft tissues. Radiation was, arguably, a much more dangerous problem and an example of difficulties with rapid adaptation of some innovations after only rudimentary and inadequate exploration of risk/benefit and value equations. An example is the proliferation of shoe-fitting fluoroscopes between about 1920 and 1970 and recently included in a listing of the "100 worst ideas of the 20th century."[31] Another example is the attempt between 1967 and 1977 to create a nuclear plutonium-238 powered TAH and LVAD system. This followed similar attempts to build nuclear-powered pacemakers. Though no clinical disasters occurred, they surely would have if this technology had become clinically available.[32] These examples serve to remind us that not all innovative approaches are wise ones to pursue. Perhaps the Batista, Myosplint, Acorn Device, and cardiomyoplasty operations (all designed to reshape a dilated left ventricle) for cardiomyopathy are additional examples innovations that failed to meet their promise.[33]

Over 50 years ago a dramatic imaging innovation, one of the first since Roentgen discovered the X-ray, was development of computed tomographic (CT) scans which in 1971 allowed visualization of a brain tumor which X-rays could not "see" because of dense skull bones.[34] Subsequent innovation included magnetic resonance imaging (MRI) and scintigraphic studies such as positron emission tomography (PET). Add to this advances in ultrasonic imaging, particularly echocardiography (with cardiac images transmitted from Space Station astronauts in orbit), and cine-angiographic procedures (which can now be combined with sophisticated analysis of vessel flow), and one sees the extraordinary changes in cardiovascular imaging that has occurred over the last century. Indeed imaging innovation has shifted modern healthcare from relying largely on physical examination skill (where quality is highly variable from clinician to clinician) to the extensive use of sophisticated (and expensive) platforms. These have been responsible for progress which now plays an essential role in diagnosis and management of patients. One example where modern-day innovation and progress is taking us that combines imaging technology with sophisticated AI application is Ultromics, a global health technology company in Great Britain that provides echocardiographic evaluation services using autonomous analytic approaches that leverage AI algorithms and cloud storage techniques during data gathering and analysis. The Ultromics team analyzed 120,000 studies done as part of various clinical trials being done at Oxford University and then tracked subsequent patient events. Outcome prediction models were created. The American Society of Echocardiography included the company as part of its 2018 "Echovation Honor Role."[35] Additional advances we can expect include greater use of three-dimensional image projections with better integration into the EMR and holographic and printed three-dimensional models to help plan complicated operative and percutaneous intervention procedures (or combinations).

Virtual/augmented reality education platforms are rapidly appearing and have dramatically changed education paradigms. The COVID-19 pandemic and lockdown spurred innovation as most education endeavors moved from in-person to remotely delivered synchronous and asynchronous curriculums. We've learned much about the advantages and disadvantages, as well as benefits and detriments of this approach but we need more data about this approach. Innovation in this arena has been informed by studies of outcomes compared to in-person efforts. Human anatomy courses are an excellent example of this paradigm shift, though not without controversy.[36–39] Since the work of Vesalius, da Vinci, and other Renaissance cognoscenti, healthcare professionals have been taught human anatomy in dissection laboratories of one sort or another for over 500 years. More recently the work was done utilizing formaldehyde embalmed cadavers, and some have argued this has become more of a right-of-passage, than an exceptional learning experience. It wasn't just the stench and caustic nature of the embalmed cadaver which prompted stringent Occupational Safety and Health

Association (OSHA) regulations, that was the problem, but the change in the way the tissues felt and looked as they lost their color and became turgid. Use of unembalmed cadavers helped solve some of these problems.[40] Also, for a medical student, or other healthcare professional needing insight into normal as much as pathologic anatomy, most cadavers were old (even ancient) and riddled with disease (though this could still could provide one aspect of a valued education experience). Finally, procuring enough cadavers for countless anatomy classes around the world has become challenging. Advances in virtual and augmented reality imaging has allowed for changes that are sweeping worldwide and impacting the way anatomy curriculums are delivered. Cadaveric anatomy education can be anticipated to decrease as these innovative virtual platforms continue to improve.

Advances in regenerative and precision (molecular-genetic and "omic") medicine are additional examples of emerging innovations that are showing promise and expected to change the way we prevent, diagnose, and care for patients with CVD.[41–43] Because of the low regenerative potential of cardiac myocytes that are responsible for powering the heart as a pump and the epidemic of atherosclerotic CHD and cardiomyopathy causing substantial morbidity and mortality, regenerative approaches to healing the heart are attractive. Pluriopotent stem cells which theoretically can be induced to differentiate into myocytes or embryonic cardiac stem cells and have potential as cell-replacement therapies. They can also be a way to model CVD and screen pharmacological agents for adverse myocardial side effects. Progress has been slow, however, and unlike cell therapies for some cancers and inflammatory diseases, stem cell therapies for CVD are still largely theoretic and experimental, requiring a great deal of investigational work with basic, preclinical, and clinical trials. However no listing of emerging innovation and progress in preventing, diagnosing, and treating CVD would be complete without mention of embryonic stem, pluripotent stem, induced pleuripotent stem, and skeletal myoblast cells.

Precision medicine seems further along than regenerative medicine but still is evolving conceptually. The concept will likely drive further innovation and progress. It is an integrative approach to CVD (and other disease) prevention and treatment. It relies on newer understanding of personalized data that is rooted in an individual's genetics and lifestyle as well as exposures to a variety of environmental factors. The "omics" of transcriptomics, epigenomics, proteomics, metabolomics, and microbiomics are nodes that create networks that are complex intermingled systems ultimately driving human health and disease. It is reasonable to consider CVD in the context of "network medicine."[44] Leopold and Loscalzo posited "that precision medicine is poised to become the next great revolution in the practice of medicine, as well as in the maintenance of cardiovascular health and the prevention and cure of cardiovascular disease."

Finally, emerging innovation and advances will continue in the arenas of MCSD development, arrhythmia management, percutaneous coronary and peripheral disease interventions, and structural implants (valves, vascular stents, medication delivery systems, monitoring devices, etc.). Some paradigm shifts will be subtle while others dramatic.

Though some might not think the evolution of the business of medicine as innovation, it most surely is, though not fitting nicely into the category of a tool, device, medication, or procedure. Nonetheless a variety of healthcare delivery systems are undergoing rapid change. One example is the shift in health care financing from an episode-based paradigm to one that is incentive and outcomes-based and sometimes referred to as Value-Based Care. This is, indeed, a creative innovation that will, hopefully, change the way we approach CVD prevention, diagnosis, and care. One might speculate that this innovation is the most dramatic of all on the Table 39.3 list.[45]

Threats to innovation and progress

Table 39.4 summarizes some of the threats to innovation and progress in the domain of CVD prevention, diagnosis, and treatment. It is not meant to be a totally inclusive listing but rather to prompt consideration of more serious challenges and potential difficulties innovators will encounter. At the top of the list is the financial and administrative support entrepreneurial efforts require. This is not a trivial challenge. There are financial, infrastructure, intellectual, skill set, education, and training program insufficiencies generally and costs associated with these features that drive technological advances can be massive and daunting.

Because many of the emerging innovations detailed in Table 39.3 involve massive databases and sophisticated internet networks, the lack of adequate cybersecurity is concerning. Because data is going to be more heavily leveraged in the future to drive CVD healthcare innovation the vulnerability of the worldwide web is frightening. Prevention of ransomware attacks that paralyze networks, nefarious programming of monitoring and therapeutic devices, and techniques to ensure privacy, security, and confidentiality of patient data are essential but not optimal at the present moment. Competition has been an important tenant of entrepreneurial efforts but destructive competition that includes corporate spying and other criminal acts pitting one group against another will stifle progress. Data capacity and the ability to manage it is a threat, particularly when AI is being used. Moore's law, for example, explains the limitation in the ability of integrated circuit transistors to support future digital data management requirements. This issue is another target for

TABLE 39.4 Threats to innovation and progress focused on CVD prevention and treatment.

- Lack of support for entrepreneurial efforts that drive innovation and progress
  - Financial, infrastructure, intellectual, education, and training insufficiencies
  - Costs associated with technological advancements
  - Lack of technological infrastructure in underserved areas
- Cybersecurity in a heavily leveraged, networked, and vulnerable worldwide web
  - Ransomware paralyzing networks
  - Nefarious programing of devices
  - Privacy, security, and confidentiality of patient data generally
- Destructive competition in innovation and development environments
- Data management capacity and speed
  - Moore's law—limitation in the ability of integrated circuits to support digital data management
  - Maintaining data accuracy and ease of use
- External device (iPhone, Apple Watch, etc.) interference with implantable cardiac devices
- Inappropriate and overuse of remote patient monitoring
- Lack of clinical trials creating data to inform caregivers about outcomes in the "modern" era of innovative technology
  - Too few efforts creating standards of care
  - Too few "quality" metrics created
  - Insufficient clinical studies of performing virtual physical exams
  - Unknown risks of using virtual technology
- Disagreements about healthcare payments in a virtual world
- Differing telehealth practice expectations among States
- Lack of Multistate telehealth licensure

Source: Modified from American Heart Association recommendations for the implementation of telehealth in CVD and stroke care. A policy statement. *Circulation*. 2017;135:e24–e44.

innovation.[46] IBM has built a new supercomputing system that addresses some of the computer challenges mentioned. Quantum computing relies on Qubits rather than binary bits and bytes. This system will revolutionize the speed and intensity of analysis of massive data sets housed in cloud and hybrid-cloud arrangements.[47,48] Of course more innovative methods of ensuring data accuracy, ease of use, and security are desperately needed. And as use of external electronic devices increases (iPhone, Apple watches, AirPods, charging stations, etc.) there is the danger of them interfering with implantable cardiac devices such as pacemakers and cardioverter-defibrillators. Yet another target for innovation.[49]

Lacking good guidelines for use of remote patient monitoring devices could lead to either overuse or underuse of these tools. This can be a problem more generally with several of the innovation opportunities discussed and detailed on Table 39.3. Although some regulatory systems (the FDA as an example overseeing development of cardiac implants) govern the use of many devices, other arenas, such as use of "big data" to develop diagnostic and therapeutic algorithms, have less regulatory oversight in some senses. Furthermore, the lack of sophisticated clinical trials in "real-world" settings that can create high-quality outcomes data to inform caregivers in the modern era of innovative technology is troubling. There have been too few efforts creating standards of care and too few quality measures have been created with these devices in mind. As an example, insufficient clinical studies have been performed to inform us about the true utility of virtual healthcare as well as its dangers. We have little information about how to perform a physical examination virtually, though some tools have been developed. Physicians training in times past might remember the iconic textbook by DeGowin and DeGowin, *Bedside Diagnostic Examination*.[50] We need a "DeGowin and DeGowin" for virtual patient encounters generally, and as well, better characterization of the risks of extensive use of virtual technology.

Other risks to further innovation include disagreements about healthcare payments in a virtual world, differing telehealth practice expectations among various States, and multistate telehealth licensure.[51]

One institution's approach to data stewardship in the age of healthcare innovation and progress

Because healthcare data is at the center of many contemporary innovations, stewardship and oversight of this asset is essential to address (Table 39.5). This has become even more important in view of recent breaches in trust that have occurred and been well-publicized.[52] To protect our patients, research subjects and ourselves, the Cleveland Clinic has developed guiding principles for data management and oversight. The Clinic was founded 100 years ago in 1921.

TABLE 39.5 An enterprise wide approach to data utilization.

Cleveland Clinic is committed to ensuring innovation will be managed with sensitivity and accountability to protect our patients, the communities we serve, and the organization.
We Intend To: • Collect, use, and analyze data and biospecimens to improve the lives of those we care for, educate those who serve, generate new knowledge in research and discovery, and address health and healthcare disparities. • Protect data by using state-of-the art data privacy monitoring tools and services (we will require the same of our partners). • Partner with organizations and institutions to collectively use data for clinical, scientific, and technological advancements • Educate our caregivers and hold them accountable to appropriate data use practices • Improve patients' access to their personal healthcare data, and educate them on how to use their data to monitor and improve their health and wellness • Protect patient data and privacy rights and responsibilities, while operating with transparency. • Comply with applicable laws and regulations for use of patient data • Comply with data breach reporting notification requirements • Not allow our collaborators, partners or caregivers, to unjustly deny or discriminate their services to our patients due to data obtained from or about them • Reinvest value generated by data-related innovations into our mission for patient care, prioritizing research, education, and discovery

During its evolution data in its many dimensions have been the foundation for operating our academic medical center. Indeed, data support activities that allow us to achieve our mission, which is singular but contains three interdependent pillars: providing care for life, researching for health, and educating those who serve. The same mission is at hand for healthcare systems generally, even when the focus is primarily on patient care and not academics (remembering that even then elements of research and education are present and important). Recent controversy has arisen about how healthcare institutions have managed, used, and shared data acquired from their patients and during day-to-day operations. Regulatory and court battles have ensued. Cleveland Clinic has confronted the challenge of data collection, collation, analysis, and protection, utilizing the principles delineated on Table 39.5. Driven in particular by recent technological advances in data science, it became necessary to determine the best approach to data (including specimen) stewardship. Clearly the importance of data gathering is an integral part of patient and population healthcare management, medical discovery, and education. Collaborations with medical schools, hospitals, biomedical research institutions, and industry have always been important but can present challenges. Because of that we codified our approach to collaboration and innovation. Since patients trust us to provide the best care, we, in particular, understand that their personal health information is sacred. Indeed, we must be good stewards of this data while using it for operations, healthcare, education, research, and innovation. Commitments have been made to several universal principles of a patient's right to privacy when it comes to their personal data and putting patients first.

Summary and conclusions

To summarize and end it is appropriate to remember the opening and closing lines of T.S. Eliot's poem, East Coker.[53] That work was about the rise and decline of an English countryside village emblematic of the industrial revolution around the world where villages prospered then declined, evolved, and prospered again. "In my beginning is my end" and then "In my end is my beginning." The stories iterate and are at play today. It is innovation that drives the renaissance with paradigm shifts making the words "in my end is my beginning" a mantra. And so taking words from the beginning of this chapter we note that our work is, in large part, preventing, diagnosing, and treating CVD. Advancement of technology with innovation for future paradigm shifts are essential if we are to meet our Oslerian goals. This requires new strategies. New strategies demand innovation. The problematic and devastating CVD epidemic must be halted. This can only be done after dreaming up new paradigm shifts. Many targets and approaches have been presented. It is now time to do more work.

References

1. Life Expectancy. https://www.google.com/search?q = life + expectancy + in + us&rlz = 1C1CHZL_enUS783US783&oq = life + expectence + &aqs = chrome.1.69i57j0i13l9.7570j0j15&sourceid = chrome&ie = UTF-8.

2. Kuhn TS. *The Structure of Scientific Revolutions*. University of Chicago Press; 1962.
3. Millard C. *Destany of the Republic: A Tale of Madness, Medicine and the Murder of a President*. Random House; 2011.
4. Brieger GH. A portrait of surgery: surgery in America 1875–1889. *Surg Clinics North America*. 1987;67:1181–1216.
5. Ravitch MA. *A Century of Surgery: 1880–1980*. Philadelphia: Lippencott; 1981.
6. Walt Disney. If you can dream it. https://www.go2nextlevels.com/if-you-can-dream-it-you-can-do-it-always-remember-that-this-whole-thing-was-started-with-a-dream-and-a-mouse-%E2%80%95-walt-disney-company/.
7. AHA. 2021 Heart disease and stroke statistics update at-a-glance. *Circulation*. 2021;143:e254–e743.
8. Maddox TM, Jannuzi J, Allen LA, et al. 2021 Update to the 2017 ACC expert consensus decision pathway for optimization of heart failure treatment: answers to 10 pivotal issues about heart failure with reduced ejection fraction: a report of the American College of Cardiology Solution Set Oversight Committee. *J Am Coll Cardiol*. 2021;77(6):772–810.
9. Kolff WK. *Artificial Organs*. New York: John Wiley & Sons; 1976.
10. Miller CA. *A Time for All Things: The Life of Michael E. DeBakey*. Oxford: Oxford University Press; 2020.
11. McKellar S. *Artificial Hearts: The Allure and Ambivalence of a Controversial Medical Technology*. Baltimore: Johns Hopkins University Press; 2018.
12. Newsweek. How technology will heal your heart; June 2001.
13. Monlina EJ, Shah P, Kierman MS, et al. The society of thoracic surgeons intermacs 2021 annual report. *Ann Thorac Surg*. 2021;111:778–792.
14. Brookings Innovation Centers. Available from: https://www.brookings.edu/blog/metropolitan-revolution/2017/09/08/12-principles-guiding-innovation-districts-2/.
15. McChrystal S. *Team of Teams: New Rules of Engagement for a Complex World*. New York: Portfolio/Penguin; 2015.
16. EPIC Slicer Dicer (Flint lead poisoning). https://medcitynews.com/2016/epic-flint-water/.
17. EPIC Slicer Dicer (Flint lead poisoning). https://madison.com/wsj/news/local/health-med-fit/flint-doctor-used-epic-systems-records/.
18. Blockchain. https://burniegroup.com/wp-content/uploads/2017/05/Blockchain-infographic-1024x793.png.
19. Maeng DD, Starr AE, Tomcavage JF, Sciandra J, Salek D, Griffith D. Can telemonitoring reduce hospitalization and cost of care? A health plan's experience in managing patients with heart failure. *Pop Health Management*. 2014;17(6):340–344. Available from: https://doi.org/10.1089/pop.2013.0107.
20. Peri MG, Shankar MN, Daniels MJ, et al. Effect of telehealth extended care for maintenance of weight loss in rural US communities: a randomized clinical trial. *JAMA Network Open*. 2020;3(6):e206764. Available from: https://doi.org/10.1001/jamanetworkopen.2020.6764.
21. Sood A, Watts SA, Johnson JK, Hirth S, Aron DC. Telemedicine consultation for patients with diabetes mellitus: a cluster randomized controlled trial. *J Telemed and Telecare*. 2018;24(6):385–391. Available from: https://doi.org/10.1177/1357633X17704346.
22. eHospital. https://consultqd.clevelandclinic.org/ehospital-program-enhances-care-medical-icus/.
23. Peloton rehabilitation. https://wellinsiders.com/i-tried-peloton-to-come-back-from-an-injury/; https://pelotonpt.net.
24. Mobile stroke unit. https://my.clevelandclinic.org/health/treatments/17242-mobile-stroke-unit.
25. Han SH, Kim KO, Shon HS. System framework for cardiovascular disease prediction based on big data technology. *Symmetry*. 2017;9:273. Available from: https://doi.org/10.3390/sym9120293.
26. ChaJohnson KW, Soto JT, Glicksberg BD, et al. Artificial intelligence in cardiology. *J Am Coll Cardiol*. 2018;71(23):2668–2679.
27. Karunathilake SP, Ganegoda GU. Secondary prevention of cardiovascular diseases and application of technology for diagnosis. *BioMed International*. 2018; https//doi.org/10.1155/2018/5767864.
28. Amazon Alexa. How do I save a life? https://heart.org/en/affiliates/alexa-how-do-i-save-a-life.
29. Feldman DI, Robison WT, Pacor JM, et al. Harnessing mHealth technologies to increase physical activity and prevent cardiovascular disease. *Clin Cardiol*. 2018;41:985–991.
30. Park LG, Howie-Esquivel J, Dracup K. A quantitative systematic review of the efficacy of mobile phone interventions to improve medication adherence. *J Adv Nurs*. 2014;70(9):1932–1953.
31. Worst ideas of the century. https://www.anvari.org/fun/Political/100_Worst_Ideas_of_the_Century.html; https://content.time.com/time/magazine/article/0,9171,991230,00.html.
32. McKellar S. Atomic hearts: a decade of US government-sponsored development. *Phys Today*. 2016;69(5):38. Available from: https://doi.org/10.1063//PT0.3.3168.
33. Dryfus G. The Batista procedure. *Heart*. 2001;85:1–2.
34. Howell JD. The CT scan after 50 years—continuity and change. *N Engl J Med*. 2021;385(2):104–105.
35. Ultromics. AI enabled cardiovascular imaging solutions. https:/http://www.ultromics.com.
36. https://3dforscience.com/hololens-medical-medical-education/.
37. Moro C, Stromberga Z, Raikos A, Stirling A. The effectiveness of virtual and augmented reality in health sciences and medical anatomy. *Anat Sci Educ*. 2017;10:549–559.
38. Iwanaga J, Loukas M, Dumont AS, Tubbs RS. A review of anatomy education during and after the COVID-19 pandemic: revisiting traditional and modern methods to achieve future innovation. *Clin Anat*. 2021;34:108–114.
39. Evans DJR, Bay BH, Wilson TD, Smith CF, Lachman N, Pawlina W. *AnnSci Educ*. 2020;34:279–283.
40. Drake RL, McBride JM, Pawlina W. An update on the status of anatomical sciences education in United States medical schools. *Anat Sci Educ*. 2014;7:312–325.
41. Soeder J. Regenerative medicine to the rescue: new cell therapies offer healing and hope where other approaches can't reach. https://magazine.clevelandclinic.org.

42. Samak M, Hinkel R. Stem cells in cardiovascular medicine: historical overview and future prospects. *Cells*. 2019;8:1530–1557.
43. Leopold JA, Loscalzo J. *Circ Res*. 2018;122:1302–1315.
44. Loscalzo J, Barabasi AL, Silverman EK. *Network Medicine: Complex Systems in Human Disease and Therapeutics*. Cambridge, MA and London: Harvard University Press; 2017.
45. DeCamp M, Sulmassey LS. Ethical and professionalism implications of physician employment and health care business practices: a policy paper from the American College of Physicians. *Ann Intern Med*. 2021;174:844–851. Available from: https://doi.org/10.7326/M20-7093.
46. Shalf J. The future of computing beyond Moore's law. *Philos Trans A Math Phys Eng Sci*. 2020;378(2166):20190061. Available from: https://doi.org/10.1098/rsta.2019.0061. Epub January 20, 2020.
47. Schramm S. Data analysis meets quantum physics. *Nature*. 2017;550:338–340.
48. Castelvecchi D. IBM's quantum cloud computer goes commercial. *Nature*. 2017;543:159. Available from: https://www.nature.com/articles/nature.2017.21585.
49. Apple lists all its products that could interfere with cardiac devices. https://www.techspot.com/news/90218-apple-lists-all-products-could-interfere-cardiac-devices/.
50. DeGowin EL, Degowan RL. *Bedside Diagnostic Examination*. 2nd ed. London: The Macmillan Company; 1969.
51. AHA. Recommendations for the implementation of telehealth in CVD and stroke care. A policy statement. *Circulation*. 2017;135:e24–e44.
52. Google defends use of patient data on Capitol Hill among scrutiny of Ascension deal. https://www.fiercehealthcare.com/tech/senators-pressing-ascension-google-data-deal-as-tech-giant-defends-its-use-patient-records.
53. Young JB. East Coker: in my beginning is my end and in my end is my beginning. *Methodist Debakey Cardiovasc J*. 2019;15(3):231–232.

Index

Note: Page numbers followed by "*f*" and "*t*" refer to figures and tables, respectively.

E

F

G

H

L

M

N

Q

R

S

U

V

W